ADVANCES IN
ATOMIC PHYSICS
An Overview

ADVANCES IN ATOMIC PHYSICS

An Overview

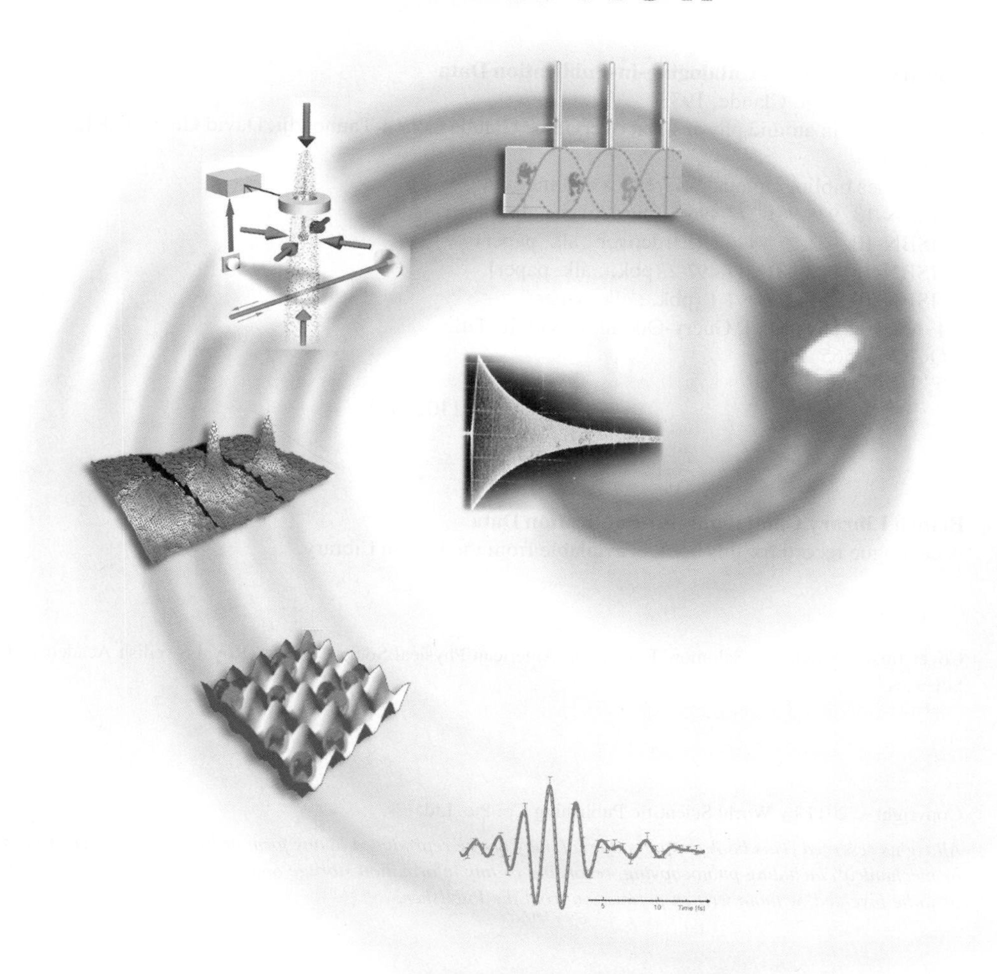

Claude Cohen-Tannoudji

Collège de France & Laboratoire Kastler Brossel, France

David Guéry-Odelin

Laboratoire Collisions Agrégats Réactivité, France

World Scientific

NEW JERSEY • LONDON • SINGAPORE • BEIJING • SHANGHAI • HONG KONG • TAIPEI • CHENNAI

Published by

World Scientific Publishing Co. Pte. Ltd.
5 Toh Tuck Link, Singapore 596224
USA office: 27 Warren Street, Suite 401-402, Hackensack, NJ 07601
UK office: 57 Shelton Street, Covent Garden, London WC2H 9HE

Library of Congress Cataloging-in-Publication Data
Cohen-Tannoudji, Claude, 1933–
Advances in atomic physics : an overview / Claude Cohen-Tannoudji, David Guéry-Odelin.
p. cm.
Includes bibliographical references and index.
ISBN-13 978-981-277-496-5 (hardcover : alk. paper)
ISBN-10 981-277-496-3 (hardcover : alk. paper)
ISBN-13 978-981-277-497-2 (pbk. : alk. paper)
ISBN-10 981-277-497-1 (pbk. : alk. paper)
1. Nuclear physics. I. Guéry-Odelin, David. II. Title.
QC776.C565 2011
539.7--dc23
2011028321

British Library Cataloguing-in-Publication Data
A catalogue record for this book is available from the British Library.

Cover Images Credit: C. Salomon, E. Cornell, American Physical Society and the Royal Swedish Academy of Sciences.

Printed in Singapore by Mainland Press Pte Ltd.

Foreword

I had the privilege of spending the academic year 1989–1990 as a visiting scientist in Claude Cohen-Tannoudji's research group at the École Normale Supérieure in Paris. Laser cooling was still in its early stages and the atmosphere at ENS was electric. Nevertheless, one of my great pleasures was to actually leave the laboratory in Rue Lhomond in order to take the short walk across the 5th Arrondissement to the Collège de France, where I attended Claude's weekly lectures on atomic physics. I recall thinking how fortunate I was to attend what were surely among the most clearly presented and thoughtfully constructed lectures ever given on the subject. The Collége de France requires that each year's lectures be on a fresh subject and I was frankly envious of my Parisian friends who could attend every year and learn first hand how Claude understood some new topic in atomic physics. This volume, by Claude Cohen-Tannoudji and his younger colleague David Guéry-Odelin, brings the spirit and scope of those legendary lectures to the printed page.

Even in a book as substantial as this, the authors must make choices about what to include. In the selection of topics, in the choice of historic experiments to illustrate those topics, and in the choice of theoretical viewpoints to explain the science, Cohen-Tannoudji and Guéry-Odelin display consummate good scientific taste. The topics tie together key themes in atomic physics, like coherence and control, and highlight not just the famous experiments but those that best teach the concepts. The choice of references to other texts, reviews, and the primary literature, also displays wonderful scientific taste. Most importantly, the theoretical treatments have that characteristic Cohen-Tannoudji quality of clarity grounded in physical intuition. A particularly appealing aspect of the theoretical discussion is the authors' commitment to providing multiple, alternative explanations of the same physical phenomenon. Such multiple physical models are invaluable for generating the insights that lead to new directions in research, as well as in simply enhancing our pleasure at understanding the physical world.

The scope of this volume is remarkable, spanning the history of 20th century and early 21st century atomic physics and covering a range of topics that encompass the most rapidly progressing areas of research today. From the Hanle effect in the early 20th century (a pre-quantum mechanics manifestation of atomic coherence) to the latest experiments on quantum degenerate Bose and Fermi gases as model systems for the study of many-body physics, *Advances in atomic physics: an overview* delivers even more than its title suggests.

For young students just beginning an education in atomic physics to experienced researchers like myself who have lived through many of the exciting and still ongoing developments recounted here, and for all those in between, this book presents an inviting feast. Bon appétit!

William D. Phillips
Gaithersburg, July 2011

Contents

Atom-photon interactions: a simple system for studying higher order effects 191

Exploring quantum interferences with few atoms and photons 405

Acknowledgments

We have been lucky to work in the atomic physics school created in France by Alfred Kastler and Jean Brossel and to benefit from their inspiring guidance.

We are extremely grateful for the invaluable help of several colleagues and friends who read carefully some parts of the manuscript and suggested many significant improvements: Marie-Anne Bouchiat, Yvan Castin, Frédéric Chevy, Roland Combescot, Jean Dalibard, Thierry Lahaye, Franck Laloë, Alfred Maquet, Guthrie Partridge, Christophe Salomon.

We would like also to acknowledge stimulating and fruitful discussions with Ennio Arimondo, Alain Aspect, François Bardou, François Biraben, Jean-Philippe Bouchaud, Maxime Dahan, Pierre Desbiolles, Jacques Dupont-Roc, Claude Fabre, Gilbert Grynberg, Serge Haroche, Paul Indelicato, Michèle Leduc, Pierre Lemonde, Christophe Mora, Jakob Reichel, Serge Reynaud, Chris Westbrook.

Part of the material contained in this book was already presented by one of us (CCT) during the lectures given at the Collège de France in Paris between 1973 and 2004.[1] We thank the Collège de France for this wonderful opportunity given to its professors. We also thank the City University of Hong Kong, and in particular Professors Michel Van Hove, Philippe Ciarlet and David Tong, who invited us to spend three two-week stays in CityU, during which part of the material of this book was presented in a set of lectures.

We also express our gratitude to Guthrie Partridge who edited the english of our manuscript and the "Institut Francilien de Recherches sur les Atomes Froids" (IFRAF) which provided financial help for this work.

We finally thank our colleagues all over the world with whom we enjoyed very stimulating discussions during scientific conferences and visits and who helped us to discover the exciting achievements of our discipline.

[1]The text of these lectures may be found (in French) at the following Web site: http://www.phys.ens.fr/cours/college-de-france/

Chapter 1

General introduction

Purpose of this book

The evolution of atomic physics during the last few decades has been spectacular. In the 1950's, many people believed that this discipline had already reached most of its objectives. The structure of atoms was well understood. Higher resolutions in atomic and molecular spectra were certainly possible, but no spectacular development was expected that could stimulate new ideas and open up new research directions. In fact, this pessimistic vision turned out to be completely wrong. Indeed, atomic physics has completely renewed itself during the last six decades, many times and in many directions. Here we mention a few of the developments that have occurred during this period in order to illustrate how impressive the renewal of this discipline has been: optical pumping methods, realization of laser light sources with unprecedented performances, nonlinear and time resolved spectroscopies, laser cooling and trapping, control of atomic systems at the single particle (electron, atom or photon) level, Bose-Einstein condensates and degenerate Fermi gases, cavity quantum electrodynamics, femtosecond and attosecond pulses, quantum information, detection of parity violation in atoms, etc. In parallel with these developments, interesting and fruitful connections have been established with many other fields of physics and chemistry, opening the way to new synergies and new possibilities of cross fertilizations.

It is clear that these spectacular advances of atomic, molecular and optical physics (AMO) require increasingly complex experimental and theoretical skills. In order to be competitive, a young researcher entering this field must often specialize in a narrow area of AMO. It is difficult for him or her to have a global view of the field, to understand how the evolution of ideas has allowed breakthroughs and opened totally new perspectives. The purpose of this book is to provide the reader with a better perception of the evolution of AMO by putting research performed at different times into perspective, and by trying to point out a few important and general concepts that can be frequently revisited from different points of view. Our hope is to give him or her a better confidence for exploring new problems where this global view of AMO can be fruitful.

Organization of the book

The choice of topics covered in this book is necessarily biased by our better knowledge of the fields in which we have worked. It is impossible in a single book to cover all the advances of atomic physics during six decades! This does not mean that we establish any hierarchy between these topics. We only hope that our general interpretation of the impressive progress of atomic physics also applies to the fields that we do not address.

The book starts (Chap. 2) with a review of the general background that is needed for reading the other chapters. It introduces the atomic and field variables, the different possible classical and quantum descriptions of atom-field interactions, the basic conservation laws, which are a useful guideline for understanding how one can manipulate atoms with light, the basic ingredients of the description of magnetic resonance, including the Rabi oscillation, the Ramsey fringes and the Einstein coefficients describing the absorption and emission rates.

The other chapters of the book are grouped in 8 parts that can be read more or less independently and whose contents will be now briefly analyzed. A detailed index at the end of the book shall help the reader to easily locate topics of interest. Despite its length, the bibliography cannot be expected to be complete and we apologize for any omission of important references.

Part 1 - Advances in spectroscopy

We begin with a review of optical pumping methods (Chap. 3), which completely renewed radiofrequency and microwave spectroscopy in the middle of the last century. Large population differences between the Zeeman sublevels of atoms can be obtained by optical pumping, allowing a very sensitive optical detection of magnetic resonance in both excited and ground atomic states. Atoms can also be prepared by these methods in linear superpositions of Zeeman sublevels, and we show that this leads to interesting applications described in Chap. 4. Optical spectroscopy was revolutionized by the advent of lasers. In Chap. 5, we analyze the properties of the fluorescence light emitted by a two-level atom excited by a resonant laser beam. Finally, in Chap. 6, we describe several high resolution spectroscopic methods that use nonlinear effects (saturated absorption, two-photon transitions), confinement (Lamb-Dicke effect), or other schemes (shelving method), to get very narrow optical lines.

The field covered in this part 1 is a case study that allows us to introduce several new theoretical approaches that have proven to be very useful for atomic physics, including optical Bloch equations, the dressed atom approach and the quantum jump description of dissipative processes. Moreover, these approaches also establish a link between the concepts of magnetic resonance and those of quantum optics.

Part 2 - Perturbations of atomic levels by light

The light used in optical pumping experiments for polarizing atoms is also a source of perturbations for the atoms and the fields. In Chap. 7, we show that atomic levels are shifted and broadened by light, and that in turn, the propagation of the light is modified. In fact, light shifts, which are a source of perturbations for high resolution spectroscopy, were found to give rise to new interesting applications when laser sources became available. We show that they can be used to trap atoms in potential wells with different shapes and dimensionalities and to detect photons in a cavity without destroying them. In Chap. 8, we study how an atom is perturbed by a high frequency excitation whose frequency ω is much higher than the atomic resonant frequencies. We give a simple semiclassical interpretation of these perturbations in terms of the vibration of the atomic electron in the high frequency incident wave.

Even in the absence of exciting light, the atom is perturbed by its interaction with the radiation field which is then in the vacuum state. These perturbations (Lamb shift, spin anomaly $g-2$) are called radiative corrections. We show that the approach followed in part 2 provides new physical insights into radiative corrections produced by comparing the perturbations due to a non-zero applied field to those obtained when the electromagnetic field is in its vacuum state.

Part 3 - Multiphoton processes

An atom can make a transition between two states by absorbing several photons. In Chap. 9, we describe these multiphoton transitions between two discrete states, first in the radiofrequency domain, where they were first discovered in experiments using optically pumped atoms, then in the optical domain with laser sources. Particular attention is given to two-photon stimulated Raman processes, where the atom absorbs one photon and emits another photon in a stimulated way. Recent important applications of these processes will be described. When the multiphoton transition connects the ground state to a state in the continuum, the process is called multiphoton ionization. In Chap. 10, we describe the main features of multiphoton ionization and we show that interesting new effects appear when the light intensity is so high that the laser electric field becomes comparable to the Coulomb field binding the atomic electron. We show that the interpretation of these effects led to dramatic advances in our ability to produce ultrashort pulses of UV and X radiation, in the attosecond range (1 as $= 10^{-18}$ s).

The research field covered in this part 3 is a new example showing how physical processes discovered in the early days of optical methods have progressively evolved to be now at the heart of an important frontier of laser physics, which will be described in more detail in part 8.

Part 4 - Control of atomic motion. Cooling and trapping

The exchange of linear momentum between atoms and photons give rise to radiative forces acting on atoms and allows a control of their motion. In Chap. 11, we first present a calculation and a physical interpretation of these radiative forces based on different approaches. As a result of the Doppler shift, the radiative forces exerted by the laser waves on the atom acquire a velocity dependence which can, in certain conditions, act as a friction force which reduces the atomic velocity. In Chap. 12, we describe the Doppler laser cooling mechanism for two-level atoms, which was the first cooling scheme proposed for trapped ions as well as for neutral atoms. When the measurement of the temperatures of laser cooled atoms became more precise at the end of the 1980's thanks to time-of-flight techniques, it was discovered that other cooling mechanisms were taking place, or could be implemented, that lead to temperatures much lower than those obtained with Doppler cooling. In Chap. 13, we describe two of these mechanisms: low intensity Sisyphus cooling based on a combination of optical pumping and light shifts that leads to a situation in which the moving atom runs up potential hills more frequently than down; subrecoil cooling, which allows atoms to be cooled below the single photon recoil momentum. Finally, in Chap. 14, we discuss the problem of trapping charged particles as well as neutral atoms, which allows a full control of the external degrees of freedom, both in momentum and position space i.e. in phase space.

The research field covered in part 4 shows how the better understanding of atom-photon interactions progressively acquired during the last decades has allowed the emergence of a new very active research field, ultracold atom physics.

Part 5 - Ultracold interactions and their control

The developments described in part 4 raise the question of the description of interactions between ultracold atoms. In Chap. 15, we analyze this problem from both microscopic and macroscopic points of views. We first introduce the scattering length, a parameter which describes most of the properties of elastic collisions between ultracold atoms. We also introduce a few regularized delta potentials that lead to simpler calculations and show how the idea of a mean field can be derived by studying how the propagation of an atom can be modified by its interaction with a background ultracold gas. One of the most important developments of atomic physics in the field of ultracold atoms has been the discovery that atom-atom interactions can be controlled by using so-called Feshbach resonances. In Chap. 16, we present a detailed description of these resonances, their physical interpretation, and their use for producing ultracold molecules.

With Feshbach resonances, one can say that atomic physics has reached a full control of atomic parameters: spin polarization with optical pumping, velocity with laser cooling, position with trapping, atom-atom interactions with Feshbach resonances.

Part 6 - Atomic interferometry. Entangled states

The concept of linear superpositions of states, leading to quantum interference effects, is central in quantum mechanics. Many examples may be found in atomic physics. The pioneering experiments using optical methods involved only internal atomic states and are described in part 2. In part 6, we describe quantum interference effects that also involve external degrees of freedom (of the center of mass) and which become easier to observe with the large de Broglie wavelengths obtained by laser cooling. In Chap. 17, we show that several well known experiments in wave optics, like Young's double slit interferences, can be extended to atomic de Broglie waves. Furthermore, by playing with the atomic internal variables, one can realize atomic interferometers, where the two paths of the interferometer differ, not only by the external variables of the center of mass, but also by the atomic internal variables. In Chap. 18, we describe these interferometers and their important applications. Atomic clocks with ultracold atoms are presently the most accurate time frequency standards, and atomic interferometers provide the most precise measurements of inertial fields.

Finally, in Chap. 19, we consider linear superpositions of states of two subsystems 1 and 2. When such linear superpositions cannot be written as a product of a state of system 1 by a state of system 2, the two systems are in an *entangled state* that exhibits quantum correlations, impossible to understand with the concepts of classical physics. Different types of entangled states can now be prepared and are playing an important role in discussions concerning the quantum non-separability (violation of Bell's inequalities), the measurement process, the decoherence due to the coupling with the environment and the possibility to use quantum correlations to transmit and process information.

The topics covered in part 6 show that atomic physics is an ideal playground for deepening the understanding of quantum concepts. Important examples of linear superpositions of states have been analyzed in a long series of experiments of increasing complexity performed over the last six decades. These have given rise to a wealth of applications.

Part 7 - Quantum gases

Quantum gases presently receive a lot of attention. In part 7, we give a brief historical review of the developments which led to the emergence of this field and a description of a few important results that have been obtained. In Chap. 20, we begin by analyzing what is meant by quantum gas by comparing three characteristic lengths: the thermal de Broglie wavelength λ_T of the atoms of the gas, the mean distance between atoms d and the range r_0 of the atom-atom interactions. We review the various quantum regimes that can appear depending on the relative values of these three characteristic lengths.

We briefly describe the pioneering work of Albert Einstein who predicted in 1924 the phenomenon of *Bose-Einstein condensation* (BEC) appearing in this regime. In Chap. 21, we then discuss the long series of theoretical and experimental works that led to the observation of BEC in a gas, 70 years after Einstein's prediction. In this chapter, we restrict ourselves to the mean-field description of Bose-Einstein condensates where each bosonic atom in the condensate is considered to evolve in the mean field produced by the other atoms. In Chap. 22, we introduce the Gross-Pitaevskii equation, which describes this situation and which explains quantitatively most of the static and dynamic properties of weakly interacting condensates at very low temperatures. In Chap. 23, we introduce the correlation functions of the atomic field operators in quantum gases and show that they allow a clear understanding of the coherence properties of condensates. Finally, in Chap. 24, we analyze from an experimental point of view another spectacular quantum macroscopic property of quantum fluids, their superfluidity. We discuss the frictionless propagation of a probe particle in the condensate, due to the impossibility for this particle to create an elementary excitation in the fluid when its velocity is lower than a certain threshold. We also show how quantized vortices appear in the condensate when the trap containing this condensate is rotated at a high enough speed.

The history of gaseous BEC is a beautiful example of a long term fundamental research endeavor that required the combination of several experimental and theoretical contributions to make a major scientific discovery possible.

Part 8 - A few frontiers of atomic physics

Another important feature of atomic physics, which is well illustrated by the example of quantum gases, is its ability to stimulate fruitful connections with other fields, such as condensed matter physics, particle physics or laser physics. A few examples of these connections are given in the next part devoted to the frontiers of atomic physics:

- Tests of fundamental theories
  Several examples of such tests using the high precision of atomic physics experiments are already given in the previous parts. In Chap. 25, we focus on the tests of fundamental symmetries. We show how it has been possible to observe for the first time a parity violation in atoms and how the results obtained in this way for the parameters of the standard model complement the information deduced from high energy experiments.
- Strongly interacting many-body systems
  In Chap. 26, we describe a few effects that appear in quantum gases when interactions are too strong to allow a mean-field description, and which present interesting connections with other similar effects (Josephson effect, superfluid- Mott insulator transition, BEC-BCS crossover).
- Extreme light
  In Chap. 27, we focus on examples that show how the interplay between basic and applied research has allowed the realization, in table-top experimental set-ups, of laser sources with unprecedented performances. We first describe how the understanding of ionization of atoms in intense laser fields, described in Chap. 10, has led to the realization of ultrashort laser pulses, in the attosecond range. These developments would not have been possible without the development of intense femtosecond pulses based on several techniques, such as mode locking or chirped pulsed amplification, that are briefly described.

Chapter 2

General background

2.1 Introduction

In this chapter, we review and comment upon a few basic results regarding atom-photon interactions that will be useful for the discussions presented in the subsequent chapters of this book.

In Sec. 2.2, we begin by listing various observables of two interacting systems: the atom and the electromagnetic field. For the atom, a clear distinction must be made between the external variables, which characterize the motion of the center of mass of the atom, and the internal variables, which describe the internal dynamics of the atom in its rest frame. With recently developed methods for cooling atoms to very low temperatures, situations can now be achieved where the de Broglie wavelength of the center of mass wavefunction becomes large compared to the other characteristic lengths (range of the atom-atom interactions, mean distance between atoms, laser wavelength). A quantum treatment of external variables is essential for interpreting the new situations that can be explored in these conditions. We also briefly review, in Sec. 2.2, the various possible semiclassical and quantum descriptions of the field variables.

Conservation laws in atom-photon interactions are another important topic and are discussed in Sec. 2.3. They are a useful guide for understanding the important advances of atomic physics of the last few decades made through our ability to manipulate the internal and external degrees of freedom of atoms. The conservation of angular momentum explains the transfer of angular momentum from polarized photons to the internal degrees of freedom of atoms and is at the basis of the optical methods (double resonance and optical pumping) that are described in detail in Chap. 3. More recently, it has been shown that special types of laser beams, called Laguerre-Gaussian beams, also carry an "orbital" angular momentum that can be transferred to the external degrees of freedom of ultracold atoms. The conservation of linear momentum in absorption-spontaneous emission or absorption-stimulated emission cycles explains how a laser beam can exert forces on an atom and modify its velocity. These radiative forces are now used for cooling and trapping atoms, and are at the origin of the most spectacular recent developments that use ultracold atoms to open new research fields and realize new states of matter, such as gaseous Bose-Einstein condensates.

The next two sections introduce, with simple examples, the two main types of time evolution that can be observed for an atom interacting with the electromagnetic field.

In Sec. 2.4, we first analyze the simple problem of a two-level atom interacting with a coherent monochromatic field. This allows the introduction of a basic quantity, the so-called *Rabi frequency*, that characterizes the strength of the interaction. The Rabi frequency was first introduced for a spin 1/2 interacting with a static magnetic field $\vec{B}_0$ and a circularly polarized RF field perpendicular to $\vec{B}_0$. At resonance, when the frequency ω of the RF field coincides with the Larmor precession frequency of the spin around $\vec{B}_0$, the spin oscillates between its two energy states $+1/2$ and $-1/2$ at the Rabi frequency Ω which is proportional to the product of the spin magnetic moment and the RF field amplitude. In this case, it is also possible to give a very simple geometric interpretation of transient effects induced by $\pi/2$ and π RF pulses. This geometric interpretation also establishes a clear connection between linear superpositions of the two spin states and the existence of a spin transverse magnetization, perpendicular to the static field. When monochromatic laser sources became available, it was soon realized that a great similarity exists between the behavior of spins interacting with RF fields and the behavior of atoms interacting with coherent monochromatic fields. All the basic concepts, first introduced in the RF domain (Rabi oscillation, transient effects, Bloch equations) were extended to the optical domain. In Sec. 2.4, we show that the Hamiltonian describing the dynamics of the system has the same structure in both cases. The existence of linear superpositions of states in the optical domain can be related to a non-vanishing value of the electric dipole moment for the corresponding atomic transitions. In the perturbative limit, we also study the transition amplitude between the two states of the two-level system induced by a weak RF or light pulse of duration τ. We show that the transition amplitude is important in a frequency interval on the order of $1/\tau$ around the resonance, which illustrates the time-frequency uncertainty relation. It is then straightforward to extend this calculation to a sequence of two coherent pulses with the same duration τ, separated by a time interval T. In this case, the transition amplitude exhibits narrow fringes with a period $2\pi/T$ inside the previous broad profile of width $1/\tau$. These fringes, named "Ramsey fringes" after the name of their inventor, Norman Ramsey, were first investigated in the RF domain. They have been extended to the optical domain and are now systematically used in all high precision spectroscopic and interferometric measurements, as this will be shown in Chap. 18. In Chap. 19, we will discuss also the extension of the concept of quantum interference to more general situations such as multi-particle quantum interference and entangled states.

In Sec. 2.5, we finally consider the case of a two-level atom interacting with a broadband field. We first use a semiclassical approach for which the field is described as a fluctuating classical field. We show that the evolution of the atom is no longer a coherent oscillation between the excited state e and the ground state g, but is instead an irreversible transition from one state to the other occurring with a certain rate. This treatment could be used to get the B_{eg} and B_{ge} coefficients introduced by Einstein for describing the absorption and stimulated emission rates of an atom interacting with a black body field. Next, we show that the same rates can be obtained from a quantum description that gives, in addition, the A coefficient which describes the spontaneous emission rate from e to g. The quantum treatment is here essential because the A coefficient cannot be derived from a semiclassical theory.

2.2 The two interacting systems: atom and field

2.2.1 *External and internal atomic variables*

The separation between these two types of variables is well known for a one-electron atom like hydrogen. Let $\vec{r}_e$ and $\vec{r}_N$ be the positions of the electron and the nucleus, with mass m_e and m_N, and momenta $\vec{p}_e$ and $\vec{p}_N$. We also introduce the new variables:

$$\begin{cases} \vec{R} = \dfrac{m_e \vec{r}_e + m_N \vec{r}_N}{M}, \\ \vec{r} = \vec{r}_e - \vec{r}_N, \end{cases} \qquad \begin{cases} \vec{P} = \vec{p}_e + \vec{p}_N, \\ \dfrac{\vec{p}}{m} = \dfrac{\vec{p}_e}{m_e} - \dfrac{\vec{p}_N}{m_N}, \end{cases} \tag{2.1}$$

where $M = m_e + m_N$ is the total mass and $m = m_e m_N / M$ the *reduced mass*. The Hamiltonian H of the two particles interacting by a potential V depending only on their relative position reads

$$H = \frac{\vec{p_e}^{\,2}}{2m_e} + \frac{\vec{p_N}^{\,2}}{2m_N} + V\left(\vec{r}_e - \vec{r}_N\right), \tag{2.2}$$

and can be rewritten with the center of mass coordinates in the form:

$$H = \frac{\vec{P}^{\,2}}{2M} + \frac{\vec{p}^{\,2}}{2m} + V\left(\vec{r}\right). \tag{2.3}$$

The new variables $\vec{R}$, $\vec{P}$, $\vec{r}$, $\vec{p}$ obey the commutation relations:

$$[r_i, p_j] = i\hbar\, \delta_{ij}, \qquad [R_i, P_j] = i\hbar \delta_{ij}, \qquad i, j = x, y, z. \tag{2.4}$$

The variables $\vec{R}$ and $\vec{P}$ are called the center of mass variables or *external* variables. The variables $\vec{r}$ and $\vec{p}$ are called *internal* variables. This separation can be easily extended to many-electron atoms where the interactions depend only on the relative positions.[1] The external variables are the variables of the center of mass and describe the motion of the atom as a whole. The internal variables describe the relative motion of the various constituents of the atom.

[1] See for example the appendix of [van Enk (1994)], which is also reprinted in [Allen *et al.* (2003)].

2.2.2 *Classical versus quantum treatments of atomic variables*

Internal variables must obviously be treated quantum mechanically because of the quantization of various atomic observables such as the internal energy or the angular momentum of the atom. A basis of states in the Hilbert space of internal variables can, for example, be labelled by a set of quantum numbers such as E (energy), J (total angular momentum), M_J (z-component of the total angular momentum).

For external variables, a classical treatment is often used when the de Broglie wavelength $\lambda_{\text{dB}} = h/Mv$ of the center of mass of the atom is very small compared to the wavelength of the field. At room temperature, λ_{dB} is on the order of a fraction of nanometer so that the atomic wave packet is very well localized in the light wave and moves like a classical particle.[2] Very often, the atomic mass M is considered as infinite and the atom is assumed to remain at rest at a given point.

Thanks to laser cooling techniques, temperatures can now be lowered to a few 10^{-9} K and λ_{dB} can reach values on the order of several microns. In this case, a quantum treatment of the center of mass dynamics is essential for understanding spectacular effects such as interference of atomic de Broglie waves or Bose-Einstein condensation of atomic gases. We will make extensive use of such a quantum treatment in several chapters of this book.

2.2.3 *Classical description of field variables*

In a great number of cases where the atom is interacting with a classical field, it is justified to use a semiclassical description of atom-field interactions. This field can be either a coherent monochromatic field, or a broadband incoherent field.

The first case was originally addressed in the context of magnetic resonance experiments, which played a crucial role in the renaissance of atomic physics after World War II. Later on, it was revisited with the experimental studies on the behavior of atoms excited by a coherent single mode laser field. For studying the evolution of atoms under the influence of monochromatic excitation and relaxation processes, one usually uses master equations for the atomic density matrix, called Bloch equations in the RF domain, and *optical Bloch equations* in the optical domain. In this book, we will show how these equations are important for understanding resonance fluorescence and the radiative forces exerted by laser beams on atoms.

The second case, atom interactions with broadband incoherent fields, corresponds to usual thermal or discharge light sources. The field is conveniently described as a fluctuating field described by its time correlation functions. This description is similar to that used for analyzing relaxation processes induced by a random perturbation. It allows for a simple introduction of the notion of transition

[2] Because of the wave packet spreading, one could think that this approximation does not hold for all times. In fact, the relevant characteristic length of an atomic wave packet is not its spatial extent, but the coherence length, which is a constant of the motion for a free particle. Furthermore, this coherence length decreases in time in the presence of dissipative processes like photon scattering. We will discuss these problems in more details in Chap. 17.

rates that characterize an irreversible behavior of the atom, in sharp contrast to the reversible Rabi oscillation that occurs when it interacts with a coherent monochromatic field.

These semiclassical approaches are sufficient for understanding most experiments, but cannot explain basic processes like spontaneous emission, which is due to the coupling of the atom with the quantum vacuum field, or more subtle effects like photon correlations observed on the fluorescence light emitted by a single atom. A brief summary of the basic ingredients of field quantization is presented in the next subsection.

2.2.4 *Quantum description of field variables*

We first recall the definition of a field mode for a classical free field (without interactions with particles). Suppose that this field is enclosed in a cubic box of volume L^3 with periodic boundary conditions. The field can be expanded on a set of plane transverse waves with wave vector $\vec{k}_i$ and polarization $\vec{\varepsilon}_i$ perpendicular to $\vec{k}_i$. The periodic boundary conditions imply that the three components of $\vec{k}_i$ are multiple integers of $2\pi/L$. In the following, we use the index i for characterizing the set $\vec{k}_i, \vec{\varepsilon}_i$. The coefficients α_i of the expansion of the field in this basis of plane waves are called *normal variables* because they evolve independently from each other: $\alpha_i(t) = \alpha_i(0)\exp(-i\omega_i t)$ where $\omega_i = ck_i$. The Hamiltonian of the classical field can be shown to be a sum of independent harmonic oscillators, one for each mode i with frequency $\omega_i = ck_i$.

The field quantization is achieved by replacing the normal variables α_i and α_i^* of the harmonic oscillator i by the annihilation and creation operators $\hat{a}_i$ and $\hat{a}_i^\dagger$ of the corresponding quantum harmonic oscillator.[3] These obey the commutation relations:

$$\left[\hat{a}_i, \hat{a}_j^\dagger\right] = \delta_{ij}. \tag{2.5}$$

For example, the operator describing the vector potential can be shown to be given by [Cohen-Tannoudji *et al.* (1989)]:

$$\hat{\vec{A}}(\vec{r}) = \sum_i \sqrt{\frac{\hbar}{2\varepsilon_0\omega_i L^3}} \left[\hat{a}_i\vec{\varepsilon}_i \exp\left(i\vec{k}_i\cdot\vec{r}\right) + \hat{a}_i^\dagger\vec{\varepsilon}_i^{\,*} \exp\left(-i\vec{k}_i\cdot\vec{r}\right)\right]. \tag{2.6}$$

The coefficient $(\hbar/2\varepsilon_0\omega_i L^3)^{1/2}$ is introduced in Eq. (2.6) in order to have a simple expression for the radiation field Hamiltonian:

$$\hat{H}_R = \sum_i \frac{\hbar\omega_i}{2}\left(\hat{a}_i^\dagger\hat{a}_i + \hat{a}_i\hat{a}_i^\dagger\right) = \sum_i \hbar\omega_i\left(\hat{a}_i^\dagger\hat{a}_i + \frac{1}{2}\right). \tag{2.7}$$

[3] We insert here a symbol hat on $\hat{a}_i$ and $\hat{a}_i^\dagger$ to make it clear that these quantities are operators and not c-numbers. However, very often, when there is no ambiguity, we will skip the hat in order to keep the notation simpler.

The basis of plane waves $\vec{\varphi}_i(\vec{r}) = L^{-3/2}\vec{\varepsilon}_i \exp\left(-i\vec{k}_i \cdot \vec{r}\right)$ is an orthonormal basis, and is well adapted for describing the momentum of the radiation field:

$$\hat{\vec{P}}_R = \sum_i \hbar\vec{k}_i \hat{a}_i^\dagger \hat{a}_i. \tag{2.8}$$

Each photon of mode i has a well defined energy $\hbar\omega_i$ and momentum $\hbar\vec{k}_i$. By introducing orthonormal linear combinations of plane waves $\vec{\varphi}_i(\vec{r})$ for each value of ω_i, one can construct another orthonormal basis of monochromatic waves $\vec{\chi}_n(\vec{r})$

$$\vec{\chi}_n(\vec{r}) = \sum_{i \,\text{with}\, \omega_i=\omega_n} U_{ni}\, \vec{\varphi}_i(\vec{r}) \tag{2.9}$$

more adapted to other physical properties of the field, such as the angular momentum. From Eqs. (2.9) and (2.6) and using the unitarity of the matrix U_{ni}, one can derive the mode expansion of the field operators in the new basis:

$$\hat{\vec{A}}(\vec{r}) = \sum_n \sqrt{\frac{\hbar}{2\varepsilon_0\omega_n}} \left[\vec{\chi}_n(\vec{r})\hat{b}_n + \vec{\chi}_n^*(\vec{r})\hat{b}_n^\dagger\right], \tag{2.10}$$

where $\hat{b}_n$ (resp. $\hat{b}_n^\dagger$), which is a linear combination of the $\hat{a}_i$'s (resp. $\hat{a}_i^\dagger$'s), destroys (resp. creates) a photon in the mode $\vec{\chi}_n(\vec{r})$. Each photon of this mode has a well defined energy $\hbar\omega_n$ and a well defined value of another physical quantity that is diagonal in the new basis.

A particularly important state of the field is the vacuum state, in which all modes are empty in all possible bases. An atom in an excited state is coupled to the vacuum state. It is this coupling that explains how it can spontaneously emit a photon. This process cannot be explained with a semiclassical approach. One can also show that in the vacuum state, all field operators have a zero average value and a non-zero variance, the same as the ground state of an harmonic oscillator for which the average value but not the variance of the position operator $\hat{X}$ vanishes. The vacuum state thus contains a fluctuating field whose power spectral density can be shown to be equal to $\hbar\omega_i/2$ per mode i. These fluctuations are called *vacuum fluctuations* and can be invoked for explaining radiative corrections such as the Lamb shift whose discovery and interpretation have played a crucial role in the development of quantum electrodynamics. We will return to these problems in Chap. 8.

2.2.5 *Atom-field interaction Hamiltonian in the long wavelength approximation*

For the discussion of conservation laws presented in the next section, it will be useful to recall the structure of the atom-field interaction Hamiltonian H_I. More detailed presentations can be found in [Cohen-Tannoudji *et al.* (1989)] and in the Appendix of [Cohen-Tannoudji *et al.* (1992b)].

In the Coulomb gauge, H_I is given by:

$$H_I = -\sum_j \frac{q_j}{m_j} \vec{p}_j \cdot \vec{A}(\vec{r}_j) + \sum_j \frac{q_j^2}{2m_j} \vec{A}^2(\vec{r}_j), \tag{2.11}$$

where q_j, m_j, $\vec{r}_j$, $\vec{p}_j$ are the charge, mass, position and momentum of particle j and $\vec{A}$ is the vector potential of the field.[4] If we consider single photon transitions, the last term of Eq. (2.11), quadratic in $\vec{A}$, can be ignored. For a neutral one electron atom, the index j can take only two values $j = 1, 2$, and we have $q_1 = -q_2 = q$, so that:

$$H_I \simeq -\frac{q_1}{m_1} \vec{p}_1 \cdot \vec{A}(\vec{r}_1) - \frac{q_2}{m_2} \vec{p}_2 \cdot \vec{A}(\vec{r}_2) = -q \left[\frac{\vec{p}_1}{m_1} \cdot \vec{A}(\vec{r}_1) - \frac{\vec{p}_2}{m_2} \cdot \vec{A}(\vec{r}_2) \right]. \tag{2.12}$$

The atom-field interactions considered in this book correspond to situations where the wavelength of the field, λ_L, is very large compared to the size of the atom, a. The long wavelength approximation consists of neglecting the variation of the field over distances on the order of a. We can thus replace the vector potential at $\vec{r}_j$ by its value taken at the center of mass position $\vec{R}$:

$$\vec{A}(\vec{r}_j) = \vec{A}(\vec{R} + \vec{r}_j - \vec{R}) \simeq \vec{A}(\vec{R}). \tag{2.13}$$

Inserting Eq. (2.13) into Eq. (2.12) leads to:

$$H_I \simeq -q\vec{A}(\vec{R}) \cdot \left[\frac{\vec{p}_1}{m_1} - \frac{\vec{p}_2}{m_2} \right] = -\frac{q}{m} \vec{A}(\vec{R}) \cdot \vec{p}, \tag{2.14}$$

where we have used the definitions (2.1) of the internal momentum $\vec{p}$ and of the reduced mass m. It thus appears that the external variables appear only in the vector potential, whereas the internal variables appear only as an internal operator $\vec{p}$ multiplying the vector potential $\vec{A}(\vec{R})$ taken at the center of mass position. This simple result can be readily extended to a multi-electron neutral atom.[5]

Let us finally mention that, in the long wavelength approximation, one can perform a unitary transformation on the atom-field Hamiltonian. This leads to a more familiar form of H_I:

$$H_I \simeq -\vec{D} \cdot \vec{E}(\vec{R}) \tag{2.15}$$

describing the interaction of the electric dipole moment $\vec{D}$ of the atom (which is an internal operator) with the electric field $\vec{E}(\vec{R})$ taken at the center of mass position and given by an expansion similar to Eq. (2.6) or (2.10) (see [Cohen-Tannoudji *et al.* (1989)]).

Remark

When the coherence length of the center of mass wave packet is very large, a mistake that is frequently made is to think that the electric dipole approximation for the atom-field interaction Hamiltonian is no longer valid. This approximation

[4] There are also spin-dependent terms in H_I that can be analyzed in the same way as those discussed here.

[5] See for example [van Enk (1994)].

is based on the smallness of the distance between the electron and the nucleus compared to the wavelength λ_L of the light field. It involves the internal variable $\vec{r} = \vec{r}_e - \vec{r}_N$, not the external variable $\vec{R}$. The center of mass can be delocalized in a large volume, with a non-zero probability amplitude to be at two distant points. For each of these possible positions of the center of mass, however, the size of the atom characterized by the electron-nucleus distance r is very small compared to λ_L, so that the electric dipole approximation is always valid.

2.2.6 *Elementary interaction processes*

The interaction Hamiltonian (2.14) describes elementary processes in which the atom goes from the initial internal and external states $|\psi_{\text{in}}^{\text{int}}\rangle \otimes |\psi_{\text{in}}^{\text{ext}}\rangle$ to the final internal and external states $|\psi_{\text{fin}}^{\text{int}}\rangle \otimes |\psi_{\text{fin}}^{\text{ext}}\rangle$ by absorbing a photon i, with an amplitude proportional to:

$$\langle\psi_{\text{fin}}^{\text{int}}|\, \vec{\varepsilon}_i \cdot \vec{p}\, |\psi_{\text{in}}^{\text{int}}\rangle \, \langle\psi_{\text{fin}}^{\text{ext}}|\, \exp(i\, \vec{k}_i \cdot \vec{R}\,)\, |\psi_{\text{in}}^{\text{ext}}\rangle \, \langle n_i - 1|\, \hat{a}_i \,|n_i\rangle \tag{2.16}$$

where n_i is the initial number of photons in the mode i and we have taken the expansion of $\vec{A}$ in plane waves. A similar expression could be obtained by using the expansion of $\vec{A}$ in the basis $\vec{\chi}_m$, where the central matrix element of (2.16) would be replaced by $\langle\psi_{\text{fin}}^{\text{ext}}|\, \vec{\chi}_m(\vec{R}\,)\, |\psi_{\text{in}}^{\text{ext}}\rangle$. The matrix element involving the field operator $\hat{a}_i$ is equal to $\sqrt{n_i}$, so that the probability of absorbing a photon is proportional to the number n_i of the incident photons.

The amplitude of an emission process starting from the same initial atomic states and with the same number n_i of photons is given by an expression similar to (2.16), where the atomic operators $\vec{\varepsilon}_i \cdot \vec{p}$ and $\exp(i\, \vec{k}_i \cdot \vec{R}\,)$ are replaced by their hermitian conjugates and where the field matrix element is replaced by $\langle n_i + 1|\, \hat{a}_i^\dagger \,|n_i\rangle = \sqrt{n_i + 1}$. The probability of the emission process is thus proportional to $n_i + 1$. If $n_i = 0$, the process corresponds to the spontaneous emission of a photon in the mode i which is initially empty. If $n_i \neq 0$, we have, in addition, the stimulated emission of a photon in the mode i with a probability proportional to the number n_i of incident photons stimulating this emission.

We turn now to the analysis of the two atomic matrix elements appearing in (2.16) and show how it is possible to deduce from them the basic conservation laws and selection rules of atomic transitions.

2.3 Basic conservation laws

2.3.1 *Conservation of the total linear momentum*

(i) *Case of free atoms*

We first consider an atom moving freely. Its center of mass is not submitted to an external potential, and one can take plane waves with momenta $\vec{p}_{\text{in}}$ and $\vec{p}_{\text{fin}}$ and with energies $p_{\text{in}}^2/2M$ and $p_{\text{fin}}^2/2M$ for the initial and final states of the center of mass. Consider an atom, initially in a lower internal state g, with a center of mass momentum $\vec{p}_{\text{in}}$, and suppose that this atom absorbs a photon with energy $\hbar\omega$ and momentum $\hbar\vec{k}$. The external atomic matrix element of (2.16) differs from zero only if the final center of mass momentum obeys:

$$\vec{p}_{\text{fin}} = \vec{p}_{\text{in}} + \hbar\vec{k}. \tag{2.17}$$

This equation expresses the conservation of the total linear momentum.

The total energy of the initial state is equal to:

$$E_{\text{in}} = \hbar\omega + E_g + \frac{p_{\text{in}}^2}{2m} \tag{2.18}$$

whereas the total energy of the final state is

$$E_{\text{fin}} = E_e + \frac{p_{\text{fin}}^2}{2m}. \tag{2.19}$$

Because of the conservation of total energy, these two energies must be equal. This implies, using Eq. (2.17) and taking $E_e - E_g = \hbar\omega_0$:

$$\hbar\omega = \hbar\omega_0 + \frac{\hbar\vec{k}\cdot\vec{p}_{\text{in}}}{M} + \frac{\hbar^2k^2}{2M}. \tag{2.20}$$

This can be also written as

$$\omega = \omega_0 + \vec{k}\cdot\vec{v}_{\text{in}} + \frac{E_{\text{rec}}}{\hbar}, \tag{2.21}$$

where $\vec{v}_{\text{in}} = \vec{p}_{\text{in}}/M$ is the initial velocity of the atomic center of mass, and where

$$E_{\text{rec}} = \frac{\hbar^2k^2}{2M} = \hbar\omega_R \tag{2.22}$$

is the recoil kinetic energy of an atom initially at rest after the absorption of a photon. This simple calculation shows that in a resonant absorption process the frequency ω of the absorbed photon differs from the atomic frequency ω_0 by the *Doppler shift*[6] $\vec{k}\cdot\vec{v}_{\text{in}}$ and the *recoil shift* $\omega_R = E_{\text{rec}}/\hbar$.

If we consider an ensemble of atoms in thermal equilibrium at the temperature T, their velocities $\vec{v}_{\text{in}}$ are distributed according to the Maxwell-Boltzmann law. This gives rise to a distribution of the Doppler shifts having a Gaussian shape of width $\Delta\omega_D \simeq k_L(k_BT/M)^{1/2}$. The absorption curve therefore has the shape of a Gaussian of width $\Delta\omega_D$ centered at $\omega_0 + \omega_R$ (see right part of Fig. 2.1).

A similar calculation can be made for the emission process in which an atom, initially in an internal excited state e and with center of mass linear momentum $\vec{p}_{\text{in}}$, goes to the lower state g by emitting a photon with energy $\hbar\omega$ and momentum $\hbar\vec{k}$. The equation corresponding to Eq. (2.21) is now replaced by:

$$\omega = \omega_0 + \vec{k}\cdot\vec{v}_{\text{in}} - \frac{E_{\text{rec}}}{\hbar}. \tag{2.23}$$

[6]In a relativistic treatment, there is also a second order correction proportional to v^2/c^2 where c is the speed of light. For thermal atoms, this contribution is negligible.

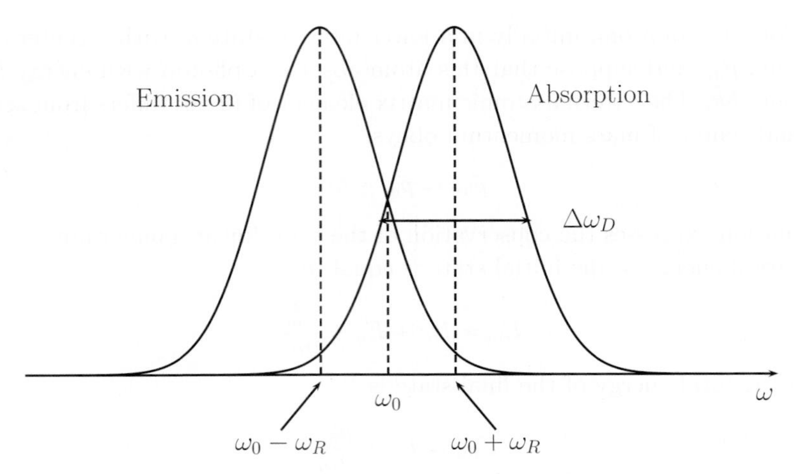

Fig. 2.1 Right: absorption curve centered at $\omega + \omega_R$ with a width $\Delta\omega_D$. Left: emission curve centered at $\omega - \omega_R$ with a width $\Delta\omega_D$.

The change of sign of the recoil shift with respect to (2.21) is easy to understand. Suppose that the atom is initially at rest so that there is no Doppler shift. In the absorption process, ω must be larger than ω_0 because the energy $\hbar\omega$ of the absorbed photon must not only increase the internal energy of the atom by an amount $\hbar\omega_0$, it must also communicate the recoil energy to the center of mass. In the emission process the internal energy $\hbar\omega_0$ lost by the atom is transformed into the energy $\hbar\omega$ of the emitted photon plus the recoil energy, leaving ω smaller than ω_0. The absorption and the emission curves are therefore two Gaussian curves with the same width $\Delta\omega_D$ centered at $\omega_0 + \omega_R$ and $\omega_0 - \omega_R$, respectively (see Fig. 2.1).

The recoil shift varies as k^2, whereas the Doppler width varies as k. The importance of the recoil shift therefore increases with the frequency of the transition. In the optical domain and at room temperatures, ω_R is generally very small compared to $\Delta\omega_D$, so that the two absorption and emission curves of Fig. 2.1 have a very large overlap and are not easily resolved.[7] In the gamma ray domain, the situation is reversed: the two curves are separated by an interval larger than their width, such that a gamma ray emitted by a nucleus in e cannot be absorbed by a similar nucleus in g. This prediction is valid for free atoms, though important changes occur when the atom is confined (see below).

In the previous discussion, the exchange of momentum between the atom and the field appears as a "bad" process that limits the resolution of spectroscopic measurements through the Doppler broadening and the recoil shift to which it gives rise. Several ingenious methods have been invented to circumvent the Doppler

[7]Even if the Doppler width is eliminated by nonlinear spectroscopy methods, which will be described in Chap. 6, the broadening of the transition by the natural width Γ of e due to spontaneous emission remains. For most optical transitions, Γ is much larger than ω_R so that the recoil doublet of Fig. 2.1 is not resolved except for very weak transitions (see Sec. 6.2.3 of Chap. 6).

broadening (without changing the atomic velocities). As it often happens in physics, a limiting process can be turned into an advantage. These exchanges of momentum can be used to exert large radiative forces on atoms that reduce their velocity spread. This effectively amounts to cooling the atoms and leads to a completely negligible Doppler broadening. This so-called *laser cooling*[8] method has given rise to a wealth of spectacular developments that will be described in Chaps. 13, 14, 21, and 26.

(ii) *Case of atoms trapped in an external potential*

The external potential is supposed to be independent of the internal state of the atom. We consider a process where the absorption of a photon of momentum $\hbar\vec{k}$ brings the atom from g to e and transfers its center of mass from the initial bound state $\psi_{\rm in}^{\rm ext}(\vec{R}\,)$ to the final one $\psi_{\rm fin}^{\rm ext}(\vec{R}\,)$. According to Eq. (2.16), the amplitude of this process is proportional to:

$$\left\langle \psi_{\rm fin}^{\rm ext} \right| \exp(i\vec{k}\cdot\vec{R}\,) \left| \psi_{\rm in}^{\rm ext} \right\rangle . \tag{2.24}$$

The operator $\exp(i\vec{k}\cdot\vec{R}\,)$ is a translation operator in momentum space by an amount $\hbar\vec{k}$. Equation (2.24) thus expresses that the amplitude to go from a bound initial state to a bound final state is simply the overlap of the final bound wavefunction with the initial bound wavefunction translated by an amount $\hbar\vec{k}$ in momentum space. If the center of mass is localized in the binding potential over distances small compared to the wavelength $\lambda_L = 2\pi/k$, the width of the wavefunctions in momentum space is much larger than $\hbar k$. The translated state $\exp(i\vec{k}\cdot\vec{R}\,)\,|\psi_{\rm in}^{\rm ext}\rangle$ is then nearly equal to $|\psi_{\rm in}^{\rm ext}\rangle$ and the amplitude (2.24) reduces to $\langle\psi_{\rm fin}^{\rm ext} \mid \psi_{\rm in}^{\rm ext}\rangle$, which is very small if the initial and final bound states are not the same. The most important transitions are therefore the zero-phonon lines i.e. those for which the initial and final states have the same vibrational quantum number. Physically, this is due to the fact that, if the binding is strong enough, the momentum of the photon, $\hbar k$, will be imparted to the whole system creating the external potential. The corresponding recoil kinetic energy is negligible for this macroscopic system. Such a suppression of the recoil shift lies at the heart of the Lamb-Dicke and Mössbauer effects that will be described in more detail in Chap. 6.

2.3.2 *Conservation of the total angular momentum*

(i) *Selection rules for the internal angular momentum*

Consider first a situation in which the external variables can be described classically with the center of mass at rest at $\vec{r} = \vec{0}$. We can replace the operator $\vec{R}$ in Eq. (2.14) by the c-number vector $\vec{0}$. The interaction Hamiltonian H_I acts only on internal variables through the internal operator $\vec{\varepsilon}\cdot\vec{p}$. According to the Wigner-Eckart theorem, the matrix elements of all vector operators between e and g are proportional. We can thus consider here the matrix elements of $\vec{\varepsilon}\cdot\vec{D}$, where $\vec{D}$ is the electric dipole

[8] See Chap. 12.

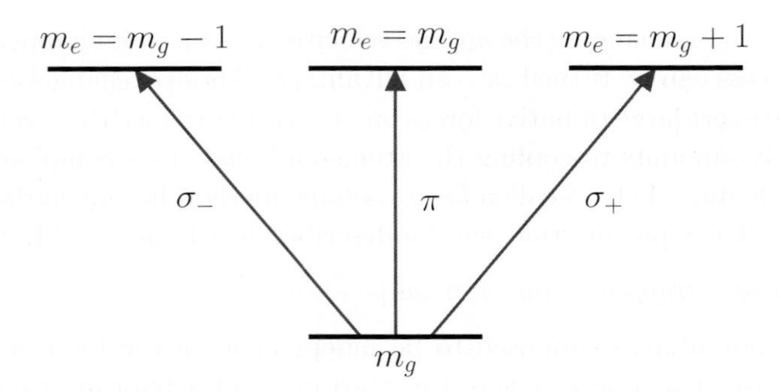

Fig. 2.2 Polarization selection rules for an electric dipole transition. The magnetic quantum number m changes by $+1$ for a σ_+-polarized excitation, by -1 for a σ_--polarized excitation and by 0 for a π-polarized excitation.

moment of the atom, and examine under what conditions the matrix elements of this operator are different from zero. First, suppose that the atomic transition connects a ground state g with an orbital angular momentum $\ell_g = 0$ to an excited state e with an orbital angular momentum $\ell_e = 1$. If the polarization of the field is σ_+, we have $\vec{\varepsilon} \propto \vec{\varepsilon}_x + i\vec{\varepsilon}_y$, so that $\vec{\varepsilon} \cdot \vec{D} \propto D_x + iD_y = q(x + iy) = qr \sin\theta \exp(i\varphi)$, where x, y, z are the coordinates of the relative particle, and θ and φ are its polar angles. Acting on the ground state $m_g = 0$ which has no φ-dependence, H_I thus creates an excited state wavefunction varying as $\exp(i\varphi)$, i.e. an excited sublevel $m_e = +1$. A similar calculation shows that a σ_--polarized excitation leads to a $m_e = -1$ excited sublevel. Finally, for a π-polarization, we have $\vec{\varepsilon} = \vec{\varepsilon}_z$, so that H_I contains the component $D_z = qz$ of $\vec{D}$, which has no φ-dependence and can excite only the $m_e = 0$ sublevel. For a $\ell_g = 0 \leftrightarrow \ell_e = 1$ transition, we therefore have the polarization selection rules

$$\sigma_+ \to m_e - m_g = +1, \qquad \sigma_- \to m_e - m_g = -1, \qquad \pi \to m_e - m_g = 0. \quad (2.25)$$

When the spins of the electrons and of the nucleus are taken into account, the internal atomic states are labelled by the total angular momentum quantum numbers (J, M). Using the Wigner-Eckart theorem and the fact that $D_x \pm iD_y$ and D_z are the spherical components of a vector operator, one can readily extend the result (2.25) and obtain the polarization selection rules sketched in Fig. 2.2. In the next chapter, the possibility to selectively excite (or detect) a particular component of an optical transition will be shown to play an important role in the manipulation of the internal degrees of freedom of atoms.

(ii) *Selection rules for the external angular momentum*

Is it possible to use certain types of laser beams to prepare external states with well-defined angular momentum for an atom? This is an interesting question since ultracold atomic gases can contain *quantized vortices* that can be described to a good approximation as a macroscopic number of atoms with the same wavefunction that has a well-defined angular momentum $m\hbar$, with $m = \pm 1, \pm 2, \ldots$ with respect to a given axis. The center of mass is delocalized over a large volume and rotates around an axis. This rotation of the center of mass adds to the internal rotation of the atom described by its internal angular momentum. We know how to control the internal angular momentum by choosing the light polarization. Can one choose a spatial dependence of the light field such that it can create quantized vortices in an ultracold atomic gas?

In this perspective, the most convenient laser beams seem to be the so-called *Laguerre-Gaussian* (LG) beams. These beams are paraxial beams propagating along the z-axis, with a wave vector k and with an azimuthal phase dependence $\exp(im\varphi)$ with $m = \pm 1, \pm 2, \ldots$ This gives rise to a non-zero angular momentum of the light beam along its direction of propagation. The phase singularity on the beam axis results in a zero intensity on this axis. The intensity profile is cylindrically symmetrical around this axis and has an annular shape. The LG beams are characterized by three quantum numbers: the wave number k, the number p of radial nodes of the mode function and the integer m characterizing its azimuthal phase dependence. They can be generated from usual Gaussian beams by various techniques. One frequently used possibility consists of using the diffraction of a Gaussian beam from numerically computed holograms. More details about LG beams and the various types of angular momentum that can be considered for a light beam can be found in [Allen *et al.* (2003)] which contains a collection of reprints on these topics.

The LG modes define a possible basis set for the expansion of the vector potential, which is convenient for studying the exchange of angular momentum between the light field and the atomic external degrees of freedom. We suppose that the light polarization $\vec{\varepsilon}$ is uniform in the beam which results in a uniform transfer of internal angular momentum. We thus use the mode functions $\vec{\varepsilon} F_{k,p,m}(\vec{r})$ of a LG beam k, p, m for the mode functions $\vec{\chi}_n(\vec{r})$ in Eq. (2.10). If we replace $\vec{A}(\vec{R})$ by this expansion in the interaction Hamiltonian (2.14), we find that H_I describes elementary processes in which a photon is absorbed under the effect of $\hat{b}_{k,p,m}$, the internal state is changed from m_g to m_e under the effect of the internal operator $\vec{\varepsilon} \cdot \vec{p}/m$ and the initial external state $\psi_{\text{in}}^{\text{ext}}(\vec{R})$ is transformed into a new external state given by

$$F_{k,p,m}(\vec{R})\, \psi_{\text{in}}^{\text{ext}}(\vec{R}). \tag{2.26}$$

This equation shows that the phase of the mode $F_{k,p,m}$ is imprinted on the initial external wave function, $\psi_{\text{in}}^{\text{ext}}(\vec{R})$ when a photon is absorbed.

Suppose that the initial external state has a zero angular momentum with respect to the z-axis. The wavefunction $\psi_{\text{in}}^{\text{ext}}(\vec{R})$ thus does not depend on the azimuthal angle φ. The absorption of a LG photon k, p, m multiplies it by $F_{k,p,m}(\vec{R})$, as shown in (2.26), so that it acquires an azimuthal phase dependence $\exp(im\varphi)$. Since the external angular momentum along the z-axis is given by the operator $L_z = (\hbar/i)\partial/\partial\varphi$, one concludes that the absorption of a LG photon k, p, m confers an angular momentum $m\hbar$ to the atom. This method has been recently used for creating a vortex in a trapped Bose-Einstein condensate [Andersen *et al.* (2006)].[9] The efficiency with which the final rotating state is created in the trapped condensate depends, of course, on the mode matching between the initial and final external wavefunctions in the trap and the mode function of the LG beam (overlap integral of (2.26) with the final rotating wavefunction). Since the waist of the LG beam is on the order of a few tens of microns, this clearly shows that the external wavefunctions must have a very large spatial extension, corresponding to ultra low temperatures.

Remark

What happens if the initial external wavefunction is a very well localized wave packet? The transformation of this wavefunction following the absorption of a LG photon is still given by Eq. (2.26), but the azimuthal dependence is no longer given by $\exp(im\varphi)$ because $\psi_{\text{in}}^{\text{ext}}(\vec{R})$ has a very well-localized azimuthal phase. The angular momentum of (2.26) is then no longer well-defined. Using the commutator of $L_z = (\hbar/i)\partial/\partial\varphi$ with $\exp(im\varphi)$, $[L_z, \exp(im\phi)] = m\,\hbar$, one can easily show that though the average value of L_z increases by $m\hbar$ after the absorption of the LG photon, this absorption does not lead to a well defined angular momentum state. The mean transfer $m\hbar$ of angular momentum can be understood in the following way. The combination of the $\exp(ikz)$ dependence along the z-axis of a LG beam with the $\exp(im\varphi)$ azimuthal phase dependence gives rise to helical phase fronts, resulting in a Poynting vector that is no longer parallel to the direction of propagation of the beam, but instead has an added transverse component. It is the transfer of momentum from this transverse component of the Poynting vector that increases the average value of the external atomic angular momentum.

2.4 Two-level atom interacting with a coherent monochromatic field. The Rabi oscillation

2.4.1 *A simple case: magnetic resonance of a spin 1/2*

We consider a spin $S = 1/2$ in a static field $\vec{B}_0$ parallel to the z-axis. Its magnetic moment $\vec{M}$ is proportional to $\vec{S}$: $\vec{M} = \gamma\vec{S}$ where γ is the gyromagnetic ratio. The interaction with $\vec{B}_0$ is described by the Hamiltonian

$$H = -\vec{M} \cdot \vec{B}_0 = \omega_0 S_z \tag{2.27}$$

where $\omega_0 = -\gamma B_0$ is the Larmor frequency that describes the precession of the spin around $\vec{B}_0$. The energies of the two spin states $\pm 1/2$ are equal to $E_\pm = \pm\hbar\omega_0/2$. In addition, the spin interacts with a radiofrequency field $2\vec{B}_1 \cos\omega t$ parallel to

[9] The transfer of angular momentum uses a slightly different scheme. Instead of transferring the atom to an excited internal state, which would decay radiatively and result in a destruction of the spatial coherence of the external wavefunction, one uses a stimulated Raman process induced by two beams, one Gaussian beam, which does not carry angular momentum, and one LG beam, which does. The atom ends in a stable internal ground state while gaining an external angular momentum $m\hbar$.

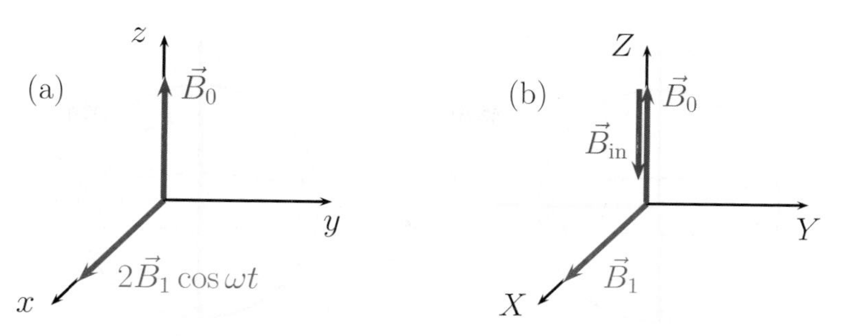

Fig. 2.3 (a) Static field $\vec{B}_0$ and radiofrequency field $2\vec{B}_1 \cos \omega t$ with which the spin 1/2 interacts. (b) In the rotating frame $OXYZ$, the spin interacts with two static fields, $\vec{B}_0 - \vec{B}_{\text{in}}$ aligned along the Z-axis and $\vec{B}_1$ aligned along the X-axis.

the x-axis and oscillating at a frequency ω close to ω_0 (see Fig. 2.3(a)) [Abragam (1961)].

The radiofrequency field can be decomposed into two components rotating clockwise and counterclockwise with respect to the z-axis, each with an amplitude B_1. Near resonance ($\omega \simeq \omega_0$), only the component rotating in the same direction as the Larmor precession has an important effect. The so-called *Rotating Wave Approximation* (RWA) amounts to keeping only this component in the equation of motion.[10] In the frame $OXYZ$ rotating with the main component, the RF field is fixed along the X-axis and has an amplitude B_1. Along the Z-axis, there is an additional inertial field $\vec{B}_{\text{in}}$ due to the rotation with an amplitude such that the Larmor precession frequency around $\vec{B}_{\text{in}}$ is equal to $-\omega$. To summarize, in the rotating frame, the Larmor precession frequency around Z is equal to $\omega_0 - \omega$ whereas the Larmor precession frequency around X is equal to $\Omega_1 = -\gamma B_1$.

At resonance ($\omega = \omega_0$), the total field along Z is equal to 0. The spin undergoes a Larmor precession around X at the frequency Ω_1, which is called the "Rabi frequency". This precession describes reversible oscillations between the two spin states $\pm 1/2$ due to absorption and stimulated emission processes.

In order to extend these results to any two-level system, it is useful to first write the Hamiltonian of the spin, in the laboratory frame $Oxyz$:

$$H = \omega_0 S_z + 2\Omega_1 \cos \omega t S_x, \tag{2.28}$$

which can also be written in matrix form

$$H = \frac{\hbar}{2} \begin{pmatrix} \omega_0 & 2\Omega_1 \cos \omega t \\ 2\Omega_1 \cos \omega t & -\omega_0 \end{pmatrix}, \tag{2.29}$$

and then in the rotating frame $OXYZ$ with the rotating wave approximation

$$\tilde{H} = \frac{\hbar}{2} \begin{pmatrix} \omega_0 - \omega & \Omega_1 \\ \Omega_1 & \omega - \omega_0 \end{pmatrix}. \tag{2.30}$$

[10]This approximation discards higher order effects such as the Bloch-Siegert shift (see, for example, Complement A_{VI} of Chap. VI of [Cohen-Tannoudji *et al.* (1992b)]).

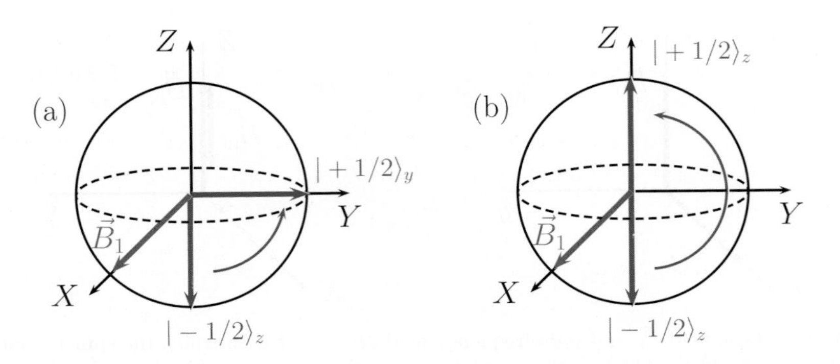

Fig. 2.4 (a) $\pi/2$ pulse. (b) π pulse.

It will be also important for the following discussions to introduce the idea of $\pi/2$ and π pulses. Suppose that we are at resonance ($\omega = \omega_0$) and that we start with a spin pointing along the negative direction of the Z-axis, i.e. in the eigenstate $|-1/2\rangle_Z$ of S_Z (see Fig. 2.4(a)). If we apply the RF field during a time τ such that $\Omega_1\tau = \pi/2$, the spin will rotate by an angle equal to $\pi/2$ around the X-axis in the OYZ plane and will end up in the state pointing along the positive direction of the Y-axis, i.e. in the eigenstate $|+1/2\rangle_Y$ of S_Y:

$$|-1/2\rangle_Z \underset{\pi/2\,\text{pulse}}{\rightarrow} |+1/2\rangle_Y = \frac{-i}{\sqrt{2}}\left(|+1/2\rangle_Z + i\,|-1/2\rangle_Z\right) \tag{2.31}$$

which is a linear superposition of the eigenstates $|\pm 1/2\rangle_Z$ of S_z. One could also apply a π pulse, i.e. apply the RF field during a time τ such that $\Omega_1\tau = \pi$. The spin then rotates by an angle equal to π around the X-axis in the OYZ plane and ends up in the state pointing along the positive direction of the Z-axis, i.e. in the eigenstate $|+1/2\rangle_Z$ of S_Z (see Fig. 2.4(b)):

$$|-1/2\rangle_Z \underset{\pi\,\text{pulse}}{\rightarrow} -i\,|+1/2\rangle_Z\,. \tag{2.32}$$

These $\pi/2$ and π pulses are very useful for preparing linear superpositions of $|+1/2\rangle_Z$ and $|-1/2\rangle_Z$ and for transforming one eigenstate of S_Z into the other one.

We will end this section by pointing out the importance of linear superpositions of states. The Larmor precession of the spin, starting from the state $|-1/2\rangle_Z$ or $|+1/2\rangle_Z$, in the vector sum $\vec{B}_0 + \vec{B}_{\text{in}} + \vec{B}_1$ of the three fields represented in Fig. 2.3(b), brings it in a state that points along a direction (with polar angles θ and φ) that does not generally coincide with the z-axis. This is due to the fact that the RF field $\vec{B}_1$ is perpendicular to the z-axis. When it interacts with a RF field, the spin is therefore put into a state

$$|\psi\rangle = \cos\left(\frac{\theta}{2}\right) e^{-i\varphi/2}\,|+1/2\rangle_Z + \sin\left(\frac{\theta}{2}\right) e^{+i\varphi/2}\,|-1/2\rangle_Z\,, \tag{2.33}$$

which is a *linear superposition* of the two states $\pm 1/2$. For example, after a $\pi/2$

pulse, the spin is put into the state (2.31) corresponding to[11] $\theta = \varphi = \pi/2$. These superpositions of states are important because they are at the origin of quantum interference effects that will be mentioned throughout this book. The simple example of a spin 1/2 studied here clearly shows that linear superpositions of states appear when the spin magnetic moment has transverse components with respect to the direction of the static field, which is the quantization axis for defining the energy levels of the spin. This connection between linear superpositions of states and the existence of a transverse magnetization remains valid for angular momenta other than 1/2. In Chap. 4, we will review a few interference effects related to the existence of transverse quantities in a atomic gas.

Effect of a $\pi/2$ pulse induced by a single photon in a resonant cavity

It is possible nowadays to build cavities with very high Q-factors. This allows one to neglect all damping processes and prepare states of the field in this cavity that have a single photon. Suppose that one mode of this cavity is resonant for the frequency ω_0 of a two-state system and that one can neglect the effect of all other modes. If we prepare the total {atom + field} system in the state $|-, N = 1\rangle$ (atom in the $|-\rangle$ state in the presence of one photon), this state is resonantly coupled to the state $|+, N = 0\rangle$ (atom in the $|+\rangle$ state with zero photon). If Ω is the Rabi frequency characterizing the coupling between these two states and if the atom-field interaction lasts a time τ such that $\Omega\tau = \pi/2$, a straightforward calculation shows that at the end of the pulse the system is put into the state:

$$|\psi\rangle = \frac{-i}{\sqrt{2}} \left(|+, N = 0\rangle + i\,|-, N = 1\rangle\right). \tag{2.34}$$

This result can be considered as giving the result of a $\pi/2$ pulse in a case where a quantum treatment of the field is needed. It also shows that the two states which are linearly superposed in (2.34) are not atomic states, as this was the case in (2.31), but tensor products of atomic and field states. The state (2.34) cannot be written as a product of a field state with an atomic state. It is an *entangled state*. Such entangled states of two systems, which often appear when these two systems interact, are a fundamental feature of quantum mechanics. They exhibit very peculiar quantum correlations, which play an essential role in the new field of quantum information. We will come back to the properties of entangled states in Chap. 19.

2.4.2 *Extension to any two-level atomic system*

We consider now a two-level system $\{e, g\}$ that is not necessarily a spin 1/2. For example, g can be the ground state of an atom and e an excited state. The energy splitting between the two states is equal to $\hbar\omega_0$, so that the atomic Hamiltonian

[11] Only the relative phase of the two coefficients of the expansion (2.31) or (2.33) is important. Multiplying (2.33) by the global phase factor $\exp(-i\varphi/2)$ transforms (2.33) into (2.31) when $\theta = \varphi = \pi/2$.

can be written as

$$H_A = \frac{\hbar\omega_0}{2}(|e\rangle\langle e| - |g\rangle\langle g|). \tag{2.35}$$

This atom interacts with a classical light field $\vec{E}_L = E_0\vec{e}_z \cos\omega_L t$ polarized along the z-axis with frequency ω_L and amplitude E_0. The atom-field interaction Hamiltonian V_{AL} can be written as $V_{AL} = -\vec{D}\cdot\vec{E}_L = -E_0 D_z \cos\omega_L t$, where $\vec{D}$ is the electric dipole moment of the atom with matrix element $\langle e| D_z |g\rangle = D_{eg}$, which can always be made real with a proper choice of the relative phases of e and g. Introducing the Rabi frequency Ω defined by $\hbar\Omega = -D_{eg}E_0$, we get:

$$V_{AL} = \hbar\Omega\,(|e\rangle\,\langle g| + |g\rangle\,\langle e|)\cos\omega_L t. \tag{2.36}$$

If we write the Schrödinger equation for the coefficients $c_g(t)$ and $c_e(t)$ of the expansion of the state vector in the basis $\{e, g\}$ and if we make the change of variables

$$\begin{cases} c_e(t) = b_e(t)e^{-i\omega_L t/2}, \\ c_g(t) = b_g(t)e^{+i\omega_L t/2}, \end{cases} \tag{2.37}$$

which is the equivalent of the transformation to the rotating frame for the spin 1/2 problem, we get:

$$\begin{cases} i\dot{b}_e(t) = \dfrac{\omega_0 - \omega_L}{2} b_e(t) + \dfrac{\Omega}{2}\left(1 + e^{2i\omega_L t}\right) b_g(t), \\ i\dot{b}_g(t) = \dfrac{\Omega}{2}\left(1 + e^{-2i\omega_L t}\right) b_e(t) - \dfrac{\omega_0 - \omega_L}{2} b_g(t). \end{cases} \tag{2.38}$$

The rapidly oscillating terms $\exp(\pm 2i\omega_L t)$ have a negligible effect near resonance. Neglecting them is the equivalent of the rotating wave approximation for the spin 1/2 problem. One then finds that the evolution in the new representation is governed by the same Hamiltonian as the spin Hamiltonian in the rotating frame $OXYZ$ given above in Eq. (2.30). One concludes that any two-level system $\{e, g\}$ can be described as a "fictitious" spin 1/2 when it is driven by a resonant electromagnetic wave coupling e and g: it oscillates between e and g at the Rabi frequency Ω.

We saw in the previous subsection that, for a real spin 1/2, linear superpositions of states are related to the existence of a transverse magnetization. What is the corresponding physical quantity that is related to linear superpositions of the two states e and g of an optical transition? One has to look for a physical observable that only has non-diagonal elements between e and g, so that its average value is non-zero only when the state of the atom is a linear superposition of e and g. This is the case for the electric dipole moment D, which is an odd operator, having non-zero matrix elements only between the two states e and g of opposite parities. When the two-level atom is driven by a coherent laser beam, this beam induces a dipole moment that oscillates with a certain phase with respect to the field.

2.4.3 *Perturbative limit*

In this subsection, we suppose that the field amplitude is sufficiently small to allow a perturbative treatment of the atom-field coupling H_I. We consider the case of a two-level system with two energy levels $|\pm\rangle$ with energy $\pm\hbar\omega_0/2$, interacting with a quasi-resonant field of frequency ω_L. The strength of the coupling is characterized by a Rabi frequency $\Omega(t)$, which could depend on time if the field is applied in the form of a pulse. We suppose that at a certain initial time t_i, the system is prepared in the state $|-\rangle$ and we calculate, to first order in $\Omega(t)$, the probability amplitude of finding it in the other state $|+\rangle$ at a final time t_f.

The evolution operator $U(t_f, t_i)$ between t_i and t_f is given to first order in $\Omega(t)$ by the following expression:

$$U(t_f, t_i) = U_0(t_f, t_i) + \int_{t_i}^{t_f} U_0(t_f, t)\, H_I(t)\, U_0(t, t_i)\, \mathrm{d}t, \tag{2.39}$$

where

$$U_0(t_2, t_1) = e^{-iH_0(t_2-t_1)/\hbar} \quad \text{with} \quad H_0 = \frac{\hbar\omega_0}{2}\left(|+\rangle\langle+| - |-\rangle\langle-|\right), \tag{2.40}$$

$$\text{and} \quad H_I(t) = \frac{\hbar\Omega(t)}{2}\left(|+\rangle\langle-|\, e^{-i\omega_L t} + |-\rangle\langle+|\, e^{+i\omega_L t}\right). \tag{2.41}$$

From these three equations one easily deduces:

$$\begin{aligned}\langle+|\, U(t_f, t_i)\, |-\rangle &= \int_{t_i}^{t_f} e^{-i\omega_0(t_f-t)/2}\frac{\Omega(t)}{2} e^{-i\omega_L t} e^{+i\omega_0(t-t_i)/2} \mathrm{d}t, \\ &= \frac{1}{2} e^{-i\omega_0(t_f+t_i)/2} \int_{t_i}^{t_f} \Omega(t)\, e^{i(\omega_0-\omega_L)t}\, \mathrm{d}t. \end{aligned} \tag{2.42}$$

Suppose that the field is applied in the form of a single pulse of duration τ around the time t_a (see Fig. 2.5(a)). In the integral of Eq. (2.42), one can make the change of variables from t to $t' = t - t_a$ so that the integral over t' runs from $t_i - t_a$ to $t_f - t_a$. If $t_a - t_i$ and $t_f - t_a$ are large compared to the width τ of the pulse, one can extend the limits of the integral to $\pm\infty$. The integral over t' is then proportional to the Fourier transform $g(\omega)$ of $\Omega(t_a + t')$ taken in $\omega_0 - \omega_L$. One gets:

$$\langle+|\, U(t_f, t_i)\, |-\rangle = \frac{1}{2} e^{-i\omega_0(t_f+t_i)/2} e^{i(\omega_0-\omega_L)t_a} g(\omega_0 - \omega_L), \tag{2.43}$$

where

$$g(\omega) = \int_{-\infty}^{+\infty} \Omega(t_a + t')\, e^{i\,\omega\, t'}\, \mathrm{d}t'. \tag{2.44}$$

Since $\Omega(t_a + t')\, e^{i\,\omega\, t'}$ is maximum for $t' = 0$ and has a width τ, $g(\omega)$ is maximum for $\omega = 0$ and has a width on the order of $1/\tau$. As a function of $\omega_L - \omega_0$, the transition probability from the state $|-\rangle$ to the state $|+\rangle$, which is given by:

$$P_{-\to+}(\omega_0 - \omega_L) = |\langle+|\, U(t_f, t_i)\, |-\rangle|^2 = \frac{1}{4}\, |g(\omega_0 - \omega_L)|^2 \tag{2.45}$$

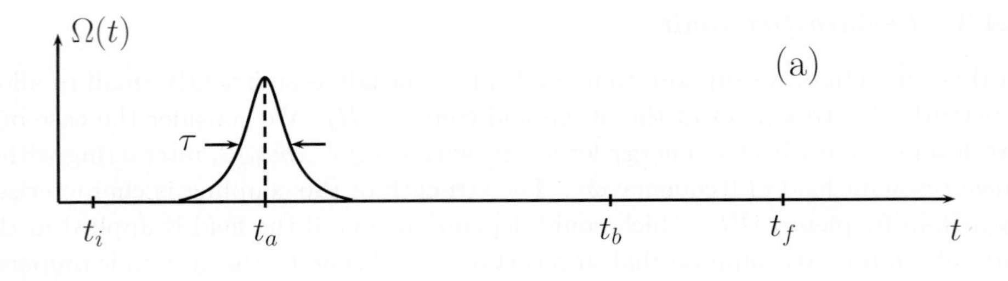

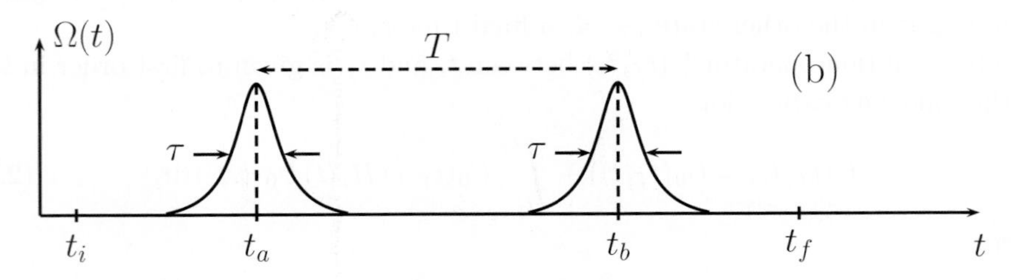

Fig. 2.5 (a) Single pulse of width τ applied at time t_a. (b) Sequence of two identical coherent pulses with the same widths applied at two different times t_a and t_b, and separated by an interval T.

exhibits resonant variations centered on $\omega_0 - \omega_L = 0$, with a width on the order of $1/\tau$ (see Fig. 2.6(a)). The smaller τ, the larger the resonance. This is an illustration of the time-frequency uncertainty relation.

Now suppose that the field is applied in the form of two coherent pulses[12] of the same duration τ at times t_a and t_b with $t_b - t_a = T \gg \tau$ (see Fig. 2.5(b)). Since we are making a first order calculation of the transition amplitude from $|-\rangle$ to $|+\rangle$, we must consider only one interaction with the field, either with the first pulse or with the second one. The total amplitude is thus the sum of the two amplitudes associated with each pulse and we have just to add to (2.43) the same expression with t_a replaced by t_b. One gets:

$$\langle + | U(t_f, t_i) | - \rangle = \frac{1}{2} e^{-i\omega_0(t_f+t_i)/2} g(\omega_0 - \omega_L) \left[e^{i(\omega_0-\omega_L)t_a} + e^{i(\omega_0-\omega_L)t_b} \right]. \quad (2.46)$$

We thus add two amplitudes which have the same modulus and a phase difference equal to $(\omega_0 - \omega_L)(t_b - t_a) = (\omega_0 - \omega_L)T$. If $(\omega_0 - \omega_L)T = 2n\pi$, with $n = 1, 2, 3, \ldots$, the interference is constructive; if $(\omega_0 - \omega_L)T = (2n+1)\pi$, the interference is destructive. The transition probability

$$\begin{aligned} P_{-\to+}(\omega_0 - \omega_L; T) &= \frac{1}{4} |g(\omega_0 - \omega_L)|^2 \left| 1 + e^{i(\omega_0-\omega_L)T} \right|^2 \\ &= \frac{1}{2} |g(\omega_0 - \omega_L)|^2 \left[1 + \cos(\omega_0 - \omega_L)T \right] \end{aligned} \quad (2.47)$$

[12]Figure 2.5 actually gives the envelope of the field. The two pulses can be considered to be two distinct portions of the same carrier that oscillates at the frequency $\omega_L \gg 1/\tau$. This is what makes the two pulses coherent.

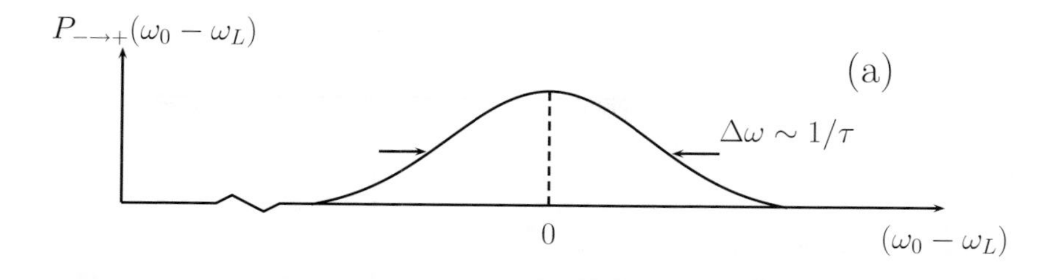

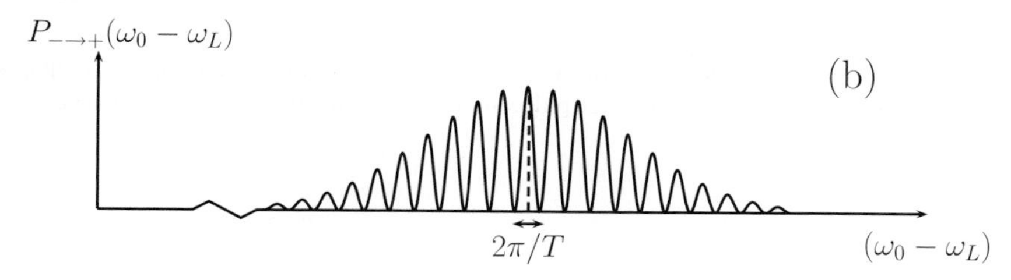

Fig. 2.6 (a) Transition probability $P_{-\to+}(\omega_0 - \omega_L)$ as a function of $\omega_0 - \omega_L$ for the single pulse excitation of Fig. 2.5(a). (b) Same quantity for the two-pulse excitation of Fig. 2.5(b).

exhibits interference fringes with a fringe spacing $2\pi/T$ inside the broad profile of width $1/\tau$ obtained with a single pulse (see Fig. 2.6(b)). These are the famous Ramsey fringes that allow an improvement of the resolution of the determination of ω_0 by a factor $T/\tau \gg 1$ [Ramsey (1950)]. There is a close analogy between these fringes and Young's double-slit interference fringes in optics, which appear with a period determined by the distance between the two slits within the broad diffraction profile observed with a single slit.

2.4.4 *Two physical pictures for Ramsey fringes*

(i) *Interference between two different paths*

This interpretation closely follows the calculation presented in the previous subsection. Starting from $|-\rangle$, the atom can end up in $|+\rangle$ either during the first pulse at time t_a, or during the second pulse at time t_b. As shown in Fig. 2.7, there are therefore two possible paths leading from the same initial state $|-\rangle$ to the same final state $|+\rangle$. The corresponding amplitudes interfere. This interpretation of quantum interference effects is very general and is very useful for interpreting interferences involving several particles. In each path, one can have several interaction processes involving several photons or several atoms. Examples of such situations will be discussed in Chap. 19 (Hanbury Brown and Twiss effect).

In Fig. 2.7, one plots the time along the horizontal axis and the internal atomic state along the vertical axis. In Chap. 18 we will see similar diagrams when we plot two space coordinates along the two axes, and where the two paths that interfere

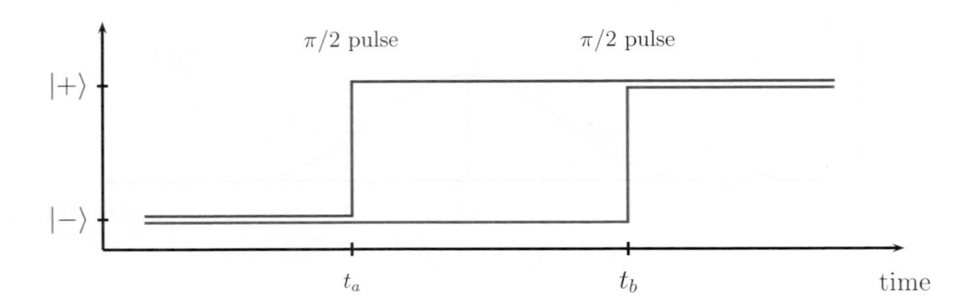

Fig. 2.7 Two possible paths leading from the initial state $|-\rangle$ to the final state $|+\rangle$. The atom can absorb a photon either during the first pulse at time t_a, or during the second pulse at time t_b.

are two different spatial paths followed by the atomic center of mass. Such schemes correspond to a generalization of Ramsey interferometers involving matter waves and called *Ramsey-Bordé interferometers*. They are currently used for measuring inertial fields.

(ii) *Interpretation in terms of linear superpositions of states*

The first pulse creates a linear superposition of states and a corresponding transverse magnetization. While the atom travels between the two cavities, it is not coupled to the RF field and its magnetization rotates freely around the z-axis at the Larmor frequency ω_0. The RF field inside the two cavities that are supplied by the same source continues to evolve at the same frequency ω. When the atom enters the second cavity, its magnetization and the RF field have accumulated a relative phase shift $(\omega_0 - \omega)T$. If $(\omega_0 - \omega)T$ is equal to 2π, the transverse magnetization and the RF field find themselves with the same relative phase that they had after the first pulse. The coherent action of the RF field continues as if the length of the first pulse was doubled and the transition amplitude at the end of the second pulse is doubled. If $(\omega_0 - \omega)T = \pi$, the relative phase of the RF field with respect to the transverse magnetization has changed its sign and the rotation of the spin around the RF field is reversed, thereby returning the spin to its initial state as shown in Fig. 2.8. In this way, we interpret the first zero and the second maximum of the curve of Fig. 2.6(b).

The advantage of this picture is that it can be extended to the case where a perturbative treatment would not be possible for the two $\pi/2$ pulses. In order to consider only the rotation of the spin around $\vec{B}_1$ during the two $\pi/2$ pulses, one should assume that $|(\omega_0 - \omega)|\tau \ll 1$, and that the time separation T between the two pulses is sufficiently long so that $(\omega_0 - \omega)T$ can reach values on the order of π.

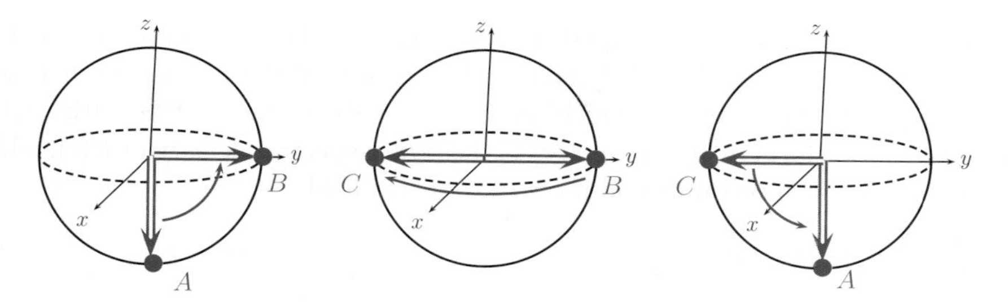

Fig. 2.8 Bloch sphere representation of a given Ramsey sequence: a two-level atom prepared in the ground state $|-\rangle$ (A) undergoes a $\pi/2$ pulse and ends up in a superposition of state $-i(|+\rangle + i|-\rangle)/\sqrt{2}$ (B); the superposition evolves freely for an adjustable time during which the Bloch vector that represents the system turns into the equatorial plane; another $\pi/2$ pulse is finally applied. If, at the start of the second pulse, the atom state corresponds to a Bloch vector pointing along C (which corresponds to $(\omega_0 - \omega)T = \pi$) the atom comes back to the ground state A after the sequence of two pulses, so that $P_{-\to+}(\omega_0 - \omega_L; T) = 0$.

2.5 Two-level atom interacting with a broadband field. Absorption and emission rates

2.5.1 *Absorption rate deduced from a semiclassical treatment of the field*

We now suppose that the field is no longer applied in the form of pulses and that it excites a two-level system.[13] Its intensity is constant and low enough to allow a first order perturbative treatment. We also suppose that its spectral width $\Delta\omega$ around its central frequency ω_L is large compared to all other broadenings of the atomic transition (natural width due to spontaneous emission, relaxation processes). We start from the atom in the state g at an initial time $t_i = 0$ and want to calculate how the probability $P_{g\to e}(t)$ to find it in the state e at a later time t depends on t.

We describe the field as a classical field $E(t) = \mathcal{E}(t)\exp(-i\omega_L t)$+c.c. of frequency ω_L, with a fluctuating complex amplitude $\mathcal{E}(t)$ that includes both amplitude and phase fluctuations. The random function of time $E(t)$ is characterized by its time correlation functions. In the lowest order treatment presented here, we need only the first order correlation function

$$g^{(1)}(t', t'') = \overline{\mathcal{E}^*(t'')\mathcal{E}(t')}e^{-i\omega_L(t'-t'')}, \tag{2.48}$$

which is the average of the product of two field values[14] taken at two different times over all possible realizations of the random field. In addition, we suppose that the random field is stationary, which means that $g^{(1)}(t', t'')$ only depends on $t' - t'' = \tau$:

$$g^{(1)}(t', t'') = \overline{\mathcal{E}^*(t'')\mathcal{E}(t')}e^{-i\omega_L(t'-t'')} = g^{(1)}(\tau) \quad \text{with} \quad \tau = t' - t''. \tag{2.49}$$

[13] We consider an optical transition and not a RF transition because at the end of this section we want to address the problem of the spontaneous emission rate which is negiligible in the RF domain.

[14] The relevant correlation for calculating the absorption rate is the average value of the product of the positive frequency part of $E(t')$ (varying as $\exp(-i\omega_L t')$) with the negative frequency part of $E^*(t'')$ (varying as $\exp(i\omega_L t'')$).

The correlation function $g^{(1)}(\tau)$ tends to zero when τ is larger than a certain characteristic time τ_C, which is called the correlation time of the fluctuating field and which represents the time after which the field loses the memory of its initial value. The Fourier transform $\tilde{g}^{(1)}(\omega)$ of $g^{(1)}(\tau)$ is the power spectral density of $E(t)$, which is nothing but the spectral distribution $I(\omega)$ of this field

$$\int_{-\infty}^{+\infty} g^{(1)}(\tau)\, e^{i\,\omega\,\tau}\, d\tau = \int_{-\infty}^{+\infty} \overline{\mathcal{E}^*(t'')\mathcal{E}(t')}\, e^{i(\omega-\omega_L)(t'-t'')} d\tau = I\,(\omega)\,. \tag{2.50}$$

This result is an illustration of the Wiener-Khinchin theorem. It follows that the correlation time τ_C is nothing but the inverse of the spectral width $\Delta\omega$ of the field:

$$\tau_C \simeq \frac{1}{\Delta\omega}. \tag{2.51}$$

Since $\Delta\omega$ is assumed to be large compared to all characteristic widths of the problem, it follows that τ_C is small compared to all characteristic times.

To calculate $P_{g\to e}(t)$, we come back to the transition amplitude given by Eq. (2.42) in which we replace $|-\rangle$ by $|g\rangle$, $|+\rangle$ by $|e\rangle$, t_i by 0, t_f by t and $\Omega(t)$ by $D_{eg}\mathcal{E}(t')$. We multiply this amplitude by its complex conjugate and take the average of $\mathcal{E}^*(t'')\mathcal{E}(t')$ over all possible realizations of the random field, which finally leads to:

$$P_{g\to e}(t) = \frac{1}{4}\left|D_{eg}\right|^2 \int_0^t \mathrm{d}t' \int_0^t \mathrm{d}t''\, \overline{\mathcal{E}^*(t'')\mathcal{E}(t')}\, e^{i(\omega_0-\omega_L)(t'-t'')}. \tag{2.52}$$

The upper limit t of the integrals appearing in (2.42) can be chosen to be very large compared to the correlation time τ_C of the fluctuating field, while being short enough to have $P_{g\to e}(t) \ll 1$, as this is required for a perturbative treatment. If we make the change of variables $\{t', t''\} \to \{t', \tau = t' - t''\}$, the limits of the double integral become

$$\int_0^t \mathrm{d}t' \int_0^t \mathrm{d}t'' \to \int_0^t \mathrm{d}t' \int_{t'-\tau}^{t'} \mathrm{d}\tau. \tag{2.53}$$

Except when t' is very small, in an interval of width τ_C near 0, the lower and upper limits of the integral over τ are very large compared to τ_C, which is the range of values of τ over which the correlation function appearing in (2.52) is not negligible. We can thus extend these upper and lower limits to $\pm\infty$, which gives, using (2.50)

$$P_{g\to e}(t) = \frac{1}{4}\left|D_{eg}\right|^2 \int_0^t \mathrm{d}t' \int_{-\infty}^{+\infty} \overline{\mathcal{E}^*(t'')\mathcal{E}(t')}\, e^{i(\omega_0-\omega_L)\tau} \mathrm{d}\tau = \left[\frac{1}{4}\left|D_{eg}\right|^2 I(\omega_0)\right] t. \tag{2.54}$$

We find that the transition probability increases linearly in time, which allows us to introduce an absorption rate $W_{g\to e}$ that is independent of time and proportional to both the field spectral intensity at the frequency ω_0 of the atomic transition and the square of the dipole matrix element of the atomic dipole moment between e and g:

$$W_{g\to e} = \frac{P_{g\to e}(t)}{t} = \frac{1}{4}\left|D_{eg}\right|^2 I(\omega_0). \tag{2.55}$$

This behavior corresponds to an irreversible departure from g due to the absorption process, and is in a sharp contrast with the reversible Rabi oscillation that occurs with a coherent monochromatic field. It is reminiscent of the absorption rates first introduced by Einstein for an atom interacting with the black body field and proportional to $I(\omega_0)$ and to a B coefficient, which corresponds in the simple model studied here to the term $\frac{1}{4}|D_{eg}|^2$ appearing in (2.55). A similar treatment could be applied to calculate $P_{e\to g}(t)$ and to deduce the stimulated emission rate $W_{e\to g}$.

2.5.2 *Physical discussion. Relaxation time and correlation time*

The inverse of the absorption rate $1/W_{e\to g}$ is the characteristic time describing the damping of the probability that the atom remains in g when it is excited by the incoming light. It is a *relaxation time* T_R that must be clearly distinguished from the correlation time of the field, which is much shorter. T_R is the mean time after which an absorption process has a great chance to occur, whereas τ_C is the duration of the transient phenomena associated with the absorption process when this process occurs. We will often return to the importance of the distinction between these two characteristic times in the subsequent chapters of this book.

One can also use the following picture. The stationary incoherent light beam with a spectral width $\Delta\omega$ can be represented as an ensemble of one-photon wave packets, with a frequency width $\Delta\omega$, randomly distributed in time with a uniform constant density depending on the intensity. Each of these wave packets passes through the atom during a time $1/\Delta\omega = \tau_C$. Most of them do not give rise to an absorption process, so the mean time separating two absorption processes by the atom is T_R. When an absorption process occurs, however, the excitation of the atom by the wave packet that produces this absorption lasts the time τ_C during which the wave packet passes through the atom.

2.5.3 *Sketch of a quantum treatment of the absorption process*

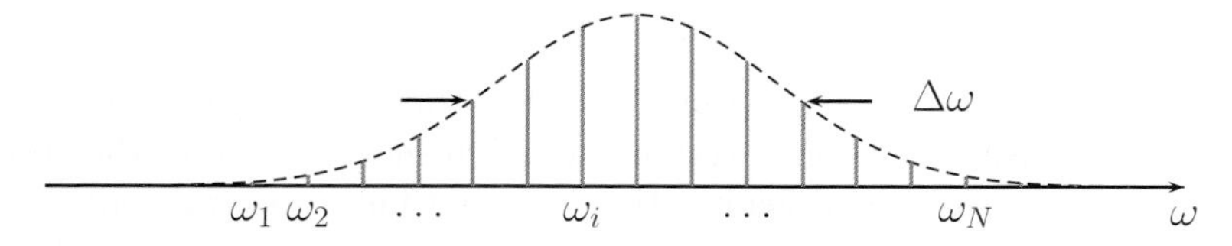

Fig. 2.9 Distribution of the number of photons in the modes of the quantum field exciting the atom.

We now use a quantum description for the field that excites the atom. This field contains photons in several modes with parallel wave vectors and closely spaced frequencies ω_1, ω_2, $\ldots\omega_i$, $\ldots$ ω_N covering a large spectral bandwidth $\Delta\omega$. We suppose that there are n_1 photons in mode 1, $\ldots n_i$ photons in mode i, $\ldots n_N$

photons in mode N. The variations with ω_i of n_i are represented in Fig. 2.9.

The initial state of the system {atom + field} can be written

$$|\psi_{\text{in}}\rangle = |g; n_1, n_2, \ldots, n_i, \ldots n_N\rangle. \tag{2.56}$$

This initial state is coupled by the absorption process to a great number of states where the atom has been transferred to the excited state e and one photon has disappeared from one of the field modes

$$|\psi_{\text{in}}\rangle \Leftrightarrow \{|e; n_1, n_2, \ldots, n_i - 1, \ldots n_N\rangle\} \quad \text{with } i = 1, 2, \ldots N. \tag{2.57}$$

These final states have closely spaced energies and can be considered to form a continuum of final states with an energy width equal to $\Delta E = \hbar\Delta\omega$.

If ΔE is large enough, one can apply Fermi's golden rule and introduce an absorption rate $W_{g\to e}$ proportional to the light intensity $I(\omega_0)$ at the atomic frequency ω_0. Similar considerations allow one to introduce a stimulated emission rate $W_{e\to g}$. We thus get the same result as the previous semiclassical treatment.

2.5.4 *Extension to spontaneous emission*

In the absence of incoming light, the light field within the semiclassical approximation is equal to zero and there is no interaction. An atom initially in the excited state e would remain there forever.

In a quantum description, all modes are in the vacuum state $|0_i\rangle$ and the initial state of the {atom + field} system is

$$|\psi_{\text{in}}\rangle = |e; \{0_i\}\rangle. \tag{2.58}$$

This state is coupled to a continuum of final states where the atom is in g and where one photon has appeared in one mode i

$$|e; \{0_i\}\rangle \Leftrightarrow \{|g; 0_1, 0_2, \ldots, 1_i, \ldots\rangle\}. \tag{2.59}$$

This continuum has a very large spectral width so that the correlation time associated with spontaneous emission is very short, much shorter than for the absorption process associated with a light beam with a large but finite spectral width. One can apply Fermi's golden rule and predict a spontaneous radiative decay of the excited state with a rate $W^{\text{spont}}_{e\to g}$, which is nothing but the Einstein's A coefficient. One can show that A scales as ω_0^3 where ω_0 is the atomic frequency.[15] This explains why spontaneous emission can be ignored in the radio-frequency and microwave domains even though it plays a key role in the optical domain.

[15] This ω_0^3 dependence comes from two contributions: the square of the modulus of the coupling matrix element, which varies as ω_0, and the ω_0^2 dependence of the density of final states.

2.6 Two-level atom interacting with a coherent monochromatic field in the presence of damping

In Sec. 2.4, we neglected all damping process, including the damping rate A of e due to spontaneous emission. More precisely, the results derived in Sec. 2.4 are valid only if the Rabi frequency Ω that characterizes the coupling with the monochromatic field is large compared to all damping rates. The damping processes during the Rabi oscillation between g and e can then be neglected.

We will show in Chap. 7, using an effective Hamiltonian approach, that if Ω is very weak compared to the damping rates the absorption process from g to e can be described by a rate as in the case of a broadband excitation. When the only damping process is spontaneous emission, this behavior can be understood with the following qualitative argument. Because of its natural width A, the excited state can be considered to be a continuum with a density of states $1/A$, corresponding to one state in an interval A. By applying Fermi's golden rule, one can then predict a transition rate from g to e equal to the square Ω^2 of the coupling matrix element Ω times the density of states $1/A$, which gives an absorption rate that scales as Ω^2/A.

A semiclassical point of view can be also taken for understanding stimulated emission. First suppose that pumping processes in the absence of the monochromatic field introduce a steady state inversion of populations between e and g, that is the population σ_{ee} of the upper state e is larger than the population σ_{gg} of the lower state g. When the monochromatic field E is applied, one can calculate the steady state value of the mean dipole moment induced by this field to the lowest order in E from optical Bloch equations. Damping processes are essential for the establishment of a steady state. One can then deduce that the field radiated in the forward direction by this dipole is proportional to $\sigma_{ee} - \sigma_{gg}$. If one considers an ensemble of such dipoles distributed uniformly in a plane perpendicular to the direction of the incident light beam, and if one calculates the total field radiated at resonance by these dipoles in the forward direction,[16] one finds that this total radiated field has the same phase as the incident field when $\sigma_{ee} > \sigma_{gg}$, and a π-phase shift when $\sigma_{ee} < \sigma_{gg}$. The two fields add in the first case, and subtract in the second case. This simple picture thus provides a clear understanding of the coherent amplification of light by stimulated emission in a population inverted medium. This basic mechanism of laser light sources contrasts with the absorption of light in a normal medium.

....

[16] See Chap. 31 of [Feynman (1963)].

We will conclude this chapter by once more emphasizing the importance of using different approaches to understand the same physical phenomenon. This allows one to grasp different physical pictures that complement one another, and provides a better understanding of the physical process involved. For example, in this chapter we have interpreted Ramsey fringes in terms of interfering paths, as well as in terms of Bloch vectors. We have shown that a semiclassical description provides a better understanding of the coherent character of the amplification by stimulated emission. On the other hand, a quantum description is essential for understanding the laser linewidth. Even in the cases where a semiclassical description of atom-field interactions is sufficient, a quantum treatment of these interactions can bring new insight into the physical mechanisms. Considering the {atom + quantum field} system as a global system described by a time independent Hamiltonian leads to a clearer identification of the various elementary interaction processes (absorption and stimulated emission, spontaneous emission) and suggests the physical picture of an atom "dressed" by the photons with which it interacts. This dressed atom approach will be presented in more detail in other chapters of this book, and will be used in the analysis of several important physical effects.

PART 1

Light: a source of information on atoms

Introduction

Spectroscopy is a basic source of information on atoms. By measuring the frequency spectrum of the electromagnetic radiation which is emitted or absorbed by atoms, one can determine the Bohr frequencies $\omega_{ij} = (E_i - E_j)/\hbar$ of the atomic transitions connecting two levels i and j. Comparison of the measured frequencies with those predicted from quantum theories of the atomic structure provides a precise test of these theories. Measuring the spectrum of the light emitted by a medium also allows the identification of the atoms or molecules that it contains, since different atoms have different spectra. Most of the information we have on the universe around us comes from the observation of the light emitted or absorbed by stellar or planetary atmospheres, by molecules in the interstellar space, or by galaxies moving away from us at a velocity which can be deduced from the shift of frequency of the light that we receive from them.

Various types of spectroscopy have been developed, depending on the frequency range in which the measured Bohr frequencies fall: radiofrequency and microwave, optical, X-ray, γ-ray. They give information on different types of atomic quantities: atomic g-factors, fine and hyperfine structures, radiative lifetimes, optical transitions, inner shell transitions. The progress of spectroscopy is essentially determined by the development of new technologies and new sources of radiation (radar, masers, lasers, synchrotron radiation, etc.) and by the invention of new methods of investigation, a few examples of which will be given in this book.

It is impossible here to review all the developments of spectroscopy that have occurred during the last decades. We have chosen to focus on two examples that illustrate how a series of advances in atomic physics extending over several decades can be put in perspective.

The first example concerns radiofrequency and microwave spectroscopy. The development of such spectroscopy in atomic beam experiments was pioneered by Isidor Rabi in Columbia at the end of the 1930's and marked the renaissance of atomic physics through the discovery of radiative corrections such as the Lamb shift and the spin anomaly. Quantum Electrodynamics was invented for explaining these effects. A few years later, the radio electric detection of magnetic resonance in liquid and solid samples by the groups of Edwin Purcell and Felix Bloch marked the birth of a new method of investigation of condensed matter which now plays an essential role in chemistry, biology and medicine. We will focus here, in Chap. 3, on the optical methods of radiofrequency and microwave spectroscopy which were initiated in the 1950's by Alfred Kastler and Jean Brossel. These methods combine magnetic resonance with an optical excitation of the atoms. They allow one to polarize atoms and detect magnetic resonance by a modification of the absorbed or emitted light. They made it possible to detect nuclear magnetic resonance in gaseous samples with an unprecedented accuracy and sensitivity. We choose to discuss them because they constitute the first example of manipulation of atoms by light. Using the conservation of the total angular momentum, one can control an internal atomic degree of freedom (spin polarization). A few decades later, another conservation law, the conservation of the total linear momentum, was used to control the external atomic degrees of freedom (position and velocity of the centre of mass). This is the new research field of laser cooling and trapping, which will be described in part 4 of this book. The second reason of our choice of optical methods is that they have allowed the discovery of several interesting effects:

- light shifts which are described in Chap. 7, and which, later on, turned out to play an important role in several cooling and trapping mechanisms
- multiphoton transitions that are described in Chap. 9
- new types of resonances associated with linear superpositions of Zeeman sublevels discovered in optical pumping experiments performed in the sixties. These studies clearly demonstrated the need of a density matrix description of the atoms and the importance of the off-diagonal elements of this matrix. A review of these resonances and of their physical interpretation is presented in Chap. 4. These pioneering studies provide a good introduction to the general field of quantum interference which will be studied in more detail in Chaps. 17–19.

Our second example deals with optical spectroscopy. Chapter 5 is devoted to a description of the fluorescence light emitted by an atom excited by a resonant laser beam. This example clearly shows the impact of lasers for the development of atomic physics. The monochromaticity and the coherence of these light sources made it possible to extend effects which before were only observable in magnetic resonance experiments, like the Rabi oscillation between the excited and the ground states, to optical transitions. Spontaneous emission, which is negligible in the RF domain, is a basic process in the optical domain and including it in the description of resonance fluorescence led to the development of new theoretical approaches such as the optical Bloch equations and the dressed atom approach. New types of detection signals have been also investigated such as time correlations between the fluorescence photons successively emitted by an atom excited by a resonant laser beam. The interpretation of these signals in the case of single atom resonance fluorescence stimulated new theoretical descriptions in terms of quantum jumps and Quantum Monte Carlo Wave Function simulations. The development of laser spectroscopy clearly shows how an improvement of experimental investigations can stimulate new theoretical ideas.

The last few decades have been also marked by spectacular advances in our ability to improve the resolution of spectroscopic measurements. Chapter 6 is devoted to a brief review of a few of these advances prior to the development of laser cooling which will be described later on in this book. New methods for circumventing the Doppler effect and the recoil effect, which are important in the optical domain, will be described. These include the saturated absorption method, the two-photon

Doppler-free spectroscopy and the Lamb-Dicke effect. We will also introduce the shelving method of Hans Dehmelt, which allows a very efficient detection of a very narrow optical transition on a single trapped ion.

Chapter 3

Optical methods

3.1 Introduction

To understand the impact of optical methods on the development of radio-frequency and microwave spectroscopy, it is useful to briefly review the state of the art in this domain before the advent of these methods.

The pioneering experiments of the group of Isidor Rabi at the end of the 1930's were performed using atomic beams [Rabi and Cohen (1934)] and state selection devices of the type introduced by Otto Stern and Walther Gerlach [Gerlach and Stern (1924)]. In such devices, atoms follow different trajectories that depend on their internal spin state. It therefore follows that spin flips induced by a resonant RF or microwave field result in a change of an atom's trajectory that can be measured by focussing the atoms onto a detector [Ramsey (1956)].

The next important step was the extension of magnetic resonance to liquid or solid samples in which the number of atoms was much larger than that in an atomic beam. Two different groups succeeded in detecting magnetic resonance by purely radioelectric methods, with each of them giving a different interpretation of their results. The group of Felix Bloch observed the phenomenon of nuclear induction using two perpendicular radio-frequency (RF) coils, one for producing the RF field, the other one for detecting the voltage induced by the rotating magnetization of the sample appearing at resonance [Bloch *et al.* (1946)]. The group of Edwin Purcell observed a resonant absorption and interpreted this result through transitions between two quantized energy levels of the atomic system [Purcell *et al.* (1946)]. Considered first as due to two different phenomena, these two types of signals were soon realized to have the same physical origin, magnetic resonance.

The radioelectric detection of magnetic resonance is possible only if the two states a and b involved in the radio-frequency or microwave transitions have unequal populations, since the detection signal is proportional to the population difference $N_a - N_b$. In the absence of state selection, this population difference comes only from the Boltzmann factor characterizing the thermal equilibrium, at the temperature

T, of the sample,

$$\frac{N_b}{N_a} = \exp\left(-\frac{E_b - E_a}{k_B T}\right) \tag{3.1}$$

where k_B is the Boltzmann constant. For energy splittings $E_b - E_a$ in the radiofrequency or microwave domain, and for temperatures that are not too low,[1]

$$E_b - E_a \ll k_B T \tag{3.2}$$

so that

$$N_a - N_b \simeq \frac{N_0}{2}\frac{E_b - E_a}{k_B T}, \tag{3.3}$$

where $N_0 = N_a + N_b$ is the total population of the two levels a and b. Because of the smallness of the factor $(E_b - E_a)/k_B T$, an appreciable signal can be obtained only if N_0 is very large, which explains why radioelectric detection of magnetic resonance is essentially limited to liquid or solid samples.

Even so, radioelectric detection of magnetic resonance in dense systems has given rise to many applications. Because of the interactions between different spins, the resonance spectra exhibit fine structures that give information on the relative spatial positions of the spins, and appear as powerful tools for determining the structure of complex molecules. Another example of application is magnetic resonance imaging (MRI) that allows one to obtain the spatial repartitions of protons in a living organism, and has become a very useful imaging technique in medicine.

On the other hand, if one is interested in very precise measurements of atomic structures, it is better to detect magnetic resonance in dilute systems such as atomic gases, where atom-atom interactions are much weaker or, at least, can be taken into account more quantitatively. For this measurement, radioelectric detection methods are not sensitive enough, and while atomic beam deflection experiments are more efficient since they do not rely on the Boltzmann factor, they are limited to atomic ground states, or to long-lived metastable states. One may wonder how to overcome these limitations and to extend magnetic resonance experiments to gaseous samples and to short-lived atomic states. In this chapter, we show that optical methods provide a solution to this problem.

We begin in Sec. 3.2 by describing the double resonance method, which uses the excitation of atoms with polarized light and the observation of the re-emitted light with a proper polarization for detecting magnetic resonance in atomic excited states. Next, we show in Sec. 3.3 how the transfer of angular momentum from polarized photons to atoms in optical pumping cycles can lead to very large spin polarization in atomic ground states. The first experiments using optical pumping are briefly described in Sec. 3.4. In Sec. 3.5, we explain how the optical pumping light, which acts only on the orbital angular momentum of the electrons, can produce large degrees of polarization of the nuclear spins. The applications of optical pumping are overviewed in Sec. 3.6. Finally, in Sec. 3.7, we summarize the important and original features of optical methods.

[1]At room temperature, $k_B T/h = 6000$ GHz, which is very large compared to radiofrequency and microwave frequencies.

3.2 Double resonance

In 1949, Francis Bitter suggested for the first time the extension of the techniques of radio-frequency spectroscopy to the excited states of atoms [Bitter (1949)], though in the end, it was shown that the method he proposed could not work [Pryce (1950)]. The same year, however, Jean Brossel and Alfred Kastler proposed a double resonance method using simultaneously resonant optical and radio-frequency excitations to accomplish this goal on mercury atoms [Brossel and Kastler (1949)].

3.2.1 *Principle of the method*

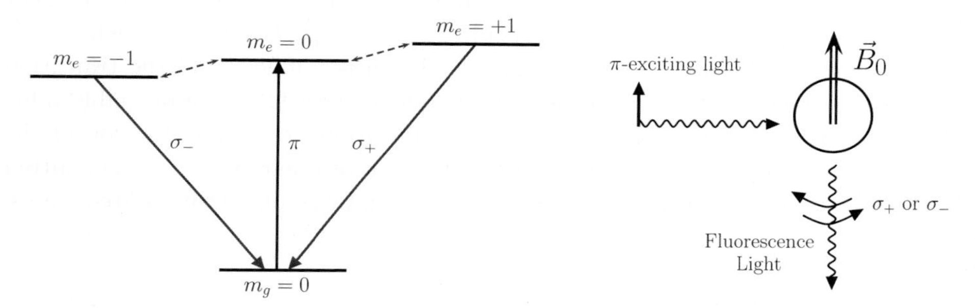

Fig. 3.1 Right: energy level diagram showing the Zeeman structure of mercury-198 ($6^1S_0 \longrightarrow 6^3P_1$ transition). The excited state has a finite lifetime $\Gamma^{-1} = 155$ ns. Left: double resonance scheme: a constant magnetic field splits the excited state e into three Zeeman sublevels, a resonant π-polarized light is applied, exciting selectively the sublevel $m_e = 0$. A radio-frequency field drives the transitions between $m_e = 0$ and $m_e = \pm 1$ which results in the spontaneous emission of circularly polarized light.

Mercury atoms have all available electron shells filled, and their isotopes with even atomic mass number have zero nuclear spin. Those isotopes have therefore a very simple low-lying electronic level diagram: the ground state has a zero angular momentum $J_g = 0$ and the excited state an angular momentum $J_e = 1$. In the presence of a magnetic field, the ground state remains unchanged while the excited state is split into three Zeeman sublevels with magnetic quantum number $m_e = -1$, $m_e = 0$ and $m_e = 1$ (see Fig. 3.1). The energy of each state reads:

$$E = E_0 + g m_e \mu_B B, \tag{3.4}$$

where E_0 is the energy of the excited state in zero field, and μ_B the Bohr magneton. The electron's Landé g-factor depends on the quantum numbers of the particle state and is on the order unity. The selection rule associated with the conservation of angular momentum allows the selective excitation of Zeeman sublevels depending on the polarization of the resonant light. Excitation with π-polarized light leads solely to the sublevel $m_e = 0$ of the excited state, whereas excitation with circular polarization σ^+ or σ^- leads, respectively, to the sublevels $m_e = 1$ or $m_e = -1$.[2] It must be emphasized that this selective excitation based on the polarization of the resonant light is independent of the spectral width of the light source, since it relies on the conservation law of angular momentum.

The emitted light also reflects the polarization of the excitation, except when the excited atom is perturbed during the short lifetime of the excited state. In fact, the proposal of Jean Brossel and Alfred Kastler is based upon this principle, and in particular, consists of modifying the population of the excited state by driving magnetic resonance transitions between excited Zeeman sublevels. For this purpose, a radio-frequency field is applied perpendicularly to the magnetic field B_0 that lifts the degeneracy of excited Zeeman sublevels. A resonant population transfer occurs when the frequency of the radio-frequency field coincides with the Larmor frequency corresponding to the Zeeman splitting $\omega_0 = g\mu_B B_0/\hbar \equiv \gamma B_0$, where γ is the gyromagnetic factor. If this resonant condition is fulfilled, and the transition probability induced by the oscillating radio-frequency field has an appreciable value within the lifetime of the excited state, a modification of the polarization of the emitted light is observed. In other words, the state of polarization of the emitted light can be used as an optical detection of the magnetic resonance in excited states.

3.2.2 *Predicted shape for the double resonance curve*

The calculation of the magnetic resonance line shape is carried out in the following manner. The excitation of the atom by the π-polarized incident light can be considered as instantaneous because of the broad spectral width of the light. Atoms are therefore prepared to the Zeeman sublevel $m_e = 0$, with a constant rate of n_0 atoms per unit of time. During the time spent into the excited state, the atom evolves under the effect of the radio-frequency field B_1 and its state becomes a linear superposition of the three sublevels $m_e = -1, 0, 1$. In the rotating field frame associated with the radio-frequency field, the evolution of the internal degrees of freedom of the excited state corresponds to the rotation of a spin 1 vector around the sum of the various fields $B_0\vec{e}_Z$, $B_1\vec{e}_X$ and $-\omega J_z\vec{e}_Z$, where ω is the angular frequency of the radio-frequency field.[3] The rotation matrices for a spin 1 are readily derived from spin 1/2 rotation matrices. This formulation permits to calculate the Majorana probability $P(J = 1; 0 \to 1; t)$ for an atom initially in the Zeeman sublevel $m_e = 0$ to be transferred after a time t in the sublevel $m_e = 1$ [Majorana (1932)].

In practice, the probability that the atom stays in the excited state decreases exponentially with a time constant given by the radiative lifetime $\tau_R = 1/\Gamma$. In the steady state regime, the total number N_{+1} of excited atoms in the sublevel $m_e = 1$

[2] See Sec. 2.3 of Chap. 2.
[3] See Fig. 2.3 of Chap. 2.

is therefore given by:

$$N_{+1} = n_0 \int_0^\infty P(J=1; 0 \to 1; t) e^{-\Gamma t} \mathrm{d}t, \tag{3.5}$$

which can be shown to be equal to [Brossel and Bitter (1952)]

$$N_{+1} = \frac{n_0 \Omega^2}{2} \frac{4\delta^2 + \Gamma^2 + \Omega^2}{(\delta^2 + \Gamma^2 + \Omega^2)(4\delta^2 + \Gamma^2 + 4\Omega^2)}, \tag{3.6}$$

where $\Omega = \gamma B_1$ is the Rabi frequency and $\delta = \omega - \omega_0$ is the detuning between the angular frequency ω of the radio-frequency field and the Zeeman splitting ω_0 between the sublevels.

The double resonance curve contains lots of useful information for the atomic physicist. The square of their half width calculated from the peak height at resonance ($\delta = 0$) varies linearly with B_1^2, and tends to $4\Gamma^2$ when B_1 vanishes. It is therefore possible to derive the natural width Γ of the excited state from the data taken at different values of the amplitude B_1 of the radio-frequency magnetic field. In addition, the gyromagnetic factor that enters the expression of $\omega_0 = \gamma B_0$ is readily obtained from the resonance condition.

3.2.3 *Experimental results*

In practice, the emitted light with a polarization different from the exciting light is collected as a function of the static field B_0 for various radio-frequency field amplitudes. Though its components are simple, such an experiment can yield very accurate results whose quantitative understanding is clearly confirmed by the perfect agreement between theoretical curves and experimental data [Brossel and Bitter (1952)] (see Fig. 3.2). In these data, the radiative broadening is exemplified with the clear increase of the width of the experimental resonance curves with the radio-frequency amplitude. In addition, the central dip that emerges for large amplitudes of the radio-frequency field, signals an effect commonly called the *Majorana reversal*. In the next section, we will develop a perturbative approach that is quite appropriate for enlightening the underlying physics of this effect.

3.2.4 *Interpretation of the Majorana reversal*

It is instructive to expand the expression (3.6) for N_{+1} in powers of Ω:

$$N_{+1} = \frac{n_0}{2} \left[\frac{\Omega^2}{\delta^2 + \Gamma^2} - \frac{\Omega^4}{(\delta^2 + \Gamma^2)^2} - \frac{3\Omega^4}{(\delta^2 + \Gamma^2)(4\delta^2 + \Gamma^2)} + \ldots \right]. \tag{3.7}$$

The lowest order term in Ω^2 gives rise to a Lorentzian curve of full width at half maximum of 2Γ. The first term of the next order in Ω^4 accounts for radiative broadening. The second term in Ω^4 of Eq. (3.7) corresponds to a narrower curve with respect to the Lorentzian curve because of the factor $1/(4\delta^2 + \Gamma^2)$. In addition, this term has a negative sign: it is responsible for the dip structure observed in Fig. 3.2.

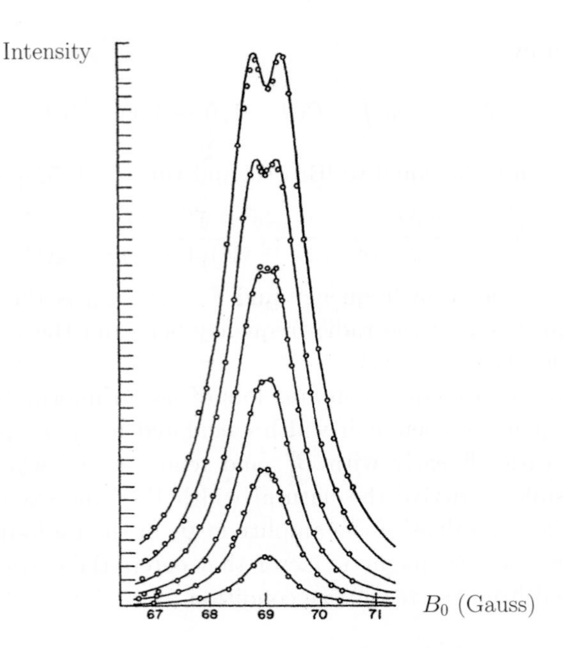

Fig. 3.2 Double resonance curves. The intensity of the emitted light with a polarization different from the exciting light is plotted as a function of the applied static field B_0 for various radio-frequency field amplitudes at a fixed RF frequency. The open-circle points are the experimental results. The solid lines are the theoretical predictions of Eq. (3.6). Figure adapted from [Brossel and Bitter (1952)]. Copyright: American Physical Society.

This expansion suggests an interpretation in terms of quantum interferences involving in total 4 photons.

The terms in Ω^4 reveal the quantum interferences between a one photon process (amplitude $\propto \Omega$) connecting state $m_e = 0$ to state $m_e = 1$, and a three-photon process (amplitude $\propto \Omega^3$) connecting the same states (see Fig. 3.3). Two diagrams can be envisioned for the latter phenomenon: (i) $m_e = 0 \rightarrow m_e = 1 \rightarrow m_e = 0 \rightarrow m_e = 1$ (Fig. 3.3(b)) and (ii) $m_e = 0 \rightarrow m_e = -1 \rightarrow m_e = 0 \rightarrow m_e = 1$ (Fig. 3.3(c)). The one photon diagram as well as the three photon diagram (i) (Fig. 3.3(b)) only involve two Zeeman sublevels and are also present for an excited state of angular momentum $J_e = 1/2$. They account for radiative broadening. For an atom with an excited state of angular momentum $J_e = 1$, there is a third Zeeman sublevel and a new diagram (ii) (Fig. 3.3(c)) appears giving rise to the Majorana reversal effect.

3.3 Optical pumping [Kastler (1950)]

In the previous section, we have emphasized, in the case of a $J_g = 0 \rightarrow J_e = 1$ transition, the role played by polarization selection rules in addressing specifically an excited atomic state with a well-defined magnetic number. Here, we shall consider

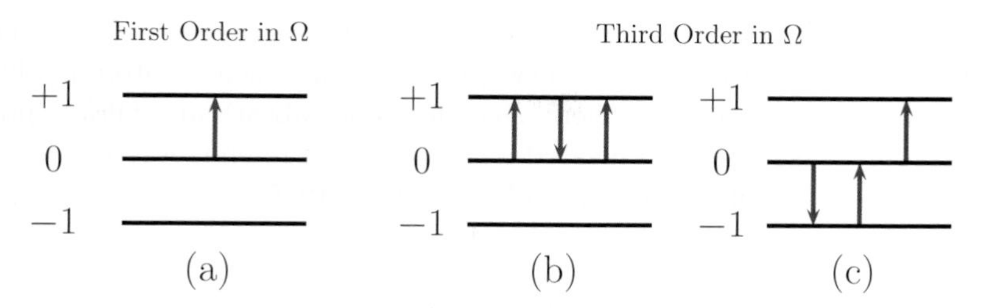

Fig. 3.3 4-photon quantum interferences for a three-level system: Both radiative broadening and Majorana reversal result from the interference between a one photon process and a three photon process. Radiative broadening involves processes (a) and (b), while Majorana reversal involves processes (a) and (c).

the simplest example of an atom having a ground state angular momentum $J_g = 1/2$ and an excited state angular momentum $J_e = 1/2$, and explore the new physics that arises with a degenerate ground state.

3.3.1 *Principle of the method for a* $J_g = 1/2 \to J_e = 1/2$ *transition*

The schematic of the experimental setup we consider is depicted in Fig. 3.4. A magnetic field is applied on a cell containing the atoms of interest. A right-circular polarized photon with a direction of propagation parallel to the magnetic field, selectively excites the transition $|g, m_g = -1/2\rangle \longrightarrow |e, m_e = 1/2\rangle$. Assuming that no disorientation by collisions occurs during the lifetime of the excited state, the atom fluoresces from state $|e, m_e = 1/2\rangle$ to either of the two states $|g, m_g = \pm 1/2\rangle$ with probabilities given by the square of the corresponding Clebsch Gordan

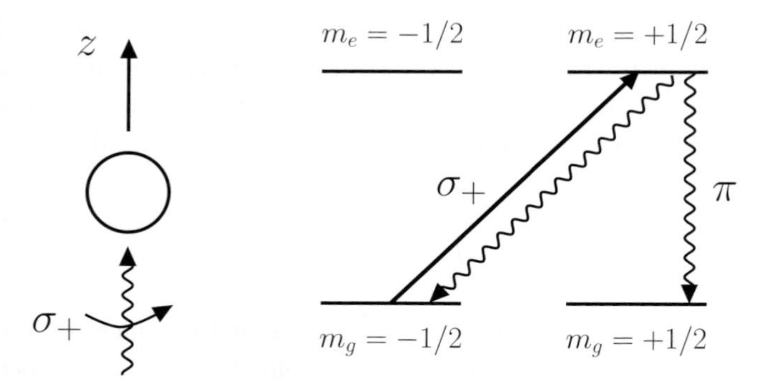

Fig. 3.4 Principle of optical pumping for a $J_g = 1/2 \to J_e = 1/2$ transition. The absorption of σ^+ light only allows transition from state $|g, m_g = -1/2\rangle$ to $|e, m_e = 1/2\rangle$. Once in this excited state, the atom decays to the ground states $|g, m_g = \pm 1/2\rangle$ with probabilities given by the square of the corresponding Clebsch Gordan coefficients. An atom pumped into the level $|g, m_g = 1/2\rangle$ is no longer coupled to the incident σ^+-polarized light since there is no σ_+ transition starting from this state. This optical pumping technique therefore enables one to control and probe the populations in the different ground state sublevels.

coefficients. If, after this absorption-spontaneous emission cycle, the atom ends up in the state $|g, -1/2\rangle$, it can start a new cycle, whereas when an atom finally lands in the $|g, +1/2\rangle$ state it remains there, since it cannot absorb another σ^+ polarized photon. The net result of the repeated absorption and re-emission is a "pumping" of the atoms to the ground state $|g, +1/2\rangle$ [Kastler (1950)].

3.3.2 *Angular momentum balance*

The process by which the pumping operates implies that a σ_+ photon disappears and a π photon is emitted. The field loses one unit $+\hbar$ of angular momentum along the magnetic field axis, while the internal state of the atom is changed from the state $|g, -1/2\rangle$ to the state $|g, +1/2\rangle$ (after passing through the excited state $|e, +1/2\rangle$), which corresponds to a gain of one unit $\hbar$ of angular momentum. Conservation of the total { atom + field } angular momentum results in a transfer of angular momentum from the field to the atom for each absorption-spontaneous emission process.

The oriented atoms in the fully-pumped state can be disoriented mainly through collisions with the cell walls, which tend to equalize the populations of the two ground state sublevels. If, however, the efficiency of optical pumping is large compared to the relaxation processes, a degree of atomic polarization close to 100% can be reached at room temperature with a relatively low magnetic field on the order of few Gauss.

3.3.3 *Double role of light*

Experimentally, the intensity of the light transmitted through the cell is reduced due to absorption. As the state $|g, -1/2\rangle$ becomes more and more emptied, the gas becomes transparent to the resonant optical radiation with polarization σ_+.

By measuring the amount of absorbed light L_A, we obtain an optical signal directly proportional to the number of atoms $N_{-1/2}$ in the sublevel $-1/2$ of the ground state g:

$$L_A \propto N_{-1/2}. \tag{3.8}$$

One can also measure the amount of fluorescence light L_F, which is directly proportional to L_A since this light is re-emitted by atoms that have absorbed a photon. By measuring L_A or L_F, it is therefore possible to optically detect any variation of $N_{-1/2}$ due to collisions or to magnetic resonance transition between $|g, -1/2\rangle$ and $|g, +1/2\rangle$.

This clearly shows that, in optical pumping experiment, light plays a double role: (i) it *pumps* an atomic sample by changing the population of the ground state sublevels and (ii) it *probes* the polarization of the ground state.

3.4 First experiments on optical pumping

So far, we have presented the principle of optical pumping using the simple example of an atom with angular momentum 1/2 in both the ground and the excited states. Optical pumping experiments have indeed been performed with atoms having such a level scheme. This is, for instance, the case of mercury-199 discussed hereafter in Sec. 3.5.2. In fact, the first proposal of optical pumping [Kastler (1950)] was presented for alkali atoms, but the principle of the method is basically the same as for a $J_g = 1/2 \rightarrow J_e = 1/2$ transition. The first demonstration of optical pumping was performed on an atomic beam of sodium [Brossel *et al.* (1952)]. The method was then applied to an atomic vapor contained in a glass cell [Cagnac (1961)]. For a review of the early experiments in optical pumping, see [Cohen-Tannoudji and Kastler (1966)].

Alkali atoms have two ground state hyperfine levels. Optical pumping can work not only inside a given ground state hyperfine level but also between two different hyperfine levels. If the optical lines starting from these two levels are resolved optically (hyperfine splitting large compared to the spectral width of the light beam), atoms can be selectively removed from a hyperfine level by a properly tuned light beam and transferred to the other hyperfine level in absorption-spontaneous emission cycles. Such a scheme is referred to as hyperfine pumping.

The finite time of interaction between the light and the atoms in experiments performed with atomic beams limits the number of atoms per unit time that are involved, which therefore limits the signal. Using a cell, the interaction time is only limited by the time-of-flight of the atoms from one wall of the cell to the other. However, the atoms can lose their polarization by colliding with the walls of the container. To overcome this limitation, the gas to be polarized is often mixed with a buffer gas that has no magnetic substates in its ground electronic state, and that is therefore unable to absorb or donate the small amount of energy required for magnetic substate transitions. The buffer gas shields the atoms of interest from colliding with one another, and slows their diffusion towards the walls [Brossel *et al.* (1955); Bender (1956); Hartmann *et al.* (1958)]. In addition, a suitable coating of the walls can substantially slow down relaxation processes [Robinson *et al.* (1958)].

3.5 How can optical pumping polarize atomic nuclei?

3.5.1 *Using hyperfine coupling with polarized electronic spins*

In the interaction Hamiltonian describing the atom-field coupling, the electric field of the incident light wave interacts with the electric dipole moment of the atoms and does not act in any way on the nuclear spin. Under these conditions, one can thus wonder how nuclear spins can become polarized by optical pumping. In this section, we address this question by showing how nuclei can become polarized

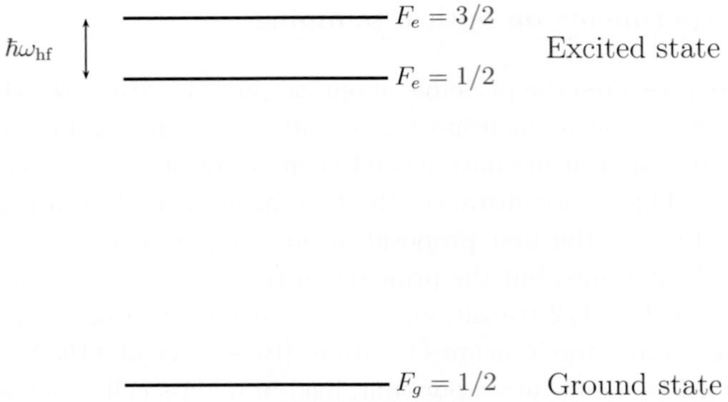

$\hbar\omega_{\rm hf}$ $F_e = 3/2$ Excited state
$F_e = 1/2$

$F_g = 1/2$ Ground state

Fig. 3.5 Hyperfine structure of an atom with nuclear spin $I = 1/2$ and electronic angular momentum $J_g = 0$ for the ground state and $J_e = 1$ for the excited state. The hyperfine coupling gives rise to a single hyperfine level for the ground state $F = I = 1/2$, and to two hyperfine levels $F = 1/2$ and $F = 3/2$ for the excited state. $\hbar\omega_{\rm hf}$ is the energy splitting between the two hyperfine levels of the excited state.

through hyperfine coupling with the polarized electrons.

We consider the case of an atom with a nuclear spin $I = 1/2$ excited on an optical electronic transition connecting a ground state $J_g = 0$ to an excited state $J_e = 1$, where J is the electronic angular momentum. The hyperfine coupling between I and J gives rise to a single hyperfine level $F = I = 1/2$ in the ground state, and to two hyperfine levels in the excited state, $F = 1/2$ and $F = 3/2$, separated by an energy splitting $\hbar\omega_{\rm hf}$ which is the hyperfine structure in the excited state e (see Fig. 3.5). The ground state hyperfine level $F = I = 1/2$ has two Zeeman sublevels $m_I = \pm 1/2$ that correspond to the two possible orientations of the nuclear spin.

First, we consider the case of a broad line excitation where the linewidth Δ of the exciting light is much larger than the hyperfine structure in the excited state e:

$$\Delta \gg \omega_{\rm hf}. \tag{3.9}$$

Let us start from the state $M_I = -1/2$ in g and suppose that the exciting light has a σ_+ polarization. The absorption of a σ_+ photon lasts a time $1/\Delta$, which is the correlation time of the exciting light.[4] Equation (3.9) entails that this time is very short, so short that the effect of the hyperfine coupling between I and J is negligible during this time:

$$\omega_{\rm hf} \times \frac{1}{\Delta} \ll 1. \tag{3.10}$$

Just after the excitation process, M_I is unchanged and the atom is thus in the state $M_J = +1, M_I = -1/2$ (see the single arrow of Fig. 3.6 where the atomic states are represented in the M_J, M_I basis).

The hyperfine interaction commutes with the total angular momentum $F = I + J$, and thus does not change the value of $M_F = M_J + M_I$. During the radiative

[4]See Sec. 2.5.2 of Chap. 2.

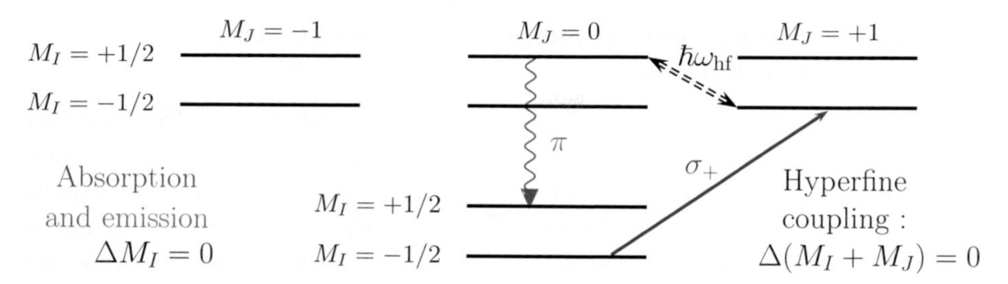

Fig. 3.6 Atomic states represented in the M_J, M_I basis. An atom initially in the Zeeman sublevel $M_I = -1/2$ in the ground state reaches the state $M_J = 1$ $M_I = -1/2$ by absorption of a σ_+ photon. The hyperfine interaction couples the $M_J = 1$ $M_I = -1/2$ state to the $M_J = 0$ $M_I = 1/2$ state, and thus, the nuclear spin changes its orientation because of the hyperfine interaction.

lifetime $1/\Gamma$ of e, hyperfine interactions couple the state $M_J = +1, M_I = -1/2$ to the state $M_J = 0, M_I = +1/2$ since it has the same value of M_F (double arrow of Fig. 3.6). The orientation of the nuclear spin can thus be changed thanks to its interaction with the polarized electron.

After a time on the order of $1/\Gamma$, the atom falls back to the ground state in a spontaneous emission process, which lasts so short a time that the nuclear spin remains unchanged during the process. If the state $M_J = 0, M_I = +1/2$ has been appreciably populated during the time spent in the excited state e, i.e. if

$$\omega_{\rm hf} \times \frac{1}{\Gamma} \gg 1 \qquad \Longrightarrow \qquad \omega_{\rm hf} \gg \Gamma \tag{3.11}$$

an appreciable population of the state $M_I = +1/2$ can be obtained in the ground state after the spontaneous transition $M_J = 0, M_I = +1/2 \to M_I = +1/2$ (vertical wavy arrow of Fig. 3.6). It thus clearly appears that, with a broad line excitation, optical pumping can polarize the nuclear spins, indirectly by their interaction with the polarized electrons, provided that hyperfine coupling lasts a long enough time.

These effects have been studied theoretically and experimentally by Jean-Claude Lehmann in his Ph.D. work [Lehmann (1967)]. It has been shown that Cd nuclei can be polarized by optical pumping on the transition $5^1\mathrm{S}_0 \leftrightarrow 5^1\mathrm{P}_1$ (in the excited state $5^1\mathrm{P}_1$, $\omega_{\rm hf}$ is equal to 3Γ), whereas Zn nuclei cannot be polarized by optical pumping on the transition $4^1\mathrm{S}_0 \leftrightarrow 4^1\mathrm{P}_1$, since in the excited state $4^1\mathrm{P}_1$, $\omega_{\rm hf}$ is much smaller than Γ.

One may wonder what happens if the linewidth Δ of the exciting light is smaller than $\omega_{\rm hf}$, while $\omega_{\rm hf}$ and Δ are larger than Γ, such that the incoming light can selectively excite one of the two hyperfine components $F_g = 1/2 \to F_e = 1/2$ or $F_g = 1/2 \to F_e = 3/2$ of the optical line:

$$\omega_{\rm hf} \gg \Delta \gg \Gamma. \tag{3.12}$$

In this case, the excitation process that brings the atom in the excited state lasts long enough to allow hyperfine coupling to produce an appreciable effect during the excitation process itself. As a result, the nuclear spin can change during the excitation process, thereby making it more appropriate to use the basis $|F, M_F\rangle$ for describing the various Zeeman sublevels in e and g.

3.5.2 *First example: optical pumping experiments with mercury-199 atoms*

In contrast to the even isotopes discussed earlier, the odd mercury isotopes have a non-vanishing nuclear spin I. For example, the isotope mercury-199 has a nuclear spin equal to $I = 1/2$. The magnetism of its ground state $6^1\mathrm{S}_0$ is consequently nuclear and not electronic. In the excited state $6^3\mathrm{P}_1$, the electronic angular momentum $J = 1$ is coupled to the nuclear spin I, resulting in an hyperfine structure of the excited state with two levels, $F_e = 1/2$ and $F_e = 3/2$. The hyperfine splitting in this case is very large ($\sim$ 22 GHz) compared to linewidth Δ ($\sim 2\pi\times$ 1.3 GHz). It turns out that there is a coincidence between the frequency of the optical line of the even isotope mercury-204 and the frequency of the transition $6^1\mathrm{S}_0, F_g = 1/2 \to 6^3\mathrm{P}_1, F_e = 1/2$ of the odd isotope mercury-199, so that one can selectively excite this transition of mercury-199 with a lamp filled with mercury-204. In this way, one realizes a situation identical to the simple level scheme of Fig. 3.4, already used to explain the principle of optical pumping. Not surprisingly one arrives at the same conclusion: in absorption-spontaneous emission cycles angular momentum can be transferred from the polarized light beam to the atom, leading to a spin polarization in g. The interesting point here is that, since $J_g = 0$, this polarization is purely nuclear, meaning that optical methods allow one to detect nuclear magnetic resonance in gaseous samples for which usual radio electric detection methods would be totally inefficient.

During his Ph.D., Bernard Cagnac studied the nuclear magnetic resonance of these atoms in their ground state. The resonance was measured optically at room temperature and in a magnetic field of a few Gauss. This provides a very high-precision measurement of the nuclear magnetic moment with an accuracy better than 10^{-5} [Cagnac (1961)].

The dynamics of optical pumping was also investigated. Figure 3.7(a), which gives the time evolution of the fluorescence light L_F, proportional to $N_{-1/2}$, shows the establishment of the steady state value of $N_{-1/2}$ resulting from the competition between optical pumping and relaxation processes. If the pumping beam is switched off during a finite time τ, the atomic populations relax in the dark towards their equilibrium value in the absence of pumping and this relaxation is analyzed by switching back on the pumping beam (see Fig. 3.7(b)). Repeating this experiment for various values of τ gives the sets of curves shown in Fig. 3.7(c) whose starting points lie on the curve describing the evolution of $N_{-1/2}$ due to relaxation processes.

3.5.3 *Second example: combining optical pumping with metastability exchange collisions for helium-3*

The ground electronic state of the helium atom is a $1^1\mathrm{S}_0$ state with zero electronic angular momentum $J_g = 0$. The nucleus of helium-3 has a nuclear spin $I = 1/2$, so that the ground state is a state $F = I = 1/2$, as in mercury-199. One can wonder if

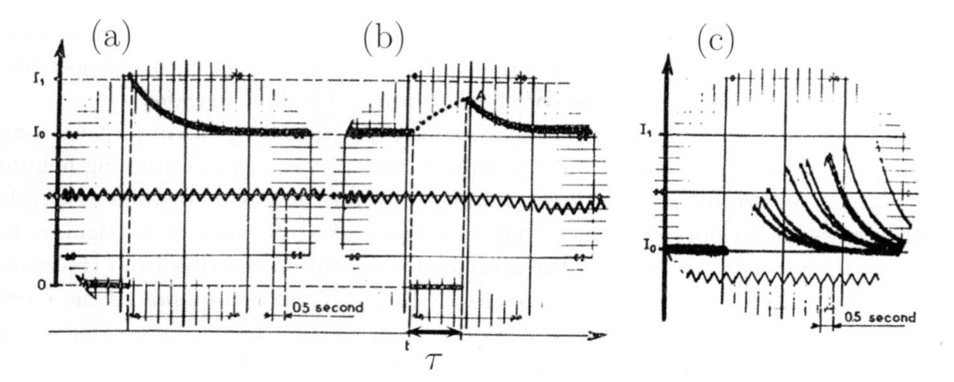

Fig. 3.7 Dynamics of optical pumping. Evolution of the fluorescence light L_F, proportional to $N_{-1/2}$ for various time-dependent pumping lights. (a) When the pumping beam is suddenly switched on, $N_{-1/2}$ decreases from its thermal equilibrium value to a steady-sate value resulting from the competition between optical pumping and thermal relaxation processes. (b) If the pumping beam is then switched off during a time τ, $N_{-1/2}$ relaxes in the dark towards its thermal equilibrium value (dotted line). When the pumping beam is switched on again, one gets a new pumping curve whose starting point gives the value $N_{-1/2}$ after a relaxation phase in the dark of duration τ. (c) Repeating the same sequence for various values of τ gives a series of transient pumping curves whose starting points yield the full thermal relaxation curve of the atoms. Figure extracted from [Cagnac (1961)]. Copyright: EDP Sciences.

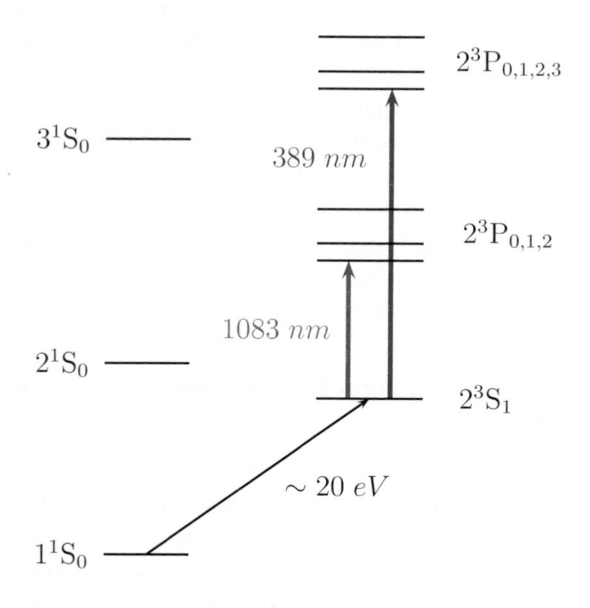

Fig. 3.8 Energy level diagram for helium atom.

it would be possible to polarize helium-3 nuclei by optical pumping. Figure 3.8 gives the energy levels of the helium atom and shows that all excited states are located more than 20 eV above the ground state. There are no convenient available light sources for exciting the transitions starting from the ground state, and an optical pumping analogous to the one considered in the previous subsection is not possible. We describe now another method for circumventing this difficulty.

The first excited state of helium is a metastable state $2^3\mathrm{S}_1$ having a very long lifetime, on the order of two hours. By using a discharge in a cell containing helium atoms, one can produce a significant number of atoms in this metastable state. It is then possible to optically pump the atoms in this state, and in particular, to polarize their nuclei, by exciting the atoms on a transition starting from the state $2^3\mathrm{S}_1$ and going to higher excited states like $2^3\mathrm{P}_{0,1,2}$. This corresponds to the more convenient wavelength of 1083 nm. Can one then transfer the nuclear orientation obtained in this way from the metastable state to the ground state?

This can be achieved by using *metastability exchange collisions*. During a collision between an atom A in the metastable state and another one B in the ground state $1^1\mathrm{S}_0$, the electronic excitation can be transferred from A to B:

$$(A,\ 2^3\mathrm{S}_1)\ +\ (B,\ 1^1\mathrm{S}_0)\ \Rightarrow\ (A,\ 1^1\mathrm{S}_0)\ +\ (B,\ 2^3\mathrm{S}_1). \qquad (3.13)$$

The duration of this collision is so short that the nuclear spin cannot evolve during this time, such that A remains polarized after it has given its electronic excitation to B. This method, first proposed and demonstrated in 1963 [Colegrove *et al.* (1963)], is very efficient and with the use of lasers for optical pumping of the $2^3\mathrm{S}_1$ state, can now lead to very high degrees of nuclear polarization in the ground state, on the order of 80 %.

Let us also mention another method for polarizing helium-3 nuclei which uses the spin dipole interactions that occurs during collisions between the nuclear spins of helium-3 and the electronic spins of alkali atoms polarized by optical pumping [Bouchiat *et al.* (1960)].

Such methods for the production of polarized helium-3 have facilitated its use in several applications, such as polarized quantum fluids, polarized targets for nuclear physics, and magnetometers. We now describe a recent application in medicine.

3.5.4 *A new application: magnetic resonance imaging of the lung cavities*

Optical pumping has recently found an interesting application for imaging the human body. We describe here experiments using polarized helium-3 atoms.[5] To begin, one prepares a sample of polarized gaseous helium-3, using optical pumping with a laser. This gas is inhaled by patients which allows magnetic resonance images of the cavities in their lungs to be obtained (see Fig. 3.9). Current proton-based MRI only provides information on solid or liquid parts of the human body such as the muscles, the brain, or the blood (left part of Fig. 3.9). The use of helium with high degrees of spin polarization provides MRI signals strong enough to be detected,

[5] The first magnetic resonance images of lungs with optically pumped noble gases was demonstrated in an excised mouse lung inflated with polarized xenon-129 atoms [Albert *et al.* (1994)].

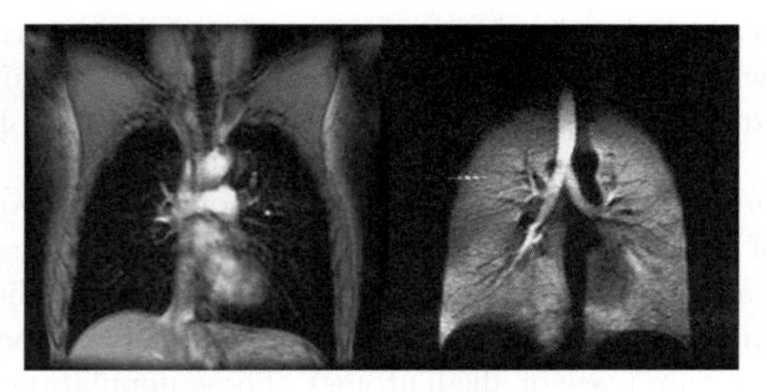

Fig. 3.9 Magnetic resonance imaging (MRI) of the chest of a patient. Left: image obtained with ordinary proton-based resonance. Right: image obtained with gaseous helium-based resonance. The patient has inhaled a mixture of air and helium-3, and the latter has been polarized using optical pumping. Figure taken from [Johnson (1998)]. Copyright: Institute of Physics.

even with a system as dilute as a gas. These signals allow internal spaces of the body, like the cavities of the lung, to be visualized at unprecedented resolutions (right part of Fig. 3.9). The use of this polarized gas provides a promising tool important for improving our understanding of lung physiology and function (see [Johnson (1998); Möller (2002)] and references therein).

3.6 Brief survey of the main applications of optical methods

The examples given in this chapter show how it is possible to measure radiative lifetimes, g-factors, nuclear magnetic moments, hyperfine structures. Because measurements are performed on very dilute gases, they are less perturbed by collisions and can be much more precise.

- *Spectroscopic applications.* When atoms are not in a beam, but in a cell, they can lose their polarization by colliding with the walls of the container. To overcome this limitation, the gas to be polarized is often mixed with a buffer gas that shields the atoms of interest from colliding with one another, and slows their diffusion towards the walls. Suitable coating of the walls of the cell can also slow down relaxation processes [Robinson *et al.* (1958)]. The pressure broadening and the pressure shift of the magnetic resonance curves when the buffer gas pressure is increased give information on the interaction potential between the optically pumped atoms and the buffer gas atoms.

- *Dynamical studies.* Atoms are removed from thermal equilibrium by the pumping beam. If the beam is switched off, they return to thermal equilibrium and their relaxation process can be followed by a probe beam. An example of such a dynamical study is given in Fig. 3.7. Such experiments can be considered to be precursors of pump-probe experiments performed later on with lasers, where a first pump laser perturbs an atomic medium whose state is subsequently monitored with a second probe laser.

- *Population inversion.* Through proper choice of the polarization or frequency of the pumping light, it is possible to concentrate atoms in the highest magnetic sublevel of an hyperfine state in an applied magnetic field, or in the highest hyperfine level when the atom has a hyperfine structure in the ground state (case of alkali atoms). These population inversions can be used to achieved masers (microwave amplification by stimulated emission radiation) oscillating at the hyperfine frequency or at the Larmor frequency [Gordon *et al.* (1954)].

- *Study of atom-field interactions.* The high precision that can be obtained with optical methods allows several physical effects associated with atom-field interactions to be investigated.

  A first example is the discovery of light shifts which are energy displacements of the atomic ground state sublevels proportional to the intensity of a quasi resonant light irradiation. They will be described in more detail in Chap. 7. With the usual weak intensity of thermal sources, these light shifts are small, on the order of a few Hz. However, they can still be detected by a shift of the magnetic resonance curves in the ground state which are very narrow. With the higher intensity of laser light, light shifts can be much larger, and it turns out that they play an important role in schemes that have been recently developed for cooling atoms to very low temperatures and for trapping them in various types of potential wells produced by spatially dependent light shifts.

  A second example concerns multiphoton radio-frequency transitions. We will see in Chap. 9 that an atom can go from one Zeeman sublevel to another one by absorbing several radio-frequency photons. Optical methods have allowed these radio-frequency multiphoton resonances, and in particular their radiative broadening and shifts, to be observed and studied in great detail. This has paved the way to the study of optical multiphoton resonances.

  We should also mention that the dressed atom approach for understanding the behavior of atoms in intense fields was first developed for atoms interacting with intense radio-frequency fields. Most of the new effects predicted by this approach, such as the modification of the Landé factor of a spin dressed by an intense high frequency RF field (described in Chap. 8) have been investigated with optical methods.

- *New effects related to linear superposition of atomic internal states.* It is during the development of optical methods that the importance of linear superposition of states in atomic physics was fully realized, showing that the description of the ensemble of atoms in terms of populations of the energy levels was insufficient and that it was necessary to describe them with a density matrix with diagonal as well as off-diagonal elements. The next chapter will be devoted to the description of the new physical effects involving linear superpositions of states and how they have considerably enriched the power of optical methods. These studies can be considered to be precursors of modern developments involving quantum interference and dealing with atomic interferometry and entangled states.

- *Tests of fundamental physics.* Any new high precision measurement allows new more refined tests of the fundamental laws of physics to be performed. We just mention here the first demonstration of parity violation in atoms, which will be described in more details in Chap. 25, and whose principle is clearly inspired by optical methods. The idea is to excite an atom with a configuration of static fields and light beams that is invariant under a mirror reflection with respect to a plane Π. One looks at the polarization of the light re-emitted by the atom. If this polarization of the light is not invariant under the same mirror reflection, it means that atom-field interactions are not invariant under a space reflection, and thus violate parity. It is in this way that parity violation due to weak neutral current interactions (due to exchanges of Z_0 bosons between the electron and the nucleus) was first demonstrated in 1982 [Bouchiat *et al.* (1982)], following the initial proposal of [Bouchiat and Bouchiat (1974a,b)].

- *Practical applications*

  - Magnetometers
    The measurement of magnetic field is performed *in situ* by measuring the resonance frequency $\omega_0 = \gamma B_0$ [Malnar and Mosnier (1961); Bloom (1962)]. This technique has been used in several modern experiments, for example, in experiments looking for the existence of an electric dipole moments of atoms.
  - Atomic clocks
    The difference of population between the two hyperfine ground state levels of rubidium or cesium is achieved by hyperfine pumping instead of Stern and Gerlach type state selecting devices. It allows one to construct much more compact atomic clocks.
  - Masers
    They rely on population inversions achieved by optical pumping Zeeman or hyperfine sublevels.

3.7 Concluding remarks

To conclude this chapter, we would like to summarize the key features of optical pumping:

(i) *Large population differences*

The transfer of angular momentum from polarized photons to atoms can concentrate atoms into a Zeeman sublevel, or, at least, produce population differences much larger than those given by the Boltzmann factor for thermal equilibrium at room temperature. Optical pumping thus opens the possibility for detecting magnetic resonance on dilute systems such as atomic vapors.

(ii) *High sensitivity*

The population changes due to magnetic resonance (or relaxation processes) are detected by a change of the absorbed or fluorescence light. This optical detection is much more sensitive than usual radioelectric detection of magnetic resonance because it is much easier to detect an optical photon since it carries a much higher energy than a radio-frequency photon.

The combination of this high sensitivity with the large population difference explains why it is possible to easily detect, at room temperatures and in low magnetic fields, nuclear magnetic resonances in atomic ground states whose magnetism comes only from the nuclear spins. A radioelectric detection would measure the radio-frequency energy absorbed or emitted during a transition between Zeeman sublevels, which is extremely small for a nuclear spin. By contrast the change of angular momentum in a Zeeman transition is as important for a nuclear spin as for an electronic spin and the probability of absorbing an optical photon only depends on the angular momentum of the Zeeman sublevel.

The combination of optical detection and radio-frequency or microwave frequency radiation thus makes it possible to observe transitions between Zeeman levels of hyperfine states in weak magnetic fields, where the spacing between neighboring Zeeman states is less than 10^{-8} eV ! The optical pumping scheme is an invaluable tool for determining the characteristics of a given atom. Conversely, it provides a very efficient method for measuring very weak magnetic fields.

(iii) *First example of manipulation of atoms by light*

Optical pumping is one of the first examples of the possibility of manipulating atoms by light: transfer of angular momentum between polarized photons and atoms allows one to control the populations of the internal atomic states and to detect their changes. In his original paper, Alfred

Kastler also suggested that it could be possible to cool the external degrees of freedom using this technique. The idea of the suggested scheme called "effet luminofrigorifique" was first to cool the internal degrees of freedom of atoms by optically pumping them into the lowest energy state and relying on collisions for equilibrating internal and translational temperatures.[6]

In fact, more efficient methods for manipulating external degrees of freedom of atoms were demonstrated a few decades later with laser cooling, which relies on the exchange of linear momentum between photons and atoms. Chapters 11–13 of this book will be devoted to the description of these methods. Let us just mention here that, by reducing considerably the atomic velocities, laser cooling allows high resolution optical spectroscopy[7] to be performed on atoms without any Doppler broadening.

[6]The feasibility of this scheme was experimentally demonstrated in 2006 with chromium-52 atoms using inelastic dipolar collisions to couple the motional and the spin degrees of freedom [Fattori *et al.* (2006)].

[7]Other methods for high resolution spectroscopy such as the Doppler free methods based on nonlinear effects are described in Chap. 6.

Chapter 4

Linear superpositions of internal atomic states

4.1 Introduction

In the description of the experiments presented in the previous chapter the emphasis has been put on the populations of the atomic Zeeman sublevels. It has been shown that the excitation of atoms with polarized light can lead to large population differences between the Zeeman sublevels in the excited state e or in the ground state g. Similarly, by monitoring the polarized light emitted or absorbed by the atom, one can optically detect any change of these population differences induced by resonant RF transitions or by relaxation processes.

In fact, the most general state of an atom cannot be described just by giving the populations of its energy states $|E_n\rangle$, which are also the probabilities π_n of occupation of these states. An atom can be in a linear superposition of the states $|E_n\rangle$:

$$|\psi\rangle = \sum_n c_n \, |E_n\rangle. \tag{4.1}$$

This happens for example when it interacts with a resonant RF field. The predictions of the measurements of an observable $\hat{G}$ performed on an atom in the state (4.1) depend not only on the populations $|c_n|^2$ of the states E_n, but also on the crossed terms $c_n c_{n'}^*$ with $n \neq n'$, which are complex numbers whose arguments depend on the relative phase between c_n and $c_{n'}$:

$$\langle \hat{G} \rangle = \sum_{n,n'} G_{n'n} c_n c_{n'}^*. \tag{4.2}$$

More generally, the state of the atom is not necessarily a pure state like (4.1), but a statistical mixture of such states described by a density operator $\hat{\sigma}$, having not only diagonal elements $\langle E_n | \, \hat{\sigma} \, | E_n \rangle = \pi_n$, but also off diagonal elements $\langle E_n | \, \hat{\sigma} \, | E_{n'} \rangle = \sigma_{n\,n'}$ with $n \neq n'$, often called *atomic coherences*, between the states $|E_n\rangle$ and $|E_{n'}\rangle$. By this denomination, one means that, if the populations $|c_n|^2$ and $|c_{n'}|^2$ are the same for all the states of the statistical mixture, the relative phase between c_n and $c_{n'}$ does not vary randomly from one state of the statistical mixture to another one:

$$\langle E_n | \, \hat{\sigma} \, | E_{n'} \rangle = \overline{c_n c_{n'}^*} \neq 0, \tag{4.3}$$

where the bar means the average over the states of the statistical mixture. For the sake of convenience, we will call $\sigma_{n\,n'}$ a *Zeeman coherence* when $|E_n\rangle$ and $|E_{n'}\rangle$ are two Zeeman sublevels belonging to the same hyperfine level, a *hyperfine coherence* when $|E_n\rangle$ and $|E_{n'}\rangle$ are two Zeeman sublevels belonging to two different hyperfine levels, and *optical coherence* when one of the two states $|E_n\rangle$ or $|E_{n'}\rangle$ belong to the ground atomic state whereas the other one belongs to the excited state.

In this chapter, we will review a few physical effects associated with the existence of atomic coherences and which have played an important role in the development of optical methods. They give new physical insight into quantum interference effects. We will address a few questions such as:

- How to prepare these atomic coherences?
- How to detect them?
- How do they evolve?
- What are the new physical effects to which they give rise?
- What are the possible applications?

We will restrict in this chapter to atomic coherences between internal atomic states, assuming that the center of mass of the atom can be treated classically as a heavy particle at rest. Except in Sec. 4.6, we will consider only Zeeman or hyperfine coherences and the light excitation will be supposed to be broadband. In Sec. 4.6, we will consider an atom excited by coherent monochromatic laser beams, a situation where optical coherences also play an important role.

4.2 First experimental evidence of the importance of atomic coherences

A very surprising result was obtained in 1956 when several sets of double resonance curves analogous to those shown in Fig. 3.2 of the previous chapter were recorded for increasing values of the temperature of the cell [Guiochon-Bouchiat *et al.* (1956, 1957)]. It clearly appeared that the width of the resonances were decreasing when the temperature increases (see Fig. 4.1). One would rather expect a broadening of the curves due to the increase of the collisions between atoms when the vapor density increases.

The interpretation of this result was readily found. When an excited atom spontaneously emits a photon, this photon can be reabsorbed by another atom. If the size of the cell is large enough, this process can be important for vapor pressures such that the collisions between atoms are still negligible. The photon emitted by the first atom keeps some information of the Zeeman coherence introduced by the RF field in the excited state of this atom and transfers to the excited state of the second atom part of this information. After averaging over all its possible positions, the second atom keeps a memory of the Zeeman coherence of the first atom. Everything happens as if the Zeeman coherence introduced by the RF field was living a

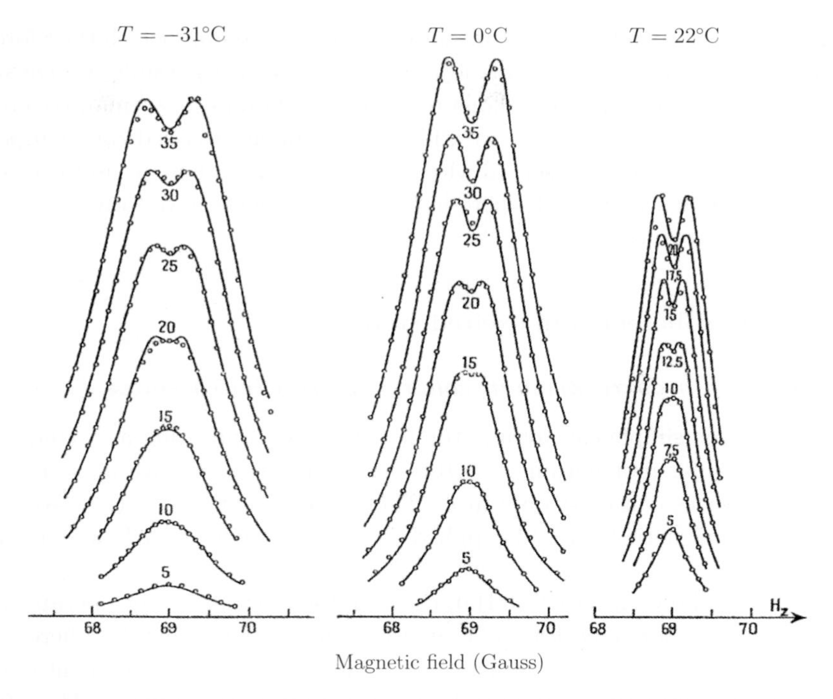

Fig. 4.1 Sets of experimental double resonance curves for increasing values of the temperature T of the cell [Guiochon-Bouchiat *et al.* (1956)]. For each value of T the intensity of the emitted light with a polarization different from the exciting light is plotted as a function of the applied static field B_0 for various radio-frequency field amplitudes at a fixed RF frequency. The open-circle points are the experimental results. The solid line is the theoretical result of Eq. (3.6). The width of the curves decreases when the temperature increases. Figure extracted from [Guiochon-Bouchiat *et al.* (1956)]. Copyright: Elsevier Masson SAS.

time longer than the radiative lifetime of a single isolated atom. As a result, there is an increase of the effective lifetime of this coherence due to coherent multiple scattering and this is why the width of the curves decreases. This interpretation was confirmed by several experimental tests and by a quantitative theory of the effect [Barrat (1959a,b,c)].

As it often happens in the development of a research field, the importance of a new concept does not necessarily appear in the simplest possible case. The narrowing of double resonance curves due to coherent multiple scattering is a rather subtle illustration of the importance of atomic coherences. There are much simpler examples of situations where atomic coherences play an important role. They will be reviewed in the subsequent sections of this chapter. Most of these examples concern Zeeman coherences and broad line optical excitations, because it is then possible to derive simple results concerning the preparation and the detection of these Zeeman coherences, and their equation of motion. Furthermore, it is possible to relate Zeeman coherences to transverse atomic physical quantities, for example, a magnetization perpendicular to the static magnetic field, and which therefore precess around this field [Cohen-Tannoudji (1962)]. It is then possible to give simple physical interpretations of the various resonant effects due to Zeeman coherences. In the last section of this chapter, we will describe an effect called *coherent population trapping* which can be observed with two narrow line coherent laser excitations of a Λ-type atomic system and where hyperfine coherences and optical coherences play an important role.

4.3 Zeeman coherences in excited states

4.3.1 *How to prepare Zeeman coherences in excited states e?*

Consider first the simple case of an atomic transition connecting a ground state g with a zero angular momentum $J_g = 0$ to an excited state e with $J_e = 1$, so that Zeeman coherences can exist only in e. The quantization axis is the z-axis along which a static magnetic field $\vec{B}_0$ is applied, the energy sublevels $|M_e\rangle$ in e being the eigenstates of J_z.

In the atom-photon interaction Hamiltonian, the internal atomic operator multiplying the annihilation operator a of a photon is proportional to $\vec{\varepsilon}\cdot\vec{D}$, where $\vec{\varepsilon}$ is the polarization vector of the exciting light and $\vec{D}$ the electric dipole moment operator of the atom. The probability amplitude that the atom in the sublevel $|M_g = 0\rangle$ of g absorbs a photon and goes to the sublevel M_e of e is proportional to $\langle M_e|\,\vec{\varepsilon}\cdot\vec{D}\,|0\rangle$. If the polarization $\vec{\varepsilon}$ is a linear superposition of the three basic polarization vectors, $\vec{\varepsilon}_+$, $\vec{\varepsilon}_-$, $\vec{\varepsilon}_z$, corresponding to σ_+, $\vec{\sigma}_-$, $\vec{\pi}$ polarizations, respectively:

$$\vec{\varepsilon} = c_+\vec{\varepsilon}_+ + c_-\vec{\varepsilon}_- + c_z\vec{\varepsilon}_z, \tag{4.4}$$

the atom is put, after the absorption of a photon in the state:[1]

$$|\psi\rangle = c_+\,|M_e = +1\rangle + c_-\,|M_e = -1\rangle + c_0\,|M_e = 0\rangle\,. \tag{4.5}$$

To prepare in e a linear superpositions of Zeeman sublevels, one must therefore excite the atoms with a polarization which is not one of the three basic polarizations σ_+, σ_- or π, but a linear superposition of at least two of these three polarizations.

4.3.2 *Physical interpretation*

The spin angular momentum of the photon is transferred to the atom during the absorption process. What should be the polarization of the photon, which characterizes its spin angular momentum (for an electric dipole transition), in order to get Zeeman coherences in e?

[1] In writing (4.5), we use the fact that the Clebsch Gordan coefficients for a $J_g = 0 \leftrightarrow J_e = 1$ transition are all equal so that $\langle M_e = +1|\,\vec{\varepsilon}_+\cdot\vec{D}\,|0\rangle = \langle M_e = -1|\,\vec{\varepsilon}_-\cdot\vec{D}\,|0\rangle = \langle M_e = 0|\,\vec{\varepsilon}_z\cdot\vec{D}\,|0\rangle$.

To answer this question, one must first look for the atomic observables in e whose average values depend on the Zeeman coherences (see Eq. (4.2)). The z-component J_z along the quantization axis of the atomic angular momentum is diagonal in the $\{|M_e\rangle\}$ basis and its mean value depends only on the populations $\langle M_e|\,\hat{\sigma}\,|M_e\rangle$. On the other hand, the transverse components of $\vec{J}$, like J_x, J_y, $J_\pm = J_x \pm iJ_y$, obey selection rules $\Delta M_e = \pm 1$ and have therefore mean values depending on the Zeeman coherences $\langle M_e \pm 1|\,\hat{\sigma}\,|M_e\rangle$. Similarly, quadratic functions of the transverse components of $\vec{J}$, like J_x^2, J_y^2, or $J_\pm^2$ depend on the Zeeman coherences $\langle M_e \pm 2|\,\hat{\sigma}\,|M_e\rangle$. In order to have Zeeman coherences in e, one must therefore introduce an angular momentum in e having transverse components with respect to the quantization axis determined by the applied static magnetic field. Because of the conservation of the total angular momentum of the { atom + photon } system, this transverse angular momentum must come from the absorbed photon. The spin angular momentum of the photon must therefore have components transverse with respect to the quantization axis z which is equivalent to say that its polarization $\vec{\varepsilon}$ must be a linear superposition of the three basic polarization vectors, $\vec{\varepsilon}_+$, $\vec{\varepsilon}_-$, $\vec{\varepsilon}_z$.

Remark

More generally, one can show that the density operator $\hat{\sigma}_e$ in the excited state e after the absorption of a photon of polarization $\vec{\varepsilon}$ is given by:

$$\hat{\sigma}_e \propto (\vec{\varepsilon}\cdot\vec{D})\,\hat{\sigma}_g\,(\vec{\varepsilon}^{\,*}\cdot\vec{D}) \tag{4.6}$$

where $\hat{\sigma}_g = |M_g = 0\rangle\,\langle M_g = 0|$ is the density operator in g, which, in the simple case $J_g = 0$ considered here, is a scalar proportional to the unit operator. Equation (4.6) then shows that $\hat{\sigma}_e$ is proportional to a product of two vector operators and can thus, according to Wigner-Eckart theorem, be expanded in a basis of irreducible tensor operators $T_q^{(k)}$, with $k = 0, 1, 2$ and $q = -k, -k+1, \ldots k-1, k$.[2] The three operators $T_q^{(1)}$, with $q = -1, 0, +1$ are the three components of a vector operator called *orientation*. The five operators $T_q^{(2)}$, with $q = -2, -1, 0, +1, +2$ are the five components of a quadrupole operator called *alignment*.

These operators must have the same symmetry properties as the polarization $\vec{\varepsilon}$ with respect to rotations and reflections. Suppose that the exciting light beam is σ_+ polarized and propagates along a direction $\vec{u}$ which is transverse with respect to the direction of the quantization axis. Then, only the operators $T_q^{(1)}$ appear in the expansion of $\hat{\sigma}_e$, describing an orientation, pointing along the direction of $\vec{u}$ and having the same symmetries as a magnetic field. If the exciting light beam has a linear polarization parallel to $\vec{u}$, only the operators $T_q^{(2)}$ appear in the expansion of $\hat{\sigma}_e$, describing an alignment, having the same symmetries as an electric field or an ellipsoid invariant by rotation around the $\vec{u}$ axis.

[2]See for example [Messiah (2003); Cohen-Tannoudji *et al.* (1977)].

4.3.3 *How to detect Zeeman coherences in e?*

Consider an atom which is in a linear superposition of the three Zeeman sublevels of e:

$$|\psi\rangle = \sum_{M_e=-1,0,+1} c_{M_e} |M_e\rangle \,. \tag{4.7}$$

The probability amplitude $\mathcal{A}$ that it emits a photon with polarization $\vec{\varepsilon}$ is the sum of the products of the amplitudes c_{M_e} to be in $|M_e\rangle$ by the amplitudes $\langle 0|\, \vec{\varepsilon}^{\,*} \cdot \vec{D}\, |M_e\rangle$ to emit a photon $\vec{\varepsilon}$ from $|M_e\rangle$:

$$\mathcal{A} \propto \sum_{M_e} c_{M_e} \langle 0|\, \vec{\varepsilon}^{\,*} \cdot \vec{D}\, |M_e\rangle \,. \tag{4.8}$$

The intensity $I(\vec{\varepsilon})$ of the light emitted by the atom in the state (4.7) is proportional to the square of the modulus of the amplitude (4.8):

$$I(\vec{\varepsilon}) \propto |\mathcal{A}|^2 \propto \sum_{M_e, M_e'} c_{M_e} c^*_{M_e'} \langle M_e'|\, \vec{\varepsilon} \cdot \vec{D}\, |0\rangle \langle 0|\, \vec{\varepsilon}^{\,*} \cdot \vec{D}\, |M_e\rangle \,. \tag{4.9}$$

It depends on the Zeeman coherences in e, $\sigma_{M_e, M_e'} = c_{M_e} c^*_{M_e'}$, with $M_e \neq M_e'$ only if $\langle M_e'|\, \vec{\varepsilon} \cdot \vec{D}\, |0\rangle \langle 0|\, \vec{\varepsilon}^{\,*} \cdot \vec{D}\, |M_e\rangle \neq 0$ with $M_e \neq M_e'$. Like for the excitation process, one concludes that the light emitted with a polarization $\vec{\varepsilon}$ is a signal proportional to the Zeeman coherences in e only if $\vec{\varepsilon}$ is a linear superposition of the three basic polarization vectors, $\vec{\varepsilon}_+$, $\vec{\varepsilon}_-$ and $\vec{\varepsilon}_z$.

4.3.4 *Equation of motion of Zeeman coherences in e*

We suppose here that the optical excitation is broadline, with a spectral width Δ large compared to the Larmor precession frequency ω_L around the static magnetic field $\vec{B}_0$. One can then describe the excitation by a rate which, for the Zeeman coherence $\sigma_{M_e M_e'}$, is given by[3]

$$\left(\frac{\mathrm{d}\sigma_{M_e M_e'}}{\mathrm{d}t}\right)_{\text{exc}} = W \sigma_{gg}, \tag{4.10}$$

where σ_{gg} is the population of the non-degenerate ground state, and where W is a coefficient proportional to the light intensity and to $\langle M_e|\, \vec{\varepsilon} \cdot \vec{D}\, |0\rangle \langle 0|\, \vec{\varepsilon}^{\,*} \cdot \vec{D}\, |M_e'\rangle$. For low intensity of the exciting light, one can neglect the depletion of the ground state and replace σ_{gg} by 1 in Eq. (4.10).

The evolution of the Zeeman coherence due to the Larmor precession in the static field $\vec{B}_0$ is described by:

$$\left(\frac{\mathrm{d}\sigma_{M_e M_e'}}{\mathrm{d}t}\right)_{\text{Larmor}} = -\frac{i}{\hbar}\left(E_{M_e} - E_{M_e'}\right) \sigma_{M_e M_e'} = -i\omega_0 \left(M_e - M_e'\right) \sigma_{M_e M_e'} \,. \tag{4.11}$$

[3]To keep the notation simple we do not write explcitly the M_e and M_e' dependence of the rate W.

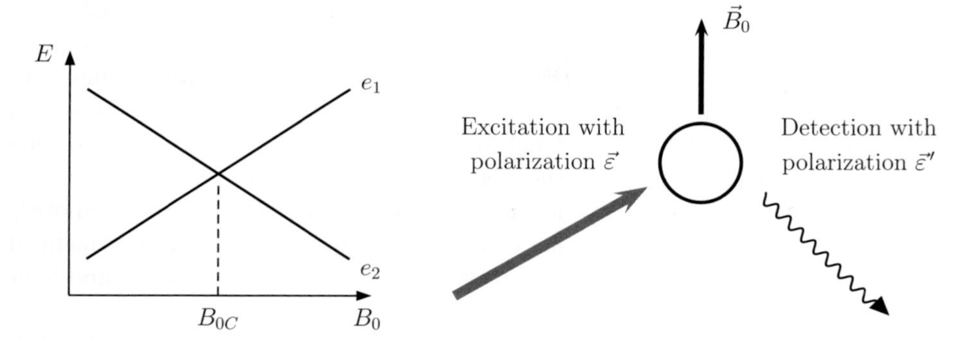

Fig. 4.2 Left part: when the static field B_0 is scanned, two sublevels e_1 and e_2 of the excited state cross when $B_0 = B_{0C}$. Right part: atoms are excited with a light beam having a polarization $\vec{\varepsilon}$ and one detects the fluorescence light emitted with a polarization $\vec{\varepsilon}'$.

Finally, there is the damping due to spontaneous emission which is responsible for a decay of the coherence:

$$\left(\frac{\mathrm{d}\sigma_{M_e M_e'}}{\mathrm{d}t}\right)_{\text{spont.em.}} = -\Gamma\,\sigma_{M_e M_e'} \tag{4.12}$$

where Γ is the natural width of e.

Adding the three previous rates (with $\sigma_{gg} \simeq 1$) leads finally to:

$$\left(\frac{\mathrm{d}\sigma_{M_e M_e'}}{\mathrm{d}t}\right) = W - i\,(M_e - M_e')\,\omega_0\sigma_{M_e M_e'} - \Gamma\,\sigma_{M_e M_e'}. \tag{4.13}$$

4.3.5 *Level crossing resonances in the excited state e*

Consider the following experiment sketched in the right part of Fig. 4.2. Atoms are excited with a light beam having a constant intensity and a polarization $\vec{\varepsilon}$. One detects the fluorescence light emitted with a polarization $\vec{\varepsilon}'$. Both polarizations $\vec{\varepsilon}$ and $\vec{\varepsilon}'$ are linear superpositions of the polarizations $\vec{\varepsilon}_1$ and $\vec{\varepsilon}_2$ associated with the transitions $g \leftrightarrow e_1$ and $g \leftrightarrow e_2$, so that a Zeeman coherence σ_{12} can be created by the exciting beam and detected by monitoring the $\vec{\varepsilon}'$-polarized fluorescence light. One sweeps the static magnetic field around a value B_{0C} for which the two sublevels e_1 and e_2 cross (left part of Fig. 4.2). Since we consider here an excited state with $J_e = 1$, the three Zeeman sublevels cross in zero magnetic field and $B_{0C} = 0$. How does the intensity of the fluorescence light vary when B_0 is scanned around B_{0C}?

The answer to this question can be readily obtained by using the equation of motion (4.13) for $\sigma_{12} = \langle e_1|\,\hat{\sigma}\,|e_2\rangle$. Since the intensity of the exciting light is constant, W is time independent and the equation of evolution of σ_{12} has a steady-state solution:

$$\sigma_{12}^{\text{st}} = \frac{W}{\Gamma + i\,\omega_{12}}, \tag{4.14}$$

where $\omega_{12} = (E_{e_1} - E_{e_2})/\hbar$. The detection signal contains a term proportional to the real part of σ_{12}, equal to $\Gamma\, W/(\Gamma^2 + \omega_{12}^2)$, and a term proportional to the imaginary part, $\Gamma\, \omega_{12}/(\Gamma^2 + \omega_{12}^2)$. One therefore concludes that the fluorescence intensity exhibits resonant variations when B_0 is scanned around 0 in an interval determined by the natural width Γ of e, and not by the spectral width Δ of the exciting light or by the Doppler width of the optical line. Like for the double resonance and optical pumping experiments described in the previous chapter, this is another example of an experiment where one can make a high resolution measurement of an atomic parameter, which is here the natural width Γ of e, with broadband light sources.

This zero field level crossing resonance was observed for the first time by Wilhelm Hanle and it is referred to as the *Hanle effect* [Hanle (1924, 1926)]. The geometrical interpretation of Zeeman coherences given in Sec. 4.3.2 above leads to a simple physical picture for the Hanle effect. Because of the transverse character of the polarization of the exciting light, a transverse polarization, and thus a transverse magnetization, are introduced in the excited state e. Once created in a very short time $1/\Delta$, the transverse magnetization is submitted to two competing processes: Larmor precession at the frequency ω_{12} around the z-axis along which B_0 is applied and damping with a rate Γ due to spontaneous emission. If $|\omega_{12}| \ll \Gamma$, the angle of precession remains negligible during the lifetime Γ^{-1} of e. A large transverse magnetization can build up in steady state in e, giving rise to large signal in the fluorescence light detected with a polarization $\vec{\varepsilon}\,'$ sensitive to this transverse magnetization. If $|\omega_{12}| \gg \Gamma$, the transverse magnetization makes several turns during Γ^{-1} and its average value vanishes in steady state, giving a zero contribution to the detection signal. Note the important difference with the behavior of the populations σ_{11} and σ_{22} which describe a longitudinal magnetization not affected by the Larmor precession (since $\omega_{11} = \omega_{22} = 0$). The steady state value of this longitudinal magnetization is large and does not depend on B_0.

Similar level crossing resonances have also been observed around the non-zero magnetic field values where two Zeeman sublevels originating from two different fine structure levels cross. This is called the *Franken effect* [Colegrove *et al.* (1959)].

Interpretation of the level crossing resonance in terms of interfering paths

There are two different paths for going from the same initial state $g, \vec{\varepsilon}$ to the same final state $g, \vec{\varepsilon}\,'$, the first one passing through e_1, the second one through e_2 (see Fig. 4.3). They are both open because $\vec{\varepsilon}$ and $\vec{\varepsilon}\,'$ are both linear superpositions of the polarizations $\vec{\varepsilon}_1$ and $\vec{\varepsilon}_2$ of the transitions $g \to e_1$ and $g \to e_2$. Near the crossing point where $E_{e_1} = E_{e_2}$, the two scattering amplitudes can be simultaneously resonant and large and one can show that the level crossing resonance is due to the interference between them.

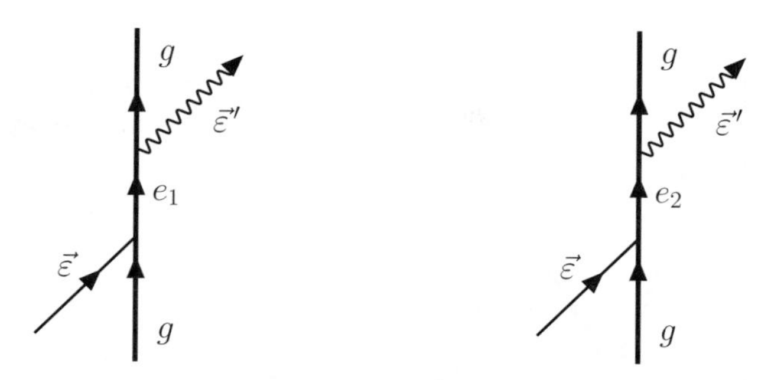

Fig. 4.3 Two possible paths corresponding to the absorption of a photon with polarization $\vec{\varepsilon}$ followed by the emission of a photon with polarization $\vec{\varepsilon}'$. The two paths differ by the intermediate state e_1 or e_2.

4.3.6 *Pulsed excitation. Quantum beats*

Suppose now that the exciting light, instead of being constant, is applied in the form of a short pulse, lasting a time τ short compared to all other times of the problem (Larmor period, radiative lifetime). One can then replace, in the equation of motion (4.13) of σ_{12}, W by $W(t) = W_0\delta(t)$. This equation then becomes:

$$\frac{\mathrm{d}\sigma_{12}}{\mathrm{d}t} = W_0\,\delta(t) - i\,\omega_{12}\sigma_{12} - \Gamma\,\sigma_{12}, \tag{4.15}$$

and its solution is:

$$\sigma_{12}(t) = W_0\;\theta(t)\,e^{-i\,\omega_{12}\,t}\,e^{-\Gamma\,t} \tag{4.16}$$

where $\theta(t)$ is the Heaviside function.

The excitation prepares the coherence σ_{12} in a very short time. This coherence then evolves freely at the frequency ω_{12} and is damped with a rate Γ. This evolution is detected as a damped oscillation of the intensity of the fluorescence light which contains a term proportional to σ_{12}. There is a certain analogy between this effect and the sound beats which are heard after a short percussion of two different guitar strings.

The first observation of quantum beats was made in 1964 [Dodd *et al.* (1964); Aleksandrov (1964)]. In [Dodd *et al.* (1964)], the experiment was realized on cadmium atoms and the modulation frequency was in the MHz range (see Fig. 4.4). The development of ultrashort laser pulses[4] has given to this method an increasing importance and opened new research fields, like femtochemistry. A femtosecond pulse (lasting a few 10^{-15} sec) excites a molecule in a coherent superposition of vibrational states. The corresponding wave packet oscillates in the molecular potential well at the vibration frequency, and a second laser pulse detects the oscillating wave packet at a later time. One can in this way detect in real time the vibration of a molecule [Zewail (2003)].

[4]See Chap. 27.

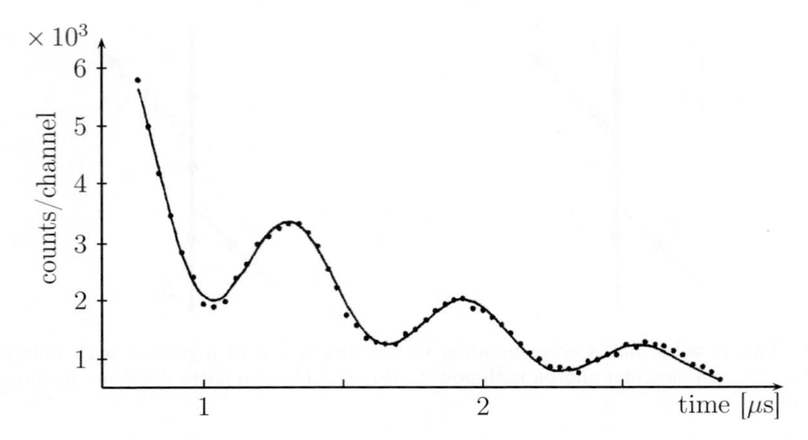

Fig. 4.4 Quantum beats observed with Cd atoms. Figure adapted from [Dodd *et al.* (1964)]. Copyright: Institute of Physics.

4.3.7 *Excitation with modulated light*

Suppose now that the intensity of the excitation is modulated at the frequency Ω. One must then replace, in the equation of motion (4.13) of σ_{12}, W by $W(t) = W_0 \exp(-i\Omega t) + \text{c.c.}$, which gives for the equation of motion of σ_{12}:

$$\frac{\mathrm{d}\sigma_{12}}{\mathrm{d}t} = W_0 \exp(-i\Omega t) - i\,\omega_{12}\sigma_{12} - \Gamma\sigma_{12}, \tag{4.17}$$

that has a driven solution oscillating at frequency Ω:

$$\sigma_{12} = \frac{W_0}{\Gamma + i\,(\omega_{12} - \Omega)} \exp(-i\Omega t). \tag{4.18}$$

The intensity of the fluorescence light contains a component modulated at frequency Ω, which exhibits resonant variations when the frequency Ω of the excitation is scanned around ω_{12} in an interval of width Γ. The first observation of this resonant effect was done in 1964 [Corney and Series (1964)].

4.3.8 *Modulation of the fluorescence light in a double resonance experiment. Light beats*

In a double resonance experiment, the RF field produces Zeeman coherences. The Rabi precession around the RF field gives rise to a transverse magnetization. In steady-state, this transverse magnetization is static in the rotating frame[5] which rotates around the z- axis at the frequency ω of the RF field. Consequently, in the laboratory frame, we have a transverse magnetization rotating around the z-axis at the frequency ω. We thus expect that the fluorescence light, emitted with a proper polarization, should exhibit modulations at ω (and at its harmonic 2ω coming from the coherences $\Delta M_e = \pm 2$ if any). These modulations were observed for the first

[5]See Chap. 2.

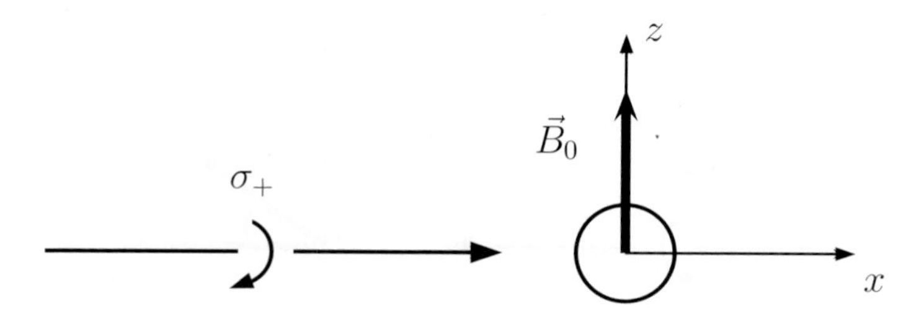

Fig. 4.5 Scheme of a transverse optical pumping experiment.

time in 1959 [Dodd *et al.* (1959)]. They were called *light beats* because the two light waves emitted from e_1 and e_2, and can be considered as keeping an information of the phase coherence between e_1 and e_2 if $\sigma_{12} \neq 0$ and can thus beat. One of the first density matrix treatments of Zeeman coherences for interpreting light beats was developed in 1961 [Barrat (1961)].

4.4 Zeeman coherences in atomic ground states

All the results described in the previous section concerning Zeeman coherences in excited states e can be easily extended to ground states g. We will only discuss here a few of them leading to interesting applications.

4.4.1 *Hanle effect in atomic ground states*

For the sake of simplicity, we will consider here the simple case of a $J_g = 1/2 \leftrightarrow J_e = 1/2$ because it is then easy to explain how it is possible to prepare and to detect Zeeman coherences in g. Extension to more complicated cases is straight forward.

Zeeman coherences can be readily introduced in g by performing a transverse optical pumping of the atoms. Figure 4.5 gives the scheme of such an experiment. Atoms are put in a static magnetic field B_0 parallel to the z-axis and are excited by a σ_+-polarized resonant light beam propagating along the x-axis. As explained in Sec. 3.3 of Chap. 3, optical pumping cycles (absorption of one incident photon followed by a spontaneous emission) tend to orient the atoms in the eigenstate $|M_g = +1/2\rangle_x$ of J_x with eigenvalue $+\hbar/2$:

$$|M_g = +1/2\rangle_x = \frac{1}{\sqrt{2}} \left[|M_g = +1/2\rangle_z + |M_g = -1/2\rangle_z \right] \tag{4.19}$$

which is a linear superposition of the two eigenstates[6] of J_z. Transverse optical

[6]This result is valid only if B_0 is low enough for allowing one to neglect the Larmor precession of the atomic magnetization during the time Γ^{-1} spent in the excited states e. We will see later on that the Hanle resonance in the ground states g is much narrower than the Hanle resonance in e so that the values of B_0 which are explored here are very low.

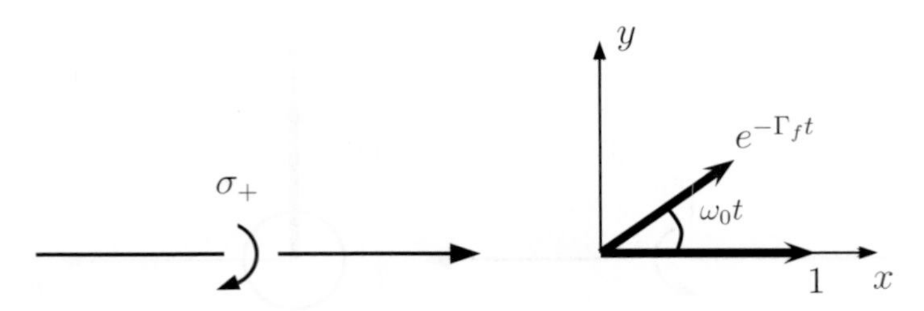

Fig. 4.6 Competition between the Larmor precession at frequency ω_0 and the damping due to spontaneous emission for an atom initially pumped along the x-axis.

pumping can thus introduce Zeeman coherences in g. Also, for a $J_g = 1/2 \leftrightarrow J_e = 1/2$ transition, the σ_+-polarized incident photons propagating along the x-axis are absorbed only by the atoms in the eigenstate $|M_g = -1/2\rangle_x$ of J_x. By monitoring the absorption of the pumping beam, we thus get a signal depending on the Zeeman coherence in g.

It is possible to give a simple geometrical interpretation of the Hanle resonance in g (see Fig. 4.6). In the plane xy perpendicular to B_0, there is a competition between

(i) optical pumping which tends to orient the spins along the x-axis,

(ii) the Larmor precession in g which makes them rotating at the Larmor frequency ω_0 in the xy plane.

A spin magnetic moment prepared at $t = 0$ along the x-axis with a length 1 (in appropriate units) is rotated by an angle $\omega_0 t$ after a time t and is reduced by a factor $\exp(-\Gamma_g t)$ where Γ_g is the relaxation rate in g.

If $\omega_0 \gg \Gamma_g$, at a given time t, the spins pumped before t are isotropically distributed in the xy plane with a resultant equal to zero, so that $\langle J_x \rangle = \langle J_y \rangle = 0$. If $\omega_0 \leq \Gamma_g$, at a given time t, the spins form a span starting from the x-axis with a small opening, so that $\langle J_x \rangle$ ($\langle J_y \rangle$) varies with ω_0 as an absorption (dispersion) curve with a width $\delta\omega_0 \simeq \Gamma_g$.

The interest of Hanle resonances in g is that they are very narrow because Γ_g is much smaller than Γ_e. They can be thus used for detecting very small variations of B_0. Let us give a few orders of magnitude. For an electronic angular momentum, ω_0 is on the order of 1 MHz/G. If Γ_g is on the order of 1 sec^{-1}, the width δB_0 of the resonance, given by $\delta\omega_0 \simeq \Gamma_g$ is on the order of 10^{-6} G. The left part of Fig. 4.7 shows an example of a dispersion shaped Hanle resonance detected in the ground state of rubidium-87 atoms [Dupont-Roc *et al.* (1969); Cohen-Tannoudji *et al.* (1970)]. The width of the resonance is on the order of 1 μG. If one applies square variations of B_0 around $B_0 = 0$, the signal to noise ratio is high enough for allowing one to detect field variations as small as 5×10^{-10} G, as shown in the right part of Fig. 4.7. Using such a high sensitivity, it has been possible to detect the static magnetic field produced at a macroscopic distance ($\simeq$ 10 cm) by a gaseous sample of polarized helium-3 nuclei (about 5×10^{15} nuclei in a volume of 100 cm^3) [Cohen-Tannoudji *et al.* (1969)].

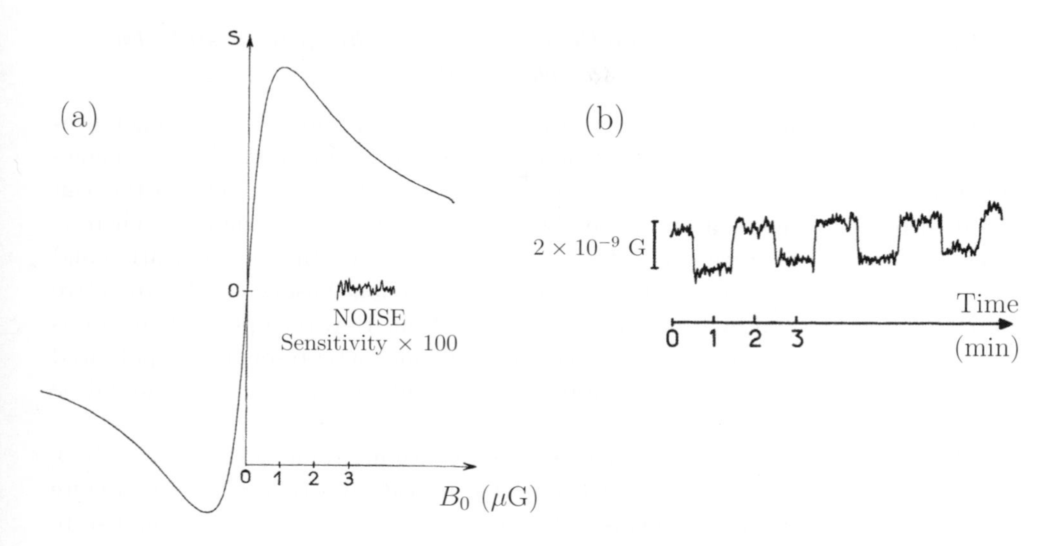

Fig. 4.7 Left: dispersion shaped Hanle resonance in the ground state of rubidium-87 atoms. Right: variations of the signal when square variations of B_0 equal to 2×10^{-9} G are applied. Figure extracted from [Cohen-Tannoudji *et al.* (1970)]. Copyright: EDP Sciences.

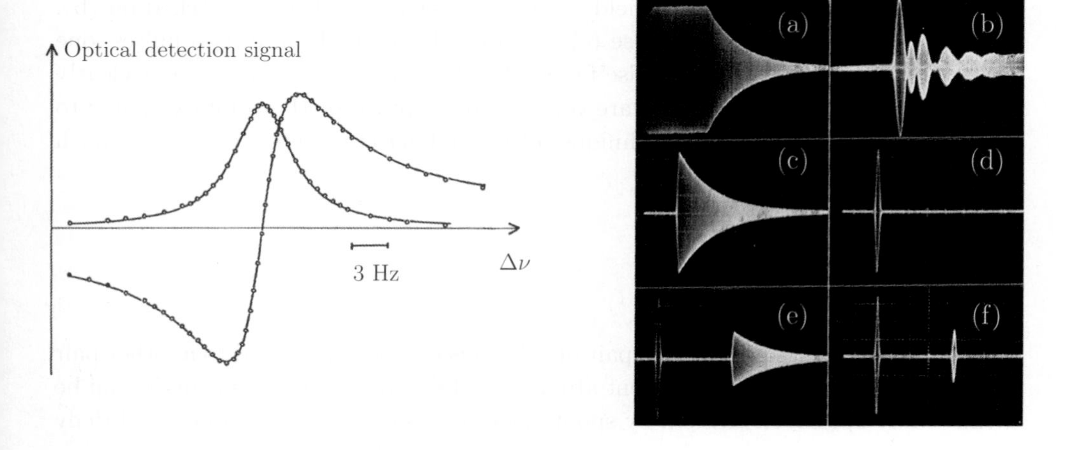

Fig. 4.8 Optical detection signals of the nuclear magnetic resonance in the ground state of mercury-199 atoms, based on the modulation of the absorbed light. Left part: signals recorded in steady state with a lock-in detection. Right part: signals recorded in a transient regime, when the RF field is suddenly switched off (a), switched on (b), applied in the form of a $\pi/2$ pulse (c), π pulse (d), in the form of two pulses, $\pi, \pi/2$ (e), π, π (f). Figure taken from [Cohen-Tannoudji (1962)]. Copyright: EDP Sciences.

4.4.2 *Detection of the magnetic resonance in the ground state by the modulation of the absorbed light*

Even if they are not prepared directly by the optical excitation, Zeeman coherences in the ground state g can appear when a resonant RF field is applied. The magnetization rotates around the z-axis at the frequency ω of the RF field and this can be detected by a modulation of the absorbed light at the frequency ω. There is a certain analogy between this optical detection of the rotating magnetization and the usual radioelectric detection of the magnetic resonance using the electromotive force induced in a coil by this rotating magnetization. But, the optical detection is much more sensitive and allows a detection of the magnetic resonance of polarized nuclei in a gaseous sample, in conditions where a radio electric detection would be extremely difficult.

The first optical detection of magnetic resonance in atomic ground states by a modulation of the absorbed light was reported in [Bell and Bloom (1957)]. Figure 4.8 shows similar detection signals of the magnetic resonance in the ground state of mercury-199 atoms [Cohen-Tannoudji (1962)]. The angular momentum in this ground state is purely nuclear ($J = 0, I = 1/2$), so that we are dealing here with a nuclear magnetic resonance. The signal to noise ratio is however very good. The left part of the figure shows the signal observed in steady-state with a lock-in detection which allows the detection of the components of the resonance that varies as absorption and dispersion curves. The right part shows the signal detected in transient regime, when the RF field is suddenly switched off (a), switched on (b), applied in the form of a $\pi/2$ pulse (c), π pulse (d), in the form of two pulses, one π and a $\pi/2$ one (e), and a π pulse followed by another π pulse (f). It thus clearly appears that Zeeman coherences are very useful in optical methods for extending to gaseous samples all the usual techniques of the radioelectric detection, with a much better signal to noise ratio.

4.5 Transfer of coherences

A coherence prepared between a pair of sublevels can be transferred to another pair of sublevels belonging to a different atom or to the same atom. This transfer can be achieved by optical excitation, by spontaneous emission or by collisions. We briefly review in this subsection a few examples of such transfers.

(i) *Transfers between two atoms mediated by a spontaneous photon*

The narrowing of double resonance curves due to coherent multiple scattering described above in Sec. 4.2 is an example of such transfer. The Zeeman coherence created by the RF field in the excited state of an atom A can be transferred to the excited state of another atom B by a photon which is emitted by A and reabsorbed by B. In *geometrical* terms, one can say that a transverse magnetization appears in the excited state of B whose direction is correlated with the the one of A, when one averages over all possible directions of emission of the photon which is exchanged by the two atoms. Because of the isotropy of spontaneous emission, the exchanged photon cannot create by itself a transverse magnetization, and the only privileged direction which remains after the excitation of B is the initial direction of the transverse magnetization in the excited state of A.

Shift of the double resonance curves due to coherent multiple scattering

Because of the propagation time τ_{prop} of the photon between A and B, the photon which is emitted at time t by A arrives in B at time $t+\tau_{\text{prop}}$, bringing information on the transverse magnetization of A at time t. There is therefore a phase shift between the orientation of the magnetization which is transferred and the one which would remain in A if A was living a longer time. Another source of phase shift occurring during the transfer is the change of polarization of the photon due to the dispersion of the medium between A and B (producing for example a Faraday rotation). These phase shifts, accumulated during several exchanges of photons whose number increases with the vapor density, produce a change of the effective Larmor frequency in e, and consequently a shift of the double resonance curve when the vapor pressure increases. These effects have been observed [Omont (1964)].

(ii) *Transfer between two atoms mediated by collision*

Another example of transfer of coherence between two atoms, this time due to collisions, may be found in [Haroche and Cohen-Tannoudji (1970)]. A transverse magnetization is introduced in the ground state of optically pumped cesium atoms by a $\pi/2$ resonant RF pulse. This transverse magnetization is transferred to the ground state of rubidium atoms by spin exchange collisions and detected by the modulation of the absorption of a properly polarized probe beam exciting the rubidium atoms. In fact, the transfer cannot lead to a build up of the transverse magnetization in rubidium if the Larmor frequencies of the two atoms are different. It is however possible to make these two frequencies equal by "dressing" the magnetic moments of the two atoms by a high frequency RF field. It can be indeed shown[7] that the interaction with a high frequency RF field changes the g-factor of an atom by an amount depending on the amplitude B_1 of the RF field. For certain values of the amplitude of B_1, the two Landé factors of cesium and rubidium become equal and one actually observes that the transfer of coherence is large for these values of B_1. For more details, see [Haroche and Cohen-Tannoudji (1970)].

(iii) *Transfers between two levels of the same atom*

When the angular momenta J_e and J_g of e and g are different from 0, Zeeman coherences can exist in both states. A Zeeman coherence in g can be transferred to e by absorption of a photon. A Zeeman coherence in e can be transferred to g by

[7] See Chap. 8.

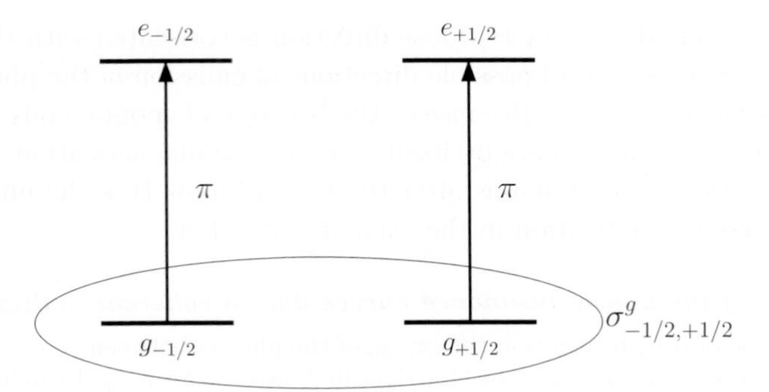

Fig. 4.9 A π-polarized excitation can transfer a Zeeman coherence in a ground state with $J_g = 1/2$ to an excited state with $J_e = 1/2$.

spontaneous emission.[8]

A Zeeman coherence which already exists in g can be transferred to e even if the polarization of the exciting light is not transverse with respect to B_0. If there is already a privileged direction for the transverse magnetization in g, it is not necessary that the absorbed photon introduces by itself a transverse magnetization. For example, for a $J_g = 1/2 \leftrightarrow J_e = 1/2$ transition, one can show that the rate of preparation of the Zeeman coherence $\sigma^e_{-1/2,+1/2} = \langle e_{-1/2} | \hat{\sigma} | e_{+1/2} \rangle$ for a π-polarized optical excitation is given by:

$$\frac{\mathrm{d}}{\mathrm{d}t} \langle e_{-1/2} | \hat{\sigma} | e_{+1/2} \rangle = W \langle e_{-1/2} | \vec{\varepsilon}_z \cdot \vec{D} | g_{-1/2} \rangle \langle g_{-1/2} | \hat{\sigma} | g_{+1/2} \rangle \langle g_{+1/2} | \vec{\varepsilon}^{\,*}_z \cdot \vec{D} | e_{+1/2} \rangle \tag{4.20}$$

where W is a coefficient proportional to the light intensity. Two π-polarized excitation amplitudes connect the two Zeeman sublevels in g between which there is a Zeeman coherence to two Zeeman sublevels in e. A Zeeman coherence between the two sublevels $|e_{-1/2}\rangle$ and $|e_{+1/2}\rangle$ thus appears (see Fig. 4.9). Of course, one can also transfer to e a coherence already existing in g with a polarization which is transverse with respect to B_0.

Consider now the transfer by spontaneous emission. If one does not observe the direction and the polarization of the emitted photon, one must average the rate of transfer over all possible directions and polarizations of this photon. It is then possible to show that one must add the rates of independent transfers involving two emission amplitudes which can be σ_+, σ_- or π polarized. The average over the directions and the polarizations of the emitted photon suppresses all possible contributions of a transverse privileged direction coming from this photon. For example, for a $J_g = 1/2 \leftrightarrow J_e = 3/2$ transition, the coherence $\sigma^g_{-1/2,+1/2}$ in g can be fed from $\sigma^e_{+1/2,+3/2}$ by two σ_+ emission amplitudes (see Fig. 4.10). Similar figures could be drawn for representing the feeding of $\sigma^g_{-1/2,+1/2}$ from $\sigma^e_{-1/2,+1/2}$ by two π emission amplitudes and from $\sigma^e_{-3/2,-1/2}$ by two σ_- emission amplitudes.

[8] It could be also transferred by stimulated emission, but we suppose here that the light intensity is low, so that stimulated emission remains negligible compared to spontaneous emission.

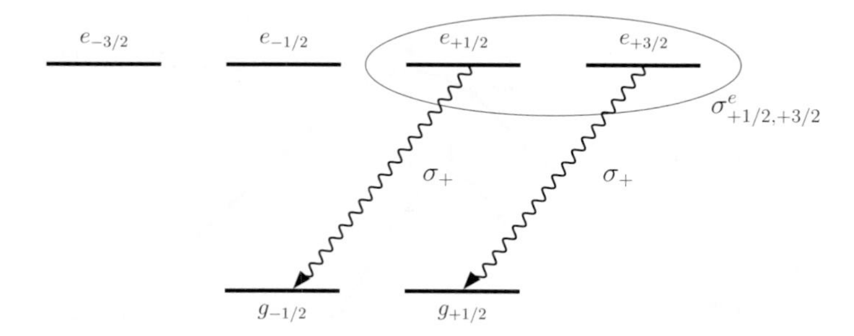

Fig. 4.10 Feeding of the coherence $\sigma^g_{-1/2,+1/2}$ in g from $\sigma^e_{+1/2,+3/2}$ in e by two σ_+ emission amplitudes.

The previous results can be used for deriving the equations of motion of the atomic density matrix in an optical pumping experiment. For a broadline excitation, one can show that the density operators σ_e in e and σ_g in g obey coupled rate equations. The rate of variation $\mathrm{d}\sigma_e/\mathrm{d}t$ of σ_e is the sum of three terms: the rate of excitation from σ_g, the precession in the static field B_0 and in an applied RF field B_1 if any and the damping due to spontaneous emission. Similarly, the rate of variation $\mathrm{d}\sigma_g/\mathrm{d}t$ of σ_g is the sum of the departure rate from g due to absorption processes, the feeding rate from σ_e due to spontaneous emission and the precession in B_0 and B_1.[9]

We will just mention here an interesting prediction of these equations. When a magnetic resonance is performed in the ground state of optically pumped atoms, the Zeeman coherence introduced in g can be transferred to e by absorption of photons and then fall back in g by spontaneous emission. Zeeman coherences thus circulate along the optical pumping cycle. They do not disappear completely after one absorption. The transverse magnetization removed from g by the absorption process can come back in g with a certain memory of its initial direction. However, during the time Γ^{-1} spent in e, this magnetization precesses with the Larmor frequency ω_0^e of e which is generally different from the Larmor frequency ω_0^g of g. The transverse magnetization of an atom having experienced an optical pumping cycle has thus acquired a phase shift with respect to the magnetization of an atom having not absorbed a photon. If optical pumping cycles are repeated with a certain rate proportional to the light intensity, the accumulation of these phase shifts will induce a change of the effective Larmor frequency in g, and thus a shift of the magnetic resonance in g, proportional to the light intensity. In addition, the magnetic resonance curve will be broadened due to the shortening of the lifetime of g that follows from absorption processes. This effect was experimentally observed [Cohen-Tannoudji (1961b)] and the results found in excellent agreement with the theoretical predictions. Figure 4.11 shows the magnetic resonance curve in the ground state

[9]For a derivation of these equations, see [Barrat and Cohen-Tannoudji (1961)]; see also [Cohen-Tannoudji (1962)] for a physical interpretation of these equations and for a description of the experiments which have been performed for testing their predictions.

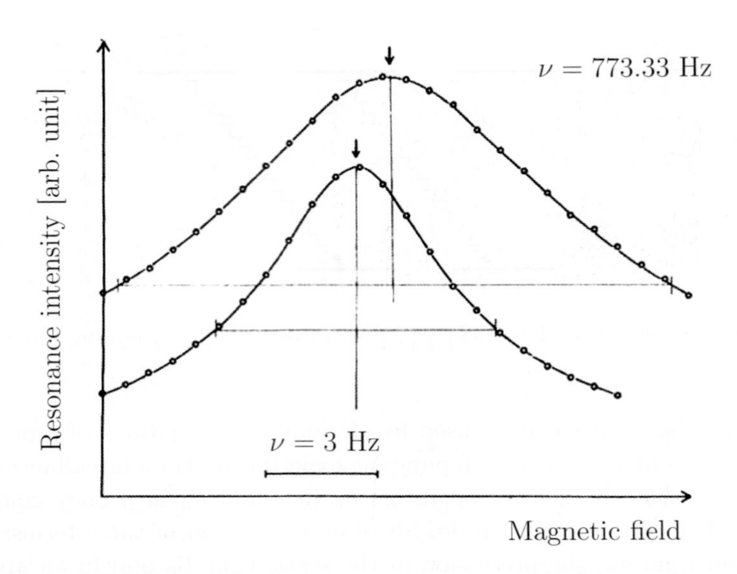

Fig. 4.11 Shift and broadening of the magnetic resonance curve in the ground state of mercury-199 atoms due to the circulation of the Zeeman coherence along the optical pumping cycle. Figure adapted from [Cohen-Tannoudji (1962)]. Copyright: EDP Sciences.

($J = 0$, $I = 1/2$) of mercury-199 atoms excited by a light beam resonant for the transition connecting the ground state to the hyperfine level $F = 1/2$ of the $6^3\mathrm{P}_1$ level. The two curves correspond to different light intensities. The broadening and the shift of the curves when the light intensity increases clearly appear.

Note that this light shift is totally different from another type of light shifts, discussed in Chap. 7. The shift discussed here is due to resonant absorption processes inducing a circulation of the coherence. The light shifts described in Chap. 7 are due to virtual non-resonant absorption and re-emission processes, and can be considered as ac Stark shifts.

Transfers of coherence in systems with a series of equidistant energy levels

Several systems have equidistant energy levels, like an harmonic oscillator or an atom "dressed" by photons.[10] Transfers of coherence between a pair of levels and another pair of levels having the same energy splitting are then very important. The equality of the evolution frequencies allows the coherence to be preserved from any phase shift. For example, the coherence between the levels $n+1$ and n of a one-dimensional harmonic oscillator is not completely destroyed in a spontaneous emission process. It is partially transferred to the pair of levels n and $n-1$.[11] Neglecting these transfers would lead to the wrong prediction that the damping rate of the oscillator due to spontaneous emission is proportional to its energy. Similarly, it is essential to take into account these transfers of coherence in the

[10] See Sec. 5.4 of Chap. 5.
[11] See [Cohen-Tannoudji *et al.* (1992b)], Complement B_{IV}.

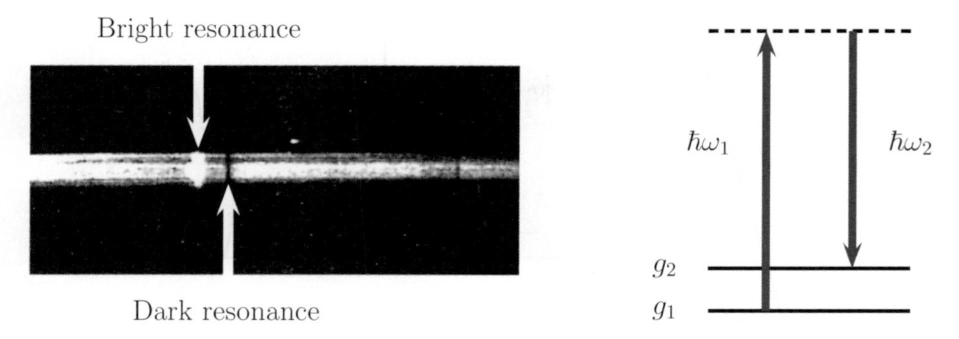

Fig. 4.12 Bright and dark resonances observed in the fluorescence light emitted by a sodium vapor put in a spatially inhomogeneous magnetic field. The dark resonance appears at a point where the frequency difference $\omega_1 - \omega_2$ between two laser modes matches the frequency splitting between two Zeeman sublevels g_1 and g_2. Figure adapted from [Alzetta *et al.* (1976)], reproduced with kind permission of Società Italiana di Fisica.

energy diagram of the dressed atom for obtaining the correct values of the widths of the components of the Mollow fluorescence triplet [12] described in Chap. 5.

4.6 Dark resonances. Coherent population trapping

Most experiments described in the previous sections were performed with broadband discharge light sources. We describe now an experiment using two coherent laser beams exciting two transitions starting from two lower sublevels and ending at the same excited sublevel (Λ-type atomic system) and giving evidence for a quenching of the absorption of light due to destructive interference.

4.6.1 *Discovery of dark resonances*

Dark resonances were discovered in 1976 in Pisa [Alzetta *et al.* (1976)]. A cell containing an optically pumped sodium vapor was put in a spatially inhomogeneous magnetic field along the z-axis. The splitting between the Zeeman sublevels is thus z-dependent. If one applies a radiofrequency field with frequency ω_{RF}, it induces resonant transitions between two Zeeman sublevels g_1 and g_2 only at the point z where $E_{g_2} - E_{g_1} = \hbar\omega_{RF}$. The population difference between g_1 and g_2 is modified by the RF resonant transitions and this results in a modification of the fluorescence light (generally an increase) at the points where the resonance condition is fulfilled. One observes a series of bright lines in the fluorescence emitted along the path of the laser beam in the cell (see Fig. 4.12), forming a spatially resolved RF spectrum.

The surprising result of this experiment was the disappearance of the fluorescence at certain points along the path of the laser beam, forming dark lines, also called *dark resonances* (see Fig. 4.12), which remain present if the RF field is switched off, provided however that the laser beam is spectrally multimode. It was

[12] See [Cohen-Tannoudji *et al.* (1992b)], Chap. VI.D.

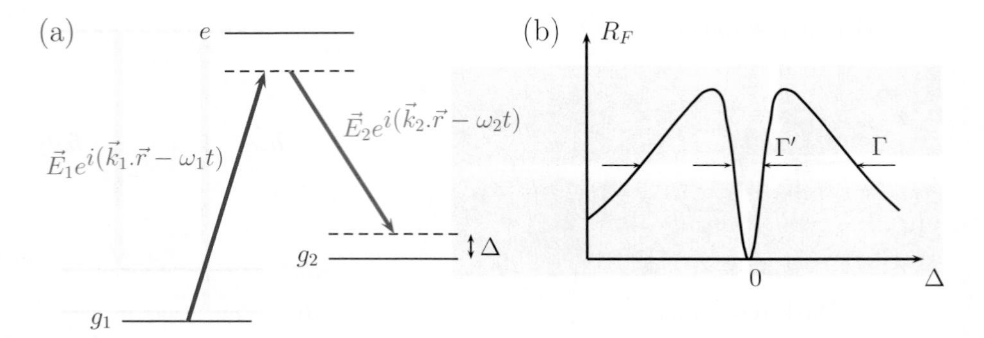

Fig. 4.13 (a) Three-level atom g_1, g_2, e interacting with two laser waves exciting the two transitions $g_1 \to e$ and $g_2 \to e$, respectively. (b) Variation of the fluorescence rate R_F with the detuning Δ from the Raman resonance condition. This curve corresponds to the case where the laser ω_1 is resonant for the transition $g_1 \leftrightarrow e$.

readily found that dark resonances appear for values of z such that the frequency difference between two laser modes is equal to the frequency splitting between two Zeeman sublevels g_1 and g_2 belonging to the two hyperfine states of the sodium atom:

$$E_{g_2} - E_{g_1} = \hbar\omega_1 - \hbar\omega_2. \tag{4.21}$$

This equation is a resonance condition for the stimulated Raman transitions between g_1 and g_2.[13] One photon is absorbed in one laser mode, another emitted in a stimulated way in another mode, the atom going from one Zeeman sublevel to another one with conservation of the total energy.

4.6.2 *First theoretical treatment of dark resonances*

The first theoretical treatment of dark resonances [Arimondo and Orriols (1976)] was using optical Bloch equations (OBE). A three-level atom $\{g_1, g_2, e\}$, with two lower sublevels g_1 and g_2 and one upper sublevel e, interacts with two laser beams: $\vec{E}_1 \exp(i\,\vec{k}_1 \cdot \vec{r} - \omega_1\, t)$ exciting only the transition $g_1 \leftrightarrow e$ and $\vec{E}_2 \exp(i\,\vec{k}_2 \cdot \vec{r} - \omega_2\, t)$ exciting only the transition $g_2 \leftrightarrow e$ (see Fig. 4.13(a)). Within the rotating wave approximation, OBE become a set of first order coupled differential equations with time-independent coefficients which can be solved exactly. One finds that, when the detuning Δ from the resonance Raman condition is equal to 0, the population σ_{ee} of the excited state e vanishes: there is no fluorescence. Simultaneously, the off diagonal element $\sigma_{g_1 g_2}$ of the atomic density matrix σ between g_1 and g_2 takes a large value. This means that atoms are put in a linear combination of the two lower states g_1 and g_2 which does not absorb light, hence the denomination *coherent population trapping* (CPT) given to this effect.

The variations with $\Delta = (\omega_1 - \omega_2) - (E_{g_2} - E_{g_1})/\hbar$ (see Fig. 4.13(a)) of the

[13]See Sec. 9.4.3 of Chap. 9.

fluoresence rate R_F, which is proportional to the population σ_{ee} of e, are represented in Fig. 4.13(b), in the case where the laser ω_1 is resonant for the transition $g_1 \leftrightarrow e$. R_F vanishes for $\Delta = 0$ and then increases, forming a narrow dip. The width Γ' of this dip is determined by the relaxation time in the ground state and is much smaller than the width Γ of the excited state.

The OBE approach gives a quantitative description of the dark resonances but the physical reason why condition (4.21) is essential for the quenching of fluorescence does not appear clearly. We present in the following section a simple physical interpretation of this Raman resonance condition. Let us also mention that dark resonances can be simply related to the radiative cascade[14] of the atom "dressed" by the two types of photons ω_1 and ω_2. The radiative cascade of the dressed atom is also discussed in Sec. 5.4 of Chap. 5.

4.6.3 *Interpretation of the Raman resonance condition*

(i) *Expression of the dark state at time $t = 0$*

Consider an atom in the state:

$$|\psi(t=0)\rangle = c_1 |g_1\rangle + c_2 |g_2\rangle , \tag{4.22}$$

$c_1(c_2)$ is the amplitude for the atom to be in g_1 (g_2). Let us introduce the Rabi frequencies characterizing the interaction with the two laser fields:

$$\Omega_i(t=0) = -\vec{D}_{eg_i} \cdot \vec{E}_i/\hbar, \qquad i = 1, 2 \tag{4.23}$$

where $\vec{D}_{eg_i}$ is the matrix element of the dipole moment operator between e and g_i. The amplitude to have an absorption process from g_i to e is equal to the amplitude c_i for the atom to be in g_i times the amplitude to absorb a photon from g_i which is proportional to Ω_i. If the amplitude c_1 and c_2 appearing in Eq. (4.22) are such that:

$$c_1 \Omega_1 + c_2 \Omega_2 = 0 \tag{4.24}$$

the two absorption amplitudes interfere destructively and the atom cannot be excited. A state (4.22) obeying condition (4.24) is called a dark state.

(ii) *If a state is dark at time $t = 0$, does it remain dark at a later time?*

We suppose that the state remains dark as time goes on and we try to find the condition for this to happen. If the state remains dark, this means that there is no interaction between the atom and the laser fields and the time evolution of the atom is a free evolution as if the laser fields were switched off. The coefficients c_i of the state (4.22) acquire a phase factor $\exp(-i\, E_i t/\hbar)$ where E_i is the unperturbed energy of the state g_i, without any light shift. The laser fields E_i acquire a phase

[14]See for example [Dalibard *et al.* (1987)].

factor $\exp(-i\,\omega_i t)$ due to their frequencies ω_i. The total absorption amplitude thus becomes:

$$c_1 \exp(-iE_1\,t/\hbar)\Omega_1 \exp(-i\omega_1 t) + c_2 \exp(-iE_2\,t/\hbar)\Omega_2 \exp(-i\omega_2 t). \tag{4.25}$$

It remains equal to 0, like in $t = 0$, only if the two phase factors multiplying $c_1\Omega_1$ and $c_2\Omega_2$ are the same, i.e. if:

$$(E_1/\hbar) + \omega_1 = (E_2/\hbar) + \omega_2 \quad \Leftrightarrow \quad E_2 - E_1 = \hbar\,(\omega_1 - \omega_2)\,. \tag{4.26}$$

One recovers the Raman resonance condition (4.21) and understands now why the dark resonance appears only when this condition is fulfilled. If it is not fulfilled, a state which is dark at a given time does not remain dark at a later time and can then absorb light. This approach shows also clearly that the resonance Raman condition involves the unperturbed atomic states without any light shifts. In Eq. (4.21), the atomic energies E_1 and E_2 are unperturbed energies.

Note finally that the dark resonance exists only if the two laser fields are coherent. If their phases change in a random way, the phase difference between the two absorption amplitudes appearing in Eq. (4.25) changes also in a random way so that the destructive interference cannot hold for all times. If the frequency splitting between g_1 and g_2 falls in the radiofrequency or in the microwave domains, a convenient way for generating the two frequencies ω_1 and ω_2 is to start with a single laser beam and to modulate its amplitude or its frequency at a frequency near $\omega_1 - \omega_2$, in order to generate sidebands. Alternatively, one can split a laser beam in two parts and use acousto-optic modulators to adjust the frequency difference between the two beams. An important advantage of these methods is that, even if the laser has a frequency jitter, this jitter is the same for the central frequency and the sidebands so that it cancels out in the difference $\omega_1 - \omega_2$. The position of the dark resonance is thus insensitive to the frequency jitter, which explains the great interest of dark resonances for high resolution spectroscopy.[15]

4.6.4 *A few applications of dark resonances*

(i) *Electromagnetically induced transparency (EIT)*

Consider an atomic vapor of three-level atoms $\{e, g_1, g_2\}$, having a great optical depth for a laser ω_1 exciting the transition $e \leftrightarrow g_1$, so that this laser beam is fully absorbed by the vapor. If one adds a second laser beam coherent with the first one and exciting the transition $e \leftrightarrow g_2$, with a frequency ω_2 such that the resonance Raman condition between unperturbed states is fulfilled, the absorption coefficient vanishes because of the dark resonance and the vapor becomes transparent.[16]

[15] For a review, see [Arimondo (1996)].
[16] See [Harris (1997)] and references therein.

(ii) *Slow light*

With the narrow dip in the absorption coefficient due to the dark resonance is associated a rapid variation of the refractive index of the vapor in the vicinity of the dip. This can change dramatically the group velocity v_g of light in the vapor. Using dark resonances for reducing v_g is one of the methods which have been used to produce slow light [Hau *et al.* (1999)].

(iii) *Stimulated Raman adiabatic passage (STIRAP)*

In a dressed atom description of coherent population trapping,[17] the dark state is a linear combination of states labelled by an atomic quantum number, g_1, g_2 or e, and by the numbers N_1 and N_2 of photons in the two modes:

$$|\psi_D\rangle = c_1\,\Omega_1\;|g_1, N_1+1, N_2\rangle + c_2\,\Omega_2\,|g_2, N_1, N_2+1\rangle\,. \tag{4.27}$$

This state is not coupled by the interaction Hamiltonian V to the state $|e, N_1, N_2\rangle$ and is an eigenstate of the unperturbed Hamiltonian H_0 (the two states appearing in (4.27) are two degenerate states of H_0, which corresponds to the resonance Raman condition between unperturbed states). The energy separation between the dark state and the other dressed states (eigenstates of the total Hamiltonian $H_0 + V$) is on the order of the Rabi frequencies Ω_1 and Ω_2 and is supposed large enough. If one sweeps slowly enough Ω_1 and Ω_2, i.e. the intensities of the lasers, the system remains in the dark state corresponding to the instantaneous values of Ω_1 and Ω_2.

Suppose that one starts in a state for which $\Omega_1 = 0$ and $\Omega_2 \neq 0$, the atom being in g_1. This state is a dark state because an atom in g_1 cannot absorb a photon ω_2. If one slowly decreases Ω_2 and increases Ω_1 until Ω_2 vanishes, the dark state will evolve adiabatically to a dark state for which $\Omega_1 \neq 0$ and $\Omega_2 = 0$. The system will have been transferred adiabatically from g_1 to g_2. Note that this way of changing Ω_1 and Ω_2 is at first sight counterintuitive for transferring the atom from g_1 to g_2. One would rather think that laser ω_1 should be first used to transfer the atom from g_1 to e, the laser ω_2 being then used to transfer by stimulated emission the atom from e to g_2. In fact, we do not use in STIRAP any real absorption or stimulated emission process. One follows adiabatically a dark state where the excited state e is never populated. The advantage of STIRAP, first introduced in molecular physics [Kuklinski *et al.* (1989)], is to avoid any spontaneous emission process from e, which would populate rovibrational states other than g_2 where one wants to transfer molecules selectively.[18]

(iv) *Velocity selective coherent population trapping*

Up to now, we have implicitly supposed that the atom is at rest. If the atom is moving with a velocity $\vec{v}$, the laser frequencies are shifted by amounts equal to $\vec{k}_1 \cdot \vec{v}$ and $\vec{k}_2 \cdot \vec{v}$, respectively. In a Raman stimulated process these two Doppler shifts add with a negative sign. The resonance Raman condition remains fulfilled only if

[17] See for example [Dalibard *et al.* (1987)].

[18] For a review on STIRAP, see also [Vitanov *et al.* (2001)].

the two Doppler shifts are equal. If $\vec{k}_1 \cdot \vec{v} \neq \vec{k}_2 \cdot \vec{v}$, a state which is dark when $\vec{v} = \vec{0}$ is no longer dark when the atom is moving. An atom at rest cannot absorb light. But, as soon as it is moving, photon absorption processes can take place. Coherent population trapping becomes velocity selective. Because of the random changes of atomic velocity due to the random recoils communicated by the photons which are spontaneously emitted in random directions, the atoms perform a random walk in velocity space. They can thus fall in a small region near $v = 0$ where they remain trapped and accumulate. This is the principle of a laser cooling method which has been used to cool atoms for the first time below the single photon recoil limit [Aspect *et al.* (1989); Cohen-Tannoudji (1992a)]. This method will be described in more detail in Chap. 13.

4.7 Conclusion

In this chapter, we have described several original effects that have been observed in the early days of optical pumping. They are all related to the existence of linear superpositions of internal atomic states and most of them can be observed with broadband light sources since they rely only on the polarization of the light beam. The key point is that this polarization must give rise to an atomic angular momentum perpendicular to the static magnetic field when the photon is absorbed by the atom. The polarization of the emitted photon must also allow a detection of a transverse angular momentum in the excited state of the atom. It is possible in this way to give a clear physical interpretation of most of the effects described in this chapter and to understand the differences between these effects and those involving only populations: transverse angular momentum precesses around the static magnetic field whereas longitudinal angular momentum remains fixed.

The quantum interference effects described in this chapter are not only interesting for the applications to which they gave rise: measurement of radiative lifetime from the width of level crossing resonances, detection of very weak magnetic fields, high signal to noise ratio in the detection of magnetic resonance by modulation of the absorbed or emitted light, quantum beats, electromagnetically induced transparency, slow light, STIRAP, ... These early studies attracted the attention on the importance of linear superpositions of states and paved the way for the development of new investigations of quantum interference effects which will be described in other chapters of the book:

- In Chap. 17, we will focus on the external degrees of freedom, ignoring the internal degrees of freedom of the atom, and considering the off diagonal elements of the density operator between two different points in space $\sigma_{\vec{r}\vec{r}'} = \langle\vec{r}|\,\hat{\sigma}\,|\vec{r}'\rangle$, called *spatial coherences*. This will allow us to describe the interference effects observed with atomic de Broglie waves and to point out the analogies and differences with the interference effects observed with optical waves.

- In Chap. 18, we will discuss the new effects which can be observed when both internal and external degrees of freedom are involved in an experiment. It will be possible in this way to describe Ramsey-type interferometers which play an important role in high precision measurements.
- Finally, we will show in Chap. 19 how it is possible to generalize the notion of quantum interference to more complex situations involving entangled states of two systems and multi-particle interferometry, situations which are now playing a central role in the development of quantum information. In this way, we hope to give in this book a survey of the spectacular developments which have occurred during the last few decades in our understanding of quantum interference and to put in perspective the various approaches to these problems which have been explored.

Chapter 5

Resonance fluorescence

5.1 Introduction

This chapter is devoted to the description of the fluorescence light emitted by a two-level atom excited by a resonant or quasi resonant monochromatic laser beam. Experimental investigations of this system started at the end of the 1960's, when tunable laser sources became available in the laboratories, and led to the discovery of several new effects, such as the Mollow fluorescence triplet or photon antibunching. The interpretation of such effects raised a series of new questions. How does the behavior of the system change when the laser intensity increases and when a perturbative treatment of the atom-laser interaction is no longer valid? How can one interpret the detection signals observed on a single atom, in a single experimental realization, when ensemble averages are meaningless?

Our motivation here is not to present an exhaustive review of this research field. We would rather like to show how a simple problem, a two-level atom in a resonant optical laser field, has played the role of a test bed for the development of quantum optics. In particular, by stimulating the development of new theoretical approaches, this problem has provided a better insight into the underlying physical mechanisms.

The type of experiment we want to analyze is sketched in Fig. 5.1. A monochromatic laser beam of frequency, ω_L, excites an atomic beam at a right angle (so that there is no Doppler effect). In addition, the excitation is resonant or quasi resonant (ω_L is very close to the atomic frequency ω_A). The fluorescence light that is spontaneously reemitted by the atoms is collected by a detector.

Various types of experimental signals can be considered:

- **Total intensity.** The total intensity I_{tot} of the fluorescence light is measured by a broadband detector. The laser frequency ω_L is scanned around ω_A and I_{tot} is plotted versus ω_L.
- **Fluorescence spectrum.** The laser frequency ω_L is fixed and the frequency spectrum $S(\omega)$ of the fluorescence light is measured with a narrow band detector.
- **Photon correlations.** The laser frequency ω_L is fixed and one measures

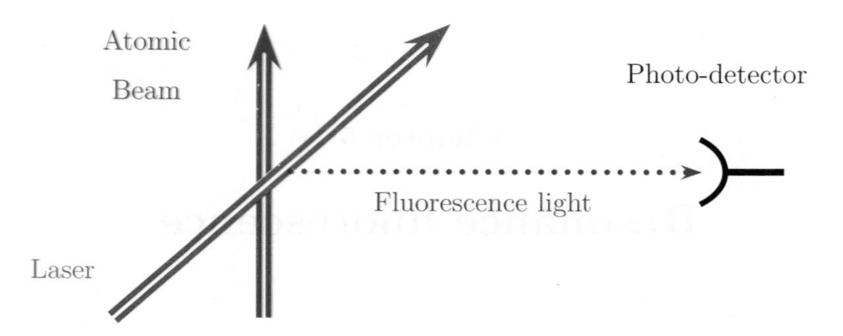

Fig. 5.1 Sketch of the experimental setup corresponding to the experiments analyzed in this chapter. A monochromatic laser beam, with frequency ω_L equal or very close to the atomic frequency ω_A, excites an atomic beam at right angle (in order to avoid Doppler shifts). The fluorescence light, made of spontaneously reemitted photons by the excited atoms, is collected by a photo-detector.

with a broadband detector the probability $G_2(t, t+\tau)$ to detect one photon at time t and another one at time $t+\tau$. In steady state, $G_2(t, t+\tau)$ does not depend on t, and $G_2(\tau)$ is plotted versus τ.

We begin in Sec. 5.2 by a description of resonance fluorescence in the low intensity limit where a perturbative treatment of the atom-laser interaction is possible. To interpret the new phenomena which appear at high laser intensity, several approaches are possible. We briefly introduce optical Bloch equations in Sec. 5.3 and the "dressed atom" approach in Sec. 5.4. The dressed atom approach provides simple physical pictures for the time evolution of the system which are applied to the interpretation of various detection signals: photon correlations in Sec. 5.5 and fluorescence spectrum in Sec. 5.6.

5.2 Low intensity limit. Perturbative approach

5.2.1 *Lowest order process*

This section is devoted to the lowest order description of absorption-emission processes of a two-level atom illuminated by a quasi-resonant laser of frequency ω_L. We neglect the Doppler effect and the recoil energy. Within these assumptions, the external atomic quantum numbers can be ignored.

The physical quantity of interest is the resonant scattering amplitude of the impinging photon. In an elementary absorption-spontaneous emission process, three states play a privileged role: (i) the initial state $|\psi_{\text{init}}\rangle = |g, N\omega_L\rangle$ where the atom is in the ground state in the presence of N laser photons; (ii) the intermediate state $|\psi_{\text{interm}}\rangle = |e, (N-1)\omega_L\rangle$ onto which the system is projected just after the absorption of one ω_L photon; (iii) the final state $|\psi_{\text{fin}}\rangle = |g, (N-1)\omega_L, \omega_F\rangle$ reached when the atom falls back from the excited state to the ground state by spontaneously

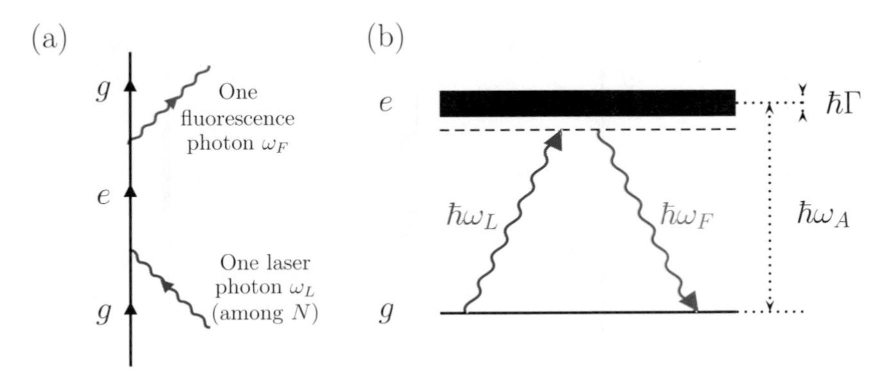

Fig. 5.2 Diagrammatic representation of an elementary absorption-spontaneous emission cycle between two discrete internal states g and e. (a) The evolution in time of the internal state is represented by a vertical line. The wavy entering line corresponds to the absorption process while the outgoing one represents the spontaneous emission. (b) The internal states are represented by horizontal lines. The finite lifetime of the excited state is displayed by the width of the state. The wavy lines have the same meaning for both diagrams.

emitting a photon of frequency ω_F. This elementary cycle is diagrammatically represented in two different manners in Fig. 5.2.

5.2.2 *Resonant scattering amplitude*

The resonant scattering amplitude is obtained from the matrix element of the evolution operator of the coupled atom-laser system during the interaction time T: $A = \langle \psi_{\text{fin}} | U(T) | \psi_{\text{init}} \rangle$. Its calculation can be done within the second order perturbation theory:[1]

$$A = -2\pi i \delta^{(T)}(E_{\text{fin}} - E_{\text{init}}) \frac{\langle \psi_{\text{fin}} | V_{AR} | \psi_{\text{interm}} \rangle \langle \psi_{\text{interm}} | V_{AL} | \psi_{\text{init}} \rangle}{E_{\text{init}} - E_{\text{interm}}}. \tag{5.1}$$

This amplitude describes the succession of two elementary processes: (i) the absorption of one photon from the laser, which brings the state of the system from $|\psi_{\text{init}}\rangle$ to $|\psi_{\text{interm}}\rangle$, is taken into account by the term proportional to the Rabi frequency, $\hbar\Omega = 2\langle \psi_{\text{interm}} | V_{AL} | \psi_{\text{init}} \rangle$, where V_{AL} is the atom-laser interaction Hamiltonian, and (ii) the spontaneous emission of a photon through the coupling term V_{AR} with the vacuum field reservoir which brings the system from the intermediate state to the final state $|\psi_{\text{fin}}\rangle$. The delta function $\delta^{(T)}$ expresses the conservation of energy with an uncertainty $\hbar/T$ due to the finite duration of the excitation. The energy differences entering the expression of A are given by

$$\begin{aligned} E_{\text{init}} - E_{\text{interm}} &= E_g + N\hbar\omega_L - E_e - (N-1)\hbar\omega_L = \hbar(\omega_L - \omega_A) = \hbar\delta, \\ E_{\text{init}} - E_{\text{fin}} &= E_g + N\hbar\omega_L - E_g - (N-1)\hbar\omega_L - \hbar\omega_F = \hbar(\omega_L - \omega_F), \end{aligned} \tag{5.2}$$

[1] See [Cohen-Tannoudji *et al.* (1992b)], complement A_I.

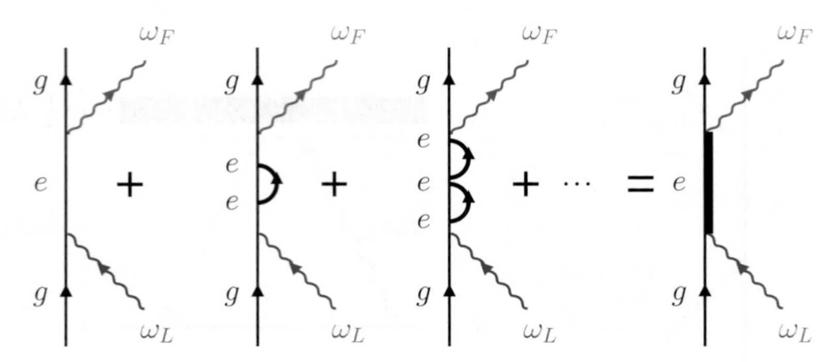

Fig. 5.3 The resonant scattering amplitude does not diverge as soon as a non-perturbative treatment of the coupling V_{AR} with the vacuum field reservoir is carried out. The scattering amplitude is then a sum over all processes where a photon is spontaneously emitted and then reabsorbed as diagrammatically illustrated in this figure.

where the detuning $\delta = \omega_L - \omega_A$ accounts for the frequency mismatch between the laser and the atomic frequencies. We can thus rewrite the resonant scattering amplitude as a function of the frequencies ω_A, ω_L, ω_F and Ω:

$$A \propto \delta^{(T)}(\omega_L - \omega_F)\frac{\Omega}{\omega_L - \omega_A}. \tag{5.3}$$

When ω_L becomes equal to ω_A, the amplitude of the resonant scattering diverges! The solution to this divergence requires us to take into account all the amplitudes where the atom, in the excited state e, emits and reabsorbs a photon an arbitrary number of times (see Fig. 5.3). Thus, we have to deal with a non-perturbative treatment of V_{AR}, though, at this stage, we neglect processes where several interactions with the incident light beam occur, and therefore, keep a perturbative treatment of the coupling term V_{AL} between the laser and the atom. Such an approach is valid for very low intensity of the light beam.

The summation over all the emission-reabsorption processes introduces a "renormalized" propagator[2] for the excited state e obtained by replacing its energy E_e by $E_e - i\hbar\Gamma/2$. It is represented by a heavy gray line in the last diagram of Fig. 5.3. This result simply explains the radiative instability of the excited state e, since $E_e \longrightarrow E_e - i\hbar\Gamma/2$ gives the exponential decay of the probability amplitude to remain in the excited state after a time t: $\exp(-iE_e t/\hbar) \longrightarrow \exp(-iE_e t/\hbar)\exp(-\Gamma t/2)$. The new energy denominator that enters the resonant scattering amplitude becomes

$$E_{\text{init}} - E_{\text{interm}} = E_g + \hbar\omega_L - \left(E_e - i\hbar\frac{\Gamma}{2}\right) = \hbar\left(\omega_L - \omega_A + i\frac{\Gamma}{2}\right), \tag{5.4}$$

and no longer vanishes for $\omega_L = \omega_A$. This non-perturbative treatment of the coupling term V_{AR} yields the following expression for the resonant scattering amplitude:

$$A(\omega_L) \propto \delta^{(T)}(\omega_L - \omega_F)\frac{\Omega}{\omega_L - \omega_A + i(\Gamma/2)}. \tag{5.5}$$

[2]See [Cohen-Tannoudji *et al.* (1992b)], Complement B_{III}.

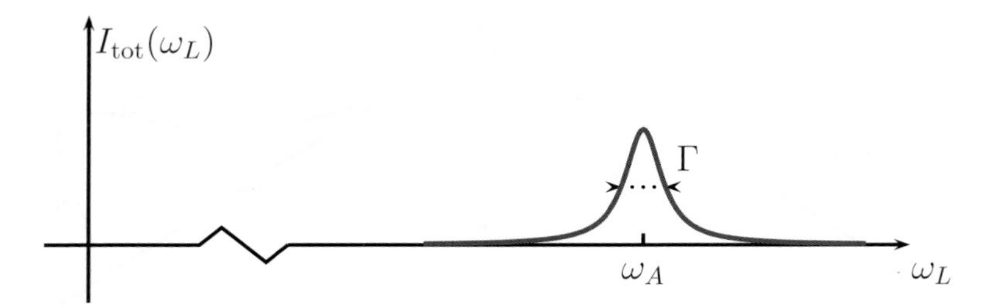

Fig. 5.4 At the lowest order, the scattered light close to resonance is monochromatic and has the same frequency as the incident one. Its intensity varies with the detuning $\omega_L - \omega_A$ as a Lorentzian of width Γ.

This amplitude remains valid in the perturbative limit for which

$$\epsilon \equiv \frac{\Omega}{\sqrt{\delta^2 + (\Gamma^2/4)}} \ll 1. \tag{5.6}$$

Equation (5.5) can also be interpreted in semiclassical terms by considering the atom as a classical oscillator of eigenfrequency ω_A damped with a time constant Γ^{-1}. If this oscillator is excited at a monochromatic frequency ω_L, its "response" will exhibit a resonant behavior over an interval of width Γ about $\omega_L = \omega_A$.

The preceding analogy allows us to understand in a simple way the main features of the scattered light close to resonance in the low intensity limit. For a fixed frequency ω_L of the impinging laser, the spectral distribution of the radiated photons ω_F is a delta function centered at ω_L. If the incident light is monochromatic, the scattered light is also monochromatic with the same frequency (Rayleigh scattering). In addition, the total fluorescence intensity I_{tot} ($\propto |A|^2 \propto \Omega^2$) is proportional to the laser intensity.

Alternatively, by using a broadband detector, one can measure the total intensity I_{tot} of the scattered light versus the incident laser frequency ω_L:

$$I_{\mathrm{tot}}(\omega_L) \sim \int |A|^2 \mathrm{d}\omega_F \sim \frac{\Omega^2}{(\omega_L - \omega_A)^2 + (\Gamma^2/4)}. \tag{5.7}$$

We find out that the total intensity varies with the detuning from resonance $\omega_L - \omega_A$ as a Lorentzian of width Γ. This behavior allows one to infer the atomic frequency ω_A with a resolution given by the spectral width of the excited level Γ (see Fig. 5.4). If the incident radiation contains photons with all frequencies, forming a white continuous spectrum near $\omega_L = \omega_A$, each individual photon of frequency ω_L is scattered elastically with an efficiency given by $I_{\mathrm{tot}}(\omega_L)$.

5.2.3 *Scattering of a light wave packet*

Similar considerations allow one to understand the scattering of a light wave packet. Let us rewrite the wave function of the incident pulse as an integral over the Fourier frequencies:

$$\psi_{\mathrm{inc}}(s) \sim \int \mathrm{d}\omega g(\omega) \exp(-i\omega s), \tag{5.8}$$

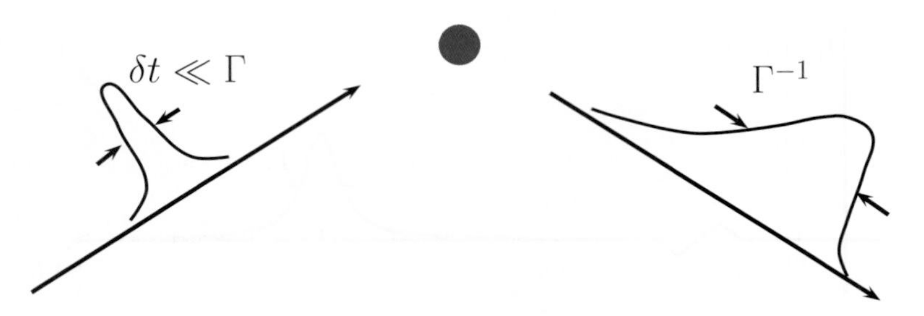

Fig. 5.5 The envelop of an incident short wave packet ($\Delta t \ll \Gamma^{-1}$) is broadened after being scattered.

where $s = t - r/c$, c being the speed of light. Each component ω of the incident wave packet is scattered with an amplitude $A(\omega)$ given by Eq. (5.5) so that

$$\psi_{\text{scat}}(s) \sim \int d\omega g(\omega) \frac{1}{\omega - \omega_A + i\Gamma/2} \exp(-i\omega s). \tag{5.9}$$

The scattered packet $\psi_{\text{scat}}(s)$ has a time dependence given by the convolution product of the Fourier transform of $g(\omega)$, proportional to the incident wave packet $\psi_{\text{inc}}(s)$, by the Fourier transform of the scattering amplitude $A(\omega) = [\omega - \omega_A + i\Gamma/2]^{-1}$, proportional to $\exp(-i\omega s - \Gamma s/2)\theta(s)$ where $\theta(s)$ is the Heaviside function.

Two limiting cases can be distinguished depending on the relative value of the width $\delta\omega$ of the Fourier transform $g(\omega)$ of the incident wave packet compared to the frequency width Γ of the excited state:

- *Narrow line excitation:* $\delta\omega \ll \Gamma$. The wave packet passes through the atom during a very long time $\delta t \sim 1/\delta\omega \gg \Gamma^{-1}$, so that the atomic dipole reaches a steady state and oscillates at the same frequency as the laser driving field. As a corollary, the scattered wave packet has the same shape as the incident one.
- *Broad line excitation:* $\delta\omega \gg \Gamma$. The wave packet passes through the atom during a very short time $\delta t \sim 1/\delta\omega \ll \Gamma^{-1}$. After a rising edge analogous to the one of the incident wave packet, the scattered wave packet decays exponentially with a time constant Γ^{-1} and a frequency ω_A. Resonance scattering is then decomposed into two steps: (i) a percussion excitation (ii) followed by a free exponential decay (see Fig. 5.5).

5.2.4 *First higher order processes*

So far, we have restricted ourselves to very low incident intensity. We have consequently neglected the processes where several interactions with the incident light beam occur. The simplest example of a nonlinear scattering process involves the absorption of two laser photons ω_L (see Fig. 5.6). The amplitude for such a second order process is proportional to Ω^2 to be contrasted with the first order elastic

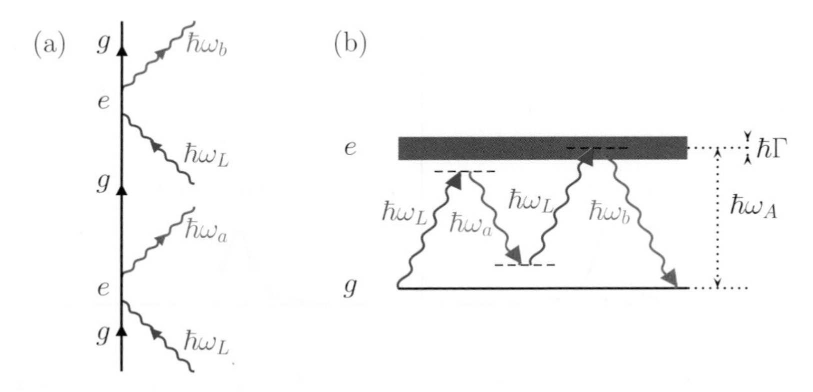

Fig. 5.6 Two different diagrammatic representations (see caption of Fig. 5.2) of the same quasi-resonant scattering process at the second order in Ω: two incident ω_L photons are absorbed and two photons with frequencies ω_a and ω_b are emitted. When such nonlinear processes are taken into account, the light scattered by an excited atom near resonance is no longer necessarily monochromatic.

process proportional to Ω. The conservation of the total energy implies that the frequencies ω_a and ω_b of the emitted photons fulfill the equality $2\omega_L = \omega_a + \omega_b$ and not $\omega_a = \omega_b = \omega_L$.

The amplitude associated with such a nonlinear process exhibits a resonance when the energy attained after the absorption of the second photon ω_L varies about the energy of the excited state, within an energy interval of width $\hbar\Gamma$. It follows that ω_b is centered about ω_A in an interval of width Γ, and consequently the frequency ω_a over an interval symmetric with respect to ω_L, thus distributed about $2\omega_L - \omega_A$. Thereby, at high intensities, the scattering processes involving several photons of the incident light beam give rise to inelastic scattering . The corresponding perturbative fluorescence spectrum, assuming that $|\omega_L - \omega_A| > \Gamma$, is represented in Fig. 5.7, showing a characteristic triplet structure. The central line corresponds to the elastic scattering contribution, and the sidebands account for the inelastic scattering which becomes all the more important as the excitation grows stronger.

One could consider developing higher and higher order diagrams for understanding the behavior of atoms in strong resonant fields. However, the perturbation series does not converge, and the calculation becomes more difficult the nearer ω_L comes to the atomic frequency ω_A. In the case $\delta = 0$ and $\Omega \gg \Gamma$, the parameter $\epsilon = \Omega/\sqrt{\delta^2 + (\Gamma^2/4)}$ for the perturbative expansion is larger than one, and the previous treatment is no longer valid.

5.3 Optical Bloch equations

A first non-perturbative treatment of the interaction with the laser field can be given using optical Bloch equations (OBE). These equations give the rate of variation of

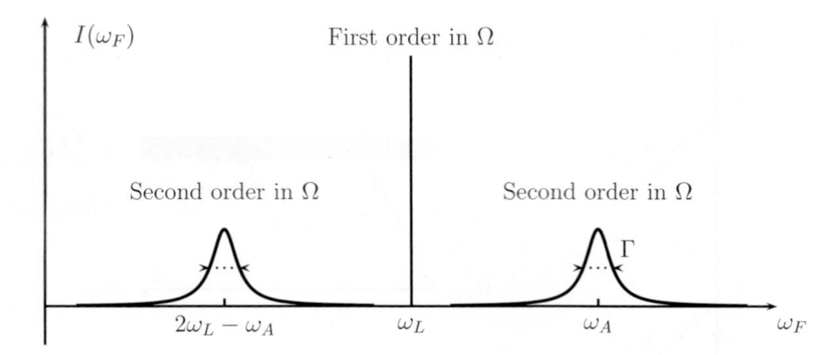

Fig. 5.7 Fluorescence spectrum (perturbative results). The delta function at $\omega_F = \omega_L$ is the elastic component given by the lowest order diagram (Fig. 5.2). It is proportional to the intensity I of the incident light. The two Lorentz curves, centered at ω_A and $2\omega_L - \omega_A$, are the inelastic components resulting from the nonlinear scattering processes of Fig. 5.6. The total area below these curves is proportional to I^2.

the atomic density matrix σ_A, including the effect of the interaction with the laser light and the damping due to spontaneous emission. They can be solved for any value of the Rabi frequency Ω, and in this way, provide a non-perturbative description of the atomic state. From these equations, one can also calculate the correlation functions of the atomic dipole moment using the quantum regression theorem. From this result, the properties of the fluorescence light can be deduced. This approach will not be presented here, but is explained in great detail in [Mollow (1969); Cohen-Tannoudji (1977)] and, Chap. V and Complement A_V of [Cohen-Tannoudji *et al.* (1992b)]. We will come back to optical Bloch equations and give their steady-state solution in Chap. 11, which is devoted to the study of radiative forces. In this section, we will focus on the terms describing the damping due to spontaneous emission since we will need the corresponding equations in the following.

In the OBE approach, the laser field is treated as a classical external field $\vec{\mathcal{E}}_L \cos\omega_L t$, so that the Hamiltonian H in the absence of spontaneous emission is given by:

$$H = H_A + V_{AL}(t), \tag{5.10}$$

where

$$H_A = \hbar\omega_A \left|e\right\rangle \left\langle e\right| \qquad \text{and} \qquad V_{AL}(t) = -\vec{D}\cdot\vec{\mathcal{E}}_L \cos\omega_L t, \tag{5.11}$$

$\vec{D}$ being the atomic dipole moment operator.

The equations of motion of the atomic matrix density σ_A have the following form:

$$\frac{\mathrm{d}\sigma_A}{\mathrm{d}t} = -\frac{i}{\hbar}\left[H, \sigma_A\right] + \left(\frac{\mathrm{d}\sigma_A}{\mathrm{d}t}\right)_{\text{spont}}, \tag{5.12}$$

where the first term is a commutator giving the rate of variation due to H and where the last term describes the damping due to spontaneous emission. This damping is produced by the coupling of the atom with the reservoir of the empty modes of the quantum radiation field. One can show[3] that the corresponding rate of variation of the matrix elements of σ_A is given by:

$$\left(\frac{d\sigma_{ee}}{dt}\right)_{\text{spont}} = -\Gamma\sigma_{ee}, \quad \left(\frac{d\sigma_{eg}}{dt}\right)_{\text{spont}} = -\frac{\Gamma}{2}\sigma_{eg}, \quad \left(\frac{d\sigma_{gg}}{dt}\right)_{\text{spont}} = \Gamma\sigma_{ee}. \quad (5.13)$$

The interpretation of these equations is straightforward. The first one describes how the population of the excited state e decays because of spontaneous emission with a rate Γ; the damping of the coherence σ_{eg} between e and g appears in the second equation with a rate twice smaller, which is the half sum of the widths of e (equal to Γ) and of g (equal to zero); finally, the third one describes how atoms are transferred from e to g by spontaneous emission with a rate Γ.

Strictly speaking, Eqs. (5.13) describe the damping due to spontaneous emission calculated in the absence of any laser excitation. Adding *independently* in Eq. (5.12) the rates of variation of σ_A due to the laser excitation and to spontaneous emission, respectively, as each of these two processes was acting alone, is therefore an approximation which is valid for the following reason. The duration of an elementary spontaneous emission process[4] is extremely short. It scales as the correlation time τ_C of vacuum fluctuations, which are the fluctuations of the reservoir R of the empty modes of the radiation field.[5] This time, which is smaller than an optical period $2\pi/\omega_A$, is very short compared to the inverse of the Rabi frequency Ω characterizing the atom-laser coupling so that one can neglect the atom-laser coupling during an elementary spontaneous emission process and use the damping coefficients calculated in the absence of laser.

Equations (5.13) can also be written as an operator equation:

$$\left(\frac{d\sigma_A}{dt}\right)_{\text{spont}} = -\frac{\Gamma}{2}\left(\mathcal{S}_+\mathcal{S}_-\sigma_A + \sigma_A\mathcal{S}_+\mathcal{S}_-\right) + \Gamma\mathcal{S}_-\sigma_A\mathcal{S}_+ \quad (5.14)$$

where

$$\mathcal{S}_+ = |e\rangle\langle g| \text{ and } \mathcal{S}_- = |g\rangle\langle e|. \quad (5.15)$$

The first term of the right-hand side of Eq. (5.14) is an anticommutator, which can be also considered as describing the evolution of σ_A due to an effective anti-Hermitian Hamiltonian[6] $-i\hbar\,(\Gamma/2)\,|e\rangle\langle e|$ introducing the radiative instability of e. The last term of the right-hand side of Eq. (5.14) is a transfer term. Master equations having the structure of Eq. (5.14) are called "Lindblad equations" [Lindblad (1976)].

Optical Bloch equations are a very powerful tool in quantum optics and provide quantitative expressions for the various signals which can be observed in resonance

[3]See for example Chap. IV of [Cohen-Tannoudji *et al.* (1992b)].

[4]This duration should not be confused with the radiative lifetime $\tau_R = 1/\Gamma$ which is the mean time that an atom spends in the excited state e before spontaneously emitting a photon.

[5]See Chap. IV of [Cohen-Tannoudji *et al.* (1992b)].

[6]The commutator of σ_A with an anti-Hermitian operator O is taken equal to: $[O, \sigma_A] = O\sigma_A - \sigma_A O^\dagger = O\sigma_A + \sigma_A O$. This insures that the rate of variation of σ_A is Hermitian.

fluorescence. However, their physical interpretation is not straightforward since the classical treatment of the laser field does not allow a clear identification of the elementary processes of absorption, stimulated emission and spontaneous emission. In the next section, we present another non-perturbative treatment, the dressed atom approach, from which physical pictures can be more easily extracted.

5.4 The dressed atom approach

The dressed atom approach was first developed to interpret various effects observed in optical pumping experiments performed on atoms interacting with strong resonant or non-resonant radio-frequency fields. A clear interpretation of these effects can be given by studying the properties of the whole system {atom + RF photons} in interaction, the so-called atom "dressed" by RF quanta.[7] Later on, this approach was generalized to the optical domain for analyzing the behavior of atoms interacting with intense monochromatic laser fields. The new ingredient one has to include in the theory is spontaneous emission,[8] which was negligible in the RF domain. This latter effect, as detailed in Sec. 5.4.5, gives rise to a radiative cascade in the { atom + photon } energy-level ladder. The analysis of this radiative cascade provides a simple interpretation of the statistical properties of the sequence of fluorescence photons spontaneously emitted by the atom, and is a simple introduction to the stochastic description of quantum dissipative processes in terms of quantum jumps.

5.4.1 *The interacting systems*

The dressed atom approach consists of considering, in a first step, the atom and the laser photons as a single isolated quantum system described by a time independent Hamiltonian (system inside the ellipse of Fig. 5.8). Near resonance, the energy levels of such a system, called dressed states, can be easily calculated and form a ladder of energy levels. In a second step, we consider the coupling of the dressed atom with the reservoir of the empty modes of the quantized radiation field which introduces a damping of the system and gives rise to photons spontaneously emitted by the dressed atom. In this approach, the sequence of fluorescence photons emitted by the atom excited by the resonant laser field appears as photons spontaneously emitted by the dressed atom cascading downwards its ladder of energy levels.

The two-level atom with a ground state $|g\rangle$ and an excited state $|e\rangle$ considered here is described by the Hamiltonian H_A:

$$H_A\,|e\rangle = \hbar\omega_A\,|e\rangle\,, \qquad H_A\,|g\rangle = 0. \tag{5.16}$$

[7] See [Cohen-Tannoudji and Haroche (1966a)] and complement A_{VI} of [Cohen-Tannoudji *et al.* (1992b)].

[8] See [Cohen-Tannoudji and Reynaud (1977)] and Chap. VI of [Cohen-Tannoudji *et al.* (1992b)].

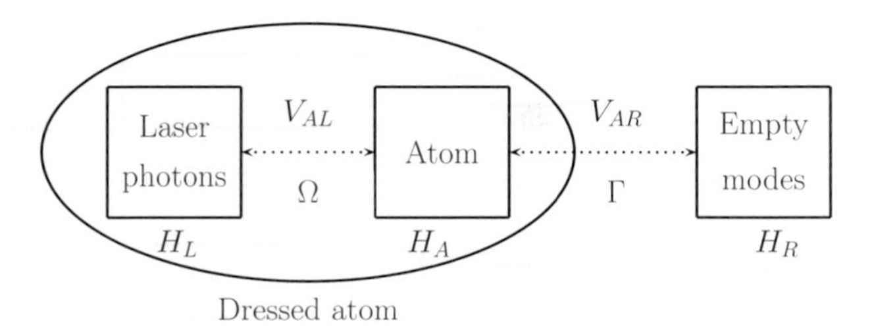

Fig. 5.8 The various systems considered in the dressed atom approach.

We give now a quantum description of the laser field. The laser photons are the photons of a single mode L of a cavity whose Hamiltonian H_L can be written:

$$H_L = \hbar\omega_L a^\dagger a, \tag{5.17}$$

$a^\dagger$ and a being the creation and annihilation operators of a laser photon. The atom-laser interaction Hamiltonian is denoted V_{AL}, so that the total Hamiltonian of the dressed atom is equal to:

$$H_{LA} = H_A + H_L + V_{AL}. \tag{5.18}$$

We call H_R the Hamiltonian of the empty modes of the radiation field and V_{AR} the coupling between the atom A and the vacuum field reservoir R.

5.4.2 *Uncoupled states of the atom-laser system*

The uncoupled states of the atom-laser system correspond to $V_{AL} = 0$ and are the eigenstates of $H_A + H_L$. They are labelled by two quantum numbers: e or g for the internal degree of freedom of the atom, and N for the number of laser photons. Their eigenvalues are given by:

$$(H_A + H_L)|e, N\rangle = \hbar(\omega_A + N\omega_L)|e, N\rangle, \tag{5.19}$$

$$(H_A + H_L)|g, N+1\rangle = (N+1)\hbar\omega_L|g, N+1\rangle. \tag{5.20}$$

Near resonance, the two states $|g, N+1\rangle$ and $|e, N\rangle$ have very close energies. Their energy difference is equal to:

$$E_{g,N+1} - E_{e,N} = \hbar\left(\omega_L - \omega_A\right) = \hbar\delta, \tag{5.21}$$

where $\delta = \omega_L - \omega_A$ is the detuning between the laser frequency ω_L and the atomic frequency ω_A. The two states are degenerate if $\delta = 0$. If $\delta > 0$, $|g, N+1\rangle$ is above $|e, N\rangle$. If $\delta < 0$, $|g, N+1\rangle$ is below $|e, N\rangle$. The uncoupled states are gathered in two-dimensional manifolds $\ldots \mathcal{E}_N, \mathcal{E}_{N+1}, \mathcal{E}_{N+2}, \ldots$ separated by a distance $\hbar\omega_L$, the splitting inside each manifold being equal to $\hbar\delta$ (see left part of Fig. 5.9). Note however that the state $|g, 0\rangle$ (not represented in Fig. 5.9) is isolated.

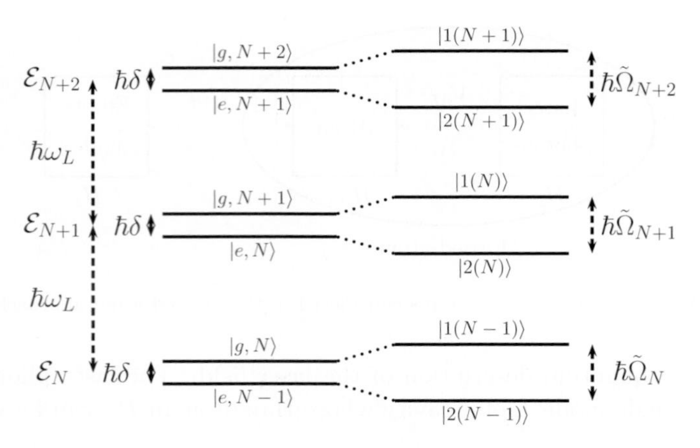

Fig. 5.9 Three adjacent manifolds of uncoupled states of the atom-laser system (left part of the figure) and corresponding dressed states (right part).

5.4.3 *Effect of the coupling. Dressed states*

The coupling V_{AL} between A and L is proportional to the atomic dipole moment $\vec{D}$ and to the laser electric field $\vec{E}_L$ which is a linear combination of the annihilation and creation operators a and $a^\dagger$ of a laser photon. The only non-zero matrix element of the odd operator $\vec{D}$ (in a space reflection) is between $|e\rangle$ and $|g\rangle$. The only non-zero matrix elements of $\vec{E}_L$ are between $|N\rangle$ and $|N+1\rangle$ or $|N-1\rangle$ and are proportional to $\sqrt{N+1}$ or $\sqrt{N}$. The non-zero matrix element of V_{AL} in $\mathcal{E}_{N+1}$ is thus:

$$\langle e, N|V_{AL}|g, N+1\rangle \propto \sqrt{N+1}. \tag{5.22}$$

This coupling expresses the fact that the atom in the ground state g and in the presence of $N+1$ photons can absorb a photon and reach the excited state e, while the number of photons decreases by one unit. We introduce Ω_0 the *vacuum Rabi frequency*,

$$\frac{\hbar\Omega_0}{2} = \langle e, 0|V_{AL}|g, 1\rangle, \tag{5.23}$$

so that we can write

$$\langle e, N|V_{AL}|g, N+1\rangle = \frac{\hbar\Omega_{N+1}}{2} \qquad \text{with} \qquad \Omega_{N+1} = \Omega_0\sqrt{N+1}. \tag{5.24}$$

Note that the coupling V_{AL} has non-zero matrix elements, not only within each manifold $\mathcal{E}_N$, but also between different manifolds. For example, $|g, N+1\rangle$ is coupled not only to $|e, N\rangle$, but also to $|e, N+2\rangle$ (the atom goes from g to e by stimulated emission of a laser photon). This coupling is, however, highly non-resonant and we will neglect it in comparison with the resonant coupling within a manifold. This approximation is the equivalent of the rotating wave approximation described in Chap. 2, Sec. 2.4. In other words, our treatment is valid to zeroth order in Ω_{N+1}/ω_L, but we impose no conditions on the relative values of Ω_{N+1}, δ and Γ.

As a result of the coupling (5.24), the two uncoupled states $|g, N+1\rangle$ and $|e, N\rangle$ transform into two perturbed states $|1(N)\rangle$ and $|2(N)\rangle$, also called "dressed states", which are linear combinations of $|g, N+1\rangle$ and $|e, N\rangle$. The wave functions and the energies of these two dressed states are obtained by diagonalizing the 2×2 matrix:

$$\hbar \begin{pmatrix} \delta & \Omega_{N+1}/2 \\ \Omega_{N+1}/2 & 0 \end{pmatrix}. \tag{5.25}$$

Their energy splitting (see right part of Fig. 5.9) is equal to $\hbar\tilde{\Omega}_{N+1}$ where

$$\tilde{\Omega}_{N+1} = \sqrt{\Omega_{N+1}^2 + \delta^2}. \tag{5.26}$$

Consider for example the case $\delta = 0$ (resonant excitation). The two dressed states $|1(N)\rangle$ and $|2(N)\rangle$ are then the linear symmetric and antisymmetric combinations of $|g, N+1\rangle$ and $|e, N\rangle$ and their splitting is equal to $\hbar\Omega_{N+1}$. If one starts at time $t = 0$ from the state $|g, N+1\rangle$, the system will oscillate back and forth between $|g, N+1\rangle$ and $|e, N\rangle$ at the angular frequency Ω_{N+1}, in a series of reversible absorption and stimulated emission processes.[9] This justifies the denomination Rabi frequency given to Ω_{N+1}.

5.4.4 *Two different situations*

The dressed atom approach described in this section can be applied to the interpretation of two types of experiments: (i) atom in a real cavity and interacting with the field contained in this cavity; (ii) atom in free space interacting with an incident laser beam. For each of these two situations, we explain here how to define the Rabi frequencies introduced in the previous subsection.

(i) *Atom in a real cavity. Cavity Quantum Electrodynamics*

In this case, the quantum number N represents the number of photons in the real cavity. One must, in particular, consider the case $N = 0$ (empty cavity) and the case of a small number of photons $N = 1, 2, 3, \ldots$ The lowest manifold $\mathcal{E}_0$ is one dimensional and contains a single state $|g, 0\rangle$. The first excited manifold $\mathcal{E}_1$ contains the two states $|g, 1\rangle$ and $|e, 0\rangle$, the second excited manifold $\mathcal{E}_2$ contains the two states $|g, 2\rangle$ and $|e, 1\rangle$, and so on. The uncoupled states and the dressed states in the first four manifolds are represented in Fig. 5.10 in the case of a resonant excitation ($\delta = 0$). The splitting between the dressed states in the manifolds $\mathcal{E}_N$ increases as $\Omega_0\sqrt{N}$. If the state of the field in the cavity is the superposition of states with different values of N, the Rabi oscillation between e and g, will therefore involve several Rabi frequencies. Such an effect has been experimentally observed [Brune *et al.* (1996a)].

[9] In Sec. 7.2 of Chap. 7, we will present a more precise treatment of this problem, including the radiative instability of the excited state e due to spontaneous emission and described by the finite energy linewidth $\hbar\Gamma$. We will show that the Rabi oscillation appears only when the Rabi frequency Ω_{N+1} is large enough compared to Γ.

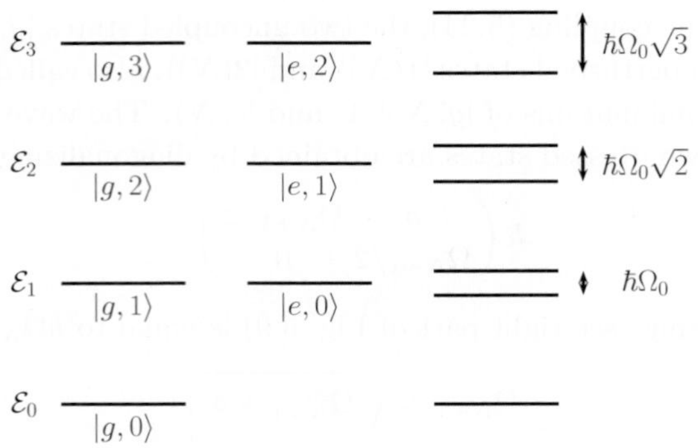

$\mathcal{E}_3$ $|g,3\rangle$ $|e,2\rangle$ $\hbar\Omega_0\sqrt{3}$

$\mathcal{E}_2$ $|g,2\rangle$ $|e,1\rangle$ $\hbar\Omega_0\sqrt{2}$

$\mathcal{E}_1$ $|g,1\rangle$ $|e,0\rangle$ $\hbar\Omega_0$

$\mathcal{E}_0$ $|g,0\rangle$

Fig. 5.10 The first four manifolds of the atom-laser system in a real cavity for a resonant excitation.

This representation of the atom-photons states in a cavity constitutes the so-called Jaynes Cummings model [Jaynes and Cummings (1963)]. It is frequently used in the research field called *Cavity Quantum Electrodynamic*.[10] We will use it in Chap. 7 for studying the phase shift of the field due to its interaction with the atom.

Inhibition of spontaneous emission

If the atomic frequency differs too much from all the cavity mode frequencies, an excited atom cannot deposit its internal energy into the cavity so that spontaneous emission is quenched. More generally, the cavity modifies the boundary conditions for the electromagnetic field. As a result, the vacuum field frequency spectrum is modified which can change the characteristics of spontaneous emission [Kleppner and Haroche (1989)].

(ii) *Atom in free space*

In this case, it is convenient to model the propagating laser field as a single mode field in a fictitious ring cavity (see Fig. 5.11). The cavity will be supposed to be very large, so that the modification of spontaneous emission due to cavity effects can be neglected. In addition, the mode of the cavity and the number of photons will be chosen in such a way that the field "seen" by the atom in the fictitious cavity has the same spatial dependence and the same amplitude as in the real experiment.

It is then clear that the number of photons, N, appearing in the atom-photon states introduced above has no real physical meaning. We can take its mean value very large, and similarly the volume V of the cavity mode, the ratio $\langle N\rangle/V$ remaining fixed and equal to the energy density in the real experiment. If the state of the single mode field is a coherent state, or close to a coherent state, the dispersion δN of N around $\langle N\rangle$ is very small in relative value

$$\frac{\delta N}{\langle N\rangle} = \frac{1}{\sqrt{\langle N\rangle}} \ll 1. \tag{5.27}$$

[10] For a review, see for example [Haroche and Raimond (2006)].

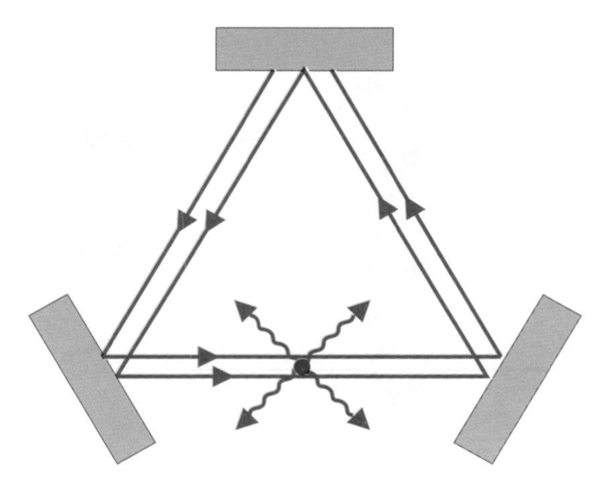

Fig. 5.11 Model used for the laser beam described as a single mode field in a fictitious ring cavity.

In this case, we can neglect the variation with N of the Rabi frequencies Ω_N, and replace them by a single Rabi frequency which we will denote Ω:

$$\Omega_N = \Omega_0\sqrt{N} \simeq \Omega_0\sqrt{\langle N\rangle} \equiv \Omega. \tag{5.28}$$

Similarly, we can replace the generalized Rabi frequencies $\tilde{\Omega}_N = \sqrt{\Omega_N^2 + \delta^2}$ giving the splitting between the dressed states of $\mathcal{E}_N$ by a N-independent frequency:

$$\tilde{\Omega} = \sqrt{\Omega^2 + \delta^2}. \tag{5.29}$$

The Rabi frequencies Ω and $\tilde{\Omega}$ that we introduce here are the quantum equivalent of the corresponding Rabi frequencies defined in a semiclassical treatment of the laser field.[11] The fact that the N-dependence of $\tilde{\Omega}_N$ can be neglected allows us to consider the energy diagram of the dressed states as periodic in the relevant range of values of N.

5.4.5 *Radiative cascade in the basis of uncoupled states*

The next step of the dressed atom approach consists of taking into account the coupling V_{AR} between the atom A and the reservoir R of the empty modes of the quantized radiation field, thus giving rise to photons spontaneously emitted in these modes. Each fluorescence photon appears in a scattering process where a laser photon disappears and a fluorescence photon appears in a mode different from the laser mode. Since a laser photon disappears, this process corresponds to a radiative transition of the dressed atom from a manifold $\mathcal{E}_N$ to the adjacent lower manifold $\mathcal{E}_{N-1}$. The next fluorescence photon corresponds to a radiative transition from $\mathcal{E}_{N-1}$ to $\mathcal{E}_{N-2}$, and so on. This leads us to the picture of the dressed atom cascading downwards through its energy diagram.

[11]See Chap. 2.

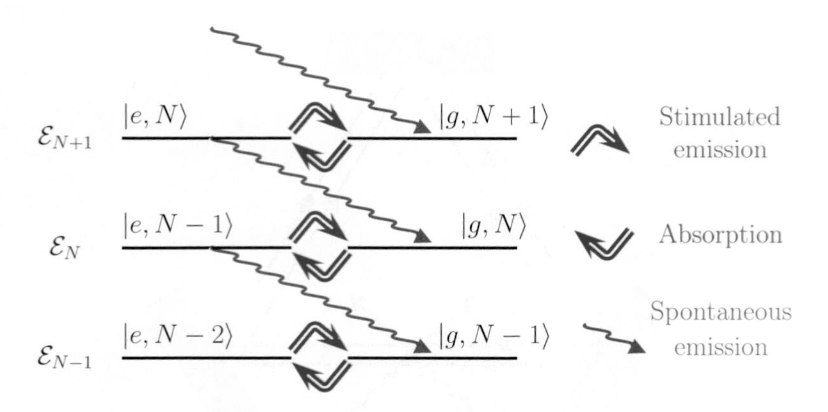

Fig. 5.12 Radiative cascade of the dressed atom.

The quantitative analysis of this cascade is based on the master equation that describes the damping of the dressed atom $A + L$ due to spontaneous emission and generalizes the master equation (5.14) given in Sec. 5.3 for A to $A + L$. As in Sec. 5.3, the short value of the correlation time of vacuum fluctuations allows us to independently add the rates of variation of the dressed atom density matrix σ_{AL} due to the atom-laser coupling and to spontaneous emission, respectively, calculated as if each process was acting alone. This leads us to the following master equation for $A + L$:

$$\begin{aligned}\frac{\mathrm{d}\,\sigma_{AL}}{\mathrm{d}\,t} &= \frac{1}{i\,\hbar}\left[H_A + H_L + V_{AL},\, \sigma_{AL}\right] \\ &\quad - \frac{\Gamma}{2}\left(\mathcal{S}_+\mathcal{S}_-\sigma_{AL} + \sigma_{AL}\mathcal{S}_+\mathcal{S}_-\right) + \Gamma\mathcal{S}_-\sigma_{AL}\mathcal{S}_+ .\end{aligned} \tag{5.30}$$

As already mentioned above and in Sec. 5.3, each elementary spontaneous emission process lasts a very short time during which one can neglect the atom-laser coupling. It is thus more appropriate to describe the radiative cascade in the uncoupled state basis (see Fig. 5.12).

Just after a spontaneous emission process, the atom is projected onto the ground state $|g\rangle$, while the laser photons remain spectator. The combined {atom + photon} system enters in the state $|g\rangle$ of a given manifold, for example, the state $|g, N+1\rangle$ of $\mathcal{E}_{N+1}$. It then evolves between $|g, N+1\rangle$ and $|e, N\rangle$ within $\mathcal{E}_{N+1}$ by absorption and stimulated emission processes. At a certain time, the atom will leave $\mathcal{E}_{N+1}$ by a spontaneous emission jump from $|e, N\rangle$ to $|g, N\rangle$, and enter the next lower manifold $\mathcal{E}_N$. And so on.

A more detailed analysis of this cascade can be given using the master equation (5.30) for studying the rate of variation of the restriction σ_N of the dressed atom density matrix within $\mathcal{E}_N$:

$$\sigma_N = P_N \sigma_A P_N \text{ with } P_N = |g, N\rangle\langle g, N| + |e, N-1\rangle\langle e, N-1| . \tag{5.31}$$

The restrictions within $\mathcal{E}_N$ of the various terms appearing in the right-hand side of

Eq. (5.30) are easily calculated:

$$P_N\left(H_A+H_L+V_{AL}\right)P_N=(N-1)\hbar\omega_L+\hbar\omega_A+\hbar\begin{pmatrix}\delta & \Omega/2\\ \Omega/2 & 0\end{pmatrix} \tag{5.32}$$

$$P_N\mathcal{S}_+\mathcal{S}_-P_N=|e,N-1\rangle\langle e,N-1| \tag{5.33}$$

$$P_N\mathcal{S}_-\sigma_{AL}\mathcal{S}_+P_N=|g,N\rangle\langle g,N|\langle e,N|\sigma_{N+1}|e,N\rangle\,. \tag{5.34}$$

Using Eqs. (5.32) and (5.33), one gets the effective Hamiltonian H_{eff}^N describing the reduced evolution within $\mathcal{E}_N$. Within a constant term this non-Hermitian Hamiltonian is equal to:

$$H_{\text{eff}}^N=\hbar\begin{pmatrix}\delta & \Omega/2\\ \Omega/2 & -i\Gamma/2\end{pmatrix}. \tag{5.35}$$

By taking into account the contribution of the last term of Eq. (5.30) and using Eq. (5.34), we finally get for the rate of variation of σ_N:

$$\frac{\mathrm{d}\sigma_N(t)}{\mathrm{d}t}=\frac{1}{i\hbar}\left[H_{\text{eff}}^N,\sigma_N(t)\right]+\Gamma\pi_{e,N}(t)\,|g,N\rangle\langle g,N| \tag{5.36}$$

where

$$\pi_{e,N}(t)=\langle e,N|\,\sigma_{N+1}(t)\,|e,N\rangle \tag{5.37}$$

is the probability to find at time t the dressed atom in the state $|e,N\rangle$ of $\mathcal{E}_{N+1}$.

Equation (5.36) is a linear homogeneous differential equation with a source term (last term in the right-hand side). Its solution can be written:

$$\begin{aligned}\sigma_N(t)&=\Gamma\int_{t_0}^{t}\mathrm{d}t'\,\pi_{e,N}(t')\exp[-i\,H_{\text{eff}}^N(t-t')]\,|g,N\rangle\langle g,N|\exp[+i\,\left(H_{\text{eff}}^N\right)^{\dagger}(t-t')]\\&=\Gamma\int_{t_0}^{t}\mathrm{d}t'\,\pi_{e,N}(t')\,|\psi_N(t-t')\rangle\langle\psi_N(t-t')|\end{aligned} \tag{5.38}$$

where

$$|\psi_N(t-t')\rangle=\exp[-i\,H_{\text{eff}}^N(t-t')]\,|g,N\rangle\,. \tag{5.39}$$

We have supposed that t_0 is sufficiently far in the past of t so that the solution of the homogeneous equation is damped to 0 at time t. This form of $\sigma_N(t)$ clearly shows that the system enters $\mathcal{E}_N$ at a certain time t' in the state $|g,N\rangle$ with a rate $\Gamma\pi_{e,N}(t')$, coming from the state $|e,N\rangle$. Its evolution within $\mathcal{E}_N$ is then governed by the effective Hamiltonian H_{eff}^N given in Eq. (5.35). The non-Hermitian character of H_{eff}^N is due to the fact that the system can leave $\mathcal{E}_N$ by a spontaneous emission jump that brings it from $|e,N-1\rangle$ to the state $|g,N-1\rangle$ of $\mathcal{E}_{N-1}$ (the norm is not conserved in $\mathcal{E}_N$). For an atom having entered the manifold $\mathcal{E}_N$ at time t', the rate of departure from $\mathcal{E}_N$ at time t is given according to Eq. (5.39) by:

$$\Gamma\left|\langle e,N-1\,|\psi_N(t-t')\rangle\right|^2=\Gamma\left|\langle e,N-1|\exp[-i\,H_{\text{eff}}^N(t-t')]\,|g,N\rangle\right|^2. \tag{5.40}$$

The integral over t' in Eq. (5.38) means that we have to sum over all possible "histories" corresponding to different times of entries t' in $\mathcal{E}_N$. Similarly, to describe the subsequent evolution in $\mathcal{E}_{N-1}$, we would have to sum over all possible departures times t from $\mathcal{E}_N$ given by the probability distribution of Eq. (5.40). And so on.

The radiative cascade thus appears as (see Fig. 5.13):

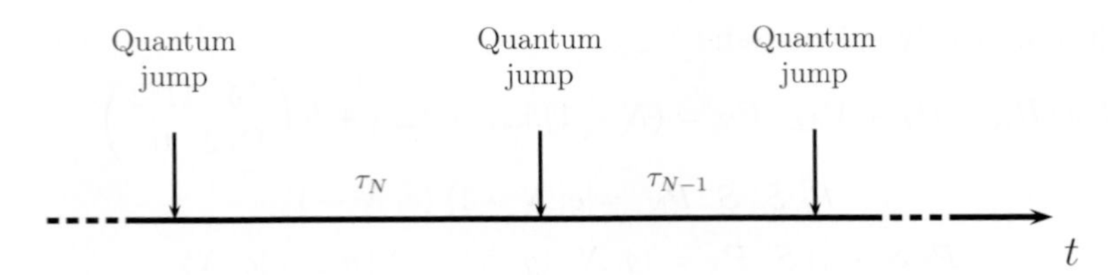

Fig. 5.13 The dynamics of an atom illuminated by a resonant light is made of a succession of coherent evolution periods within a manifold $\mathcal{E}_N$ for a duration τ_N, followed by quantum jumps towards the manifold $\mathcal{E}_{N-1}$.

- a sequence of spontaneous quantum jumps $|e, N+1\rangle \to |g, N+1\rangle$, $|e, N\rangle \to |g, N\rangle$, $|e, N-1\rangle \to |g, N-1\rangle$, ... occurring at random times
- separated by coherent evolution periods describing absorption and stimulated emission processes and governed by an effective hamiltonian.

It is worth noting that the temporal analysis made in this section is valid whatever is the value of the Rabi frequency Ω with respect to the linewidth Γ.

5.4.6 *A new description of quantum dissipative processes*

The analysis of the dressed atom radiative cascade given in the previous subsection describes the time evolution of the system in terms of several possible histories that differ by the times at which quantum jumps from one manifold to the next lower one occurred. For each of these histories, however, the state of the system is described as a pure state between two jumps. For example, the system enters $\mathcal{E}_N$ in the pure state $|g, N\rangle$ and remains in the pure state given by Eq. (5.39) before leaving $\mathcal{E}_N$. The quantum state of the system in a dissipative process is usually described by a density matrix because of its correlations with the reservoir responsible for the dissipation, which prevents a description of the system by a pure state. Moreover, the usual description in terms of a density operator, like the one given by the master equation (5.36), concerns ensemble averages. The analysis of the previous subsection shows that one can extract the idea of individual histories, for a single realization of the ensemble average described by this master equation. For each of these histories, we have a wave function description of the system, with a random change of the wave function from time to time. In this sense, we can use this new description for making predictions on the time evolution of a single quantum system in a single experimental realization. Moreover, summing the predictions for each history over all possible histories gives the same results as the master equation.

In this way, the radiative cascade of the dressed atom provides a simple introduction to new descriptions of dissipative processes that rely on the idea of stochastic wave functions and which presently attract a lot of interest. There are various slightly different schemes [Dalibard *et al.* (1992); Carmichael (1993); Dum *et al.* (1992); Gisin and Percival (1992)], all of which share two basic ideas to be contrasted with the traditional approach of optical Bloch equations: (i) atoms are no longer described by a density matrix, but rather by an ensemble of wave functions; (ii) the time evolution of these wave functions is not a continuous deterministic process; it is a sequence of coherent evolution periods governed by an effective non-Hermitian Hamiltonian and interrupted by instantaneous quantum jumps occurring at random times. There is a certain analogy between this new description of quantum dissipative processes and a classical random walk. Indeed, between two quantum jumps, the atomic wave function follows a coherent deterministic evolution like a Brownian particle moving freely between two collisions.

Such a general stochastic wave function approach can be shown to be mathematically equivalent to optical Bloch equations and, more generally, to systems described by a master equation of the Lindblad type. In the following chapters, we will present a few examples of interpretations of physical effects inspired by this approach.

5.5 Photon correlations. The quantum jump approach

We come back to the physical arrangement of Fig. 5.1, supposing that the detector monitoring the fluorescence light is a broadband detector with a very short response time. Its output is a random sequence of pulses. How can one analyze the statistical properties of this sequence of pulses?

The usual description of such an experiment uses the optical Bloch equations and the quantum regression theorem for calculating the second order correlation function[12] $G_2(t, t+\tau)$ giving the probability to detect one photon at time t and another one at time $t+\tau$. In spite of its remarkable efficiency for the computation of most signals, this method does not provide a simple understanding of the statistical properties of the physical processes. In this section, we show how the picture of the radiative cascade of the $\{$ atom + photon $\}$ system in the uncoupled basis can give a deeper physical insight into the time evolution of the system when it is observed with a broadband detector having a good time resolution.

5.5.1 *The waiting time distribution*

The analysis of Sec. 5.4.6 naturally leads us to introduce the probability distribution $W(\tau)$, called the "delay function" or the "waiting time distribution," of the time intervals τ between two *successive* spontaneous jumps. In the next subsection, we will establish the connection between $W(\tau)$ and the second order correlation function $G_2(\tau)$. The time interval τ is also the time spent τ_N by the system in a given manifold $\mathcal{E}_N$ between the time of entry in this manifold and the time of exit

[12]One can show that this correlation function is also proportional to $\langle \mathcal{S}_+(t)\mathcal{S}_+(t+\tau)\mathcal{S}_-(t+\tau)\mathcal{S}_-(t)\rangle$ where $\mathcal{S}_+$ and $\mathcal{S}_-$ are defined in Eq. (5.15) (see for example Sec. V.D of [Cohen-Tannoudji *et al.* (1992b)]).

(see Fig. 5.13). We have already given in Eq. (5.40) the rate of departure at time t of an atom having entered the manifold $\mathcal{E}_N$ at an earlier time t'. Putting $t - t' = \tau$, we get from this equation:

$$W(\tau) = \Gamma |\langle e, N-1| e^{-iH_{\text{eff}}\tau/\hbar} |g, N\rangle|^2. \tag{5.41}$$

5.5.2 *From the waiting time distribution to the second order correlation function*

The steady state second order correlation function $G_2(\tau)$ gives the probability distribution of the time intervals τ between two spontaneous jumps, where the second one is not necessarily the next jump after the first one. It can be the first jump with a probability $W(\tau)$, the second one, with a probability distribution given by the convolution product $W \otimes W$, etc ... The calculation of $G_2(\tau)$ from the knowledge of the delay function W is thus readily performed using Laplace transforms [Kim *et al.* (1987); Reynaud *et al.* (1988)]:

$$\tilde{G}_2(s) = \tilde{W}(s) + \tilde{W}^2(s) + \cdots = \frac{\tilde{W}(s)}{1 - \tilde{W}(s)}. \tag{5.42}$$

For the two-level atom, the explicit calculation of $W(\tau)$ requires the diagonalization of the effective hamiltonian given by Eq. (5.35). One finds for a resonant excitation ($\delta = 0$) and a sufficiently large intensity ($\Omega > \Gamma$):

$$W(\tau) = \Gamma \frac{\Omega^2}{\lambda^2} \left(\sin^2 \frac{\lambda \tau}{2} \right) e^{-\Gamma\tau/2} \qquad \text{with} \qquad \lambda^2 = \Omega^2 - \frac{\Gamma^2}{4}. \tag{5.43}$$

and, using Eq. (5.42),

$$G_2(\tau) = 1 - e^{-3\Gamma\tau/4} \left(\cos(\tilde{\lambda}\tau) + \frac{3\Gamma}{4\tilde{\lambda}} \sin(\tilde{\lambda}\tau) \right) \qquad \text{with} \qquad \tilde{\lambda}^2 = \Omega^2 - \frac{\Gamma^2}{16}. \tag{5.44}$$

5.5.3 *Photon antibunching*

It appears on Eq. (5.41) that $W(\tau) \longrightarrow 0$ when $\tau \longrightarrow 0$. The same result holds for $G_2(\tau)$. This has a clear physical meaning: As the atom is in the ground state $|g, N\rangle$ of a given manifold just after a spontaneous quantum jump, it cannot immediately emit a photon from this state, and instead must first absorb a laser photon to be put in an excited state $|e, N-1\rangle$, from which it can emit a photon. This takes some time, and explains why the waiting time distribution $W(\tau)$ and the second order correlation function $G_2(\tau)$ vanish when $\tau \longrightarrow 0$, exhibiting what is called a "photon antibunching." Such an effect only appears on the light emitted by a single atom. Indeed, two photons spontaneously emitted by two different atoms can be detected at the same time.

The first experiment investigating the statistical properties of the fluorescence light [Kimble *et al.* (1977)] emitted by a single atom was performed as illustrated in Fig. 5.1: a dilute beam was excited at right angle by a resonant laser beam, and

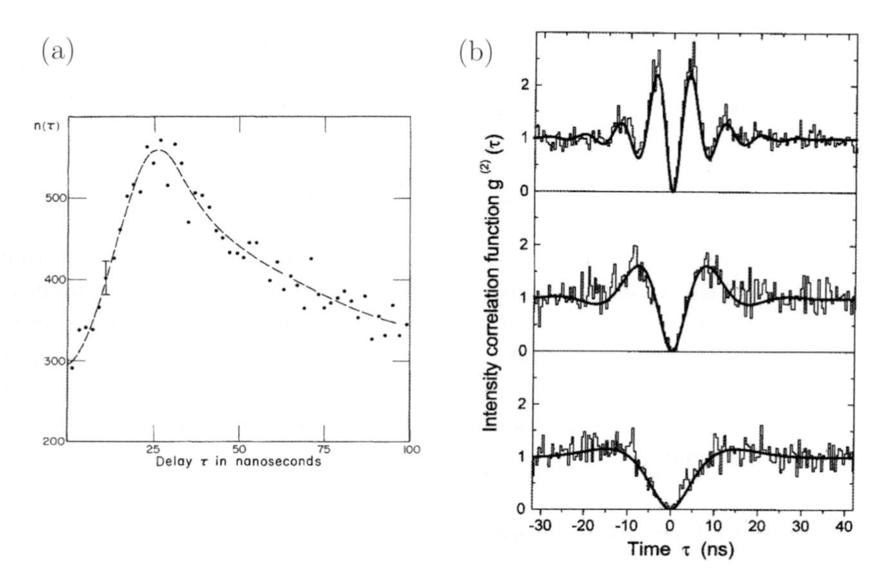

Fig. 5.14 (a) Experimental observation of antibunching from [Kimble *et al.* (1977)]. The number of recorded pulse pairs $n(\tau)$ is plotted as a function of the time delay τ in nanosecond. The growth of $n(\tau)$ from $\tau = 0$ clearly demonstrates the antibunching effect. The corresponding setup is schematically represented in Fig. 5.1. Copyright: American Physical Society. (b) Intensity correlation measurements for a single $^{24}Mg^{+}$ ion for three different detunings and intensities of the incident field from [Höffges *et al.* (1997)]. The Rabi frequency increases from the bottom figure to the top one. One clearly sees the Rabi oscillation at high intensities for the correlation measurements. Copyright: Taylor and Francis publisher.

the fluorescence light was collected by a photo-detector. The fluorescence signal clearly exhibits a dip, but does not completely vanish for $\tau = 0$ (see Fig. 5.14(a)). This reduced contrast is due to the difficulty in such an experiment to avoid the presence of several atoms in the excitation volume, whereas antibunching is a one-atom phenomenon and its signature is rapidly washed out in the computation of the correlation function when an increasing number of atoms contributes to the signal.

More recently, intensity correlation measurements for a single trapped ion $^{24}Mg^{+}$ were carried out (see Fig. 5.14(b)). The results clearly show the antibunching effect with a vanishing probability for the consecutive detection of two atoms at time $\tau = 0$ and the oscillations of the correlation signal for increasing values of the incident field [Höffges *et al.* (1997)].

The interest of photon antibunching is that it is a pure quantum effect, which cannot occur with classical fields. In contrast, a classical stochastic field gives rise to a bunching effect in the photocurrent as first demonstrated by Hanbury Brown and Twiss [Hanbury Brown and Twiss (1956)]. For such a classical field, the time correlation function $\overline{I(t)I(t+\tau)}$ of the classical intensity $I(t)$ is a non-increasing function of τ near $\tau = 0$ because $\overline{I(t)I(t)} \geq \overline{I(t)I(t+\tau)}$. This can be easily demonstrated by using a Schwarz's inequality. The fact that the measured correlation signal shown in Fig. 5.14 is an increasing function of τ near $\tau = 0$ demonstrates that the light spontaneously emitted by an atom cannot be simulated as a classical field.

5.6 Fluorescence triplet at high laser intensities

In this section, we analyze the spectral distribution of fluorescence light that is measured by a narrow band detector. With such an accurate spectrometer, one does not measure the time at which the fluorescence photons arrive but rather their energy, or in other words, the frequency of the light that is emitted by the oscillating dipole moment of the atom.[13] The dressed atom approach is particularly well suited for identifying the oscillation frequencies of the atomic dipole moment in the limit of large Rabi frequency.

5.6.1 *Limit of large Rabi frequency*

We suppose in this section that the Rabi frequency Ω is so large that the perturbative treatment of Sec. 5.2 cannot be applied. More precisely, we suppose that

$$\epsilon = \frac{\Omega}{\sqrt{\delta^2 + (\Gamma^2/4)}} \gg 1. \tag{5.45}$$

The dimensionless parameter ϵ can no longer be considered as a perturbative expansion parameter as this was the case in Sec. 5.2.

Condition (5.45) means that the splitting $\tilde{\Omega}$ between the dressed states, which is larger than Ω, is large compared to the width of these states, which is on the order of Γ. It is then appropriate to use the dressed state basis for writing the master equation (5.30). This basis includes the effect of the laser-atom coupling to all orders, and one can then make approximations when the effect of the small damping terms of Eq. (5.30), proportional to Γ, is taken into account.

5.6.2 *Mollow fluorescence triplet*

If, in a first step, one neglects the damping terms proportional to Γ, one finds that the only frequencies which can appear in the evolution of the atomic dipole moment $\vec{D}$, and thus in the fluorescence spectrum, are the allowed Bohr frequencies of the dressed atom. These frequencies correspond to pairs of dressed levels between which $\vec{D}$ has a non-zero matrix element.

In the uncoupled basis, the atomic dipole, which cannot change N, can only connect the two levels $|e, N\rangle$ and $|g, N\rangle$ of two adjacent manifolds $\mathcal{E}_{N+1}$ and $\mathcal{E}_N$, and we have $\langle e, N|\vec{D}|g, N\rangle = \langle e|\vec{D}|g\rangle \neq 0$. The two dressed states $|1(N)\rangle$ and

[13] More precisely, one can show that the fluorescence spectrum is proportional to the Fourier transform of the correlation function $\langle \mathcal{S}_+(t+\tau)\mathcal{S}_-(t)\rangle$ where $\mathcal{S}_+$ and $\mathcal{S}_-$ are defined in (5.15) (see for example Sec. V.D of [Cohen-Tannoudji *et al.* (1992b)].).

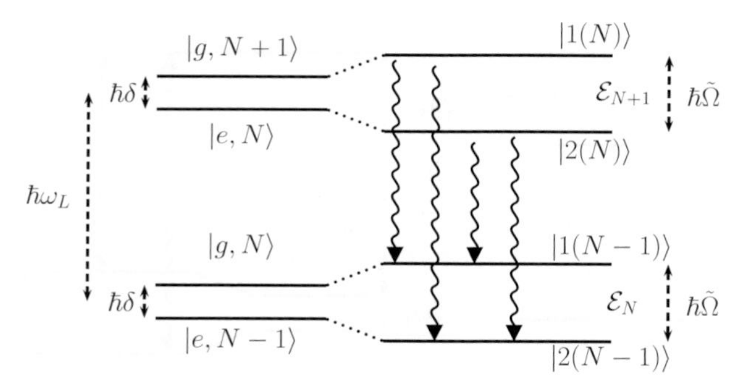

Fig. 5.15 The four allowed transitions between the two dressed states of two adjacent manifolds are represented by the wavy arrows involving photons of frequencies $\omega_L - \tilde{\Omega}$, ω_L and $\omega_L + \tilde{\Omega}$.

$|2(N)\rangle$ of the manifold $\mathcal{E}_{N+1}$ are contaminated by $|e, N\rangle$, while the states $|1(N-1)\rangle$ and $|2(N-1)\rangle$ of $\mathcal{E}_N$ are contaminated by $|g, N\rangle$. This means that the four transitions connecting the two dressed states of two adjacent manifolds are allowed (see Fig. 5.15). We deduce the following frequencies of the spontaneously emitted lines:

transition $|2(N)\rangle \longrightarrow |1(N-1)\rangle$: frequency $\omega_L - \tilde{\Omega}$,
transition $|2(N)\rangle \longrightarrow |2(N-1)\rangle$: frequency ω_L,
transition $|1(N)\rangle \longrightarrow |1(N-1)\rangle$: frequency ω_L,
transition $|1(N)\rangle \longrightarrow |2(N-1)\rangle$: frequency $\omega_L + \tilde{\Omega}$.

Consequently, the fluorescence spectrum, commonly called the Mollow fluorescence triplet [Mollow (1969)], has a structure made of a central line at ω_L and two sidebands at $\omega_L \pm \tilde{\Omega}$. At resonance ($\delta = 0$) and for $\Omega \gg \Gamma$, we get three well separated lines centered at ω_L, $\omega_L + \Omega$ and $\omega_L - \Omega$. Let us emphasize that this result could not be obtained with the perturbative approach in Sec. 5.2, since the expansion parameter ϵ, which is equal to Ω/Γ when $\delta = 0$, is then very large compared to unity. However, in the limit, $\delta \gg \Omega \gg \Gamma$, where the expansion parameter ϵ is small with respect to unity, both formalisms give the same result.

Interpretation of another effect: the Autler-Townes doublet [Autler and Townes (1955)]

The dressed atom approach can be also applied to the interpretation of another physical effect. We consider a three-level atom (see Fig. 5.16(a)). An intense and resonant laser is driving the transition $|g\rangle \longrightarrow |e_1\rangle$ and a very weak probe is scanned around the frequency of the transition $|e_1\rangle \longrightarrow |e_2\rangle$. We assume that the frequencies ω_0 and ω_0' of these transitions are sufficiently different so that the level $|e_2, N\rangle$ is not affected by the intense laser with N photons and a frequency $\omega_L = \omega_0$. The observed absorption line for the probe turns out to be split into two components. This effect is readily explained within the dressed atom formalism. Indeed, under the excitation by the intense laser on the transition $|g\rangle \longrightarrow |e_1\rangle$, the states $|g, N+1\rangle$ and $|e_1, N\rangle$ give rise to the two dressed states $|1(N)\rangle$ and $|2(N)\rangle$ separated by $\hbar\tilde{\Omega}$. Because the two dressed states are contaminated by

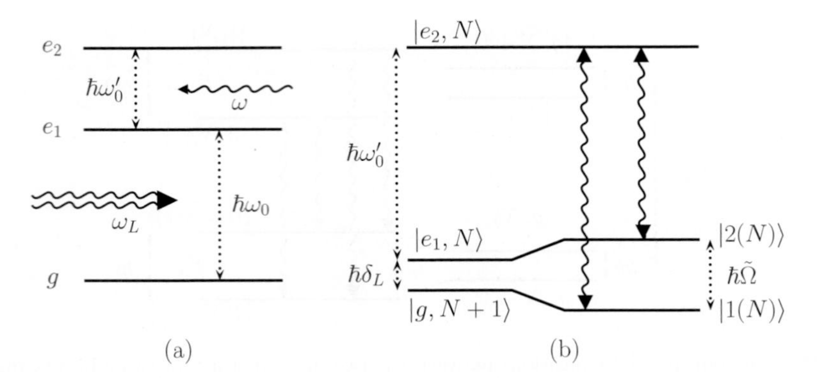

Fig. 5.16 (a) A three-level atom in presence of an intense laser resonant with the transition $|g\rangle \longrightarrow |e_1\rangle$. The transition $|e_1\rangle \longrightarrow |e_2\rangle$ is probed by a very weak laser. (b) Unperturbed levels (left-hand part) and dressed levels (right-hand part). The dressed state basis gives a simple interpretation of the doublet structure (Autler-Townes effect) observed when the frequency of the probe is scanned around ω_0'.

the state $|e_1, N\rangle$, two transitions appear: (i) $|1(N)\rangle \longrightarrow |e_2, N\rangle$ with a frequency $\omega_0' - (\tilde{\Omega} + \delta_L)/2$ and (ii) $|2(N)\rangle \longrightarrow |e_2, N\rangle$ with a frequency $\omega_0' + (\tilde{\Omega} - \delta_L)/2$, as illustrated by the wavy vectors in Fig. 5.16(b).

5.6.3 *Widths and weights of the components of the Mollow triplet*

To calculate the widths and the weights of the various components of the fluorescence triplet, one must solve the full master equation in the dressed state basis while keeping the damping terms proportional to Γ. A "secular approximation" for the damping terms describing spontaneous emission can then be made, using the fact that the splitting between dressed states is larger than their widths. This approximation consists of neglecting the couplings proportional to Γ between the populations of the dressed states and the coherences (off-diagonal elements of σ_{AL} between dressed states), or between coherences evolving at different frequencies. This approximation is possible because a coupling proportional to Γ between two density matrix elements evolving at frequencies ω_1 and ω_2 cannot have an appreciable effect if $|\omega_1 - \omega_2| \gg \Gamma$. The error made in neglecting these non-secular couplings is on the order of $\Gamma/|\omega_1 - \omega_2|$.

We will not present the calculations here, but refer the reader to Chap. VI of [Cohen-Tannoudji *et al.* (1992b)] for more details. Let us just mention that the secular equations of motion of the dressed state populations are rate equations involving only populations with a very simple meaning: the population of a dressed state in $\mathcal{E}_N$ is fed by radiative transitions from the two dressed states of $\mathcal{E}_{N+1}$ and is emptied by radiative transitions towards the dressed states of $\mathcal{E}_{N-1}$. The weights of the three components of the Mollow triplet can be calculated from the solution of the rate equations for the dressed state populations and from the transition rates between the pair of dressed states corresponding to a certain component of the triplet. The widths of the components of the triplet are obtained by solving the secular equations of motion of the coherences evolving at the corresponding frequency.

5.6.4 *Time correlations between the photons emitted in the two sidebands of the fluorescence triplet*

We now consider experiments where the detectors monitoring the fluorescence light are put behind frequency filters that select photons having a frequency corresponding to a given component of the fluorescence triplet. We still suppose that these three components are well resolved with a splitting $\tilde{\Omega}$ large compared to their width Γ. The bandwidth $\Delta\omega$ of the filters obeys the condition:

$$\Gamma \ll \Delta\omega \ll \tilde{\Omega}. \tag{5.46}$$

They are selective enough for detecting only a given component of the triplet while keeping a time resolution better than the atomic damping time $1/\Gamma$. We suppose that we have two such detectors monitoring the two sidebands at $\omega_L - \tilde{\Omega}$ and $\omega_L + \tilde{\Omega}$ respectively and we try to see if there are time correlations between the outputs of these two detectors.

Before addressing this question, it will be useful to come back to the radiative cascade described in Sec. 5.4.5 and to analyze it in the basis of dressed states. A spontaneous jump of the dressed atom projects it in the state $|g, N\rangle$ of a given manifold $\mathcal{E}_{N+1}$. This state is a linear superposition of the two dressed states $|1(N)\rangle$ and $|2(N)\rangle$, and is characterized by a vanishing emission rate which explains the photon antibunching effect. The probability to emit a photon afterwards is modulated in time because of the Rabi oscillation between $|g, N\rangle$ and $|e, N-1\rangle$ which is weakly damped in the limit $\tilde{\Omega} \gg \Gamma$. In the dressed state basis, this modulation appears as a quantum beat signal, related to the coherence between $|1(N)\rangle$ and $|2(N)\rangle$, and gives rise to two possible paths between these two states and a final state which is reached after the spontaneous jump (see Fig. 5.17(a)). The use of narrow band detectors considered in this section discriminates one of the two frequencies that corresponds to its bandpass. Therefore, the quantum beat signal is suppressed much like the interference pattern of a Young's two-slit experiment is washed out by a "which-path" information. The conclusion of this analysis is that, with frequency filters obeying condition (5.46), the radiative cascade should take place only between dressed states, without involving any linear superposition of dressed states as in Fig. 5.17(a).

An example of such a cascade between dressed states is shown in Fig. 5.17(b). After emission of a photon in the sideband $\omega_L + \tilde{\Omega}$, the atom is in a dressed state of type 2. It cannot emit another photon at the same frequency, so it must go to a state of type 1. This is only possible by emission of a photon in the sideband $\omega_L - \tilde{\Omega}$. This

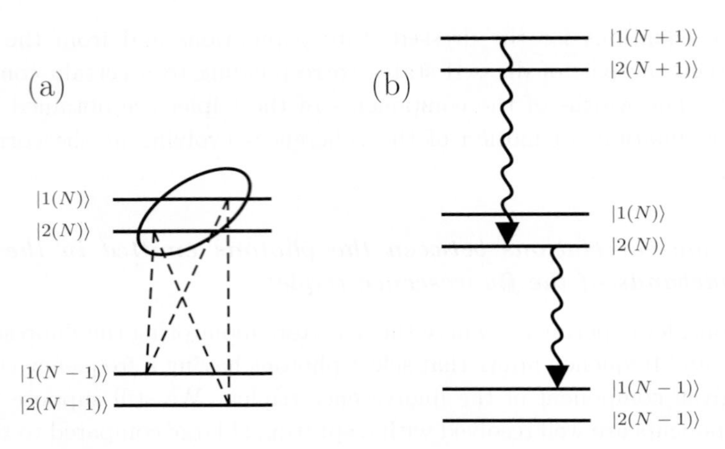

Fig. 5.17 Dressed { atom + photon } energy-level ladder. (a) The coherence between $|1(N)\rangle$ and $|2(N)\rangle$ gives rise to two possible paths between these two states and a given final state. A narrow band detector used to analyze the fluorescence photon selects one of these paths, and therefore destroys the interference between the two paths. (b) After emission of a photon in the sideband $\omega_L + \tilde{\Omega}$, the atom is in a dressed state of type 2. It cannot emit another photon at the same frequency without joining first a state of type 1 by emitting a photon in the sideband $\omega_L - \tilde{\Omega}$. It follows that emissions in the two sidebands of the triplet are necessarily alternated in time.

shows that emissions in the two sidebands of the triplet are necessarily alternated in time.[14] While the existence of such correlations is easily predicted in this approach based on the radiative cascade of the dressed atom, the analysis would be less obvious in the optical Bloch equations approach. More details on the theory of these correlations can be found in [Cohen-Tannoudji and Reynaud (1979); Apanasevich and Kilin (1979)]. They have also been observed experimentally [Aspect *et al.* (1980a)].

5.7 Conclusion

In this chapter, we have presented a detailed analysis of resonance fluorescence of a two-level atom interacting with a monochromatic laser field. This has given us the opportunity to introduce and describe several important physical effects, such as resonance scattering, Mollow fluorescence triplet, photon correlations, photon antibunching. To interpret these effects, we have used several different approaches, including the perturbative expansion of scattering amplitudes, optical Bloch equations, the dressed atom approach and the description of dissipative processes in terms of quantum jumps and stochastic wave functions.

The treatments presented here for two-level atoms could be easily extended to more complex systems. For example, the delay function approach will be used in

[14]Between two emissions in the sidebands, there can be any number of emissions in the central component at ω_L since these emissions do not change the type 1 or 2 of the dressed states.

Chap. 6 for interpreting the intermittent fluorescence that occurs for a three-level atom with two excited states that have very different lifetimes being driven by two resonant laser fields. Another important effect, coherent population trapping, occurs in a three-level atom with two ground state sublevels, and is particularly well adapted for a quantum trajectory analysis. It will be discussed in subsequent chapters.

We will conclude this chapter with a remark concerning the quantum description of the laser field in the analysis of resonance fluorescence: This quantum description is not essential. When the laser field is in a coherent state, one can prove that a semiclassical description and a quantum description of the effect of the laser excitation are fully equivalent. We think, however, that a quantum description of the laser field, and the introduction of combined atom-laser photon states, is more fruitful, since it provides more insight into the elementary physical processes of absorption, stimulated emission and spontaneous emission and suggests new descriptions of dissipative processes in terms of quantum jumps and stochastic wave functions.

Chapter 6

Advances in high resolution spectroscopy

6.1 Introduction

The resolution of a spectroscopic measurement depends on the precision with which the central frequency of an atomic line can be determined. This precision is higher when the width of the line is smaller. To increase the resolution of spectroscopy, one must therefore reduce the sources of line broadening as much as possible. In the optical domain, the width of the lines for thermal atoms at room temperature or higher is determined mainly by the Doppler effect.[1] In Secs. 6.2, 6.3 and 6.4 of this chapter, we review a few methods that have been developed for circumventing the Doppler effect as well as other limiting phenomena like the recoil shift.[2]

Once the Doppler effect is eliminated, one is left with the natural width Γ of the line which is equal to the inverse of the radiative lifetime of the excited state. It is not easy to control spontaneous emission and to increase this radiative lifetime, though by putting the atom into a cavity, one can change the radiative lifetime,[3] at the risk of introducing cavity shifts of the line. To get very narrow optical lines for frequency standards, it is therefore better to use transitions connecting the ground state to a very long lived excited state. The difficulty is then the very low fluorescence rate (proportional to Γ) leading to very low signal to noise ratios. In Secs. 6.5 and 6.6, we present the shelving method and the quantum logic method which are elegant solutions for solving this problem. Finally, in Sec. 6.7, we address the problem of high precision frequency measurement for which spectacular progress has been recently achieved with the use of frequency combs.

Before presenting these various methods, we recall a few results and orders of magnitude about the Doppler effect and the recoil shift for free atoms and for atoms trapped in an external potential.

---

[1] We will come back in Chaps. 12 and 13 on laser cooling methods which lead to atomic velocities so low that the Doppler effect can be ignored.

[2] See Chap. 2.

[3] See [Kleppner and Haroche (1989)] and references therein.

Doppler effect and recoil shift for free atoms. Orders of magnitude

The momentum $\hbar\vec{k}$ transferred to an atom by the emission or absorption of a photon changes its initial momentum $\vec{p}_i$, and thus the corresponding kinetic energy. The variation of this kinetic energy at the end of the emission or absorption process is the sum of a term independent of $\vec{p}_i$, which is the recoil energy $E_R = \hbar^2k^2/2M$, and a term $\hbar\vec{k}\cdot\vec{p}_i/M$, which is the Doppler shift.[4]

In a gas at thermal equilibrium at temperature T, the velocities are distributed according to the Maxwell-Boltzmann law and the Doppler shift varies from one atom to another. The absorption or emission lines are therefore broadened. The corresponding width $\Delta\nu_D$, which depends on the standard deviation of v, is called the Doppler width and reads

$$\frac{\Delta\nu_D}{\nu} \simeq \frac{\sqrt{\Delta v^2}}{c} \simeq \frac{\sqrt{k_BT/M}}{c} \sim 10^{-6}. \tag{6.1}$$

For an optical line, $\Delta\nu_D$ is on the order of 10^9 Hz. This width is generally very large compared to the natural width of the line Γ, typically on the order of 10^7 Hz. The Doppler width is therefore a serious limitation for high resolution optical spectroscopy. The recoil shift in the optical domain is on the order of 10^5 Hz, and can generally be ignored.[5] It can be however detected on weakly allowed optical lines, with small values of Γ, when the Doppler broadening of the line is eliminated (see Sec. 6.2.3).

Since the frequency of a hyperfine transition is much smaller, $\Delta\nu_D$ is much smaller, on the order of 10^4 Hz. In the ground state, however, the relaxation time can become very long, such that the intrinsic width of the hyperfine lines may be smaller than the Doppler width (this is no longer true for the RF transitions between Zeeman sublevels). Here also, the Doppler width is a serious limitation for high resolution spectroscopy.

Atom trapped in an external potential

We consider now an atom, or an ion, trapped in an external potential. We suppose that this potential acts only on the center of mass variables and is independent of the internal variables: it is essentially the same for an atom in the ground state g or in an excited state e. This is clearly the case for an ion since the trapping force only depends on the charge position and velocity of the ion.

The motion of the atomic center of mass is now characterized by a discrete set of energy levels φ_v with energies E_v ($v = 0, 1, 2, 3\ldots$) which are respectively the eigenstates and eigenvalues of the center of mass Hamiltonian. When the atom absorbs (or emits) a photon and goes from the ground state g to an excited state e, its external state also changes, so that the absorption (or emission) spectrum is a discrete spectrum of lines with frequencies given by:

$$E_{v',v} = \hbar\omega_{v'v} = E_e - E_g + E_{v'} - E_v. \tag{6.2}$$

[4] See Chap. 2.

[5] This is no longer true in the gamma ray domain (see Sec. 2.3.1 of Chap. 2).

A few questions can then be asked: What are the relative intensities of these various lines? How does the atom absorb (or lose) the momentum of the photon? Is it possible to obtain recoil free absorption or emission lines, corresponding to $v = v'$? We will see in Sec. 6.4 that this is possible in the commonly-called Lamb-Dicke regime, which can be considered as a precursor of the Mössbauer effect.

6.2 Saturated absorption

Nonlinear effects can give rise to narrow Doppler free structures in various types of lineshapes observed for atomic gases. The first example of such a structure is the so-called Lamb-dip observed on the output of a gas laser [Lamb (1964); Brewer *et al.* (1969)]. The gain curve of such a laser has a Doppler width determined by the temperature of the gas. The standing wave within the cavity burns two holes in the velocity distribution corresponding to the saturation of two velocity classes interacting separately with the two counter-propagating waves forming the standing wave. When the cavity frequency is tuned to the center of the Doppler profile, these two holes coincide leading to a higher saturation and resulting in a reduced output power of the laser. Therefore, when the frequency of the cavity is swept, the output power exhibits a narrow hole with a width scaling as the natural width. This is called the Lamb-dip.

The saturation absorption technique described hereafter is reminiscent of the Lamb-dip effect applied to a passive medium rather than to an active medium such as a gas laser.

6.2.1 *Principle of the method*

The saturated absorption method uses the nonlinear coupling of an atom with two laser waves of the same frequency that are propagating in opposite directions (see Fig. 6.1) [Bordé (1970); Hänsch *et al.* (1971a,b); Ouhayoun and Bordé (1972); Baklanov and Chebotayev (1972); Haroche and Hartmann (1972); Hall (1973)].

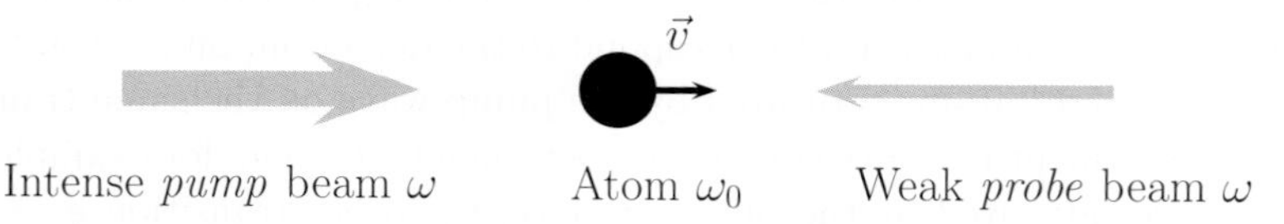

Fig. 6.1 An atomic gas interacts with two counter propagating laser beams with the same frequency ω: one intense *pump* beam and one weak *probe* beam. The atom interacts simultaneously with the two beams, only when its velocity component v along the direction of the laser beams is equal to zero within Γ. This leads to a reduction of the absorption of the probe beam when ω is scanned in an interval of width Γ about the atomic frequency ω_0.

The pump beam "saturates" the atoms (i.e. reduces the population difference between the excited state e and the ground state g), when their velocity v_1 is such

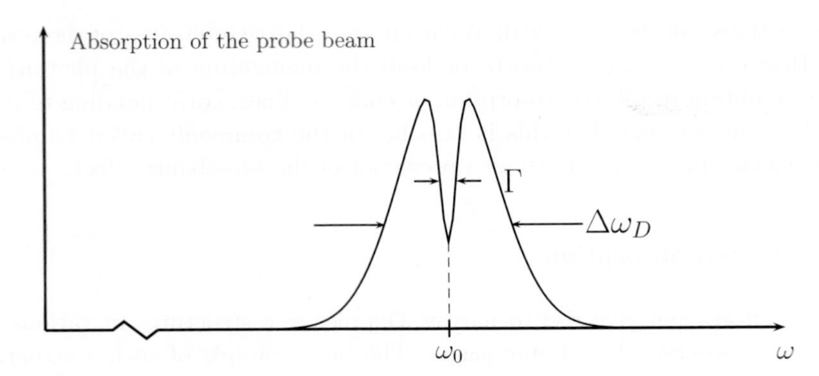

Fig. 6.2 The shape of the narrow saturation dip of width Γ observed on the absorption of the probe beam in the laser configuration of Fig. 6.1 as the frequency ω is scanned about the frequency ω_0 of the atomic transition.

that $\omega - kv_1 = \omega_0$ within Γ (resonance condition). It creates a hole in the velocity distribution of atoms in g centered at $v_1 = (\omega - \omega_0)/k$ with a width $\delta v \sim \Gamma/k$. The probe beam detects atoms with a velocity v_2 such that $\omega + kv_2 = \omega_0$ within Γ. This gives $v_2 = -(\omega - \omega_0)/k$. In general, the two velocity classes are different. If we want the atoms to interact simultaneously with the two beams, we must have $v_1 = v_2$ within Γ, which imposes to ω to be equal to ω_0 within Γ. In this case, the absorption of the probe beam is reduced, since the pump beam reduces the population difference between e and g. We conclude that the curve giving the absorption of the probe beam versus ω must exhibit a narrow dip around $\omega = \omega_0$ within the broad Doppler profile of width $\Delta\omega_D$ (see Fig. 6.2).

6.2.2 *Crossover resonances*

We now suppose that the excited state has a structure corresponding to two excited sublevels e_1 and e_2, and we denote ω_{01} and ω_{02} as the frequencies of the two transitions connecting the ground state g to the excited states e_1 and e_2.

Like in the previous case, there are saturated absorption resonances when ω is scanned around ω_{01} and ω_{02} that correspond to the detection, on a given transition $g \to e_1$ or $g \to e_2$, of atoms saturated by the pump wave on the same transition.

In addition, the probe wave can now detect, on a transition, for example $g \to e_1$, atoms that are saturated by the pump wave on the other transition $g \to e_2$. The corresponding resonance conditions can be written:

$$\omega + kv = \omega_{01} \qquad \text{and} \qquad \omega - kv = \omega_{02}, \tag{6.3}$$

and their solution is:

$$kv = (\omega_{01} - \omega_{02})/2 \qquad \text{and thus} \qquad \omega = (\omega_{01} + \omega_{02})/2. \tag{6.4}$$

We conclude that a new saturated absorption resonance appears at $\omega = (\omega_{01} + \omega_{02})/2$, half way between the resonances at $\omega = \omega_{01}$ and $\omega = \omega_{02}$. This resonance is called a *crossover* resonance.

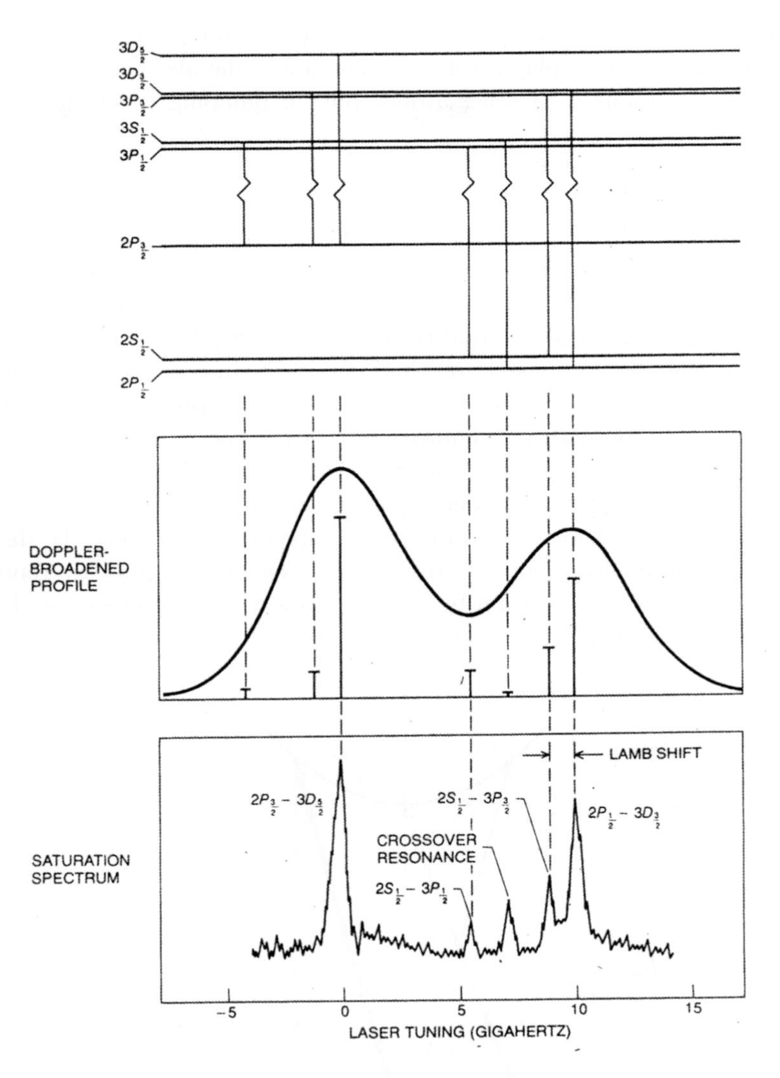

Fig. 6.3 Optical spectra obtained on the Balmer α line of hydrogen. The saturation absorption spectrum allows the Lamb shift to be optically resolved. Figure extracted from [Series *et al.* (1979)]. Copyright: Scientific American.

To illustrate the previous considerations, we give an example of a saturated absorption signal obtained on the Balmer α line of hydrogen (see Fig. 6.3) [Hänsch *et al.* (1972)]. The upper part of the figure gives the fine structure energy levels in the lower levels 2s and 2p and the upper levels 3s, 3p and 3d. The Doppler broadened profile shows the fine structure splitting in the $n = 2$ level, but the Lamb shift splitting between $2S_{1/2}$ and $2P_{1/2}$ is not resolved. The saturation absorption spectrum allowed this splitting to be resolved optically for the first time.[6]

Note that other effects play a role and may affect the absorption profile. Such effects include collisions with other atoms, optical pumping effects for multilevel atoms, ...

6.2.3 *Recoil doublet*

So far, the saturation absorption method has been described without taking into account the recoil shift. We now show that the recoil shift gives rise to a doublet in the saturation spectrum provided that the recoil energy E_R is larger than the natural width $\hbar\Gamma$. It will be convenient for this analysis to use the energy-momentum diagrams introduced by Christian Bordé [Bordé (1976); Bordé (1977)]. In these diagrams, the variation with the momentum p of the energy E of an atomic state is represented for each internal state, g or e, by a parabola $E = p^2/2M$ centered on the E-axis. We thus get two parabolas separated by a distance $\hbar\omega_0$. Since the pump and probe beams are propagating in opposite directions, and since the dispersion relations for these photons can be written $E = \pm cp$, the photons are represented in the energy-momentum diagram by straight lines with a slope $+c$ for the pump photons and $-c$ for the probe photons.

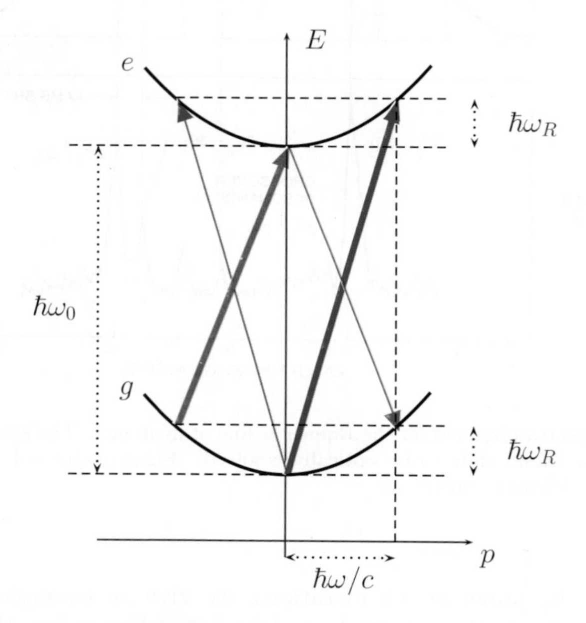

Fig. 6.4 Energy-momentum diagram showing how an atom can interact simultaneously with the two counter propagating pump (thick arrow) and probe (thin arrow) beams when the momentum of the atom, either in the ground state g or in the excited state e, is equal to zero (see text).

[6]The splitting between the two excited states $3P_{3/2}$ and $3D_{3/2}$ is negligible compared to the Lamb shift.

Consider first an atom in the ground state g with $p = 0$. The absorption of a pump photon is represented in the energy-momentum diagram of Fig. 6.4 by a straight line with a slope $+c$ joining the point $E = E_g, p = 0$ of the g-parabola to the point $p = \hbar\omega/c = \hbar k$ of the e-parabola which has an energy $E = E_e + \hbar\omega_R$ where $\hbar\omega_R = E_R$ is the recoil energy. The angular frequency of this photon is thus equal to $\omega_0 + \omega_R$. It is clear in Fig. 6.4 that this atom in g with $p = 0$ can also absorb a probe photon having the same frequency, with the corresponding transition being represented by a straight line with the same length and with a slope $-c$. We conclude that an atom in g with $p = 0$ can interact simultaneously with the pump and probe waves when $\omega = \omega_0 + \omega_R$, where the hole created in g at $p = 0$ diminishes the absorption of the probe. This simultaneous interaction with the two waves occurs only for $p = 0$, since another point of the g-parabola with $p \neq 0$, cannot be joined to two different points of the e-parabola with two arrows having the same length and opposite slopes.

Another possible simultaneous interaction with the two waves occurs when the final state of the transition is the state $p = 0$ of the e-parabola. One can see in Fig. 6.4 that the absorption of a pump photon with frequency $\omega = \omega_0 - \omega_R$ takes the atom from the state $p = -\hbar k$ of the g-parabola to the state $p = 0$ of the e-parabola, creating an excess of population in the e-state at $p = 0$. Atoms put in this state $p = 0$ in the excited state e can interact resonantly with the probe beam at frequency $\omega = \omega_0 - \omega_R$ which brings them by stimulated emission to the state $p = +\hbar k$ of the g-parabola (see Fig. 6.4). This stimulated emission reduces the absorption of the probe beam. As before, there are no other $p \neq 0$ states in the e-parabola that can lead to a simultaneous resonant interaction with the two beams.

One therefore concludes that when ω is scanned around ω_0, the absorption of the probe beam should exhibit two narrow dips with a width Γ centered at $\omega = \omega_0 + \omega_R$ and $\omega = \omega_0 - \omega_R$, respectively. This so-called recoil doublet can only be observed when ω_R is larger than Γ. Such a condition is generally difficult to fulfill, except if one takes a weakly allowed transition with a small value of Γ. This is the case for the transition of CH_4 at 3.39 μm, which has allowed the first observation of the recoil doublet in the infrared domain [Hall *et al.* (1976)]. Figure 6.5 clearly shows the recoil doublets appearing on the three hyperfine components of this transition.

6.3 Two-photon Doppler-free spectroscopy

6.3.1 *Principle of the method*

This Doppler-free method relies on a two-photon absorption process in which the atom goes from the ground state g to the excited state e by absorbing two photons of the same frequency ω propagating in opposite directions. The momenta of the two photons are opposite such that the total momentum absorbed by the atom in

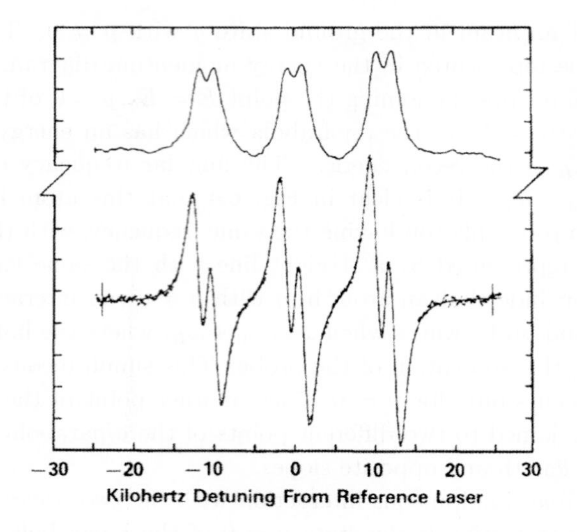

Fig. 6.5 Upper curve: recoil doublets appearing on the three hyperfine components of the transition of CH_4 at 3.39 μm. The splitting frequency interval is $2E_R/h = 2.15$ kHz. Lower curve: derivative spectrum of the three hyperfine lines. Figure extracted from [Hall *et al.* (1976)]. Copyright: American Physical Society.

the two-photon absorption process is equal to zero for any initial momentum p of the atom. For example, if the atom starts from the state $p = 0$ in g, it ends up in the state $p = 0$ in e (see Fig. 6.6). There is no Doppler effect and no recoil shift and the center of the resonance is located at $\omega_0 = 2\omega$. This method was suggested independently by the groups of Venyamin Chebotayev [Vasilenko *et al.* (1970)] and Bernard Cagnac [Cagnac *et al.* (1973)]. Experiments confirming these predictions were performed by several groups [Biraben *et al.* (1974); Levenson and Bloembergen (1974); Hänsch *et al.* (1975)].

This idea can be extended to p-photon processes with $p > 2$ if the wave vectors of the p waves obey the condition

$$\vec{k}_1 + \vec{k}_2 + \cdots + \vec{k}_p = \vec{0}, \tag{6.5}$$

such that the total momentum transferred to the atom when it absorbs p photons, one from each wave, is equal to zero.

6.3.2 *Examples of results*

Figure 6.7 shows an example of a Doppler-free two-photon spectrum obtained on the 3s $\to$ 5s transition of sodium [Cagnac (1975)]. When ω is scanned around $\omega_0/2$, one gets a narrow resonance[7] with a width Γ, centered at $\omega = \omega_0/2$. Note that, since the electric dipole matrix elements do not change the electron and nuclear spin

[7] The atom can also absorb two photons propagating along the same direction, a process which gives rise to a Doppler broadened background below the Doppler-free line.

states, the two-photon transition can be shown to obey the selection rule $\Delta F = 0$ since the connected states have the same angular momentum.

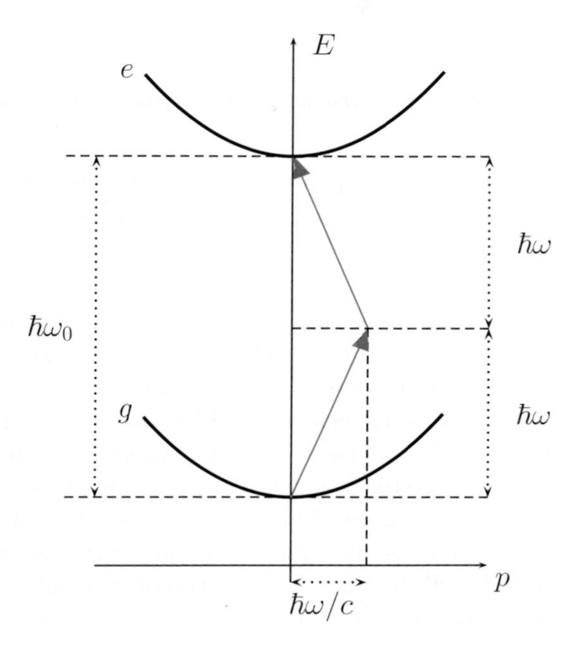

Fig. 6.6 Principle of Doppler-free two-photon spectroscopy. The total momentum transferred to the atom when it absorbs two counter propagating photons with the same frequency is equal to zero, so that there is no Doppler shift and no recoil shift.

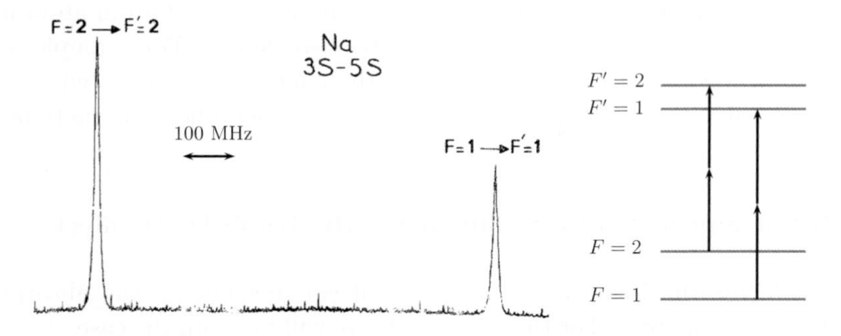

Fig. 6.7 Doppler-free two-photon spectrum obtained on the 3s→ 5s transition of sodium. F is the hyperfine quantum number in 3s, F' the hyperfine quantum number in 5s [Cagnac (1975)]. The hyperfine structure of the transition is clearly resolved.

The 1s → 2s transition of hydrogen is particularly interesting because of the very long lifetime (on the order of one second) of the metastable state 2s. The two-photon line is extremely narrow, and its observation[8] has allowed for dramatic advances in the determination of the Rydberg constant [Hänsch (2006); Mohr *et al.* (2008)].

6.3.3 *Comparison between saturated absorption and two-photon spectroscopy*

Two-photon Doppler-free spectroscopy offers a general technique for studying the structure of excited levels which is complementary to the saturated absorption technique:

- It permits one to probe levels with the same parity as the ground state, while the saturated absorption technique is restricted to resonant levels with a parity opposite to that of the ground state.
- The saturated absorption technique addresses a small class of atoms for which the Doppler effect of the two counter-propagating beams is the same, while two-photon spectroscopy relies on the compensation of the Doppler effect for all velocity classes. This occurs for all atoms regardless of their thermal velocity. This is an important point for two-photon absorption, a weak second order effect, since the spectroscopic signal is magnified by the large number of atoms that contribute.

To conclude this section, it should be pointed out that, in general, light interacting with an atom in a multiphoton process produces shifts of the energy levels. These light shifts, to be described in more detail in Chap. 7, must be carefully taken into account if one wants to determine the unperturbed atomic frequency from the center of the Doppler-free two-photon line. In the presence of atom-atom interactions, the two-photon Doppler free line can be also shifted. For example, gaseous Bose-Einstein condensate of spin-polarized hydrogen have been detected in this way through a density dependent shift of this line due to mean-field interactions.[9]

6.4 Recoil suppressed by confinement: the Lamb-Dicke effect

We now address the issue of the influence of confinement on the absorption or emission lines of an atom. For this discussion, we will focus on the case of a trapped ion. The important point is that the external potential acts essentially only on the

[8] The atoms transferred from 1s to 2s by the two-photon Doppler free transition are detected by applying a static electric field which contaminates the state 2s by the nearby state 2p. This gives rise to a quenching of the metastability of the state 2s and to the emission of ultraviolet Lyman α photons, which can be easily detected.

[9] See Chap. 21.

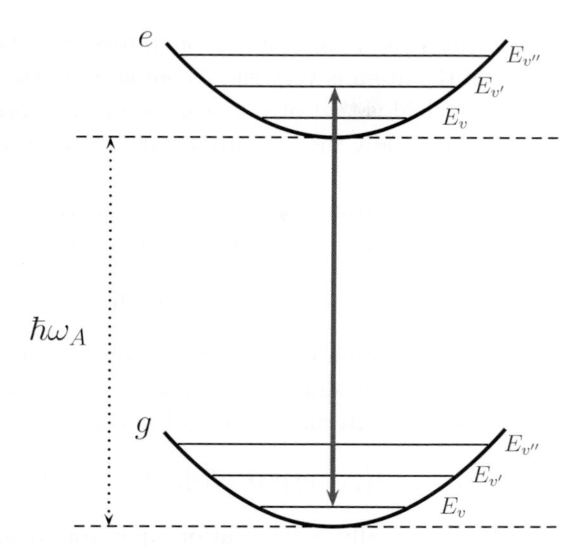

Fig. 6.8 As the trapping of an ion is ensured through its charge, the trapping potential does not depend on the internal excited e or ground state g.

charge of the ion, and not on the internal atomic variables.[10] The splitting between the two internal states g and e of a given optical transition of the ion is nearly independent of the trapping potential.

In Fig. 6.8, we represent the two identical external potentials for the ground state g and the excited state e, along with their identical ladder of vibrational levels $E_v, E_{v'}, E_{v''} \ldots$. The frequencies of the lines for absorption or emission are equal to: $\hbar\omega_A + E_{v'} - E_v$, where $\hbar\omega_A$ is the energy of the atomic transition, and $E_{v'} - E_v$ is the difference between two vibrational energies. In the following sections, we detail the characteristics of these lines.

6.4.1 *Intensities of the vibrational lines*

The intensities of the vibrational lines are related to the matrix elements of the atom-laser interaction Hamiltonian H_{I} between the pair of levels involved in the transition. The laser field has an amplitude $\mathcal{E}_0$, a polarization $\vec{\varepsilon}$, and a wave vector $\vec{k}$. For the atom, the external variable is the center of mass position $\vec{R}$. The internal atomic variables are coupled to the field through the dipole moment $\vec{D}$. The expression of H_{I} reads:

$$H_{\mathrm{I}} = -\vec{D} \cdot \vec{E}_L(\vec{R}) = -\mathcal{E}_0 \vec{\varepsilon} \cdot \vec{D} \left(\hat{a} \exp(i\vec{k} \cdot \vec{R}) + \hat{a}^\dagger \exp(-i\vec{k} \cdot \vec{R}) \right).$$

[10] Note that depending on the application, one may have to take into account a residual differential Stark effect between the ground state and the excited state that originates from a non-perfect cancellation of the electric static field at the position of the particle. In addition, thermal electric fields may also play a role.

where $\hat{a}$ and $\hat{a}^\dagger$ are the creation and annihilation operators of a photon of the laser mode $(\vec{k}, \vec{\varepsilon})$. As the size of the atom is very small compared to the wavelength, the relevant value of the electric field is that at the position of the center of mass of the atom. In this Hamiltonian, it is only the operators $\exp(\pm i\vec{k}\cdot\vec{R})$ that depend on the external variable $\vec{R}$.

Let us consider the transition from the state $|g, \varphi_v\rangle$ to the state $|e, \varphi_{v'}\rangle$. The corresponding matrix elements of H_I can be factorized in the form:

$$\langle e, \varphi_{v'}, N-1|H_\text{I}|g, \varphi_v, N\rangle \propto \langle e|\vec{D}\cdot\vec{\varepsilon}|g\rangle\langle\varphi_{v'}|\exp(i\vec{k}\cdot\vec{R})|\varphi_v\rangle\langle N-1|\hat{a}|N\rangle. \qquad (6.6)$$

The intensity $I_{v',v}$ of the transition $v \to v'$ between the vibrational levels v' and v has a v' and v dependence that is simply proportional to the square modulus of the matrix element involving the external degrees of freedom:

$$I_{v',v} = |\langle\varphi_{v'}|\exp(i\vec{k}\cdot\vec{R})|\varphi_v\rangle|^2. \qquad (6.7)$$

The operator $\exp(i\vec{k}\cdot\vec{R})$ is actually a translation operator in momentum space by an amount $\hbar\vec{k}$. The intensity of the line $v \to v'$ is thus proportional to the modulus squared of the scalar product of the vibrational wave function $|\varphi_{v'}\rangle$ by the vibrational wave function $|\varphi_v\rangle$ translated in momentum space by an amount $\hbar\vec{k}$.

Sum rules

The line intensities $I_{v',v}$ obey the normalization sum rule:

$$\sum_{v'} I_{v',v} = \sum_{v'}\langle\varphi_v|\exp(-i\vec{k}\cdot\vec{R})|\varphi_{v'}\rangle\langle\varphi_{v'}|\exp(i\vec{k}\cdot\vec{R})|\varphi_v\rangle = 1. \qquad (6.8)$$

Another interesting sum rule is:

$$\langle E\rangle = \sum_{v'} I_{v',v}(E_{v'} - E_v) = \frac{\hbar^2 k^2}{2M} = E_\text{rec}. \qquad (6.9)$$

It expresses that the mean energy gained by the ion absorbing a photon and going from one vibrational level v to any other one v' is equal to the recoil energy. To prove the result of Eq. (6.9), one writes the left-hand side of the equation as

$$\sum_{v'}\langle\varphi_v|\left[\exp(-i\vec{k}\cdot\vec{R}), H_\text{ext}\right]|\varphi_{v'}\rangle\langle\varphi_{v'}|\exp(i\vec{k}\cdot\vec{R})|\varphi_v\rangle$$

where H_ext is the Hamiltonian of the center of mass. The closure relation over $|\varphi_{v'}\rangle$ leads to the following average value in $|\varphi_v\rangle$

$$\langle\varphi_v|\exp(-i\vec{k}\cdot\vec{R})H_\text{ext}\exp(+i\vec{k}\cdot\vec{R}) - H_\text{ext}|\varphi_v\rangle = E_\text{rec} - \langle\varphi_v|\hbar\vec{k}\cdot\vec{P}/M|\varphi_v\rangle = E_\text{rec}.$$

We have used the fact that the only term in H_ext that does not commute with the translation operator $\exp(i\vec{k}\cdot\vec{R})$ is $\vec{P}^2/2M$ and the well defined parity of $|\varphi_v\rangle$. This leads to a zero average value of $\vec{P}$ in $|\varphi_v\rangle$.

6.4.2 *Influence of the localization of the ion*

The importance of the localization of the ion on absorption or emission lines can be easily understood from Eq. (6.7). The localization criterion is readily obtained by comparing the width Δx of the spatial distribution of the center of mass to the laser wavelength $\lambda = 2\pi/k$. The atom is well localized when the spatial distribution is very narrow compared to the laser wavelength: $\Delta x \ll \lambda$. This inequality has a direct consequence on the line intensities. Indeed, it can be recast in the momentum space:

$$\Delta x \ll \lambda \Longrightarrow \Delta p = \frac{\hbar}{\Delta x} \gg \frac{\hbar}{\lambda} = \frac{\hbar k}{2\pi}, \tag{6.10}$$

and means that the momentum dispersion of the center of mass is much larger than the photon recoil momentum. The wave functions $|\varphi_v\rangle$ and $\exp(i\vec{k}\cdot\vec{R})|\varphi_v\rangle$ are very close in p-space since the latter is equal to the former just translated by an amount $\hbar\vec{k}$ much smaller than its width $\hbar/\Delta x$. Consequently, $\exp(i\vec{k}\cdot\vec{R})|\varphi_v\rangle$ is quasi orthogonal to $|\varphi_{v'}\rangle$ if $v \neq v'$, so that $I_{v',v\neq v'} \ll I_{v',v'}$.

We conclude that when an atom is well localized in a trapping potential ($\Delta x \ll \lambda$), the most intense line in absorption or emission is the line at the atomic frequency ω_A without change of the external state. In practice, this means the suppression of any Doppler effect or recoil shift. This effect is known as the Lamb-Dicke effect (for reasons which will be explained in Sec. 6.4.4).

Finally, one may ask where the momentum of the absorbed photon goes? In this case, the answer is that the presence of the trap breaks the translational invariance of the problem, and it is the trap itself that absorbs the photon's momentum. As the trap has a macroscopic mass, the translational velocity it acquires and its recoil energy are negligibly small.

6.4.3 *Case of a harmonic potential*

In the case of a harmonic confining potential with a vibrational angular frequency ω_v, more precise calculations can be done. The spatial extension of the vibrational wave functions scales as $R_0 = \sqrt{\hbar/m\omega_v}$ and the localization condition can be written $kR_0 = 2\pi R_0/\lambda \ll 1$. The dimensionless parameter

$$\eta = kR_0 = \sqrt{\frac{\hbar k^2}{m\omega_v}} = \sqrt{\frac{2E_{\text{rec}}}{\hbar\omega_v}}, \tag{6.11}$$

where E_{rec} is the recoil energy, is called the Lamb-Dicke parameter. The Lamb-Dicke criterion is readily expressed as $\eta \ll 1$. The strong confinement can also be interpreted as a strong binding, since $\eta \ll 1 \Longrightarrow E_{\text{rec}} \ll \hbar\omega_v$: the recoil energy is small compared to the spacing between the vibrational levels.

The line intensities can also be readily calculated for this particular confinement. The position $\hat{R}$ of the atomic center of mass can be written $R_0(\hat{a} + \hat{a}^\dagger)$ and the exponential $\exp(ik\hat{R}) = \exp(ikR_0(\hat{a} + \hat{a}^\dagger))$ appearing in the expression of the interaction Hamiltonian must be expanded to order p in $kR_0 = \eta$ to find a term connecting two vibrational levels v and $v + p$. It follows that the "0 -phonon" lines $v \to v$ are the most intense ones if $\eta \ll 1$, and that the "p -phonon" lines $v \to v \pm p$ have an intensity that scales as $\propto \eta^p$.

6.4.4 *Historical perspective*

Though the suppression of the recoil and of the Doppler effect due to the confinement of the emitter or absorber is easy to understand for the single trapped ion system considered in this section, it was first discovered during the investigation of more complex situations.

The first work considering the influence of the confinement on absorption and emission line shapes [Lamb (1939)] dealt with the resonant absorption of slow neutrons (not photons!) by the nuclei of atoms bound in a crystal lattice. In 1939, Willis Lamb did calculations showing that for weak crystal binding, the curve giving the absorption rate of neutrons versus their energy has the same form as for free atoms and exhibits the recoil shift, whereas when the binding is strong, the curve has a more complex shape and contains a narrow recoil-free component. The following interpretation can be given for this narrow structure: For strong binding, the crystal as a whole recoils with a momentum equal to the momentum of the absorbed neutron. As a consequence of the macroscopic mass of the crystal, the corresponding recoil kinetic energy of the crystal is negligible such that there is no recoil shift of the absorption curve.

The theoretical work of Lamb was somewhat premature inasmuch as it could not be observed in the context for which it was developed for. Nearly twenty years later, in 1958, Rudolf Mössbauer applied Lamb's theory to the analogous problem of the resonance absorption of gamma radiation by a crystal. Because of their large frequency, gamma ray photons give rise to large recoil shifts which have opposite signs for absorption and emission.[11] These shifts are so large that the absorption and emission lines do not overlap, so that the photon emitted by an excited nucleus cannot be absorbed by a similar non-excited nucleus. Mössbauer showed that, under accessible experimental conditions, there exists a high probability of recoil-free nuclear transitions with no simultaneous change of the lattice state, giving rise to what is called now the Mössbauer effect [Mössbauer (1958a); Mössbauer (1958b); Mössbauer (1959)].

Meanwhile, in 1953, Robert Dicke published a very short paper considering the effect of collisions on the Doppler width of spectral lines [Dicke (1953)]. In fact, the model studied by Dicke was developed for the hyperfine lines in the ground state of alkali atoms diffusing in a buffer gas. Since, in this gas, the electronic and nuclear spins of the atoms are not affected by the the electrostatic interactions appearing during a collision, it is possible to consider that collisions change the

[11] See Fig. 2.1 of Chap. 2.

velocity of an atom without changing its internal state. Dicke was considering a very simple model where the radiating atom is confined in a one-dimensional well of width a, moving back and forth between the two walls. Classically, the wave emitted by the atom is frequency modulated by the Doppler shift associated with the back and forth motion of the atom, and one finds that the central component of the Fourier spectrum at the atomic frequency is predominant when a is small compared to the wavelength λ of the transition. Similar conclusions are reached in a quantum treatment where the transition probabilities between two external states of the atom in the well are calculated. In other words, strong confinement ($a \ll \lambda$) again gives rise to recoil-free absorption and emission processes.

In fact, rather than for atoms diffusing in a buffer gas, Dicke's model is more appropriate for atoms contained in a cell without buffer gas whose dimensions a are smaller than the wavelength of the hyperfine transition. The atoms fly freely between the walls of the cell, the collisions being assumed not to perturb the internal spin states of the hyperfine transition. The microwave can be either a running wave, which corresponds to the case considered by Dicke, or a standing wave, as exemplified in many types of atomic frequency standards. In this case, the phase of the wave is constant but its amplitude varies in space. This gives rise to an amplitude modulation for the moving atom, where, as before the central component is predominant when $a \ll \lambda$. The case of atoms diffusing in a buffer gas has been also studied. In particular, Anatole Abragam [Abragam (1964)], established an interesting connection between the narrowing of the line[12] when the mean free path ℓ between collisions becomes smaller than the wavelength λ and the motional narrowing effect in magnetic resonance.

Finally, as trapping techniques for charged particles were developed, trapped ions emerged as the simplest situation where Dicke's model could be directly applied. Indeed, in this case, there is only one binding frequency. This is to be contrasted with the problems developed by Lamb for the resonance capture of neutrons by bound atoms in a crystal, or similarly by Mössbauer for the absorption of gamma radiation by a crystal. They are much more complex since the whole spectrum of binding frequencies has to be taken into account, leading in practice to more complicated absorption lineshapes. In recognition of the pioneering work of Lamb and Dicke, the regime of strong localization, or strong binding, is now commonly referred to as the *Lamb-Dicke regime*.

6.5 The shelving method

We have seen in the previous sections that the Doppler and recoil effects can be both circumvented by nonlinear effects and suppressed by confinement. Metrology seems then ultimately limited by the natural width of the lines. Is it possible to

[12] For an atom diffusing in a gas, the lineshape can be shown to be a Lorentzian with a width $\Delta\nu \sim \Delta\nu_D\, \ell/\lambda \ll \Delta\nu_D$, where $\Delta\nu_D$ is the Doppler width and ℓ the mean free path.

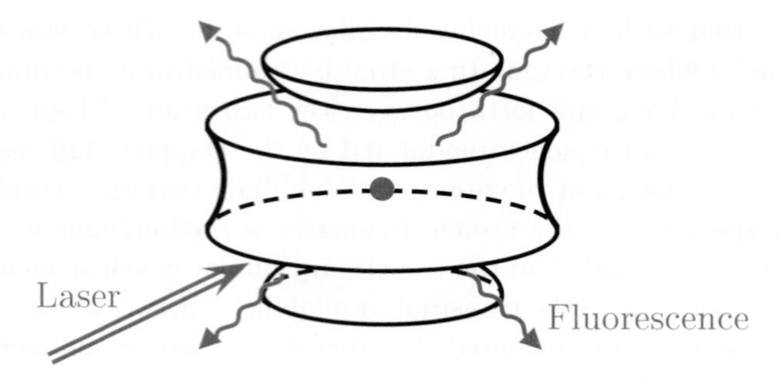

Fig. 6.9 The fluorescence of a trapped ion is driven by a laser beam. The spontaneously emitted light that goes to the free spaces between the electrodes can be collected by a photodetector.

overcome this limitation? Two main methods have been explored:

- Spontaneous emission can be controlled and even suppressed by putting the atom in a cavity [Purcell *et al.* (1946); Kleppner (1981)]. The cavity modifies the spectrum of the quantum vacuum field, and in particular its spectral density at the atomic frequency. This effect has been observed in the microwave [Hulet *et al.* (1985); Gabrielse and Dehmelt (1985)] and optical domains [Jhe *et al.* (1987)]. Note however that it is, in general, impossible to avoid cavity shifts.
- A very elegant method, originally proposed by Hans Dehmelt [Dehmelt (1975)], is known as the shelving method. It is based on a three-level atom with two excited states having very different lifetimes. We now explain this method in more detail.

6.5.1 *Single ion spectroscopy*

During the last few decades, spectacular advances have been achieved in ion physics. Methods, which will be described in Chaps. 12–14, have been developed for trapping and cooling ions with a remarkable efficiency. It is possible now to trap a single ion for days and observe the fluorescence light that it emits when it is excited by a resonant laser beam (see Fig. 6.9). The first experiment demonstrating oprical detection of a trapped ion is described in [Neuhauser *et al.* (1980)]. Let us give a few orders of magnitude for the rate of emitted photons: If the laser saturates the atomic transition, the population of the excited state is equal to $1/2$ and the ion emits about $\Gamma/2$ photons per second. For a typical radiative lifetime on the order of 10^{-8} sec, and even given a typical detection efficiency of only 10^{-4} (typical solid angle), the photodetector collects 10^4 photons per second. This number of photons can easily be detected, not only by a standard photodetector, but even by the naked eye[13]!

[13]This does not mean, strictly speaking, that we observe the ion itself. Instead, we observe the light that it emits. The ion cannot thus be optically observed with a spatial resolution better than the laser wavelength.

This research field has opened new frontiers for ultra high resolution spectroscopy and metrology. It has many advantages: (i) the first and second order Doppler effect are essentially suppressed since the trapped ion is cooled, (ii) there is no more transit time broadening because the ion can be trapped for hours, and (iii) there are no collisions since there is a single particle in the trap.

In the absence of Doppler effect, the width of the curve giving the total intensity of the fluorescence light versus the laser frequency ω_L is (at sufficiently low intensity) given by the natural width Γ of the excited state e. For this reason, one would definitely prefer to use narrow lines (small Γ) for frequency standards and high resolution spectroscopy. However, working with a forbidden transition dramatically decreases the number of detected photons and results in a very small signal to noise, which ultimately limits the resolution. These two contradictory requirements have been satisfied by the *electron shelving* method, proposed by Hans Dehmelt, which allows the detection of very weak transitions on a single trapped ion through a very sensitive double resonance scheme [Dehmelt (1975)].

6.5.2 *Intermittent fluorescence*

Hans Dehmelt proposed the exploitation of a three-level configuration with two excited states and two corresponding transitions, that we call for convenience, "blue" and "red". These transitions connect the lower state g to the two excited states with a V-configuration, where the strong blue transition $g \to e_B$ has a very broad linewidth Γ_B, and the weak red transition $g \to e_R$ has a very narrow linewidth $\Gamma_R \ll \Gamma_B$ (see Fig. 6.10). Two lasers simultaneously drive the two transitions such that the blue laser is used to cool the ion and detect the fluorescence, and the red transition is used for high resolution spectroscopy.

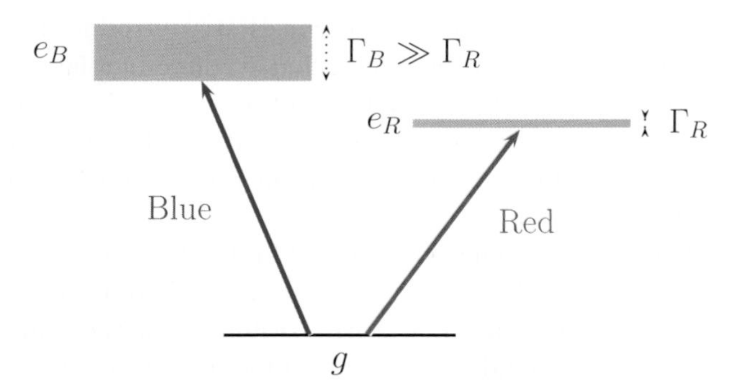

Fig. 6.10 A three-level atom with two excited states. The strong transition $g \longrightarrow e_B$ (blue arrow) with a large linewidth Γ_B is used to cool the ion. The weak transition (red arrow) with a much smaller linewidth $\Gamma_R \ll \Gamma_B$ is used for high resolution spectroscopy.

When the atom absorbs a red photon, it is "shelved" on the excited metastable state e_R. It can remain there for seconds. The ion placed on this shelf is no longer

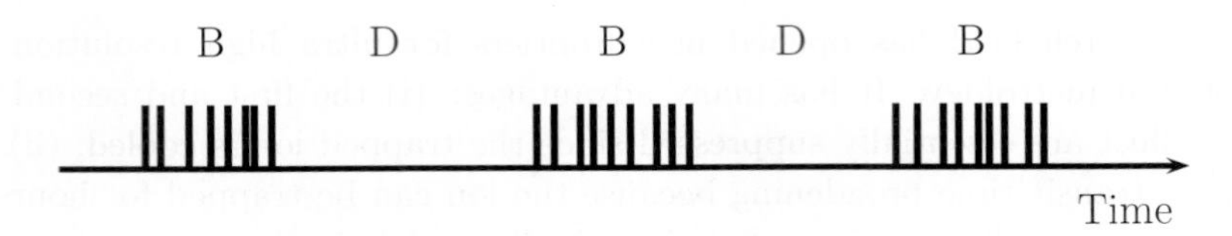

Fig. 6.11 Telegraph-like fluorescence signal collected by a photodetector recording the light emitted by a trapped ion with a three-level electronic structure. The output signal of the detector exhibits an intermittent fluorescence (alternance of period of brightness (B) and darkness (D)).

available for the blue laser and the blue fluorescence stops. In this way, one detects the absorption of one red photon by the absence of a huge number, on the order of Γ_B/Γ_R, of blue photons, since the absorbed red photon switches off the intense blue fluorescence for a typical time of the order of $\tau_R = (\Gamma_R)^{-1}$. The sequence of pulses given by the broadband photodetector recording the fluorescence light is therefore expected to exhibit periods of brightness, with closely spaced pulses, alternating with periods of darkness corresponding to the periods of shelving. The output of the detector should look like a telegraph signal, as suggested by Richard Cook and Jeff Kimble [Cook and Kimble (1985)] (see Fig. 6.11). Note that the random telegraph signal radiated by the strong transition (see Fig. 6.11) collected by the detector provides a direct monitor of the occupation of the electronic state e_R.

6.5.3 *Properties of the detected signal*

The dressed atom approach developed in Chap. 5 for analyzing the radiative cascade of the dressed atom is very useful for formulating a quantitative description of the shelving method. The waiting time distribution $W(\tau)$, introduced in Sec. 5.5.1 of Chap. 5 and extended to a three-level atom, is indeed directly related to the repartition of delays τ between two successive jumps of the atom and thus provides simple evidence for the existence of periods of darkness. It provides quantitative predictions for the average duration and the repetition rate of those periods [Dalibard and Cohen-Tannoudji (1986)].

We will now describe the calculation of $W(\tau)$ for the three-level atom introduced above [Dalibard and Cohen-Tannoudji (1986)]. Immediately after the detection of a first fluorescence photon at time t, the system is in the state $|g, N_B, N_R\rangle$, i.e. the atom is in the ground state in the presence of N_B blue photons and N_R red photons. Neglecting antiresonant terms, this state is only coupled by the laser-atom interaction Hamiltonian to the two other states $|e_B, N_B - 1, N_R\rangle$ and $|e_R, N_B, N_R - 1\rangle$. The atom absorbs a blue or red photon and jumps from the ground state $|g\rangle$ to one of the two excited states $|e_B\rangle$ or $|e_R\rangle$. These three states form a nearly degenerate three-dimensional manifold $\mathcal{E}(N_B, N_R)$ made of the three states $\{|g, N_B, N_R\rangle, |e_B, N_B - 1, N_R\rangle, |e_R, N_B, N_R - 1\rangle\}$ from which the atom can escape only by emitting a second fluorescence photon. The detection of this photon then projects the atom in a lower manifold, and so on.

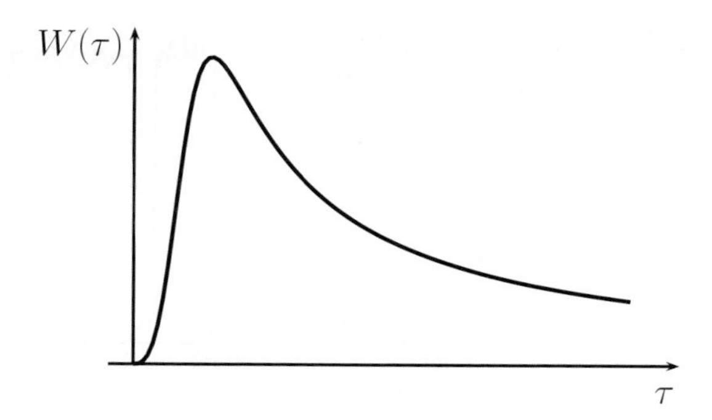

Fig. 6.12 The waiting time distribution for a level scheme depicted in Fig. 6.10. It exhibits a long time tail revealing the presence of long periods of darkness.

The evolution within the manifold $\mathcal{E}(N_B, N_R)$ is governed by the effective Hamiltonian :

$$H_{\text{eff}} = \hbar \begin{pmatrix} 0 & \Omega_B/2 & \Omega_R/2 \\ \Omega_B/2 & -\delta_B - i\Gamma_B/2 & 0 \\ \Omega_R/2 & 0 & -\delta_R - i\Gamma_R/2 \end{pmatrix}, \tag{6.12}$$

where δ_B (resp. δ_R) is the detuning from resonance on the strong (resp. weak) transition, and Ω_B (resp. Ω_R) the corresponding Rabi frequency. The waiting time distribution (or delay function) between two successive emissions corresponds to the probability to leave the manifold $\mathcal{E}(N_B, N_R)$ at time τ and is given by:

$$\begin{aligned} W(\tau) = \Gamma_B |\langle e_B, N_B - 1, N_R| \exp\{-iH_{\text{eff}}\tau/\hbar\} |g, N_B, N_R\rangle|^2 + \\ \Gamma_R |\langle e_R, N_B, N_R - 1| \exp\{-iH_{\text{eff}}\tau/\hbar\} |g, N_B, N_R\rangle|^2. \end{aligned}$$

The calculation reveals the existence of a long tail of the function $W(\tau)$ (see Fig. 6.12). This is due to the fact that one of the three eigenvalues of H_{eff} has a very small imaginary part (because $\Gamma_R \ll \Gamma_B$, $\Omega_R \ll \Omega_B$). After one quantum jump, it can thus take a very long time before observing the next jump. We recover the telegraph-like signal with a sequence of pulses made of periods of darkness (without any pulse) between periods of brightness (with many closely spaced pulses) as illustrated in Fig. 6.11. Alternatively, the waiting time distribution can be calculated using a simulation of the individual photon emissions [Zoller *et al.* (1987)].

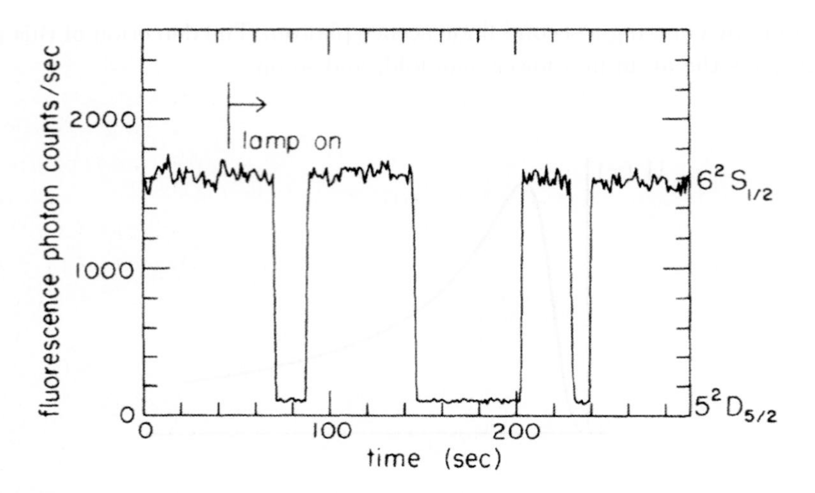

Fig. 6.13 Alternance of period of brightness and darkness in the collected fluorescence of a trapped Ba^+ ion. The suppression of the fluorescence for long periods of time (> 30 seconds) occurs when the atom is in the excited state of the weak transition (see Fig. 6.10). Figure extracted from [Nagourney *et al.* (1986)]. Copyright: American Physical Society.

6.5.4 *Observation of quantum jumps*

The first direct observations of quantum jumps were performed on a single laser-cooled Ba^+ ion contained in a radio-frequency trap[14] by the groups of Hans Dehmelt [Nagourney *et al.* (1986)] and Peter Toschek [Sauter *et al.* (1986)]. These experiments provided a validation of the shelving method. The fluorescence from the strong transition state was suppressed for long time periods (> 30 seconds), after which the fluorescence reappeared as shown in Fig. 6.13. The issue of the signal-to-noise ratio limitation on a weak transition was consequently circumvented.[15] A detector such as the eye is thus sufficient to directly observe quantum jumps!

In conclusion, experiments using the shelving method are interesting examples of experiments performed on a single quantum system. Moreover, for analyzing the results observed in a single experimental realization, stochastic wave functions approaches, like those based on the waiting time distribution for simulating the random sequence of quantum jumps of the system, appear as the most appropriate theoretical tool.

[14]See Chap. 14.

[15]In a high resolution experiment, excitations with the blue and red lasers are alternated in time to avoid light shifts due to the blue laser during the periods of excitation of the red transition by the red laser.

6.6 Quantum logic spectroscopy

The shelving method described in the previous section works only for an ion having two transitions sharing a common level, one narrow transition for spectroscopy and one broad transition for cooling and detection. It may happen that an ion with a very narrow transition, that could be used as a very precise frequency standard, does not possess another strong transition falling in a convenient frequency range and which could be easily used for cooling and detection. For example, the $^{27}\mathrm{Al}^+$ ion has a very weak transition at $\lambda = 267$ nm connecting the ground state $^1\mathrm{S}_0$, $F = 5/2$ to the excited state $^3\mathrm{P}_1$, $F = 7/2$ with a Q factor, $Q = \nu/\Delta\nu$, as high as 2×10^{17}. The wavelength of this transition can be produced by a frequency doubled dye laser, but the strong transition connecting the ground state $^1\mathrm{S}_0$, $F = 5/2$ to the excited state $^1\mathrm{P}_1$, which could be used for cooling and detection, has a wavelength $\lambda = 167$ nm not easy to produce. In this section, we show how it is possible to solve this difficulty by taking two ions in the trap [Schmidt *et al.* (2005)]. The first one, called the *spectroscopy ion* (S), for example the $^{27}\mathrm{Al}^+$ ion, has the narrow transition interesting for metrology. The quantum state of this ion produced by its coherent coupling with a laser tuned near its resonance frequency is mapped by a quantum logic operation onto another ion, called the *logic ion* (L), which has convenient transitions for cooling and for detecting the state of L, which is by construction identical to the state of S. The cooling of S is obtained by its coupling with L. The corresponding thermalization of the two ions at the same temperature is called *sympathetic cooling*.

The mapping of the state of S onto L is achieved through one of their normal modes of vibration in the trap. These normal modes result from the Coulomb coupling of the two ions which do not move independently in the trap. Consider one of these modes with frequency ω_v. By exciting S with a laser tuned at the red sideband $\omega_{\mathrm{spec}} - \omega_v$, where ω_{spec} is the resonance frequency of S, one can selectively excite the first vibrational state $n = 1$ of this normal mode. The internal state of L is not changed, but both ions have their external motion changed since the quantum number of their normal mode of vibration has changed. A similar operation can be performed by exciting L with a laser tuned at the red sideband $\omega_{\mathrm{logic}} - \omega_v$, where ω_{logic} is the resonance frequency of L.

Figure 6.14 shows the various steps of the mapping. The states written at the bottom of the figure are the states of the whole system at each step.

In the initial state, each ion is in its lower state $|\downarrow\rangle_S$ for S, $|\downarrow\rangle_L$ for L. The normal mode of vibration is in the ground state $|0\rangle_m$ (step (a)).

A laser field with a frequency close to ω_{spec} puts the ion S in a linear superposition of the lower $|\downarrow\rangle_S$ and upper $|\uparrow\rangle_S$ states with coefficients α and β. The L ion and the normal mode of vibration remain unchanged (step (b)).

A π-pulse is applied to S at the frequency $\omega_{\mathrm{spec}} - \omega_v$ of the red sideband of this ion. The internal state of S is changed from $|\uparrow\rangle_S$ to $|\downarrow\rangle_S$ and the quantum

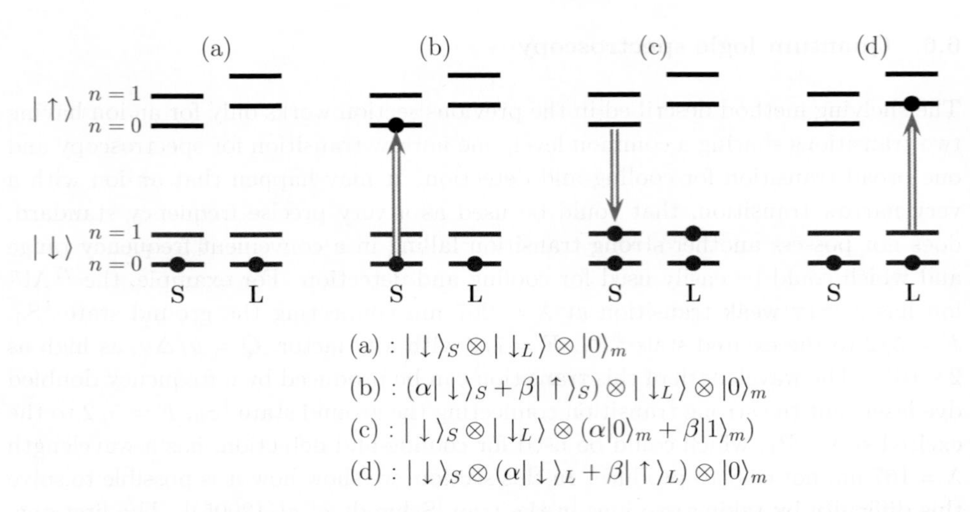

$$(a): |\downarrow\rangle_S \otimes |\downarrow_L\rangle \otimes |0\rangle_m$$

$$(b): (\alpha|\downarrow\rangle_S + \beta|\uparrow\rangle_S) \otimes |\downarrow_L\rangle \otimes |0\rangle_m$$

$$(c): |\downarrow\rangle_S \otimes |\downarrow_L\rangle \otimes (\alpha|0\rangle_m + \beta|1\rangle_m)$$

$$(d): |\downarrow\rangle_S \otimes (\alpha|\downarrow\rangle_L + \beta|\uparrow\rangle_L) \otimes |0\rangle_m$$

Fig. 6.14 Different steps of the quantum logic operation mapping the state of the spectroscopy ion onto the logic ion.

number of the vibration mode is changed from 0 to 1. The internal state of L is unchanged (step (c)). The expression (c) of the state of the total system shows that both ions are in their lower states whereas the state of the normal mode is in a linear superposition of the two states $n = 0$ and $n = 1$ with the same coefficients α and β obtained in the step (b). In other words, the state of S has been copied on the normal mode of vibration.

A π-pulse is applied to L at the frequency $\omega_{\text{logic}} - \omega_v$ of the red sideband of this ion. The internal state of L is changed from $|\downarrow\rangle_L$ to $|\uparrow\rangle_L$ and the quantum number of the vibration mode is changed from 1 to 0. The internal state of S is unchanged (step (d)). The expression (d) of the state of the total system shows that S and the normal mode of vibration are in their lowest states whereas the state of L is in a linear superposition of the two states $|\downarrow\rangle_L$ and $|\uparrow\rangle_L$with the same coefficients α and β obtained in the step (b). In other words, the state of S has been mapped onto L.

This method has been successfully demonstrated by taking for S the $^{27}\text{Al}^+$ ion mentioned above and for L a $^9\text{Be}^+$ ion [Schmidt *et al.* (2005)]. The measurement of L consists of determining $|\alpha|^2$ and $|\beta|^2$ by monitoring the fluorescence of L which has not the same intensity depending whether L is in $|\downarrow\rangle_L$ or $|\uparrow\rangle_L$. The most precise optical atomic clocks built by the NIST-Boulder group use this quantum logic method. They are used to test fundamental effects such as gravitational red shift predicted by general relativity or a possible variations of fundamental constants.[16]

[16]See Chap. 18.

6.7 Frequency measurement with frequency combs

In the previous sections, we have described several methods for obtaining very narrow atomic lines. When the frequency of a laser is locked to the center of the atomic line, it still remains to determine its wavelength or its frequency. High precision wavelength measurements are not easy. It is much more convenient to measure frequencies, but they must be expressed in terms of the frequency of the clock transition used for defining the unit of time. This clock transition is a hyperfine transition of the cesium atom in the microwace domain. One must therefore find a way to link the microwave to the optical domains. For several decades, this was achieved by using very complex frequency chains based on harmonic generation. An important breakthrough in this research field was the introduction of *frequency combs* for laser-based precision spectroscopy for which John Hall and Theodor Hänsch got the Nobel Prize in 2005 [Hall (2006); Hänsch (2006)]. This method has dramatically simplified and improved the accuracy of frequency metrology.[17]

A frequency comb consists of the superposition of many continuous and coherent laser modes, equidistant in frequency space. This technology was made possible by progress in mode locked lasers,[18] which deliver short light pulses. In these systems, laser light is repeatedly reflected within a mirrored cavity and the longitudinal modes are synchronized such that the peaks of the different modes coincide at regular intervals, evenly spaced in time. The peaks sum up to form a solitary short pulse that contains many different frequencies. An attenuated copy of this pulse escapes at one of the mirrors of the cavity that is partially transparent. This frequency spectrum is close to the Fourier limit with a spectral width equal to the inverse of the pulse duration: a 10 fs pulse therefore has a 100 THz bandwidth centered around the optical carrier frequency, typically 800 nm ($\omega_c/2\pi \simeq 375$ THz). The group velocity v_g determines the time needed for a cavity round trip: $T = 2L/v_g$, and thus the angular frequency of the repetition rate is $\omega_r = 2\pi T^{-1}$. In practice, the repetition rate ranges from 4 MHz to 2 GHz and can thus be easily compared to the microwave reference of the cesium clock. The frequencies ω_n of the combs are given by

$$\omega_n = n\omega_r + \omega_{CE}, \tag{6.13}$$

where ω_{CE} is a carrier-envelope frequency offset common to all modes, by convention $0 \leq \omega_{CE} \leq \omega_r$. When $\omega_{CE} \neq 0$, the maxima of the wave do not coincide with that of the pulse envelope as shown in Fig. 6.15. The offset originates from the mismatch between the group v_g and the phase v_p velocities:

$$\omega_{CE} T = \Delta\varphi \qquad \text{where} \qquad \Delta\varphi = \omega_c \left(\frac{2L}{v_g} - \frac{2L}{v_p} \right) \tag{6.14}$$

[17] For a review see [Udem and Riehle (2007)].
[18] See Chap. 27.

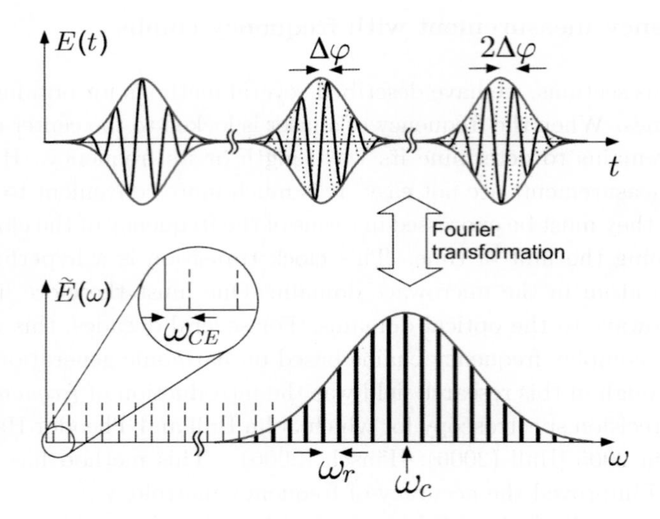

Fig. 6.15 Representation in time and frequency spaces of the electric field of a pulsed femtosecond laser with a carrier frequency ω_c. A pulse-to-pulse phase shift $\Delta\varphi$ is associated with a carrier-envelope frequency offset ω_{CE}. The mode spacing is dictated by the repetition rate ω_r. Figure reproduced from [Udem and Riehle (2007)] with kind permission of Società Italiana di Fisica. Copyright: Italian Physical Society.

is the pulse-to-pulse phase shift between the carrier envelop and the carrier, also called carrier envelope phase (see Fig. 6.15). The use of the frequency comb for frequency measurement requires the servo control of both ω_r and ω_{CE}. The control of the repetition rate can be readily performed by mounting one of the laser cavity mirrors on a piezo electric transducer. Other methods have been used that depend on the mode-locked laser characteristics. The measurement of ω_{CE} is more delicate, and is carried out by the self-referencing technique referred to as the $f - 2f$ interferometer. The first step consists of increasing the spectral width of the pulse train so that the highest frequencies are more than twice as high as the lowest ones. To generate such an octave spanning comb, the pulse train is broadened through its injection into a photonic crystal fiber, where self-phase modulation generates a frequency chirp. The enlarged pulse in frequency space is no longer Fourier limited, but the mode spacing remains unchanged. The carrier-envelope frequency offset is measured by beating the frequency doubled red wing of the comb $2\omega_n$ with the blue side of the comb at ω_{2n}. The beat notes have a frequency:

$$2\omega_n - \omega_{2n} = 2(n\omega_r + \omega_{CE}) - (2n\omega_r + \omega_{CE}) = \omega_{CE}, \tag{6.15}$$

and are used to determine the offset frequency[19] ω_{CE}. The measurement of an unknown frequency ω_L of a cw laser with a stabilized frequency comb also requires the generation of a beat note. This is realized by superimposing both beams onto

[19]We will see in Chap. 27 that the measure and the control of ω_{CE} are essential for generating ultrashort attosecond pulses.

a beam splitter and collecting the interference signal on a photodetector. The frequency ω_L of the laser is related to the frequency ω_b of the beat note by

$$\omega_L = n\omega_r \pm \omega_{CE} \pm \omega_b. \tag{6.16}$$

Depending on the precision to which the optical frequency ω_L is known, different methods have been developped for extracting the value of n and of the signs in Eq. (6.16). One can use, for example, different values for the repetition rate.

Finally, frequency combs enable the comparison of an optical frequency with a microwave reference, as well as the direct comparison between two optical frequencies.

6.8 Conclusion

In this chapter, we have reviewed a few methods that have been invented for improving the resolution of spectroscopic measurements.

Doppler free lines can be obtained through the use of nonlinear effects, such as saturation spectroscopy, or two-photon transitions using counter propagating laser beams.

Another interesting method for avoiding the Doppler broadening and the recoil shift due to the momentum transfers between photons and atoms during the transition is to confine the radiating system in a small region with a spatial extent smaller than the wavelength of the transition. Along these lines, a series of developments extending over several decades have led to the realization of the Lamb-Dicke regime for trapped ions.

When the Doppler broadening of spectral lines is circumvented, very weak transitions with a very small natural width are most attractive. Fortunately, even though the weak signal to noise ratio associated with optical detection raises serious problems, they can be solved by using other detection schemes, such as the shelving method. An extension of the shelving method to ions that do not have the appropriate level configuration can be realized by using quantum logic operation.

Finally, an elegant solution to the problem of the accurate measurement of optical frequencies has been provided by the use of frequency combs. This technique allows one to connect the microwave transition of cesium atoms used to define the unit of time to any optical frequency.

It is amusing to note that all the methods described in this chapter try to avoid or circumvent the perturbations associated with the exchanges of momentum between atoms and photons. As it often occurs in physics, however, a perturbation can be turned into an advantage. The most recent developments in atomic physics, which have occurred during the last two decades, use the exchanges of momentum between atoms and photons for cooling and trapping atoms and thereby eliminating, in a drastic way, first and second order Doppler effects. These methods are opening new avenues for ultra high resolution spectroscopy, in particular with atomic clocks.[20]

[20]See Chap. 18.

PART 2

Atom-photon interactions: a source of perturbations for atoms which can be useful

Introduction

When an atom interacts with an electromagnetic field, both systems are perturbed. The energy levels of the atom can be shifted and broadened by light. These so-called light shifts and light broadenings must be carefully taken into account in high resolution spectroscopy since they change the atomic frequencies to be measured. The field is also perturbed through a modification of its speed of propagation described by an index of refraction, which is complex if a damping of the field amplitude is present.

In the early sixties the great sensitivity of optical methods made it possible to discover light shifts and light broadenings, and to study their properties in great detail. The purpose of part 2 is to describe these perturbations, to interpret them, and to give their main characteristics. It turns out that these perturbations can be also used in a positive way. Light shifts have recently been used to achieve laser traps, mirrors for atoms, optical lattices, and spin dependent coherent transport.[1] We will briefly review a few of these applications. We will also show how the interpretation of the perturbation of an atom due to its interaction with an incident field can give new physical insights into the radiative corrections of quantum electrodynamics due to the interaction of this atom with the quantum vacuum field.

We will restrict ourselves to two limiting cases leading to simple discussions.

(i) Case of a quasi resonant optical excitation considered in Chap. 7.

The light frequency ω_L is close to the frequency ω_A of an atomic transition $g \leftrightarrow e$. It is sufficiently far from resonance for all other atomic frequencies so that one can keep only the two levels g and e in the description of the atomic system.[2] In the dressed atom approach, the manifolds of the closely spaced levels of the atom-photon system have a finite dimension, which allows simpler calculations.

(ii) Case of a high frequency excitation considered in Chap. 8.

The light frequency ω_L is large compared to the frequency of all atomic transitions. Using a second order treatment of the atom-field coupling, we will show how the slow motion of the atomic electron is perturbed by the high frequency vibration of this electron in the incident wave leading to a new effective mass for the electron and to new electric and magnetic form factors.

Note that the distinction between low frequency, resonant, and high frequency excitation is artificial since all situations can occur in a given physical problem. For example, when an atom interacts with an intense infrared laser with frequency ω_L, the excitation appears to have a low frequency for the ground state g since ω_L is in general very small compared to the frequencies of the transitions connecting g to the first excited states. The light shift of g is thus negative, corresponding to a Stark shift of g in the quasi static laser field. On the other hand, for the highly excited

[1] "Sisyphus cooling" is another application of light shifts which will be discussed in Chap. 13.
[2] These levels may have however several Zeeman sublevels.

Rydberg states the excitation appears to have a high frequency since ω_L is large compared to the frequency splitting between adjacent Rydberg states. We will see in Chap. 8 that all Rydberg states are light shifted by the same positive amount, which is in fact the mean kinetic energy of vibration of the electron in the incident laser field. It follows that the energy of ionization of the atom, which corresponds to the energy difference between the ground state and the beginning of the ionization continuum is increased by the interaction with the laser. Furthermore, resonant excitation of an excited discrete state can occur when the energy difference between this excited state and the ground state is a multiple integer of the photon energy. This resonant multiphoton excitation of a discrete excited state can considerably enhance the multiphoton ionization rate of the atom. We will discuss situations of this type in Chap. 10.

Chapter 7

Perturbations due to a quasi resonant optical excitation

7.1 Introduction

In this chapter, we describe the perturbations experienced by a two-level atom when it interacts with an incident light beam whose frequency ω_L is close to the atomic transition frequency ω_A. We will use the dressed atom approach already introduced in Sec. 5.4 of Chap. 5 for studying the properties of the fluorescence light emitted by the atom. The emphasis is put here, not on the properties of the emitted light, but on the modifications of the atomic energy levels.

The advantages of using a quasi-resonant excitation of an atomic transition is that we can keep only the two levels g and e of this transition. The relevant manifolds $\mathcal{E}(N)$ of dressed states are then two-dimensional and the effective Hamiltonian H_{eff}, which describes the reduced evolution within $\mathcal{E}(N)$, can be exactly diagonalized.

We suppose here that g is the ground state and that e is an excited state with a natural width Γ equal to the spontaneous emission rate from this state. Including Γ in H_{eff} will allow a discussion of the effect of dissipation in the perturbations of atoms induced by light. In Sec. 7.2, we will study, in particular, the evolution of the behavior of the system when the Rabi frequency Ω which characterizes the light intensity increases. Various regimes can be observed depending on the relative values of Ω, Γ, and the detuning $\delta = \omega_L - \omega_A$ between the light and the atomic frequencies. At low intensity, the dressed states are close to the unperturbed states that describe the atom in g or in e in the presence of a certain number of photons. Their complex energies describe two types of effects: first a shift of the ground state called *light shift* and due to the interaction with the incident light; second, an instability of the ground state due to its light-induced contamination by the unstable excited state, and which can also be interpreted as a photon absorption rate by an atom in g. At high intensity, the two dressed states become entangled atom-photon states separated by an energy splitting $\hbar\Omega$, and can no longer be interpreted as states describing an atom in g or in e perturbed by the incident light. Starting from g, the atom will oscillate back and forth between g and e at the frequency Ω, performing what is called a "Rabi oscillation". This analysis provides a simple example of a continuous transition between the two extreme types of time evolution, the Weisskopf-Wigner exponential decay, which corresponds here to the irreversible departure rate from g by absorption of a photon and the Rabi oscillation.

Another advantage of including the dissipation in the description of the system is that dissipation plays an essential role in the various laser cooling mechanisms that will be described in Chaps. 12 and 13. This is due to the fact that light shifts and light broadenings are correlated, since they both result from the interaction with the incident light. One of the most efficient laser cooling mechanism, "Sisyphus cooling," is based on these correlations.

The perturbations induced by the atom-light interactions are not restricted to the atom. They also concern the incident light. In Sec. 7.3 we study the perturbations of the light field in a cavity due to the presence of an atom in this cavity, and show that the perturbations of the light field can be classified, as for the atom, into two categories: first, a modification of the frequency of the field, a "reactive" effect, analogous to the light shift that can be related to the real part of the index of refraction; and second, a damping of the field, a "dissipative" effect, analogous to the light broadening of atomic levels, that is related to the imaginary part of the index of refraction.

It is also instructive to try to give a semiclassical interpretation of these various results, which is possible in the low intensity limit. We will present the principle of such a semiclassical approach in Sec. 7.2.5 by considering the atomic dipole moment induced by the incident field and related to this field by a *dynamical polarizability*. The absorption rate and the light shift can be interpreted by considering the energy couplings between the incident field and the induced dipole moment. The interference between the incident field and the field radiated in the forward direction by the induced dipole moment explains the damping and the phase shift of the transmitted field.

Light shifts were first observed in the early 1960's in optical pumping experiments. Since lasers were not available at that time, and the light intensity delivered by discharge lamps was not very high, the light shifts were on the order of 1 Hz. It was possible, however, to detect these small light shifts by using the fact that the different Zeeman sublevels in g are, in general, shifted by different amounts depending on the light polarization. These differential light shifts produce a displacement of the magnetic resonance curves in g that can be detected since, due to the very long lifetime of g, these resonance curves are very narrow. These experiments will be briefly described in Sec. 7.4.

Light shifts from lasers can be much larger, and represent an important source of perturbations for atomic clocks and high precision measurements of atomic frequencies. In these cases, one can try to eliminate them by extrapolating the measured atomic frequencies to zero light intensity. On the other hand, as is often the case in physics, it turns out that a perturbation can be turned into an advantage. If the detuning δ between the light and the atomic frequencies is large enough, light broadenings are very small and can be ignored. Position dependent light shifts in a laser field whose intensity varies in space then appear as an external potential for the atom in g. Depending on the sign of δ, one can create potential wells with different geometries in which sufficiently cold atoms can be trapped, or potential barriers that can reflect them like a mirror. Some applications of these ideas will be described in Sec. 7.5. Light shifts are also useful for manipulating fields, for preparing new interesting states of the field in the cavity, or for detecting photons in this cavity without destroying them. A brief review of these recent applications is given in Sec. 7.6.

7.2 Light shift, light broadening and Rabi oscillation

7.2.1 *Effective Hamiltonian*

In the following, we use the dressed atom approach introduced in Sec. 5.4 of Chap. 5. Consider an atom A with two levels e and g that interacts with a light field L of frequency ω_L close to the frequency $\omega_A = (E_e - E_g)/\hbar$ of the $e \leftrightarrow g$ transition. We suppose that g is the ground state and that e is an excited state with a natural width Γ. The detuning between the light and the atomic frequencies is $\delta = \omega_L - \omega_A$.

In the absence of coupling V_{AL} between A and L, the uncoupled states of the atom-photon system gather in two-dimensional manifolds $\mathcal{E}(N+1) = \{|g, N+1\rangle; |e, N\rangle\}$ where N is the number of photons (left part of Fig. 7.1). The atom in g can absorb one photon and go to e, so that the two states $|g, N+1\rangle$ and $|e, N\rangle$ are coupled. We have:

$$\langle e, N| \, V_{AL} \, |g, N+1\rangle = \frac{\hbar \, \Omega_{N+1}}{2}, \tag{7.1}$$

where Ω_{N+1} is a Rabi frequency. Here we consider an experiment in free space, so that we can neglect the N-dependence of Ω_{N+1} and replace Ω_{N+1} by Ω (second situation considered in Sec. 5.4.4 of Chap. 5).

We have already mentioned in Sec. 5.4.5 of Chap. 5 that the reduced evolution

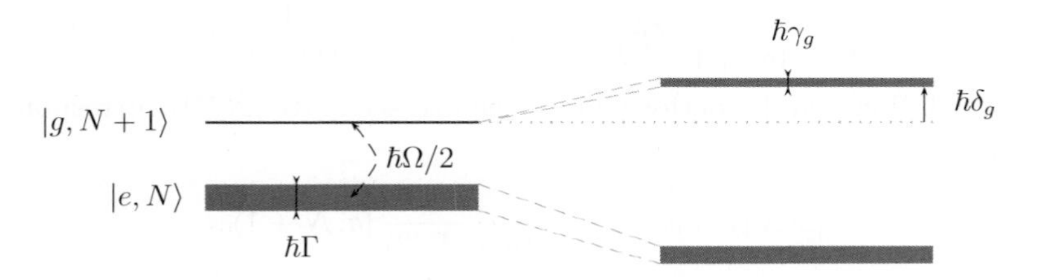

Fig. 7.1 Uncoupled states (left part of the figure) and dressed states (right part) in the manifold $\mathcal{E}(N+1)$. Due to the atom-photon coupling V_{AL} described by the Rabi frequency Ω, the state $|g, N+1\rangle$ is shifted by an amount $\hbar\delta_g$ which is the light shift of the ground state and broadened by an amount $\hbar\gamma_g$ which is the light broadening.

within $\mathcal{E}(N+1)$ is described by an effective Hamiltonian

$$H_{\text{eff}} = \hbar \begin{pmatrix} \delta & \Omega/2 \\ \Omega/2 & -i\Gamma/2 \end{pmatrix}, \tag{7.2}$$

obtained by adding an imaginary term $-i\Gamma/2$, which accounts for the radiative instability of e due to spontaneous emission, to the energy of the excited state e.[1] We now discuss the eigenvalues and the eigenstates of H_{eff} in certain limits. They describe the dressed states of the system, including the effect of the radiative damping described by Γ.

7.2.2 *Weak coupling limit. Light shift and light broadening*

Suppose first that the coupling $\Omega/2$ between the two uncoupled states $|g, N+1\rangle$ and $|e, N\rangle$ is small compared to the difference between the unperturbed energies of these two states. Since the energy of $|e, N\rangle$ is complex, we take the modulus of their energy difference:

$$\Omega \ll \left|\delta + i\frac{\Gamma}{2}\right|, \tag{7.3}$$

or, equivalently: $\Omega \ll \Gamma$ or $\Omega \ll |\delta|$. This limit corresponds to weak light intensities or large detunings.

It is then possible to apply perturbation theory for obtaining the expressions of the two dressed states $|1(N)\rangle$ and $|2(N)\rangle$ of $\mathcal{E}(N+1)$. If $\delta > 0$, $|g, N+1\rangle$ is above $|e, N\rangle$ and transforms into $|1(N)\rangle$ when the coupling is introduced. To first order in V_{AL} we get:

$$|1(N)\rangle = |g, N+1\rangle + \frac{\Omega}{2\,(\delta + i\Gamma/2)}\,|e, N\rangle\,, \tag{7.4}$$

and the energy shift of $|1(N)\rangle$ with respect to $|g, N+1\rangle$ is given to second order in V_{AL} by:

$$\delta E_{1N} = \hbar\frac{\Omega^2}{4\,(\delta + i\Gamma/2)} = \hbar\delta_g - i\hbar\frac{\gamma_g}{2}, \tag{7.5}$$

where:

$$\delta_g = \frac{\delta}{4\delta^2 + \Gamma^2}\Omega^2, \qquad \text{and} \qquad \gamma_g = \frac{\Gamma}{4\delta^2 + \Gamma^2}\Omega^2. \tag{7.6}$$

Similar calculations can be performed for the dressed state $|2(N)\rangle$ corresponding to $|e, N\rangle$ and give:

$$|2(N)\rangle = |e, N\rangle - \frac{\Omega}{2\,(\delta + i\Gamma/2)}\,|g, N+1\rangle\,, \tag{7.7}$$

and:

$$\delta E_{2N} = -\hbar\delta_g + i\hbar\frac{\gamma_g}{2}. \tag{7.8}$$

[1] See [Cohen-Tannoudji *et al.* (1992b)], Sec. III.C.3 and Sec. VI.C.2.

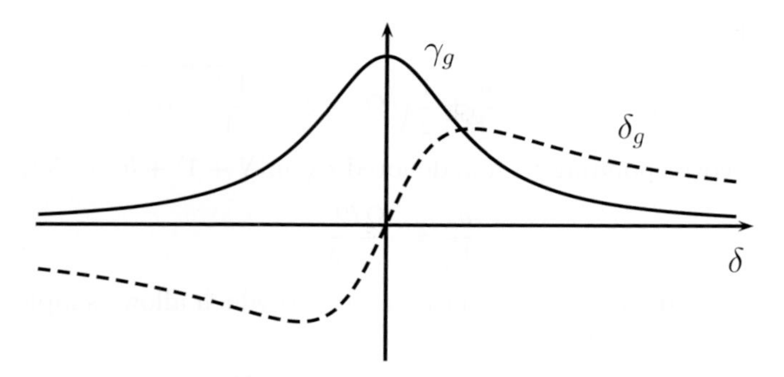

Fig. 7.2 Variations with the detuning δ of the light shift δ_g (dashed line) and of the light broadening γ_g (solid line).

The main component of the dressed state $|1(N)\rangle$ is $|g, N+1\rangle$, the last term of Eq. (7.4) representing a small contamination of $|g, N+1\rangle$ by $|e, N\rangle$. The dressed state $|1(N)\rangle$ can thus be considered as representing the ground state perturbed by the coupling. From Eq. (7.5), we deduce that $\hbar\delta_g$ is the shift of the ground state due to the incident light, also called *light shift*, whereas $\hbar\gamma_g$ is the *light broadening*. Similarly, Eqs. (7.7) and (7.8) show that the excited state e undergoes a light shift $-\hbar\delta_g$ and that its width is reduced from $\hbar\Gamma$ to $\hbar(\Gamma - \gamma_g)$.

These results are represented on the right part of Fig. 7.1. The two light shifts of $|g, N+1\rangle$ and $|e, N\rangle$ are equal and opposite. Because of the non diagonal coupling between them, the two unperturbed states repel each other. The light broadening $\hbar\gamma_g$ of $|g, N+1\rangle$ is due to the contamination of $|g, N+1\rangle$ by $|e, N\rangle$ induced by V_{AL}, which transfers to g a small part of the radiative instability of e. In the presence of light, an atom in g does not remain there for ever. It has a certain probability per unit time γ_g to leave this state by absorption of a photon. One can thus also interpret γ_g as a photon absorption rate from g.

According to Eq. (7.6), both δ_g and γ_g are proportional to Ω^2. The light shift and the light broadening are thus proportional to the light intensity. As a function of the detuning δ, the light shift δ_g varies as a dispersion curve whereas the light broadening γ_g varies as an absorption curve (see Fig. 7.2).

In the limit of large detunings ($|\delta| \gg \Gamma$), we have, according to Eq. (7.6):

$$|\delta_g| = \frac{\Omega^2}{4|\delta|} \quad \gg \quad \gamma_g = \frac{\Omega^2\Gamma}{4\delta^2} = |\delta_g| \frac{\Gamma}{|\delta|}. \tag{7.9}$$

The light shift is then much larger than the line broadening.

7.2.3 *High coupling limit. Rabi oscillation*

To discuss this limit, it will be useful to calculate exactly the eigenvalues and the eigenstates of H_{eff}. The eigenvalues λ are given by the second degree equation:

$$(\delta - \lambda)\left(-i\frac{\Gamma}{2} - \lambda\right) - \frac{\Omega^2}{4} = \lambda^2 + \lambda\left(i\frac{\Gamma}{2} - \delta\right) - \left(i\frac{\Gamma\delta}{2} + \frac{\Omega^2}{4}\right) = 0, \tag{7.10}$$

and are equal to:

$$\lambda_{\pm} = -i\frac{\Gamma}{4} + \frac{\delta}{2} \pm \frac{1}{2}\sqrt{\Omega^2 + \delta^2 - \frac{\Gamma^2}{4} + i\Gamma\delta}. \tag{7.11}$$

The eigenstate corresponding to λ is denoted $a_\lambda \left|g, N+1\right\rangle + b_\lambda \left|e, N\right\rangle$, with

$$\frac{a_\lambda}{b_\lambda} = \frac{\Omega/2}{\lambda - \delta}. \tag{7.12}$$

We will focus here on the resonant case ($\delta = 0$) which allows simple discussions. When $\delta = 0$, Eq. (7.11) gives:

$$\lambda_{\pm} = -i\frac{\Gamma}{4} \pm \frac{1}{2}\sqrt{\Omega^2 - \frac{\Gamma^2}{4}}. \tag{7.13}$$

It clearly appears that, as long as $\Omega < \Gamma/2$, the two eigenvalues are complex and given by:

$$\lambda_{\pm} = -i\left[\frac{\Gamma}{4} \mp \frac{1}{2}\sqrt{\frac{\Gamma^2}{4} - \Omega^2}\right]. \tag{7.14}$$

When $\Omega \ll \Gamma/2$, one of these eigenvalues, λ_+, is equal to $-i\Omega^2/2\Gamma$, i.e. to $-i\gamma_g/2$, where γ_g is given by Eq. (7.6) with $\delta = 0$. The other one, λ_-, is equal to $-i(\Gamma - \gamma_g)/2$. We note also that, according to Eq. (7.12), $|a_{\lambda+}/b_{\lambda+}| = \Gamma/\Omega \gg 1$, which means that the main component of the dressed state corresponding to λ_+ is $\left|g, N+1\right\rangle$. We thus recover all the results derived in the previous subsection in the low coupling limit.

When Ω increases, while still remaining smaller than $\Gamma/2$, λ_+ increases and λ_- decreases, their sum $(\lambda_+ + \lambda_-)$ remaining constant and equal to $-i\Gamma/2$. The two eigenvalues become equal to $-i\Gamma/4$ when Ω reaches the value $\Gamma/2$.

The high intensity regime begins when Ω becomes larger than $\Gamma/2$. The last term of Eq. (7.13) becomes then real. The two eigenvalues λ_+ and λ_- have opposite real parts. Their imaginary parts are both equal to $-i\Gamma/4$. When $\Omega \gg \Gamma$, the two real parts become equal to $\pm\Omega/2$ and the frequency splitting between the two dressed states is equal to the Rabi frequency Ω. Equation (7.12) gives $a_{\lambda_\pm} = \pm b_{\lambda_\pm}$, which means that the two dressed states are then equal to the symmetric and antisymmetric linear combinations of $\left|g, N+1\right\rangle$ and $\left|e, N\right\rangle$:

$$\frac{1}{\sqrt{2}}\left[\left|g, N+1\right\rangle \pm \left|e, N\right\rangle\right]. \tag{7.15}$$

It is then no longer possible to consider that the two dressed states represent $\left|g, N+1\right\rangle$ and $\left|e, N\right\rangle$ slightly perturbed by the coupling. The two dressed states are *entangled* states of the atom-photon system. In these states, which have interesting applications for quantum information, the atom-photon system must be considered as a whole inseparable system exhibiting quantum correlations. Suppose that the system starts at time $t = 0$ from $\left|g, N+1\right\rangle$, which is a linear superposition of the two dressed states. These two dressed states then evolve with damped evolution phase factors $\exp(\mp i\Omega t/2)\exp(-\Gamma t/4)$, so that the system oscillates back and forth between $|g, N+1\rangle$ and $|e, N\rangle$ at the Rabi frequency Ω, the probability to be in one of the unperturbed states being damped at a rate $\Gamma/2$.

All the previous considerations can be easily extended to the non-resonant case. The high coupling limit corresponds to

$$\Omega \gg \left|\delta + i\frac{\Gamma}{2}\right|. \tag{7.16}$$

and the first term, Ω^2, in the square root of Eq. (7.11) is dominant. If, in addition, we suppose that $|\delta| \gg \Gamma$, we can neglect the last two terms in this square root. We thus get:

$$\lambda_\pm = -i\frac{\Gamma}{4} + \frac{\delta}{2} \pm \frac{\tilde{\Omega}}{2}, \tag{7.17}$$

where $\tilde{\Omega}^2 = \Omega^2 + \delta^2$. The frequency of the oscillation between g and e is now the effective Rabi frequency $\tilde{\Omega}$.

7.2.4 *Absorption rate versus Rabi oscillation*

The previous analysis shows that, when an atom in the ground state g is excited by a resonant monochromatic field to an excited atomic state e, we have not necessarily a Rabi oscillation between g and e. If e is unstable and if the light intensity is low enough, we have an irreversible departure rate from g with a rate γ_g proportional to the light intensity and which can be interpreted as a photon absorption rate. A Rabi oscillation appears only when the light intensity is high enough and this oscillation is damped. To interpret this result, we must consider that the interaction V_{AL} couples the discrete stable state g to a "broad" state e, which can be considered as a continuum of width Γ. If the coupling strength Ω is small compared to the width of the continuum Γ, one can apply Fermi golden rule and find a transition rate γ_g from g to e given (within numerical factors) by the square Ω^2 of the coupling times the density of final states, which is here $1/\Gamma$(one state in an interval Γ). This gives $\gamma_g \simeq \Omega^2/\Gamma$, in agreement with the results found above for γ_g (see Eq. (7.6) with $\delta = 0$). If the coupling strength Ω is large compared to Γ, it is no longer correct to apply Fermi golden rule and we find a damped Rabi oscillation.

We refer the reader to Complement CIII of [Cohen-Tannoudji *et al.* (1992b)] for a study of the effect of the coupling of a discrete state with a continuum of finite width when the coupling is varied.

7.2.5 *Semiclassical interpretation in the weak coupling limit*

In the weak coupling limit, the response of the atom to the field is linear in the field amplitude and a semiclassical interpretation of the various effects discussed in this chapter can be given in terms of this induced electric dipole moment.

The atomic dipole driven by the incident monochromatic field has a component in phase with the field and a component in quadrature related to the field by a dynamic polarizability $\alpha(\omega_L)$.

The component of the dipole moment in quadrature with the field absorbs energy from the field. It varies with the detuning δ as an absorption curve. It is this component which is responsible for the absorption rate associated with the light broadening of g. The component of the dipole moment in phase with the field gives rise to a polarization energy. It varies with the detuning δ as a dispersion curve. It is this component which is responsible for the light shift.[2] This effect is analogous to the Stark effect describing the interaction of a static electric field with the static dipole that it induces. The light shift δ_g is often called for that reason "dynamical Stark shift" or "ac-Stark shift".

Consider now the field radiated in the forward direction by the induced dipole moment. It is well known that there is a phase shift equal to $\pi/2$ between the radiating dipole moment and the field radiated at long distance in the forward direction.[3] The field radiated by the component of the dipole moment *in quadrature* with the incident field will thus have in the forward direction a phase shift equal to π with the incident field. Its interference with the incident field will thus result in a reduction of the amplitude of the total transmitted field. This is related to the imaginary part of the index of refraction. The field radiated by the component of the dipole moment *in phase* with the incident field will have in the forward direction a phase shift equal to $\pi/2$ with the incident field. Its interference with the incident field gives rise in the complex plane to a small rotation of the total transmitted field which can be interpreted as a phase shift associated with the real part of the index of refraction.

7.2.6 *Generalization to a non-resonant excitation*

A perturbative expression of the light shift δ_g of g (to order 2 in the field amplitude $\mathcal{E}_L$) can be given when the light frequency ω_L is far from the frequencies ω_A of all transitions $g \rightarrow e$ starting from g, so that all dissipative effects can be ignored. This expression generalizes the one obtained in this chapter for a quasi-resonant excitation (ω_L close to the frequency ω_A of one particular transition and far from all others).

All allowed transitions (for which $\langle e|\hat{D}|g\rangle = D_{eg} \neq 0$) must then be taken into account, and not only a single one. One must also include the effect of processes where the atom goes from g to a higher state e, not only by absorbing a photon ω_L, but also by emitting in a stimulated way a photon ω_L. The energy defect is equal to $E_g-(E_e-\hbar\omega_L) = \hbar(\omega_L-\omega_A)$ in the first case and to $E_g-(E_e+\hbar\omega_L) = -\hbar(\omega_L+\omega_A)$ in the second one. When ω_L is close to ω_A, one can neglect the second process which

[2]The connection between the light shift and the polarization energy of the atomic dipole moment driven by the light field has been first pointed out in [Pancharatnam (1966)].

[3]See also Sec. 15.6 of Chap. 15 on forward scattering in quantum scattering theory.

gives a perturbative contribution to δ_g containing a very large energy denominator. Neglecting this process corresponds to the *rotating wave approximation*.[4] In the general case where $(\omega_L - \omega_A)$ is not negligible compared to ω_L, both terms are of the same order and must be kept.

A second order treatment of the atom-field coupling $-\hat{D}\mathcal{E}_L \cos\omega_L t$ leads to the following expression for the light shift δ_g:

$$\delta_g = \frac{1}{4\hbar^2} \sum_e \left[\frac{|D_{eg}|^2 \mathcal{E}_L^2}{\omega_L - \omega_A} - \frac{|D_{eg}|^2 \mathcal{E}_L^2}{\omega_L + \omega_A} \right]. \tag{7.18}$$

When $\omega_L \to 0$, this expression coincides with the expression giving the dc Stark shift of g. When $(\omega_L - \omega_A)$ is very small compared to ω_L, while being large compared to Γ_e, one gets:

$$\delta_g \simeq \frac{\Omega_{eg}^2}{4(\omega_L - \omega_A)}, \tag{7.19}$$

where $\Omega_{eg} = D_{eg}\mathcal{E}_L/\hbar$ is the Rabi frequency of the transition $e \to g$, a result which coincides with the expression (7.9) derived in this chapter. When the laser frequency is high $(\omega_L \gg \omega_A)$, it is possible to expand the expression (7.18) of δ_g in a power series of ω_A/ω_L. Calculations of this type will be presented in the next chapter.

Remark

It is interesting to relate the expression (7.18) of the light shift of g to the dynamic polarizability $\alpha(\omega_L)$ of this state which describes the linear response of the mean dipole moment of the atom to the laser excitation:

$$\left\langle \hat{D}(t) \right\rangle = \alpha(\omega_L)\, \mathcal{E}_L \cos\omega_L t. \tag{7.20}$$

A standard linear response theory applied to the interaction Hamiltonian $-\hat{D}\mathcal{E}_L \cos\omega_L t$ leads to the following expression of $\alpha(\omega_L)$:

$$\alpha(\omega_L) = -\frac{1}{\hbar} \sum_e \left[\frac{|D_{eg}|^2}{\omega_L - \omega_A} - \frac{|D_{eg}|^2}{\omega_L + \omega_A} \right]. \tag{7.21}$$

The light shift δ_g can be re-expressed as half[5] the time average of the interaction energy between the mean induced dipole moment and the laser electric field:

$$\hbar\delta_g = -\frac{1}{2}\overline{\left\langle \hat{D}(t) \right\rangle \mathcal{E}_L \cos\omega_L t}. \tag{7.22}$$

Inserting Eq. (7.20) into (7.22) and using the expression (7.21) of $\alpha(\omega_L)$ leads to the expression (7.18) of δ_g, which can be also written:

$$\hbar\delta_g = -\frac{1}{4}\alpha(\omega_L)\, \mathcal{E}_L^2. \tag{7.23}$$

The light shift can therefore be interpreted as the polarization energy of the atom in the non-resonant laser field.

[4]See Sec. 2.4 of Chap. 2.

[5]The factor 1/2 in Eq. (7.22) is the analogue of the factor 1/2 appearing in the calculation of the polarization energy of a dielectric.

7.2.7 *Case of a degenerate ground state*

The previous subsections deal with non-degenerate ground states, more precisely with atoms having a ground state with zero angular momentum excited by a light beam having a polarization such that this ground state is coupled only to one particular Zeeman sublevel of the excited state e. For example, a π-polarization couples only $J_g = 0, M_g = 0$ to $J_e, M_e = 0$.[6]

These calculations can be easily generalized to atoms having a degenerate ground state with $J_g \neq 0$. If $\vec{D}$ is the atomic dipole moment and $\vec{\varepsilon}_L$ the polarization of the laser field, one finds that the effective Hamiltonian describing the perturbation of the ground state manifold by the light beam has matrix elements given by [Cohen-Tannoudji (1962)] :

$$\langle M_g | H_{\text{eff}} | M'_g \rangle \propto \sum_{M_e} \langle M_g | \vec{\varepsilon}_L{}^* \cdot \vec{D} | M_e \rangle \langle M_e | \vec{\varepsilon}_L \cdot \vec{D} | M'_g \rangle. \tag{7.24}$$

The matrix given in Eq. (7.24) is Hermitian so that its eigenvectors $|g_\alpha\rangle$ are orthogonal and its eigenvalues λ_α are real. The states $|g_\alpha\rangle$ have well-defined light shifts and absorption rates proportional to $\lambda_\alpha \delta_g$ and $\lambda_\alpha \gamma_g$, where δ_g and γ_g are the light shift and absorption rate for a two-level atom.

In the matrix (7.24), a vector operator $\vec{\varepsilon}_L \cdot \vec{D}$ appears twice. This matrix can thus be expanded into irreducible tensor operators $T_q^{(k)}$ with $k = 0, 1, 2$ and $q = -k, -k+1, ...k-1, k$. The term $k = 0$ describes a global displacement of the ground state manifold. The three terms $k = 1$ describe a perturbation having the same structure as the one associated with a static magnetic field. The five terms $k = 2$ describe a perturbation similar to the one associated with a static electric field. For a discussion of the equivalence of light shifts with fictitious static magnetic and electric fields, see for example [Cohen-Tannoudji and Dupont-Roc (1972)] which contains also a description of experiments demonstrating this equivalence.

Light shifts in the presence of fine and hyperfine structures

Suppose that the ground state has a zero orbital angular momentum ($L = 0$), so that its degeneracy comes only from the existence of a non-zero electronic spin S or nuclear spin I, or both. The ground and excited states can then have fine and hyperfine structures. Suppose that the light frequency is close to the frequency of one particular component $F_g \leftrightarrow F_e$ of the optical line, where F is the hyperfine quantum number, and sufficiently far from all other components for allowing one to ignore these other components. One can then consider only the transition $F_g \leftrightarrow F_e$ for calculating the light shifts and one finds, in general, that there are in F_g different internal states having different light shifts. The matrix (7.24) is not just proportional to the unit matrix. This is the case for the experiments described in Sec. 7.4 below.

Suppose now that the detuning δ of the exciting light is much larger than all fine and hyperfine structures of the optical line, so that all these hyperfine components contribute to the light shift. One can then show that all sublevels in the ground state undergo the same light shift. The ground state manifold is globally

[6] See Sec. 2.3.2 of Chap. 2.

displaced. A first mathematical argument for understanding this result is to note that the fine and hyperfine structures should in principle appear in the energy denominators of the second order expression of the light shifts. But, if these structures are small compared to the energy detuning $\hbar\delta$, one can neglect them and replace all energy denominators by $\hbar\delta$. The effective Hamiltonian describing the light perturbation is then proportional to $(\vec{\varepsilon}_L^* \cdot \vec{D})(\vec{\varepsilon}_L \cdot \vec{D})/\hbar\delta$, which is a pure electronic orbital operator, and which acts as the unit matrix for electronic and nuclear spin variables. A second more physical argument is to note that the second order expression of light shifts can be interpreted as describing the effect of virtual absorptions and re-emissions of photons by the atom, these virtual transitions lasting a time on the order of $\hbar/\hbar\delta = 1/\delta$. If the detuning δ is very large compared to the fine and hyperfine structures, this time is too short for allowing any coupling involving the spins. These spins thus remain "spectator" during the virtual transition and cannot influence the magnitude of the light shift which remains independent of their orientation.

7.3 Perturbation of the field. Dispersion and absorption

7.3.1 *Atom in a cavity*

To study the perturbation of the field, we consider an atom placed in a real cavity. This cavity is supposed to be perfect, which means that the typical time scale for losses is long compared to all other relevant times of the experiment. The corresponding experiments can now be realized and belong to the research field commonly called "Cavity Quantum Electrodynamics".[7]

The field in the cavity is supposed to be initially in a coherent state. One may wonder to which extent the amplitude and the phase of the electromagnetic field inside the cavity are perturbed by the atom-field interaction if an atom, whose frequency ω_A is close to the frequency ω of the cavity field, is introduced in the cavity.

To answer quantitatively to this question, it is necessary to introduce the explicit N-dependence of the Rabi frequency appearing in the expression of the effective Hamiltonian H_{eff} governing the reduced evolution in the manifold $\mathcal{E}(N+1)$. In the perturbative limit, which is realized when N is sufficiently small, the eigenstates of H_{eff} are very close to the uncoupled states $|g, N+1\rangle$ and $|e, N\rangle$. The state $|g, N+1\rangle$ undergoes a shift δ_g and a broadening γ_g, that depend on N. They will be denoted in the following δ_N and γ_N:

$$\delta_N = (N+1)\Omega_0^2 \frac{\delta}{4\delta^2 + \Gamma^2}, \qquad \text{and} \qquad \gamma_N = (N+1)\Omega_0^2 \frac{\Gamma}{4\delta^2 + \Gamma^2}, \tag{7.25}$$

where Ω_0 is the vacuum Rabi frequency defined by Eq. (5.23) of Sec. 5.4.3 of Chap. 5. We remind that the state $|e, N\rangle$ undergoes a shift with an opposite sign $-\delta_N$.

[7]See [Haroche and Raimond (2006)] for a review.

7.3.2 *Frequency shift of the field due to the atom*

The separation between the states $|g, N+1\rangle$ (or $|e, N\rangle$) displaced by light can be viewed as a renormalization of the field's frequency. Indeed, the splitting between dressed states of two successive manifolds is no more given by $\hbar\omega$ but by $\hbar(\omega \pm \delta_0)$ depending on the internal state of the atom: $E_{g,N+1} - E_{g,N} = \hbar(\omega + \delta_0)$ and $E_{e,N} - E_{e,N-1} = \hbar(\omega - \delta_0)$. The field-atom interaction thus displaces the frequency of the intracavity field by opposite amounts depending on the internal state of the atom (see Fig. 7.3). If this interaction lasts for a time T, the oscillation of the field accumulates a phase shift ϕ with respect to the free oscillation in the absence of interaction: $\phi = +\delta_0 T$ (resp. $\phi = -\delta_0 T$) if the atom is in the ground (resp. excited) state.

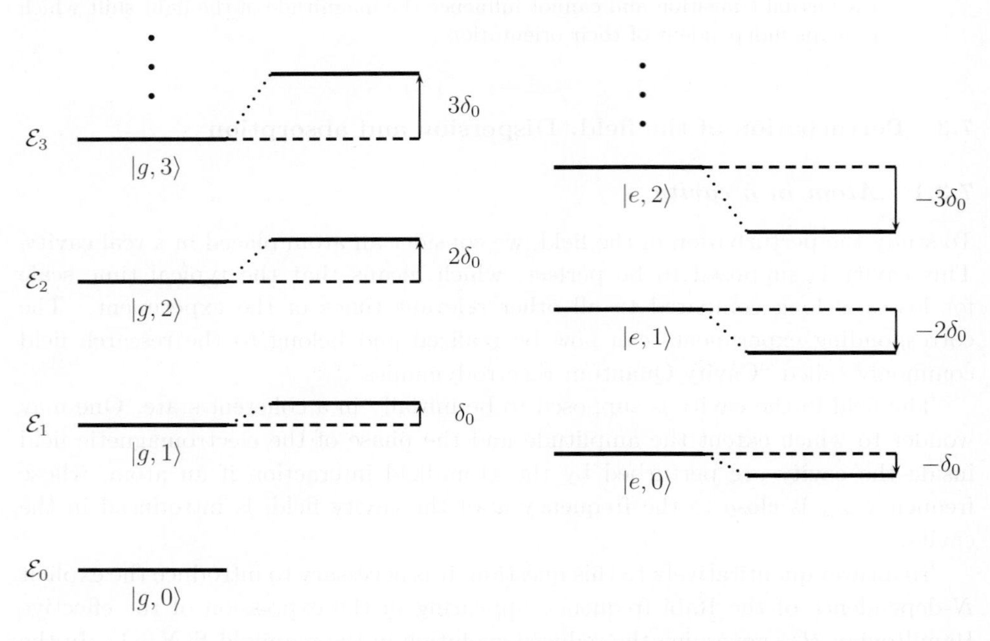

Fig. 7.3 Light shifts of the states $|g, N+1\rangle$ and $|e, N\rangle$ for low values of N. Because of the N-dependence of the light shifts, the splitting between the dressed states corresponding to g (resp. e) is changed from ω to $\omega + \delta_0$ (resp. from ω to $\omega - \delta_0$). The interaction of the intra-cavity field with the atom displaces the frequency of this field by opposite amounts depending on the internal state of the atom.

It is worthwhile to make an analogy with another phenomenon involving atom-light interaction. Indeed, a light beam passing through an atomic medium of length L undergoes a phase shift proportional to L and to the real part of the index of refraction. Close to an atomic resonance, the phase shifts have an opposite sign depending on the side of the resonance the light frequency is. In other words, the real part of the index of refraction varies like a dispersion curve with the frequency of the field. This is known as the "anomalous dispersion". The effect studied for an intracavity field is of the same nature despite the fact that it involves a single atom. One readily shows in this latter context that the phase shift is accumulated in time, and not in space by contrast with the previous example, and varies as a dispersion curve with the field frequency:

$$\phi = \frac{\omega - \omega_A}{4(\omega - \omega_A)^2 + \Gamma^2}\Omega_0^2 T. \tag{7.26}$$

The wavelength of a single mode field is determined by the geometry of the cavity. The change of the resonance frequency in the presence of an atom inside the cavity can thus be interpreted as an index of refraction change depending on the internal state of the atom.

7.3.3 *Damping of the field*

We now address the effect of the width γ_N of the dressed state. The initial state $|\psi(0)\rangle$ corresponding to an atom in the ground state in the presence of a field in a coherent state $|\alpha\rangle$ can be expanded on the uncoupled states basis:

$$|\psi(0)\rangle = |g\rangle \otimes |\alpha\rangle = \sum_{N=0}^{\infty} \frac{\alpha^N}{\sqrt{N!}} e^{-|\alpha|^2/2} |g, N\rangle. \tag{7.27}$$

Within the perturbative limit, each state $|g, N\rangle$ evolves with an eigenenergy $\tilde{E}_N$:

$$\tilde{E}_N = N\hbar\omega + \hbar\delta_{N-1} - i\hbar\gamma_{N-1}/2 = N\hbar(\omega + \delta_0 - i\gamma_0/2). \tag{7.28}$$

After a time t, the initial state becomes:

$$\begin{aligned} |\psi(t)\rangle &= \sum_{N=0}^{\infty} \frac{\alpha^N}{\sqrt{N!}} e^{-|\alpha|^2/2} \exp[-iN(\omega + \delta_0 - i\gamma_0/2)t] |g, N\rangle \\ &\propto |g\rangle \otimes |\alpha \exp[-i(\omega + \delta_0 - i\gamma_0/2)t]\rangle. \end{aligned} \tag{7.29}$$

The field is still described by a coherent state, evolving at the frequency $\omega + \delta_0$, and with an amplitude damped at a rate $\gamma_0/2$, where γ_0 is given by:

$$\gamma_0 = \Omega_0^2 \frac{\Gamma}{4(\omega - \omega_A)^2 + \Gamma^2}. \tag{7.30}$$

This damping rate varies with the detuning as an absorption curve. The analogy presented in the previous subsection is still valid. Indeed, the amplitude of a quasi-resonant light after crossing an atomic medium is damped by an amount which varies as an absorption curve when the frequency of this incident beam is varied around the atomic resonance frequency. This is reflected by the absorption curve shape of the imaginary part of the index of refraction of the medium.

7.4 Experimental observation of light shifts

7.4.1 *Principle of the experiment*

The first experimental studies of light shifts were performed in the early 1960's, at a time when laser sources were not yet available in the laboratories. These experiments used discharge lamps whose intensity was not very high, so that the corresponding light shifts were not larger than a few Hz. It was thus hopeless to observe these light shifts directly by a displacement of optical lines. The detection scheme was based on a displacement of the magnetic resonance lines in the ground state of an atom [Cohen-Tannoudji (1961a)].

Consider, for example, a transition connecting a $J_g = 1/2$ atomic ground state having two Zeeman sublevels $m_g = \pm 1/2$ to an excited state $J_e = 1/2$ having also two Zeeman sublevels $m_e = \pm 1/2$ (see left part of Fig. 7.4). Because of the polarization selection rules, a σ_+-polarized non-resonant light displaces only the ground state sublevel $m_g = -1/2$, whereas a σ_--polarized non-resonant light displaces only the ground state sublevel $m_g = +1/2$ (see right part of Fig. 7.4). The Zeeman splitting between the ground state sublevels produced by a static magnetic field is increased in the first case, decreased in the second one. The detuning δ of the non-resonant shifting light is supposed to be positive (so that the light shifts of $m_g = \pm 1/2$ are positive) and also much larger than the Zeeman splitting so that the two light shifts have the same magnitude. It thus clearly appears that the magnetic resonance curve in g should be light-shifted by opposite amounts depending on whether the shifting light is σ_+ or σ_- polarized. Since magnetic resonance curves in the ground state are very narrow because of the long lifetime of g, one can hope to be able to detect very small light shifts of these magnetic resonance curves.

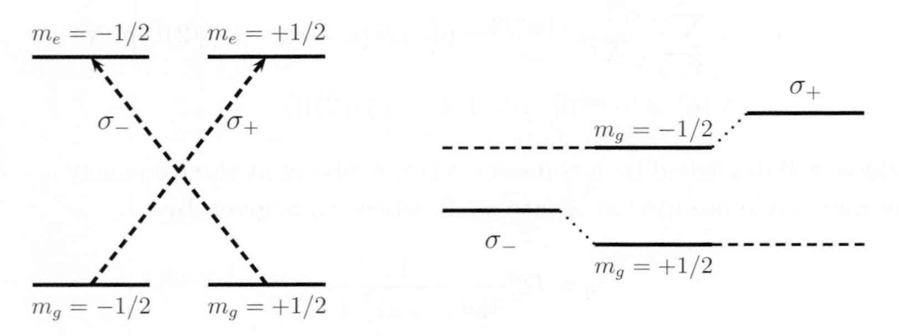

Fig. 7.4 Left: transition $J_g = 1/2 \to J_e = 1/2$. Right: a σ_+ (resp. σ_-)-polarized non-resonant light beam shifts selectively the ground state sublevel $m_g = -1/2$ (resp. $m_g = +1/2$). The Zeeman splitting between the two Zeeman sublevels in a static magnetic field is changed by opposite amounts depending on whether the shifting light is σ_+ or σ_- polarized.

7.4.2 *Examples of results*

This idea has been implemented in experiments performed on mercury-199 atoms [Cohen-Tannoudji (1961a)]. A first resonant σ_+-polarized beam (see upper part of Fig. 7.5) excites the transition $(6^1\mathrm{S}_0, F = 1/2) \to (6^3\mathrm{P}_1, F = 1/2)$ of the atoms and optically pumps them, as already explained in Sec. 3.3 of Chap. 3. The lamp used for this beam is actually a lamp filled with the isotope mercury-204 whose optical frequency coincides with the frequency of the transition $(6^1\mathrm{S}_0, F = 1/2) \to (6^3\mathrm{P}_1, F = 1/2)$ of mercury-199. Very narrow magnetic resonance curves can then be observed with these optically pumped atoms. A second perturbing light beam that originates from a mercury source filled with another isotope (mercury-201) in order to have a non-resonant excitation ($\delta \neq 0$) is then added. One observes that, in the presence of this second beam, the magnetic resonance curve is light shifted and that the light shifts have opposite signs for a σ_+ and a σ_- polarization of this second beam (lower part of Fig. 7.5).

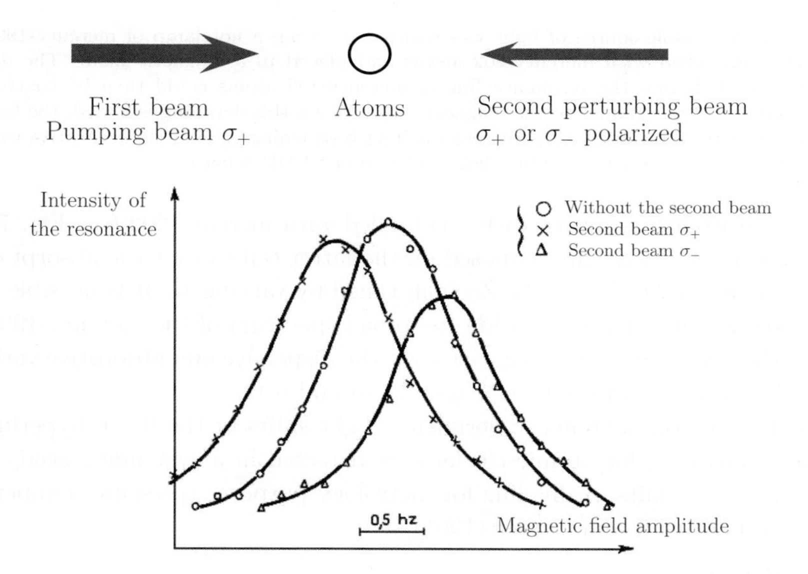

Fig. 7.5 Upper part: mercury atoms (mercury-199) are illuminated by two beams, a resonant σ_+-polarized beam which optically pumps the atoms and which allows an optical detection of the magnetic resonance in the ground state of these atoms, a non-resonant beam which produces light shifts. Lower part: magnetic resonance curves observed in the absence of the non-resonant beam and in the presence of this beam with a σ_+ or σ_- polarization. The light shifts of the magnetic resonance curve have opposite signs for a σ_+ and a σ_- polarization of this non-resonant beam. Adapted from [Cohen-Tannoudji (1961a)]. Copyright: Elsevier Masson SAS.

To investigate the variations with the detuning δ of the light shift δ_g and of the light broadening γ_g, a tunable source would be needed. It was generated in the following manner [Cohen-Tannoudji (1962)]. A hole was burnt in the very broad spectral width of a very hot lamp filled with mercury-199 atoms by filtering the

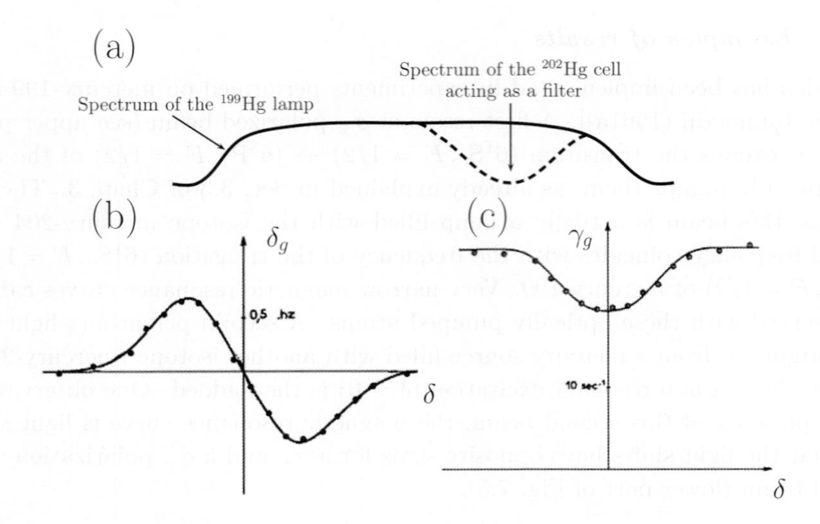

Fig. 7.6 (a) A tunable source of light was realized by using a hot lamp of mercury-199 atoms filtered by a cell filled with mercury-202 atoms and placed in a magnetic field. The detuning δ between this hole and the resonance line of mercury-199 atoms could then be controlled by Zeeman tuning the hole thanks to the magnetic field. When this detuning is varied, the light shift δ_g exhibits dispersive variations (b) whereas the light broadening γ_g exhibits absorptive variations (c). Adapted from [Cohen-Tannoudji (1962)]. Copyright: EDP Sciences.

light coming from this source with a cell filled with mercury-202 (see Fig. 7.6(a)). A magnetic field B was superimposed on the latter cell so that the absorption line of this mercury-202 cell could be Zeeman tuned by varying B. It is possible in this way to position at will a hole inside the broad spectrum of the mercury-199 lamp. One can then observe, with a negative sign, the dispersive and absorptive variations of δ_g and γ_g when δ is varied (see Figs. 7.6(b) and (c)).

Shortly after the mercury experiments, light shifts of the $0-0$ hyperfine line of alkali atoms used for atomic clocks were detected in a systematic study of the various causes of shifts of this line for metrology purposes (pressure, temperature, light intensity)[Arditi and Carver (1961)].

Remark

The calculations presented in this chapter suppose that the perturbing light is monochromatic. The light used in the experiments described in this section is not monochromatic. It is a thermal incoherent radiation with a central frequency ω_0 and a spectral width $\Delta\omega$. In a second order perturbation treatment, one can add independently the light broadenings and the light shifts produced by the various frequency components. One can show that γ_g and δ_g are then proportional to:[8]

$$\gamma_g \propto \int \mathrm{d}\omega \, I(\omega) \frac{\Gamma}{\Gamma^2 + 4\delta^2}, \quad \text{and} \quad \delta_g \propto \int \mathrm{d}\omega \, I(\omega) \frac{\delta}{\Gamma^2 + 4\delta^2}, \tag{7.31}$$

where $I(\omega)$ is the spectral density of the exciting radiation and $\delta = \omega - \omega_A$. If

[8]See, for example, [Cohen-Tannoudji (1962)].

$\Delta\omega \gg \Gamma$, the curves giving the variations of γ_g and δ_g with $\omega_0 - \omega_A$ are then expected to have a width on the order of the spectral width $\Delta\omega$ of the exciting radiation rather than Γ as it is the case for a monochromatic excitation. This is what is observed in Fig. 7.6.

7.5 Using light shifts for manipulating atoms

Light shifts are a perturbation for high resolution spectroscopy (atomic clocks, measurement of the Rydberg constant, ...) since they change the atomic frequencies that one tries to measure. An extrapolation to zero light intensity of the position of the resonances is then required in order to extract the unperturbed frequency.

However, with a high intensity delivered by laser sources, light shifts turned out to be useful for creating potential wells, potential barriers for neutral atoms, spatially periodic arrays of potential wells whose parameters can be fully controlled. Indeed, they are proportional to the light intensity, and can therefore be space dependent if the laser intensity is spatially inhomogeneous. The depth of these wells or the height of these barriers are small whilst large enough for trapping and reflecting ultracold atoms obtained by laser cooling techniques. We review in this section a few applications of these optical potentials using light shifts. A more detailed presentation is provided in Chap. 14.

7.5.1 *Laser traps*

The simplest laser trap, usually termed dipole trap, is made of a red detuned ($\omega_L < \omega_A$) focused laser as shown in Fig. 7.7. The light shift δE_g of the ground state g is then negative so that the laser creates a potential well for the atom in the ground state with a depth which is proportional to the peak intensity of the beam. In a semiclassical picture, the dipole trap results from the interaction between the dipole induced by the laser and the laser electric field. In such a dipole well, atoms are trapped if they have a sufficiently low kinetic energy, or in other words if they are cold enough [Chu *et al.* (1986)].

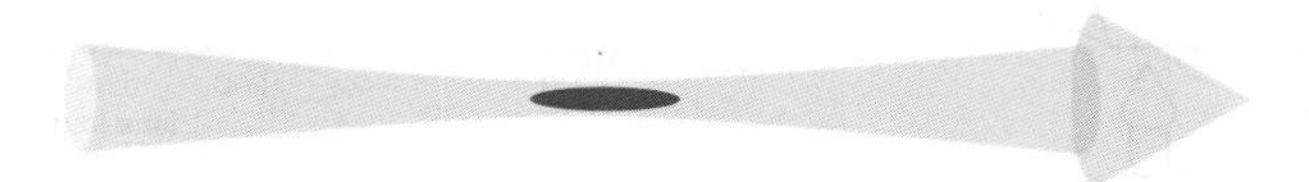

Fig. 7.7 The simplest dipole trap is made of a red detuned ($\omega_L < \omega_A$) focused laser beam. It realizes an attractive potential well in which neutral atoms can be trapped if they are slow enough.

If the laser frequency is far enough from the atomic resonances, far compared to the hyperfine and fine structure, one can show that all ground state levels are displaced by the same amount.[9] In this limit, the trap characteristics are no longer

[9] See remark at the end of Sec. 7.2.7.

dependent on the laser polarization.

There are many other types of traps using radiation pressure forces of polarized waves and magnetic field gradients. They will be described in Chap. 14.

7.5.2 *Atomic mirrors*

For a positive detuning of the laser wave with respect to the atomic frequency, the dipole force repels the atoms from the high intensity region. It is thus possible to create a potential barrier on which the atoms can be elastically reflected. Following a suggestion by Richard Cook and Richard Hill [Cook and Hill (1982)], several groups have realized a mirror for atoms with an evanescent wave formed by internal reflection of the laser beam at the glass surface (Fig. 7.8(a)). The intensity of the evanescent wave decreases exponentially with the distance to the surface on a typical distance on the order of $\lambda/2\pi$ where λ is the laser wavelength.

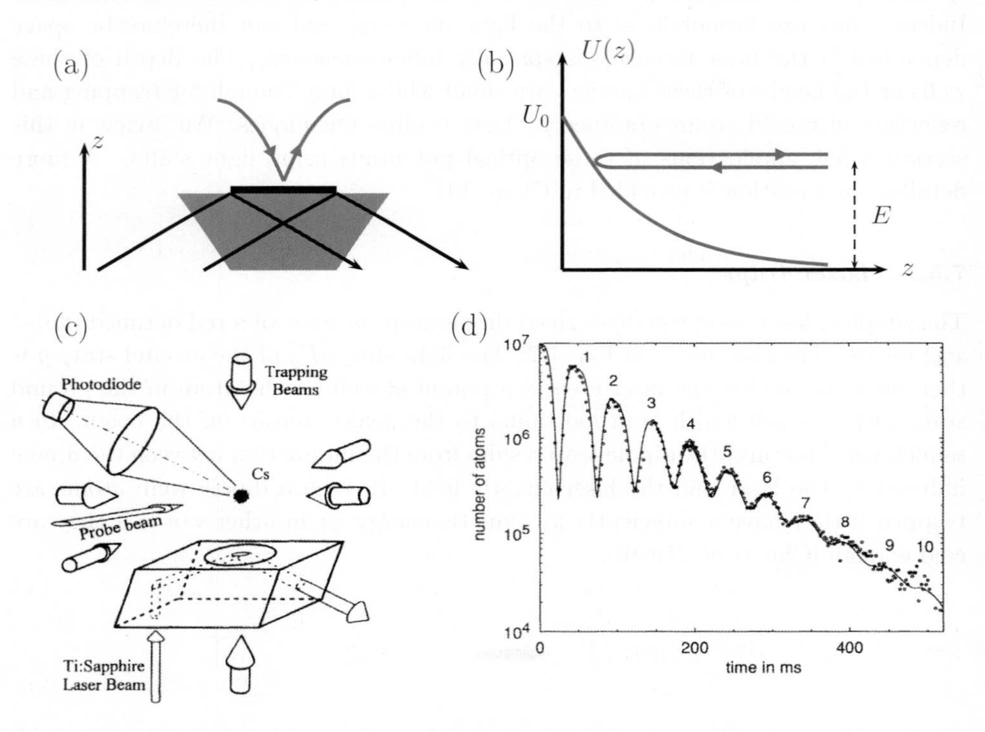

Fig. 7.8 (a) A repulsive potential barrier is realized with a blue-detuned evanescent wave at a glass surface. (b) Atoms with an incident kinetic energy E lower than the height U_0 of the potential barrier are elastically reflected. (c) A gravitational cavity is realized using a concave mirror in order to stabilize the transverse degrees of freedom. (d) Up to ten successive bounces have been experimentally observed with such a curved mirror [Aminoff *et al.* (1993)]. Figure adapted from [Aminoff *et al.* (1993)]. Copyright: American Physical Society.

The incident atoms arrive on the vacuum side and feel the repulsive dipole force as they enter the evanescent wave (see Fig. 7.8(b)). If their incident kinetic energy is smaller than the height of the light induced potential barrier, atoms turn back before touching the glass. They are reflected. In practice, with a laser intensity of 1 Watt focused on a surface of the order of 1 mm^2, an atom can be reflected if the component of its velocity normal to the mirror is lower than a few meters per second. This scale of velocities corresponds to a temperature on the order of few mK. Such atomic mirrors are therefore well suited for manipulating laser cooled atoms[10] while the prism is at room temperature. Actually, the repulsive evanescent barrier is not coupled to this temperature reservoir, and behaves like a perfect potential barrier.

The atoms can undergo multiple bounces as soon as their transverse degrees of freedom are under control. This has been performed by using a concave evanescent mirror such that the classical trajectories close to the vertical axis are stable[11] (see Fig. 7.8(d)). More than ten successive bounces have been observed thanks to this trick [Aminoff *et al.* (1993)]. One can view this system as a cavity where one wall is provided by the evanescent mirror and the second wall is made by the gravity. Other examples of gravito-optical trap based using blue detuned light have been demonstrated. In [Milner *et al.* (2001)], such a trap has been realized using an optical wedge made of two intersecting blue-detuned light sheets (see Fig. 7.9(a)).

Gravitational cavities made with evanescent wave at the surface of a prism have been used to investigate atom optics in the time domain [Arndt *et al.* (1996)]. Indeed the intensity of the evanescent wave can be readily changed in time: an intensity modulation induces a phase modulation of atomic de Broglie waves, an abrupt time aperture is also achievable and yields diffraction in time domain. Those topics are addressed in more detail in Sec. 17.7 of Chap. 17.

7.5.3 *Blue detuned traps: a few examples*

A blue-detuned dipole trap consists in surrounding a spatial region with blue "walls" made of blue detuned sheets of light. As an example, an optical box is readily realized by intersecting three pairs of Gaussian blue detuned laser sheets as illustrated in Fig. 7.9(b). Hollow blue-detuned laser beams provide radial confinement and can be used to guide or in combination with plugging beams to trap atoms [Kuga *et al.* (1997); Bongs *et al.* (2001)]. Evanescent-wave guiding of atoms in hollow optical fibers has also been demonstrated [Renn *et al.* (1996)].

The shaping of blue detuned walls can also be achieved by scanning rapidly the position of a blue-detuned laser beam. This technique offers a very flexible tool to realize different geometries. For instance, the integrable and chaotic motion of ultracold rubidium atoms were investigated by confining atoms with different

[10] See Fig. 7.8(c) and Chap. 12.

[11] The quantum mechanical modes of such a cavity made of a mirror placed in the earth gravitational field have been calculated in [Wallis *et al.* (1992)]. The quantum modes of a trap made of a flat mirror plus the gravitational field have been directly observed with neutrons in 2002 [Nesvizhevsky *et al.* (2002)].

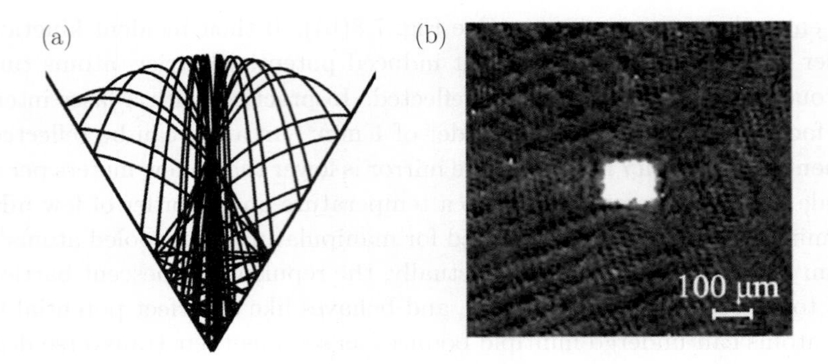

Fig. 7.9 (a) Optical wedge made of blue detuned light to store atoms and study their dynamics [Milner *et al.* (2001)]. (b) Repulsitve "box-like" potential made of three pairs of blue-detuned sheets of light. Ref. [Price *et al.* (2008)] reports the loading of 1.5×10^5 atoms in this optical box. Figures extracted from [Milner *et al.* (2001); Price *et al.* (2008)]. Copyright: American Physical Society.

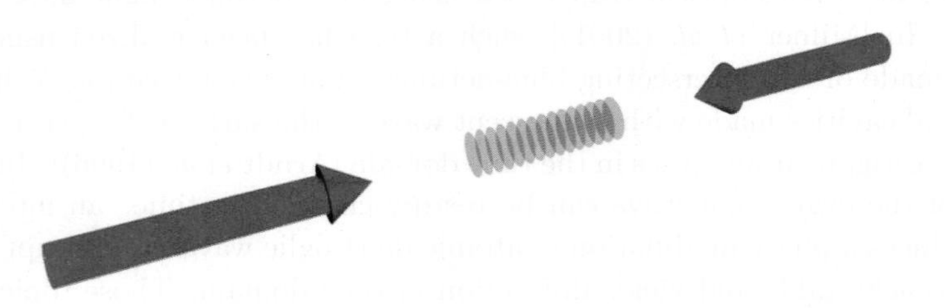

Fig. 7.10 Periodic array of potential wells for atoms created by the light shifts associated with a detuned laser standing wave.

billiard-shaped optical dipole potentials [Friedman *et al.* (2001)].

The great advantage of traps with blue-detuned light is that atoms are essentially stored in a place without light. This minimizes unwanted effects such as spontaneous emission, ... This is exemplified in [Davidson *et al.* (1995)] where a superposition of magnetic field insensitive states could be observed during a few seconds, while the coherence time of the same superposition of states held in a red-detuned dipole trap with comparable depth was 300 times shorter. Traps based on blue-detuned light are therefore well-suited for precision spectroscopy and fundamental tests.

7.5.4 *Optical lattices*

Periodic lattices confining ultracold atoms can be obtained by introducing a spatial periodic modulation of far-detuned light intensity. The simplest configuration of such a scheme involves two coherent off-resonance counter-propagating laser beams. As a result of the interference, the intensity is spatially modulated with a period equal to $\lambda/2$ where λ is the laser wavelength (see Fig. 7.10). Depending on the sign of the detuning Δ, the atoms accumulate at the nodes (blue detuning) or the antinodes (red detuning) of the standing wave.[12]

Another interesting application of optical lattices is the possibility to move the lattice by using two different frequencies ω_1 and ω_2 for the two counter-propagating laser waves forming a one-dimensional standing wave. For instance, two coherent and counter-propagating beams with the same linear polarization and a frequency mistmatch produce a total electric field of the form:

$$\begin{aligned} \vec{E}(z,t) &= \vec{E}_0(e^{ik_1z-i\omega_1t} + e^{-ik_2z+i\omega_2t}) + \text{c.c.} \\ &= 4\vec{E}_0 \cos\left(k(z-\bar{v}t)\right)\cos\left(\frac{\omega_1+\omega_2}{2}t\right), \end{aligned} \tag{7.32}$$

where $k = (k_1+k_2)/2$ and $\bar{v} = (\omega_1-\omega_2)/2k$. Therefore, the interference pattern moves at the velocity $\bar{v}$. This technique has been used in many different contexts: to launch packets of cold atoms through microwave cavities for atomic clocks [Riis *et al.* (1990); Clairon *et al.* (1991)], to realize a continuous beam of cold atoms [Weyers *et al.* (1997); Cren *et al.* (2002)], ... By increasing one of the two frequencies linearly in time, one gets a standing wave whose velocity is increasing in time. In the rest frame of this accelerating standing wave an atom feels a constant inertial force in addition to the spatially periodic force due to the lattice. The quantum motion of this atom submitted to these two forces is completely different from the motion of a classical particle. It exhibits oscillations, called Bloch oscillations, more easily observed in this case than for an electron moving in a crystal and accelerated by a constant electric field.[13]

By changing the intensity of the lasers forming the optical lattice, one can control the tunnelling rate of atoms between adjacent potential wells and explore the transition between a regime where atoms are confined in the potential wells and a regime where they are delocalized over the whole lattice. This possibility has been recently used for demonstrating the superfluid Mott insulator transition for ultracold bosons placed in an optical lattice [Greiner *et al.* (2002)] and will be described in more detail in Chap. 26.

7.5.5 *Internal state dependent optical lattices*

When the atom has several ground state Zeeman sublevels, interesting effects appear when the optical lattice in which these atoms are put is not the same for the different sublevels. This occurs for a not too large detuning $\delta = \omega_L - \omega_A$. Consider for example an atom with a transition connecting a $J_g = 1/2$ atomic ground state to an excited state $J_e = 1/2$ and suppose that this atom interacts with a one-dimensional set of coherent counterpropagating beams with equal intensities and linear polarizations $\vec{e}_1$ and $\vec{e}_2$ making an angle θ (see Fig. 7.11):

---

[12] See also Sec. 14.4.3 of Chap. 14.

[13] See [Ben Dahan *et al.* (1996)] and Sec. 17.6 of Chap. 17.

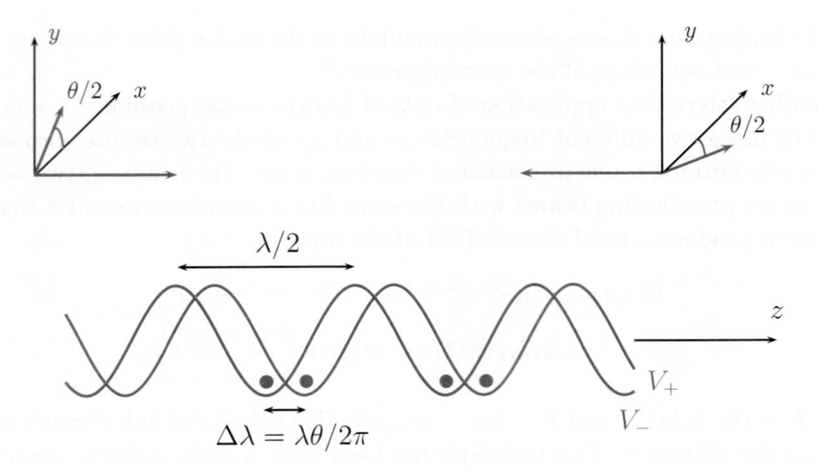

Fig. 7.11 Interfering, counter-propagating beams having linear polarization vectors making an angle θ provide a laser field which is the sum of two σ_+ and σ_- standing waves spatially shifted from one another by an amount : $\Delta z = \lambda\theta/2\pi$. The different Zeeman sublevels are shifted differently in light fields with different polarizations. Two optical lattices that depend on the internal state result from this configuration.

$$\vec{E}(z,t) = (\vec{e}_1 e^{ikz} + \vec{e}_2 e^{-ikz})e^{-i\omega t} + \text{c.c.}, \tag{7.33}$$

with

$$\vec{e}_1 = \cos(\theta/2)\vec{e}_x + \sin(\theta/2)\vec{e}_y, \quad \text{and} \quad \vec{e}_2 = \cos(\theta/2)\vec{e}_x - \sin(\theta/2)\vec{e}_y. \tag{7.34}$$

The expression (7.33) for the electric field is usefully recast in terms of the right and left circular polarizations:

$$\vec{E}(z,t) = \sqrt{2}[-\vec{e}_+ \cos(kz - \theta/2) + \vec{e}_- \cos(kz + \theta/2)]e^{-i\omega t} + \text{c.c.} \tag{7.35}$$

with

$$\vec{e}_+ = -\frac{\vec{e}_x + i\vec{e}_y}{\sqrt{2}}, \quad \text{and} \quad \vec{e}_- = \frac{\vec{e}_x - i\vec{e}_y}{\sqrt{2}}. \tag{7.36}$$

The interference of the two counterpropagating beams thus produces a superposition of two standing waves with polarization σ_+ and σ_- shifted from one another by an amount Δz depending on θ: $\Delta z = \lambda\theta/2\pi$. The intensities of each standing wave are proportional respectively to I_+ and I_-:

$$\begin{aligned} I_+(z,\theta) &= 2\cos^2(kz - \theta/2) = 1 - \cos(2kz - \theta), \\ I_-(z,\theta) &= 2\cos^2(kz + \theta/2) = 1 - \cos(2kz + \theta). \end{aligned} \tag{7.37}$$

Because of the polarization selection rules already described above in Sec. 7.4, atoms in $M_g = -1/2$ (resp. $M_g = +1/2$) interact only with the σ_+ (resp. σ_-) wave and feel the potential $I_+(z,\theta)$ (resp. $I_-(z,\theta)$). In other words, the laser

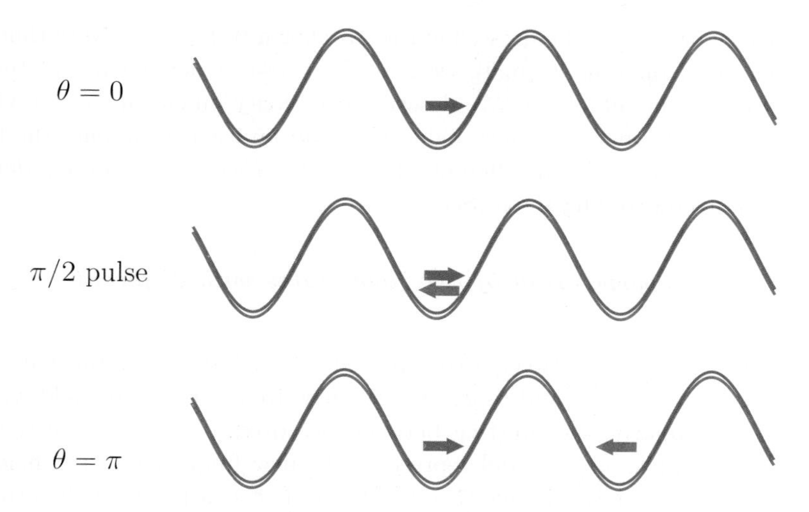

Fig. 7.12 Illustration of the coherent transport of neutral atoms using spin-dependent optical lattice potentials (see text).

configuration shown in Fig. 7.11 results in an internal state-dependent potential. The two potentials $V_+(z,\theta)$ and $V_-(z,\theta)$ coincide for $\theta = 0$ and $\theta = \pi$. They have opposite variations for[14] $\theta = \pi/2$ (see Fig. 7.11).

7.5.6 *Coherent transport*

By changing the polarization angle θ one can thereby control the separation between the two potentials. This simple idea opens new possibilities for quantum state engineering. Suppose that the angle θ is chosen initially equal to zero ($\theta = 0$), so that the two spin-dependent potentials $V_+(z)$ and $V_-(z)$ coincide (see Fig. 7.12). An atom in the ground state $|-\rangle$ is trapped in one well and submitted to a $\pi/2$ microwave pulse which puts it in a coherent superposition of both internal states: $(|-\rangle + |+\rangle)/\sqrt{2}$. The angle θ is then adjusted to the value $\theta = \pi$. The spatial wave packet of the atom is consequently split into two components moving in opposite directions. Indeed, the two potentials $V_+(z)$ and $V_-(z)$ are moved by $\lambda/2$ and the two states $|-\rangle$ and $|+\rangle$, which follow these potentials, become trapped in different potential wells. This coherent transport opens the way to interesting interference experiments [Mandel *et al.* (2003)].

7.6 Using light shifts for manipulating fields

In this section, as in Sec. 7.3, we suppose that the atom is placed in a cavity. This cavity is detuned from resonance so that the effect of the atom-field coupling is

[14]This latter configuration plays an important role for the "Sisyphus cooling" mechanism addressed in Chap. 13.

essentially a light shift of the dressed states, as shown in Fig. 7.3. Note that, since the number N of photons is small, we must keep the N dependence of the Rabi frequency in each manifold $\mathcal{E}_N$. We describe two cavity quantum electrodynamics (QED) experiments that show how light shifts can be used to prepare the field in a linear superposition of two different phase states (Sec. 7.6.1), or for detecting photons without destroying them (Sec. 7.6.2).

7.6.1 *Linear superposition of two field states with different phases*

The frequency splitting between two adjacent dressed states of the left part of Fig. 7.3 is equal to $\omega + \delta_0$, and represents the new frequency of the field when the atom is in g. Similarly, the splitting between two dressed states of the right part of Fig. 7.3 is equal to $\omega - \delta_0$, and represents the new frequency of the field when the atom is in e. As shown in Sec. 7.3, it follows that as a result of the atom-field interaction, the field acquires a phase shift that depends on the atomic state: $+\phi$ if the atom is in the ground state $|g\rangle$, and $-\phi$ if the atom is in the excited state $|e\rangle$.

This can be exploited to generate Schrödinger cat states of the field [Brune *et al.* (1996b)]. Suppose that the atom that enters the cavity is in a superposition of the states $|g\rangle$ and $|e\rangle$: $(|g\rangle + |e\rangle)/\sqrt{2}$. The state of the total system $\{\,\text{atom+field}\,\}$ at the end of the interaction, when the atom leaves the cavity, is given by $(|g,+\phi\rangle + |e,-\phi\rangle)/\sqrt{2}$, that is the field gets entangled with the atom. The two states $+\phi$ and $-\phi$ can be "mesoscopically" different. Applying an extra $\pi/2$ pulse to the atom transforms $|e\rangle$ and $|g\rangle$ into linear superpositions of $|e\rangle$ and $|g\rangle$:

$$\left(\frac{|g,+\phi\rangle + |e,-\phi\rangle}{\sqrt{2}}\right) \xrightarrow{\pi/2} \left(\frac{|g\rangle + |e\rangle}{\sqrt{2}} \otimes |+\phi\rangle + \frac{|g\rangle - |e\rangle}{\sqrt{2}} \otimes |-\phi\rangle\right). \tag{7.38}$$

The detection of the atom in $|e\rangle$ or $|g\rangle$ then projects the field into a linear superposition of two states with opposite phases(see Fig. 7.13):

$$\frac{1}{\sqrt{2}}(|+\phi\rangle \pm |-\phi\rangle). \tag{7.39}$$

One can then study the evolution of this superposition state under the effect of the dissipation that originates in this context from the cavity losses, and not from spontaneous emission from the excited state. This makes it possible to study the issue of decoherence by studying the damping rate of the coherence between the states $|+\phi\rangle$ and $|-\phi\rangle$ as a function of their "distance" [Haroche and Raimond (2006)].

7.6.2 *Non-destructive detection of photons*

Now consider the modification of the frequency splitting between two states with the same value of N, where one state belongs to the left ladder and the other belongs to the right ladder of Fig. 7.3. For example, the splitting between $|e,N\rangle$ and $|g,N\rangle$,

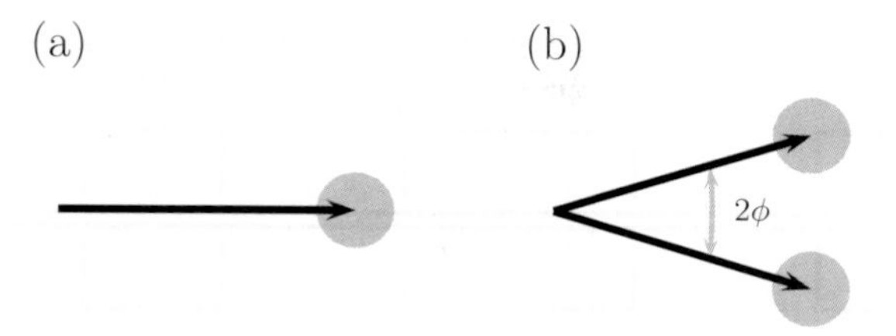

Fig. 7.13 (a) Pictorial representation in phase space of a coherent field. The length and the direction of the arrow accounts for the amplitude and phase of the field. The disk describes the quantum uncertainties of the field. (b) Components of the field for the entangled state of Eq. (7.39).

equal to ω in the absence of coupling, becomes equal to $\omega - \delta_0$ if $N = 0$, and $\omega - 3\delta_0$ if $N = 1$. There is therefore a N-dependent change of the atomic frequency. If the atom enters the cavity in a linear superposition of e and g, the atomic coherence between e and g is phase shifted by an amount $\chi(N)$, that depends on N, when it leaves the cavity. The phase shift can be used to determine N without changing it, since in the non-resonant cavity, there is no possible absorption of photons.

More precisely, suppose that the cavity is at a very low temperature and contains only 0 or 1 photon. Before entering the cavity, the atom, initially in g, is prepared in a linear superposition of e and g by a $\pi/2$ pulse in a first auxiliary cavity (left part of Fig. 7.14). The phase shift $\chi(N)$ induced by the atom field coupling when the atom crosses the cavity containing 0 or 1 photon is probed with a $\pi/2$ pulse in a second auxiliary cavity (right part of Fig. 7.14), and gives rise to Ramsey fringes. Suppose that the relative phase of the two fields inducing the two $\pi/2$ pulses is such that, if $N = 0$, the atom exits the second auxiliary cavity in g. Then, if the detuning and interaction time are adjusted in such a way that $\chi(1) - \chi(0) = \pi$, the atom exits the second auxiliary cavity in e if $N = 1$. Measuring the state of the atom when it exits the second auxiliary cavity thus allows a non-destructive detection of the number of photons in the cavity. If the cavity damping time is long enough, one can send hundreds of atoms through the cavity, to probe the presence of a photon in the cavity and follow its fate from its birth due to thermal fluctuations to its death due to a damping process. Such an experiment has been performed recently and is described in more detail in [Gleyzes *et al.* (2007)]. It has been also extended to value of N larger than 1 [Guerlin *et al.* (2007)].

7.7 Conclusion

In this chapter, we have shown that a quasi-resonant excitation of an atomic transition produces a displacement and a broadening of the energy levels of this transition. These light shifts and light broadenings are reactive and dissipative effects which are the equivalent, for the atom, of the dispersion and absorption of a light beam passing through an atomic vapor. In subsequent chapters, we will see that a similar

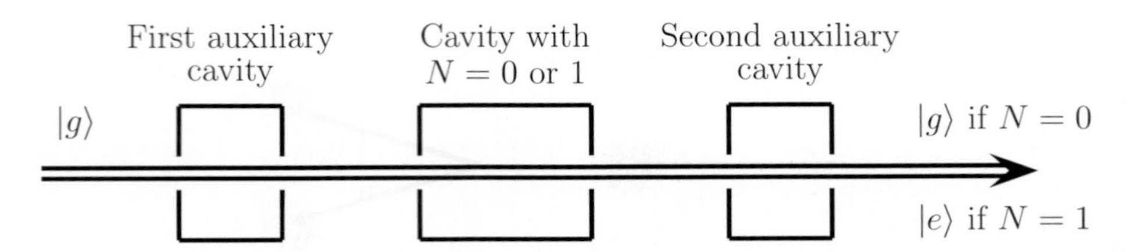

Fig. 7.14 Scheme of the experiment allowing the non-destructive detection of a photon in a non-resonant cavity.

distinction can be introduced concerning the evolution of the external degrees of freedom of an atom interacting with a light beam. Two types of radiative forces exerted by the light beam can be identified: a radiation pressure force and a dipole force, associated respectively with dissipative effects (resonant scattering of photons) and reactive effects (spatial gradients of light shifts).

When the atom has several ground state sublevels, these sublevels undergo in general different light shifts depending on the light polarization. It follows that the splitting between the Zeeman sublevels is changed by light in a polarization dependent way, resulting in a shift of the magnetic resonance curves in the atomic ground state g. This light shift can be easily detected, even if it is smaller than the natural width of the excited state e, because magnetic resonance curves in g are very narrow, as a consequence of the long lifetime of g. It is in this way that light shifts have been observed for the first time, using discharge light sources much less intense than laser sources.

When laser sources became available, much larger light shifts could be observed and great care had to be taken for evaluating the perturbation that they introduced in high precision measurements. It appeared however that this perturbation associated with light shifts can be considered from a different point of view, as a convenient way of realizing potential wells and potential barriers, of different types and dimensionalities, or for manipulating the state of a cavity field. Furthermore, dissipative effects and reactive effects, like optical pumping and light shifts, can conspire to give a very efficient cooling mechanism, such as Sisyphus cooling that will be described in Chap. 13. These various developments provide an interesting example of a situation which often happens in physics. A few decades after their discovery, physical effects can be revisited in a different perspective, opening the way to new unexpected applications.

Chapter 8

Perturbations due to a high frequency excitation

8.1 Introduction

The purpose of this chapter is to point out similarities and differences that exist between the perturbation of atomic systems due to a high frequency excitation, and the radiative corrections of quantum electrodynamics (QED) resulting from the interaction of a charged particle with the vacuum field and with its self field. Through this approach, we hope to provide more physical insight into radiative corrections. By high frequency fields we mean fields with a frequency ω large compared to the atomic frequencies ω_A. Expansions in power series of ω_A/ω can then be done, leading to explicit expressions for the corrections due to the coupling with the field. We suppose however that ω is sufficiently low for allowing a non-relativistic treatment of atom-field interactions.[1] For an atomic electron, these two conditions on ω can be written:

$$\omega_A \ll \omega \ll mc^2/\hbar, \tag{8.1}$$

where m is the electron mass.

A first well-known example of radiative correction is the *Lamb shift*. In 1947, precise measurements demonstrated the existence of a small energy shift between the $2s_{1/2}$ and $2p_{1/2}$ levels of hydrogen that was not predicted by Dirac equation [Lamb and Retherford (1947)]. Another example is provided by the electron spin anomaly $g-2$. The g-factor of the electron was found to be not equal to 2, as predicted by Dirac equation, but slightly larger [Kush and Foley (1948)].[2] The theoretical interpretation of these effects was the starting point of QED and all modern quantum field theories [Schwinger (1948); Itzykson and Zuber (1986)].[3] To the lowest order in the fine structure constant α, which characterizes the strength of electromagnetic interactions, these corrections can be interpreted as being due to virtual emissions and re-absorptions of photons by the electron.

[1] Lowest order relativistic corrections will be however kept in the calculations, described by terms containing $1/c^2$ factors, where c is the speed of light.

[2] For these experimental discoveries, Willis Lamb and Polykarp Kusch were awarded the Physics Nobel Prize in 1955.

[3] Richard Feynman, Julian Schwinger and Sin-Itiro Tomonaga were awarded the Physics Nobel Prize in 1965 for their pioneering work on quantum electrodynamics.

On the other hand, the development of RF and optical spectroscopies led to the discovery of perturbations due to a high frequency incident optical or RF field. Atomic levels can be shifted by light. We will also see in this chapter that the interaction with a high frequency RF field can change the g-factor of a neutral atom. To the lowest order in the atom-field coupling, these perturbations can be interpreted in terms of virtual absorptions and stimulated emissions of photons by the atom.

It is tempting to try to interpret the radiative corrections of QED as being due to a fluctuating quantum vacuum field having a zero average value and a power spectral density equal to $\hbar\omega/2$ per mode ω. Theodore Welton gave a simple interpretation of the Lamb shift based on this idea [Welton (1948)]. However, the same picture does not work for the spin anomaly $g-2$. The predicted sign of the anomaly based on this picture is wrong. Several authors have pointed out the difficulty of giving a simple interpretation of the spin anomaly [Feynman (1962); Weisskopf (1975)], and several attempts have been made for finding explanations based on relativistic effects [Koba (1949); Huang (1952); Lai *et al.* (1974)], without, however, giving convincing arguments.

In this chapter, we show that the interaction with vacuum fluctuations is not sufficient for interpreting radiative corrections and that one must also take into account the interaction of the charged particle with its self field, the so-called radiation reaction. The approach that we follow is to focus on the interaction of the atomic system with a given high frequency mode ω of the radiation field containing N photons. The interaction is treated quantum mechanically at the lowest order, where perturbations of the atomic system appear. We find two types of terms: those proportional to N called "stimulated" corrections, because they are proportional to the intensity of the incident field; and those independent of N called "spontaneous" corrections, because they remain present, as spontaneous emission, when the mode is empty ($N = 0$). A comparison between the two types of terms allows one to get a better understanding of spontaneous corrections.

We begin in Sec. 8.2 with a very simple problem: a spin 1/2 coupled to a high frequency RF field. The fact that the atomic system has only two states leads to very simple calculations that show that the g-factor of the spin is reduced (and can even be cancelled) by the interaction with the RF field. Experiments confirming these predictions are briefly described. The same conclusion holds when the mode is empty: the g-factor of the spin is always reduced and this effect is entirely due to vacuum fluctuations.

The system studied in Sec. 8.2 is a neutral spin 1/2 system. The addition of charge raises other questions: Should one take into account the vibration of the electron's charge in vacuum fluctuations? Could the coupled vibrations of the charge and spin lead to a positive correction for g? In Sec. 8.3, we present an effective Hamiltonian approach, valid to order 2 in the electron charge, which includes the first relativistic corrections containing $1/c^2$ factors. This approach shows how the slow motion of the charge and spin of a weakly bound electron are modified by the interaction with a high frequency mode ω. These modifications can be described in terms of a new mass, a new magnetic moment and electric and magnetic form factors. We first review a few stimulated corrections that can all be interpreted semiclassically in terms of the vibration of the charge and spin induced by the incident field. We then turn to spontaneous corrections. The important point is that the terms independent of N are not equal to the terms proportional to N, with N replaced by 1/2. This means that radiative corrections are not just corrections due to vacuum fluctuations. In addition, there are other effects that are interpreted as being due to radiation reaction (interaction of the electron with its self field).

This finally leads, in Sec. 8.4, to a simple interpretation of radiative corrections. We first review a few spontaneous corrections, distinguishing those that are due to vacuum fluctuations from those that are due to radiation reaction. In this way, we show that the Lamb shift is primarily due to vacuum fluctuations, thereby confirming the interpretation given by Welton.

For the spin anomaly, we start from the remark that g is experimentally defined by the ratio between the Larmor frequency ω_L of the electron spin in a static field, and the cyclotron frequency ω_C of the electron charge in the same field: $g/2 = \omega_L/\omega_C$. Both frequencies ω_L and ω_C are reduced, but ω_C is reduced more than ω_L, essentially by radiation reaction. This explains why g becomes larger than 2.

8.2 Spin 1/2 coupled to a high frequency RF field

In this section, we consider a spin 1/2 with a gyromagnetic ratio γ coupled to a static magnetic field $\vec{B}_0$ parallel to the z-axis and an RF field $\vec{B}_1 \cos\omega t$ with frequency ω parallel to the x-axis (see Fig. 8.1). We denote

$$\omega_L = -\gamma B_0 \tag{8.2}$$

the *bare* Larmor precession frequency around $\vec{B}_0$, and

$$\omega_1 = -\gamma B_1 \tag{8.3}$$

the Larmor precession frequency around $\vec{B}_1$. The frequency ω_1 is in fact the Rabi frequency characterizing the atom-RF field coupling.[4] The high frequency condition for the RF field reads:

$$\omega \gg \omega_L. \tag{8.4}$$

We want to study how the Larmor precession around $\vec{B}_0$ is modified by the interaction with the RF field. The resulting modified Larmor frequency is denoted $\bar{\omega}_L$ in the following. As this section is restricted to a *neutral* spin 1/2 particle, the radiative corrections to the g-factor are only due to the modification of the Larmor frequency $\omega_L \to \bar{\omega}_L$.

[4]It is denoted here ω_1, and not Ω, as in other chapters of this book, because the notation ω_1 is generally used in all references mentioned in this chapter.

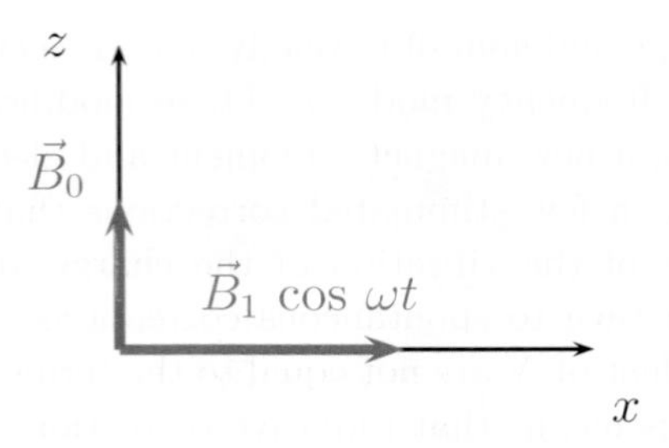

Fig. 8.1 Static and RF fields interacting with the spin 1/2.

8.2.1 *Hamiltonian*

For studying the modification of the Larmor frequency of the spin due to its coupling to the RF field, a semiclassical treatment would be sufficient. It is based on the Hamiltonian:

$$\hat{H} = \hbar\omega_L \hat{S}_z + \hbar\omega_1 \hat{S}_x \cos\omega t, \tag{8.5}$$

where the $\hat{S}_i$'s are the components of the spin ($i = x, y, z$).

In fact, as we will see it later on, a quantum treatment is more appropriate for obtaining physical insights into radiative corrections, so we will therefore use the Hamiltonian:

$$\hat{H} = \hbar\omega_L \hat{S}_z + \hbar\omega \hat{a}^\dagger \hat{a} + \lambda \hat{S}_x (\hat{a} + \hat{a}^\dagger), \tag{8.6}$$

where $\hat{a}^\dagger(\hat{a})$ creates (destroys) one RF photon, and λ is a coupling constant. The term $\hat{a}^\dagger + \hat{a}$ comes from the expression of the RF field operator in the magnetic dipole approximation.

8.2.2 *Perturbative treatment of the coupling*

We first give a perturbative treatment of the coupling. The central part of Fig. 8.2 represents the unperturbed states of the spin-RF photons system. The states gather in two-dimensional manifolds $\{|+, N-1\rangle, |-, N-1\rangle\}, \{|+, N\rangle, |-, N\rangle\}, \{|+, N+1\rangle, |-, N+1\rangle\}$ with a frequency splitting inside each manifold equal to ω_L, the frequency splitting between adjacent manifolds being equal to ω. The interaction Hamiltonian

$$\hat{V} = \lambda \hat{S}_x (\hat{a} + \hat{a}^\dagger) = \frac{\lambda}{2}(\hat{S}_+ + \hat{S}_-)(\hat{a} + \hat{a}^\dagger) \tag{8.7}$$

only has off-diagonal elements between the two spin states $|+\rangle$ and $|-\rangle$ and obeys the selection rule $\Delta N = \pm 1$ for the RF photon states. The level $|+, N\rangle$ is coupled to $|-, N+1\rangle$ and to $|-, N-1\rangle$ (dotted arrows in the left part of Fig. 8.2) with matrix elements equal to:

$$\langle -, N+1| \hat{V} |+, N\rangle = (\lambda\sqrt{N+1})/2, \qquad \langle -, N-1| \hat{V} |+, N\rangle = (\lambda\sqrt{N})/2. \tag{8.8}$$

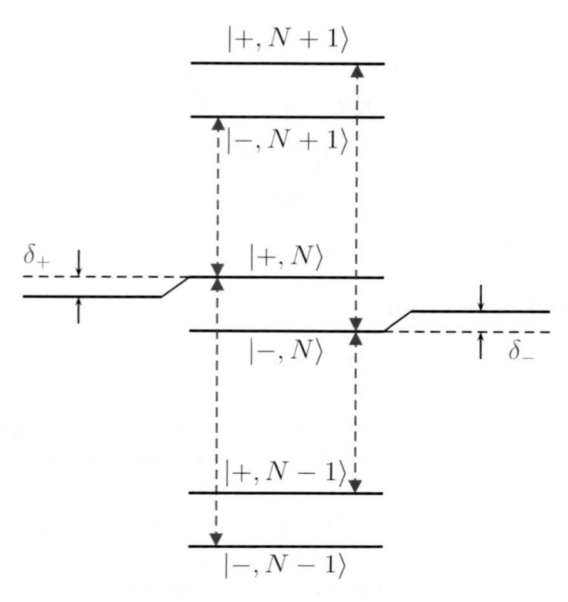

Fig. 8.2 Central part: unperturbed states of the spin-RF photons system. They are gathered in two-dimensional manifolds $\{|+, N-1\rangle, |-, N-1\rangle\}, \{|+, N\rangle, |-, N\rangle\}, \{|+, N+1\rangle, |-, N+1\rangle\}$ with a frequency splitting inside each manifold equal to ω_L. The arrow connects levels that are coupled by the interaction hamiltonian $\hat{V}$ that yields energy shifts of the levels (sides part of the figure).

Note that we keep the exact N-dependence of the matrix elements without replacing $\sqrt{N+1}$ and $\sqrt{N}$ by $\sqrt{\langle N \rangle}$. The motivation is to keep the exact N-dependence of the corrections in order to distinguish those due to the incident field, which depend on N, from those that remain when $N = 0$, which represent the contribution of the mode ω to radiative corrections. The coupling with $|-, N+1\rangle$ shifts the level $|+, N\rangle$ downwards whereas the coupling with $|-, N-1\rangle$ shifts it upwards. But, the shift due to $|-, N+1\rangle$ predominates because this level is closer to $|+, N\rangle$ than $|-, N-1\rangle$, so that the net shift δ_+ of $|+, N\rangle$ is negative. More precisely, we get:

$$\delta_+ = \frac{\lambda^2}{4}\left[-\frac{N+1}{\omega-\omega_L} + \frac{N}{\omega+\omega_L}\right] = \frac{\lambda^2}{4}\left[-N\frac{2\omega_L}{\omega^2-\omega_L^2} - \frac{1}{\omega-\omega_L}\right]. \tag{8.9}$$

Similar calculations show that the net shift δ_- of $|-, N\rangle$ due to the couplings with $|+, N+1\rangle$ and $|+, N-1\rangle$ (dotted arrows in the right part of Fig. 8.2) is equal to

$$\delta_- = \frac{\lambda^2}{4}\left[\frac{N}{\omega-\omega_L} - \frac{N+1}{\omega+\omega_L}\right] = \frac{\lambda^2}{4}\left[+N\frac{2\omega_L}{\omega^2-\omega_L^2} - \frac{1}{\omega+\omega_L}\right]. \tag{8.10}$$

Finally, we conclude that the Larmor frequency ω_L, the frequency splitting between $|+, N\rangle$ and $|-, N\rangle$, is changed by the coupling with the RF field and becomes:

$$\omega_L \to \bar{\omega}_L = \omega_L + \delta_+ - \delta_- \simeq \omega_L\left[1 - \frac{\lambda^2}{\omega^2}\left(N+\frac{1}{2}\right)\right]. \tag{8.11}$$

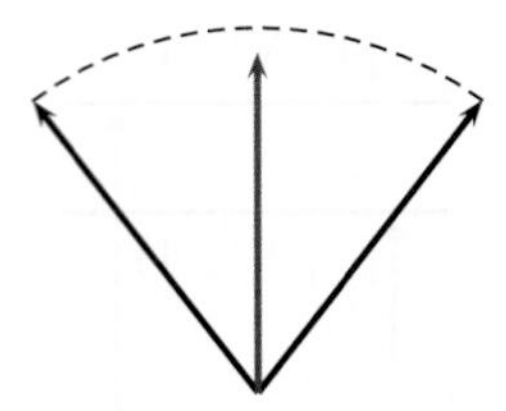

Fig. 8.3 Angular vibration of the spin in the RF field. Its static component has a reduced length.

8.2.3 *Stimulated corrections*

The term proportional to N in Eq. (8.11) represents the corrections due to the incident RF field, proportional in this perturbative treatment to the intensity of the field. They have a simple semiclassical interpretation. Suppose first that the static magnetic field is equal to zero. The spin precesses around the time-dependent RF field, keeping the same length during this precession. The coupling with the RF field thus induces an angular vibration of the spin at high frequency ω. The static component of the spin obtained after averaging over a period $2\pi/\omega$ of the field necessarily has a smaller length than that in the absence of vibration (see Fig. 8.3). It is this static component that has a non-zero average interaction with a small applied static field through an effectively reduced magnetic moment.

Non-perturbative treatment of the interaction with the incident field

The modified value $\bar{\omega}_L$ of the Larmor frequency can be calculated to all orders in the amplitude of the incident field [Cohen-Tannoudji and Haroche (1966b)]. One finds:

$$\bar{\omega}_L = \omega_L J_0\left(\frac{\omega_1}{\omega}\right), \tag{8.12}$$

where J_0 is the Bessel function of order 0 and ω_1 is the Rabi frequency. Even though the first investigation of this effect [Cohen-Tannoudji and Haroche (1966b)] was done using a dressed atom approach, this result has also a simple semiclassical interpretation. If the static field is first taken equal to zero, the precession angular frequency around the RF field, equal to $\omega_1 \cos\omega t$, is frequency modulated and the angle of the spin in the plane perpendicular to the RF field is given by a Fourier series, whose static component is equal to $J_0(\omega_1/\omega)$. The coupling of this static component with a small applied static field then leads to the perturbed Larmor frequency given in Eq. (8.12). Because of the interaction with the high frequency field, the g-factor of the spin always decreases since $|J_0(\omega_1/\omega)| \leq 1$ for all values of ω_1/ω and vanishes for all values of ω_1/ω corresponding to the zeros of J_0.

This modification of the Larmor frequency due to the interaction with the high frequency RF field has been experimentally observed on the Zeeman hyperfine spectrum of the ground state of rubidium-87 and hydrogen [Haroche *et al.* (1970)]. When ω_1/ω increases, the perturbed Larmor frequency decreases and vanishes for certain values of ω_1/ω. The variation of ω_L/ω with ω_1/ω follows the predicted $J_0(\omega_1/\omega)$ law (see Fig. 8.4).[5]

[5]See Sec. 26.2.7 of Chap. 26 where the same Hamiltonian leading to similar effects appears for a Bose-Einstein condensate in a time-dependent double well potential.

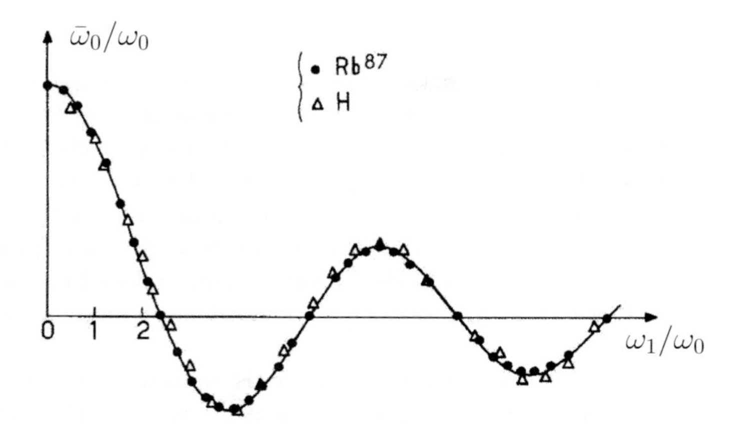

Fig. 8.4 Measured value of the modified Larmor frequency for rubidium and hydrogen. The theoretical curve given by Eq. (8.12) is in excellent agreement with the experimental data. Figure adapted from [Haroche *et al.* (1970)]. Copyright: American Physical Society.

8.2.4 *Radiative corrections*

Even if the mode ω is empty ($N = 0$), there is a correction of the Larmor frequency described by the term $1/2$ of Eq. (8.11). We obtain this term because we have kept the exact N-dependence of the matrix elements of $\hat{a}$ and $\hat{a}^\dagger$ in the pertubative treatment of Sec. 8.2.2. This term describes the contribution of the mode ω to the radiative corrections, and has the same form as the term proportional to N, with N replaced by $1/2$. It can thus be interpreted in the same way by considering the spin vibrating in a fluctuating field having a spectral power density equal to $\hbar\omega/2$. For a neutral spin, vacuum fluctuations are thus sufficient for understanding the modification of the g-factor due to radiative corrections in the mode ω.

In a second order calculation of the effect of the coupling, like that done here, the contributions of the various modes can be added in an independent way.[6] There are no crossed terms between different modes. Starting from the vacuum state of the field, the only possible process is the emission of a photon in a mode, followed by the reabsorption of this photon. The full radiative correction to second order in the coupling constant λ is thus the integral over ω of the radiative corrections due to the various modes. The contribution of non-relativistic modes ($\hbar\omega \ll mc^2$) taken into account in this treatment is, according to Eq. (8.11), obviously negative, and so the coupling with these modes cannot explain the spin anomaly a defined by the equation

$$\bar{g} = g\,(1 + a)\,, \tag{8.13}$$

relating the perturbed $\bar{g}$-factor to the unperturbed g and which is predicted by QED to be equal to $+\alpha/2\pi$, where α is the fine structure constant.

[6] This is why we have not included terms describing the coupling of the spin with the other modes of the field in the quantum Hamiltonian (8.6).

We are then led to consider two possibilities:

- The main contribution to the spin anomaly comes from the relativistic modes and cannot be explained by the treatment given here. In fact, this point of view has been investigated by several authors [Koba (1949); Weisskopf (1975); Lai *et al.* (1974)] who tried to relate the electron magnetic moment to the so-called "Zitterbewegung" appearing in the relativistic dynamics of the Dirac electron [Huang (1952)] and to interpret the spin anomaly as a modification of the Zitterbewegung induced by vacuum fluctuations. However, no precise calculation was done and no convincing result was obtained.
- The contribution of non-relativistic modes is important, but the treatment given in this section concerns only a neutral system. The electron is a charged particle and it is thus necessary to extend this treatment to a charged particle to see if the coupling of the charge to the vacuum field, combined with the coupling of the spin, can give rise to a positive anomaly [Grotch and Kazes (1977)]. This is what is done in the next section.

8.3 Weakly bound electron coupled to a high frequency field

In this section, we show how to extend the treatment of Sec. 8.2 to a charged particle such as an electron that has both a charge q and a spin magnetic moment. We suppose that the evolution frequencies in the absence of the perturbing field are low compared to the frequency ω of the field.

8.3.1 *Effective Hamiltonian describing the modifications of the dynamical properties of the electron*

The first step of this analysis is the derivation of an effective Hamiltonian describing how the slow motion of the electron is, to order 2 in q, perturbed by the field.

Figure 8.5 represents manifolds $\mathcal{E}_N$ of electron energy levels in the presence of N photons. The distance $\hbar\omega$ between adjacent manifolds is large compared to the splitting within each manifold because of the high frequency assumption for the field. The situation is similar to that of Fig. 8.2 except that the manifolds now have an infinite dimension (instead of 2).

The atom-field interaction Hamiltonian, derived from the electron kinetic energy $(\vec{p} - q\vec{A})^2/2m$ and the magnetic interactions of the spin, contains two types of terms: terms linear in q, that we denote $\hat{V}_1$ and terms quadratic in q that we denote $\hat{V}_2$. The terms linear in q are proportional to the potential vector $\vec{A}$ (and thus to the magnetic field $\vec{B}$) which are linear superpositions of the creation and annihilation operators $\hat{a}^\dagger$ and $\hat{a}$, and which connect $\mathcal{E}_N$ to $\mathcal{E}_{N+1}$ and $\mathcal{E}_{N-1}$ with matrix elements proportional to $\sqrt{N+1}$ and $\sqrt{N}$, respectively (see arrows in Fig. 8.5). The terms

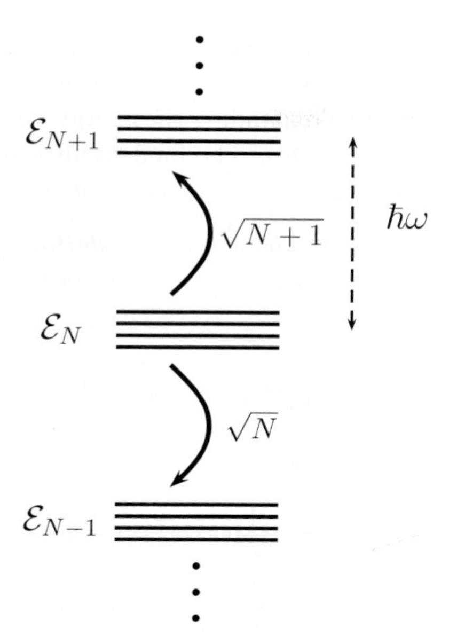

Fig. 8.5 Manifolds $\mathcal{E}_N$ of electron energy levels in the presence of N photons. The splitting within each manifold is small compared to $\hbar\omega$ because of the high frequency assumption for the field.

quadratic in q are proportional to $\vec{A}^2$, and contain terms proportional to $\hat{a}\hat{a}^\dagger$ and $\hat{a}^\dagger\hat{a}$, which operate only within $\mathcal{E}_N$ with matrix elements proportional to $N+1$ and N, respectively. These quadratic terms also couple $\mathcal{E}_N$ to $\mathcal{E}_{N\pm 2}$, but these couplings can be neglected in a calculation to order 2 in q. All terms in the interaction Hamiltonian, as well as in the electron Hamiltonian, include the first relativistic corrections up to order $1/c^2$.

The idea of the calculation sketched in this section is to derive an effective Hamiltonian acting within $\mathcal{E}_N$ and giving the electron energies within this manifold perturbed by the interaction with the high frequency field inducing virtual transitions $\mathcal{E}_N \to \mathcal{E}_{N\pm 1} \to \mathcal{E}_N$. The calculation is limited to order 2 in q, i.e. to order 1 in the fine structure constant $\alpha = q^2/4\pi\epsilon_0\hbar c$. This effective Hamiltonian is obtained by performing a unitary transformation that suppresses the couplings between $\mathcal{E}_N$ and $\mathcal{E}_{N\pm 1}$ to order 1 in q. The restriction within $\mathcal{E}_N$ of the transformed Hamiltonian then gives the correct perturbed energies within $\mathcal{E}_N$ to order 2 in q and to order 1 in $1/c^2$. Reference [Avan *et al.* (1976)] gives the details of this calculation. One finds for the matrix elements of this effective Hamiltonian between two states of $\mathcal{E}_N$:

$$\begin{aligned}
&\langle N,\alpha|\,\hat{H}^N_{\text{eff}}\,|N,\beta\rangle = \langle N,\alpha|\,\hat{H}^e\,|N,\beta\rangle + N\hbar\omega + \langle N,\alpha|\,\hat{V}_2\,|N,\beta\rangle \\
&+\frac{1}{2}\sum_{\gamma,N'\neq N}\left(\frac{1}{E_{N,\alpha}-E_{N',\gamma}}+\frac{1}{E_{N,\beta}-E_{N',\gamma}}\right)\langle N,\alpha|\,\hat{V}_1\,|N',\gamma\rangle\,\langle N',\gamma|\,\hat{V}_1\,|N,\beta\rangle\,,
\end{aligned} \tag{8.14}$$

where $\hat{H}^e$ is the electronic Hamiltonian. The second line of Eq. (8.14) looks like a second order perturbation shift, but since the two unperturbed states $|N,\alpha\rangle$ and

$|N,\beta\rangle$ do not have the same unperturbed energy, one gets the half sum of two energy denominators corresponding to each of these unperturbed energies.[7] This expression can be transformed by expanding the energy denominators appearing in Eq. (8.14) in powers of $(E_\alpha - E_\gamma)/\hbar\omega$. The final result is an explicit expression of the effective Hamiltonian in terms of purely electronic operators.

Note. A few more details about this calculation

The expansion of the first energy denominator of the second line of Eq. (8.14) gives, with $N' = N + 1$:

$$\frac{1}{E_{N,\alpha} - E_{N+1,\gamma}} = \frac{1}{E_\alpha - E_\gamma - \hbar\omega} = -\frac{1}{\hbar\omega} - \left(\frac{1}{\hbar\omega}\right)^2 (E_\alpha - E_\gamma) - \ldots \tag{8.15}$$

leading to the following expression

$$\begin{aligned}\frac{\langle N,\alpha|\hat{V}_1|N,\gamma+1\rangle}{E_{N,\alpha} - E_{N+1,\gamma}} &= -\frac{1}{\hbar\omega}\langle N,\alpha|\hat{V}_1|N+1,\gamma\rangle \\ &\quad - \left(\frac{1}{\hbar\omega}\right)^2 \sqrt{N+1}\langle\alpha|\hat{V}_1^e|\gamma\rangle(E_\alpha - E_\gamma) - \ldots\end{aligned} \tag{8.16}$$

where $\hat{V}_1^e$ is the electron part of $\hat{V}_1$. The energy difference $E_\alpha - E_\gamma$ can be included in a commutator

$$\langle\alpha|\,\hat{V}_1^e\,|\gamma\rangle\,(E_\alpha - E_\gamma) = \langle\alpha|\left[\hat{H}^e, \hat{V}_1^e\right]|\gamma\rangle\,. \tag{8.17}$$

When Eq. (8.16) is multiplied by $\langle N+1,\gamma|\,\hat{V}_1\,|N,\beta\rangle$ and summed over γ, the closure relation for γ leads finally to the following result

$$-\frac{N+1}{\hbar\omega}\,\langle\alpha|\,(\hat{V}_1^e)^2\,|\beta\rangle - \frac{N+1}{\hbar^2\omega^2}\,\langle\alpha|\,\left[\hat{H}^e, \hat{V}_1^e\right]\,\hat{V}_1^e\,|\beta\rangle\,, \tag{8.18}$$

which is the matrix element between two electronic states of an electronic operator.

8.3.2 *Stimulated effects*

The final result of the previous calculation is:

$$H_{\text{eff}}^N = (N+1)R + NS = N(R+S) + R, \tag{8.19}$$

where R and S are electronic operators. We will not give the explicit expression of R and S, which can be found in [Avan *et al.* (1976)]. We will just discuss the physical meaning of these terms, beginning with the term proportional to N which represents the stimulated corrections induced by the N incident photons. The last term, R, independent of N will be discussed in the next subsection.

[7] This peculiarity did not exist for the simple example of Sec. 8.2. The reason is that in this case the selection rules obeyed by the interaction Hamiltonian prevent any second order off diagonal coupling between the two spin states $|N,+\rangle$ and $|N,-\rangle$.

Let us first give a few characteristic energies that appear in the expression of R and S:

mc^2: rest mass energy of the electron
$\hbar\omega$: energy of the photon
E_B: binding energy of the electron in the static potentials, which, as a consequence of the high frequency assumption, obeys the relation:

$$E_B \ll \hbar\omega. \tag{8.20}$$

$\mathcal{E}_V$: kinetic energy of vibration of the electron in the incoming high frequency field E, given by:

$$\mathcal{E}_V = \frac{q^2\langle E^2\rangle}{2m\omega^2}. \tag{8.21}$$

δx: amplitude of the vibration motion of the electron obeying the following relations:

$$\langle \delta x\rangle = 0, \qquad \langle \delta x^2\rangle = \frac{q^2\langle E^2\rangle}{m^2\omega^4}. \tag{8.22}$$

The term $N(R+S)$ contains several stimulated corrections that can all be interpreted semiclassically in terms of the vibration of the charge and the spin of the electron in the incident wave. We will describe four of them here.

(1) *Global displacement of energy levels.* All levels of $\mathcal{E}_N$ are globally displaced by an amount $\mathcal{E}_V$ equal to the vibration kinetic energy. For an atomic electron, this happens for the Rydberg states whose splitting is small compared to $\hbar\omega$ since the high frequency assumption is valid for these states (see Fig. 8.6). Note however that this assumption is not valid for the ground state g for which $\hbar\omega$ can be smaller than the energy splitting between g and the first excited states. The shift of g, generally different from the shift $\mathcal{E}_V$ of the Rydberg states, leads to a displacement of the ionization threshold.[8]

(2) *Mass correction δm.* The mass correction δm associated with the vibration energy $\mathcal{E}_V$: $\delta m = \mathcal{E}_V/c^2$ introduces corrections to the kinetic energy of the slow motion of the electron:

$$\frac{\vec{\pi}_0^2}{2m} \longrightarrow \frac{\vec{\pi}_0^2}{2m} - \frac{\delta m}{m}\frac{\vec{\pi}_0^2}{2m} \simeq \frac{\vec{\pi}_0^2}{2(m+\delta m)}, \tag{8.23}$$

where $\vec{\pi}_0 = \vec{p}_0 - q\vec{A}_0$, $q\vec{A}_0$ is the static vector potential that corresponds to a static magnetic field when this field exists.

(3) *Corrections to the potential energy $\delta\phi_0$.* The vibrating electron averages the static potential ϕ_0 over the spatial extent δx of its vibration

$$\phi_0 \longrightarrow \phi_0 + \delta\phi_0 \tag{8.24}$$

$$\delta\phi_0 = \frac{q^2\langle E^2\rangle}{2m^2\omega^4}(\vec{\varepsilon}.\vec{\nabla}).(\vec{\varepsilon}^*.\vec{\nabla})\phi_0 \tag{8.25}$$

where $\vec{\varepsilon}$ is polarization of the incident field.

[8]This effect has been observed experimentally, see Chap. 10.

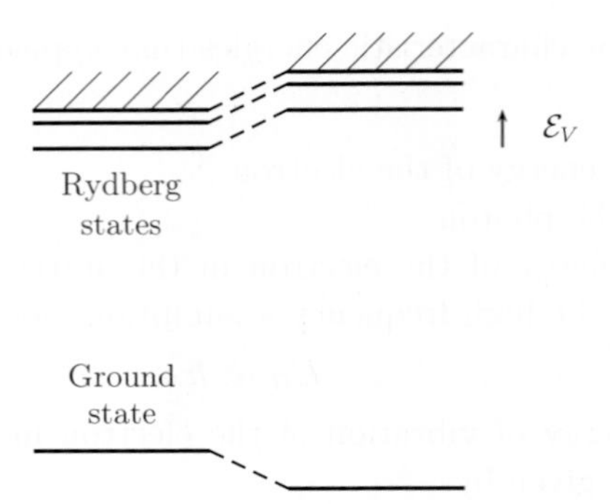

Fig. 8.6 The high frequency assumption is valid for the Rydberg states of an atomic electron resulting in a global displacement of their levels by an amount of energy equal to the vibrational kinetic energy.

(4) *Corrections to the spin magnetic moment*

- There are corrections which are analogous to those considered above for a spin 1/2 vibrating in a high frequency RF field.
- There are also terms describing the coupling of the electron spin with the motional magnetic field "seen" by the electron charge vibrating in the electric field of the incident wave. All these new corrections related to the charge also reduce the electron g-factor.

8.3.3 *Spontaneous effects. Vacuum fluctuations and radiation reaction*

The last term R of Eq. (8.19) does not vanish, as the first one does when $N = 0$. It describes the radiative corrections associated with virtual emissions and re-absorptions of photons by the electron in the initially empty mode ω. This term is obtained here because the exact N-dependence of the matrix elements of $\hat{a}^\dagger$ and $\hat{a}$ have been kept, as in Sec. 8.3.2. There is however an important difference with the result obtained in Sec. 8.3.2. The spontaneous corrections R are not just proportional to the stimulated corrections $N(R+S)$ with N replaced by 1/2. They can be written:

$$R = \frac{1}{2}(R+S) + \frac{1}{2}(R-S). \tag{8.26}$$

The first term of Eq. (8.26) represents the effect of vacuum fluctuations. What is the physical meaning of the second one?

To understand the physical origin of the mechanisms that are involved in addition to the coupling with vacuum fluctuations, it is useful to consider the evolution of the field variables in the Heisenberg picture. Since the total Hamiltonian is a quadratic function of $\hat{a}^\dagger$ and $\hat{a}$, one can easily show that the equation of evolution of $\hat{a}$ is a linear differential equation with a source term depending on electron variables. Its solution is thus the sum of two terms: a solution of the homogeneous equation, and a forced solution driven by the motion of the electron charge and spin. The first term describes the vacuum field $\vec{E}_{\text{vac}}$, which would exist in the absence of the electron, while the second one is the source field $\vec{E}_{\text{source}}$ produced by the electron itself. The equations of motion of electron variables depend on the total field. When the total field $\vec{E}_{\text{tot}}$ is replaced by $\vec{E}_{\text{vac}} + \vec{E}_{\text{source}}$, the interaction of the electron with $\vec{E}_{\text{vac}}$ describes the effect of vacuum fluctuations, whereas the interaction of the electron with $\vec{E}_{\text{source}}$ describes the effect of radiation reaction, i.e. the reaction on the electron of its self field.

There is a difficulty in this approach. The electron variables commute with the total field variables. So, while any order can be taken in the product of electron variables and total field variables, this is not true separately for $\vec{E}_{\text{vac}}$ and $\vec{E}_{\text{source}}$, which do not commute with electron variables as their sum does. It therefore seems that there is an ambiguity in the identification of the two effects, since their relative contribution can be changed by changing the order of electron and total field variables, which can be arbitrarily chosen in the equations of evolution of electron variables. A solution to this ambiguity is presented in [Dalibard *et al.* (1982)]. If one imposes the constraint that the two rates of variation due to vacuum fluctuations and radiation reaction be physical, i.e. be separately Hermitian and contain only Hermitian operators, the only possible order between electron and total field variables that is physically acceptable is the completely symmetrical order. This order leads to results identical to those obtained here from an effective Hamiltonian approach: the effect of vacuum fluctuations is described by $(R+S)/2$ and the effect of radiation reaction by $(R-S)/2$. Here we will focus on the effective Hamiltonian approach, described in detail in Ref. [Dupont-Roc *et al.* (1978)], and, in the next section, we will summarize the results obtained with this approach.

8.4 New insights into radiative corrections

8.4.1 *Examples of spontaneous corrections*

In this subsection, we first review a few examples of corrections contained in the term R of Eq. (8.19), adding a subscript "vf" to those due to vacuum fluctuations and a subscript "rr" to those due to radiation reaction. Since the effects of the various modes ω can be added independently, we will average the corrections over all possible directions of the wave vector $\vec{k}$ and polarization $\vec{\varepsilon}$, while keeping ω fixed. This restores the spherical symmetry, and amounts to evaluating the mean contribution of a mode in the "shell" of frequency ω.

(1) *Overall shift of all levels (for which the high frequency assumption is valid)*

$$\delta E_{\text{vf}} = \mathcal{E}_V^0 = \frac{q^2}{2m\omega^2}\frac{\hbar\omega}{2\epsilon_0 L^3} \tag{8.27}$$

where L^3 is the quantization volume. The shift (8.27), which is due to vacuum fluctuations, is the equivalent of kinetic energy of vibration of Eq. (8.21) with $\langle E^2 \rangle$ replaced by half the variance of the vacuum field and represents the kinetic energy of vibration of the electron in the vacuum fluctuations of the shell ω.

(2) *Mass corrections*

As in Eq. (8.23), these appear in the corrections to the kinetic energy of the slow motion of the electron.

$$\frac{\vec{\pi}_0^2}{2m} \longrightarrow \frac{\vec{\pi}_0^2}{2m} - \frac{\delta m_{\rm vf}}{m}\frac{\vec{\pi}_0^2}{2m} - \frac{\delta m_{\rm rr}}{m}\frac{\vec{\pi}_0^2}{2m} \simeq \frac{\vec{\pi}_0^2}{2(m+\delta m)}, \tag{8.28}$$

where

$$\delta m_{\rm vf} = \frac{\mathcal{E}_V^0}{m\,c^2}, \qquad \delta m_{\rm rr} = \frac{4}{3}\frac{\mathcal{E}_V^0}{\hbar\,\omega}. \tag{8.29}$$

The first mass correction $\delta m_{\rm vf}$ is analogous to the one introduced in Sec. 8.3.2. The second one, $\delta m_{\rm rr}$ is due to radiation reaction, and represents the inertia of the self field that the electron carries with it when it moves. Note that $\delta m_{\rm rr}$ is much higher than $\delta m_{\rm vf}$ by a factor on the order of $mc^2/\hbar\omega$, and that it does not depend on $\hbar$ since $\mathcal{E}_V^0$ is also proportional to $\hbar$ in the second equation (8.29). This mass correction due to radiation reaction is therefore a classical effect.

(3) *Electric form factor*

As in Sec. 8.3.2, this correction represents the effect of the spatial average of the electrostatic potential over the spatial extent $\delta x_{\rm vf}$ of the vibration of the electron induced by vacuum fluctuations. Because of the angular average over the wave vector of the mode, we get a Laplacian rather that a second order spatial derivative along the polarization of the field.

$$\phi_0 \to \phi_0 + (\delta\phi_0)_{\rm vf}, \tag{8.30}$$

where

$$(\delta\phi_0)_{\rm vf} = \frac{\mathcal{E}_V^0}{3m\omega^2}\Delta\phi_0 = \frac{1}{6}\langle \delta x_{\rm vf}^2 \rangle \Delta\phi_0. \tag{8.31}$$

(4) *Magnetic couplings of the spin*

The interaction of the electron spin with a static magnetic field $\vec{B}_0$ is modified as follows:

$$-\frac{q\hbar}{2m}\vec{\sigma}.\vec{B}_0 \to -\frac{q\hbar}{2m}\vec{\sigma}\cdot\vec{B}_0\left(1 - \frac{5}{3}\frac{\mathcal{E}_V^0}{m\,c^2}\right). \tag{8.32}$$

This correction is entirely due to vacuum fluctuations and reduces the coupling with $\vec{B}_0$. There are no contributions of radiation reaction. This is due to the fact that, in the non-relativistic domain, a magnetic moment is less coupled to the radiation field than a charge. The inertia of the self field of the charge slows the motion of the charge more efficiently than the Larmor precession of the magnetic moment.

Note also that the correction appearing in Eq. (8.32) includes not only the effect of vacuum fluctuations on the spin, analogous to the correction calculated in

Sec. 8.2, but also crossed effects between the vibrations of the charge and the spin (the factor 5/3 is the sum of a factor 2/3 due to the first effect, and of a factor 1 for the second). Taking into account the charge of the particle in its response to vacuum fluctuations further decreases its magnetic coupling to $\vec{B}_0$.

8.4.2 *Interpretation of the Lamb shift*

We suppose that there is no static magnetic field ($\vec{A}_0 = \vec{0}$), so that $\vec{\pi}_0 = \vec{p}$. The effective Hamiltonian describing radiative corrections for a hydrogen atom in the manifold $\mathcal{E}_0$ is thus:

$$H^0_{\text{eff}} = \frac{\vec{p}^2}{2m}\left(1 - \frac{\mathcal{E}^0_{\text{v}}}{mc^2} - \frac{4}{3}\,\frac{\mathcal{E}^0_{\text{v}}}{\hbar\omega}\right) + q\,\phi_0 + q\,\frac{\mathcal{E}^0_{\text{v}}}{3m\omega^2}\Delta\phi_0 + H_{\text{fs}}, \tag{8.33}$$

where H_{fs} is the fine structure atomic Hamiltonian and $\phi_0 = -q/(4\pi\epsilon_0 r)$ is the electrostatic potential due to the proton with charge $-q$, where q is the charge of the electron. Is the degeneracy between the two states $2\text{s}_{1/2}$ and $2\text{p}_{1/2}$ predicted by the Dirac equation removed by the radiative corrections appearing in Eq. (8.33) and proportional to $\mathcal{E}^0_{\text{v}}$?

None of the mass corrections due to vacuum fluctuations or radiation reaction (first term of Eq. (8.33)) can remove this degeneracy since the corresponding change of kinetic energy is the same for the $2\text{s}_{1/2}$ and $2\text{p}_{1/2}$ states which have the same average kinetic energy.

We now consider the modification of the Coulomb potential ϕ_0 due to the vibration of the electron in vacuum fluctuations. Since $\Delta(1/r) = -4\pi\delta(\vec{r})$, the third term of Eq. (8.33) can be written:

$$\frac{q^2}{\varepsilon_0}\;\frac{\mathcal{E}^0_{\text{v}}}{3m\,\omega^2}\delta(\vec{r}). \tag{8.34}$$

This correction, which is positive, affects only the $2\text{s}_{1/2}$ state, for which the electron wave function is non-zero at $\vec{r} = \vec{0}$ and shifts it upwards. The vibration of the electron in vacuum fluctuations changes the Coulomb potential "seen" by the electron in the state $2\text{s}_{1/2}$ and can explain the Lamb shift, as first pointed out by Welton [Welton (1948)].

To complete the calculation, one should now integrate the result given in Eq. (8.34) over all modes, which involves the angular averaging already done, as well as a summation over ω weighted by the mode density $2 \times 4\pi \times (L/2\pi)^3 \times k^2\,dk$. This gives for the shift of the $2s_{1/2}$ state:

$$\delta E\left(2\text{s}_{1/2}\right) = \frac{\alpha}{3\pi}\left(\frac{\hbar}{mc}\right)^2 \langle 2\text{s}_{1/2}|\,\frac{q^2}{\varepsilon_0}\delta(\vec{r})\,|2\text{s}_{1/2}\rangle \int_{\omega_m}^{\omega_M}\frac{d\omega}{\omega}. \tag{8.35}$$

Our calculation is valid for a frequency ω large compared to the binding energy E_B of the electron. So, we must introduce a lower cut off ω_m in the integral over ω of Eq. (8.35) with $\omega_m > \omega_B$. Because our calculation is non-relativistic, we must also introduce a cut off at ω_M with $\hbar\omega_M < mc^2$. Using these values to calculate the integral over ω, one finds the following order of magnitude for the Lamb shift from Eq. (8.35):

$$\delta E\left(2s_{1/2}\right) \simeq -\frac{\alpha^3}{3\pi} R_y \ln\left(\frac{\omega_M}{\omega_m}\right),$$

where $R_y = mc^2\alpha^2/2$ is the Rydberg constant. If we use the minimum value of $\omega_m \simeq R_y/\hbar$, and the maximum value of $\omega_M = mc^2/\hbar$, we find $\delta E\left(2s_{1/2}\right)/h \simeq 969$ MHz. The first non-relativistic calculation of the Lamb shift made by Hans Bethe and his colleagues yielded 1051.4 MHz [Bethe *et al.* (1950)]. Recent calculations for all effects contributing to the $2s_{1/2} - 2p_{1/2}$ Lamb shift result in a value 1057.833 (4) MHz, to be compared to the best experimental value 1057.845 (3) MHz [Eides *et al.* (2001)]. It is notable that the simple calculation performed in this section leads to a good order of magnitude of the Lamb shift.

8.4.3 *Interpretation of the spin anomaly $g-2$*

We now consider an electron in a uniform static field $\vec{B}_0$. There is no electrostatic potential and the vector potential $\vec{A}_0$ may be chosen to be equal to $\vec{A}_0(\vec{r}) = (\vec{B}_0 \times \vec{r})/2$.

In the absence of radiative corrections, the electron Hamiltonian is given by:

$$H_e = \frac{\vec{\pi}_0^2}{2m} - \frac{q\hbar}{2m}\vec{\sigma}.\vec{B}_0. \tag{8.36}$$

The energy levels of the orbital part of H_e are the Landau levels labelled by a quantum number $n = 0, 1, 2, ..$ with energies given by:

$$E_n = \left(n + \frac{1}{2}\right)\hbar\omega_C \quad \text{with} \quad \omega_C = \frac{|q|}{m}B_0. \tag{8.37}$$

The frequency ω_C is the cyclotron frequency. The spin part of H_e has two energy levels $|+\rangle$ and $|-\rangle$ separated by an energy interval $\hbar\omega_L$, where ω_L is the Larmor frequency given by:

$$\omega_L = g\frac{|q|}{2m}B_0. \tag{8.38}$$

The g-factor appearing in this equation is equal to 2 such that the Larmor and cyclotron frequencies are equal:

$$g = 2 \quad \rightarrow \quad \omega_C = \omega_L. \tag{8.39}$$

How are these results modified by radiative corrections?

The effective Hamiltonian describing the dynamics of the electron in the presence of radiative corrections is equal to:

$$H_{\text{eff}}^0 = \frac{\vec{\pi}_0^2}{2m}\left(1 - \frac{\mathcal{E}_{\text{v}}^0}{mc^2} - \frac{4}{3}\frac{\mathcal{E}_{\text{v}}^0}{\hbar\omega}\right) - \frac{q\hbar}{2m}\vec{\sigma}\cdot\vec{B}_0\left(1 - \frac{5}{3}\frac{\mathcal{E}_{\text{v}}^0}{mc^2}\right). \tag{8.40}$$

It clearly appears that the cyclotron and Larmor frequencies are both modified and reduced according to:

$$\omega_C \quad \to \quad \bar{\omega}_C = \omega_C \left[1 \; - \; \frac{4}{3}\frac{\mathcal{E}_{\rm V}^0}{\hbar\,\omega} \; - \; \frac{\mathcal{E}_{\rm V}^0}{m\,c^2} \right], \tag{8.41}$$

$$\omega_L \quad \to \quad \bar{\omega}_L = \omega_L \left[1 \; - \; 0 \; - \; \frac{5}{3}\frac{\mathcal{E}_{\rm V}^0}{m\,c^2} \right]. \tag{8.42}$$

The first correction appearing in the brackets of the right-hand sides of these two equations is the contribution of radiation reaction, and the second one is the contribution of vacuum fluctuations. The contribution of radiation reaction is the largest since $\hbar\omega \ll mc^2$, and appears for ω_C but not for ω_L. As already explained in Sec. 8.4.1, this is due to the fact that, in the non-relativistic domain, a charge is more coupled than a magnetic moment to its self field.

We now try to deduce the corrected value $\bar{g}$ of the g-factor from the previous results. It is defined by:

$$\bar{\omega}_L = \bar{g}\frac{|q|}{2\,\bar{m}}B_0, \tag{8.43}$$

where $\bar{m}$ is the corrected (or renormalized) mass appearing in the definition of the Bohr magneton. This corrected mass also appears in the corrected kinetic energy $\vec{\pi}_0^2/2\bar{m}$, and, thus, in the corrected cyclotron frequency

$$\bar{\omega}_C = \frac{|q|}{\bar{m}}B_0. \tag{8.44}$$

From these two equations, one deduces that:[9]

$$\frac{\bar{g}}{2} = \frac{\bar{\omega}_L}{\bar{\omega}_C}. \tag{8.45}$$

Since the cyclotron frequency is reduced more than the Larmor frequency, it is clear than $\bar{g}$ will be larger than 2. More precisely, one deduces from the previous equations that:

$$\frac{\bar{g}}{2} = \frac{1-(5/3)(\mathcal{E}_V^0/mc^2)}{1-(4/3)(\mathcal{E}_V^0/\hbar\omega)-(\mathcal{E}_V^0/mc^2)} \simeq 1+\frac{4}{3}\frac{\mathcal{E}_V^0}{\hbar\omega} - \frac{2}{3}\frac{\mathcal{E}_V^0}{mc^2}. \tag{8.46}$$

The first correction in $\mathcal{E}_V^0/\hbar\omega$ is positive and due to radiation reaction. It is much larger than the second one, which is negative and due to vacuum fluctuations.

To summarize, the positive sign of $g-2$ can be explained in the non-relativistic limit by radiation reaction, which slows down the cyclotron motion of the charge more efficiently than the Larmor precession of the spin.

One can finally ask if the main contribution to $g-2$ could not come from the relativistic modes of the radiation field for which $\hbar\omega > mc^2$, in which case the physical pictures developed here would be of little interest. The previous calculations can

[9] In fact, the precise measurements of the g-factor of a single trapped electron consist of measuring the Larmor and cyclotron frequencies of the electron in the same static field B_0 and taking their ratio [Brown and Gabrielse (1986)].

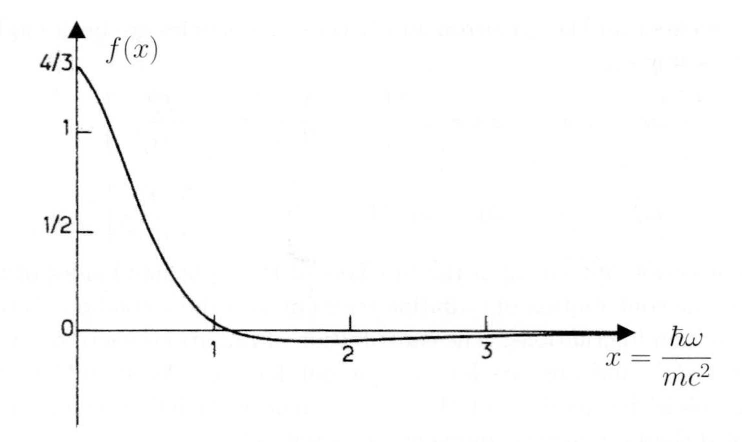

Fig. 8.7 Function $f(\hbar\omega/mc^2)$, which accounts for the contribution of the modes of frequency ω to the spin anomaly (see Eq. (8.47)). The main contribution clearly comes from the non-relativistic domain. Figure adapted from [Dupont-Roc and Cohen-Tannoudji (1984)]. Copyright: Elsevier.

be extended to answer this question [Dupont-Roc and Cohen-Tannoudji (1984)]. Starting from the full relativistic Hamiltonian for coupled Dirac and Maxwell fields, one can derive an effective Hamiltonian giving the energy levels of a non-relativistic electron (i.e. in a frame where this electron is moving slowly) that includes the relativistic contributions of virtual emissions and re-absorptions of photons of *any* frequency ω, as well as electron-positron pair creations. One finally finds [Dupont-Roc and Cohen-Tannoudji (1984)]:

$$a_e = \frac{g-2}{2} = \frac{\alpha}{\pi}\int_0^{\hbar\omega_M/mc^2} f(x)dx = \frac{\alpha}{2\pi}, \tag{8.47}$$

where the function $f(x)$, represented in Fig. 8.7, gives the x dependence of the contribution of the various modes to the spin anomaly. This approach gives the correct result of QED to first order in α, and clearly confirms that the main contribution comes from the non-relativistic domain.[10] The physical interpretation derived from the non-relativistic calculation given above is thus justified.

8.5 Conclusion

The modifications of the dynamics of a weakly bound electron due to its coupling with a high frequency non-relativistic mode ω of the radiation field can be detailed explicitly in the form of a single-particle effective Hamiltonian.

The terms of this Hamiltonian that are proportional to the number N of photons in the mode ω describe stimulated corrections proportional to the intensity of

[10] Actually, the contribution of the domain $\hbar\omega > mc^2$ is small and negative.

the incident field, and can all be interpreted semiclassically as resulting from the vibration of the charge and spin of the electron in the incident field. Among these corrections, there is a global shift of the Rydberg states of an atomic electron that is equal to the vibration kinetic energy of the electron in the incident field, and in general, different from the light shift of the strongly bound ground state. This produces a change of the ionization threshold of the atom that can play an important role in the multiphoton ionization of this atom.[11]

The terms of the effective Hamiltonian that are independent of N describe spontaneous corrections that remain present when the mode ω is empty. They represent the contributions of this mode, to lowest order in the interaction, to the radiative corrections. A term to term comparison between stimulated and spontaneous corrections shows that two types of effects are involved in spontaneous corrections: the vibration of the electron charge and spin due to vacuum fluctuations, and the radiation reaction of the charge.

The radiation reaction of the charge is due to the inertia of its self field. It also exists in classical radiation theory and the corresponding mass change of the electron is independent of $\hbar$. Radiation reaction is much smaller for the spin in the non-relativistic limit where magnetic effects are small. Vacuum fluctuations are a pure quantum effect, and do not exist for a classical field. However, once their existence is admitted, their effect on the vibration of the charge and spin can be interpreted semiclassically as stimulated corrections induced by a fluctuating field with a power spectral density equal to $\hbar\omega/2$ per mode.

The Lamb shift is simply interpreted as the averaging of the Coulomb potential by the electron charge vibrating in vacuum fluctuations. The $g-2$ anomaly is essentially due to radiation reaction which, for an electron in a static magnetic field, slows down the cyclotron frequency of the charge more than the Larmor frequency of the spin. These conclusions for $g-2$ are not changed by a relativistic generalization of the effective Hamiltonian approach, which shows, to lowest order in α, that the contribution of non-relativistic modes to $g-2$ is predominant.

Note, finally, that the effective Hamiltonian approach followed in this chapter should not be considered as a new efficient method for calculating radiative corrections. Actually, the most efficient methods are the fully covariant perturbation methods of QED. For example, new measurements of $g-2$ on a single trapped electron [Odom *et al.* (2006)] are so precise that in order to compare the experimental results to theoretical predictions, one must take into account Feynman diagrams of order 4 in the fine structure constant α [Gabrielse *et al.* (2006)].

Calculating these corrections with the non-manifestly covariant "old fashioned" perturbation method of this chapter would be hopeless! Fortunately, our motivation here was not to calculate radiative corrections to any order in α, but instead to understand the physical mechanisms responsible for them in the non-relativistic

[11]See Chap. 10.

limit and at the lowest order in α where they appear. To do this, we considered an electron moving with a velocity v small compared to c in a certain reference frame, and tried to understand how its motion was changed when it interacted with an incident field and with the vacuum field of a given non-relativistic mode.

PART 3

Atom-photon interactions: a simple system for studying higher order effects

Introduction

In part 3, we review some higher order effects appearing in atom-photon interactions. We are interested in them because they provide clear examples of situations that require going beyond a lowest-order treatment of the interaction, and even, in certain cases, finding a new description of the observed phenomena that is not based on a perturbative approach.

Higher order effects involving the absorption or the emission of several photons by the same atom can be easily observed only if the amplitude of the electromagnetic field exciting the atoms is high enough. It is therefore not surprising that they were first observed in the RF or microwave domains, where it is relatively easy to generate electromagnetic fields with an amplitude such that the corresponding Rabi frequency Ω, characterizing the strength of the atom-field coupling, is not small compared to the frequency ω_A of the atomic transition. In some cases, Ω can even be on the same order or larger than ω_A. In Chap. 9 we describe the first experiments that have clearly demonstrated the existence of RF multiphoton processes using the optical methods described in Part 1 of this book. The interesting characteristic of RF multiphoton processes is that they concern atomic spin systems with a finite number of levels efficiently coupled to the RF field. The radiative shift and the radiative broadening of multiphoton resonances can then be easily calculated, especially with a dressed atom approach using a time independent Hamiltonian.

When laser sources with a sufficiently high intensity became available, they allowed multiphoton processes to also be observed in the optical domain. In Chap. 9, we describe a few of these processes taking place between discrete atomic states. It is then necessary to take into account effects like the Doppler shift and spontaneous emission, which can be ignored in the RF domain. Two-photon processes, like stimulated Raman effect, where one photon is absorbed and another one is emitted in a stimulated way, will be also described in detail. These offer interesting possibilities for cooling atoms and ions, for probing the elementary excitations of a Bose-Einstein condensate, and for selectively transferring molecules from one state to another (e.g. STIRAP mechanism that is described in Chap. 4).

Multiphoton processes in the optical range can also be described in terms of nonlinear atomic susceptibilities. When the laser intensity is high enough, the induced atomic dipole moment is no longer simply proportional to the driving laser field amplitude E. Nonlinear terms, proportional to E^2, E^3, $\ldots$ appear in the induced dipole moment, and give rise to emissions at harmonic frequencies $2\omega_L$, $3\omega_L$, ..., where ω_L is the laser frequency. If two lasers with frequencies ω_{L1} and ω_{L2} excite the atom, they can generate sum and difference frequencies $\omega_{L1} \pm \omega_{L2}$ in the atomic response. These effects open a large range of applications in nonlinear optics for generating new frequencies, for coupling and guiding different waves, and for correcting phase distortions. Their quantitative analysis requires solving two types of problems:

- calculation of the individual nonlinear response of each atom,
- calculation of the propagation of the various new waves generated by the various atoms in order to see if they can interfere constructively.

The approach usually followed in nonlinear optics is first to calculate the individual induced dipole moments of each atom using optical Bloch equations, and then to take these dipole moments as source terms in the Maxwell equations describing the propagation of the field. Such semiclassical equations, where atoms are described quantum mechanically and the field classically, are sometimes called *Maxwell-Bloch equations* or *Maxwell-Schrödinger equations.* In this part, we will not address the problem of the field propagation, but will rather focus on the description of multiphoton processes for a single atom. The interested reader can find more detailed presentations of nonlinear optics in [Shen (1984); Bloembergen (1982, 1996); Scully and Zubairy (1997)].

When the final atomic state of a multiphoton process belongs to the ionization continuum of the atom, the atom is ionized at the end of the process. The first description of the photoelectric effect goes back to 1905, when Einstein introduced the concept of *light quantum* for the first time. Einstein focussed on one-photon ionization occurring only if the energy $h\nu$ of the photon is larger than the ionization energy E_I of the atom. In Chap. 10, we describe experiments performed in the 1960's, in which ionization can be produced even if $h\nu < E_I$, through a multiphoton process that can provide enough energy to ionize the atom. We also describe new physical effects, such as *above threshold ionization* (ATI) and *high harmonic generation* (HHG) that were observed when laser sources with much higher intensity became available. These effects exhibit features that cannot be understood with a perturbative treatment. They signal the onset of a new ionization mechanism based on the tunnel effect. When the laser field is on the same order as the Coulomb field binding the atomic electron to the ion core, the Coulomb barrier preventing the electron from escaping acquires a finite width and the electron can tunnel through this barrier. The interesting feature of this mechanism is that once the electron has been released, it is accelerated by the oscillating laser field and can come back near the ion and *re-collide* with it. In Chap. 10, we describe a few interesting consequences of this re-collision. It turns out that they are at the origin of recent developments in the generation of very short *attosecond* UV pulses, which will be described in more detail in Chap. 27.

Chapter 9

Multiphoton processes between discrete states

9.1 Introduction

A p-photon resonant process (with $p > 1$) between two discrete states a and b, with $E_b > E_a$, is a process where the atom, starting from state a in the presence of N photons, goes to state b, with the number of photons decreasing from N to $N - p$. The initial state $|a, N\rangle$ and the final state $|b, N - p\rangle$ of the transition must have the same total energy and the same total linear and angular momentum. Since the interaction Hamiltonian $\hat{V}$ can change the number of photons by only one unit, higher order terms in the perturbation expansion of the transition amplitude, where $\hat{V}$ appears p times, are required for explaining the absorption of p photons.

The first theoretical investigation of multiphoton processes[1] was done by Maria Göppert Mayer in 1931 [Göppert Mayer (1931)]. It took about twenty years before experimental evidence of multiphoton resonances was obtained. As explained in the introduction of Part 3, this was due to the weak intensity of the electromagnetic sources available at that time. The development of radiofrequency and microwave sources after World War II, and then of laser sources during the 1960's, changed the situation.

The first indication of two-photon resonances appeared through unexpected structures in the electric RF spectrum of RbF molecules observed in a molecular beam experiment [Hughes and Grabner (1950a,b)]. The authors tried to explain a non-predicted group of lines in terms of two-photon transitions. Much clearer evidence of RF multiphoton resonances between the Zeeman sublevels of optically pumped alkali atoms was obtained shortly afterwards during the PhD works of Bernard Cagnac and Jacques-Michel Winter. In Sec. 9.2, the corresponding experiments will be described in detail and it will be shown how their results can be understood with simple arguments based on the conservation of energy and angular momentum. When the intensity of the RF field increases, multiphoton resonances

[1] In fact, in his 1905 paper on light quanta, Einstein envisions the possibility that several light quanta could be involved in a deviation from the "Stokes rule" according to which the emitted light from a fluorescence substance has an upper bound frequency determined by the incident excitation light frequency (see [Arons and Peppard (1965)]).

are shifted and broadened. In Sec. 9.3, we present a quantitative analysis of these radiative shifts and broadenings using a dressed atom approach.

Multiphoton resonances also exist in the optical domain. In Sec. 9.4, we describe the first observations of resonances of this type made possible in the 1960's by the availability of laser sources with high enough intensities. In this case also, radiative shifts exist and can affect the precision of the measurements if they are not taken into account. In this chapter, we only consider multiphoton resonances between discrete states[2] and show how the Doppler effect depends on the relative directions of propagation of the various absorbed photons. Stimulated Raman processes, involving the absorption of one photon and the stimulated emission of another one are particularly interesting. They can connect two sublevels of the ground state with long lifetimes and the corresponding coupling can be described by an effective Hamiltonian. We show that stimulated Raman resonances can have a very high selectivity in velocity that explains why they are playing an important role in recent developments of atomic and molecular physics. Finally, at the end of Sec. 9.4, we present a simple interpretation of the momentum conservation equation that is usually presented as a necessary condition for realizing phase-matching in nonlinear optics. This condition expresses that the total momentum transferred to each atom in a multiphoton process must be equal to zero. Otherwise, the waves scattered by the various atoms will correspond to orthogonal final states of these atoms, preventing them to interfere.

9.2 Radiofrequency multiphoton processes

9.2.1 *Multiphoton RF transitions between two Zeeman sublevels m_F and $m_F + 2$*

Consider three Zeeman sublevels m_F, $m_F + 1$ and $m_F + 2$ belonging to a certain hyperfine level F of an alkali atom. If the static magnetic field B_0 is high enough, these three sublevels are not equidistant because of the Paschen-Back effect.

Figure 9.1(a) shows the one-photon transition between m_F and m_F+1 occurring for a frequency ω_1 of the RF field and the one-photon transition between $m_F + 1$ and $m_F + 2$ occurring for a frequency ω_2:

$$E_{m_F+1} - E_{m_F} = \hbar\omega_1, \qquad\qquad E_{m_F+2} - E_{m_F+1} = \hbar\omega_2. \tag{9.1}$$

The atom can also go directly from m_F to $m_F + 2$ by absorbing two photons ω if the condition

$$E_{m_F+2} - E_{m_F} = 2\hbar\omega, \tag{9.2}$$

expressing the conservation of the total energy is fulfilled. The conservation of the total angular momentum also requires that the two RF photons must each have an

[2]The case where the final atomic state of the multiphoton transition belongs to the ionization continuum will be studied in Chap. 10.

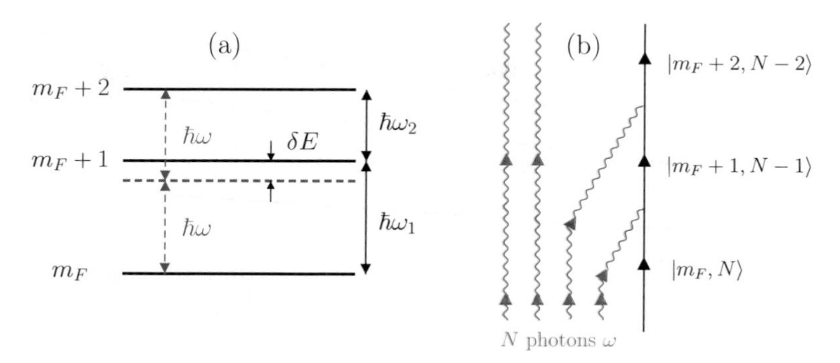

Fig. 9.1 One and two-photon transitions between three non-equidistant Zeeman sublevels. (b) Another diagrammatic representation of the two-photon process between m_F and $m_F + 2$.

angular momentum equal to $+\hbar$ along the quantization axis in order to provide the gain of angular momentum of the atom going from m_F to $m_F + 2$

$$(m_F + 2)\hbar - m_F\hbar = 2\hbar.$$

They must therefore have a σ_+ polarization.[3]

In Fig. 9.1(b), another representation of the two-photon process by a Feynman diagram is given. In the initial state $|m_F, N\rangle$, the atom is in the state m_F in the presence of N photons ω. After the absorption of one photon ω, the system goes to the intermediate state $|m_F + 1, N - 1\rangle$ where the atom is in the state $m_F + 1$ in the presence of $N - 1$ photons. Finally, after the absorption of the second photon ω, the system ends in the final state $|m_F + 2, N - 2\rangle$ where the atom is in the state $m_F + 2$ in the presence of $N - 2$ photons. The energies of the initial and final states are equal. They differ from the energy of the intermediate state by an amount $\delta E = [E_{m_F} + N\hbar\omega] - [E_{m_F+1} + (N - 1)\hbar\omega] = \hbar(\omega - \omega_1)$. This energy mismatch clearly appears in Fig. 9.1(a): it is equal to the energy difference between the solid line $m_F + 1$ and the dashed line.

Remark

A transition that does not conserve the total energy of the system is called a virtual transition. It cannot last a time longer than $\hbar/\delta E$ where δE is the energy defect. The dashed line of Fig. 9.1(a) is sometimes called a "virtual level", but it is, actually, an energy level of the total system { atom+photon }. Solid lines of Fig. 9.1(a) represent the energy levels of an atom in m_F, $m_F + 1$, $m_F + 2$ in the presence of $N - 1$ photons, the dashed line represents the energy level of the dressed atom in m_F in the presence of N photons.

The one and two-photon processes described above correspond to a situation in which the static magnetic field B_0 is fixed and where the frequency ω of the RF field is varied. Alternatively, one can also keep the frequency ω of the RF field fixed and vary B_0. This is represented in Fig. 9.2. The two-photon resonance

[3]See Sec. 2.3.2 of Chap. 2.

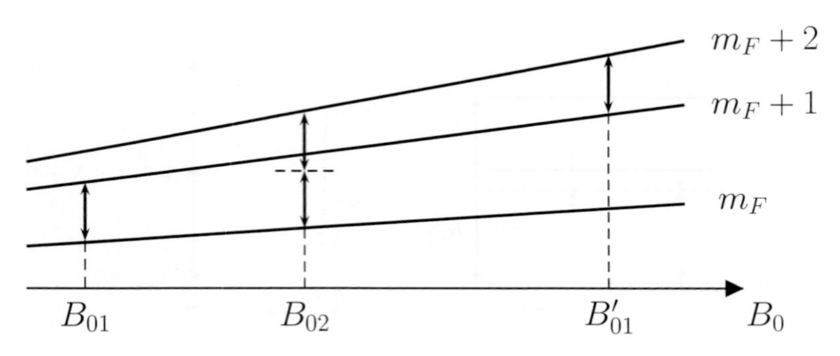

Fig. 9.2 When the static field B_0 is scanned, for a fixed frequency ω of the RF field, the one and two-photon resonances of Fig. 9.1 appear at three different values of B_0.

$m_F \longrightarrow m_F + 2$ appears at a value B_{02} of the field which is intermediate between the values B_{01} and B'_{01} of the field where the one-photon resonances $m_F \longrightarrow m_F+1$ and $m_F + 1 \longrightarrow m_F + 2$ appear.

9.2.2 *Experimental observation on sodium atoms*

The hyperfine level $F = 2$ in the ground state of a sodium atom has five Zeeman sublevels $m_F = -2, -1, 0, +1, +2$ which are not equidistant because of the Paschen-Back effect (see left part of Fig. 9.3). For a fixed value of the frequency ω of the RF field, one predicts one, two and three-photon resonances that should appear for different values of the static field B_0: four one-photon resonances $-2 \to -1, -1 \to 0, 0 \to +1, +1 \to +2$, denoted A, B, C, D, respectively; three two-photon resonances $-2 \to -0, -1 \to +1, 0 \to +2$, denoted a, b, c; two three-photon resonances $-2 \to +1, -1 \to +2$, denoted α, β.

These resonances have been experimentally observed in optical pumping experiments performed on sodium atoms [Brossel *et al.* (1953, 1954)]. Optical pumping creates population differences between the Zeeman sublevels, necessary for observing magnetic resonance signals. When the value of the static field is swept around the value corresponding to a one-photon or multiphoton resonance, the populations of the Zeeman sublevels are changed by the RF transitions. This induces a change in the intensity of the fluorescence light, thereby allowing an optical detection of the resonance. The right part of Fig. 9.3 shows examples of experimental curves. For low RF power (upper curve), one sees only the four one-photon resonances A, B, C, D. For a medium power (intermediate curve), the three two-photon resonances a, b, c appear between the four one-photon resonances. Finally, for high RF power (lower curve), the two three-photon resonances α, β appear.[4]

[4]In principle, one should also have a four-photon resonance $-2 \to +2$, but this resonance has not been observed, probably because of insufficient RF power.

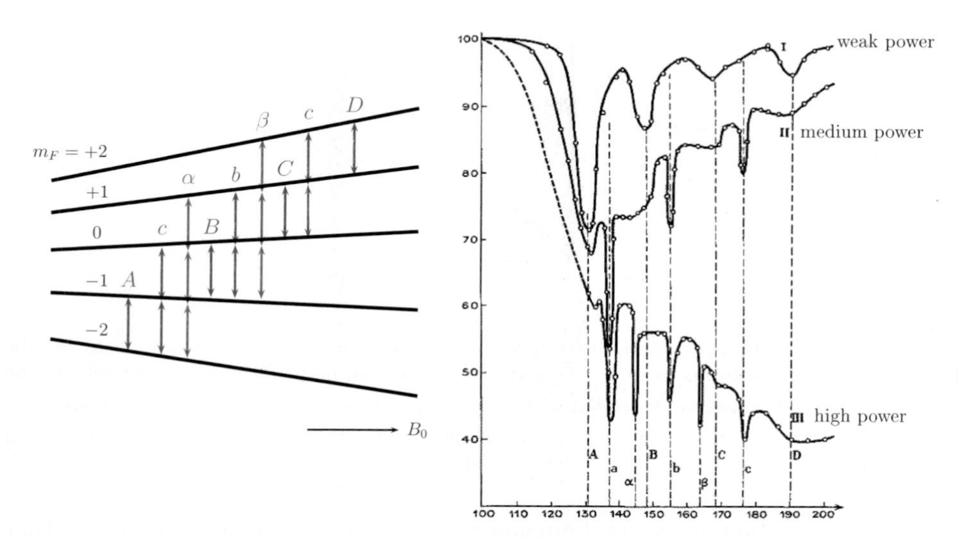

Fig. 9.3 (a) Various one, two and three-photon resonances appearing between the Zeeman sublevels of the hyperfine level $F = 2$ of the ground state of a sodium atom. The frequency ω of the RF field is fixed and the static field B_0 is scanned. (b) Three different experimental spectra of RF resonances obtained with increasing RF powers. At low power, only the one-photon resonances appear. For intermediate power, one sees also the two-photon resonances. The three-photon resonances appear only at high power. Figure adapted from [Brossel *et al.* (1954)]. Copyright: EDP Sciences.

9.2.3 *Multiphoton resonances between two Zeeman sublevels m_F and $m_F + 1$*

RF multiphoton resonances can also be observed between two Zeeman sublevels m_F and $m_F + 1$. Let $\hbar\omega_0$ be the Zeeman energy splitting between $m_F + 1$ and m_F. When $\hbar\omega_0$ is varied while keeping a fixed value ω of the frequency of the RF field, one expects to see a one-photon resonance for $\omega_0 = \omega$, a two-photon resonance for $\omega_0 = 2\omega$, a three-photon resonance for $\omega_0 = 3\omega$, a four-photon resonance for $\omega_0 = 4\omega$, and so on. These resonances are represented in Fig. 9.4.

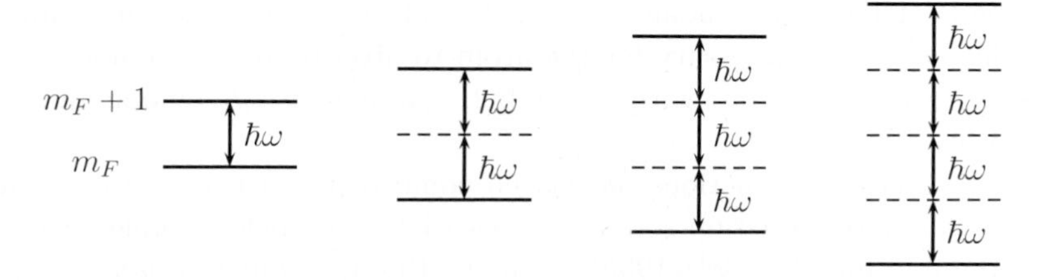

Fig. 9.4 One, two, three and four-photon resonances between two Zeeman sublevels m_F and $m_F + 1$.

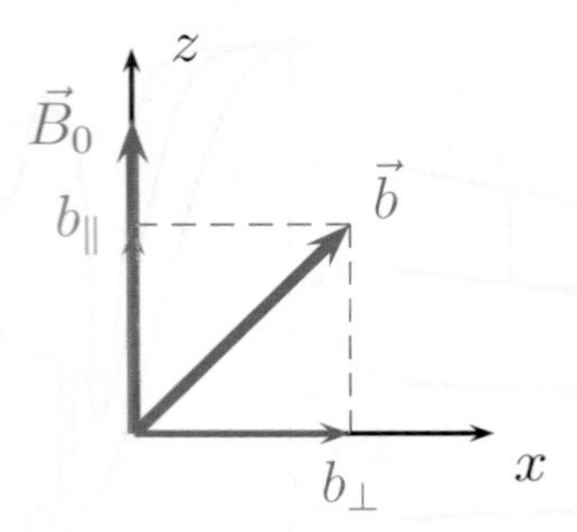

Fig. 9.5 When the linearly polarized RF field $\vec{b}$ is oblique with respect to the static field $\vec{B}_0$, the RF field is the sum of two components $\vec{b}_{\parallel}$ and $\vec{b}_{\perp}$, parallel and perpendicular, respectively, to the static field B_0. The photons associated with $\vec{b}_{\parallel}$ are π-polarized and have an angular momentum $J_z = 0$ with respect to the z-axis. The photons associated with $\vec{b}_{\perp}$ are σ-polarized and have an angular momentum $J_z = +\hbar$ or $J_z = -\hbar$ with respect to the z-axis.

The conservation of the total angular momentum in the transition implies that the sum of the angular momenta of the absorbed RF photons is equal to the angular momentum $+\hbar$ gained by the atom going from m_F to $m_F + 1$. For the one-photon resonance, the absorbed photon must have a σ_+ polarization ($J_z = +\hbar$). For the two-photon resonance, one of the two photons must have a σ_+ polarization, and the other one a π polarization ($J_z = 0$). For the three-photon resonance, we have two possibilities: one photon with a σ_+ polarization, and the other two with a π polarization; two photons with a σ_+ polarization, the third with a σ_- polarization ($J_z = -\hbar$); for the four-photon resonance, we also have two possibilities: one photon with a σ_+ polarization and the other three photons with a π polarization; two photons with a σ_+ polarization, the third one with a σ_- polarization, and the fourth with a π polarization.

The polarization of the RF field can easily be adjusted by changing the angle between the static field $\vec{B}_0$ and the RF field $\vec{b}$ (see Fig. 9.5). The photons associated with the component $\vec{b}_{\parallel}$ of the RF field $\vec{b}$ parallel to the static field $\vec{B}_0$ are π-polarized and have an angular momentum $J_z = 0$. The photons associated with the component $\vec{b}_{\perp}$ of the RF field $\vec{b}$ perpendicular to $\vec{B}_0$ are σ-polarized and have an angular momentum $J_z = +\hbar$ or $-\hbar$. If the RF field $\vec{b}$ is oblique with respect to $\vec{B}_0$, all the values of $J_z(0, +\hbar, -\hbar)$ are possible for the RF photons. If the RF field $\vec{b}$ is perpendicular to $\vec{B}_0$ (σ-polarization), the only possible values of J_z are $+\hbar$ or $-\hbar$. In this case, it is necessary for the atom to absorb an odd number $2n + 1$ of photons: $(n + 1)\sigma_+$ and $n\sigma_-$ with $n = 0, 1, 2, \ldots$ to increase J_z by one unit $+\hbar$ in a transition $m_F \longrightarrow m_F + 1$.

These theoretical predictions have been confirmed experimentally. Figure 9.6 shows resonances observed on sodium atoms with a RF field oblique with respect to $\vec{B}_0$ [Margerie and Brossel (1955); Winter (1955)]. Odd resonances ($\omega_0 = \omega$, $\omega_0 = 3\omega$) as well as even resonances ($\omega_0 = 2\omega$, $\omega_0 = 4\omega$) are observed when the RF power increases. Figure 9.7 shows resonances observed on mercury-199 atoms with

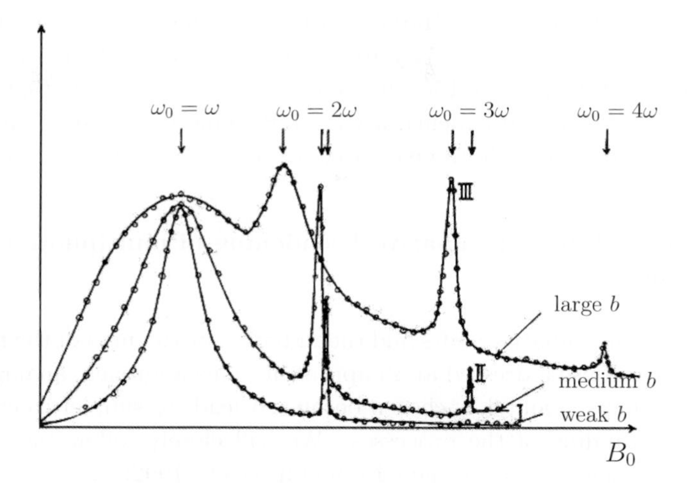

Fig. 9.6 Three experimental spectra of RF resonances $m_F \to m_F + 1$ observed in the ground state of sodium for increasing values of the RF power. When the RF power increases, the resonances are shifted to lower values of B_0 and they become broader. The four-photon resonance appears only at high power. Adapted from [Margerie and Brossel (1955)]. Copyright: Elsevier Masson SAS.

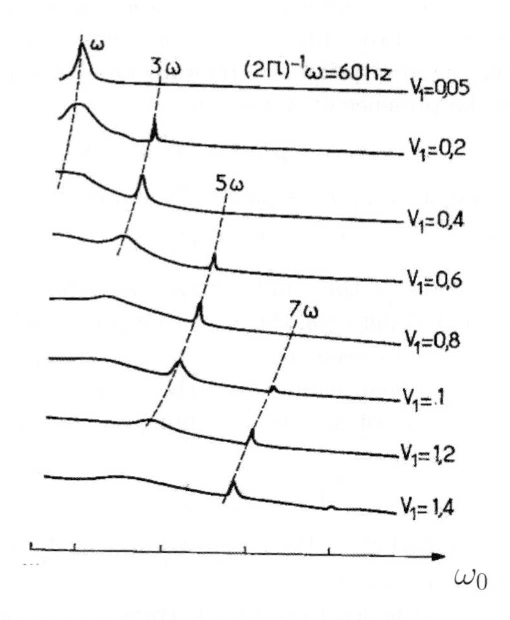

Fig. 9.7 Eight experimental spectra of RF resonances observed in the ground state of mercury-199 atoms for increasing values of the RF power and for a σ-polarized RF field. V_1 is the RF voltage. Only odd resonances $\omega_0 = \omega, 3\omega, 5\omega, 7\omega, \ldots$ appear. Clearly such resonances exhibit a radiative shift and a radiative broadening as V_1 increases [Haroche (1971)]. Courtesy of S. Haroche.

an RF field perpendicular to $\vec{B}_0$ [Haroche (1971)]. In this case, only odd resonances ($\omega_0 = \omega$, $\omega_0 = 3\omega, \omega_0 = 5\omega$, $\omega_0 = 7\omega$) are observed when the RF power increases.

The experimental curves of Figs. 9.6 and 9.7 also clearly show that the various resonances are shifted and broadened when the RF power increases. We show in the next section how it is possible to give a quantitative interpretation of these effects.

9.3 Radiative shift and radiative broadening of multiphoton resonances

For interpreting the radiative shifts and the radiative broadening of the multiphoton resonances, we will use a dressed atom approach. This approach, analogous to that introduced in Chaps. 5 and 6, is chosen because it leads to simpler calculations and a better understanding of the processes. We will closely follow the presentation given in Complement A_{VI} of [Cohen-Tannoudji *et al.* (1992b)].

9.3.1 *Energy levels of the atom+RF photons system. Transition amplitude*

Consider first the initial and final states of a p-photon transition: the state α, corresponding to the atom in the state a in the presence of $N+p$ photons, and the state β, reached after the absorption of p photons and corresponding to the atom in another state b in the presence of N photons

$$|\alpha\rangle = |a, N+p\rangle, \qquad |\beta\rangle = |b, N\rangle. \tag{9.3}$$

Energy conservation implies that $E_a + p\hbar\omega = E_b$ and consequently that:

$$E_\alpha = E_\beta = E_0. \tag{9.4}$$

The initial and final states of the total system have the same energy E_0 when the frequency ω of the RF field (or the static magnetic field B_0) has the value corresponding to the p-photon resonance.

The interaction Hamiltonian cannot directly connect the two states α and β when $p > 1$. The system must go through intermediate levels $|\gamma\rangle = |c, N'\rangle$. In general,

$$E_\gamma \neq E_\alpha, E_\beta. \tag{9.5}$$

There is therefore a manifold $\mathcal{E}_0 = \{|\alpha\rangle, |\beta\rangle\}$ of states of the total system that are well isolated from the other states[5] $|\gamma\rangle$.

The p-photon transition is described by the transition amplitude

$$\langle b, N|\hat{U}(t)|a, N+p\rangle = \langle\beta|\hat{U}(t)|\alpha\rangle, \tag{9.6}$$

[5]It may happen that other states $|\gamma\rangle$ have an energy E_γ close to E_α and E_β. This is the case, for example, when the three atomic levels of Fig. 9.1(a) are equidistant. The three states $|m_F, N+2\rangle$, $|m_F+1, N+1\rangle$ and $|m_F+2, N\rangle$ then have the same energy when $\hbar\omega$ is equal to the spacing between the three sublevels. The manifold $\mathcal{E}_0$ must then be enlarged to include the state $|\gamma\rangle$.

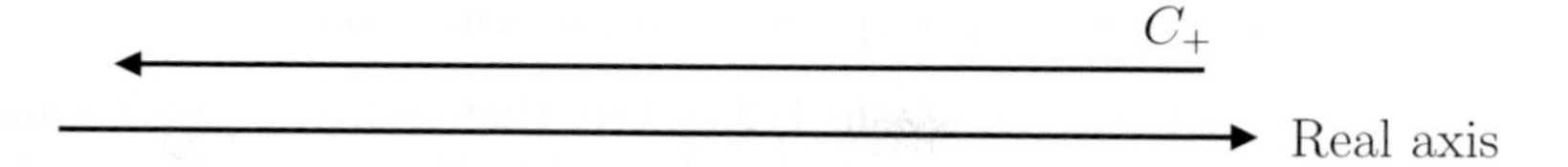

Fig. 9.8 Contour of integration C_+ in the complex plane.

where $\hat{U}(t)$ is the evolution operator. For $t > 0$, one can show that:

$$\hat{U}_{\beta\alpha}(t) = \frac{1}{2\pi i}\int_{C_+} \mathrm{d}z e^{-izt/\hbar}\hat{G}_{\beta\alpha}(z), \quad \text{where} \quad \hat{G}(z) = \frac{1}{z - \hat{H}} \tag{9.7}$$

is the resolvent operator and where C_+ is the contour of integration in the complex plane represented in Fig. 9.8 by a line situated immediately above the real axis and followed from right to left.[6]

The subspace $\mathcal{E}_0 = \{|\alpha\rangle, |\beta\rangle\}$ is well isolated. It will be therefore appropriate to calculate the restriction of $\hat{U}(t)$, and consequently of $\hat{G}(z)$, within $\mathcal{E}_0$:

$$\hat{P}\hat{G}(z)\hat{P}, \qquad \hat{P} = |\alpha\rangle\langle\alpha| + |\beta\rangle\langle\beta|. \tag{9.8}$$

A simple calculation, presented in Chap. III of [Cohen-Tannoudji *et al.* (1992b)], gives

$$\hat{P}\hat{G}(z)\hat{P} = \frac{\hat{P}}{z - \hat{P}\hat{H}_0\hat{P} - \hat{P}\hat{R}(z)\hat{P}} \tag{9.9}$$

where

$$\hat{R} = \hat{V} + \hat{V}\frac{\hat{Q}}{z - \hat{Q}\hat{H}_0\hat{Q} - \hat{Q}\hat{V}\hat{Q}}\hat{V}, \qquad \text{and} \qquad \hat{Q} = 1 - \hat{P}. \tag{9.10}$$

In a perturbative treatment, one can expand $\hat{R}(z)$ in powers of $\hat{V}$:

$$\hat{R} = \hat{V} + \hat{V}\frac{\hat{Q}}{z - \hat{H}_0}\hat{V} + \hat{V}\frac{\hat{Q}}{z - \hat{H}_0}\hat{V}\frac{\hat{Q}}{z - \hat{H}_0}\hat{V} + \ldots \tag{9.11}$$

The matrix elements of $\hat{R}(z)$ then appear as products of matrix elements of $\hat{V}$ with energy denominators related to intermediate states that do not belong to $\mathcal{E}_0$ and are far from the states of $\mathcal{E}_0$.

If one could neglect the z-dependence of $\hat{R}(z)$, $\hat{P}\hat{H}_0\hat{P} + \hat{P}\hat{R}(z)\hat{P}$ would have the meaning of an effective Hamiltonian acting within $\mathcal{E}_0$. It is actually possible to make an approximation leading to this interpretation of $\hat{R}$. $\hat{P}\hat{G}(z)\hat{P}$ has poles located at the eigenvalues of $\hat{H}$, and thus takes large values in the neighborhood of the energies of the unperturbed states α and β of the multiphoton transition. These energies are equal to $E_\alpha = E_\beta = E_0$. All other eigenvalues E_γ of $\hat{H}_0$ are very far, so that $\hat{P}\hat{R}(z)\hat{P}$ varies very slowly with z in the neighborhood of $z = E_0 + i\epsilon$, since all denominators, on the order of $E_0 - E_\gamma$, are very large. Thus, one can replace $\hat{P}\hat{R}(z)\hat{P}$ by $\hat{P}\hat{R}(E_0 + i\epsilon)\hat{P} = \hat{\hat{R}}$, in the expression of $\hat{P}\hat{G}(z)\hat{P}$, which gives

$$\hat{P}\hat{G}(z)\hat{P} = \frac{\hat{P}}{z - \hat{P}\hat{H}_0\hat{P} - \hat{\hat{R}}}. \tag{9.12}$$

[6]See Chap. III of [Cohen-Tannoudji *et al.* (1992b)].

9.3.2 *Pure single photon resonance. Simple anticrossing*

Consider first the simple case of a spin 1/2 system that has only two Zeeman sublevels $\pm 1/2$ and interacts with a σ_+-polarized RF field. The interaction Hamiltonian $\hat{V}$ has then a single non-zero matrix element:

$$\langle +1/2, N|\hat{V}| - 1/2, N+1\rangle = V_{\beta\alpha} = \hbar\Omega/2 \tag{9.13}$$

where Ω is the Rabi frequency. In the subspace $\{|\beta = +1/2, N\rangle; |\alpha = -1/2, N+1\rangle\}$, the evolution is exactly described by the Hamiltonian

$$\begin{pmatrix} E_\beta & \hbar\Omega/2 \\ \hbar\Omega/2 & E_\alpha \end{pmatrix}. \tag{9.14}$$

Suppose that E_α and E_β are varied linearly by scanning a static magnetic field B_0. How does the transition probability to go from state $|\alpha\rangle$ to $|\beta\rangle$ after a time t vary with $E_\beta - E_\alpha = \hbar\omega_{\beta\alpha}$?

The dashed lines of Fig. 9.9 represent the unperturbed energies E_α and E_β, which vary linearly with B_0 and cross when $B_0 = B_{0c}$. When the coupling between these two states is taken into account, the perturbed energies, given here by the energies of the dressed states of the { atom+RF photons } system, are obtained by diagonalizing the matrix (9.14). It is well known that the perturbed states "anticross", and give rise to a hyperbola with the unperturbed states as asymptots. The minimum distance of $2V_{\beta\alpha} = \hbar\Omega$ between the two branches of the hyperbola is reached when the two unperturbed levels cross, i.e. when $\omega_{\beta\alpha} = 0$.

From the expression of the perturbed states in terms of the unperturbed states, one can calculate the probability $P_{\beta\alpha}(t)$ to find the system in the state β at time t

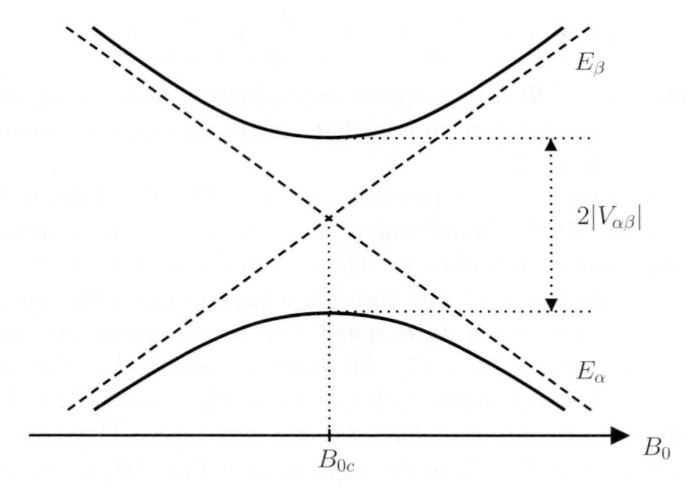

Fig. 9.9 Simple anticrossing. The two dotted lines represent the energies of the unperturbed levels α and β varying linearly with B_0. They cross when $B_0 = B_{0c}$. The interaction Hamiltonian $\hat{V}$ couples only these two states. The two perturbed states form an hyperbola with the unperturbed states as asymptots, with a minimum distance equal to $2|V_{\alpha\beta}|$.

if it starts from the state α at time $t = 0$. A simple calculation gives:

$$P_{\beta\alpha}(t) = \frac{\Omega^2}{\omega_{\beta\alpha}^2 + \Omega^2} \sin^2 \left[\sqrt{\omega_{\beta\alpha}^2 + \Omega^2} \frac{t}{2} \right]. \tag{9.15}$$

In many cases, the system is prepared in the state α with a constant rate and then decays with a rate Γ. One can show that the steady-state population, $\bar{\Pi}_\beta$, of β is proportional to

$$\int_0^\infty \Gamma e^{-\Gamma t} P_{\beta\alpha}(t) \mathrm{d}t = \frac{1}{2} \frac{\Omega^2}{\Omega^2 + \Gamma^2 + \omega_{\beta\alpha}^2}. \tag{9.16}$$

When B_0 is varied, $\bar{\Pi}_\beta$ exhibits resonant variations that have the shape of a Lorentzian curve centered about the anticrossing with a radiative broadening (Ω) equal to the minimum distance between the two branches of the anticrossing.

9.3.3 *Higher order anticrossing for a p-photon resonance ($p > 1$)*

Contrary to the previous case, the evolution can no longer be described exactly. Instead, if the two levels α and β involved in the p-photon resonance form a well isolated manifold, one can use the approximation introduced at the end of Sec. 9.3.1 that replaces z by $E_0 + i\epsilon$ in the expression of $\hat{R}(z)$, which is then taken equal to $\bar{R} = \hat{R}(E_0 + i\epsilon)$. With this approximation, the evolution is described by the effective Hamiltonian

$$\begin{pmatrix} E_{\beta\beta} + \bar{R}_{\beta\beta} & \bar{R}_{\beta\alpha} \\ \bar{R}_{\alpha\beta} & E_{\alpha\alpha} + \bar{R}_{\alpha\alpha} \end{pmatrix}. \tag{9.17}$$

The comparison of Eqs. (9.14) and (9.17) leads to the following conclusions:

- The unperturbed energies E_α and E_β are corrected by the terms $\bar{R}_{\alpha\alpha}$ and $\bar{R}_{\beta\beta}$. These terms are of order 2 in the interaction Hamiltonian $\hat{V}$, and represent the radiative shifts of the states α and β due to non-resonant couplings with the states γ other than α and β. The dotted straight lines of Fig. 9.10 represent the energies of the unperturbed levels α and β. The straight full lines represent the energies of these levels shifted by the quantities $\bar{R}_{\alpha\alpha}$ and $\bar{R}_{\beta\beta}$. The resonance occurs now at the crossing of these shifted levels, i.e. at the value of the field for which:

$$E_{\alpha\alpha} + \bar{R}_{\alpha\alpha} = E_{\beta\beta} + \bar{R}_{\beta\beta}. \tag{9.18}$$

- The non-diagonal coupling $\bar{R}_{\alpha\beta}$ between the two levels that cross is now at least of order p in $\hat{V}$, and is equal to a sum of products of p matrix elements of $\hat{V}$ and of $p - 1$ energy denominators $1/(E_0 - E_\gamma)$, where $\gamma \neq \alpha, \beta$.
- Comparing with the results of the previous section, one concludes that a p-photon resonance has the shape of a Lorentzian curve. The center of this curve undergoes a radiative shift determined by $\bar{R}_{\alpha\alpha}$ and $\bar{R}_{\beta\beta}$, which are of order 2 in the RF field amplitude. The radiative broadening of the resonance is determined by $\bar{R}_{\alpha\beta}$ and is thus of order p in the RF field amplitude.

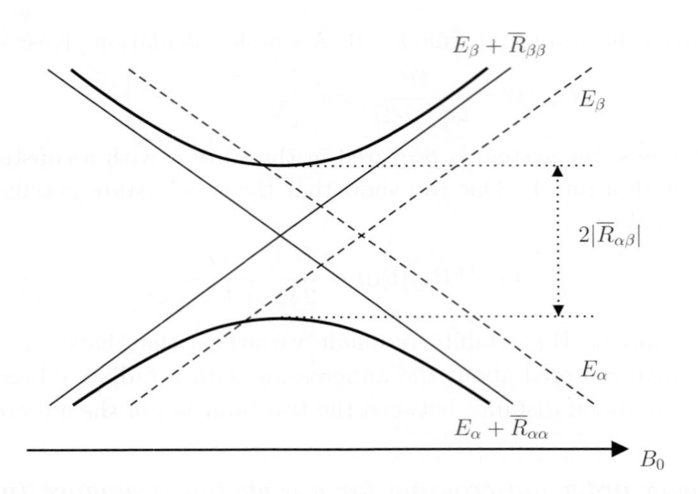

Fig. 9.10 Higher order anticrossing. The dotted straight lines represent the energies of the two unperturbed states α and β. Because of couplings to far distant states, these two states undergo radiative shifts $\bar{R}_{\alpha\alpha}$ and $\bar{R}_{\beta\beta}$. The crossing between these two shifted states, represented by straight full lines, is displaced. The two states are also coupled by a higher order term $\bar{R}_{\alpha\beta}$ and form a hyperbola with a minimum distance between its two branches $2|\bar{R}_{\alpha\beta}|$. For a p-photon resonance, $\bar{R}_{\alpha\alpha}$ and $\bar{R}_{\beta\beta}$ are of order 2 in the RF field amplitude, and $\bar{R}_{\alpha\beta}$ is of order p.

9.3.4 *Application to the case of a spin 1/2 coupled to a σ-polarized RF field*

The previous predictions can be illustrated with the simple case of a spin 1/2 interacting with:

- a static field parallel to the z-axis that produces a Zeeman splitting $\hbar\omega_0$ between the two sublevels $m_F = \pm 1/2$, where ω_0 is the Larmor frequency proportional to the static field B_0
- a linearly polarized RF field with frequency ω parallel to the x-axis (σ-polarization).

We have already considered such a situation in Chap. 8 (see Fig. 8.1 of this chapter).

The following interpretation of RF multiphoton resonance for a spin 1/2 was first given in [Cohen-Tannoudji and Haroche (1966a)]. The interaction Hamiltonian $\hat{V}$ is proportional to the x-component $\hat{S}_x$ of the spin. It obeys the selection rule $\Delta m_F = \pm 1$. It is also proportional to $\hat{a} + \hat{a}^\dagger$, so that it obeys also the selection rule $\Delta N = \pm 1$, where N is the number of RF photons. We will now review a few new effects that can appear with a σ-polarized RF field, in comparison with the case of σ_+-polarized field considered in Sec. 9.3.2.

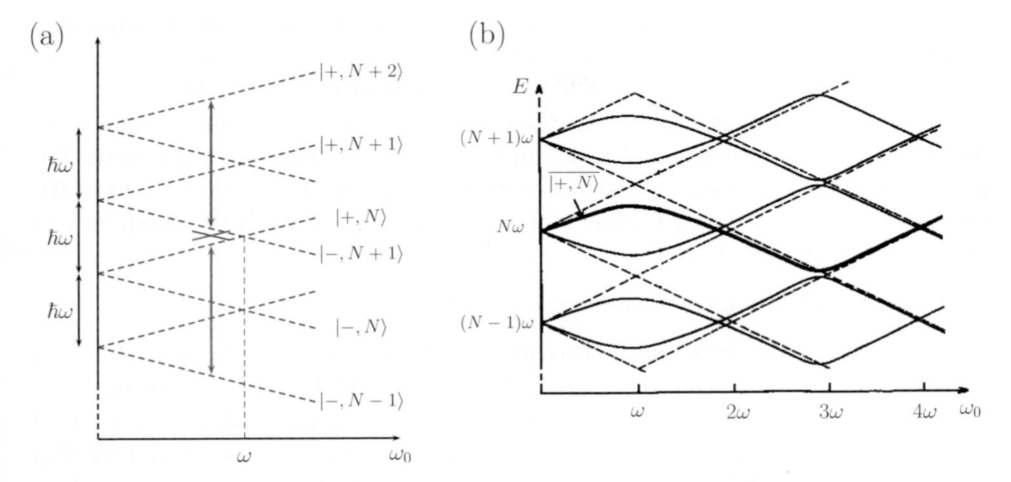

Fig. 9.11 Bloch-Siegert shift and multiphoton resonances for a spin 1/2 interacting with a σ-polarized RF field. (a) Near the crossing $\omega_0 = \omega$ corresponding to the one-photon resonance, the level $|+, N\rangle$ which is non-resonantly coupled to $|-, N-1\rangle$ is shifted upwards, whereas the level $|-, N+1\rangle$ which is non-resonantly coupled to $|+, N+2\rangle$ is shifted downwards. The anticrossing, which results from the resonant coupling between $|+, N\rangle$ and $|-, N+1\rangle$, is thus shifted towards lower static field values. This is the Bloch-Siegert shift. (b) The odd crossings between unperturbed states, $\omega_0 = \omega, 3\omega, \ldots$ become anticrossings describing multiphoton resonances when the coupling that exists between them is taken into account. There is no coupling between the non-perturbed levels that cross at an even crossing $\omega_0 = 0, 2\omega, 4\omega \ldots$, so these crossings remain crossings. Figure adapted from [Landre *et al.* (1970)]. Copyright: Elsevier Masson SAS.

Bloch-Siegert shift for the one-photon resonance

The dotted straight lines of Fig. 9.11 represent the unperturbed energies of the {atom+RF photons} states versus $\hbar\omega_0$. They form an ensemble of two straight lines with a slope ± 1 starting from points of the ordinate axis located at $N\hbar\omega$.

Consider the two energy levels $|-, N+1\rangle$ and $|+, N\rangle$ which cross when $\omega = \omega_0$. When the RF field is σ_+-polarized, these two states are coupled together, but no other coupling of each of these two states to another state exists. This is the case studied in Sec. 9.3.2, and is a consequence of the conservation of the total angular momentum: the photons associated with a σ_+-polarized RF field have an angular momentum $J_z = +\hbar$, and an atom in the $-1/2$ state can only absorb a photon and go to the state $+1/2$. In this case, the two diagonal matrix elements $\bar{R}_{\alpha\alpha}$ and $\bar{R}_{\beta\beta}$ introduced in Sec. 9.3.3 are equal to zero.

A σ-polarized field can be decomposed into the sum of σ_+ and σ_--polarized RF fields that correspond to photons having an angular momentum $+\hbar$ and $-\hbar$. When the RF field is σ-polarized, the $|-, N+1\rangle$ state is now coupled, not only to $|+, N\rangle$, but also to $|+, N+2\rangle$, by a (non-resonant) stimulated emission of a σ_- photon. Similarly, the state $|+, N\rangle$ is coupled, not only to $|-, N+1\rangle$, but also to $|-, N-1\rangle$ by a (non-resonant) absorption of a σ_- photon (see vertical arrows of the left part of Fig. 9.11). These extra couplings produce a negative displacement of $|-, N+1\rangle$

and a positive displacement of $|+, N\rangle$ described by the two diagonal elements $\bar{R}_{\alpha\alpha}$ and $\bar{R}_{\beta\beta}$ which are now no longer equal to zero.[7] The crossing between $|-, N+1\rangle$ and $|+, N\rangle$ is thus displaced to the left (see left part of Fig. 9.11).

We conclude from the previous discussion that the one-photon resonance induced by a σ-polarized RF field undergoes a radiative shift proportional to the RF power. When the resonance is monitored by sweeping the static field for a fixed RF frequency, the shift is towards lower static field values. It is called the Bloch-Siegert shift [Bloch and Siegert (1940)].

Odd spectrum of multiphoton resonances

Consider the crossing between the two unperturbed levels $|-, N+1\rangle$ and $|+, N-2\rangle$, occurring for $\omega_0 = 3\omega$ (see right part of Fig. 9.11). While the two crossing levels cannot be directly coupled by $\hat{V}$, because they correspond to $\Delta N = 3$, each of them is coupled to distant levels. This produces radiative shifts $\bar{R}_{\alpha\alpha}$ and $\bar{R}_{\beta\beta}$ proportional to the RF power. These can be easily calculated,[8] and one can show that they displace the crossing towards lower values of ω_0. In addition, the two crossing levels can be connected by a third order chain involving two non-resonant intermediate states $|+, N\rangle$ and $|-, N-1\rangle$:

$$|-, N+1\rangle \to |+, N\rangle \to |-, N-1\rangle \to |+, N-2\rangle.$$

An off-diagonal coupling $\bar{R}_{\alpha\beta}$, of order 3 in the RF field amplitude, thus exists between the two crossing levels and gives rise to a third order anticrossing (see right part of Fig. 9.11). We conclude that there exists a three-photon resonance appearing for $\omega_0 = 3\omega$ and that undergoes a radiative shift proportional to the square of the RF field amplitude towards low static field values and a radiative broadening proportional to the third power of the RF field amplitude.

These conclusions can readily be extended to all odd crossings of unperturbed levels occurring for $\omega_0 = (2p+1)\omega$, with $p = 1, 2, 3, \ldots$. They explain the $(2p+1)$-photon resonances appearing at these values of ω_0, their radiative shifts, proportional to the square of the RF field amplitude, and their radiative broadenings, proportional to the $(2p+1)$th power of the RF field amplitude. In this way, one can give a quantitative account of the experimental results shown in Fig. 9.7.

Even spectrum of crossings

Even crossings such as the crossing between $|-, N+1\rangle$ and $|+, N-1\rangle$, occurring for $\omega_0 = 2\omega$, also appear in Fig. 9.11. Radiative shifts exist for these crossing levels because of couplings to distant states, and give rise to a shift of the crossing towards low field values. But there is no off-diagonal coupling because of the conservation of the total angular momentum. To change N by 2, two interactions with the RF

[7] These matrix elements, as well as the analogous terms appearing for the multiphoton resonances considered below, are explicitly calculated in Complement A_{VI} of [Cohen-Tannoudji *et al.* (1992b)].
[8] See Complement A_{VI} of [Cohen-Tannoudji *et al.* (1992b)].

field are needed. The change of angular momentum of the field can be equal to 0 or $\pm 2\hbar$ but this is not the angular momentum $+\hbar$ corresponding to the atomic transition $-1/2 \to +1/2$. Similar arguments can be given for all even crossings $\omega_0 = 2p\omega$, with $p = 1, 2, 3, \ldots$. For all these crossings, the off-diagonal coupling $\bar{R}_{\alpha\beta}$ is equal to zero. The even crossings remain crossings and do not transform into anticrossings corresponding to multiphoton resonances.

Level crossing resonances

In Chap. 4, we have seen that, for free atoms, level crossing resonances can be observed near the values of the static magnetic field $\vec{B}_0$ where two Zeeman sublevels g_1 and g_2 cross. These resonances are related to off-diagonal elements $\langle g_2|\sigma|g_1\rangle$ of the atomic density matrix σ between the two sublevels, the so-called Zeeman coherences. Zeeman coherences can be introduced by the optical excitation if the light polarization is transverse with respect to the static field, and can build up if their evolution frequency $\omega_{21} = (E_{g_2} - E_{g_1})/\hbar$ is small enough, i.e. near the crossing point. Physically, the Zeeman coherences reflect the existence of a transverse magnetization $\langle J_\pm \rangle$ resulting from a transfer of angular momentum from the optical photons to the atoms in a direction perpendicular to $\vec{B}_0$.

In the problem considered in this chapter, the two levels which cross, for example near $\omega_0 = 2\omega$, are the two dressed states $\overline{|-, N+1\rangle}$ and $\overline{|+, N-1\rangle}$ (see Fig. 9.11). A transverse optical pumping can introduce a coherence between the two dressed states if the transverse angular momentum $J_\pm = J_x \pm iJ_y$ has a non-zero matrix element between them. In the absence of atom-RF coupling, $\langle -, N+1|J_\pm|+, N-1\rangle = 0$ because the atomic operator $J_\pm$ cannot change the number of photons. In the presence of the coupling, $\overline{|-, N+1\rangle}$ is contaminated by $|+, N\rangle$, and $\overline{|+, N-1\rangle}$ is contaminated by $|-, N\rangle$, so that $J_\pm$ can connect the two perturbed states containing admixtures with the same number of photons and different spin states. If the detection signal uses also a transverse polarization, it is sensitive to the coherence $\overline{\langle -, N+1|\sigma|+, N-1\rangle}$ and can reflect its resonant variations when ω_0 is scanned around 2ω. Level crossing resonances of the dressed atom have been observed near $\omega_0 = 2\omega$ and $\omega_0 = 4\omega$ [Cohen-Tannoudji and Haroche (1965)]. Their advantage is that they do not broaden appreciably when the RF power increases. Their radiative shifts can be measured more precisely than the radiative shifts of multiphoton resonances, allowing a more rigorous test of the theory presented in this section.

Deformation of the dressed-atom energy diagram for high RF power

Consider the dressed state $\overline{|+, N\rangle}$ represented by the heavy line of the right part of Fig. 9.11. A numerical diagonalization of the dressed atom Hamiltonian, performed by keeping a finite number of unperturbed states, allows one to follow the deformation of this dressed state even when the RF power increases to values so high that a perturbative calculation is no longer valid. Figure 9.12 taken from [Landre *et al.* (1970)] gives the shape of this dressed state for increasing values of Ω/ω where

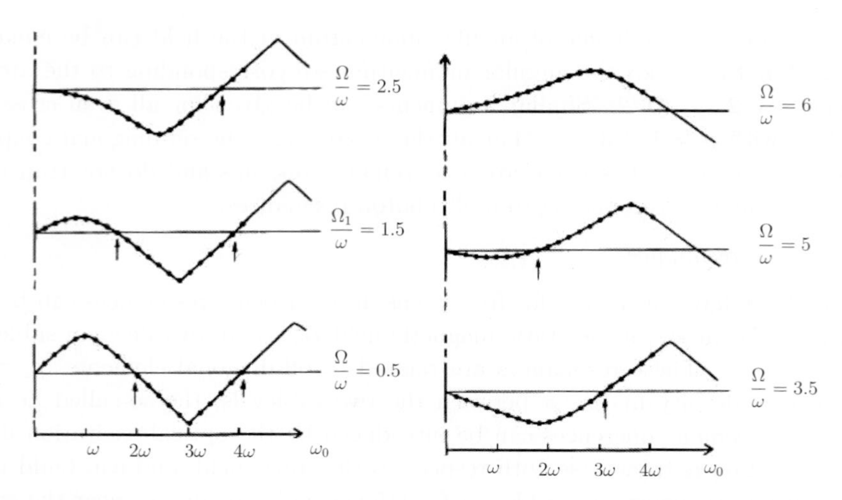

Fig. 9.12 Deformation of the dressed state represented by a heavy line in Fig. 9.11.b for increasing values of the RF power. Each curve corresponds to a value of Ω/ω, where Ω is the Rabi frequency. Figure adapted from [Landre *et al.* (1970)]. Copyright: Elsevier Masson SAS.

Ω is the Rabi frequency.[9] This deformation can be understood qualitatively. As the RF power increases, the strength of the first anticrossing initially appearing at $\omega_0 = \omega$ increases. This results in an increase of the distance between the two branches of the corresponding hyperbola, and thus a decrease of the amplitude of the first oscillation of the dressed state energy versus ω_0 appearing in Fig. 9.12. At the same time, the position of the first even crossing appearing at $\omega_0 = 2\omega$ moves to the left. This produces a decrease of the slope of the dressed state near $\omega_0 = 0$, which is none other than the decrease of the g-factor of a spin 1/2 interacting with a high frequency σ-polarized RF field studied in Chap. 8. When the shift of the first even crossing is so high that it arrives at $\omega_0 = 0$, the slope of the dressed state near $\omega_0 = 0$ is equal to zero. This happens for a value of $\Omega \simeq 2.4\omega$ corresponding to the first zero of the Bessel function $J_0(\Omega/\omega)$ introduced in Chap. 8 (see Eq. (8.12)). Then the slope of the dressed state is negative. Its absolute value increases, passes through a maximum and decreases until the shift of the even crossing, initially appearing at $\omega_0 = 4\omega$, is so high that it arrives at $\omega_0 = 0$, and gives rise to a second cancellation of the g, which corresponds to the second zero of $J_0(\Omega/\omega)$, and so on. We can thus understand the deformation of the energy diagram described by the various curves of Fig. 9.12 and relate it to other physical effects studied in Chap. 8.

[9] The points are experimental measurements performed with a transverse modulated optical pumping method described in Chap. 4.

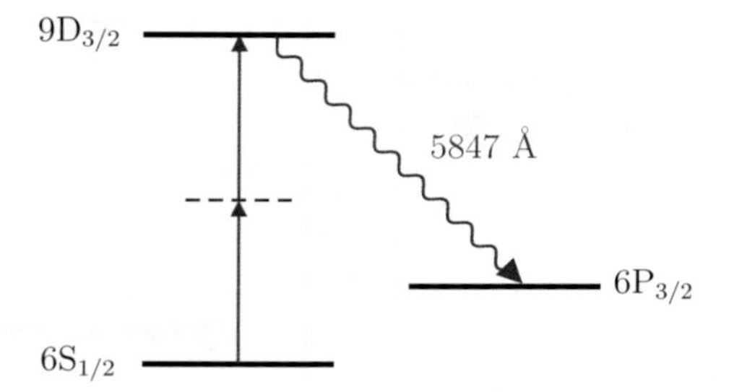

Fig. 9.13 Two-photon transition $6S_{1/2} - 9D_{3/2}$ of the cesium atom detected by the fluorescence light emitted in the transition $9D_{3/2} - 6P_{3/2}$.

9.4 Optical multiphoton processes between discrete states

9.4.1 *Introduction*

Multiphoton processes in the optical domain could not be observed before the advent of lasers because the intensity of the available light sources was not high enough. Among the first successful experiments, were the two-photon excitation in CaF_2 crystals by W. Kaiser and C. Garrett [Kaiser and Garrett (1961)], and the observation by I. Abella [Abella (1962)] of a two-photon excitation of the transition $6S_{1/2} - 9D_{3/2}$ in a cesium vapor and its detection by fluorescence at 5847 Å (see Fig. 9.13).

New features for optical multiphoton processes must be mentioned in comparison with what happens in the RF domain:

- Exchanges of linear momentum between atoms and photons give rise to appreciable Doppler shift while these shifts are generally negligible in the RF domain. Laser configurations can be found for cancelling the total linear momentum absorbed by the atom, as is the case for the two-photon Doppler-free spectroscopy discussed in Chap. 6.
- Spontaneous emission is no longer negligible in the optical domain, and spontaneously emitted photons can appear in a multiphoton process.
- There is a greater variety of atomic states that can be involved. For example, multiphoton processes involving continuum states can appear. They will be discussed in the next chapter.

9.4.2 *Radiative shift of Doppler-free two-photon resonances*

Like RF multiphoton resonances, optical multiphoton resonances can undergo radiative shifts that can be calculated by similar methods. These radiative shifts (which can be also considered as light shifts) perturb the atomic frequencies that one often wants to measure very precisely. This is particularly important for Doppler-free two-photon resonances, which are very narrow, and which are used for:

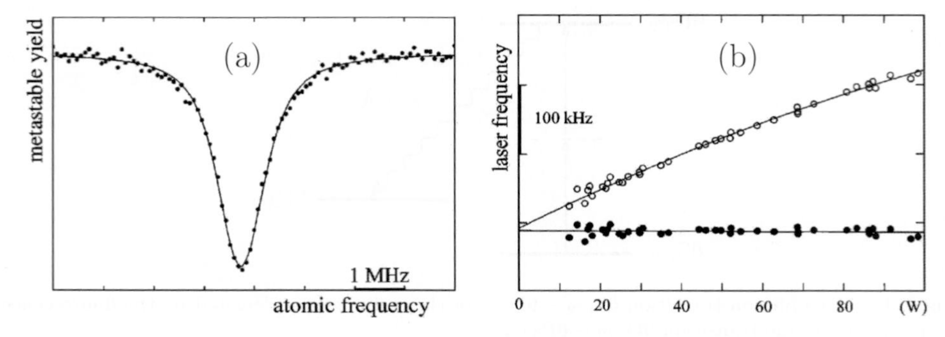

Fig. 9.14 (a) Two-photon Doppler-free transition $2S_{1/2}(F=1)-8S_{5/2}$ of hydrogen. (b) Radiative shift of the $2S_{1/2}-8D_{5/2}$ transition of deuterium when the light power increases (∘). The shift is large compared to the experimental uncertainties. Low part of the graph, same data but with the line position corrected from light shifts (•). Figure extracted from [de Beauvoir *et al.* (1997)]. Copyright: American Physical Society.

- creating frequency standards,
- measuring fundamental constants (such as the Rydberg constant Ry, or the fine structure constant α).

If these radiative shifts vary in space due to a spatial inhomogeneity of the laser intensity, they can also introduce a broadening of the resonance line. One must therefore carefully take them into account in several studies (e.g. optically pumped atomic clocks, transition 1s-2s of hydrogen). The position of the resonance must be extrapolated to zero laser intensity.

For example, Fig. 9.14 shows the radiative shift of the two-photon Doppler-free transitions 2S-8S/8D of hydrogen [de Beauvoir *et al.* (1997)]. It clearly appears that the shift is large. However, as shown in the lower part of the figure giving the corrected position of the line, which appears not to depend any longer on the light intensity, it is well understood theoretically.

9.4.3 *Stimulated Raman processes*

Stimulated Raman processes are other examples of two-photon processes where one photon is absorbed and a second one is emitted in a stimulated way. Figure 9.15(a) shows a process where the atom starting from level g_1 absorbs a photon ω_1 and goes to g_2 by stimulated emission of a photon ω_2. Figure 9.15(b) shows the reverse process.

9.4.3.1 *Differences with spontaneous Raman processes*

If the emitted photon is emitted by spontaneous emission and not by stimulated emission, the two-photon process is called spontaneous Raman effect. It is shown in Fig. 9.16, where the spontaneously emitted photon is represented by a wavy line. Depending on whether the spontaneously emitted photon has a frequency larger or

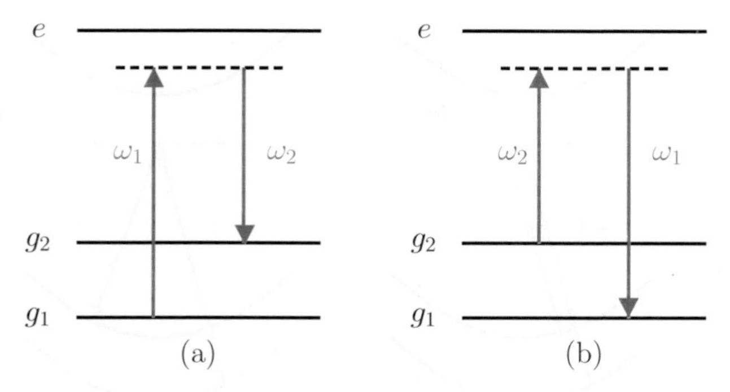

Fig. 9.15 (a) The atom goes from g_1 to g_2 by absorption of a photon ω_1 and stimulated emission of a photon ω_2. (b) Reverse process.

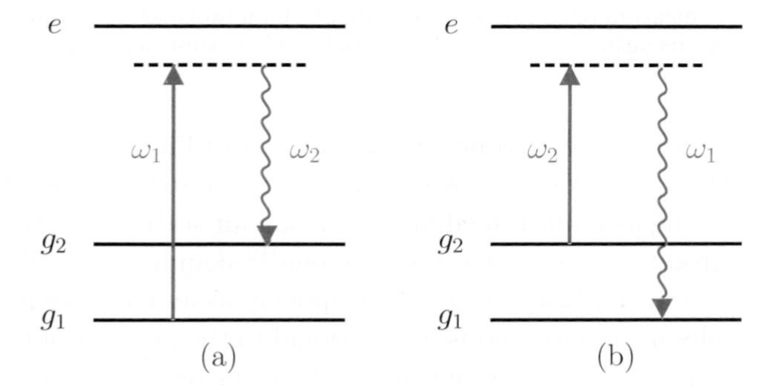

Fig. 9.16 (a) Spontaneous Raman Stokes process. (b) Spontaneous Raman anti-Stokes process.

smaller than the absorbed photon, the process is called Raman Stokes (Fig. 9.16(a)) or Raman anti-Stokes (Fig. 9.16(b)).

9.4.3.2 *Doppler effect*

The Doppler effect for stimulated Raman process can be easily understood by using energy-momentum diagrams analogous to those introduced in Chap. 6 for Doppler free two-photon absorption processes. Figure 9.17 is analogous to Fig. 6.6 of Chap. 6 and illustrates the influence of the Doppler effect for stimulated Raman processes. Depending whether the two photons propagate in the same direction (Fig. 9.17(a)) or in opposite directions (Fig. 9.17(b)), the Doppler effects subtract or add.

9.4.3.3 *Applications*

Stimulated Raman processes using laser light have a wealth of applications in chemistry, for chemical analysis, and in nonlinear optics, for generating new frequencies. Here, we will focus on applications in atomic physics and consider the case where g_1 and g_2 are two closely spaced ground state sublevels with a very long lifetime.

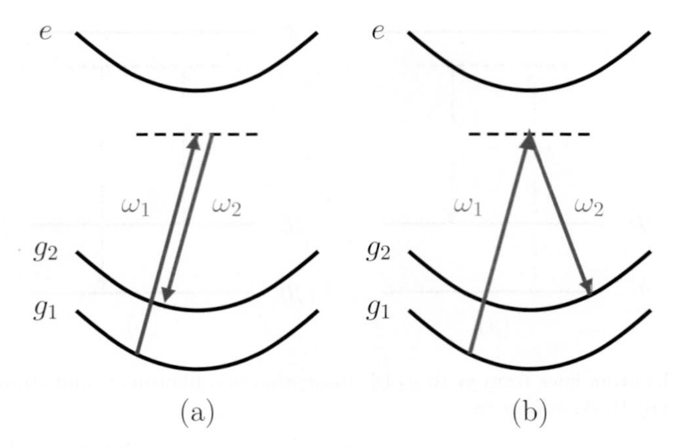

Fig. 9.17 Energy-momentum diagrams for stimulated Raman processes for co-propagating beams (a) and for counter-propagating beams (b). The Doppler effects subtract in the first case, and add in the second.

We will suppose that the two frequencies ω_1 and ω_2 of Fig. 9.15 are very far from resonance with the transitions $g_1 \to e$ and $g_2 \to e$ connecting g_1 and g_2 to any excited state e serving as an intermediate non-resonant state in the Raman process. One can then introduce an effective non-diagonal[10] coupling $\bar{R}_{12}$ between g_1 and g_2 that can be written as $\hbar\Omega_{\text{eff}}/2$ where Ω_{eff} appears as an effective Rabi frequency that directly couples g_1 and g_2, and is proportional to the product of the amplitudes of the two fields ω_1 and ω_2. If the detunings Δ_e from resonance of the transitions $g_1 \to e$ and $g_2 \to e$ are very large compared to the natural widths Γ_e of the levels e, one can neglect any spontaneous emission process during the Raman process and the effective Hamiltonian describing the reduced evolution within the manifold $\{g_1, g_2\}$ is Hermitian: There is no dissipation.

If the contribution of a particular level e is predominant in the sum over intermediate states, and if Ω_{1e} and Ω_{2e} are the Rabi frequencies of the two transitions $g_1 \to e$ and $g_2 \to e$, respectively, one can show that:

$$\bar{R}_{12} = \frac{\hbar\Omega_{\text{eff}}}{2} = \hbar\frac{\Omega_{1e}\Omega_{2e}}{4\Delta_e} \qquad \Rightarrow \qquad \Omega_{\text{eff}} = \frac{\Omega_{1e}\Omega_{2e}}{2\Delta_e}. \tag{9.19}$$

The widths of levels g_1 and g_2 due to their contamination by e are smaller than Ω_{eff} by a factor on the order of Γ_e/Δ_e. We have also supposed that Δ_e is large compared to the splitting between g_1 and g_2.

We thus get a situation formally equivalent to that of a two-level system $\{g_1, g_2\}$ excited by a monochromatic excitation at frequency[11] $\omega_1 - \omega_2$, with an effective Rabi

[10] There are also diagonal terms $\bar{R}_{11}$ and $\bar{R}_{22}$ describing radiative shifts of g_1 and g_2. We suppose that these shifts are re-included in the unperturbed energies of these states.

[11] Note that the laser at frequency ω_2 is usually obtained by an acousto-optic or electro-optic technique from the laser at frequency ω_1, so that the jitter of the laser disappears in the subtraction $\omega_1 - \omega_2$.

frequency Ω_{eff}. In this sense, one could think that it would be simpler to directly excite the transition $g_1 \to g_2$ with a resonant microwave field. The advantage of using a stimulated Raman process, however, is that one can transfer a considerably higher momentum to the atom if the two laser beams at frequencies ω_1 and ω_2 are counter-propagating. In this case, the transfer of momentum is on the order of $\hbar(\vec{k}_1 - \vec{k}_2) \simeq 2\hbar\vec{k}_1$, since $\vec{k}_2 \simeq -\vec{k}_1$, whereas it is negligible for a microwave transition. A consequence of this large momentum transfer in a stimulated Raman process, and of the long lifetime of the ground state sublevels, is that the velocity selectivity of the transfer of atoms from g_1 to g_2 is extremely high. The velocity spread δv of the transferred atoms is given by $2k_1\delta v \simeq \gamma$, where γ is the width of g_1 and g_2, which is essentially determined by the transit time of atoms in the excitation zone, and which is much smaller than the natural width of an excited atomic state. For values of γ on the order of 10^3 Hz and for laser wavelengths on the order of 1 μm, one gets values of δv smaller than 1 mm.s^{-1}. This can be very useful for certain atomic interferometry experiments that require the preparation of atomic wave packets in different internal atomic states with very long spatial coherence lengths.[12] Examples of such experiments will be described in Chap. 18. This high selectivity in velocity can be also used to measure the momentum distribution of an ensemble of atoms with a very high resolution. For example, this method, known as Bragg scattering, has been used to measure the momentum distribution of atoms in a Bose-Einstein condensate.[13]

Stimulated Raman transitions are also used to imprint the phase of the lasers onto the atomic wave function. This will be illustrated in Chap. 18 where this phase imprinting is used to measure the free fall of atoms in a gravimeter. Another application stressed in Sec. 2.3.2 of Chap. 2 is the generation of vortices by imprinting the phase pattern of a Laguerre Gauss beam on a Bose-Eintein condensate.

Stimulated Raman processes are also very useful for laser cooling of ions. In this case, the two states $|g_1 = v, m\rangle$ and $|g_2 = v', m'\rangle$ describe the ion in two different vibrational states v and v' in the trapping potential and in two different internal states m and m' with long lifetimes. As a result, the transition $g_1 \to g_2$ has a very narrow linewidth and is well resolved from the other vibrational transitions. This allows a resolved sideband excitation of the ion that can be used to cool it to the lowest vibrational state. More detail regarding this process will be given in Chap. 13.

Note finally that the configuration of Fig. 9.15 with two lower and one upper atomic states driven by two coherent laser fields at frequency ω_1 and ω_2 in a Λ-type configuration was already introduced in Sec. 4.6 of Chap. 4 in connection with coherent population trapping. Stimulated Raman resonances are thus closely con-

[12]Note that this velocity selective excitation process is not a cooling, but a selection since the velocity distribution is not changed.

[13]The two states g_1 and g_2 are then two different energy momentum states of the atom in the same internal state and the stimulated Raman process can be also interpreted as a stimulated Rayleigh process (see end of this subsection).

nected to other effects studied in this book, like dark resonances, Stimulated Raman Adiabatic Passage (STIRAP), allowing one to selectively transfer molecules from one state to another [Bergmann *et al.* (1998)], subrecoil laser cooling by Velocity Selective Coherent Population Trapping (VSCPT) which will be described in more detail in Chap. 13.

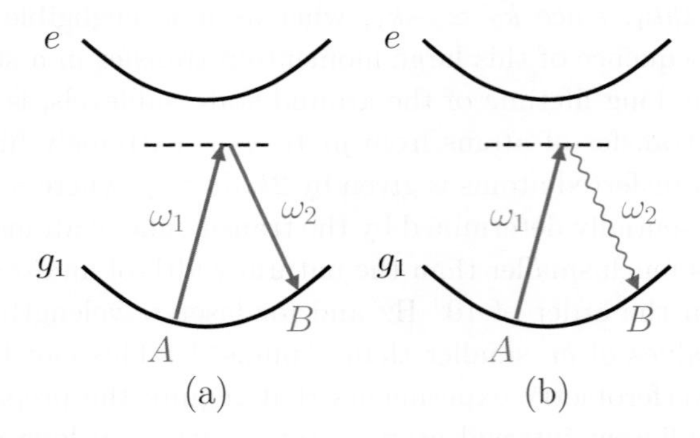

Fig. 9.18 Stimulated (a) and spontaneous (b) Rayleigh processes. A photon ω_1 is absorbed and a photon ω_2 propagating in the opposite direction is emitted in a stimulated (a) or spontaneous way (b) while the atom remains in the same internal state g_1.

Stimulated Rayleigh processes

One can also consider two-photon processes analogous to the previous ones but where the initial and final atomic internal states are the same. They correspond to Rayleigh scattering. Figure 9.18(a) represents such a process in the case where the two photons propagate in opposite directions. If the photon ω_2 is emitted spontaneously (Fig. 9.18(b)), the process is called spontaneous Rayleigh scattering. The conservation of the total energy and total momentum clearly appears in Fig. 9.18 since the initial and final atomic states A and B belong to the parabola of the energy momentum states of the atom. If the two photons ω_1 and ω_2 of Fig.9.18(a) are two counterpropagating photons of a standing wave, they have the same frequency ω and the initial and final atomic states A and B must have opposite momenta (see Fig. 9.19). The process is often called Bragg scattering. It can be interpreted as the diffraction of the initial atomic de Broglie wave by the periodic optical potential created by the non-resonant standing wave and is analogous to the diffraction of a light wave by a material grating. The respective roles of light and matter are interchanged. We will come back in Chap. 17 to the diffraction of atomic de Broglie waves by standing waves. Even if ω_1 and ω_2 are not equal, stimulated Rayleigh processes like those of Fig. 9.18(a) are called Bragg scattering. Because of their high selectivity in velocity, they are currently used now to measure the dispersion law of the elementary excitations of a Bose-Einstein condensate.[14]

Analogy with Compton scattering. If the parabola of Fig. 9.18(a) giving the energy momentum states of an atom is replaced by the corresponding parabola giving, in the non-relativistic limit, the energy momentum states of a free electron, the processes represented in Fig. 9.18 correspond to stimulated and spontaneous Compton scattering processes where a photon is scattered by a free electron with conservation of the total

[14]See Chap. 24.

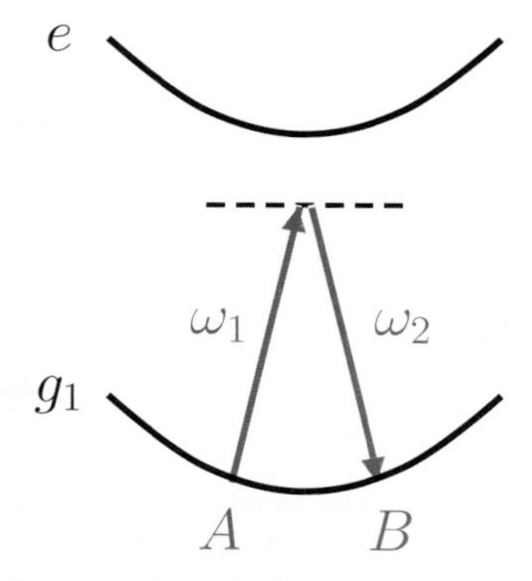

Fig. 9.19 Stimulated Rayleigh processes in a laser standing wave. We have then $\omega = \omega_1 = \omega_2$ and the initial and final atomic states A and B have opposite momenta. The scattering process can be considered as the Bragg diffraction of the atomic de Broglie wave by the optical periodic potential created by the non-resonant standing wave.

energy and momentum. The equivalent for electrons of the process represented in Fig. 9.19 was suggested by Piotr Kapitza and Paul Dirac and was the first proposal to diffract an electronic de Broglie wave by a periodic light structure [Kapitza and Dirac (1933)].

9.4.4 *Phase matching condition. Application to degenerate four-wave mixing*

In a multiphoton process during which a certain number of photons are absorbed and a certain number emitted, the net momentum gained by the atom is equal to:[15]

$$\delta\vec{p} = \hbar\left[\left(\sum \vec{k}_i\right)_{\text{absorbed}} - \left(\sum \vec{k}_j\right)_{\text{emitted}}\right]. \tag{9.20}$$

Consider two atoms A and B, with initial momenta $\vec{p}_A$ and $\vec{p}_B$. If atom A is involved in the multiphoton process, its momentum will increase by an amount equal to $\delta\vec{p}$. A similar result holds for B. The final momenta of the two atoms will be $(\vec{p}_A + \delta\vec{p}, \vec{p}_B)$ in the first case, $(\vec{p}_A, \vec{p}_B + \delta\vec{p})$ in the second one. The two amplitudes associated with the two multiphoton processes involving either A or B can interfere only if the final states of the two atoms are the same, which implies:

$$\delta\vec{p} = \vec{0} \quad \Rightarrow \quad \left(\sum \vec{k}_i\right)_{\text{absorbed}} = \left(\sum \vec{k}_j\right)_{\text{emitted}}. \tag{9.21}$$

Condition $\delta\vec{p} = \vec{0}$, called phase matching condition, insures that there is no which-path information preventing the waves generated by the various atoms of the medium to interfere constructively. We now apply these ideas to a nonlinear optical process called *degenerate four-wave mixing*. Consider three waves 1, 2, 3 of the same frequency ω propagating in an atomic medium with waves 1 and 2

[15] Strictly speaking, the index of refraction of the medium containing the two atoms has to be taken into account.

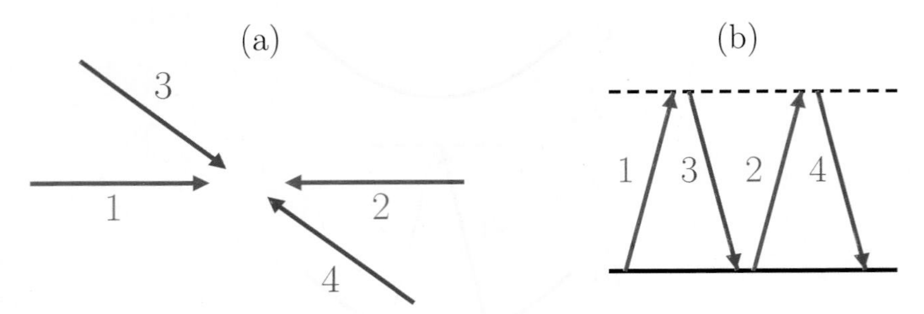

Fig. 9.20 (a) Three waves 1, 2, 3 having the same frequency ω propagate in an atomic medium. Waves 1 and 2 have opposite directions of propagation. A nonlinear process gives rise to a fourth wave 4 propagating in a direction opposite to wave 3. (b) The nonlinear process consists of the absorption of a photon from wave 1, the stimulated emission of a photon in wave 3, the absorption of a photon from wave 2 and, finally the emission of a photon in the new wave 4. The phase matching condition implies that waves 3 and 4 propagate in opposite directions

in opposite directions (see Fig. 9.20(a)). A sequence of absorption and stimulated emission processes gives rise to a new wave 4 (see Fig. 9.20(b)). Conservation of energy implies that wave 4 must have the same frequency as the three other waves. According to Eq. (9.21), the waves 4 emitted by the various atoms of the medium can interfere only if $\vec{k}_1 + \vec{k}_2 = \vec{k}_3 + \vec{k}_4$, which implies $\vec{k}_4 = -\vec{k}_3$ since $\vec{k}_1 = -\vec{k}_2$. The new wave 4 must therefore propagate in a direction opposite to wave 3.[16]

It is also instructive to study the space and time dependence of the dipole moments induced in the medium by the waves 1, 2 and 3, whose positive frequency component (varying as $\exp(-i\omega t)$) is given by the product of the positive frequency components of the absorbed waves 1 and 2

$$\mathcal{E}_1 \exp[i(\vec{k}_1 \cdot \vec{r} - \omega\, t)] \times \mathcal{E}_2 \exp[i(\vec{k}_2 \cdot \vec{r} - \omega\, t)] = \mathcal{E}_1 \mathcal{E}_2 \exp(-2i\omega t)$$

(since $\vec{k}_1 = -\vec{k}_2$) multiplied by the negative frequency component $\mathcal{E}_3^* \exp[-i(\vec{k}_3 \cdot \vec{r} - \omega t - \varphi)]$ of the emitted wave 3, where φ is the phase of wave 3. By a proper choice of the origin of time, one can always suppose $\mathcal{E}_1$ and $\mathcal{E}_2$ to be real. The product of these three components added to its complex conjugate is the source of wave 4, and gives the space time dependence of this wave 4:

$$\text{Wave}\, 4 \propto \mathcal{E}_1 \mathcal{E}_2 \mathcal{E}_3^* \exp[-i(\vec{k}_3 \cdot \vec{r} + \omega t - \varphi)] + \mathcal{E}_1 \mathcal{E}_2 \mathcal{E}_3 \exp[i(\vec{k}_3 \cdot \vec{r} + \omega t - \varphi)]. \quad (9.22)$$

It is interesting to compare this expression with the space time dependence of wave 3:

$$\text{Wave}\, 3 \propto \mathcal{E}_3 \exp[i(\vec{k}_3 \cdot \vec{r} - \omega t - \varphi)] + \mathcal{E}_3^* \exp[-i(\vec{k}_3 \cdot \vec{r} - \omega t - \varphi)]. \quad (9.23)$$

The amplitude of the positive frequency part of wave 3 is multiplied by $\mathcal{E}_3 \exp(-i\varphi)$ whereas the amplitude of the positive frequency part of wave 4 is multiplied by the complex conjugate of this factor, $\mathcal{E}_3^* \exp(+i\varphi)$. This is why wave 4 is called phase

[16] Another interpretation of the condition $\vec{k}_1 + \vec{k}_2 = \vec{k}_3 + \vec{k}_4$ can be given in terms of Bragg diffraction of wave 2 by the intensity grating resulting from the interference of the two waves 1 and 3.

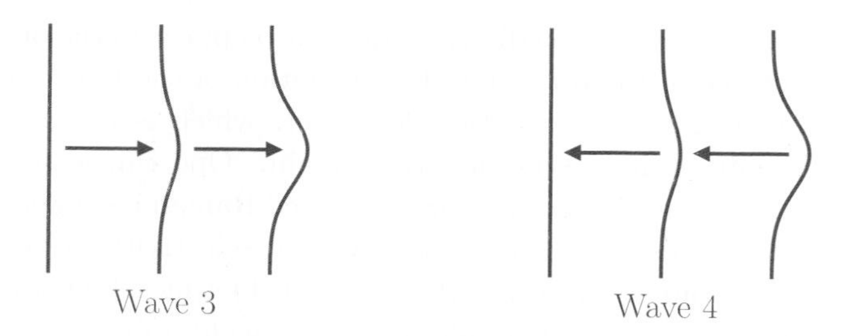

Fig. 9.21 The time evolution of wave 4 is the time reversal of the evolution of wave 3.

conjugate of wave 3. The two equations (9.22) and (9.23) also show that wave 4 is obtained from wave 3 by changing t into $-t$ (time reversal). If wave 3 undergoes phase distortions while it propagates in the medium, the wave 4 obtained by four wave mixing will have these distortions corrected (see Fig. 9.21).

9.5 Conclusion

In this chapter, we have described several examples of multiphoton resonances between discrete states that provide clear examples of situations that require to go beyond a first order treatment of atom-photon interactions.

These multiphoton resonances have been first observed in the RF domain in experiments using the high sensitivity of optical methods described in Chap. 3. The polarization of the RF field, which determines the angular momentum of the RF photons, can be easily changed experimentally and the positions of the multiphoton resonances can be clearly interpreted by considering how the atom can go from one Zeeman sublevel to another by absorbing a certain number of photons with well-defined energies and angular momenta.

In the description of these resonances and in the calculation of their radiative shifts and broadenings, we have frequently used a quantum treatment of the RF field. Obviously, the quantum nature of the RF field is not essential and a purely classical description would have been sufficient. A quantum treatment of the RF field, however, leads to simpler calculations of the radiative shifts and broadenings because the Hamiltonian of the total system of $\{$ atom+RF photons $\}$ is time independent. It is then possible to introduce true energy levels, corresponding to the energy levels of the atom "dressed" by RF photons, and to use time independent perturbation theory. From this point of view, multiphoton resonances appear at the position of "level anticrossings" of the dressed atom energy levels. We have shown that consideration of the dressed atom energy diagram provides a global interpretation of various effects appearing in RF spectroscopy (multiphoton resonances, level crossing resonances, radiative shifts and broadenings, modification of the g-factor), even in a high intensity regime where perturbation theory cannot be applied.

Multiphoton processes between discrete states also appear in the optical domain. By a convenient choice of the direction of propagation of the laser beams exciting the atom, one can compensate the Doppler effect, which is no longer negligible in the optical domain as it was in the RF domain. One can also enhance it in certain configurations that can be used for stimulated Raman resonances and realize situations where these resonances have a very high selectivity in velocity if they connect two ground state sublevels with very long lifetimes. This opens several interesting perspectives for the manipulation of ultracold atoms and molecules. A few of them will be described in subsequent chapters.

Chapter 10

Photoionization of atoms in intense laser fields

10.1 Introduction

Up to now, we have only considered single-photon or multiphoton transitions taking place between two atomic discrete states. In this chapter, we suppose that the final state of the atom belongs to a continuum starting at an energy E_I above the atomic ground state. E_I is the *ionization energy* of the atom, i.e. the minimum energy required for removing an electron from the Coulomb potential well of the nucleus. Such a process is called *photoionization*.

The left part of Fig. 10.1 represents a single-photon ionization process. Conservation of energy requires that the energy of the continuum state reached after the absorption of the photon, which is nothing but the kinetic energy of the photoelectron once it is far from the parent ion, is given by the equation:

$$h\nu = E_I + \frac{1}{2}mv^2. \tag{10.1}$$

An equation like (10.1) was first given by Albert Einstein in his famous 1905 paper on light quanta [Einstein (1905)]. He was applying his theory of light quanta introduced in this paper to the photoelectric effect with metals [1] and predicting that the kinetic energy of the photo-electrons was increasing linearly with the frequency of light and not with the light intensity. Einstein's ideas were not accepted by his contemporaries. It took more than ten years before Robert Millikan, who did not believe initially in Einstein's law, obtained in 1916 experimental results showing that it was correct and that it did provide a precise value of the Planck constant h. The denomination of photons for these light quanta was given only in 1926 by Gilbert Lewis [Lewis (1926)].

Before the advent of intense light sources, only single-photon ionization processes were essentially considered. Actually, atoms can be photoionized, not only by absorbing a single photon with an energy $h\nu$ larger than E_I/h, but also by absorbing several photons of energy $h\nu' < E_I$ such that the sum of the energy of those photons is larger than the ionization energy E_I. The right part of Fig. 10.1

[1] The equivalent of the ionization energy considered above is the extraction work which must be provided to extract the electron from the metal.

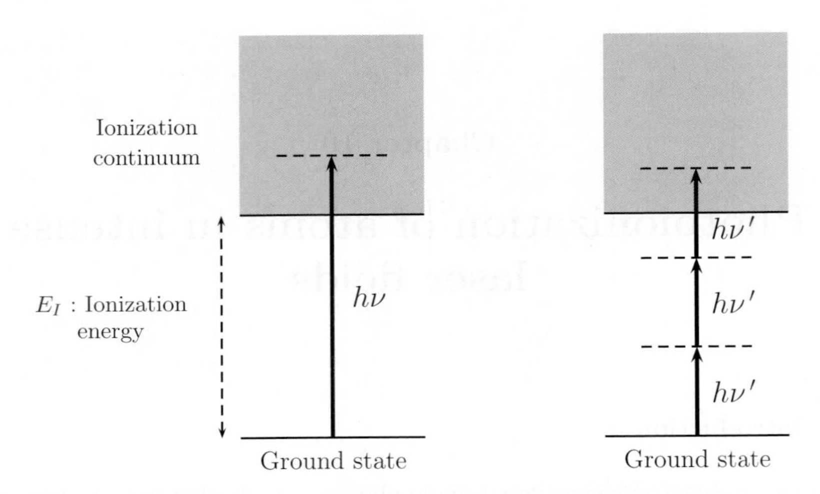

Fig. 10.1 Left: single-photon ionization. Right: three-photon ionization.

represents a three-photon ionization process, where the light frequency ν' is such that $h\nu' < E_I$, $2h\nu' < E_I$, but $3h\nu' > E_I$. The generalization of Eq. (10.1) for a three-photon ionization is:

$$3h\nu' = E_I + \frac{1}{2}mv^2. \tag{10.2}$$

Multiphoton ionization was first observed in the 1960's with Q-switched lasers[2]: 7-photon ionization of xenon in Russia [Voronov and Delone (1966)]; multiphoton ionization of rare gases, including helium, by the Saclay group in France [Agostini *et al.* (1968)]. In Sec. 10.2, we review a few properties of these processes.

In a multiphoton ionization process, the atom can absorb more photons than the minimum number required to reach the ionization threshold. It then reaches higher states in the continuum. This means that the kinetic energy of the photoelectrons can increase with the light intensity, but this increase is only by discrete amounts equal to an integer number of $h\nu$. These processes are called *Above Threshold Ionization* (ATI) and will be described in Sec. 10.3.

In parallel with ATI, other processes can take place, like harmonic generation, consisting of the transformation of an odd number $2n+1$ of photons $h\nu$ into a single photon of energy $(2n+1)h\nu$. These processes will be briefly described in Sec. 10.4. We will see that, after an initial decrease of the harmonic intensity with its order n, a plateau appears allowing the observation of higher order harmonics.

For sufficiently high laser intensities, ATI and harmonic generation (which then contains harmonics with large orders) exhibit features which cannot be explained by a perturbative treatment. These features characterize a new regime of photoionization where the electron leaves the atomic ion, not by absorbing a certain

[2]See Sec. 27.3.1 of Chap. 27.

number of photons, but by tunneling through a potential barrier resulting from the superposition of the Coulomb potential of the ion and of a Stark potential due to the laser electric field. This tunnel ionization regime will be described in Sec. 10.5, using a semiclassical description of the electron motion in the laser field once it has been released from the ion. The possibility that the electron comes back in the vicinity of the ion and recollides with it gives rise to very interesting effects, like the generation of extremely short bursts of extreme UV, XUV or X-ray photons.

This better understanding of tunnel ionization has stimulated spectacular advances in our ability to generate, to measure and to control attosecond pulses of XUV radiation (1 as $= 10^{-18}$ s), and to apply them to the investigation of ultra-short electron dynamics in atoms and molecules. We review a few of these recent advances in Chap. 27.

10.2 Multiphoton ionization

10.2.1 *Parameters influencing the multiphoton ionization rate*

Laser intensity

In a perturbative treatment, the K-photon ionization rate W is proportional to the K-th power I^K of the laser intensity I. In a log-log plot, the curve giving the variation of W with I should thus be a straight line with a slope K. This is actually what is observed with laser intensities on the order of 10^{13} W·cm^{-2} (see Fig. 10.2). For example, the number of photons required for ionizing xenon, krypton and argon with a Neodynium laser are K =11, 13, 14, respectively. One finds that these numbers are actually the slopes of the three straight lines of Fig. 10.2 giving W versus I for these three atoms [Mainfray and Manus (1978)].

Coherence of the laser field

Two laser fields with the same average intensity $\langle I \rangle$ do not in general give the same rate W because $\langle I^K \rangle \neq \langle I \rangle^K$. For example, several experiments performed in Saclay have shown that multiphoton ionization rates are larger with a multimode laser in comparison with a cw laser having the same average intensity.

Intermediate resonance

It is possible that a discrete state e, with a width Γ and an energy E_e is very close to $E_g + K'\hbar\omega$ with $K' < K$. The K-photon ionization process has in this case a quasi-resonant intermediate state. The propagator

$$\frac{1}{E_g + K'\hbar\omega - E_e + i\hbar\Gamma/2} \tag{10.3}$$

relative to the intermediate state e can become very large.

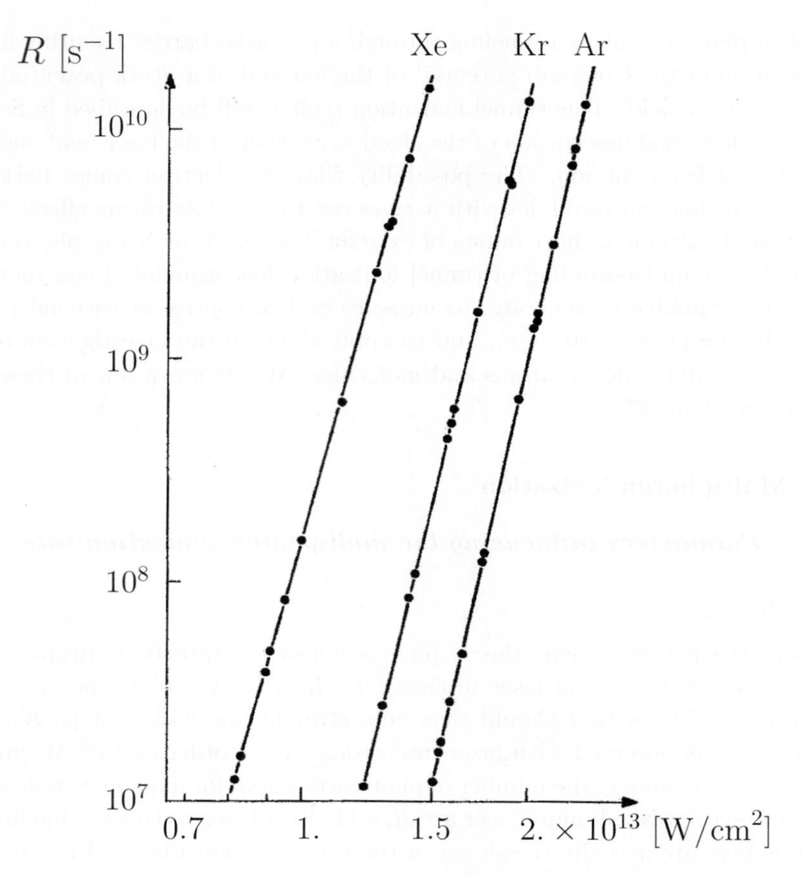

Fig. 10.2 Variations in a log-log scale of the photoionization rate, R, versus the Neodymium laser intensity for three rare gas atoms: xenon, krypton and argon. Figure adapted from [Mainfray and Manus (1978)]. Copyright: EDP Sciences.

Figure 10.3 shows an example of intermediate resonance in the case $K = 2$, $K' = 1$. When the light intensity varies, the light shifts of the intermediate state vary and this changes the detuning from resonance to the intermediate state. The variation of the ionization rate with the laser intensity is, in this case, more complex than the power law dependence mentioned above.

Polarization of the laser

The polarization selection rules combined with the parity selection rules (electric dipole transitions) lead to different multiphoton ionization paths depending whether the laser polarization is σ_+ or π [Klarsfeld and Maquet (1972)]. Resonances can appear on one path and not on the other, making the ionization rate polarization-dependent.

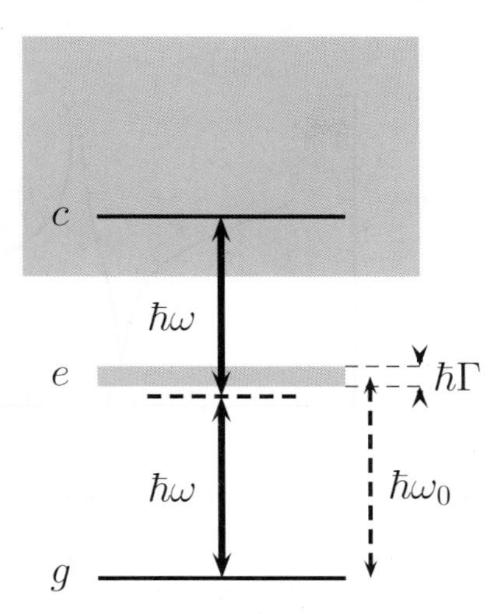

Fig. 10.3 Quasi-resonant intermediate state in a two-photon ionization process.

10.2.2 *Quantum interference effects in multiphoton ionization*

Consider the case of a two-photon ionization process. Figure 10.4(a) shows how an atom can go from the ground state g to a state c of the continuum by absorbing two photons $\hbar\omega$. Two excited discrete states e_1 and e_2 are also represented in this figure. When ω varies, resonances occur when ω coincides with one of the frequencies ω_{10} and ω_{20} corresponding to the transitions $g \to e_1$ and $g \to e_2$. When ω lies between ω_{10} and ω_{20}, the two energy denominators associated with the two intermediate states e_1 and e_2 have opposite signs and the corresponding amplitudes interfere with a minus sign.

These predictions are confirmed by a theoretical calculation of the two-photon ionization rate of cesium atoms (see Fig. 10.4(b)) showing clearly the maxima corresponding to the resonant intermediate excitation of the np states with $n = 7, 8, 9$ and the minima between two resonances.[3]

The comparison of the position and the depth of these minima with the theoretical predictions allows one to test the accuracy of the calculations of the energies and matrix elements appearing in the perturbative expression of the transition amplitude. More interestingly, the possibility to use these quantum interference effects for blocking a physical or chemical process in one sense and to orient it in another sense is called *coherent control* and is presently the subject of several theoretical [Brumer and Shapiro (1986); Tannor *et al.* (1986)] and experimental investigations [Chen *et al.* (1990); Potter *et al.* (1992); Zhu *et al.* (1995)].

[3] Experiments have been also performed to demonstrate the existence of these minima (see [Morellec *et al.* (1980)]).

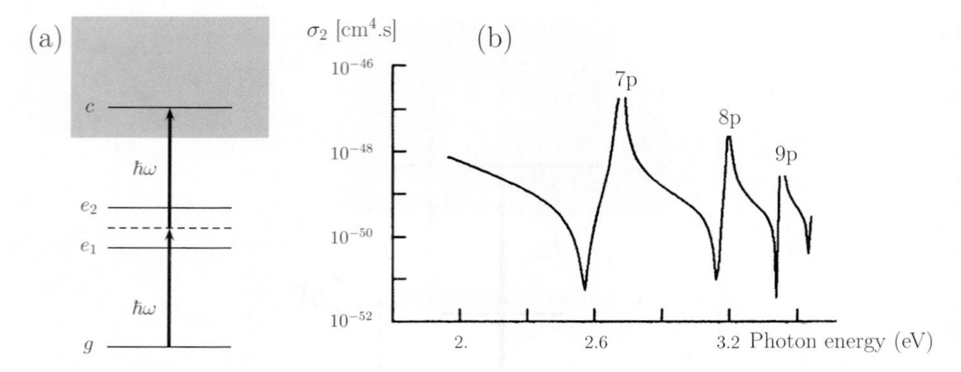

Fig. 10.4 (a) Two-photon ionization process. Different transition amplitudes have to be considered differing by the intermediate atomic state e_1 or e_2 that appears after the absorption of the first photon. (b) Calculated two-photon ionization of cesium atoms. The resonances associated with the intermediate states 7p, 8p, 9p, clearly appear as well as the minima of ionization probability between two resonances. Figure taken from [Barry Bebb (1966)]. Copyright: American Physical Society.

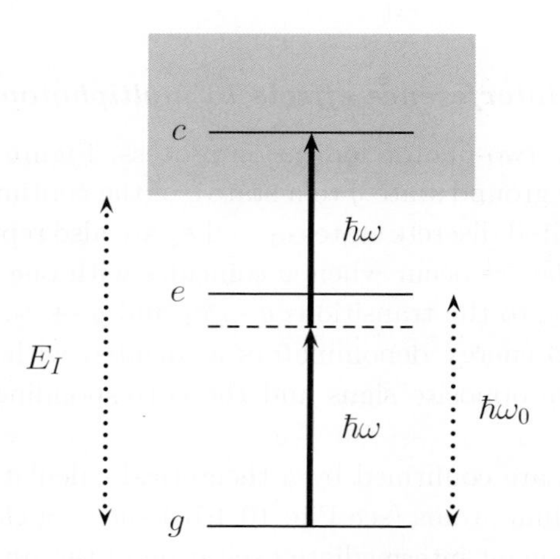

Fig. 10.5 Two-photon ionization near a resonance with an intermediate state e. The transition amplitudes where the intermediate state is e or any other intermediate state interfere, and give rise to an asymmetric profile in the two-photon ionization rate.

10.2.3 *Asymmetric line profiles in resonant multiphoton ionization*

Figure 10.5 shows a two-photon ionization process where ω is close to the frequency ω_0 of an atomic transition $g \to e$. The atom can go from the ground state g to a continuum state c

- either through the intermediate state e. If ω is close to ω_0, the amplitude varies resonantly when ω is scanned.

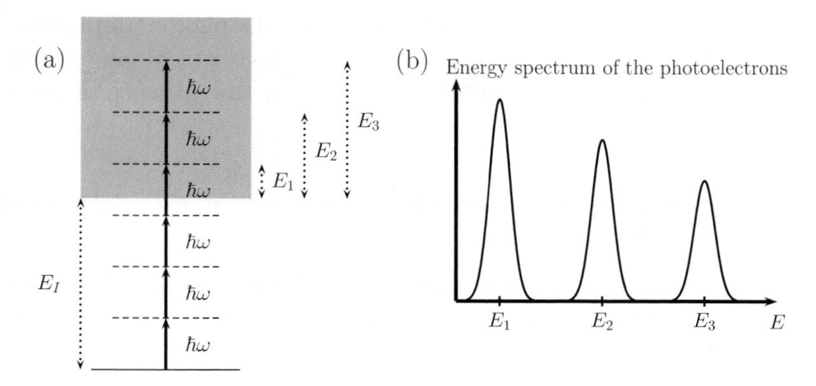

Fig. 10.6 (a) Above threshold ionization. Once brought in the ionization continuum, the photoelectron continues to absorb photons reaching states with kinetic energies E_1, E_2, E_3, ...(b) The energy spectrum of the photoelectrons consists of equidistant peaks $E_3 - E_2 = E_2 - E_1 = \hbar\omega$.

- or through any other non-resonant intermediate state. The corresponding amplitude varies smoothly when ω is scanned about ω_0.

The interference between these two amplitudes gives rise to asymmetric profiles reminiscent of the Fano profiles observed in the photoionization of an atom near a autoionizing state [Armstrong (1975); Dixit and Lambropoulos (1979)].

10.3 Above threshold ionization (ATI)

10.3.1 *Multiphoton transitions between states of the continuum*

Figure 10.6(a) describes a four-photon ionization process bringing the atom in a state of the continuum with a kinetic energy E_1 given by $E_1 = 4\hbar\omega - E_I$. It can also absorb (directly or sequentially) an extra photon and reach a state of the continuum with a kinetic energy E_2 given by $E_2 = 5\hbar\omega - E_I$, or two extra photons allowing it to reach another state of the continuum with a kinetic energy E_3 given by $E_3 = 6\hbar\omega - E_I$. And so on. If one looks at the energy spectrum of the photo-electrons, one thus expects to see a series of equidistant peaks separated by $h\nu = \hbar\omega$ and located at $E_1, E_2, E_3, \ldots$, as shown in Fig. 10.6(b). This phenomenon is called Above Threshold Ionization (ATI) and was observed for the first time in 1979 [Agostini *et al.* (1979)].

ATI clearly shows that, once photoionized, the electron energy continues to increase when the light intensity increases, contrary to what was believed in the first interpretations of the photoelectric effect. However, this increase is not continuous, it occurs by discrete steps given by an integer of the photon energy $\hbar\omega$.

At first sight, one could think that ATI is impossible because a free electron cannot absorb a photon (the total energy and the total momentum could not be conserved). But the photoionized electron is not really free. It still feels the Coulomb field of the ion which can absorb the recoil following the photon absorption. What happens in ATI is somewhat analogous to inverse "bremsstrahlung," which is the acceleration of electrons by absorption of photons in the presence of an external field.

10.3.2 *Consequences of the oscillatory motion of the electron in the laser field*

We have seen in Chap. 8 that a weakly bound electron oscillates in a laser field $E_0 \cos \omega t$ at the frequency ω of this field. The corresponding kinetic energy, $\mathcal{E}_V$, called "ponderomotive energy", is given by

$$\mathcal{E}_V = \frac{q^2 E_0^2}{4m\omega^2}, \tag{10.4}$$

where q is the electron charge and E_0 the laser field amplitude. $\mathcal{E}_V$ is proportional to the laser intensity I and to the square λ^2 of the laser wavelength.

With intensities on the order of 10^{13} W.cm^{-2}, and for laser wavelengths on the order of 1 μm, $\mathcal{E}_V$ is on the order of 1 eV. The light shift of the ground state g can be shown to be smaller and will be neglected in the following. The ionization energy E_I of the atom and also the energy of the Rydberg states thus increase by an amount $\mathcal{E}_V$ that is not negligible compared to E_I.[4] Figure 10.7 shows how this change of the ionization energy E_I varies along a line crossing the laser beam. This introduces a decrease of the kinetic energy of the photoelectron depending on the place x where it is produced. At point x, this kinetic energy is equal to

$$E_C^{(1)}(x) = N\hbar\omega - E_I - \mathcal{E}_V(x). \tag{10.5}$$

It is smaller than the kinetic energy in the absence of oscillatory motion:

$$E_C^{(2)} = N\hbar\omega - E_I. \tag{10.6}$$

In fact, the electron is not detected inside the laser beam, but once it is outside. The ponderomotive energy $\mathcal{E}_V(x)$ depends on the position x of the electron inside the laser beam and appears as an effective potential energy for its motion giving rise to a force

$$\vec{F}(x) = -\vec{\nabla}\mathcal{E}_V(x) \tag{10.7}$$

called *ponderomotive force*. Under the effect of this force, the electron is accelerated during its exit from the laser beam, and its energy increases by an amount equal to the ponderomotive energy $\mathcal{E}_V(x)$ at the point x where it has been created. It thus seems that the oscillatory motion has no influence on the energy spectrum of the photoelectrons detected outside the laser.

This is however true only if the laser pulse lasts a time sufficiently long for allowing the photoelectron to feel this force during the whole exit time. This leads us to consider two extreme situations:

[4] See Fig. 8.6 of Chap. 8.

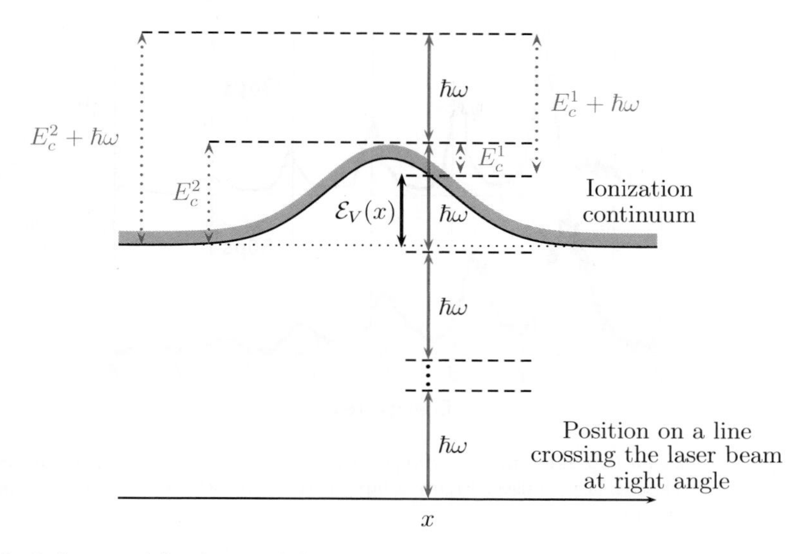

Fig. 10.7 Influence of the change of the ionization energy due to the oscillation of the electron in the laser field. The kinetic energy $E_C^{(1)}$ of the photoelectron produced by a multiphoton ionization process is reduced in comparison with its value $E_C^{(2)}$ in the absence of shift of the ionization energy.

Long laser pulses. The laser field does not change during the exit of the electron from the laser beam. The slow motion of the electron has the time to regain the energy $\mathcal{E}_V(x)$ taken by the fast oscillatory motion at the point x where the photoelectron has been created. The ionization spectrum above threshold is thus the same as that one would calculate ignoring $\mathcal{E}_V$ (nevertheless low energy peaks are suppressed).

Short laser pulses. The laser field vanishes before the photoelectron could be accelerated. All the peaks of the ATI spectrum have an energy diminished by an amount $\mathcal{E}_V$. Actually, this is even, for a given peak duration, a technique used to measure the laser intensity.

This dependence of the ATI spectrum on the pulse duration has been experimentally observed [Agostini *et al.* (1987)] and is shown in Fig. 10.8. The peaks of the ATI spectrum are shifted to lower energies when the duration of the laser pulses decreases. With a sufficiently high resolution, it is even possible to observe the signature of the excitation of the Rydberg states as a fine structure in the ATI peaks [Freeman *et al.* (1987)].

10.3.3 *Evidence for non-perturbative effects*

When the laser intensity is increased to values on the order of or larger than 10^{14} W·cm^{-2}, such that the ponderomotive energy $\mathcal{E}_V$ becomes comparable or larger than the ionization energy, the ATI spectrum extends over broader energy intervals. "Plateaus" appear in the curve giving the intensity of the peaks of the ATI spectrum versus the photon order, showing that the intensity of the peaks decreases much

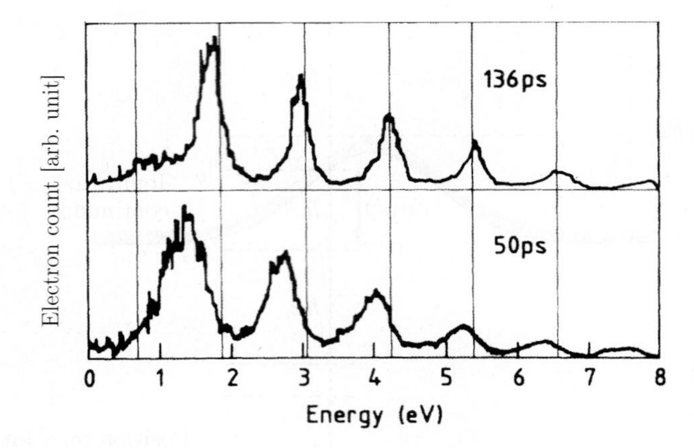

Fig. 10.8 ATI spectra for two different pulse durations, 136 ps and 50 ps. The spectrum is shifted to lower energies for shorter laser pulses. Figure adapted from [Agostini *et al.* (1987)]. Copyright: American Physical Society.

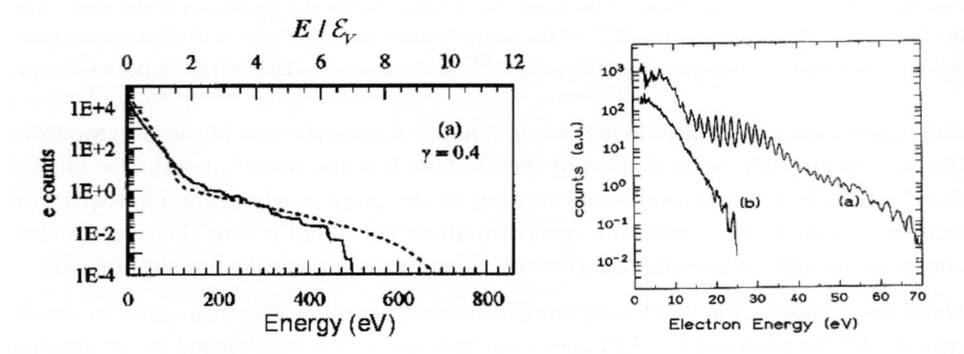

Fig. 10.9 Left: ATI spectrum of helium atoms for a 780 nm excitation at 12×10^{14} W·cm^{-2}. The decrease of the electron counts exhibits a change of slope for an energy E larger than $2\mathcal{E}_V$ giving rise to a plateau which extends up to about 8 $\mathcal{E}_V$. γ is the Keldysh parameter introduced in Sec. 10.5. The dotted line is the theoretical prediction of the semiclassical model presented in Sec. 10.5. Figure taken from [Walker *et al.* (1996)]. Copyright: American Physical Society. Right: Effect of the laser polarization for the ATI spectra of argon atoms at an intensity of 2.4×10^{14} W·cm^{-2}. The plateau clearly appears for a linearly polarized excitation (a), but not for a circularly polarized one (b). Figure taken from [Paulus *et al.* (1994)]. Copyright: American Physical Society.

more slowly than what would be expected from perturbation theory. The left part of Fig. 10.9, taken from [Walker *et al.* (1996)], clearly shows such a behavior observed with helium atoms. The right part of this figure, taken from [Paulus *et al.* (1994)], shows that the plateaus only appear for a linearly polarized laser excitation. We will show later on (see Sec. 10.5) that, at these laser intensities, ionization is no longer a multiphoton process, but rather a tunnel ionization, and we will present a semiclassical model explaining the results shown in Fig. 10.9.[5]

[5]For a review of the classical and quantum features of ATI, see [Becker *et al.* (2002)].

10.4 Harmonic generation

10.4.1 *Physical interpretation*

Harmonic generation is a nonlinear process where N photons with energy $\hbar\omega$ disappear and are replaced by a photon with energy $N\hbar\omega$. Figure 10.10(left) shows for example the generation of the fifth harmonic. A semiclassical interpretation of this effect can be given by considering the mean induced dipole moment $\langle D(t)\rangle$ of the atom driven by the laser field. At very low laser intensity, $\langle D(t)\rangle$ is proportional to the laser field and its motion is harmonic. At higher laser intensities, its motion is still periodic but no longer purely harmonic because the binding Coulomb potential of the electron is not harmonic: $\langle D(t)\rangle$ has Fourier components at integer multiples of ω explaining why harmonics of the laser frequency appear in the light emitted by the dipole. In a perturbative treatment of this problem, the nth order Fourier component of $\langle D(t)\rangle$ is related to E^n by a nonlinear susceptibility χ_n. One can then easily predict that only odd harmonics of ω can appear in the emitted light as a consequence of the reflection symmetry. The change of sign of $\langle D\rangle$ and E in such a symmetry operation implies that only odd powers of E can appear in the expansion of $\langle D\rangle$ in powers of E.

The first evidence for harmonic generation was obtained in 1961 by Peter Franken and collaborators [Franken *et al.* (1961)] with the observation of a fre-

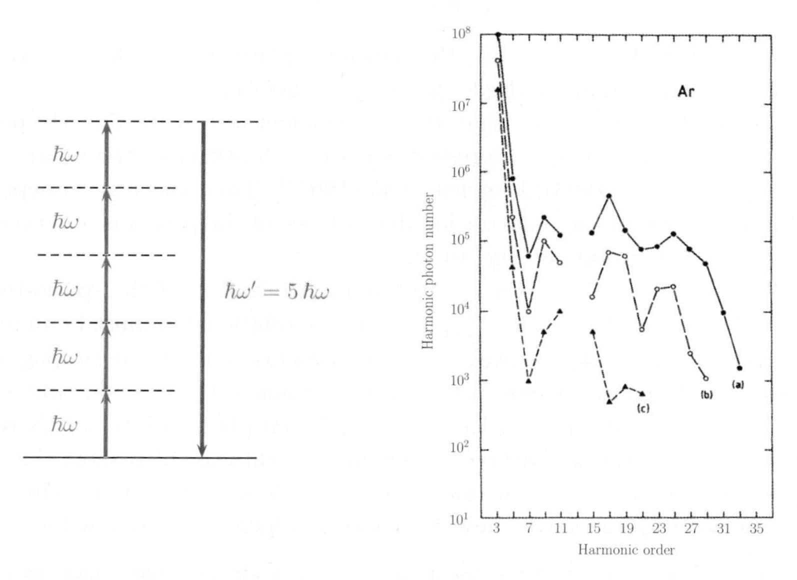

Fig. 10.10 Left: harmonic generation process where 5 photons $\hbar\omega$ disappear and are replaced by a photon with energy $5\hbar\omega$. Right: variations with the harmonic order of the efficiency of harmonic generation for Ar atoms and for several laser intensities (a) 3×10^{13} W·cm^{-2}, (b) 2.2×10^{13} W·cm^{-2}, (c) 1.6×10^{13} W·cm^{-2}. Figure adapted from [Li *et al.* (1989)]. Copyright: American Physical Society.

quency doubling of a ruby laser passing through a quartz crystal.[6] A few years later, in 1967, third harmonic generation has been observed in a gas [New and Ward (1967)].

10.4.2 *High order harmonic generation (HHG). Evidence for non-perturbative effects*

With the development of laser sources providing intense and short pulses it has become possible to observe harmonic generation in rare gases with a high order n [McPherson *et al.* (1987); Ferray *et al.* (1988); Li *et al.* (1989)].

Figure 10.10(right) gives an example of results obtained in one of the first experiments of this type [Li *et al.* (1989)]. The intensity of the various harmonics is plotted versus the harmonic order for argon gas and for several intensities. After a rapid decrease of the efficiency of harmonic generation when the harmonic order increases, this efficiency, for sufficiently lage intensity, reaches a plateau before dropping to very low values beyond a certain cut-off. The existence of this plateau, similar to those observed for ATI spectra, cannot be explained with a perturbative treatment. It gives a second evidence for the onset of new physical processes occurring at high laser intensities (see Sec. 10.5). The measured cut-off energy is found to obey an empirical law given by:

$$E_{\text{cut-off}} \simeq 3.2\, \mathcal{E}_V + E_I. \tag{10.8}$$

With sufficiently high values of $\mathcal{E}_V$, the harmonic plateau can extend to very high orders, larger than several hundreds [Seres *et al.* (2004)].

It must be also emphasized that HHG is efficient only with linearly polarized laser excitation and strongly suppressed when the ellipticity of the polarization is increased [Budil *et al.* (1993); Dietrich *et al.* (1994)]. This polarization dependence of HHG is analogous to the polarization dependence of the plateaus observed in the ATI spectra (see right part of Fig. 10.9).

A critical issue of HHG is the control and optimization of the spatio-temporal characteristics of high harmonics. A proper phase relationship must be maintained between the fundamental laser wave and the harmonics along the propagation direction so that the contributions of the various atoms of the gas add coherently.[7]

HHG sources of radiation, which convert infrared photons into soft X-ray photons, have very interesting features (coherence, brightness, ultrashort duration of the pulses) which make them complementary of synchrotron radiation sources. They play also an essential role in the new mechanisms which are used now for analyzing

[6] In a crystal the reflection symmetry disappears and this is why even harmonics can appear in harmonic generation.

[7] We refer the reader to [Salières *et al.* (1999)] giving a review of studies dealing with these problems prior to 2000 and to [Kapteyn *et al.* (2005)] for a review of more recent experiments realizing phase matching by guiding the light in gas-filled, hollow-core wave guides, or in spatially periodic structures ("quasi-phase matching").

physical phenomena with a sub-femtosecond time scale, on the order of one hundred attoseconds (1 as $= 10^{-18}$ s). We will come back to these achievements in Chap. 27.

10.5 Tunnel ionization and recollision

In this section, we present a new description of strong field photoionization initiated by the work of Leonid Keldysh in 1964 [Keldysh (1965)] and describing ionization in terms of a tunneling of the electron through the Coulomb barrier. A breakthrough in the understanding of the effects associated with tunnel ionization occurred in the early 1990's when it was realized that, once freed, the electron continues to interact with the laser field and can return to the vicinity of the parent ion and "recollide" with it [Schafer *et al.* (1993); Corkum (1993)]. We show here how a simple quasiclassical model of this recollision can explain most of the intriguing features of ATI and HHG mentioned in the previous sections. Most importantly, we show also that this recollision can give rise to extremely short bursts of soft X-ray photons, perfectly synchronized with the incident laser wave, and opening the way to what is called now *attosecond science*.

10.5.1 *The breakdown of perturbation theory*

It is easy to understand why perturbative treatments are no longer valid when the laser intensity reaches values on the order of 10^{14} W/cm^2. The laser field is then on the order of 2×10^{10} V/m and becomes comparable to the Coulomb field, since the Coulomb field created by a proton at a distance equal to twice the Bohr radius is on the order of 5×10^{11} V/m. The perturbation due to the laser field is no longer small compared to the binding energy of the electron and the linear potential $-qEx$ generated by the laser electric field E can lower the Coulomb potential sufficiently for allowing the electron to leave the parent ion by tunneling through the potential barrier (see Fig. 10.11). This ionization mechanism is called *tunnel ionization.*

In fact, the observation of this new regime of ionization, and all the non-perturbative effects described in the previous two sections, has been made possible only thanks to the realization of short femtosecond laser pulses.[8] Short laser pulses are important for two reasons:

- First, they make it relatively easy to obtain high laser intensities. Concentrating a small light energy, on the order of 10^{-4} J, on a time interval of a few 10^{-15} s, and on a beam waist on the order of 50 μm, leads to intensities in the range of few 10^{15} W/cm^2.

[8] Spectacular progress has been achieved in this research field when the dye laser systems used in the 1980's were replaced by the femtosecond mode-locked titanium sapphire (Ti:S) lasers, providing much shorter pulse durations as short as 6 fs, that is, two optical cycles (see Chap. 27).

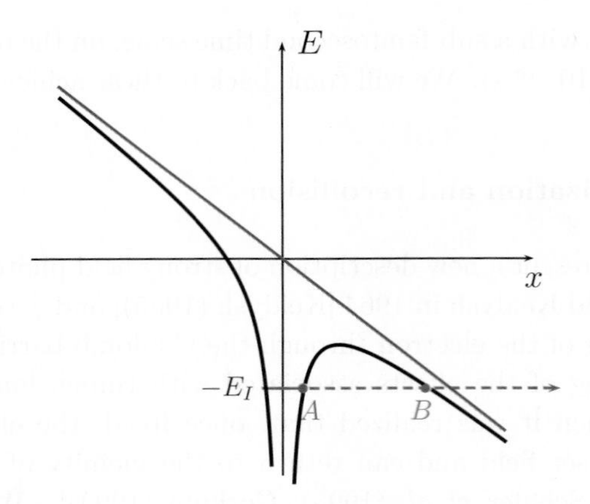

Fig. 10.11 Tunnel ionization. The electron leaves the ion by tunneling through the potential barrier formed by the sum of the Coulomb potential of the ion and the linear potential $-qEx$ due to the laser electric field E, supposed linearly polarized along the x-axis.

- If the pulse is short enough, the atom has no time to be ionized during the rising edge of the pulse. It can survive perturbative ionization and feel the maximum intensity of the pulse where the new regime of tunnel ionization takes place.

10.5.2 *Keldysh parameter*

The description of tunneling given in the previous subsection implicitly supposes that the electric field is static. In fact, the laser electric field oscillates at the frequency ω and the slope of the linear potential $-qEx$ changes its sign every half period π/ω. If, at a certain time, the slope is negative, as shown in Fig. 10.11, and if the electron has not enough time during half a period to tunnel along the positive direction of the x-axis, the sign of the slope will have changed and the electron will be pushed back to the left before having time to tunnel. It is therefore expected that tunnel ionization can occur only for low enough frequencies.

This problem has been addressed by Leonid Keldysh [Keldysh (1965)] (see also [Perelomov *et al* (1966, 1967); Ammosov *et al.* (1986)]). The tunneling rate is calculated for an oscillating field and averaged over one period. It is found that, when a certain parameter γ called the *Keldysh parameter* is small enough, the average tunneling rate is equal to the average of the instantaneous tunneling rate calculated at each time t for a fixed value $E(t)$ of the laser field (quasi-static approximation). The Keldysh parameter is given by:

$$\gamma = \sqrt{\frac{E_I}{2\,\mathcal{E}_V}}. \tag{10.9}$$

Tunnel ionization is the main physical process when $\gamma \ll 1$, i.e. when the ponderomotive energy $\mathcal{E}_V$ is much larger than the ionization energy E_I. Multiphoton ionization is a correct description when $\gamma \gg 1$.

Equation (10.9) can also be written as:[9]

$$\gamma = \omega \frac{\sqrt{2mE_I}}{qE_0} = \omega\tau_{\text{tunnel}} \tag{10.10}$$

where

$$\tau_{\text{tunnel}} = \sqrt{2mE_I}/qE_0 \tag{10.11}$$

appears as a characteristic time for tunneling,[10] since the quasi-static condition $\gamma = \omega\tau_{\text{tunnel}} \ll 1$ expresses that the variations of the field amplitude can be neglected during τ_{tunnel}.

We finally give a few orders of magnitude of $\mathcal{E}_V$. A laser pulse at a wavelength λ=1 μm and with an intensity of 10^{15} W/cm^2 gives a ponderomotive energy of 93 eV, which is large compared to the ionization energies of rare gases, ranging from 24.6 eV for helium to 12 eV for xenon. This shows that it is possible to have at the same time laser fields on the order or larger than the Coulomb field and oscillatory energies larger than the ionization energy leading to small Keldysh parameters and therefore to quasi-static tunneling.

10.5.3 *Two-step quantum-classical model*

The tunneling of the electron through the potential barrier of Fig. 10.11 is a pure quantum effect. Once the electron has left the ion, it continues to interact with the laser field. One model, which explains many features of strong field ionization, consists of describing this second step in a classical way, by considering the classical motion of the released electron driven by the laser field. We will see later on that the electron can gain a considerable energy during this second step, much larger than the initial energy just after the first tunneling step. In the tunneling model, the initial velocity of the classical trajectory is equal to zero: $v_{\text{in}} = 0$. One also neglects the effect of the Coulomb potential of the ion on the motion of the released electron. This two-step model has been developed by Ken Kulander, and Paul Corkum [Schafer *et al.* (1993); Corkum (1993)] and will be shortly described hereafter.

We will suppose first that the laser field $\vec{E}(t) = E_0\,\vec{\varepsilon}_x \sin\omega t$ is linearly polarized in the x-direction. The equation of motion of the electron (with a negative charge $q = -|q|$ and a mass m) along the x-axis, $m\,\ddot{x}(t) = -|q|E_0 \sin\omega t$, is readily integrated with the initial conditions $x_{\text{in}}(t_0) = 0$, $v_{\text{in}}(t_0) = 0$ where t_0 is the time at which the electron is leaving the ion:

$$\dot{x}(t) = v(t) = \frac{|q|E_0}{m\omega}\cos\omega t - \frac{|q|E_0}{m\omega}\cos\omega t_0, \tag{10.12}$$

$$x(t) = \frac{|q|E_0}{m\omega^2}\left[\sin\omega t - \sin\omega t_0\right] - \frac{|q|E_0}{m\omega}(t - t_0)\cos\omega t_0. \tag{10.13}$$

[9] $\gamma \sim 1$ for an incident intensity $\sim 10^{14}$ W.cm^{-2}.

[10] In fact, if L is the distance between the two points A and B of Fig. 10.11, it is easily shown that τ_{tunnel} scales as $L/\bar{v}$ where $\bar{v}$ is a characteristic velocity of the electron in the bound state of energy $-E_I$, defined by $m\bar{v}^2 \simeq E_I$.

Equation (10.12) shows that the velocity of the electron is the sum of an oscillatory component at the laser frequency ω, due to the forced oscillation in the laser field, and a constant drift velocity

$$v_{\text{drift}} = -\frac{|q|E_0}{m\omega}\cos\omega t_0. \tag{10.14}$$

If the laser pulse is very short, it passes over the electron during a time so short that the spatial gradient of the light intensity cannot produce an appreciable change of the electron velocity. At the end of the pulse the velocity of the electron reduces to the drift velocity and the kinetic energy of the electron is only due to the drift motion, and not to the oscillatory motion:

$$E_{\text{kin}} = \frac{m}{2}v_{\text{drift}}^2 = \frac{q^2E_0^2}{2m^2\omega^2}\cos^2\omega t_0 = 2\mathcal{E}_V\cos^2\omega t_0. \tag{10.15}$$

This result shows that, after the tunneling, the energy of the ejected electron can gain from the laser field an energy ranging between 0 and $2\mathcal{E}_V$. One understands in this way why the ATI spectrum extends over an energy interval on the order of $2\mathcal{E}_V$ before the onset of the plateau in Fig. 10.9.

For the following, it will be useful to introduce a graphic construction giving the position and the velocity of the ejected electron interacting with the laser field (see Fig. 10.12). The sinusoid of this figure is the curve $(|q|E_0/m\omega^2)\sin\omega t$, the straight dotted line the tangent to the sinusoid at the time t_0 where the tunneling occurs and which is close to the value where the laser field is maximum (in absolute value). From Eqs. (10.13) and (10.14), one can easily show that the slope of this tangent is equal to $-v_{\text{drift}}$ and that the abscissa of the electron along the x-axis at time t is given by the difference of the ordinates of the points M and N, while its velocity is given by the difference of the slopes of the two curves at M and N.

Figure 10.12 clearly shows also that the dotted straight line can intersect the sinusoid at a second point B at a time t_1. This means that the ejected electron can be accelerated by the laser field in a certain direction, reverse its velocity when the direction of the laser electric field changes, and arrives at the position of the ion at time t_1.[11] It can then recollide with the ion. This process can give rise to interesting phenomena that are analyzed in the next subsection.

Similar calculations can be done with a circularly polarized laser excitation. The classical trajectory is no longer confined to the x-axis and takes place in the xy-plane. The important point is that it never comes back in the vicinity of the ion, so that recollision does not occur in this case. The kinetic energy gained by the released electron is found to be equal to $\mathcal{E}_V$.

The velocity and the position of the photoelectron after its birth time t_0 have been calculated in Eqs. (10.12) and (10.13) assuming that the laser field has a constant amplitude E_0. These calculations can be easily generalized to the case of

[11] For other values of t_0, several intersections between the two curves can appear, implying that the released electron can come back several times in the vicinity of the ion.

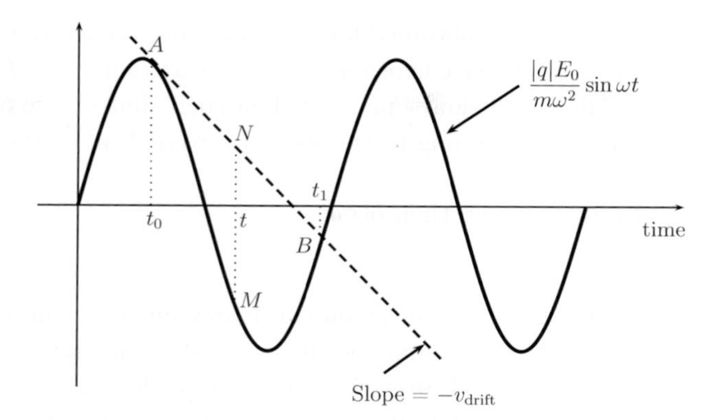

Fig. 10.12 Graphic construction giving the position and the velocity of the ejected electron interacting with the laser field. The sinusoid is the curve $(|q|E_0/m\omega^2)\sin\omega t$, the straight dotted line the tangent to the sinusoid at the time t_0 where the tunneling occurs. The slope of this tangent is equal to $-v_{\text{drift}}$. At a given time t, the abscissa of the electron along the x-axis is given by the difference of the ordinates of the points M and N, its velocity by the difference of the slopes of the two curves at M and N.

a few-cycle laser pulse for which the variation of E_0 cannot be ignored. The velocity of the photoelectron at time t is given by:

$$\dot{x}(t) = v(t) = -\frac{|q|}{m}\int_{t_0}^{t} E(t')\,\mathrm{d}\,t = -\frac{|q|}{m}\left[A(t_0) - A(t)\right] \tag{10.16}$$

where $A(t)$ is the vector potential related to $E(t)$ by $E(t) = -\partial A(t)/\partial t$. At the end of the pulse, $A(t) = 0$, so that the velocity of the photoelectron has increased by an amount

$$\Delta v = -\frac{|q|}{m}A(t_0) \tag{10.17}$$

depending only on the value $A(t_0)$ of the potential vector of the laser field at its birth time. Similar calculations can be done for the position $x(t)$ of the photoelectron. One finds:

$$x(t) = -\frac{|q|}{m}\left[(t - t_0)A(t_0) - \int_{t_0}^{t} A(t')\,\mathrm{d}\,t'\right]. \tag{10.18}$$

10.5.4 *Recollision*

When the electron returns to the ion at time t_1, it has a kinetic energy determined by its total velocity $v(t_1)$ given by (10.13), and not just by $-v_{\text{drift}}$, because it is still in the laser beam. The velocity $v(t_1)$ is given by the difference of the slopes of the two curves intersecting in B and is clearly not equal to zero, as this was the case in A. For different tunneling times t_0, we have different return times t_1 and consequently, different velocities $v(t_1)$. It is possible to calculate the maximum value of $v(t_1)$, and consequently, the maximum kinetic energy that the electron can have when it returns in the neighborhood of the ion. One finds:

$$E_{\text{kin}}^{\text{max}} = 3.17\ \mathcal{E}_V. \tag{10.19}$$

This maximum kinetic energy is obtained for a release time t_0 such that $\omega t_0 \simeq 108°$, slightly larger than the value t_M corresponding to an extremum of E in absolute value ($\omega t_M = 90°$). The recollision time t_1 is then such that $\omega t_1 \simeq 342°$, slightly smaller than the value corresponding to the second zero of E after the release time (see Fig. 10.12).

Several physical processes can then occur.

(i) *Double ionization*

If $\mathcal{E}_V$ is large enough, the returning electron can knock out a second electron from the ion in an inelastic collision giving rise to a double ionization. This process occurs a fraction of optical period after the tunneling of the first electron and can be considered to be the main contribution to non-sequential double ionization.

(ii) *High harmonic generation*

When the electron comes back near the ion, it can recombine with the ion and end in the atomic ground state of the atom while emitting radiation. The energy of this radiation is equal to the kinetic energy of the returning electron plus the ionization energy E_I of the atom. The energy of the emitted radiation has thus an upper bound, or a cut-off, given by:

$$E_{\text{cut-off}} = E_{\text{kin}}^{\text{max}} + E_I = 3.17\ \mathcal{E}_V + E_I. \tag{10.20}$$

This result for the cut-off energy of the high harmonic spectrum, derived here from the recollision model, explains the empirical law (10.8) fitting the observations. One understands also why high harmonic generation cannot be observed with a circularly polarized excitation for which recollision cannot occur.

Note also that, at each half period of the oscillating field, the direction of E changes from the positive direction of the x-axis to the negative one, giving rise to tunneling electrons moving in opposite directions each half period. The emission of light pulses in the recombination process is thus periodic with a repetition frequency 2ω, which explains why the harmonic frequencies form a comb with a spacing equal to 2ω.

(iii) *High energy part of the ATI spectrum*

The returning electron can also scatter from the ion. It becomes dephased from its harmonic motion and this can give rise to an absorption of energy from the laser field analogous to the collision induced RF heating in Paul traps.[12] Suppose, for example, that the electron is back scattered at time t_1, which means that its velocity $v(t_1)$ along the x-axis changes into $-v(t_1)$. Replacing t by t_1 in (10.12), one gets for the velocity of the electron, just after the elastic collision:

$$v\left(t_1+\varepsilon\right) = -\frac{|q|\,E_0}{m\omega}\cos\omega t_1 + \frac{|q|\,E_0}{m\omega}\cos\omega t_0. \tag{10.21}$$

[12] See Chap. 14.

To obtain the new drift velocity $v_{\text{drift}}^{\text{new}}$ after the elastic collision, we write:

$$v(t_1 + \varepsilon) = \frac{|q|\, E_0}{m\omega} \cos \omega t_1 + v_{\text{drift}}^{\text{new}}, \tag{10.22}$$

which gives, combined with Eq. (10.21):

$$v_{\text{drift}}^{\text{new}} = -2\frac{|q|\, E_0}{m\omega} \cos \omega t_1 + \frac{|q|\, E_0}{m\omega} \cos \omega t_0. \tag{10.23}$$

From this equation, one can calculate the maximum value of the new drift velocity and one finds that the maximum kinetic energy of the electron after the laser pulse can reach $10\mathcal{E}_V$ if a recollision process occurs. This explains the higher energy plateaus appearing in the ATI spectra of Fig. 10.9 and the absence of these plateaus when the laser polarization is circular.

10.5.5 *Full quantum treatments*

The two-step quantum-classical description of tunnel ionization can be incorporated in full quantum descriptions that we will not describe here [Lewenstein *et al.* (1995)]. Using a Feynman's path integral approach, it has been in particular possible to introduce "quantum orbits" which generalize the classical orbits discussed above and which allow quantum interference between different orbits to be taken into account [Salières *et al.* (2001)].

The quantum nature of the re-scattering process opens interesting perspectives. The returning electron wave can be considered as a quantum probe whose diffraction by the atomic core can give important information on the structure of this core. The tunneling process can be actually considered as a beam splitter in an interferometer giving rise to two coherent waves, the wave of the released electron, and the wave of the remaining atomic core. The released electron is then brought back and interferes with the core wave. For more details concerning these developments, see [Wagner *et al.* (2006)].

10.6 Conclusion

In this chapter, we have reviewed several interesting mechanisms occurring in the photoionization of atoms. The atom can be photoionized even if the energy $h\nu$ of the photons is smaller than the ionization energy E_I of the atoms, but the mechanism of this ionization changes when the laser intensity is increased.

Of course, this intensity has to be high enough for ionizing the atom when single photon ionization is not possible. But, as long as the amplitude of the laser field remains small compared to the Coulomb field binding the atomic electron to the ion core, a perturbative treatment is possible, leading to the description of the ionization as a multiphoton process where p photons (with $p > 1$) bring the energy $ph\nu$ required to excite the atom from the ground state to the continuum, p being the smaller integer for which $ph\nu > E_I$. The observed number of photoelectrons then varies with the laser intensity I_L as I_L^p, as predicted by perturbation theory for a higher order process. This power law can be modified when intermediate resonances appear in the multiphoton process, giving rise to enhancement of the ionization rate, to destructive interferences, to asymmetric line shapes, but all these effects can be understood with a perturbative treatment of the atom-laser interaction.

When the laser intensity reaches values where the laser field is no longer negligible compared to the Coulomb field, new interesting effects appear like above-threshold ionization and high harmonic generation which exhibit features that cannot be explained with a perturbative treatment. The ponderomotive energy $\mathcal{E}_V$ of the electron in the laser field becomes also comparable or larger than the ionization energy E_I. We have shown that, in this regime of laser intensities, the electron is no longer ionized by a multiphoton process, but by tunneling through the potential barrier resulting from the sum of the Coulomb potential of the ion and the Stark potential produced by the quasi-static laser field. Furthermore, once it has left the ion by this tunneling process, the electron can be accelerated by the laser field, come back near the ion and recollide with it. We have shown that this recollision picture can explain all the non-perturbative features found in ATI and HHG. In particular, the effect of the polarization of the laser field can be clearly understood: the trajectory of the electron can come back in the vicinity of the ion only for a linearly polarized field. This explains the great difference of the observed effects when the laser polarization is changed.

All these effects are very interesting by themselves because they provide evidence for qualitatively new physical mechanisms in atomic physics. As it happens often in physics, a discovery of new mechanisms generally opens new perspectives for interesting applications. HHG provides new sources of XUV radiation by converting infrared photons in XUV photons. The XUV pulses generated by the released electron when it recollides with the parent ion are also extremely short, as short as 100 attoseconds. This is due to the fact that tunnel ionization occurs only near the crests of the laser field in a time window smaller than the period of the optical field. The time resolution of the physical processes is no longer determined by the temporal width of the envelope of the light pulse, as this is the case in usual nonlinear optics, but by a fraction of the optical cycle of the carrier. This spectacular gain in time resolution is now being used to investigate ultrashort electron dynamics in atomic and molecular processes. Furthermore, the de Broglie wavelength of the recolliding electron is on the order of the size of the parent ion so that its diffraction pattern can give structural information on this ion, adding spatial resolution to time resolution. This technique is of particular interest with molecules. A few of these developments will be described in Chap. 27. It is clear that this research field will give rise to exciting new developments.

PART 4

Atom-photon interactions: a tool for controlling and manipulating atomic motion

Introduction

Light can be used to manipulate the external degrees of freedom of an atom, i.e. its velocity and its position. Spectacular developments have occurred in this research field during the last few decades. It is now possible to cool atoms to extremely low temperatures, on the order of a few nanokelvin, and to store them in various types of traps. The realization of ultracold atomic gases has opened the way to spectacular developments, such as atomic clocks with an extremely high accuracy and stability, interferometers with atomic de Broglie waves, gaseous Bose-Einstein condensates, and quantum degenerate Fermi gases.

In this part, we focus on the basic mechanisms used to control and manipulate atomic motion while leaving applications of ultracold atomic gases to be reviewed in subsequent parts of this book. Several review articles have been written on this subject.[1]

In Chap. 3, we presented an example of manipulation of atoms by light. The principle of optical pumping is to polarize atoms using the exchanges of angular momentum between polarized light and atoms. The physical mechanism at the origin of the forces exerted by light on atoms is the transfer of linear momentum of photons to atoms in absorption-spontaneous emission cycles or absorption-stimulated emission cycles. We analyze these processes in Chap. 11 by calculating the mean radiative force exerted on a two-level atom initially at rest at a given point of the laser beam. The concept of a mean radiative force at a given point seems, at first glance, incompatible with Heisenberg relations. It is meaningful only if one can construct atomic wave packets sufficiently well localized in position and momentum. We show that this is possible if the energy scale $\hbar\Gamma$ associated with the natural width of the excited state e is large compared to the recoil kinetic energy $E_{\text{rec}} = \hbar^2k^2/2m$ of an atom emitting or absorbing a photon. This large natural width condition also insures that external variables, such as the position and the velocity of the atomic center of mass, evolve at a much smaller rate than internal variables, such as the mean electric dipole moment. In Chap. 11, we show that under these conditions there are two types of radiative forces: dissipative (or radiation pressure) forces and reactive (or dipole) forces. We give a physical interpretation of each of these forces.

The apparent laser frequency experienced by a moving atom is Doppler-shifted, which modifies the rate of exchange of linear momentum with the light. In Chap. 12, we analyze the Doppler induced velocity-dependent forces experienced by a two-level atom moving in a laser plane wave or in a laser standing wave and show that these forces can introduce a friction that damps the atomic velocity. This is the principle of laser Doppler cooling, which works for free atoms as well as for ions moving in a trap. The lower limit for the temperature that can be obtained by Doppler cooling,

[1] See, for example, the Nobel lectures of Steve Chu, Claude Cohen-Tannoudji and Williams Phillips [Chu (1998); Cohen-Tannoudji (1998b); Phillips (1998)], courses given at summer schools [Cohen-Tannoudji (1992a); Arimondo *et al.* (1992)] and books [Metcalf and Van Der Straten (2001); Letokhov (2007)].

called the Doppler limit, results from a competition between the cooling effect of light and the heating effect of the fluctuations introduced by the random character of spontaneous emission. A simple treatment of this competition, inspired by the analogy with the theory of Brownian motion, provides an estimate of this Doppler limit. New cooling mechanisms appear when the atom moves in a high intensity laser standing wave, and we interpret these with the dressed atom approach introduced in Chap. 5 which describes resonance fluorescence in intense laser fields. A key idea in this analysis is the time lag with which the fast internal variables follow the evolution of the slow external variables. It is this small correction to the adiabatic following approximation that is at the origin of the friction. It is also here that the idea of "Sisyphus cooling" appears for the first time in this research field.

When precise methods such as time-of-flight were implemented for measuring the temperature of laser cooled atomic gases, it was discovered that their temperatures were much lower than those predicted by the theory of Doppler cooling by almost two orders of magnitude! This was a rather good surprise and showed that other cooling mechanisms, more efficient than Doppler cooling, were operating. We discuss these in Chap. 13 and we show that they result from a new and interesting combination of physical effects such as optical pumping and light shifts that had been discovered several years before, and which are described in other chapters of this book.[2] The new cooling mechanisms, which exist only for atoms with several Zeeman sublevels in the ground state and moving in a laser configuration exhibiting polarization gradients, result from the fact that the atom moves uphill more frequently than downhill in a spatially modulated optical potential. This is a low intensity version of the Sisyphus cooling effect introduced in Chap. 12 for a two-level atom moving in an intense laser standing wave. In Chap. 13, we show that temperatures obtained through low intensity Sisyphus cooling have a lower limit on the order of a few recoil temperatures T_{rec}, where T_{rec} is defined by $k_B T_{\text{rec}} = E_{\text{rec}}$. The origin of this limit, called the "recoil limit", is easy to understand. It comes from the fact that the random direction along which photons are spontaneously emitted by the atom cannot be controlled. This gives rise to an atomic momentum spread Δp at least equal to $\hbar k$, and to a disordered kinetic energy at least equal to E_{rec}.

It turns out that this seemingly fundamental limit can also be overcome, and we describe in the second part of Chap. 13 the principle of the so-called "sub-recoil" cooling methods that have led to temperatures lower than T_{rec}. The idea is to quench the absorption and subsequent re-emission of light for atoms with a very low velocity. This is accomplished by introducing a trap in momentum space into which the atoms can fall and accumulate during their random walk. Two different methods for producing these "dark states" are described: velocity selective coherent population trapping and Raman cooling. Understanding the long time limit of these cooling methods raises interesting theoretical problems, and we briefly review a few

[2] See Chaps. 3 and 7.

advances in this direction by establishing connections between sub-recoil cooling and the theory of random walks dominated by a few rare events (the so-called "Lévy flights").

Finally, in Chap. 14, we describe the various methods that have been developed for trapping either charged particles, like ions, or neutral atoms. Traps for ions such as the Penning trap or the Paul trap, which couple to the charge, are much deeper than traps for neutral atoms, which use the coupling of the magnetic or electric atomic dipole moment with magnetic field or electric field gradients. As a result, neutral atoms must be pre-cooled before being trapped. Various schemes have been invented for circumventing the constraints introduced by Maxwell equations and for introducing configurations of fields that lead to a minimum of effective potential energy. Artificial magnetic fields have been also recently realized that give rise to effective Lorentz force for neutral atoms. All these schemes will be briefly reviewed.

We will conclude this introduction with two remarks:
(i) It can seem paradoxical that laser light can cool atoms to extremely low temperatures since lasers can also be associated with the idea of fusion following the deposition of a large amount of energy in a small volume of matter. In fact, laser cooling does not use high power lasers that can fully ionize atoms and heat the plasma that is produced in this way. Instead, laser cooling experiments use lasers with powers on the order of a few tens of milliwatts, which is far too low to ionize the atom by multiphoton processes, but large enough to saturate an atomic transition, i.e. for appreciably populating the excited state e of this transition when the frequency of the laser is tuned near resonance. The atom can then undergo several absorption-emission cycles per second, with a rate on the order of the inverse of the radiative lifetime of e, so that the corresponding rate of transfer of linear momentum from the laser is high enough to give rise to a large radiation pressure force that can change the atomic velocity in a very short time.
(ii) In view of the number of applications that have followed the development of laser cooling and trapping methods, one wonders why the control of atomic motion by light was not investigated sooner. Most physical concepts concerning the exchanges of linear momentum between atoms and light were already known, and the order of magnitude of the radiative forces could have been easily estimated. The reason for this delay is technological: For laser cooling, one needs monochromatic laser light that can be tuned near resonance with an atomic transition and whose spectral width is smaller than the natural width of the transition. A prerequisite for the development of this research field has been the availability of such monochromatic laser sources.[3] Furthermore, a given experiment generally needs several laser wavelengths for cooling, trapping, and repumping the atoms. It is only with the development of compact, relatively cheap and easy to use laser sources, such as laser diodes, that laser cooling and trapping experiments became practical to build. Finally, one should not forget the importance of the combination of laser cooling

[3] A few manifestations of radiation pressure were already observed, such as in comet tails.

with other methods, such as evaporative cooling, which has allowed the observation of Bose-Einstein condensation, or Feshbach resonances, which provide a simple way to control atom-atom interactions in ultracold atomic gases.

Chapter 11

Radiative forces exerted on a two-level atom at rest

11.1 Introduction

This chapter is devoted to an analysis of the radiative forces experienced by a two-level atom interacting with a quasi-resonant laser wave. Here we will suppose that the atom is initially at rest in the laser wave, while in the next chapter we will study the modification of these forces when the atom is moving with a velocity v. Before explaining how these forces can be calculated (Sec. 11.2) and interpreted (Secs. 11.3 and 11.4), it is useful to give a few orders of magnitude and a few general considerations about the concept of a mean force at a given point.

11.1.1 *Order of magnitude of the force*

The basic mechanism for controlling atomic motion with light is the transfer of the momentum of a laser photon to the atom. When the atom absorbs the photon, the momentum $\vec{P}$ of the atomic center of mass is changed by $\hbar\vec{k}_L$. An important energy scale in this problem is the recoil kinetic energy

$$E_{\text{rec}} = \frac{\hbar^2 k_L^2}{2M}, \tag{11.1}$$

associated with the transfer of this momentum.

After the absorption of a photon, the atom is in the excited state e, but does not remain there for ever. After a certain time, the average value of which is the radiative lifetime τ_R of the excited state e, equal to the inverse of the natural width Γ of e, it falls back in the ground state g by spontaneous emission of a photon.[1] The momentum lost by the atom in the emission process is equal to zero on average, since the spontaneous photon can be emitted with equal probability in two opposite directions. It follows that after an absorption-spontaneous emission cycle, the mean velocity change of the atom δv is due only to the absorption process, and is equal

[1] Here, we consider the effect of absorption-spontaneous emission cycles. We will study later on the effect of absorption-stimulated emission cycles.

to the recoil velocity $v_{\rm rec}$:

$$\delta v = v_{\rm rec} = \frac{\hbar k_L}{M}. \tag{11.2}$$

Let $\mathrm{d}\langle N\rangle/\mathrm{d}t$ be the mean number of fluorescence cycles per second. Since the mean time spent in the excited state e is on the order of τ_R, $\mathrm{d}\langle N\rangle/\mathrm{d}t$ is at most equal to $1/\tau_R = \Gamma$. We conclude that the mean velocity change per second is given approximately by:

$$\frac{\delta v}{\delta t} = \frac{\mathrm{d}\langle N\rangle}{\mathrm{d}t} v_{\rm rec} \simeq \frac{\hbar k_L}{M\tau_R} = \frac{\hbar k_L \Gamma}{M}, \tag{11.3}$$

and consequently that the mean radiative force is equal to:

$$\mathcal{F} = M\frac{\delta v}{\delta t} \simeq \hbar k_L \Gamma. \tag{11.4}$$

This result for the order of magnitude of the mean radiative force $\mathcal{F}$ will be confirmed by more precise calculations in the following.

11.1.2 *Characteristic times*

The calculations of the mean radiative forces presented in this chapter can be simplified with approximations that take into account the existence of different time scales in the problem.

There are actually two types of atomic variables evolving at different speeds: external variables, such as the position $\vec{R}$ and the momentum $\vec{P}$ of the atomic center of mass, which characterize the motion of the atom as a whole; and internal variables such as the atomic dipole moment $\vec{D}$, which characterize the state of the atom in its rest frame.

The typical time of evolution of the internal variables is the radiative lifetime $\tau_R = 1/\Gamma$ of the excited state e. When parameters of the laser beam, such as the frequency or the amplitude, are changed, the mean electric dipole moment and the populations of e and g change with a time constant τ_R. The characteristic time $T_{\rm int}$ of internal variables is thus equal to:

$$T_{\rm int} \simeq \tau_R = 1/\Gamma. \tag{11.5}$$

The characteristic time $T_{\rm ext}$ of external variables is the time it takes for the velocity v of the center of mass to change appreciably under the effect of the mean force $\mathcal{F}$. This change $\delta v = (\mathcal{F}/M)T_{\rm ext}$ will be appreciable if it changes the interaction with the laser in a significative way, i.e. if the change of the Doppler shift associated with δv is on the order of the natural width Γ of the atomic transition. This gives:

$$k_L \frac{\mathcal{F}}{M} T_{\rm ext} \sim \Gamma. \tag{11.6}$$

Using the expression (11.4) of the mean force, we get:

$$T_{\rm ext} \sim \frac{M}{\hbar k_L^2} \sim \frac{\hbar}{E_{\rm rec}}. \tag{11.7}$$

For most allowed optical transitions, $\hbar\Gamma \gg E_{\text{rec}}$, so that:

$$T_{\text{ext}} = \frac{\hbar}{E_{\text{rec}}} \gg T_{\text{int}} = \frac{1}{\Gamma}. \tag{11.8}$$

In the regime of large natural linewidths, internal variables are thus much faster than external variables. It will therefore be possible to consider that internal variables follow in a quasi adiabatic way the slow motion of external variables.

The following table summarizes the value for the recoil velocity $v_{\text{rec}} = \hbar k_L/M$ and the ratio $T_{\text{ext}}/T_{\text{int}} = M\Gamma/\hbar k_L^2$ for different alkali atoms:

Element	λ_L (nm)	$\Gamma/2\pi$ (MHz)	v_{rec} ($\text{m}\cdot\text{s}^{-1}$)	$T_{\text{ext}}/T_{\text{int}}$	E_{rec}/k_B (μK)
H	121.5	100	3.3	7.4	645
Li	671	5.9	0.085	94	3
Na	589	9.8	0.03	394	1.2
K	766	6.1	0.013	740	0.4
Rb	780	6	0.006	1600	0.18
Cs	852	5.2	0.0035	2500	0.1

11.1.3 *Validity of the concept of a mean force at a given point*

The idea of a mean force $\mathcal{F}(x)$ (due to transfer of momentum) acting on an atom at x is valid only if one can construct atomic wave packets with sufficiently small spreads Δx and Δv, without any violation of Heisenberg relations.

Consider first the constraint on Δx. The atom can be considered as being well localized in a laser wave if:

$$\Delta x \ll \lambda_L \qquad \Leftrightarrow \qquad \Delta x \ll 1/k_L. \tag{11.9}$$

Next consider the constraint on Δv. The wave packet can be considered as being well localized in velocity space if the spread of the Doppler shifts $k_L\Delta v$ associated with the velocity spread Δv is small compared to the natural width Γ of the excited state e. Using $\Delta v = \Delta p/M$, where Δp is the momentum spread gives:

$$k_L\Delta p/M \ll \Gamma. \tag{11.10}$$

Multiplying (11.9) by (11.10) leads to:

$$\Delta x\Delta p \ll M\Gamma/k_L^2, \tag{11.11}$$

which is compatible with Heisenberg relations

$$\Delta x\Delta p \geq \hbar \tag{11.12}$$

only if:

$$M\Gamma/k_L^2 \gg \hbar. \tag{11.13}$$

One recovers the large linewidth condition $\hbar\Gamma \gg E_{\text{rec}}$, which is thus essential for two approximations:

(1) The possibility to introduce two different time scales for internal and external variables.
(2) The possibility to define a mean force $\mathcal{F}(x)$ at a given point x.

Remark about the spreading of the wave packet
If Δx is much smaller than λ_L at a given time, this is no longer true at a later time because of the increase of Δx with t. Can this spreading change the conclusions of the above discussion? In fact, the exchange of momentum between the atom and the field, at the origin of laser cooling, transforms the initial wave packet into a statistical mixture of wave packets. The relevant length scale of the problem is no longer the position spread Δx, but rather the *coherence length*[2] :$\xi = \hbar/\Delta p$. The localization condition in position space (11.9) should thus be replaced by:

$$\xi = \hbar/\Delta p \ll 1/k_L \qquad \Leftrightarrow \qquad \Delta p \gg \hbar k_L. \tag{11.14}$$

Most laser cooling schemes obey this condition (except those studied in Chap. 13). The compatibility between Eqs. (11.10) and (11.14) leads again to the large linewidth condition (11.13).

11.2 Calculation of the mean radiative force

The purpose of this section is to derive the mean radiative force $\vec{\mathcal{F}}(\vec{r})$ exerted on an atom at rest by quasi-resonant light [Ashkin (1978); Cook (1979); Gordon and Ashkin (1980); Cohen-Tannoudji (1992a); Cohen-Tannoudji *et al.* (1992b)]. Consequently, we will explicitly take into account the external degrees of freedom of the atom.

11.2.1 *Principle of the calculation*

The approach followed in this section is based on the Heisenberg equations that give the rate of change of the associated variables. In particular, the Heisenberg equation that gives the rate of variation of the momentum operator $\hat{\vec{P}}$ of the atomic center of mass can be identified as the force operator $\hat{\vec{F}}$:

$$\hat{\vec{F}} = \frac{\mathrm{d}\hat{\vec{P}}}{\mathrm{d}t} = \frac{1}{i\hbar}[\hat{\vec{P}}, \hat{H}], \tag{11.15}$$

where $\hat{H}$ is the Hamiltonian. This force operator depends on multiple variables, and in particular, the position operator $\hat{\vec{R}}$ of the center of mass. When we take the average value of the force operator $\langle\hat{\vec{F}}\rangle = \vec{\mathcal{F}}$, we can replace the position operator by its mean value $\hat{\vec{R}} \to \langle\hat{\vec{R}}\rangle = \vec{R}$ because the wave packet is well localized in space. It is therefore possible in this way to define a mean force $\vec{\mathcal{F}}(\vec{R})$ at the position $\vec{R}$ of the center of the atomic wave packet.

[2]See Chap. 17 and references therein.

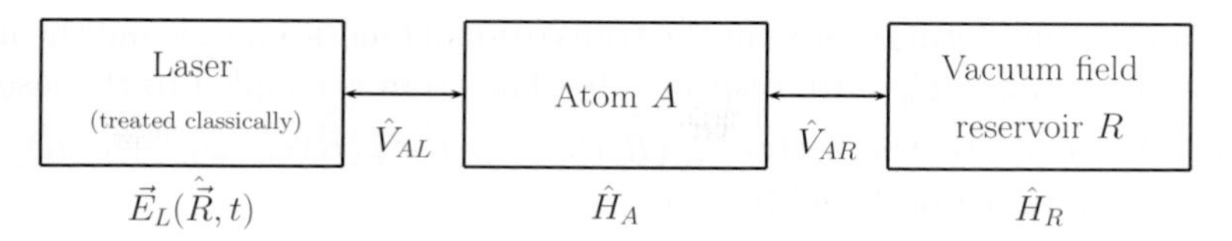

Fig. 11.1 Various subsystems considered in this chapter: the laser field $\vec{E}_L(\hat{\vec{R}},t)$ treated classically, the atom A with Hamiltonian $\hat{H}_A$, and the reservoir with Hamiltonian $\hat{H}_R$. $\hat{V}_{AL}$ and $\hat{V}_{AR}$ account for the coupling of the atom with the laser field and the reservoir of empty electromagnetic modes, respectively.

The next step is then to write the force operator as:

$$\hat{\vec{F}} = \langle\hat{\vec{F}}\rangle + \delta\hat{\vec{F}} = \vec{\mathcal{F}} + \delta\hat{\vec{F}}, \tag{11.16}$$

where $\delta\hat{\vec{F}} = \hat{\vec{F}} - \langle\hat{\vec{F}}\rangle$ is the fluctuating part of the force, analogous to the Langevin force in the theory of Brownian motion. By calculating the correlation function of this fluctuating part, one can determine the momentum diffusion coefficient of the atoms. Here we will just describe the physical mechanisms responsible for the fluctuations of the radiative force.[3]

Radiative forces can also be studied in the Schrödinger point of view. The principle of this approach is to start from the master equation describing the evolution of the atomic density matrix in the Wigner representation. The advantage of this representation is that it describes the atomic external state by a quasi probability distribution quite similar to a classical phase space distribution.[4] By expanding the Wigner function in powers of $\hbar k_L/\Delta p$, it is possible to derive a Fokker-Planck equation for this function analogous to the corresponding equation for Brownian motion, with a drift term describing the effect of the mean force and a momentum diffusion coefficient describing the fluctuations of the force [Dalibard and Cohen-Tannoudji (1985a)].

11.2.2 *Hamiltonian and the rotating wave approximation*

Figure 11.1 summarizes the different subsystems along with their respective Hamiltonian and coupling terms:

$$\hat{H} = \hat{H}_A + \hat{H}_R + \hat{V}_{AR} + \hat{V}_{AL}. \tag{11.17}$$

The two-level atom A is characterized by the Hamiltonian $\hat{H}_A$ that takes into account both the internal and external degrees of freedom:

$$\hat{H}_A = \hat{H}_A^{\text{ext}} + \hat{H}_A^{\text{int}} = \frac{\hat{\vec{P}}^2}{2M} + \hbar\omega_A|e\rangle\langle e|. \tag{11.18}$$

[3] More details can be found in references [Gordon and Ashkin (1980); Cohen-Tannoudji (1992a); Cohen-Tannoudji *et al.* (1992b)].

[4] Note however that the Wigner function is not a probability distribution since it can take negative values.

The first and second terms represent the translational kinetic energy and the internal energy (we have taken $E_g = 0$), respectively. The atom is coupled to the laser field, described classically by the vector $\vec{E}_L(\hat{\vec{R}}, t) = \vec{e}_L(\hat{\vec{R}})\mathcal{E}_L(\hat{\vec{R}}) \cos[\omega_L t + \phi(\hat{\vec{R}})]$ by the electric dipole interaction Hamiltonian:

$$\hat{V}_{AL} = -\hat{\vec{D}} \cdot \vec{E}_L(\hat{\vec{R}}, t), \tag{11.19}$$

where $\hat{\vec{D}} = \vec{d}_{eg}(|e\rangle\langle g| + |g\rangle\langle e|)$ denotes the dipole moment operator, and where the laser electric field $\vec{E}_L(\hat{\vec{R}}, t)$ is evaluated at the position operator of the center of mass. The atom is also coupled to the "reservoir" of initially empty modes of the electromagnetic field responsible for spontaneous emission of fluorescence photons:

$$V_{AR} = -\hat{\vec{D}} \cdot \vec{E}(\hat{\vec{R}}). \tag{11.20}$$

We also define the position dependent Rabi frequency by: $\hbar\Omega(\hat{\vec{R}}) = -\mathcal{E}_L(\hat{\vec{R}})\vec{d}_{eg} \cdot \vec{e}_L(\hat{\vec{R}})$. As we are interested only in quasi-resonant radiative processes, the atom-field coupling can be recast with our notations in the form:

$$\hat{V}_{AL} = \frac{\hbar\Omega(\hat{\vec{R}})}{2}\left[e^{-i\phi(\hat{\vec{R}})-i\omega_L t}|e\rangle\langle g| + \text{h.c.}\right], \tag{11.21}$$

where we have neglected the "antiresonant" terms $e^{-i\omega_L t}|g\rangle\langle e|$ and $e^{i\omega_L t}|e\rangle\langle g|$ since they have eigenfrequencies very different from the resonant processes. This approximation, known as the rotating wave approximation, is also carried out for the coupling term $\hat{V}_{AR}$ with the free radiation field.

11.2.3 *Heisenberg equations for the external variables. Force operator*

Consider the Heisenberg equations of motion for the center of mass position operator $\hat{\vec{R}}$ and its conjugate operator, the atomic momentum operator $\hat{\vec{P}}$:

$$\frac{\mathrm{d}\hat{\vec{R}}}{\mathrm{d}t} = \frac{1}{i\hbar}[\hat{\vec{R}}, \hat{H}] = \frac{\partial \hat{H}}{\partial \hat{\vec{P}}} = \frac{\hat{\vec{P}}}{M}, \tag{11.22}$$

$$\frac{\mathrm{d}\hat{\vec{P}}}{\mathrm{d}t} = \frac{1}{i\hbar}[\hat{\vec{P}}, \hat{H}] = -\frac{\partial \hat{H}}{\partial \hat{\vec{R}}} = -\vec{\nabla}\hat{V}_{AL}(\hat{\vec{R}}, t) - \vec{\nabla}\hat{V}_{AR}(\hat{\vec{R}}, t) = \hat{\vec{F}}(\hat{\vec{R}}). \tag{11.23}$$

In the semiclassical treatment followed here, we are interested in the average value $\vec{\mathcal{F}}$ of the operator $\hat{\vec{F}}$:

$$\vec{\mathcal{F}} = \langle\hat{\vec{F}}\rangle = -\langle\vec{\nabla}\hat{V}_{AL}(\hat{\vec{R}}, t)\rangle - \langle\vec{\nabla}\hat{V}_{AR}(\hat{\vec{R}}, t)\rangle. \tag{11.24}$$

One can show that the contribution of $\hat{V}_{AR}$ to the mean radiative force is equal to zero [Cohen-Tannoudji *et al.* (1992b)]. Physically, this is due to the fact that the vacuum field has a zero average value and that the field radiated by the dipole has no gradient at the position of the atom. Note however that the coupling $\hat{V}_{AR}$ with the reservoir R introduces damping terms, which play a major role in the master equation for the internal degrees of freedom. The expression for the mean radiative force then reads:

$$\vec{\mathcal{F}} = -\langle \vec{\nabla} \hat{V}_{AL}(\hat{\vec{R}}, t) \rangle + 0 = \sum_{i=x,y,z} \langle \hat{D}_i \vec{\nabla} E_{Li}(\hat{\vec{R}}, t) \rangle. \quad (11.25)$$

11.2.4 *Approximations. Mean radiative force*

In order to derive the explicit expression of the mean radiative force, we will introduce two approximations:

(i) As a result of the large mass of the atom, the de Broglie wavelength of the atom $\lambda_{\text{dB}} = h/Mv$ is in general much smaller than the optical wavelength λ_L, i.e. the typical length scale of the driving field. It is therefore possible to construct atomic wave packets with very small dimensions compared with λ_L. For such wave packets, it is legitimate to replace the position operator $\hat{\vec{R}}$ of the center of mass by its mean value $\vec{R}$. As a consequence, the gradient of the laser field becomes a c-number that can be taken out of the average. The mean force at $\vec{R}$ is thus equal to:

$$\vec{\mathcal{F}}(\vec{R}, t) = -\langle \vec{\nabla} \hat{V}_{AL}(\vec{R}, t) \rangle = + \sum_{i=x,y,z} \langle \hat{D}_i \rangle \vec{\nabla} E_{Li}(\vec{R}, t). \quad (11.26)$$

(ii) As already emphasized in the introduction of the chapter, the internal degrees of freedom of the atom evolve on a much shorter time scale than the external ones. We will consequently neglect the variations of the slow external variables during the time needed for the fast internal variables to reach an equilibrium. If we suppose that the atom is initially at rest, we can thus consider that the mean dipole moment appearing in Eq. (11.26) is equal to its steady state value $\langle \hat{D}_i \rangle \longrightarrow \langle \hat{D}_i \rangle_{\text{st}}$.

11.2.5 *The two types of mean radiative forces: dissipative and reactive*

From Eq. (11.26), we deduce that the contribution of the internal degrees of freedom to the mean force $\vec{\mathcal{F}}$ appears through the averages $\langle \hat{D}_i \rangle$. In practice, the determination of the steady state value of each dipole moment component requires the knowledge of the steady state solution for the density matrix $\hat{\sigma}$ for the internal degrees of freedom. The contribution of the external degrees of freedom to the mean force $\vec{\mathcal{F}}$ originates from the gradient of the laser electric field, and, as a consequence, has two parts: one proportional to the gradient of the phase of the laser field, and another proportional to the gradient of the amplitude of the laser field.

Remark on polarization gradients

If the polarization $\vec{e}_L(\vec{r})$ of the laser wave varies in space, one may wonder if polarization gradients should not also appear in the expression of the radiative forces in addition to the phase and amplitude gradients. In fact, it is not the amplitude $\mathcal{E}_L(\vec{r})$ that appears in the coupling Hamiltonian $\hat{V}_{AL}$ (and subsequently in its gradient) that gives the expression of the radiative force. It is the Rabi frequency $\Omega(\vec{r})$ (see Eq. (11.21)) which contains two functions of $\vec{r}$: $\mathcal{E}_L(\vec{r})$ and $\vec{d}_{eg} \cdot \vec{e}_L(\vec{r})$. For the case of a two-level atom considered in this chapter, there is a single dipole transition matrix element $\vec{d}_{eg}$ that has a fixed direction $\vec{u}$ in space. Only the component $\vec{u}.\vec{e}_L(\vec{r})$ of $\vec{e}_L(\vec{r})$ along this direction $\vec{u}$ appears in the Rabi frequency $\Omega(\vec{r})$. If $\vec{e}_L(\vec{r})$ varies in space, polarization gradients contribute to the spatial variation of the Rabi frequency and their effect can be taken into account by redefining an effective amplitude $\mathcal{E}_{\text{eff}}(\vec{r})$ equal to the product of $\mathcal{E}(\vec{r})$ by $\vec{d}_{eg} \cdot \vec{e}_L(\vec{r})$. Radiative forces can then be expressed in terms of the phase gradient and the gradient of the effective amplitude. For an atom with several Zeeman sublevels in e and g, there are several Zeeman components of the optical line that can be excited with relative efficiencies depending on $\vec{e}_L(\vec{r})$ and that can then vary in space in the presence of polarization gradients. The light shifts of the various Zeeman sublevels in g and the optical pumping rates between them can then vary in space, giving rise to new interesting laser cooling mechanisms to be studied in Chap. 13.

We suppose an atom at rest at $\vec{R}$ and we choose the origin of space and time such that $\vec{R} = \vec{0}$ and $\phi(\vec{0}) = 0$. Using the steady state solution of the Bloch equations (see Note below for detailed calculations) we readily derive the expression for the force experienced by the atom illuminated by a quasi-resonant light:

$$\vec{\mathcal{F}}\left(\vec{R}, t\right) = -\operatorname{Tr}\left\{\hat{\sigma}_{st}\,\vec{\nabla}\hat{V}_{AL}(\vec{R}, t)\right\} = \vec{\mathcal{F}}_{\text{dissip}} + \vec{\mathcal{F}}_{\text{react}} \tag{11.27}$$

with

$$\vec{\mathcal{F}}_{\text{dissip}} = -\hbar\,\Omega v_{\text{st}}\,\vec{\nabla}\phi\Big|_{\vec{R}=\vec{0}} \quad \text{and} \quad \vec{\mathcal{F}}_{\text{react}} = -\hbar\,\Omega u_{\text{st}}\,\frac{\vec{\nabla}\Omega}{\Omega}\Big|_{\vec{R}=\vec{0}}, \tag{11.28}$$

where u_{st} and v_{st} are the steady state components of the atomic dipole moment in phase and in quadrature with the laser field, respectively (see Eq. 11.33). One finds

$$u_{\text{st}} = \frac{\delta}{\Omega}\frac{s}{1+s}, \qquad v_{\text{st}} = \frac{\Gamma}{2\Omega}\frac{s}{1+s}, \tag{11.29}$$

where

$$s = \frac{\Omega^2}{2}\frac{1}{\delta^2 + \Gamma^2/4}, \tag{11.30}$$

is the saturation parameter.

The radiative force is thus the sum of two forces:

- a "dissipative force" $\vec{\mathcal{F}}_{\text{dissip}} = -\,\hbar\,\Omega\,v_{\text{st}}\;\vec{\nabla}\phi\Big|_{\vec{R}=\vec{0}}$, proportional to the dissipative component v_{st} of the atomic dipole moment in quadrature with the laser field and to the phase gradient of the laser field.

- a "reactive force" $\vec{\mathcal{F}}_{\text{react}} = \hbar\, u_{\text{st}}\, \vec{\nabla}\Omega\Big|_{\vec{R}=\vec{0}}$ proportional to the reactive component u_{st} of the atomic dipole moment in phase with the laser field and to the amplitude gradient of the laser field.

Note. *Optical Bloch equations and their steady-state solution*

The optical Bloch equations have already been briefly discussed in Sec. 5.3 of Chap. 5. In this note, we work out their steady state solution. A more extensive study can be found in [Cohen-Tannoudji *et al.* (1992b)]. These equations give the rate of variation of the density matrix for the atomic internal degrees of freedom:

$$\frac{\mathrm{d}\hat{\sigma}}{\mathrm{d}t} = \frac{1}{i\hbar}\left[\hat{H}_A + \hat{V}_{AL}\,,\hat{\sigma}\right] + \left(\frac{\mathrm{d}\hat{\sigma}}{\mathrm{d}t}\right)_{\text{spont}}, \tag{11.31}$$

where the damping terms due to the spontaneous emission have the following expression:

$$\left(\frac{\mathrm{d}\sigma_{ee}}{\mathrm{d}t}\right)_{\text{spont}} = -\left(\frac{\mathrm{d}\sigma_{gg}}{\mathrm{d}t}\right)_{\text{spont}} = -\,\Gamma\sigma_{ee}, \quad \left(\frac{\mathrm{d}\sigma_{eg}}{\mathrm{d}t}\right)_{\text{spont}} = -\,\frac{\Gamma}{2}\sigma_{eg}.$$

As already pointed out, the evolution of the external degrees of freedom depend only on the steady state solution $\hat{\sigma}^{\text{st}}$ of the master equation. To derive its expression, it is convenient to rewrite Eq. (11.31) in terms of the Bloch vector $\vec{A}$ of components (u, v, w), related to the elements of the density matrix $\hat{\sigma}$ in the following manner:

$$u = \frac{\sigma_{ge}e^{-i\omega_L t} + \sigma_{eg}e^{i\omega_L t}}{2}, \qquad v = \frac{\sigma_{ge}e^{-i\omega_L t} - \sigma_{eg}e^{i\omega_L t}}{2i}, \qquad w = \frac{\sigma_{ee} - \sigma_{gg}}{2}. \tag{11.32}$$

The use of these variables permits one to recast optical Bloch equations in terms of a set of three coupled linear differential equations with time-independent coefficients. Physically, this transformation corresponds to the rewriting of the Bloch equations in the rotating frame (rotating wave approximation).

The third component w of the Bloch vector represents half the difference between the populations of the two levels g and e. The physical interpretation of u and v is readily obtained by calculating the mean value of the dipole operator:

$$\langle\hat{\vec{D}}\rangle = \vec{d}_{eg}\left(\sigma_{ge} + \sigma_{eg}\right) = 2\,\vec{d}_{eg}\mathrm{Re}\left[\sigma_{ge}\right] = 2\,\vec{d}_{eg}\left(u\cos\omega_L t - v\sin\omega_L t\right). \tag{11.33}$$

The components u and v of the Bloch vector are thus proportional to the components of $\langle\hat{\vec{D}}\rangle$ in phase and in advance quadrature with the incident field $\vec{E}_L(\vec{R},t)$, respectively.

The optical Bloch equations (11.31) are readily recast in the following set of three equations:

$$\frac{\mathrm{d}\vec{A}}{\mathrm{d}t} = \mathcal{B}\vec{A} - \vec{A}_0, \text{ with } \mathcal{B} = \begin{pmatrix} -\Gamma/2 & \delta & 0 \\ -\delta & -\Gamma/2 & -\Omega \\ 0 & \Omega & -\Gamma \end{pmatrix} \text{ and } \vec{A}_0 = \begin{pmatrix} 0 \\ 0 \\ \Gamma/2 \end{pmatrix}. \tag{11.34}$$

Since the atom can be considered to remain at rest at the origin during the characteristic time Γ^{-1} for the evolution of the internal variables, the Rabi frequency Ω does not depend on time, and the optical Bloch equations are then a set of three linear differential equations with time independent coefficients. Their steady state solution is given by $\vec{A}_{\rm st} = \mathcal{B}^{-1}\vec{A}_0$. One finds

$$u_{\rm st} = \frac{\delta}{\Omega}\frac{s}{1+s}, \qquad v_{\rm st} = \frac{\Gamma}{2\Omega}\frac{s}{1+s}, \qquad w_{\rm st} = -\frac{1}{2\,(1+s)}, \tag{11.35}$$

where $s = (\Omega^2/2)/(\delta^2 + \Gamma^2/4)$ is the saturation parameter. The steady state solutions for $u_{\rm st}$, $v_{\rm st}$ and $w_{\rm st}$ are not perturbative with respect to the driving field since the square of the Rabi frequency Ω^2 enters their denominator.

The component $v_{\rm st}$ in quadrature of the dipole and the population of the upper state ($\sigma_{ee}^{\rm st} = w_{\rm st} + 1/2$) vary as an absorption curve as a function of the detuning $\delta = \omega_L - \omega_A$, centered at $\delta = 0$ and with a half-width equal to $[\Gamma^2/4 + \Omega^2/2]^{1/2}$. The component $v_{\rm st}$ is proportional to the average power absorbed by the atom, and corresponds to the "dissipative" response of the dipole. The component $u_{\rm st}$ in phase with the driving field varies as a dispersion curve, and corresponds to the "reactive" response of the dipole.

When $s = 1$, one easily finds from the definition of w and from the normalization of $\hat{\sigma}$ ($\sigma_{ee} + \sigma_{gg} = 1$) that $\sigma_{gg} = 3/4$ and $\sigma_{ee} = 1/4$. We usually introduce the saturation intensity of a laser $I_{\rm sat}$ as the intensity for which $s = 1$ at resonance ($\delta = 0$), i.e. for which the population of the excited state e becomes appreciable. For rubidium atoms, one finds that $I_{\rm sat} = 1.6$ mW / cm^2, a value easily accessible with standard laser diodes.

11.3 Dissipative force

In this section, we analyze in greater detail the physics of the dissipative force along with the early fundamental applications of radiation pressure.

11.3.1 *Theoretical results*

We will focus on the simplest case where the laser wave is a plane wave with wave vector $\vec{k}_L$: $\vec{E}_L(\vec{r},t) = \vec{e}_L\mathcal{E}_L\cos[\omega_L t - \vec{k}_L\cdot\vec{r}]$. This field is characterized by a constant amplitude and by an $\vec{r}$-dependent phase $\phi(\vec{r}) = -\vec{k}_L\cdot\vec{r}$ that gives rise to a phase gradient $\vec{\nabla}\phi = -\vec{k}_L$. For such a wave, the reactive force vanishes (because $\vec{\nabla}\mathcal{E}_L = 0$), and the radiative force is given entirely by its dissipative component:

$$\vec{\mathcal{F}}_{\rm dissip} = -\hbar\,\Omega\,v_{\rm st}\;\vec{\nabla}\phi\Big|_{\vec{R}=\vec{0}} = \hbar\,\vec{k}_L\,\Gamma\,\frac{\Omega^2/4}{\delta^2 + (\Gamma^2/4) + (\Omega^2/2)}. \tag{11.36}$$

At low laser intensity ($\Omega \ll |\delta|$ or Γ), the force is proportional to Ω^2 and thus to the intensity. At high intensity ($\Omega \gg \delta$ and Γ), the force saturates to a maximum value

$$\vec{\mathcal{F}}_{\rm max} = \hbar\,\vec{k}_L\,\Gamma/2, \tag{11.37}$$

that is independent of the intensity. The average force reaches a maximum when the light is resonant with the atomic transition ($\delta = 0$), and varies with δ as a Lorentzian absorption curve centered at $\omega_L = \omega_0$ with a width $\Delta\omega = \sqrt{(\Gamma^2/4) + (\Omega^2/2)}$.

11.3.2 *Physical interpretation*

The physical interpretation of Eq. (11.36) is straightforward when one expresses the Bloch vector component $v_{\rm st}$ in terms of the steady state population $\sigma_{ee}^{\rm st}$ of the excited state. This relation is deduced from the third optical Bloch equation in the steady state regime: $\Omega\, v_{\rm st} = \Gamma w_{\rm st} + \Gamma/2 = \Gamma \sigma_{ee}^{\rm st}$. The dissipative force can thus be rewritten in the form:

$$\vec{\mathcal{F}}_{\rm dissip} = \hbar\, \vec{k}_L\, \Gamma\, \sigma_{ee}^{\rm st} = \hbar\, \vec{k}_L \left\langle \frac{{\rm d}N_{\rm spont}}{{\rm d}t} \right\rangle. \tag{11.38}$$

We have used the fact that $\Gamma\sigma_{ee}^{\rm st}$ is nothing but the number of spontaneous emission processes per second, or in other words, the number of fluorescence cycles per second. At each of these cycles the atom gains the momentum $\hbar\, \vec{k}_L$. It must be stressed that only the absorbed laser photon contributes to the momentum change of the atom since the spontaneous photon can be emitted with equal probabilities in opposite directions, so that the corresponding momentum change is zero on average.

The dissipative force appears as the mean momentum transfer per second from the laser to the atom in photon scattering processes. For that reason, it is also called the "scattering force". If the intensity associated with the plane wave is sufficiently high, stimulated emissions also occur, though this latter process does not contribute to the force exerted by the light. Indeed, the momentum acquired during the absorption process is regained by the incident beam through stimulated emission, and thus the dissipative force results exclusively from spontaneous emission processes, which cause a photon to disappear from the incident beam.

It is also instructive to calculate the energy absorbed from the field by the atom. Between t and $t + {\rm d}t$, the driving field $\vec{E}_L(\vec{r} = \vec{0}, t)$ carries out the following elementary work on the atom: $\langle {\rm d}W \rangle = \vec{E}_L \cdot {\rm d}\langle \hat{\vec{D}}_{\rm st} \rangle$. We deduce the expression for the mean absorbed power averaged over an optical period:

$$\overline{\left\langle \frac{{\rm d}\, W}{{\rm d}t} \right\rangle} = -\omega_L\, \vec{E}_L \cdot \vec{d}_{eg}\, (v_{\rm st}) = \hbar\, \omega_L\, \Omega\, v_{\rm st}. \tag{11.39}$$

As expected, the energy is absorbed from the field by the component $v_{\rm st}$ in advance quadrature of the mean dipole moment. The mean number of photons absorbed per second is thus given by

$$\left\langle \frac{{\rm d}N_{\rm abs}}{{\rm d}t} \right\rangle = \frac{1}{\hbar\, \omega_L} \overline{\left\langle \frac{{\rm d}W}{{\rm d}t} \right\rangle} = \Omega\, v_{\rm st}. \tag{11.40}$$

The dissipative force can thus also be rewritten as:

$$\vec{\mathcal{F}}_{\rm dissip} = \hbar\, \vec{k}_L \left\langle \frac{{\rm d}N_{\rm abs}}{{\rm d}t} \right\rangle. \tag{11.41}$$

The two expressions (11.38) and (11.41) are equivalent since in the steady state regime the number of photons spontaneously emitted per second is equal to the number of photons absorbed per second.

11.3.3 *Application to the deflection and to the slowing down of an atomic beam*

To see how large the scattering force can be, let us calculate the maximum acceleration (or deceleration) $a_{\max}$ that can be communicated to a given atom with radiation pressure:

$$a_{\max} = \frac{\mathcal{F}_{\text{dissip}}^{\max}}{M} = \left(\frac{\hbar k_L}{M}\right)\frac{\Gamma}{2} = \frac{\Gamma}{2}v_{\text{rec}}. \tag{11.42}$$

At each cycle, the atomic velocity changes by an amount equal to the recoil velocity v_{rec}. This velocity change, on the order of 10^{-2} m·s^{-1} in the optical range, is very weak, but in principle it can be accumulated a large number of times $\Gamma/2 \simeq 10^8$ per second. We infer the following order of magnitude: $a_{\max} = v_{\text{rec}} \times (\Gamma/2) \simeq 10^{-2}\text{m·s}^{-1} \times 10^8\,\text{s}^{-1} = 10^6\,\text{m}\cdot\text{s}^{-2} \simeq 10^5\,g$, where g is the acceleration due to gravity. This is a very large value that clearly shows the potential of resonant laser light for acting on atoms.

In 1933, Otto Frisch did the first experiment to directly show the momentum transferred to an atom by the absorption of a photon [Frisch (1933)]. In this experiment, light from a sodium lamp deflected a beam of sodium atoms. However, since the light beam was very far from saturating the atomic transition, it yielded a small deflection of the beam. In the sixties and early seventies, Arthur Ashkin [Ashkin (1970a,b)] and several scientists in Soviet Union [Askar'yan (1962); Kazantsev (1972); Letokhov (1968)] recognized the potential of intense, narrow-band laser light for manipulating atoms in various ways, such as accelerating them or trapping them in optical potential wells. Experiments of deflection of atomic beams were then revisited with resonant lasers in 1972 [Picqué and Vialle (1972); Schieder *et al.* (1972)]. Due to the much higher spectral intensities available with resonant laser light, an atom can absorb many photons at a high rate resulting in larger deflections. Figure 11.2(a) represents the typical sketch of such an experiment. A resonant laser beam excites an atomic beam at a right angle. The deviation and the broadening of the position distribution of the deflected atoms are measured with a detector as illustrated in Fig. 11.2(b).

The mechanical action of resonant light also offers the possibility of very efficiently stopping an atomic beam emerging from an oven [Balykin *et al.* (1979); Phillips and Metcalf (1982); Phillips and Prodan (1983); Ertmer *et al.* (1985)]. Consider an atom moving at a velocity $v_0 = 10^3$ m/s. A resonant and counter-propagating beam can decelerate this atom down to zero velocity in a time equal to $T = v_0/a_{\max} \simeq 10^{-3}$ s, over a distance of only $L = (v_0)^2/2a_{\max} = 0.5$ m.

The first experiments of this type were carried out with lasers at fixed optical frequency. In the velocity distribution of the atomic beam, only atoms with a velocity allowing them to be in resonance with the laser could interact with it. The laser initially decelerates these atoms, but when their velocity change is large enough, they are Doppler shifted out of resonance and the deceleration stops. One thus expects to get a hole in the velocity profile of the beam corresponding to

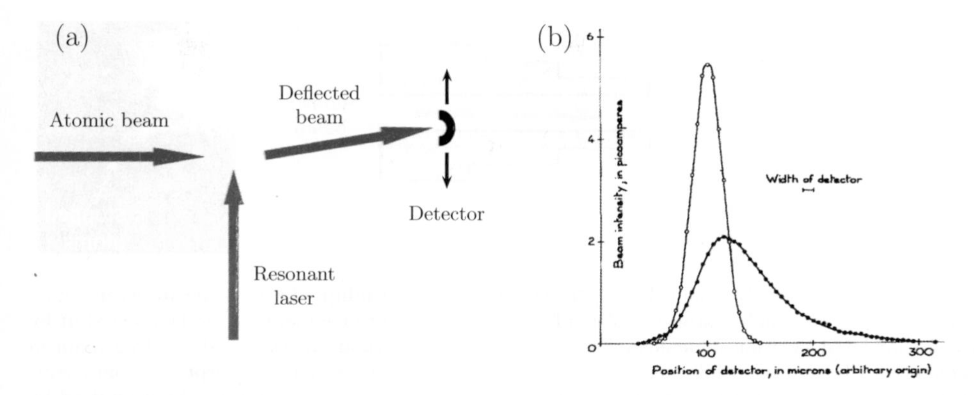

Fig. 11.2 (a) Sketch of an experiment demonstrating the deflection of an atomic beam by resonant laser light. A resonant laser beam, perpendicular to the atomic beam, exerts a radiation pressure force that deviates the beam. (b) Experimental results from [Picqué and Vialle (1972)]. The narrow curve gives the position distribution of the deflected atoms in the absence of the pushing laser. In the presence of the laser, this distribution is shifted and broadened. Copyright: Institute of Physics.

the resonant velocities and a narrow peak on the low velocity side of this hole corresponding to atoms that have been pushed to lower velocities (see Fig. 11.3). The class of atoms affected by the counter-propagating laser has been both slowed down and cooled, through the narrow peak still consists of rather fast atoms.

One solution to prevent atoms from coming out of resonance as they are decelerated was proposed in 1976 by Letokhov, Minogin, and Pavlik [Letokhov *et al.* (1976)]. They suggested chirping the frequency of the laser so as to keep it in resonance with a much larger class of atomic velocities. The Moscow group applied the technique to decelerate an atomic beam [Balykin *et al.* (1979)], but without

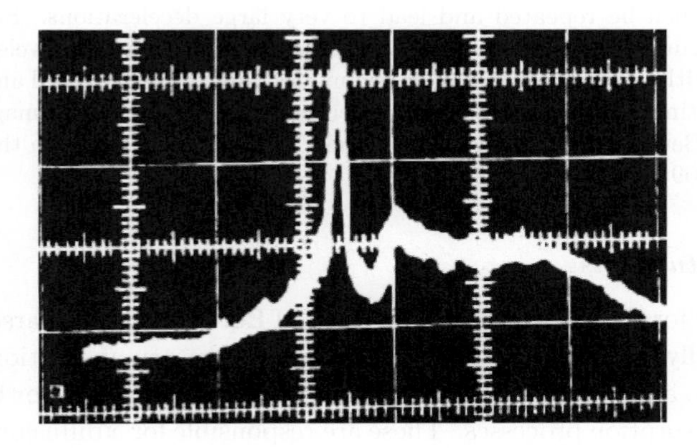

Fig. 11.3 Velocity distribution of an atomic beam in the presence of a resonant and counter-propagating beam: experimental results [Balykin *et al.* (1979)]. The laser drills a hole in the velocity distribution and the corresponding atoms are accumulated in a narrow peak at a lower velocity: The laser slows down these atoms and cools them (narrower distribution). Courtesy JETP Letters.

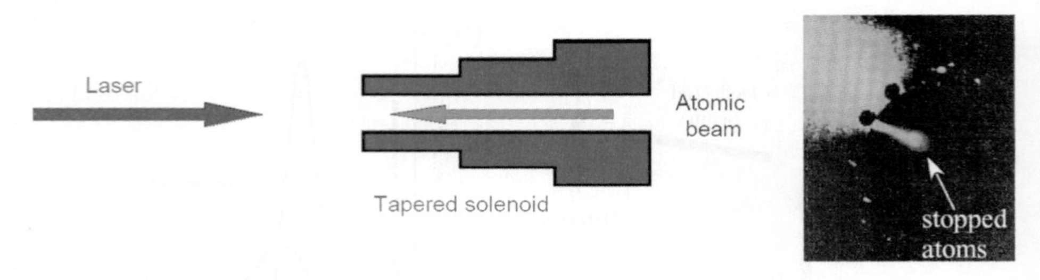

Fig. 11.4 Principle of a Zeeman slower [Prodan *et al.* (1982); Phillips (1998)]. The atoms emerging from an oven go through a tapered solenoid which allows one to compensate the Doppler shift by a well-calculated position dependent Zeeman shift. This method allows the atoms to remain in resonance with the laser while they are slowed down by the succession of absorption-spontaneous emission cycles. The photo on the right part of the figure shows the fluorescence light emitted by the atoms during the stopping process. Courtesy of W. Phillips.

clear success [Balykin (1980)]. A convincing and significant deceleration and cooling of an atomic beam with this chirp-cooling was obtained in 1983 by John Prodan and William Phillips [Phillips and Prodan (1983)]. Later it was shown that it is even possible to bring the atoms to rest using this technique [Ertmer *et al.* (1985)]. The other standard technique for decelerating beams is realized through the use of a "Zeeman slower"[Prodan *et al.* (1982); Phillips (1998)]. This method consists of keeping the atoms on resonance by compensating the Doppler shift due to the deceleration of the atoms with a position dependent Zeeman shift produced by a magnetic field gradient along the direction of the laser (see Fig. 11.4).

Remark

The large value of the radiation pressure force considered in this section is due to the fact that the atomic transition excited by the laser is closed. Several fluorescence cycles can then be repeated and lead to very large decelerations. Such a situation does not occur for molecules. Atomic transitions can also have a wavelength not easily accessible with tunable lasers. Other techniques have been suggested and implemented for decelerating a supersonic beam of heteropolar molecules or paramagnetic atoms or molecules [Bethlem *et al.* (1999); Vanhaecke *et al.* (2007)]. See also the review paper [Raizen (2009)] and references therein.

11.3.4 *Fluctuations*

The expression for the dissipative force $\vec{\mathcal{F}}_{\text{dissip}}$ of Eq. (11.36) is a coarsed-grained average. Physically, this force results from many elementary absorption-spontaneous emission cycles, and is consequently accompanied by fluctuations for both the emission and the absorption processes. These are responsible for a diffusion of the atomic momentum.

We distinguish two contributions to the momentum diffusion driven by the fluctuations of the force [Gordon and Ashkin (1980)]. The first one originates from the fluctuations of the recoil momentum transferred by the fluorescence photons that are emitted in random directions. The second one is associated with the fluctuations in the number of fluorescence cycles occurring in a given time interval.

These diffusion mechanisms in velocity space are responsible for a heating of the translational degrees of freedom. It is thus important to quantitatively evaluate these effects since they ultimately dictate, as detailed in the following chapter, the limit of Doppler laser cooling.

For each photon spontaneously emitted in a given direction, the atom recoils in the opposite direction. As the emission occurs in a random direction, the atom experiences a random walk of step size $\hbar k_L$ in momentum space. The number of steps per second is given by the absorption rate $\mathrm{d}N/\mathrm{d}t = \Gamma\sigma_{ee}^{\mathrm{st}}$. The corresponding Brownian motion in momentum space gives rise to a diffusion of the atomic momentum. For a time interval Δt, the momentum dispersion increases according to

$$\overline{\Delta p^2} = \hbar^2 k_L^2 \left(\frac{\mathrm{d}N}{\mathrm{d}t}\right) \Delta t = 2D_{\mathrm{spont}}\Delta t, \tag{11.43}$$

where the momentum diffusion coefficient D_{spont} reads

$$D_{\mathrm{spont}} = \frac{1}{2}\hbar^2 k_L^2\, \Gamma\, \sigma_{ee}^{\mathrm{st}} = \frac{1}{4}\hbar^2 k_L^2\, \Gamma \frac{s}{1+s}. \tag{11.44}$$

Note that at sufficiently high intensity, the diffusion coefficient D_{spont} saturates to $\hbar^2 k_L^2 \Gamma/4$. This simply reflects the saturation of the absorption rate.

The fluctuations of the number of absorbed photons ΔN_{abs} during a time interval Δt also leads to an atomic diffusion along the laser beam axis. The momentum absorbed from the laser beam is: $\Delta p = \Delta N_{\mathrm{abs}}\, \hbar k_L$, and the corresponding variance of the momentum is related to the variance of the number of absorbed photons in the following manner:

$$\overline{\Delta p^2} - \left(\overline{\Delta p}\right)^2 = \hbar^2 k_L^2 \left[\overline{\Delta N_{\mathrm{abs}}^2} - \left(\overline{\Delta N_{\mathrm{abs}}}\right)^2\right] \equiv \overline{\Delta N_{\mathrm{abs}}}\,(1+Q)\,, \tag{11.45}$$

where Q is a correction to Poisson statistics that can be deduced from the theory of resonance fluorescence. At low intensity, $Q < 0$ and reveals a sub-Poissonian probability distribution in resonance fluorescence [Short and Mandel (1983); Stenholm (1986)]. Since $\overline{\Delta N_{\mathrm{abs}}} = \Gamma\, \sigma_{ee}^{\mathrm{st}}\, \Delta t$ one finds:

$$\overline{\Delta p^2} - \left(\overline{\Delta p}\right)^2 = 2D_{\mathrm{abs}}\, \Delta\, t \quad \text{where} \quad D_{\mathrm{abs}} = \frac{1}{4}\hbar^2 k_L^2\, \Gamma\, \frac{s}{1+s}\,(1+Q)\,. \tag{11.46}$$

In conclusion, we find that the momentum diffusion coefficient D_{abs} due to the fluctuations of the number of absorbed photons is of the same order of magnitude as D_{spont}.

11.4 Reactive force

In this section, we work out the different expressions for the mean value of the reactive force $\vec{\mathcal{F}}_{\mathrm{react}}$. We also present a physical interpretation of this force and of its fluctuations using the dressed-atom approach.

11.4.1 *Theoretical results*

Using Eqs. (11.28) and (11.35), we readily derive the following expression for the reactive force:

$$\vec{\mathcal{F}}_{\text{react}} = -\,\hbar\,u_{\text{st}}\;\vec{\nabla}\Omega\Big|_{\vec{R}=\vec{0}} = -\frac{\hbar\,\delta}{4}\,\frac{\vec{\nabla}\,\Omega^2}{\delta^2+(\Gamma^2/4)+(\Omega^2/2)}, \tag{11.47}$$

$\vec{\mathcal{F}}_{\text{react}}$ varies with the detuning $\delta = \omega_L - \omega_A$ as a dispersion curve having the same width as the dissipative force $\vec{\mathcal{F}}_{\text{dissip}}$. For $\omega_L > \omega_A$ ($\delta > 0$, blue detuning), the force repels the atoms away from the high intensity regions. For $\omega_L < \omega_A$ ($\delta < 0$, red detuning), the force attracts the atoms toward regions of high laser intensities. The electric polarizability is thus positive below resonance and negative above resonance. This is in agreement with the classical response of an electron elastically bound to the positive core of an atom. The charge oscillates in phase with the external force if the frequency is below resonance, and with an opposite phase if the frequency is above.

The value of the detuning δ that maximizes $|\vec{\mathcal{F}}_{\text{react}}|$ depends upon the intensity through the Rabi frequency Ω. This intensity-dependent optimal detuning is on the order of Ω. We conclude that $|\vec{\mathcal{F}}_{\text{react}}|_{\max} \sim \hbar|\vec{\nabla}\Omega|$. The reactive force thus increases with the intensity of the light, in contrast to the dissipative force that saturates at large Ω.

Finally, the reactive force derives from a potential

$$\vec{\mathcal{F}}_{\text{react}} = -\vec{\nabla}\,U, \quad \text{where} \quad U = \frac{\hbar\,\delta}{2}\ln\left[1+\frac{\Omega^2/2}{\delta^2+(\Gamma^2/4)}\right]. \tag{11.48}$$

The reactive force, also called the "dipole force", appears as a conservative force that is intimately related to the coherent absorption stimulated emission processes. If $\omega_L < \omega_A$, the maximum optical potential depth is on the order of $0.3\;|\hbar\,\Omega_{\max}|$ and therefore increases indefinitely with the light intensity.

11.4.2 *Physical interpretation*

A plane wave has no amplitude gradient. To have a non-zero amplitude gradient, and therefore a non-zero reactive force, one has to superimpose several plane waves with different wave vectors.

To ease the interpretation of the reactive force, let us describe the laser wave as a superposition of several plane waves with the same frequency $\omega_i = \omega_L$, $i = 1, 2, 3$, ... but with different wave vector directions $\vec{k}_i$.

The atom can absorb one photon in the wave i and emit, in a stimulated way, one photon in another wave $j \neq i$. In such a cycle, no energy is absorbed from the laser wave since $\hbar\omega_i = \hbar\omega_j = \hbar\omega_L$. However, as $\hbar\vec{k}_i \neq \hbar\vec{k}_j$, the atomic momentum changes by an amount $\hbar(\vec{k}_i - \vec{k}_j)$. The reactive force thus appears as a force due to the "redistribution" of photons between the various plane waves forming the laser wave. This is why it is also called "redistribution force".

The reactive force involves absorption-stimulated emission with Ω as a typical rate to be compared with the dissipative force, which involves absorption spontaneous emission cycles with a rate ultimately limited by the width Γ of the excited state. This explains why the reactive force increases indefinitely with the light intensity in contrast to the dissipative force that saturates.

One may wonder what determines the sense of the redistribution $i \to j$ or $j \to i$? The response is the relative phase between the two waves i and j at the atom location. This important point is illustrated by considering the simplest case leading to a non-vanishing dipole force: a laser wave consisting of two counter-propagating plane waves. The electric fields E_1 and E_2 are represented in the complex plane in Fig. 11.5 at a point where they are in quadrature. The component u that accounts for the reactive response of the dipole is in phase with the total field E and does not absorb energy from E. Let u_1 and u_2 be its components in phase with E_1 and E_2. The component u_2, which is in phase with E_2 does not absorb energy from E_2. However, u_2 is in advance quadrature with E_1, so it absorbs (from E_1) an energy proportional to $E_1 u_2$. Similarly, the component u_1, which is in phase with E_1, does not exchange energy with E_1, but it is in retarded quadrature with E_2, and thus emits an energy proportional to $E_2 u_1$ into this field. Since $|E_1||u_2| = |E_2||u_1|$, the total energy does not change, but the direction of this coherent redistribution is determined by the relative phase of E_1 and E_2.

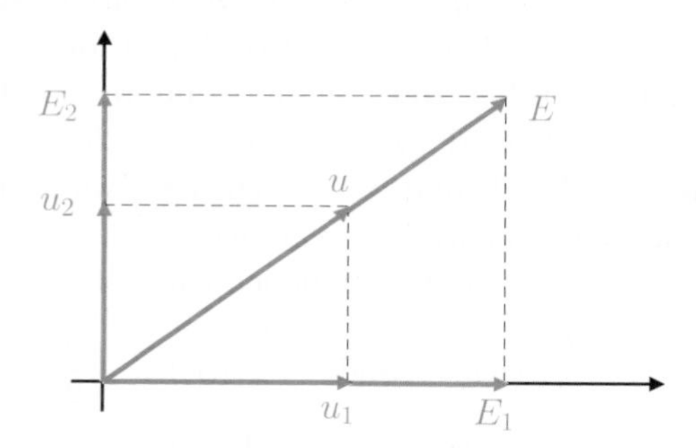

Fig. 11.5 Representation in the complex plane of the field E that results from the superposition of two fields E_1 and E_2 at a point where they are in quadrature. u, u_1 and u_2 are the dipole components in phase respectively with E, E_1 and E_2.

11.4.3 *Dressed atom interpretation*

The picture of the dressed atom cascading down its energy diagram leads to a simple interpretation of the mean value and fluctuations of the dipole force exerted on an atom initially at rest at $\vec{r}$ [Dalibard and Cohen-Tannoudji (1985b)].

As analyzed in Sec. 5.4 of Chap. 5, the dressed atom approach relies on a quantum description of the laser field that we suppose to be in a coherent state of a single mode. In the absence of coupling, the eigenstates of the Hamiltonian $H_A + H_L$ of

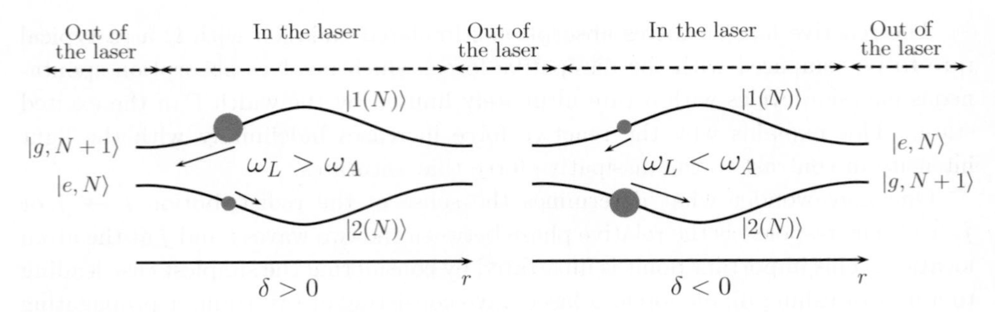

Fig. 11.6 Dressed atom-laser states at a given point $\vec{r}$. The atom-laser coupling produces a position dependent splitting $\Omega(\vec{r})$ between the two dressed states of a given manifold. We have represented the variation across the laser beam of the dressed-atom energy levels for a positive detuning (left part) and for a negative detuning (right part). The energy splitting and the wave functions both depend on the position $\vec{r}$. Out of the laser beam, the energy levels connect with the uncoupled atom-laser states. The dipole force is the average of the two-dressed-state position-dependent forces weighted by the steady state populations of these states, represented by the filled circles. The sign of the mean force is determined by the most populated level, which is less contaminated by the unstable state $|e, N\rangle$ and which thus connects to $|g, N+1\rangle$ out of the laser beam region.

the { atom+laser field } system are labelled by two quantum numbers: the number N of laser photons and the internal atomic number g for the ground state and e for the excited state. The states $|g, N+1\rangle$ and $|e, N\rangle$ are close to each other near resonance (when $|\delta_L| \ll \omega_A$), and generate a two-dimensional manifold. Because of the off-diagonal coupling resulting from the dipole interaction between them, these eigenstates give rise to two dressed states, $|1(N)\rangle$ and $|2(N)\rangle$ which are linear coherent superpositions of $|g, N+1\rangle$ and $|e, N\rangle$. The eigenvalues repel each other as a result of the coupling and are separated by an energy[5] $\hbar\tilde{\Omega}(\vec{r}) = \hbar\,(\Omega^2(\vec{r}) + \delta^2)^{1/2}$. The dressed states vary in space, and the $\vec{r}$-dependence of their energies is given by:

$$E_1(\vec{r}) = \frac{\hbar}{2}\tilde{\Omega}(\vec{r}) \qquad \text{and} \qquad E_2(\vec{r}) = -\frac{\hbar}{2}\tilde{\Omega}(\vec{r}). \tag{11.49}$$

Their separation increases when the position-dependent Rabi frequency $\Omega(\vec{r})$ increases in the laser beam region (see Fig. 11.6).

On each dressed state, the dressed atom feels a gradient force that does not have the same sign on the levels $|1(N)\rangle$ and $|2(N)\rangle$. If the dressed atom is initially in a dressed state of type 1, it undergoes a force $\vec{F_1} = -\vec{\nabla}E_1(\vec{r}) = -\hbar\vec{\nabla}\tilde{\Omega}(\vec{r})/2$. By spontaneous emission of a "long" photon $\hbar(\omega_L + \tilde{\Omega})$, which occurs at a random time, it jumps into a dressed state of type 2 where the force has a value opposite to the one of the dressed state of type 1: $\vec{F_2} = -\vec{F_1}$. A subsequent emission of a "short" photon $\hbar(\omega_L - \tilde{\Omega})$ brings back the dressed atom in a dressed state of type 1, and the force changes its sign again. We conclude that the instantaneous force experienced by the atoms switches back and forth between two opposite values at random time intervals. As shown in [Dalibard and Cohen-Tannoudji (1985b)], this

[5]See Fig. 5.9 of Chap. 5.

physical picture leads to a simple and quantitative understanding of the mean value and fluctuations of the reactive force. Here we give a brief outline of this derivation.

Mean value of the reactive force

The mean value of the reactive force is the average of the two dressed-state position-dependent forces weighted by the respective steady state populations $\Pi_1^{\rm st}$ and $\Pi_1^{\rm st}$ of these two dressed states:

$$\vec{F} = \Pi_1^{\rm st}\vec{F}_1 + \Pi_2^{\rm st}\vec{F}_2. \tag{11.50}$$

The populations $\Pi_1^{\rm st}$ and $\Pi_2^{\rm st}$ are proportional to the mean times spent in levels of type 1 and 2. When $\delta = \omega_L - \omega_A > 0$ (left part of Fig. 11.6), the state $|1(N)\rangle$ is less contaminated by $|e, N\rangle$ than $|2(N)\rangle$ because it connects to the stable state $|g, N+1\rangle$ out of the laser beam (see left part of Fig. 11.6). It is therefore more populated and the mean dipole force expels the atom out of the laser beam region. When $\delta = \omega_L - \omega_A < 0$ (right part of Fig. 11.6), the conclusion is reversed: the mean dipole force attracts the atom towards the high-intensity region. This interpretation explains the change of sign of the force with the sign of the detuning $\delta = \omega_L - \omega_A$.

Fluctuations of the reactive force

The fluctuations of the instantaneous force, shown in Fig. 11.7, clearly explain the fluctuations of the reactive force and are responsible for a diffusion of atomic momentum described by a diffusion coefficient $D_{\rm react}$ given by:

$$D_{\rm react} = \int_0^\infty {\rm d}t \left[\left\langle \vec{F}(t).\vec{F}(t+\tau)\right\rangle - \vec{\mathcal{F}}^2_{\rm react}\right], \tag{11.51}$$

where $\vec{F}$ is the two-valued instantaneous dipole force with mean value $\vec{\mathcal{F}}_{\rm react}$. The calculation of $D_{\rm react}$ can be found in [Dalibard and Cohen-Tannoudji (1985b)]. Its scaling and order of magnitude is readily obtained. At resonance, the mean times spent in levels of type 1 and 2 are equal to $2/\Gamma$ because they are equally contaminated by the unstable state e. The mean reactive force is thus equal to zero. The instantaneous forces F_1 and F_2 scale as $\pm\hbar k_L\Omega$ since the length scale[6] of the position-dependent Rabi frequency is given by k_L^{-1}. According to Eq. (11.51), the diffusion coefficient is thus on the order of F^2 times the correlation time of the force, on the order of $2/\Gamma$. At resonance, it scales as:

$$D_{\rm react} \sim \frac{(\hbar\vec{\nabla}\Omega)^2}{2\Gamma} \sim \frac{\hbar^2 k_L^2 \Omega^2}{\Gamma} \sim \hbar^2 k_L^2 \Gamma \left(\frac{\Omega}{\Gamma}\right)^2. \tag{11.52}$$

The most important feature of this result is that $D_{\rm react}$ increases when the light intensity increases and does not saturate. This is to be contrasted with the diffusion coefficient of radiation pressure that saturates to values of the order of $\hbar^2 k_L^2 \Gamma$.

[6]This estimate refers to a laser standing wave for which the intensity changes over a scale determined by the wavelength λ_L. In a focussed laser beam, the length scale to be considered is either the waist or the Rayleigh length.

Physically, the large value of D_{react} is due to the fact that, between two successive spontaneous emission processes, the atomic momentum increases as $\vec{\nabla}\tilde{\Omega}$ and is therefore not limited when Ω increases.

Very far from resonance, the dressed atom spends most of its time in the state that contains the larger admixture of the stable state g, and the diffusion coefficient becomes negligible. This explains the interest of the reactive force for achieving nearly conservative far detuned optical traps.[7]

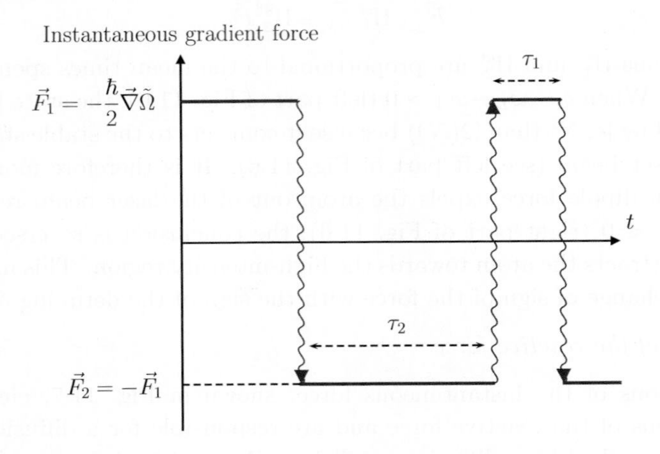

Fig. 11.7 The instantaneous reactive force switches back and forth between the two dressed-state-dependent forces $\vec{F}_1 = -\hbar\vec{\nabla}\tilde{\Omega}/2$ and $\vec{F}_2 = -\vec{F}_1 = \hbar\vec{\nabla}\tilde{\Omega}/2$. The lengths of time τ_1 and τ_2 spent in each dressed state between two successive jumps are random.

11.5 Conclusion

When the natural width $\hbar\Gamma$ of the excited state of a two-level atom is large compared to the recoil energy E_{rec}, the concept of a mean radiative force at a given point is valid.

An atom at rest at a given point experiences two types of mean radiative forces:

(1) A *dissipative force* due to transfer of momentum in scattering processes (absorption followed by spontaneous emission). This force is also called "scattering force" or "radiation pressure force". It varies with the detuning as a Lorentz absorption curve centered at $\omega_L = \omega_A$. It saturates at high intensity at the value $\hbar k_L \Gamma/2$.

(2) A *reactive force* due to redistribution processes between the various plane waves forming the laser wave (absorption of one photon in a plane wave followed by stimulated emission in another plane wave). This force can also be interpreted as a force associated with a spatial gradient of dressed state energies. For that

[7]See Chap. 14.

reason, it is also called "gradient force" or "dipole force". It varies with the detuning as a dispersion curve, does not saturate at high intensity and derives from a potential. Near resonance, its fluctuations can be much larger than the fluctuations of the dissipative force.

The distinction between the reactive and the dissipative responses of an atom to an electromagnetic excitation thus appears as a useful guideline in the study of atom-photon interactions. For the internal degrees of freedom of the atom, these two types of responses give rise to a shift and a broadening of the atomic levels, respectively, not only for the interaction with the vacuum field (Lamb shift and natural linewidth), but also for the interaction with an incident field (light shifts and power broadenings). For the propagation of light, they give rise to a dispersion and an absorption described by the real and imaginary parts of the index of refraction, respectively. The results derived in this chapter show that a similar distinction also exists for the external degrees of freedom of the atom, which are driven by two types of forces with different physical interpretations: a reactive force and a dissipative force.

Chapter 12

Laser cooling of two-level atoms

12.1 Introduction

The radiative forces described in the previous chapter for a two-level atom initially at rest can be also used for changing the velocity distribution of an ensemble of moving atoms. This distribution is in general a bell-shaped curve characterized by two important parameters: the mean velocity $\langle v \rangle$ and the velocity dispersion Δv around the mean value $\langle v \rangle$. The interaction with a laser beam can modify both of them. It can reduce or increase $\langle v \rangle$, an effect called slowing down or acceleration. It can also reduce or increase the velocity spread Δv, which characterizes the disordered motion of the atoms and which is related to the temperature of the sample. The smaller Δv, the lower the temperature.[1] Strictly speaking, an equilibrium temperature can be defined only if the velocity distribution is Gaussian (Maxwell Boltzmann distribution). If this is not the case, one can always define an effective temperature from the width at $1/e$ of the velocity distribution. The decrease of Δv induced by the interaction with laser light is called "laser cooling". This chapter is devoted to the description of a few laser cooling mechanisms, in the simple case of two-level atoms. Laser cooling of atoms deserves a special attention since it has opened a wealth of important developments in atomic physics during the last decades. A few of them will be described in the subsequent chapters.

In order to change Δv, the radiative force must depend on v. A force independent of v would just translate the whole velocity distribution without changing its shape. This is the case, for example, for ions with a charge q and a mass m submitted to the electrostatic force produced by a static electric field E during a time T. Each velocity is changed by the same amount qET/m. The simplest way to introduce a velocity dependence in the radiative force exerted by a laser beam with frequency ω_L on atoms with frequency ω_A is to use the Doppler effect, which changes the detuning $\delta = \omega_L - \omega_A$ from resonance by an amount proportional to the velocity v of the atom. Since the radiative force depends on the detuning δ, the velocity dependence of the detuning induces a velocity dependence of the force.

[1] Suppose for example that $\Delta v = 0$. In the reference frame moving with the mean velocity $\langle v \rangle$, all atoms are at rest, a situation which obviously corresponds to a zero temperature.

The first proposals of laser cooling for trapped ions [Wineland and Dehmelt (1975)] and neutral atoms [Hänsch and Schawlow (1975)] were actually based on the Doppler effect and called, for that reason, *laser Doppler cooling.* They are first presented in Sec. 12.2 in the simplest possible case of a red detuned laser plane wave. Near $v = 0$, the mean force $\mathcal{F}(v)$ can be written as $\mathcal{F}(v) = \mathcal{F}_0 - \alpha v + \ldots$. The first term $\mathcal{F}_0$ is the velocity-independent part of the force. The linear term $-\alpha v$ (where the coefficient α is positive for a red detuning) can be interpreted as a friction force that would damp the atomic velocity to zero if the force $\mathcal{F}_0$ was equal to zero. A few possible ways of eliminating this force $\mathcal{F}_0$ are briefly reviewed, and provide a simple introduction to the various schemes that have been proposed for Doppler cooling.

For neutral atoms, the idea for eliminating $\mathcal{F}_0$ consists of using two counter-propagating waves instead of a single one. This scheme, which was first suggested in [Hänsch and Schawlow (1975)], is analyzed in Sec. 12.3. It is shown that it is valid only in the low intensity limit and if the total radiative force is spatially averaged. One can then independently add the radiation pressure forces of the two waves and ignore the interference effects between them. The limits of laser Doppler cooling are also investigated. These result from the competition between the damping due to the friction force and the heating due to the fluctuations of the radiative forces, and which are described by the momentum diffusion coefficients introduced in the previous chapter.

At higher laser intensities, the interference effects between the two counter-propagating laser waves give rise to higher order corrections which can no longer be ignored and are evaluated in Sec. 12.4 by solving the optical Bloch equations for a moving atom. At low velocities, the velocity dependence of the spatially averaged force is shown to result from the fact that the internal variables do not follow the slow motion of external variables in a perfectly adiabatic way. Instead, there is a small time lag, linear in the atomic velocity v. The result of this analysis is that the force is linear in v, and corresponds to a damping force at low intensity for a red detuning, to an anti-damping force[2] at high intensity. The conclusions are reversed for blue detuning, where the force is anti-damping at low intensity and damping at high intensity. A physical interpretation of this surprising result is given in Sec. 12.5 using the dressed atom approach.

Trying to understand this result in simple physical terms has led to the discovery of the Sisyphus mechanism. In the high intensity regime considered here, the main ingredient is that the energies of the dressed states in the intense laser standing wave resulting from the interference between the two counter-propagating waves are spatially modulated. At high intensity and for blue detuning, cooling results from the fact that the moving atom rolls up potential hills more frequently than down. A similar effect, appearing at low intensity for an atom having several

[2]By "anti-damping force" we refer to a force which communicates an acceleration increasing linearly with velocity.

Zeeman sublevels in the ground state, was identified a few years later and played an important role in the development of laser cooling, since it led to temperatures much lower than the Doppler limit found in Sec. 12.3. This "low intensity Sisyphus cooling", operating for multilevel atoms, will be analyzed in the following chapter.

12.2 Doppler-induced friction force

In the previous chapter, an expression for the mean radiative force $\vec{\mathcal{F}}$ exerted by light on a two-level atom initially at rest was derived in terms of the amplitude and phase gradients of the laser field and the reactive and dissipative components of the atomic dipole moment. In this chapter, we suppose that the atom is moving so that its center of mass position evolves with time according to $\vec{R} = \vec{v}t$. Since external variables change on a time scale much longer than that of internal variables, it is possible to neglect the variation of the atom velocity during the characteristic time Γ^{-1} of evolution of the internal variables. The velocity is thus considered constant for the calculation of the force. This time dependence of $\vec{R}$ must be included in the expression of the force and in the optical Bloch equations which have to be solved for finding the reactive and dissipative components of the atomic dipole moment.

12.2.1 *Doppler effect in a red detuned laser plane wave*

This section is devoted to the analysis of the very simple situation of a moving atom that is irradiated by a counter-propagating red detuned ($\omega_L < \omega_A$) plane wave with wave vector $-\vec{k}_L$ and frequency ω_L (see Fig. 12.1(a)). In this case, since the field amplitude and the Rabi frequency are not position dependent, the reactive force cancels out, and the mean radiative force reduces to its dissipative component.

Since the atom is moving against the laser beam, the frequency of the light in the rest frame of the atom is Doppler shifted toward resonance and the radiation pressure is correspondingly larger. The electric field experienced by the moving atom is given by:

$$\vec{E}_L\left(\vec{R},t\right) = \vec{e}_L\mathcal{E}_L\cos\left[\omega_L t + \vec{k}_L\cdot\vec{R}\right] = \vec{e}_L\mathcal{E}_L\cos\left((\omega_L + \vec{k}_L\cdot\vec{v})t\right). \tag{12.1}$$

Since the Rabi frequency remains the same for the moving atom, we conclude that the optical Bloch equations essentially keep the same form as those for an atom at rest, provided that ω_L is replaced by the Doppler shifted value $\omega_L + \vec{k}_L\cdot\vec{v}$.

Since the laser plane wave is propagating along the negative direction of the x-axis, $-\vec{k}_L = -k_L\vec{e}_x$ (with $k_L > 0$) and the radiation pressure force $\mathcal{F}$ is along the x-axis and negative (see Fig. 12.1(b)). Replacing ω_L by $\omega_L + \vec{k}_L\cdot\vec{v} = \omega_L + k_L v$ in Eq. (11.36) of Chap. 11 leads to the following expression for $\mathcal{F}(v)$:

$$\mathcal{F}(v) = -\hbar\, k_L\,\Gamma\,\frac{\Omega^2/4}{\left(\omega_L + k_L v - \omega_A\right)^2 + (\Gamma^2/4) + (\Omega^2/2)}. \tag{12.2}$$

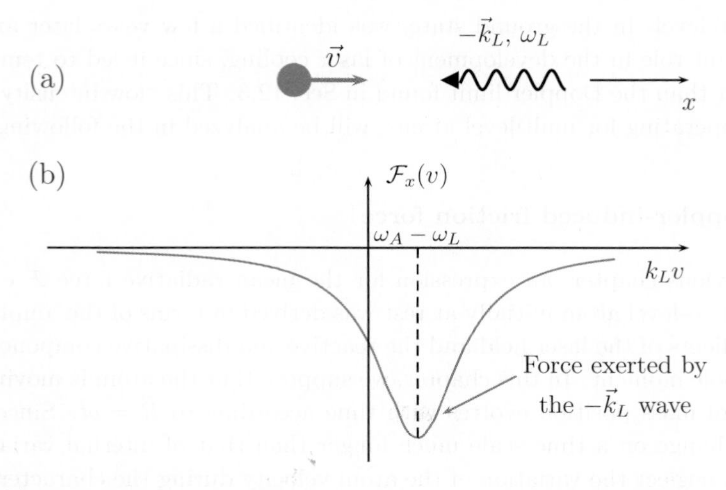

Fig. 12.1 (a) Atom moving with a velocity v along the x-axis and interacting with a counter-propagating laser plane wave with wave vector $-\vec{k}_L = -k_L\vec{e}_x$ (with $k_L > 0$) and frequency ω_L. (b) Variations with $k_L v$ of the radiation pressure force $\mathcal{F}$ experienced by the atom.

This force is represented in Fig. 12.1(b) as a function of $k_L v$. It has the shape of a Lorentz absorption curve and reaches its maximum absolute value when the Doppler shift compensates the detuning, i.e. when $k_L v = \omega_A - \omega_L$.

12.2.2 *Low velocity behavior of the force*

A power series expansion of $\mathcal{F}$ near $v = 0$ gives:

$$\mathcal{F} = \mathcal{F}_0 - \alpha\, v + \ldots \tag{12.3}$$

where $\mathcal{F}_0 = \mathcal{F}(0)$. The slope of the curve of Fig. 12.1(b) near $v = 0$ is negative, so that the coefficient α of Eq. (12.3) is positive and can be interpreted as a friction coefficient. Its expression is:

$$\alpha = \hbar\, k_L^2 \frac{s}{(1+s)^2} \frac{(-\delta)\,\Gamma}{\delta^2 + (\Gamma^2/4)}, \tag{12.4}$$

where $s = (\Omega^2/2)/(\delta^2 + \Gamma^2/4)$ is the saturation parameter.

A laser cooling effect, i.e. a decrease of the velocity dispersion, is obtained with a pure friction force. In order to compensante for the constant force $\mathcal{F}_0$, one can imagine using an external force that is independent of the velocity v. A first possibility (which has not been implemented) would be to take advantage of gravity and to use the radiation pressure force associated with $\mathcal{F}_0$ to levitate atoms while cooling them with the friction force term. The next two subsections review other possibilities that have been explored and which have led to successful laser cooling of trapped ions and neutral atoms.

12.2.3 *Idea of Doppler cooling for trapped ions*

For charged particles such as trapped ions, the idea is to use the trapping force to compensate $\mathcal{F}_0$. Let us give a very simple semiclassical description of laser cooling of trapped ions that was first proposed in [Wineland and Dehmelt (1975)].

Consider an ion oscillating along the x-axis with an oscillation frequency ω_v much smaller than the natural width Γ of the excited state of the ion, and suppose that this ion is excited by a red detuned laser beam propagating along the negative direction of the x-axis as shown in Fig. 12.1(a). Condition $\omega_v \ll \Gamma$ means that the ion can undergo several fluorescence cycles before its velocity changes appreciably due to its oscillatory motion so that, at each step of this motion, one can apply the previous results for the force, as if the ion was moving with a constant velocity. Because of the Doppler shift, the ion gets closer to resonance with the laser beam and feels a large force opposite to its velocity during the half period of the oscillation where its velocity is positive. During the other half period, where the velocity is negative, the ion gets farther from resonance and therefore the force in the same direction as the velocity is weaker. As a result of this asymmetry, the decrease of velocity during the first half period is greater than the increase of velocity during the second half period, and the oscillatory motion is damped. The velocity-independent force $\mathcal{F}_0$ simply results in a displacement of the equilibrium position of the trapped ion to a place where the trapping force compensates $\mathcal{F}_0$.

A quantum treatment of the motion of the ion in the limit $\omega_v \ll \Gamma$ leads to the same conclusions [Wineland and Itano (1979)]. The opposite limit, $\omega_v \gg \Gamma$, is also very interesting since it allows ions to be cooled in the lowest vibrational level of the trap. This will be described in Chap. 13.

12.2.4 *Idea of Doppler cooling for neutral atoms*

For neutral atoms, the idea presented in [Hänsch and Schawlow (1975)] consists of adding a second laser plane wave with the same intensity and frequency propagating in the opposite direction (see Fig. 12.2(a)), so that the two velocity-independent forces exerted by the two waves cancel out. The force exerted by the $+\vec{k}_L$ wave is positive and maximum for $k_L v = \omega_L - \omega_A$ (see Fig. 12.2(b)) and can be obtained from the force exerted by the $-\vec{k}_L$ wave through symmetry when the forces exerted by the two waves are added independently. The resultant total force, represented in Fig. 12.2(b), is an antisymmetric function of $\vec{v}$ with a friction coefficient twice as large as that for one beam.

The idea of laser cooling with the configuration of Fig. 12.2 is very simple. For an atom at rest, there is no Doppler effect and the two radiation pressure forces cancel out. If the atom is moving with a velocity v, the counter-propagating wave gets closer to resonance because of the Doppler shift and exerts a stronger force than the co-propagating wave, which appears farther from resonance. The net force is opposite to v and proportional to v for v small enough. Laser Doppler cooling thus

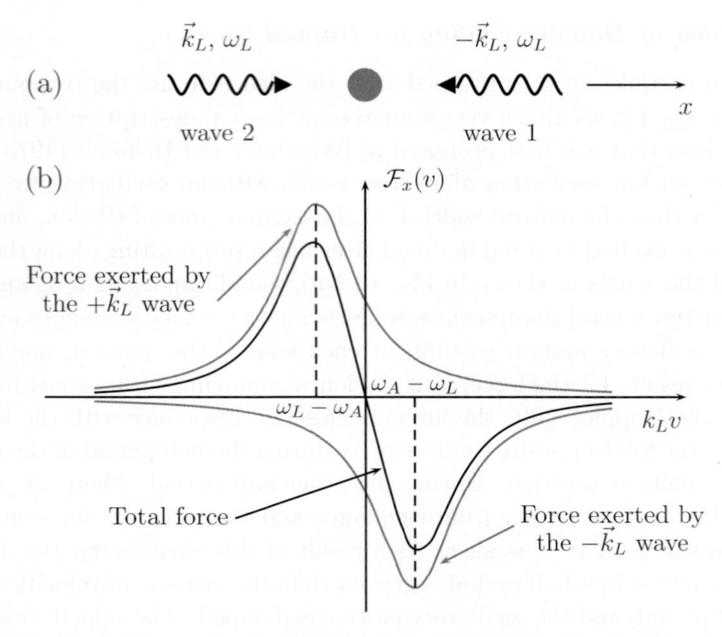

Fig. 12.2 Laser Doppler cooling scheme. (a) An atom moving with velocity v in a weak standing wave of frequency ω_L. In the frame of the atom, the frequencies of the two travelling waves are Doppler shifted, such that the atom experiences counter-propagating travelling waves with different frequencies $\omega_\pm = \omega \pm k_L v$. (b) The resulting radiation pressure force can be considered, in the weak intensity limit, as the sum of the forces from each individual travelling-wave component of the standing wave.

appears as a Doppler induced imbalance between two opposite radiation pressure forces.

In the previous reasoning there are however a number of intriguing questions. Firstly, the two counter-propagating laser waves are coherent and interfere. One may wonder to which extent one can consider that each beam acts independently on the atom as if it was alone. Actually, the two counter-propagating waves give rise to a standing wave that has a constant phase. So there is no phase gradient, there are only amplitude gradients. It thus seems that, in this laser configuration, the force more likely has a reactive character than a dissipative one.

In the next two sections, we will give a more precise description of Doppler cooling for neutral atoms by revisiting the optical Bloch equations for the $u(t)$, $v(t)$ and $w(t)$ components of the Bloch vector of the internal state density matrix and by explicitly taking into account the periodic character of the Rabi frequency associated with the standing wave. In Sec. 12.4, we consider the weak intensity limit ($\Omega \ll \Gamma$ or $|\delta|$), and show that interference effects can be ignored if the force is spatially averaged. Next, in Sec. 12.5, we consider the high intensity limit ($\Omega \gg \Gamma$ or $|\delta|$).

12.3 Two-level atom moving in a weak standing wave. Doppler cooling

12.3.1 *Perturbative approach for calculating the force*

Rather than using optical Bloch equations, it will be simpler to use the expression (11.26) of Chap. 11 for the force $\vec{\mathcal{F}} = \sum_{i=x,y,z} \langle D_i \rangle_{\text{st}} \vec{\nabla} E_{Li}(\vec{R}, t)$, where $\langle D_i \rangle_{\text{st}}$ is the mean dipole. In the low intensity regime, the mean dipole moment can be calculated to the lowest order in the field amplitude E_{Li}, so that $\langle D_i \rangle_{\text{st}}$ appears as the sum of the mean first order dipoles induced by wave 1 of wave vector $-\vec{k}_L$ and wave 2 of wave vector $+\vec{k}_L$. This leads to the following expression for the mean force:

$$\vec{\mathcal{F}} = \sum_{i=x,y,z} \left(\langle D_i \rangle_{\text{st}}^{(1)} \vec{\nabla} E_{Li}^{(1)} + \langle D_i \rangle_{\text{st}}^{(2)} \vec{\nabla} E_{Li}^{(2)} \right.$$
$$\left. + \langle D_i \rangle_{\text{st}}^{(2)} \vec{\nabla} E_{Li}^{(1)} + \langle D_i \rangle_{\text{st}}^{(1)} \vec{\nabla} E_{Li}^{(2)} \right), \tag{12.5}$$

containing four types of terms:

- the mean dipole $\langle D_i \rangle_{\text{st}}^{(1)}$ induced by wave (1) interacting with the gradient $\vec{\nabla} E_{Li}^{(1)}$ of the field of wave (1) plus another similar term with $(1) \to (2)$.
- the mean dipole $\langle D_i \rangle_{\text{st}}^{(1)}$ induced by wave (1) interacting with the gradient $\vec{\nabla} E_{Li}^{(2)}$ of the field of wave (2) plus another similar term with $(1) \leftrightarrow (2)$.

The two "square" terms ($\langle D_i \rangle^{(j)} \vec{\nabla} E_{Li}^{(j)}$, $j = 1, 2$) represent the radiation pressure forces exerted by each wave, independent of the other one. The "crossed" terms ($\langle D_i \rangle^{(j)} \vec{\nabla} E_{Li}^{(k)}$, $j = 1, 2$, $k \neq j$) represent interference effects between the two waves, which have been ignored in the previous discussion.

Let us analyze the contribution of these "crossed" terms in the rest frame of the atom. The two waves $+\vec{k}_L$ and $-\vec{k}_L$ undergo opposite Doppler shifts $+k_L v$ and $-k_L v$. The mean dipole induced by wave 1 of wave vector $-\vec{k}_L$ varies as $e^{i(\omega_L - k_L v)t}$. The gradient of the field of wave 2 of wave vector $+\vec{k}_L$ varies as $e^{i(\omega_L + k_L v)t}$. The product of the first term by the complex conjugate of the second gives a beat note equal to $e^{-2ik_L vt} = e^{-2ik_L R}$. A similar argument can be used for the other crossed term. The interference terms are thus spatially modulated with a spatial period $\lambda_L/2$ and therefore have a zero spatial average value.

In conclusion, the force exerted on a two-level atom moving in a weak intensity standing wave is not simply the sum of the two radiation pressure forces exerted by the two counter-propagating waves. There are interference terms that only vanish if one considers the spatially averaged force. It is only for this spatially averaged force, and in the low intensity regime, that the simple picture given in Sec. 12.2 is valid.

12.3.2 *Friction coefficient for a red-detuned weak standing wave*

The previous discussion shows that the friction coefficient in a weak intensity red-detuned standing wave is equal to twice the friction coefficient found for a single running wave (see Eq. (12.4)). In the weak intensity regime, the saturation parameter s is very small with respect to unity ($s \ll 1$), so that $s/(1+s)^2$ can be replaced by s. This gives finally:

$$\vec{\mathcal{F}} \simeq -\alpha\, v \quad \text{with} \quad \alpha = 2\hbar k_L^2 s \frac{-\delta\,\Gamma}{\delta^2 + (\Gamma^2/4)}. \tag{12.6}$$

The maximum value of the friction coefficient is obtained for $\delta = -\Gamma/2$ and is equal to $\alpha_{\max} = 2\hbar k_L^2 s$. The shortest damping time is given by $\tau_{\min} = M/\alpha = M/2\hbar k_L^2 s$. For rubidium atoms and a saturation parameter $s = 0.1$, one finds $\tau_{\min} = 100\,\mu$s.

As shown in Fig. 12.2, there is a finite velocity capture range around zero velocity for which the radiative forces of the two red-detuned and counter-propagating waves act as an effective damping force. For $\delta = -\Gamma/2$, the variation with $k_L v$ of the total force shows that the velocity interval v_{capt} over which the force is appreciable is given by:

$$k_L v_{\text{capt}} \sim \Gamma \quad \Longrightarrow \quad v_{\text{capt}} \sim \Gamma/k_L. \tag{12.7}$$

The following table summarizes examples of values for the capture velocity for different alkali atoms:

Element	Li	Na	K	Rb	Cs
v_{capt} (m·s^{-1})	0.25	0.17	0.21	0.21	0.23

12.3.3 *Momentum-energy balance. Entropy balance*

It is worth having a closer look at momentum, energy and entropy balance in laser Doppler cooling. Momentum balance is clear to understand since momentum is transferred to atoms when more photons are absorbed from one wave than from the other one.

In order to understand energy balance, consider the case of an atom moving in the x direction with a velocity $\vec{v}_x = v\vec{e}_x$.

(a) If the atom absorbs a photon from the $-\vec{k}_L$ wave, it re-emits a photon $\omega_L + 2k_L v$ in the $+x$ direction, and ω_L in the $-x$ direction. Indeed, in the low intensity limit and according to the resonance fluorescence study of Chap. 5, the atom re-emits a photon at the same frequency as the one it absorbs in a frame where it is at rest. On average, the energy of the re-emitted photon is larger than the energy of the absorbed one, even after angular integration over the direction of the re-emitted photon.

(b) If the atom absorbs a photon from the $+\vec{k}_L$ wave, it re-emits a photon $\omega_L - 2k_L v$ in the $-x$ direction and ω_L in the $+x$ direction. On the average the energy of the re-emitted photon is smaller than the energy of the absorbed one.

If processes (a) and (b) have the same rate, then on average there is no change of energy of the radiation field, and thus of the atom. Since the laser is red detuned, however, process (a) is more frequent than process (b) when the atom velocity is positive ($v > 0$), so that the energy of the re-emitted photon is on average larger than the energy of the absorbed one. We conclude that the energy of the radiation field increases while the energy of the atoms decreases.

Cooling a cloud of atoms results in a decrease of the entropy of the atoms. However, one must not forget that to achieve Doppler laser cooling, photons are absorbed from the laser beam. In this process, photons from the laser beam, which has a low entropy, are transformed into fluorescence photons emitted in all possible directions. The fluorescence field is a disordered system with a high entropy. The entropy of the radiation field thus increases while the entropy of the atoms decreases. In this context, the second law of thermodynamics states that the entropy for the total system atom plus field increases [van Enk and Nienhuis (1992)].

12.3.4 *Limits of Doppler cooling. Lowest temperature*

The limits of Doppler cooling for a two-level atom were studied in the late seventies [Hänsch and Schawlow (1975); Letokhov *et al.* (1976); Wineland and Itano (1979); Gordon and Ashkin (1980)]. This subsection is devoted to a semiclassical discussion that yields an estimate of the minimum achievable temperature. The quantum theory of one-dimensional laser cooling of free atoms using a transition from a $J = 0$ ground state to a $J = 1$ excited state can be found in [Castin *et al.* (1989)].

The damping of the atomic velocity, obtained within the semiclassical approach, yields a damping of its momentum dispersion:

$$\frac{\mathrm{d}p}{\mathrm{d}t} = -\frac{\alpha}{M}p \implies \left.\frac{\mathrm{d}(\Delta p)^2}{\mathrm{d}t}\right|_{\text{cooling}} = -2\gamma\,(\Delta p)^2\,, \tag{12.8}$$

with $\gamma = \alpha/M = 2\hbar k_L^2 s/M$. In Chap. 11, we have calculated the diffusion coefficient D due to the fluctuations of the dissipative force and found that it is related to the fluctuations of the number of photons absorbed in each wave, and the fluctuations of the momentum carried away by the fluorescence photons emitted in all possible directions. Taking into account those two contributions, one finds that for two identical counter-propagating waves in the weak intensity regime,

$$D = \hbar^2 k_L^2 \Gamma s, \tag{12.9}$$

where s is the saturation parameter of each beam. The corresponding diffusion acts as a heating mechanism:

$$\left.\frac{\mathrm{d}(\Delta p)^2}{\mathrm{d}t}\right|_{\text{heating}} = 2D. \tag{12.10}$$

In steady state, $(\Delta p)^2_{\text{st}}$ results from the trade-off between heating and cooling mechanism:

$$\left.\frac{\mathrm{d}(\Delta p)^2}{\mathrm{d}t}\right|_{\text{heating}} + \left.\frac{\mathrm{d}(\Delta p)^2}{\mathrm{d}t}\right|_{\text{cooling}} = 0. \tag{12.11}$$

We find that $(\Delta p)^2_{\rm st}$ scales as $M\hbar\Gamma$. We infer an effective equilibrium temperature $T_{\rm eq}$ on the order of:

$$k_B T_{\rm eq} \sim (\Delta p)^2_{\rm st}/M \sim \hbar\Gamma. \tag{12.12}$$

Doppler cooling can be readily generalized to three dimensions by using three mutually orthogonal, intersecting pairs of counter-propagating laser beams. In the intersection region, there is a laser cooling damping force in all directions. A calculation taking into account the angular diagram of emission, for any value of the detuning $\delta < 0$, and under the assumption of weak intensities ($s \ll 1$), yields an equilibrium temperature[3]

$$k_B T_{\rm eq} = \frac{\hbar\Gamma}{4}\left(\frac{2\,|\delta|}{\Gamma} + \frac{\Gamma}{2\,|\delta|}\right). \tag{12.13}$$

The minimum temperature is obtained for $\delta = -\Gamma/2$ and is equal to:

$$k_B T_{\rm min} = \hbar\Gamma/2. \tag{12.14}$$

The corresponding value of the velocity dispersion is given by $M\left(\Delta v\right)^2_{\rm min} \simeq \hbar\Gamma \;\rightarrow\; \left(\Delta v\right)_{\rm min} \simeq (\hbar\Gamma/M)^{1/2}$. The first estimations of $T_{\rm min}$ made in [Hänsch and Schawlow (1975)], along with the proposition of Doppler cooling for neutral atoms, were not correct since they mixed up $(\Delta v)_{\rm min}$ and $(\Delta v)_{\rm capt} \sim \Gamma/k_L$. In fact $(\Delta v)_{\rm min}/(\Delta v)_{\rm capt} \sim (\hbar\Gamma/M)^{1/2}/(\Gamma/k_L) \sim (\hbar\Gamma/E_{\rm rec})^{1/2} \ll 1$. The first correct estimate of the limits of Doppler cooling was made by the Moscow group in 1976 [Letokhov *et al.* (1976)]. For rubidium atoms, one finds $T_{\rm min} = 140\,\mu$K, and for sodium atoms $T_{\rm min} = 240\,\mu$K.

In fact, when precise methods for measuring temperature were developed, it was found that the measured temperatures were much lower than these values and were not varying with the various parameters as predicted by Eq. (12.13). This was the indication that other cooling mechanisms were operating. These will be studied in the Chap. 13, where it will be shown that they appear because rubidium and sodium are not simply two-level atoms, but atoms with several Zeeman sublevels in the ground state.

Lowest temperatures for laser cooling of trapped ions

For trapped atoms, the theoretical limit of Doppler cooling is found to be the same as that for free atoms when the natural width Γ of the atomic transition used for laser cooling is much larger than the oscillation frequency $\omega \ll \Gamma$ of the atoms in the trap [Wineland and Itano (1979)].

The first experiments on Doppler cooling of trapped ions were carried out on Ba^+ and Mg^+ in the NIST Boulder group [Wineland *et al.* (1978)] and in the Heidelberg group [Neuhauser *et al.* (1978)], respectively. In subsequent experiments, physicists laser cooled a sample of ions down to a few mK. The group of Hans Dehmelt reported the measurement of an effective temperature of 5^{+15}_{-5} mK for a single magnesium ion, to be compared with the prediction of Eq. (12.13) that yields 1 mK, given that the

[3] See W. Phillips lecture in [Arimondo *et al.* (1992)].

natural width of the magnesium transition used for laser cooling is 43 MHz [Nagourney *et al.* (1983)]. For this measurement, the temperature was deduced from the Doppler broadening of spectral lines, which is not a very sensitive method. A large improvement was made, however, by probing on another transition having a linewidth much less than the motional oscillation frequency of an ion in the trap [Janik *et al.* (1985)].

12.3.5 *Consistency of the various approximations*

Only the linear term in v has been kept in the expansion of the force in powers of $k_L v/\Gamma$, i.e. in powers of $v/(\Delta v)_{\text{capt}}$. The velocity spread of the cooled atoms is on the order of $(\Delta v)_{\min}$ and we have found that $(\Delta v)_{\min} \ll (\Delta v)_{\text{capt}}$. The cooled atoms therefore have a velocity that falls in the linear domain of the force.

The coherence length of those atoms is equal to:

$$\frac{\hbar}{M\,(\Delta v)_{\min}} \simeq \left(\frac{\hbar}{M\Gamma}\right)^{1/2} = \left(\frac{\hbar k_L^2}{M\Gamma}\right)^{1/2} \frac{1}{k_L} \simeq \lambda_L \left(\frac{E_{\text{rec}}}{\hbar\Gamma}\right)^{1/2} \ll \lambda_L. \tag{12.15}$$

The atomic wave packets are thus well localized in the laser wave, which validates our semiclassical approach.

Finally, for $\delta = -\Gamma/2$, the depth of the optical potential of the standing wave is on the order of $\hbar\Gamma s$. Since $s \ll 1$, this depth is very small compared to the kinetic energy of the cooled atoms. Therefore, the cooled atoms are not trapped in the optical potential wells of the laser standing wave.

12.3.6 *Spatial diffusion. Optical molasses*

The diffusion in momentum space has its counterpart in position space. A detailed calculation of the corresponding diffusion coefficient can be found in [Lett *et al.* (1989)]. Here, we give an estimate of this coefficient using a qualitative approach. Let us consider an atom embedded in a three-dimensional Doppler-cooling configuration. We are interested in the position distribution of this atom as a function of time. According to the equation of motion, the atom "loses" the memory of its initial velocity after the damping time $\tau = M/\alpha$, which gives the characteristic time for the evolution of external degrees of freedom. The spatial diffusion can be described by a random walk with a step size $\ell \simeq \tau\Delta v$. The random walk model gives $\langle r^2\rangle = 2\mathcal{N}\ell^2$, where $\mathcal{N}$ is the number of steps. After a time T, this number of steps is $\mathcal{N} = T/\tau$. One finally gets:

$$\langle r^2\rangle = 2D_x T \quad \text{with} \quad D_x = (\Delta v)^2\tau. \tag{12.16}$$

The same random walk model leads for the momentum diffusion coefficient D

$$D \simeq \frac{(\Delta p)^2}{2\tau} = \frac{M^2(\Delta v)^2}{2\tau}. \tag{12.17}$$

When the friction increases, τ decreases. Equations (12.16) and (12.17) show that the ratio D_x/D decreases. Atoms are thus more and more embedded in viscous medium which was called for that reason *optical molasses*.

The previous results can be obtained through a more rigorous analysis based on Fokker-Planck-Kramers equation for the phase space distribution of the atomic cloud [Hodapp *et al.* (1995)]. Using the friction coefficient given by Eq. (12.6) and the momentum diffusion coefficient of Eq. (12.9), one obtains the following expression for the spatial-diffusion coefficient ($|\delta| > \Gamma$):

$$D_x = \frac{\Gamma}{k_L^2}\left(\frac{\delta}{\Gamma}\right)^2 \frac{I_{\mathrm{sat}}}{I}. \tag{12.18}$$

With $\delta = 3\Gamma$, and $I = I_{\mathrm{sat}}$, one finds for rubidium atoms $D_x = 0.5\ \mathrm{cm^2 s^{-1}}$ which is a very low rate of expansion of the quadratic size of the atomic cloud. The first experimental realization of optical molasses was made at Bell laboratories in 1985 by the group led by Steve Chu and Arthur Ashkin [Chu *et al.* (1985); Chu (1998)].

12.4 Beyond the perturbative approach

So far, we have restricted our analysis of Doppler cooling to laser configurations that consist of two low intensity counter-propagating beams, for which a perturbative approach is sufficient to describe the dynamics. The forces experienced by an atom in an intense-standing wave involve new mechanisms that depend on the velocity of the atom and the intensity of the field. Multiphoton processes become important, and can give rise to very large forces. This section is devoted to an investigation of this regime with a non-perturbative approach based on optical Bloch equations. New effects will be identified, including cooling in an intense blue-detuned standing wave.

12.4.1 *Optical Bloch equations for a moving atom*

As already pointed out, the interference between the two counter-propagating waves results in a sinusoidal variation of the Rabi frequency: $\Omega(x) = \Omega_0 \sin(k_L x)$. Therefore, the Bloch vector $\vec{A}$ for the internal degrees of freedom of components (u, v, w) evolves according to a position-dependent Bloch matrix $\mathcal{B}(x)$:[4]

$$\frac{\mathrm{d}\vec{A}}{\mathrm{d}t} = \mathcal{B}(x)\vec{A} - \vec{A}_0. \tag{12.19}$$

If the atom is at rest at x, the Bloch vector reaches a steady state $\vec{A}_{\mathrm{st}}(x) = \mathcal{B}^{-1}(x)\vec{A}_0$. Equation (12.19) can thus be recast in the useful form:

$$\frac{\mathrm{d}\vec{A}(x)}{\mathrm{d}t} = \mathcal{B}(x)[\vec{A}(x) - \vec{A}_{\mathrm{st}}(x)]. \tag{12.20}$$

[4]See note on Bloch equations in Chap. 11, page 255.

12.4.2 *Time lag of internal variables*

One can qualitatively consider that the term $\mathcal{B}(x)$ acts as a damping term. The typical timescale for this damping is on the order of $\tau = \Gamma^{-1}$, much faster than the timescale for the evolution of external degrees of freedom. This hierarchy of timescales permits one to show from Eq. (12.20) that when the atom is moving, its internal state at x is not the steady state that it would reach if it was at rest at x. It is the steady state corresponding to the position occupied at an earlier time $t-\tau$, or equivalently at an earlier position $x - v\tau$, where v is the atom velocity that can be assumed to be constant on the time required for internal variables to reach their steady-state (see Note below):

$$\vec{A}\left[x\left(t\right)\right] \simeq \vec{A}_{\mathrm{st}}\left[x\left(t-\tau\right)\right] = \vec{A}_{\mathrm{st}}\left[x(t) - v\tau\right] \tag{12.21}$$

There is therefore a time lag for the internal variables when the atom is moving in the standing wave. This is an important point that will be used in the following to interpret some special features of laser-cooling.

Time lag in the presence of damping. A simple one-dimensional model

To introduce the idea of time lag in a simple way, it is convenient to consider the one-dimensional version of Eq. (12.20):

$$\frac{\mathrm{d}\chi\left(x\right)}{\mathrm{d}t} = -\Gamma\left[\chi\left(x\right) - \chi_{\mathrm{st}}\left(x\right)\right] \qquad \text{where} \qquad x = vt. \tag{12.22}$$

If x is fixed, $\chi\left(x\right)$ tends to $\chi_{\mathrm{st}}\left(x\right)$ with a time constant $\tau = \Gamma^{-1}$. If x varies, the solution of Eq. (12.22) can be written:

$$\chi\left(t\right) = \chi\left(t_0\right)\exp\left[-\Gamma\left(t-t_0\right)\right] + \Gamma\int_{t_0}^{t}\mathrm{d}t'\exp\left[-\Gamma\left(t-t'\right)\right]\chi_{\mathrm{st}}\left(t'\right). \tag{12.23}$$

For $(t - t_0) \gg \Gamma^{-1}$, the first term vanishes. Since external variables evolve slowly during Γ^{-1}, one can write:

$$\begin{aligned}\chi_{\mathrm{st}}(t') = \chi_{\mathrm{st}}\left[x(t')\right] &\simeq \chi_{\mathrm{st}}\left[x(t) - v(t'-t)\right] \\ &\simeq \chi_{\mathrm{st}}\left[x(t)\right] - (t'-t)v(\partial/\partial x)\chi_{\mathrm{st}}\left[x(t)\right].\end{aligned} \tag{12.24}$$

Inserting this expression into the last term of Eq. (12.23) leads to:

$$\chi[x(t)] = \chi_{\mathrm{st}}\left[x(t)\right] - \Gamma^{-1}v(\partial/\partial x)\chi_{\mathrm{st}}\left[x(t)\right] \tag{12.25}$$

$$\simeq \chi_{\mathrm{st}}\left[x(t) - v\tau\right], \tag{12.26}$$

where $\tau = \Gamma^{-1}$. It thus clearly appears that when the atom is moving, its state at x is not the steady state that it would reach if it was at rest at x. Instead, it is the steady state corresponding to the position occupied at an earlier time $t - \tau$, or equivalently at an earlier position $x - v\tau$.

12.4.3 *Low velocity limit ($k_L v \ll \Gamma$)*

In the forced regime, the components of the Bloch vector $\vec{A}$ are periodic functions of x or equivalently of t, since $x = vt$. One can thus rewrite the time evolution of $\vec{A}$ as a differential equation for the spatial coordinate x:

$$v\frac{\mathrm{d}\vec{A}}{\mathrm{d}x} = \mathcal{B}(x)\left[\vec{A}(x) - \vec{A}_{\mathrm{st}}(x)\right]. \tag{12.27}$$

In the low velocity domain, $v \ll \Gamma/k_L = (\Delta v)_{\mathrm{capt}}$, one can search for a solution of Eq. (12.27) by expanding the Bloch vector in powers of the velocity v: $\vec{A} = \vec{A}^{(0)} + v\vec{A}^{(1)} + \cdots$. At zero order in v, one recovers the steady-state solution for a motionless atom:

$$0 = \mathcal{B}(x)\left[\vec{A}^{(0)}(x) - \vec{A}_{\mathrm{st}}(x)\right] \quad \Rightarrow \quad \vec{A}^{(0)}(x) = \vec{A}_{\mathrm{st}}(x) = \mathcal{B}^{-1}(x)\vec{A}_0. \tag{12.28}$$

At first order in v, one finds

$$v\frac{\mathrm{d}\vec{A}^{(0)}}{\mathrm{d}x} = v\frac{\mathrm{d}\vec{A}_{\mathrm{st}}}{\mathrm{d}x} = \mathcal{B}(x)v\vec{A}^{(1)}(x) \;\Rightarrow\; \vec{A}^{(1)}(x) = \mathcal{B}^{-1}(x)\frac{\mathrm{d}}{\mathrm{d}x}\left(\mathcal{B}^{-1}(x)\vec{A}_0\right). \tag{12.29}$$

Since the laser standing wave has no phase gradient, the radiative force depends only on the u-component of the Bloch vector. Thus, the force linear in v may be calculated from the u-component of $\vec{A}^{(1)}$. After a spatial averaging over one optical wavelength, this leads to an explicit expression for the friction coefficient appearing in the spatially averaged force $\langle\mathcal{F}\rangle = -\alpha v$. We will not give this explicit expression of α, which can be found for example in [Gordon and Ashkin (1980)], but instead summarize the results of this calculation. At low intensity, one recovers the perturbative result of Eq. (12.6) that describes a net friction force for a red-detuned and weak standing wave. With the non-perturbative approach based on optical-Bloch equations, it is also possible to obtain results in the high intensity regime ($I \gg I_{\mathrm{sat}}$). The surprising result in this limit is that one finds a friction $\langle\mathcal{F}\rangle = -\alpha v$ (with $\alpha > 0$) only for a blue detuning. In this limit, the expression of α is given by:

$$\alpha \simeq \frac{3}{4\sqrt{2}}\hbar k_L^2 \frac{\Omega}{\Gamma}\frac{\delta}{|\delta|}. \tag{12.30}$$

The underlying physics will be revealed by the dressed-atom approach detailed in Sec. 12.5, and will be related to the spatial modulation of the dressed state energies and to the time lag between internal and external variables.

12.4.4 *Higher velocities*

Optical Bloch equations are valid at high intensity and for any velocity. In the presence of a standing wave, the Bloch vector is a periodic function of x and its components can be expanded in a Fourier series of x. The coefficients of these expansions obey recurrence relations, which can be solved in terms of continued fractions. This approach is very convenient for numerical calculations [Kyröla and Stenholm (1977); Minogin and Serimaa (1979)]. Moreover, if the continued fraction is stopped at a certain order, one gets a ratio of two polynomials that is more precise than a perturbative calculation performed at the same order.

Examples of velocity dependence of the spatially averaged force are depicted in Fig. 12.3(a) for $\delta = -10\Gamma$ and for increasing intensities. At high intensity, resonances appear in the spatially averaged force at certain velocities. These velocity-tuned multiphoton resonances are commonly called "Doppleron" resonances.

The resonances result from processes in which an atom makes a resonant transition from g to e by absorbing $N+1$ photons from one travelling wave and re-emitting N photons in a stimulated manner into the other travelling wave. The resonance condition is

$$(N+1)(\omega_L - k_L v) - N(\omega_L + k_L v) = \omega_A \quad \Rightarrow \quad \omega_L - \omega_A = (2N+1)k_L v. \quad (12.31)$$

The atom finally returns from e to g by spontaneously emitting a photon. An example of such a multiphoton resonance (corresponding to $2N+1 = 3$) is shown in Fig. 12.3(b). Doppleron resonances can give rise to very large accelerations, as experimentally reported in [Bigelow and Prentiss (1990); Tollett *et al.* (1990)] on atomic beams.

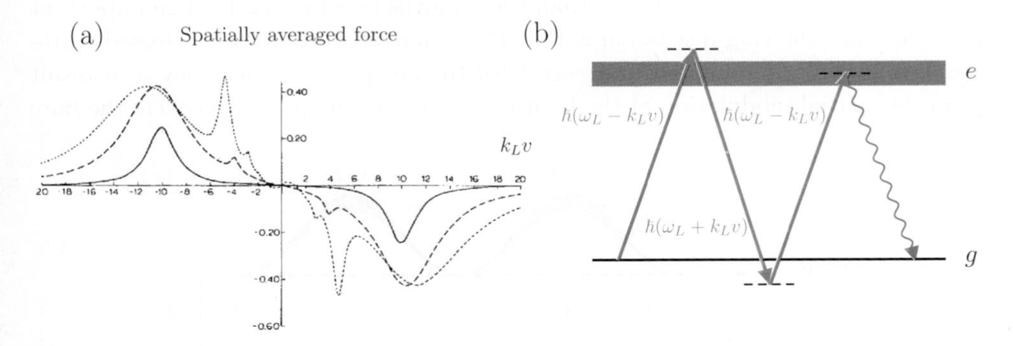

Fig. 12.3 (a) Results of the calculation of the force experienced by an atom in an intense standing wave versus its velocity using a continued-fraction solution from [Minogin and Serimaa (1979)]. The curve displays sharp resonances referred to as Dopplerons for an increasing intensity ($\delta = -10\Gamma$: solid line $I = I_{\text{sat}}$, dashed line $I = 9I_{\text{sat}}$, dotted $I = 25I_{\text{sat}}$). Figure adapted from [Minogin and Serimaa (1979)]. Copyright: Institute of Physics. (b) Multiphoton stimulated Raman transitions involving the absorption and stimulated emission of photons from both traveling waves induce resonant transitions from g to e for certain velocities $|\delta| = nk_L v$, where n is an odd integer. The atom then returns from e to g by spontaneous emission of a photon. The figure illustrates the case $n = 3$. The photons absorbed or emitted in a stimulated way are represented by full arrows, the photon spontaneously emitted by a wavy arrow.

12.5 Dressed atom approach to atomic motion in an intense standing wave. Blue cooling

The purpose of this section is to provide a physical picture, based on the dressed atom approach, of the friction mechanism that occurs, according to the Bloch equation analysis, for an atom moving in an intense blue-detuned one-dimensional standing wave [Dalibard and Cohen-Tannoudji (1985b)]. In the following, we restrict ourselves to the study of atomic motion along the direction of propagation of the laser beams, referred to as the x-axis.

12.5.1 *Energy and radiative widths of the dressed states*

The atom-laser uncoupled basis is made, as already described in Chap. 5, by an infinite set of two-dimensional manifolds $\mathcal{E}_{N+1} = \{|g, N+1\rangle, |e, N\rangle\}$. The state $|g, N+1\rangle$ is represented in Fig. 12.4 above the state $|e, N\rangle$, since we consider blue-detuned light ($\delta > 0$). Since the interference between the two counter-propagating lasers forms a standing wave, the Rabi frequency is spatially modulated: $\Omega(x) = \Omega_0 \sin(k_L x)$. The nodes of the standing wave are thus located at $x = 0$, $x = \lambda/2$, $x = \lambda$, ..., and the antinodes are located at $x = \lambda/4$, $x = 3\lambda/4$, ... The splitting between the dressed states $|1(N)\rangle$ and $|2(N)\rangle$ reads:

$$\hbar\tilde{\Omega}(x) = \hbar[(\omega_L - \omega_0)^2 + \Omega_0^2 \sin^2(k_L x)]^{1/2}. \tag{12.32}$$

This splitting is maximum at the antinodes and minimum at the nodes. Since we consider the high intensity limit, $\Omega \gg \Gamma$, $|\delta|$, the dressed states in the standing wave are separated by a splitting larger than their widths (see Fig. 12.4). The important point for the following discussion is that the radiative widths of the dressed states are also spatially modulated and *correlated* to the splitting since they also result from the spatial modulation of the light intensity: at a node, one recovers the bare

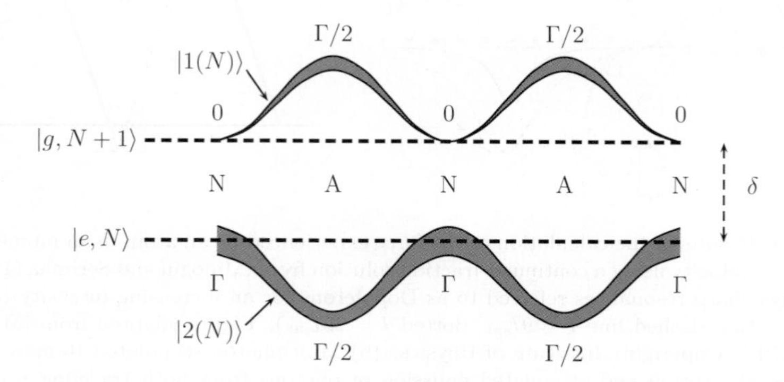

Fig. 12.4 The dressed states of an atom interacting with an intense blue-detuned standing wave ($\Omega \gg \Gamma$, δ) are separated by a splitting larger than their widths. The splitting is maximum at the antinodes A and minimum at the nodes N. The radiative width of each dressed state increases on an ascending branch and decreases on a descending branch.

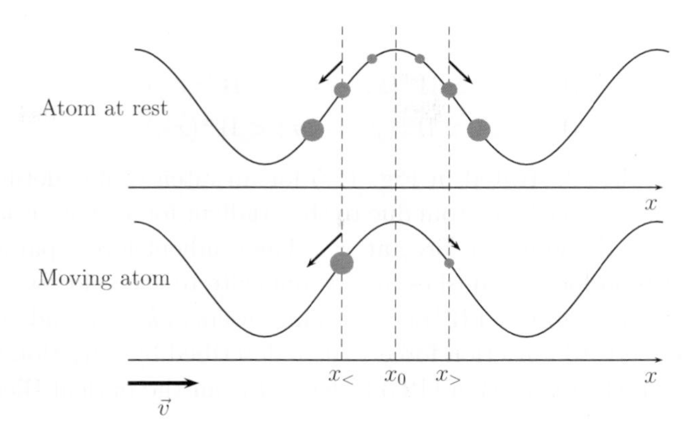

Fig. 12.5 For an atom at rest, the populations of one dressed state at two positions symmetric with respect to a maximum of the blue detuned standing wave are equal. The corresponding gradient forces are thus equal and opposite. This is no longer true for a moving atom since the time lag in the atom response leads to an asymmetry between the populations. This asymmetry yields a non-vanishing spatial average force that acts as a friction force on the moving atom.

atom regime: $|1(N)\rangle = |g, N+1\rangle$, with a width $\Gamma_1 = 0$ and $|2(N)\rangle = |e, N\rangle$ with the width of the excited state $\Gamma_2 = \Gamma$; at an antinode, under the large intensity assumption, $\Omega \gg \Gamma$, $|\delta|$, and both dressed states are equally contaminated by the excited state $|e, N\rangle$, resulting in widths $\Gamma_1 = \Gamma_2 = \Gamma/2$. In summary, the radiative widths of the dressed states always increase (decrease) on an ascending (descending) branch of the light potential.

12.5.2 *Friction mechanism*

As already emphasized in Chap. 5, in the high intensity limit the dressed state populations evolve independently of the off diagonal elements of the dressed atom density matrix and obey rate equations. The larger the radiative width, the smaller the steady state population. As a consequence, $\Pi^{\mathrm{st}}(x)$ decreases as a function of x on an ascending branch, and increases on a descending branch. When the atom is at rest, the populations of one dressed state at two positions $x_<$ and $x_>$, symmetric with respect to a maximum x_0 of the standing wave, are equal: $\Pi^{\mathrm{st}}(x_<) = \Pi^{\mathrm{st}}(x_>)$. The corresponding gradient forces are thus equal and opposite, and according to Eq. (11.50), the spatially averaged dipole force vanishes. Because of the time lag between the populations at x and the steady state populations, the dressed state populations at two symmetric points are no longer equal and the gradient force no longer has a zero spatial average value when the atom is moving.

In this subsection, we suppose first that the atom has a small velocity, so that during the time lag $\tau = \Gamma^{-1}$, it moves over a distance $v\tau$ that is small compared to the laser wavelength ($k_L v \ll \Gamma$). It is then possible to expand the dressed state populations in powers of $k_L v/\Gamma$. For an atom with a positive velocity along the

x-axis:

$$\Pi(x_<, v) \simeq \Pi^{\rm st}(x_< - v\tau) > \Pi^{\rm st}(x_<),$$
$$\Pi(x_>, v) \simeq \Pi^{\rm st}(x_> - v\tau) < \Pi^{\rm st}(x_>). \qquad (12.33)$$

This is schematically illustrated in Fig. 12.5 for an intense blue-detuned standing wave. The slowing down of the atom due to the gradient force at $x_<$ is larger that its acceleration due to the gradient force at $x_>$. The gradient force spatially averaged over a wavelength no longer vanishes, and is opposite to the velocity v. Expanding in Eq. (12.33) $\Pi^{\rm st}(x_< - v\tau)$ and $\Pi^{\rm st}(x_> - v\tau)$ in powers of $k_L v/\Gamma$ and averaging over a spatial period leads to a friction force $-\alpha_{\rm bd} v$ described by a friction coefficient α, in agreement with the expression (12.30) derived from the optical Bloch equations approach.

It is instructive to compare this friction coefficient $\alpha_{\rm bd}$, obtained for an atom in a blue-detuned intense standing wave, to the friction coefficient $\alpha_{\rm rd}$ obtained in laser cooling with a red-detuned weak standing wave (see Eq. (12.6)). According to Eq. (12.30), $\alpha_{\rm bd}$ scales as $\hbar k_L^2 \Omega/\Gamma$ whereas, according to Eq. (12.6), $\alpha_{\rm rd}$ scales as $\hbar k_L^2 s$ where s is the saturation parameter. For blue cooling, the standing wave is intense, so that $\Omega/\Gamma \gg 1$. For red Doppler cooling, the standing wave is weak, so that $s \ll 1$. We conclude that:

$$\frac{\alpha_{\rm rd}}{\alpha_{\rm bd}} \sim \frac{\hbar k_L^2 s}{\hbar k_L^2 \Omega/\Gamma} \ll 1, \qquad (12.34)$$

Blue cooling is therefore much more efficient than Doppler cooling for damping the atomic velocity.[5]

12.5.3 *High intensity Sisyphus cooling*

We now suppose that the atom has a velocity large enough that it can travel over several laser wavelengths during the lifetime in the excited state Γ^{-1}, but low enough that one can neglect non-adiabatic transitions between dressed states. This will allow us to give an interpretation of blue cooling in terms of a Sisyphus effect.

In both dressed states, the probability to leave this state by spontaneous emission is larger at the top of the potential hills (see Fig. 12.4). For a blue detuning, the transitions from a state of type 1 to a state of type 2, corresponding to the upper sideband of the fluorescence spectrum, occur preferentially at the antinodes of the standing wave, where the contamination of the state 1 by the excited state $|e\rangle$ is the largest. By contrast, for the lower sideband, the transitions from states of type 2 to states of type 1 occur preferentially at the nodes, where the dressed state coincides with $|e, N\rangle$. Therefore, the moving atom experiences on average more "uphills"

[5]However, the fluctuations of the gradient forces involved in blue cooling are much larger than the fluctuations of the weak intensity radiation pressure forces involved in Doppler cooling [Dalibard and Cohen-Tannoudji (1985b)]. The momentum diffusion coefficient for blue cooling increases more rapidly with Ω/Γ than the friction coefficient, so that the equilibrium temperatures achievable with blue cooling are much higher.

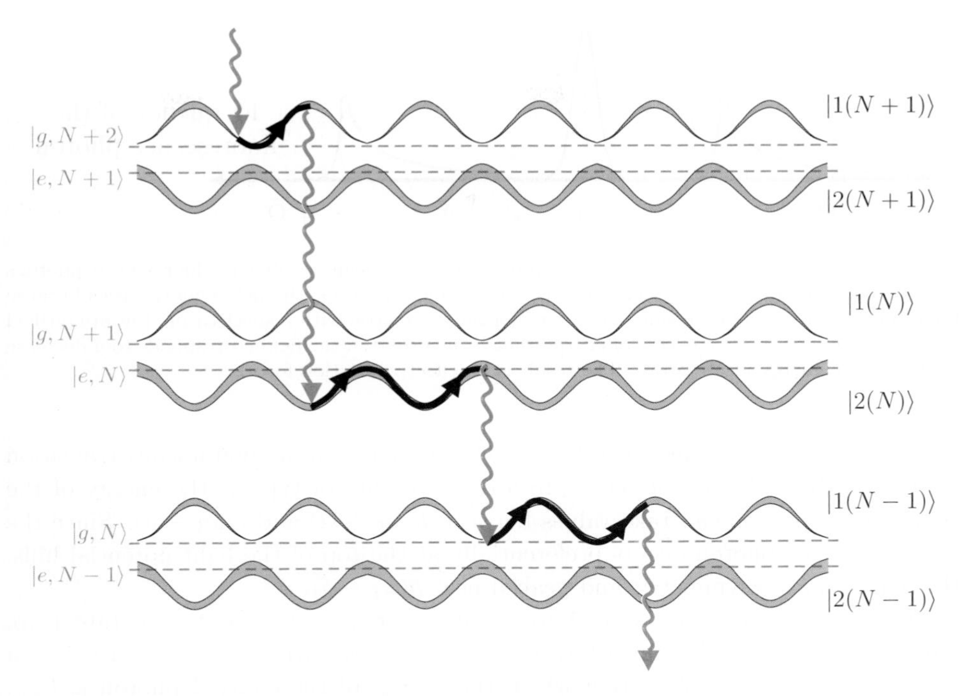

Fig. 12.6 One-dimensional high intensity blue-detuned Sisyphus cooling. For a two-level atom, the spatial modulation of the laser intensity of the standing wave results in correlated spatial modulations of the dressed states splitting and of their widths. Because of these correlations, the departure rate by spontaneous emission from a dressed state is maximum at the top of a potential hill and a moving atom runs up potential hills more frequently than down.

than "downhills" in the dressed-atom energy diagram (see Fig. 12.6). Like Sisyphus in Greek mythology who was always rolling a stone up a slope, the atom runs up potential hills more frequently than down [Dalibard and Cohen-Tannoudji (1985b)]. This is the reason why this cooling mechanism is commonly referred to as Sisyphus cooling. When the atom climbs a potential hill, its kinetic energy is transformed into potential energy. Dissipation then occurs because the spontaneously emitted photon has an energy higher than the absorbed laser photon. After each Sisyphus cycle, the total energy E of the atom decreases by an amount on the order of $\hbar\tilde{\Omega}$, where $\hbar\tilde{\Omega}$ is the depth of the optical potential wells.

The momentum balance in blue cooling is easy to understand. When the atom climbs a potential hill, it absorbs photons from the counter-propagating beam and re-emits them in a stimulated manner into the other beam. This redistribution of laser photons tends to decrease the atomic momentum.

The energy balance is more subtle. In a spontaneous transition between a dressed state of type 1 to a dressed state of type 2, the energy of the spontaneous photon can take values between $\hbar(\omega_L + \tilde{\Omega})$ and $\hbar(\omega_L + \delta)$. Since the emission of the photon occurs preferentially at the top of the potential hills, the line shape

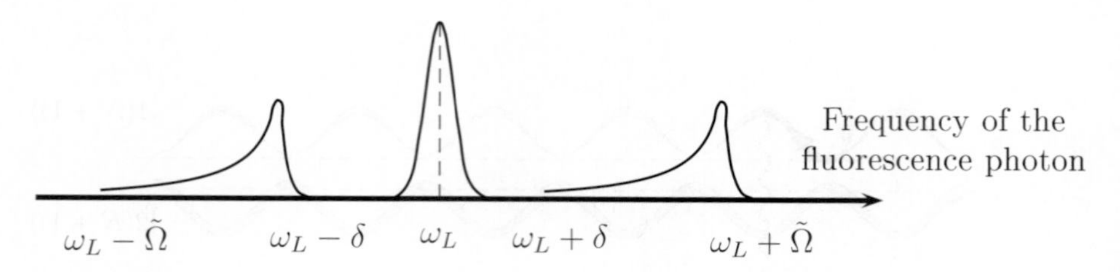

Fig. 12.7 For an atom moving in a standing wave, the mean number of fluorescence photons emitted per unit time in the two sidebands is the same. However, the field energy varies because the centers of gravity of each sideband are no longer symmetric with respect to ω_L, but are pushed to higher frequencies. The excess of energy that is deposited in the field corresponds to a decrease of the kinetic energy of the atom. The result is an effective friction force.

is asymmetric and peaked near $\hbar(\omega_L + \tilde{\Omega})$. Similarly, in a spontaneous transition between a dressed state of type 2 to a dressed state of type 1, the energy of the spontaneous photon can take values between $\hbar(\omega_L - \tilde{\Omega})$ and $\hbar(\omega_L - \delta)$. Since the emission of the photon occurs preferentially at the top of the light potential hills, the line shape is asymmetric and peaked near $\hbar(\omega_L - \delta)$.

The transitions from a state 1 to a state 2, or from a state 2 to a state 1 are alternated in time, with an arbitrary number of transitions from a state i to a state i, with $i = 1$ or 2, during which the energy of the emitted photon is $\hbar\omega_L$. The two sidebands of the fluorescence triplet correspond to the same number of photons, but are asymmetric, i.e. pushed towards high frequencies according to our analysis (see Fig. 12.7). After the emission of a pair of photons in the two sidebands, the energy of the field has varied by a positive amount on the order of: $\hbar(\omega_L + \Omega) + \hbar(\omega_L - \delta) - 2\hbar\omega_L = \hbar(\Omega - \delta) > 0$, since we consider the high intensity regime $\Omega \gg \delta$. The conclusion is that the energy of the field increases while the energy of the atom decreases.

12.5.4 *Experimental results*

Blue Sisyphus cooling has been investigated experimentally in [Aspect *et al.* (1986)]. This experiment was carried out on a cesium atomic beam. The transverse density profile of the beam was analyzed by scanning the position of a tungsten hot-wire detector, first in the absence of laser (curve (a) of Fig. 12.8). The atomic beam was then irradiated by an intense standing wave at right angle. For a blue detuning $\delta = +6\Gamma$, and a maximum Rabi frequency of 50Γ, a strong collimation of the beam was observed that yields a width five times smaller than that of the unperturbed atomic beam (curve (b) of Fig. 12.8). This clear evidence for transverse cooling is to be contrasted with the strong decollimation observed in the same experimental conditions but with an opposite detuning $\delta = -6\Gamma$ (curve (c) of Fig. 12.8).

Another effect which has been observed on an atomic beam crossing at right angle an off-resonance intense standing laser wave was the channeling of atoms through

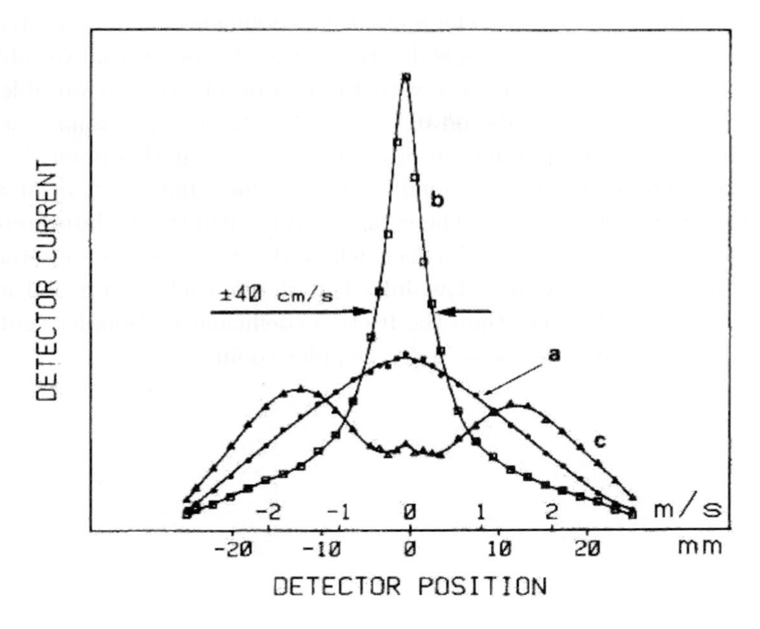

Fig. 12.8 The transverse density of a thermal beam of cesium-133 atoms is inferred by plotting the hot-wire detector current as a function of its transverse position, in the absence and in the presence of a transverse standing wave. (a) Signal observed in the absence of a transverse standing wave. (b) Signal observed in the presence of an intense and blue-detuned standing wave ($\delta/2\pi = 30$ MHz). The width is five times narrower than that of the unperturbed beam, which clearly demonstrates the cooling effect. (c) Same measurements with a red detuning ($\delta/2\pi = -30$ MHz), yielding a decollimation of the beam. Figure extracted from [Aspect *et al.* (1986)]. Copyright: American Physical Society.

the potential wells [Salomon *et al.* (1987)]. This was one of the first examples of the possibility of a sub wavelength spatial confinement of atoms with off-resonance laser light.

12.6 Conclusion

A two-level atom moving in a weak intensity red-detuned standing wave feels a force whose spatial average is a friction force. The corresponding cooling mechanism, called laser Doppler cooling, can be interpreted as resulting from a Doppler induced imbalance between two opposite radiation pressure forces. The friction coefficient is on the order of $\hbar k_L^2 s$, where s is the saturation parameter, assumed to be very small compared to unity. The minimum temperature that can be obtained by Doppler cooling is given by $k_B T_{\text{min}} = \hbar\Gamma/2$. The corresponding velocity spread Δv_{min} is smaller than the velocity capture range Δv_{capt} by a factor $(E_{\text{rec}}/\hbar\Gamma) \ll 1$. Similar results are obtained for Doppler cooling of trapped ions, where a single running wave can be used for cooling the ions.

When the standing wave has a high intensity, cooling occurs not for a red, but for a blue detuning. This blue cooling is due to the time lag of internal variables which do not follow, in a perfect adiabatic way, the motion of external variables. When the atom is moving, the population of a dressed state, at a given point x, at time t, is not the steady state population of an atom at rest at this point, but instead the steady state population corresponding to an earlier time $t - \tau$, or at an earlier position $x - v\tau$, where $\tau = \Gamma^{-1}$. The cooling mechanism can be interpreted in the dressed atom basis as a "Sisyphus" effect, where the atom moving in potential hills runs uphill more frequently than downhill. The friction coefficient is on the order of $\hbar k_L^2(\Omega/\Gamma)$ and is much larger than the friction coefficient of Doppler cooling. The velocity capture range is the same as for Doppler cooling.

Chapter 13

Sub-Doppler cooling. Sub-recoil cooling

13.1 Introduction

In the previous chapter, several theoretical predictions were presented concerning Doppler cooling of a two-level atom, the simple cooling scheme based on a Doppler induced imbalance between opposite radiation pressure forces. We summarize them here:

- Doppler cooling works at low laser intensity (the saturation parameter s is much smaller than 1).
- the minimum temperature is equal to:

$$k_B T_{\min} = \hbar \Gamma / 2, \tag{13.1}$$

  where Γ is the natural width of the excited state e.
- This minimum is reached when:

$$\delta = \omega_L - \omega_A = -\Gamma/2, \tag{13.2}$$

- The friction coefficient is given by:

$$\alpha = \hbar \, k_L^2 s, \tag{13.3}$$

  and the velocity capture range is on the order of

$$(\Delta \, v)_{\mathrm{capt}} \simeq \Gamma / k_L. \tag{13.4}$$

To test these predictions experimentally, a precise method for measuring the temperature of a very dilute gaseous sample was needed. This became possible with the development of time-of-flight methods [Lett *et al.* (1988)], whose principle is sketched in Fig. 13.1. Atoms stuck in an optical molasses are suddenly released to fall in the gravity field when the laser beams of the molasses are switched off. One measures the time at which the atoms cross a laser beam located below the molasses by recording the fluorescence photons that are emitted as they cross the probing zone. The distribution of the times of flight depends on:

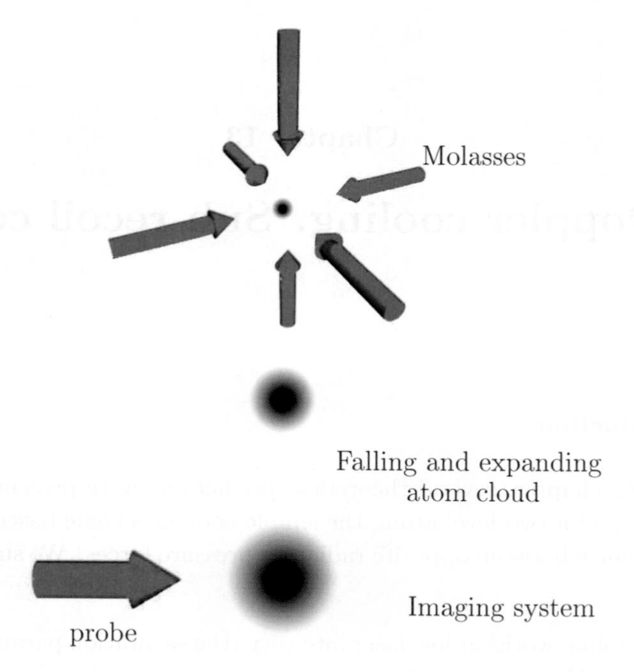

Fig. 13.1 Principle of the time-of-flight method for measuring the temperature of a cold gaseous sample. The distribution of the times taken by the atoms to fly from an optical molasses, which is suddenly switched off, to a detection laser beam is measured and fitted with a theoretical distribution that assumes a certain initial velocity distribution.

- the acceleration due to gravity g,
- the initial position distribution of the atoms (which can be deduced from a photograph of the molasses),
- the initial velocity distribution (which is determined by the temperature).

The fit between the observed time-of-flight signal and the theoretical prediction assuming a Gaussian velocity distribution corresponding to a certain temperature provides a precise estimate of the temperature, with an accuracy on the order of 5%.

The first experimental results obtained by this method brought a great surprise. The measured temperatures were not at all in agreement with the predictions deduced from a quantitative theory of Doppler cooling. Figure 13.2, taken from [Lett *et al.* (1988)] gives the variation of the temperature with the detuning and shows that the temperatures are much lower than expected and do not pass through a minimum when the detuning is increased.

This means that other mechanisms, different from Doppler cooling, are playing a role in optical molasses. A first motivation of this chapter is to explain the physical mechanisms at the origin of this "sub-Doppler" cooling (cooling below the Doppler limit).

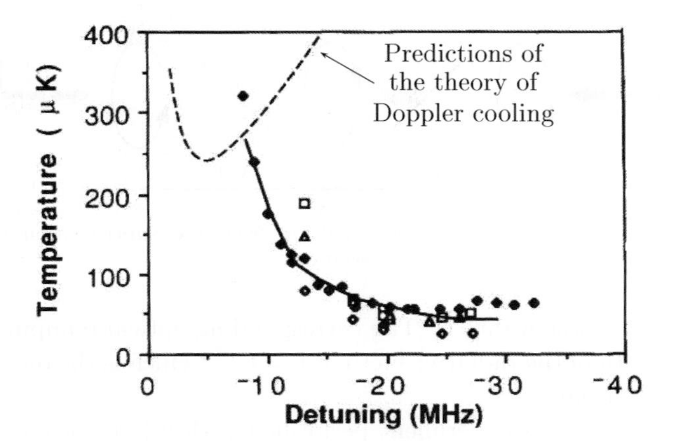

Fig. 13.2 Example of experimental results obtained by the time-of-flight method. The variations with the detuning of the measured temperatures (data points and solid line) are not in agreement with the prediction deduced from the theory of Doppler cooling (dashed curve). The temperatures are much lower than expected and do not pass through a minimum when the detuning is increased. Figure adapted from [Lett *et al.* (1988)]. Copyright: American Physical Society.

In this chapter, we also address another fundamental limit, the so-called recoil limit. In most laser cooling schemes, fluorescence cycles never stop. Since the random recoil $\hbar k$ communicated to the atom by the spontaneously emitted photon cannot be controlled, it seems impossible to reduce the atomic momentum spread below a value corresponding to a single photon momentum $\hbar k$. Another motivation of this chapter is to show that this fundamental limit can also be overcome, and that one can obtain atomic momentum spreads smaller than $\hbar k$. The corresponding cooling schemes are called "sub-recoil" (cooling below the single photon recoil limit)

$$(\Delta p)_{\text{sub-recoil}} < \hbar k_L \quad \text{or, equivalently} \quad (\Delta v)_{\text{sub-recoil}} < \frac{\hbar k_L}{M}. \tag{13.5}$$

13.2 Sub-Doppler cooling

13.2.1 *The basic ingredients of sub-Doppler cooling*

The new mechanisms that will be described in this section use two important features that were not included in the description of laser cooling given in Chap. 12.

(1) The temperature measurements described above are performed on alkali atoms, which have several Zeeman sublevels in the ground state. This introduces new internal times. Atoms can absorb one photon from one Zeeman sublevel and fall, by spontaneous emission of a photon, into another Zeeman sublevel. This *optical pumping* process has been described in Chap. 3. At low intensity, the photon absorption rate γ_g from the ground state can be much smaller than the

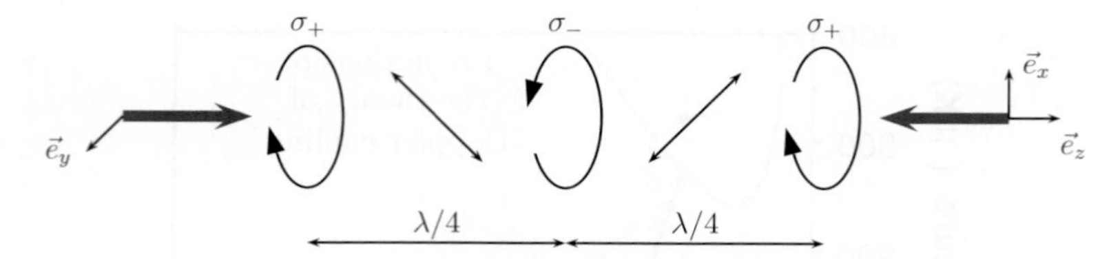

Fig. 13.3 Polarization of the total field resulting from the superposition of two counter-propagating laser fields with orthogonal polarizations.

spontaneous emission rate Γ. The corresponding optical pumping time $1/\gamma_g$ is much longer than the radiative lifetime $\tau = \Gamma^{-1}$, which is the only internal time for two-level atoms.

(2) In a 3D laser cooling experiment performed with 6 laser beams, the laser polarization cannot be uniform in space and so there are necessarily spatial polarization gradients that need to be taken into account.

13.2.2 *Laser configuration and atomic transition*

In order to have a situation exhibiting polarization gradients, we consider a 1D laser configuration where the two counter-propagating laser beams have orthogonal linear polarizations. Figure 13.3 represents how the polarization of the total field varies along the direction z of the two beams. The polarization of the total field is alternatively σ_+ and σ_- in planes separated by a distance $\lambda/4$. In between, it is linear at $\pm 45°$ of the polarization axis or elliptical. Such a laser configuration is called the "lin $\perp$ lin" laser configuration.

For the atomic transition, we must consider an atom with several ground state sublevels. We take the simplest possible case of a $J_g = 1/2$ ground state having only two Zeeman sublevels. Since most laser cooling experiments use optical transitions $J_g \to J_e = J_g + 1$, we will consider a transition $J_g = 1/2 \to J_e = 3/2$. As in Doppler cooling, we suppose $\delta = \omega_L - \omega_A < 0$ (red detuning).

13.2.3 *Light shifts and optical pumping for an atom at rest*

First consider an atom at rest at a location where the polarization of the total field is σ_+. Figure 13.4(a) shows the two σ_+ transitions $\Delta m = m_e - m_g = +1$ which are excited by the σ_+-polarized field. Atoms are optically pumped from $g_{-1/2}$ to $g_{+1/2}$ by the absorption of a σ_+ photon and the spontaneous emission of a π photon.[1] In steady-state, the population $\Pi_{g+1/2}$ of $g_{+1/2}$ is equal to 1:

$$\Pi_{g_{+1/2}} = 1 \text{ and } \Pi_{g_{-1/2}} = 0. \tag{13.6}$$

Since the laser is not exactly resonant, the two ground state sublevels are light-

[1] Atoms in $g_{+1/2}$ can also absorb a σ_+ photon but they fall back by spontaneous emission in the same state.

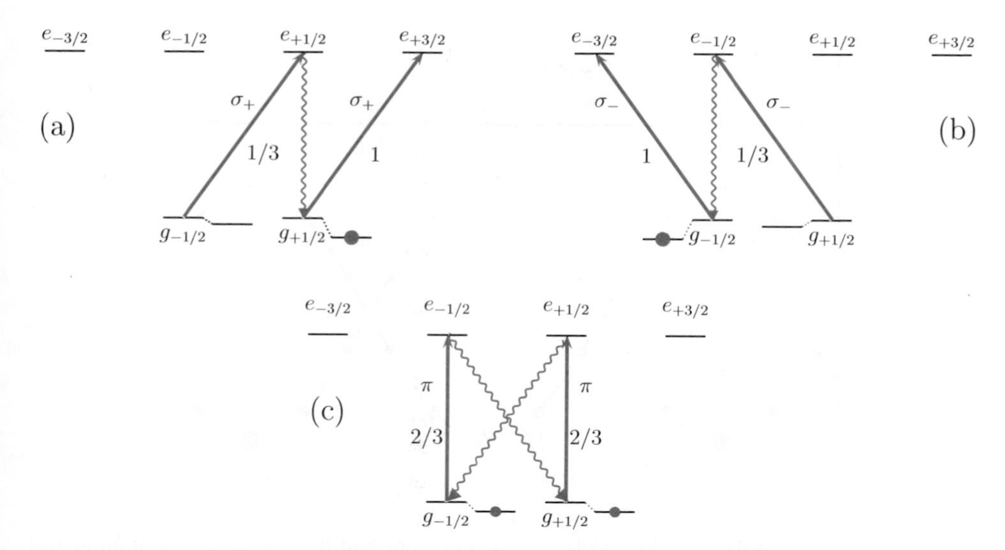

Fig. 13.4 Steady-state populations and light shifts for an atom at rest at a location where the polarization of the total field is (a) σ_+, or (b) σ_-, or (c) π.

shifted.[2] For the red detuning considered here, these light shifts are negative. Because of the 3:1 ratio of the squares of the Clebsch Gordan coefficients of the two σ_+ transitions starting from $g_{-1/2}$ and $g_{+1/2}$, the light shift of $g_{+1/2}$ is three times larger than the light shift of $g_{-1/2}$. We thus have:

$$\delta_{g_{+1/2}} = \delta_g \qquad \text{and} \qquad \delta_{g_{-1/2}} = \delta_g/3, \tag{13.7}$$

where δ_g is the light shift of the ground state for a $J_g = 0 \to J_e = 1$ having a Clebsch Gordan coefficient equal to 1.

Similar considerations for an atom at rest at a location where the polarization of the total field is σ_- lead to the conclusion that the atom is optically pumped to $g_{-1/2}$ (see Fig. 13.4(b)):

$$\Pi_{g_{+1/2}} = 0 \qquad \text{and} \qquad \Pi_{g_{-1/2}} = 1, \tag{13.8}$$

and that the light shift of $g_{-1/2}$ is three times larger than the light shift of $g_{+1/2}$:

$$\delta_{g_{+1/2}} = \delta_g/3 \qquad \text{and} \qquad \delta_{g_{-1/2}} = \delta_g. \tag{13.9}$$

Finally consider an atom at rest at a place where the polarization of the total field is π (see Fig. 13.4(c)). Optical pumping is symmetric, so that:

$$\Pi_{g_{+1/2}} = \Pi_{g_{-1/2}} = 1/2. \tag{13.10}$$

The light shifts of $g_{-1/2}$ and $g_{+1/2}$ are equal:

$$\delta_{g_{+1/2}} = \delta_{g_{-1/2}} = 2\delta_g/3. \tag{13.11}$$

[2] See Chap. 3.

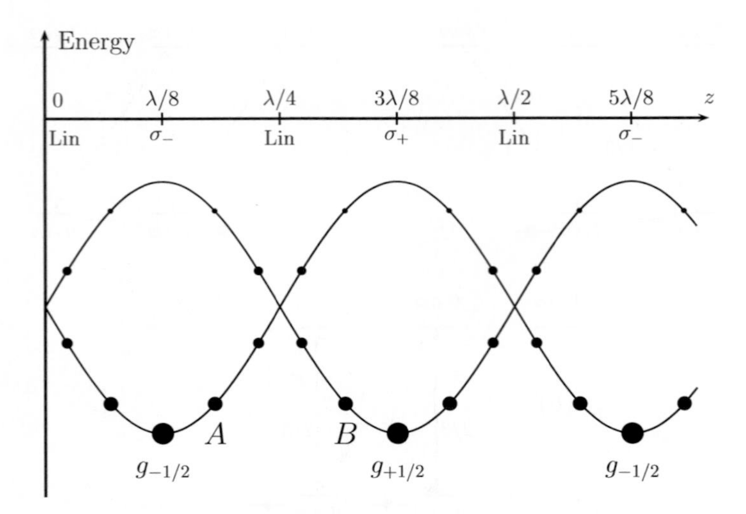

Fig. 13.5 Variations with z of the steady-state populations and light shifts for an atom at rest at z.

All these results are summarized in Fig. 13.5 which gives the variations with z of the steady-state populations and light shifts for an atom at rest at z. The light shifts of the two ground state sublevels are spatially modulated. The modulations are out of phase. The steady-state populations for an atom at rest are also spatially modulated and correlated with the light shifts. As with the high intensity Sisyphus cooling studied in Chap. 12, the steady-state populations decrease on an ascending branch, and increase on a descending branch.

13.2.4 *Low intensity Sisyphus cooling for a moving atom*

Suppose now that the atom is moving along the z axis and that it travels over a distance on the order of the laser wavelength λ_L during the optical pumping time (on the order of $1/\gamma_g$). In this case, if it starts from the bottom of a valley in a given Zeeman sublevel, it has the time to reach the top of the hill (see Fig. 13.6) where it has a large probability to be optically pumped in the other sublevel, i.e. in the bottom of a valley, and so on. As in the high intensity Sisyphus cooling studied in Chap. 12, the moving atom is running uphill more frequently than downhill. This new Sisyphus effect is now due to correlations between the spatial modulations of light shifts and optical pumping rates. A detailed description of this effect is given in [Dalibard and Cohen-Tannoudji (1989)], where another laser configuration, consisting of two counter-propagating laser waves with orthogonal circular polarizations is also studied. Similar conclusions are described in [Ungar *et al.* (1989)].

A simple argument can be given for evaluating the equilibrium temperature that can be obtained by this cooling mechanism. Each optical pumping cycle represented in Fig. 13.6 gives rise to a fluorescence photon with an energy that is higher than

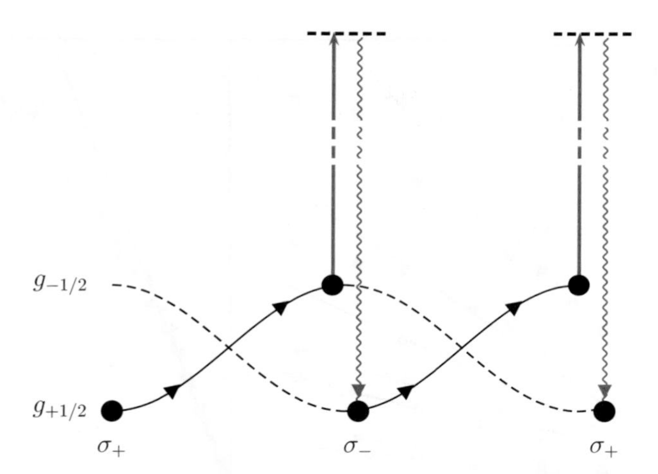

Fig. 13.6 Principle of low intensity Sisyphus cooling. If the optical pumping time is sufficiently long, an atom, initially in the $g_{+1/2}$ Zeeman sublevel, has the time to go from the bottom of the valley to the top of the hill where it has a large probability of being optically pumped into the other sublevel $g_{-1/2}$ at the bottom of the corresponding valley. From there, the cycle can be repeated.

the energy of the absorbed photon, by an amount on the order of the light shift $\hbar\delta_g$. After each fluorescence cycle, the energy of the radiation field increases by an amount on the order of $\hbar\delta_g$ and, consequently, the energy of the atom decreases by the same amount. The energy of the atom thus decreases step by step until its total energy is so low that it gets trapped in the optical potential wells associated with the spatially modulated light shifts. The equilibrium temperature of sub-Doppler cooling is therefore expected to be given by:

$$k_B T_{\text{sub-Doppler}} \sim \hbar\delta_g = \frac{\hbar\Omega^2\,|\delta|}{4\delta^2 + \Gamma^2} \tag{13.12}$$

where we have used the expression of the light shift δ_g derived in Chap. 7 (see Eq. (7.6)). We have assumed here a low intensity. If the detuning is large, we get:

$$|\delta| \gg \Gamma \quad \Rightarrow \quad k_B T_{\text{sub-Doppler}} \sim \frac{\hbar\Omega^2}{4\,|\delta|} \propto \frac{\text{Laser intensity}}{\text{Detuning}}. \tag{13.13}$$

These qualitative predictions are confirmed by a more quantitative treatment [Dalibard and Cohen-Tannoudji (1989)]. At low intensity, the light shift of the ground state is much smaller than the natural width Γ of the excited state. In this way, one understands how it is possible to obtain temperatures about two orders of magnitude lower than the Doppler limit, which scales as $\hbar\Gamma$:

$$k_B T_{\text{sub-Doppler}} \sim \hbar\delta_g \ll k_B T_{\text{Doppler}} \sim \hbar\Gamma. \tag{13.14}$$

Experiments concerning the equilibrium temperature have been performed to check these predictions [Salomon *et al.* (1990)]. The left part of Fig. 13.7 shows the

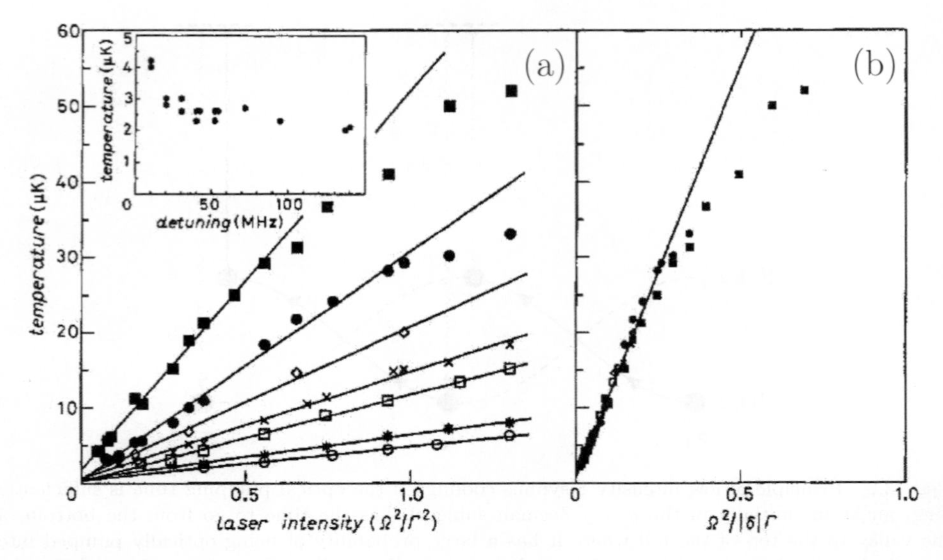

Fig. 13.7 (a) Variation of the measured temperature of a laser cooled cesium cloud with laser intensity for various values of the detuning: $|\delta|/2\pi=$ 10 MHz (■), 20 MHz (•), 30 MHz (◇), 40 MHz (×), 54 MHz (□), 95 MHz (⋆), 140 MHz (○). The inset shows the variation of the temperature with detuning for a given value of the laser intensity. (b) Variations of the measured temperatures plotted versus the light shift. For low light shifts, the temperature is a linear function of the light shift. Figure adapted from [Salomon *et al.* (1990)]. Copyright: EDP Sciences.

variation of the measured temperature of a laser cooled cesium cloud as a function of the light intensity (proportional to Ω^2/Γ^2, where Ω is the Rabi frequency), for various values of the detuning δ (the values of $|\delta|/2\pi$ for the various experimental points are indicated in the caption of the figure). The inset shows the variations of the temperature with the detuning for a given value of the intensity. These results show that the temperature of sub-Doppler cooling is proportional to the laser intensity. If one plots the measured temperatures as a function of the light shift, which is proportional to $\Omega^2/|\delta|\Gamma$, all points collapse onto the same curve which behaves linearly for low values of the light shift. This shows that the equilibrium temperature scales as $\hbar\delta_g$ for low values of the light shift, in agreement with the theoretical prediction given by Eq. (13.14).

13.2.5 *Characteristics of the friction force. Qualitative discussion*

We have already mentioned in the previous subsection that the force experienced by the atom will be maximum for a velocity $v = v_{\text{capt}}$ corresponding to an atom going from the bottom of a valley to the top of the next hill during a time γ_g^{-1}. We conclude that the velocity capture range is given by the following equation:

$$v_{\text{capt}}\,\gamma_g^{-1} \sim \lambda_L \quad \leftrightarrow \quad k_L v_{\text{capt}} \sim \gamma_g. \tag{13.15}$$

Comparing with the results derived in Chap. 12 for the velocity capture range of Doppler cooling for a two-level atom (see Eq. (13.4)),[3] we see from Eq. (13.15) that the velocity capture range of low intensity Sisyphus cooling is much smaller, by a factor on the order of $\gamma_g/\Gamma \ll 1$ as illustrated in Fig. 13.8.

Using a similar reasoning to that presented in Sec. 12.5.3 of Chap. 12 for high intensity Sisyphus cooling, we now try to give an order of magnitude of the friction coefficient of low intensity Sisyphus cooling.

In Fig. 13.5, the mean force exerted on an atom at rest at position A has the same modulus but opposite direction as the force exerted on an atom at B, symmetric to A with respect to the position $z = \lambda/4$. This results from the same steady-state population of the corresponding sublevels $g_{-1/2}$ and $g_{+1/2}$. This force $\mathcal{F}_0$ is proportional to the energy gradient, which is itself on the order of the variation of light shift energy $\hbar\delta_g$ over a distance on the order of λ_L:

$$\mathcal{F}_0 \sim \frac{\partial E}{\partial R} \sim \frac{\hbar\,\delta_g}{\lambda_L} \sim \hbar\,k_L\,\delta_g. \tag{13.16}$$

For an atom moving with a velocity v smaller than v_{capt}, the steady-state population at A and B are different because of time lag in the atomic response, as already stressed in Sec. 12.5.2 of Chap. 12. The difference in populations is given by:

$$\Delta\Pi = \Pi\,(A) - \Pi\,(B) \sim 2\left(\frac{\partial\Pi^{\text{st}}}{\partial R}\right) v\,\gamma_g^{-1} \sim k_L v\,\gamma_g^{-1}, \tag{13.17}$$

where we have used the fact that the steady-state populations of $g_{-1/2}$ and $g_{+1/2}$ vary by an amount on the order of one in a distance on the order of λ_L, such that $\partial\Pi^{\text{st}}/\partial R \sim, 1/\lambda_L \sim k_L$. We deduce the expression for the friction force:

$$\mathcal{F} \sim \mathcal{F}_0\Delta\Pi \sim \left(\frac{\partial E}{\partial R}\right)\Delta\Pi \sim (\hbar\,\delta_g\,k_L)\,k_L v\,\gamma_g^{-1} \sim \hbar\,k_L^2\frac{\delta_g}{\gamma_g}v, \tag{13.18}$$

from which we get the friction coefficient :

$$\alpha \sim \hbar\,k_L^2\frac{\delta_g}{\gamma_g} \sim \hbar\,k_L^2\frac{|\delta|}{\Gamma}, \qquad \text{if} \qquad |\delta| \gg \Gamma. \tag{13.19}$$

This qualitative derivation of the friction coefficient is in agreement with the more quantitative calculations given in [Dalibard and Cohen-Tannoudji (1989)].

The friction coefficient of low intensity Sisyphus cooling given in Eq. (13.19) is much larger than the friction coefficient of Doppler cooling, which is on the order of $\hbar\,k_L^2 s$, where the saturation parameter s must be smaller than one.[4] However the friction force of low intensity Sisyphus cooling acts on a much smaller velocity interval than Doppler cooling, and in fact, both forces are useful (see Fig. 13.8). Doppler cooling drags atoms from a relatively large velocity interval towards the region where Sisyphus cooling operates, thereby dramatically enhancing the number of atoms affected by the sub-Doppler mechanism.

[3] The same velocity capture range is found for high intensity Sisyphus cooling.
[4] See Sec. 12.3.2 of Chap. 12.

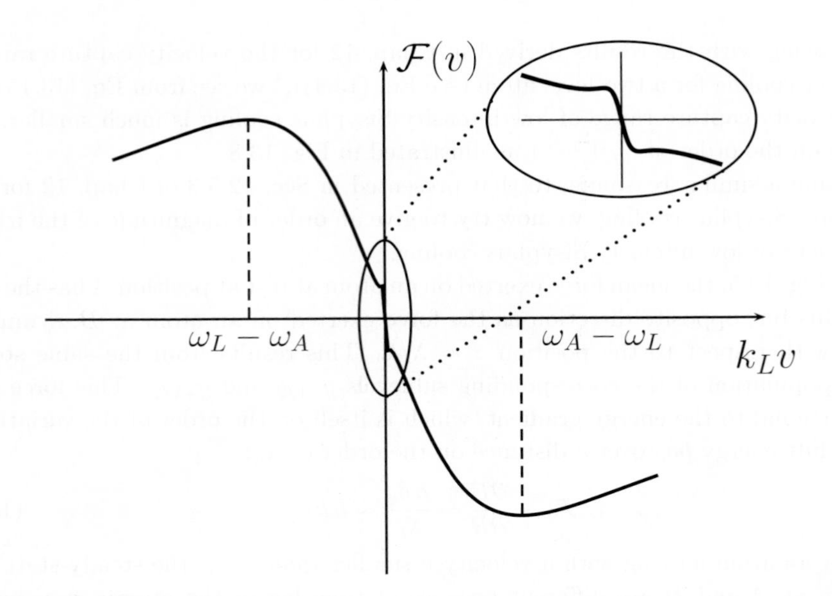

Fig. 13.8 Variation with $k_L v$ of the radiative force $\mathcal{F}$ experienced by an atom moving at a velocity v. Compared to Doppler cooling, a much larger friction coefficient appears in a smaller velocity interval than for Doppler cooling as a result of the low intensity Sisyphus cooling mechanism.

The fact that the Sisyphus friction coefficient is highest at low intensity, when the spatial gradient of the light shifts is small, seems at first paradoxical. This can be understood by noting that while the light shifts decrease, the time lag γ_g^{-1}(optical pumping time) increases.

13.2.6 *Quantum limits of sub-Doppler cooling*

The previous analysis shows that the equilibrium temperature is proportional to the laser intensity I_L: the lower I_L, the lower T. One cannot decrease I_L indefinitely however, and sub-Doppler cooling must certainly become inefficient at very low intensity. In fact, the previous analysis ignores the fact that the photon emitted spontaneously in an optical pumping cycle imparts a recoil momentum $\hbar k_L$ to the atom. This increases the kinetic energy of the atom. Cooling is possible only if the decrease of the total atomic energy due to the Sisyphus effect, on the order of $\hbar\delta_g$, is larger than the increase of the kinetic energy, on the order of E_{rec}, due to recoil associated with the spontaneously emitted photon.

When I_L is decreased, one thus expects that the equilibrium temperature T should decrease linearly with I_L, pass through a minimum on the order of a few recoil energies E_{rec}, and then increase:

$$k_B T_{\min} \sim \text{a few } E_{\text{rec}}. \tag{13.20}$$

This qualitative prediction is confirmed by a full quantum calculation based on a numerical integration of the quantum equations of motion [Castin *et al.* (1991)].

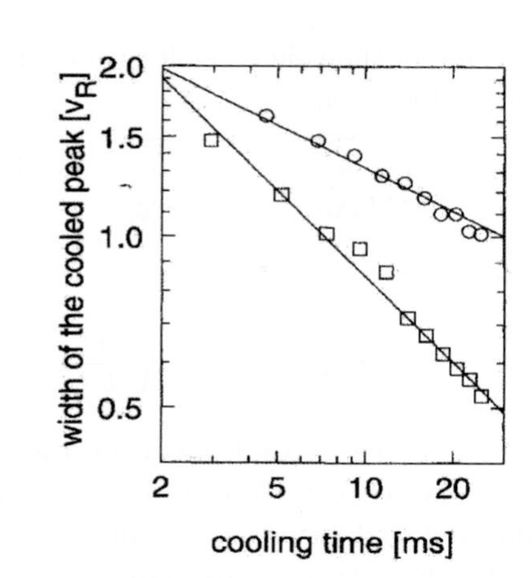

Fig. 13.17 Variations with the cooling time θ of the width δv_θ of the cooled atoms (in log-log scale). Circles : Blackman pulses ($\kappa = 4$) leading to a straight line with a slope $-1/4$. Squares: square pulses ($\kappa = 2$) leading to a straight line with a slope $-1/2$. Figure extracted from [Reichel *et al.* (1995)]. Copyright: American Physical Society.

Moreover, it is very difficult to predict if the weight of the area below these narrow peaks, which is related to the proportion of cooled atoms, tends to a finite value or to zero.

Other approaches have been tried for investigating these problems, and in particular, quantum Monte Carlo simulations using the waiting time distribution.[10] For example, for 1D-VSCPT of metastable helium, the dressed atom approach has been used for calculating the distribution of the time intervals between two successive spontaneous emissions of photons by the atom [Cohen-Tannoudji *et al.* (1992c); Bardou *et al.* (1994a)]. Between two spontaneous emissions, the atom evolves in the family of three states $\{|e,p\rangle; |g_1, p - \hbar k\rangle; |g_2, p + \hbar k\rangle\}$ coupled by absorption and stimulated emission processes. After each spontaneous emission process, whose probability of occurrence can be calculated by diagonalizing a 3×3 matrix, p changes in a random way and the system enters a new family of three states labelled by the new value of p, and so on. Figure 13.18 shows the random time evolution of p obtained from such a simulation. It clearly suggests an anomalous random walk along the time axis dominated by a few rare events occurring when p becomes close to zero. In fact, there are about 4000 time intervals in the simulation of Fig. 13.18(a), of which a single one lasts a time that is on the order of three quarters of the total interaction time. The zoom shown in Fig. 13.18(b) exhibits a similar structure and demonstrates a self-similarity of the random walk at all scales.

All these results suggest the existence of a close connection between sub-recoil cooling and random processes that are dominated by a few rare events and for which Gaussian statistics cannot be applied. Instead, a completely new approach, using Lévy statistics, has been introduced for analyzing the long time limit of sub-recoil

[10]See Sec. 5.5.1 of Chap. 5.

so that the order of magnitude of the coherence length of the atomic wave packets describing the atomic center of mass is:

$$\xi \sim \frac{\hbar}{M\,\Delta v_{\text{eq}}} \sim \sqrt{\frac{\hbar}{M\,|\delta_g|}}. \tag{13.22}$$

Let us compare now ξ and λ_L:

$$\frac{\xi}{\lambda_L} \sim k_L\xi \sim \sqrt{\frac{\hbar\,k_L^2}{|\delta_g|\,M}} \sim \sqrt{\frac{E_{\text{rec}}}{\hbar\,|\delta_g|}}. \tag{13.23}$$

Near the minimum of the curve of Fig. 13.9, $\bar{E}_K$ is on the order of $50E_R$, so that $\Delta p \sim 7\hbar k_L$ and $\xi/\lambda_L \sim 1/7$, which is no longer very small compared to 1. A full quantum treatment of internal and external variables, as the one leading to the curve of Fig. 13.9, seems therefore more appropriate.

Rather than numerically solving the full quantum equations of motion, one can instead calculate the energy levels of the center of mass motion in the spatially periodic potential wells of Fig. 13.6. This is what is done in [Castin and Dalibard (1991)]. As in solid state physics, this leads to a description of atomic motion in terms of Bloch states and energy bands.[5] If the separation between energy bands is large compared to $\hbar\gamma_g$, a secular approximation can be made in the equations of motion of the atomic density matrix, that leads to rate equations that only couple the populations of the various energy bands. Low intensity Sisyphus cooling then appears as an optical pumping process that concentrates the atoms into the lowest energy bands. In fact, these lowest energy bands are very narrow, due to the smallness of the tunneling rates of the atoms between adjacent potential wells. These bands can be also considered as describing vibrational energy levels in the potential wells. Experiments, using either stimulated Raman spectroscopy [Verkerk *et al.* (1992)] or spontaneous Raman spectroscopy [Jessen *et al.* (1992)], have been performed to demonstrate the existence of these quantized vibrational levels.

13.3 Sub-recoil cooling

13.3.1 *Physical mechanism*

To achieve a momentum spread Δp smaller than the photon momentum, one must prevent atoms with very small velocity v (smaller than the recoil velocity v_{rec}) from absorbing light, since this would lead to a recoil in a random direction and a velocity v_{rec} due to the spontaneously re-emitted photon. This is achieved by realizing a situation where the fluorescence rate R depends on v and vanishes for $v = 0$ (see Fig. 13.10(a)). Atoms in the so-called "dark states" with zero velocity (or with v very close to 0) no longer absorb photons and are protected from the "bad effects" of light that are associated with the random recoils due to the spontaneous emission

[5] See also Sec. 17.6.1 of Chap. 17.

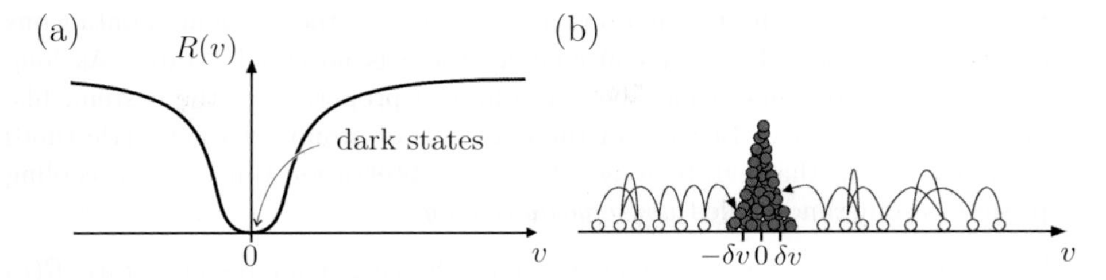

Fig. 13.10 Principle of sub-recoil laser cooling. (a) The fluorescence rate $R(v)$ depends on v and vanishes for $v = 0$. (b) The atoms perform a random walk in v-space and accumulate in a small interval around $v = 0$ where they remain trapped. Courtesy of F. Bardou.

processes that follow the absorption of photons, while atoms with non-zero velocity can absorb and re-emit photons. The velocity changes associated with the random recoils due to spontaneously emitted photons can make them fall into the region close to $v = 0$, where they remain trapped and accumulate (see Fig. 13.10(b)).

It must be realized that the width δv of the velocity distribution obtained by this method is not the width of the dip that appears in the variations with v of the fluorescence rate $R(v)$ (see Fig. 13.10(a)). The width δv of the velocity distribution is the width δv_θ of the interval around $v = 0$ in which the atoms can remain trapped during the interaction time θ. The longer θ, the narrower δv_θ. Let us evaluate the order of magnitude of δv_θ when $R(v)$ increases quadratically with v near $v = 0$. Atoms in this interval will not absorb photons during the time θ, i.e. will not jump out of this interval during θ, if $\theta\, R(\delta v_\theta) < 1$. We conclude that, if $R(v)$ varies as v^2,

$$\delta v_\theta \propto 1/\sqrt{\theta}. \tag{13.24}$$

The width of the velocity distribution obtained after an interaction time θ is thus expected to decrease indefinitely as $1/\sqrt{\theta}$. For θ long enough, we can have $\delta p = M\delta v < \hbar k$, which explains why this cooling scheme is called sub-recoil cooling.

All these qualitative predictions are confirmed by a more quantitative treatment based on Lévy statistics.[6] This sub-recoil cooling scheme differs significantly from the other cooling schemes described in Chap. 12 and in the beginning of this chapter. Two important differences should be noted:

- Doppler cooling and Sisyphus cooling are based on a friction force that damps the atomic velocity. Here, the cooling is not based on a friction force,[7] but rather on an inhomogeneous random walk in v-space that slows down and vanishes when $v \to 0$.
- In Doppler cooling and Sisyphus cooling, the system reaches a steady state as a result of the competition between the cooling introduced by the friction and

[6] See Sec. 13.3.4 and [Bardou *et al.* (2002)] for a review.

[7] The fact that the cooling process considered here does not rely on a friction force does not mean, of course, that it would not benefit from the presence of a friction force pushing the atoms towards the origin of v-space and improving the accumulation process. Sub-recoil cooling schemes including friction forces have been also investigated [Marte *et al.* (1994); Shahriar *et al.* (1993)].

the heating due to the fluctuations associated with the random spontaneous emission processes. In sub-recoil cooling, there is no steady state. As long as the interaction time θ may be, the physical properties of the system, like δv_θ, continue to vary. Because of the existence of atomic characteristic times (trapping times) that can be longer than the observation time, such a cooling process is sometimes called *non-ergodic cooling.*

Two methods have been implemented for achieving a fluorescence rate $R(v)$ varying with v as the curve of Fig. 13.10(a).

- The first starts from a quantum interference effect, coherent population trapping, which blocks the absorption of light and makes it velocity dependent [Aspect *et al.* (1988, 1989)]. This method is called "Velocity Selective Coherent Population Trapping" (VSCPT).
- The second uses velocity selective Raman processes to achieve a velocity selective excitation of the atoms [Kasevich and Chu (1992a)].

13.3.2 *Velocity selective coherent population trapping (VSCPT)*

Coherent population trapping (CPT). Dark states.

We first briefly summarize the important features of coherent population trapping, which have been already described in Sec. 4.6 of Chap. 4 as an example of a quantum interference effect associated with linear superpositions of internal atomic states. Figure 4.13 of Chap. 4 shows a three-level atom with two lower sublevels g_1, g_2 and one excited sublevel e. Two laser waves, $E_1 \exp[i(\vec{k}_1 \cdot \vec{r} - \omega_1 t)]$ + c.c. and $E_2 \exp[i(\vec{k}_2 \cdot \vec{r} - \omega_2 t)]$+c.c., excite the two transitions $g_1 \to e$ and $g_2 \to e$, respectively. Let Δ be the detuning from the Raman resonance condition:

$$\hbar\Delta = \hbar\omega_1 - \hbar\omega_2 + (E_{g_1} - E_{g_2}). \tag{13.25}$$

This detuning Δ can be adjusted by varying the splitting between g_1 and g_2, while holding ω_1 and ω_2 fixed, or by sweeping ω_2, with ω_1 fixed. We have shown in Chap. 4 that the variations with Δ of the intensity of the fluorescence light emitted by the atom (fluorescence rate R_F) have the shape represented in Fig. 4.13(b) of Chap. 4: R_F vanishes for $\Delta = 0$ and then increases, forming a narrow dip. This shape is similar to the one represented in Fig. 13.10(a), with v replaced by Δ.

The interpretation of this effect involves a quantum destructive interference. The atom in g_1 can absorb a photon ω_1 and go from g_1 to e, with an amplitude A_1. It can absorb a photon ω_2 and go from g_2 to e, with an amplitude A_2. But, if the atom has been put in a linear superposition of g_1 and g_2 :

$$|\psi_D\rangle = c_1|g_1\rangle + c_2|g_2\rangle, \tag{13.26}$$

such that

$$c_1 A_1 + c_2 A_2 = 0, \tag{13.27}$$

the two absorption amplitudes $c_1 A_1$ and $c_2 A_2$ interfere destructively and the absorption of light is quenched. For this reason, the state $|\psi_D\rangle$ is called a "dark state". In Chap. 4, we gave an interpretation of the resonance Raman condition (13.25), which insures that, if the state is dark at a certain time, it remains dark at all times. This is why R_F vanishes for $\Delta = 0$. If $\Delta \neq 0$, a state that is dark at a given time does not remain dark, and starts to absorb light at a later time. One understands in this way the increase of R_F with Δ around $\Delta = 0$.

How to make CPT velocity selective

Now suppose that the atom is moving with a velocity $\vec{v}$. Is it possible to achieve a situation where the detuning Δ from the resonance Raman condition is proportional to v? The curve represented in Fig. 4.13(b) of Chap. 4 would then represent a velocity dependent fluorescence rate that has the shape assumed in Fig. 13.10(a), necessary to achieve sub-recoil cooling.

Let us first use a semiclassical treatment of atomic motion where the atomic center of mass is treated as a classical particle moving with a velocity $\vec{v}$. Suppose that the resonance Raman condition is fulfilled when $\vec{v} = \vec{0}$. When $\vec{v} \neq \vec{0}$, the two laser frequencies are Doppler shifted in the atom rest frame by amounts equal to $\vec{k}_1 \cdot \vec{v}$ and $\vec{k}_2 \cdot \vec{v}$, respectively. They subtract in the Raman process, thereby changing the detuning Δ, which becomes $\Delta + (\vec{k}_1 \cdot \vec{v} - \vec{k}_2 \cdot \vec{v})$. This is different from Δ if $\vec{k}_1 \cdot \vec{v} \neq \vec{k}_2 \cdot \vec{v}$. If the splitting between g_1 and g_2 is very small compared to ω_1 and ω_2, $|\vec{k}_1|$ and $|\vec{k}_2|$ are nearly equal, and coherent population trapping is velocity selective only if $\vec{k}_1$ and $\vec{k}_2$ are not parallel (for example, opposite).

Of course, a semiclassical treatment of atomic motion is highly questionable in the sub-recoil limit, when $\Delta p < \hbar k$, since the coherence length $\xi = \hbar/\Delta p$ of the atomic wave packets becomes greater than the wavelength λ_L of the lasers. Under these conditions, the atom can no longer be considered to be localized in these waves

$$\Delta p < \hbar k \to \xi = \frac{h}{\Delta p} > \frac{2\pi}{k} = \lambda_L . \tag{13.28}$$

On the other hand, the two absorption amplitudes from g_1 and g_2 can interfere only if they reach the same final state $|e, \vec{p}\rangle$ (atom in the state e with momentum $\vec{p}$). Because of momentum conservation, these two amplitudes must start from $|g_1, \vec{p} - \hbar\vec{k}_1\rangle$ and $|g_2, \vec{p} - \hbar\vec{k}_2\rangle$ and the energies of these states must include not only the internal energy E_{g_i} but also the external kinetic energy $(\vec{p} - \hbar\vec{k}_i)^2/2m$. We conclude that the dark state is a superposition of two states differing not only by the internal state g_i, but also by the external state $\vec{p} - \hbar\vec{k}_i$:

$$|\psi\rangle = c_1|g_1, \vec{p} - \hbar\vec{k}_1\rangle + c_2|g_2, \vec{p} - \hbar\vec{k}_2\rangle, \tag{13.29}$$

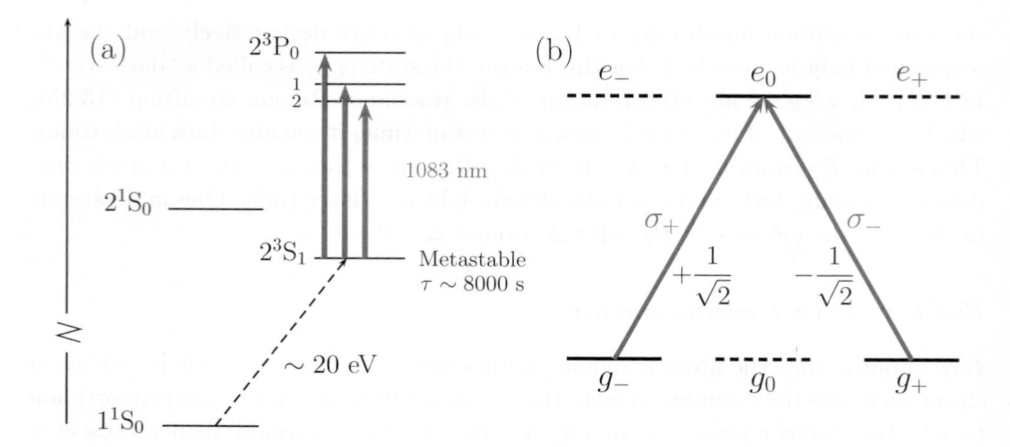

Fig. 13.11 (a) A few energy levels of the helium atom. (b) Transition $g = 2^3S_1 \to e = 2^3P_1$ excited by two counter-propagating laser beams with polarizations σ_+ and σ_-, respectively. The number near the transitions are the Clebsch Gordan coefficients.

with (Raman resonance condition including external energy):

$$E_{g_1} + \frac{(\vec{p} - \hbar\vec{k}_1)^2}{2m} + \hbar\omega_1 = E_{g_2} + \frac{(\vec{p} - \hbar\vec{k}_2)^2}{2m} + \hbar\omega_2. \tag{13.30}$$

Experimental study of VSCPT with metastable helium atoms

The feasibility of cooling atoms to temperatures below the recoil limit by VSCPT has been demonstrated in one-dimensional experiments performed on helium atoms that have been excited, by an electric discharge, to the metastable state 2^3S_1, which has a very long lifetime ($\sim 8000s$) [Aspect *et al.* (1988)]. Laser beams excite these metastable atoms with a frequency close to one of the three atomic transitions $2^3S_1 \to 2^3P_J$ ($J = 0, 1, 2$) corresponding to a wavelength $\lambda_L \sim 1083$ nm (see Fig. 13.11(a)).

The transition $2^3S_1 \to 2^3P_1$ is particularly interesting (see Fig. 13.11(b)). If one uses two counter-propagating laser beams with polarizations σ_+ and σ_-, respectively, atoms are optically pumped out of g_0, and cannot come back in g_0 because of the value 0 of the Clebsch Gordan coefficient $e_0 \leftrightarrow g_0$. In this way, a pure three-level system $\{g_-, g_+, e_0\}$ is realized in which each laser beam excites only one transition, $\{g_- \to e_0\}$ for the σ_+ beam, and $\{g_+ \to e_0\}$ for the σ_- beam.

The principle of the experiment is sketched in Fig. 13.12(a). A metastable helium beam crosses, at a right angle, two counter-propagating laser beams that have the same frequency ($\omega_1 = \omega_2$), and σ_+ and σ_- polarizations. The transition $2^3S_1 \to 2^3P_1$ is thereby excited. Since the static magnetic field is equal to 0,

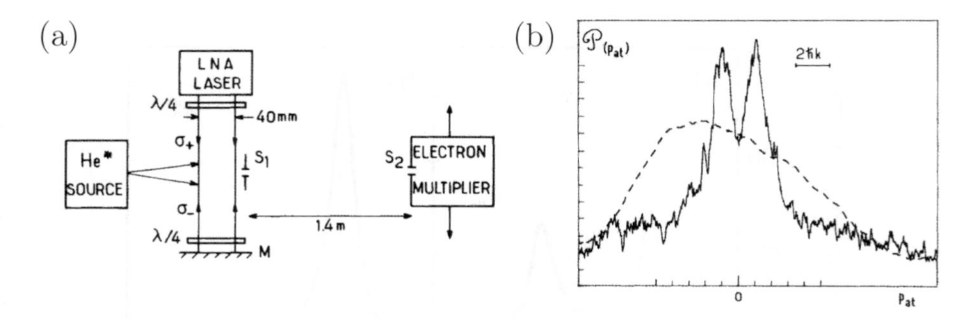

Fig. 13.12 (a) Sketch of the experimental set up. (b) Measured momentum distribution exhibiting two peaks separated by a distance $2\hbar k$ well above the momentum distribution in the absence of lasers (dashed line). Figure adapted from [Aspect *et al.* (1988)]. Copyright: American Physical Society.

$E_{g+} = E_{g-}$, and the Raman resonance condition (13.30) gives $p = 0$ with the dark state:[8]

$$|\psi\rangle = \frac{1}{\sqrt{2}}\left(|g_-\rangle \otimes |-\hbar\vec{k}\,\rangle + |g_+\rangle \otimes |+\hbar\vec{k}\,\rangle\right). \tag{13.31}$$

This dark state is an entangled internal-external state with a double peak momentum distribution, where the two peaks are centered at $+\hbar k$ and $-\hbar k$.

Figure 13.12(b) shows the transverse momentum distribution measured by an electron multiplier that is displaced in a direction parallel to the laser beams. Two peaks separated by a distance $2\hbar k$ clearly appear well above the momentum distribution in the absence of lasers, which shows that there is a cooling and not just a velocity selection. The fact that the two peaks are well resolved shows also that their width is smaller than $\hbar k$, which is a signature of sub-recoil cooling.

Significant improvements have been made in this experiment by replacing the atomic beam with a magneto-optically trapped and cooled sample [Bardou *et al.* (1994b)]. When the trap is switched off, two VSCPT laser beams with opposite directions and σ_+ and σ_- polarizations are applied during a time θ that is sufficiently short that the effect of gravity field is negligible. The atoms then fall onto a multi channel detector that can detect individual atoms. The interaction time in this new experiment is much longer than in the previous (1 ms instead of 30 μs). This leads to narrower peaks and a better signal to noise ratio, as shown in Fig. 13.13(a).

It is also possible to obtain velocity distributions with a single peak by adiabatically changing the relative intensity of the two lasers. By doing this, one changes the coefficients c_1 and c_2 of the dark state (13.26) which allows the transfer of the total population of the two peaks into a single peak (see Fig. 13.13(b)). A similar idea, used in the "Stimulated Raman Adiabatic Passage" (STIRAP) method

[8]The + sign in Eq. (13.31) comes from the fact that the two Clebsch Gordan coefficients of the two transitions $\{g_- \to e_0\}$ and $\{g_+ \to e_0\}$ are equal and opposite (see Fig. 13.11(b)). It is therefore a + sign in Eq. (13.31) that leads to a destructive interference between the two absorption amplitudes $\{g_- \to e_0\}$ and $\{g_+ \to e_0\}$.

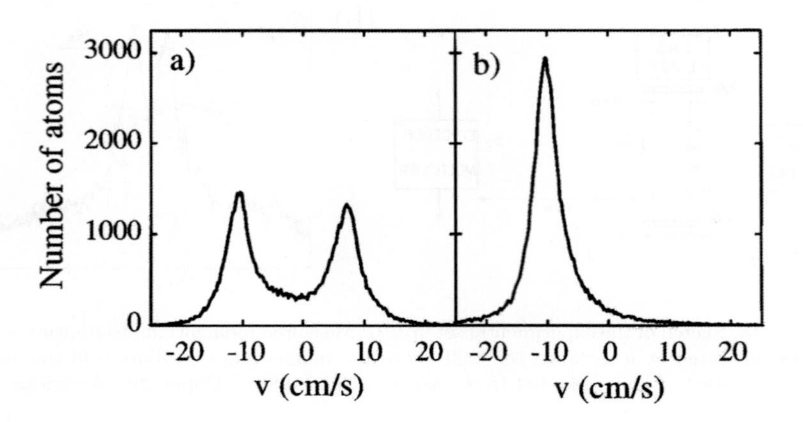

Fig. 13.13 (a) Velocity distribution of the atoms along the cooling axis. Atoms have been trapped in a magneto-optical trap, subjected to 1D VSCPT cooling during a time $\theta = 1$ ms after the switching off of the trap and subsequently detected on a multichannel plate. (b) The same distribution after a 100 % efficient adiabatic transfer in one of the two peaks states. Figure taken from [Kulin *et al.* (1997)]. Copyright: American Physical Society

in molecular physics, allows one to transfer all molecules from one level to another one, while remaining in a dark state [Bergmann *et al.* (1998)].[9]

Sub-recoil cooling of metastable helium atoms has been extended to two and three dimensions [Lawall *et al.* (1994, 1995)] using four and six VSCPT laser beams, respectively. One can show that, for a $J_g = 1 \to J_e = 1$ transition, the dark state is a superposition of four or six atomic wave packets whose momentum and polarization [Olshanii and Minogin (1991, 1992)] are the same as those of the four or six laser beams used to cool them. Figure 13.14 shows for example the four sub-recoil peaks obtained in a two-dimensional VSCPT experiment [Lawall *et al.* (1994)] and the single peak obtained after adiabatically switching off three of the four laser beams.

13.3.3 *Sub-recoil Raman cooling*

We have seen in Chap. 9 that stimulated Raman transitions induced by two counter-propagating laser beams have a very high selectivity in velocity. Sub-recoil Raman cooling [Kasevich and Chu (1992a)] uses such stimulated Raman transitions to transfer sodium atoms with positive velocity ($v > 0$) from the lower hyperfine level F_1 to the upper one F_2 and change their velocity by an amount $-2\hbar k/M$ (see left part of Fig. 13.15). By optical pumping with another beam, atoms are transferred back from F_2 to F_1 and in the process undergo a random momentum recoil (due to the spontaneous photon) varying between $+\hbar\vec{k}$ and $\hbar\vec{k}$ (see right part of Fig. 13.15). Atoms are thus put back in F_1 with a velocity smaller than their initial velocity. This cycle is repeated for atoms with $v < 0$ whose velocity is changed by a positive

[9] See also Sec. 9.4.3 of Chap. 9.

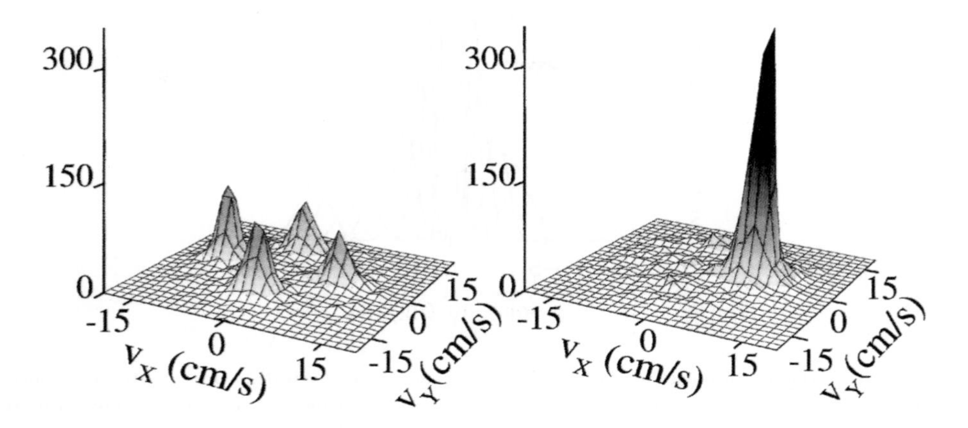

Fig. 13.14 Left: four peak momentum distribution of metastable helium atoms obtained in a 2D-VSCPT experiment performed with four laser beams. Right: single peak obtained after adiabatically switching off three of the four laser beams. Figure adapted from [Kulin *et al.* (1997)]. Copyright: American Physical Society.

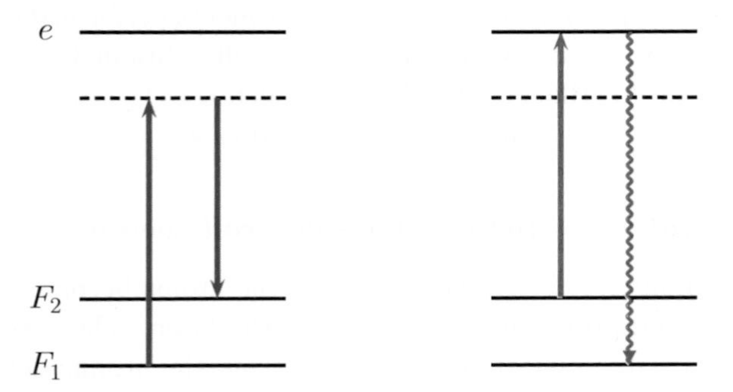

Fig. 13.15 Stimulated Raman pulses with two counter-propagating laser beams (left), and optical pumping pulses (right) used for compressing the velocity distribution towards $v = 0$.

amount on the order of $+\hbar\vec{k}/M$. In this way, the velocity distribution is compressed towards $v = 0$.

One does not, however, excite atoms that have a velocity in a small interval δv on the order of $\hbar k/M$ around $v = 0$. This interval plays the role of a trap in velocity space: because of the random changes of momentum associated with absorption-spontaneous emission cycles, atoms have a certain probability to end up in this zone from which they cannot escape. This sequence of pulses finally gives the desired shape for $R_F(v)$ (see Fig. 13.16).

One of the advantages of sub-recoil Raman cooling is that one can choose an appropriate shape for the stimulated Raman pulses and in this way change the exponent κ of the power law v-dependence describing the increase of the jump

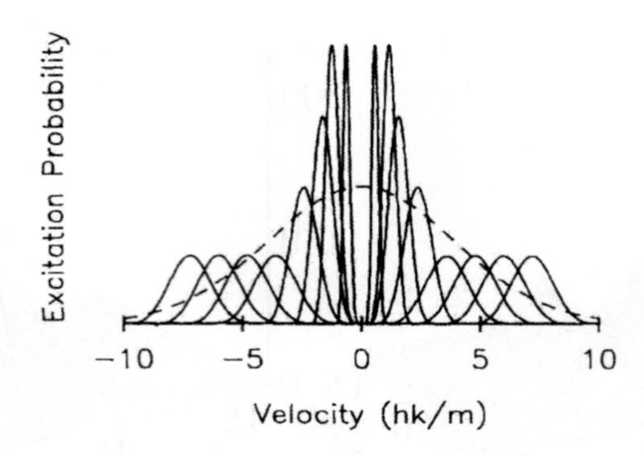

Fig. 13.16 Sequence of pulses leading to a v-dependence of the excitation probability (and consequently of the fluorescence rate) having the desired shape for achieving sub-recoil cooling. Figure from [Kasevich and Chu (1992a)]. Copyright: American Physical Society.

rate $R_F(v)$ around $v = 0$. The same argument as the one leading to Eq. (13.24) given above shows that the width δv_θ of the velocity distribution obtained after an interaction time θ should decrease as $\theta^{-1/\kappa}$. This prediction has been checked experimentally in sub-recoil Raman cooling experiments performed on cesium atoms [Reichel *et al.* (1995)]. Figure 13.17 shows the results obtained with square pulses, for which $\kappa = 2$, and for Blackman pulses for which $\kappa = 4$; δv_θ decreases with θ as $\theta^{-1/2}$ in the first case, and as $\theta^{-1/4}$ in the second case.

13.3.4 *Quantitative predictions for sub-recoil cooling*

In the sub-recoil limit, the coherence length of the atoms becomes larger than the laser wavelength, and atoms are delocalized in the laser. The external degrees of freedom must be treated quantum-mechanically and the atomic density matrix σ_A must have two types of quantum numbers, internal (i, j) and external (p, p'):

$$\langle i, p|\sigma_A|j, p'\rangle.$$

The equations of motion of these matrix elements are called "generalized optical Bloch equations (GOBE)". They are more complex than usual Bloch equations, where atomic motion is treated semiclassically and which involve only internal quantum numbers.

Because of their complexity, it is in general not possible to get analytical solutions of the GOBE. Furthermore, sub-recoil cooling has no steady-state solution so it is very difficult (except in a very small number of particular cases [Alekseev and Krylova (1992, 1996)]) to make simple predictions concerning the long time limit where these mechanisms are most interesting. A numerical integration of the GOBE also raises serious practical problems for long times. The momentum distribution contains a series of very narrow peaks, analogous to delta functions, and so requires a very large number of discrete values for its description in a numerical calculation.

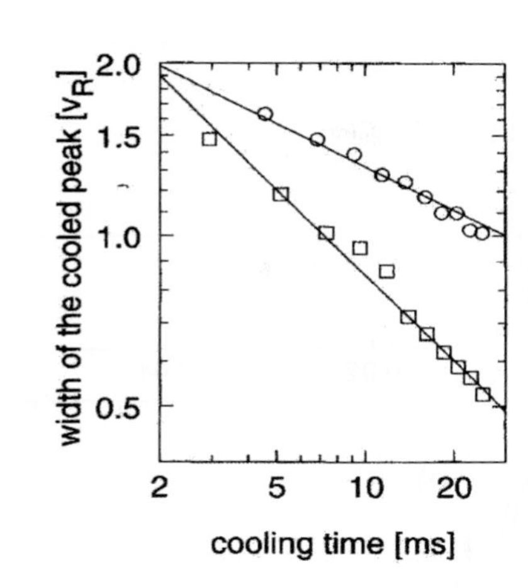

Fig. 13.17 Variations with the cooling time θ of the width δv_θ of the cooled atoms (in log-log scale). Circles : Blackman pulses ($\kappa = 4$) leading to a straight line with a slope $-1/4$. Squares: square pulses ($\kappa = 2$) leading to a straight line with a slope $-1/2$. Figure extracted from [Reichel *et al.* (1995)]. Copyright: American Physical Society.

Moreover, it is very difficult to predict if the weight of the area below these narrow peaks, which is related to the proportion of cooled atoms, tends to a finite value or to zero.

Other approaches have been tried for investigating these problems, and in particular, quantum Monte Carlo simulations using the waiting time distribution.[10] For example, for 1D-VSCPT of metastable helium, the dressed atom approach has been used for calculating the distribution of the time intervals between two successive spontaneous emissions of photons by the atom [Cohen-Tannoudji *et al.* (1992c); Bardou *et al.* (1994a)]. Between two spontaneous emissions, the atom evolves in the family of three states $\{|e, p\rangle; |g_1, p - \hbar k\rangle; |g_2, p + \hbar k\rangle\}$ coupled by absorption and stimulated emission processes. After each spontaneous emission process, whose probability of occurrence can be calculated by diagonalizing a 3×3 matrix, p changes in a random way and the system enters a new family of three states labelled by the new value of p, and so on. Figure 13.18 shows the random time evolution of p obtained from such a simulation. It clearly suggests an anomalous random walk along the time axis dominated by a few rare events occurring when p becomes close to zero. In fact, there are about 4000 time intervals in the simulation of Fig. 13.18(a), of which a single one lasts a time that is on the order of three quarters of the total interaction time. The zoom shown in Fig. 13.18(b) exhibits a similar structure and demonstrates a self-similarity of the random walk at all scales.

All these results suggest the existence of a close connection between sub-recoil cooling and random processes that are dominated by a few rare events and for which Gaussian statistics cannot be applied. Instead, a completely new approach, using Lévy statistics, has been introduced for analyzing the long time limit of sub-recoil

[10]See Sec. 5.5.1 of Chap. 5.

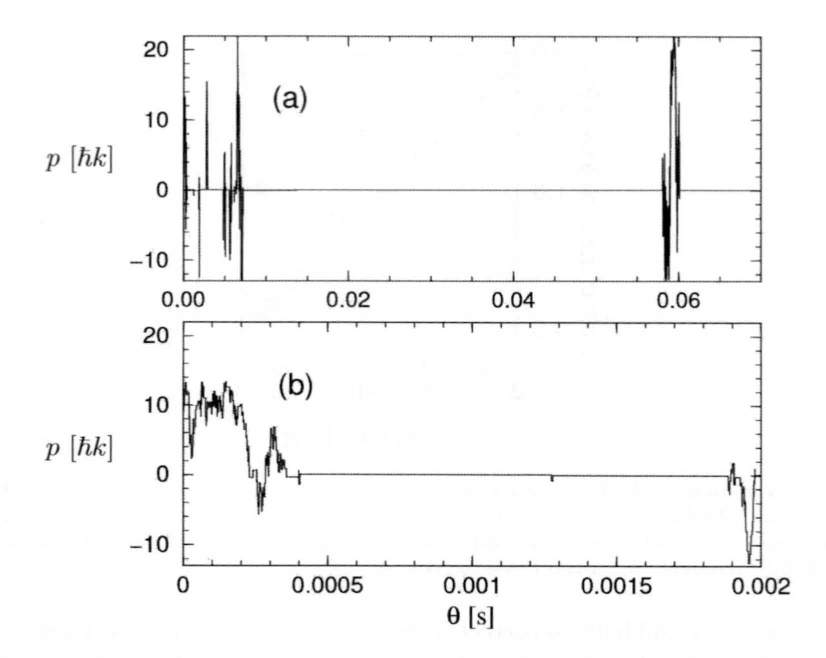

Fig. 13.18 (a) Monte-Carlo simulation of 1D-VSCPT of metastable helium atoms. Each vertical discontinuity corresponds to a spontaneous emission jump during which the atomic momentum changes abruptly. The random walk is clearly dominated by a few rare events. (b) Zoomed part of the sequence. Courtesy of F. Bardou.

cooling [Bardou *et al.* (1994a)]. It is described in detail in [Bardou *et al.* (2002)]. Let us just mention here that it has provided quantitative analytical predictions, that have been checked experimentally, for the main features of the cooled atoms such as their momentum distribution [Saubamea *et al.* (1999)].

13.4 Resolved sideband cooling of trapped ions

Laser cooling of trapped ions was briefly described in Chap. 12 in the limit where the natural width Γ of the excited state is large compared to the vibration frequency ω_v of the ions in the trap. In this limit, the motion of the ion during the characteristic time $1/\Gamma$ of internal variable is negligible compared to the amplitude of the vibrational motion in the trap so that the ion can be considered free during this time. The cooling mechanism is then identical to the Doppler cooling mechanism for free atoms, and the lowest temperature that can be attained is on the same order, scaling as $\hbar\Gamma/k_B$.

Several mechanisms are possible for going below this limit. One of them is Sisyphus cooling and is discussed theoretically in [Wineland *et al.* (1992a)]. We will focus here on another method, called *resolved sideband cooling*, that allows ions

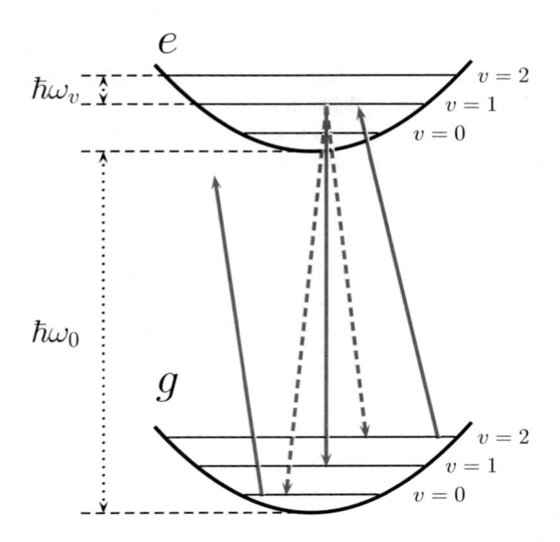

Fig. 13.19 Principle of resolved sideband cooling.

to be accumulated in the lowest vibrational level of the trap [Wineland and Itano (1979)].

Sideband resolved cooling works when the vibration frequency ω_v is large compared to the natural width Γ of the excited state

$$\omega_v \gg \Gamma, \tag{13.32}$$

and to the recoil frequency $E_{\text{rec}}/\hbar$ (Lamb-Dicke limit)

$$\omega_v \gg E_{\text{rec}}/\hbar. \tag{13.33}$$

The principle of this method is sketched in Fig. 13.19. Condition (13.32) means that the various vibrational levels in e and g are well resolved, so that it is possible to selectively excite the transitions $|g, v\rangle \to |e, v-1\rangle$ ($\Delta v = -1$) by a laser excitation at frequency $\omega_A - \omega_v$, where ω_A is the frequency of the atomic transition. Condition (13.33) means that the most probable re-emission lines are: $|e, v-1\rangle \to |g, v-1\rangle$ with $\Delta v = 0$.[11] After an absorption-emission cycle, the vibrational quantum number in g essentially decreases by one unit, and the atoms progressively accumulate in $|g, v = 0\rangle$, where they can no longer exit by an excitation at $\omega_A - \omega_v$. There is a certain analogy with dark state cooling of neutral atoms described in the previous section in that one puts the ion into the lowest vibrational state, which appears as a trap from which it can no longer absorb photons.[12]

[11]See Sec. 6.4 of Chap. 6.

[12]Note however that atoms cannot be entirely transferred into the lowest vibrational level because of the off-resonant excitation of the transitions ($\Delta v = 0$) by the laser excitation at $\omega_A - \omega_v$. One can show that the population of the excited vibrational levels scales as $(\Gamma/\omega_v)^2$ [Wineland and Itano (1979)].

How to achieve $\omega_v \gg \Gamma$***?***

Condition (13.32) is not easy to achieve, since using weakly allowed transitions with a small value of Γ leads to weak excitation rates and very long cooling times. An elegant solution for solving this difficulty is to use stimulated Raman transitions [Monroe *et al.* (1995)].[13] The role of levels g and e is played by two different hyperfine levels F_1 and F_2 of the ground state of the ion which have a very small width because relaxation times are long in the ground state. In this way, it is easy to achieve the condition (13.32). As with any ground state level, these levels have a vibrational structure, and condition (13.33) is also easy to achieve. Stimulated Raman transitions allow one to transfer ions from $|F_1, v\rangle$ to $|F_2, v-1\rangle$ with a high selectivity. By optical pumping with a laser resonant with a transition starting from[14] F_2, ions then preferentially fall back in $|F_1, v-1\rangle$. In this way, with a redefinition of e and g, one achieves the cooling scheme of Fig. 13.19. This method is now currently used in experiments using ultracold trapped ions.

13.5 Conclusion

To conclude this chapter, we summarize a few important results that have been established.

- The existence of several Zeeman sublevels in the ground state g is not just a source of complications. It gives rise to interesting new effects and allows important improvements of laser cooling.
- Low intensity Sisyphus effect is due to a correlation between the spatial modulations of light shifts (reactive effects) and dissipative effects (optical pumping rates).
- Sisyphus temperatures scale as light shifts $\hbar\delta_g$ and not as $\hbar\Gamma$. They can be much lower than the Doppler limit, but remain larger than the recoil limit. The velocity capture range is much smaller than that of Doppler cooling.
- One of the key ingredients of low intensity Sisyphus cooling is the appearance of long internal times associated with optical pumping. These optical pumping times τ_p are inversely proportional to the light intensity I_L and can become longer than external times.
- The friction in low intensity Sisyphus cooling remains very large when $I_L \to 0$ because the decrease of light shifts is compensated for by the lengthening of the time lag τ_p between internal and external variables.
- The recoil limit can be overcome by introducing a trap in momentum space around $p = 0$. During their random walk in momentum space, atoms fall and accumulate into velocity dark states, that cannot absorb light. These dark states can be achieved by velocity selective quantum interference effects or by velocity selective Raman processes.

---

[13] See Sec. 9.4.3 of Chap. 9.

[14] With this resonant excitation, one can confer an adjustable width to level F_2.

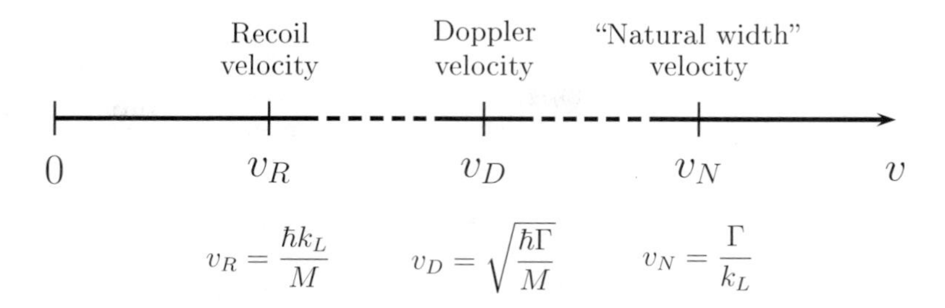

Fig. 13.20 A few characteristic velocities associated with the various cooling schemes.

- There is no lower limit to the effective temperature which can be obtained by this method. The longer the interaction time, the lower the temperature.
- Sub-recoil cooling is based on anomalous random walks dominated by a few rare events. Lévy statistics is an appropriate theoretical framework for analyzing this cooling mechanism and provides quantitative results in the long time limit.

We will end this conclusion by showing in Fig. 13.20 a few characteristic velocities, given by the previous analysis, that clearly show the successive improvements of laser cooling techniques. The first is the recoil velocity, $v_R = \hbar k_L/M$. The second, $v_D = \sqrt{\hbar\Gamma/M}$, is such that $Mv_D^2 = \hbar\Gamma$ and thus gives the velocity dispersion that can be achieved by Doppler cooling. The last velocity, $v_N = \Gamma/k_L$, satisfies $k_L v_N = \Gamma$, which means that the Doppler effect associated with v_N is equal to the natural width Γ. It gives therefore the velocity spread of the atoms that can be efficiently excited by a monochromatic light. It is easy to check that v_N/v_D and v_D/v_R are both equal to $\sqrt{\hbar\Gamma/E_R}$, where $E_R = \hbar^2 k_L^2/2M$ is the recoil energy. For most allowed transitions, we have $\hbar\Gamma \gg E_R$, so that $v_R \ll v_D \ll v_N$ (the v-scale in Fig. 13.20 is not linear). This shows the advantage of laser cooling. Laser spectroscopy with a monochromatic laser beam gives lines whose width, expressed in velocity units, cannot be smaller than v_N. Doppler cooling reduces the velocity dispersion of the atoms to a much lower value, though it cannot be smaller than v_D. Sisyphus cooling reduces this lower limit to a few v_R. Finally, sub-recoil cooling allows one to go below v_R.

Chapter 14

Trapping of particles

14.1 Introduction

Trapping particles by combinations of electric and magnetic fields allows one to make measurements that would be difficult or impossible to perform by other techniques. Indeed, trapping makes it possible to prevent any contact between the particle and the thermal walls. In this way, one can obtain well-controlled environments over long periods of time, which, according to the Heisenberg principle, allows very high accuracy measurements of their properties. Therefore, trapping is a very efficient technique for achieving high resolution spectroscopy, and for increasing precision in measurements of many fundamental physical quantities such as masses, magnetic moments, g-factors ... We must keep in mind, however, that metrology measurements require avoiding or minimizing the perturbation of the internal atomic frequencies by the trapping mechanism.

Trapping many atomic particles also provides good conditions for investigating collective effects such as crystallization for ions, or many-body quantum phases for neutral atoms. Indeed, traps are crucial for reaching high spatial densities and high phase space densities, and they have played a key role in the realization of Bose-Einstein condensation for neutral atoms.

A wide variety of trapping geometries allows one to realize simple model systems which can provide a better understanding of more complex situations found in other fields. For instance, achieving tight confinement in one or two dimensions allows the investigation of the properties of low dimensional quantum systems. Spatially periodic arrays of traps provide simple models that can simulate situations encountered in solid state physics.

Trapping requires physical mechanisms that provide restoring forces. In addition, pre-cooling of particles may be needed to make the kinetic energy of the particles smaller than the trap depth.

This chapter gives an overview of trapping techniques used for low energy particles, and is organized as follows. First, we address the possibility of trapping charged particles with electromagnetic fields. While constraints resulting from Maxwell's equations prevent charged particles from being trapped in a pure electrostatic trap, two standard kinds of trap, the Penning and the Paul traps, have been developed to overcome this difficulty. These are described in Sec. 14.2. It is worth noting that the corresponding trapping mechanisms for ions do not rely on their internal structure, which is therefore not perturbed and can thus be measured accurately by other techniques.

The other sections are devoted to neutral particles for which the Lorentz force is zero, with an emphasis on neutral atoms. Conservative trapping potentials are based on the coupling of the magnetic or electric dipole moment of the atom with gradients of magnetic or electric fields. For instance, the interaction energy of the magnetic dipole $\vec{\mu}$ of an atom depends on the relative orientation between this dipole and the magnetic field $\vec{B}$, and is given by $-\vec{\mu}\cdot\vec{B}$. It acts as a potential energy for the motion of the center of mass of the atomic particle, or in other terms for its external degrees of freedom. Such a trapping mechanism necessarily perturbs the internal atomic energy levels.

The strength of the force exerted by electromagnetic fields on a dipole is much weaker than that exerted on a charged particle: the electrostatic force exerted on an electron in a field of 10^4 V·m^{-1} is on the order of 1.6×10^{-15} N, to be compared with the force exerted on a magnetic moment equal to the Bohr magneton in a typical gradient of 10 T·m^{-1} which gives 10^{-22} N. As a consequence, the typical depth of a magnetic trap for neutral atoms is on the order of few tens of mK, and it turns out that it is essential to implement a pre-cooling stage to load them efficiently.

We discuss separately the cases of magnetic (Sec. 14.3) and electric (Sec. 14.4) trapping, since the physical concepts involved are quite different: there is a state dependent permanent dipole moment in the first case, and an induced moment in the latter. In Sec. 14.5, we explain how artificial orbital magnetism can be realized for neutral atoms using either a rotating trap or gauge potentials related to the existence of a Berry phase. Situations can thus be achieved in which a neutral atom experiences an artificial Lorentz force like a charged particle in a magnetic field. The last section (Sec. 14.6) is devoted to the magneto-optical trap, which is based on the radiation pressure force and constitutes the starting point of most experiments using cold neutral atoms.

14.2 Trapping of charged particles

14.2.1 *The Earnshaw theorem*

In the presence of an electrostatic field $\vec{E}(\vec{r})$, a charge q experiences the electrostatic force $\vec{F}=q\vec{E}$. This particle, however, can never be subjected to a restoring force in all space directions due to the Gauss equation for electrostatics $\vec{\nabla}\cdot\vec{E}=0$, which entails $\vec{\nabla}\cdot\vec{F}=0$. This result is known as the Earnshaw theorem [Jackson (1975)]. In the following, we present two possibilities for circumventing this theorem, therefore achieving three-dimensional confinement. The first uses a combination of electric

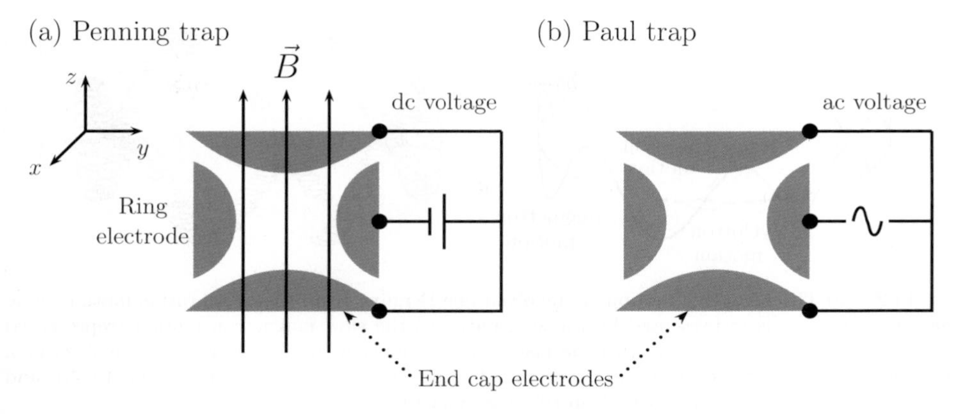

Fig. 14.1 Basic electrode configuration (cross-section) for (a) Penning trap made with a static electric field and a magnetic field and (b) Paul trap where oscillating electric fields are used.

and magnetic fields that lead to a dynamical stabilization of the charged particle. The second takes advantage of the time dependence of the applied electric field obtained by modulating it at the frequency $\omega/2\pi$. The kinetic energy of the particle vibrating at $\omega/2\pi$ appears in certain limits as an effective potential for the particle and provides a restoring force. While these traps are the most commonly used, other techniques have also been developed.[1]

14.2.2 *The Penning trap*

The Penning trap is formed by the superposition of a quadrupole electrostatic potential $V(\vec{r})$:

$$U\left(x,y,z\right) = qV\left(x,y,z\right) = \frac{1}{2}m\omega^2\left[z^2 - \frac{x^2+y^2}{2}\right], \tag{14.1}$$

and a uniform magnetic field $\vec{B}_0 = \vec{B}_0\,\vec{e}_z$ parallel to the symmetry axis (see Fig. 14.1). The electrostatic potential ensures trapping along the z direction, but expels the particle in the perpendicular xy plane. The axial magnetic field generates a rotation of the transverse velocity vector in the plane xy that counteracts the expelling effect of the electrostatic potential if the magnetic field is sufficiently large. This results in a dynamical stablilization of the particle's trajectories.

The Newton equations of motion for the particle are linear in position $\vec{r}$ and velocity $\vec{v}$ with time independent coefficients, and therefore can be solved exactly.[2] There are stable solutions when the cyclotron frequency ω_C is larger than $\sqrt{2}\,\omega$, with a motion of the particle being, as shown in Fig. 14.2, the superposition of:

[1]See [Major *et al.* (2005)] and references therein.
[2]The quantum equations of motion can also be solved exactly [Major *et al.* (2005)].

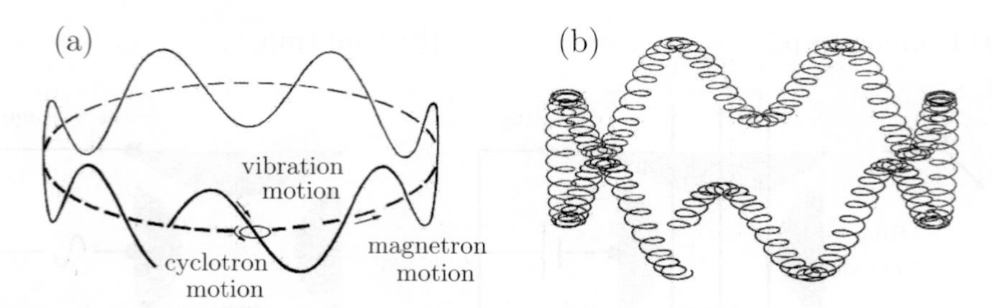

Fig. 14.2 (a) Three kinds of motion are involved in a Penning trap: (i) the vibration motion along the z symmetry axis (represented by the sinusoid), (ii) the slow magnetron motion (represented by the large dashed circle), and (iii) the fast cyclotron motion (represented by the small circle). (b) Representation of a typical trajectory in a Penning trap. Figure adapted from [Brown and Gabrielse (1986)]. Copyright: American Physical Society.

- a *vibration* motion along the z-axis of angular frequency ω.
- a *cyclotron* motion perturbed by the quadrupole electric field with a characteristic frequency ω'_C:

$$\omega'_C = \frac{\omega_C}{2} + \frac{1}{2}\sqrt{\omega_C^2 - 2\omega^2}, \tag{14.2}$$

- A *magnetron* motion describing a drift of the particle in the crossed electric and magnetic fields with a characteristic frequency ω_m:

$$\omega_m = \frac{\omega_C}{2} - \frac{1}{2}\sqrt{\omega_C^2 - 2\omega^2}. \tag{14.3}$$

For $B_0 = 0.1$ T, $U_0 = 10$ V, and a distance between the electrodes of a few mm, one finds the following typical orders of magnitude: $\omega'_C/2\pi \sim$ a few GHz, $\omega/2\pi \sim$ 100 MHz, and $\omega_m/2\pi \sim 1$ MHz. Note that the measurement of the ratio of the cyclotron frequencies of two different species of ions can be used as a mass spectrometer.

The displacement of the charged particle from its equilibrium position induces a current in an electrical circuit connected to the electrodes. This effect can be used for detection [Dehmelt (1968)]. The axial motion is particularly interesting since it lies in the radiofrequency range, which is easily detected. This oscillation may also be driven by applying an alternating voltage, and the resulting driven oscillation can be observed by phase-sensitive detection. An example of such an experimental signal is depicted in Fig. 14.3. It reveals the trapping of seven electrons initially, and the removal of electrons one by one achieved by increasing the driving as a function of time. This figure shows that it is possible to trap and detect a single electron!

If trap electrodes are connected to an external circuit, the charged particles dissipate their energy by inducing current in the circuit and come to equilibrium at the temperature of the external circuit [Church and Dehmelt (1969)]. A more quantitative study shows that the corresponding cooling, referred to as resistive cooling, is most effective for charged particles with a large charge to mass ratio

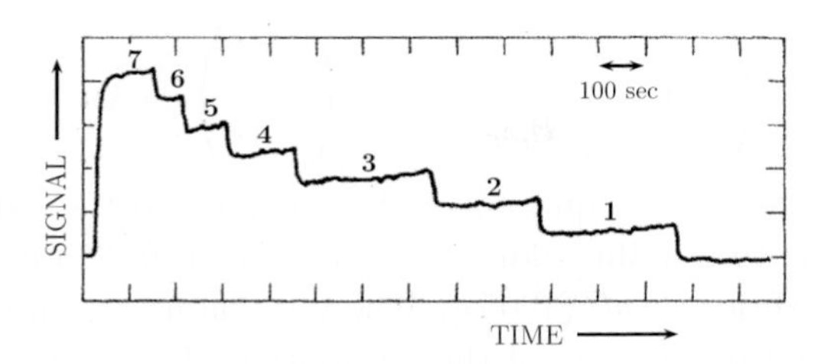

Fig. 14.3 The forced-oscillation signal as a function of time. The trap is initially loaded with seven electrons. The driving frequency yields a loss of the electrons one by one. Figure adapted from [Wineland *et al.* (1973)]. Copyright: American Physical Society.

[Itano *et al.* (1995)]. It is particularly efficient for electrons and most commonly used in Penning traps. We emphasize that this technique, in contrast to laser cooling techniques for trapped ions, does not require knowledge of the internal level structure of the ions nor the narrow band radiation sources matched to the level spacing.

14.2.3 *The Paul trap*

The Penning trap implies the use of a relatively large magnetic field, which induces large Zeeman shifts of the energy levels of the trapped ions. This can constitute a drawback for some applications.

In the absence of magnetic field, the Earnshaw theorem, which states the impossibility of trapping a charged particle by a purely electrostatic potential, can be circumvented by using a rapidly oscillating electric field, as suggested by Wolfgang Paul. The basic idea[3] is to drive a micromotion of the particle with the oscillating field [Paul (1990)]. The farther the particle is from the central point, the larger is its micromotion amplitude. As explained below, the kinetic energy of the micromotion plays the role of the potential energy for the slow motion of the particle.

Using the same geometry as the Penning trap, but with the charge of the conductors oscillating in time with a frequency $\Omega/2\pi$, a particle of charge q and mass m in the trap experiences the following time-dependent potential energy for its external degrees of freedom:

$$U\left(x, y, z, t\right) = qV(x, y, z)\cos\Omega t = \frac{1}{2}m\omega^2\left[z^2 - \frac{x^2+y^2}{2}\right]\cos\Omega t, \qquad (14.4)$$

where the parameter ω is imposed by the applied voltage, the geometry and the position of the electrodes. According to Newton's law, the equation of motion reads:

[3]Wolfgang Paul and Hans Dehmelt won the physics Nobel prize in 1989 for the development of charged particle trap techniques [Paul (1990); Dehmelt (1990)].

$$\begin{pmatrix} \ddot{x} \\ \ddot{y} \\ \ddot{z} \end{pmatrix} = -\frac{1}{m}\vec{\nabla}U = \frac{\omega^2}{2}\begin{pmatrix} x \\ y \\ -2z \end{pmatrix} \cos \Omega t. \tag{14.5}$$

This set of linear differential equations, which belong to the category of Mathieu equations, gives rise to stability domains that can be used for trapping particles [Campbell (1955); Major *et al.* (2005)]. These domains depend on the ratio ω/Ω but not on the initial parameters of the ion motion. It is worth pointing out that the stability domain depends on the charge to mass ratio of the trapped particle. Such configurations have been used extensively to design very accurate mass spectrometers.

It is instructive, assuming $\Omega \gg \omega$, to work out an approximate solution of the equations of motion (14.5). In this limit, the motion of the particle can be decomposed into a fast oscillation at frequency Ω (the micro-motion) superimposed with a motion that has a slower characteristic frequency [Landau and Lifshitz (1982)]:

$$\vec{r}(t) = \vec{r}_{\text{fast}}(t) + \vec{r}_{\text{slow}}(t). \tag{14.6}$$

The equations of motion for the slow variables are obtained by averaging the micromotion in time over the period of the fast motion $2\pi/\Omega$. One finds that the slow motion is derived from the three-dimensional confining effective potential:

$$U_{\text{eff}}\,(x,y,z) = \frac{m\omega^4}{16\Omega^2}\left[x^2_{\text{slow}} + y^2_{\text{slow}} + 4z^2_{\text{slow}}\right]. \tag{14.7}$$

This is a notable result, since confinement is now ensured. The energy stored in the micromotion is of the same order as that stored in the slow motion which is readily shown by calculating the average kinetic energy of the micromotion over the period of the fast motion: $\langle m\, v^2_{\text{fast}}/2\rangle = U_{\text{eff}}$. The effective potential is confining for both positive and negative charged particles since it does not depend on the sign of the charge. The consistency of the above treatment is ensured by choosing a field amplitude such that the oscillation frequency of the particle in the potential U_{eff} is indeed much smaller than the fast frequency Ω: $\omega_{\text{slow}} \simeq \omega^2/\Omega \ll \omega \ll \Omega$.

Once trapped, the ion can remain in the trap for several hours. However, if the micromotion is interrupted by collisions with sufficiently heavy particles, an energy transfer from the fast motion to the slow motion causes a heating of the particle. This effect is referred to as radiofrequency heating [Major and Dehmelt (1968)].

It is possible to build very compact traps with high frequencies: The trap used in [Monroe *et al.* (1995)] uses a ring electrode of radius 170 μm, a sinusoïdal voltage amplitude of $V_0 = 600$ V and a frequency of oscillation $\Omega/2\pi = 230$ MHz. The "slow" frequencies are on the order of $\omega/2\pi \simeq 30$ MHz.

The trap geometry is flexible. For instance, by using four bars to generate a potential of the form:

$$U(x,y,z,t) = \frac{1}{2}m\omega^2(x^2 - y^2)\cos\Omega t, \tag{14.8}$$

one can build a linear Paul trap. The string of ions of Fig. 14.4(a) has been produced in such a trap.

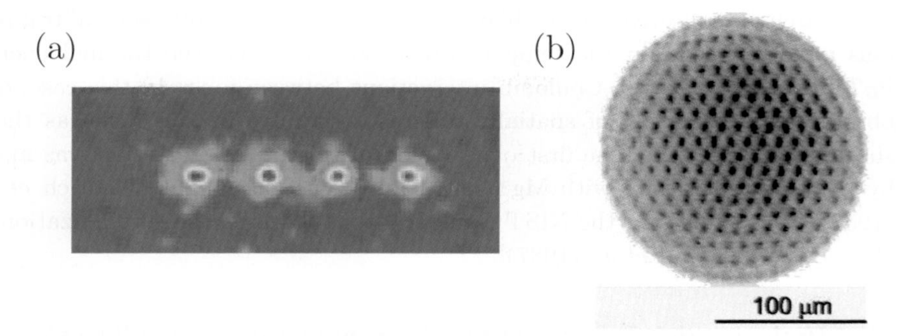

Fig. 14.4 Coulomb crystals: (a) four atoms trapped in a linear Paul trap (courtesy of R. Blatt) (b) Three-dimensional long-range ordered structures of about 2300 $^{40}Ca^+$ ions confined in a linear Paul trap (view along the trap axis). Figure adapted from [Mortensen *et al.* (2006)]. Copyright: American Physical Society.

14.2.4 *Cooling of the trapped ions*

Ions are usually produced in the trap by electron bombardment of neutral atoms, and as a result, their temperature is high. They remain trapped, however, since the trap depth is large. For example, with a typical voltage $V_0 \simeq 600$ V applied between the electrodes, the trap depth is on the order of $qV_0/k_B = 10^6$ K!

Several methods can be used to reduce the temperature of the ions once they are trapped, which is essential for applications. An efficient primary cooling technique is to use thermal contact with a residual vapor (at a pressure on the order of $\sim 10^{-4}$ mbar) of helium at room temperature (25 meV). The helium gas serves as a reservoir of low energy atoms with which the trapped ions can thermalize. This is known as the buffer gas cooling technique.

Efficient laser cooling techniques can then be used to reduce the temperature of the ions to much lower values [Stenholm (1986); Wineland and Itano (1987)]. Laser cooling with broad lines ($\hbar\Gamma \gg E_{\text{rec}}$) has been described in Sec. 12.2.3 of Chap. 12 and results in temperatures in the range of 0.5 mK. It is even possible, as discussed in Sec. 13.4 of Chap. 13, to bring the ions in the lowest vibrational state of the trap by using laser cooling with narrow lines ($\hbar\Gamma \ll E_{\text{rec}}$) [Wineland and Itano (1979)].

When laser sources at the wavelength needed for cooling an ion A^+ are not available, it can be mixed with another one B^+ that can be easily laser cooled. This scheme, which uses thermal contact between A^+ and B^+ to cool A^+ is called *sympathetic cooling*.

Ions confined to a small region of space at very low temperatures enable several interesting applications that can be classified into two categories:

- High precision measurements are performed with a very small number of ions, possibly a single ion, in order to minimize the perturbations due to ion-ion interactions. These applications, which can be extended to unstable radioactive elements [Bollen *et al.* (1992)], are briefly reviewed in the next subsection.

- The search for collective effects, like crystallization, that involve several trapped ions and appear when the temperature is so small that the thermal energy $k_B T$ is smaller that the Coulomb interactions between ions. In this case, one observes the formation of spatially ordered structures of ions, such as those shown in Fig. 14.4.b. The first observation of this phase transition was made by the Garching group with Mg^+ ions cooled with laser light [Diedrich *et al.* (1987)]. The same year the NIST Boulder group observed the crystallization of Hg^+ ions [Wineland *et al.* (1987)].

14.2.5 *High precision measurements performed with ultracold trapped ions*

A first category of experiments consists of measuring atomic parameters like atomic masses or gyromagnetic ratios. Mass spectrometry using the comparison between cyclotron frequencies has been already mentioned above. Measuring the ratio between the cyclotron frequency of a charged particle and the Larmor frequency that characterizes the precession of its magnetic moment provides a measurement of the gyromagnetic ratio of the particle.[4] In this way, the group of Hans Dehmelt measured the g factor of the electron with a remarkable precision by making measurements on a single electron stored for several months [Gabrielse *et al.* (1985); Van Dyck *et al.* (1986)]. More recently, the team led by Gerald Gabrielse at Harvard University realized a measurement of this quantity with an unprecedent accuracy through a careful and refined analysis of the dynamics of a single electron in a Penning trap [Hanneke *et al.* (2008)]. This later experiment gave the following value for the electron g factor:

$$g/2 = 1.00115965218073(28)[0.28 \text{ ppt}]. \tag{14.9}$$

High precision measurements with trapped ions also provide stringent tests of fundamental theories. For example, the small difference between the measured value of the g factor of the electron and the value predicted by Dirac equation, equal to 2, is a radiative correction due to quantum electrodynamics effects (QED). The comparison between the most recent experimental value given in (14.9) and refined QED calculations[5] provides a test of QED at the 10^{-12} level of accuracy [Aoyama *et al.* (2008)]. Note that this new measurement of g combined with QED theory determines the fine structure constant with an unprecedented accuracy:

$$\alpha^{-1} = 1.137035999084(51)[0.37 \text{ ppb}]. \tag{14.10}$$

By comparing the Penning trap cyclotron frequencies for a single electron and a single positron [Van Dyck *et al.* (1987)], or for a single proton and a single antipro-

[4] See Chap. 8.

[5] To match the precision of the measurement, the theory must include radiative corrections up to the fourth order in α of QED perturbation theory as well as hadronic and weak interaction contributions.

ton, tests of the symmetry between matter and anti-matter can be performed with a very high precision [Dehmelt (1981)].

Laser cooled trapped ions can also be used as frequency standards for atomic clocks. Microwave clocks using hyperfine transitions of a single $^{199}\mathrm{Hg}^+$ ion [Berkeland *et al.* (1998)] or a single $^{9}\mathrm{Be}^+$ ion [Bollinger *et al.* (1985)] have been achieved [Prestage *et al.* (2001)]. Optical clocks using optical transitions of laser cooled trapped ions are also very promising [Oskay *et al.* (2006); Rosenband *et al.* (2008)].[6]

Let us finally mention the interest in trapped ions for quantum information. The coupling of two trapped ions to the vibration modes in the trap can be used to put these ions in an entangled state, as first suggested by Ignacio Cirac and Peter Zoller [Cirac and Zoller (1995)].[7] This scheme is currently applied and pursued by several groups in the world for realizing quantum gates, quantum bits, ... [Leibfried *et al.* (2003)].

14.3 Magnetic traps

For neutral atoms, trapping mechanisms rely on the coupling of a permanent or induced dipole moment to electromagnetic field gradients. In this section, we give a brief overview of magnetic trapping for polarized atoms.

14.3.1 *Introduction*

The action of magnetic field gradients on individual atoms was first investigated by the pioneering experimental work of Otto Stern and Walther Gerlach in 1921. In their experiment, a beam of silver atoms passed through a region with a large magnetic gradient that resulted in the splitting of the beam into two components. A part of the beam was deflected towards the region of large magnetic field, while the other one was deflected in the opposite direction.

This experiment demonstrated the quantization of angular momentum. The energy levels of an atom with angular momentum J in a magnetic field B are discrete and given by

$$E(m_J) = g\mu_B m_J B, \tag{14.11}$$

where g is the gyromagnetic factor, μ_B the Bohr magneton, m_J the quantum number along the field axis, and B the modulus of the magnetic field. For instance, silver atoms have a spin $J = 1/2$, and the projection of the magnetic moment along the axis of the magnetic field can take only two values, either parallel or anti-parallel to $\vec{B}$. Atoms with $\vec{\mu}$ parallel to $\vec{B}$ (energy $-\mu B$) are attracted by the high field region, and atoms with $\vec{\mu}$ anti-parallel to $\vec{B}$ are deflected towards the low field region. Equation (14.11) is valid as long as the magnetic moment $\vec{\mu}$ adjusts adiabatically

[6] See Sec. 18.6 of Chap. 18.
[7] See Sec. 19.4.2 of Chap. 19.

to the local direction of the magnetic field $\vec{B}$ during the displacement of the atom in the trap. In this case, the atomic motion is determined by the gradient of the modulus of the magnetic field. Note that, in free space, only local minima of the magnetic field modulus are achievable, and therefore only so-called "low field seekers" can be trapped by static magnetic fields. This result, known as the Wing's theorem, is dictated by Maxwell's equations [Wing (1984)]. In the following, the shaping of this minimum is addressed through the presentation of different trap geometries that provide a minimum of $|\vec{B}|$.

The magnetic force exerted on an atom by a magnetic field gradient of ~ 10 T/m, is hundreds of times stronger than gravity. Magnetic confinement was first demonstrated by Wolfgang Paul and his group in the late seventies using neutrons, despite their small magnetic moment [Paul (1990)]. In almost all experiments carried out with neutral atoms, pre-cooling is performed with a magneto-optical trap (see Sec. 14.6) in order to efficiently load the atoms into the magnetic trap.

14.3.2 *Quadrupole trap and Majorana losses*

The simplest magnetic trap is composed of two circular coils with currents flowing in opposite senses ("anti-Helmholtz" like configuration) as illustrated in Fig. 14.5. William Phillips and his team observed the first magnetically trapped atomic gas in 1985 with this magnetic configuration [Migdall *et al.* (1985)]. The convenient aspect of this trap is its straightforward implementation. This is also the reason that this simple geometry is used in many experiments for the transport of magnetically trapped cold atoms over large distances (few tens of centimeters) into ultra-high-vacuum chambers with large optical access. This can be accomplished by either using a set of coils with time-varying currents [Greiner *et al.* (2001)] or by physically moving a pair of coils [Lewandowski *et al.* (2003); Nakagawa *et al.* (2005)].

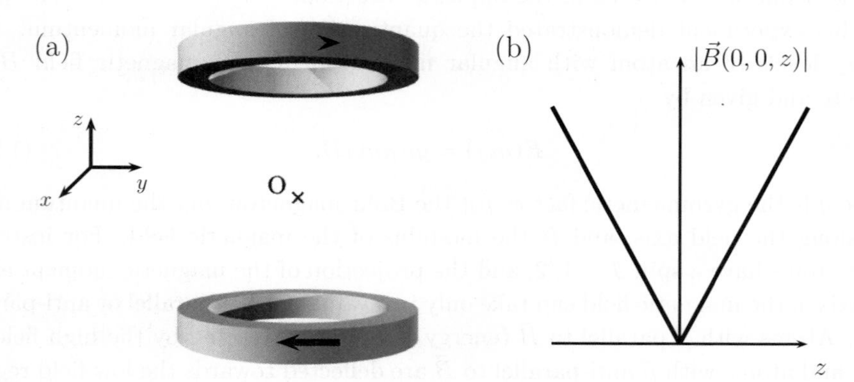

Fig. 14.5 (a) Two coils configuration to generate a three-dimensional quadrupole trap. (b) Modulus of the magnetic field along the symmetry axis. The magnetic field vanishes at O and increases linearly with the distance z from this point.

For a three-dimensional quadrupole configuration, the field vanishes only at a point O. The expansion of the magnetic field around this point for distances sufficiently small compared to the radius of the coils, takes the following form (see Fig. 14.5) :

$$\vec{B}(x,y,z) = \begin{pmatrix} b'x \\ b'y \\ -2b'z \end{pmatrix}, \tag{14.12}$$

assuming that the coils have the same symmetry axis z. This local expansion of the magnetic field components fulfills the Maxwell's relation for static fields $\vec{\nabla} \cdot \vec{B} = 0$ and $\vec{\nabla} \times \vec{B} = \vec{0}$, and according to Eq. (14.12), the modulus of the magnetic field increases linearly with the distance from O in its neighborhood: $\left|\vec{B}\right| = b'\sqrt{x^2 + y^2 + 4z^2}$.

An atom moving near the center may experience rapid variations of the direction of the field. In the case that these variations are on a timescale shorter than ω_L^{-1} where $\omega_L = \mu B/\hbar$ is the Larmor frequency, the magnetic moment of the atom cannot adiabatically follow the magnetic field variations, and the direction of the magnetic moment with respect to the local direction of the magnetic field can be reversed, thereby transferring the atom to a non-trapping state. As a result, atoms are expelled from the trap. This spin flip mechanism is commonly referred to as "Majorana losses" [Majorana (1932)]. Such loss increases with the atomic density and is enhanced for low mass atoms [Petrich *et al.* (1995)].

The addition of a static magnetic field is not sufficient to counteract this loss mechanism, as this will just displace the position of the zero. However, more elaborate magnetic configurations using static magnetic fields in which the modulus of the magnetic field no longer vanishes (Ioffe-Pritchard trap), or time-dependent magnetic fields (TOP traps) can be envisioned. These two important magnetic traps are detailed in the next subsections. Alternatively, one can superimpose a blue detuned optical "plug" on the center of the trap to prevent atoms from approaching the center O of the quadrupole configuration. This method was used in the group of Wolfgang Ketterle at MIT for the first realization of Bose-Einstein condensation of sodium atoms [Davis *et al.* (1995a)].

14.3.3 *Ioffe-Pritchard trap*

The Ioffe configuration, used initially for plasma confinement [Gott *et al.* (1962)], was proposed for neutral atoms by David Pritchard [Pritchard (1983)] and first demonstrated a few years later [Bagnato *et al.* (1987a)]. It is now one of the most commonly used magnetic traps for neutral atoms.

This magnetostatic trap provides confinement around a non-vanishing magnetic field $\vec{B}_0$ to inhibit Majorana losses and can be generated by superimposing the magnetic field ($\vec{B}_1$) of two pinch coils with current running in the same direction to the one ($\vec{B}_2$) of four bars in two-dimensional quadrupole configuration (see Fig. 14.6).

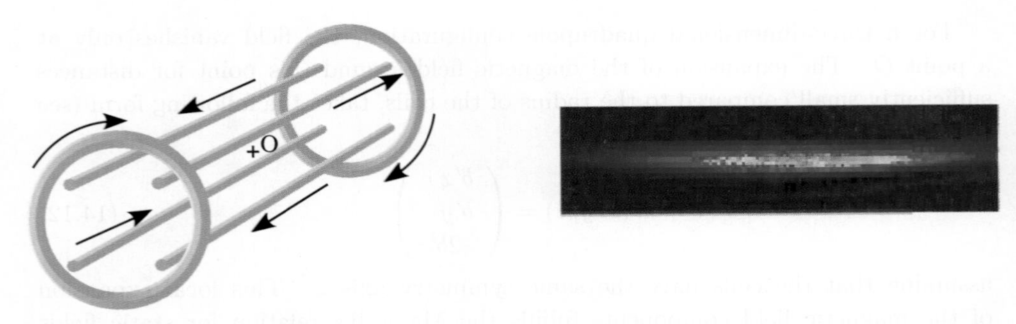

Fig. 14.6 Left: Ioffe-Pritchard configuration consist of four bars with opposite currents in neighbouring bars to provide the transverse confinement, and pinch coils to provide axial confinement. Right: absorption image of a cloud of cold atoms trapped in a Ioffe-Pritchard trap.

Near the center O of the trap, the field can be expanded as

$$\vec{B}(\vec{r}) = \vec{B}_1(\vec{r}) + \vec{B}_2(\vec{r}) = B_0\vec{e}_z + \frac{b''}{4}\begin{pmatrix} -2xz \\ -2yz \\ 2z^2 - x^2 - y^2 \end{pmatrix} + b'\begin{pmatrix} x \\ -y \\ 0 \end{pmatrix}. \quad (14.13)$$

One immediately checks that $\vec{B}(\vec{r})$ obeys the Maxwell equations for static fields: $\vec{\nabla}.\vec{B} = 0$ and $\vec{\nabla} \times \vec{B} = \vec{0}$.

Typically, the bias field B_0 ranges from 0.1 to 1 mT, the gradient b' from 1 to 10 $\mathrm{T.m^{-1}}$ and the curvature b'' from 10 to 100 $\mathrm{T{\cdot}m^{-2}}$. This ensures that the Larmor frequency that characterizes the evolution of the magnetic moment $\vec{\mu}$ remains larger than 1 MHz. Since the oscillation frequencies of the atom in the trap are usually much smaller (a few hundreds Hz), the magnetic moment $\vec{\mu}$ adjusts adiabatically to the local direction of the magnetic field $\vec{B}$ during the displacement of the atom in the trap. As a consequence, an atom initially prepared in a low field seeking state will remain in this state. At low temperatures ($k_BT \ll \mu B_0$), the potential $U(\vec{r})$ experienced by the atoms is that of a three-dimensional anisotropic harmonic oscillator:

$$U(\vec{r}) = \mu|\vec{B}(\vec{r})| \simeq \mu B_0 + \frac{1}{2}m\omega_\perp^2(x^2 + y^2) + \frac{1}{2}m\omega_z^2 z^2, \quad (14.14)$$

where $\omega_\perp^2 = \mu(b'^2/2B_0 - b''/4)/m$ and $\omega_z^2 = \mu b''/2m$ are respectively the transverse and longitudinal angular frequencies. A confinement for the transverse degrees of freedom is obtained only if the parameters B_0, b' and b'' fulfil the inequality $b'^2 \geq b''B_0/2$.

Many implementations of this magnetic configuration are feasible and described in the following references: the Cloverleaf trap [Mewes *et al.* (1996)], the Baseball configuration [Myatt *et al.* (1997)], the three identical coil configuration [Söding *et al.* (1998)], the QUIC trap [Esslinger *et al.* (1998)], the Four-Dee magnetic bottle [Hau *et al.* (1998)] ... The Ioffe-Pritchard configuration has also been realized with permanent magnets [Tollett *et al.* (1995)] and iron-core electromagnets [Desruelle *et al.* (1998)].

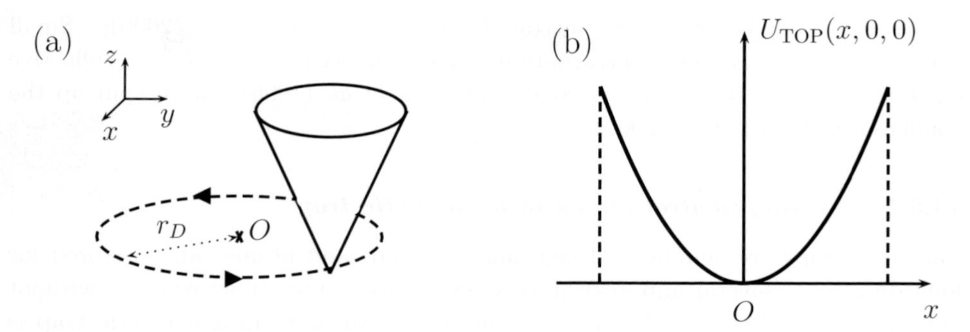

Fig. 14.7 TOP trap configuration. (a) The magnetic field generated by a three-dimensional quadrupole configuration is rotated on a circle of radius r_D sufficiently fast with respect to the trapping oscillation period. (b) This yields an effective harmonic potential for the atoms, with a finite depth given by the potential energy at r_D.

14.3.4 *Time-averaged orbiting potential (TOP)*

To avoid the losses due to the zero field of the three-dimensional quadrupole, the group led by Eric Cornell in Boulder added a rotating magnetic field in the plane perpendicular to the symmetry axis of the quadrupole. This generates a time dependent magnetic field of the form [Petrich *et al.* (1995)]:

$$\vec{B}(\vec{r},t) = \begin{pmatrix} b'x \\ b'y \\ -2b'z \end{pmatrix} + \begin{pmatrix} B_0\cos(\Omega t) \\ B_0\sin(\Omega t) \\ 0 \end{pmatrix}. \tag{14.15}$$

This configuration permits (i) the rotation of the zero of the quadrupole magnetic field on a circle of radius $r_D = B_0/b'$, and (ii) by a proper choice of the rotation frequency Ω, the generation of an essentially harmonic confinement with a characteristic angular frequency $\omega_{\rm trap}$, for the atoms inside the volume delimited by the circle. The frequency Ω is chosen to be high compared to the typical orbiting frequencies $\omega_{\rm trap}$ of atoms, and low compared to the Larmor frequency ω_L:

$$\omega_{\rm trap}\ (\sim 100\ \text{Hz}\ ) \ll \Omega\ (\sim 5\ \text{kHz}\ ) \ll \omega_L\ (\geq 1\ \text{MHz})\,. \tag{14.16}$$

Given this hierarchy of frequencies, atoms essentially experience the mean value of the instantaneous magnetic field averaged over a period $2\pi/\Omega$. They thus evolve in the effective potential

$$U_{\rm TOP}(\vec{r}) = \frac{2\pi}{\Omega}\int_0^{2\pi/\Omega} \mu|\vec{B}(\vec{r},t)|\mathrm{d}t \simeq \mu B_0 + \frac{\mu b^2}{4B_0}(x^2+y^2+8z^2). \tag{14.17}$$

Therefore, the confining potential is harmonic. Due to Majorana flops for atoms crossing the circle of radius r_D (referred to as "the circle of death"), the depth of this potential is given by $U_{\rm TOP}(r_D) = \mu B_0/4$.

From Eq. (14.17), we deduce that the aspect ratio at thermal equilibrium is $e = (\langle x^2\rangle/\langle z^2\rangle)^{1/2} = 2\sqrt{2}$. However, by adding other time dependent magnetic fields it is possible to change the geometry. For instance, an axial modulation provides for an independent control of the trap frequencies [Hodby *et al.* (2000)]. Small time-dependent transverse perturbations have also been used to excite collective excitations of the Bose-Einstein condensate [Jin *et al.* (1996)], or to spin up the condensate [Haljan *et al.* (2001)].

14.3.5 *Loading neutral atoms in a magnetic trap*

Magnetic traps are simple to design and to build, and atoms can be stored for long durations (several minutes) at very low temperatures (microkelvin), without any appreciable heating. In fact, the lifetime of an atom in a magnetic trap is essentially determined by the quality of the vacuum in the chamber containing the trap, since given the relatively low trap depth, a collision with a molecule from the background gas can eject the atom from the trap.

In order to be transferred into a magnetic trap, atoms have to be polarized in the weak-field-seeker state. In addition, for Bose-Einstein condensation experiments, special care may be taken to "mode-match" the trapping parameters in order to maintain, as constant as possible, the phase space density of the cloud when atoms are transferred in the magnetic trap. To facilitate evaporative cooling,[8] the cloud can then be adiabatically compressed in order to increase the elastic collision rate.

We finally emphasize that many other magnetostatic configurations for trapping and guiding atoms are achievable [Bergeman *et al.* (1987)]. However, there is a fundamental restriction in the use of magnetic traps, due to the fact that the trapping mechanism is obtained only for weak field seeking internal states. In dipole traps, realized with far-detuned light and presented in the next section, this restriction is removed, and the dependence of the strength of the trapping upon the internal state can be adjusted, a prospect which offers many interesting possibilities.

14.4 Electric dipole traps

14.4.1 *Induced dipole moment*

14.4.1.1 *Static induced dipole moment*

When an atom is placed in a static electric field $\vec{E}$, it acquires a dipole moment $\vec{D} = \alpha_0 \vec{E}$, where α_0 is the static polarizability. For simplicity, we shall assume that α_0 is a scalar, although it may also have a tensorial part.

The potential energy of an atom in this field is $W = -\alpha_0 E^2/2$ and the corresponding force is $\vec{F} = -\vec{\nabla} W = \alpha_0 \vec{\nabla}(E^2)/2$. For an atom in its ground state g, α_0 is positive and the atom is attracted to the regions where the electric field is the largest. The potential energy W is nothing but the Stark shift of the relevant atomic internal state induced by the electric field. Here we assume a linear response of the atom with respect to the field, such that the shift is readily calcu-

[8] See Sec. 21.3.3 of Chap. 21.

lated through second order perturbation theory. Note that this assumption, which consists of neglecting saturation effects, is valid as long as the applied electric field is small compared to the inner field of the atom created by the nucleus.

14.4.1.2 *Dynamic induced dipole moment*

This description can be generalized to the case of a time-dependent electric field. Consider a field oscillating with the angular frequency ω_L. The static polarizability must then be replaced by the dynamic polarizability $\alpha(\omega_L)$.

As already discussed in Chap. 11, dipole confinement occurs when the induced atomic dipole is placed in a spatially inhomogeneous light intensity region. The use of the dipole force turns out to be the only way to provide a conservative trapping potential for atoms with light, since their permanent dipole moment is null, as a consequence of the reflection symmetry of the physical interactions.[9]

When ω_L is much smaller than the relevant atomic Bohr frequencies $\omega_{\rm A}$, one recovers the static limit for the polarization: $\alpha(\omega_L) \simeq \alpha_0$. This is the case in many experiments in which ground state alkali atoms are manipulated (Bohr frequencies $\omega_{\rm A} \sim 3 \times 10^{15}$ s^{-1}) with very far detuned laser light, such as the radiation from a CO_2 laser ($\omega_L \sim 2 \times 10^{14}$ s^{-1}).

14.4.1.3 *Near resonance induced dipole moment*

A very important practical case is that of an atom in its internal ground state that is irradiated with a laser wave of frequency ω_L close to, ω_A, the atomic Bohr frequency corresponding to the transition $g \leftrightarrow e$. The force exerted by light on an atom is the sum of a reactive force proportional to the amplitude gradient of the laser electric field and a dissipative force proportional to the phase gradient.[10] To achieve a nearly conservative trap based on the reactive force of light, one needs to minimize spontaneous emission processes. This is realized by using a large detuning compared to the linewidth of the atomic transition $|\delta| = |\omega_L - \omega_A| \gg \Gamma$.

The trapping potential can be designed to be independent of the particular magnetic sublevel by using sufficiently far-detuned light for the dipole trap. Conversely, one can make the trapping potential dependent on sublevels, in a controlled manner, through an appropriate choice of both the polarization and light detuning. This possibility was exploited for the coherent transport of atoms.[11]

The expression for the dipole potential is obtained through the dressed atom formalism in which the energy levels of the { atom+laser } system are considered. One finds two types of dressed states that connect either to the ground state or the excited state when the laser intensity tends to zero. The spontaneous emission processes cause random jumps between the two types of dressed states, with an instantaneous force oscillating back and forth between the two opposite values

[9] For a review see [Grimm *et al.* (2000)].
[10] See Chap. 11.
[11] See Sec. 7.5.6 of Chap. 7.

associated with the two dressed states. The dressed atom picture provides a simple interpretation of the mean value $\vec{\mathcal{F}}_{\text{react}} = -\vec{\nabla} U$ and of the fluctuations of the dipole force.[12]

The potential energy for the reactive or dipole force is given by

$$U = \frac{\hbar\left(\omega_L - \omega_A\right)}{2} \text{Log}\left[1 + \frac{\Omega^2/2}{\left(\omega_L - \omega_A\right)^2 + \left(\Gamma^2/4\right)}\right] \simeq \frac{\Omega^2}{4(\omega_L - \omega_A)}, \qquad (14.18)$$

where Ω is the Rabi frequency. The last equality holds if $|\omega_L - \omega_A| \gg \Omega\,,\,\Gamma$. In this regime, the rate of spontaneous emission $\gamma_g = \Gamma\Omega^2/4\left(\omega_L - \omega_A\right)^2$ (see Eq. (7.6) of Chap. 7) fulfills the inequality $|U/\hbar| \gg \gamma_g$.

The sign of the light shift, and hence the direction of the dipole force, depends on the sign of the detuning from resonance $\delta = \omega_L - \omega_A$. When δ is negative (red detuning), the result is qualitatively the same as for a static field: the atom is attracted to the region of large laser intensities. U is a potential well in which atoms can be trapped (position dependent light shifts). On the contrary, when δ is positive (blue detuning), the force exerted on the atoms tends to push them away from high intensity regions.

14.4.2 *Application of dipole forces to trapping*

14.4.2.1 *Historical perspective*

Historically, Gurgen Ashotovich Askar'yan was the first to point out the possible use of the ponderomotive force for plasmas and atoms [Askar'yan (1962)]. One-dimensional trapping using standing waves was proposed for neutral atoms by Vladilen Letokhov [Letokhov (1968)]. A few years later, Arthur Ashkin addressed the use of the dipole force to trap neutral atoms in three dimensions within the two-level atom approximation [Ashkin (1978)]. The first evidence for dipole force with neutral atoms was achieved the same year on an atomic beam [Bjorkholm *et al.* (1978)], and the optical trapping of atoms was observed in the same group a few years later [Chu *et al.* (1986)].

This section recalls the basics of dipole traps for neutral atoms and gives some of their applications. As explained through many examples, a large variety of dipole trap geometries can be realized.

14.4.2.2 *The simplest optical trap*

The first observation of optically trapped atoms was made using a single, strongly focused, Gaussian laser beam tuned about 10^4 natural linewidths below resonance. Atoms were cooled below 10^{-3} K in an "optical molasses"[13] before being captured by the dipole-force optical trap. In this pioneering experiment, about 500 sodium atoms were optically trapped [Chu *et al.* (1986)].

[12] See Sec. 11.4.3 of Chap. 11.
[13] See Sec. 12.3.6 of Chap. 12.

Conceptually, the single Gaussian beam is the simplest configuration. Let us denote the axial direction of the beam z and the radial coordinate r. If the thermal energy of the trapped cloud is small compared to the depth, atoms experience a harmonic confining potential:

$$U(r,z) \simeq U_0 \left[-1 + \frac{2r^2}{w_0^2} + \frac{z^2}{z_r^2} \right] = -U_0 + \frac{1}{2} m\omega_\perp^2 r^2 + \frac{1}{2} m\omega_z^2 z^2, \tag{14.19}$$

where w_0 is the beam waist of the Gaussian beam, $z_r = \pi w_0^2/\lambda$ its Rayleigh length, and λ the wavelength of the dipole laser. The dipole trap generated by a single focussed beam is therefore very anisotropic since $\omega_z/\omega_\perp \sim \lambda/w_0 \ll 1$.

Multi-level alkali atoms

For an alkali atom, the depth U_0 of the dipole potential takes the following simple form when the detuning is large compared to the hyperfine splitting [Grimm *et al.* (2000); Kuppens *et al.* (2000)]:

$$\begin{aligned} U_0 = \frac{\pi c^2 \Gamma I_0}{2\omega_A^3} \Bigg[&\left(\frac{1}{\delta_{D_1}^{(-)}} + \frac{1}{\delta_{D_1}^{(+)}} + \frac{2}{\delta_{D_2}^{(-)}} + \frac{2}{\delta_{D_2}^{(+)}} \right) \\ &- g_F m_F \sqrt{1-\epsilon^2} \left(\frac{1}{\delta_{D_1}^{(-)}} + \frac{1}{\delta_{D_1}^{(+)}} - \frac{1}{\delta_{D_2}^{(-)}} - \frac{1}{\delta_{D_2}^{(+)}} \right) \Bigg], \end{aligned} \tag{14.20}$$

where Γ denotes the natural linewidth, I_0 is the peak intensity, $g_F = [F(F+1) + S(S+1) - I(I+1)]/[F(F+1)]$ the Landé factor and $\delta_{D_i}^{(\pm)} = \omega_L \pm \omega_{D_i}$, where D_i refers to the D_1 and D_2 lines of the alkali atom. The ellipticity ϵ is defined by writing the polarization vector of the light in the form

$$\vec{\epsilon} = \frac{1}{\sqrt{2}} (\vec{e}_x \sqrt{1+\epsilon} + i \vec{e}_y \sqrt{1-\epsilon}). \tag{14.21}$$

The term that depends upon the ellipticity ϵ of the light describes the so-called optical Zeeman splitting [Cohen-Tannoudji and Dupont-Roc (1972)].

14.4.2.3 *Transport of atoms with moving laser traps*

The transport of cold atoms or even Bose-Einstein condensates can be accomplished with a high degree of control and over large distances by trapping the atomic cloud in the focus of a red-detuned laser and translating its location [Gustavon *et al.* (2001); Couvert *et al.* (2008a)]. The dipole trap used for such a transport is also called "optical tweezers". In this way, it is possible to realize a continuous source of Bose-Einstein condensate by periodically replenishing a condensate held in an optical dipole trap with new condensates delivered by optical tweezers [Chikkatur *et al.* (2002)]. It is worth noting that in contrast to magnetic transport (see Sec. 14.3.2), atoms can be transported with optical tweezers regardless of their Zeeman substates.

Optical manipulation of microspheres and biological particles systems and tweezers: a widely used tool

The focus of a laser with a frequency that is low compared to the resonance frequencies of the constituent molecules of an object can be used to trap the object. For instance, in 1986 the group of Arthur Ashkin reported the optical trapping of a micron size sphere [Ashkin *et al.* (1986)]. These dielectric particles are far more polarizable than atoms, and therefore are simpler to trap. Ashkin proposed these experiments as a proof of principle that atoms could be optically trapped.

Soon after, the same group observed the laser trapping of bacteria and viruses [Ashkin and Dziedzic (1987)], and even of live e-coli bacteria, for hours without damage, despite the large laser intensity used for the trapping, on the order of few MW/cm^2 [Ashkin *et al.* (1987)]. Optical tweezers also enable the microscopic manipulation of biological molecules. For instance, by attaching a particle that can be optically trapped to a molecule of interest, it is possible to stretch molecules such as DNA and study their elastic response [Smith *et al.* (1996)], or their relaxation towards equilibrium [Perkins *et al.* (1994b)]. Optical tweezers were also used to investigate polymer dynamics [Perkins *et al.* (1994a)] and have provided direct evidence for several key assumptions in the reptation model developed by Pierre Gilles de Gennes, Sam Edwards, and Masao Doi.

For particles in the Mie size regime, where the diameter of the particle is large compared to the light wavelength, the momentum transfer from the incident light to the particle can be interpreted in terms of the refraction of the beam when the light rays cross the dielectric object [Ashkin *et al.* (1986); Foot (2005)].

14.4.2.4 *Crossed dipole traps*

Two beams of orthogonal polarizations with approximately the same waist and power can be crossed to create tight confinement in all directions, as first experimentally demonstrated in [Adams *et al.* (1995)]. Alternatively, a very tight axial confinement is readily obtained by retro-reflecting the laser beam, thereby yielding a standing wave.

Radio-frequency and microwave traps

The use of spatially inhomogenous dressed states to trap atoms has been successfully applied to the radio-frequency or microwave domain [Agosta and Silvera (1989); Spreeuw *et al.* (1994)]. In order to give insights into such traps, let us consider an atom with an angular momentum $J = 1$ in the presence of an inhomogeneous magnetic field. The potential energy curves for the different Zeeman levels around a minimum of the magnetic field modulus are represented in the left part of Fig. 14.8. In the presence of the radiofrequency or microwave field, the uncoupled states $\{|+1, N+2\rangle, |0, N+1\rangle, |-1, N\rangle\}$ have energy curves that cross where the radio-frequency or microwave field resonantly connects the Zeeman levels (dashed curves in the right part of Fig. 14.8). As a result of the atom-field interaction, the coupled states exhibited anti-crossings so that one of the three dressed states gives rise to a trap with a finite depth, whereas another one gives rise to two traps located at the crossing position of the uncoupled states. The latter traps enable the confinement of atoms on a non-planar surface. The use of adiabatic potentials for trapping atoms in new geometries was proposed in [Zobay and Garraway (2001)], and first realized with thermal atoms [Colombe *et al.* (2004)] and then with ultracold or condensed atoms [Morizot *et al.*

(2006); Hofferberth *et al.* (2006)]. This kind of trap turns out to be very flexible since one can shape the trap geometry by using different magnetic fields and radio-frequency field configurations.

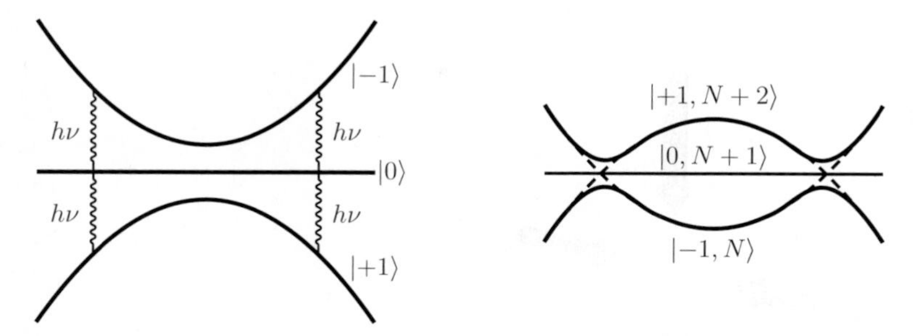

Fig. 14.8 Potential energy curves for different Zeeman levels associated with an angular momentum $J = 1$. Left: the level $|m = -1\rangle$ exhibits a minimum at the location of the minimum of the magnetic field modulus. Conversely the level $|m = +1\rangle$ has a maximum. The former can be used as a trap while the latter always expels atoms. Right: by superimposing a microwave or radio-frequency, two dressed { atom + field } states can be used to trap atoms: one where the magnetic field modulus is minimum, and the other at the position for which the radio-frequency or microwave connects resonantly the Zeeman levels. This latter kind of trap enables one to confine atoms in non-planar geometries defined by $\mu||B(\vec{r})|| = h\nu$.

14.4.3 *Optical lattices*

Optical lattices, already briefly presented in Sec. 7.5.4 of Chap. 7, were initially used for manipulating ultracold atoms in the early nineties by the group of Gilbert Grynberg at ENS (Paris) [Verkerk *et al.* (1992)] and by the group of William Phillips at NIST (Gaithersburg) [Jessen *et al.* (1992)]. They are formed from the periodic modulation of the light intensity in a laser standing wave. The space-dependent light shift acts as a potential for the atomic external degree of freedom. Depending on the sign of the detuning δ, the atoms accumulate at the nodes (for blue-detuned light) or the antinodes (for red-detuned light) of the standing wave.

An analogy is quite often made with a crystal. However, there are at least three important differences with a true crystal that one should keep in mind:

(i) The spatial order does not result from interactions between atoms, but rather from an external potential created by light.
(ii) The orders of magnitude are completely different for the spatial period: Angstrom for the crystal, and micrometer for the optical lattice. Atoms are thus very far apart and do not interact very much.
(iii) Optical lattices are very flexible since their characteristics can be varied by changing the parameters of the laser standing wave.

Optical lattices allow for many kinds of cold atom manipulations. They have led

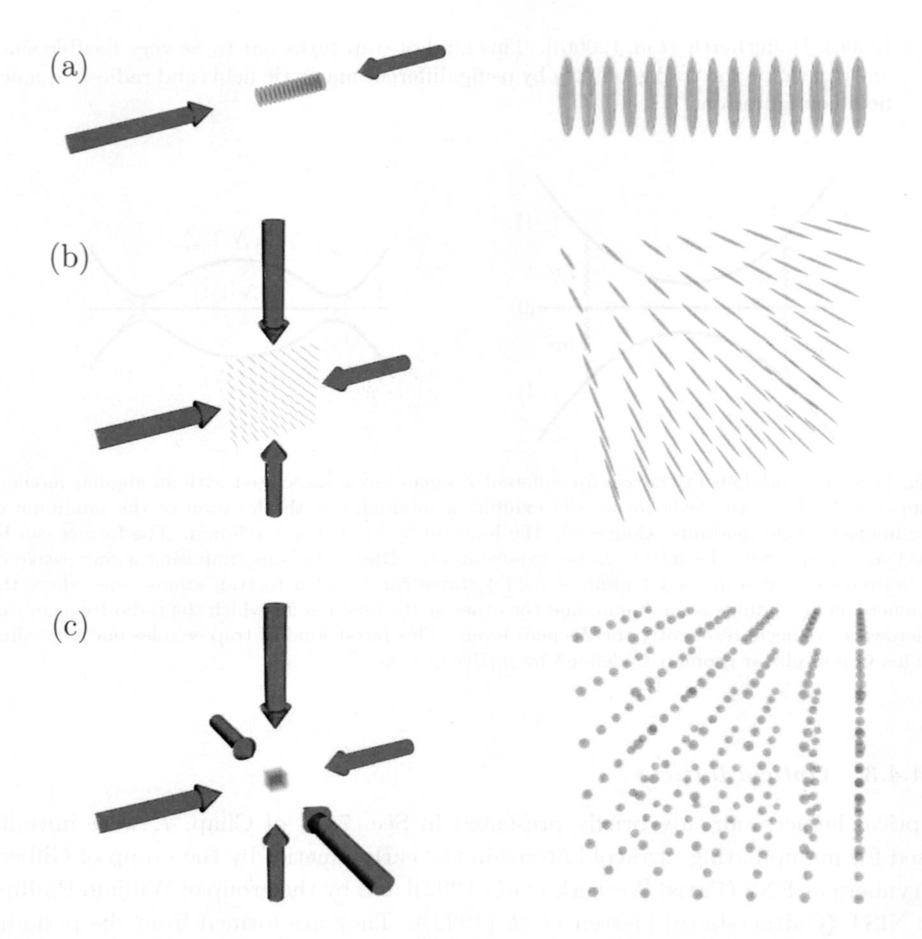

Fig. 14.9 Using the interference pattern of (a) one standing wave, (b) two perpendicular standing waves, or (c) three mutually orthogonal standing waves, it is possible to generate arrays of traps with a large variety of geometries: (a) pancake, (b) cigar shape or (c) spherical.

to several spectacular developments, both from theoretical and experimental points of view. In the following, we summarize some of their main applications:

- *Dimensionality.* The simplest optical lattice is realized by reflecting the laser beam back onto itself. The potential experienced by the atoms varies along the laser beam direction. This configuration is used to produce arrays of two-dimensional traps (see Fig. 14.9(a)). A configuration made of two standing waves along two orthogonal directions yields an array of elongated tubes of light. This configuration with red-detuned light is used to realize arrays of one-dimensional traps (see Fig. 14.9(b)). With three mutually orthogonal standing wave, one has a three dimensional simple cubic array of three-dimensional traps (see Fig. 14.9(c)).

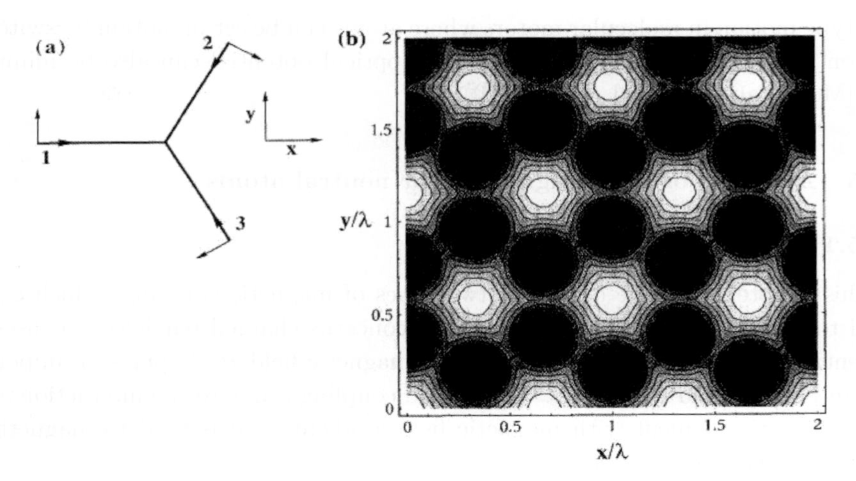

Fig. 14.10 (a) Three coplanar beams of equal intensity with wave vectors at 120 degrees angle from each other. (b) The optical potential wells have their minima on a hexagonal lattice. Figure extracted from [Grynberg *et al.* (1993)]. Copyright: American Physical Society.

- *Geometry.* One interesting feature of optical lattices is that a wide range of possibilities for the laser beam directions and polarization is available. As a first example, the distance between two wells can be controlled by changing the angle between the two counter-propagating beams. More elaborate geometries such as a hexagonal lattice as illustrated in Fig. 14.10 [Grynberg *et al.* (1993)] or a face-centered cubic lattice can be easily generated.
- *Moving molasses.* This application has already been discussed in Sec. 7.5.4 of Chap. 7. It consists of introducing a slight frequency difference $\omega \pm \delta\omega$ between the two beams used to generate the standing wave. As a result, the interference pattern moves at the velocity $\bar{v} = \delta\omega/k$. This technique enables a packet of atoms to be set in motion while remaining cool, and is used, for instance, in cold atomic clocks [Riis *et al.* (1990); Clairon *et al.* (1991)].
- *Variable potential depths.* By changing the intensity of the standing wave, the tunnelling between adjacent wells can be controlled. This offers the possibility of investigating the transition from a regime where atoms are localized in potential wells to a regime where they are delocalized over the whole lattice. This possibility has been recently employed to investigate the superfluid Mott-insulator transition for bosons [Greiner *et al.* (2002)] and fermions [Jördens *et al.* (2008); Schneider *et al.* (2008)].[14]
- *Time-dependent optical lattices* constitute test beds for the study of the quantum counterpart of classically chaotic systems such as the delta-kicked rotor, where dynamical localization has been observed [Casati *et al.* (1979); Moore *et al.* (1994); Ringot *et al.* (2000)]. Other complex systems, such as the ratchet

[14]See Sec. 26.3 of Chap. 26.

type motion in molecular motors where atoms can be set in motion by switching on and off a periodic and asymmetric optical potential can also be mimicked [Mennerat-Robilliard *et al.* (1999)].

14.5 Artificial orbital magnetism for neutral atoms

14.5.1 *Introduction*

In this chapter, we have described two types of magnetic couplings which can be used to trap particles. The first one only concerns charged particles and uses the Lorentz force $q\vec{v} \times \vec{B}$ in the presence of a magnetic field $\vec{B}$. It plays an important role in Penning and Paul traps. The second coupling is due to the interaction of the spin magnetic moment with magnetic field gradients and is used to magnetically trap polarized neutral atoms.

The Landau energy levels of an electron in a uniform magnetic field originate from the first kind of coupling between charged particle and a magnetic field. This energy level spectrum plays an essential role in several effects of condensed matter physics such as quantum Hall effects. We will see in Chap. 26 that ultracold quantum gases of neutral atoms can be used as simple systems simulating condensed matter problems with strong interactions. It is therefore interesting to see whether artificial Lorentz forces can be engineered for neutral atoms. This would make it possible to simulate in a controlled manner quantum Hall type effects with neutral quantum gases.

In this section, we describe two methods to achieve this goal: the rotation of quantum gases and the use of geometric potentials induced by external fields.

14.5.2 *Rotating a harmonically trapped quantum gas*

There is a close analogy between the Hamiltonian experienced by a charged particle in the presence of a magnetic field and the Hamiltonian H' of a neutral atom in a rotating frame. Consider a neutral atom of mass m confined in a harmonic potential that is static in a rotating frame associated with the constant rotation vector $\vec{\Omega} = \Omega\hat{z}$. In this frame, the Hamiltonian which governs its evolution reads

$$H' = H - \Omega L_z = \frac{(\vec{p} - m\vec{\Omega} \times \vec{r})^2}{2m} + \frac{1}{2}m(\omega_\perp^2 - \Omega^2)(x^2 + y^2) + \frac{1}{2}m\omega_\parallel^2 z^2, \quad (14.22)$$

where $\omega_\perp$ ($\omega_\parallel$) is the trap angular frequency in the transverse (longitudinal) xy (z) plane (direction). The kinetic energy term of Eq. (14.22) is analogous to the one of a particle of charge q and mass m that experiences a vector potential $\vec{A} = m\vec{\Omega} \times \vec{r}/q$ which is with a uniform magnetic field $\vec{B} = 2m\vec{\Omega}/q$. The radial confining potential is reduced by the centrifugal potential ($\omega_\perp^2 \longrightarrow (\omega_\perp^2 - \Omega^2)$). The Coriolis force experienced by the neutral atom in the rotating frame appears as equivalent to the Lorentz force on a charged particle. As Ω approaches the trap frequency $\omega_\perp$, the centrifugal potential balances the transverse trapping potential, the system flattens out and becomes effectively two-dimensional. The one-body energy spectrum of (14.22) becomes thus reminiscent of the Landau levels of a charged particle in a uniform magnetic field. It is organized in a succession of Landau levels separated by an energy tending to $2\hbar\omega_\perp$ when $\Omega \to \omega_\perp$. Each Landau level is made of energy states with a splitting equal to $(\omega_\perp - \Omega)$. In the limit $\Omega = \omega_\perp$, the one-body Hamiltonian ground state becomes macroscopically degenerate. Beyond the one-body problem, the determination of the ground state of a rapidly rotating interacting gas is closely related to that which appears in fractional quantum Hall effect. The combined effect of interactions and quantum statistics eventually determines the features of the many-body ground state. Such rotating systems therefore allows one to study artificial orbital magnetism in quantum gases.[15]

14.5.3 *Artificial gauge potential from adiabatic evolution*

Alternatively, one can realize artificial gauge potentials acting on neutral atoms using, for instance, atom-laser interactions.[16] Such potentials are produced in the laboratory frame and do not impose any constraint on the symmetry properties of the initial Hamiltonian. In such schemes, one uses the internal structure of the atoms and the position dependent coupling between the external field and atoms. If the internal atomic state is initially prepared in a given dressed state $|\chi_1(\vec{r})\rangle$ of energy $E_1(\vec{r})$ of the atom-laser system and if the atom moves slowly enough for the adiabatic theorem to hold, the atom remains in state $|\chi_1(\vec{r})\rangle$. Writing the state of the system as $\sum_i \psi_i(\vec{r},t)|\chi_i(\vec{r})\rangle$, one can show that the wave function $\psi_1(\vec{r},t)$ associated with this dressed state experiences the energy $E_1(\vec{r})$ which plays the role of an potential energy for the center of mass motion, and two other contributions originating from the Berry's phase accumulated by the moving atom: a vector potential $\vec{A}(\vec{r})$ giving rise to an effective magnetic field $\vec{B}(\vec{r}) = \vec{\nabla} \times \vec{A}(\vec{r})$ and a scalar potential $W(\vec{r})$.

Let us consider in the following the simple example of a two-level atom interacting with a laser field. Using the rotating wave approximation, the corresponding dressed atom Hamiltonian in a given two-dimensional manifold is time-independent and reads

$$H = \frac{\vec{p}}{2m} + V(\vec{r}), \tag{14.23}$$

where we have introduced the dressed states at a given position $\vec{r}$, $|\chi_1(\vec{r})\rangle$ and $|\chi_2(\vec{r})\rangle$, and the potential $V(\vec{r}) = E_1(\vec{r})|\chi_1(\vec{r})\rangle\langle\chi_1(\vec{r})| + E_2(\vec{r})|\chi_2(\vec{r})\rangle\langle\chi_2(\vec{r})|$. The expansion of the spinor wave function on this basis reads here $|\Psi(\vec{r},t)\rangle = \psi_1(\vec{r},t)|\chi_1(\vec{r})\rangle + \psi_2(\vec{r},t)|\chi_2(\vec{r})\rangle$. If the atom is initially prepared in the dressed

[15]See [Bretin *et al.* (2004); Schweikhard *et al.* (2004); Gemelke *et al.* (2010); Cooper (2008); Fetter (2009)] and references therein.
[16]See [Dalibard *et al.* (2010)] and references therein.

state $|\chi_1(\vec{r})\rangle$ and moves sufficiently slowly, the wave function $\psi_2(\vec{r},t)$ remains negligible at all time, and the wave function $\psi_1(\vec{r},t)$ evolves according to the effective Hamiltonian

$$H = \frac{[\vec{p} - \vec{A}(\vec{r})]^2}{2m} + E_1(\vec{r}) + W(\vec{r}), \tag{14.24}$$

where $\vec{A}(\vec{r}) = i\hbar\langle\chi_1|\vec{\nabla}\chi_1\rangle$ and $W(\vec{r}) = \hbar^2|\langle\chi_2|\vec{\nabla}\chi_1\rangle|^2/2m$. The Berry phase is nothing but the circulation of the vector potential on a closed contour $\mathcal{C}$:

$$\gamma(\mathcal{C}) = \frac{1}{\hbar}\oint_{\mathcal{C}} \vec{A}\mathrm{d}\vec{\ell}.$$

Remark

The dressed state $|\chi_1(\vec{r})\rangle$ can always be expressed in terms of the two bare internal atomic states g_1 and g_2

$$|\chi_1(\vec{r})\rangle = \cos\left(\frac{\beta(\vec{r})}{2}\right)|g_1\rangle + e^{i\alpha(\vec{r})}\sin\left(\frac{\beta(\vec{r})}{2}\right)|g_2\rangle. \tag{14.25}$$

where $\alpha(\vec{r})$ and $\beta(\vec{r})$ are the two angles which define the orientation of the Bloch vector associated with the dressed states. The vector potential can be calculated explicitly

$$\vec{A}(\vec{r}) = \frac{\hbar}{2}(\cos(\beta(\vec{r})) - 1)\vec{\nabla}\alpha, \tag{14.26}$$

and depends only on $\beta(\vec{r})$ and on the gradient of the phase $\alpha(\vec{r})$. Note that the potential vector vanishes in a standing wave.

The interpretation of the geometric scalar and vector potentials goes as follows [Cheneau *et al.* (2008)]. The mean value of the force operator $\langle\vec{F}\rangle = -\langle\vec{\nabla}V\rangle$ in the adiabatic state $|\chi_1(\vec{r})\rangle$ is equal to $-\vec{\nabla}E_1(\vec{r})$ which accounts for the mean dipole force [Dalibard and Cohen-Tannoudji (1985b)]. However, we do not obtain an explicit signature of the scalar potential W in this expression. This originates from the fact that $|\chi_1(\vec{r})\rangle$ is an eigenstate of the potential $V(\vec{r})$ but not of the force operator $\vec{F} = -\vec{\nabla}V$. As a result $\langle\vec{F}^2\rangle$ is not equal to $\langle\vec{F}\rangle^2$ which means that the force acting on the atom has quantum fluctuations about its average value. These fluctuations of the radiative force cause a micromotion whose kinetic energy plays the role of a ponderomotive potential for the slow motion. This potential is nothing but the scalar potential $W(\vec{r})$.[17] The origin of the Lorentz force is due to the motion-induced contamination of the state $|\chi_1(\vec{r})\rangle$ by the other state $|\chi_2(\vec{r})\rangle$ when the atom is set in motion. This is reminiscent of the semiclassical calculation of the velocity-dependent radiative forces.[18] The Lorentz force cannot, however, lead to any cooling, since it is perpendicular to $\vec{v}$ and does not performed any mechanical work.

Such artificial gauge potentials have been implemented in an experiment performed at NIST (Gaithersburg) to produce an optically synthesized magnetic field.

[17] This effect is analogous to that studied in Sec. 14.2.3 for the Paul trap.

[18] See Chap. 12.

The observed signature of that field was the appearance of vortices in a Bose-Einstein condensate [Lin *et al.* (2009)]. The same team has generated a synthetic electric field through the time dependence of an effective vector potential [Lin *et al.* (2011)].

Choosing a suitable atomic level structure, a wide variety of Abelian or non-Abelian gauge fields can be generated by an appropriate engineering of external fields in both bulk or discrete systems where atoms are trapped in an optical lattice. This novel tool opens new possibilities for producing and studying strongly interacting many-body states.

14.6 Magneto-optical trap (MOT)

In the dipole traps for neutral atoms presented so far, it is necessary to take special care to avoid spontaneous emission in order to produce a truly conservative trap.

The volume of these traps, however, is rather small since the laser light has to be tightly focused in order to reach the high intensities required to have large depths for the traps.

Another possibility for achieving traps with a larger volume and a large depth is to use the radiation pressure force. In the previous chapter, we already have explained how the large deceleration provided by this force can be used to decelerate atoms in short distances (Zeeman slower). The velocity dependence of the force associated with the Doppler effect is also exploited in optical molasses to realize a friction force. However, optical molasses do not constitute a trap but rather just a viscous medium. In order to produce a trapping force, it is necessary to have a position dependent force.

This is realized in the so-called magneto-optical trap, where an inhomogeneous magnetic field ($B_z = b'z$) modifies the force through the Zeeman effect. The principle of the trap takes advantage of both the linear and angular momenta carried by the photons. For simplicity, its principle is presented for a one-dimensional configuration, as first suggested by Jean Dalibard in 1986, and we assume the angular momenta of the ground state g and the excited state e involved in the trapping to be respectively $J_g = 0$ and $J_e = 1$ (Fig. 14.11(a)). The two counter-propagating waves are red-detuned by the same amount ($\delta = \omega_L - \omega_A$) and have opposite circular polarizations. They are therefore in resonance with the atom at different places. At the center of the trap O, the magnetic field is zero: by symmetry the two radiation pressure forces at this location have the same magnitude and opposite directions, so that an atom at O feels no net force. For an atom placed at the left of O, the laser wave coming from the left, which is σ_+-polarized, is closer to resonance with the allowed transition $g \leftrightarrow e, m = +1$ compared to the one experienced by an atom at O. The radiation pressure created by this wave is therefore increased with respect to its value at O. Conversely, the radiation pressure force created by the wave coming from the right is decreased with respect to its value at O. Indeed the wave is σ_-

polarized and it is further from resonance with the transition $g \leftrightarrow e, m = -1$ than it is at O. Therefore, the net force for an atom to the left of O is pointing towards O. For an atom located to the right of O, the reverse phenomenon occurs: the radiation pressure force created by the wave coming from the right now dominates, such that the resulting force also points towards O. One therefore achieves a stable trapping around O.

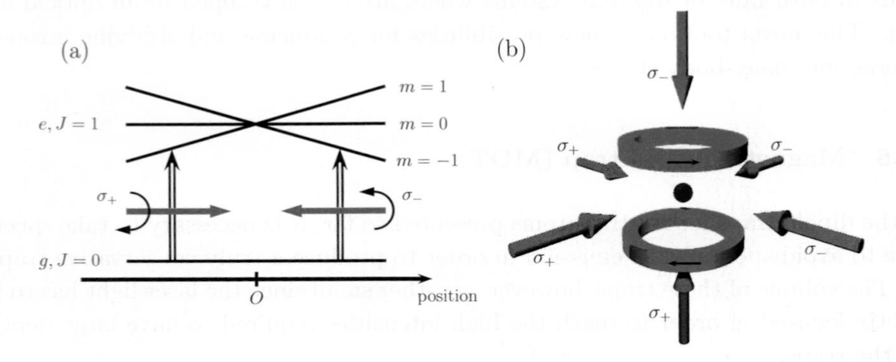

Fig. 14.11 (a) Principle of the magneto-optical trap (MOT) using an optical transition from a ground state with zero angular momentum to an excited state with an angular momentum $J = 1$. The two counter-propagating laser waves, are σ_+ and σ_- polarized, have the same intensity and the same frequency. The balance between the two radiation pressure forces is broken by the presence of a magnetic field. A gradient of magnetic field creates a situation where atoms feel a restoring force towards the zero magnetic field position. (b) Regular three-dimensional configuration of the magneto-optical trap. The gradient is provided by a pair of coils placed in anti-Helmholtz like configuration.

In one dimension, the explicit calculation of the restoring force is carried out in the same manner as in Chap. 12 for the frictional force in a one-dimensional molasses configuration, taking into account the Zeeman shift. An atom placed in a red-detuned weak standing wave in the presence of a magnetic gradient thus experiences a force that can be expanded around O in the form:

$$\mathcal{F}_z \simeq -\alpha\, v_z - \kappa z \quad \text{with} \quad \kappa = 2k_L \mu b' s \frac{-\delta\,\Gamma}{\delta^2 + (\Gamma^2/4)}, \tag{14.27}$$

where μ is the magnetic moment of the excited state and s the saturation parameter. In addition to the friction term, a restoring force arises from the Zeeman shift contribution. The non-zero value of the red-detuning also provides cooling of the trapped atoms, along the lines already discussed in Chap. 12.

Such a scheme can be extended to three dimensions. The magnetic gradient is usually made with a pair of coils in anti-Helmholtz configuration (see Fig. 14.11(b)). The aspect ratio of the trap is governed by the magnetic gradient strength along the different space directions. For instance, it is possible to generate an elongated trap along the z-axis by imposing a two-dimensional magnetic quadrupole field $\vec{B}(b'x, -b'y, 0)$.

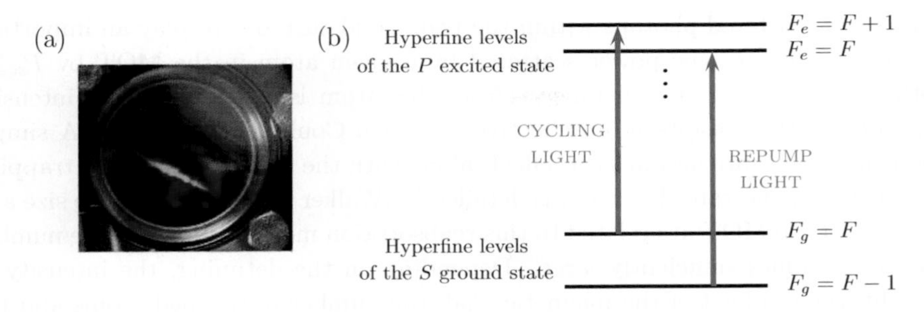

Fig. 14.12 (a) Picture of an elongated magneto-optical trap. The geometry is dictated by the gradient of the magnetic field. In this case, a two-dimensional quadrupole field was used. (b) Level scheme of alkali atoms. A repumping beam is required to cycle back atoms that are pumped into the lower hyperfine level through absorption-spontaneous emission cycles.

In practice, the level structure of atoms is usually not that of a two-level atom. The generalization of the principle presented above to multi-level atoms requires some care. Indeed, the Zeeman splitting of the ground and excited states plays an important role. For alkali atoms, the cycling transition corresponds to the transition $F_g = F \to F_e = F + 1$ (see Fig. 14.12(b)). However, the cycling light can also populate the $F_e = F$ state. As a result of such an off resonance excitation, atoms can be transferred into the lower hyperfine state. In this case, they are no longer coupled to the cycling light. To counteract this process, a repumping beam that connects the lower hyperfine level to one of the Zeeman states of the excited state is usually required.

The first demonstration of the magneto-optical trap was made in the context of a collaboration between the groups of MIT and Bell Labs [Raab *et al.* (1987)]. The remarkable feature of this trap is its robustness. In addition, it has a relatively large velocity capture range and it can be used for trapping atoms in a cell filled with a low pressure vapor, as shown by the JILA Boulder group [Monroe *et al.* (1990)]. The typical capture velocity is on the order of 20 $\mathrm{m.s^{-1}}$ for rubidium atoms, the capture volume depends on the size of the laser beam.

It is worth calculating the expected size of the MOT from the expression (14.27) of the restoring force. Assuming that the cloud is at thermal equilibrium, the quadratic size is given by the equipartition theorem[19] $\langle z^2 \rangle \simeq k_B T/\kappa$. Using the expression (14.27) with the optimal detuning for Doppler cooling ($\delta = -\Gamma/2$), a saturation parameter equal to 1/4, and the typical values for the magnetic moment taken equal to one Bohr magneton and a magnetic gradient of $b' = 0.1$ T/m, we find a size of few tens of micrometers.

Actually, in most experiments, the observed size of the MOT is on the order of a few millimeters to a few centimeters in the case of large laser beams. Indeed, when the number of atoms in the trap is sufficiently large (typically above 10^4), reab-

[19]Strictly speaking, one cannot apply the equipartition theorem since a MOT is not an isolated system in thermal equilibrium. However, it is useful to work out a simple estimate.

sorptions of scattered photons within the trap cloud turn out to play an important role. If one denotes the power scattered by a given atom in the MOT by P_{scatt}, another atom placed at a distance r from this atom is subjected to an intensity $P_{\text{scatt}}/(4\pi r^2)$ that results in an effective repulsive Coulomb type force. A simple model that takes into account this effect, along with the attenuation of the trapping beam as they penetrate the cloud, is detailed in [Walker *et al.* (1990)]. The size and the shape of the MOT are related to this reabsorption mechanism when the number of atoms becomes sufficiently large. Depending on the detuning, the intensity of the light, the gradient of the magnetic field, the number of trapped atoms and the alignment of the beams, a large variety of collective effects have been observed in MOT clouds. These inlude flat density profiles, orbital modes [Walker *et al.* (1990)], and self-sustained oscillations [Labeyrie *et al.* (2006)], ...

The nonlinear variation of the restoring force with position, the reabsorption of scattered photons, and the occurrence of light assisted collisions [Sesko *et al.* (1989)] ultimately limit the typical maximum achievable atomic density to the range of a few 10^{10} atoms/cm^3 for alkali atoms, which in turn limits the density in phase space [Townsend *et al.* (1995)]. Despite elegant methods developed to counteract these limitations, such as the dark MOT technique [Ketterle *et al.* (1993)], it turns out to be impossible within the MOT to directly reach the quantum degenerate regime, for which the mean inter-particle distance is on the order of the thermal de Broglie wavelength. However, the MOT provides a large number of atoms that can be loaded into a non-dissipative trap.

The temperature reached in a MOT ranges from few μK to 1 mK depending on the atomic species and the MOT parameters. Let us emphasize that the MOT has become the workhorse of cold atom physics.

14.7 Conclusion

In conclusion, one can say that spectacular advances have been realized in our ability to trap atoms. Several different trapping mechanisms have been demonstrated, for ions as well as for neutral atoms that enable the removal of any contact between the trapped particles and thermal walls, which is essential for keeping laser cooled atoms at low temperatures. The dream of keeping a single electron, a single ion, or a single atom confined to a small region of space for a long time in order to make precise measurements has become a reality. Trapping a particle and moving it with optical tweezers can be considered as a real manipulation of particles. Trapping is usually preceded by cooling, which is essential for neutral atoms because traps for neutral atoms are shallow. Cooling increases the density of an atomic cloud in velocity space, whereas trapping increases the density in position space. Combining cooling and trapping leads to higher phase space densities, which are essential for reaching the threshold for Bose-Einstein condensation. The possibility of realizing traps, or array of traps such as optical lattices, with different geometries, and different dimensionalities, is also very interesting. It provides simple systems that can mimic more complex situations, usually found in condensed matter. In Sec. 14.5, we have described another example of connection between atomic and solid state physics provided by the possibility of simulating quantum Hall type effects with ultracold atoms interacting with artificial magnetic fields.

Once the atoms are trapped, atom-atom interactions begin to play an important role in the dynamics of the gas. In the next two chapters, we address the problem of two-body collisions between ultracold atoms. First, we show that at very low temperatures, the effect of collisions is described by a single parameter, the scattering length. We also show that it is possible to control the scattering length with the so-called Feshbach resonances. By varying a static magnetic field, one can achieve repulsive or attractive interactions between atoms, or even suppress these interactions and obtain an ideal gas. One can considerably increase the strength of the interactions and reach a regime of strongly correlated atoms. Finally, it now appears that one can gain a full control of all the degrees of freedom of an atom: its spin orientation with optical pumping, its velocity with laser cooling, its position with trapping, its interactions with its neighbours with Feshbach resonances. This explains the recent spectacular advances of atomic physics.

PART 5

Ultracold interactions and their control

Introduction

The laser cooling and trapping techniques described in the previous part allow one to prepare gaseous samples of atoms at very low temperatures. Understanding how these atoms interact is essential for several reasons. First, elastic collisions restore thermal equilibrium and play a key role in evaporative cooling techniques described in Chap. 21 which are essential for reaching the temperatures below which Bose-Einstein condensation can be observed in an ultracold bosonic gas. Second, it was realized during the last two decades that atom-atom interactions can be fully controlled with the use of the so-called *Feshbach resonances.* These interactions can be made attractive, repulsive, very large or very small just by adjusting a static magnetic field. Feshbach resonances are very interesting from the perspective of using ultracold quantum gases as simple model systems for a better understanding of more complex situations involving strongly interacting particles. The purpose of Part 5 is to introduce the basic concepts that allow a quantum description of ultracold interactions and of their control.

We begin in Chap. 15 by describing how elastic collisions at very low temperatures can be quantum mechanically described by a single parameter in most cases, the *scattering length* a. We will show later on, in Chap. 22, that most physical properties of Bose-Einstein condensates in the mean-field approximation depend only on this single parameter. It is also important to understand why the sign of the scattering length determines the attractive or repulsive character of atom-atom interactions. A simple argument is presented for understanding this point.

Another important problem addressed in Chap. 15 concerns the model potentials very often used for describing ultracold collisions. Using the exact interaction potential generally leads to overly difficult calculations. In dilute systems, only the asymptotic part of the scattering wave function is relevant for describing the behavior of the gas and this asymptotic part only depends on the scattering length a. Calculations can therefore be greatly simplified by replacing the exact potential by model potentials with a simpler mathematical form, provided that they lead to the same value for the scattering length. Furthermore, these model potentials are chosen in such a way that they lead to the good value of a in the Born approximation, a property which is important for the mean field description of gaseous condensates presented in Chap. 22.

The properties of forward scattering are also briefly described. Using a Gaussian beam model, it is possible to give a simple description of the interference between the incident atomic wave and the wave scattered in the forward direction that is responsible for the absorption and the phase shift of the incident beam. The appearance of a mean-field energy for an atom crossing a slab of scatterers is also a first introduction to the idea of mean field energy of an atom in a condensate, a topic which will be studied in more detail in Chap. 22.

Chapter 16 is entirely devoted to Feshbach resonances. These resonances belong to the general class of resonances resulting from the coupling of a discrete state to a continuum. In the problem considered here, we have two colliding atoms in a certain spin configuration that, during the collision, pass through another resonant intermediate state in which they are in a molecular bound state with a different spin configuration. We develop the analogy between this situation and the resonant scattering of a photon by an atom, which allows us to derive the general form of the resonant scattering amplitude. This result is then confirmed by a more quantitative treatment using a two-channel collision model. The Feshbach resonance is easily tuned by scanning a static magnetic field since the two relevant states, the incident collision state (belonging to a continuum) and the intermediate molecular bound state, correspond to different spin configurations with different magnetic moments. Their respective positions thus change when the magnetic field is varied. The result of the calculation is that in the vicinity of a Feshbach resonance, the strength, and even the sign of the two-body scattering length changes. Indeed, when the magnetic field is scanned through the resonance, a goes from $+\infty$ to $-\infty$ or vice versa.

Special attention is also given to the bound states of the system. In the two-channel model, one finds that there is a bound state in the region $a > 0$ whose binding energy tends to zero near the resonant value of the magnetic field. In this region, the energy and the wave function of the bound state are universal: they depend only on the scattering length a, and not on the details of the interaction potential. By sweeping the magnetic field through the resonance from region $a < 0$ to $a > 0$, one can transform a pair of ultracold atoms into an ultracold molecule.

In the region $a < 0$, there is no bound state of two atoms, but bound states of three bosonic atoms, with universal properties, are predicted to appear. In Chap. 16, we briefly describe experiments that used ultracold atoms and Feshbach resonances to demonstrate for the first time the existence of these trimers, called *Efimov states*. This shows that Feshbach resonances are not only useful for controlling ultracold interactions and achieving strongly interacting systems that require descriptions that go beyond the mean field approximation. They also allow the investigation of exotic states, such as Efimov trimers or tetramers, that appear in the study of few body quantum systems.

Chapter 15

Two-body interactions at low temperatures

15.1 Introduction

The purpose of this chapter is to give a brief review of the concepts that are needed for a quantitative description of interactions in dilute atomic gases. In these systems, two-body interactions dominate because of the assumption of diluteness, and can be described in terms of collisions through quantum scattering theory (Sec. 15.2). In this chapter, we only focus on elastic collision processes.

In a trapped atomic sample, elastic collisions restore thermal equilibrium. They are also responsible for the success of evaporative cooling, the only technique that, up to now, has enabled the achievement of Bose-Einstein condensation in dilute gases.[1] At very low temperature, the description of degenerate quantum gases becomes simple and depends only on a very small number of collision parameters. In particular, we will introduce the essential concept of scattering length a and provide a simple interpretation of its sign (Sec. 15.3).

In view of the application of scattering theory to quantum gases, we introduce two model potentials to take atomic interactions into account in a simple manner: the pseudo-potential (Sec. 15.4) and the delta potential truncated in momentum space (Sec. 15.5). These model potentials are designed to give the same scattering length as the exact potential while allowing simple perturbative calculations. In this way, we avoid, for instance, several difficulties associated with the use of the exact interaction potential.

In a scattering process, the interference between the incident wave and the wave scattered in the forward direction plays an important role for explaining the absorption and the phase shift of the outgoing wave in the forward direction. In Sec. 15.6, we use a Gaussian beam model for the incident wave that allows for a detailed analysis of this interference effect. As an application, we consider such a Gaussian beam crossing a slab of independant scatterers and we explain in this way its attenuation (dissipative effect) and its dephasing (reactive effect) which can be interpreted as due to an index of refraction associated with an effective mean field potential.

[1] Inelastic processes are also important since they ultimately limit the achievable atomic spatial densities.

15.2 Quantum scattering: a brief reminder

Let us consider two spinless atoms or two atoms in the same spin state, with the same mass m, interacting through a two-body interaction potential $V(\vec{r}_1 - \vec{r}_2)$. The Hamiltonian of this two-body system reads

$$H = \frac{p_1^2}{2m} + \frac{p_2^2}{2m} + V(\vec{r}_1 - \vec{r}_2). \tag{15.1}$$

A sketch of the atom-atom interaction potential $V(r)$ is given in Fig. 15.1. At small separations, the interaction is strongly repulsive because of the overlap of the electronic clouds of each atom, and weakly attractive at large distances because of van der Waals interactions. As an example, the typical depth of the singlet interaction potential of two rubidium-87 atoms is around a few thousands of Kelvin, in temperature units.

As in classical mechanics, the Hamiltonian (15.1) takes a simple form when using the center-of-mass and relative variables:

$$H = \frac{P_G^2}{2M} + \frac{p^2}{2\mu} + V(\vec{r}), \tag{15.2}$$

with $M = 2m$, $\mu = m/2$ and

$$\vec{R}_G = (\vec{r}_1 + \vec{r}_2)/2, \qquad \vec{r} = \vec{r}_1 - \vec{r}_2,$$
$$\vec{P}_G = \vec{p}_1 + \vec{p}_2, \qquad \vec{p} = (\vec{p}_1 - \vec{p}_2)/2.$$

From the expression (15.2), we deduce that the center of mass, described by $\vec{R}_G$ and $\vec{P}_G$, is decoupled from the relative motion, and evolves as a free particle of mass

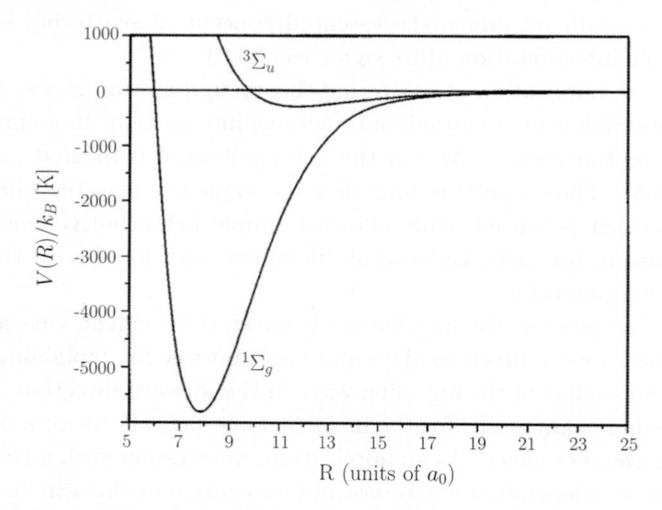

Fig. 15.1 Atom-atom interaction potential for two rubidium-87 atoms as a function of their relative distance r (in units of $a_0 \simeq 0.53$ Å): repulsive at short distances (electronic cloud repulsion) and attractive at large distances (van der Waals interactions). The singlet (triplet) potential accomodates 123 (39) bound states. Figure adapted from [Weiner *et al.* (1999)]. Copyright: American Physical Society.

$M = 2m$. The collision dynamics is thus entirely contained in the relative motion. It corresponds to the scattering, by the interaction potential $V(\vec{r})$, of a fictitious particle with position $\vec{r}$, momentum $\vec{p}$ and a *reduced mass* $\mu = m/2$. The two-body wave function of the scattering problem with (15.2) as a Hamiltonian can therefore be searched in the form

$$\Psi(\vec{r}_1; \vec{r}_2) = e^{i\vec{P}_G \cdot \vec{R}_G/\hbar} \psi(\vec{r}), \tag{15.3}$$

where $\psi(\vec{r})$ accounts for the relative motion. In the quantum treatment of scattering, the relative energy in the center of mass of the colliding particles belongs to the continuum of the reduced Hamiltonian for the relative motion, and the important quantity to characterize the interaction is the scattering amplitude presented in the following section.

15.2.1 *Scattering amplitude*

We will assume that the two-body potential V tends to zero when the relative distance between atoms goes to infinity. In the following, we also consider that the interaction potential has a finite range[2] b.

Determining the scattering properties requires the knowledge of the eigenstates of the Hamiltonian of the relative motion with a well-defined incident positive energy $E = \hbar^2 k^2/2\mu$. These are determined by solving the Schrödinger equation:

$$\left(-\frac{\hbar^2}{2\mu} \Delta + V(\vec{r}) \right) \psi(\vec{r}) = E\psi(\vec{r}) \Longrightarrow \left(\Delta + k^2 \right) \psi(\vec{r}) = \frac{2\mu}{\hbar^2} V(\vec{r}) \psi(\vec{r}). \tag{15.4}$$

This equation can be solved formally using the outgoing Green's function $G^+(\vec{r})$ solution of the equation $\left(\Delta + k^2 \right) G^+(\vec{r}) = \delta(\vec{r})$. The general solution of the linear equation (15.4) then takes the simple general form

$$\psi(\vec{r}) = \varphi_0(\vec{r}) + \frac{2\mu}{\hbar^2} \int \mathrm{d}^3\vec{r}' G^+(\vec{r} - \vec{r}') V(\vec{r}') \psi(\vec{r}'), \tag{15.5}$$

where $\varphi_0(\vec{r})$ is a solution of Eq. (15.4) without the right-hand side: $\left(\Delta + k^2 \right) \varphi_0(\vec{r}) = 0$. The integral equation (15.5), referred to as the Lippmann-Schwinger equation, provides an implicit determination of the scattering states.

Operatorial notation

Equation (15.4) can be written formally as

$$(E - \hat{T}) |\psi\rangle = \hat{V} |\psi\rangle, \tag{15.6}$$

where $\hat{T} = \hat{\vec{p}}^2/2\mu$ is the kinetic energy operator, and $\hat{V}$ the potential energy operator. The Lippmann-Schwinger equation (15.5) can then be recast in the simple form

$$|\psi\rangle = |\varphi_0\rangle + \hat{G}^+(E) \hat{U} |\psi\rangle, \tag{15.7}$$

[2] Several of our conclusions in the finite range case hold for power-law potentials, scaling as r^{-n}, provided that $n > 3$ [Landau and Lifshitz (1996)]. As an example, the van der Waals interactions between atoms, that decrease as C_6/r^6 for large r, yield a typical potential range given by $b = (2\mu C_6/\hbar^2)^{1/4}$ [Gribakin and Flambaum (1993)]. This range is the distance at which the van der Waals interaction equals the confinement energy in a volume b^3.

In this equation, $\hat{U} = (2\mu/\hbar^2)\hat{V}$, $\hat{G}^+(E) = 1/(E - \hat{T} + i\epsilon)$ denotes the outgoing Green's function operator (ϵ is an infinitely small positive number), and $|\varphi_0\rangle$ fulfills $\hat{T}|\varphi_0\rangle = E|\varphi_0\rangle$.

The precise expression for the Green's function $G^+(\vec{r})$ depends on the dimensionality and on the confinement or boundary conditions. In three dimensions[3] and in a box with periodic boundary conditions, the Green's function that corresponds to an outgoing wave, and thus to the scattered wave, reads:

$$G^+(\vec{r}) = -\frac{1}{4\pi}\frac{e^{ik|\vec{r}|}}{|\vec{r}|}. \tag{15.8}$$

Using Eqs. (15.5) and (15.8), the asymptotic form of the wave function for the relative motion is at large r ($|\vec{r}| \gg b$)

$$\psi_{\vec{k}_i}(\vec{r}) \sim e^{i\vec{k}_i\cdot\vec{r}} + f(k, \vec{\kappa}_i, \vec{n})\frac{e^{ikr}}{r}, \tag{15.9}$$

where we have introduced the incident plane wave solution $\varphi_0(\vec{r}) \sim e^{i\vec{k}_i\cdot\vec{r}}$ with a wave vector $\vec{k}_i$ and used the expansion $|\vec{r} - \vec{r}'| \simeq r - \vec{r}'\cdot\vec{n}$ with $\vec{n} = \vec{r}/r$. The direction of the incident plane wave is defined by the unit vector $\vec{\kappa}_i = \vec{k}_i/k$ and $f(k, \vec{\kappa}_i, \vec{n})$ is the *scattering amplitude* in the direction defined by the unit vector $\vec{n}$. At a given point $\vec{r}$ within the asymptotic limit, the scattering amplitude, f, does not depend on the distance r and is given by

$$f(k, \vec{\kappa}, \vec{n}) = -\frac{\mu}{2\pi\hbar^2}\int \mathrm{d}^3\vec{r}' e^{-i\vec{k}_f\cdot\vec{r}'} V(\vec{r}')\psi(\vec{r}'), \tag{15.10}$$

where we have introduced the wave vector $\vec{k}_f = k\vec{n}$ that corresponds to a given direction of the scattered spherical outgoing wave. The scattering amplitude depends only on the energy of the incident particle through k, the incident direction $\vec{\kappa}_i$, and the polar angles θ and φ of the final wave vector $\vec{k}_f$ with respect to $\vec{k}_i$. From the implicit expression (15.10) of the scattering state, one clearly understands that the value of the wave function out of the potential range is related to those of the same wave function inside the scattering region.

Note that if $kb \ll 1$, one can replace the exponential in the integral of (15.10) by 1. The scattering amplitude then no longer depends on the direction $\vec{n}$ of the final wave vector $\vec{k}_f$ and is spherically symmetric even if $V(\vec{r})$ is not.

Born approximation

In the expression (15.10) for the scattering amplitude, the potential V appears explicitly. To lowest order in V, one can replace the wave function $\psi(\vec{r}')$ by the zeroth order solution $\exp(i\vec{k}_i\cdot\vec{r}')$ of the Schrödinger equation:

$$f(k, \vec{\kappa}_i, \vec{n}) \simeq -\frac{\mu}{2\pi\hbar^2}\int \mathrm{d}^3\vec{r}' e^{-i(\vec{k}_i - \vec{k}_f)\cdot\vec{r}'} V(\vec{r}'). \tag{15.11}$$

[3]The quantum scattering theory can also be investigated with a similar approach in lower dimensional spaces (see [Pricoupenko *et al.* (2003)] and references therein). Experimentally, the scattering properties in reduced dimensionality are studied using gases that are strongly confined along one or two dimensions.

This expression is known as the Born approximation and is valid for a weak interaction potential. The scattering amplitude appears, within this approximation, as the spatial Fourier transform of the potential.

15.2.2 *Scattering cross section*

For distinguishable particles, the expressions for the differential and total cross-sections are directly related to the modulus of the scattering amplitude [Cohen-Tannoudji *et al.* (1977)]:

$$\frac{\mathrm{d}\sigma}{\mathrm{d}\Omega} = |f(k,\vec{\kappa}_i,\vec{n})|^2, \qquad \text{and} \qquad \sigma(k,\vec{\kappa}_i) = \int |f(k,\vec{\kappa}_i,\vec{n})|^2 \mathrm{d}^2 n. \tag{15.12}$$

In the following, we assume that the interaction potential is spherically symmetric[4] $V(\vec{r}) = V(r)$. Under this assumption, the scattering amplitude depends only on the angle θ between the two unit vectors $\vec{\kappa}_i$ and $\vec{n}$, and we use from now on the notation $f(k,\theta) = f(k,\vec{\kappa}_i,\vec{n})$.

For indistinguishable particles in the same spin state,[5] the two scattering diagrams of Fig. 15.2, that correspond to scattering amplitudes $f(k,\theta)$ and $f(k,\pi-\theta)$, respectively, cannot be distinguished because of the overlapping of the colliding wave packets in the collision zone. The exact two-body spatial wave functions must be either symmetric (bosons) or antisymmetric (fermions) with respect to the interchange of the coordinates $\vec{r}_1$ and $\vec{r}_2$ of the two atoms, $\Psi(\vec{r}_1;\vec{r}_2) = \varepsilon\Psi(\vec{r}_2;\vec{r}_1)$, with $\varepsilon = +1$ for bosons and $\varepsilon = -1$ for fermions, which implies the equality $\psi(\vec{r}) = \varepsilon\psi(-\vec{r})$ for the relative motion wavefunction. As a consequence, one can show that the scattering amplitude $f(k,\theta)$ should be replaced by $f(k,\theta)+\varepsilon f(k,\pi-\theta)$ *but* with θ restricted to the interval $0 \leq \theta \leq \pi/2$ [Schiff (1969)]. The differential and total cross-sections for identical particles thus read:

$$\frac{\mathrm{d}\sigma}{\mathrm{d}\Omega} = |f(k,\theta) + \varepsilon f(k,\pi-\theta)|^2, \qquad \text{and} \qquad \sigma(k) = \int_0^{\pi/2} \frac{\mathrm{d}\sigma}{\mathrm{d}\Omega} 2\pi \sin\theta \mathrm{d}\theta. \tag{15.13}$$

We can therefore calculate the scattering amplitude for identical particles as if they were distinguishable particles, and afterwards use the relations (15.13) to infer their collision properties.

15.2.3 *Partial wave expansion*

To carry out a more detailed study, we expand the incident and scattered wave functions on the spherical-wave basis:

$$\psi(\vec{r}) = \sum_{\ell=0}^{\infty} \sum_{m=-\ell}^{m=\ell} Y_\ell^m(\theta,\varphi) \frac{u_{k,\ell,m}(r)}{r}, \tag{15.14}$$

[4]This assumption would not be valid for gases where the dipole-dipole interaction is not neglibible such as in chromium-52 gases [Lahaye *et al.* (2009)], or in gases of polar heteronuclear molecules.

[5]Here, we assume in addition that the interaction potential does not depend on the spins.

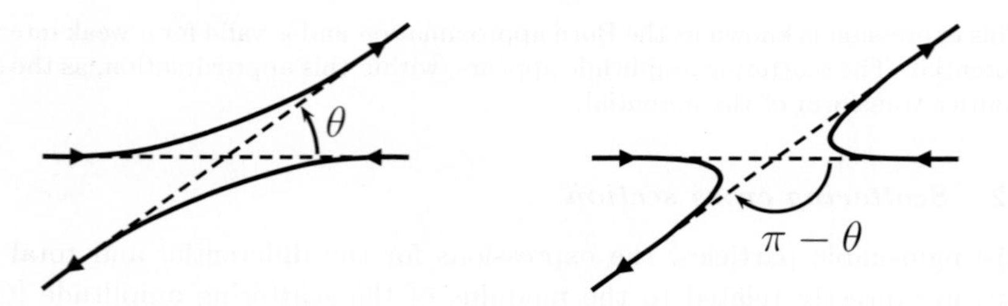

Fig. 15.2 Two scattering processes yielding the same final state for identical particles.

where φ is the azimuthal angle around the z axis, defined as the incident wave function direction, and $Y_\ell^m(\theta,\varphi)$ are the spherical harmonic functions. As explained below, the lower the incident energy, the lower the number of partial waves that have to be taken into account to describe the collision. This is the reason that this expansion plays a crucial role in the studies of the collisional properties of ultracold gases.

The index m turns out to be equal to zero because of the spherical symmetry of the potential, and will be omitted in the following. The only spherical harmonics that contribute to the scattering are thus $Y_\ell^0(\theta,\varphi) \propto P_\ell(\cos\theta)$, where $P_\ell(x)$ are the Legendre polynomials. Inserting the expansion (15.14) into the Schrödinger equation (15.4), we obtain an effective one-dimensional Schrödinger equation for the determination of each radial wave function $u_{k,\ell}(r)$:

$$\left[\frac{\mathrm{d}^2}{\mathrm{d}r^2} + k^2 - \frac{\ell(\ell+1)}{r^2} - \frac{2\mu}{\hbar^2}V(r)\right] u_{k,\ell}(r) = 0, \tag{15.15}$$

where an extra term containing the centrifugal barrier $\hbar^2\ell(\ell+1)/2\mu r^2$ adds to the interaction potential.

To solve this set of equations, we need boundary conditions. We shall assume that $u_{k,\ell}(0) = 0$ and therefore that $u_{k,\ell}(r)/r$ is regular at $r = 0$. The asymptotic form at large distances ($kr \gg 1$) is obtained by expanding the incident and the scattered wave functions on the spherical-wave basis:

$$e^{ikz} \sim \frac{1}{2ikr}\sum_{\ell=0}^{\infty}(2\ell+1)P_\ell(\cos\theta)\left[(-1)^{\ell+1}e^{-ikr} + e^{ikr}\right], \tag{15.16}$$

$$\psi(\vec{r}) \sim \frac{1}{2ikr}\sum_{\ell=0}^{\infty}(2\ell+1)P_\ell(\cos\theta)\left[(-1)^{\ell+1}e^{-ikr} + e^{2i\delta_\ell}e^{ikr}\right], \tag{15.17}$$

where the phase shifts δ_ℓ are real and depend on the incident wave vector modulus k. The coefficients $e^{2i\delta_\ell}$ have a modulus equal to one which ensures the flux conservation for each partial wave. The asymptotic form of the radial wave functions for $kr \gg 1$ is therefore given by

$$u_{k,\ell}(r) \sim (-1)^{\ell+1}e^{-ikr} + e^{2i\delta_\ell}e^{ikr} \sim \sin\left(kr - \ell\frac{\pi}{2} + \delta_\ell\right). \tag{15.18}$$

In the asymptotic limit (large r), the radial wave function $u_{k,\ell}(r)$ behaves in the same manner as a free wave ($V = 0$), except for a possible phase shift for the reflected outgoing wave. Combining Eqs. (15.16) and (15.17) with the definition of the scattering amplitude (15.9), we obtain the scattering amplitude and the cross section in the form of a partial wave expansion:

$$f(k,\theta) = \frac{1}{2ik}\sum_{\ell=0}^{\infty}(2\ell+1)(e^{2i\delta_\ell}-1)P_\ell(\cos\theta), \qquad \text{and} \qquad \frac{d\sigma}{d\Omega} = |f(k,\theta)|^2. \quad (15.19)$$

This is the major result concerning the description of pure elastic scattering in the center-of-mass frame.

The total cross-section can also be expressed in terms of the imaginary part of the forward scattering amplitude $f(k,\theta=0)$ [Sakurai (1994)]:

$$\sigma_{\text{tot}} = \frac{4\pi}{k^2}\text{Im}\left[f(k,\theta=0)\right]. \quad (15.20)$$

This relationship is called the optical theorem, and its physical interpretation will be discussed below in Sec. 15.6.

The expression for the total cross-section is given by

$$\sigma(k) = \sum_{\ell=0}^{\infty}\sigma_\ell(k), \qquad \text{with} \qquad \sigma_\ell(k) = \frac{4\pi}{k^2}(2\ell+1)\sin^2\delta_\ell(k). \quad (15.21)$$

In contrast with the differential cross section, there are no interferences between partial waves for the total cross-section. They cancel out in the angular integration over the scattering directions.

Using Eq. (15.13) and the properties of Legendre polynomials, we deduce that the expression for the scattering cross section of polarized bosons (fermions) contains only even (odd) values of ℓ:

$$\text{bosons}: \sigma_B(k) = \frac{8\pi}{k^2}\sum_{\ell \text{ even}}^{\infty}(2\ell+1)\sin^2\delta_\ell(k), \quad (15.22)$$

$$\text{fermions}: \sigma_F(k) = \frac{8\pi}{k^2}\sum_{\ell \text{ odd}}^{\infty}(2\ell+1)\sin^2\delta_\ell(k). \quad (15.23)$$

An example of total cross-section for different partial waves as a function of the kinetic energy of the relative particle is provided in Fig. 15.3 for two colliding metastable helium-4 atom (boson) in the state 2^3S_1.

An important difference between the total cross-section and the differential cross section is that the latter involves interferences between partial waves (see Eq. (15.19)). As an example, the s and d partial-wave interference along with the energy dependent phase shifts have been directly observed by monitoring the elastic collision of two ultracold samples of rubidium-87 atoms with an adjustable relative energy [Thomas *et al.* (2004); Buggle *et al.* (2004)]. Furthermore, the differential cross section also enters the collision integral of the Boltzmann equation, so that the thermalization and the transport properties of a non-degenerate gas also involve interferences between the different partial waves [Anderlini and Guéry-Odelin (2006)].

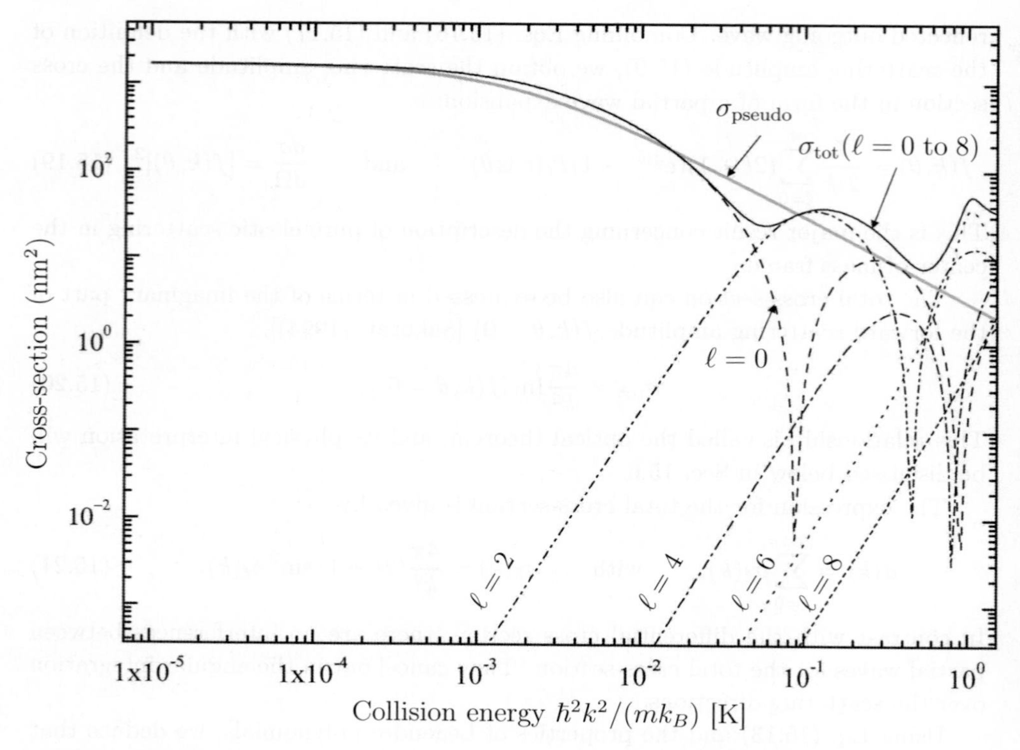

Fig. 15.3 Total cross-section for the partial waves $\ell = 0, 2, 4, 6, 8$ of two colliding metastable helium-4 atom (boson) in the state 2^3S_1 as a function of the incident kinetic energy expressed in Kelvin (dashed and/or dotted lines). These quantities have been calculated from the two-body potential ${}^5\Sigma_g^+$ given in [Stärck and Meyer (1994)]. The sum of the contributions of the five lowest partial waves is represented by the solid line. The gray line is the pseudo-potential prediction (see Sec. 15.4) that agrees well with the exact calculation for a low relative incident energy. Figure adapted from [Léonard (2003)]. Courtesy of J. Léonard.

15.3 Scattering length

15.3.1 *Low-energy limit*

As already stressed, the key point in the partial wave expansion is that for a given incident energy, only a finite number of partial waves contributes to the cross section.

This can be readily understood by a classical argument. Let us denote p the incident momentum, and δ the impact parameter. The angular momentum is therefore $|\vec{r} \times \vec{p}| = p\delta$, and corresponds to a partial wave number $\ell = p\delta/\hbar$. The interaction has a significant effect if the closest approach distance, δ, is smaller than the range b of the potential:

$$\delta \ll b \implies \ell \ll \ell_{\max} = \frac{pb}{\hbar} = \frac{2\pi b}{\lambda_{\text{dB}}} \tag{15.24}$$

where $\lambda_{\text{dB}} = h/p$ is de Broglie wavelength associated with the relative particle of

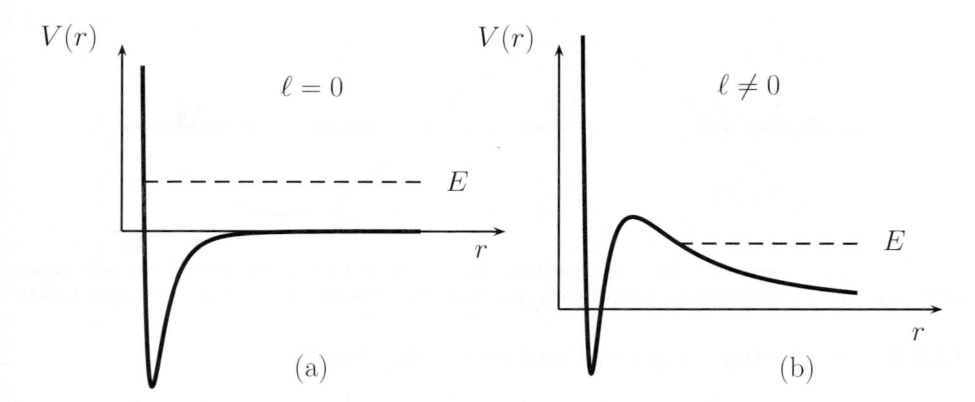

Fig. 15.4 Effective potential for the one-dimensional radial Schrödinger equation: (a) $\ell = 0$, s-wave scattering ; (b) $\ell > 0$, the centrifugal barrier sums up to the interatomic potential.

momentum p. We thus find an upper bound for the number of partial waves that are involved in the collision process.

Quantum mechanically, the phase shift of a partial wave ℓ larger than ℓ_{max} becomes negligibly small. Indeed, the effective potential experienced by the radial partial wave $u_{k,\ell}$ when $\ell \neq 0$ is the interatomic potential superimposed with a centrifugal barrier (see Fig. 15.4). A particle with a relative energy much lower than the barrier height does not feel the short distance details of the interacting potential since it is reflected:[6]

$$\frac{\hbar^2 \ell(\ell+1)}{mb^2} \gg \frac{p^2}{2m} \implies \frac{\lambda_{\mathrm{dB}}^2 \ell^2}{4\pi^2 b^2} \gg 1 \implies \ell \gg \frac{2\pi b}{\lambda_{\mathrm{dB}}}. \tag{15.25}$$

We consequently expect that, when $b \ll \lambda_{\mathrm{dB}}$, the scattering amplitudes vanish for all partial waves except $\ell = 0$ in the low-energy limit since there is no energy barrier in this latter case. More quantitatively, one finds the following scaling of the contribution of the partial wave $\ell \neq 0$ when the relative wave vector tends to zero [Joachain (1983); Landau and Lifshitz (1996)]:

$$\sigma_{\ell \neq 0} \propto k^{4\ell} \underset{k \to 0}{\longrightarrow} 0. \tag{15.26}$$

As a consequence, for bosons, the s-wave contribution governs the scattering properties in the low-energy limit:

$$f(k \to 0) \simeq \frac{e^{i\delta_0(k)}}{k} \sin \delta_0(k). \tag{15.27}$$

Since only odd partial waves contribute to the cross section between identical fermions, at low temperature a polarized fermionic gas behaves as an ideal gas.

[6]This is no longer true in the presence of a shape resonance which occurs when the energy of the incoming wave is resonant with a quasi-bound state of the effective potential [Landau and Lifshitz (1996)].

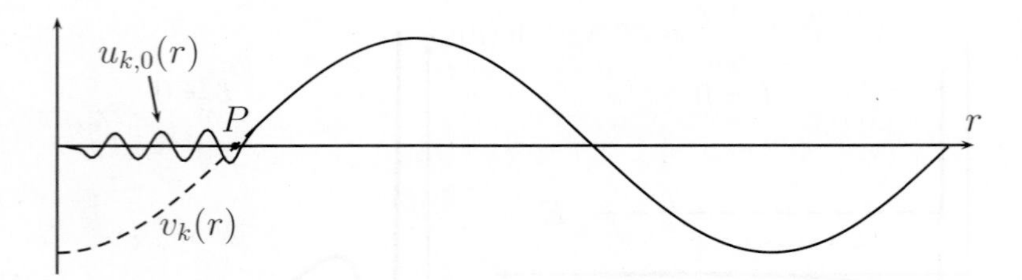

Fig. 15.5 $u_{k,0}(r)$ s-wave radial wave function, and v_k the extension for all r of the asymptotic form of $u_{k,0}(r)$. The scattering length corresponds to the limit of the abscissa of P when $k \to 0$.

15.3.2 *Scattering amplitude and scattering length*

In the low-energy limit, only the $\ell = 0$ partial wave has to be taken into account. From Eq. (15.15), we deduce that the radial wave function $u_{k,0}$ is a solution of

$$\left[\frac{\mathrm{d}^2}{\mathrm{d}r^2} + k^2 - \frac{2\mu}{\hbar^2}V(r)\right] u_{k,0}(r) = 0. \tag{15.28}$$

For distances r large compared to the potential range, $u_{k,0}$ scales as $\sin[kr + \delta_0(k)]$. Let us denote $v_k(r)$ as the extension for all r of the asymptotic form of $u_{k,0}(r)$: $v_k(r) = A\sin[kr + \delta_0(k)]$, and P as the point of intersection of $v_k(r)$ with the r-axis that is the closest to the origin (see Fig. 15.5). The expansion of $v_k(r)$ about $kr = 0$ that corresponds to the low-energy limit reads:

$$v_k(r) \underset{kr\to 0}{\longrightarrow} A[\sin\delta_0(k) + kr\cos\delta_0(k)] \propto (r-a), \tag{15.29}$$

where we have introduced the *scattering length* a:

$$a = -\lim_{k\to 0}\frac{\tan\delta_0(k)}{k}, \quad \text{or equivalently} \quad \delta_0(k) \underset{k\to 0}{\longrightarrow} -ka. \tag{15.30}$$

Using in Eq. (15.27) the expansion of $e^{i\delta_0(k)}$ and $\sin\delta_0(k)$ in powers of ka, yields $f(k\to 0) \simeq -a(1-ika)$. This result provides us with a simple geometrical interpretation of the scattering length as the limit of the abscissa of P when $k \to 0$. According to Eqs. (15.27) and (15.30), the first term in the expansion of the scattering amplitude at low-energy is simply related to the scattering length:[7]

$$f(k\to 0) = -a. \tag{15.31}$$

Using Eqs. (15.21) and (15.31), one obtains the low-energy expression for the s-wave scattering cross section for distinguishable particles

$$\sigma_0(k) = \frac{4\pi}{k^2}\sin^2\delta_0(k) \underset{k\to 0}{\longrightarrow} 4\pi a^2 < \frac{4\pi}{k^2}. \tag{15.32}$$

[7] A more refined analysis yields the following expansion at low energy:

$$f(k\to 0) = \frac{-a}{1 + ika - r_e k^2 a/2},$$

where r_e is the *effective range*, another important length that depends on the details of the interaction potential.

From Eq. (15.22), we deduce the low-energy limit of the cross section for identical bosons $\sigma_B(k \to 0) = 8\pi a^2$.

Let us give some examples of values for the scattering length of some commonly used alkali atoms:[8] 0.0635 nm for hydrogen [Jamieson *et al.* (1992)], -114 ± 13 nm for lithium-6[Abraham *et al.* (1997)], 2.93 ± 0.07 nm for sodium-23 [van Abeelen and Verhaar (1999)], 5.24 nm for rubidium-87 [Marte *et al.* (2002)], and 127 ± 6 nm for cesium-133 [Leo *et al.* (2000)].

15.3.3 *Square potential and resonances*

In this section, we determine the scattering length for square potential barriers and wells since these simple potentials contain many features that are valid for any potential. The width of the square potential is denoted by b and the height or depth by V_0 (see Fig. 15.6). Outside the potential range $r > b$, the s-wave radial wave function $u_0(r)$ satisfies

$$\frac{\mathrm{d}^2 u_0}{\mathrm{d}r^2} = 0 \qquad \text{for} \qquad r > b. \tag{15.33}$$

The solution of Eq. (15.33) is simply $u(r) = C_2(r - a)$, where C_2 is a constant. In the region where the potential is non-zero ($r < b$), one has to solve

$$\frac{\mathrm{d}^2 u_0}{\mathrm{d}r^2} = \pm k_0^2 u_0, \tag{15.34}$$

where $k_0 = (2\mu|V_0|/\hbar^2)^{1/2}$. The + sign in Eq. (15.34) refers to the potential barrier and the $-$ sign to the potential well.

Inside the barrier, the solution is therefore of the form $u_0(r) = C_1 \sinh(k_0 r)$. The continuity of the logarithmic derivative of u yields the following explicit form of the scattering length:

$$a = b - \frac{1}{k_0}\tanh(k_0 b). \tag{15.35}$$

We conclude that the scattering length is always positive and smaller than the range of the potential (see right part of Fig. 15.6(a)). In the limit $V_0 \to \infty$, the scattering length tends to the hard sphere core radius b. More generally, the scattering length is positive for any wholly repulsive potential.

In contrast, depending on the shape of the potential, an attractive potential may give either a negative or a positive value for the scattering length. This is illustrated by considering the scattering by a square well potential. A reasoning similar to the one above yields

$$a = b - \frac{1}{k_0}\tan(k_0 b). \tag{15.36}$$

In the right part of Fig. 15.6(b) the scattering length is represented as a function of the dimensionless quantity $k_0 b$. In contrast to the case of a barrier, the scattering

[8] We give here only the value of the scattering length for the triplet potential for which the electron and the nucleus are fully polarized.

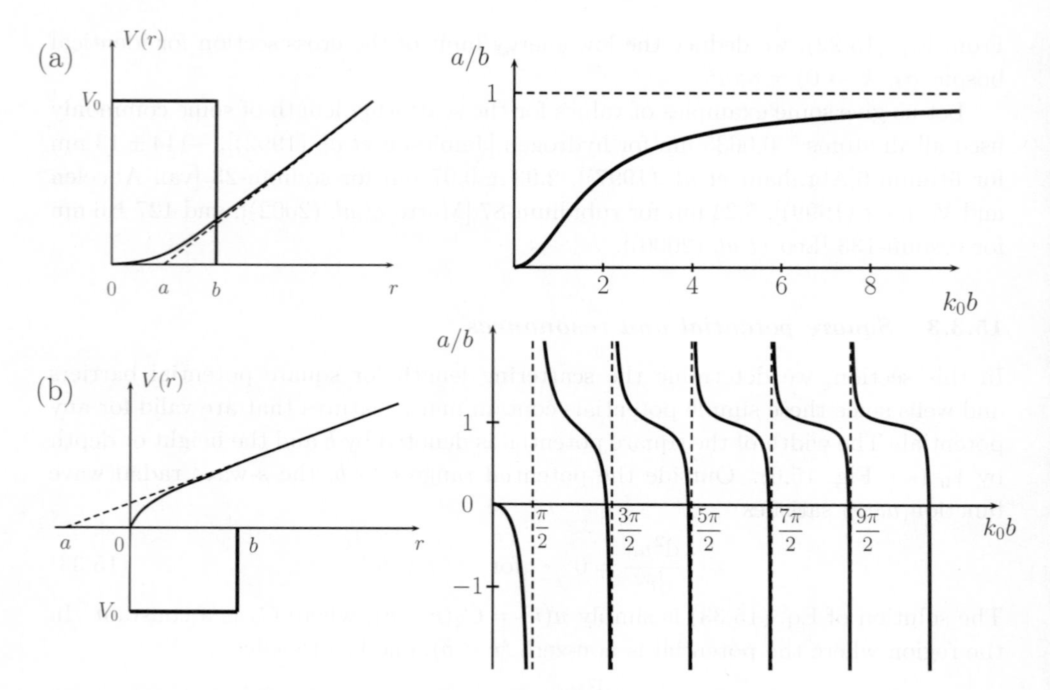

Fig. 15.6 (a) Square potential barrier. Left: the line $v(r)$ (dashed line), which is the asymptote of the s-wave radial wave function $u(r)$ (solid line), intersects the horizontal axis at a point whose abscissa is the scattering length. Right: the scattering length normalized to the width b as a function of k_0b where $k_0 = (2\mu|V_0|/\hbar^2)^{1/2}$. (b) Square potential well, same notations as for (a).

length can take any positive or negative value. When the depth of the potential increases, divergences of the scattering length a occur for values of the depth V_0 such that $k_0b = (2n+1)\pi/2$ where n is an integer. Each divergence corresponds to the appearance of a new bound state in the potential well. This remarkable feature turns out to be valid for any potential well, and is known as the Levinson's theorem. The divergences of the scattering length are referred to as *zero-energy resonances*.

How to experimentally determine the scattering length?

The fact that the scattering length strongly depends on the details of the interatomic potential makes *ab initio* calculations particularly difficult. The experimental determination of the scattering length is therefore usually required and is used in practice to improve the knowledge on the potential.

A first method consists of putting a gas of cold atoms in an out-of-equilibrium state and following the system as it recovers its thermal equilibium state [Monroe *et al.* (1993)]. The observable is a global quantity, such as the size of the cloud. These thermalization measurements are delicate to interpret since they require a precise knowledge of the link between the differential cross-section or the collision rate and the thermalization time. When complemented by a study of the cross section over a large range of temperature while remaining in the s-wave regime, such experiments can provide information on the sign of the scattering length. Indeed, for negative scattering length, the s-wave phase shift $\delta_0(k)$ increases with the wave vector k at low temperature according to the expansion $\delta_0(k) = -ka$ and becomes positive for larger values [Dalibard (1999)]. As a result, there is a value k^* of k such that $\delta_0(k^*) = 0$ and thus $\sigma_0(k^*) \propto \sin^2 \delta_0(k^*) = 0$. The cancellation of the cross section inhibits the thermalization in a range of temperature $\sim \hbar^2(k^*)^2/2mk_B$ and the cooling processes relying on elastic collisions such as evaporative cooling. This "hole" in the cross section for negative scattering length has been observed with rubidium-85, lithium-7 and neon-20 atoms [Burke *et al.* (1998); Schreck (2003); Spoden *et al.* (2005)]. Vice-versa, a detailed study of the cross section as a function of the energy may enable one to rule out the negative sign for the scattering length [Schmidt *et al.* (2003); Spoden *et al.* (2005)].

Alternatively, the evolution of the shape of a Bose-Einstein condensate in a time-of-flight experiment can be exploited to determine the scattering length [Cornish *et al.* (2000)]. Indeed, when the scattering length is positive, the radius of the condensate in the Thomas-Fermi regime scales as[9] $(Na)^{1/5}$. However, the precise determination of the value of a requires a very accurate measurement of the number of atoms.

The most accurate method by far is photoassociation. This spectroscopic method consists of determining the position of the molecular bound states, from which one can infer information about the scattering properties.[10]

15.3.4 *Effective interactions and the sign of the scattering length*

The scattering length determines how the long range behavior of the wave functions is modified by the interactions. To understand how the sign of a is related to the sign of the effective interactions, it will be useful to consider a particle enclosed in a spherical box with radius R so that we have the boundary condition $u_0(R) = 0$ leading to a discrete energy spectrum.

In the absence of interactions ($V = 0$), a possible set of normalized eigenstates (along with the corresponding eigenvalues) of the 3D Schrödinger equation are:

$$\psi_N^{(0)}(r) = \frac{u_N^{(0)}(r)}{r} = \left(\frac{1}{2\pi R}\right)^{1/2} \frac{\sin(N\pi n r/R)}{r}, \qquad \text{and} \qquad E_N^{(0)} = \frac{\hbar^2}{2\mu}\frac{N^2\pi^2}{R^2}, \tag{15.37}$$

where N is a non-zero positive integer.

Figure 15.7(a) represents for example the radial wave function $u_1^{(0)}(r)$ for $N = 1$ and $V = 0$. It coincides with its extension $v_1^{(0)}(r)$ and is a sinusoid with a half-wavelength equal to R and a wave number $k = \pi/R$. In the presence of interactions, assuming that the scattering potential is centered at $r = 0$, the extension $v_1(r)$ of $u_1(r)$ intersects the horizontal axis at a point of abscissa a (dotted lines of Fig. 15.7(b) and Fig. 15.7(c) corresponding to $a > 0$ and $a < 0$, respectively). This function $v_1(r)$ is therefore a sinusoid with a half-wavelength equal to $R - a$ and a wave number $k = \pi/(R - a)$. For $a \ll r < R$, the radial wave function $u_1(r)$ coincides with its extension $v_1(r)$ and the total energy is equal to the kinetic

[9] See Chap. 22.

[10] See Chap. 16, [Weiner *et al.* (1999)] and references therein.

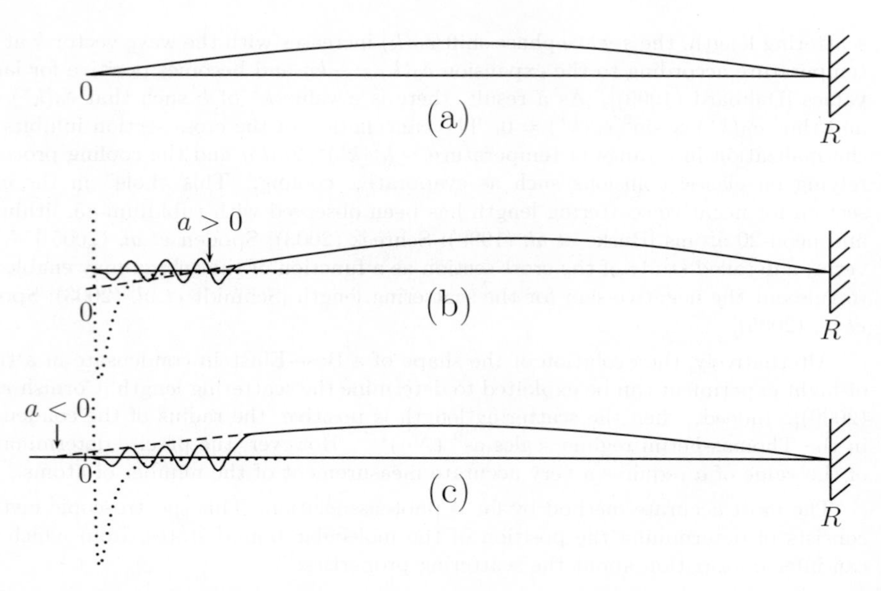

Fig. 15.7 Representation of the s-wave radial wave function in a spherical box of radius R with strict boundary conditions (solid line): (a) in the absence of interactions, (b) with a positive scattering length and (c) with a negative scattering length. The dashed line is the extrapolating function $v_k(r)$ for all r of the asymptotic form of the radial wave. Its intersection with the axis gives the value and the sign of the scattering length.

energy since $V = 0$ in this region. The kinetic energy is determined by the wave number k and is therefore equal to $E_1 = \hbar^2\pi^2/2\mu(R-a)^2$, which is larger than the unperturbed energy $E_1^{(0)} = \hbar^2\pi^2/2\mu R^2$ if $a > 0$, and smaller if $a < 0$.

The same argument can be extended to any value of N. For $V = 0$, $u_N^{(0)}(r)$ and $v_N^{(0)}(r)$ coincide and are sinusoids with a half-wavelength R/N. For $V \neq 0$, $v_N(r)$ is a sinusoid with a half-wavelength equal to $(R-a)/N$ and wave number $N\pi/(R-a)$ which coincides with $v_N^{(0)}(r)$ for $a \ll r$. In this zone, the total energy coincides with the kinetic energy $E_N = N^2\hbar^2\pi^2/2\mu(R-a)^2$, which is larger than the unperturbed energy $E_N^{(0)} = N^2\hbar^2\pi^2/2\mu R^2$ if $a > 0$, and smaller if $a < 0$.

We conclude that the energy difference introduced by the interaction is equal to:

$$\delta E_N = E_N - E_N^{(0)} = \frac{\hbar^2}{2\mu}(\pi N)^2\left(\frac{1}{(R-a)^2} - \frac{1}{R^2}\right) \simeq \frac{\hbar^2\pi^2 N^2}{\mu R^3}a. \tag{15.38}$$

The effective interactions are thus repulsive ($\delta E_N > 0$) for positive scattering length and attractive for negative scattering length.

15.4 Pseudo-potential

15.4.1 *Motivation for introducing this pseudo-potential*

As already emphasized, the exact interaction potential V is, in general, very difficult to calculate. In addition, a small error in the evaluation of the potential V can introduce a very large error for the scattering length because of the existence of resonances associated with a large number of bound states (see Fig. 15.1). Furthermore, in general the Born approximation cannot be applied to the exact potential because it is strongly repulsive at short distances, attractive at intermediate distances and has many bound molecular states.[11] This makes it difficult to use this potential for a mean-field description of ultracold quantum gases.

In fact, in the low-energy limit the macroscopic properties of dilute quantum gases no longer depend on the details of the scattering potential, but simply on the s-wave scattering length a. If the objective is not to calculate a, but instead to study the macroscopic properties of the gas, one can consider the scattering length as a given experimental parameter. It is convenient to introduce another potential, the pseudo-potential $V_{\text{pseudo}}(r)$, simpler to use than $V(r)$, that fulfills the following two conditions:

(i) it must give the same scattering length as the exact potential,
(ii) it must allow a Born approximation treatment.

15.4.2 *Localized pseudo-potential giving the correct scattering length*

Calculations will be simpler if we can choose a localized pseudo-potential proportional to a delta function, of the form $g\delta(\vec{r})$, where g is a constant. The question is whether or not it will be possible to choose the constant g such that the corresponding scattering length is equal to a.

In fact, since this potential is equal to 0 for $r \neq 0$, the function[12] $\tilde{u}_k(r)$ appearing in the s-wave radial wave function $\tilde{R}_k(r) = \tilde{u}_k(r)/r$ obeys the equation $\mathrm{d}^2\tilde{u}_k(r)/\mathrm{d}r^2 + k^2\tilde{u}_k(r) = 0$ for $r \neq 0$ and is a sinusoid of wave number k. Since this function must have the same asymptotic behavior for large r as the corresponding function for the exact potential, it must coincide with the extension $v_k(r)$ of the function $u_{k,0}(r)$ that corresponds to the real potential:

However, a difficulty then appears. According to Eq. (15.30), for $k = 0$ one must have:

$$\tilde{u}_0(r) = v_0(r) = B(r - a) \tag{15.39}$$

[11] For a square potential, the Born approximation is valid when there is no bound states [Dalibard (1999)].

[12] We add a tilde to all the wave functions associated with the pseudo-potential in order to clearly distinguish them from the wave functions associated with the real potential.

where B is a constant. The radial wave function $\tilde{R}_0(r) = \tilde{u}_0(r)/r$ has a $1/r$ divergence at $r = 0$, and the product $\delta(\vec{r})\tilde{R}_0(r)$ that appears in the Schrödinger equation is singular at $r = 0$ and has no meaning. Serious mathematical difficulties thus appear when one tries to use a potential proportional to $\delta(\vec{r})$.[13]

It is actually more appropriate to look for another boundary condition to be satisfied by $\tilde{u}_0(r)$ at $r = 0$ in order to get the correct value for a. From Eq. (15.39), one has:

$$\frac{1}{a} = -\left.\frac{\mathrm{d}\tilde{u}_0(r)/\mathrm{d}r}{\tilde{u}_0(r)}\right|_{r=0} = -\frac{\tilde{u}_0'(0)}{\tilde{u}_0(0)}. \tag{15.40}$$

The pseudo-potential must therefore be such that it leads to a logarithmic derivative of $\tilde{u}_0(r)$ equal to $-a$ at $r = 0$. We show that this is possible if we take the following expression for the pseudo-potential

$$V_{\text{pseudo}}(\vec{r}) = g\delta(\vec{r})\frac{\mathrm{d}}{\mathrm{d}r}(r\cdot) \tag{15.41}$$

where g is a constant to be determined. The operator $[(\mathrm{d}/\mathrm{d}r)(r\cdot)]$ put after the delta function regularizes the action of this delta function when it acts on functions that behave as $1/r$ near $r = 0$

$$V_{\text{pseudo}}(\vec{r})\frac{\tilde{u}_0(r)}{r} = g\delta(\vec{r})\tilde{u}_0'(0). \tag{15.42}$$

Note, however, that for functions $f(r)$ that are regular at $r = 0$, V_{pseudo} has the same action as $g\delta(\vec{r})$:

$$V_{\text{pseudo}}(\vec{r})f(r) = g\delta(\vec{r})f(0), \quad \text{if } f(r) \text{ is regular at } r = 0. \tag{15.43}$$

To see if the expression (15.41) leads to the correct value for the scattering length, and to find the corresponding value of the constant g, we now write the Schrödinger equation obeyed by the radial wave function $\tilde{R}_0(r) = \tilde{u}_0(r)/r$ for $k = 0$, where $\tilde{u}_0(r) = v_0(r)$ is given by Eq. (15.39). We get

$$-\frac{\hbar^2}{2\mu}\Delta\left(1 - \frac{a}{r}\right) + g\delta(\vec{r})\frac{\mathrm{d}}{\mathrm{d}r}r\left(1 - \frac{a}{r}\right) = 0. \tag{15.44}$$

Using $\Delta(1/r) = -4\pi\delta(\vec{r})$ and Eq. (15.42), Eq. (15.44) gives the explicit expression for the coupling constant g appearing in the expression (15.41) of the pseudo-potential:

$$g = \frac{4\pi\hbar^2 a}{2\mu} = \frac{4\pi\hbar^2 a}{m}. \tag{15.45}$$

[13]These difficulties can be circumvented if the delta function is replaced by a narrow function $\delta_{k_c}(\vec{r})$ localized at the origin and still satisfying $\int \mathrm{d}^3 r \delta_{k_c}(\vec{r}) = 1$, but with a small nonzero width $1/k_c$ (see Sec. 15.5).

15.4.3 *Scattering amplitude. Validity of the Born approximation*

We suppose now that the wave number k is different from zero, though small enough to allow s-wave scattering to dominate ($ka \ll 1$). By writing a Schrödinger equation analogous to (15.44) for $k \neq 0$, we determine the phase shift $\delta_0(k)$ of the s-wave and then the scattering amplitude f. Finally, we show that the expression of f, calculated to first order in V_{pseudo}, is correct, which proves the validity of the Born approximation applied to V_{pseudo}.

For $k \neq 0$, the Schrödinger equation becomes:

$$-\frac{\hbar^2}{2\mu}\Delta\left(\frac{\tilde{u}_k(r)}{r}\right) + V_{\text{pseudo}}(\vec{r})\frac{\tilde{u}_k(r)}{r} = \frac{\hbar^2 k^2}{2\mu}\frac{\tilde{u}_k(r)}{r}. \tag{15.46}$$

To calculate the action of the Laplacian on $\tilde{u}_k(r)/r$, it is convenient to write:

$$\Delta\left(\frac{\tilde{u}_k(r)}{r}\right) = \Delta\left[\frac{\tilde{u}_k(0)}{r} + \frac{\tilde{u}_k(r) - \tilde{u}_k(0)}{r}\right] = -4\pi\tilde{u}_k(0)\delta(\vec{r}) + \frac{1}{r}\frac{\mathrm{d}^2\tilde{u}_k(r)}{\mathrm{d}r^2}. \tag{15.47}$$

where we have used the expression of the Laplacian in spherical coordinates. When using Eq. (15.42) for calculating the second term of Eq. (15.46), the Schrödinger equation becomes:

$$-\frac{\hbar^2}{2\mu}\left[-4\pi\tilde{u}_k(0)\delta(\vec{r}) + \frac{1}{r}\tilde{u}_k''(r)\right] + g\delta(\vec{r})\tilde{u}_k'(0) = \frac{\hbar^2 k^2}{2\mu}\frac{\tilde{u}_k(r)}{r}. \tag{15.48}$$

The cancellation of the term independent of $\delta(\vec{r})$ yields:

$$\tilde{u}_k''(r) + k^2\tilde{u}_k(r) = 0 \tag{15.49}$$

whose solution is a sinusoid

$$\tilde{u}_k(r) \propto \sin\left[kr + \delta_0(k)\right]. \tag{15.50}$$

The s-wave phase shift $\delta_0(k)$ can be determined by cancelling the coefficient of $\delta(\vec{r})$ in Eq. (15.48). This gives:

$$\frac{\tilde{u}_k'(0)}{\tilde{u}_k(0)} = k\cot\delta_0(k) = -\frac{2\pi\hbar^2}{g\mu} = -\frac{1}{a}, \tag{15.51}$$

which satisfies Eq. (15.40). This relation between a and $\delta_0(k)$ can finally be used to express the s-wave scattering amplitude given in Eq. (15.27) in terms of a. Using Eqs. (15.27) and (15.51), we can write:

$$f_0(k) = \frac{\sin\delta_0(k)}{ke^{-i\delta_0(k)}} = \frac{1}{k\left[\cot\delta_0(k) - i\right]} = -\frac{a}{1 + ika}. \tag{15.52}$$

Since $ka \ll 1$, the expansion of $f_0(k)$ in powers of ka is convergent:[14]

$$f_0(k) = -a + ika^2 + \cdots \tag{15.53}$$

Since the pseudo-potential is proportional to g, and thus to a, it follows that the expansion (15.53) can also be considered to be an expansion in powers of V_{pseudo}

[14]We recover the result mentioned on page 360.

whose lowest order term is predominant, and gives the correct value for the scattering length $f_0(k \to 0) = -a$. The two conditions mentioned at the beginning of this subsection are thus fulfilled by the pseudo-potential given in Eq. (15.41) with the relation (15.45) between the coupling constant g and the scattering length a.

From the expression of the s-wave scattering amplitude (15.52), we deduce the total cross-section for scattering of identical bosons by the pseudo-potential as a function of the incident wave vector k for the relative motion:

$$\sigma(k) = \frac{8\pi a^2}{1 + k^2 a^2}. \tag{15.54}$$

The low-energy limit ($k|a| \ll 1$) gives the well-known result $\sigma(k \to 0) \to 8\pi a^2$. Close to a resonance such as those discussed for the square potential well on page 362, one can have $k|a| \gg 1$ and $kb \ll 1$ (validity domain of s-wave scattering) and the cross section can reach the so-called *unitary limit*:

$$\sigma(k) \underset{k|a|\gg 1}{\longrightarrow} \frac{8\pi}{k^2}, \tag{15.55}$$

where it does not depend on the scattering length anymore.

Application to a particle in a spherical box

The interpretation of the sign of the scattering length described in Sec. 15.3.4 was based on the modification, induced by the interaction, of the asymptotic behavior of the energy eigenstate wave functions of a particle in a box of radius R. We have calculated the energy shifts δE_N of the discrete energy levels that are produced by a potential characterized by a given scattering length a (see Eq. (15.38)). It is not possible to get the same result for an arbitrary potential with the same scattering length using perturbation theory. For instance, a hard core potential ($V = \infty$ for $r < a$) cannot obviously be treated perturbatively, however the corresponding pseudo-potential allows perturbative calculations. The expression of δE_N, to first order in V_{pseudo}, is obtained by taking the diagonal matrix element of V_{pseudo} in the normalized zeroth order state $\psi_N^{(0)}$ (see Eq. (15.37)). A straight forward calculation gives:

$$\delta E_N = \langle \psi_N^{(0)} | V_{\text{pseudo}} | \psi_N^{(0)} \rangle = \frac{\hbar^2 \pi^2 N^2}{\mu R^3} a, \tag{15.56}$$

which coincides with the result (15.38) given above.

15.4.4 *Bound state of the pseudo-potential for a positive scattering length*

The pseudo-potential also exhibits a bound state for $a > 0$. This is readily shown by repeating the previous calculation devoted to scattering states with the positive energy $\hbar^2 k^2/2\mu$ replaced by a negative one $-\hbar^2 \kappa^2/2\mu$. The Schrödinger equation that we have to solve for the radial wave function is obtained by cancelling the term independent of $\delta(\vec{r})$ in Eq. (15.48)

$$u_0'' - \kappa^2 u_0(r) = 0. \tag{15.57}$$

The cancellation of the coefficient of $\delta(\vec{r})$ yields the boundary condition at $r = 0$: $u_0'(0)/u_0(0) = -a^{-1}$. The physical solution that corresponds to a bound state must give a wave function that tends to zero at large distance, and we thus get

$$\psi_{\text{bound}}(\vec{r}) = \frac{1}{\sqrt{2\pi a}} \frac{e^{-r/a}}{r}, \tag{15.58}$$

with an energy

$$E_{\text{bound}} = -\frac{\hbar^2}{2\mu a^2}. \tag{15.59}$$

As explained in the next chapter, the existence of this bound state plays an important role in the formation of cold molecules from two species of ultracold gases.

15.5 Delta potential truncated in momentum space

15.5.1 *Expression of the potential*

Rather than using the pseudo-potential (15.41), a frequently followed procedure is to use a delta potential

$$V_\delta(\vec{r}) = g_0 \delta(\vec{r}) \tag{15.60}$$

where the new coupling constant g_0 is chosen so that the delta potential leads to the correct value of the scattering length a, which is a measurable quantity. This condition will give the equation that relates g_0 to a.

Mathematical difficulties, such as divergent expressions associated with the zero range of $\delta(\vec{r})$, appear when one uses a true delta function. One can circumvent these difficulties by using a function $\delta_{k_c}(\vec{r})$, that satisfies $\int \mathrm{d}^3 r \delta_{k_c}(\vec{r}) = 1$, but with a finite range $1/k_c$ around the origin, i.e. with a cut-off k_c in its Fourier transform:[15]

$$\tilde{V}(\vec{k}) = \int \mathrm{d}^3 r e^{-i\vec{k}\cdot\vec{r}} V(\vec{r}) = g_0 \int \mathrm{d}^3 r \delta_{k_c}(\vec{r}) e^{-i\vec{k}\cdot\vec{r}} = g_0 \eta(\vec{k}) \tag{15.61}$$

where $\eta(k) = 1$ when $k \ll k_c$ and $\eta(k) \to 0$ when $k \gg k_c$.

15.5.2 *Determination of the new coupling constant*

To calculate the relation between g_0 and a, we start from the Lippmann-Schwinger expression of the scattering state

$$|\psi_i^+\rangle = |\vec{k}_i\rangle + \frac{1}{E_i - H_0 + i\varepsilon} V |\psi_i^+\rangle \tag{15.62}$$

and from the equation relating the S and T matrices [Messiah (2003); Roman (1965)]

$$S_{ji} = \delta_{\vec{k}_j \vec{k}_i} - 2\pi i \delta_{E_j E_I} T_{ji} \tag{15.63}$$

[15] Another solution for solving this difficulty is to use a *lattice model* [Castin (2008)] where the spatial coordinates $\vec{r}$ of the particles are discretized on a grid of step d. The Dirac function is then replaced by a Kronecker symbol and V by $(g_0/d^3)\sum \delta_{\vec{r}_i \vec{r}_j}$. The step d of the grid plays the role of the inverse of the cut-off k_c considered here.

where

$$T_{ji} = \langle \vec{k}_j | V | \psi_i^+ \rangle = \langle \vec{k}_j | V | \vec{k}_i \rangle + \langle \vec{k}_j | \frac{1}{E_i - H_0 + i\varepsilon} V | \psi_i^+ \rangle. \tag{15.64}$$

One can show that the matrix element of T between the initial and final states of the scattering process is proportional to the scattering amplitude [Joachain (1983)]: $T_{ji} = \alpha f_{ji}$. The coefficient of proportionality is easily calculated by comparing the Born expressions of f_{ji} calculated from Eqs. (15.64) and (15.11). One gets

$$T_{ji}^{\text{Born}} = \langle \vec{k}_j | V | \vec{k}_i \rangle = \frac{2\pi\hbar^2}{\mu L^3} f_{ji} \Longrightarrow \alpha = -\frac{2\pi\hbar^2}{\mu L^3}. \tag{15.65}$$

Replacing T_{ji} by αf_{ji} in Eq. (15.64) and inserting the closure relation $\sum |\vec{k}_k\rangle\langle\vec{k}_k| = \mathbb{1}$ in the last term of the right-hand side of Eq. (15.64), one gets:

$$-\frac{2\pi\hbar^2}{\mu L^3} f_{ji} = V_{ji} + \sum_k \frac{V_{jk}}{E_i - E_k + i\varepsilon} \langle \vec{k}_k | V | \psi_i^+ \rangle. \tag{15.66}$$

In principle, $\langle \vec{k}_k | V | \psi_i^+ \rangle$ is proportional to f_{ki} only in the energy shell, i.e. if $E_k = E_i$. But, the energy denominator $1/(E_i - E_k + i\varepsilon)$ of Eq. (15.64) enhances the contribution of the terms $E_i = E_k$, so that we will make an approximation by replacing $\langle \vec{k}_k | V | \psi_i^+ \rangle$ by αf_{ki}, which leads to an integral equation for the scattering amplitude.

We now replace V by the truncated delta potential and consider the low energy limit $k_i = k_j = k \to 0$. In this limit, the scattering amplitudes f_{ji} and f_{ki} tend to $-a$. The first matrix element of V in Eq. (15.66) can be replaced by g_0/L^3 and the second by $g_0\eta(k)/L^3$, which gives

$$\frac{2\pi\hbar^2}{\mu L^3} a = \frac{g_0}{L^3} + \sum_k \frac{g_0\eta(k)/L^3}{-E_k + i\varepsilon} a \frac{2\pi\hbar^2}{\mu L^3} \tag{15.67}$$

which can also be written

$$\frac{1}{g_0} = \frac{m}{4\pi\hbar^2 a} - \frac{m}{\hbar^2} \frac{1}{L^3} \sum_k \frac{\eta(k)}{k^2}, \tag{15.68}$$

where we have replaced the reduced mass μ by $m/2$ and $E_k = \hbar^2 k^2/2\mu$ by $\hbar^2 k^2/m$.

15.5.3 *Comparison with the pseudo-potential*

It is interesting to compare the coupling constants g_0 and g, associated with the truncated delta potential and the pseudo-potential, respectively. From Eqs. (15.45) and (15.68), one deduces

$$g_0 = \frac{g}{1 - \dfrac{4\pi a}{L^3} \displaystyle\sum_{\vec{k}} \frac{\eta(k)}{k^2}}. \tag{15.69}$$

The sum in (15.69) is proportional to the cut-off k_c in k so that the denominator of (15.69) can be written Cak_c where C is a constant of order 1 that depends on the exact shape of $\eta(k)$. Finally, one gets:

$$g_0 = \frac{g}{1 - Cak_c} = \frac{4\pi\hbar^2}{m}\frac{a}{1 - Cak_c}. \tag{15.70}$$

If $ak_c \ll 1$, one can expand the fraction of (15.70) in a power series of ak_c.

Contrary to $g = 4\pi\hbar^2 a/m$, g_0 is not simply proportional to a. A lowest order treatment in g gives a correction proportional to a. This is not the case for g_0. This is an advantage of V_{pseudo}. However the calculations with V_{pseudo} are not always easy and this is why one often prefers to use the truncated delta potential. In Chaps. 24 and 26, we will give examples of calculations performed with a truncated delta potential that, after replacement of g_0 by its expression (15.70) in terms of a, lead to finite results that are independent of the precise value of the cut-off k_c.

15.6 Forward scattering

Forward scattering plays an essential role in various physical phenomena. For example, the interference of the incident wave, with the wave scattered in the forward direction explains the attenuation of the incident wave and provides a clear interpretation of the optical theorem. In optics, the interference of the incident light wave and the light scattered by an atomic medium where populations are inverted (upper states more populated than lower states) explains how the incident wave can be amplified by stimulated emission, and why this amplification occurs only in the forward direction.

The analysis of forward scattering is usually presented for an incident plane wave. In this section, we will follow a different description, that is more appropriate for a real experiment in which the incident wave has a finite transverse spread. The incident wave is taken to be a Gaussian wave, and is scattered by a potential localized near the focus of the incident beam. After a brief reminder on the important parameters of a Gaussian beam, we will study the structure of the incident and scattered waves in the far-field zone. The analysis of the interference between these two waves will provide new physical insights into the role played by forward scattering in collision physics.

15.6.1 *Gaussian incident wave and scattered wave*

(i) *Beam waist, Rayleigh length and divergence angle*

Reminder of Gaussian beams

Consider a Gaussian wave $\varphi(x, y, z)$ with a unity peak amplitude that propagates along the positive z direction. Within the paraxial approximation, that is, for a relatively small divergence, a Gaussian beam that propagates in free space remains Gaussian with parameters that evolve in space

$$\varphi(x, y, z) = \frac{w_0}{w(z)} e^{ikz+i\chi(z)} e^{ik\rho^2/2R(z)} e^{-\rho^2/w(z)} \tag{15.71}$$

where $\lambda = 2\pi/k$ is the wavelength of the incident wave, z_R the Rayleigh length (see below), $\chi(z)$ the Gouy phase, $R(z) = z + z_R^2/z$ the radius of curvature of the wavefronts, w_0 the *beam waist* and $w(z) = w_0(1 + z^2/z_R^2)^{1/2}$ the beam radius, which varies along the propagation direction. The *Rayleigh length*, z_R, defined as

$$z_R = \frac{\pi w_0^2}{\lambda} = \frac{k w_0^2}{2}, \tag{15.72}$$

determines the length over which the beam propagates without significantly diverging. The *Gouy phase* is defined as $\tan \chi(z) = -z/z_R$ and describes the phase shift experienced by a converging wave when it passes through its focus.

Figure 15.8 represents a Gaussian wave propagating along the z-axis. Near the focus O, the mathematical form of the wave is given in cylindrical coordinates by:

$$e^{-\rho^2/w_0^2} e^{ikz} \tag{15.73}$$

where ρ is the distance from the z-axis.

The angular divergence of the beam characterized by the angle θ_0 applies for $z > z_R$, and is due to the diffraction associated with the finite transverse spread of the beam:

$$\theta_0 = \frac{\lambda}{\pi w_0} = \frac{2}{k w_0}. \tag{15.74}$$

The solid angle subtended by the cone of angle θ_0 is equal to:

$$\delta\Omega_0 = 2\pi(1 - \cos\theta_0) \simeq \pi\theta_0^2 \simeq \frac{4\pi}{k^2 w_0^2}. \tag{15.75}$$

We assume that the scattering center is at a position $|z| \ll z_R$ such that the incident Gaussian wave can be considered to be a plane wave at the scatterer position.

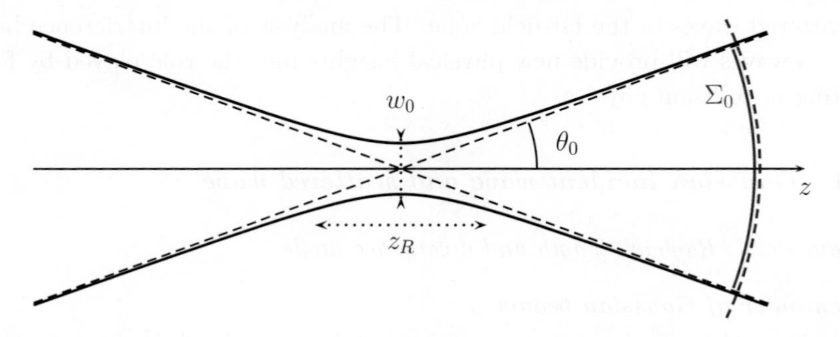

Fig. 15.8 Physical parameters of a Gaussian wave.

(ii) *Incident wave in the far-field zone ($z \gg z_R$)*

When $z \gg z_R$, the z-dependence of the field for $\rho = 0$ is given by:

$$\psi_{\text{inc}}(z \gg z_R, \rho = 0) = z_R \frac{e^{ikz}}{z} e^{i\chi(z)}. \tag{15.76}$$

In the far-field zone, the Gouy phase tends to $\chi(z \to \infty) = -\pi/2$.

When $z \gg z_R$, the wavefronts of the incident wave become spherical and it is better to use spherical coordinates of a point M: r, θ where r is the distance between O and M and θ is the angle between OM and the z-axis (there in no dependence on the azimuthal angle φ because of the cylindrical symmetry). Starting from the value (15.76) on the z-axis, the incident field $\psi_{\text{inc}}(r, \theta)$ decreases as $\exp(-\theta^2/\theta_0^2)$ on the sphere of radius r when θ increases, and vanishes when θ becomes larger than the diffraction angle θ_0. More precisely, the expression of the incident field in the far-field region is found to be equal to:

$$\psi(r \gg z_R, \theta) \simeq -iz_R \frac{e^{ikr}}{r} e^{-\theta^2/\theta_0^2}. \tag{15.77}$$

The amplitude of the incident field is thus concentrated on the surface Σ_0 of the intersection of the sphere or radius r with the cone of solid angle θ_0 as represented in blue in Fig. 15.8:

(iii) *Scattered wave in the far-field zone*

Near the focus, where the scatterer is located, the incident wave has planar wavefronts. We can thus use the asymptotic form (15.9) of the scattered wave in the far-field zone. Furthermore, the angle θ_0 below which this scattered wave can interfere with the incident wave is taken to be small enough that the θ-dependence of the scattering amplitude $f(\theta)$ can be neglected and replaced by $f(0)$. This leads to the following form for the scattering wave in the far-field zone:

$$\psi_{\text{scat}}(r \gg z_R, \theta) = f(0) \frac{e^{ikr}}{r}. \tag{15.78}$$

This wave is represented by the red dotted line of Fig. 15.8. If the scatterer is out of axis ($\rho \neq 0$), one has to take into account the amplitude reduction of the incident wave by a factor $\exp(-\rho^2/w_0^2)$.

15.6.2 *Interference of the incident and scattered waves in the far-field zone*

(i) *Modification of the amplitude and phase of the incident wave*

In the far-field zone, the wave fronts of the two waves coincide and can interfere. However, the incident wave has an appreciable amplitude only within the diffraction cone. It follows that interference is only possible inside the region defined by this cone.

In this angular region, the superposition of the two waves is obtained by adding (15.77) and (15.78) and is given by:

$$\psi_{\text{tot}}(r \gg z_R, \theta) = -iz_R \frac{e^{ikr}}{r}\left[e^{-\theta^2/\theta_0^2} + i\frac{f(0)}{z_R}\right]. \tag{15.79}$$

The effect of the scattering in the far-field zone is thus to reduce the amplitude of the incident wave by an amount proportional to $\text{Im}[f(0)/z_R]$ and to introduce the phase shift by $\text{Re}[f(0)/z_R]$. The importance of the Gouy phase clearly appears here. It is because of the $-i$ factor introduced by this phase in (15.77) that $\text{Im}f(0)$ and not $\text{Re}f(0)$ appears in the reduction of the amplitude of the incident wave.

(ii) *Connection with the optical theorem*

The scattering amplitude $f(0)$ is smaller than the beam waist, which is itself smaller than the Rayleigh length. The ratio $\text{Im}[f(0)/z_R]$ is thus much smaller than 1. To first order in this ratio, the decrease of the probability density of the total wave due to the interference between the incident and scattered waves is equal to

$$\delta\rho_{\text{interf}}(r, \theta) = |\psi_{\text{tot}}|^2 - |\psi_{\text{inc}}|^2 = -2\frac{z_R \text{Im}f(0)}{r^2} e^{-\theta^2/\theta_0^2}.$$

The interference between the incident and scattered waves is thus responsible for the decrease of the flux of the probability current of the incident wave through the surface Σ_0 of Fig. 15.8 given by:

$$\delta\phi_{\text{interf}} = J_i \int_0^\pi \delta\rho_{\text{interf}}(r, \theta) r^2 2\pi \sin\theta \text{d}\theta, \tag{15.80}$$

where $J_i = \hbar k/\mu$ is the incident probability current. Since $\theta_0 \ll 1$, $\sin\theta$ in Eq. (15.80) can be replaced by θ. The integral over θ is then straightforward. Using $z_R\theta_0^2 = 2/k$ one finds:

$$\delta\phi_{\text{interf}} = -\frac{\hbar k}{\mu}\frac{4\pi}{k}\text{Im}f(0). \tag{15.81}$$

The decrease of the flux of the probability current of the incident wave is due to the scattering in all directions produced by the potential $V(r)$, and is described by the total scattering cross section σ_{tot}. The corresponding flux due to scattering is equal to the flux of the probability current of the incident wave (15.73) evaluated near the focus O through a surface equal to σ_{tot}:

$$\delta\phi_{\text{scat}} = \frac{\hbar k}{\mu}\sigma_{\text{tot}}. \tag{15.82}$$

Equating $-\delta\phi_{\text{interf}}$ and $\delta\phi_{\text{scat}}$ leads to:

$$\sigma_{\text{tot}} = 4\pi\frac{\text{Im}f(0)}{k} \tag{15.83}$$

which is nothing but the optical theorem. The advantage of this approach is that it clearly exhibits how the incident wave is attenuated and why this attenuation is restricted to the diffraction cone.

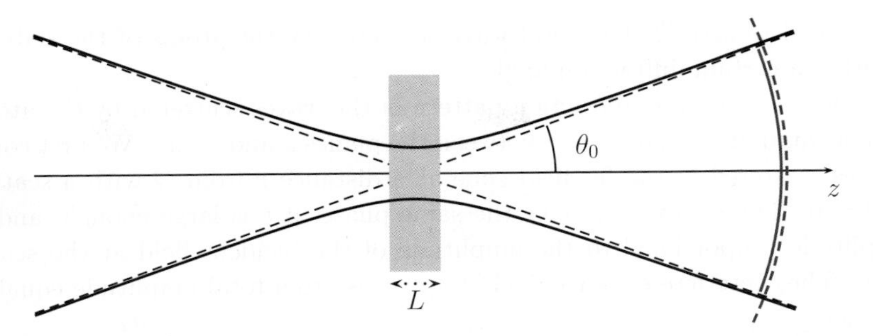

Fig. 15.9 Gaussian wave crossing a slab of thickness L containing atoms with a spatial density n.

15.6.3 *Phase shift of the incident wave and mean field energy*

In order to get a better understanding of the phase shift associated with $\mathrm{Re}f(0)$, we now consider a situation in which a slab of thickness $L \ll z_R$ perpendicular to the z-axis, containing identical atoms with a spatial density n, is put in the focal plane of the incident Gaussian beam (see Fig. 15.9). We suppose that there are no interactions between the atoms of the slab, but that they scatter the incident wave. We try to understand how this scattering modifies the incident wave in the far-field zone.

Each atom in the slab gives rise to a scattering wave, and the first question to answer is whether or not these waves interfere. An important feature of forward scattering must be taken into account: in forward scattering, the momentum of the incident particle does not change so that no momentum is transferred to the particle of the slab that scatters the incident particle. Therefore, after an individual scattering process, one cannot determine which particle of the slab is responsible for the scattering by measuring its momentum change. Even though, a transverse momentum transfer does exist if the scattering angle is not identical to zero while being limited within the diffraction zone, this momentum transfer is generally much smaller than the transverse momentum spread of the scatterer due to its transverse spatial localization within the beam. We suppose here that the transverse coherence length ξ of the scatterer is small $(1/\xi \gg k\theta_0)$. It follows that the amplitudes of the waves scattered by the different atoms of the slab must be added. We will also suppose that the scatterers are initially at rest and have a mass much larger than the mass of the incident particle, such that we can consider them to be fixed.

Since we suppose $L \ll z_R$ all the scattered waves have the same spherical wavefronts in the far-field zone. However, their amplitudes depend on the transverse position of the scatterer in the focal plane due to the transverse ρ-dependence of the amplitude of the incident wave (see Eq. (15.73)). For a given value of z, their relative phases also depend on their distance from the axis if the scattering angle

θ is not equal to zero.[16] The total wave scattered by the atoms of the slab exists only within a certain diffraction angle.

We now calculate the diffraction pattern of the waves scattered by the atoms of the slab located at position x, y, z between the planes z and z+dz. We first consider the waves scattered in the far-field zone at a distance r from O with a scattering angle $\theta = 0$. These waves all have the same phase, if r is large enough, and have an amplitude proportional to the amplitude of the incident field at the scatterer position. They interfere constructively to give rise to a total amplitude equal to:

$$n\mathrm{d}z \int_{-\infty}^{\infty} \mathrm{d}x \int_{-\infty}^{\infty} \mathrm{d}y e^{-(x^2+y^2)/w_0^2} e^{ikz} \frac{e^{ik(r-z)}}{r-z} f(0) = \simeq n\mathrm{d}z\pi w_0^2 \frac{e^{ikr}}{r} f(0). \quad (15.84)$$

If the scattering angle is not equal to zero and lies in the plane xOz, the phases of the different waves are no longer the same and the double integral of (15.84) has to be replaced by:

$$n\mathrm{d}z \int_{-\infty}^{\infty} \mathrm{d}x e^{ikx \sin\theta} \int_{-\infty}^{\infty} \mathrm{d}y e^{-(x^2+y^2)/w_0^2} \simeq n\mathrm{d}z\pi w_0^2 e^{-\theta^2/\theta_0^2}. \quad (15.85)$$

In the integral over x, we have replaced $\sin\theta$ by θ since θ is small and used (15.74). The integral over z gives an extra factor L. Adding the expression obtained for the total scattered wave to the expression (15.77) of the incident wave, and using Eq. (15.72), we finally get

$$\psi_{\text{tot}}(r \gg z_R, \theta) \simeq -i\frac{z_R}{r} e^{-\theta^2/\theta_0^2} e^{ikr} \left(1 + 2i\pi nL\frac{f(0)}{k}\right), \quad (15.86)$$

for the total wave in the far-field zone.

It thus appears that the total scattered wave has the same θ-dependence as the incident wave in the far-field zone, which was expected since this dependence comes from the same diffraction mechanism. If we write:

$$1 + 2i\pi nL\frac{f(0)}{k} \simeq e^{i\delta kL}, \quad (15.87)$$

where

$$\delta k = 2\pi n\frac{f(0)}{k} = \frac{2\pi n}{k}\mathrm{Re}f(0) + \frac{2i\pi n}{k}\mathrm{Im}f(0), \quad (15.88)$$

we see that the effect of the slab on the incident wave is just to replace $\exp(ikr)$ by $\exp[i(k+\delta k)r]$. Everything happens as if the wave number k of the incident wave is changed by an amount equal to δk inside the slab. The effect of the slab is described by an index of refraction with real and imaginary parts.

The imaginary part of δk describes the attenuation of the incident wave in the forward direction due to the scattering in other directions. The intensity of the incident wave decreases as $\exp(-L/\ell)$, where the extinction length ℓ is given by:

$$\ell^{-1} = \frac{4\pi n}{k}\mathrm{Im}f(0) = n\sigma_{\text{tot}}, \quad (15.89)$$

[16] We assume that the solid angle $\delta\Omega_0$ of the Gaussian beam is large compared to the solid angle under which the waist is seen at a distance r.

and where we have used the optical theorem (15.83). This result is what is expected for a particle moving in a medium containing n scatterers per unit volume with a cross section σ_{tot} each.

The real part of δk is associated with a change of the kinetic energy inside the slab equal to:

$$\delta E_{\text{kin}} = \frac{\hbar^2}{2m} 2k\delta k = \frac{2\pi\hbar^2 n}{m} \text{Re} f(0). \tag{15.90}$$

Note that m is the mass of the incident particle, since we suppose here that the scatterers are fixed. The total energy being constant, we deduce that the potential energy of the incident particle within the slab is changed by an amount

$$\delta E_{\text{pot}} = -\delta E_{\text{kin}} = -\frac{2\pi\hbar^2 n}{m} \text{Re} f(0). \tag{15.91}$$

This change of the potential energy can be interpreted as resulting from the interaction of the incident particle with the mean field created by the particles of the slab.

It is interesting to replace in the previous expressions $f(0)$ by its expression in terms of the scattering length a valid at low temperature. From $f(0) = -a(1 - ika)$, we deduce:

$$\text{Re} f(0) = -a \quad \text{and} \quad \text{Im} f(0) = ka^2. \tag{15.92}$$

The attenuation of the incident beam depends on a^2 and cannot be used to determine the sign of a. The mean field interaction

$$\delta E_{\text{pot}} = -\delta E_{\text{kin}} = \frac{2\pi\hbar^2 na}{m} \tag{15.93}$$

depends on a and has the same sign. For a positive scattering length, the particle slows down because of the repulsion exerted by the particles of the slab. Conversely, the particle is accelerated in the presence of attractive interactions ($a < 0$). Expressions similar to (15.93) with an extra factor of 2 due to indistinguishability are found in the mean field description of Bose-Einstein condensates.

15.7 Conclusion

In this chapter, we have shown that, at very low energy, elastic collisions between atoms are entirely characterized by a reduced number of scalar quantities, the scattering length and the effective range, whose precise value depends crucially upon the details of the interatomic potential. In addition, we have presented a graphic interpretation of the fact that interactions are attractive for $a < 0$ and repulsive for $a > 0$.

For dilute atomic gases, atoms are far apart and collisional physics is dominated by two-body interactions and governed by the asymptotic behavior of the wave function that describes the relative motion of the colliding atoms. In this context and in the low energy limit, it is possible to replace the real potential by model potentials that have the same scattering length as the real potential and allow simpler calculations. We have introduced two models of this type, the pseudo-potential and the delta potential truncated in momentum space. These model potentials will be used in Chap. 22 that is devoted to the mean-field description of Bose-Einstein condensates and in Chap. 26 to account for interactions between two-species fermionic atoms in the BCS theory (weak attractive forces).

In this chapter, we have also introduced a simple Gaussian beam model of the incident wave which provides a simple introduction of the concept of mean-field that plays an important role in the description of dilute Bose-Einstein condensate.

A fascinating feature of cold atoms physics is the possibility of controlling the scattering length and therefore the strength of the interactions. The next chapter will be devoted to Feshbach resonances that allow a simple way to achieve this control.

Chapter 16

Controlling atom-atom interactions

16.1 Introduction

In the previous chapter, we saw that collisions between ultracold atoms can be described in good approximation by a single parameter, the scattering length a. Depending on the sign of a, the effective interactions can be repulsive ($a > 0$) or attractive ($a < 0$). When they vanish ($a = 0$) the gas behaves as an ideal gas. A very important breakthrough occurred in this field when it was realized that a can be changed using a static magnetic field. The physical mechanism responsible for this variation of a is the so-called *Feshbach resonance* that appears when, during a collision process, the two colliding atoms pass through a resonant intermediate state where they are bound. There is a certain analogy between this effect and the resonant scattering of a photon by an atom when the frequency ω_i of the incident photon is close to the frequency ω_A of an atomic transition connecting the initial state of the atom (which is generally the ground state g) to an excited state e. It is well known[1] that the scattering amplitude then exhibits a resonant behavior associated with an energy denominator $\omega_L - \omega_A + i(\Gamma/2)$, where Γ is the natural width of the excited state. The Feshbach resonance is similarly associated with an energy denominator $E_{\text{coll}} - E_{\text{bound}} + i(\Gamma_{\text{bound}}/2)$, where E_{coll} is the energy of the initial state of the two atoms in the collision channel and E_{bound} and Γ_{bound} are the energy and the width of the intermediate bound state appearing in the collision process.[2] At very low temperatures, the scattering amplitude is proportional to the scattering length a. Since the magnetic moments of the two-atom system are generally not the same in the initial collision state and the intermediate bound state, $E_{\text{coll}} - E_{\text{bound}}$ can be tuned by varying a static magnetic field B. As a result, the scattering length a exhibits resonant variations when B is swept around values corresponding to the vanishing of $E_{\text{coll}} - E_{\text{bound}}$.

[1] See Chap. 5.

[2] Feshbach resonances are an example of phenomena that arise from the resonant coupling of a discrete state to a continuum. Resonances of this type appear in nuclear and atomic physics. They have been studied by renowned physicists, like Ugo Fano and Herman Feshbach. This is why they are sometimes called *Fano-Feshbach resonances* (see for example [Chin *et al.* (2010)] and the historical remarks of section I.C of this reference).

This chapter is devoted to a description of Feshbach resonances and of some of their applications. We first introduce the notion of collision channel in the simple case of two colliding alkali atoms (Sec. 16.2). We also describe the microscopic atom-atom interactions that give rise to the potential curves in the various channels as well as the couplings between different channels. The most rigorous way of calculating the scattering lengths is then to solve a set of coupled differential equations derived from the Schrödinger equation called *coupled channel equations*. The remaining part of the chapter deals with the simple case in which only two channels are involved: the "open" channel containing the initial collision state and the "closed" channel containing the intermediate resonant bound state. In Sec. 16.3, we develop the analogy with the resonant scattering of a photon by an atom in more detail by using a simple diagrammatic approach. A more precise calculation is then presented in the case of the two-channel model. In Sec. 16.4, we first calculate the scattering states of the two-channel Hamiltonian and we show that the scattering length contains a term varying as $(B - B_0)^{-1}$, where B_0 is the resonant value of the magnetic field.[3] We analyze the various features of the corresponding Feshbach resonance.

Feshbach resonances are not only useful for manipulating the scattering length and for controlling atom-atom interactions, but also for producing bound states with remarkable properties. When the scattering length is positive, the two-channel Hamiltonian has a bound state whose wave function and binding energy are calculated in Sec. 16.5. In particular, we show that in the vicinity of the Feshbach resonance, this bound state has universal properties that depend only on the scattering length and not on the microscopic details of the interaction potential. By varying the static field from a region where a is negative to a region where a is positive one can transform a pair of two ultracold colliding atoms into an ultracold molecule. In Sec. 16.6, we briefly review various methods for converting atom pairs into molecules including optical Feshbach resonances and photoassociation.[4]

16.2 Collision channels

16.2.1 *Microscopic interactions*

Let us consider two alkali atoms, ignoring in a first step the nuclear spin degrees of freedom. Figure 16.1 gives the variations with r of the Born Oppenheimer potentials that give the interaction energy of the two atoms separated by a distance r. There are two potential curves depending on the quantum number $S = 1$ (triplet state) or $S = 0$ (singlet state) of the total spin $\vec{S} = \vec{S}_1 + \vec{S}_2$, where $\vec{S}_1$ and $\vec{S}_2$ are the spins of the valence electrons of the two atoms. The antisymmetry of the wave

[3]We will see that the width of the intermediate resonant bound state is negligible when the two colliding atoms are ultracold.

[4]For the quantitative analysis of Feshbach resonances, we will follow closely the presentation of [Köhler *et al.* (2006)]. More details can also be found in this reference concerning ultracold molecules formed by sweeping a magnetic field near a Feshbach resonance.

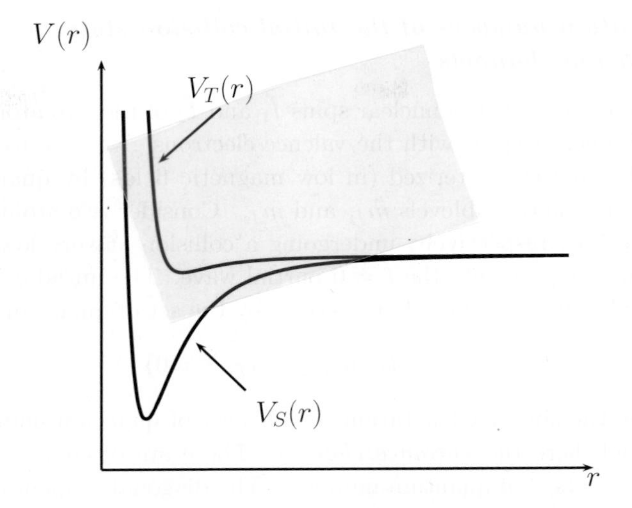

Fig. 16.1 Born Oppenheimer potentials of two alkali atoms versus the distance r between the two atoms. The potential is not the same in the triplet state ($S = 1$) and in the singlet state ($S = 0$).

function of the two electrons requires that for $S = 1$ (symmetric spin state) the orbital wave function must be antisymmetric, and vice versa for $S = 0$. It follows that, at short distances, the electrostatic repulsion is not the same in the triplet and in the singlet state. This explains the splitting of the two Born Oppenheimer potentials. At large distance, the splitting between the two curves tends to zero since both potentials are dominated by the attractive van der Waals interaction that varies as $-1/r^6$ where r is the relative distance between the two atoms. Let P_S and P_T be the projection operators onto the singlet and triplet manifolds. The electrostatic interaction between the two atoms can be written as

$$V_{\text{el}}(r) = V_S(r)P_S + V_T(r)P_T. \tag{16.1}$$

Using the expressions of P_S and P_T in terms of $\vec{S}_1$ and $\vec{S}_2$, one gets:

$$V_{\text{el}}(r) = \frac{1}{4}V_S(r) + \frac{3}{4}V_T(r) + \frac{\vec{S}_1 \cdot \vec{S}_2}{2\hbar^2}\left[V_T(r) - V_S(r)\right]. \tag{16.2}$$

This interaction depends only on the distance r between the two atoms and remains invariant under a rotation of the molecular axis. It thus commutes with the relative angular momentum $\vec{L}$ of the two atoms and cannot connect two states with different quantum numbers ℓ of this angular momentum.

There are also magnetic interactions V_{ss} between the two atoms that result from the dipole-dipole interactions between the two spin magnetic moments which are much smaller than V_{el}. Contrary to V_{el}, V_{ss} depends on the relative orientation of the molecular axis and the spin direction and can connect two states with different values of ℓ.

Finally, the total interaction Hamiltonian is equal to

$$V^{\text{int}} = V_{\text{el}} + V_{ss}. \tag{16.3}$$

16.2.2 *Quantum numbers of the initial collision state. Collision channels*

We now take into account the nuclear spins $\vec{I}_1$ and $\vec{I}_2$ of the two atoms. The interaction of these nuclear spins with the valence electrons gives rise to two hyperfine states for each atom characterized (in low magnetic fields) by quantum numbers f_1 and f_2 with magnetic sublevels m_{f_1} and m_{f_2}. Consider two atoms in the states f_1, m_{f_1} and f_2, m_{f_2}, respectively, undergoing a collision at very low temperature, so that one can consider only the $\ell = 0$ partial wave. The initial collision state of the two ultracold atoms is thus characterized by the set of quantum numbers

$$\alpha \quad : \quad \{f_1, m_{f_1}, f_2, m_{f_2}, \ell = 0\} \tag{16.4}$$

summarized by the simplified notation α. This set of quantum numbers defines a collision channel, here the *entrance channel*. There are of course other channels defined by other sets β of quantum numbers. The diagonal element of V^{int} in each channel provides a potential energy for each channel. But V^{int} also has off diagonal elements between different channels that give rise to coupled channel equations.

16.2.3 *Coupled channel equations*

The eigenstates of the total Hamiltonian with eigenvalues E can be written:

$$|\psi\rangle = \sum_{\alpha} \psi_\alpha(\vec{r})|\alpha\rangle \tag{16.5}$$

where $\psi_\alpha(\vec{r})$ is the wave function in channel α whose radial part is of the form

$$\frac{F_\alpha(r, E)}{r}. \tag{16.6}$$

The coupled equations of motion of the F_α's deduced from Schrödinger equation have the following form:

$$\frac{\partial^2}{\partial r^2} F_\alpha(r, E) + \frac{2\mu}{\hbar^2} \sum_{\beta} \left[E\, \delta_{\alpha\beta} - V_{\alpha\beta}(r)\right] F_\beta(r, E) = 0, \tag{16.7}$$

where

$$V_{\alpha\beta}(r) = \left[E_{f_i, m_{f_1}} + E_{f_2, m_{f_2}} + \frac{\ell(\ell+1)\hbar^2}{2\mu r^2}\right] \delta_{\alpha\beta} + V^{\text{int}}_{\alpha\beta}(r). \tag{16.8}$$

Solving these coupled differential equations numerically gives the asymptotic behavior of F_α for large r, from which one can determine the phase shift δ_0 and, consequently, the scattering length a in the entrance channel.

To get more insight into Feshbach resonances, we now consider a simplified model that takes only two modes into account: that containing the entering collision state and that containing the intermediate resonant bound state.

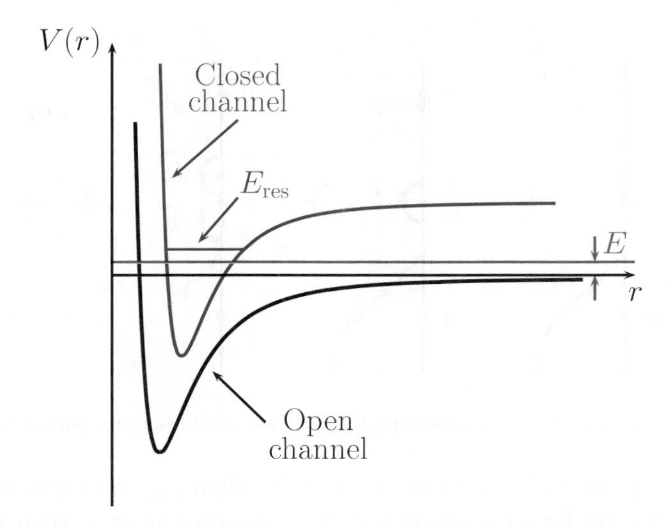

Fig. 16.2 Two-channel model: open channel containing the entering collision state with a very small energy E; closed channel, whose threshold is larger than E, containing a bound state whose energy $E_{\rm res}$ is very close to E.

16.2.4 *Two-channel model*

Figure 16.2 represents the two channels that are kept in the model. The colliding ultracold atoms, with a very small energy E, belong to the entrance collision channel, called the *open channel*. The other channel contains the resonant bound state $\varphi_{\rm res}$ whose energy $E_{\rm res}$ is very close to E. This channel is called *closed* because its dissociation threshold is above E, so that the two colliding atoms cannot belong to the continuum of this channel. By sweeping a static magnetic field, the open and the closed channel potentials move with respect to each other so that the difference between $E_{\rm res}$ and E is varied.

The state of the system can be written as a vector whose components refer to each channel:

$$\begin{pmatrix} |\varphi_{\rm op}\rangle \\ |\varphi_{\rm cl}\rangle \end{pmatrix} \tag{16.9}$$

where the indices op and cl refer to the spin state in each channel. In this two-state basis, the two-channel Hamiltonian H_2 is represented by:

$$H_2 = \begin{pmatrix} H_{\rm op} & W \\ W & H_{\rm cl} \end{pmatrix} \tag{16.10}$$

where

$$H_{\rm op} = T + V_{\rm op} \quad \text{and} \quad H_{\rm cl} = T + V_{\rm cl} \tag{16.11}$$

with $T = -\hbar^2\Delta/(2\mu)$ the kinetic energy operator. In Eq. (16.11), $V_{\rm op}$ and $V_{\rm cl}$ are the interaction potentials operators in the open and closed channels, respectively. W describes the coupling between the two channels.

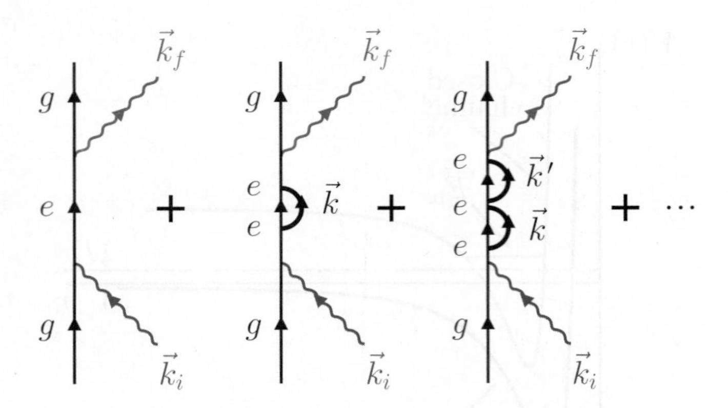

Fig. 16.3 Feynman diagrams representing the resonant scattering of a photon by an atom.

We will also neglect all eigenstates of $H_{\rm cl}$ other than $\varphi_{\rm res}$ since near the resonance ($E_{\rm res} \simeq 0$), they are far from resonance and, compared to $\varphi_{\rm res}$, their contribution is negligible. In this *single resonance approximation*, we can thus use the following simple expression for the closed channel Hamiltonian:

$$H_{\rm cl} = E_{\rm res} \left|\varphi_{\rm res}\right\rangle \left\langle\varphi_{\rm res}\right| . \tag{16.12}$$

In Secs. 16.4 and 16.5, we will show that this model is sufficiently simple to allow one to diagonalize the Hamiltonian H_2 and to get analytical expressions for the main physical quantities of interest:

- scattering length,
- energy and wave function of the bound state of the two-channel Hamiltonian when this bound state exists.

Before doing these calculations, we first present, in the next section, a simple physical interpretation of Feshbach resonances.

16.3 Qualitative discussion. Analogy between Feshbach resonances and resonant light scattering

Let us first recall the elementary processes involved in the calculation of the resonant light scattering amplitude.[5] The first diagram of Fig. 16.3 describes the lowest order scattering process. The impinging photon with wave vector $\vec{k}_i$ and energy $\hbar\omega_i = \hbar c k_i$ is absorbed by the atom which goes from the ground state g to an excited state e located at an energy $\hbar\omega_A$ above g. The atom then falls back to g by spontaneously emitting the scattered photon $(\vec{k}_f, \omega_f)$. The energy mismatch in the intermediate state is $\delta E = \hbar(\omega_i - \omega_A)$, so that the corresponding virtual transition lasts a time on the order of $\hbar/|\delta E| = 1/|\omega_i - \omega_A|$, that becomes longer

[5]For a more detailed discussion see Sec. 5.2.2 of Chap. 5.

and longer when ω_i tends to ω_A. The propagator of the intermediate state of the scattering process is proportional to $1/(\omega_i - \omega_A)$. This is the origin of the resonant enhancement of the scattering amplitude. The other Feynman diagrams represent higher order processes where the atom virtually emits and reabsorbs one or more photons before emitting the scattered photon $(\vec{k}_f, \omega_f)$. Summing the amplitudes corresponding to all these processes leads to a *renormalized propagator* for the excited state that is obtained by replacing its energy $\hbar\omega_A$ by $\hbar\left[\omega_A + \Delta_0 - i(\Gamma_0/2)\right]$, where $\hbar\Delta_0$ and $\hbar\Gamma_0$ are the energy shift and the energy width of e, respectively, due to the coupling of the discrete state e with the continuum of states where the atom is in g in the presence of a photon. The scattering process can also take place with intermediate states e' other than e, that are far from resonance when ω_i is close to ω_A. These give rise to a non-resonant background amplitude that is nearly constant when ω_i is scanned around ω_A. Finally, this analysis predicts a photon scattering amplitude of the form

$$A(\vec{k}_i \to \vec{k}_f) = A_{\text{non-resonant}} + A_{\text{resonant}} \tag{16.13}$$

where

$$A_{\text{resonant}} \propto \frac{1}{\omega_i - \omega_A - \Delta_0 + i(\Gamma_0/2)}. \tag{16.14}$$

A similar analysis can be applied to the scattering of two ultracold atoms with an intermediate bound resonant state in a closed channel (see Fig. 16.4). The first diagram at the left of this figure represents a process where the two ultracold atoms in the entrance collision state φ_{coll} combine into the bound state φ_{res} of the closed channel with energy E_{res} before dissociating into the final collision state. The propagator of this state varies as $1/(E - E_{\text{res}})$ and is very large when E_{res} is close to E. The other diagrams of Fig. 16.4 represent higher order processes where the intermediate bound state virtually dissociates and recombines a certain number of times before undergoing the final dissociation into the final collision state. As in the previous example, one expects that the resonant scattering amplitude varies as:

$$A_{\text{resonant}} \propto \frac{1}{E - E_{\text{res}} - \hbar\Delta_0 + i\hbar(\Gamma_0/2)} \tag{16.15}$$

where $\hbar\Delta_0$ and $\hbar\Gamma_0$ are the energy shift and the energy width of the bound state φ_{res}, respectively, that result from the coupling of this discrete state in the closed channel to the continuum of collision states in the open channel. The new feature here is that the density of states in the continuum of collision states is very small for the energy E of the initial state because the two colliding atoms are ultracold. It follows that the decay rate Γ_0 of φ_{res}, which according to the Fermi golden rule, is proportional to the density of states of the continuum at an energy equal to E_{res}, is negligible when E_{res} is close to $E \simeq 0$. One can thus replace Γ_0 by 0 in Eq. (16.15). Finally, one expects that the scattering length a, which is proportional to the scattering amplitude when $E \to 0$, has the form:

$$a = a_{\text{non-resonant}} + a_{\text{resonant}} \tag{16.16}$$

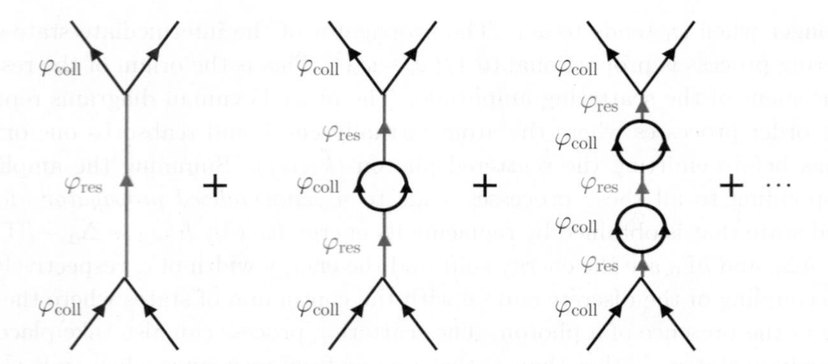

Fig. 16.4 Feynman diagrams representing the resonant scattering of two ultracold atoms.

where $a_{\text{non resonant}}$ is a background non resonant scattering length corresponding to scattering processes without intermediate association into a bound intermediate state, and where:

$$a_{\text{resonant}} \propto \frac{1}{B - B_0} \tag{16.17}$$

where B_0 is the value of the static field for which $E_{\rm res} + \hbar\Delta_0 = E \simeq 0$. Such a prediction is confirmed by the quantitative calculations presented in the next section.

16.4 Scattering states of the two-channel Hamiltonian

In this section, we first calculate the eigenstates of the Hamiltonian H_2 with a positive energy E. These form a continuum of scattering states of the two atoms, taking into account the effect of the coupling W between the two channels. From their asymptotic behavior when $E \to 0$, we deduce the scattering length a and analyze its variations with the static field.

16.4.1 *Calculation of the dressed scattering states*

Supposing that the state given by Eq. (16.9) is an eigenstate of the Hamiltonian (16.10) with an eigenvalue E that leads to a system of two coupled equations:

$$H_{\rm op}|\varphi_{\rm op}\rangle + W|\varphi_{\rm cl}\rangle = E|\varphi_{\rm op}\rangle, \tag{16.18}$$

$$W|\varphi_{\rm op}\rangle + H_{\rm cl}|\varphi_{\rm cl}\rangle = E|\varphi_{\rm cl}\rangle. \tag{16.19}$$

The two components of the scattering state corresponding to an incoming state with wave vector $\vec{k}$ will be denoted $\varphi_{\rm op}^{\vec{k}}$ and $\varphi_{\rm cl}^{\vec{k}}$. The first equation of (16.19) can then be written:

$$(E - H_{\rm op})\,|\varphi_{\rm op}^{\vec{k}}\rangle = W|\varphi_{\rm cl}^{\vec{k}}\rangle. \tag{16.20}$$

Its solution is the sum of a solution of the equation with the right-hand side set to zero and a solution of the full equation with the right-hand side considered as a source term:

$$|\varphi_{\rm op}^{\vec{k}}\rangle = |\varphi_{\vec{k}}^{+}\rangle + G_{\rm op}^{+}(E)\, W|\varphi_{\rm cl}^{\vec{k}}\rangle \quad \text{where} \quad G_{\rm op}^{+}(E) = \frac{1}{E - H_{\rm op} + i\,\varepsilon}. \tag{16.21}$$

In Eq. (16.21), $G_{\rm op}^{+}(E)$ is a Green function of $H_{\rm op}$. The term $+i\varepsilon$, where ε is a positive number that tends to zero, insures that the second term of Eq. (16.21) has the asymptotic behavior of an outgoing scattering state for $r \to \infty$. The first term of Eq. (16.21) involves only $H_{\rm op}$, and is chosen as an outgoing scattering state of $H_{\rm op}$ in order to get the good behavior for $r \to \infty$.[6]

$$\varphi_{\vec{k}}^{+}(\vec{r}) = \frac{1}{(2\pi)^{3/2}} \left[{\rm e}^{i\,\vec{k}.\vec{r}} + \frac{1}{E - T + i\varepsilon} V_{\rm op}\, \varphi_{\vec{k}}^{+}(\vec{r}) \right]. \tag{16.22}$$

The second equation of (16.19) can then be written:

$$(E - H_{\rm cl})\, |\varphi_{\rm cl}^{\vec{k}}\rangle = W|\varphi_{\rm op}^{\vec{k}}\rangle. \tag{16.23}$$

Its solution can be expressed in terms of the Green function of $H_{\rm cl}$:

$$|\varphi_{\rm cl}^{\vec{k}}\rangle = G_{\rm cl}(E)\, W|\varphi_{\rm op}^{\vec{k}}\rangle \quad \text{where} \quad G_{\rm cl}(E) = (E - H_{\rm cl})^{-1}. \tag{16.24}$$

Using the single resonance approximation (16.12), one gets:

$$|\varphi_{\rm cl}^{\vec{k}}\rangle = |\varphi_{\rm res}\rangle \frac{\langle \varphi_{\rm res}|W|\varphi_{\rm op}^{\vec{k}}\rangle}{E - E_{\rm res}}. \tag{16.25}$$

The two components $|\varphi_{\rm op}^{\vec{k}}\rangle$ and $|\varphi_{\rm cl}^{\vec{k}}\rangle$ of the scattering states of H_2 are sometimes called *dressed states* because they include the effect of the coupling W between the two channels. The eigenstates $|\varphi_{\vec{k}}^{+}\rangle$ and $|\varphi_{\rm res}\rangle$ of $H_{\rm op}$ and $H_{\rm cl}$ are called *bare states.*

Inserting Eq. (16.25) into Eq. (16.21) gives:

$$|\varphi_{\rm op}^{\vec{k}}\rangle = |\varphi_{\vec{k}}^{+}\rangle + G_{\rm op}^{+}(E)\, W|\varphi_{\rm res}\rangle \frac{\langle \varphi_{\rm res}|W|\varphi_{\rm op}^{\vec{k}}\rangle}{E - E_{\rm res}}. \tag{16.26}$$

In order to eliminate $|\varphi_{\rm op}^{\vec{k}}\rangle$ in the right-hand side, we multiply both sides of Eq. (16.21) by $\langle \varphi_{\rm res}|\, W$, which gives:

$$\frac{\langle \varphi_{\rm res}|W|\varphi_{\rm op}^{\vec{k}}\rangle}{E - E_{\rm res}} = \frac{\langle \varphi_{\rm res}|W|\varphi_{\vec{k}}^{+}\rangle}{E - E_{\rm res} - \langle \varphi_{\rm res}|W\, G_{\rm op}^{+}(E)\, W|\varphi_{\rm res}\rangle}. \tag{16.27}$$

Inserting Eq. (16.27) into Eq. (16.26), we finally get:

$$|\varphi_{\rm op}^{\vec{k}}\rangle = |\varphi_{\vec{k}}^{+}\rangle + G_{\rm op}^{+}(E)\, \frac{W|\varphi_{\rm res}\rangle\langle \varphi_{\rm res}|W}{E - E_{\rm res} - \langle \varphi_{\rm res}|W\, G_{\rm op}^{+}(E)\, W|\varphi_{\rm res}\rangle} |\varphi_{\vec{k}}^{+}\rangle. \tag{16.28}$$

Only bare states appear in the right-hand side of Eq. (16.28).

[6] The factor $(2\pi)^{-3/2}$ is introduced so that the outgoing scattering states are normalized $\langle \varphi_{\vec{k}}^{+}|\varphi_{\vec{k}'}^{+}\rangle = \delta(\vec{k} - \vec{k}')$.

Connection with the two-potential scattering

Equation (16.28) can be rewritten in a more suggestive way. If we introduce the effective coupling $V_{\rm eff}$ defined by:

$$V_{\rm eff} = W \frac{|\varphi_{\rm res}\rangle\langle\varphi_{\rm res}|}{E - E_{\rm res} - \langle\varphi_{\rm res}|W\, G^+_{\rm op}(E)\, W|\varphi_{\rm res}\rangle} W \tag{16.29}$$

we get, by inserting Eq. (16.29) into Eq. (16.28):

$$|\varphi^{\vec{k}}_{\rm op}\rangle = |\varphi^+_{\vec{k}}\rangle + \frac{1}{E - H_{\rm op} + i\,\varepsilon} V_{\rm eff}|\varphi^+_{\vec{k}}\rangle. \tag{16.30}$$

Like $V_{\rm op}$, $V_{\rm eff}$ acts only inside the open channel space. It describes the effect of virtual transitions to the closed channel subspace. The two-channel scattering problem can thus be reformulated in terms of a single-channel scattering problem (in the open channel) which describes the scattering produced by $V_{\rm eff}$ of waves distorted by $V_{\rm op}$ through a generalized Lippmann-Schwinger equation used in two-potential scattering problems.[7] In other words, the two incident particles interact with two potentials: $V_{\rm op}$ which acts in the entering channel and $V_{\rm eff}$ which accounts for the virtual transition and that corresponds to the scattering of waves distorded by the potential $V_{\rm op}$.

16.4.2 *Existence of a resonance in the scattering amplitude*

16.4.2.1 *Position and width of the resonance*

The second term of Eq. (16.28) is the most interesting since it gives the effects that arise from the coupling W. Its contribution to the scattering amplitude becomes large if its denominator vanishes, i.e. if:

$$E = E_{\rm res} + \langle\varphi_{\rm res}|\, W\, G^+_{\rm op}(E)\, W\, |\varphi_{\rm res}\rangle\,. \tag{16.31}$$

When E is close to 0, the last term of Eq. (16.31) is equal to:

$$\langle\varphi_{\rm res}|W\, G^+_{\rm op}(0)\, W|\varphi_{\rm res}\rangle = \sum_{\vec{k}} \frac{|\langle\varphi_{\rm res}|W|\varphi^+_{\vec{k}}\rangle|^2}{-E_{\vec{k}} + i\,\varepsilon} = \hbar\Delta_0. \tag{16.32}$$

Its interpretation is clear: It gives the shift $\hbar\Delta_0$ of $|\varphi_{\rm res}\rangle$ due to the second order coupling induced by W between $|\varphi_{\rm res}\rangle$ and the continuum of $H_{\rm op}$. We thus predict that the scattering amplitude, and so the scattering length, will be maximum (in absolute value), not when $E_{\rm res}$ is close to 0, but when the shifted energy of $|\varphi_{\rm res}\rangle$

$$\tilde{E}_{\rm res} = E_{\rm res} + \hbar\,\Delta_0 \tag{16.33}$$

is close to the energy $E \simeq 0$ of the incoming state.

Strictly speaking, the Green function $G^+_{\rm op}(E) = \left(E - E^{\rm op}_{\vec{k}} + i\,\varepsilon\right)^{-1}$ appearing in Eq. (16.31) is equal to:

$$\frac{1}{E - E^{\rm op}_{\vec{k}} + i\,\varepsilon} = \mathcal{P}\left(\frac{1}{E - E^{\rm op}_{\vec{k}}}\right) - i\,\pi\,\delta\left(E - E^{\rm op}_{\vec{k}}\right) \tag{16.34}$$

[7] See for example Chap. 17 of [Joachain (1983)].

where $\mathcal{P}$ means principal part. Because of the last term of Eq. (16.34), Eq. (16.32) should also contain an imaginary term that describes the damping of $|\varphi_{\rm res}\rangle$ due to its coupling with the continuum of $H_{\rm op}$ induced by W. However, we are considering here the limit of ultracold collisions ($E \to 0$) where the density of states of the continuum of $H_{\rm op}$ vanishes, which means that the damping of $|\varphi_{\rm res}\rangle$ can be ignored in the limit $E \to 0$.

All the qualitative predictions of the previous section concerning the position and the width of the resonance of the scattering amplitude are thus confirmed.

16.4.2.2 *Resonant value B_0 of the magnetic field*

Assuming that the spin configurations of the two channels have different magnetic moments, the energies of the states in these channels vary differently when a static magnetic field B is applied and scanned. If ξ is the difference of the magnetic moments in the two channels, the difference between the energies of two states belonging to the channels varies linearly with B, with a slope ξ. If we take the energy of the dissociation threshold of the open channel as the zero of energy, the energy $E_{\rm res}$ of $\varphi_{\rm res}$ is equal to:

$$E_{\rm res} = \xi\,(B - B_{\rm res}) \tag{16.35}$$

where $B_{\rm res}$ is the magnetic field for which $E_{\rm res}$ is degenerate with the energy $E \simeq 0$ of the ultracold collision state. In fact, the position of the resonance is given, not by the zero of $E_{\rm res}$, but by the zero of $\tilde{E}_{\rm res}$

$$\tilde{E}_{\rm res} = E_{\rm res} + \hbar\Delta_0 = \xi\,(B - B_0)\,. \tag{16.36}$$

This equation gives the correct value of the field, B_0, at which one expects a divergence of the scattering amplitude. Figure 16.5 gives the variations of $E_{\rm res}$ and $\tilde{E}_{\rm res}$ with the static magnetic field B. We suppose here that $\xi < 0$. Since, according to Eq. (16.32), Δ_0 is also negative, B_0 is smaller than $B_{\rm res}$.

16.4.3 *Asymptotic behavior of the dressed scattering states*

Only the asymptotic behavior of the open channel component of the dressed scattering state (16.28) is relevant because the closed channel component, proportional to $|\varphi_{\rm res}\rangle$, tends to zero much more rapidly. We expect the asymptotic behavior of $|\varphi_{\rm op}^{\vec{k}}\rangle$ to be of the form:

$$\varphi_{\rm op}^{\vec{k}}(\vec{r}) \underset{r\to\infty}{\simeq} \frac{1}{(2\pi)^{3/2}}\left[\mathrm{e}^{i\,\vec{k}\cdot\vec{r}} + f(k,\vec{n})\,\frac{\mathrm{e}^{i\,k\,r}}{r}\right] \quad \text{where } \vec{n} = \frac{\vec{r}}{r}. \tag{16.37}$$

In the limit $\vec{k} \to 0$, the scattering amplitude becomes spherically symmetric and gives the scattering length that we want to calculate:

$$f(k,\vec{n}) \underset{k\to 0}{\to} -a. \tag{16.38}$$

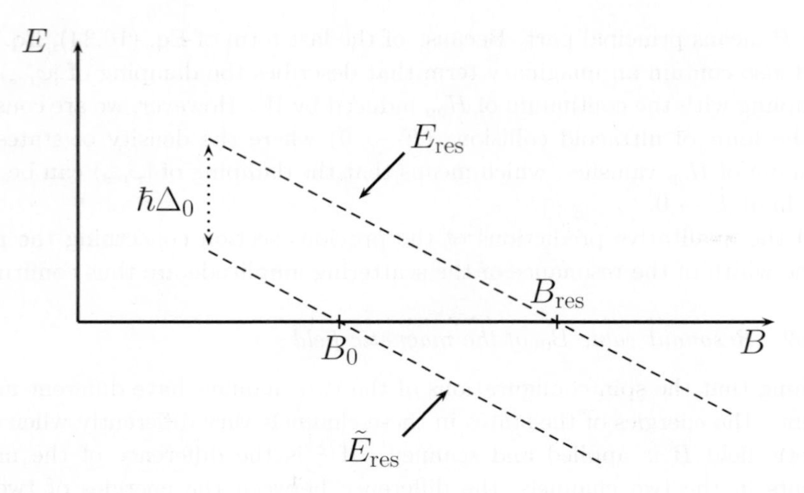

Fig. 16.5 Variations of $E_{\rm res}$ and $\tilde{E}_{\rm res}$ with the static magnetic field B.

16.4.3.1 *Background scattering length*

The first term of Eq. (16.28) describes the scattering in the open channel without coupling to the closed channel. Its asymptotic behavior gives the scattering length $a_{\rm op}$ in the open channel with $W = 0$. This scattering length is often called the *background* scattering length

$$a_{\rm op} = a_{\rm bg}. \tag{16.39}$$

16.4.3.2 *Effect of the inter-channel coupling on the asymptotic behavior*

The second term of Eq. (16.28) describes the effect of the coupling to the closed channel. Using Eqs. (16.31) and (16.33), we can rewrite (when $E \simeq 0$) Eq. (16.26) in the following way:

$$|\varphi_{\rm op}^{\vec{k}}\rangle = |\varphi_{\vec{k}}^{+}\rangle + G_{\rm op}^{+}(E)\,\frac{W|\varphi_{\rm res}\rangle\langle\varphi_{\rm res}|W}{E-\tilde{E}_{\rm res}}|\varphi_{\vec{k}}^{+}\rangle. \tag{16.40}$$

The contribution of the inter-channel coupling to the asymptotic behavior of the dressed scattering state is given by the large r dependence of the last term of this equation:

$$\langle\vec{r}|G_{\rm op}^{+}(E)\,\frac{W|\varphi_{\rm res}\rangle\langle\varphi_{\rm res}|W}{E-\tilde{E}_{\rm res}}|\varphi_{\vec{k}}^{+}\rangle = \int \mathrm{d}^3r'\langle\vec{r}|G_{\rm op}^{+}(E)|\vec{r}\,'\rangle\langle\vec{r}\,'|\,\frac{W|\varphi_{\rm res}\rangle\langle\varphi_{\rm res}|W}{E-\tilde{E}_{\rm res}}|\varphi_{\vec{k}}^{+}\rangle. \tag{16.41}$$

We thus need to know the asymptotic behavior of the Green's function of $H_{\rm op}$ for large r:

$$G_{\rm op}^{+}\left(E,\vec{r},\vec{r}\,'\right) = \langle\vec{r}|\,\frac{1}{E-H_{\rm op}+i\,\varepsilon}\,|\vec{r}\,'\rangle\,. \tag{16.42}$$

One can show that this Green function can be expressed in terms of the incoming scattering states of $H_{\rm op}$:[8]

$$G^+_{\rm op}(E,\vec{r},\vec{r}\,') \underset{r\to\infty}{\simeq} -\frac{{\rm e}^{ikr}}{r}\frac{2\mu}{\hbar^2}\sqrt{\frac{\pi}{2}}\left[\varphi^-_{k\,\vec{n}}(\vec{r}\,')\right]^* \qquad \vec{n}=\vec{r}/r \tag{16.43}$$

where $\varphi^-_{k\,\vec{n}}$ is the incoming scattering state of $H_{\rm op}$ leading to an outgoing state $k\,\vec{n}$. Using $\left[\varphi^-_{k\,\vec{n}}(\vec{r}\,')\right]^* = \langle\varphi^-_{k\,\vec{n}}|\vec{r}\,'\rangle$ and the closure relation for $\vec{r}\,'$, we get for the asymptotic behavior of the right-hand side of Eq. (16.41):

$$-\frac{{\rm e}^{ikr}}{r}\frac{2\mu}{\hbar^2}\sqrt{\frac{\pi}{2}}\frac{\langle\varphi^-_{k\,\vec{n}}|W|\varphi_{\rm res}\rangle\langle\varphi_{\rm res}|W|\varphi^+_{\vec{k}}\rangle}{E-\tilde{E}_{\rm res}}. \tag{16.44}$$

In the limit $k\to 0$, this expression can be written:

$$-\frac{1}{r}\frac{2\mu}{\hbar^2}\sqrt{\frac{\pi}{2}}\frac{|\langle\varphi^+_{\vec{0}}|W|\varphi_{\rm res}\rangle|^2}{0-\tilde{E}_{\rm res}} = +\frac{1}{r}\frac{2\mu}{\hbar^2}2\pi^2\frac{|\langle\varphi^+_{\vec{0}}|W|\varphi_{\rm res}\rangle|^2}{\xi(B-B_0)}. \tag{16.45}$$

16.4.4 *Scattering length. Feshbach resonance*

The asymptotic behavior of the first term of Eq. (16.28) gives the background scattering length. We must now add the contribution of the second term we have just calculated in Eq. (16.45). To get the asymptotic behavior of the scattering amplitude $f(k,\vec{n})$ appearing in Eq. (16.37), we have to multiply (16.45) by $(2\pi)^{3/2}$ because of the normalization coefficient $(2\pi)^{-3/2}$ in Eq. (16.37). We thus get for the total scattering length:

$$a = a_{\rm bg} - \frac{2\mu}{\hbar^2}2\pi^2\frac{|\langle\varphi^+_{\vec{0}}|W|\varphi_{\rm res}\rangle|^2}{\xi(B-B_0)} = a_{\rm bg}\left(1-\frac{\Delta B}{B-B_0}\right) \tag{16.46}$$

where

$$\Delta B = \frac{2\mu}{\hbar^2}2\pi^2\frac{|\langle\varphi^+_{\vec{0}}|W|\varphi_{\rm res}\rangle|^2}{\xi\, a_{\rm bg}}. \tag{16.47}$$

The simple expression (16.46) of the variation of a with B was given in [Moerdijk *et al.* (1995)].

As predicted from the simple discussion of Sec. 16.3, we find that in the two-channel model the scattering length is a sum of two terms: one background term which does not vary around B_0, and another term that varies as $1/(B-B_0)$. The variations of the scattering length with the static field are represented in Fig. 16.6. They clearly exhibit the main results derived in this chapter:

- The scattering length diverges when $B=B_0$,
- Its sign changes when B is scanned around B_0,
- It vanishes when $B-B_0=\Delta B$.

[8] To demonstrate (16.43), one can use the approach detailed in Exercise 4 of Chap. XIX of Ref. [Messiah (2003)].

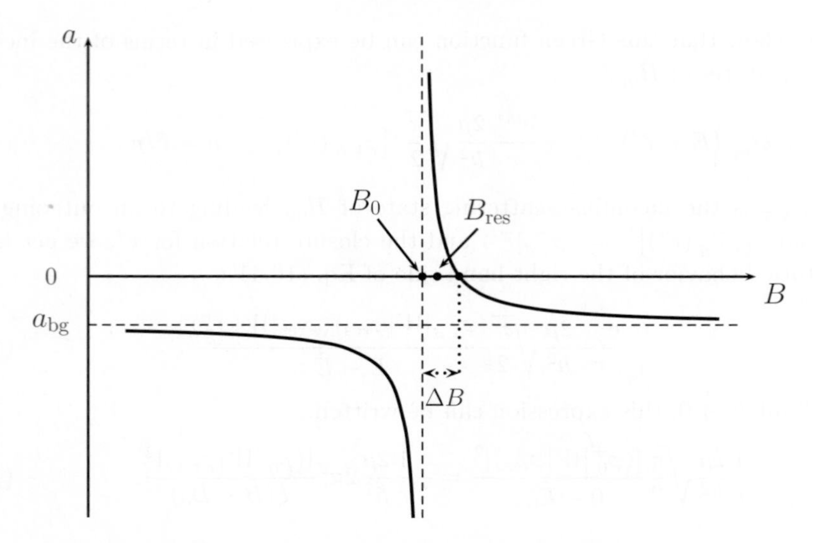

Fig. 16.6 Variations of the scattering length a with the magnetic field B. The figure corresponds to two colliding rubidium-85 atoms each in the state $f = 2, m_f = -2$. In this case, we have $a_{bg} < 0$ and $\xi < 0$.

It is also evident in Fig. 16.6 that the width of the Feshbach resonance is related to the quantity ΔB. The larger ΔB, the broader the resonance. According to Eq. (16.47), ΔB is proportional to the square of the modulus of the matrix element of the coupling W between the discrete state and the continuum and inversely proportional to the background scattering length.

The first experiment reporting the observation of Feshbach resonances with clouds of cold atoms was done at MIT in the group of Wolfgang Ketterle [Inouye *et al.* (1998)]. The experiment was performed on a Bose-Einstein condensate of sodium atoms confined in a dipole trap. A uniform external magnetic fields was varied in the range where the theoretical prediction of the resonance position was made. The dispersive variation of the scattering length, by a factor over ten, was observed by measuring the interaction energy of the condensate through a time-of-flight experiment. In addition, enhanced inelastic processes that resulted in trap loss were observed in the vicinity of the resonance.[9] Soon after, a Feshbach resonance of rubidium-85 atoms was observed by Zeeman-resolved photoassociation spectroscopy [Courteille *et al.* (1998)], and by a dramatic increase of the collision rate in the vicinity of the resonance [Roberts *et al.* (1998)].

[9]Near the resonance, the duration of the collision process increases which favors inelastic output channels.

16.5 Bound states of the two-channel Hamiltonian

In this section, we show that the two-channel Hamiltonian has a single bound state when the static magnetic field B has a value corresponding to a positive scattering length. We study the B-dependence of the energy and wave function of this bound state. In the following, we denote such a bound state as:

$$\begin{pmatrix} |\varphi^b_{\rm op}\rangle \\ |\varphi^b_{\rm cl}\rangle \end{pmatrix}, \tag{16.48}$$

where $|\varphi^b_{\rm op}\rangle$ and $|\varphi^b_{\rm cl}\rangle$ are the components of the bound state in the open and closed channels, respectively. Expressing that the state (16.48) is an eigenstate of the Hamiltonian (16.10) with eigenvalue E_b, we get the following two equations:

$$H_{\rm op}|\varphi^b_{\rm op}\rangle + W|\varphi^b_{\rm cl}\rangle = E_b|\varphi^b_{\rm op}\rangle, \tag{16.49}$$

$$W|\varphi^b_{\rm op}\rangle + H_{\rm cl}|\varphi^b_{\rm cl}\rangle = E_b|\varphi^b_{\rm cl}\rangle. \tag{16.50}$$

16.5.1 *Calculation of the energy of the bound state*

To solve the set of Eqs. (16.50), we can use the Green's function of $H_{\rm op}$ and $H_{\rm cl}$ without the $i\varepsilon$ term, since E_b is negative (below the threshold of $V_{\rm op}$). This gives:

$$\left|\varphi^b_{\rm op}\right\rangle = G_{\rm op}(E_b)\, W\, \left|\varphi^b_{\rm cl}\right\rangle \quad \text{and} \quad \left|\varphi^b_{\rm cl}\right\rangle = G_{\rm cl}(E_b)\, W\, \left|\varphi^b_{\rm op}\right\rangle. \tag{16.51}$$

Using the single resonance approximation for $G_{\rm cl}$

$$G_{\rm cl}(E_b) = \frac{|\varphi_{\rm res}\rangle\,\langle\varphi_{\rm res}|}{E_b - E_{\rm res}} \tag{16.52}$$

in the second equation of (16.51) shows that $|\varphi^b_{\rm cl}\rangle$ is proportional to $|\varphi_{\rm res}\rangle$, so that:

$$\begin{pmatrix} |\varphi^b_{\rm op}\rangle \\ |\varphi^b_{\rm cl}\rangle \end{pmatrix} = \frac{1}{N_b}\begin{pmatrix} G_{\rm op}(E_b)\, W|\varphi_{\rm res}\rangle \\ |\varphi_{\rm res}\rangle \end{pmatrix} \tag{16.53}$$

where N_b is a normalization factor given by:

$$N_b = \sqrt{1 + \langle\varphi_{\rm res}|W\, G^2_{\rm op}(E_b)W|\varphi_{\rm res}\rangle}. \tag{16.54}$$

From Eqs. (16.51) and (16.52) and the results of the previous section, one can derive an implicit equation for E_b:

$$E_b = \tilde{E}_{\rm res} - (2\mu)^2 E_b \int {\rm d}^3k \frac{|\langle\varphi_{\rm res}|W|\varphi^+_{\vec{k}}\rangle|^2}{\hbar^2k^2(\hbar^2k^2 + 2\mu|E_b|)}. \tag{16.55}$$

Demonstration of Eq. (16.55)

Inserting Eq. (16.52) into the second equation of (16.51) gives:

$$|\varphi^b_{\rm cl}\rangle = \frac{1}{E_b - E_{\rm res}}|\varphi_{\rm res}\rangle\langle\varphi_{\rm res}|W|\varphi^{\rm b}_{\rm op}\rangle \tag{16.56}$$

which, inserted into Eq. (16.49) leads to:

$$|\varphi^{\rm b}_{\rm op}\rangle = \frac{1}{E_b - E_{\rm res}} G_{\rm op}(E_b)\, W|\varphi_{\rm res}\rangle\langle\varphi_{\rm res}|W|\varphi^{\rm b}_{\rm op}\rangle. \tag{16.57}$$

As for Eq.(16.26), we can eliminate the dressed state $|\varphi_{\text{op}}^b\rangle$ by multiplying both sides of Eq. (16.57) by $\langle\varphi_{\text{res}}|W$. This gives:

$$E_b - E_{\text{res}} = \langle\varphi_{\text{res}}|\, W\, G_{\text{op}}(E_b)\, W\, |\varphi_{\text{res}}\rangle\,. \tag{16.58}$$

Using the identity

$$G_{\text{op}}(E_b) = G_{\text{op}}(0) - E_b G_{\text{op}}(0) G_{\text{op}}(E_b). \tag{16.59}$$

we can rewrite (16.58) as:

$$E_b = E_{\text{res}} + \langle\varphi_{\text{res}}|\, W G_{\text{op}}(0) W\, |\varphi_{\text{res}}\rangle - E_b\, \langle\varphi_{\text{res}}|\, W G_{\text{op}}(0) G_{\text{op}}(E_b) W\, |\varphi_{\text{res}}\rangle \tag{16.60}$$

The second term of the right-hand side of Eq. (16.60) is the shift $\hbar\Delta_0$ of φ_{res}. Adding it to E_{res}, we get $\tilde{E}_{\text{res}}$, so that Eq. (16.60) can be rewritten:

$$E_{\text{b}} = \tilde{E}_{\text{res}} - E_b\, \langle\varphi_{\text{res}}|\, W\, G_{\text{op}}(0)\, G_{\text{op}}(E_b)\, W\, |\varphi_{\text{res}}\rangle\,. \tag{16.61}$$

To go further, we introduce the spectral decomposition of $G_{\text{op}}(z)$:

$$G_{\text{op}}(z) = \int \mathrm{d}^3k \frac{|\varphi_{\vec{k}}^{+}\rangle\langle\varphi_{\vec{k}}^{+}|}{z - \hbar^2k^2/2\mu} + G_{\text{op}}^{\text{b}}(z). \tag{16.62}$$

The last term of Eq. (16.62) gives the contribution of the bound states of H_{op}. We suppose here that their energy is far below $E = 0$, so that we can ignore this term. Inserting Eq. (16.62) into Eq. (16.61) finally leads to Eq. (16.55).

We now try to solve the implicit Eq. (16.55). We first note that E_b is negative, so that the right-hand side of Eq. (16.55) is larger than $\tilde{E}_{\text{res}}$. Therefore, a negative solution of (16.55) for E_b only exists if $\tilde{E}_{\text{res}} < 0$. According to Fig. 16.5, this is possible only if $B > B_0$, which corresponds to a positive scattering length. There is no bound state for the two-channel Hamiltonian when $a < 0$.

To calculate the integral of Eq. (16.55), we introduce the new variable:

$$u = \frac{\hbar\, k}{\sqrt{2\mu\, |E_b|}} \tag{16.63}$$

which allows one to rewrite, after angular integration, the integral of Eq. (16.55) as:

$$\frac{1}{\hbar^3} \frac{4\pi}{\sqrt{2\mu|E_b|}} \int_0^\infty \mathrm{d}u \frac{|\langle\varphi_{\text{res}}|W|\varphi_{\vec{k}}^{+}\rangle|^2}{(u^2+1)}. \tag{16.64}$$

Let k_0 be the width of $|\langle\varphi_{\text{res}}|W|\varphi_{\vec{k}}^{+}\rangle|^2$ considered as a function of k. This defines a value u_0 of u

$$u_0 = \frac{\hbar\, k_0}{\sqrt{2\mu\, |E_b|}} \tag{16.65}$$

that characterizes the width in u of the numerator of the integrand of Eq. (16.64).

Two different limits can then be considered depending on whether u_0 is much larger or much smaller than 1.

First limit: $u_0 \gg 1 \quad \Leftrightarrow \quad |E_b| \ll \hbar^2 k_0^2/2\mu$

The denominator of the integral of Eq. (16.64) varies more rapidly with u than the numerator which can be replaced by its value for $\vec{k} = \vec{0}$. Equation (16.64) can thus be approximated by:

$$\frac{1}{\hbar^3}\frac{4\pi}{\sqrt{2\mu|E_b|}}|\langle\varphi_{\text{res}}|W|\varphi_{\vec{0}}^+\rangle|^2 \underbrace{\int_0^\infty \frac{\mathrm{d}u}{u^2+1}}_{=\pi/2}. \tag{16.66}$$

Replacing the integral of Eq. (16.55) by Eq. (16.66) yields:

$$E_b = \tilde{E}_{\text{res}} + \sqrt{|E_b|}\frac{2\pi^2(2\mu)^{3/2}}{\hbar^3}|\langle\varphi_{\text{res}}|W|\varphi_{\vec{0}}^+\rangle|^2. \tag{16.67}$$

We can re-express $\left|\langle\varphi_{\text{res}}|\, W\, |\varphi_{\vec{0}}^+\rangle\right|^2$ in terms of ΔB thanks to (16.47) and $\tilde{E}_{\text{res}} < 0$ in terms of $\xi(B - B_0)$ thanks to Eq. (16.36). When B is larger but close to B_0, one can neglect the l.h.s of Eq. (16.67), and obtain an approximate value for E_b:

$$E_b \simeq -\frac{\hbar^2}{2\mu a^2} \propto (B - B_0)^2. \tag{16.68}$$

Second limit: $u_0 \ll 1 \quad \Leftrightarrow \quad |E_b| \gg \hbar^2 k_0^2/2\mu$

The numerator of the integral of Eq. (16.64) varies more rapidly with u than the denominator, so that we can neglect the term in u^2 in the denominator. In fact, this approximation amounts to neglecting $\hbar^2 k^2$ compared to $2\mu|E_b|$ in the denominator of the integral of Eq. (16.55) and allows us to transform Eq. (16.55) into:

$$\begin{aligned} E_b &= \tilde{E}_{\text{res}} + (2\mu)^2 \int \mathrm{d}^3k \frac{|\langle\varphi_{\text{res}}|W|\varphi_{\vec{k}}^+\rangle|^2}{2\mu\,\hbar^2 k^2}, \\ &= \tilde{E}_{\text{res}} + \int \mathrm{d}^3k \frac{|\langle\varphi_{\text{res}}|W|\varphi_{\vec{k}}^+\rangle|^2}{\hbar^2 k^2/2\mu}, \\ &= \tilde{E}_{\text{res}} - \hbar\Delta_0 = E_{\text{res}} = \xi\left(B - B_{\text{res}}\right). \end{aligned} \tag{16.69}$$

We have used the expression (16.32) of $\hbar\Delta_0$ and Eq. (16.35).

The results of this calculation are summarized in Fig. 16.7. The two-channel Hamiltonian has no bound state when $\tilde{E}_{\text{res}} > 0$. This is easy to understand. When the bound state of the open channel, shifted by its coupling with the continuum of the open channel, has a shifted energy $\tilde{E}_{\text{res}}$ which is still positive, it remains resonantly coupled to the continuum and can decay into this continuum, so that it is no longer a discrete state but a metastable one. From Eqs. (16.36), (16.46) and (16.47), it is easy to check that condition $\tilde{E}_{\text{res}} > 0$ is equivalent to $a > 0$ and to $B - B_0 > 0$ (< 0) when $\xi < 0$ (> 0).

When $B - B_0$ is very small compared to $B_{\text{res}} - B_0$, $|E_b|$ increases quadratically with $B - B_0$. When $E_{\text{res}} \to -\infty$, the coupling of the bound state of the closed channel to the continuum of the open channel becomes negligible and the bound state of the two-channel Hamiltonian coincides with the bound state of the open channel, so that E_b coincides with E_{res}.

16.5.2 *Wave function of the bound state*

From Eqs. (16.53) and (16.54), it is possible to calculate the relative weight of the closed channel component $\varphi_{\rm cl}^b$ in the (normalized) bound state wave function of the Hamiltonian H_2. One gets:

$$\langle\varphi_{\rm cl}^b|\varphi_{\rm cl}^b\rangle = \frac{1}{N_b^2} \quad \text{with} \quad N_b^2 = 1 + \langle\varphi_{\rm res}|\, W\, G_{\rm op}^2(E_b) W\, |\varphi_{\rm res}\rangle\,. \tag{16.70}$$

Using

$$G_{\rm op}(E_b) = \frac{1}{E_b - H_{\rm op}} \quad \Rightarrow \quad \frac{\partial}{\partial E_b} G_{\rm op}(E_b) = -\frac{1}{\left(E_b - H_{\rm op}\right)^2} = -G_{\rm op}^2(E_b). \tag{16.71}$$

we can rewrite the second equation of (16.70) as:

$$N_b^2 = 1 - \frac{\partial}{\partial E_b} \langle\varphi_{\rm res}|\, W\, G_{\rm op}(E_b) W\, |\varphi_{\rm res}\rangle\,. \tag{16.72}$$

Combining this equation with Eqs. (16.35) and (16.58), we find for the normalization coefficient N_b the simple expression

$$\frac{1}{N_b^2} = \frac{1}{\xi}\frac{\partial E_b}{\partial B}. \tag{16.73}$$

This result can be expressed in simple terms. For a given value of B, the weight of the closed channel component in the wave function of the dressed bound state is just given by the slope of the curve giving $E_b(B)$ versus B, divided by the slope ξ of the asymptote of this curve (see Fig. 16.7). When the bound state of the two-channel Hamiltonian appears near $B = B_0$ in the region $a > 0$, the slope of the curve $E_b(B)$ is equal to 0 and the weight of the closed channel component in its wave function is negligible. The dressed bound state is essentially a linear combination of states of the continuum of $H_{\rm op}$. For larger values of B, near the asymptote of $E_b(B)$, this weight tends to 1 and the dressed bound state coincides with the bare one.

The previous conclusion means that, near the Feshbach resonance, the coupling with the closed channel can be neglected for calculating the wave function of the bound state and that we can thus look for the eigenfunction of $H_{\rm op}$ with an eigenvalue $-\hbar^2/2\mu a^2$. The asymptotic behavior of this wave function (at distances larger than the range of $V_{\rm op}$) can be obtained by solving the one-dimensional radial Schrödinger equation for $u_0(r)$ with $V_{\rm op} = 0$:

$$-\frac{\hbar^2}{2\mu}\frac{{\rm d}^2 u_0(r)}{{\rm d}\, r^2} = -\frac{\hbar^2}{2\mu a^2} u_0(r). \tag{16.74}$$

The three-dimensional wave function of the dressed bound state thus behaves asymptotically as:

$$\frac{\exp(-r/a)}{r}. \tag{16.75}$$

Its spatial extent is given by the scattering length a.

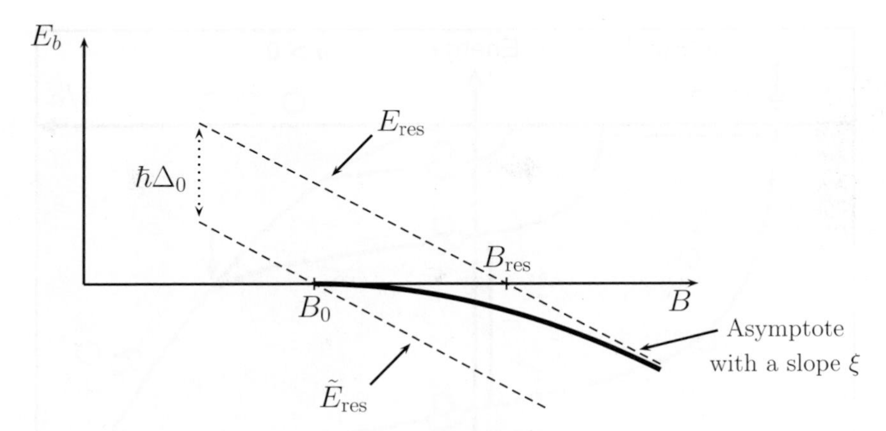

Fig. 16.7 Energy E_b of the bound state of the two-channel Hamiltonian versus the magnetic field B. As in Fig. 16.5, we suppose here that $\xi < 0$. The bound state only exists for $\tilde{E}_{res} < 0$, which is equivalent to $B > B_0$ when ξ is negative (and also to $a > 0$ for all signs of ξ). After a quadratic increase of $|E_b|$ with $B - B_0$, E_b tends to the unshifted energy E_{res} of the bound state of the closed channel.

16.5.3 *Halo states*

Equations (16.68) and (16.75) show that near the Feshbach resonance the properties of the bound state are universal: its wave function and its energy depend only on a single quantity, the scattering length a, and not on the details of the interaction potential giving rise to this scattering length. Near the Feshbach resonance, a is much larger than the range of the atom-atom interaction potential, and the wave function of the Feshbach molecule (also referred to as Feshbach dimer) extends far into the classically forbidden region. These giant molecules are also called *halo states.* Examples of halo states are found in other fields, like the deuteron in nuclear physics and the helium molecules.[10] Feshbach dimers are particularly interesting because one can change their size by just sweeping a magnetic field.

Efimov states

Quantum halo states also exist for three body systems. In 1970, Vitaly Efimov predicted the existence of giant bound states of three identical bosons [Efimov (1970, 1971)]. After this theoretical prediction, a long period of more than three decades elapsed before homonuclear Efimov trimers were experimentally observed in ultracold cesium, potassium and lithium atoms [Kraemer *et al.* (2006); Zaccanti *et al.* (2009); Gross *et al.* (2009); Pollack *et al.* (2009)]. Figure 16.8 reproduces the energy of these trimers T versus the inverse of the scattering length $1/a$ (green curves of the figure). There are an infinite series of such curves. Two successive Efimov states appear at values of a differing by a universal scaling factor x equal to 22.7 (the figure is not at scale). Their binding energies differ by x^2. In the region $a < 0$, there are no bound states.

Efimov states were detected as resonances in the three-body decay rates near a Fes-

[10]For a review on quantum halos see [Jensen *et al.* (2004)] and references therein.

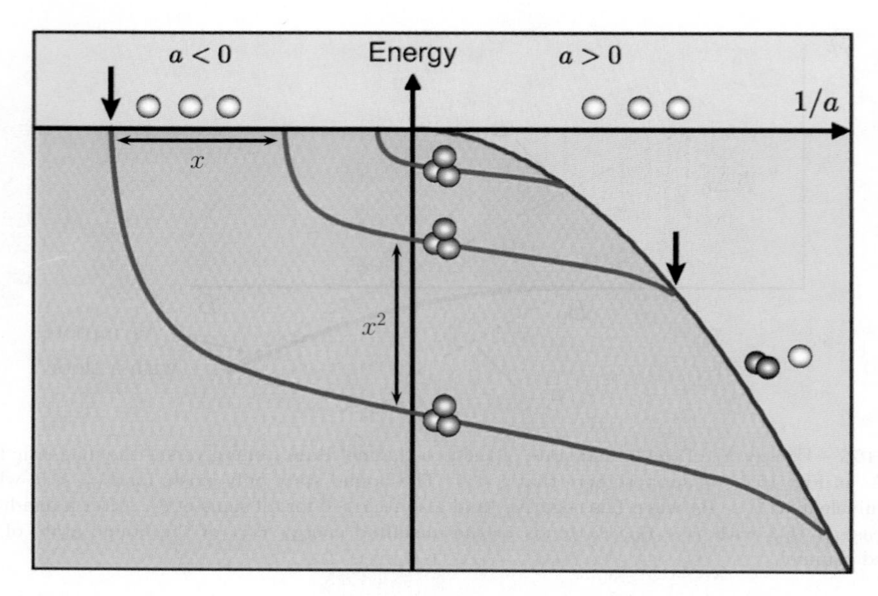

Fig. 16.8 Energy of the Efimov trimer states versus the inverse of the scattering length (green curves). A series of families of states associated with the different green curves is predicted with a scaling factor $x = 22.7$ for the relative sizes of two successive trimers and x^2 for their binding energies. The heavy black line in the right part of the figure is the energy of the dimer predicted to exist for $a > 0$. Figure adapted from [Ferlaino and Grimm (2010)]. Copyright: American Physical Society.

hbach resonance. Suppose that $|a|$ is slowly decreased in the region $a < 0$. When a reaches the value corresponding to the abscissa of the starting point of the lowest green curve of Fig. 16.8, three atoms A can combine and form an Efimov trimer: $A + A + A \to T$. This opens decay channels into deeply bound dimer states and appears as resonant enhancement of the atom loss near the resonant value of a. Note that we have already shown that there are no dimers D corresponding to a bound state of two atoms A in the region $a < 0$. There are, however, Efimov trimers. This is why these Efimov states are sometimes called *Borromean* states since they can stay together without any two-body binding.

In the region $a > 0$, the heavy black line represents the energy of the dimers discussed in Sec. 16.5. The lowest green curve intersects this heavy black line for a certain value of a. If a is decreased in the region $a > 0$ and reaches this value, a dimer D can combine with an atom A and form a trimer T: $D + A \to T$. This results in resonant atom-dimer relaxation loss rates and can be used to detect the right end of the trimer curve.

More recently, the next family of Efimov states (second green curve of Fig. 16.8) has been detected in ultracold potassium atoms [Zaccanti *et al.* (2009)]. The sizes of the trimers of the two first families have been found to scale with the expected factor $x = 22.7$. Interspecies trimers KRbRb and KKRb formed with potassium and rubidium atoms have also been observed [Barontini *et al.* (2009)].

16.6 Producing ultracold molecules

A great amount of attention is presently being paid to the production of ultracold molecules because important developments are expected in this research field, in fundamental and applied physics as well as in quantum chemistry. In this chapter, we will not review all these developments[11] though a few of them will be described in subsequent chapters, in connection with the study of quantum degenerate gases. Our main concern in this chapter, devoted to the control of atom-atom interactions, is to show how ultracold molecules can be produced from ultracold atoms by controlling their interactions with a magnetic field (magnetic tuning of a Feshbach resonance) or an optical field (photoassociation).

16.6.1 *Magnetic tuning of a Feshbach resonance*

The principle of this method is shown in Fig. 16.9. If the magnetic field B is slowly swept through the Feshbach resonance from the region $a < 0$ to the region $a > 0$, the system of the two ultracold atoms, initially in a very low energy state, remains adiabatically in the lowest energy state, which, in the region $a > 0$, becomes the molecular bound state studied in the previous section.[12]

Once they are formed, these Feshbach molecules, with a spatial extent on the order of a, can undergo three-body inelastic collisions with the background ultracold gas which transfer them to deep molecular bound states with a spatial extent on the order of the potential range, $b \ll a$. The released energy in this process is shared between the molecules and the remaining atom.

It turns out that Feshbach molecules formed from two fermionic atoms in two different spin states are much less sensitive to these inelastic processes than those formed with bosonic atoms. Such a difference comes from the Pauli principle. In a collision between a Feshbach molecule formed from two fermions in different spin states and a fermionic atom from the background ultracold gas, two of the three fermions are necessarily in the same spin state. The inelastic collision rate is actually determined by the probability to have these three fermions in the volume b^3 associated with the deeply bound molecular state. Pauli principle dramatically reduces this probability because it forbids two identical fermions to be close enough to one another. As a result, the inelastic collision rate is reduced by a factor proportional to a power of $b/a \ll 1$ [Petrov *et al.* (2005)].

We will see in Chap. 26 that this method of sweeping the magnetic field from a region where $a < 0$ to a region where $a > 0$ has been extensively used to produce ultracold molecules from an ultracold sample of a two-component atomic Fermi gas. A sufficiently large number of such molecules can be formed, to reach the threshold

---

[11] We refer the interested reader to recent review papers [Heinzen (1999); Pillet *et al.* (2003); Carr and Ye (2009); Bell and Softley (2009); Dulieu and Gabbabini (2009)].

[12] For a review on ultracold Feshbach molecules, see [Köhler *et al.* (2006); Chin *et al.* (2010); Ferlaino *et al.* (2009)].

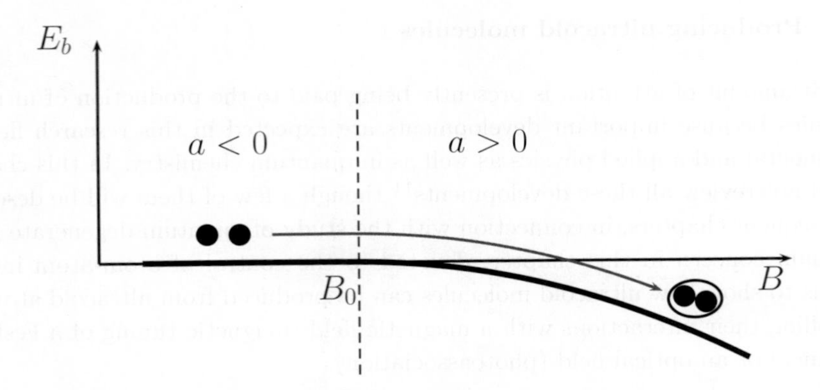

Fig. 16.9 If the magnetic field B is swept slowly through the Feshbach resonance from the region $a < 0$ to the region $a > 0$, a pair of two ultracold atoms is transformed into an ultracold molecule.

for Bose-Einstein condensation of the molecules (that obey the bosonic statistics since they are made of an even number of fermions).

Generally, their binding energy is very weak and they are well above the ground state of the molecular potential curve of the open channel, though STIRAP techniques[13] can be applied to transfer them with a very high efficiency to this ground state. The JILA group has recently been able to produce a high phase-space density of polar KRb molecules in the rovibrational ground state [Ni *et al.* (2008)]. Polar molecules are very interesting because their electric dipole-dipole interactions are anisotropic and have a long range [Lahaye *et al.* (2009)].

Other methods for producing Feshbach molecules

These methods consist of starting with cold atoms in the $a > 0$ region and inducing transition between the scattering states of two atoms and the weakly bound state. To ensure conservation of energy and momentum a third partner is required. It can be

- either a photon from a resonant electromagnetic field which stimulates the transition [Thompson *et al.* (2005); Gaebler *et al.* (2007); Papp and Wieman (2006); Weber *et al.* (2008)],
- or another atom participating in the three-body inelastic process which preferentially populates the least bound state [Jochim *et al.* (2003b); Zwierlein *et al.* (2003)].

16.6.2 *Photoassociation of ultracold atoms*

The simplest photoassociation process, represented in Fig. 16.10(a), puts the pair of atoms $A + A$ into a bound molecular state of an excited electronic potential $A + A^*$ by absorption of a photon ω_1. The transition considered here connects a state of the continuum of the entrance channel to a discrete state of another excited potential. The energy distribution of the pair of atoms depends on the temperature and has a certain width. Consequently the width of the absorption line when ω_1 is scanned is

[13] See page 307 in Chap. 13.

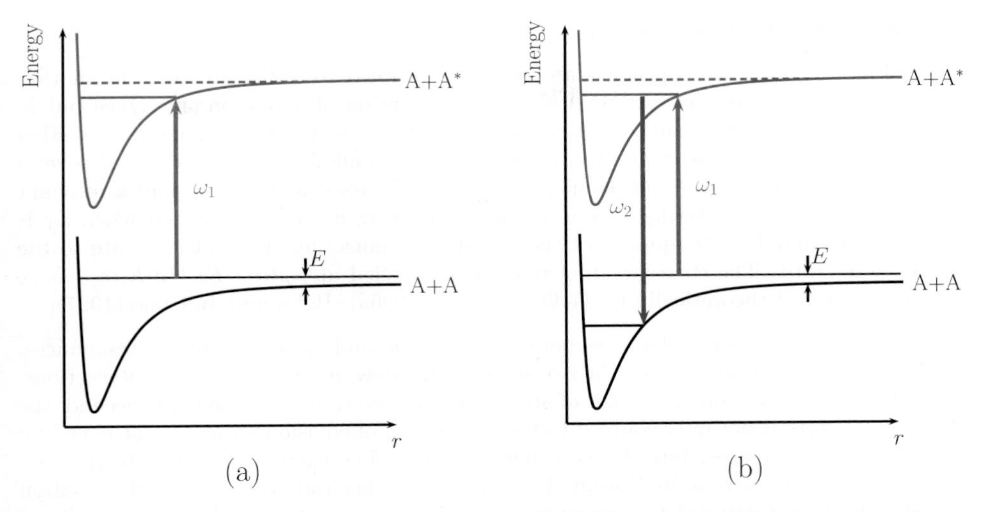

Fig. 16.10 (a) one-color photoassociation. A pair of two ultracold atoms $A + A$ absorbs a photon ω_1 and is put in a bound molecular state of an excited electronic potential $A + A^*$. (b) two-color photoassociation. A two-photon stimulated Raman transition (absorption of a photon ω_1 and stimulated emission of a photon ω_2 transfers the pair of atoms $A + A$ into a bound molecular state of the same potential $A + A$).

in general larger than for a transition between two discrete states. With ultracold atoms, however, the width of the energy distribution of the initial collision state is very small and the broadening of the photo-absorption spectrum is negligible.

In general, the discrete molecular state reached after the absorption of the photon is not stable radiatively since one of the two atoms, A^*, is in an excited electronic state and can decay by spontaneous emission. The ultracold molecules obtained in this way thus have a short lifetime.

Ultracold molecules in the lower electronic potential, with a much longer lifetime, can be obtained in stimulated Raman processes (two-color photoassociation process represented in Fig. 16.10(b)).[14] If the two lasers ω_1 and ω_2 are coherent, the configuration involved in the two-photon stimulated Raman process is a Λ-configuration that gives rise to coherent population trapping and dark states.[15] The two atoms can be trapped in a linear superposition of a collision state and a bound molecular state. The dark resonance obtained when $\omega_1 - \omega_2$ is scanned is very narrow. It allows a very precise determination of the binding energies of the molecular states in the lower electronic potential $A + A$, and in particular of the least bound state, from which one can deduce a very precise value of the scattering length. This method has been applied to measure the scattering length of two metastable helium-4 atoms in the 2^3S_1 state with an accuracy two orders of magnitude better that all previous determinations [Moal *et al.* (2006)].

[14] For a review of production of ultracold molecules by photoassociation, see Refs. [Heinzen (1999); Pillet *et al.* (2003)].

[15] See Chaps. 4 and 13.

Optical Feshbach resonances

In Fig. 16.10(a), the collision state $A + A$ is coupled to a discrete state in another channel relative to the system $A + A^*$ by absorption of a photon ω_1. Once put in this state, the system can come back to the initial state $A + A$ by stimulated emission of a photon ω_1. This cycle absorption-stimulated emission can be repeated several times. The situation is thus similar to the one discussed in Sec. 16.3 and a resonant behavior of the scattering amplitude and scattering length is expected when ω_1 is scanned around the frequency of the transition connecting the collision state to the discrete state. The corresponding resonance is called an *optical Feshbach resonance* and is studied theoretically in [Fedichev *et al.* (1996a); Bohn and Julienne (1997)].

There are important differences between magnetic and optical Feshbach resonances. First, the transition between the two states is due now, not to atom-atom interactions, but to the interaction of the pair of atoms with an external radiation field. Second, the discrete state is not only coupled to the continuum of collision states, as for magnetic Feshbach resonances, but also to other channels. The excited bound state $A + A^*$, which contains one excited atom A^*, can decay either radiatively or by dissociation (it is no longer coupled to a quasi zero energy continuum state where the density of states vanishes). The factor $i\Gamma_0/2$ must now be kept in Eq. (16.15).

Despite the limitations that result from the instability of the discrete state, optical Feshbach resonances have a few advantages: They can be used for diamagnetic atoms for which magnetic tuning is not available. Light fields can be switched on and off and swept more rapidly than magnetic fields. They can be also more tightly focussed, allowing different groups of atoms to be addressed selectively [Yamazaki *et al.* (2010)]. Optical Feshbach resonances have been experimentally observed with rubidium-87 atoms [Theis *et al.* (2004)]. The principle of optical coupling between the collision state and a discrete state can be extended to the two-photon Raman coupling of Fig. 16.10(b), and has been also demonstrated with rubidium-87 atoms [Thalhammer *et al.* (2005)].

16.7 Conclusion

With Feshbach resonances, an unprecedented control of ultracold atoms has been reached. Not only can atoms be cooled to very low temperatures, confined in traps with all possible geometries, but now their interactions can be changed at will by simply tuning a magnetic field. These interactions can be cancelled, their sign can be changed, and their strength can reach very high values at the center of the resonance.

In this chapter, the emphasis has been placed on the elementary collision process, though a primary interest of Feshbach resonances is that they open new exciting possibilities for investigating the fundamental behavior of quantum degenerate bosonic and fermionic gases, and for exploring new regimes with strong interactions that require descriptions beyond mean field. This explains why they play a central role in the experiments trying to simulate, with quantum gases, the behavior of more complex systems found in other fields.

Feshbach resonances are also very attractive for preparing ultracold homo-and hetero nuclear molecules and for exploring the quantum properties of exotic few body systems like Efimov trimers. Several investigations are presently devoted to the formation of tetramers near a Feshbach resonance. These resonances are thus also very interesting for the study of few body quantum systems [Greene (2010)].

PART 6

Exploring quantum interferences with few atoms and photons

Introduction

In Chap. 4, we described several physical effects associated with the existence of linear superpositions of internal states. The properties of the light absorbed or emitted by an atom in a linear superposition of two Zeeman sublevels, $c_1|\varphi_1\rangle + c_2|\varphi_2\rangle$, depends, not only on the probabilities of occupation $|c_1|^2$ and $|c_2|^2$ of the two sublevels, but also on the Zeeman coherences $c_1c_2^*$ and $c_2c_1^*$. These quantum interference terms give rise to interesting new phenomena such as level crossing resonances, quantum beats and coherent population trapping. The physical interpretation of these effects and their applications were discussed in detail in Chap. 4. In this part, we show that quantum interference effects show up in other contexts involving the external degrees of freedom of atoms or composite systems such as two atoms, two photons, or one atom and one photon.

In the experiments described in Chap. 4, the temperature of the atomic gas is on the order of a few hundred K, so that the thermal de Broglie wavelength, λ_{T}, of these atoms is a fraction of Angstrom. The coherence length of the atomic wave packets is then so small that atomic motion can be described classically in most cases.[1] The development of laser cooling techniques, described in part 4, has changed the situation. The coherence length of the atomic wave packets can reach values on the order of a few micrometers and a quantum description of atomic motion becomes necessary. The off-diagonal elements $\langle\vec{r}|\sigma|\vec{r}'\rangle$ of the atomic density operator σ between two different positions $\vec{r}$ and $\vec{r}'$, the so-called *spatial coherences*, play an important role. In Chap. 17, we describe a few atom optics experiments that extend the well known interference phenomena observable with optical waves to atomic de Broglie waves. In order to keep the discussion as simple as possible, we consider atoms in a single internal state, for example in a non-degenerate ground state. To illustrate the quantum nature of atomic motion, we also describe a few physical effects that cannot be understood classically, such as the Bloch oscillations observable on atoms moving in the spatially periodic potential of an optical lattice, and the frequency modulation of atomic de Broglie waves reflected by a vibrating mirror.

Introducing the internal degrees of atoms in atom optics experiments[2] introduces a wealth of interesting physical effects, some of which are described in Chap. 18. For example, a two-level atom $\{g, e\}$ crossing a resonant laser wave during a time interval such that it feels a $\pi/2$ pulse exits the laser wave in a state which is a linear superposition of $|g, \vec{p}\rangle$ and $|e, \vec{p}+\hbar k\rangle$, where $\vec{p}$ is the initial momentum of the atom, and $\hbar\vec{k}$ the momentum of the absorbed laser photon. The crossing of the laser wave is thus equivalent to the crossing of a beam splitter that changes not only the

[1] The use of supersonic beams can, however, lead to smaller velocity spreads and longer coherence lengths that give rise to observable interference effects between atomic de Broglie waves. A few examples of experiments of this type will be described in Chap. 17.

[2] The equivalent of internal degrees of freedom in usual interference experiments with light waves is the polarization of the photon.

direction of the momentum, but also the internal state. In Chap. 18, we describe atomic interferometers that use several beam splitters of this type to produce atomic wave packets which follow different coherent paths starting from the same point and ending at a different point, where they recombine and interfere. We show that these experiments can be considered to be the extension to the optical domain of the Ramsey method of the separate oscillatory fields which has been so successful in the radio-frequency and microwave domains. We also describe some applications of atomic interferometry to high precision measurements: measurement of time (atomic clocks with ultracold atoms); measurement of inertial fields (gravimeters, gyrometers).

In the interference effects described in Chaps. 17 and 18, a single atom is involved. In Chap. 19, we consider two quantum systems 1 and 2 and the so-called *entangled* states $|\psi_{12}\rangle$ of the global system $\{1+2\}$ that cannot be written as a product of a state $|\varphi_1\rangle$ of 1 by a state $|\chi_2\rangle$ of 2: $|\psi_{12}\rangle \neq |\varphi_1\rangle \otimes |\chi_2\rangle$. Entangled states are another important example of linear superpositions of states: they are linear superpositions of at least two product states of 1 and 2. A well known example of entangled state is the singlet state of two spins 1/2: $(|1+\rangle \otimes |2,-\rangle - |1-\rangle \otimes |2,+\rangle)/\sqrt{2}$. In Chap. 19, we discuss in detail the quantum correlations that exist between two systems when they are in an entangled state, as well as a few important properties associated with entanglement that illustrate the non-intuitive features of quantum mechanics, such as non-separability and complementarity. We also show that entanglement plays an essential role in the quantum measurement process, and is therefore central in quantum mechanics. It turns out that it is also at the basis of new methods being explored for transmitting and processing information, which shows that abstract concepts can give rise to interesting practical applications.

Chapter 17

Interference of atomic de Broglie waves

17.1 Introduction

Wave-particle duality for particles with non-zero mass was first postulated by Louis de Broglie [de Broglie (1923)]. He introduced the idea that every particle of mass m and momentum p, equal to mv in the non-relativistic limit, has associated with it a wave whose wavelength, now called the *de Broglie wavelength*, is given by:

$$\lambda_{\mathrm{dB}} = \frac{h}{p} \underset{v \ll c}{\simeq} \frac{h}{mv}. \tag{17.1}$$

The first evidence for the wave nature of massive particles was obtained by Clinton Davisson and Lester Germer with the diffraction of electrons by a crystal [Davisson and Germer (1927)]. This experiment can be considered as the transposition to electronic waves of the Bragg diffraction of X-rays by a crystal. Another diffraction experiment of an electron beam was performed the same year by George Paget Thomson.[1] Immanuel Esterman and Otto Stern realized the first atom diffraction experiment when they observed the diffraction of helium atoms from the surface of a LiF crystal [Esterman and Stern (1930)].

Diffraction and interference phenomena were also observed with neutrons. This field started in the mid seventies [Bonse and Rauch (1979)]. Because of the electric neutrality of neutrons and the smallness of their magnetic moment, matter-wave experiments with neutrons are well suited for high-precision experimental studies. These experiments provide important information on the structure and dynamics of various types of media. They also allow the study of interesting quantum effects such as the change of sign of the wave function of a spin $1/2$ in a 2π rotation, or the phase shifts associated with topological effects or inertial fields.[2]

The diffraction of helium atoms by a crystal, mentioned above, proved that not only elementary particles, like electrons and neutrons, but also composite particles such as atoms[3] behave as waves. Since atoms, unlike electrons, carry no charge, and

[1] C. J. Davisson and G. P. Thomson were awarded the 1937 physics Nobel prize for these works.

[2] For a review of the experimental techniques used in this field and the various results that have been obtained, see [Rauch and Werner (2000)].

[3] We suppose of course that the energies involved in the experiment are low enough so that the atom keeps its identity and is not broken into smaller constituents.

unlike neutrons, do not penetrate through matter, new techniques were required to provide optical elements for atoms. A new field devoted to the study of matter waves propagation, reflection, refraction, diffraction and interference has therefore emerged. New devices for manipulating atomic de Broglie waves (mirrors, beam splitters, diffracting structures) have been imagined and realized. In this respect, the trapping and cooling methods described in Part 4 have been extremely useful, and are at the origin of important differences between atom optics and optics with light waves. For example, laser cooling is very useful for reducing the angular spread of an atomic beam without using collimating slits. Atom optics can be dissipative and the "etendue" of the beam is no longer necessarily conserved, like in optics.

For the sake of simplicity, in this chapter we will focus on the external degrees of freedom of the atom associated with its center of mass, and ignore its internal degrees of freedom. This simplification is valid if we consider an atom in a non-degenerate ground state g with zero angular momentum ($J_g = 0$). New effects that appear in the interference of atomic de Broglie waves which are associated with the existence of several internal states will be investigated in Chap. 18.

17.2 De Broglie waves versus optical waves

17.2.1 *Dispersion relations. Position and momentum distributions*

The first difference between de Broglie waves and optical waves[4] lies in their dispersion relations, which relate the energy $E = \hbar\omega$ and the wave vector $\vec{k}$. This relation is linear for photons $\omega_L = k_L c$, but quadratic for particles with non-zero mass $\omega = \hbar k^2/2m$.

Consider a one-dimensional packet of light waves:

$$E(x,t) = \int \mathrm{d}k_L\, g(k_L) e^{i(k_L x - \omega_L t)} = \int \mathrm{d}k_L\, g(k_L) e^{ik_L(x-ct)}. \tag{17.2}$$

Because of the linear dispersion relation $\omega_L = ck_L$, the light field $E(x,t)$ depends only on $x - ct$ and propagates without any deformation at velocity c. This is to be contrasted with the well known phenomenon in quantum mechanics of wave packet spreading that results from the quadratic relation between ω and k for a particle with non-zero mass. Two distributions can be associated with an atomic wave packet: the position distribution $R(\vec{r},t) = \langle \vec{r}|\hat{\sigma}(t)|\vec{r}\rangle$, and the momentum distribution $P(\vec{p},t) = \langle \vec{p}|\hat{\sigma}(t)|\vec{p}\rangle$, where $\hat{\sigma}(t)$ is the atomic density operator. For a free particle, the momentum $\hat{p}$ is a constant of the motion, so that:

$$P(\vec{p},t) = P(\vec{p},0). \tag{17.3}$$

Therefore, the momentum distribution does not change in time though the position distribution does change due to the wave packet spreading.

[4] We denote the angular frequency and the wave vector of light waves by ω_L and $\vec{k}_L$, and their counterpart for matter waves by ω and $\vec{k}$.

It is worth noting that $P(\vec{p},t)$ is *not* the Fourier transform of $R(\vec{r},t)$. This is readily shown by considering the pure case for which the density matrix takes the simple form $\hat{\sigma}(t) = |\Psi(t)\rangle\langle\Psi(t)|$. Although the wave function in momentum space $\tilde{\Psi}(\vec{p},t) = \langle\vec{p}\,|\Psi(t)\rangle$ is, by definition, the Fourier transform of the wave function in position space $\Psi(\vec{r},t) = \langle\vec{r}\,|\Psi(t)\rangle$, there is no such relation between $R(\vec{r},t) = |\Psi(\vec{r},t)|^2$ and $P(\vec{p},t) = |\tilde{\Psi}(\vec{p},t)|^2$.

17.2.2 *Spatial coherences. Coherence length*

In Chap. 4, we discussed the quantum interference effects associated with linear superpositions of different internal states E, E' in terms of the off-diagonal elements $\langle E'|\,\hat{\sigma}(t)\,|E\rangle$ of the atomic density operator. In the present chapter, where we discuss quantum interference effects associated with the center of mass variables, it seems natural to focus on the off-diagonal elements $\langle\vec{r}\,|\sigma(t)|\vec{r}\,'\rangle$ of $\hat{\sigma}(t)$ between two localized states $|\vec{r}\,\rangle$ and $|\vec{r}\,'\rangle$. We call these off-diagonal elements *spatial coherences*.

$$\hat{\sigma}\left(\vec{r},\vec{r}\,',t\right) = \langle\vec{r}\,|\,\hat{\sigma}\left(t\right)|\vec{r}\,'\rangle\,. \tag{17.4}$$

When the distance $|\vec{r}-\vec{r}\,'|$ between the two points $\vec{r}$ and $\vec{r}\,'$ becomes very large, one expects that $\hat{\sigma}\left(\vec{r},\vec{r}\,',t\right)$ tends to zero. Except in the case of a de Broglie plane wave with a perfectly well defined wave vector, there is no well defined phase relation between the values of the atomic wave function at two different points separated by a sufficiently large distance. The concept of coherence length is thus associated with a characteristic distance ξ, beyond which the spatial coherence tends to zero. This length plays a crucial role in atomic interferometry. Generally, the two waves that are interfering in a given experiment are obtained by splitting an incident wave into two parts and then recombining them. If the length difference between the two different paths is larger than the coherence length of the incident wave, there is no phase relation between the two waves at the place where they are recombined, and the interference disappears.

At this stage, a frequent mistake that is made is to relate the coherence length ξ to the width Δr of the position distribution $R(\vec{r}\,)$. This would imply that the coherence length increases with time, since the spreading of the wave packet is responsible for an increase of Δr with time. This prediction is obviously absurd.

When trying to precisely calculate the coherence length, one is faced with the difficulty that $\sigma\left(\vec{r},\vec{r}\,',t\right)$ does not only depend on $\vec{r}-\vec{r}\,'$, but on both $\vec{r}$ and $\vec{r}\,'$. This difficulty can be circumvented by considering the sum of all spatial coherences $\sigma\left(\vec{r},\vec{r}\,',t\right)$ with a fixed value of $\vec{a} = \vec{r}-\vec{r}\,'$. We call this sum $G(\vec{a}\,)$ the global spatial coherence for a spatial separation $\vec{a}$:

$$G(\vec{a}\,) = \int \mathrm{d}^3 r\langle\vec{r}\,|\hat{\sigma}|\vec{r}+\vec{a}\,\rangle. \tag{17.5}$$

In the case of a pure state $|\Psi\rangle$, $G(\vec{a}\,)$ is given by

$$G(\vec{a}\,) = \int \mathrm{d}^3 r\langle\vec{r}\,|\Psi\rangle\langle\Psi|\vec{r}+\vec{a}\,\rangle = \int \mathrm{d}^3 r\Psi(\vec{r}\,)\Psi^*(\vec{r}+\vec{a}\,), \tag{17.6}$$

and corresponds to the overlapping of the wave packet described by $\Psi(\vec{r})$ and that described by $\Psi^*(\vec{r})$, translated by $-\vec{a}$.

From the definition (17.5) of the global spatial coherence $G(\vec{a})$, it seems appropriate to define the coherence length ξ as the width of $G(\vec{a})$. We now show that $G(\vec{a})$ is simply related to the momentum distribution $P(\vec{p})$ and not to the position distribution $R(\vec{r})$. The closure relation for the eigenstates of the momentum operator gives

$$\begin{aligned} G(\vec{a}) &= \int \mathrm{d}^3 r \, \langle \vec{r} \, |\hat{\sigma}| \, \vec{r} + \vec{a} \rangle \qquad (17.7) \\ &= \int \mathrm{d}^3 r \mathrm{d}^3 p \, \mathrm{d}^3 p' \, \langle \vec{r} \mid \vec{p} \rangle \, \langle \vec{p} \, |\hat{\sigma}| \, \vec{p}' \rangle \, \langle \vec{p}' \mid \vec{r} + \vec{a} \rangle . \end{aligned}$$

Inserting $\langle \vec{r} \, | \vec{p} \rangle = (2\pi\hbar)^{-3/2} \exp{(i \vec{p} \cdot \vec{r} / \hbar)}$ and a similar equation for $\langle \vec{p}' | \vec{r} + \vec{a} \rangle$ into this equation, and using

$$\frac{1}{(2\pi\hbar)^3} \int \mathrm{d}^3 r \, \exp{[i \, (\vec{p} - \vec{p}') \cdot \vec{r} / \hbar]} = \delta \, (\vec{p} - \vec{p}') \, ,$$

yields

$$G(\vec{a}) = \int \mathrm{d}^3 p \; P(\vec{p}) \, \exp{(-i \, \vec{p} \cdot \vec{a} / \hbar)} \, . \qquad (17.8)$$

This shows that the global spatial coherence $G(\vec{a})$ is simply the Fourier transform of the momentum distribution $P(\vec{p})$. We conclude that $G(\vec{a})$, and consequently its width ξ, are time independent like $P(\vec{p})$, and that the coherence length ξ is given by:[5]

$$\xi \simeq \frac{\hbar}{\Delta p} . \qquad (17.9)$$

For a particle of mass m at thermal equilibrium at a temperature T, the coherence length is commonly called the *thermal de Broglie wavelength* and is given by:

$$\lambda_T = \frac{h}{\sqrt{2\pi m k_B T}} . \qquad (17.10)$$

Measuring the momentum distribution of cold atoms

The usual method to infer the momentum distribution of a sample of cold atoms is through a time-of-flight experiment in which the size of the cloud is monitored as a function of time. However, at very low temperature the time required to obtain a significant expansion may be too long as atoms fall under gravity. A more efficient method described in [Saubamea *et al.* (1997, 1999)] uses a controlled splitting of the wave packet to determine $G(\vec{a})$ by interferometric means. From the Fourier transform $P(\vec{p})$ of $G(\vec{a})$, one then infers the momentum dispersion. This is reminiscent of the method used in classical optics to infer the spectral width $\delta\omega_L$ of a very narrow spectral line. This width can be more easily deduced from the time correlation function $G(\tau) = \int E^*(t+\tau)E(t)\mathrm{d}t$ of the light field, which is the Fourier transform of the spectral intensity distribution $I(\omega_L)$.

[5] One could initially think that Heisenberg uncertainty relations imply a relation similar to Eq. (17.9) between Δr and Δp, however this is true only for a minimum uncertainty wave packet before it starts to spread.

17.2.3 *Fragility of spatial coherences*

All processes that increase the momentum dispersion Δp contribute to a decrease of the coherence length. Let us comment on two important examples:

- Spontaneous emission occurs in a random direction. As a result, the atomic momentum dispersion increases by a quantity on the order of $\hbar k_L$. Even if the coherence length is initially infinite (i.e. $\Delta p = 0$), ξ becomes smaller than $\hbar/(\hbar k_L) = \lambda_L/(2\pi)$ after a single spontaneous emission! This is the reason that spontaneous emission must be strictly avoided in all devices that use light for manipulating atoms in atomic interferometers (see Sec. 17.5).
- An atom that undergoes collisions with other atoms is also subjected to a momentum diffusion that increases its momentum dispersion and therefore reduces its coherence length.

Decoherence time and damping time

Consider an atom in a linear superposition of two wave packets separated by a distance $2a$ large compared to their width σ. The wave function of the atom reads:

$$\psi(x) = \frac{1}{\sqrt{2}}[\varphi(x-a) - \varphi(x+a)], \tag{17.11}$$

and the corresponding momentum distribution is $P_0(p) = 2|\tilde{\varphi}(p)|^2 \sin^2(pa/\hbar)$ where $\tilde{\varphi}(p)$ denotes the Fourier transform of $\varphi(x)$. In momentum space, the coherence between the two wave packets gives rise to an interference pattern with a fringe period $\delta p = \pi\hbar/a$.

Let us now take into account the environment of the atom within a semiclassical description.[6] We assume that the atom undergoes elastic collisions with the other atoms of the gas. As a result, the atom undergoes a random walk and its momentum diffuses. Starting with an initial well-defined momentum, the atom acquires a momentum dispersion Δp after a time Δt given by $(\Delta p)^2 = 2D\Delta t$, where D is the momentum diffusion coefficient. At time Δt, the momentum distribution is thus the convolution product of $P_0(p)$ and a function of width Δp. When Δp increases and becomes very large compared to the fringe period $\pi\hbar/a$, the interference pattern of the momentum distribution is blurred. We can therefore introduce a decoherence time defined by $T_R = \pi^2\hbar^2/(2Da^2)$. For a time evolution t such that $t \gg T_R$, the coherence between the two packets has been destroyed, and the state of the atom is a statistical mixture of the two wave packets.

It is worth comparing the decoherence rate T_R^{-1} to the damping rate of the external degrees of freedom. As a result of the collisions, the mean momentum of the atom obeys:

$$\frac{\mathrm{d}\langle p\rangle}{\mathrm{d}t} = -\gamma\langle p\rangle, \tag{17.12}$$

where γ is the friction coefficient. According to the fluctuation dissipation theorem, the diffusion coefficient D and the damping rate γ are related by the formula $D/\gamma =$

[6] A quantum treatment of this problem is proposed in [Unruh and Zurek (1989)].

$3mk_BT$. We deduce the following expression for the decoherence rate

$$\frac{1}{T_R} = \frac{3\gamma}{2\pi}\left(\frac{a}{\lambda_{\mathrm{T}}}\right)^2, \tag{17.13}$$

where λ_{T} is the thermal de Broglie wavelength which is on the order of 10^{-11} m at $T = 300$ K. The distance between the two wave packets is supposed to be large compared to their width, so that the ratio a/λ_{T} is very large. We conclude that any superposition of two wave packets separated by a distance much larger than λ_{T} is destroyed at a rate much larger than the damping time. Therefore, such states cannot be kept for long times.

17.3 Young's two-slit interferences with atoms

Young's experiment is a textbook example of interferometry, and can be considered as the fundamental gedanken experiment that exhibits the wave properties of matter. Interferences emerge as a result of the existence of two indistinguishable possible paths. The first experimental observation of Young's experiment using particles with non-zero mass was made in 1961, with electrons, by Claus Jönsson [Jönsson (1961, 1974)]. In this section, we address its realization with atoms. Before describing a few of these experiments, however, we first point out the important parameters that characterize the atomic source, whose control is essential.

17.3.1 *Important parameters of Young's double-slit interferometer*

Three quantities that characterize the atoms of the beam sent through the two slits have to be considered (see Fig. 17.1):

- the mean velocity $\langle v \rangle$,
- the longitudinal velocity dispersion $\Delta v_{\parallel}$,
- the transverse velocity dispersion $\Delta v_{\perp}$.

A transverse collimation is first required in order to obtain a transverse coherence length on the order or larger than the slits separation. This condition, essential for insuring that the two waves emerging from the two slits are coherent,[7] is readily achieved by a transverse filtering of the atomic beam emitted by an oven. The improved transverse coherence of the source is thus obtained at the expense of a drastic reduction of the flux. When possible, a better solution is to introduce a transverse laser cooling of the beam, to improve the collimation without any loss of atoms. In either case, the velocity distribution of the atomic source is such that: $\Delta v_{\perp} \ll \langle v \rangle$.

[7]A similar requirement is well known in optics and is usually analyzed with the Cittert-Zernike theorem which relates the transverse spatial coherence of the light beam to the diffraction of the aperture along the beam [Cittert (1934); Zernike (1938)].

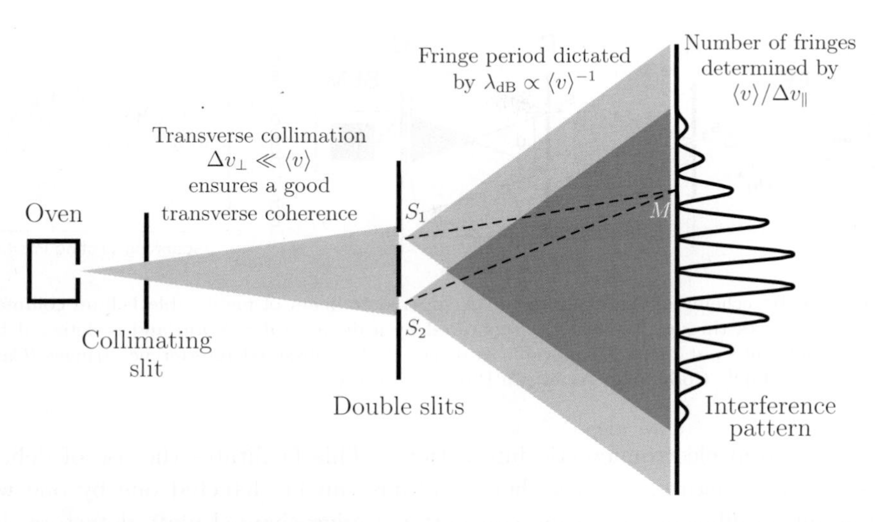

Fig. 17.1 Sketch of a typical Young's two-slit atom interferometer experiment: atoms from an oven are collimated by slits to improve the transverse coherence of the atomic source. This filtered beam impinges a screen with two slits separated by a distance smaller than the transverse coherence length. The pattern made by the atoms and collected on a screen placed far away from the two slit screen exhibits interference fringes.

The fringe period is proportional to the ratio between the de Broglie wavelength $\lambda_{\mathrm{dB}} \propto 1/\langle v \rangle$ and the distance between the slits. Reducing the mean velocity $\langle v \rangle$ thus increases the fringe period and makes them easier to observe on the screen.

The number of fringes that can be observed on the screen depends on the longitudinal coherence length $\xi_{\parallel} = \hbar/\Delta v_{\parallel}$. To understand this result, consider a point M on the screen (see Fig. 17.1) such that the length difference between the two paths S_1M and S_2M connecting M to the two slits S_1 and S_2 is equal to an integer number n of de Broglie wavelengths λ_{dB}. This point M thus belongs to the nth fringe (the central fringe corresponds to $n = 0$). The two paths arriving at M with a length difference equal to $n\lambda_{\mathrm{dB}}$ can interfere only if this length difference is on the order or smaller than the longitudinal coherence length $\xi_{\parallel}$. This clearly shows that the maximum number of observable fringes is given by the ratio $\xi_{\parallel}/\lambda_{\mathrm{dB}}$. To increase this number, one must therefore increase $\xi_{\parallel}$, i.e. decrease $\Delta v_{\parallel}$. The reason supersonic beams were used in the pioneering experiments performed with atoms was that they ensure a large ratio $\langle v \rangle / \Delta v_{\parallel}$.

17.3.2 *Young's double-slit interferences with supersonic beams*

The first Young's double-slit interferometer with atoms was demonstrated using a supersonic beam of metastable helium and micrometer size slits [Carnal and Mlynek (1991)]. The choice of metastable helium was made for three main reasons: (i) the low mass favors a relatively large de Broglie wavelength, (ii) the lowest excited state of rare gas is a metastable level with zero angular momentum, and is therefore

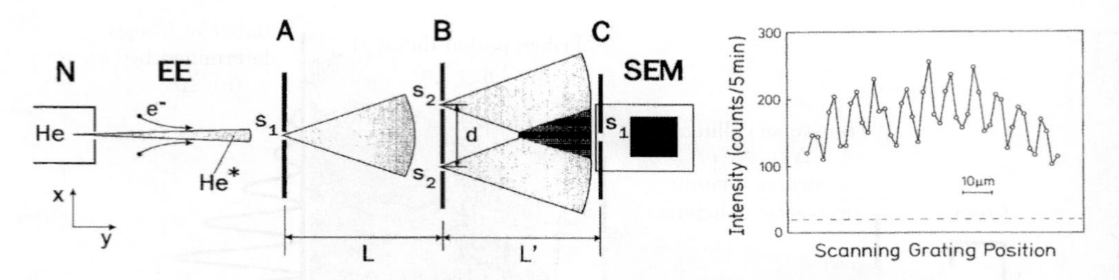

Fig. 17.2 Left: scheme of the experiment. A supersonic beam of metastable helium collimated by a slit s_1 passes through two slits s_2 separated by a distance $d = 8$ μm and is detected by a microchannel plate put behind a movable grating. Right: observed interference fringes [Carnal and Mlynek (1991)]. Copyright: American Physical Society.

nearly free from electromagnetic interactions. This facilitates the use of delicate transmission structure, and (iii) helium atoms can be detected one by one with high quantum efficiency by using conventional microchannel plate detectors since the metastable states have an energy of several electron-volts compared to the true ground state. Figure 17.2 gives a scheme of the experimental set up and an example of the interference fringes that were observed.

17.3.3 *Young's double-slit interferences with cold atoms*

We now describe a two-slit interference experiment performed by the group of Fujio Shimizu in Tokyo with cold metastable neon atoms [Shimizu *et al.* (1992)]. In this experiment, the low longitudinal mean velocity gives a large de Broglie wavelength and a large fringe spacing and makes it possible to directly see the interference pattern without scanning a grating as was done in the previous experiment. Using metastable atoms also gives the possibility to detect the atoms one by one as they arrive on the screen and to follow the build up of the interference pattern.[8]

The experiment is done as follows. Metastable neon atoms are laser cooled in a magneto-optical trap. They are released from a very small region of the trap by an optical transfer to another untrapped state.[9] The released atoms fall because of gravity and cross a screen with two slits located a few centimeters below before reaching a multichannel plate where they are detected one by one. Figure 17.3 shows an example of the interference pattern that can be observed in real time as each atom gives a localized impact on the screen. Even though they arrive in an apparently random way and the location of each impact cannot be predicted with certainty, a spatial structure clearly emerges as the number of events increases. The

[8]This fascinating regime for double-slit experiments where one sends particles through the double-slit one by one was achieved for the first time with photons in 1909 by Geoffrey Taylor [Taylor (1909)], and with electrons by Akira Tonomura and co-workers at Hitachi in 1989 [Tonomura *et al.* (1989)]. In the experiment with cold atoms described here, the fringe spacing is much larger, and therefore the interference pattern is much easier to observe.

[9]For more details, see [Shimizu *et al.* (1992)].

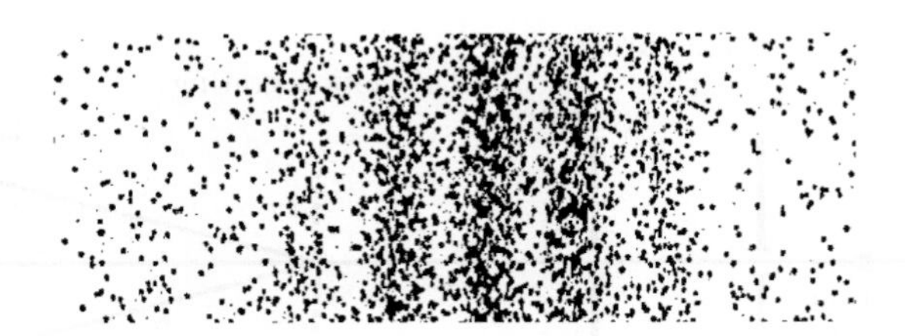

Fig. 17.3 Interference pattern observed in the screen for the two-slit experiment performed with cold metastable neon atoms. Each point corresponds to a localized impact of a neon atom. The spatial repartition of these impacts is modulated and exhibits interference fringes. The horizontal size of the picture is ~ 2 mm. Courtesy of F. Shimizu.

spatial density of impacts is modulated and exhibits a fringe structure with high density impact zones separated by low density zones.

This experiment provides a very clear and beautiful textbook illustration of wave particle duality for atoms. Each atom is a particle giving rise to a localized impact, but at the same time, there is a wave associated with this atom that passes through the two slits and gives rise to an interference structure in the spatial density of impacts. This wave is a linear superposition of two waves diffracted by each slit. It is not a concrete physical wave like, for example, sound waves, it is a *probability amplitude wave* that allows one to calculate the probability of finding the particle at a given point. This is the interpretation of the wave function in quantum mechanics formulated by Max Born in 1926. The connection between the particle aspect and the wave aspect is probabilistic. The square of the modulus of the wave function at a given point gives the probability to find the particle at this point.

17.3.4 *Can one determine which slit the atom passes through?*

Since the early days of quantum mechanics, famous physicists like Albert Einstein and Niels Bohr have debated about the possibility of an experiment that allows one to determine which slit the particle passes through in a Young's two-slit interference experiment.[10] In the fifth Solvay meeting in 1927, Einstein proposed making the screen with the two slits free to move in the plane perpendicular to the direction of the incoming particles. Because of the conservation of the total momentum, the screen gains a certain momentum δp if the particle passes through the upper slit before arriving in the center of the detection screen, and the opposite momentum $-\delta p$ if the particle passes through the lower slit. By measuring the momentum of the screen, one could thus determine the path of the particle. Niels Bohr answered that the screen must be described quantum mechanically. To be able to determine the momentum transferred by the particle, one should know its initial momentum with an accuracy better than δp. But then, because of the uncertainty relations, the position spread δx of the screen cannot be ignored and is easily calculated to

[10]This type of question is also called "which-path information" or in German "welcher weg".

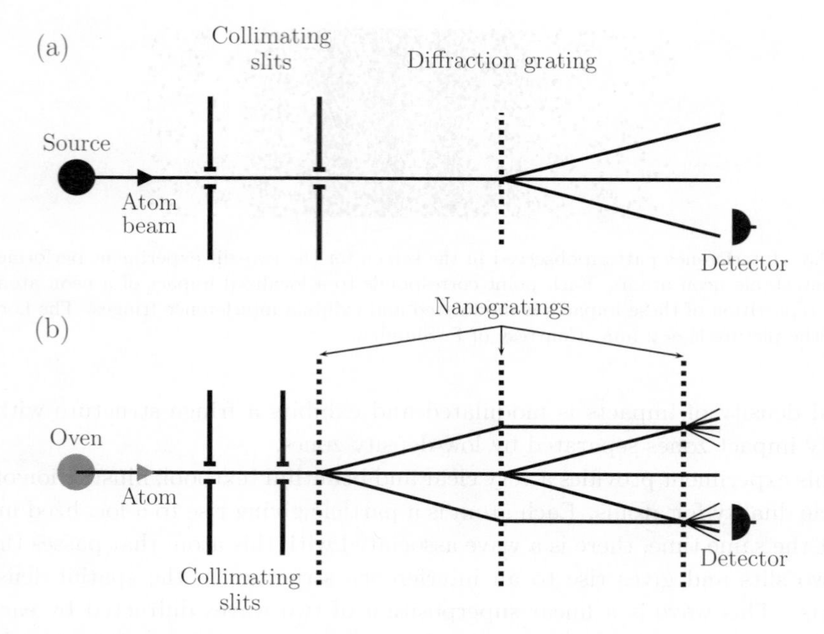

Fig. 17.4 (a) A typical setup for atom diffraction has four main parts: the source, the collimation, the diffraction grating and the detector. (b) Three grating Mach-Zehnder atom interferometer [Keith *et al.* (1991)].

be larger than the fringe spacing so that the interference pattern is washed out. In other words, this attempt to determine through which slit the particle passes destroys the interference pattern. The previous argument can be recast in a more general form, using the modern language of entangled states.[11]

17.4 Diffraction of atoms by material structures

The double-slit interferometer can be considered to be an example of atomic beam manipulation. With the advent of nano-lithography techniques, material gratings have been used for atom diffraction[12] and atom interferometry [Keith *et al.* (1988, 1991)].

The principle of a diffraction experiment using a material grating is sketched in Fig. 17.4(a). It consists of an atom source, a collimation stage, a diffraction grating and a detector. The first collimation slit is used to achieve a transverse coherence length larger than a few grating periods. In order to clearly resolve the diffraction peaks, a second slit may be added to ensure that the beam width is smaller than the separation between the different diffraction orders. The grating separates the

[11]See Sec. 19.7 of Chap. 19.

[12]More elaborate structures such as Fresnel zone plates have also been designed and used to deflect and focus an atomic beam [Carnal *et al.* (1991b); Ekstrom *et al.* (1992)].

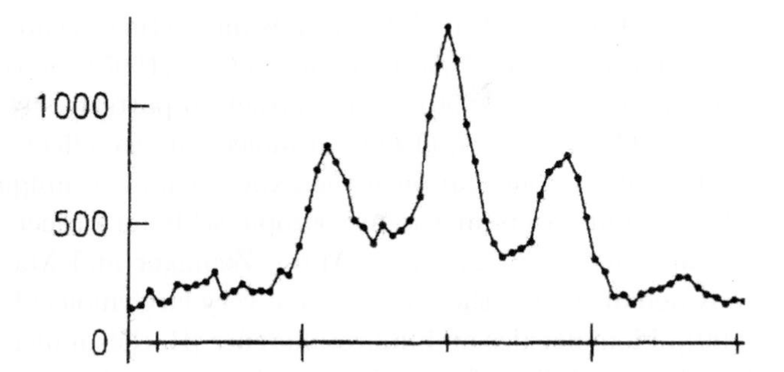

Fig. 17.5 Atomic diffraction grating pattern observed with sodium atom. Figure extracted from [Keith *et al.* (1988)]. Copyright: American Physical Society.

incident wave plane in several plane waves that are coherent with respect to each other. If the distance between two successive slits is equal to d, the angle of the nth order diffraction peak is simply given by:

$$\theta_n \simeq n\frac{\lambda_{\mathrm{dB}}}{d}, \tag{17.14}$$

where $\lambda_{\mathrm{dB}} = h/(m\langle v\rangle)$. The angular separation between the different orders thus depends on $\langle v\rangle$. In order to have well resolved diffraction peaks, a fairly monochromatic beam, like a supersonic beam, is required.

The first observation of atomic diffraction from nanofabricated structure was done by the group of David Pritchard at MIT using a supersonic sodium beam [Keith *et al.* (1988)] (see Fig. 17.5). The typical orders of magnitude were the following: the de Broglie wavelength was 17 pm, the distance between the slits was 200 nm and their width 100 nm. Similar experiments have been carried out with helium atoms with a de Broglie wavelength on the order of 1 Å passing through a gold grating with a periodicity of 0.5 μm [Carnal *et al.* (1991a)].

By using three equidistant gratings, one can realize a Mach-Zehnder interferometer for atoms as shown in Fig. 17.4(b) and described in more detail in [Keith *et al.* (1991)]. The first grating diffracts the incident atomic beam into the diverging orders $-1, 0$ and $+1$. The 0th and the 1st diffraction order are diffracted through the second grating placed downstream. This grating diffracts a portion of each of the two incident beams toward each other. The recombination of the -1st and +1st diffraction order respectively from the second grating occurs on the third grating. By a proper choice of the grating parameters and of the incoming atomic de Broglie wavelength, the two paths of the interferometer can be sufficiently far apart in the region between the two gratings to allow for a partial isolation of one from another.

By placing a metal foil between the two arms and applying an electric field to one arm, David Pritchard's group obtained a very accurate measurement of the sodium electric polarizability [Ekstrom *et al.* (1995)]. A similar apparatus using laser standing waves (see next section) as a beam splitter has been used to measure the electric polarizability of lithium atoms [Miffre *et al.* (2006)].

Using material structures for atom diffraction is interesting because experiments can be extended to molecules like Na_2 [Chapman *et al.* (1995)] or He_2 [Luo *et al.* (1996); Schöllkopf and Toennies (1996)]. The diffraction pattern is sensitive to van der Waals interaction [Grisenti *et al.* (1999)], a molecular size effect.

One may wonder how far one can push the experimental techniques to observe clear evidence of quantum effects at the mesoscopic scale for objects of increasing size, mass and complexity? The group of Anton Zeilinger and Markus Arndt in Vienna has already demonstrated the diffraction of very large molecules, such as C_{60} [Arndt *et al.* (1999)], biomolecules and fluorofullerenes [Hackermüller *et al.* (2003)]. Such studies offer the possibility of investigating the frontier between classical and quantum behavior.

Material structures have disadvantages as well. They are associated with the loss of atoms, due to sticking, or bouncing off of the structure, and result in a relatively small fraction of diffracted atoms. In Sec. 17.5, beam splitters based on non resonant laser standing waves will be described. These are species and state selective, and turn out to be very flexible.

Atom holograms

Holography can be defined as a technique to reconstruct an object by interferometrically manipulating a wave front. Fujio Shimizu and coworkers have developed different holographic techniques to draw an arbitrary pattern of atoms on a two-dimensional surface. With a setup similar to the one used for observing Young fringes with cold atoms, they have demonstrated the transmission [Morinaga *et al.* (1996)] and reflection amplitude [Shimizu and Fujita (2002)] holography, and phase holography [Fujita *et al.* (2000)].

For example, the hologram used for transmission amplitude holography consists of 1024 × 1024 cells with square holes of 500 nm size that diffract the incident matter waves. The position of holes (see Fig. 17.6(a)) is calculated in two steps: in the first step, one calculates the complex transmission amplitude function $t(x, y)$ of the hologram that generates the intended wave front; in the second step, one approximates the transmission function by a binary pattern. An example of such an hologram is presented in Fig. 17.6(a) along with the image produced on the MCP [Morinaga *et al.* (1996)].

17.5 Diffraction by laser standing waves

17.5.1 *New features compared to the diffraction by material gratings*

Non-resonant laser standing waves give rise to spatially periodic and conservative optical potentials. Optical lattices, described in Chaps. 7 and 14, are an important example of these structures.

In this section, we consider two well-collimated counter-propagating laser beams that give rise to a standing wave along the z-axis (see Fig. 17.7(a)). The common

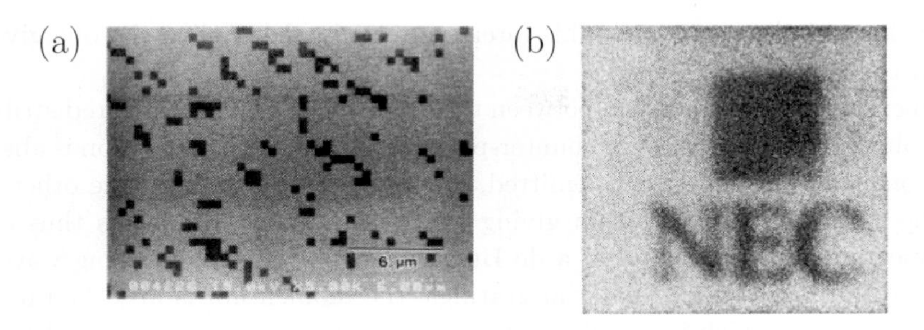

Fig. 17.6 (a) Transmission mask hologram made of square holes (black) of 500 nm size. (b) Corresponding reconstructed pattern (NEC) obtained experimentally (the square is formed by the nondiffracted waves). Figure extracted from [Morinaga *et al.* (1996)]. Copyright: American Physical Society.

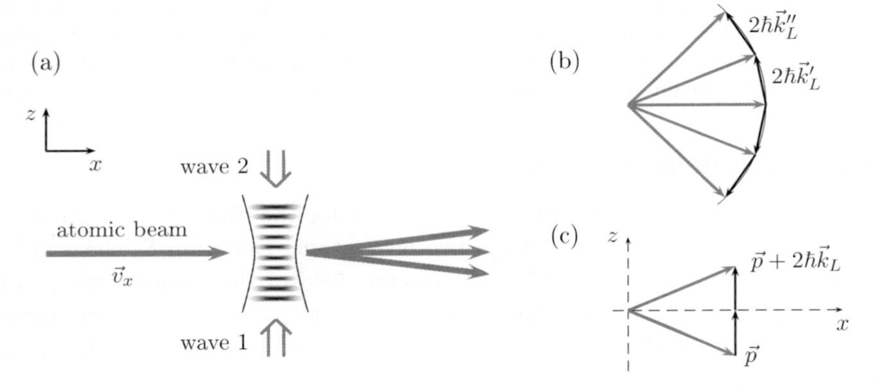

Fig. 17.7 (a) The experiment consists of measuring the momentum distribution of an atomic beam that has crossed a perpendicular standing wave during a finite time duration. (b) Raman-Nath regime where many wave vectors are involved in the absorption-stimulated absorption process. (c) Bragg scattering condition: the standing wave is no longer perpendicular to the atomic beam. A small angle is required to fulfill the momentum conservation.

frequency of the two laser beams is sufficiently far from the resonance[13] of the atoms of an atomic beam propagating along the x-axis that no spontaneous emission can take place during the transient time $T = w_0/v_x$ during which the atoms, with velocity v_x, pass through the laser beam of waist w_0.

We describe in this section the diffraction of the atomic de Broglie wave crossing at right angle the periodic optical potential associated with the laser standing wave.

Two points of view can be developed for understanding the diffraction pattern:

(i) A constructive interference appears in certain directions of the waves diffracted

[13]The internal ground state g of a two-level atom is the same before and after the diffraction process, so that only external degrees of freedom seem to be involved. Actually, virtual transitions between the ground state g and the excited state e are responsible for the spatially periodic light shift of the state g that gives rise to the periodic optical potential.

by the periodic structure. This interpretation is identical to the one given for material gratings.

(ii) An exchange of momentum between the atom and the field due to redistribution of photons between the two counter-propagating waves. One photon is absorbed in one wave and another is emitted, in a stimulated way, into the other wave. The exchange of momentum giving rise to the diffraction peaks thus clearly appears in the diffraction of a de Broglie wave by a laser standing wave. For the experiments using material gratings, the momentum is absorbed or given by the mask, which is usually considered as a classical macroscopic object.

Kapitza-Dirac effect

The diffraction of de Broglie waves by light was actually first proposed for electrons by Kapitza and Dirac in 1933 [Kapitza and Dirac (1933)]. The mechanism was a redistribution of photons between the two waves by stimulated Compton scattering: one photon of a given wave is Compton scattered by the electron in the other wave. Altshuler and co-workers then proposed an extension of this idea to the diffraction of atoms, pointing out that the diffraction probability would be dramatically enhanced if the light frequency is not too far from the atomic frequency [Altshuler *et al.* (1966)].

17.5.2 *Light-atom momentum exchange*

For sake of simplicity, the theoretical framework for the diffraction of atoms by a laser standing wave is presented for a two-level atom. Denoting $\mathcal{E}_0$ the amplitude of each travelling wave of Fig. 17.7, the electric field of the standing wave is $\mathcal{E} = 2\mathcal{E}_0 \cos(\omega_L t)\cos(k_L z)$. Each beam is assumed to have the same waist, equal to w_0. The interaction Hamiltonian between atoms and light in the rotating wave approximation is given by:

$$V_{AL} = \frac{\hbar\Omega}{2}(e^{ik_L z} + e^{-ik_L z})\left[e^{i\omega_L t}|g\rangle\langle e| + e^{-i\omega_L t}|e\rangle\langle g|\right], \tag{17.15}$$

where $\Omega = -D_{eg}\mathcal{E}_0/\hbar$ is the Rabi frequency associated with each beam, and D_{eg} the matrix element of the dipole operator $\hat{D}_{eg}$ between the ground state g and the excited state e. The atom Hamiltonian is:

$$H_A = \hbar\omega_A|e\rangle\langle e| + \frac{\vec{P}^2}{2M}, \tag{17.16}$$

where the external degrees of freedom are explicitly taken into account. Since we describe the diffraction of a de Broglie wave, the atomic motion has to be treated quantum mechanically. In Eq. (17.15), z is an operator, and the translation operator $e^{\pm ik_L z}$ can be expanded in the momentum space according to

$$e^{\pm ik_L z} = \sum_p |\vec{p} \pm \hbar\vec{k}_L\rangle\langle\vec{p}|. \tag{17.17}$$

The interaction Hamiltonian V_{AL} couples the state $|g,\vec{p}\rangle$ to $|e,\vec{p} \pm \hbar\vec{k}_L\rangle$, and $|e,\vec{p} \pm \hbar\vec{k}_L\rangle$ also to $|g,\vec{p} \pm 2\hbar\vec{k}_L\rangle$, ... The initial state $|g,\vec{p}\rangle$ is therefore coupled to $|g,\vec{p} + 2n\hbar\vec{k}_L\rangle$ with $n = \pm 1, \pm 2, \ldots$ The momentum gain $2n\hbar\vec{k}_L$ corresponds to a redistribution of photons between the travelling waves that forms the standing wave: n photons are absorbed from one wave and re-emitted in a stimulated way into the other one.

In the process that connects $|g, \vec{p}\rangle$ to $|g, \vec{p}+2n\hbar\vec{k}_L\rangle$, the photons have the same energy $\hbar\omega_L$. Conservation of energy within the system {atom+field}, means that the kinetic energy remains unchanged i.e.

$$|\vec{p}| = |\vec{p} + 2n\hbar\vec{k}_L|. \tag{17.18}$$

If the initial atomic momentum is along the x-axis, and if the photon momentum is along the z-axis, it is impossible to fulfill the criterium of energy conservation. However, if the photon momentum has a component along the x-axis, there are always solutions as shown in Fig. 17.7(b). Actually, the finite waist of the beam implies a wave vector dispersion along the x-axis of extension on the order of $\hbar/w_0$. This dispersion is significant when a sufficiently small waist is used. The corresponding regime, referred to as the Raman-Nath regime, is detailed in the next section.

When the beam waist is too large, the photon momentum dispersion along the x-axis is too small to allow processes like those of Fig. 17.7(b) to occur. However, the criterium (17.18) can be fulfilled if the atomic beam is slightly tilted with respect to the normal direction of the grating. Two photon processes directly coupling the states $\vec{p}$ and $\vec{p} + 2\hbar\vec{k}_L$ can thus occur, thereby yielding a symmetric change of the momentum vector with respect to the x-axis as shown in Fig. 17.7(c). The splitting of the wave function is based on energy and momentum transfer, in units of $2\hbar\vec{k}_L$, from the light field to the atom through the redistribution of photons between the two traveling waves forming the standing waves. This regime is known as the Bragg regime, and will be also detailed.

17.5.3 *Raman-Nath regime*

In the following, we perform the analysis of the Raman-Nath regime[14] with a far off resonance standing wave. In this regime, the waist of the laser is very small and the displacement of the atom along the standing wave direction during the crossing time T is small with respect to the light wavelength λ_L. Under this assumption, the kinetic energy term $\hat{P}_z^2/2M$ can be neglected, and all remaining terms of the Hamiltonian commute with the position operator $\hat{z}$. The atom enters the interaction region in the state $|z_0\rangle \otimes |g\rangle$. The edges of the laser beam profile are assumed to be sufficiently smooth to allow an adiabatic entry and exit of the atoms. As a result of the interaction between the atom and the far off resonant light of the standing wave, the state after the interaction acquires a phase shift that is proportional to the local light intensity experienced by the atoms:

$$|\chi_{\rm in}\rangle = |z_0\rangle \otimes |g\rangle \rightarrow |\chi_{\rm fin}\rangle = |z_0\rangle \otimes |g\rangle e^{-i\delta\phi(z)}, \tag{17.19}$$

[14]Venkata Raman and Nagendra Nath have identified this regime by analyzing the diffraction of light by ultra-sonic waves in the years 1935–1937 (see [Raman and Nath (1936)] and references therein).

where $\delta\phi(z_0) = \int_{-\infty}^{\infty} \mathrm{d}t \delta E_g(z_0, x(t))/\hbar$ is the time integral of the position dependent light shift of the ground state of the atom crossing the laser beam at a fixed value of z. The phase shift $\delta\phi(z)$ is readily calculated from the second-order perturbation theory:

$$\delta\phi(z) = \int_{-\infty}^{\infty} \frac{\Omega^2[x(t)]\cos^2 k_L z}{\delta}\mathrm{d}t = \int_{-\infty}^{\infty} \frac{\Omega^2[x(t)]}{\delta}(1 - \cos 2k_L z)\mathrm{d}t, \tag{17.20}$$

The standing wave therefore acts as *a sinusoidal phase grating.* The expansion of Eq. (17.19) clearly exhibits the transfer of transverse momentum, quantized in units of $2\hbar k_L$, and reveals the underlying coherent process:

$$|\chi_{\mathrm{fin}}\rangle = |\chi_{\mathrm{in}}\rangle \exp\left(-i\frac{\gamma}{2\delta}\right) \sum_{n=0}^{\infty} (-1)^n J_{2n}\left(\frac{\gamma}{2\delta}\right) e^{-2ink_L z}, \tag{17.21}$$

where $\gamma = \int_{-\infty}^{\infty} \Omega^2[x(t)]dt$, and J_{2n} is the Bessel function of order $2n$. There is no loss of atoms through their interaction with the standing waves. This is a key advantage of phase grating compared to material grating where many atoms are lost since they stick on the structures as previously discussed.

The wave interpretation of the quantized transfers of momentum is straightforward. The interaction of an incoming atom with a standing wave perpendicular to its propagation direction is nothing but the diffraction of a de Broglie wave with wavelength $\lambda_{\mathrm{dB}} = h/p_x$ by a periodic structure with a spatial period d. According to (17.14), the angles of the diffracted waves are given by

$$\theta_n = \frac{p_z}{p_x} = \frac{n\lambda_{\mathrm{dB}}}{d} \Longrightarrow p_z = \frac{nh}{d}. \tag{17.22}$$

In the case of an off-resonant standing wave, $d = \lambda_L/2$, where $\lambda_L = 2\pi/k_L$ is the wavelength of the laser light used for the standing wave. p_z is thus equal to $2n\hbar k_L$ in agreement with Eq. (17.21).

The diffraction of atoms by a near-resonant standing wave in the Raman-Nath regime has been observed by the group of David Pritchard at MIT [Moskowitz *et al.* (1983); Gould *et al.* (1986)] (see Fig. 17.8).

Remark

If spontaneous emissions occurs, the global picture based on coherent interaction processes is deeply modified. Random momentum changes due to the spontaneously emitted photons occur, giving rise to a diffusion of the photon momentum. The transition between the diffractive and diffusive regimes is described in [Tanguy *et al.* (1984)].

17.5.4 *Bragg regime*

We have seen in Sec. 17.5.2 that the diffraction process directly couples the states $|g, p_z = -\hbar k_L\rangle$ and $|g, p_z = +\hbar k_L\rangle$ in the Bragg regime. Figure 17.9 represents those states and a few neighbouring states of the total system { atom+field }. The states $|g, -\hbar k_L\rangle$ and $|g, +\hbar k_L\rangle$ are coupled through the excited state $|e, 0\rangle$ by a

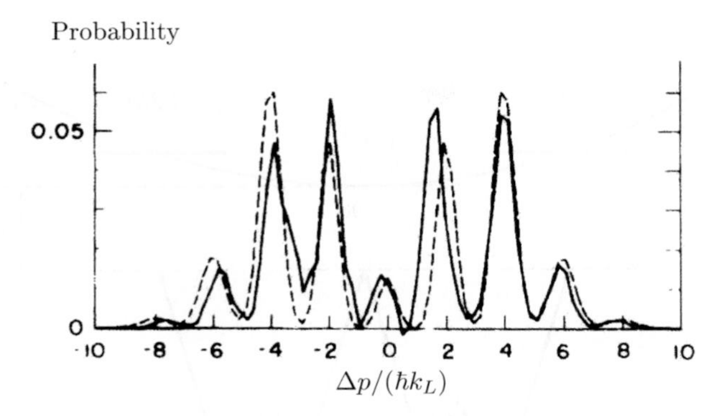

Fig. 17.8 Raman-Nath multiple diffraction pattern obtained with atoms. The solid line is the experimental measurement and the dashed line is the expected signal, according to the experimental parameters. Figure extracted from [Gould *et al.* (1986)]. Copyright: American Physical Soceity.

stimulated Raman process: absorption of a photon from one of the travelling waves of the standing wave to reach virtually the state $|e,0\rangle$, followed by the stimulated emission of a photon in the other travelling wave. The energy mismatch in the intermediate state is

$$E_{|g,-\hbar\vec{k}_L\rangle} + \hbar\omega_L - E_{|e,\vec{0}\rangle} = \hbar(\omega_L - \omega_A) + \frac{\hbar^2 k_L^2}{2m} = \hbar(\delta + \omega_R). \tag{17.23}$$

The effective coupling between the two degenerate states $|g,-\hbar k_L\rangle$ and $|g,+\hbar k_L\rangle$ is therefore given by:

$$\frac{\hbar\Omega_{\rm eff}}{2} = \frac{(\hbar\Omega/2)^2}{\hbar(\delta+\omega_R)}. \tag{17.24}$$

The corresponding Rabi oscillation with the effective Rabi frequency $\Omega_{\rm eff}$ can be used to generate for instance $\pi/2$ or π pulses. Note that many other states, as illustrated in Fig. 17.9, are involved. For instance, the state $|g,+\hbar k_L\rangle$ is also coupled to $|g,+3\hbar k_L\rangle$ through the intermediate state $|e,+2\hbar k_L\rangle$. The energy mismatch between $|g,+\hbar k_L\rangle$ and $|g,+3\hbar k_L\rangle$ is $(3\hbar k_L)^2/2m - (\hbar k_L)^2/2m = 8\hbar\omega_R$. Similarly, $|g,-\hbar k_L\rangle$ is also coupled to $|g,-3\hbar k_L\rangle$.

The Bragg diffraction of atoms by a laser standing wave with a large waist has been observed for the first time in 1988 by the group of David Pritchard [Martin *et al.* (1988)].

Bragg's scattering is used for implementing Mach-Zehnder interferometers[15] (see also Sec. 17.4) as illustrated in Fig. 17.10. An atomic beam passes through three successive and equidistant standing waves. The effective Rabi frequency for the first standing wave is adjusted in such a manner to achieve a $\pi/2$ pulse which acts

[15] Mach-Zehnder interferometers for non-zero mass particles were first demonstrated with electrons using electron diffraction from three very thin metallic crystals [Marton *et al.* (1953)]. Concerning neutrons, interferometers are essentially based on successive Bragg reflection on three gratings cut in the same silicon crystal, and was first realized by H. Rauch and coworkers in 1974 [Rauch *et al.* (1974)].

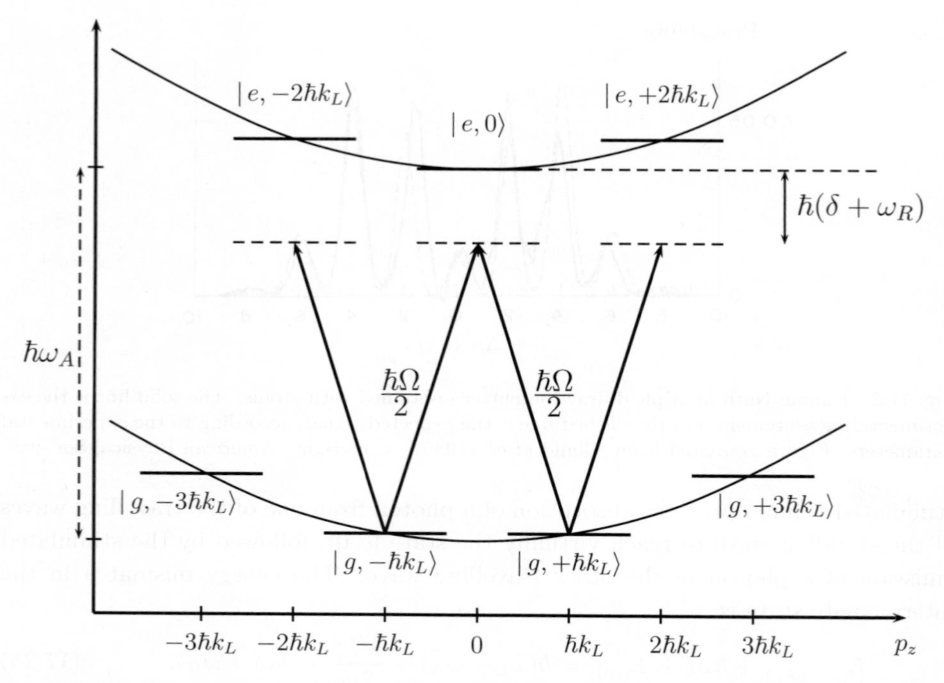

Fig. 17.9 Diagram representing the combined internal and external states for an atom placed in a standing wave made from two counterpropagating coherent waves 1 and 2. Off-resonant Bragg diffraction corresponds to a stimulated Raman process. By absorbing a photon from wave 1, the atom starting from $|g,-\hbar k_L\rangle$ reaches (virtually) the state $|e,0\rangle$, from which it reemits, in a stimulated manner, a photon in wave 2, to go to the state $|g,+\hbar k_L\rangle$.

as a 50%-50% beam splitter. The second zone, with a π pulse, is equivalent to a mirror. As a result the two coherent beams recombine at the location of the third standing wave identical to the first one. Such an interferometer that uses light as beam splitter requires very good control of the angle θ.

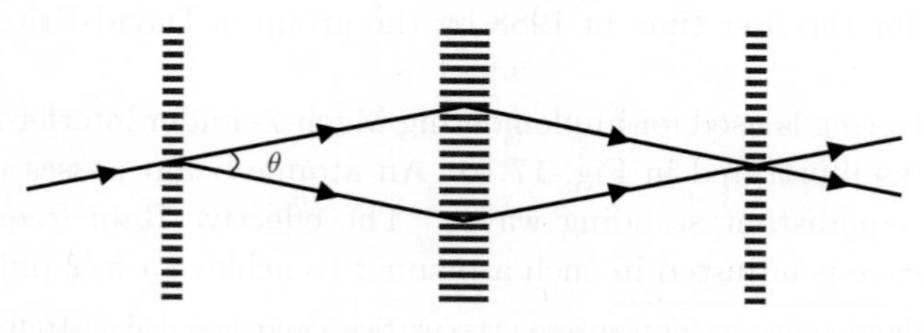

Fig. 17.10 Succession of three equidistant standing waves to realize a Mach-Zehnder interferometer for atoms.

17.6 Bloch oscillations

So far, we have studied the evolution of atomic de Broglie waves crossing a perpendicular optical potential. In this subsection, we address another interesting configuration where the atoms move along the optical potential.

With this configuration, one can observe a physical effect, usually studied in solid state physics, called *Bloch oscillations*. Consider an electron moving in a crystal. It feels the periodic potential created by the ions of the crystal. If, in addition, it is submitted to an homogeneous electric field, quantum mechanics predicts that it will execute an oscillatory motion rather than an accelerated one. In this subsection, we show that this striking quantum prediction can be more easily observed with atoms moving in a non-resonant laser standing wave and submitted in addition to a constant inertial force.

Let us start by reviewing the basics of the quantum treatment of an electron in a one-dimensional crystal. The application to cold atoms will be analyzed afterwards.

17.6.1 *Review on the quantum treatment of a particle in a periodic potential*

A particle of mass m that experiences a periodic potential of spatial period d: $V(x) = V_0 \sin^2(\pi x/d)$, is governed by the Hamiltonian H_0:

$$H_0 = \frac{p^2}{2m} + V(x). \tag{17.25}$$

The energy levels $E_n(q)$ are labelled by a band index n and a quasi-momentum $\hbar q$. The corresponding eigenvectors $|n, q\rangle$ are commonly called *Bloch states*. The available energy bands are separated by energy gaps (see Fig. 17.11). As a consequence of the periodicity of the potential, the eigenstates and eigenvalues are $2\pi/d$ periodic in quasi-momentum q. By convention, the interval $[-\pi/d, \pi/d]$ is referred to as the first Brillouin zone. It is worth noting that the eigenstates are delocalized over the whole lattice. The mean velocity of the electron in a band of index n is given by:

$$\langle v\rangle_n = \frac{1}{\hbar}\frac{dE_n(q)}{dq}. \tag{17.26}$$

The dispersion relation in each band is not quadratic. Therefore, the quasi-momentum $\hbar q$ differs from the mean momentum $m\langle v\rangle_n$.

In free space, the mean momentum of an electron subjected to a uniform electric field increases linearly with time. This is to be contrasted with an electron in a periodic potential. In the presence of an electric field $\vec{E} = E\vec{e}_x$, the electron motion is governed by the Hamiltonian:

$$H = \frac{p^2}{2m} + V(x) - Fx, \tag{17.27}$$

where the force F is given by $F = (-e)E$. One can show that the solution of the Schrödinger equation with the Hamiltonian H is simply given by a Bloch state

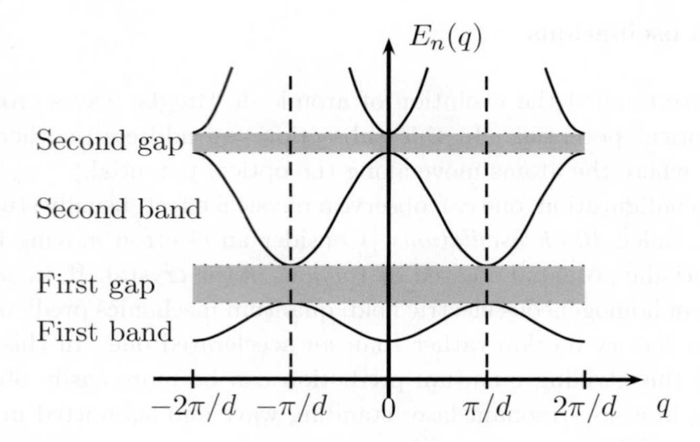

Fig. 17.11 Band structure $E_n(q)$ for a particle in a periodic potential $V(x) = V_0 \sin^2(\pi x/d)$. Under the influence of a weak uniform force F, the quasi-momentum $\hbar q(t)$ scans the q-axis at a uniform speed F, so that a particle prepared in the fundamental band and remaining in this band executes periodic motion, referred to as Bloch oscillations.

$|n, q(t)\rangle$ with a time-dependent quasi-momentum $\hbar q(t) = \hbar q(0) + Ft$, which means that $q(t)$ scans the q-axis at a uniform speed $F/\hbar$ [Ashcroft and Mermin (1976)]. The particle remains in the same band if this speed is not too large, so that non-adiabatic transitions between different bands can be neglected. Instead of increasing linearly with time as in free space, the mean velocity $\langle v \rangle_n$ thus oscillates with a period:

$$\tau_B = \frac{2\pi/d}{|F|/\hbar} = \frac{h}{|F|d}. \tag{17.28}$$

17.6.2 *Implementation with cold atoms*

Bloch oscillations have been investigated with atoms in laser standing waves [Ben Dahan *et al.* (1996)]. To mimic the role of the static electric field, the atom is put in an accelerated standing wave that is realized through a tunable frequency difference $\delta\nu(t)$ between the two counter-propagating laser travelling waves.

A constant frequency difference $\delta\nu(t) = \delta\nu_0$ yields a standing wave in global translation at a constant velocity proportional to $\delta\nu_0$ as already discussed in Sec. 7.5.4 of Chap. 7 and in Sec. 14.4.3 of Chap. 14. A linear variation in time of $\delta\nu$ results in a constant acceleration a, and therefore a constant inertial force $F = -ma$ is exerted on the atom in the frame attached to the standing wave. The atom is therefore governed by a Hamiltonian of the form (17.27) in this frame.

One may wonder how to prepare an initial Bloch state with well defined quasi-momentum q. As already emphasized, the Bloch state wave function is delocalized over the whole lattice. In practice, this means that

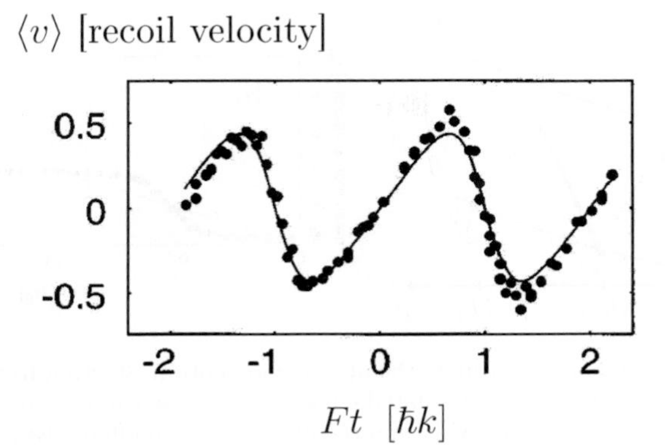

Fig. 17.12 Experimental data revealing the Bloch oscillations of cold cesium atoms in a one-dimensional optical lattice. The measured mean atomic velocity, in units of the recoil velocity $\hbar k_L/m$, is plotted versus time, in units of $\hbar k_L/F$, where F is the inertial force experienced by the atoms in the accelerating standing wave. Figure extracted from [Ben Dahan *et al.* (1996)]. Copyright: American Physical Society.

(i) the coherence length ξ of the wave packet should be larger than the lattice period $\lambda_L/2$,
(ii) the momentum spread along the standing wave axis $\delta p = \hbar/\xi$ should be smaller than $\hbar/\lambda_L \sim \hbar k_L$.

This latter requirement means that the atomic momentum spread should be smaller that the photon momentum, or in other words that sub-recoil cooling techniques should be used. For instance, the experiment of [Ben Dahan *et al.* (1996)] relied on one-dimensional Raman cooling.[16]

Examples of experimental results are shown in Fig. 17.12. The measured mean velocity $\langle v\rangle$ (in units of the recoil velocity $\hbar k_L/m$), plotted versus time t (in units of $\hbar k_L/F$), clearly exhibits an oscillatory behavior. The observed asymmetry of the variations of $\langle v\rangle$ is due to the change of the curvature of $E_n(q)$ when q varies from 0 to π/d (the effective mass m^* of the particle is not the same in the center of the band and in the edges).

Bloch oscillations are easier to observe for cold atoms in an optical lattice than for electrons in a crystal due to the fact that the Bloch period is inversely proportional to the period d of the periodic potential, and is therefore three orders of magnitude longer for a real crystal than for an optical standing wave. Bloch oscillations are very difficult to observe in natural crystals because of dissipative effects that occur on a much shorter timescale than the Bloch period. They have, however, been observed on semiconductor superlattices [Waschke *et al.* (1993)].

[16]See Sec. 13.3.3 of Chap. 13.

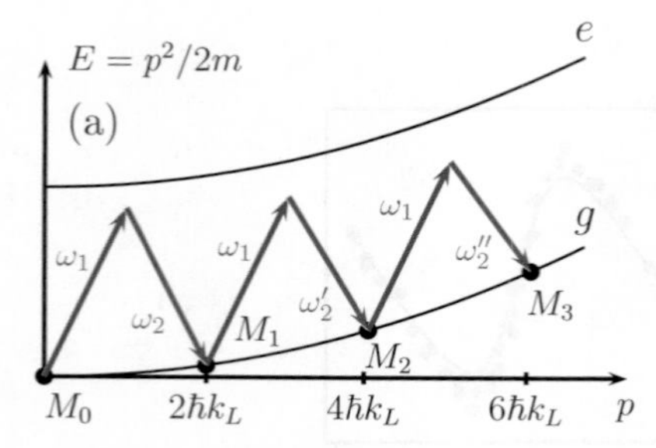

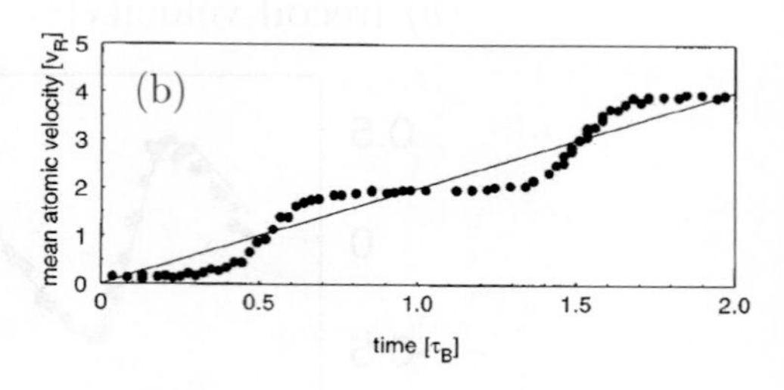

Fig. 17.13 (a) In the laboratory frame, the atomic momentum, starting from $p = 0$, increases by steps $2\hbar k_L$ in resonant absorption-stimulated emission cycles, where the atom absorbs one photon in the wave with the higher frequency ω_1 and re-emits it in a stimulated way in the other wave ω_2. (b) Experimental measurement of the mean atomic velocity in the laboratory frame as a function of time. Figure extracted from [Peik *et al.* (1997)]. Copyright: American Physical Society.

17.6.3 *Physical interpretations*

Consider an atomic plane wave, initially prepared with momentum $p \simeq 0$ in the fundamental band. This initial state is delocalized over the whole optical lattice. Its momentum p increases linearly with time according to Newton's law because of the external force until it reaches a critical value that satisfies the Bragg reflection condition: $k_{\text{at}} = \pi j/d$, where k_{at} is the atomic wave vector, and j is an integer. At this stage, the wave is Bragg reflected by the periodic potential and its momentum is reversed. The further evolution consists in the repetition of that process: an acceleration by the force followed by a Bragg reflection.

In the following, we propose a different physical picture of Bloch oscillations, where the analysis is performed in the laboratory frame.

In the absence of spontaneous emission, the atomic momentum can change by units of $\hbar\vec{k}_1 - \hbar\vec{k}_2 \simeq 2\hbar\vec{k}_L$ through the absorption of a photon from one wave and the stimulated emission of a photon into the other wave, as depicted in Fig. 17.13(a). The possible energy-momentum states of atoms after their interaction with the standing wave are therefore discrete points $M_j(p = 2j\hbar k_L, E = p^2/2m = 4j^2 E_{\text{rec}})$ with $j = 0, 1, 2, 3 \ldots$, on the momentum energy parabola of the free particle, since atoms have been initially prepared with a momentum spread much smaller than $\hbar k_L$. The gain in kinetic energy is provided by the frequency difference between the two laser waves: the atoms are accelerated in the direction of the laser beam with the higher frequency ω_1 by absorbing photons from it and re-emitting photons with a lower frequency ω_2 into the other beam. The transition from M_j to M_{j+1} is resonant for an angular frequency difference $\delta\omega_j = 4(2j+1)E_{\text{rec}}/\hbar$. These absorption-stimulated emission cycles start with an atom initially at rest. The different resonant values of $\delta\omega_j$ are thus encountered sequentially when the frequency ω_2 is lowered, and accompanied with a gain of atomic momentum equal to $2\hbar k_L$. One can show that, for a linear increase of the angular frequency difference $\delta\omega$, the mean atomic velocity increases by $2\hbar k_L/m$ during each Bloch period. Therefore, the Bloch oscillations in the laboratory frame appear as a periodic deviation of the mean velocity around the linear increase in time at as illustrated in Fig. 17.13(b).

Bloch oscillations is thus an efficient method for increasing the momentum by units of $2\hbar k_L$ in a short time scale given by the Bloch period. This feature is extensively used in precision measurement.[17]

The Wannier-Stark ladders

As already discussed, an atom placed in a periodic potential has an energy spectrum made of allowed bands separated by gaps (see Fig. 17.11). One may wonder how this energy spectrum is modified in the presence of an accelerating field E. The answer was given in the early 1960s by Gregory Wannier through his study of the behavior of electrons in a solid subjected to an electric field. When one adds to the periodic potential a linear potential describing the effect of the field E, the periodic potential is tilted. The discrete levels in adjacent potential wells no longer have the same energy and the tunnel coupling between them is no longer resonant. One can show that the continuum of each band obtained for $E = 0$ splits in a series of discrete levels separated by $\Delta E = h/\tau_B$ thereby forming a so-called *Wannier-Stark ladder*. The Bloch oscillations can be reinterpreted within this formalism, and their frequency related to the energy splitting between the discrete states of the ladder. The group of Mark Raizen has performed spectroscopic measurements on the energy levels of these ladders using cold atoms in a one-dimensional lattice [Niu *et al.* (1996); Wilkinson *et al.* (1996)].

17.7 Diffraction of atomic de Broglie waves by time-dependent structures

So far, we have considered interference and diffraction of atomic de Broglie waves moving through static material slits and gratings. The observed phenomena can be analyzed using time-independent Schrödinger equation. In this section, we describe new phenomena that can be observed with diffracting structures that are modulated in time.

The generation of wave diffraction from a time-dependent boundary condition was first discussed theoretically by Marcos Moshinsky [Moshinsky (1952)]. The first experiments directly demonstrating diffraction in time for matter-waves were carried out with ultra-cold neutrons reflected from vibrating mirrors [Felber *et al.* (1990); Hamilton *et al.* (1987)]. In this section, we describe experiments performed at Ecole Normale Supérieure by the group of Jean Dalibard where the phase modulation [Steane *et al.* (1995)], time diffraction and interference [Szriftgiser *et al.* (1996)] of atomic de Broglie waves were investigated using cold atoms bouncing on an intensity modulated blue detuned evanescent light wave at the surface of a glass prism [Cook and Hill (1982)].

[17] See Sec. 18.5.3 of Chap. 18.

17.7.1 *Phase modulation of atomic de Broglie waves*

Atomic mirrors using evanescent waves[18] can mimic a vibrating mirror if the evanescent wave intensity is modulated. The distance from the mirror at which a classical particle would be reflected changes when the light intensity varies. If atoms prepared with a well-defined total energy are reflected from such a vibrating mirror, the associated matter waves are phase modulated. This effect is well known for light waves. As a consequence of this phase modulation, frequency sidebands appear for the reflected de Broglie waves at integer multiples of the angular frequency Ω of the vibrating mirror. If E_i is the incident kinetic energy of the atoms, one expects to get reflected atoms with a kinetic energy E_f having several components $E_f = E_i + n\hbar\Omega$, with $n = 0, \pm 1, \pm 2, \ldots$ One can also consider that the incoming particle "sees" a modulated time-dependent perturbation that induces transitions corresponding to energy changes equal to $n\hbar\Omega$. For a detailed theory on the intensity of the various sidebands, see [Henkel *et al.* (1994)].

To perform this experiment, a magneto-optical trap placed 3.3 mm above the prism surface is used as a source of atoms. The laser cooled sample provides a small value of $\langle v \rangle$. Since the kinetic energy of the atoms is smaller than the height of the light potential barrier associated with the evanescent wave, it is possible for them to bounce off of it. The experiment proceeds as follows [Steane *et al.* (1995)]. The mirror is applied during a first pulse in order to select a fraction of the atom distribution in the MOT with a well-defined total energy E_i that corresponds in practice to a velocity of 0.25 m$\cdot$s^{-1} at the reflecting spot with a velocity spread of less than 1 %. This subset of atoms bounces on the mirror and when it enters back into the mirror zone, the intensity of the evanescent wave is modulated at an angular frequency Ω. A time of flight measurement with a probe beam is performed to extract the energy spectrum of the atoms that have bounced on the vibrating mirror.

If the atoms were to behave like classical point-like particles, they would bounce at different times on the mirror and acquire different energies from this mirror depending on its velocity at the time of the interaction, such that their final energy distribution would be continuous and Gaussian. Experimental results are clearly different from this classical prediction, as illustrated in Fig. 17.14. The observed time of flight spectrum contains a central peak corresponding to particles having bounced with an energy $E_f = E_i$, plus various sidebands corresponding to particles having bounced with an energy $E_n = E_i + n\hbar\Omega$. This result clearly shows the quantum nature of the atomic motion. Measuring precisely, the energy of the bouncing atoms could provide a direct measurement of the Planck constant $\hbar$ since Ω is known.

Note that the maximum frequency at which efficient transfer can occur is determined by the interaction time between the atom and the oscillating potential.

[18] See Sec. 7.5.2 of Chap. 7.

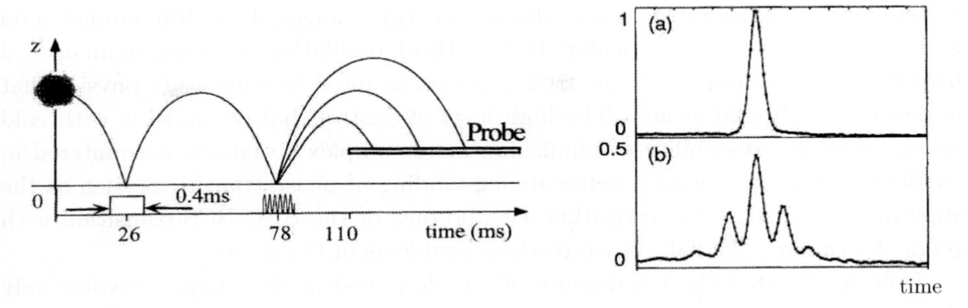

Fig. 17.14 Phase modulation of atomic de Broglie waves. Left: experimental scheme. Right: time-of-flight experimental results without (a) and with (b) modulation. Figure adapted from [Steane *et al.* (1995)]. Copyright: American Physical Society.

This phase modulator for de Broglie waves is the temporal equivalent of the spatial periodic phase gratings presented in Sec. 17.5.

17.7.2 *Atomic wave diffraction and interference using temporal slits*

Alternatively, by switching on and off the light that is totally reflected in the prism, the evanescent mirror can be used as a time-dependent aperture as envisioned by Moshinsky in the context of neutron optics [Moshinsky (1952); Gerasimov and Kazarnovskii (1976); Felber *et al.* (1990)]. According to the time-energy uncertainty relation, one expects that a beam with a well-defined energy that is chopped with a short pulse has an energy distribution that is broadened by a time diffraction effect. One can also apply the mirror during two short pulses. The two waves reflected by these two pulses can then overlap and give rise to interference fringes in the time domain. These time diffraction and interference effects have been observed with atomic de Broglie waves associated with cold atoms [Szriftgiser *et al.* (1996)].

17.8 Conclusion

The experiments described in this chapter clearly demonstrate the wave nature of atomic motion. Atomic de Broglie waves are diffracted and interfere like light waves. There are, however, important differences between the two types of waves. The dispersion relation is not the same, and atoms interact with each other, which is not the case for photons.

Atom optics with de Broglie waves offers interesting possibilities. Laser cooling can be used to increase the coherence length of atomic beams without losing atoms. The evolution of atomic de Broglie waves in non-resonant optical waves can be clearly interpreted in terms of momentum exchanges between atoms and photons in stimulated Raman processes. Quantum effects that can be observed with atomic waves, such as Bloch oscillations or discrete energy changes of an atom bouncing on a vibrating mirror, are spectacular. In fact, Bloch oscillations of atoms in an optical lattice is a first example of a physical situation inspired by solid state physics that is revisited with cold atoms. The high level of control that is available with cold atoms offers the possibility of mimicking more complex situations encountered in condensed matter, to gain a better understanding of phase transitions such as the superfluid-Mott insulator transition with bosons, or the BEC-BCS transition with pairs of fermions. We will return to these problems in Chap. 26.

Note finally that the interference effects described in this chapter involve only the atomic external degrees of freedom. They are associated with the off diagonal elements of the atomic density matrix between two different points in space. These spatial coherences are the counterpart of the Zeeman coherences considered in Chap. 4 which involve only the internal degrees of freedom. In the next chapter, we will review several new interesting interference effects where both internal and external degrees of freedom play important roles.

Chapter 18

Ramsey fringes and atomic interferometry

18.1 Introduction

The first experiments to use magnetic resonance spectroscopy to investigate the structure of the ground state of atoms and molecules were carried out at Columbia university in the laboratory of Isidor Rabi using molecular beam techniques described in detail in [Ramsey (1956)]. A great breakthrough in this field was the invention of the method of separate oscillatory fields [Ramsey (1949, 1990)], for which Norman Ramsey got the Nobel Prize in Physics in 1989. We have already described this method in Sec. 2.4 of Chap. 2 and given several interpretations of the Ramsey fringes: analogy in the time domain with the Young two-slit interference fringes in the space domain; interference between two different paths where the RF transition occurs during either the first or second pulse; creation of a linear superposition of states by the first pulse and interrogation of this superposition of states by a second pulse after a free evolution period.

The principle of an experiment using two Ramsey zones is sketched in Fig. 18.1. A state selecting device, like that used in a Stern-Gerlach experiment, selects the atoms in one of the two levels, g, of the transition. The atomic beam passes through two cavities that are fed by the same RF or microwave source, to ensure that the fields in the two cavities are coherent. The beam emerging from the two cavities passes through a second state selecting device that filters the atoms in the other state, e. The number of atoms N_e transferred in this state is measured with a detector. The variations of N_e with the frequency of the RF or microwave field exhibit Ramsey fringes with a width that scales as $1/T$, where T is the transit time between the two Ramsey zones, within a broad diffraction profile whose width scales as the inverse of the transit time, τ, through each cavity. By using two zones instead of one, the resolution of the experiment can be improved by two orders of magnitude since T/τ can reach values on the order of 100.

Since their invention, Ramsey fringes have been revisited several times. Combined with the most recent developments of atomic physics, which provide a much better control of the internal and external degrees of freedom of atoms, they are at the basis of new atomic interferometry methods that allow more refined tests of the

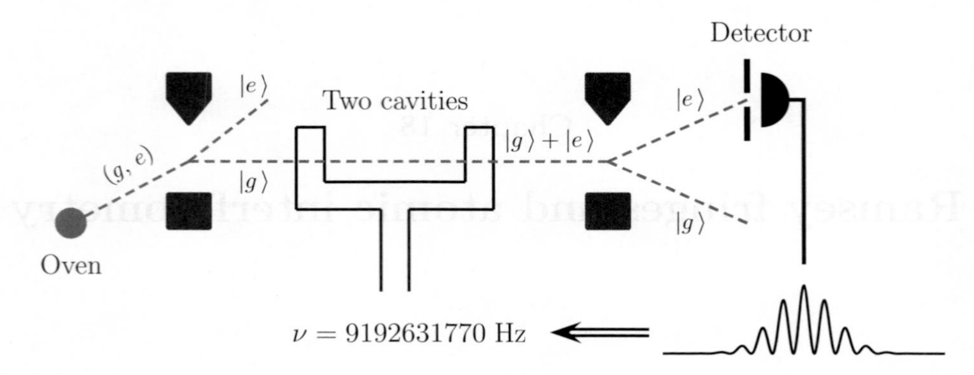

Fig. 18.1 Principle of an atomic clock with molecular beam using two Ramsey zones.

basic physical laws and a much higher precision in the measurement of fundamental constants.

A first straightforward idea to improve the resolution of the measurements is to reduce the velocity of the atoms by laser cooling techniques in order to increase the transit time T between the two Ramsey zones. In Sec. 18.2, we describe the spectacular improvement of microwave cesium clocks made possible by the use of cold atoms, either on earth, with the so-called *atomic fountains*, or in space, with zero gravity flights. We also show that the relative stability and accuracy of these clocks opens new possibilities for more accurately testing the gravitational red shift predicted by general relativity.

The development of coherent monochromatic laser sources has naturally led physicists to consider the possibility of extending the Ramsey scheme from the RF and microwave domains, where it was initially used, to the optical domain. In Sec. 18.3, we analyze the new features that arise in the optical domain due to the fact that the photon momentum can no longer be ignored such that atomic external degrees of freedom must be treated quantum mechanically. We show that an atomic wave packet that crosses a laser beam splits into two coherent wave packets whose centers follow different trajectories. The crossing of a laser beam therefore has an equivalent effect as a coherent beam splitter. By using several beam splitters of this type, atomic interferometers can be realized in which the center of the atomic wave packet follows several different coherent paths starting from the same point and recombining at another point where they interfere. Two atomic interferometers of this type will be described, the atomic Mach-Zehnder interferometer and the Ramsey-Bordé interferometer.

If the atoms propagate at different altitudes in the two arms of the atomic interferometer, the interference signal will be sensitive to the acceleration of gravity and will allow one to measure it. If the platform supporting the interferometer is rotating with an angular speed Ω_{rot}, the interference signal will be sensitive to Ω_{rot}. This effect is the equivalent, for de Broglie waves, of the Sagnac effect for light waves. An important issue is to calculate the phase shift due to the inertial gravitational and rotational fields along the two arms of the interferometer. In Sec. 18.4, we summarize the main lines of a path integral method leading to a simple calculation of these phase shifts when the one-particle Lagrangian of the system is quadratic with the position and momentum coordinates. In Sec. 18.5, we address some applications of atomic interferometry for the measurement of gravitational fields and rotational inertial fields. We also discuss the measurement of the ratio h/M, where h is the Planck constant and M the atom mass, and its importance for a determination of the fine constant structure α.

18.2 Microwave atomic clocks with cold atoms

18.2.1 *Principle of an atomic clock*

An atomic clock is an oscillator whose frequency is locked to the frequency of an atomic transition. In 1967, the unit of time, the second, was defined by an international convention as *the duration of 9 192 631 770 periods of the radiation corresponding to the transition between the two hyperfine levels of the ground state of the cesium-133 atom.* The principle of a cesium atomic clock is sketched in Fig. 18.2. An oscillator of frequency ω_0 generates a microwave field that scans the transition connecting the two ground state hyperfine levels $F = 3$ and $F = 4$ of the cesium atom. The resonance is centered at ω_A and has a width $\Delta\omega$. With a correction servo loop, the frequency ω_0 is locked to the central frequency ω_A of the cesium atom. It is clear that the precision of the clock will improve as the width $\Delta\omega$ of the atomic resonance is reduced. Usual atomic clocks with thermal atoms use an apparatus similar to that shown in Fig. 18.1. Atoms with a velocity on the order of 100 m.s^{-1} cross two Ramsey cavities separated by a distance L on the order of 0.5 m, leading to a transit time between the two cavities on the order of 5 ms and to a width $\Delta\omega/2\pi$ of the resonance on the order of 100 Hz.

18.2.2 *Atomic fountains*

A first idea for increasing the transit time T between the two Ramsey zones would be to slow down the atoms with radiation pressure of a counter-propagating laser beam. In this case, however, the atoms fall due to gravity and so it is necessary to use a vertical configuration.[1] With a laser pulse,[2] a cloud of cold cesium atoms

[1] This fountain configuration in which atoms pass through the cavity on the way up and once again on the way down, was first proposed in 1953 by Jerrold Zacharias [Ramsey (1956)]. Despite many efforts by Zacharias and his collaborators, the fountain experiment failed. This was due to the too small number of ultraslow atoms in the beam, which is far below the theoretical prediction due to atom-atom scattering in the collimation region, where the beam is not in an equilibrium state.

[2] In order to avoid heating of the atoms by the momentum diffusion associated with radiation pressure, it is better to use a *moving molasses* to throw the atoms upwards. The atoms are

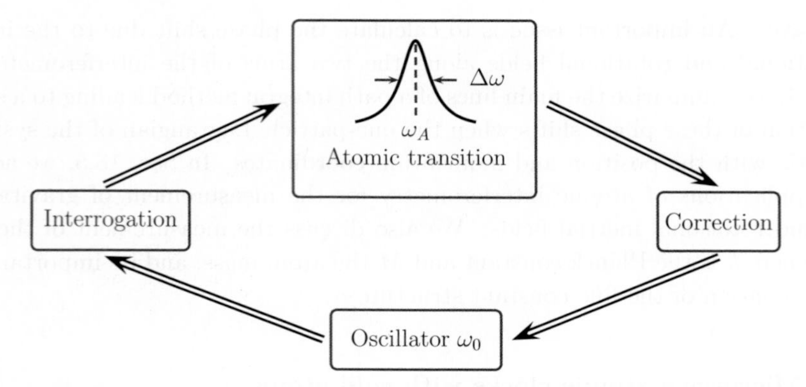

Fig. 18.2 Principle of an atomic clock. The oscillator generating the microwave field used to scan the hyperfine transition in the ground state of cesium atoms has its frequency ω_0 locked by a correction servo loop to the central frequency ω_A of the atomic transition.

is sent upwards through the microwave cavity. This cloud crosses the cavity, once on the way up and once on the way down (see Fig. 18.3(a)). In this way, a kind of fountain is formed where the drops of water are replaced by atoms. Each atom feels two coherent interactions with the microwave field, separated in time by an interval corresponding to the difference of times T at which it crosses the cavity on the way up and on the way down. If H is the height of the fountain, a simple calculation gives:

$$\frac{1}{2}g\,(T/2)^2 = H \quad \Rightarrow \quad T = 2\sqrt{2H/g}. \tag{18.1}$$

If the fountain has a height of $H = 30$ cm, we get from Eq. (18.1) with $g \simeq 10$ m.s^{-2}, $T = 0.5$ s, which represents an improvement by two orders of magnitude over the transit time of clocks using thermal atoms.

18.2.3 *Performances of atomic fountains*

Atomic fountains were realized with sodium atoms by the group of Steve Chu at Stanford University [Gibble and Chu (1993)] and with cesium atoms by the group of Christophe Salomon and André Clairon at the Observatoire de Paris (BNM/SYRTE) [Lea *et al.* (1994)]. Figure 18.3(b) gives an example of the Ramsey fringes that can be obtained with a fountain of $H = 30$ cm. The signal to noise ratio is very good, on the order of 2000, and the width of the central fringe is on the order of 1 Hz [Lemonde *et al.* (2001)].

The relative stability of cesium clocks using atomic fountains is on the order of 1.6×10^{-16} for an integration time of 5×10^4 s. Their relative accuracy is on the

cooled below the Doppler limit in a standing wave. This standing wave is set in motion during a certain time by chirping the frequency of one of the two counter propagating laser beams, so that atoms remain cooled during the launching phase (see [Riis *et al.* (1990); Clairon *et al.* (1991)] and Sec. 7.5.4 of Chap. 7).

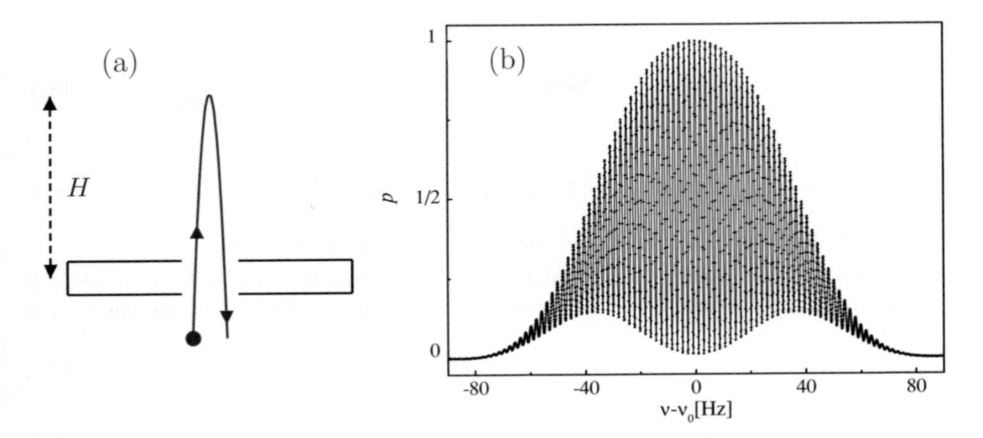

Fig. 18.3 (a) Principle of an atomic fountain. (b) Example of Ramsey fringes obtained with an atomic fountain. Courtesy of C. Salomon.

order of 3×10^{-16}. To give an idea of the meaning of these numbers, let us just mention that a relative accuracy of 10^{-16} corresponds to an error smaller than 1 s in a period of 300 millions years!

To reach this performance, a certain number of effects must be carefully controlled. One of them is the density shift of the atomic line due to atom-atom collisions in the cloud of atoms. To reduce the importance of this effect, the density of atoms in the cloud must be lowered, and an extrapolation of the position of the atomic line at zero density must be performed. It turns out that the collisional shift is at least 50 times smaller for rubidium than for cesium [Sortais *et al.* (2000)]. From this point of view, we can therefore say, a posteriori, that it would have been better to choose rubidium rather than cesium for the definition of the second, even if its smaller hyperfine structure leads to a smaller Q-factor of the atomic resonance.

Quantum projection noise

One important source of noise for atomic clocks is the *quantum projection noise* due to the discrete character of the measured signal. Consider an atom prepared in the state $|g\rangle$ and that crosses the two cavities in Fig. 18.1. Its state becomes a linear superposition of the ground state $|g\rangle$ and the excited state $|e\rangle$:

$$|\psi\rangle = \alpha\,|e\rangle + \beta\,|g\rangle\,. \tag{18.2}$$

Ramsey fringes appear in the probability $|\alpha|^2$ to find it in state $|e\rangle$, but the state selecting device, which detects only atoms in the state $|e\rangle$, does not directly measure $|\alpha|^2$. It gives a click with a probability $|\alpha|^2$ and no click with a probability $|\beta|^2$. The signal S given by the detector is thus equal to 1 if the atom is in state $|e\rangle$, to 0 if the atom is in $|g\rangle$. It can be written as the projector into $|e\rangle$

$$S = \Pi_e = |e\rangle\,\langle e|\,, \tag{18.3}$$

and is characterized by a mean value

$$\langle S\rangle = \langle\psi|\,\Pi_e\,|\psi\rangle = |\alpha|^2\,, \tag{18.4}$$

and a variance

$$(\Delta S)^2 = \langle\psi|\,\Pi_e^2\,|\psi\rangle - (\langle\psi|\,\Pi_e\,|\psi\rangle)^2 = |\alpha|^2 - |\alpha|^4 = |\alpha|^2\left(1-|\alpha|^2\right). \tag{18.5}$$

To derive Eq. (18.5), we have used the projector property $\Pi_e^2 = \Pi_e$.

The previous calculation shows that the signal S_N recorded by the detector when N independent atoms have crossed the two cavities, i.e. the number of clicks, has a mean value

$$\langle S_N\rangle = N\,|\alpha|^2\,, \tag{18.6}$$

and a variance

$$(\Delta S_N)^2 = N(|\alpha|^2 - |\alpha|^4) = N\,|\alpha|^2\left(1-|\alpha|^2\right). \tag{18.7}$$

The fact that a single measurement can give only a discrete result, 0 or 1, gives rise to a dispersion of the number of clicks around the mean value, and consequently to a fundamental noise, called the quantum projection noise. Equations (18.6) and (18.7) show that the signal to noise ratio is proportional to $N^{1/2}$.

This noise can become appreciable when the other sources of noise (technical noise of the oscillator, shot noise) become negligible. The $N^{1/2}$-dependence of the signal to noise ratio has been experimentally observed [Itano *et al.* (1993); Santarelli *et al.* (1999)] (see Fig. 18.4). To improve it, one must increase the density of the beam, but then, the collisional shift increases. Proposals to reduce the quantum projection noise have been made that consist of using atomic squeezed states where the N atoms are prepared in a correlated state [Wineland *et al.* (1992b); Kitagawa and Ueda (1993); Sørensen and Mølmer (2001)]. Recent experiments have confirmed the possibility of reducing the quantum projection noise using squeezing [Appel *et al.* (2009); Riedel *et al.* (2010)].

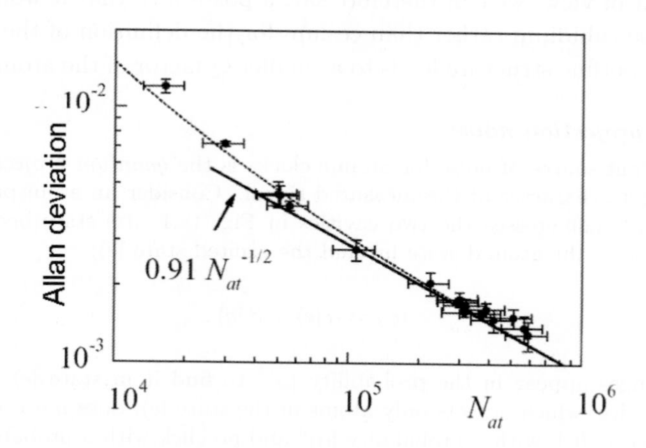

Fig. 18.4 Allan deviation in normalized units versus the number N_{at} of atoms. The Allan deviation is a convenient characterization of the dispersion of the results given by the detector. Figure extracted from [Santarelli *et al.* (1999)]. Copyright: American Physical Society.

18.2.4 *Cold atoms clocks in space*

According to Eq. (18.1), the time interval between the two interactions in the microwave cavity increases only as $H^{1/2}$ when the height H of the fountain is increased. It is therefore not realistic to try to improve the performances of the fountain by increasing its height. A more straightforward way to avoid the acceleration due to gravity is to put the clock in a satellite, so that in the frame of the satellite the atoms are not falling and keep the very low velocity obtained with laser cooling.

As a preliminary test of the feasibility of such experiments, cold atom clocks have been put in a plane making parabolic flights. The pilot turns off the engines when the plane is in an ascending phase. The plane then follows a free fall parabolic trajectory. The pilot switches on the engines when the plane is in the descending phase to start a new ascending phase and a new free fall parabolic trajectory. Each parabolic phase lasts about 20 s, during which time the effective gravity field in the plane is reduced, allowing the various experimental groups in the plane to perform zero gravity experiments. The performances of the cold atom clocks have been improved in these conditions by a factor 10, in comparison with the atomic fountains on earth [Lemonde *et al.* (1998)].

With the success of these zero gravity experiments, it is now planned to put a cold atom clock in the International Space Station (ISS), together with a Hydrogen maser. The project is called ACES (Atomic Clock ensemble in Space).The signal delivered by the cold atom clock in the ISS would be used to synchronize all atomic clocks on earth and in the satellites of the Global Positioning System (GPS) and Galileo system. Comparison between the frequency of the clock in the ISS and the earth clocks would also allow a very precise test of general relativity, as explained in the next section.

18.2.5 *Tests of general relativity*

According to the theory of general relativity, two clocks in different gravitational potentials do not have the same frequency. This effect is referred to as the gravitational red shift. For example, two clocks at two different altitudes z and $z + \delta z$ in the earth field have a relative frequency difference given by:

$$\frac{\omega_A\,(z+dz) - \omega_A\,(z)}{\omega_A\,(z)} = \frac{\delta\,\omega_A}{\omega_A} = \frac{g\,\delta z}{c^2}. \tag{18.8}$$

If we take $\delta z = 1$ m, we get, using $g = 10$ m·s^{-2} and $c = 10^8$ m·s^{-1} : $\delta\,\omega_A/\omega_A = 10^{-16}$. This means that the accuracy of cold atom clocks is now high enough to detect an altitude difference of 1 m between two clocks![3] A space clock at an altitude of 400 km and a terrestrial clock would have frequencies differing by 4×10^{-11} meaning that it should be possible to test this prediction of general relativity with

[3] In this Sec. 18.2, we focus on microwave atomic clocks. Recent progress in optical atomic clocks (see Sec. 18.6) allows now the detection of the differential gravitational red shift for two clocks whose height differs only by 33 cm [Chou *et al.* (2010b)].

a precision 50 times higher than all previous tests.[4] Another possible application of cold atom clocks could be the determination of the geoid surface, where the gravitational potential has a given value. Another possible application of cold atom clocks could be the determination of the geoid surface, where the gravitational potential has a given value.

18.3 Extension of Ramsey fringes to the optical domain

The simplest idea for producing two interaction zones with optical fields is to take two parallel coherent laser beams and send atoms across them. Consider first, as in Sec. 2.4.3 of Chap. 2, the perturbative limit where the atom-field interaction is treated to the lowest order. There are two possible paths leading the atom from g to e, depending whether it absorbs a laser photon in the first laser beam or in the second one. Can these two possible paths interfere and give rise to Ramsey fringes? In this section, we show that the change of momentum of the atom when it absorbs a photon causes a spatial separation of the two atomic wave packets corresponding to the two paths. This effect progressively quenches the interference signal, since the two final atomic external states become orthogonal. We later show how it is possible to restore the interference between the two paths by using additional laser beams.

18.3.1 *Equivalence of the crossing of a laser beam with a coherent beam splitter*

Consider an atom in its ground state $|g\rangle$, with a momentum $\vec{p}$, crossing a laser beam of wave vector $\vec{k}_L$ and of angular frequency ω_L (see Fig. 18.5(a)). The atom-laser interaction time is assumed to be short enough that spontaneous emission can be neglected. Because of momentum conservation, the interaction couples together the two states $|g, \vec{p}\rangle$ and $|e, \vec{p} + h\vec{k}_L\rangle$ through absorption-stimulated emission cycles. We are dealing here with the well known two-state problem considered in Chap. 2, whose dynamics are equivalent to those of a fictitious spin 1/2. The system oscillates between the two states at the Rabi frequency Ω. After the crossing, the atom is put in a coherent linear superposition of the two states $|g, \vec{p}\rangle$ and $|e, \vec{p} + h\vec{k}_L\rangle$. The crossing of the laser beam is thus equivalent to the effect of a coherent beam splitter, which transforms the incident de Broglie wave packet into a linear super-

[4] In a previous spatial test of the gravitational redshift, a hydrogen-maser clock was launched on a rocket to an altitude of 10^4 km and its frequency compared to a similar clock on ground [Vessot and Levine (1979)]. The redshift was tested with an accuracy of about 10^{-4}. The expected accuracy of the ACES test with cold atoms in space is significantly higher and on the order of 2×10^{-6} [Cacciapuoti and Salomon (2009)]. In fact, the first spectroscopy test of the redshift used gamma rays and the Mossbauer effect [Pound and Rebka (1960)]. The emission and absorption frequencies of ^{57}Fe nuclei were compared by putting the emitter and the absorber at altitudes differing by 22.5 m. The accuracy of the test was on the order of 10^{-2}.

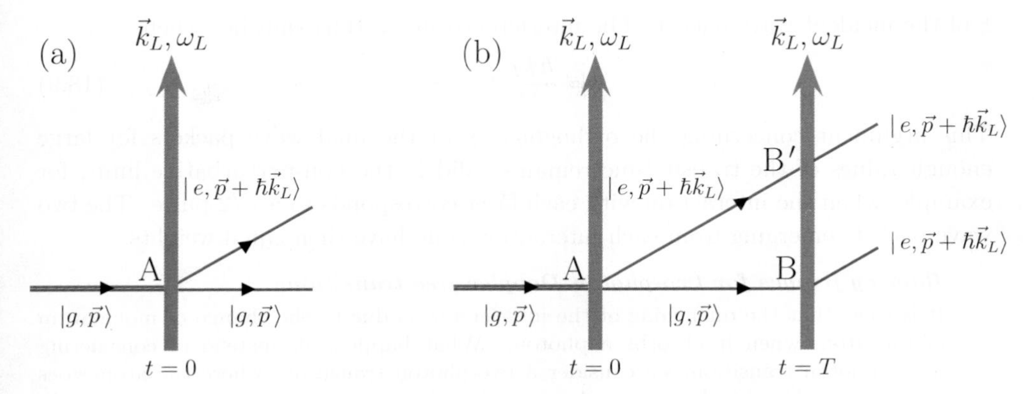

Fig. 18.5 (a) Interaction of a two-level atom of momentum $\vec{p}$, initially in its ground state, with a travelling wave propagating orthogonally to the atom direction. The interaction puts the atom in a superposition of states that entangles the external and internal degrees of freedom. (b) A naive Ramsey interferometer using two interaction zones. No interference can be observed if the separation between the final atomic wave packets exceeds their coherence length.

position of two wave packets that differ, not only by their internal state, but also by their momentum [Bordé *et al.* (1984)]. The two wave packets fly apart from each other with a relative velocity $\hbar k_L/M$. In this section, we suppose that the atom propagates freely when it is not in a laser beam, so that the various trajectories of the centers of the wave packets between two interaction zones are straight lines, along which the atom moves with a constant velocity.

18.3.2 *Spatial separation of the two final wave packets. Quenching of the interference*

When the atom enters the second laser beam (see Fig. 18.5(b)), these two wave packets are separated by a distance $d = (\hbar k_L/M)T$, where T is the transit time between the two lasers. The interaction at B with the second laser excites the atom from $|g, \vec{p}\rangle$ to $|e, \vec{p} + h\vec{k}_L\rangle$, and we get the path represented in blue in Fig. 18.5(b). Does it interfere with the other path, represented in red, where the interaction takes place at A with the first laser? The two final states correspond to the same internal state e, to the same mean momentum $|\vec{p} + h\vec{k}_L\rangle$, but to mean positions along the laser axis that differ by d. Their scalar product is nothing but the overlap integral of the incident wave packet by the same wave packet translated by an amount d, which is also the definition of the global spatial coherence at a distance d.[5] We conclude that the two final wave packets corresponding to the two paths become orthogonal when the transit time T between the two lasers is large enough that the distance d between the two wave packets becomes larger than the coherence length

[5]See Sec. 17.2.2 of Chap. 17.

ξ of the incident wave packet. The interference signal thus vanishes when:

$$d = \frac{\hbar k_L}{M} T \gg \xi. \tag{18.9}$$

This argument concerning the orthogonality of the final wave packets for large enough values of the transit time remains valid in the non-perturbative limit, for example, when the interaction with each laser corresponds to a $\pi/2$ pulse. The two wave packets emerging from each interaction zone have then equal weights.

Ramsey fringes for two-photon Doppler free transitions

It is clear that the quenching of the interference is due to the change of momentum of the atom when it absorbs a photon. What happens if, instead of considering a one-photon transition, we consider a two-photon transition, where the atom goes from g to e by absorbing two photons with the same energy, but propagating in opposite directions? The total momentum transferred to the atom is then equal to zero. We have seen in Chap. 6 that one can, in this way, suppress the velocity change of the atoms due to the two-photon absorption. In the problem considered here, the vanishing of the total momentum transferred to the atom prevents any spatial separation of the two final wave packets so that they do not become orthogonal and the interference is no longer quenched. Ramsey fringes in two-photon Doppler free transitions have indeed been observed by using two pulses of a laser standing wave separated by a time T [Salour and Cohen-Tannoudji (1977)].

18.3.3 *How to restore the interference signal?*

Coming back to one-photon transitions, we now show that a possible solution for restoring the interference signal is to use more than two laser beams in order to recombine the two final atomic wave packets.

(i) *Schemes using three travelling waves*

Suppose a third laser beam $\vec{k}_L, \omega_L$ is added at a distance L from the second in order to have three equidistant laser beams (see Fig. 18.6). The wave packet $|g, \vec{p}\rangle$ emerging at B' from the second interaction zone intersects the wave packet $|e, \vec{p} + \hbar\vec{k}_L\rangle$ emerging from B at C, and both wave packets are transformed into $|e, \vec{p}+\hbar\vec{k}_L\rangle$ after the third interaction zone. They then overlap both in momentum and position space and correspond to the same internal state, so that the interference signal is no longer quenched.

If the interaction with the second laser beam corresponds to a π-pulse, the state $|e, \vec{p} + \hbar\vec{k}_L\rangle$ is integrally transformed into $|g, \vec{p}\rangle$ at B', and vice versa at B. If in addition, the interactions at A and C correspond to $\pi/2$-pulses, the scheme of Fig. 18.6 is equivalent to a Mach-Zehnder interferometer with the first and third lasers playing the role of beam splitters, and the second laser playing the role of a mirror.

In the next section, we will show that for a free particle moving in an atomic interferometer, the phase difference between two paths with common starting and ending points is determined by the energy difference of the total system

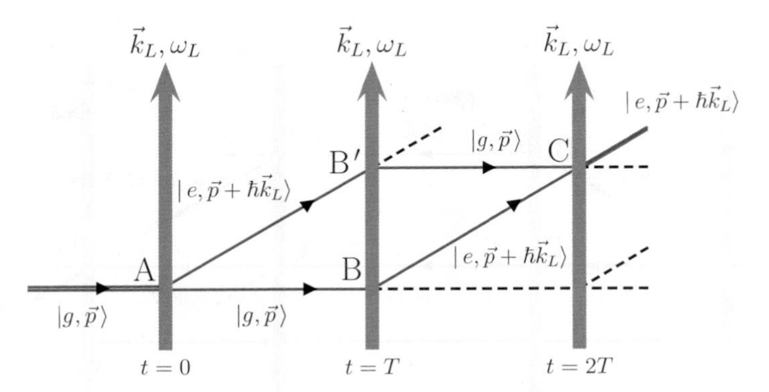

Fig. 18.6 Interferometer using three equidistant laser waves. The two paths represented in blue and red lead to two final wave packets that correspond to the same internal state and that overlap in both momentum and position space. They can therefore interfere.

{ particle+field } accumulated along these two paths. For example, the contribution $\varphi_{AB'} - \varphi_{AB}$ of the two segments AB' and AB of Fig. 18.6 to the total phase difference is equal to $-(E_{AB'} - E_{AB})T/\hbar$ where $E_{AB'}$ and E_{AB} are the energies of the total system in these two segments:

$$\begin{aligned}\varphi_{AB'} - \varphi_{AB} &= -\left[E_e - \hbar\omega_L + \frac{(\vec{p}+\hbar\vec{k}_L)^2}{2M} - E_g - \frac{\vec{p}^2}{2M}\right]\frac{T}{\hbar}, \\ &= -\left[\hbar\left(\omega_A - \omega_L\right) + \frac{\hbar^2 k_L^2}{2M} + \frac{\hbar\,\vec{p}.\vec{k}_L}{M}\right]\frac{T}{\hbar}. \end{aligned} \tag{18.10}$$

We have used the notation $\hbar\omega_A = E_e - E_g$ and subtracted $\hbar\omega_L$ from E_e because the number of photons is reduced by one unit along AB' after the absorption of a photon at A. It is clear from the symmetry of Fig. 18.6 that the two segments $B'C$ and BC give a contribution opposite to that of AB' and AB, so that the phase difference between the two blue and red interfering paths of Fig. 18.6 is equal to zero. Thus, no Ramsey fringes can appear when the laser frequency ω_L is scanned. However, such an interferometer can be sensitive to inertial fields. For example, if the paths $AB'C$ and ABC are located at different heights, the interference signal is sensitive to the acceleration g due to gravity. In the next section, we describe an experiment performed by Mark Kasevich and Steve Chu for measuring g using a three-pulse sequence $\pi/2 - \pi - \pi/2$ [Kasevich and Chu (1991, 1992b)].

(ii) *Schemes using four travelling waves. Ramsey-Bordé interferometers*

Another interesting scheme is provided by the so-called Ramsey-Bordé configuration shown in Fig. 18.7 that involves four interaction zones [Bordé *et al.* (1984)]. The sequence consists of one pair of waves propagating in the same direction followed by another pair of waves propagating in the opposite direction with the same spacing L between the two waves that make up each pair. Two paths, represented in blue and red, are shown in the figure. These paths start from the same initial state and

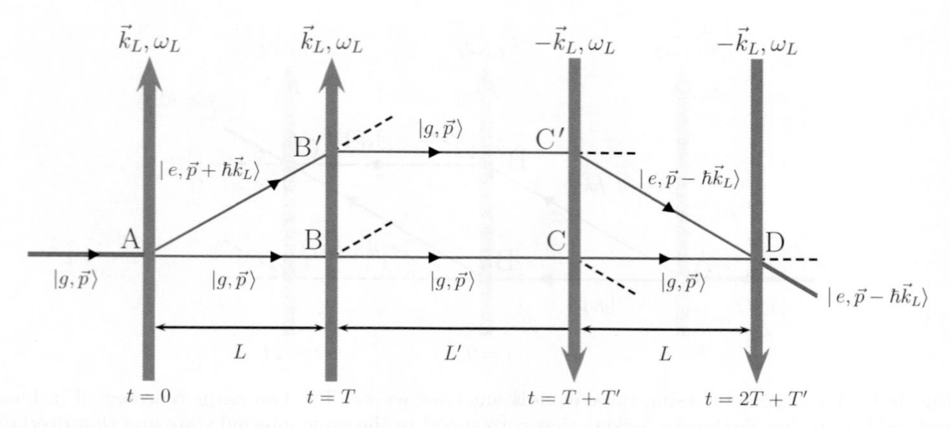

Fig. 18.7 Ramsey-Bordé interferometer using four travelling laser waves. The first pair of waves propagate along the same direction, the second pair along an opposite direction. The distance L between the two beams of each pair is the same for both pairs. The two paths represented in blue and red lead to two final wave packets corresponding to the same internal state and overlapping both in momentum and position spaces. They can thus interfere.

end in the same final state. Since the two final wave packets correspond to the same internal state and overlap both in momentum and position spaces they can interfere.

In contrast to the interferometer of Fig. 18.6, the phase difference $\Delta\phi$ between the blue and red paths is now no longer equal to zero. The contributions to $\Delta\phi$ of the segments AB' and AB is the same as that calculated for the segments AB' and AB of Fig. 18.6, and is given by Eq. (18.10). The contribution $\varphi_{C'D} - \varphi_{CD}$ of the segments $C'D$ and CD is the same, with $\vec{k}_L$ replaced by $-\vec{k}_L$, such that the terms linear in $\vec{p}$ cancel with those of Eq. (18.10). Finally, we get for $\Delta\phi$:

$$\Delta\phi = -2\left(\omega_A - \omega_L\right)T - 2\omega_R T, \tag{18.11}$$

where $\omega_R = \hbar k_L^2/2M$ is the recoil shift. When ω_L is scanned, Ramsey fringes $\cos[2(\omega_L - \omega_A - \omega_R)T]$, centered at $\omega_L = \omega_A + \omega_R$ with a fringe spacing π/T are obtained.

Another Ramsey-Bordé interferometer that leads to a final internal state e is represented in Fig. 18.8. The phase shift $\Delta\phi'$ between the blue and red paths is different from that corresponding to Fig. 18.7. The contributions of the segments AB' and AB is still given by (18.10), but the contribution of $C'D$ and CD is now given by:

$$\begin{aligned}
\varphi_{C'D} - \varphi_{CD} &= -\left[E_e - \hbar\omega_L + \frac{(\vec{p} + \hbar\vec{k}_L)^2}{2M} - E_g - \frac{(\vec{p} + 2\hbar\vec{k}_L)^2}{2M}\right]\frac{T}{\hbar}, \\
&= -\left[\hbar\left(\omega_A - \omega_L\right) - 3\frac{\hbar^2 k_L^2}{2M} - \frac{\hbar\,\vec{p}.\vec{k}_L}{M}\right]\frac{T}{\hbar}.
\end{aligned} \tag{18.12}$$

When Eqs. (18.10) and (18.12) are added, the terms linear in $\vec{k}_L$ cancel out to give

$$\Delta\phi' = -2\left(\omega_A - \omega_L\right)T + 2\omega_R T. \tag{18.13}$$

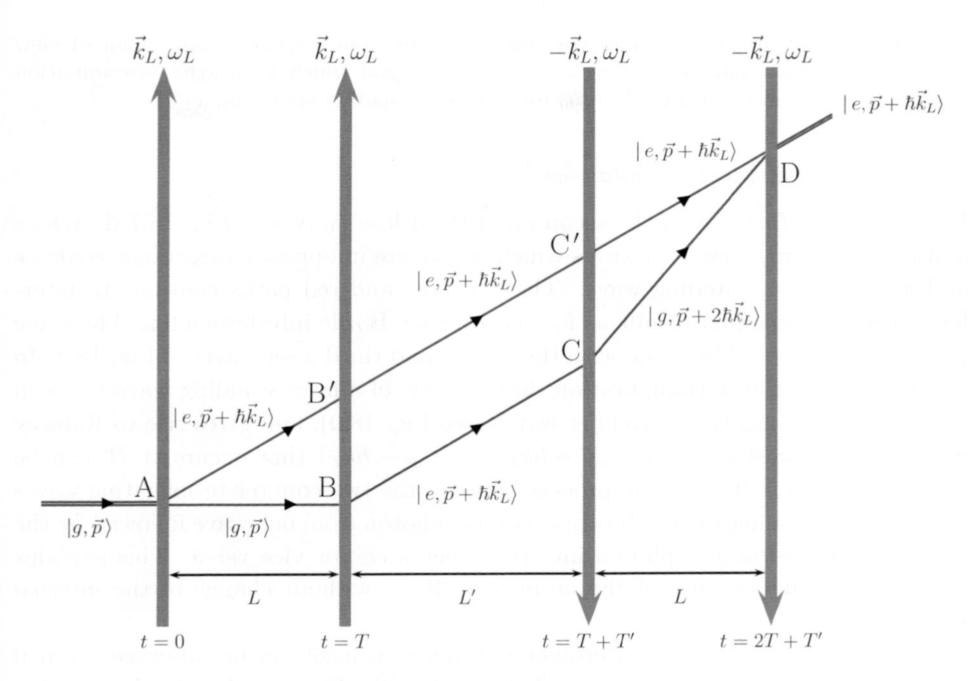

Fig. 18.8 Another Ramsey-Bordé interferometer leading to a final internal state e.

When ω_L is scanned, the Ramsey fringes $\cos[2(\omega_L - \omega_A + \omega_R)T]$ are now centered at $\omega_L = \omega_A - \omega_R$ with a fringe spacing π/T. One can easily check that there is no other Ramsey-Bordé interferometer that leads to a final internal state e.

The two final states of Figs. 18.7 and 18.8 correspond to the same internal state e, but to two different momentum states $\vec{p} - \hbar\vec{k}_L$ and $\vec{p} + \hbar\vec{k}_L$. If the detector is only sensitive to the internal state, the transition probabilities corresponding to these two final states, which are orthogonal because of the difference of momentum, must be added. The variations with ω_L of the transition probability $P(g \to e)$ is therefore the sum of two Ramsey patterns centered at $\omega_L = \omega_A \pm \omega_R$. From these patterns, a very precise measurement of the frequency ω_A of the atomic transition and the recoil shift ω_R can be deduced.

The Ramsey-Bordé interferometer was first realized with calcium atoms and was used as an inertial sensor [Riehle *et al.* (1991)]. It has also been employed with hydrogen, magnesium atoms along with I_2 molecules for high resolution spectroscopy and for fundamental studies.[6]

Symmetric versus asymmetric interferometric schemes

The scheme of Fig. 18.6 is symmetric in the sense that the atom spends the same amount of time in both e and g states along the two interfering paths. As a result, the global phase shift does not depend on $\omega_L - \omega_A$. In contrast, the Ramsey-Bordé interferometer is asymmetric and the global phase shift now explicitly depends on $\omega_L -$

[6]See [Cronin *et al.* (2009)] and references therein, and Sec. 18.5 of this chapter.

ω_A. The advantage of asymmetric configurations from a spectroscopic point of view lies in the fact that, by varying ω_L, one get a signal which allows the determination of $\omega_A \pm \omega_R$, and which could thus be used for designing an atomic clock.

(iii) *Schemes using laser standing waves*

If the distance L' between the second and third laser waves of Fig. 18.7 decreases and tends to zero, these two waves which propagate in opposite directions, coalesce and give rise to a standing wave. The two blue and red paths continue to interfere with the same phase shift as for the Ramsey-Bordé interferometer. The same transformation could be made with the second and third laser waves of Fig. 18.8. In this way we get a new configuration that consists of a laser standing wave between two counter-propagating travelling waves (see Fig. 18.9), and gives rise to Ramsey fringes. The transition from $|e, \vec{p} + \hbar\vec{k}_L\rangle$ to $|e, \vec{p} - \hbar\vec{k}_L\rangle$ that occurs at B' can be interpreted as a redistribution process between the two counter-propagating waves forming the standing wave: absorption of one photon from one wave followed by the stimulated emission of a photon into the other wave, or vice versa. This explains the change of momentum of the atom by $\pm 2\hbar\vec{k}_L$, without change of the internal state.

In the Ramsey-Bordé interferometer, Ramsey fringes can be observed, even if the interaction with each laser wave is treated to the lowest order. In the interferometer of Fig. 18.9, the interaction with the standing wave must be treated at least to second order. In this case, the observation of Ramsey fringes involves a nonlinear effect.[7] In fact, the first proposals for extending Ramsey fringes considered nonlinear effects that occur in standing waves [Baklanov *et al.* (1976)], and the first experimental observation of these fringes was obtained using a three laser configuration analogous to the one of Fig. 18.9 [Bergquist *et al.* (1977)]. We have chosen here not to follow the historical order but instead to first present the Ramsey-Bordé interferometer, both because it is simpler to explain, and because it allows the understanding, with a very simple transformation, of why Ramsey fringes can also be observed with standing waves.

18.3.4 *Other possible schemes*

In experiments that use the interferometers described in this section, it is crucial to avoid any spontaneous emission processes, which would destroy the spatial coherence. The lifetime of the excited state e must therefore be longer than the experiment's duration. This is why the first experiments using this technique were restricted to long-lived, metastable excited states. On the other hand, it would be advantageous to use optical transitions that give rise to large momentum transfers in absorption and emission processes. In this case, the large area between the two

[7] This is also the case for the optical Ramsey fringes observed in two-photon Doppler free transitions mentioned at the end of Sec. 18.3.2.

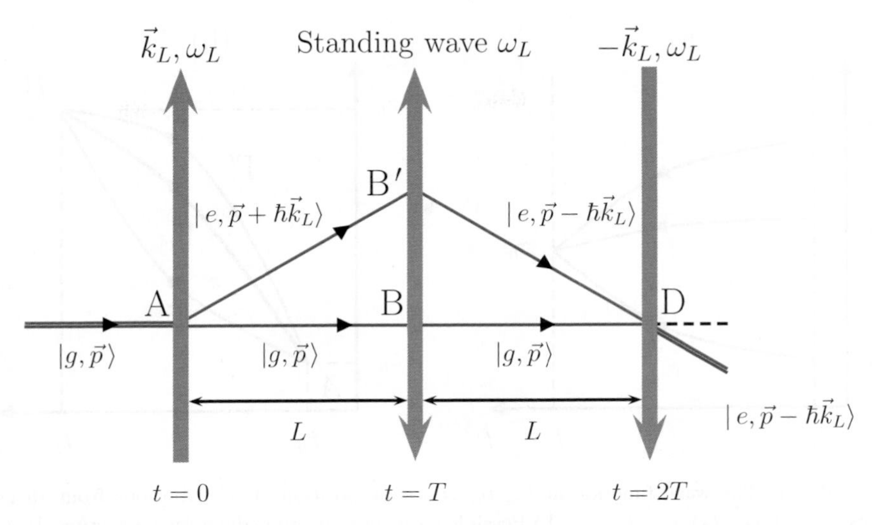

Fig. 18.9 If the distance L' between the second and third laser beams of Fig. 18.7 is decreased to zero, they combine to give a laser standing wave. The two paths represented in blue and red lead to two identical final wave packets that can interfere.

arms of the interferometer would allow a precise measurement of inertial fields.

An elegant way to circumvent the decoherence due to spontaneous emission, while at the same time achieving large momentum transfers, consists of using stimulated Raman transitions between two ground state sublevels induced by a pair of counter-propagating lasers[8] [Kasevich and Chu (1992b)]. In this manner, the two ground state levels $|g_1\rangle$ and $|g_2\rangle$ are coupled with a transfer of momentum equal to the sum of the momenta of two optical photons without populating any excited state, as long as the detuning of the lasers from optical resonance is large enough. After the crossing of the interaction zone with the two Raman lasers, an atom initially in the state $|g_1, \vec{p}\rangle$ is put in a linear combination of $|g_1, \vec{p}\rangle$ and $|g_2, \vec{p} + \hbar(\vec{k}_1 - \vec{k}_2)\rangle \simeq |g_2, \vec{p} + 2\hbar\vec{k}_1\rangle$ since $\vec{k}_1 \simeq -\vec{k}_2$.

Let us finally mention that other schemes for atom interferometry that entangle the external and internal degrees of freedom have been developed. These include the longitudinal Stern-Gerlach effect [Miniatura *et al.* (1991)], and longitudinal radio-frequency spectroscopy [Gupta *et al.* (2001)].

18.4 Calculation of the phase difference between the two arms of an atomic interferometer

At the output of an atomic interferometer, the wave function of the outgoing beam is a linear superposition of two wave functions that correspond to two possible paths that can be followed by the atoms. How can we calculate the phase difference

[8]See Sec. 9.4.3 of Chap. 9.

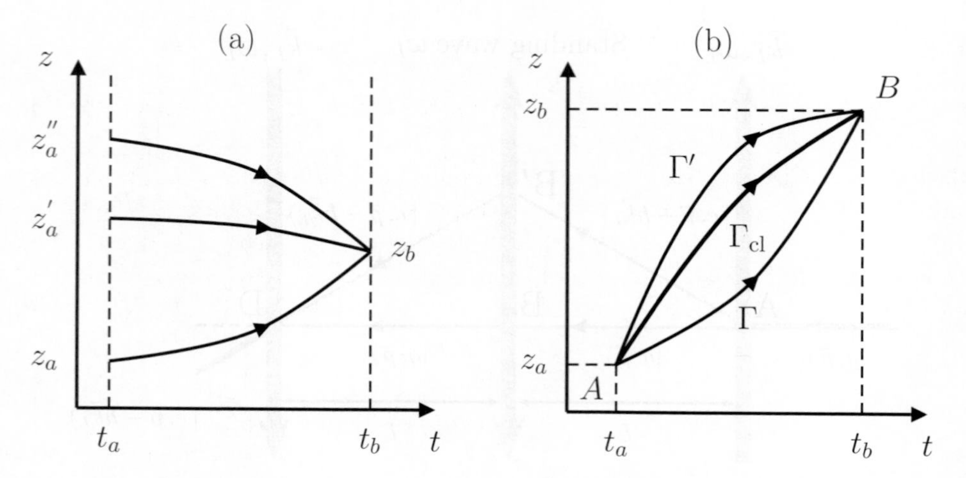

Fig. 18.10 (a) The wave function at (z_b, t_b) is a superposition of contributions from all point sources (z_a, t_a), (z_a', t_a), (z_a'', t_a), ... (b) Possible paths in space time diagram connecting $A(z_a, t_a)$ to $B(z_b, t_b)$.

between these two wave functions due to various causes (free propagation, external or inertial fields)? In this section, we present a simple way of calculating this phase difference using the Feynman path integral approach [Storey and Cohen-Tannoudji (1994)]. Here we just give the principle of these calculations. The main result is that, in a certain number of situations that are realized for most atomic interferometry experiments, it is possible to express the phase shifts of the wave functions as integrals over classical trajectories.

18.4.1 *Quantum propagator and Feynman path integral*

Quantum mechanically, the state of a system at time t_b is inferred from the state at an earlier time t_a through the evolution operator $|\psi(t_b)\rangle = U(t_b, t_a)|\psi(t_a)\rangle$. The final state projected on the position basis[9] therefore reads

$$\psi(z_b, t_b) = \langle z_b| U(t_b, t_a)|\psi(t_a)\rangle = \int_{-\infty}^{+\infty} \mathrm{d}\, z_a\, K(z_b, t_b; z_a, t_a)\, \psi(z_a, t_a)\,, \quad (18.14)$$

where we have introduced the quantum propagator $K(z_b, t_b; , z_a, t_a) \equiv \langle z_b| U(t_b, t_a)|z_a\rangle$. This quantity gives the amplitude for the particle to arrive at z_b at time t_b if it starts from z_a at time t_a. Equation (18.14) is reminiscent of the Fresnel-Huygens principle: the value of the wave function at (z_b, t_b) is the superposition of all wavelets originating from the radiating point sources (z_a, t_a), (z_a', t_a), (z_a'', t_a), ... (see Fig. 18.10(a)).

Richard Feynman gave an equivalent expression for K that can be written in

[9]To keep the notation simple, we consider here a one-dimensional problem. The extension to higher dimensions is straightforward.

the form [Feynman (1948); Feynman and Hibbs (1965)]

$$K\left(z_b,t_b;z_a,t_a\right)=K(b,a)=\mathcal{N}\sum_{\Gamma}\exp\left(i\,S_\Gamma/\hbar\right)=\int_a^b\exp(i\,S_\Gamma/\hbar)\,\mathcal{D}z(t),\quad(18.15)$$

where $\mathcal{N}$ is a normalization coefficient, and the sum is carried out over all possible paths Γ connecting (z_a,t_a) to (z_b,t_b) (see Fig. 18.10(b)) and defined by a function $z(t)$ such that $z(t_a)=z_a$ and $z(t_b)=z_b$. The sum over Γ is a functional integral in the space of functions $z(t)$. Each path contributes through a trajectory dependent phase factor $\exp(i\,S_\Gamma/\hbar)$, where S_Γ is the *action* along the path Γ, defined by:

$$S_\Gamma=\int_{t_a}^{t_b}\mathrm{d}t\,\mathcal{L}\left[z(t),\dot{z}(t)\right]\quad(18.16)$$

where

$$\mathcal{L}[z,\dot{z}]=\frac{1}{2}M\dot{z}^2-V(z)\quad(18.17)$$

is the classical *Lagrangian* of the system. In Fig. 18.10(b), Γ_{cl} is the classical path for which the action is extremal, according to the principle of least action.[10]

18.4.2 *Simple case of quadratic Lagrangians*

For a quadratic Lagrangian in the z and $\dot{z}$ variables,

$$\mathcal{L}(z,\dot{z})=a(t)^2\dot{z}^2+b(t)z\,\dot{z}+c(t)z^2+d(t)\dot{z}+e(t)z+f(t),\quad(18.18)$$

one can show that the sum over all possible paths in the expression of the quantum propagator simply reduces to the classical path

$$K\left(z_b,t_b;z_a,t_a\right)=F(t_b,t_a)\,\exp\{i\,S_{\mathrm{cl}}\left(z_b,t_b;z_a,t_a\right)/\hbar\},\quad(18.19)$$

where $S_{\mathrm{cl}}(z_b,t_b;z_a,t_a)=S\left[z_{\mathrm{cl}}(t)\right]$ and $z_{\mathrm{cl}}(t)$ is the classical path with the boundary conditions $z_{\mathrm{cl}}(t_a)=z_a$ and $z_{\mathrm{cl}}(t_b)=z_b$. The function $F(t_b,t_a)$ that appears in Eq. (18.19) does not depend on z_b and z_a.

If, in addition, the initial state is a plane wave $\psi(z_a,t_a)=\exp(i\,p_0z_a/\hbar)$, one can show that the integral in Eq. (18.14) reduces to a single contribution [Storey and Cohen-Tannoudji (1994)]: the final wave function at z_b at time t_b is obtained by considering only the *single classical trajectory* of a particle that starts at time t_a from the point z_a with momentum p_0 and passes z_b at time t_b.

$$\psi\left(z_b,t_b\right)\propto F(t_b,t_a)\psi(z_a,t_a)\exp\left\{iS_{\mathrm{cl}}\left(z_b,t_b;z_a,t_a\right)/\hbar\right\}.\quad(18.20)$$

The phase of the final wave function is therefore determined by the phase of the wave function at the trajectory's initial point and the action along the classical path. The amplitude of the final wave function is independent of position and depends only on the initial and final times.

[10] The classical limit is recovered when $S_\Gamma\gg\hbar$, a condition for which the phase factors $\exp\left(i\,S_\Gamma/\hbar\right)$ vary rapidly with Γ and interfere destructively except near the classical path Γ_{cl}, where S_Γ does not vary to first order in $(\Gamma-\Gamma_{\mathrm{cl}})$ when Γ varies about Γ_{cl}.

A useful expression of $S_{\rm cl}(z_b, t_b; z_a, t_a)$ is also given by [Landau and Lifshitz (1982)]:

$$S_{\rm cl} = \int_{\Gamma_{\rm cl}} (p{\rm d}z - H{\rm d}t) \tag{18.21}$$

where p and H are the canonical momentum and the Hamiltonian of the particle.

18.4.3 *Phase shift in the absence of external potentials and inertial fields*

In this case, the classical trajectories between two crossings with laser beams are straight lines along which the momentum p and the Hamiltonian H are constant. Figures 18.6 and 18.7 give examples of such situations. The changes of slopes of these straight lines when the atom crosses a laser beam result from the exchange of momentum between the atom and the light.

The phase shift along each path of the interferometer, for example the path $AB'C'D$ of Fig. 18.7, is obtained by adding:

- the phase shifts $\int(p{\rm d}z - H{\rm d}t)$ due to the free propagation between two crossings, where p and H are the momentum and the total (kinetic plus internal) energy of the atom, respectively.
- the phase shifts that come from the emission and absorption amplitudes of a laser photon of the atom when it crosses a laser beam. If the two-level atom $\{e, g\}$ crosses in z_1 at time t_1 a travelling wave of wave vector $\vec{k}_L$, frequency ω_L and absolute phase φ, one can show that the atomic wave function is multiplied by an amplitude [Bordé (1989)]
  - $U_{eg}\exp\{i(k_L z_1 - \omega_L t_1 - \varphi)\}$, if a photon absorption occurs,
  - $U_{ge}\exp\{-i(k_L z_1 - \omega_L t_1 - \varphi)\}$, if a photon emission occurs,
  - U_{gg} (resp. U_{ee}) in the absence of absorption (resp. emission),

  where U_{ij} with $i, j = 1$ or 2 are the transition amplitudes from j to i due to the laser excitation.

If the various phase shifts are combined, new expressions are obtained where the momentum and the energy of the total system { atom+field } explicitly appear.

Consider first the contribution of the term $\int p{\rm d}z$ of Eq. (18.21) along the segment AB' of Fig. 18.6. The momentum of the atom along this segment is $p + \hbar k_L$ and its integral (divided by $\hbar$) along AB' is equal to $(p + \hbar k_L)(z_{B'} - z_A)/\hbar$. The product of the two exponentials $\exp(ik_L z_A)$ and $\exp(-ik_L z_{B'})$ associated with the absorption of a photon at z_A and the emission of a photon at $z_{B'}$ gives rise to a phase shift equal to $-k_L(z_{B'} - z_A)$. Adding this phase shift to the one due to the free propagation, gives a total contribution of the segment AB' of $p(z_{B'} - z_A)/\hbar$, equivalent to a decrease in the atomic momentum by an amount $\hbar k_L$, which is the momentum of the absorbed photon. We finally get the integral along AB' of the initial momentum p of the total system { atom+field }.

A similar argument holds for contribution of the term $-\int H\mathrm{d}t$ of Eq. (18.21). The two exponentials $\exp(-i\omega_L t_A)$ and $\exp(+i\omega_L t_{B'})$ give a contribution to phase shift equal to $\exp\left[i\omega_L(t_{B'} - t_A)\right]$, which amounts to subtract the energy $\hbar\omega_L$ of the absorbed photon along the segment AB'

Finally, when the phase shifts due to absorption and emission of photons are taken into account, the phase difference between the two paths can be written as

$$\varphi_{AB'C'D} - \varphi_{ABCD} = \int_{AB'C'D} (p_{\mathrm{tot}}\mathrm{d}z - H_{\mathrm{tot}}\mathrm{d}t) - \int_{ABCD} (p_{\mathrm{tot}}\mathrm{d}z - H_{\mathrm{tot}}\mathrm{d}t), \quad (18.22)$$

where p_{tot} and H_{tot} are the momentum and energy of the {atom+field} system. Since as a consequence of momentum conservation the total momentum p_{tot} is the same for all segments of the two paths, only H_{tot} contributes to the phase difference.[11] This justifies the calculation presented in Sec. 18.3.

Remark

One may wonder why the argument using momentum conservation for showing that $\oint p_{\mathrm{tot}}\mathrm{d}z = 0$, cannot be transposed to the total energy. Actually, the atom field interaction can connect states with different total energies. This is due to the fact that the interaction time τ, given by the transient time of atoms through the laser beams, is finite, so that the energy is conserved within $\hbar/\tau$. We assume here that $\hbar/\tau$ is larger than all differences of energy of the states connected by the atom-field interaction. A similar approximation has already been discussed in Chap. 2 in the context of magnetic resonance pulses, where the RF field is detuned from resonance (see Sec 2.4.4).

18.4.4 *Phase shift due to external potentials and inertial fields in the perturbative limit*

When external potentials and inertial fields are present, we must add to the Lagrangian a new term $\varepsilon\mathcal{L}_1$ that describes their effect, where ε is a dimensionless parameter that accounts for their strength. We suppose that $\mathcal{L}_1$ is a quadratic function of positions and velocities. We can thus use the previous results concerning the quantum propagator.

A general and exact method for calculating the new phase shift is to determine the new classical trajectories associated with the two paths of the interferometer, which are no longer straight lines, calculate the integrals $\int(p\mathrm{d}z - H\mathrm{d}t)$ along the various paths, noting that p is no longer a constant of the motion,[12] and include the contributions of the emission and absorption amplitudes.

A simplification occurs if ε is small enough to allow an approximate perturbative calculation of the new phase shift. [Storey and Cohen-Tannoudji (1994)] showed that it is then possible to keep the unperturbed trajectories (without external potentials and inertial fields) and to calculate the integral $\varepsilon\int\mathcal{L}_1\mathrm{d}t$ along each path I

[11] We must also include the absolute phases φ of the laser beams.

[12] We suppose that $\mathcal{L}_1$ is time independent, so that H remains a constant of the motion between two interactions with the lasers.

and II of the interferometer. The phase shift between the two paths is thus given by the integral of $\varepsilon\mathcal{L}_1$ along the unperturbed closed contour $I + II$:

$$\delta\phi = \frac{\varepsilon}{\hbar}\oint_{I+II} \mathcal{L}_1 \mathrm{d}t. \tag{18.23}$$

18.5 Applications of atomic interferometry

In this section, we apply the previous formalism to explain how atom interferometers can be used to measure inertial fields and fundamental constants.

18.5.1 *Measurement of gravitational fields. Gravimeters*

We describe the first atomic gravimeter configuration,[13] which realizes the atomic analog of an optical Mach-Zehnder interferometer [Kasevich and Chu (1991)]. A cloud of cold sodium atoms, in one of the hyperfine states, g_1, of the ground state, is launched vertically using the moving molasses technique. A $\pi/2-\pi-\pi/2$ sequence of stimulated Raman pulses (see Sec. 18.3.4) between g_1 and the other hyperfine state g_2 is provided by two-counterpropagating and vertical beams, whose wavevectors and frequencies are denoted by $(\vec{k}_1, \omega_1)$ and $(\vec{k}_2, \omega_2)$ (see the inset of Fig. 18.11). The first $\pi/2$ pulse puts the atoms, originally in the state $|g_1, \vec{p}\rangle$ where $\vec{p}$ is the momentum along the vertical axis, in a superposition of $|g_1, \vec{p}\rangle$ and $|g_2, \vec{p} + \hbar(\vec{k}_1 - \vec{k}_2)\rangle \simeq |g_2, \vec{p} - 2\hbar\vec{k}_L\rangle$, since $|\vec{k}_1| \simeq |\vec{k}_2| \equiv k_L$. At time T, the applied π-pulse exchanges the momenta and internal states of the two components. The last $\pi/2$ pulse is applied when the two components spatially overlap at time $2T$ in order to recombine them coherently in the state $|g_2, \vec{p} - 2\hbar\vec{k}_L\rangle$.

Figure 18.11 shows the space-time classical paths followed by the atoms, which are straight lines in the absence of gravitational field (dotted lines) and parabolas in the presence of the gravitational field (full lines). The two paths of the interferometer recombine at B_0 in the first case, at B in the second one. Because of the free fall of the atoms in the gravitational field, C, D and B are below C_0, D_0 and B_0 and we have:

$$\overline{C_0C} = \overline{D_0D} = -gT^2/2, \qquad \overline{B_0B} = -2gT^2. \tag{18.24}$$

We begin by an exact non-perturbative calculation of the phase shift due to the gravitational field between the two waves recombined at B.

Consider first the phase shift due to the propagation along ACB for the first path, along ADB for the second. It is given by the expression:

$$\delta\phi_{\text{prop}} = \frac{1}{\hbar}\left[S_{\text{cl}}(AC) + S_{\text{cl}}(CB) - S_{\text{cl}}(AD) - S_{\text{cl}}(DB)\right] \tag{18.25}$$

[13] Recently, another gravimeter scheme using Bloch oscillations has been realized [Poli *et al.* (2011)].

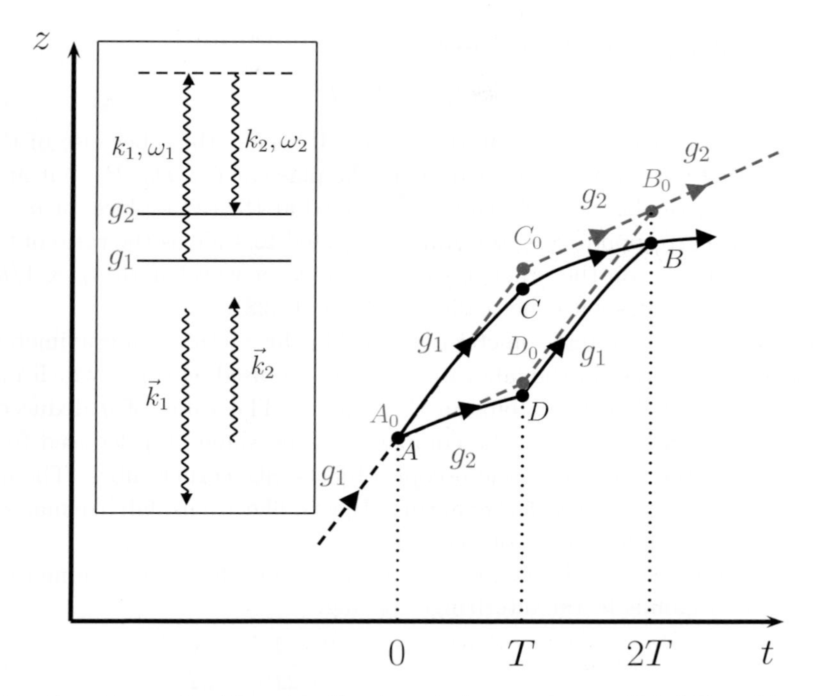

Fig. 18.11 Spacetime paths followed by the atoms in the experiment of Mark Kasevich and Steve Chu [Kasevich and Chu (1991)] in the absence of gravitational field (dashed lines) and in the presence of the gravitational field (full lines). Raman pulses occur at times 0, T and $2T$.
Inset: Stimulated Raman transitions between the two hyperfine levels g_1 and g_2 of sodium atoms are used for the coherent manipulation of the atomic wavepackets. The directions of the two counter-propagating beams are represented in the lower part of the inset.

where the classical actions along each segment of the two paths appear. One can calculate exactly the classical action corresponding to the Lagrangian $\mathcal{L}(z, \dot{z}) = M\dot{z}^2/2 - Mgz$ governing the motion of the atom in the gravitational field along the classical path joining the two points (z_a, t_a) and (z_b, t_b). One finds [Storey and Cohen-Tannoudji (1994)]:

$$S_{\text{cl}}(z_a t_a, z_b t_b) = \frac{M}{2} \frac{(z_b - z_a)^2}{t_b - t_a} - \frac{Mg}{2} (z_b + z_a)(t_b - t_a) - \frac{Mg^2}{24} (t_b - t_a)^3 . \quad (18.26)$$

Using this equation and (18.24) for calculating $\delta\phi_{\text{prop}}$ (18.25), one finds after a straightforward calculation given in [Storey and Cohen-Tannoudji (1994)]:

$$\delta\phi_{\text{prop}} = 0. \quad (18.27)$$

The previous exact result means that the free propagation of the atoms along the two arms of the interferometer does not introduce any phase shift between the two waves which are recombined at B. The phase shift can come only from the interactions with the lasers. With the rules given in Sec. 18.4.3, one can calculate this phase shift $\delta\phi_{\text{prop}}$ due to the lasers:[14]

$$\delta\phi_{\text{laser}} = 2k_L g T^2 \tag{18.28}$$

This result has a clear physical interpretation. It means that, because of the free fall, the interactions with the lasers do not take place at C_0, D_0, B_0, but at C, D, B. The phase of the lasers is therefore not probed at the same place for $g = 0$ and for $g \neq 0$. The corresponding phase shift is expected to scale as the ratio of the free fall during a time T, on the order of gT^2, by the laser wavelength $\lambda_L \propto 1/k_L$, i.e. as $k_L g T^2$, which agrees with the scaling given in (18.28).

In other words, the interferometer described in this section is a gravimeter. The lasers, which are at rest in the laboratory frame, are used as rulers which measure precisely the free fall of the atoms in this frame. The value of g deduced from the phase shift (18.28) can then be compared to the value of g deduced from the measurement of the free fall of macroscopic objects, like corner cubes. The interest of this comparison is to check that quantum objects, like atoms, fall like macroscopic objects (test of the universality of free fall).

To evaluate the sensitivity of such an interferometer, let us determine the value of δg that corresponds to the interfringe spacing:

$$2k_L \delta g T^2 = 2\pi \implies \delta g = \frac{2\pi}{k_L}\frac{1}{2T^2} = \frac{\lambda_L}{2T^2}. \tag{18.29}$$

For sodium atoms, $\lambda_L \sim 0.6\ \mu\text{m} = 6 \times 10^{-7}$ m. With a typical time between pulses of $T = 5 \times 10^{-2}$ s, $\delta g/g \simeq 10^{-5}$ for one interfringe. If one thousandth of an interfringe can be measured, a relative variation of the gravitational field of $\delta g/g \simeq 10^{-8}$ can be detected. With such a precision, the modulation of local g due to tidal forces has been observed [Peters *et al.* (2001); Mehlstäubler *et al.* (2007)].

Measurement of the gravitational constant G have also been performed using a gravity gradiometer based on atom interferometry. The change in gravitational field along one dimension is measured for different positions of a well characterized control mass [Fixler *et al.* (2007); Lamporesi *et al.* (2008)].

Remark

One can easily check that the perturbative calculation described in Sec. 18.4.4 gives results in agreement with (18.28). The phase shift can be evaluated by integrating the term $\epsilon\mathcal{L}_1 = -Mgz$ of the Lagrangian, when it can be considered as a weak perturbation, along the contour $A_0C_0B_0D_0A_0$ formed by the unperturbed trajectories (calculated with $g = 0$). According to Eq. (18.23), the phase difference between the two arms of this interferometer is:

$$\delta\phi_{\text{approx.}} = -\frac{Mg}{\hbar}\oint_{A_0C_0B_0D_0A_0} z(t)\mathrm{d}t = -\frac{Mg}{\hbar}\Sigma, \tag{18.30}$$

where Σ denotes the area of $A_0C_0B_0D_0A_0$. At time T, the two paths are separated by the algebraic distance $D_0C_0 = -2\hbar k_L T/M$. The area of the interferometer is

[14]There is also a term depending on the absolute phases of the lasers. We do not give it here because it does not depend on g and does not contribute to the phase shift due to gravity.

therefore given by $\Sigma = (2T)D_0C_0/2 = -2\hbar k_L T^2/M$, and the corresponding phase shift is $\delta\phi_{\text{grav}} = 2k_L g T^2$. In this approximation, the phase shift induced by the laser is not modified by g since the phase shift involves the trajectories for $g = 0$.

The equivalence between the two results is purely mathematical and redistributes the different contributions to the phase shifts. It should not be used to deduce physical conclusions. The use of the parallelogram $A_0C_0B_0D_0A_0$ in the perturbative calculation could suggest that the lasers do not introduce any phase shift since they appear, in the perturbative approach, to be probed at the same place for any value of g. This is obviously not the case because atoms are falling. Also, one could think that the phase shift due to the free propagation is not equal to zero and entirely due to the gravitational potential whose contribution along the two paths would be different because these paths have different altitudes. The integral of the Lagrangian $-Mgz$ along unperturbed paths calculated with $g = 0$, cannot give the correct phase shift due to propagation. According to the Feynman formulation of quantum mechanics, the correct phase shift, for a quadratic Lagrangian, uses the classical action evaluated by integrating the Lagrangian along the classical paths calculated with the *same* Lagrangian. This is what has been done in the exact calculation given above and which shows that $\delta\phi_{\text{prop}} = 0$ for any value of g.

18.5.2 *Measurement of rotational inertial fields*

In this subsection, we apply the perturbative path treatment of Sec. 18.4.4 to a free particle in a rotating frame, and determine the sensitivity of matter-wave interferometers to rotation. As shown in Fig. 18.12(a), we consider a Galilean reference frame $\mathcal{R}'$, and a frame $\mathcal{R}$ rotating at an angular velocity $\vec{\Omega} = \Omega\vec{e}_z$, where the z-axis coincides for both frames. The relation between the velocities $\vec{v}\,'$ and $\vec{v}$ in the Galilean and rotating frames respectively is given by $\vec{v}\,' = \vec{v} + \vec{\Omega} \times \vec{r}$. The Lagrangian rewritten in terms of the coordinates of the rotating frame turns out to be quadratic:

$$\mathcal{L}(\vec{r}, \vec{v}\,) = \mathcal{L}'(\vec{r}\,', \vec{v}\,') = \frac{M}{2}\vec{v}\,'^2 = \frac{M}{2}\vec{v}\,^2 + M\vec{\Omega} \cdot (\vec{r} \times \vec{v}\,) + \frac{M}{2}(\vec{\Omega} \times \vec{r}\,)^2. \qquad (18.31)$$

For a sufficiently small angular frequency Ω, the quadratic term in Ω in Eq. (18.31) can be neglected. The phase shift due to the slow rotation is calculated along the unperturbed path which is a straight line segment in real space, as illustrated in Fig. 18.12(b):

$$\delta\phi = \frac{M\vec{\Omega}}{\hbar} \cdot \int_{t_a}^{t_b} \mathrm{d}t[\vec{r}(t) \times \vec{v}(t)] = \frac{M\vec{\Omega}}{\hbar} . \int_{t_a}^{t_b} \vec{r}(t) \times \mathrm{d}\vec{r}(t) = 2\frac{M\vec{\Omega}}{\hbar}\Sigma_0, \qquad (18.32)$$

where Σ_0 is the area OAB subtended by the path $\vec{r}(t)$ with respect to the origin. We obtain from the phase shift due to the rotation

$$\delta\phi = 2\frac{M\Omega}{\hbar}\Sigma, \qquad (18.33)$$

where Σ is the area $ACBDA$ (see Fig. 18.13(a)).

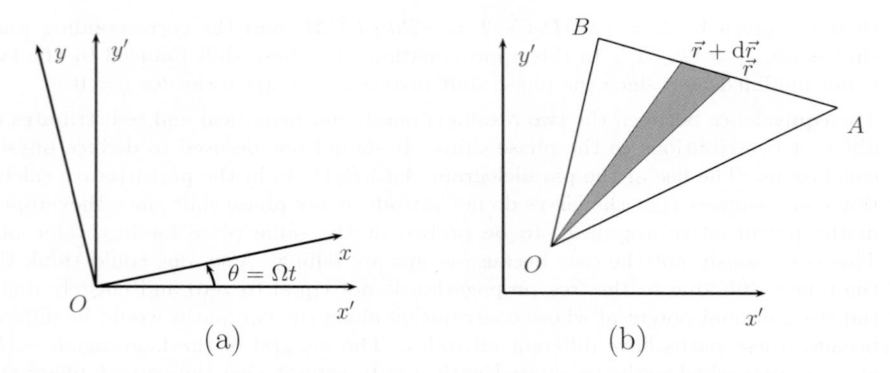

Fig. 18.12 (a) Galilean reference frame $\mathcal{R}'$ described by the coordinates x', y', z' and rotating reference frame $\mathcal{R}$ described by the coordinates x, y, z. The z-axis coincides for both frames. (b) Unperturbed path AB in the plane perpendicular to the rotation axis $\vec{\Omega} = \Omega\vec{e}_z$.

Sagnac effect for light waves and matter waves

Let us consider a circular interferometer $\mathcal{I}$ with radius ρ rotating at the angular frequency Ω with respect to a Galilean frame as depicted in Fig. 18.13(b). First, we consider an optical interferometer for light waves. Two rays emitted from the point A at $t = 0$ circulate in opposite direction around the interferometer and interfere at the beam splitter location, which is assumed to be at point B at $t = 0$. For an observer in the Galilean frame, the light ray evolving in the same sense as the interferometer $\mathcal{I}$ takes a longer time $t_+ = (\pi\rho + \rho\Omega t_+)/c$ to reach the beam splitter than the light ray rotating in the opposite sense, which needs a time $t_- = (\pi\rho - \rho\Omega t_-)/c$. The time difference $\Delta = t_+ - t_-$ is

$$\Delta t = \frac{\pi\rho}{c - \rho\Omega} - \frac{\pi\rho}{c + \rho\Omega} \simeq \frac{\pi\rho}{c}\frac{2\rho\Omega}{c} = \frac{2\Omega\Sigma}{c^2}, \tag{18.34}$$

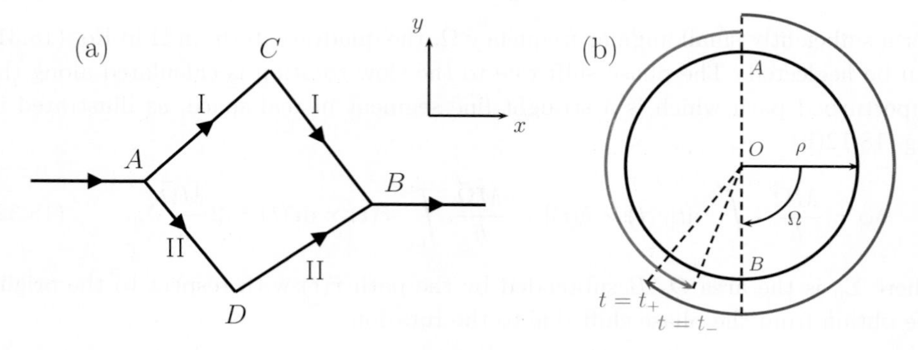

Fig. 18.13 (a) Unperturbed paths in real space describing the two arms of an atom interferometer sensitive to rotation. (b) Rotating circular optical interferometer. The beam splitter starts from the point B at time $t = 0$, and rotates at angular frequency Ω. The two rays leaving the point A at $t = 0$ circulate in opposite senses and reach the beam splitter at different times t_+ and t_-.

where $\Sigma = \pi\rho^2$ is the area of the interferometer. The so-called Sagnac phase shift is therefore [Sagnac (1913)]

$$\delta\phi_{\text{light}} = \omega_L \Delta t = \frac{2\Omega\Sigma\omega_L}{c^2}, \tag{18.35}$$

where ω_L is the light angular frequency.

If we compare this result for the Sagnac effect with light waves to the result (18.33) derived above for matter waves, we get:

$$\frac{\delta\phi_{\text{atom}}}{\delta\phi_{\text{light}}} = \frac{2M\Omega\Sigma/\hbar}{2\Sigma\Omega\omega_L/c^2} = \frac{Mc^2}{\hbar\omega_L} \gg 1. \tag{18.36}$$

From this result, we expect to increase the sensitivity by a factor of as much as 10^{11} by using matter waves instead of light. However, atomic gyrometers, using for instance Ramsey-Bordé type interferometers, have a very small surface, on the order of few mm^2, because of the small angular separation of the two paths due to photon recoil. This is to be contrasted with optical interferometers using optical fibers, which can make several turns, and thus enclose a much larger surface. In addition, a better signal-to-noise ratio can also be achieved in optical interferometers because of the higher available flux.

The first experimental observation of the Sagnac effect for atomic de Broglie waves was made in Germany at the Physikalisch-Technische Bundesanstalt (PTB) [Riehle *et al.* (1991)]. More precise atomic gyrometers reaching a sensitivity of 6×10^{-10} rad·s^{-1} for an integration time of one second have been realized by Mark Kasevich's group [Gustavon *et al.* (1997, 2000); Durfee *et al.* (2006)]. This can be compared, for instance, to the rotation speed of the earth, which is 7.29×10^{-6} rad·s^{-1}. High precision gyrometers could find practical applications in navigation (for this purpose long term stability is crucial), in geophysical studies and tests of predictions of general relativity such as the Lense-Thirring effect, which is due to the rotation of a massive body.

18.5.3 *Measurement of h/M and α*

We saw in Sec. 18.3.3 that the Ramsey-Bordé interferometer gives two sets of Ramsey fringes centered at $\omega_A \pm \omega_R$, from which one can deduce precise measurements of ω_A and ω_R. The measurement of the recoil shift $\omega_R = \hbar k_L^2/2M$ of an atom that absorbs a photon enables the determination of the ratio h/M, where M is the atom mass, since the light wave length $\lambda_L = 2\pi/k_L$ is known with accuracy.

The motivation for an accurate measurement of the recoil shift lies in the precise determination of the fine structure constant α. This constant characterizes the strength of electromagnetic interactions and plays a key role in quantum electrodynamics (QED). Its definition $\alpha = e^2/(4\pi\varepsilon_0\hbar c)$ contains several constants: the electron charge e, the speed of light in vacuum c, and the dielectric constant ε_0. Its name originates from the relativistic corrections to the hydrogen spectrum that scale as $\alpha^4 m_e c^2$, where m_e is the electron mass.

So far, the most precise measurement of α is obtained from the measurement of the electron anomalous magnetic moment $a_e = (g-2)/2$ using a trapped electron [Van Dyck *et al.* (1986); Gabrielse *et al.* (2006); Hanneke *et al.* (2008)]. Assuming the validity of QED, and equating the QED calculation for a_e to the experimental value, gives a determination of α:

$$\alpha^{-1} = 137.035999084(51)[0.37\,\mathrm{ppb}]. \tag{18.37}$$

In this approach, the obtained value of α relies on very elaborated QED calculations (involving the calculation of several hundreds of Feynman diagrams!). Other independent determinations of α that do not require QED calculations are therefore desirable, like for instance, those relying on the muonium hyperfine structure [Liu *et al.* (1999)], the quantum Hall effect [Jeffery *et al.* (1998)] ...

Atom interferometry offers the possibility of accurately determining the ratio h/M. The connection between this ratio and the fine structure constant is obtained from the following identity:

$$\alpha^2 = \frac{2R_y}{c}\frac{h}{M}\frac{M}{m_\mathrm{e}}. \tag{18.38}$$

The limiting factor is the ratio h/M since the uncertainty of the Rydberg constant R_y is 7×10^{-12} [Udem *et al.* (1997); Schwob *et al.* (1999)] and that of the mass ratio M/m_e 4.8×10^{-10} [Bradley *et al.* (1999); Mohr *et al.* (2008)]. The sensitivity of atomic interferometers can be increased via the relative momentum splitting between the two arms. Different methods have been investigated for providing large momentum transfer beam splitter using momentum transfer by extra stimulated Raman light pulses[15] [McGuirk *et al.* (2000)], multiphoton Bragg diffraction by an optical lattice as a beam splitter [Giltner and McGowan (1995); Müller *et al.* (2008)], and Bloch oscillations [Cadoret *et al.* (2008)]. The most recent work using this latter technique yields [Bouchendira *et al.* (2011)]:

$$\alpha^{-1} = 137.035999037(91). \tag{18.39}$$

Other method to measure h/M

The accurate measurement of h/M has been also achieved by non-interferometric measurement based on Bloch oscillations [Cladé *et al.* (2006a,b, 2009b)]. The method proceeds in three steps. First, a narrow and well-determined initial velocity distribution is realized thanks to a velocity selective Raman transition. Second, a transfer of a very high number of photon momenta is provided to the atoms by a succession of stimulated two-photon transitions using two counterpropagating laser beams. Each transition modifies the atomic velocity by $2v_r$, while leaving the internal state unchanged. This acceleration process can also be interpreted in terms of Bloch oscillations in the fundamental energy band of the periodic potential created by an optical standing wave.[16] Third, the velocity variation of the atoms is determined by measuring the Doppler effect using a velocity selective Raman transition whose frequency is scanned to get the whole velocity profile.

[15]See Sec. 9.4.3 of Chap. 9.
[16]See Sec. 17.6 of Chap. 17.

18.6 New perspectives opened by optical clocks

During the last few years, spectacular advances have been accomplished with the realization of optical atomic clocks, using optical, rather than microwave, transitions. The interest in using optical transitions is clear. If the width $\Delta\nu$ of the resonance is the same, the quality factor of the resonance $Q = \nu/\Delta\nu$ can be much higher because the frequency ν of the transition is higher by several orders of magnitude.

Up to now, two types of optical atomic clocks have been achieved.

- ***Single ion optical clocks.*** A single ion is trapped, cooled and a very narrow optical transition connecting the ground state of the ion to a long lived excited state is used as the clock transition. It is important to make the measurement on a single ion to avoid frequency shifts due to ion-ion interactions. The clock transition has a very low fluorescence rate because of the long lifetime of the upper state. The fluorescence rate is too weak for detecting the resonance. One rather uses the shelving method described in Sec. 6.5 of Chap. 6. To give an idea of the reachable accuracy of single ion optical clocks, let us mention the recent realization at NIST of a Al^+ optical clock with a fractional frequency inaccuracy of 8.6×10^{-18} [Chou *et al.* (2010a)].
- ***Neutral atoms in an optical lattice.*** Single ion optical clocks do not give a large signal to noise ratio because they use a single ion. Other systems using neutral atoms trapped in the potential wells of optical lattices have been studied. They give much higher signal to noise ratios because of the large number of atoms, but one has to avoid the light shifts of the clock transition produced by the light that creates the lattice. A very ingenious method for solving this difficulty is to use a wavelength for the lattice laser, called the *magic wavelength* such that the light shifts of the two levels of the clock transition are equal [Katori (2002)].

The relative accuracy of optical clocks has improved dramatically during the last decades, and now have the best performances among atomic clocks. For example, a group at NIST has measured the ratio of the clock transition frequencies of two ions, Al^+ and Hg^+, with an accuracy of a few 10^{-17} [Rosenband *et al.* (2008)]. The ratio of these two frequencies depends on the fine structure constant α, so by measuring this ratio during periods extending over several months, one can look for a possible secular variation of α predicted by certain cosmological models. For example, from their measurements, the NIST Boulder group has recently given some improved limits for such a possible variation: $\dot{\alpha}/\alpha = (-1.6 \pm 2.3) \times 10^{-17}/\text{year}$.

We conclude this chapter with Fig. 18.14 which summarizes the spectacular progresses of the relative accuracy of atomic clocks over the last decades.

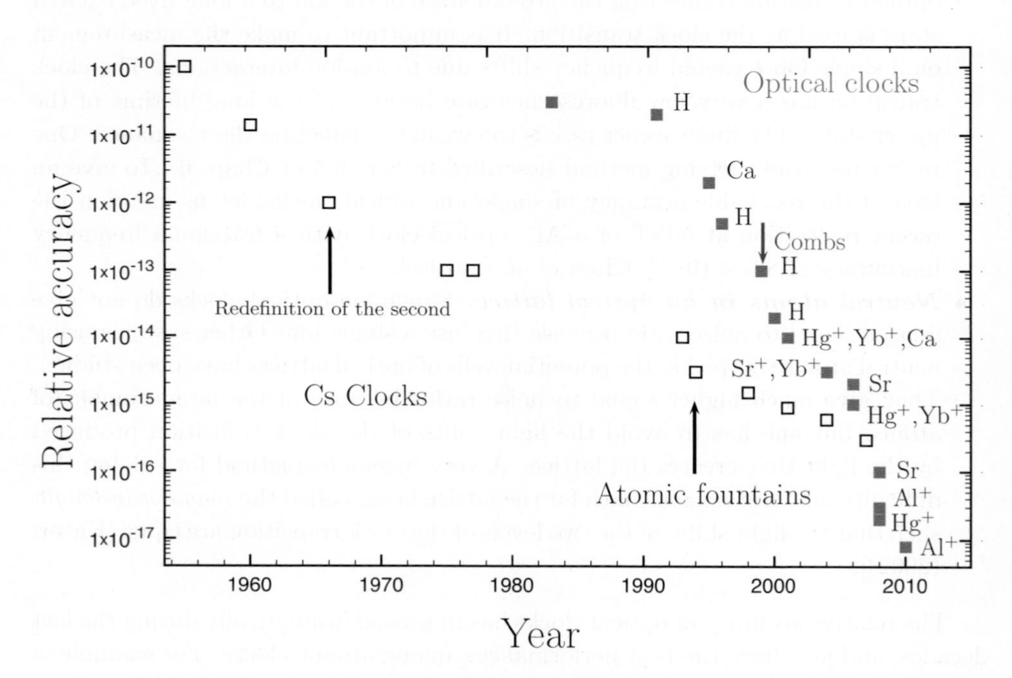

Fig. 18.14 Relative accuracy of atomic clocks as a function of time.

Chapter 19

Quantum correlations. Entangled states

19.1 Introduction

In all previous chapters dealing with quantum interference,[1] we considered experiments involving a single quantum system such as a single atom or a single photon. Interference effects were interpreted in terms of linear superpositions of states that described situations in which the system could exist in two different quantum states $|\varphi_1\rangle$ and $|\varphi_2\rangle$, with amplitudes c_1 and c_2. Certain physical predictions can then depend not only on the probabilities $|c_1|^2$ and $|c_2|^2$ to find the system in $|\varphi_1\rangle$ and $|\varphi_2\rangle$, but also on the crossed terms $c_1c_2^*$ and $c_1^*c_2$. Several interference effects have also been related to the existence of several paths that can be followed by the particle, and which lead to the interference of the various waves associated with the particle when these paths are recombined.

In fact, interference effects can also be observed in situations where several quantum systems composed of several subsytems are involved. In this chapter, we review a few situations of this type and we show that the new physical effects that can be observed, including interference effects, are related to the existence of quantum correlations between the two systems and result from the fact that the wave function of the whole system is not a product of two wave functions, one for each system, but is rather a so-called *entangled state*.

We begin in Sec. 19.2 by discussing intensity correlations in photodetection experiments. More precisely, we consider two photodetectors A and B in a light field, the first one at point $\vec{r}_A$ and the second one at point $\vec{r}_B$, and we calculate the double counting rate $w_{II}(\vec{r}_A, t_A; \vec{r}_B, t_B)$, which gives the probability to detect one photon at point $\vec{r}_A$ and at time t_A *and* another one at point $\vec{r}_B$ at time t_B. We take a simple model of the light field, where only two modes $(\vec{k}_1, \omega_1)$ and $(\vec{k}_2, \omega_2)$ of the field contain photons. If the state of the field is a product $|N_1, N_2\rangle = |N_1\rangle \otimes |N_2\rangle$ of two Fock states containing a well defined number of photons in each mode, the single counting rate $w_I(\vec{r}_A, t_A)$ that gives the probability to detect one photon at point $\vec{r}_A$ at time t_A is constant, which means that a single photodetection can

[1] Chaps. 4, 17 and 18.

occur anywhere in space at any time (a similar result holds for $w_I(\vec{r}_B, t_B)$). In contrast, one finds that the double counting rate $w_{II}(\vec{r}_A, t_A; \vec{r}_B, t_B)$ exhibits fringes, i.e. contains sinusoidal functions of $\vec{r}_A - \vec{r}_B$ and $t_A - t_B$, which means that the two photodetections are correlated. When a photodetection occurs at $\vec{r}_A, t_A$, the place and the time of the second photodetection will be correlated with it. This two-photon interference effect was first discovered by Hanbury Brown and Twiss [Hanbury Brown and Twiss (1956)]. We show that it can be interpreted as the result of an interference between two transition amplitudes: in the first, a photon of mode 1 is detected by photodetector A, and a photon of mode 2 is detected by photodetector B; in the second, a photon of mode 1 is detected by photodetector B, and a photon of mode 2 is detected by photodetector A. We also show how the first detection introduces correlations between the two modes by transforming the initial state of the field, which is a product state with no correlations, into an entangled state that depends on the first photodetection.

In view of their importance, the remaining part of this chapter is devoted to the discussion of the properties of entangled states. These are linear superpositions of two-particle states 1 and 2 that cannot be written as a product of a state of particle 1 by a state of particle 2 in any basis of the Hilbert space of the two particles. In Sec. 19.3, we show how it is possible to characterize an entangled state by using the Schmidt decomposition of the state vector of the whole system. A simple way for entangling two systems is to let them interact. In Sec. 19.4, we illustrate this idea by giving several examples of preparation of entangled states involving either one photon and one atom, two atoms, two types of degrees of freedom, or two photons.

The concept of entanglement is a central concept in quantum mechanics and has been at the origin of important debates concerning both its completeness and the interpretation of the measurement process. It is now at the heart of an important new research field called *quantum information* in which one tries to use the quantum correlations of entangled states to transmit and process information. In the last four sections of this chapter, we discuss the connections between entanglement and several important concepts such as interference (Sec. 19.5), non-locality and non-separability (Sec. 19.6), which-path information (Sec. 19.7), quantum measurement process and decoherence (Sec. 19.8).[2]

19.2 Interference effects in double counting rates

19.2.1 *Photodetection signals*

We first recall a few results concerning the expressions of single and double counting rates (for a derivation of these expressions see [Cohen-Tannoudji *et al.* (1992b)]).

[2]More detailed discussions can be found in [Bouwmeester *et al.* (2000); Laloë (2001); Haroche and Raimond (2006); Kwek *et al.* (2011)].

(i) *Single counting rates*

The probability per unit time $w_I(\vec{r}_A, t_A)$ that a photodetector at point $\vec{r}_A$ detects a photon at time t_A is proportional to:

$$w_I(\vec{r}_A, t_A) = \langle\psi| \hat{E}^{(-)}(\vec{r}_A, t_A)\hat{E}^{(+)}(\vec{r}_A, t_A) |\psi\rangle , \tag{19.1}$$

where $|\psi\rangle$ is the state of the field, and $\hat{E}^{(+)}(\hat{E}^{(-)})$ the positive (negative) frequency component of the field operator[3] $\hat{E}$ given by the mode expansion[4]:

$$\hat{E}^{(+)}(\vec{r}, t) = \sum_{\text{modes i}} \mathcal{E}_i \, \hat{a}_i \, \exp\left[i\left(\vec{k}_i.\vec{r} - \omega_i t\right)\right] = \left[\hat{E}^{(-)}(\vec{r}, t)\right]^{\dagger} . \tag{19.2}$$

In Eq. (19.2), $\hat{a}_i$ is the annihilation operator of a photon in the mode i and $\mathcal{E}_i$ is a normalization coefficient.

(ii) *Double counting rates*

The probability per unit time $w_{II}(\vec{r}_A, t_A; \vec{r}_B, t_B)$ that a photodetector at $\vec{r}_A$ detects a photon at time t_A and another photodetector at $\vec{r}_B$ detects a photon at time t_B is equal to:

$$w_{II}(\vec{r}_A, t_A; \vec{r}_B, t_B) = \langle\psi| \hat{E}^{(-)}(\vec{r}_A, t_A)\hat{E}^{(-)}(\vec{r}_B, t_B)\hat{E}^{(+)}(\vec{r}_B, t_B)\hat{E}^{(+)}(\vec{r}_A, t_A) |\psi\rangle . \tag{19.3}$$

19.2.2 *Two-mode model for the light field*

We now discuss the properties of w_I and w_{II} when only two modes $(\vec{k}_1, \omega_1)$ and $(\vec{k}_2, \omega_2)$ of the field contain photons and when the state of the field is a product of two Fock states in these two modes:

$$|\psi\rangle = |N_1\rangle \otimes |N_2\rangle = |N_1, N_2\rangle. \tag{19.4}$$

In the mode expansion (19.2) of $E^{(+)}$, we can keep only the contributions of the modes 1 and 2, since the contributions of all other empty modes vanish in w_I and w_{II}:

$$\hat{E}^{(+)}(\vec{r}, t) = \hat{E}_1^{(+)}(\vec{r}, t) + \hat{E}_2^{(+)}(\vec{r}, t) . \tag{19.5}$$

(i) *Can one observe interference fringes on w_I for such a state?*

Inserting Eq. (19.5) and its hermitian conjugate into Eq. (19.1), one finds that there are two types of terms contributing to w_I:

- square terms $\langle N_1, N_2| \hat{E}_1^{(-)} \hat{E}_1^{(+)} |N_1, N_2\rangle$ and $\langle N_1, N_2| \hat{E}_2^{(-)} \hat{E}_2^{(+)} |N_1, N_2\rangle$,
- crossed terms $\langle N_1, N_2| \hat{E}_1^{(-)} \hat{E}_2^{(+)} |N_1, N_2\rangle$ and $\langle N_1, N_2| \hat{E}_2^{(-)} \hat{E}_1^{(+)} |N_1, N_2\rangle$.

[3]See Chap. 2.
[4]For sake of simplicity, we ignore the polarization degree of freedom.

Using $\hat{a}_1 \, |N_1, N_2\rangle = \sqrt{N_1}\,|N_1 - 1, N_2\rangle$, $\hat{a}_2 \, |N_1, N_2\rangle = \sqrt{N_2}\,|N_1, N_2 - 1\rangle$, and the hermitian conjugates of these equations, one easily finds that the contributions of the crossed terms, proportional to $\langle N_1, N_2-1|N_1-1, N_2\rangle$ and $\langle N_1-1, N_2|N_1, N_2-1\rangle$ vanish. Only the square terms contribute to w_I, which is equal to:

$$w_I(\vec{r}_A, t_A) = N_1 \mathcal{E}_1^2 + N_2 \mathcal{E}_2^2. \tag{19.6}$$

There are thus no interference fringes on w_I. This result reflects the fact that in the Fock state (19.4), there is no coherence between the two fields in the modes 1 and 2.

(ii) *Can one observe interference fringes on w_{II}?*

The only non-zero matrix elements that appear in w_{II} are: $\langle\psi|\hat{a}_1^\dagger \hat{a}_1^\dagger \hat{a}_1 \hat{a}_1|\psi\rangle = N_1(N_1-1)$, $\langle\psi|\hat{a}_2^\dagger \hat{a}_2^\dagger \hat{a}_2 \hat{a}_2|\psi\rangle = N_2(N_2-1)$, and $\langle\psi|\hat{a}_1^\dagger \hat{a}_2^\dagger \hat{a}_2 \hat{a}_1|\psi\rangle$, $\langle\psi|\hat{a}_2^\dagger \hat{a}_1^\dagger \hat{a}_1 \hat{a}_2|\psi\rangle$, $\langle\psi|\hat{a}_1^\dagger \hat{a}_2^\dagger \hat{a}_1 \hat{a}_2|\psi\rangle$, $\langle\psi|\hat{a}_2^\dagger \hat{a}_1^\dagger \hat{a}_2 \hat{a}_1|\psi\rangle$, which are all equal to $N_1 N_2$. Inserting these results into Eq. (19.3), one gets:

$$\begin{aligned} w_{II}\left(\vec{r}_A, t_A; \vec{r}_B, t_B\right) &= \mathcal{E}_1^2 \mathcal{E}_2^2 \left[N_1\left(N_1-1\right) + N_2\left(N_2-1\right) + 2N_1N_2\right] \\ &+ 2N_1N_2\mathcal{E}_1^2\mathcal{E}_2^2 \,\mathrm{Re}\left(\exp\left[i\left(\vec{k}_1 - \vec{k}_2\right)\cdot\left(\vec{r}_A - \vec{r}_B\right) - i\left(\omega_1-\omega_2\right)\left(t_A - t_B\right)\right]\right). \end{aligned} \tag{19.7}$$

We have now interference fringes: the probability to detect one photon at $\vec{r}_A, t_A$ and one photon at $\vec{r}_B, t_B$ is an oscillating function of $\vec{r}_A - \vec{r}_B$ and $t_A - t_B$. The two events are correlated because even though the probability to detect one photon at $\vec{r}_A, t_A$ is uniform and does not depend on $\vec{r}_A$ and t_A, once a photon is detected at $\vec{r}_A, t_A$ the probability to detect a second photon at $\vec{r}_B, t_B$ is no longer uniform and depends on $\vec{r}_A - \vec{r}_B$ and $t_A - t_B$.

19.2.3 *What are the "objects" which interfere in w_{II}?*

To answer this question, it is useful to re-express w_{II} in another form:

$$\begin{aligned} w_{II}(\vec{r}_A, t_A; \vec{r}_B, t_B) &= \langle\psi|\, \hat{E}^{(-)}(\vec{r}_A, t_A)\hat{E}^{(-)}(\vec{r}_B, t_B)\hat{E}^{(+)}(\vec{r}_B, t_B)\hat{E}^{(+)}(\vec{r}_A, t_A)\,|\psi\rangle \\ &= \sum_f \langle\psi|\, \hat{E}^{(-)}(\vec{r}_A, t_A)\hat{E}^{(-)}(\vec{r}_B, t_B)\,|\psi_f\rangle\,\langle\psi_f| \\ &\quad \hat{E}^{(+)}(\vec{r}_B, t_B)\hat{E}^{(+)}(\vec{r}_A, t_A)\,|\psi\rangle\,, \end{aligned} \tag{19.8}$$

where we have introduced the closure relation over a complete set of states $|\psi_f\rangle$.

In fact, because of the two annihilation operators acting on the state $|\psi\rangle = |N_1, N_2\rangle$, the only states $|\psi_f\rangle$ that give a non-zero contribution in Eq. (19.8) are the states that contain two photons less than $|N_1, N_2\rangle$, i.e. $|N_1-2, N_2\rangle$, $|N_1, N_2-2\rangle$, and $|N_1-1, N_2-1\rangle$. We can thus write:

$$\begin{aligned} w_{II}(\vec{r}_A, t_A; \vec{r}_B, t_B) &= |\langle N_1-2, N_1|\hat{E}_1^{(+)}(\vec{r}_B, t_B)\hat{E}_1^{(+)}(\vec{r}_A, t_A)|N_1, N_2\rangle|^2 \\ &+ \text{same term with } 1 \leftrightarrow 2 \\ &+ |\langle N_1-1, N_2-1|\hat{E}_1^{(+)}(\vec{r}_B, t_B)\hat{E}_2^{(+)}(\vec{r}_A, t_A) + \hat{E}_2^{(+)}(\vec{r}_B, t_B)\hat{E}_1^{(+)}(\vec{r}_A, t_A)|N_1, N_2\rangle|^2 \end{aligned} \tag{19.9}$$

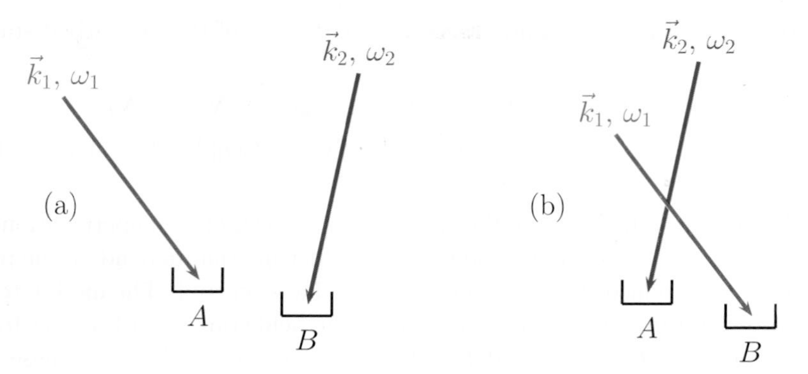

Fig. 19.1 The two paths leading from the initial state $|N_1, N_2\rangle$ to the final state where one photon in each mode has been detected [Fano (1961)]. In the first path (a), a photon $\vec{k}_1, \omega_1$ is detected by photodetector A and the second photon is detected by photodetector B, whereas in the second path (b), photon $\vec{k}_1, \omega_1$ is detected by B while photon $\vec{k}_2, \omega_2$ is detected by A. The interference fringes in w_{II} are due to the interference between these two amplitudes.

One can easily check that the sinusoidal term of Eq. (19.7) comes from the term in the last line of Eq. (19.9). This term is the modulus squared $|A_1 + A_2|^2$ of the sum of two amplitudes A_1 and A_2: the first one

$$A_1 = \langle N_1 - 1, N_2 - 1|\hat{E}_2^{(+)}(\vec{r}_B, t_B)\hat{E}_1^{(+)}(\vec{r}_A, t_A)|N_1, N_2\rangle, \tag{19.10}$$

describes a process in which a photon $(\vec{k}_1, \omega_1)$ is detected by photodetector A while a photon $(\vec{k}_2, \omega_2)$ is detected by photodetector B; the second one,

$$A_2 = \langle N_1 - 1, N_2 - 1|\hat{E}_1^{(+)}(\vec{r}_B, t_B)\hat{E}_2^{(+)}(\vec{r}_A, t_A)|N_1, N_2\rangle,$$

describes a process where the roles of the two photodetectors are interchanged (see Fig. 19.1). This discussion shows that it is not the waves in the two modes that interfere, but rather the transition amplitudes associated with different paths followed by the system.

19.2.4 *Establishment of correlations between the two modes*

A new physical insight can be obtained into the correlations exhibited by w_{II} if one writes the expression (19.3) of w_{II} in the following way:

$$w_{II}(\vec{r}_A, t_A; \vec{r}_B, t_B) = \langle\psi(\vec{r}_A, t_A)|\hat{E}^{(-)}(\vec{r}_B, t_B)E^{(+)}(\vec{r}_B, t_B)|\psi(\vec{r}_A, t_A)\rangle \tag{19.11}$$

where

$$|\psi(\vec{r}_A, t_A)\rangle = E^{(+)}(\vec{r}_A, t_A)|N_1, N_2\rangle. \tag{19.12}$$

The double counting rate at $\vec{r}_A, t_A; \vec{r}_B, t_B$ can thus be considered as the single counting rate at $\vec{r}_B, t_B$ in a field state $|\psi(\vec{r}_A, t_A)\rangle$ in which the initial field state $|N_1, N_2\rangle$ has been perturbed by the first photodetection at $\vec{r}_A, t_A$.

According to Eqs. (19.2) and (19.4), the expression of this perturbed state is:

$$|\psi(\vec{r}_A, t_A)\rangle = \mathcal{E}_1\sqrt{N_1}\exp[i(\vec{k}_1\cdot\vec{r}_A - \omega_1 t_A)]|N_1 - 1, N_2\rangle$$
$$+\mathcal{E}_2\sqrt{N_2}\exp[i(\vec{k}_2\cdot\vec{r}_A - \omega_2 t_A)]|N_1, N_2 - 1\rangle. \qquad (19.13)$$

It is no longer a product state of the two modes, but is a linear superposition of two different states of the two-mode system with coefficients that depend on the position $\vec{r}_A$ and time t_A at which the first photodetection has occurred. The modulated term in the single counting rate w_I associated with the field state (19.13) comes from the matrix elements of $E_2^{(-)}E_1^{(+)}$ and $E_1^{(-)}E_2^{(+)}$, i.e. of $\hat{a}_2^\dagger\hat{a}_1$ and $\hat{a}_1^\dagger\hat{a}_2$. These matrix elements are non-zero only between states with the same total number of photons in which N_1 increases by one unit while N_2 decreases by one unit, or vice versa. This is the case for the two states $|N_1 - 1, N_2\rangle$ and $|N_1, N_2 - 1\rangle$ appearing in the expansion (19.13). One thus understands how interference fringes, which are not present in w_I for a product state like $|N_1, N_2\rangle$, can appear with state (19.13).

The state (19.13) is an entangled state and cannot be written as a product state in any basis. In the next sections, we will come back to the quantum correlations that exist between two systems when they are in an entangled state. The previous discussion shows that the first photodetection at $\vec{r}_A, t_A$ transforms the initial product state into an entangled state. It is this entanglement that gives rise to spatial and temporal modulations for the second photodetection.

Remarks

(i) One can show that the relative phase φ between two field modes is the conjugate variable of the difference $N_1 - N_2$ of the number of photons in each mode for a fixed value of $N_1 + N_2$. In a product of two Fock states, $N_1 - N_2$ has a fixed value, so that φ is completely undetermined. The state (19.13) is a linear superposition of two states with different values of $N_1 - N_2$ but with the same value of $N_1 + N_2$. It is this uncertainty in $N_1 - N_2$ that partly removes the uncertainty on φ and which gives rise to interference fringes.

(ii) The previous calculations can be extended to experiments involving more that two photodetections. For example, in a triple photodetection experiment, the third photodetection at $\vec{r}_C, t_C$ can be considered as a single photodetection experiment in a state perturbed by the first two photodetections at $\vec{r}_A, t_A$ and $\vec{r}_B, t_B$. The place and time at which it occurs are correlated not only with $\vec{r}_A, t_A$, but also with $\vec{r}_B, t_B$. The uncertainty in $N_1 - N_2$ increases, leading to a better determination on the relative phase φ, and so on. A single run experiment involving a large number $\mathcal{N}$ of photodetections will exhibit clear interference fringes, even if it starts from a product of two Fock states with a completely undetermined relative phase. Moreover, if one performs a new single run experiment with $\mathcal{N}$ photodetections, the new system of fringes will not coincide with the first one, but will instead be centered on a different position, as if the relative phase is well defined at the end of a single run experiment, but varies randomly from one experiment to another. We will come back to these questions in Chap. 23 when we will discuss the interference between two Bose-Einstein condensates.

19.3 Entangled states

19.3.1 *Definition*

An entangled state $|\psi_{12}\rangle$ of two systems 1 and 2 is a state that cannot be written as a product of a state $|\varphi_1\rangle$ of system 1 by a state $|\chi_2\rangle$ of system 2:

$$|\psi_{12}\rangle \neq |\varphi_1\rangle \otimes |\chi_2\rangle. \tag{19.14}$$

It is a linear superposition of at least two such product states. A simple and important example of entangled state is the singlet state $|S=0, M=0\rangle$ of two spins 1/2:

$$|S=0, M=0\rangle = \frac{|1+\rangle_z \otimes |2-\rangle_z| - |1-\rangle_z \otimes |2+\rangle_z}{\sqrt{2}} \tag{19.15}$$

where $|1\pm\rangle_z$ ($|2\pm\rangle_z$) are the two eigenstates of S_{1z} (S_{2z}).

The impossibility of factorizing an entangled state into the form of a product state must hold in all bases. For example, the rotational invariance of the singlet state means that the expression (19.15) keeps the same form if, instead of using the basis of eigenstates of S_z, one uses the eigenstates of the components of the two spins in a direction defined by the polar angles θ, φ:

$$|S=0, M=0\rangle = \frac{|1+\rangle_{\theta,\varphi} \otimes |2-\rangle_{\theta,\varphi}| - |1-\rangle_{\theta,\varphi} \otimes |2+\rangle_{\theta,\varphi}}{\sqrt{2}}. \tag{19.16}$$

Let us also give an example of a state

$$\frac{|1+\rangle_z \otimes |2+\rangle_z - |1+\rangle_z \otimes |2-\rangle_z - |1-\rangle_z \otimes |2+\rangle_z + |1-\rangle_z \otimes |2-\rangle_z}{2} \tag{19.17}$$

that looks like an entangled state, but that is not. Indeed, it can be also written as $|1-\rangle_x \otimes |2-\rangle_x$ if one takes the basis of the eigenstates of S_{1x} and S_{1x}.

19.3.2 *Schmidt decomposition of an entangled state*

To determine if a state $|\psi_{1,2}\rangle$ of two systems $\mathcal{S}_1$ and $\mathcal{S}_2$ is an entangled state or not, it is useful to consider the reduced density operators ρ_1 and ρ_2 of each system. We first recall a few properties of these reduced density operators.

We suppose that the state of the whole system is a pure state described by a state vector $|\psi_{1,2}\rangle$. The density operator ρ_{12} of the whole system

$$\rho_{12} = |\psi_{1,2}\rangle\langle\psi_{1,2}| \tag{19.18}$$

is then the projector onto $|\psi_{1,2}\rangle$. If one is interested only by measurements of observables concerning $\mathcal{S}_1$ alone (or $\mathcal{S}_2$ alone), the predictions concerning the measurements of these observables can be made using the reduced density operators ρ_1 (or ρ_2) obtained by tracing ρ_{12} over the variables of $\mathcal{S}_1$ (or $\mathcal{S}_2$):

$$\rho_1 = \mathrm{Tr}_2\left(\rho_{12}\right), \qquad \rho_2 = \mathrm{Tr}_1\left(\rho_{12}\right). \tag{19.19}$$

Since density operators are semipositive definite Hermitian operators, the eigenvalues π_i of ρ_1 are real and non-negative and their sum is equal to 1 since the trace of a density operator is equal to 1: $\sum_i \pi_i = 1$. Some of the π_i's can be equal to zero. We will suppose here that the Hilbert spaces $\mathcal{H}_1$ and $\mathcal{H}_2$ of $\mathcal{S}_1$ and $\mathcal{S}_2$ have finite dimensions equal to n_1 and n_2 and, without loss of generality, that $n_1 \leq n_2$. If we take in $\mathcal{H}_1$ a basis formed by the eigenstates $|1u_i\rangle$ of ρ_1 corresponding to the eigenvalues π_i, ρ_1 is given by

$$\rho_1 = \sum_i \pi_i \,|1\,u_i\rangle\,\langle 1\,u_i|\,. \tag{19.20}$$

Coming back to $|\psi_{1,2}\rangle$, we can expand it in a basis $\{|1\,u_i\rangle \otimes |2\,w_k\rangle\}$ of the total Hilbert space $\mathcal{H}_1 \otimes \mathcal{H}_2$, containing the eigenstates $|1\,u_i\rangle$ of ρ_1 introduced above:

$$|\psi_{1,2}\rangle = \sum_i \sum_k c_{i\,k}\,|1\,u_i\rangle \otimes |2\,w_k\rangle\,. \tag{19.21}$$

This expansion can be also written:

$$|\psi_{1,2}\rangle = \sum_i \,|1\,u_i\rangle \otimes |2\,\bar{v}_i\rangle\,, \tag{19.22}$$

where we have introduced the vectors:

$$|2\,\bar{v}_i\rangle = \sum_k c_{i\,k}\,|2\,w_k\rangle\,. \tag{19.23}$$

It is convenient to introduce the basis:

$$|2\,v_i\rangle = \frac{1}{\sqrt{\pi_i}}\,|2\,\bar{v}_i\rangle\,, \quad \text{for} \quad \pi_i \neq 0 \quad \text{and} \quad |2\,v_i\rangle = |2\,\bar{v}_i\rangle \quad \text{otherwise.} \tag{19.24}$$

If one now takes the trace over the variables of $\mathcal{S}_2$ of Eq. (19.18) where $|\psi_{1,2}\rangle$ is replaced by the relation (19.22), one finds that:

$$\rho_1 = \sum_i \sum_{i'} \sqrt{\pi_i \pi_{i'}}\langle 2\,v_{i'}|2\,v_i\rangle\,|1\,u_i\rangle\langle 1\,u_{i'}|. \tag{19.25}$$

Comparing Eq. (19.25) with Eq. (19.20), one deduces that $\langle 2\,v_{i'} \mid 2\,v_i\rangle = \delta_{i\,i'}$, which means that the $|2\,v_i\rangle$'s are an orthonormal basis in $\mathcal{H}_2$. We can finally take the trace over the variables of $\mathcal{S}_1$ of the whole density matrix $\rho_{12} = |\psi_{1,2}\rangle\langle\psi_{1,2}|$, using the expression (19.22) of $|\psi_{1,2}\rangle$. This gives the following expression for ρ_2:

$$\rho_2 = \sum_i \pi_i \,|2\,v_i\rangle\,\langle 2\,v_i|\,. \tag{19.26}$$

In other words, both systems, considered alone, are described by a statistical mixture of states with the same weights π_i.

These results lead us to the following conclusion concerning the possibility of distinguishing between a product state and an entangled state. If all the π_i's appearing in the reduced density matrices of $\mathcal{S}_1$ and $\mathcal{S}_2$ are equal to zero, except one, π_0, which must be therefore equal to 1, both systems $\mathcal{S}_1$ and $\mathcal{S}_2$ are described by a pure state, $|1\,u_0\rangle$ for $\mathcal{S}_1$, $|2\,v_0\rangle$ for $\mathcal{S}_2$ and $|\psi_{1,2}\rangle$ is the product state $|1\,u_0\rangle \otimes |2\,v_0\rangle$. If several π_i's are different from zero, each system, considered alone is a statistical mixture and $|\psi_{1,2}\rangle$ is no longer a product state, but an entangled state.

19.3.3 *Information content of an entangled state*

It is interesting to try to get a more quantitative estimate of the degree of entanglement between two systems. This can be done by using the concept of statistical entropy S introduced by von Neumann, to characterize the lacking information on a system described by a density operator ρ:

$$S = -k_B \mathrm{Tr}\,(\rho \ln \rho) = -k_B \sum_i \pi_i \ln \pi_i, \tag{19.27}$$

where k_B is Boltzmann's constant and where the π_i's are the weights of the various components of the statistical mixture associated with ρ.

If the whole system $\mathcal{S}_1 + \mathcal{S}_2$ is in a pure state, so that all the π_i's are equal to 0 except one, which is equal to 1, and:

$$S_{12} = 0. \tag{19.28}$$

There is no lacking information for a quantum system described by a pure state, since this description is the best possible in quantum mechanics. If $|\psi_{1,2}\rangle$ is an entangled state, the reduced density operators ρ_1 and ρ_2 of $\mathcal{S}_1$ and $\mathcal{S}_2$ are statistical mixtures involving the same weights π_i, according to Eqs. (19.20) and (19.26). It follows that the statistical entropies S_1 and S_2 characterizing each system $\mathcal{S}_1$ and $\mathcal{S}_2$ considered alone are equal and given by:

$$S_1 = S_2 = -k_B \sum_i \pi_i \ln \pi_i > S_{12}. \tag{19.29}$$

The fact that they are larger than S_{12} is easy to understand. If we focus on $\mathcal{S}_1$ alone or on $\mathcal{S}_2$, we ignore the correlations between these two systems and we lose the corresponding information associated with these correlations and which is contained in the entangled state $|\psi_{1,2}\rangle$. Ignoring the correlations between $\mathcal{S}_1$ and $\mathcal{S}_2$ amounts to replacing ρ_{12} by $\rho_1 \otimes \rho_2$. It can be easily shown that

$$S(\rho_1 \otimes \rho_2) = S(\rho_1) + S(\rho_2) = S_1 + S_2. \tag{19.30}$$

One can thus consider that

$$\Delta S = S_1 + S_2 - S_{12} = -2k_B \sum_i \pi_i \ln \pi_i \tag{19.31}$$

characterizes the information associated with the entanglement.[5]

The quantity ΔS is maximum when all the π_i's are equal. In this case, the corresponding states $|\psi_{1,2}\rangle$ are called maximally entangled states. The singlet state of two spins 1/2 given above in Eq. (19.15) is such a maximally entangled state since the reduced density operators of each spin are statistical mixtures of $|1\pm\rangle$ and $|2\pm\rangle$ with equal weights equal to 1/2. For two spins 1/2, there are other useful maximally entangled states:

$$\frac{|1+\rangle_z \otimes |2-\rangle_z + |1-\rangle_z \otimes |2+\rangle_z}{\sqrt{2}} \quad \text{and} \quad \frac{|1+\rangle_z \otimes |2+\rangle_z \pm |1-\rangle_z \otimes |2-\rangle_z}{\sqrt{2}} \tag{19.32}$$

which are also called *Bell states.*

[5] This result remains valid even if the whole system is not in a pure state.

19.4 Preparing entangled states

Now that we have introduced a few basic properties of entangled states, we will give a few examples showing how it is possible to prepare such states by letting two systems interact, either directly or indirectly.

19.4.1 *Entanglement between one atom and one field mode*

We now use a few basic results from Chap. 2 concerning the interaction of a spin 1/2 with a monochromatic resonant radiofrequency (RF) field. In particular, we recall that the Rabi oscillation between the two spin states occurs at the Rabi frequency Ω, which characterizes the strength of the atom-field coupling. If the interaction is switched on for a time T such that $\Omega T = \pi/2$, we have a so-called $\pi/2$ pulse. In the rotating frame, the spin rotates by an angle $\pi/2$ around the RF field (which is fixed in the rotating frame) and a spin initially in the $|-\rangle$ state ends up in a linear superposition with equal weights of the two spin states $|\pm\rangle$. In a π-pulse ($\Omega T = \pi$), the state $|-\rangle$ is transformed into $|+\rangle$ and vice versa. More generally, the transformation of the spin state $|\psi\rangle$ in a pulse such that $\Omega T = \alpha$, is described by a rotation $R(\alpha) = \exp[-i\alpha S_X/\hbar]$ of angle α around the direction X of the RF field. Using $S_X = \hbar\sigma_X/2$, where σ_X is a Pauli matrix, and the fact that all even powers of σ_X are equal to 1, we get:

$$|\psi\rangle \to \exp\left(-\frac{i\alpha S_X}{\hbar}\right)|\psi\rangle = \exp\left(-\frac{i\alpha\sigma_X}{2}\right)|\psi\rangle = \left[\cos\frac{\alpha}{2} - i\sin\frac{\alpha}{2}\sigma_X\right]|\psi\rangle . \tag{19.33}$$

In most discussions of Chap. 2, the monochromatic field was treated as a classical field. However, such an approach is no longer possible in a resonant cavity when the field contains only one or zero photons. When a two-level atom e, g initially in e enters an empty resonant cavity ($N = 0$), the two states that must be considered are $|e, N = 0\rangle$ and $|g, N = 1\rangle$. These are coupled because the atom in g in the presence of one photon can absorb the photon and jump to e. The corresponding Rabi frequency Ω is then called the vacuum Rabi frequency, and is supposed to be large compared to all damping rates, such that we can neglect all these damping processes. This regime is referred to as the strong coupling regime of cavity quantum electrodynamics. We then have a two-level problem, isomorphic to a spin 1/2, with the correspondence $|+\rangle \leftrightarrow |e, N = 0\rangle$ and $|-\rangle \leftrightarrow |g, N = 1\rangle$, and we can apply (19.33) for evaluating the effect of a $\pi/2$ pulse for a system initially in $|e, N = 0\rangle$. We get:

$$|e, N = 0\rangle \underset{\pi/2 \text{ pulse}}{\to} \frac{|e, N = 0\rangle - i\,|g, N = 1\rangle}{\sqrt{2}} . \tag{19.34}$$

This is clearly a field-atom entangled state, which is now routinely prepared in cavity quantum electrodynamics experiments.[6] As shown in the next sections this state

[6] See for example [Haroche and Raimond (2006)].

can also be used to prepare entangled states of two atoms and is also considered for preparing entangled states of two field modes of two spatially separated cavities.

Absence of entanglement for the interaction with a semiclassical field
If the field with which the atom interacts is in a quasi-classical state, like a coherent state, and if one can neglect the interaction with the vacuum field (negligible spontaneous emission), one can show that it is correct to describe the field as a classical external field.[7] The idea is to perform a unitary transformation on the full quantum Hamiltonian such that, in the new representation, the atom is coupled, on the one hand, to a classical external field corresponding to the coherent state of the field, and on the other hand, to the quantum field initially in the vacuum state. If this last coupling produces a negligible effect during the duration of the experiment, it can be ignored and the evolution of the atom can be calculated as if it was interacting only with a classical external field. Coming back to the initial representation, one finds that the state of the whole system is a product state, in which the atom is in the state whose evolution has been calculated semiclassically and the field is in the same coherent state as in the initial state. This shows that there are cases where the interaction does not produce entanglement and where the field can be considered as a classical field that is not perturbed by the interaction with the atom. This result is the basis of all semiclassical treatments of radiation-matter interactions.

19.4.2 *Entanglement between two atoms*

We now describe how it is possible to entangle two atoms by sending them one after the other through a resonant cavity and by submitting them to appropriate pulses.

The first atom 1 in the upper state $|e_1\rangle$ is sent through an empty cavity containing zero photon. If the transit time through the cavity is such that the atom experiences a $\pi/2$ pulse, its state at the exit of the cavity is given by Eq. (19.34) where we add a subscript 1 to e and g.

Just before atom 2 enters the cavity, the state of the whole system { atom 1+atom 2+cavity } is thus:

$$\frac{|e_1, N=0\rangle - i\,|g_1, N=1\rangle}{\sqrt{2}} \otimes |g_2\rangle = \frac{|e_1,\, g_2, N=0\rangle - i\,|g_1,\, g_2, N=1\rangle}{\sqrt{2}}. \tag{19.35}$$

Suppose now that a second atom 2 in g enters the cavity once the first atom has left it and that the interaction time is adjusted in such a way that the second atom experiences a π pulse. How do the two components of the last term of Eq. (19.35) transform at the end of the π pulse, when atom 2 leaves the cavity? The component $|e_1, g_2, N = 0\rangle$ does not change because the atom in the lower state g_2 does not interact with the empty cavity ($N = 0$). Putting $\alpha = \pi$ in Eq. (19.33), one finds that $|g_1, g_2, N = 1\rangle$ transforms into $-i|g_1, e_2, N = 0\rangle$. Finally, when atom 2 leaves the cavity, the state of the whole system becomes:

$$\frac{|e_1,\, g_2, N=0\rangle - |g_1,\, e_2, N=0\rangle}{\sqrt{2}} = \frac{|e_1, g_2\rangle - |g_1, e_2\rangle}{\sqrt{2}} \otimes |N=0\rangle\,. \tag{19.36}$$

[7] See for example exercise 17 of [Cohen-Tannoudji *et al.* (1992b)].

The cavity returns to the empty state $N = 0$, and the two atoms are in a linear superposition of $|e_1, g_2\rangle$ and $|g_1, e_2\rangle$, i.e. an entangled state. Contrary to the case studied in the previous subsection, where the entanglement is due to a direct interaction between the atom and the field, in the case described here the two atoms do not interact directly. Each atom interacts only with the cavity which transmits information to atom 2 about its previous interaction with atom 1. The cavity is a mediator between the two atoms. As its final and initial states are identical ($N = 0$), it acts like a catalyst.

An experiment of this type has been performed on Rydberg atoms and has provided pairs of entangled atoms separated by distances on the order of 1 cm [Hagley *et al.* (1997)].

Entanglement of two trapped ions

The previous scheme for entangling two atoms through their coupling with a cavity mode is in fact inspired by another scheme proposed in 1995 by Ignacio Cirac and Peter Zoller for entangling trapped cold ions through their coupling with a coupled mode of vibration of the ions in the trap [Cirac and Zoller (1995)]. Consider two such ions $i = 1, 2$ with two internal states g_i, e_i separated by a splitting $\hbar\omega_i$.[8] Let ω_v be the frequency of the coupled vibration mode. The frequencies of the transitions of ion i are therefore equal to $\omega_i \pm n\omega_v$, where $n = 0, 1, 2, \ldots$ is the quantum number of vibration. Suppose that initially ion 1 is in the state e_1, ion 2 in the state g_2 and the vibration mode in the state $n = 0$. A laser exciting only ion 1 is tuned to the frequency $\omega_1 - \omega_v$ of the red sideband connecting the state $|e_1, n = 0\rangle$ to the state $|g_1, n = 1\rangle$. If the duration of the interaction corresponds to a $\pi/2$ pulse, the state of the total system after the pulse is:

$$\frac{1}{\sqrt{2}}\left[|e_1, 0, g_2\rangle - i\,|g_1, 1, g_2\rangle\right] \tag{19.37}$$

since ion 2 is not coupled to this first laser.

We suppose now that a second laser, interacting only with ion 2 is tuned to the frequency $\omega_2 - \omega_v$ exciting the transition of this ion connecting the state $|n = 1, g_2\rangle$ to the state $|n = 0, e_2\rangle$. The duration of the interaction corresponds now to a π pulse. Ion 1 is not affected by this second laser pulse. The state $|e_1, 0, g_2\rangle$ is not affected because there are no resonant transition at the frequency $\omega_2 - \omega_v$ starting from the state $|n = 0, g_2\rangle$. The state $|g_1, 1, g_2\rangle$ is transformed into $-i|g_1, 0, e_2\rangle$. Finally, after the two laser pulses, the state of the total system is given by:

$$\frac{1}{\sqrt{2}}\left[|e_1, 0, g_2\rangle - |g_1, 0, e_2\rangle\right] = \frac{1}{\sqrt{2}}\left[|e_1, g_2\rangle - |g_1, e_2\rangle\right] \otimes |0\rangle \tag{19.38}$$

The two ions are entangled and the vibration mode is back in the initial state $n = 0$.

---

[8] The two internal states g_i and e_i can be also two hyperfine ground states g_i and g_i' coupled by stimulated Raman transitions.

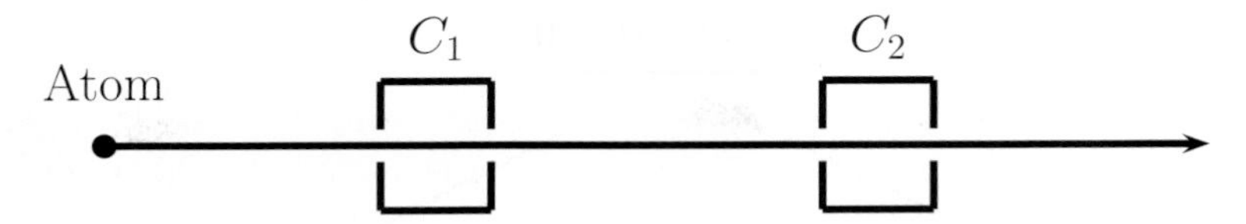

Fig. 19.2 An atom in the upper state e is sent through two cavities C_1 and C_2 that are initially empty. It experiences a $\pi/2$ pulse in C_1, and a π pulse in C_2, resulting in an entanglement of the two cavity fields.

19.4.3 *Entanglement between two separate cavity fields*

Instead of considering two atoms passing one after the other through the same cavity, we now study what happens when a single atom passes through two separated cavities (see Fig. 19.2).

Initially, the atom is in the excited state e while the two cavities are empty ($N_1 = N_2 = 0$). When the atom crosses the first cavity, it experiences a $\pi/2$ pulse. When it leaves the first cavity, and before it enters the second, the state of the whole system { atom+cavity 1+cavity 2 } is thus given by:

$$\frac{|e, N_1 = 0\rangle - i\,|g, N_1 = 1\rangle}{\sqrt{2}} \otimes |N_2 = 0\rangle \,. \tag{19.39}$$

The interaction with the second cavity corresponds to a π pulse. The same considerations as in the previous subsection show that $|g, N_2 = 0\rangle$ does not change while $|e, N_2 = 0\rangle$ transforms into $-i|g, N_2 = 1\rangle$. Finally when the atom leaves the second cavity, the state of the whole system is:

$$\begin{aligned} &-i\frac{|g, N_1 = 0, N_2 = 1\rangle + |g, N_1 = 1, N_2 = 0\rangle}{\sqrt{2}} \\ &= -i\,|g\rangle \otimes \frac{|\,N_1 = 0, N_2 = 1\rangle + |N_1 = 1, N_2 = 0\rangle}{\sqrt{2}} . \end{aligned} \tag{19.40}$$

The atom has returned to the ground state and the two cavity fields are in a linear superposition of $|N_1 = 0, N_2 = 1\rangle$ and $|N_1 = 1, N_2 = 0\rangle$, i.e. an entangled state. Here also the two cavity fields do not interact directly. The atom is a mediator between the two cavities and acts like a catalyst. This very challenging experiment has not yet been performed.

19.4.4 *Entanglement between two photons*

Pairs of entangled photons can be obtained in a radiative cascade. Figure 19.3 shows, for example, a radiative cascade $J_a = 0 \rightarrow J_b = 1 \rightarrow J_c = 0$ during which two photons with frequencies ω_1 and ω_2 are emitted. If the two photons are detected in opposite directions along the z-axis, the polarizations of these photons are necessarily either σ_+ or $\sigma-$. Furthermore it clearly appears from Fig. 19.3 that, if the ω_1 photon detected in one direction has a polarization σ_+ (σ_-), the ω_2

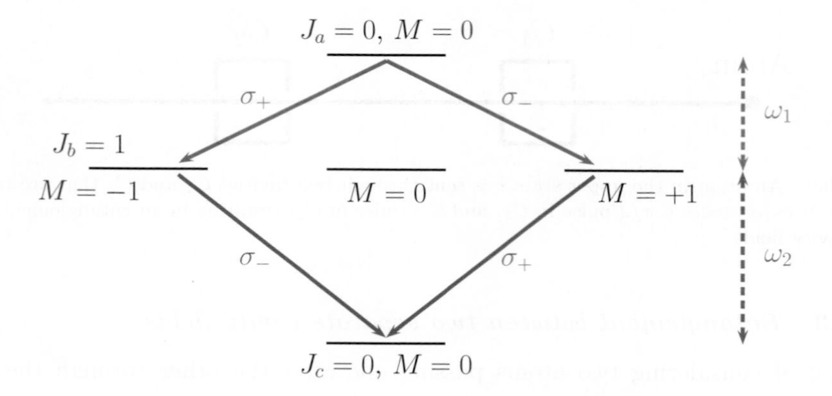

Fig. 19.3 Radiative cascade $J_a = 0 \to J_b = 1 \to J_c = 0$. If the two emitted photons with frequencies ω_1 and ω_2 are detected in opposite directions with analyzers A and B along the z-axis (see lower part of the figure), they have necessarily opposite circular polarizations, and there are two possible paths leading the atom from $J_a = 0$ to $J_c = 0$.

photon detected in the opposite direction has a polarization σ_- (σ_+). The state of the two photons emitted during the radiative cascade is thus necessarily a linear superposition of two states:

$$|\psi\rangle = \frac{1}{\sqrt{2}} \left[|\omega_1, \sigma_+\rangle \otimes |\omega_2, \sigma_-\rangle + |\omega_1, \sigma_-\rangle \otimes |\omega_2, \sigma_+\rangle \right]. \tag{19.41}$$

Instead of measuring the circular polarizations of the photons emitted in the radiative cascade, one could use analyzers to measure the linear polarizations along two orthogonal directions $\vec{e}_x$ and $\vec{e}_y$ in the plane xy perpendicular to the z-axis along which the two photons propagate in opposite directions. One can show that the state (19.41) can be also written:

$$|\psi\rangle = \frac{1}{\sqrt{2}} \left[|\omega_1, \vec{e}_x\rangle \otimes |\omega_2, \vec{e}_x\rangle + |\omega_1, \vec{e}_y\rangle \otimes |\omega_2, \vec{e}_y\rangle \right]. \tag{19.42}$$

If one photon is emitted with a linear polarization, the other photon is emitted with the same polarization. This is an entangled state of two photons that has been used to test Bell's inequalities (see Sec. 19.6).

Other, more efficient sources of entangled pairs of photons have been developed. These are based on parametric down-conversion in nonlinear crystals. In such processes, one photon ω is absorbed and two photons ω_1 and ω_2 are emitted, and the atom returns to the ground state. Energy conservation implies that $\omega = \omega_1 + \omega_2$. Because of the phase matching conditions in the nonlinear crystal, the two photons of the pair have orthogonal linear polarizations and the directions of emission of these two photons lie on two cones. If $\omega_1 = \omega_2 = \omega/2$, one can have, along the two intersections I_1 and I_2 of the two cones, photons emitted along these directions with the same frequency $\omega/2$ and with orthogonal polarizations $\vec{e}$ and $\vec{e}\,'$. A pair can thus consist of two photons with the same frequency, one emitted along I_1 with polarization $\vec{e}$, and one emitted along I_2 with polarization $\vec{e}\,'$. This pair can consist as well of two photons with the same frequency, one emitted along I_1 with polarization $\vec{e}\,'$ and one emitted along I_2 with polarization $\vec{e}$. The state of the pair is thus a linear superposition of $|I_1, \vec{e}\rangle \otimes |I_2, \vec{e}\,'\rangle$ and $|I_1, \vec{e}\,'\rangle \otimes |I_2, \vec{e}\rangle$. The interest of such entangled pairs is that the two photons are emitted in well defined directions.

19.5 Entanglement and interference

Entangled states give rise in general to interference phenomena. We have already seen in Sec. 19.2 that in a double counting experiment, the entanglement between two field modes produced by a first photodetection gives rise to interference fringes in a second photodetection. In this section, we discuss an even simpler example of the connection between entanglement and interference. We consider two identical atoms A_1 and A_2, with an excited state e and a ground state g. We suppose that they are in the entangled state:

$$|\psi\rangle = c_a\,|e_1, g_2\rangle + c_b\,|g_1, e_2\rangle \tag{19.43}$$

which describes a linear superposition of two states where one of the two atoms is in e while the other is in g. We try to understand the spontaneous emission rate Γ from this ensemble of two atoms.

The rate $\Gamma(\vec{k})$ of spontaneous emission of a photon $\vec{k}$ is, within a multiplicative constant factor, given by

$$\Gamma(\vec{k}) = \langle\psi|\,(D_1^+ \mathrm{e}^{i\,\vec{k}\cdot\vec{R}_1} + D_2^+ \mathrm{e}^{i\,\vec{k}\cdot\vec{R}_2})(D_1^- \mathrm{e}^{-i\,\vec{k}\cdot\vec{R}_1} + D_2^- \mathrm{e}^{-i\,\vec{k}\cdot\vec{R}_2})\,|\psi\rangle \tag{19.44}$$

where $\vec{R}_1$ and $\vec{R}_2$ are the positions of the centers of mass of the two atoms and where D_i^- is the lowering component of the dipole moment operator of atom i given by

$$D_i^- = |g_i\rangle\,\langle e_i| = \left(D_i^+\right)^\dagger . \tag{19.45}$$

We will suppose that the distance between the two atoms is small compared to the wavelength $\lambda = 2\pi/k$ of the radiation that they emit when going from e to g. The exponential terms of Eq. (19.44) can then be considered as equal and factored out, so that the spontaneous emission rate appears to be proportional to:

$$\begin{aligned}\Gamma &\simeq \langle\psi|\left(D_1^+ + D_2^+\right)\left(D_1^- + D_2^-\right)|\psi\rangle \\ &= \langle\psi|\,D_1^+ D_1^-\,|\psi\rangle + \langle\psi|\,D_2^+ D_2^-\,|\psi\rangle + 2\mathrm{Re}\,\langle\psi|\,D_1^+ D_2^-\,|\psi\rangle .\end{aligned} \tag{19.46}$$

The first two terms of the right-hand side of Eq. (19.46) are the average values of observables involving only one atom, and they could therefore be calculated using the reduced density operators of each atom. Using the expression (19.43) of $|\psi\rangle$, they are easily found to be equal to $|c_a|^2|D_{ge}|^2$ and $|c_b|^2|D_{ge}|^2$, respectively, where $D_{ge} = \langle g|D|e\rangle$ is the dipole matrix element. Their physical interpretation is clear. They represent the spontaneous emission rates of each atom considered alone, weighted by the probability to find this atom in the excited state. The last term involves an operator $D_1^+ D_2^-$ which acts on both atoms and is therefore sensitive to the correlations that exist between them. Its value is found to be equal to $2|D_{ge}|^2\mathrm{Re}\,(c_a^* c_b)$. It is sensitive to the coherence $c_a^* c_b$ between the two coefficients of the linear superposition (19.43). Summing all the previous results, one finds that the spontaneous emission rate Π is proportional to:

$$\Gamma \propto |D_{ge}|^2 \left(|c_a|^2 + |c_b|^2 + 2\mathrm{Re}\,(c_a^* c_b) \right). \tag{19.47}$$

It is interesting to consider two particular cases: the case of a symmetric linear combination of $|e_1, g_2\rangle$ and $|e_2, g_1\rangle$ ($c_a = c_b = 1/\sqrt{2}$), and the case of an antisymmetric one ($c_a = 1/\sqrt{2}, c_b = -1/\sqrt{2}$). We then get from Eq. (19.47):

$$|\psi\rangle = \frac{|e_1, g_2\rangle + |g_1, e_2\rangle}{\sqrt{2}} \qquad \Rightarrow \qquad \Gamma \propto 2|D_{ge}|^2,$$

$$|\psi\rangle = \frac{|e_1, g_2\rangle - |g_1, e_2\rangle}{\sqrt{2}} \qquad \Rightarrow \qquad \Gamma = 0. \tag{19.48}$$

For a statistical mixture of the two states with equal weights, one would get:

$$\text{Statistical mixture with equal weights} \qquad \Rightarrow \qquad \Gamma \propto |D_{ge}|^2. \tag{19.49}$$

The emission rate is twice as strong for the symmetric state than for the statistical mixture and is called a *superradiant state* [Dicke (1954)]. The antisymmetric state does not radiate, and is called a *subradiant state*. It clearly appears that the entanglement between the two atoms leads to a constructive interference in the first case, and to a destructive interference in the second case.

One can gain deeper physical insight into the correlations which are responsible for this constructive or destructive interference. Since the state (19.43) does not contain a linear superposition of e_1 and g_1 and also of e_2 and g_2, the mean dipole moments of each atom are equal to zero. One can consider that these dipoles oscillate with a random phase. The square terms $\langle D_1^+ D_1^- \rangle$ and $\langle D_2^+ D_2^- \rangle$ represent the average values of the square of these dipoles. The cross term $\langle D_1^+ D_2^- \rangle$ has the same sign as the square terms for the symmetric state. This means that the dipole moments oscillate in phase for the two atoms. The phase of the oscillation is random, but it is the same for each atom, so that the field radiated is larger as is the emission rate. On the contrary, for the antisymmetric state, the sign of $\langle D_1^+ D_2^- \rangle$

is opposite to the sign of the square terms. The dipole moments of the two atoms oscillate out of phase, the radiated field vanishes, and there is no emission.

19.6 Entanglement and non-separability

19.6.1 *The Einstein-Podolsky-Rosen (EPR) argument [Einstein* et al. *(1935)]*

Consider two spins 1/2 in a singlet state (19.15) and suppose that they are very far apart. This singlet state is a maximally entangled state. The reduced density operator of each spin is a multiple of the unit matrix describing a statistical mixture with equal weights of the two $|\pm\rangle$ states in any direction. All quantum information is related to correlations described by the entangled state (19.15).

Suppose that an observer A measures the component S_{1z} of spin 1 and finds $+1/2$. Then, it is sure that an observer B measuring the component S_{2z} of spin 2 will find $-1/2$ because of the correlations described by Eq. (19.15). This correlation between distant measurements led Albert Einstein, Boris Podolsky and Nathan Rosen to develop the following argument.[9] The measurement of spin 1 by A who is very far cannot influence spin 2. If one is sure of the result that B will find when measuring S_{2z}, this means that this value of S_{2z} was pre-existing before the measurement performed by B, even if B does not perform the measurement.[10] This pre-existing value of S_{2z} is what they call an *element of reality*. Attributing an element of reality to a localized entity, such as the value of S_{2z} for the spin 2 very far from the spin 1 is also known as *local realism*.

The problem then raised by EPR is that observer A could as well decide to measure S_{1x}. He would find $\pm 1/2$ and the result of the measurement of S_{2x} by B would give with certainty the opposite value, because the form of the singlet state is the same in all basis (invariance by rotation). The same reasoning as above would lead to the conclusion that the value of S_{2x} is also a pre-existing element of reality. How can one then reconcile this prediction with the fact that, in quantum mechanics, S_z and S_x are non-commuting observables that cannot have simultaneously well defined values in a quantum state?

The conclusion of EPR is that all possible correlated pairs in the singlet state cannot be considered as fully described by a single quantum mechanical wave function like (19.15). Their argument shows that the description of each pair must involve an extra random parameter, sometimes called a "hidden variable", which characterizes the correlations between the two spins in the past, at the moment

---

[9]In fact, EPR were not considering in their paper an entangled state of two spins 1/2 in the singlet state, but a pair of two particles in an entangled position momentum state [Einstein *et al.* (1935)]. It was David Bohm who introduced entangled states involving only spin degrees of freedom which allowed a simpler presentation of the EPR argument [Bohm (1951)].

[10]This point of view that considers an absence of measurement is called "counterfactual".

when this pair was created.[11] To measure the correlations between the results of the measurements of two arbitrary components of $\vec{S}_1$ by A and $\vec{S}_2$ by B, these two observers must repeat the experiment several times with pairs prepared in the same conditions and make averages of the product of the results that they get. Would it be possible to reproduce the quantum predictions derived from the wave function (19.15) by using another description, based on a random parameter λ, that varies from one pair to the next, and fixes the results that they obtain?

19.6.2 *Bell's inequalities*

Niels Bohr did not agree with the EPR argument. He maintained the standard quantum point of view, according to which, an observable cannot be considered as having a pre-existing value until it has been measured [Bohr (1935)]. Einstein was not claiming that the predictions of quantum mechanics were wrong, but was hoping that it would be possible to find a different description that gave the same predictions as quantum mechanics while remaining compatible with the idea of local realism. The debate continued for some time at a philosophical level, and both points of view seemed acceptable.

Great progress was made in 1964 when John Bell proposed inequalities that would allow one to decide unambiguously if both points of view were acceptable [Bell (1964)]. Assuming the existence of hidden variables, along with the assumption of local realism, he derived inequalities satisfied by the predictions concerning the correlations between the results of certain measurements performed by two observers on a pair of two distant systems. The conclusion was that these inequalities were, in certain cases, violated by the predictions of quantum mechanics. It thus appeared that both points of view were irreconcilable and that it could be possible to discriminate between them with an experiment.

John Clauser, Michael Horne, Abner Shimony and Richard Holt derived a Bell type inequality (CHSH inequality) for the radiative cascade $J_a = 0 \to J_b = 1 \to J_c = 0$ described in Sec. 19.4.4 [Clauser *et al.* (1969)]. They suppose that A and B measure the linear polarizations of the photons ω_1 and ω_2 in the plane xy perpendicular to the z-axis along which the two photons propagate in opposite directions. Both observers repeat their measurement several times on photon pairs emitted in the same conditions, with A sometimes measuring the polarization with a two-output analyzer oriented along a direction $\vec{a}$ to detect if the polarization is parallel or perpendicular to $\vec{a}$, and sometimes with an analyzer oriented along a direction $\vec{a}\,'$. Similarly, B uses an analyzer oriented sometimes along $\vec{b}$, and sometimes along

[11] One can give an example of correlations described by a random parameter chosen in the past. One person C takes two balls, one white and one black. He tosses a coin to choose one of the two balls, for example the black one, and he sends it to A in a box, and the other one to B also in a box. The observer A does not know the tossing made by C. He knows only that he has equal chances to receive a white ball or a black one. But, when he opens his box and finds the black ball, he knows immediately that B will find a white ball when he opens his box.

$\vec{b}'$. The hidden parameter λ has a normalized probability distribution $\rho(\lambda)$. For each value of λ, the result of the measurement by A of the polarization along $\vec{a}$ is equal to $A(\lambda, \vec{a})$. The value of $A(\lambda, \vec{a})$ is $+1$ if the polarization of the photon ω_1 is along $\vec{a}$, and -1 if the polarization is along the perpendicular direction. In a similar way, one can define the quantities $A(\lambda, \vec{a}')$, $B(\lambda, \vec{b})$, $B(\lambda, \vec{b}')$. The important point is that the result $A(\lambda, \vec{a})$ depends only on $\vec{a}$, and not on the direction of the other distant analyzer used by B. This expresses the idea of local realism, according to which the value of $A(\lambda, \vec{a})$ is an element of reality, with a well defined value for each value of λ,[12] for the photon analyzed by A. Using these assumptions one can then derive with simple algebra[13] the following inequality:

$$-2 \leq E(\vec{a}, \vec{b}) - E(\vec{a}, \vec{b}') + E(\vec{a}', \vec{b}) + E(\vec{a}', \vec{b}') \leq +2 \tag{19.50}$$

where

$$E(\vec{a}, \vec{b}) = \langle A(\lambda, \vec{a})B(\lambda, \vec{b})\rangle = \int \mathrm{d}\lambda\, \rho(\lambda)\, A(\lambda, \vec{a})B(\lambda, \vec{b}) \tag{19.51}$$

is the average value of the product of the results found by A and B when they use the analyzers along $\vec{a}$ and $\vec{b}$.

Now, the same calculations made with the quantum formalism and using the entangled state (19.15) give

$$E(\vec{a}, \vec{b}) - E(\vec{a}, \vec{b}') + E(\vec{a}', \vec{b}) + E(\vec{a}', \vec{b}') = \vec{a} \cdot \vec{b} - \vec{a} \cdot \vec{b}' + \vec{a}' \cdot \vec{b} + \vec{a}' \cdot \vec{b}'. \tag{19.52}$$

For some orientations of $\vec{a}$, $\vec{a}'$, $\vec{b}$, $\vec{b}'$, one finds that the quantity written in (19.52) can reach values equal to $2\sqrt{2}$, clearly violating the inequality (19.50).

19.6.3 *Experimental results and conclusion*

The generalization of the Bell theorem in the form of the CHSH inequality [Clauser *et al.* (1969)] contains a concrete experimental proposal inspired by the work of Carl Kocher and Eugene Commins that reported the polarization correlation of photon pairs emitted in a mercury cascade between three levels [Kocher and Commins (1967)].

In 1972, Stuart Freedman and John Clauser realized the first experiment following this proposal and using a $J_a = 0 \to J_b = 1 \to J_c = 0$ cascade in calcium atoms. They reported an experimental confirmation of the quantum predictions with a violation of the Bell's inequality by six standard deviations [Freedman and Clauser (1972)]. Two other experiments with correlated emitted photons in a similar cascade scheme were carried out in 1976 by John Clauser [Clauser (1976)], and by Edward Fry and Randall Thompson [Fry and Thompson (1976)] respectively.

[12] Of course the correlations between the results obtained by A and B can be taken into account in this description. If $\vec{a}$ and $\vec{b}$ are parallel, for certain values of λ, both $A(\lambda, \vec{a})$ and $B(\lambda, \vec{b})$ will be equal to $+1$, for other values of λ, they will be both equal to -1. This reproduces the correlations contained in the entangled state (19.15).

[13] See for example [Eberhard (1978)].

Both experiments agreed with quantum mechanics predictions. In these pioneering experiments, the detection relied on one-channel analyzers that gave a signal (equal to +1) only when the photon had a polarization parallel to the direction of the analyzer.[14]

In early eighties, another series of experiments were performed in the group of Alain Aspect using the same cascade of calcium atoms as in [Freedman and Clauser (1972)] with a certain number of significant improvements:

- The upper state of the cascade was selectively populated with a two-photon laser excitation [Aspect *et al.* (1980b)]. In this way, the coincidence rate was improved by several orders of magnitude compared to the very first experiments.
- The distance between the two detectors was large compared to the spatial extent of the wave packet due to the finite lifetime of the excited states. The two detections were therefore space-like separated [Aspect *et al.* (1981)].
- Two-channel analyzers were used to avoid an incomplete measurement inherent to one-channel polarizers as used in previous experiments [Aspect *et al.* (1982a)].
- In each channel, a rapid switch of the polarization that is measured is performed with acousto-optic modulators. The switching time is shorter than the travel time of the light between the two detectors of the two channels [Aspect *et al.* (1982b)].

In all these experiments, clear violations of Bell's inequalities, by more than 10 standard deviations, were obtained.

More recently, a great improvement of the experiment was achieved by using entangled pairs of photons obtained in parametric down conversion . As mentioned in Sec. 19.4.4, the two photons are emitted in well defined cones. At the two intersections of these two cones, the pairs of photons are described by states analogous to the state of two entangles spins. There are no solid angle limitations, and it is possible to inject the pair of photons into two fibers where they can propagate over large distances. In this way, violation of Bell's inequalities were demonstrated with a distance between the source and the measuring apparatus of more than four kilometres [Tapster *et al.* (1994); Tittel *et al.* (1998)].

In 1995, the group of Anton Zeilinger reported a violation of Bell's inequalities by 100 standard deviations in only 5 minutes, whereas all previous experiments required several hours of data recording [Kwiat *et al.* (1995)]. As in [Aspect *et al.* (1982b)], they have addressed the question of the possible exchange of signal between the detectors. However, instead of using an acousto-optic modulator that results in a periodic change of the detectors, they have used a complete random setting of the analyzers [Weihs *et al.* (1998)].

All these results give a clear confirmation of quantum mechanics and exclude any other interpretation in terms of hidden variables and local realism. It is now clear that, no matter how far they are, the two systems that appear in an entangled state

[14] See [Clauser and Shimony (1978)] and references therein.

cannot be considered to be separate entities. They form a single, non-separable entity. Such a situation characterizes what is called *quantum non-separability*.

GHZ and W states

Daniel Greenberger, Michael Horne, and Anton Zeilinger have shown that systems containing more than two correlated particles can lead to even more dramatic violations of local realism [Greenberger *et al.* (1989, 1990)]. Consider a system of three spins 1/2 in the following entangled state, referred to as the GHZ state:

$$|GHZ\rangle = \frac{1}{\sqrt{2}}(|1+, 2+, 3+\rangle - |1-, 2-, 3-\rangle) \tag{19.53}$$

where the $|\pm\rangle$ states are the eigenstates of the spins along the z-axis.

The GHZ states are eigenvectors of the following operators with the eigenvalue $+1$:

$$\begin{aligned} \sigma_{1x}\sigma_{2y}\sigma_{3y}|GHZ\rangle &= +|GHZ\rangle, \\ \sigma_{1y}\sigma_{2x}\sigma_{3y}|GHZ\rangle &= +|GHZ\rangle, \\ \sigma_{1y}\sigma_{2y}\sigma_{3x}|GHZ\rangle &= +|GHZ\rangle. \end{aligned} \tag{19.54}$$

Similarly, one finds that the state (19.53) is an eigenvector of the product operator $\sigma_{1x}\sigma_{2x}\sigma_{3x}$ with the eigenvalue -1:

$$\sigma_{1x}\sigma_{2x}\sigma_{3x}|GHZ\rangle = -|GHZ\rangle. \tag{19.55}$$

Therefore, if we consider three observers A, B and C far apart one from another that measure σ_x on spin 1 for A, 2 for B and 3 for C, the measurement of the product $\sigma_{1x}\sigma_{2x}\sigma_{3x}$ of the three x components of each spin gives -1 with a probability 1 if the inital tripartite state is described by the GHZ vector (19.53).

Arguments based on local realism, similar to those developed in Sec. 19.6.2, predict values for the results $A_x(\lambda) = \pm 1$ (resp. $A_y(\lambda) = \pm 1$) of the operator σ_x (resp. σ_y) measured on spin 1 by A, where λ accounts for the hidden parameters that are independent of the types of measurements performed by B and C. Similarly, the measurement on spin 2 (3) by B (C) of the component σ_i with $i = x, y$ is given by $B_i(\lambda) = \pm 1$ ($C_i(\lambda) = \pm 1$).

The local realism point of view should give the same result as the standard quantum point of view, which is

$$A_x(\lambda)B_y(\lambda)C_y(\lambda) = 1, \tag{19.56}$$
$$A_y(\lambda)B_x(\lambda)C_y(\lambda) = 1, \tag{19.57}$$
$$A_y(\lambda)B_y(\lambda)C_x(\lambda) = 1, \tag{19.58}$$
$$A_x(\lambda)B_x(\lambda)C_x(\lambda) = -1. \tag{19.59}$$

As $A_y^2 = B_y^2 = C_y^2 = (\pm 1)^2 = 1$, if we multiply together the first three equations, we find $A_x(\lambda)B_x(\lambda)C_x(\lambda) = 1$ which contradicts Eq. (19.59) [Mermin (1990); Laloë (2001)]! The contradiction with the EPR assumption of local realism is in this case even more dramatic than the violation of Bell inequalities since it occurs on the outcome of measurements that can be predicted with certainty. However, the experimental implementation requires the design of an apparatus that can measure the product of three spin components, which is a non-trivial task.

The state (19.53) can be generalized to N particles. Such a generalized GHZ state is nothing but a "Schrödinger cat" state, which is a superposition of two maximally different quantum states:

$$|\psi\rangle = \frac{1}{\sqrt{2}}(|1+, 2+, \ldots, N+\rangle - |1-, 2-, \ldots, N-\rangle). \qquad (19.60)$$

Such cat states are very sensitive to decoherence and are not robust against particle loss. Indeed, the measurement of any one of the N spin of the state (19.60) projects the system into a statistical mixture of $|1+, 2+, \ldots, N+\rangle$ and $|1-, 2-, \ldots, N-\rangle$ and therefore completely destroys the entanglement. This is to be contrasted with another class of maximally entangled states, the W states. For three spins this state reads

$$|W\rangle = \frac{1}{\sqrt{3}}(|1+, 2-, 3-\rangle + |1-, 2+, 3-\rangle + |1-, 2-, 3+\rangle). \qquad (19.61)$$

This W state can retain bipartite entanglement after the measurement of the state of one of the three spins. Indeed, one finds for the state (19.61) that when spin 1 is measured in state $|1-\rangle$, spins 2 and 3 are in the Bell state $(|2+, 3-\rangle + |2-, 3+\rangle)/\sqrt{2}$, (and when spin 1 is measured in the state $|1+\rangle$, spins 2 and 3 are in state $|2-, 3-\rangle$). GHZ and W states have been produced in nuclear magnetic resonance experiments [Teklemariam *et al.* (2002)], with photons [Eibl *et al.* (2004)] and with a few ions [Roos *et al.* (2004); Häffner *et al.* (2005); Leibfried *et al.* (2005)].

Intrication and separability for a statistical mixture

In the previous discussions, we have only considered the case where the global system $1 + 2$ is in a pure state. Intrication is then easy to define as a situation where this pure state can never be written in any basis as a product of a state of subsystem 1 by a state of subsystem 2. The two subsystems cannot then be considered as two separable entities and Bell's inequalities, based on local realism, are violated by the predictions of quantum mechanics. What happens if the density operator ρ_{12} of the global system is a statistical mixture?

The density operator is called *separable* if it can be written as a statistical mixture of product states $\rho_1^n \rho_2^n$ with weights π_n that are real positive numbers:

$$\rho_{12} = \sum_n \pi_n \rho_1^n \rho_2^n. \qquad (19.62)$$

In this case, separate measurements performed on each subsystem will always give results that are consistent with local realism and satisfy Bell's inequalities. Each subsystem has quantum properties but the correlations between them are classical. If the density operator can never been written as in Eq. (19.62), the global system is non-separable and the two subsystems are entangled.

As a matter of fact, it is not easy to show that a given density operator cannot be written in a separable form. However, a few simple criteria have been established. For example, Asher Peres has shown that a necessary condition for separability is that the matrix obtained from ρ_{12} by transposition of the indices of one of the two subsystems [15] has only non-negative eigenvalues [Peres (1996)]. This criterion is more sensitive than others using Bell's inequalities [Horodecki *et al.* (1995)].

[15] This new matrix is hermitian as ρ_{12} and has real eigenvalues.

19.7 Entanglement and which-path information

In Sec. 17.3.4 of Chap. 17, we discussed the possibility of determining the path followed by the particle in a Young's double slit experiment. Consider a particle flying from the source to the detection screen at the middle of the interference pattern. If the particle passes through the upper slit, its change of momentum is transferred to the double slit screen which gains this momentum in the upward direction. If the particle passes through the lower slit, the screen gains a momentum in the downward direction. So, one could think that by measuring the momentum of the screen, one could determine the path of the particle unambiguously. This conclusion is actually wrong because of the Heisenberg relations. To be able to determine the momentum δp transferred by the particle to the screen, one should know the initial momentum of the screen with an accuracy better than δp. But then, because of the uncertainty relations, the position spread δx of the screen cannot be ignored. A simple calculation shows that the interference pattern is then washed out.

This argument can be recast in a more general form, using the concept of entangled states. The initial state of the { atom+screen } system is a product state $|\psi_{\rm in}\rangle = |\varphi_A\rangle \otimes |E\rangle$ of an atomic state $|\varphi_A\rangle$ by a screen state $|E\rangle$. In the final state, the total system is in a linear superposition of two states: a first state corresponding to an atom in the state $|\varphi_A^{(1)}\rangle$, describing the state of the particle emerging from the upper slit 1 and going to the center of the detection screen, while the screen is in the state $|E_1\rangle$ equal to its initial state displaced by an amount δp; a second state corresponding to an atom in the state $|\varphi_A^{(2)}\rangle$ describing the state of the particle emerging from the lower slit 2 and going to the center of the detection screen, while the screen is in the state $|E_2\rangle$ equal to its initial state displaced by an amount $-\delta p$. The final state of the total system is thus an entangled state:

$$|\psi_{\rm fin}\rangle \propto |\varphi_A^{(1)}\rangle \otimes |E_1\rangle + |\varphi_A^{(2)}\rangle \otimes |E_2\rangle \tag{19.63}$$

that describes how each path of the atom is correlated with a state of another system which is used for determining this path.

The probability to find the atom at a point $\vec{r}$ of the detection screen is given by $\Pi(\vec{r}) = \langle\psi_{\rm fin}|\vec{r}\rangle\langle\vec{r}|\psi_{\rm fin}\rangle$, which can be also written, using Eq. (19.63):

$$\begin{aligned}\Pi(\vec{r}) = {} & |\varphi_A^{(1)}(\vec{r})|^2\langle E_1 \mid E_1\rangle + |\varphi_A^{(2)}(\vec{r})|^2\langle E_2 \mid E_2\rangle \\ & + 2\langle E_2|E_1\rangle \,\mathrm{Re}\left(\varphi_A^{(1)}(\vec{r})\varphi_A^{(2)*}(\vec{r})\right) .\end{aligned} \tag{19.64}$$

The interference fringes can come only from the last term, which contains the scalar product $\langle E_2|E_1\rangle$. If the screen is well localized in space, being for instance very massive, its momentum dispersion is relatively large, and the momentum imparted by the particle will have nearly no influence on its momentum. As a result, the scalar product will be very close to 1: $\langle E_1|E_2\rangle \simeq 1$. In the opposite limit, for which the state of the screen determines unambiguously the path of the atom, the two states E_1 and E_2 of the screen must be clearly distinct without any overlap. Their scalar product must therefore vanish, resulting in a disappearance of the interference term in Eq. (19.64).

This reasoning can be extended to any other quantum device that could be introduced for identifying the path of the atom. If the device is efficient, i.e. if its two states corresponding to the two possible paths are orthogonal, the interference fringes disappear. In the end, one cannot observe fringes and simultaneously know through which slit the atom has passed. The discussion of this section illustrates in precise terms the complementarity principle introduced by Niels Bohr stating that the wave nature and the corpuscular nature are complementary; i.e. they are only revealed with specific experimental arrangements which are not the same, depending upon if one wants to investigate the wave aspect or the corpuscular one.

19.8 Entanglement and the measurement process

19.8.1 *Von Neumann model of an ideal measurement process*

The microscopic system $\mathcal{S}$ to be measured is coupled to a measuring apparatus $\mathcal{M}$. Let $|\varphi_i\rangle$ be the eigenstates of a certain observable O of $\mathcal{S}$, with eigenvalues a_i. Suppose that $\mathcal{S}$ in the state $|\varphi_i\rangle$ interacts with $\mathcal{M}$ in the initial state $|\chi_0\rangle$. In the von Neumann model, one assumes that, after the interaction, $\mathcal{S}$ remains in the same state $|\varphi_i\rangle$ while $\mathcal{M}$ ends in the state $|\chi_i\rangle$ depending on $|\varphi_i\rangle$:

$$|\varphi_i\rangle \otimes |\chi_0\rangle \to |\varphi_i\rangle \otimes |\chi_i\rangle \,. \tag{19.65}$$

To each state of $\mathcal{S}$ corresponds a well defined state of $\mathcal{M}$. A measurement gives good results if the two final states $|\chi_i\rangle$ and $|\chi_j\rangle$ of $\mathcal{M}$, corresponding to two different states $|\varphi_i\rangle$ and $|\varphi_j\rangle$ of $\mathcal{S}$, are clearly distinguishable: $\langle\chi_i|\chi_j\rangle = \delta_{ij}$. For example the states $|\chi_i\rangle$ and $|\chi_j\rangle$ are two distinct positions of the extremity of a needle of $\mathcal{M}$.

The linearity of quantum mechanics then allows us to predict the final state $|\psi_{SM}\rangle$ of the whole system $\mathcal{S}+\mathcal{M}$ when $\mathcal{S}$ is initially in a linear superposition of the states $|\varphi_i\rangle$:

$$\left(\sum_i c_i\,|\varphi_i\rangle\right) \otimes |\chi_0\rangle \;\;\to\;\; |\psi_{SM}\rangle = \sum_i c_i\,|\varphi_i\rangle \otimes |\chi_i\rangle \tag{19.66}$$

This is clearly an entangled state. The probability of occupation $|c_i|^2$ of the state $|\varphi_i\rangle$ of S is unchanged. Looking at the state $|\chi_i\rangle$ of $\mathcal{M}$, one can deduce the value a_i of the observable O of $\mathcal{S}$. But, in addition, non-separable quantum correlations appear between $\mathcal{S}$ and $\mathcal{M}$. The entangled state (19.66) is a linear superposition of product states of $\mathcal{S}$ and $\mathcal{M}$, and not a statistical mixture of these states. In the next subsection, we discuss a few difficulties associated with this situation.

19.8.2 *Difficulty associated with macroscopic coherences*

The expression (19.66) of the total system implies the appearance, at the end of the measurement process, of linear superpositions of states where $\mathcal{M}$ is in macroscopically different states $|\chi_i\rangle$ and $|\chi_j\rangle$. This situation does not correspond to our usual perception of the macroscopic world. The strangeness of situations of this type is well illustrated by the example discussed by Erwin Schrödinger, where he considers a radioactive atom A^* put together with a cat in a box. If the atom decays to a lower state A, the emitted γ-ray triggers the release of a poison which kills the cat. During the decay, the atom is in a linear superposition of A^* and A, leading to the existence of a linear superposition of A^* in the presence of a living cat and A in the presence of a dead cat. The cat can be alive and dead *at the same time*.

19.8.3 *A possible solution: coupling of $\mathcal{M}$ with the environment*

For solving the difficulty raised by macroscopic coherences, a possible idea is to consider that $\mathcal{M}$ is not isolated, but in general coupled to a large environment $\mathcal{E}$ consisting of particles, photons, and various types of perturbing systems. One can then see if this coupling $\mathcal{M}$-$\mathcal{E}$ can prevent the appearance of macroscopic coherences and suggest a necessary condition for having a good measuring apparatus. The corresponding approach uses the concept of *decoherence*. We will not describe this approach in detail, but give only its general idea. More details may be found in the papers of Wojciech Zurek [Zurek (2003)] and in [Cohen-Tannoudji (1989); Haroche and Raimond (2006)].

As in the previous section, we focus only on two states $|\varphi_1\rangle$ and $|\varphi_2\rangle$ of $\mathcal{S}$ and the corresponding two states $|\chi_1\rangle$ and $|\chi_2\rangle$ of $\mathcal{M}$. We suppose that the coupling of $|\chi_1\rangle$ and $|\chi_2\rangle$ with the environment satisfies two important properties which make them appear as *pointer states* that any good measuring apparatus is expected to possess. We will then, in the next subsection, give a concrete example of such pointer states.

(i) We suppose first that the coupling $\mathcal{M}$-$\mathcal{E}$ does not change the occupation of $|\chi_1\rangle$ and $|\chi_2\rangle$. Mathematically, this amounts to supposing that the states $|\chi_i\rangle$ of $\mathcal{M}$ are eigenstates of the interaction Hamiltonian H_{ME}. If $\mathcal{M}$ is in the state $|\chi_1\rangle$ ($|\chi_2\rangle$), it remains in this state when it interacts with $\mathcal{E}$ initially in the state $|\xi_0\rangle$, and it becomes correlated with a state $|\xi_1\rangle$ ($|\xi_2\rangle$) of $\mathcal{E}$:

$$|\chi_1\rangle \otimes |\xi_0\rangle \to |\chi_1\rangle \otimes |\xi_1\rangle \qquad \text{and} \qquad |\chi_2\rangle \otimes |\xi_0\rangle \to |\chi_2\rangle \otimes |\xi_2\rangle \,. \tag{19.67}$$

(ii) The second assumption is that the scalar product $\langle\xi_1|\xi_2\rangle$ is very small: $|\langle\xi_1|\xi_2\rangle| \ll \langle\xi_1|\xi_1\rangle = \langle\xi_2|\xi_2\rangle = 1$.

Including the system $\mathcal{S}$ in the analysis, we conclude from the previous equations that if $\mathcal{S}$ is initially in the symmetric linear combination $(|\varphi_1\rangle + |\varphi_2\rangle)/\sqrt{2}$, the state of the total system $\mathcal{S} + \mathcal{M} + \mathcal{E}$ becomes equal to:

$$|\psi_{\text{SME}}\rangle = \frac{|\varphi_1\rangle \otimes |\chi_1\rangle \otimes |\xi_1\rangle + |\varphi_2\rangle \otimes |\chi_2\rangle \otimes |\xi_2\rangle}{\sqrt{2}} \,. \tag{19.68}$$

The important point to be noted is that the coupling H_{ME} between $\mathcal{M}$ and $\mathcal{E}$ does not alter the correlations between $\mathcal{S}$ and $\mathcal{M}$. The state $|\varphi_1\rangle$ remains correlated to $|\chi_1\rangle$ and the state $|\varphi_2\rangle$ remains correlated to $|\chi_2\rangle$.

We now come back to the difficulty associated with the linear superpositions of states of $\mathcal{S} + \mathcal{M}$. If one is interested only in $\mathcal{S} + \mathcal{M}$, one can use the reduced density operator of $\mathcal{S} + \mathcal{M}$ obtained by taking the trace over $\mathcal{E}$ of $|\psi_{\text{SME}}\rangle\langle\psi_{\text{SME}}|$. The off diagonal elements of the reduced density operator between the bipartite states $|\varphi_1\rangle \otimes |\chi_1\rangle$ and $|\varphi_2\rangle \otimes |\chi_2\rangle$ are multiplied by the scalar product $\langle\xi_2|\xi_1\rangle$, which is equal to zero according to our assumptions. The environment suppresses all interference effects between these two paths.

Finally, the conclusion of this discussion is that the existence of a basis of pointer states for $\mathcal{M}$, determined by the properties of its interaction with $\mathcal{E}$, is a necessary condition for having a good measuring apparatus. We illustrate in the next subsection these general considerations by giving a concrete example of pointer states that are stable when coupled to the environment and which can therefore keep the information about the measured result. By contrast, their linear superpositions are very rapidly transformed into statistical mixtures, which prevents the appearance of exotic macroscopic coherences.

19.8.4 *Simple example of pointer states*

We take the von Neumann model of a measuring apparatus $\mathcal{M}$ consisting of a large particle $\mathcal{P}$ coupled to the system to be measured by a Hamiltonian $H = g\hat{O}\hat{P}$ proportional to the observable $\hat{O}$ of the system and to the momentum $\hat{P}$ of $\mathcal{P}$. If one can ignore the free evolution of $\mathcal{S}$ and $\mathcal{P}$ during the interaction time, the effect of the interaction $\mathcal{S}$-$\mathcal{P}$ when $\mathcal{S}$ is in the eigenstate $|\varphi_i\rangle$ of $\hat{O}$ with eigenvalue a_i is to displace $\mathcal{P}$ by an amount proportional to a_i. One can in fact consider that the position of the big particle $\mathcal{P}$ is the needle of the apparatus measuring the value a_i of $\hat{O}$.

Let us assume that the environment in which $\mathcal{P}$ is immersed is a bath of light particles colliding with $\mathcal{P}$, which gives rise to a Brownian motion of $\mathcal{P}$. In Chap. 17, we have studied the damping of the spatial coherences of such a system. If $\mathcal{P}$ is in a linear superposition of two wave packets separated by a distance R, the coherence between the two wave packets decays with a decoherence rate larger than the velocity damping rate γ by the huge factor $(R/\lambda_{\text{dB}})^2$, where λ_{dB} is the de Broglie wavelength of $\mathcal{P}$. The mechanism of this damping is the momentum diffusion of the large particle, which erases the oscillation of its momentum distribution appearing with a period in p equal to $\pi\hbar/R$ when the large particle in a linear superposition of two wave packets separated by a distance R. When the density of the light particles forming the environment increases, the momentum diffusion coefficient D_p of the big particle increases and consequently so does the decoherence rate. It becomes more and more difficult to keep the needle of the measuring apparatus, i.e. the position of $\mathcal{P}$, in a linear superposition of two macroscopically distinguishable states. On the other hand, the spatial diffusion coefficient D_x of the big particle, which is inversely proportional to D_p according to the theory of Brownian motion, becomes smaller and smaller. The large particle is more and more stuck in the surrounding medium as this medium becomes denser. We thus have here a clear example of a situation in which the coupling of the measuring system with the environment does not appreciably change the probabilities of occupation of the pointer states, i.e. the positions of the needle, while at the same time it very rapidly destroys the coherences between two macroscopically distinguishable positions of the needle.

Another way of understanding these results is to note that most interactions between $\mathcal{P}$ and its environment are local, depending only on the position $\vec{R}$ of $\mathcal{P}$. If the free motion of $\mathcal{P}$ between two collisions could be neglected, $\vec{R}$ would be a constant of the motion and would not change such that the position state would play approximately the role of a pointer state. On the other hand, the environment is perturbed in a small region around the neighborhood of $\vec{R}$. If one considers two possible positions of $\mathcal{P}$ around $\vec{R}_1$ and $\vec{R}_2$, the two corresponding perturbed states of the environment become very rapidly orthogonal when $|\vec{R}_1 - \vec{R}_2|$ increases.

19.8.5 *The infinite chain of Von Neumann*

In fact, introducing the environment $\mathcal{E}$ in the discussion of the measurement process does not solve all the difficulties of a quantum description of this process. The coherences between two different states of $\mathcal{S}+\mathcal{M}$ are suppressed when one takes the trace of the total density matrix over the variables of $\mathcal{E}$. But Eq. (19.68) shows that we still have linear superpositions of states of $\mathcal{S}+\mathcal{M}+\mathcal{E}$, which are even more macroscopic than the states of $\mathcal{S}+\mathcal{M}$.

More generally, introducing more and more systems, $\mathcal{M}$ which is coupled to $\mathcal{E}$, $\mathcal{E}$ which is coupled to $\mathcal{E}'$, and so on, will never solve this difficulty. The whole system $\mathcal{S} + \mathcal{M} + \mathcal{E} + \mathcal{E}' + \cdots$ will be always in a linear superposition of states. This series of systems entangled with each other is called the von Neumann chain. How can one explain that, when one "observes" such a global system, one will find a well defined value a_i of the quantity which is measured, and not the others? How can one derive such a non-deterministic result of the observation from a deterministic Schrödinger equation describing the coupled evolution of a large number of entangled systems. The probability $|c_i|^2$ to find the result a_i can be readily predicted, but the fact that the observation selects a single result cannot be explained.

The Copenhagen interpretation of the quantum measurement process implies that we have to break somehow the von Neumann chain at a certain point. Before this point all the systems which are taken into account evolve according to the Schrödinger equation. But, after this point, one must consider that the observation is a classical process that selects, in a random way, one of the possible results with a certain probability that can be calculated from the previous Schrödinger equation. In other words, one cannot derive the observation of a single result from the Schrödinger equation applied to the observed system and to the ensemble of measuring apparatus. This difficulty raises a lot of questions. Where to put the point where the von Neumann chain has to be broken? In other words, where to locate the boundary between the quantum world and the classical one? Could it happen that quantum mechanics ceases to be valid at a certain macroscopic level and that it should be replaced by another theory explaining the random switching of the observing device to a certain result?

19.9 Conclusion

We will conclude this chapter with the following remark. Quantum concepts are not easy to grasp. Feynman said that "no one really understands quantum physics"! Entangled states illustrate in the most spectacular way the strangeness and the non-intuitive character of quantum mechanics, though they allow one to give a more precise meaning to concepts, such as non-local realism or complementarity, which are usually presented in a more qualitative and vague way. A very interesting feature of the evolution of this research field is that entangled states are not only useful for deepening our understanding of quantum physics. They can be experimentally produced, and gedanken experiments of the early days of quantum mechanics can now be realized in laboratoires.

We know of course that quantum mechanics is already present in practically all technical objects that we use now in our "daily lives": lasers, laptops, mobile phones, magnetic resonance imaging, etc. It was not obvious however that seemingly abstract concepts, such as entanglement, would find practical applications. As an example, quantum cryptography, based on the sharing of entangled pairs of photons, is already used as the most secure method for exchanging messages.

Entangled states are now at the heart of the development of a very promising field, called "quantum information" which exploits the quantum correlations of two entangled states for transmitting and processing information. Several groups in the world are trying to implement quantum gates which could be the building blocks of quantum computing machines of unprecedented performances. Of course, major difficulties have to be solved before reaching this goal, such as the reduction of decoherence processes to an acceptable level. But, one can reasonably hope to see an entirely new technology emerging from these investigations. Obviously, we cannot discuss all these developments in this book. We have chosen to present here an overall view of various features of quantum systems where entanglement plays an essential role.

PART 7
Degenerate quantum gases

Introduction

Quantum gases are presently one of the most active fields of atomic physics. The purpose of this part is, first, to give an historical review of the developments which led to the emergence of this field since the first prediction by Albert Einstein in 1924 of *Bose-Einstein Condensation* (BEC); second, to describe a few important properties of Bose-Einstein condensates which have been investigated both theoretically and experimentally since their first observation in 1995.

In Chap. 20, we begin by analyzing what is meant by quantum gas. The key parameter for this discussion is the thermal de Broglie wavelength λ_{T} of the atoms of the gas, which scales as $T^{-1/2}$. A first quantum regime appears when λ_{T} becomes larger than the range r_0 of the atom-atom interactions while remaining small compared to the mean distance between atoms. Collisions must then be treated quantum mechanically and interesting quantum effects due to the indistinguishability of the colliding atoms appear. We describe a few of these effects which have been investigated on helium-3 atoms polarized by optical pumping and on spin polarized hydrogen atoms: quenching of collisions between ultracold spin polarized fermions; rotation of the spins of the two colliding atoms around their vectorial sum, even in the absence of spin dependent interactions. Some of these effects continue to play an important role in ultracold atom physics.

The other important quantum regime is reached when λ_{T} becomes on the order or larger than the mean distance d between atoms. In fact, in 1924, Einstein considered a situation which corresponds to this regime. Extending to atoms the way of counting the number of microscopic states corresponding to a given macroscopic state, which was first introduced by Satyendra Bose for photons, Einstein predicted that, below a certain critical temperature T_c, a macroscopic number of atoms condenses in the ground state of the box which contains them. We briefly recall the calculation of Einstein and show that the condition $T < T_c$ is equivalent to $\lambda_{\mathrm{T}} > d$. However, the value of T_c predicted by Einstein for an ideal gas is extremely low. All gases known at that time would liquefy or even solidify at usual pressure at such low temperatures. Ignoring atom-atom interactions is therefore completely unrealistic and this is why Einstein's prediction was considered as purely academic. His calculation played however an important role by giving an exact model of phase transition allowing a clear discussion of the concept of thermodynamical limit and showing how correlations due to their indistinguishability can appear between atoms even if these atoms do not interact. In addition, Einstein's work was also invoked by Fritz London for interpreting the phase transition of liquid helium below 2.17 K.

Seven decades elapsed between the first prediction of gaseous Bose-Einstein condensation and its first observation on alkali atoms in 1995 by Eric Cornell, Carl Wieman and Wolfgang Ketterle. In Chap. 21, we describe a few important steps in this long quest for gaseous BEC. Spin polarized hydrogen atoms first appeared as the most promising system for observing this condensation since they were predicted to remain in the gas phase even at $T = 0$ K. We briefly review the pioneering works of the group of Daniel Kleppner and Tom Greytak at MIT, and the group of Jook Walraven and Ike Silvera at Amsterdam. We describe ingenious schemes which have been implemented during these studies, such as wall free confinement of neutral atoms and evaporative cooling. It turned out that it was not that easy to achieve BEC with hydrogen. With the development of laser cooling techniques, which are not easily applicable to hydrogen, alkali atoms can be cooled in the microkelvin range. Very efficient optical detection schemes can be also used for these atoms. It was finally on these atoms that BEC was first observed in 1995 on a trapped sample.[1] The condensate is actually in a metastable state living a time long enough for allowing its observation. We describe these first demonstrations of BEC on rubidium-87 and sodium-23. A few years later, BEC was finally observed on hydrogen, as well as on several other atomic species. In view of their importance for further investigations on pairs of fermionic atoms, we also briefly review experiments demonstrating quantum degeneracy in fermionic gases.

A very important issue in the study of gaseous condensates is the effect of atom-atom interactions. In Chap. 22, we address this problem within the mean-field approximation. We use a variational approach looking for the best approximation of the ground state of the N atoms system by a product of N identical 3D wave functions $\varphi(\vec{r})$. We find that the best wave function $\varphi(\vec{r})$ obeys a nonlinear Schrödinger-type equation, the Gross-Pitaevskii (GP) equation, describing the evolution of each atom in the mean field exerted on it by the $N-1$ other atoms. This equation provides a good description of the static properties of the condensate and the stability domain in the presence of attractive interactions. It also allows a characterization of the strength of the interactions by a single dimensionless parameter $\chi = Na/a_{\text{ho}}$, where a is the scattering length and a_{ho} is the width of the ground state wave function of the trapping harmonic potential. In the limit $\chi \gg 1$, called the Thomas-Fermi limit, the GP equation simplifies considerably and analytical expressions can be obtained for all relevant physical quantities. Other types of solutions of the GP equation in axisymmetric traps deserve attention such as vortex states where all atoms of the condensate have an angular momentum along the axis of the trap. We study them through the shape of the corresponding wave functions and we show that the circulation of the velocity field around the axis of the vortex is quantized.

In Chap. 22, we also introduce the time-dependent form of the GP equation which allows a quantitative analysis of the dynamical properties of the condensates, such as the ballistic expansion when the trapping potential is suddenly switched off or the normal modes of vibration. An interesting feature of the time-dependent GP equation is that it can be rewritten as a set of hydrodynamical equations for the

[1]Under standard condition of pressure, rubidium atoms are in a solid phase at such low temperatures.

spatial density of particles $\rho(\vec{r},t)$ and for the density current $\vec{j}(\vec{r},t)$. In the limit $N \gg 1$, the macroscopic occupation of the wave function $\varphi(\vec{r},t)$ by N identical bosons [2] and the hydrodynamic equations allow a classical re-interpretation of the wave function $\varphi(\vec{r},t)$ as a classical matter wave.[3]

There is a certain analogy between the solution $\varphi(\vec{r},t)$ of the GP equation in the classical limit ($N \gg 1$), and a classical Maxwell wave containing a macroscopic number of photons all in the same quantum state. But we know that, in optics, this wave description is not always valid, in particular for interpreting photon correlation signals. A correct description must quantize the electromagnetic field and use field operators whose correlation functions appear to be the good physical quantities for describing the coherence properties of the field. In Chap. 23, we show that the coherence properties of Bose-Einstein condensates can be analyzed with theoretical tools quite similar to those used in quantum optics. We introduce atomic field operators and field correlation functions, we calculate them in some particular cases and we show that there is a close analogy between thermal electromagnetic fields and thermal clouds of atoms on the one hand, and laser fields and Bose-Einstein condensates on the other hand. There are also important differences due to the fact that atoms directly interact contrary to photons. All these considerations are illustrated by the description of a few recent experiments on condensates.

In Chap. 23, we also address the important problem of the relative phase between two condensates 1 and 2. We show that, even if the two condensates are initially uncorrelated (their state is a product $|N_1\rangle \otimes |N_2\rangle$ of two Fock states), the detection of an atom, which can come either from condensate 1 or from condensate 2, puts them in an entangled state which describes correlations induced by the first detection. These correlations increase with the number of detections and one can understand in this way how a relative phase can appear between two initially independent condensates as a result of a succession of detection processes.

In Chap. 24, we present a first step beyond the mean-field treatment of the previous chapters where all atoms are supposed to be in the same state. We describe the Bogolubov approach which provides a better description of the ground state of the N atoms and of its first excited states which are the elementary excitations associated with the normal modes of vibrations of the condensate. In Chap. 22, we have shown using the time-dependent GP equation that the frequencies ω of these modes can be calculated and are discrete but the amplitudes, and thus the energies, of these waves can vary continuously, in the same way as the amplitude of a classical Maxwell wave can take all possible values. In fact, we know that the energy of a Maxwell wave becomes quantized in a quantum description of the field. The same situation occurs for a condensate. The energy of its elementary excitations is quantized and their energy spectrum can be calculated with the Bogolubov approach

---

[2] The normalization of φ is chosen such that $\int |\varphi|^2 \mathrm{d}^3 r = N$.

[3] This classical limit does not exist for fermionic atoms or for low values of the number N of bosonic atoms.

presented in Chap. 24 and which is valid for a dilute gas obeying the condition $na^3 \ll 1$ where n is the spatial density and a the scattering length.

The knowledge of the energy spectrum of elementary excitations is important for understanding the phenomenon of superfluidity. It can happen that the properties of this energy spectrum are such that an external perturbation cannot create an elementary excitation in a condensate below a certain threshold because the basic conservation laws could not be satisfied. The condition for the impossibility of such a transfer of excitation defines the Landau criterion and explains superfluid properties. First, we illustrate this idea with the example of a probe particle in an homogeneous condensate that can move with velocity v smaller than the sound velocity c without experiencing any friction. Such a spectacular effect would not exist in the absence of interactions. This shows that interactions are essential for superfluidity. The second example concerns a condensate in a bucket rotating at the angular frequency Ω. We show that if Ω is smaller than a certain critical angular velocity Ω_c, no elementary excitation can be created by the bucket. The condensate remains at rest and does not acquire any angular momentum. We also show that there is another critical angular velocity Ω_v above which the state with one vortex becomes the ground state of the system in the rotating frame. We discuss the various regimes which can be observed depending on the value of Ω compared to Ω_c and Ω_v. Gaseous condensates are very interesting for an experimental investigation of these various regimes. We describe experiments that have provided a better understanding of the mechanisms of nucleation of vortices.

Of course, it would be exciting to study the strong interaction regimes where na^3 is larger than 1 and where the Bogolubov approach is no longer sufficient. Feshbach resonances give the possibility to study these regimes experimentally by choosing the static field near the center of the resonance where $|a|$ diverges. An increasing number of theoretical and experimental works is presently devoted to these problems and reviewing them is outside the scope of this book. We will just mention in Chap. 26 a few examples of studies showing the originality of the atomic physics approach and explaining how gaseous condensates could bring a better understanding of strongly interacting systems.

Chapter 20

Emergence of quantum effects in a gas

20.1 Introduction

One of the most spectacular achievements of atomic physics during the last few years was the observation in 1995 of Bose-Einstein condensation (BEC) in an atomic gas [Anderson *et al.* (1995); Davis *et al.* (1995a)]. With laser cooling and trapping techniques, combined with evaporative cooling, it is possible to reach a regime of high phase space densities where, as predicted by Albert Einstein in 1925, a macroscopic number of atoms condenses in the ground state of the box which contains the gas. They then form what is called a *condensate* with fascinating macroscopic properties.

In the next chapter, we will describe the difficulties that had to be surmounted before BEC could be observed in a gas, 70 years after Einstein's prediction. In Sec. 20.2, we would like to show that when the temperature of the gas is lowered interesting quantum phenomena can be observed before the critical temperature for BEC is reached. These phenomena are due to indistinguishability effects in binary collisions between identical atoms and give rise to quantum features in the macroscopic properties of the gas: collisions can be quenched if the two colliding atoms are identical fermions in the same spin state; the orientations of the spins of two colliding atoms can change even in the absence of any spin-spin interactions as the two spins rotate around their vector sum. In Sec. 20.2, we will review a few of these effects and show how they can explain recently observed phenomena in ultracold gases.

Before describing the search for BEC in a gas, it will be also useful to briefly recall how Albert Einstein was led to the prediction of this effect in an ideal gas and how Bose-Einstein and Fermi-Dirac statistics were established in the early days of quantum mechanics. In Sec. 20.3, we will present the history of these developments and interpret the critical temperature derived by Einstein for BEC in terms of the thermal de Broglie wavelength λ_T of the atoms that was introduced at the same time by Louis de Broglie: BEC appears when λ_T becomes on the order of or larger than the mean distance between atoms.

Various characteristic lengths in a gas

To understand the hierarchy of the quantum effects that can be observed in a gas when the temperature is lowered, it is useful to consider the various characteristic lengths of the gas and their relative orders of magnitude. Three lengths play an important role:

- The mean distance d between atoms, related to the spatial density n by:

$$d = n^{-1/3}, \tag{20.1}$$

- The range r_0 of atom-atom interactions, which can also be characterized by the scattering length a at low enough temperatures;
- The thermal de Broglie wavelength λ_{T} given by:

$$\lambda_{\mathrm{T}} = \sqrt{\frac{2\pi\hbar^2}{mk_B T}}. \tag{20.2}$$

In a gas phase, d is much larger than r_0, since atoms interact only during collisions:

$$d \gg r_0. \tag{20.3}$$

When the temperature T is lowered, λ_{T} increases and three different regimes can be considered, depending on the value of λ_{T} compared to r_0 and d.

(1) *Classical regime:* $\lambda_{\mathrm{T}} \ll r_0 \ll d$

The atomic wave packets have a very small coherence length. Collisions involve many partial waves and can be described in terms of classical trajectories. The equilibrium populations of the energy levels of the gas can also be described by a classical Maxwell-Boltzmann distribution.

(2) *Quantum collisions:* $r_0 \ll \lambda_{\mathrm{T}} \ll d$

Collisions can no longer be treated classically. In this low energy limit, only s-waves (p-waves) are to be taken into account for identical bosons (fermions). In particular, quantum interference effects due to the indistinguishability of the colliding atoms can modify collisional cross-sections and change the directions of the spins in a coherent way even in the absence of spin-dependent interactions. The transport properties of the gas are modified, even if the equilibrium populations of the energy levels of the gas are still given by a classical Maxwell-Boltzmann distribution, since $\lambda_{\mathrm{T}} \ll d$.

(3) *Quantum degeneracy:* $r_0 \ll d \leq \lambda_{\mathrm{T}}$

The gas becomes degenerate when $d \leq \lambda_{\mathrm{T}}$. All degrees of freedom of the gas must be described quantum mechanically.

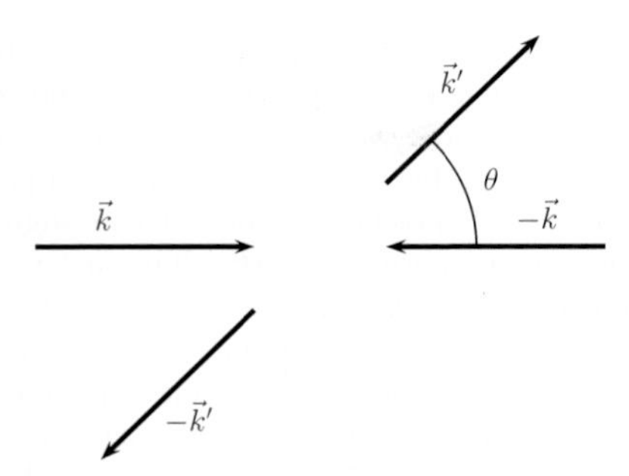

Fig. 20.1 Initial and final states of a scattering of two spinless particles in the center of mass frame. The interaction potential is chosen to be spherically symmetric, so that the transition amplitude depends only on the angle θ between $\vec{k}$ and $\vec{k}'$ (and not on the azimuthal angle φ in the plane perpendicular to $\vec{k}$).

20.2 Quantum effects in collisions

20.2.1 *S-matrix and T-matrix*

For studying the effect of a collision on the spin degrees of freedom of two identical colliding atoms, it will be simpler to use an *S-matrix* description of the collision [1] rather than the asymptotic form of the scattering wave function.

First, consider spinless particles. In the center of mass system, the S-matrix gives the transition amplitude $S_{\vec{k}'\vec{k}} = \langle \vec{k}', -\vec{k}' | S | \vec{k}, -\vec{k} \rangle$ that the system of two atoms starting from an initial state $|\vec{k}, -\vec{k}\,\rangle$, representing the two particles arriving in the collision zone with momenta $\hbar\vec{k}$ and $-\hbar\vec{k}$ ends in a final state $|\vec{k}', -\vec{k}'\rangle$, in which they fly apart from one another with momenta $\hbar\vec{k}'$ and $-\hbar\vec{k}'$ (see Fig. 20.1).

One can show that

$$S_{\vec{k}'\vec{k}} = \delta_{\vec{k}'\vec{k}} - 2\pi i \delta_{E'E} T_{\vec{k}'\vec{k}} \tag{20.4}$$

where the so-called *transition-matrix* $T_{\vec{k}'\vec{k}}$ is proportional to the usual scattering amplitude $f(\theta)$.

$$T_{\vec{k}'\vec{k}} = -Cf\,(\theta) \tag{20.5}$$

where C is a real and positive coefficient. In Eq. (20.4), $\delta_{E'E}$ expresses the conservation of energy.[2]

Remark

It is interesting to write the final two-particle wave function $\psi(\vec{r}_1, \vec{r}_2)$ of the system starting with the initial state wave function $\exp[i\vec{k}\cdot(\vec{r}_1 - \vec{r}_2)]$ that describes the initial state $|\vec{k}, -\vec{k}\rangle$:

$$\psi\,(\vec{r}_1 - \vec{r}_2) = e^{i\vec{k}\cdot(\vec{r}_1 - \vec{r}_2)} - 2\pi i \sum_{\vec{k}' \text{ with } E'=E} T_{\vec{k}'\vec{k}} e^{i\vec{k}'\cdot(\vec{r}_1 - \vec{r}_2)}. \tag{20.6}$$

[1] See Chap. XIX of [Messiah (2003)], or Chap. 4 of [Roman (1965)].

[2] In this section, we use discrete Kronecker symbols $\delta_{\vec{k}'\vec{k}}$ and $\delta_{E'E}$ with discrete indices rather than delta functions. This allows a simpler presentation. For a more rigorous treatment using wave packets, see [Lhuillier and Laloë (1982a,b)].

In the center of mass system, $\vec{r}_1 - \vec{r}_2$ is the position $\vec{r}$ of the relative particle. If we compare Eq. (20.6) with the usual asymptotic form of the scattering wave function,[3] we see that the outgoing spherical wave $f(\theta)e^{ikr}/r$ that appears in this asymptotic form can also be considered to be a sum of counter-propagating plane waves with wave vectors $\vec{k}'$ and $-\vec{k}'$ for all possible values of the angle θ, with each pair being multiplied by a coefficient proportional to $f(\theta)$. The last term on the right-hand side of (20.6) can thus be considered to be the Fourier transform of the spherical outgoing wave of the asymptotic scattering wave function. It is also interesting to note the factor i in Eq. (20.6) multiplying the T-matrix element and introducing a $\pi/2$ phase shift between the incident plane wave and the wave scattered in the forward direction, an effect already discussed in Chap. 15, using a Gaussian beam approach.

20.2.2 *Interfering scattering amplitudes for identical particles*

(i) *Scattering in a lateral direction*

When the two colliding particles are identical, the scattering amplitudes corresponding to the scattering angles θ and $\pi - \theta$ (and to the azimuthal angles φ and $\varphi + \pi$) lead to the same physical final state for the external degrees of freedom: one particle in the direction $\vec{k}'$ and one particle in the direction $-\vec{k}'$. If, as a first step, we ignore the spin quantum numbers, the two scattering amplitudes $f(\theta)$ and $f(\pi - \theta)$ corresponding to the two scattering processes represented in Fig. 20.2(a), interfere. For lateral scattering ($\theta \neq 0$), we have:

$$\langle \vec{k}', -\vec{k}' | S | \vec{k}, -\vec{k} \rangle \propto f(\theta) + \varepsilon f(\pi - \theta). \tag{20.7}$$

where $\varepsilon = +1$ if the particles are bosons, and $\varepsilon = -1$ if the particles are fermions.

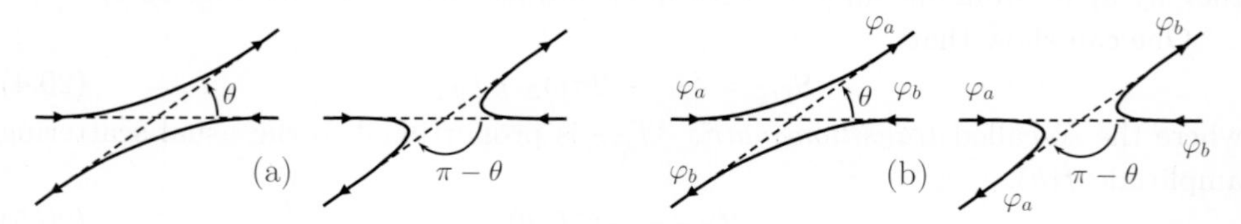

Fig. 20.2 (a) Interfering scattering amplitudes without spin quantum numbers. (b) Interfering scattering amplitudes with spin quantum numbers. If there are no spin-dependent interactions, the spin quantum numbers remain unchanged along each trajectory that represents the scattering path of each particle.

We now include the spin quantum numbers. In the initial state, the particle arriving with the momentum $\hbar\vec{k}$ has a spin state φ_a, and the particle arriving with the momentum $-\hbar\vec{k}$ has a spin state φ_b. We suppose that all interactions between the two particles are spin-independent. This is, for example, the case for two helium-3 atoms in the ground state. The electronic angular momentum J is equal to zero and the spin degrees of freedom are purely nuclear: $I = 1/2, m_I = \pm 1/2$. The magnetic interactions between the two nuclear spins are very small and

[3]See Sec. 15.2.1 of Chap. 15.

can be ignored. The spin quantum numbers thus remain unchanged along each particle's scattering trajectory (see Fig. 20.2(b)). If $\varphi_a = \varphi_b$, the two final states of Fig. 20.2(b) are the same and the two scattering amplitudes interfere as for Fig. 20.2(a). If φ_a and φ_b are orthogonal, the two final states are orthogonal and the two amplitudes no longer interfere. The spin quantum numbers φ_a and φ_b then appear as labels which make the two particles distinguishable. A very interesting situation that we will study in more detail in Sec. 20.2.4 appears when φ_a and φ_b are neither equal nor orthogonal. We will show that in this case the interference of the two scattering amplitudes gives rise to a change of the spin orientations after the collision, even if the two spins do not interact.

In the general case, the two scattering amplitudes represented in Fig. 20.2(b) must be added with the sign ε. One concludes that if we start from the initial state $|\vec{k}, \varphi_a; -\vec{k}, \varphi_b\rangle$, the system ends in the final state:

$$|\psi_{\text{fin}}\rangle \propto f(\theta)|\vec{k}', \varphi_a; -\vec{k}', \varphi_b\rangle + \varepsilon f(\pi - \theta)|\vec{k}', \varphi_b; -\vec{k}', \varphi_a\rangle. \tag{20.8}$$

More rigorous derivation

So far, we have only considered physical states not labelled with numbers 1 or 2 for the two identical particles. Normally, one should use numbers for labelling the particles and then symmetrize or antisymmetrize the state of the system. For example, the initial state considered in this section should be written:

$$\begin{aligned} |\psi_{\text{in}}(1,2)\rangle &= \frac{1+\varepsilon P_{12}}{\sqrt{2}}|1, \vec{k}, \varphi_a; 2, -\vec{k}, \varphi_b\rangle, \\ &= \frac{|1, \vec{k}, \varphi_a; 2, -\vec{k}, \varphi_b\rangle + \varepsilon|2, \vec{k}, \varphi_a; 1, -\vec{k}, \varphi_b\rangle}{\sqrt{2}}, \end{aligned} \tag{20.9}$$

where P_{12} is the permutation operator of the two particles. The final state $|\psi_{\text{fin}}(1,2)\rangle$ corresponding to one particle in the direction $\vec{k}'$ and one particle in the direction $-\vec{k}'$ is given by the action, on this initial state, of $S_\theta + S_{\pi-\theta}$, where S_θ is the component of the S-matrix corresponding to a scattering angle θ and a similar definition for $S_{\pi-\theta}$: For the lateral scattering represented in Fig. 20.2 with a scattering angle $\theta \neq 0$, we get:

$$S_\theta|\psi_{\text{in}}(1,2)\rangle \propto f(\theta)\left[|1, \vec{k}', \varphi_a; 2, -\vec{k}', \varphi_b\rangle + \varepsilon|2, \vec{k}', \varphi_a; 1, -\vec{k}', \varphi_b\rangle\right]. \tag{20.10}$$

$$\begin{aligned} S_{\pi-\theta}|\psi_{\text{in}}(1,2)\rangle &\propto f(\pi-\theta)\left[|1, -\vec{k}', \varphi_a; 2, \vec{k}', \varphi_b\rangle + \varepsilon|2, -\vec{k}', \varphi_a; 1, \vec{k}', \varphi_b\rangle\right] \\ &= \varepsilon f(\pi-\theta)\left[\varepsilon|1, -\vec{k}', \varphi_a; 2, \vec{k}', \varphi_b\rangle + |2, -\vec{k}', \varphi_a; 1, \vec{k}', \varphi_b\rangle\right] \end{aligned} \tag{20.11}$$

since $\varepsilon^2 = 1$. If we add Eqs. (20.10) and (20.11), we get a sum of two correctly symmetrized or antisymmetrized kets describing the physical state (20.8).

(ii) *Scattering in the forward direction*

For the discussions of the next sections, it will be useful to also consider the final state corresponding to forward scattering, that is, for a final state where one of the

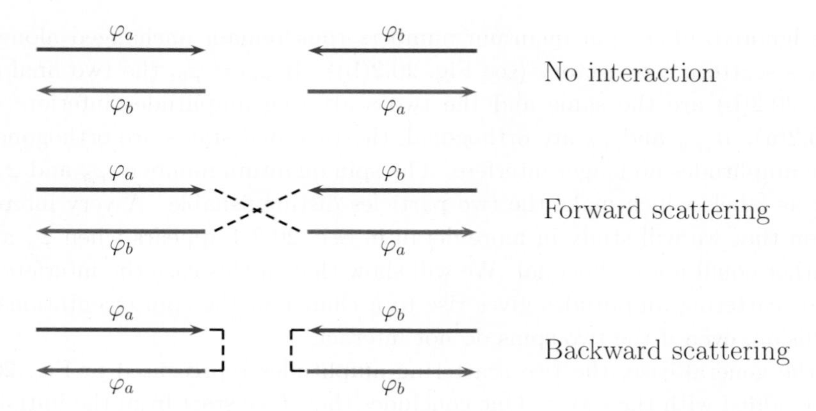

Fig. 20.3 Three different paths leading from the initial state $|\vec{k}, \varphi_a; -\vec{k}, \varphi_b\rangle$ to a final state where one of the particles ends in the state $\vec{k}$. The initial state is represented in blue, the final state in red. Three possibilities exist: no interaction, forward scattering and backward scattering transforming $|-\vec{k}, \varphi_b\rangle$ into $|\vec{k}, \varphi_b\rangle$.

particles ends in the state $\vec{k}$. Figure 20.3 shows the three possibilities leading to such a final state. For each of them, the initial physical state of the two particles is represented in blue, and the final state in red. The first possibility is that the two particles do not interact (term $\delta_{\vec{k}'\vec{k}}$ of (20.4) equal to 1 if $\vec{k} = \vec{k}'$). The second possibility is that the particle entering in the state $\vec{k}, \varphi_a$ undergoes a forward scattering. The third possibility is that the particle entering in the state $-\vec{k}, \varphi_b$ undergoes a backward scattering and ends in the state $\vec{k}, \varphi_b$. The physical final state can be written:

$$|\psi_{\text{fin}}\rangle \propto |\vec{k}, \varphi_a; -\vec{k}, \varphi_b\rangle + 2i\pi C[f(0)|\vec{k}, \varphi_a; -\vec{k}, \varphi_b\rangle + \varepsilon f(\pi)|\vec{k}, \varphi_b; -\vec{k}, \varphi_a\rangle]. \quad (20.12)$$

(iii) *From collision cross-sections to transport coefficients*

Equations (20.8) and (20.12) give the final state of a collision in a lateral direction and in the forward direction respectively, and include the interference effects due to the indistinguishability of the colliding atoms. Collision cross-sections can be deduced from these equations and used for the calculation of the collision term of a Boltzmann equation that describes the transport properties in a spin polarized gas. We will not give here the details of these calculations, which can be found in [Lhuillier and Laloë (1982a,b)].[4] As a result, one obtains the expression for various transport coefficients such as the heat conduction and the viscosity. The spin polarization appears as a new variable that enters the gas dynamics and gives rise to new effects that we describe in the next subsections.

[4]In this work, the mean free path is assumed to be small enough compared to the box size so that the hydrodynamic formalism can be applied. The first correction to ideal hydrodynamics are calculated using the Chapman-Enskog approximation which amounts to linearizing the collision operator that enters the Boltzmann equation about the hydrodynamic local equilibrium solution.

20.2.3 *Polarized Fermi gas at low temperature*

Consider a polarized Fermi gas. All atoms are in the same spin state. In all figures of the previous subsection, $\varphi_a = \varphi_b$. The final states in the two scattering processes of Fig. 20.2(b) are the same and the two scattering amplitude interfere to give a total amplitude proportional to $f(\theta) - f(\pi - \theta)$ since $\varepsilon = -1$ for fermions. At very low temperatures, $f(\theta)$, which tends to $-a$, where a is the scattering length, no longer depends on θ, so the total scattering amplitude vanishes. The particles no longer interact and the polarized Fermi gas behaves as an ideal gas.

Another way of deriving this result is to describe the collision through a partial wave expansion. Since the two colliding atoms are in the same spin state, they form a totally symmetric spin state so that the orbital wave function must be odd under a permutation of the two particles, or equivalently under a space reflection. It follows that the two particles can collide only in an ℓ-wave with ℓ odd. But, it is only in the s-wave ($\ell = 0$) that the two particles can approach at a distance sufficiently small to feel their interaction potential. In a p-wave ($\ell = 1$), their closest distance of approach is larger than the de Broglie wavelength, which in this section is taken to be much larger than the range r_0 of atom-atom interactions.

If the two colliding atoms are in orthogonal spin states, the final states in the two scattering processes of Fig. 20.2(b) are orthogonal. The two amplitudes can no longer interfere and the gas behaves as a mixture of two distinguishable species with non zero collision cross sections only for atoms in non orthogonal spin states. So, by just changing the spin polarization, for example by optical pumping, one can control atom-atom interactions and even quench them completely if the polarization is total. It must be emphasized once more that this quenching of interactions due to the spin polarization is not due to spin-dependent interactions, which are completely negligible for nuclear spins. It is due to a quantum interference between scattering amplitudes that result from the indistinguishability of identical particles and the anti-symmetrization postulate.

The control of interactions by spin polarization has dramatic consequences for the transport properties of a gas in the hydrodynamic regime. The thermal conductivity and the viscosity, which are determined by the collision cross sections, vary with the polarization of the gas. For example, if the gas is contained between two plates, and one of the two plates is heated, the temperature difference between them will be set by the transfer of heat by atoms from the hot plate to the cold plate. If collisions are rare, atoms will travel more rapidly from one plate to the other and the temperature difference is expected to be smaller than when collisions are frequent. This effect has been observed [Leduc *et al.* (1987)].

20.2.4 *Interference effects in forward and backward scattering*

In the previous subsection, we described interference effects between scattering amplitudes in a lateral direction. We show now that interesting new effects can appear as a result of quantum interference involving the incident waves. More precisely, we consider the various scattering processes represented in Fig. 20.3 in which there is, in the final state, a particle with the same value of $\vec{k}$ as one of the incident particles. If $\varphi_a = \varphi_b$, the forward and backward scattering waves both interfere with the incident wave. If φ_a and φ_b are orthogonal, only the forward scattering wave interferes with the incident wave, which gives rise to the attenuation of the incident wave, as explained in Chap. 15. What happens if φ_a and φ_b are different but not orthogonal?

To answer this question, we will try to determine the new spin state of the particle that emerges from the collision with a wave vector $\vec{k}$ and which was initially in the spin state φ_a. A similar question can be asked for the particle emerging from the collision with a wave vector $-\vec{k}$ which was initially in the spin state φ_b. We will call the direction of $+\vec{k}$ "right" and denote it $\rightarrow$, and the direction of $-\vec{k}$ "left" and denote it $\leftarrow$.

Since the final state (20.12) is not a product state, we must use a density matrix description for the final state of the two particles and take the trace over the variables of the particle with a wave vector $-\vec{k}$ when we are interested in the particle emerging with a wave vector $+\vec{k}$, and vice versa. If

$$\sigma_{\rightarrow}^{\text{in}} = |\vec{k}\rangle\langle\vec{k}| \otimes |\varphi_a\rangle\langle\varphi_a| \tag{20.13}$$

is the initial density operator (including both external and spin degrees of freedom) of the atom going to the right, the final density operator of the atom going to the right will be given by:

$$\sigma_{\rightarrow}^{\text{fin}} = \underset{\leftarrow}{\text{Tr}}\ |\psi_{\text{fin}}\rangle\ \langle\psi_{\text{fin}}| \tag{20.14}$$

where $|\psi_{\text{fin}}\rangle$ is given in (20.12) and where $\underset{\leftarrow}{\text{Tr}}$ means trace over the variables of the particle emerging to the left with a wave vector $-\vec{k}$.

In this calculation, we will keep only the terms of order 0 and 1 in the scattering amplitudes $f(0)$ and $f(\pi)$, neglecting the smaller terms of order 2. We briefly outline the results of this calculation for $\sigma_{\rightarrow}^{\text{fin}}$.

- *Order 0 in the scattering amplitudes*
  Using

$$\underset{\leftarrow}{\text{Tr}}\,|\vec{k},\varphi_a;-\vec{k},\varphi_b\rangle\langle\vec{k},\varphi_a;-\vec{k},\varphi_b| = |\vec{k},\varphi_a\rangle\langle\vec{k},\varphi_a|(\underset{\leftarrow}{\text{Tr}}\,|-\vec{k},\varphi_b\rangle\langle-\vec{k},\varphi_b|) \tag{20.15}$$

$$\text{and}\quad \underset{\leftarrow}{\text{Tr}}\,|-\vec{k},\varphi_b\rangle\langle-\vec{k},\varphi_b| = \langle-\vec{k},\varphi_b\,|-\vec{k},\varphi_b\rangle = 1, \tag{20.16}$$

  we get

$$\left(\sigma_{\rightarrow}^{\text{fin}}\right)_{\text{order}\,0} = |\vec{k}\rangle\langle\vec{k}| \otimes |\varphi_a\rangle\langle\varphi_a| \;=\; \sigma_{\rightarrow}^{\text{in}}. \tag{20.17}$$

- *Order 1 in $f(0)$*
  We have to calculate

$$\left(\sigma_{\rightarrow}^{\text{fin}}\right)_{\substack{\text{order1}\\ \text{in}f(0)}} = 2i\pi C f\,(0)\ \underset{\leftarrow}{\text{Tr}}\,|\vec{k},\varphi_a;-\vec{k},\varphi_b\rangle\langle\vec{k},\varphi_a;-\vec{k},\varphi_b| + \text{h.c.} \tag{20.18}$$

The trace is the same as in Eq. (20.15), which gives

$$\left(\sigma_{\rightarrow}^{\text{fin}}\right)_{\substack{\text{order 1}\\ \text{in } f(0)}} = -4\pi C\,\text{Im}f\,(0)\,|\vec{k}\rangle\langle\vec{k}|\,\otimes|\varphi_a\rangle\langle\varphi_a| \;=\; -4\pi C\,\text{Im}f\,(0)\;\sigma_{\rightarrow}^{\text{in}}. \quad (20.19)$$

- *Order 1 in $f(\pi)$*

 The equivalent of Eq. (20.18) is now

$$\left(\sigma_{\rightarrow}^{\text{fin}}\right)_{\substack{\text{order 1}\\ \text{in } f(\pi)}} = 2i\pi\varepsilon Cf\,(\pi)\,\underset{\leftarrow}{\text{Tr}}\,|\vec{k},\varphi_b;-\vec{k},\varphi_a\rangle\langle\vec{k},\varphi_a;-\vec{k},\varphi_b| + \text{h.c.} \qquad (20.20)$$

 Using the two equations

$$\underset{\leftarrow}{\text{Tr}}\,|\vec{k},\varphi_b;-\vec{k},\varphi_a\rangle\langle\vec{k},\varphi_a;-\vec{k},\varphi_b| = |\vec{k},\varphi_b\rangle\langle\vec{k},\varphi_a|\left(\underset{\leftarrow}{\text{Tr}}\,|-\vec{k},\varphi_a\rangle\langle-\vec{k},\varphi_b|\right)$$

$$\text{and}\quad \underset{\leftarrow}{\text{Tr}}\,|-\vec{k},\varphi_a\rangle\langle-\vec{k},\varphi_b| = \langle-\vec{k},\varphi_b|-\vec{k},\varphi_a\rangle = \langle\varphi_b|\varphi_a\rangle, \qquad (20.21)$$

 we transform Eq. (20.20) into

$$\left(\sigma_{\rightarrow}^{\text{fin}}\right)_{\substack{\text{order 1}\\ \text{in } f(\pi)}} = 2i\pi\varepsilon Cf\,(\pi)\,|\vec{k}\rangle\langle\vec{k}|\,\otimes\left[|\varphi_b\rangle\,\langle\varphi_a|\left(\langle\varphi_b\mid\varphi_a\rangle\right)\right] + \text{h.c.}. \qquad (20.22)$$

 If we insert the c-number $(\langle\varphi_b|\varphi_a\rangle)$ between $|\varphi_b\rangle$ and $\langle\varphi_a|$, we get

$$|\varphi_b\rangle\,\langle\varphi_a|\left(\langle\varphi_b\mid\varphi_a\rangle\right) = |\varphi_b\rangle\,\langle\varphi_b\mid\varphi_a\rangle\,\langle\varphi_a| = \rho_b^{\text{in}}\rho_a^{\text{in}} \qquad (20.23)$$

 where

$$\rho_a^{\text{in}} = |\varphi_a\rangle\,\langle\varphi_a| \qquad \text{and} \qquad \rho_b^{\text{in}} = |\varphi_b\rangle\,\langle\varphi_b| \qquad (20.24)$$

 are the initial spin density operators of the two atoms.[5] Finally, introducing the real and imaginary parts of $f(\pi)$ leads to

$$\left(\sigma_{\rightarrow}^{\text{fin}}\right)_{\substack{\text{order 1}\\ \text{in } f(\pi)}} = -2\pi\varepsilon C\,|\vec{k}\rangle\langle\vec{k}|\otimes\left\{\text{Im}f\,(\pi)\left[\rho_b^{\text{in}},\rho_a^{\text{in}}\right]_+ - i\text{Re}f\,(\pi)\left[\rho_b^{\text{in}},\rho_a^{\text{in}}\right]\right\} \qquad (20.25)$$

 where the notation $[A,B]_+$ means the anticommutator $AB+BA$.

If we collect all the spin density operators multiplying $|\vec{k}\rangle\langle\vec{k}|$ in Eqs. (20.17), (20.19) and (20.25), we get for the change of the spin state of the atom emerging in the $+\vec{k}$ direction:

$$\begin{aligned}\rho_a^{\text{fin}} - \rho_a^{\text{in}} &= -4\pi C\text{Im}f\,(0)\,\rho_a^{\text{in}}\\ &\quad - 2\pi\varepsilon C\left\{\text{Im}f\,(\pi)\left[\rho_b^{\text{in}},\rho_a^{\text{in}}\right]_+ - i\text{Re}f\,(\pi)\left[\rho_b^{\text{in}},\rho_a^{\text{in}}\right]\right\}\end{aligned} \qquad (20.26)$$

and a similar result for the spin state of the atom emerging in the $-\vec{k}$ direction by interchanging in (20.26) a and b.

The terms of Eq. (20.26) that are proportional to $\text{Im}f(0)$ involve interference between the incident wave and the forward scattered wave. They already exist for distinguishable atoms and do not change the spin state because they only depend on ρ_a^{in}. They describe the attenuation of the incident wave.

[5] We use the greek letter ρ for the spin density operators and the greek letter σ for the density operators including all external and spin degrees of freedom. In Eq. (20.23), ρ_a^{in} and ρ_b^{in} act in the same space, the spin space associated with the atom which emerges in the right direction $\rightarrow$.

The new terms resulting from the indistinguishability of the colliding atoms involve the interference between the incident wave and the back-scattered wave. These are proportional to the real and imaginary parts of $f(\pi)$ and are multiplied by $\varepsilon = +1$ for bosons and $\varepsilon = -1$ for fermions. They depend on both ρ_a^{in} and ρ_b^{in}, which means that the atom moving in one direction has a spin state that depends on the spin state of the atom moving in the opposite direction with which it has collided. Even though everything happens as if the two spins interact during the collision, this change of the spin states is not due to an spin-spin interaction, but rather to a quantum interference associated with their indistinguishability.

20.2.5 *Identical spin rotation effect (ISRE)*

The new terms of Eq. (20.26) that are due to the indistinguishability of the two atoms appear as a sum of a commutator and an anticommutator between ρ_a^{in} and ρ_b^{in}. We first analyze the physical content of the term proportional to the commutator. We will consider the simple case where the two atoms have a spin 1/2. In the initial density operators of the two colliding atoms $\sigma_{\rightarrow}^{\text{in}} = |\vec{k}\rangle\langle\vec{k}| \otimes \rho_a^{\text{in}}$ and $\sigma_{\leftarrow}^{\text{in}} = |-\vec{k}\rangle\langle-\vec{k}| \otimes \rho_b^{\text{in}}$, we can use

$$\rho_a^{\text{in}} = \frac{1}{2}(1 + \vec{a} \cdot \vec{\sigma}) \qquad \text{and} \qquad \rho_b^{\text{in}} = \frac{1}{2}(1 + \vec{b} \cdot \vec{\sigma}) \tag{20.27}$$

where the three components of $\vec{\sigma}$ are the Pauli matrices, and where $\vec{a}$ and $\vec{b}$ are the so-called *Bloch vectors* that describe the spin states of the two atoms. If ρ_a^{in} and ρ_b^{in} describe pure states (projectors onto $|\varphi_a\rangle$ and $|\varphi_b\rangle$), the two Bloch vectors have a unit length and $\vec{a} \cdot \vec{\sigma}/2$ and $\vec{b} \cdot \vec{\sigma}/2$ are the components of the spin along the direction of $\vec{a}$ and $\vec{b}$, respectively. If the state of one spin is a statistical mixture, the corresponding Bloch vector has a length smaller than 1. In all cases, Eq. (20.27) describe a spin that is totally polarized (for a pure case) or partially polarized (for a statistical mixture) along the direction of the Bloch vector.

It will be useful to use the following identity resulting from the commutation relations of the Pauli matrices:

$$(\vec{\sigma} \cdot \vec{a})(\vec{\sigma} \cdot \vec{b}) = \vec{a} \cdot \vec{b}\,\mathbb{1} + i\vec{\sigma} \cdot \left(\vec{a} \times \vec{b}\right) \tag{20.28}$$

where $\mathbb{1}$ is the unit operator. Using this equation, it is straightforward to get the following expression for the commutator of the two initial spin density operators:

$$\left[\rho_a^{\text{in}}, \rho_b^{\text{in}}\right] = \frac{1}{2} i\vec{\sigma}. \left(\vec{a} \times \vec{b}\right) = -\left[\rho_b^{\text{in}}, \rho_a^{\text{in}}\right]. \tag{20.29}$$

Inserting (20.29) into (20.26), we get the following expressions for the change $\delta\rho_a$ of ρ_a^{in} coming from the commutator term:

$$\delta\rho_a = \rho_a^{\text{fin}} - \rho_a^{\text{in}} = \frac{1}{2}\vec{\sigma} \cdot \delta\vec{a} = -\pi\varepsilon C \,\text{Re} f(\pi)\, \vec{\sigma} \cdot \left(\vec{a} \times \vec{b}\right) = -\delta\rho_b = -\frac{1}{2}\vec{\sigma} \cdot \delta\vec{b} \tag{20.30}$$

from which we deduce:

$$\delta\vec{a} = -\delta\vec{b} = -2\pi\varepsilon C \,\text{Re} f(\pi)\, \vec{a} \times \vec{b}. \tag{20.31}$$

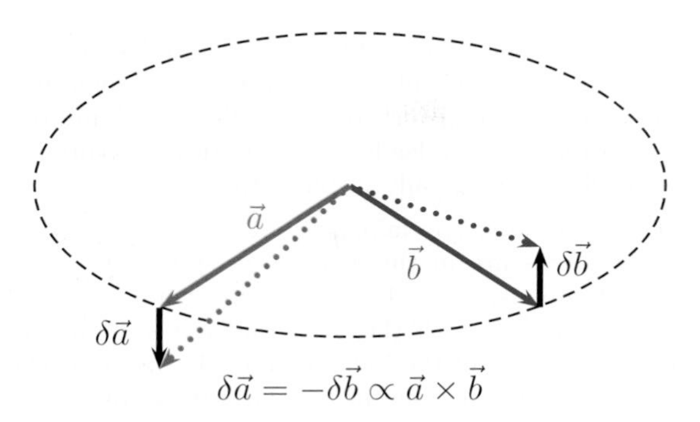

Fig. 20.4 Change of the Bloch vectors $\vec{a}$ and $\vec{b}$ of the two spins associated with the commutator of Eq. (20.26). The changes $\delta\vec{a}$ and $\delta\vec{b}$ are opposite and perpendicular to $\vec{a}$ and $\vec{b}$ and give rise to a rotation of the two Bloch vectors around their resultant.

Figure 20.4 shows the corresponding changes of the Bloch vectors $\vec{a}$ and $\vec{b}$. We suppose that the two spins are in pure states, so that $\vec{a}$ and $\vec{b}$ have equal unit lengths. The two Bloch vectors are changed by small opposite amounts $\delta\vec{a}$ and $\delta\vec{b}$ perpendicular to the plane containing $\vec{a}$ and $\vec{b}$. The final Bloch vectors represented by dotted lines are therefore derived from the initial ones, represented in full lines, by a rotation around their vector sum $\vec{a} + \vec{b}$. The same result holds if one or both of the two Bloch vectors are in a statistical mixture. The rotations experienced by a spin in the state ρ_a^{in} propagating along the direction $+\vec{k}$ in a gas containing spins in the state ρ_b^{in} add up collision by collision to give rise to large rotation angles. This effect is analogous to the Faraday rotation[6] of the polarization of a photon propagating in a polarized atomic medium, though the physical origin here is the indistinguishability of the two spins. This is why it is called *identical spin rotation effect* (ISRE). The evolution due to ISRE is unitary: the length of the Bloch vectors do not change during the rotation. This can also be understood by noting that, according to (20.26), the evolution of ρ_a^{in} in a series of collisions appears as an evolution due to an effective Hamiltonian proportional to ρ_b^{in}, or to $\rho_b^{\text{in}} + \rho_a^{\text{in}}$, since ρ_a^{in} commutes with itself. Indeed, $\rho_b^{\text{in}} + \rho_a^{\text{in}}$ is a constant of motion since the interaction Hamiltonian commutes with the total spin (rotational invariance). The two spins thus rotates around a fixed direction.

Finally, let us note that ISRE does not exist when the two atoms collide in the same spin state: $\vec{a}$ and $\vec{b}$ are then parallel and $\vec{a} \times \vec{b}$ vanishes. The same result holds if the two spins are in orthogonal states described by anti-parallel Bloch vectors.

Contribution of the anti-commutator of (20.26)

The identity (20.28) can also be used to express the anti-commutator of ρ_a^{in} and ρ_b^{in} that appears in Eq. (20.26) in terms of the Bloch vectors $\vec{a}$ and $\vec{b}$. The terms proportional

[6] However, here and in contrast with Faraday effect, there is no external magnetic field so that the time reversal symmetry remains valid.

to $\vec{a}\times\vec{b}$ cancel out, so that the changes $\delta\vec{a}$ and $\delta\vec{b}$ of $\vec{a}$ and $\vec{b}$ are in the plane containing $\vec{a}$ and $\vec{b}$. The evolution due to the anti-commutator is no longer unitary since it cannot be reproduced by an evolution governed by an effective Hamiltonian. Consequently, taking the trace over one particle leads to a statistical mixture for the other particle even if the two particles were initially in pure states.

Let us just mention here a few special cases. At very low temperatures, $f(0) = f(\pi)$. If the two colliding atoms are in the same pure spin state, $\rho_a^{\rm in}\rho_b^{\rm in} = (\rho_a^{\rm in})^2 = \rho_a^{\rm in}$. Since spins 1/2 are fermions, $\varepsilon = -1$, and the anti-commutator that appears in the second line of (20.26) cancels with the term that appears on the right-hand side of the first line. The absorption of the incident wave is thus quenched. We recover the quenching of atom-atom interactions in a polarized Fermi gas at low temperatures already mentioned in Sec. 20.2.3. If the two spins are in orthogonal pure states, $\rho_a^{\rm in}\rho_b^{\rm in} = 0$ and the contribution of the anti-commutator vanishes.

20.2.6 *A few examples of effects involving ISRE*

(i) *Spin waves*

Spatial inhomogeneities of the spin directions in a gas are generally washed out by diffusion processes. When atoms are identical, ISRE can give rise to propagating (rather than diffusive) modes of the spin direction called *spin waves*. An example of such mode is shown in Fig. 20.5. The spin polarization is the sum of a large uniform z-component and a small transverse component whose azimuthal angle in the plane perpendicular to the z-axis varies linearly with x.

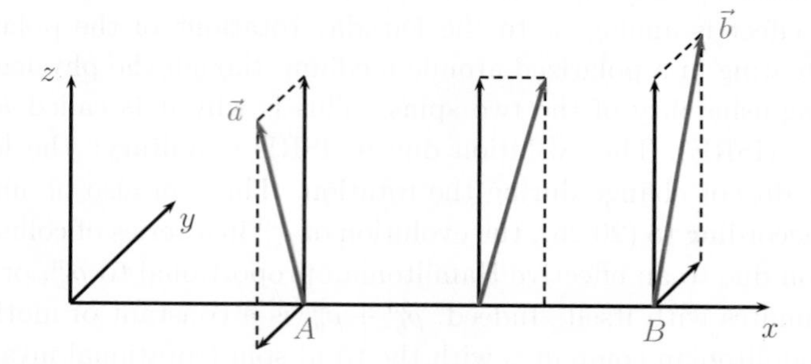

Fig. 20.5 Example of spin wave. The z-component of the spin polarization is uniform while the much smaller transverse component has an azimuthal angle in the plane perpendicular to the z-axis that varies linearly with x.

A qualitative picture can be given for understanding the role of ISRE in the appearance of propagating modes. Consider two points A and B along the x-axis of Fig. 20.5 where the local transverse magnetizations are parallel and anti-parallel to the y-axis, respectively, and two atoms at A and B, with Bloch vectors $\vec{a}$ and $\vec{b}$. Suppose that these two atoms collide along the x-axis, with the atom initially at A arriving at B, and the atom initially at B arriving at A. In the absence of

ISRE, the atom arriving at B brings a transverse magnetization opposite to the local magnetization. The same situation occurs for the atom arriving at A. It is easy to understand how this diffusive process damps the transverse magnetization. In the presence of ISRE, the spin arriving at B will have rotated around the z-axis, and its magnetization will no longer be opposite to the local magnetization. If the rotation angle is equal to π, it will even reinforce the local magnetization. The same applies to the atom starting from B and arriving at A.

We now give a brief outline of a more quantitative approach based on the concept of magnetization currents [Lhuillier and Laloë (1982a,b)]. In the absence of ISRE, a spatial gradient of transverse magnetization $\vec{M}_\perp$ gives rise to a transverse magnetization current $\vec{J}(\vec{M}_\perp)$ in a direction opposite to the gradient of $\vec{M}_\perp$:

$$\vec{J}(M_{\perp,i}) = -\alpha\vec{\nabla}M_{\perp,i}, \tag{20.32}$$

where α is a coefficient depending on the collision cross-sections without ISRE and $i = x$ or y. In the presence of ISRE, an additional current appears that is still proportional to minus the gradient of $\vec{M}_\perp$, but with a rotation around the large longitudinal magnetization M_z, which is taken to be uniform. A second current must then be added to (20.32)

$$\vec{J}(M_{\perp,i}) = -\beta M_z\vec{e}_z \times \vec{\nabla}M_{\perp,i}, \tag{20.33}$$

where β is a coefficient characterizing the importance of IRSE and where $\vec{e}_z$ is the unit vector along the z-axis. Adding the right-hand sides of (20.32) and (20.33) gives the following equation for $\vec{J}(M_+)$, where $M_+ = M_x + iM_y$:

$$\vec{J}(M_+) = \vec{J}(M_x) + i\vec{J}(M_y) = -(\alpha + i\beta M_z)\vec{\nabla}M_+. \tag{20.34}$$

On the other hand, conservation of the total spin gives the continuity equation:

$$\frac{\partial M_+}{\partial t} + \vec{\nabla}\cdot\vec{J}(M_+) = 0. \tag{20.35}$$

Eliminating $\vec{J}(M_+)$ between (20.34) and (20.35) finally gives:

$$\frac{\partial M_+}{\partial t} = (\alpha + i\beta M_z)\Delta M_+ \tag{20.36}$$

which admits solutions of the form $\exp i(\omega t - kz)$ with the following dispersion relation:

$$\omega = \beta M_z k^2 - i\alpha k^2. \tag{20.37}$$

We get a spin wave with a frequency $\omega = \beta M_z k^2$ determined by ISRE, and a damping rate $\gamma = \alpha k^2$ determined by the purely diffusive process. The quadratic dependence of the frequency with the square of the wave vector is analogous to the one of a free non-relativistic particle of effective mass $m^* = \hbar/(2\beta M_z)$. The two coefficients, α and β, can be calculated from the quantum transport equations derived in [Lhuillier and Laloë (1982a,b)]. One finds regimes of temperatures where the frequency of the wave is larger than its damping rate, which justifies the name of spin waves. Such waves have been observed in polarized hydrogen [Johnson *et al* (1984)] and in polarized helium-3 [Nacher *et al* (1984); Tastevin *et al* (1985)].

(ii) *Spin-state segregation in an ultracold trapped gas*

An experiment performed in the group of Eric Cornell at JILA in 2001 showed a curious phenomenon that they called spin-state segregation because ultracold trapped rubidium-87 atoms in different Zeeman sublevels were observed to occupy different spatial domains in the trap [Lewandowski *et al* (2002); McGuirk *et al* (2002)]. The trap was axially symmetric and elongated along the x-axis, and the temperature was higher than the critical temperature for Bose-Einstein condensation, so that the gas was non degenerate. A $\pi/2$ RF pulse put the atoms in a linear superposition of two ground state sublevels g_1 and g_2 of the ground state. The system evolved freely and a state selective optical detection gave the variations with x of the spatial densities n_1 and n_2 of atoms in g_1 and g_2, respectively. At certain times, they observed that n_1 was large near the center of the trap, while n_2 took appreciable values at the left and right extremities of the trap. After a certain time, the situation reversed: n_2 is large at the center and n_1 at the extremities. And so on. During these oscillations, $n_1 + n_2$ remained constant at each point x.

This spatial segregation was not due to a differential Stern and Gerlach effect which is too weak. In fact, it was due to ISRE. Here we will give only a qualitative interpretation. After the $\pi/2$ pulse, the fictitious spins 1/2 associated with the two-level systems $\{g_1, g_2\}$ point along the y-axis, which is perpendicular to the x-axis of the trap, and the z-axis, along which the static magnetic field B_0 is applied (see Fig. 20.6). Initially, all the spins are parallel in the xy-plane, but then tend to spread out in this plane due to the spread of Larmor frequencies resulting from spatial variations of B_0 along the x-axis. Two different atoms coming from two different points of the trap and colliding thus have different orientations of their spins so that ISRE can take place. In this rotation, each spin develops a non-zero component of its z-component, with opposite signs for the two spins, corresponding to a larger admixture of g_1 for one spin and of g_2 for the other spin. More precisely, one expects that ISRE can give rise to a standing spin wave along the x-axis where the z-component of the spin varies as $\cos(kx)\sin(\omega t)$. When $\sin(\omega t)$ is positive, S_z is positive at the center of the trap (where $\cos(kx) > 0$), giving $n_1 > n_2$ at the center while n_1 is smaller than n_2 outside the center zone, where $\cos(kx) < 0$. After half a period π/ω of the spin wave, $\sin(\omega t)$ becomes negative and the conclusions are reversed. A quantitative study of this effect can be found in [Fuchs *et al* (2002); Oktel and Levitov (2002); Williams *et al.* (2002); Fuchs *et al* (2003)].

(iii) *Long coherence times in an ultracold trapped gas*

Ultracold atomic vapors trapped on microchips are interesting because they allow the realization of miniaturized devices for precision measurements such as atomic clocks. Sequences of two $\pi/2$ microwave pulses separated by a time interval T give

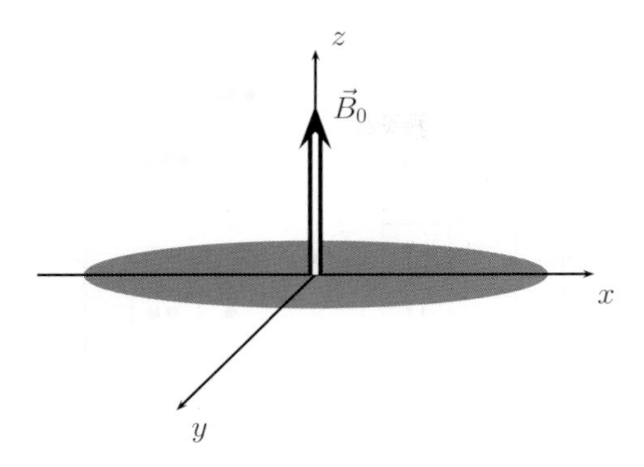

Fig. 20.6 Spatial segregation experiment: A static magnetic field perpendicular to the cylindrical symmetry axis is applied [Lewandowski *et al* (2002); McGuirk *et al* (2002)].

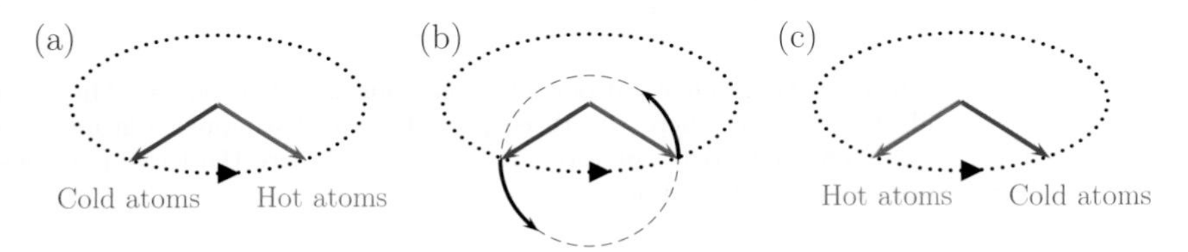

Fig. 20.7 After a $\pi/2$ pulse, the transverse magnetizations of hot atoms precess faster than that of the cold atoms and acquire a certain advance. As a result of ISRE, the two magnetizations rotate around their vector sum. If the rotation angle is equal to π, the two magnetizations are interchanged and the fast spins can catch up to the slow spins after a certain time.

rise to Ramsey fringes[7] with a width scaling as π/T. The longer T, the higher the precision. One cannot however increase T beyond the coherence time τ of the transverse magnetization. This coherence time is essentially limited by the spatial inhomogeneities of the static magnetic field experienced by the atoms during their free motion in the trap. In a recent experiment, despite a theoretical estimation of the effect of field inhomogeneities that should lead to values of τ on the order of 1 to 2 seconds, a coherence time on the order of 50 seconds was observed [Deutsch *et al.* (2010)]!

This lengthening of τ is in fact due to ISRE occurring in forward direction, as we explain it now, in a qualitative way. For this purpose, it will be convenient to divide the atoms in two classes, the "hot" atoms that explore large regions of the trap, and consequently large fields, and the "cold" atoms that remain near the center of the trap, where the field is small. Just after the first $\pi/2$ pulse, the transverse magnetizations of the two classes are parallel. But the hot atoms precess faster because they experience a larger magnetic field than the cold atoms. After a certain time, the magnetization of the hot atoms (red arrow of Fig. 20.7) will have advanced with respect to the magnetization of cold atoms (blue arrow of Fig. 20.7).

[7]See Sec. 2.4 of Chap. 2.

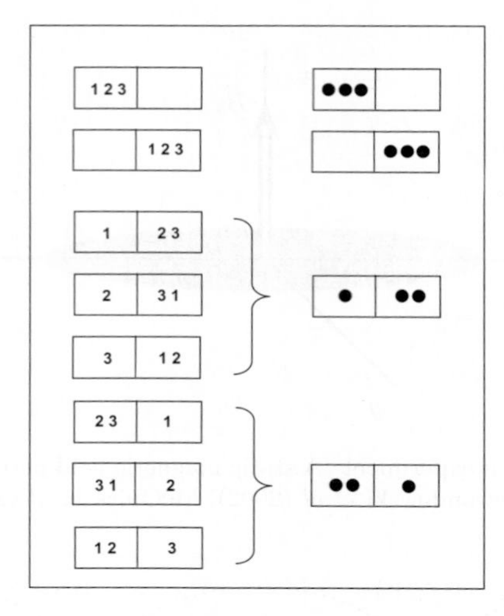

Fig. 20.8 Microscopic states of three identical particles distributed in two boxes. The left part of the figure corresponds to the case where the three particles are distinguishable and can be labelled by numbers $1, 2, 3$. The right part corresponds to the case where the three particles are indistinguishable and represented by the same dot.

But ISRE takes place every time the hot atoms cross the cloud of cold atoms which makes the two magnetizations rotate around their vector sum (dotted circle and black arrows of Fig. 20.7(b)). When the rotation angle reaches the value π, the two magnetizations are interchanged: the hot atoms become late compared to the cold atoms. Since they continue to experience the same field, hot atoms will catch up to the cold atoms after a certain time and all spins will be synchronized.[8] This effect is somewhat similar to what happens in a spin echo experiment, but here the refocusing is due to an internal effect rather than to an additional RF pulse.

20.3 The first prediction of BEC in a gas

20.3.1 *A new derivation of Planck's law for black body radiation*

In 1924, a young Indian physicist named Satyendra Nath Bose proposed a completely new derivation of Planck's law for black body radiation [Bose (1924)] . He was considering thermal radiation as a gas of *indistinguishable* particles, which leads to a change in the number of microscopic states corresponding to a macroscopic state. This can be easily understood by the simple example of three particles distributed in two boxes shown in Fig. 20.8. The number of microscopic states corresponding to a macroscopic state depends on whether the three particles are considered to be distinguishable or indistinguishable. This number is the same

[8]This presentation of the increase of the coherence time separates in time the dephasing and the refocusing of the spins. In reality, both processes act simultaneously leading to the synchronization of the spins.

when the three particles are in the same box, whereas the number of microscopic states corresponding to two particles in a box and the third one in the other box is larger when the particles are distinguishable [9] since there are three ways of choosing two numbers among $1, 2, 3$. By counting the number of microscopic states of light quanta as indistinguishable particles and maximizing the entropy of the system for a given value of the total energy, Bose re-derived Planck's law. He sent his paper to Albert Einstein who found it very interesting and helped Bose to publish it.[10]

20.3.2 *Extension of Bose statistics to atomic particles*

Shortly after, Einstein extended this way of counting the number of microscopic states to an ideal gas of identical atoms considered to indistinguishable. A difference appears in the calculations: the number of light quanta is not a fixed quantity since they can be emitted or absorbed. The maximization of the entropy was performed by Bose with the single auxiliary condition that the total energy is fixed. This introduces a single Lagrange multiplier related to the temperature. If the particles considered are atoms, their number is fixed. This introduces a second auxiliary condition and a second Lagrange multiplier, which turns out to be related to the chemical potential. In a series of two papers [Einstein (1924, 1925)] published in 1924 and 1925, Einstein derived the following expression:

$$N_i = \frac{1}{\exp[\beta\,(\varepsilon_i - \mu)]-1} \tag{20.38}$$

for the mean number n_i of atoms in an individual state of energy ε_i. In this equation, $\beta = 1/k_B T$ where T is the temperature, k_B the Boltzmann constant, and μ is the chemical potential. This distribution characterizes what is called now *Bose-Einstein statistics*. It is remarkable that it was introduced at a time when the fundamental equations of quantum mechanics were not yet established.

Instead of the chemical potential μ it is convenient to use the fugacity z, defined as

$$z = \exp(\beta\mu). \tag{20.39}$$

The fugacity is a positive number, in terms of which (20.38) can be rewritten:

$$N_i = \frac{1}{z^{-1}\exp(\beta\varepsilon_i)-1} = \frac{z}{\exp(\beta\varepsilon_i)-z}. \tag{20.40}$$

One can always take the energy of the ground state $i = 0$ as a zero of energy ($\exp(\beta\varepsilon_0) = 1$). Since N_0 cannot be negative, z must necessarily be smaller than 1, which implies the following domain of variation of z:

$$0 \leqslant z < 1. \tag{20.41}$$

[9] This shows that the relative weight of a macroscopic state in which all particles are in the same microscopic state is higher when the particles are indistinguishable.

[10] See [Bose *et al.* (1977)] for a translation in English of the original article.

To determine z, we write that the total number of particles is equal to N, which gives:

$$N = \sum_i g_i N_i = \sum_i g_i \frac{z}{\exp(\beta\varepsilon_i) - z} \tag{20.42}$$

where g_i is the degeneracy of the energy ε_i (number of microscopic states having the same energy ε_i). Inversion of Eq. (20.42) gives z as a function of N and T.

Two years after the first Einstein paper, Enrico Fermi [Fermi (1926)] and Paul Dirac [Dirac (1926)] independently realized that Bose-Einstein statistics could not be applied to electrons, which obey Pauli exclusion principle. They made a new counting of the number of microscopic states, still assuming that the particles were indistinguishable, but imposing that the occupation number of each microscopic state could not be larger than 1. In this way, they derived *Fermi-Dirac statistics* where (20.38) is replaced by:

$$N_i = \frac{1}{\exp[\beta(\varepsilon_i - \mu)] + 1}. \tag{20.43}$$

20.3.3 *The condensation phenomenon*

Consider a gas of atoms that obeys Bose-Einstein statistics contained in a cubic box with a volume $V = L^3$. Suppose that T and V are fixed and that the number of atoms in the box N is progressively increased. For very low values of N, the resolution of Eq. (20.42) gives a very small value for z: $z \ll 1$. In the denominator of (20.40), one can then neglect z in comparison with the exponential, which is always larger than 1. This gives:

$$N_i \simeq z \exp(-\beta\varepsilon_i). \tag{20.44}$$

The Bose-Einstein distribution is then close to the classical Maxwell-Boltzmann distribution. When N becomes large enough, z gets closer to the upper limit $z = 1$ of the interval given in (20.41), and dramatic new features appear. To describe them, it is convenient to single out in (20.42) the contribution N_0 of the ground state $i = 0$ taken to be non-degenerate ($g_0 = 1$), which gives, since $\varepsilon_0 = 0$:

$$N = \frac{z}{1-z} + \sum_{i \neq 0} g_i \frac{z}{\exp(\beta\varepsilon_i) - z}. \tag{20.45}$$

When z gets close to 1, $N_0 = z/(1-z)$ becomes very large, whereas the sum N_{exc} of the populations of the excited states $i \neq 0$ remains smaller than a maximum value $N_{\max}$ obtained by replacing each term in the sum appearing on the right-hand side of (20.45) by its maximum value reached for $z = 1$

$$N_{\text{exc}} = \sum_{i \neq 0} g_i \frac{z}{\exp(\beta\varepsilon_i) - z} \leq N_{\max} = \sum_{i \neq 0} \frac{g_i}{\exp(\beta\varepsilon_i) - 1}. \tag{20.46}$$

The important point, which explains the condensation phenomenon, is that, in a certain number of cases, and in particular for a gas contained in a 3D box, the value of $N_{\max}$ given in (20.46) is non-infinite. In other words, when T and V are fixed and the total number N of atoms is increased and becomes larger than $N_{\max}$, it is impossible for the excited states $i \neq 0$ to accommodate the total number of atoms because all their populations are saturated. The excess of populations, $N - N_{\max}$, must therefore be accommodated by the ground state, whose population $N_0 = z/(1-z)$ is never saturated and can reach arbitrarily high values if z is sufficiently close to 1. Once the threshold $N = N_{\max}$ is reached, every atom added to the gas has a high probability to condense in the ground state. The non-condensed atoms form a gas that Einstein called a *saturated ideal gas* because it cannot accept extra atoms. Now it is called the *thermal cloud*, whereas the atoms condensed in the ground state form what is called the *condensate*.

20.3.4 *Critical temperature*

Instead of varying N with T and V fixed, it is also possible to vary T with N and V fixed. If T is progressively decreased from a high value, the value of $N_{\max}$, which is a decreasing function of T, decreases from a large initial value until it becomes equal to N for a certain value of T called the BEC *critical temperature* T_c. Below T_c, $N_{\max}$ becomes smaller than N, and the ground state must accomodate at least $N - N_{\max}$ atoms. The critical temperature is defined for a given N as the temperature below which BEC appears.

The value of T_c is obtained by calculating the sum appearing in the expression (20.46) of $N_{\max}$ and by equating this value to N. For a 3D box, a standard calculation, sketched below, gives:

$$\frac{N}{V} = \zeta(3/2)\left[\frac{m\,k_B T_c}{2\pi\hbar^2}\right]^{3/2}, \tag{20.47}$$

where $\zeta(x)$ is the Riemann zeta function and $\zeta(3/2) \simeq 2.6124$. With the definition (20.2) of the thermal de Broglie wavelength, this equation can be rewritten:

$$n\lambda_{\mathrm{T}}^3(T_c) = \zeta(3/2) \tag{20.48}$$

where $n = N/V$ is the spatial atomic density of the gas. This equation expresses that the mean distance between the atoms is on the order of the de Broglie wavelength when $T = T_c$. This corresponds to the entrance in the quantum degenerate regime, mentioned in Sec. 20.1 of this chapter.

This wave interpretation of the critical temperature was not given in Einstein's paper. In fact, Einstein was writing his second paper on BEC when he received a copy of Louis de Broglie's thesis from Paul Langevin in which the idea of associating a wave with a particle was introduced for the first time. Einstein fully realized the importance of this idea but he did not apply it to the interpretation of the critical temperature. He rather invoked the interference of de Broglie waves to explain a term appearing in the fluctuations of the ideal gas that he found analogous to a similar term appearing in the fluctuations of thermal radiation and that he interpreted in a previous work as being due to interference of light waves [Einstein (1909)].

The values of T_c derived from (20.47) are extremely low in practice. For example, for a ultracold gas of hydrogen atoms with a density of 1.6×10^{16} atoms/cm^3, $T_c = 1$ mK; for a gas of rubidium atoms with a density of 10^{14} atoms/cm^3, $T_c = 350$ nK.[11] At these temperatures, all substances were believed to exist only in a liquid or solid state, for which Einstein's model of an ideal gas cannot be applied. Everybody, including Einstein, considered BEC as a purely academic phenomenon. It took about 70 years before methods to achieve BEC in a gas were developed and demonstrated. These will be described in the next chapter.

Derivation of the equation giving T_c

The gas is contained in a cubic box whose sides have a length L. With periodic boundary conditions the three components of the atomic momentum are quantized:

$$p_i = \frac{2\pi\hbar}{L} n_i \qquad i = x, y, z \tag{20.49}$$

where the n_i's are positive or negative integers. The corresponding energy of this state is:

$$\varepsilon_{\vec{p}} = \frac{2\pi^2\hbar^2}{mL^2}\left(n_x^2 + n_y^2 + n_z^2\right). \tag{20.50}$$

The ground state corresponds to $n_x = n_y = n_z = 0$ and has an energy $\varepsilon_0 = 0$. The energy separation between the ground state and the first excited state is equal to:

$$\delta\varepsilon = \frac{2\pi^2\hbar^2}{mL^2}. \tag{20.51}$$

Later on, we will see that $k_B T_c$ is very large compared to $\delta\varepsilon$. It is thus possible to replace the discrete sum appearing in (20.42) by an integral:

$$\sum_{\vec{p}} \to \frac{L^3}{(2\pi\hbar)^3}\int_0^\infty 4\pi^2 p^2 \mathrm{d}p. \tag{20.52}$$

Using the new dimensionless variable $x = \varepsilon/k_B T$ and the corresponding density of states $\rho(x)$, it is possible to rewrite (20.52) as:

$$\sum_{\vec{p}} \to \int_0^\infty \rho(x)\mathrm{d}x \quad \text{with} \quad \rho(x) = \frac{V}{\lambda_{\mathrm{T}}^3}\frac{2}{\sqrt{\pi}}\sqrt{x} \tag{20.53}$$

which gives the following expression for the population of the excited states [12] N_{exc}:

$$N_{\mathrm{exc}} = \int_0^\infty \rho(x)\frac{1}{z^{-1}e^x - 1}\mathrm{d}x = \frac{V}{\lambda_{\mathrm{T}}^3}\frac{2}{\sqrt{\pi}}\int_0^\infty \sqrt{x}\,\frac{ze^{-x}}{1 - ze^{-x}}\mathrm{d}x. \tag{20.54}$$

[11] In both cases, inelastic collisions limit the achievable atomic densities. This is the reason why we choose for the numerical estimate typical atomic densities that have been achieved experimentally.
[12] The density of states $\rho(x)$ vanishes for $x = 0$, so that the population of the ground state is not included in Eq. (20.54).

In the last fraction of (20.54), z and e^{-x} are smaller than 1. It is therefore possible to use the following expansion:

$$\frac{ze^{-x}}{1-ze^{-x}} = \sum_{\ell=1}^{\infty} \left(ze^{-x}\right)^{\ell} \tag{20.55}$$

which, inserted in (20.54), gives:

$$N_{\text{exc}} = \frac{V}{\lambda_{\text{T}}^3} \frac{2}{\sqrt{\pi}} \sum_{\ell=1}^{\infty} \frac{z^{\ell}}{\ell^{3/2}} \underbrace{\int_0^{\infty} \sqrt{u} e^{-u} \mathrm{d}u}_{=\sqrt{\pi}/2} = \frac{V}{\lambda_{\text{T}}^3} g_{3/2}(z) \tag{20.56}$$

where

$$g_n(z) = \sum_{\ell=1}^{\infty} \frac{z^{\ell}}{\ell^n}. \tag{20.57}$$

Equation (20.56) is usually rewritten as

$$n\lambda_{\text{T}} = g_{3/2}(z). \tag{20.58}$$

N_{exc} reaches its maximum value when $z = 1$ and is equal to the total number of atoms when $T = T_c$. Using

$$g_{3/2}(1) = \zeta(1) = \sum_{\ell=1}^{\infty} \frac{1}{\ell^{3/2}} \simeq 2.6124. \tag{20.59}$$

We finally get

$$\frac{N}{V} = \frac{g_{3/2}(1)}{\left[\lambda_{\text{T}}(T_c)\right]^3} \tag{20.60}$$

which is equivalent to (20.47).

Finally it should be emphasized that T_c given in Eq. (20.47) does not change if N and V tend to infinity with N/V and T fixed [13] (i.e. for a fixed spatial density $n = N/V$ of the gas), whereas the spacing $\delta\varepsilon = \varepsilon_1 - \varepsilon_0$ between the ground state and the first excited quantum state in the box tends to zero (see Eq. (20.51)). This shows that BEC is not a trivial thermal effect that appears when $k_B T$ becomes smaller than the spacing between the ground state and the first excited quantum state.

It is instructive to compare the population of the first excited state with that of the ground state. Let us consider the degenerate regime and assume that half of the particles are in the ground state, then the fugacity z is given by:

$$N_0 = \frac{N}{2} = \frac{1}{1-z} \quad \Longrightarrow \quad z = 1 - \frac{2}{N}. \tag{20.61}$$

[13] This limit corresponds to what is called the *thermodynamic limit*, briefly discussed in the next subsection.

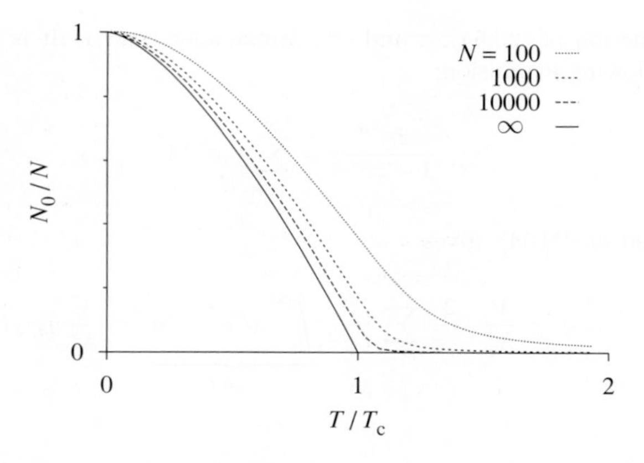

Fig. 20.9 Variations of N_0/N with T/T_c for different values of the total number N of atoms. The variations with T/T_c of the slope of the curve become more and more rapid in the neighborhood of $T = T_c$ when N increases with N/V fixed. The slope is discontinuous in the thermodynamic limit $N \to \infty$, $V \to \infty$, with T and N/V fixed. Courtesy of Werner Krauth.

As $\delta\varepsilon \ll k_B T$, the population of the first excited state can be readily evaluated

$$N_1 = \frac{z e^{-\beta\varepsilon_1}}{1 - z e^{-\beta\varepsilon_1}} \simeq \frac{1}{1-(1-2/N)(1-\delta\varepsilon/k_B T)} = \frac{1}{2/N + \varepsilon/k_B T}. \tag{20.62}$$

Since $\varepsilon/k_B T \sim \varepsilon/k_B T_c \sim N^{-2/3}$, we find that $N_1 \sim N^{2/3}$, and the ratio $N_1/N_0 \sim N^{-1/3}$ tends to zero for large N. This calculation justifies why only the ground state is singled out in the calculation of the Bose-Einstein condensation.

20.3.5 *Variation of the number N_0 of condensed atoms with the temperature. Thermodynamic limit*

When $T > T_c$, N_0 is negligible. When T decreases and reaches the value $T = T_c$, N_0 is still negligible at the thermodynamical limit and the total number N of atoms is given by (20.47). When T decreased below T_c, the number of atoms that are not condensed, which is equal to $N - N_0$, is given by the value of $N_{\max}$ corresponding to this value of T. It is given by an equation similar to (20.47) where N is replaced by $N - N_0$ and T_c is replaced by T:

$$\frac{N - N_0}{V} = 2.612 \left[\frac{m\, k_B T}{2\pi\hbar^2}\right]^{3/2} \quad \text{if} \quad T < T_c. \tag{20.63}$$

From Eqs. (20.47) and (20.63), one easily deduces:

$$\begin{array}{ll} T \geq T_c & N_0 \text{ negligible at the thermodynamical limit,} \\ T < T_c & \dfrac{N_0}{N} = 1 - \left(\dfrac{T}{T_c}\right)^{3/2}. \end{array} \tag{20.64}$$

This equation predicts that the slope of the curve that gives the variations of N_0/N with T/T_c is discontinuous in $T = T_c$. This is an artefact due to the replacement of the discrete sum in (20.45) by an integral. If the calculation is performed numerically, keeping the discrete sums, the slope of the curve remains continuous for all values of T/T_c (see Fig. 20.9). The variations, however, are the more rapid around $T = T_c$ when N is larger, with T and N/V fixed. This limit, $N \to \infty$, $V \to \infty$, with T and N/V fixed, corresponds to what is called the thermodynamic limit. It is only in this limit that BEC appears as a phase transition. In three dimensions, this transition for the ideal gas is a first order phase transition, and a second order phase transition in the presence of interactions [Huang (1988)].

20.3.6 *Influence of dimensionality*

In this subsection, we show that BEC cannot occur for a 2D ideal gas. Instead of showing that, contrary to what happens in 3D, the population of the excited states does not saturate when N increases, we calculate directly the proportion N_0/N of atoms in the ground state of the box containing the gas, and we show that N_0/N tends to zero for any temperature in the thermodynamical limit.

Consider a 2D box that consists of a square with side length equal to L and a surface $S = L^2$. The density of states is equal to:

$$\rho(\varepsilon) = \frac{mL^2}{2\pi\hbar^2}. \tag{20.65}$$

After replacement of the sums over discrete states by integrals, the total number of atoms is given by:

$$N = \int_0^\infty \frac{\rho(\varepsilon)}{e^{\beta(\varepsilon-\mu)} - 1}\,\mathrm{d}\varepsilon = \frac{mL^2}{2\pi\hbar^2}\int_0^\infty \frac{1}{e^{\beta(\varepsilon-\mu)} - 1}\,\mathrm{d}\varepsilon. \tag{20.66}$$

In the last integral of (20.66), the function of ε to be integrated is proportional to the derivative of $f(\varepsilon) = g_1(z\exp[-\beta\varepsilon]) = -\ln\{1 - \exp[-\beta(\varepsilon-\mu)]\}$.[14] The integral is thus readily calculated and gives:

$$N = \frac{mL^2}{2\pi\hbar^2\beta}\left[f(\infty) - f(0)\right] = \frac{L^2}{\lambda_\mathrm{T}^2}\ln\frac{1}{1-e^{\beta\mu}} = \frac{L^2}{\lambda_\mathrm{T}^2}\ln\frac{1}{1-z}. \tag{20.67}$$

This latter equation can be recast in terms of the surface density $n = N/L^2$ as

$$n\lambda_\mathrm{T}^2 = g_1(z). \tag{20.68}$$

This form is equivalent to Eq. (20.58) for the three-dimensional case. From (20.67), the fugacity can be calculated:

$$z = 1 - \exp\left(-n\lambda_\mathrm{T}^2\right). \tag{20.69}$$

This gives the value of the number N_0 of condensed atoms

$$N_0 = \frac{z}{1-z} = e^{n\lambda_\mathrm{T}^2} - 1 \tag{20.70}$$

[14] The Bose functions $g_n(z)$ are defined in Eq. (20.57).

and, finally, the proportion of condensed atoms:

$$\frac{N_0}{N} = \frac{e^{n\lambda_T^2} - 1}{N}. \tag{20.71}$$

In the thermodynamic limit, N and L tend to infinity with n and T fixed. It clearly appears that the numerator of (20.71) is then fixed while the denominator tends to infinity. Contrary to what happens in 3D, the proportion of condensed atoms tends to zero in the thermodynamic limit for all values of the temperature. Mathematically, this is due to the divergence of the function $g_1(z)$ when z approaches one from below which ensures that Eq. (20.68) has always a solution. A similar result is found in 1D. This shows the importance of dimensionality. For an ideal gas in a box in the thermodynamic limit, BEC can exist in 3D, but not in 2D or 1D.

20.4 Conclusion

In this chapter, we have reviewed a certain number of quantum effects that can appear in a gas when the temperature is progressively lowered.

The quantum nature of atom-atom collisions appears when the de Broglie wavelength becomes large compared to the range of the interaction potential, while remaining small compared to the mean distance between atoms. When the atoms of the gas are identical, spectacular effects, can appear due to their indistinguishability which can modify collision cross-sections and induce a spin dynamics, even in the absence of spin-spin interactions. These effects are now playing an important role in the physics of ultracold atoms by giving rise, for example, to a synchronization of the precession of the spins and a lengthening of their coherence time.

In the second part of this chapter, we described how the idea of quantum statistics and quantum degenerate gases appeared in physics at a time when quantum mechanics was not yet formulated. It is remarkable that important results, like the condensation of atoms in the lowest state of the box containing the gas, were simply derived from a new way of counting the number of microscopic states, without using wave functions and the symmetrization postulate. Of course, the description of the gas as an ideal gas was unrealistic at the temperatures predicted for the condensation. But, afterwards, one can consider now that this simple model has been useful for clarifying a certain number of important concepts. It provides an exact model of phase transition without the complexity introduced by interactions and allows a clear understanding of what is meant by thermodynamic limit and of the importance of the dimensionality.

When the condensate begins to form, a new atom added to the gas will preferentially go into the condensate. It therefore seems to be correlated with the condensed atoms even though it does not interact with them. Of course, we know now that this is due to the $\sqrt{N+1}$ factor appearing in the transition amplitude, but this was not known in 1924, and many physicists, like Erwin Schrödinger, were worried by the fact that the non-interacting atoms seemed not to be independent. This was actually the first clear example in physics of correlations resulting from quantum statistics and not from interactions. In fact, the identical spin rotation effect discussed in Sec. 20.2 of this chapter is also due to quantum statistics and appears even if the two spins do not interact.

Chapter 21

The long quest for Bose-Einstein condensation

21.1 Introduction

The paper of Albert Einstein [Einstein (1925)] that introduced the spectacular phenomenon of Bose-Einstein condensation (BEC) did not drive any experimental activity when it appeared. This is easy to understand, for at least two reasons: first, the predicted critical temperature was in a range (few mK at standard density) that was not achievable at that time and, second, the calculation was performed for an ideal gas. At such low temperatures, all known substances at that time were solid, except for helium which remains liquid under usual pressure. Neglecting interactions in a liquid or a solid is unrealistic.

A few years later, in April 1938, Fritz London suggested that the transition to superfluidity observed in liquid helium below 2.17 K could be related to BEC [London (1938)]. Soon after, Laszlo Tisza derived a two-fluid model to account for the observed transport phenomena of helium II [Tisza (1938)].[1]

It was only in 1959 that physicists, like Charles Hecht, realized that atoms, like hydrogen, can remain gaseous even at zero Kelvin provided that their electrons are spin polarized [Hecht (1959)]. In 1976, William Stwalley and Lewis Nosanow confirmed the absence of a liquid phase for polarized atomic hydrogen using improved information about their interaction potential [Stwalley and Nosanow (1976)]. Furthermore, the predicted critical temperature for spin polarized hydrogen was not too low because of the small mass of this atom. This was the starting point of an intense experimental activity to achieve BEC with hydrogen atoms by the groups of Daniel Kleppner and Thomas Greytak at MIT and Isaac Silvera and Jook Walraven at Amsterdam. Several difficulties were encountered, and are described in Secs. 21.2 and 21.3 of this chapter. Trying to solve these difficulties stimulated the emergence of new important and fruitful ideas such as wall free confinement of atoms and cooling by evaporation.

Meanwhile, spectacular progress was made in the cooling of alkali atoms to extremely low temperatures in the microkelvin range, as well as in the trapping

[1] The confirmation of this point of view came much later with neutron scattering [Sokol (1995)], and path-integral Monte-Carlo simulations [Ceperley (1995)].

techniques. It was also realized that it should be possible to keep an ensemble of alkali atoms in a gaseous phase for a time long enough to allow Bose-Einstein condensation to be observed as a metastable situation before the gas liquefied or crystallized. In Sec. 21.4, we describe the difficulties and the advantages of using alkali atoms compared to hydrogen, since, in fact it is with alkali atoms that BEC was observed for the first time in 1995. Eric Cornell, Carl Wieman and Wolfgang Ketterle [Anderson *et al.* (1995); Davis *et al.* (1995a)] received the Nobel Prize in physics in 2001 for this remarkable achievement with rubidium and sodium atoms [Cornell and Wieman (2002); Ketterle (2002)]. An evidence for quantum degenerate regime with lithium-7 atoms was also reported in 1995 by the group of Randall Hulet [Bradley *et al.* (1995)]. Three years later, BEC was finally also observed in hydrogen.

The history of gaseous BEC is a beautiful example of a long term fundamental research endeavor that required the combination of several experimental and theoretical contributions to make a major scientific discovery possible. It is also clear that this long term effort has been, and continues to be, rewarding since several new interesting methods were developed during the quest for BEC, and because now quantum degenerate gases appear as interesting models for understanding more complex situations found in other fields of physics, thanks to the ability to control the various parameters of the gas, and in particular the atom-atom interactions with Feshbach resonances. In this respect, one should also mention the importance of quantum degenerate Fermi gases, which have also been investigated after the discovery of BEC. In Sec. 21.7, we briefly describe the very first experiments performed on these systems.

21.2 First attempts on hydrogen

21.2.1 *Spin polarized hydrogen as a quantum gas*

The fact that helium does not crystallize at normal pressure and at temperatures very close to absolute zero has a quantum-mechanical origin: the interatomic attractive potential is compensated by the zero-point kinetic energy. This is the reason that such a liquid is called a quantum fluid. Because of their smaller mass, hydrogen atoms with polarized electronic spins should provide an even more extreme example of a bosonic[2] quantum fluid. In fact, spin polarized hydrogen cannot liquefy even at zero temperature, as was predicted by Charles Hecht [Hecht (1959)] and confirmed a few years later by William Stwalley and Lewis Nosanow [Stwalley and Nosanow (1976)] using more precise calculations, such as the quantum theory of corresponding states. Another advantage of hydrogen is the fact that its simplicity enables very accurate comparison of experimental results with *ab initio* calculations.

[2]The hydrogen atom consists of two tightly bound fermions (1 electron + 1 proton), and therefore behaves as a composite boson. Deuterium, which is made of an odd number of fermions (1 electron + 1 proton + 1 neutron) behaves as a composite fermion.

It is easy to understand why hydrogen atoms must be spin polarized. The interaction potential between two ground state hydrogen atoms strongly depends on the total electronic spin $S = s_1 + s_2$ of the pair of atoms. The singlet potential ($S = 0$) has many bound states and can accommodate a H_2 molecule with a ground state internuclear separation on the order of 0.75 Å and a binding energy of $\sim$ 4.5 eV. In contrast, the triplet potential ($S = 1$) does not support *any* bound states. This means that a gas of atomic hydrogen, which is generally highly unstable against recombination into H_2 molecules, can be stabilized by polarizing the atoms. Looking for BEC in such a gas is very attractive since it is predicted that the critical temperature should fall in the Kelvin range for reasonable pressures where the system is expected to behave as a nearly ideal gas.

21.2.2 *Production of a spin polarized sample at low temperature*

The first problem the experimentalists had to solve was how to create a spin polarized hydrogen gas. Atomic hydrogen is produced in an electrical discharge of molecular H_2. However, the resulting atomic gas is not spin-polarized, it is produced at high temperature, and it is short lived due to adsorption on surfaces where rapid recombinations to the molecular form occur. Using inert teflon surfaces, the recombination can be partially suppressed so that a low-density atomic hydrogen gas is achievable with a life time on the order of one second. This technique was used primarily for producing the active medium of a hydrogen maser.

The first production of a long-lived polarized atomic hydrogen gas was realized by Isaac Silvera and Jook Walraven [Silvera and Walraven (1980a)]. It was made in a cell whose walls were covered by a sub-Kelvin liquid-helium film and in the presence of a high magnetic field, on the order of 7 T. In this way, the confinement and the cooling problems were both solved. The polarization of the gas was obtained from strong magnetic field gradients (see the ground-state hyperfine energy levels of hydrogen denoted **a**, **b**, **c**, **d** in Fig. 21.1) since the dissociator was located at a low field region while the cell was in the high field region. Indeed, at sufficiently high magnetic field B and sufficiently low temperature, where the energy gap between the high and low field seekers is much larger than the thermal kinetic energy, ($T < \mu_B B/k_B$), the separation is complete. For a field of 10 T and a temperature of 0.25 K, the ratio of densities is $n(H\uparrow)/n(H\downarrow) \sim 10^{-23}$.

The electron-polarized gas made of hydrogen atoms in **a** and **b** states is not stable against recombination. Indeed, the hyperfine mixing of the state **a** allows the reactions: **a** + **b** → ortho H_2 and **a** + **a** → para H_2.[3] Such reactions require a third body to ensure the energy conservation. In principle, the third body could be another atomic hydrogen atom. However, at low enough density this is quite unlikely and recombination occurs mainly at the walls of the cell. As a result, atoms

[3] In ortho-hydrogen, the two nuclear spins are parallel, in para-hydrogen, they are anti-parallel.

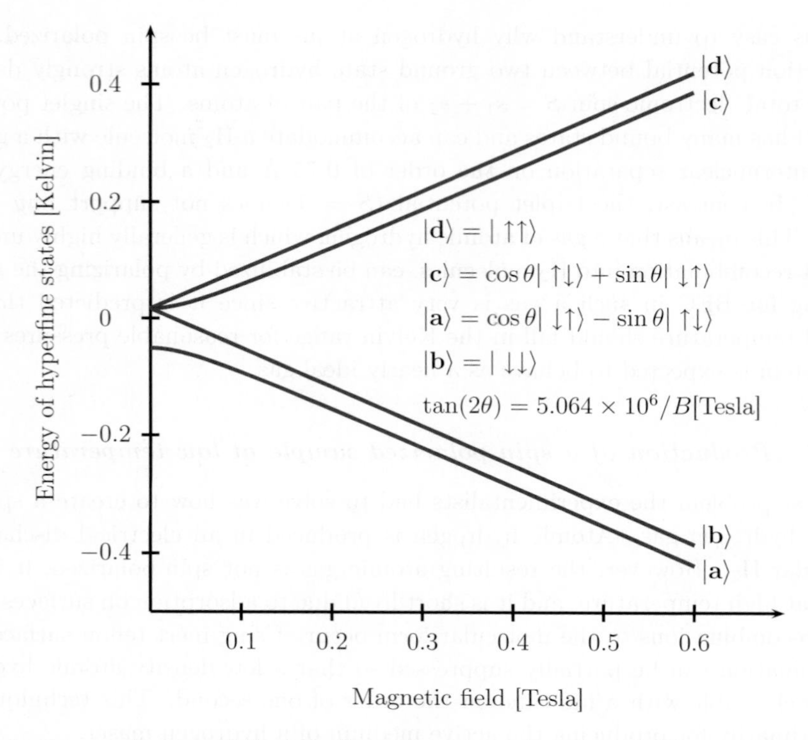

Fig. 21.1 Zeeman diagram of the hyperfine states of atomic hydrogen. The two lowest energy states, labelled **a** and **b**, are the high field seekers, and the two highest energy states, **c** and **d**, are the low field seekers. In the equations defining the states **a**, **b**, **c**, **d**, the first arrow is for the electronic spin, and the second one refers to the nuclear spin. H↑ denotes hydrogen atoms in **c** or **d** states, that is, with a magnetic moment anti-parallel with the external magnetic field and corresponding to the low energy electronic spin states, $m_S = +1/2$ because of the sign of the gyromagnetic factor with either nuclear spin.

in state **a** are progressively eliminated and the resulting gas is made of **b** atoms, for which nuclear as well as electron spins are polarized.

21.2.3 *Difficulties associated with collisions*

In the first experiment, the atomic density was low, on the order of 10^{14} atoms per cm^3, at a temperature of ~ 0.3 K. The collisions with the cold surface were crucial to ensure thermalization at cryogenic temperatures. However, atoms are adsorbed on the liquid helium surface with a binding energy E_b. Since $E_b/k_B \simeq 1$ K is larger than the temperature $T \simeq 0.3$ K of the sample, most hydrogen atoms remain adsorbed on the surface where they can recombine by inelastic processes.

The critical density for Bose-Einstein condensation at a temperature of 0.3 K is at a much higher density, on the order of 10^{20} cm^{-3}. Several groups thus tried to increase the atomic density by compressing the doubly polarized gas [Sprik *et al.* (1985); Bell *et al.* (1986)]. However, at a temperature on the order of 0.5 K, the maximum atomic densities that could be obtained were in the range of few 10^{18} atoms per cm^{-3}. It was not possible to get higher densities because of three-body recombination processes,[4] as theoretically explained in [Kagan *et al.* (1980)].

21.2.4 *Need for other methods*

Let us summarize what we have just seen. The first attempts to observe BEC with spin polarized hydrogen were based on trying to reduce the distance between particles by increasing the pressure at a fixed temperature in order to reach the Bose-Einstein threshold, for which the thermal de Broglie wavelength is on the order of the mean distance between atoms (see green line in Fig. 21.2). When it became clear that these attempts would fail, another strategy was envisioned that consisted of cooling the atoms at a fixed density (see Fig. 21.2). However, as explained above, surface adsorption and recombination at the surface becomes prohibitively large at temperatures much below 0.1 K. A series of problems were then raised:

(1) How to prevent the atoms from being adsorbed on the surface?
(2) How to describe BEC if atoms are not confined within walls?
(3) How to cool hydrogen atoms to much lower temperatures?

In the next section, we describe how these various problems were solved.

21.3 Second attempts on hydrogen

21.3.1 *Wall free confinement. Magnetic trapping*

The molecular recombination of hydrogen occurring in the liquid helium film covering the walls of the cell can be circumvented by using a wall free confinement. In 1986, Harald Hess suggested magnetically trapping hydrogen atoms as a means to implement this idea [Hess (1986)]. This was a strong motivation for the development of magnetic traps for neutral atoms.[5]

As already discussed in Sec. 14.3.1 of Chap. 14, Maxwell's equations forbid the existence of a maximum of static magnetic field modulus in free space. Thereby, only the low field seekers, **c** and **d** states, can be trapped by an appropriate magnetic field configuration (see Fig. 21.1). Spin-exchange collisions between two **c**-state atoms or between a **c**-state and a **d**-state atom lead to a rapid relaxation to **a** and **b** state atoms, which experience a repelling force at the location of the trap, and are therefore rapidly lost. The resulting rapid decrease of the **c**-state atom population results in the production of a trapped gas in the doubly-polarized **d** state. At the end of the eighties, two groups succeeded in magnetically trapping spin polarized

[4] Note that the corresponding molecular recombination requires a third body, either an atom or a surface, to conserve energy and momentum.

[5] In fact, the first magnetic trap for neutral atoms was demonstrated, not for hydrogen, but for sodium [Migdall *et al.* (1985)].

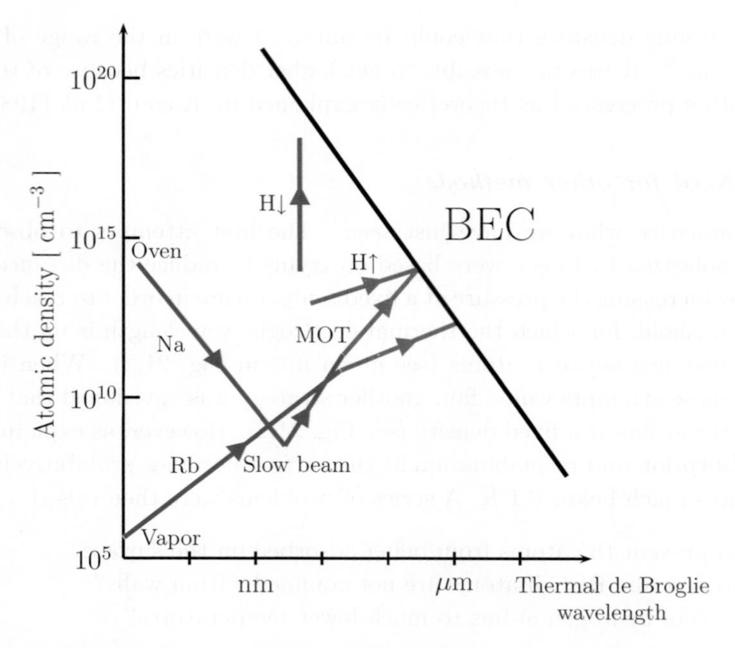

Fig. 21.2 The different routes to reach BEC. The gray line corresponds to the Bose-Einstein condensation threshold in phase space density. The first attempt on H↓ is represented by a green line. In blue, the rubidium and sodium experiment with cold atoms provided by a magneto-optical trap (MOT) fed by a vapor (Rb) or a slow beam at the exit of a Zeeman slower (Na). Red lines denote the last increase in phase space density performed in magnetic traps using evaporative cooling.

atomic hydrogen gas [Hess *et al.* (1987); van Roijen *et al.* (1988)]. Since the gas was now no longer confined between walls, but trapped in a nearly harmonic potential, it was necessary to extend Einstein's theory of BEC in a box to the case of an ideal gas trapped in a harmonic potential. This was done in [Bagnato *et al.* (1987b)] and the results will be summarized in the next Sec. 21.3.2.

The lifetime of a **d** state hydrogen gas is limited by dipolar spin relaxation. At a density of $n = 10^{14}$ cm^{-3} the decay time is about 10 seconds (and scales as n^{-1}), and the elastic collision rate is ~ 30 s^{-1} at a temperature of 1 mK. Hence, experiments with magnetically trapped low field seekers can be performed only at very low density such that the threshold temperature for Bose-Einstein condensation is below 50 μK. The need to reach such a low temperature stimulated the invention of new cooling methods which will be described in Sec. 21.3.3.

21.3.2 *Bose-Einstein condensation in a harmonic trap*

We consider here the specific example of a harmonic and isotropic trap with angular frequency ω_0, since, at very low temperature, the minimum of most traps can be approximated by a harmonic potential, and anisotropy does not fundamentally change the result.[6] The eigenenergies of the three-dimensional isotropic harmonic oscillator Hamiltonian are

$$\varepsilon_n = \hbar\omega_0 \left(n_x + n_y + n_z + \frac{3}{2} \right), \tag{21.1}$$

where n_x, n_y, n_z are positive integers. The energy of the ground state is thus $\varepsilon_{\min} = 3\hbar\omega_0/2$. Each energy state $\varepsilon_n = n\hbar\omega_0 + \varepsilon_{\min}$ where $n = n_x + n_y + n_z$ has a degeneracy g_n:

$$g_n = \frac{(n+1)(n+2)}{2}. \tag{21.2}$$

According to Eq. (20.46) of Chap. 20, the upper bound $N_{\max}$ on the number of atoms in the excited states is given by

$$N_{\max} = \sum_{n=1}^{\infty} \frac{g_n}{e^{nx} - 1}, \qquad \text{with} \qquad x = \frac{\hbar\omega_0}{k_B T}. \tag{21.3}$$

In the semiclassical limit, $\hbar\omega_0 \ll k_B T$, the discrete sum in Eq. (21.3) can be replaced by an integral and one eventually finds[7]

$$N_{\max} \simeq \left(\frac{k_B T}{\hbar\omega_0} \right)^3 g_3(1), \qquad \text{where} \qquad g_\alpha(z) = \sum_{\ell=1}^{\infty} \frac{z^\ell}{\ell^\alpha}. \tag{21.4}$$

The critical temperature T_c below which Bose-Einstein condensation takes place is simply obtained by equating $N_{\max}$ to the total number of atoms:[8]

$$k_B T_c = \hbar\omega_0 \left(\frac{N}{g_3(1)} \right)^{1/3} \simeq 0.94 \hbar\omega_0 N^{1/3}. \tag{21.5}$$

We recover the fact that the critical temperature is large compared to the energy spacing $\hbar\omega_0$ between successive levels since $N \gg 1$. Below the critical temperature, the chemical potential tends to $\varepsilon_{\min}$ so that the fugacity[9] z tends to 1. We can thus rewrite the total number of atoms as

$$N = \sum_{n=0}^{\infty} \frac{g_n}{z^{-1} e^{nx} - 1} \simeq N_0 + x^{-3} g_3(z), \quad \text{where} \qquad N_0 = \frac{z}{1-z}. \tag{21.6}$$

From this relation, we deduce the population of the ground state below the critical temperature:

$$N_0(T) = N \left[1 - \left(\frac{T}{T_c} \right)^3 \right]. \tag{21.7}$$

21.3.3 *New cooling method: evaporative cooling*

The first idea to cool atoms below 100 μK would be to use laser cooling. But laser cooling is not very efficient for hydrogen because there is no large scale separation

[6] See footnote 8 page 529.
[7] $g_3(1) \simeq 1.202$.
[8] For an anisotropic harmonic trap, one finds the same formula with ω_0 replaced by $(\omega_x \omega_y \omega_z)^{1/3}$.
[9] The zero-point energy has been taken into account in the definition of the fugacity $z = \exp[(\mu - 3\hbar\omega_0/2)/k_B T]$.

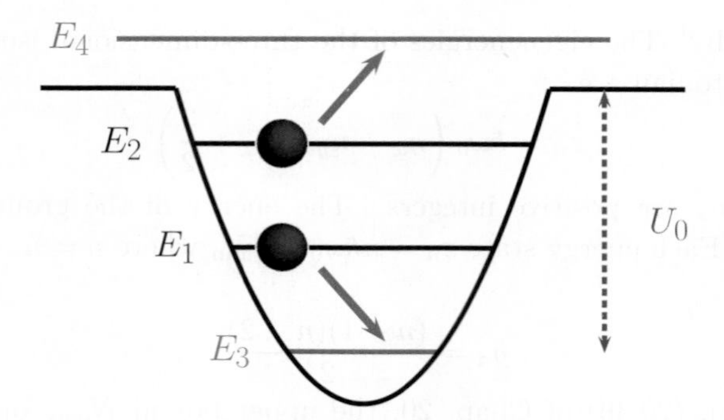

Fig. 21.3 Evaporative cooling. Consider two atoms of energy E_1 and E_2 trapped in a potential well of finite depth U_0. They undergo an elastic collision and their energies after the collision are E_3 and E_4 with $E_1 + E_2 = E_3 + E_4$. If $E_4 > U_0$, the atom with energy E_4 escapes from the well, it is evaporated. The remaining atom has a much lower energy E_3. After many processes of this type and a subsequent rethermalization, the temperature of the remaining trapped gas decreases, and its phase space density increases.

between the typical evolution times for internal and external degrees of freedom.[10] The natural width Γ of the excited state 1p is equal to $2\pi \times 100$ MHz, whereas the recoil energy E_{rec}/h is equal to $2\pi \times 13.5$ MHz. The value of the ratio $\hbar\Gamma/E_{\text{rec}}$ is not favorable for laser cooling (see Chap. 12). In addition, lasers are not so easy to operate at the wavelength $\lambda = 121.6$ nm of the Lyman α transition. This is why new cooling methods were needed.

In 1986, Harald Hess introduced the idea of evaporative cooling [Hess (1986)], which allows atoms to be cooled at temperatures far below the temperature of the surrounding walls. The mechanism of evaporative cooling is implemented on a sample trapped in a potential of non-infinite depth U_0 and works as follows: because of elastic collisions between trapped atoms, the most energetic atoms leave the trap (see Fig. 21.3 and the scaling laws comment below) and the total energy of the atoms remaining in the trap decreases. After thermalization, the temperature of the whole sample decreases, tending to an equilibrium value governed by the trap depth U_0. Even though atoms are lost, the decrease of the temperature can be so large that the phase space density increases. The initial spatial density in the trap must be large enough to ensure an elastic collision rate sufficiently large in comparison with the loss rates (due to background pressure or inelastic processes). Before starting the evaporation, the magnetic trap is loaded from a vapor of cold polarized atomic hydrogen, and the trap is compressed to increase the spatial density.

If the depth U_0 of the trap is fixed, the evaporation slows down when the temperature of the trapped sample decreases to values such that $k_B T \ll U_0$, since elastic collisions can no longer transfer enough energy to allow atoms to leave the trap. A simple idea then consists of forcing the evaporation by reducing the depth of the trapping potential [Masuhara *et al.* (1988)]. Using this technique, the MIT group

[10]See Sec. 11.1.2 of Chap. 11.

reported in 1991 the production of a hydrogen gas in the **d** state at a temperature as low as 100 μK for an on-axis atomic density of 8×10^{13} cm^{-3} [Doyle *et al.* (1991)], just at a factor 6 in phase space density $n\lambda_{\mathrm{T}}^3$ below the BEC threshold. In practice, the maximum gain in phase space density depends on the ratio of the elastic collision rate (which ensures the thermalization) over the atom loss rate. A large ratio (> 200) enables one to gain orders of magnitude for the phase space density [Ketterle and van Druten (1996); Walraven (1996); Cohen-Tannoudji (1997)].

Scaling laws of evaporative cooling

A simple model that captures the main features of this cooling mechanism can be worked out by considering sequences of discrete evaporation and re-thermalization processes occurring sequentially in time [Davis *et al.* (1995b); Cohen-Tannoudji (1997)].

Let us assume that at time $t = 0$, N atoms are trapped in a harmonic and isotropic trap. The mean total (kinetic + potential) energy of the gas is given by the equipartition theorem $E = 3Nk_BT/2 + 3Nk_BT/2 = 3Nk_BT$. The energy distribution is then truncated at an energy ηk_BT, where η is a dimensionless number on the order of 10. As η is sufficently large, only a small number of atoms dN is removed. Those atoms have a mean energy slightly higher than ηk_BT, defined as $(\eta + \kappa)k_BT$, where κ can be shown to be slightly smaller than 1 [Luiten *et al.* (1996)]. The total energy of the remaining atoms of the gas has therefore decreased by an amount dE:

$$\mathrm{d}E = \mathrm{d}N[(\eta + \kappa) - 3]k_BT. \tag{21.8}$$

The average energy amount $(\eta + \kappa)k_BT$ per evaporated atom has been taken from the other remaining atoms whose initial energy was $3(N - \mathrm{d}N)k_BT$. After thermal relaxation, the temperature of the remaining atoms has thus decreased by an amount dT that is readily calculated by energy conservation arguments:

$$3(N - \mathrm{d}N)k_BT - \mathrm{d}E = 3(N - \mathrm{d}N)k_B(T - \mathrm{d}T). \tag{21.9}$$

Combining Eqs. (21.8) and (21.9), leads to the scaling law linking the decrease of temperature to the number of remaining atoms:

$$\frac{\mathrm{d}T}{T} = \alpha\frac{\mathrm{d}N}{N} \implies \frac{T_f}{T_i} = \left(\frac{N_f}{N_i}\right)^{\alpha}, \tag{21.10}$$

where $\alpha = (\eta + \kappa)/3 - 1$. Since η is large compared to one, the parameter α is positive and larger than one. Equation (21.10) yields the following scaling laws with the atom number for all relevant quantities:[11]

$$\begin{array}{rl} \text{Temperature} & T \propto N^{\alpha}, \\ \text{Spatial volume of the gas} & V \propto N^{3\alpha/2}, \\ \text{Spatial density} & n \sim \dfrac{N}{V} \propto N^{1-3\alpha/2}, \\ \text{Phase space density} & D \propto N^{1-3\alpha}, \\ \text{Mean atom velocity} & v \sim T^{1/2} \propto N^{\alpha/2}, \\ \text{Collision rate} & \sim n\sigma v \propto N^{1-\alpha}. \end{array} \quad (21.11)$$

When the condition $\alpha > 1/3$ is fulfilled, the decrease in the number of atoms is accompanied with an increase of the phase space density and yields the possibility of reaching the Bose-Einstein threshold at the expense of a loss of atoms.

Although the sequential model that we have considered here cannot provide any information on the kinetics of evaporation, the scaling law on the collision rate shows an interesting phenomenon that occurs when $\alpha > 1$. In this regime, evaporation causes the elastic collision rate to increase, which in turn improves the efficiency of evaporation, leading to a situation where the increase of the phase space density accelerates ($\dot{D}/D > 0$). This remarkable property, referred to as the runaway regime, is confirmed by a kinetic model of evaporation [Luiten *et al.* (1996)] and explains the success of evaporative cooling.

To illustrate, let us consider typical numbers: the loss of 99 % of the atoms from an evaporation ramp that keeps a value of η on the order of 7.5 gives a reduction of the temperature by three orders of magnitude and a gain in the phase space density of seven orders of magnitude!

21.3.4 *Need for new detection method of polarized hydrogen*

In the early experiments on hydrogen, detection of atoms was based on a sensitive bolometer placed in the cell outside the trap, that measured the heat of molecular recombination (4.6 eV per event) when hydrogen was adsorbed on the wall of the cell. The number of atoms was obtained by decreasing the depth of the potential as a function of time.[12] Hydrogen atoms then progressively escaped from the trap and their number was monitored as a function of time. The energy distribution of atoms versus the trap barrier was obtained in this way.

This method, which consists of monitoring the hydrogen flux as the atoms are dumped from the trap is not appropriate for the detection of a Bose-Einstein condensate, since the condensate would be destroyed during the dumping process. Another diagnostic technique, for which atoms are not released from the trap, was therefore required.

Optical methods based on photo-absorption using one- or two-photon transitions provide an elegant way to detect *in situ* the magnetically trapped hydrogen atoms.

[11] The value of the various exponents depends on the confinement and its dimensionality, they are given here for a three-dimensional harmonical potential.

[12] This was realized by decreasing the current in one of the axial coils of the magnetic trap.

This method was initially implemented by the Amsterdam group using the absorption of a pulsed Lyman-α source [Setija *et al.* (1993)]. In this way, they could obtain information on the temperature and atomic density. However, the absorption lines were broadened by the Zeeman effect, so this group also used two-photon excitation with two different frequencies of the 1s-3s transition that has the advantage of being insensitive to magnetic fields [Pinkse *et al.* (1997)].

Similarly, the MIT group employed the two-photon Doppler-free spectroscopy of the 1s-2s transition that is essentially unperturbed by the magnetic field, which provides an excellent momentum resolution at the expense of a low excitation rate [Cesar *et al.* (1996)]. The atoms transferred to the 2s state are detected by applying an electric field, which contaminates the 2s state with the 2p state which has a short life time. The emitted Lyman-α photon from the 2p state is then detected. This latter detection method turned out to be crucial for the detection of the Bose-Einstein condensate of spin-polarized hydrogen produced by the MIT group in 1998 [Fried *et al.* (1998)], three years after its first achievement on alkali atoms [Anderson *et al.* (1995); Davis *et al.* (1995a)].

21.4 The quest for BEC for alkali atoms

21.4.1 *Difficulties associated with alkali atoms*

In the seventies, spin-polarized hydrogen gas appeared to be an ideal candidate for realizing Bose-Einstein condensation because of its unique property of remaining a gas at $T = 0$ and its relatively high condensation temperature. In addition, its small scattering length, $a = 0.0648$ nm [Jamieson *et al.* (1995)], makes the hydrogen gas behave as a nearly ideal gas at temperatures below 1 K. Hydrogen atoms are also appealing for basic studies since many of their properties can be calculated from first principles.

It turns out that the fact that spin-polarized hydrogen remains in a thermodynamically stable gaseous phase at low temperature due to quantum effects is not essential for observing BEC. Indeed, laser-cooled atoms in a wall free trap formed from either inhomogeneous magnetic fields or an appropriate dipole beam configuration[13] do not liquefy. In the absence of material surfaces, the gas to liquid transition has no support to nucleate liquid droplets, and must rely on three-body recombination to initiate this nucleation. At densities used to achieve BEC with alkali atoms, below 10^{14} atoms per cm^3, the corresponding rate remains sufficiently low and one can hope to observe BEC in a metastable state with a sufficiently long lifetime to allow precise measurements to be made.

The main difficulty is to find a domain of spatial densities in which three-body collisions remain negligible while the rate of two-body elastic collisions is high enough to ensure efficient evaporative cooling. In other words, the thermalization

[13] See Chap. 14.

time must be short compared to the lifetime of the atomic sample.[14] The limitations due to two-body inelastic collisions also have to be considered. An important inelastic process is dipolar relaxation. The low magnetic field seeker states of alkali atoms are not the lowest energy states. Dipolar relaxation in a magnetic trap can therefore yield two types of transitions: those from the upper to the lower hyperfine level, which release a relatively large energy (fraction of a Kelvin), or those between Zeeman sublevels, which release a much lower energy (tens of microKelvin for a magnetic field of a few Gauss).

Like three-body recombination processes, these two-body loss processes depend on the local density. Since in thermal equilibrium the atomic density is highest at the bottom of the trap, inelastic collisions either remove atoms with low potential energy or directly give them an increased kinetic energy by transferring internal energy to the external degrees of freedom. The corresponding depletion of atoms with less than the mean energy acts as an anti-evaporation mechanism, and thus yields a heating of the sample [Söding *et al.* (1998)].

21.4.2 *Advantages of alkali atoms*

The difficulties mentioned in the previous subsection are compensated for by several advantages that we now discuss, and which explain why BEC was first observed with alkali atoms.

A notable advantage of alkali atoms is the spectacular advances in laser cooling techniques developed in the late 1980's that make it possible to cool alkali atoms to the microkelvin regime, a range of temperatures far colder than what was achievable by conventional cryogenic methods. Due to the heavier alkali mass and the lower atomic density (ultimately limited by inelastic processes) at which the gas is prepared, the critical temperature for alkali gases is very low, in the tens of nanokelvin range, far below that of hydrogen. This, however, is not an uncircumventable obstacle thanks to the efficiency of evaporative cooling.

As already emphasized, the time required for an efficient evaporation decreases as the elastic collision rate increases.[15] In this respect, the large scattering length available for many alkali atoms (on the order of a few nm) turns out to be a serious advantage. Finally, the optical diagnostics with alkali atoms using either an absorption imaging technique or phase-contrast imaging [Ketterle *et al.* (1999)] enable the observation of the atomic sample either *in situ* or after a ballistic expansion. These techniques directly give the integral of the atomic density along a column extending through the thickness of the gas along the probe beam direction.

[14] The background pressure should not be a limitation if the experiment is performed in an ultra-high vacuum chamber (pressure below 10^{-11} mbar).

[15] This statement is valid only if the gas is not in the hydrodynamic regime [Roos *et al.* (2003)].

21.5 First observation of Bose-Einstein condensation

As explained in Chap. 13, laser cooling techniques have enabled the attainment of sub-Doppler and even sub-recoil temperatures. However the lowest temperatures obtained with these techniques were not with trapped atoms but with free dilute atomic samples. Starting with ultralow atomic densities is a problem for at least two reasons: (i) it implies an extremely low critical temperature and (ii) the time required for evaporative cooling becomes prohibitively long.

21.5.1 *Time sequence*

Nearly all BEC experiments start by loading atoms into a magneto-optical trap (MOT).[16] The phase space density $n\lambda_{\mathrm{T}}^3$ of a MOT is on the order of 10^{-7}. Both the temperature and the density turn out to be limited in the trap: the random momentum fluctuations associated with spontaneous emission limit the temperature to a factor at least ten above the recoil limit [Dalibard and Cohen-Tannoudji (1989)]; and the re-absorption process of spontaneous photons acts as an effective repulsive force that reduces the spatial density [Walker *et al.* (1990)]. Losses from light-assisted collisions also tend to limit the achievable temperature and density. Many efforts, both experimentally and theoretically, have been devoted to circumventing, at least partially, these limits. One successful scheme that was used in the first BEC experiments [Anderson *et al.* (1995); Davis *et al.* (1995a)] for increasing the spatial density was the so-called *dark spot* technique [Ketterle *et al.* (1993)]. This is a variant of the magneto-optical trap in which the repumping light does not reach a region around the center of the trap. Atoms in this volume are therefore kept in the lower hyperfine state, which does not absorb the cycling light of the MOT. The absence of re-scattering processes leads to an increase of the atomic density by one or two orders of magnitude.

Inspired by the techniques used to implement evaporative cooling of hydrogen, atoms pre-cooled in the MOT are loaded in a non-dissipative trap. There are several types of such traps already described in Chap. 14. The first demonstrations of BEC were made using magnetic traps, a TOP trap for rubidium-87 atoms [Anderson *et al.* (1995)] and a quadrupole trap plus an optical plug for sodium-23 atoms [Davis *et al.* (1995a)]. The non-dissipative trap is mode matched with the MOT cloud, that is adjusted so as to keep the phase space density approximately constant. By changing the strength of the magnetic field, the cloud is compressed to achieve high collision rates and ensure an optimal evaporation efficiency.

The implementation of evaporative cooling with magnetic traps relies primarily on radio-frequency spin flips of atoms from a trapped state to an untrapped state at a controllable location in the trap (see Fig. 21.4). This technique, proposed by David Pritchard [Pritchard *et al.* (1989)] and Jook Walraven [Hijmans *et al.*

[16] See Sec. 14.6 of Chap. 14.

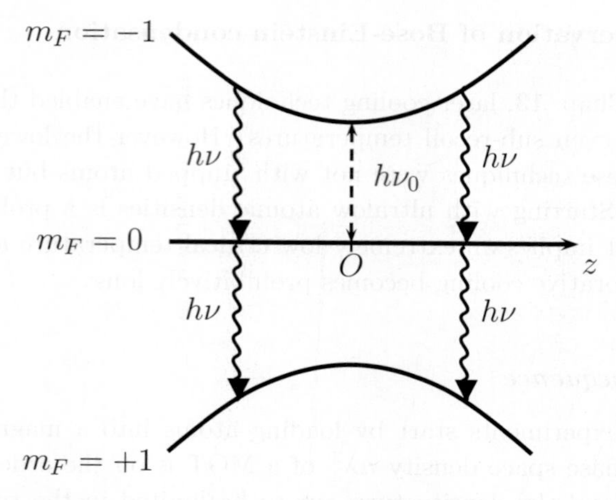

Fig. 21.4 Zeeman sublevels energies of an atom of angular momentum $F = 1$ as a function of the position in space in the presence of a minimum of magnetic field modulus located at O. A radio-frequency of frequency ν drives transitions from the trapped state $m_F = -1$ to $m_F = 0$.

(1989)], is the most commonly used.[17] It is energy selective since a given radio frequency, ν, is resonant on a surface where the modulus of the magnetic field, proportional to the magnetic potential energy experienced by the atoms, is constant: $h\nu = g_F \mu_B ||\vec{B}(\vec{r})||$. As the bottom of the trap corresponds to a non-zero magnetic field $\vec{B}(\vec{0})$, atoms that are evaporated when irradiated by a rf of frequency ν have a total energy E larger than $g_F \mu_B (||\vec{B}(\vec{r})|| - ||\vec{B}(\vec{0})||) = h(\nu - \nu_0)$. In practice, the frequency is progressively ramped from a value that removes the high tail of the energy distribution at the initial temperature to a final value close to ν_0. In this way, the evaporation is forced. The time scale for the evaporation ramp is governed by the thermalization time so as to keep the η parameter sufficiently high at all times. Similarly, for alkali atoms one can drive a transition from a trapped state of an hyperfine level to an untrapped state of the other hyperfine level.

21.5.2 *Signature of Bose-Einstein condensation*

The first observation of Bose-Einstein condensation was made on a cloud of rubidium-87 atoms at JILA by the group of Eric Cornell and Carl Wieman [Anderson *et al.* (1995)]. A few months later, the group of Wolfgang Ketterle at MIT achieved Bose-Einstein condensation of sodium-23 atoms [Davis *et al.* (1995a)]. The signature of condensation was obtained by an absorption imaging performed after releasing the atoms from the trap. Such an image reflects the density profile of

[17]The group of Eric Cornell at JILA has demonstrated another method to evaporate atoms by progressively approaching the trap with a dielectric surface [Harber *et al.* (2003)]. The high energy atoms which have a trajectory with a larger spatial extent, are therefore selectively adsorbed onto the surface. The evaporation is forced by progressively decreasing the distance between the sample and the surface.

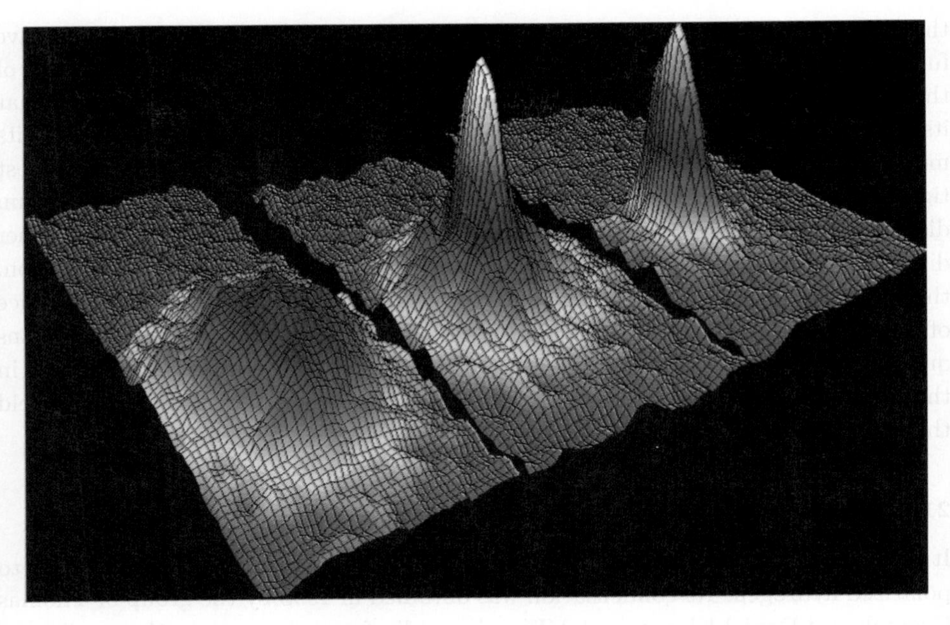

Fig. 21.5 False color images that display the velocity distribution of the cloud. Left: at a temperature above the critical temperature T_c, the cloud is spherical reflecting the Boltzmann isotropic velocity distribution. Middle: below T_c a peak emerges from the thermal cloud at zero velocity, indicating the presence of a Bose-Einstein condensate. Right: after further evaporation, a nearly pure condensate is produced. Courtesy of E. Cornell.

the cloud of atoms. In the presence of the confinement, the condensation not only occurs in momentum space but also in position space. This is directly used to prove the existence of a condensate.

Above the critical temperature, the cloud becomes spherical after a sufficiently long time-of-flight (see Fig. 21.5). Indeed, in this case one observes the isotropic momentum distribution[18] of a gas governed by the classical Boltzmann distribution. Below the critical temperature, a radical change is observed in the images (see Fig. 21.5). A dense peak emerges from the spherical thermal bath of atoms. This peak is narrower than the thermal cloud because it represents the contribution of the atoms condensed in the ground state of the trap, which have a mean square velocity smaller than the atoms of the thermal cloud that populate the excited states of the trap. This narrow peak also reflects the anisotropy of the trap, a feature that can be readily understood. Below the critical temperature, a macroscopic fraction of

[18]For a thermal cloud of initial size, Δx_0 along the x direction and Δy_0 along the y direction, the expansion in a time-of-flight experiment makes the size along the x direction to evolve as a function of time as $\Delta x(t) = [\Delta x_0^2 + (\Delta v)^2 t^2]^{1/2}$ where $\Delta v = (k_B T/m)^{1/2}$ is the velocity dispersion governed by the temperature. A similar result holds for the y direction. After an expansion time large compared to $\Delta x_0/\Delta v$ and $\Delta y_0/\Delta v$, the size of the cloud is $\Delta x(t) \simeq (\Delta v)t$ along the x-axis and $\Delta y(t) \simeq (\Delta v)t$ along the y. The initial anisotropy, $\Delta x_0/\Delta y_0$, if any, disappears in such a measurement performed after a large expansion time ($t \to \infty$) since $\Delta x(t)/\Delta y(t) \to 1$.

the atoms condenses. The atoms of the condensate share the same one-body wave function. In the absence of interactions, this wave function is the ground state of the trap. For a harmonic trap, this wave function is minimal which means that its quadratic size Δx along a given symmetry axis x of the trap is related to its momentum dispersion through the Heisenberg equality: $\Delta x \Delta p_x = \hbar/2$. The most tightly confined direction of the trap has a smaller size and thus a larger momentum dispersion. As a result, the expansion in this direction is faster than in the other directions once the trap is switched off, and during the time-of-flight expansion, the inversion of ellipticity of the condensate shape is observed. In the presence of interactions and in the dilute limit ($n|a|^3 \ll 1$), the same reasoning remains qualitatively valid.[19] A more quantitative treatment of the expansion is given in the next chapter where interactions are taken into account through a mean-field theory.

21.5.3 *Subsequent observation on hydrogen*

It is worth concluding this historical overview on the quest of BEC by coming back to polarized hydrogen. Its condensation was obtained in 1998 by the group of Thomas Greytak and Daniel Kleppner at MIT using radio-frequency evaporative cooling on a magnetically trapped sample [Fried *et al.* (1998)]. Below the critical temperature, a redshift of the two-photon line[20] by an amount proportional to the density was observed, and interpreted as being due to the formation of a condensate with a high density of weakly interacting atoms. The mean field treatment of the interactions valid in this atomic density regime effectively yields an energy shift proportional to the atomic density, as explained in the next chapter. In this context, the increase of atomic density due to the condensation in position space is detected through the interactions.

21.6 Bose-Einstein condensation of other atomic species

After the pioneering experiments on rubidium-87, sodium-23 and hydrogen, BEC was observed using several other atomic species. In this section, we briefly describe the experimental improvements that allowed these developments and the interesting features of the new condensates that have been obtained.

21.6.1 *Experimental improvements*

More systematic use has been made of optical traps and of their versatility. Contrary to magnetic traps which work only for low field seekers, optical traps can confine atoms regardless of their Zeeman sublevel and hyperfine state. As a result, the spin

[19] See Fig. 22.5 of Chap. 22.
[20] See Sec. 21.3.4.

can be used as a new degree of freedom and condensates of atoms in different Zeeman sublevels can be obtained, forming what is called a *spinor condensate*. The order parameter is then a spinor wave function, which is a vector in the internal space state [Barrett *et al.* (2001)]. Quantum fluids with a spin degree of freedom appear in other contexts such as in the superfluid behavior of helium-3 [Vollhardt and Wölfle (1990)] or in the *p*-wave superconductivity [Mackenzie and Maeno (2003)].

Forced evaporation in an optical trap

Optical traps are sometimes used in conjunction with magnetic traps: the condensate is transferred from a magnetic trap into an optical trap. The condensate can also be directly formed in an optical trap. For instance, in 2001, a Bose-Einstein condensate of rubidium-87 atoms was achieved by loading atoms pre-cooled in a MOT directly into a far detuned optical trap made of two crossing beams [Barrett *et al.* (2001)]. In this context, the evaporation is simply carried out by lowering the intensity of both beams as a function of time. The most energetic atoms leave the trap and fall under gravity. A drawback of this method is that the trap is decompressed in the course of the evaporation, which slows down the evaporation rate. However, it turns out that in practice this limitation is not usually too drastic.

Another fundamental advantage of optical traps is that they allow the use of Feshbach resonances obtained by applying a uniform static magnetic field and sweeping it without modification of the trapping potential. Feshbach resonances are very useful for tuning atom-atom interactions. This makes it easier to reach the BEC threshold by optimizing the evaporation efficiency, and is also very interesting for converting atoms into molecules, as explained in Sec. 16.6 of Chap. 16.

A new interesting type of cooling, called *sympathetic cooling*, has also been used. This method consists of cooling one species by interaction with another species that is directly cooled. This technique was proposed in 1978 for the cooling of ions [Wineland *et al.* (1978)], and first demonstrated with different isotopes of magnesium ions where one of them was directly laser cooled [Drullinger *et al.* (1980)]. In 1997 it was extended to neutral atoms, and demonstrated on a mixture of two rubidium spin species with one species cooled directly by evaporative cooling [Myatt *et al.* (1997)]. Later on, this technique was also used with two isotopes of the same species [Schreck *et al.* (2001a); Bloch *et al.* (2001); Truscott *et al.* (2001)] and also with different atomic species [Modugno *et al.* (2001)].

Let us finally mention the development of atom chip experiments. In such devices, microscopic lithographic conductors replace the customary magnetic coils used to produce magnetic trapping. The strong confinement accelerates the condensation process down to one second, which allows significant simplifications in the vacuum system.[21] Furthermore, micromanipulation by time dependent magnetic field configurations is feasible. For instance, adiabatic magnetic transport has been demonstrated [Reichel *et al.* (1999)]. Such systems have a large modularity since one can integrate many atom optical devices on them [Colombe *et al.* (2007)]. The

[21] See [Reichel (2002)] and references therein.

miniaturization therefore appears as an attractive technique for fundamental studies as well as for applications such as atom interferometry or quantum information processing.

21.6.2 *Review of new condensates*

(1) Other alkali atoms

In the condensates described in the previous section, interactions are repulsive (positive scattering length). In the presence of attractive interactions between atoms (negative scattering length,[22]) the number of atoms in the trapped condensate is limited. Above this limit, a collapse of the condensate occurs when the attractive interaction energy overcomes the zero-point kinetic energy.[23] Such a limitation on the number of condensed atoms has been observed with lithium-7 atoms [Bradley *et al.* (1997)]. The condensate had a number of atoms ranging from 650 to 1300 below the theoretical upper bound on the order of 1400, which corresponds to the triplet scattering length of lithium atoms $a = -1.45$ nm.

It has not been possible to observe BEC of cesium-133 in a magnetic trap. Cesium atoms have a negative scattering length in the low field seeking state and huge inelastic collision cross-sections. In an optical trap, atoms can be trapped in the lowest energy Zeeman sublevel $F = 3, M_F = +3$ which is stable with respect to two-body inelastic collisions. Using Feshbach resonances to optimize evaporative cooling, it was finally possible to obtain a condensate of cesium in 2002 [Weber *et al.* (2003)].

Feshbach resonances have also been useful for studying condensates of rubidium-85 atoms [Cornish *et al.* (2000)], and lithium-7 atoms [Khaykovich *et al.* (2002); Strecker *et al.* (2002)]. In the first attempt to directly cool the bosonic isotopes of potassium, some limitations in the temperature and density ranges achievable by laser cooling were observed. The bosonic isotope of potassium, ^{41}K, was later condensed by using sympathetic cooling with evaporatively cooled rubidium [Modugno *et al.* (2001)].

(2) Metastable helium

Helium atoms in the metastable 2^3S_1 state have a high internal energy, on the order of 20 eV and huge ionization Penning cross-sections. In a magnetic trap, these atoms are spin polarized and the conservation of the total spin in a collision considerably reduces these Penning cross-sections [Shlyapnikov *et al.* (1994); Fedichev *et al.* (1996b); Venturi *et al.* (1999)]. Two groups in France obtained a condensate of metastable helium in 2001 [Robert *et al.* (2001); Pereira *et al.* (2001)]. This was the first example of a condensate in which atoms were condensed in a highly internal excited state. This large internal energy

[22]See Sec. 22.5 of Chap. 22.
[23]See Sec. 22.5.3 of Chap. 22.

makes it possible to detect a single atom with both high spatial and temporal resolution when this atom falls onto a micro channel plate.

(3) Chromium
In 2005, the Bose-Einstein condensation of chromium-52 atoms was observed [Griesmaier *et al.* (2005)]. This transition metal has a valence shell in its ground state that contains six electrons with parallel spin alignment. This gives rise to a very high magnetic moment of $6\mu_B$ (μ_B is the Bohr magneton). As the dipole-dipole interaction scales with the square of the magnetic moment, its strength is 36 higher for chromium than for alkali atoms. The effects of this long-range and anisotropic interaction have been directly observed [Stuhler *et al.* (2005)], and even magnified by reducing the short-range contact interaction using a Feshbach resonance [Lahaye *et al.* (2007)].

(4) Ytterbium and rare earth atoms
Bose-Einstein condensation with ytterbium, a rare earth element, or with alkaline earth elements such as calcium or strontium are motivated by two remarkable properties that favor applications in metrology or atom interferometry: their narrow intercombination transitions and their ground state without a magnetic moment. The Kyoto group achieved BEC on different isotopes of ytterbium: ytterbium-174 [Takasu *et al.* (2003)], ytterbium-170 [Fukuhara *et al.* (2007)] and ytterbium-176 [Fukuhara *et al.* (2009)]. Concerning the alkaline earth elements, a BEC of calcium-40 was produced at the Physikalisch-Technische Bundesanstalt in Braunschweig [Kraft *et al.* (2009)], and a BEC of strontium-84 was achieved at Innsbruck and Rice University [Stellmer *et al.* (2009); Martinez de Escobar *et al.* (2009)].

Bose-Einstein condensation of photons

When a gas of bosonic atoms is put in a box and cooled down, it thermalizes by collisions (either between atoms or with the walls of the box) and the number of atoms does not change if losses are negligible. This is why, when the entropy is extremalized for finding the equilibrium state, the conservation of the atom number appears as an auxiliary condition giving rise to a Lagrange multiplier which is the chemical potential. The situation is radically different for a gas of photons. They thermalize by interaction with the walls of the cell (not by photon-photon interactions) and their number varies because they can be absorbed or emitted by the walls. When the temperature T decreases, the total number of photons decreases and tends to zero when $T \to 0$. There is no auxiliary condition for the total number of photons and no chemical potential. The lowest energy mode of the box cannot be macroscopically occupied and there is no BEC.

A recent spectacular experiment [Klaers *et al.* (2011)] has shown that it is possible to circumvent these limitations and to achieve BEC for photons.[24] Several ingredients are used for achieving this goal:

– The photon gas is put in a narrow cavity bounded by two spherical mirrors with a maximum separation of 1.5 μm, and filled with a dye. Because of the small separation of the mirrors, the longitudinal mode spacing is very large and only

[24]For a simple presentation of the experiment, see also [Wilson (2011)].

the longitudinal mode $q = 7$ can interact with the absorption and fluorescence spectrum of the dye. The dye is excited by a pump laser.

- The photons coming from the fluorescence of the dye cannot have an energy lower than the energy of the mode $q = 7$. There is a minimum cut-off of $\hbar\omega_{\text{cutoff}} = 2.1\text{eV}$. The excitation of transverse modes of the cavity can increase the energy of the photons above $\hbar\omega_{\text{cutoff}}$. The dispersion relation $\omega = f(k)$ of the photons trapped in the cavity is no longer a straight line $\omega = ck$, but a curve starting from a minimum at ω_{cutoff} and increasing with k. It looks like the dispersion relation of a massive particle with an effective mass equal to $m_{\text{eff}} = \hbar\omega_{\text{cutoff}}/c^2$, which is on the order of 10^{-35} kg, very much smaller than the mass of an ordinary atom.
- Since $\hbar\omega_{\text{cutoff}} \gg k_B T$, there is no thermal excitation of the dye. When a photon in the cavity is absorbed by the dye, it is always subsequently reemitted as a fluorescence photon. The interaction with the dye does not change the photon number, while allowing them to reach a thermal equilibrium at the temperature of the dye. This allows the realization of a number-conserving thermalization, essential for BEC.
- In order to observe BEC, one usually lowers T until the critical temperature T_c is reached. Because of the very small value of m_{eff}, the critical temperature for this gas of massive photons is much higher than for ordinary atoms, as high as the room temperature. Rather than lowering T, it is then simpler to increase the mean number of trapped photons by increasing the pump intensity.
- At low pump intensity, all transverse modes are populated. The spectrum of the light emitted by the medium extends above ω_{cutoff} and the spatial extent of the image of the cloud is large. When the pump intensity is increased, the spectrum becomes sharper around ω_{cutoff} and a very bright narrow spot becomes visible in the center of the cavity. The observed features are in good agreement with what is theoretically expected for BEC.

21.7 The first experiments on quantum degenerate Fermi gases

One may wonder to what extent the experimental techniques developed for bosonic gases can be transposed to fermionic gases. Laser cooling definitely applies for both systems, but evaporative cooling requires some care when transposed to fermionic gases, as explained below. Furthermore, the situation for fermionic gases at ultralow temperature is radically different from that of bosons, as schematically depicted in Fig. 21.6. Indeed, as the temperature is decreased there is no phase transition for an ideal Fermi gas such as the macroscopic occupation of a single state for bosons, but rather the progressive occupation of the lowest single particle states with at most one particle per state (Pauli exclusion principle). As a direct consequence, if one adds a fermion to a trapped ideal gas of N polarized fermions at zero temperature, it will occupy a level of energy $\geq E_F$. The size of the gas thus increases. This effect has been experimentally observed and is a manifestation of the so-called Fermi pressure [Truscott *et al.* (2001); Schreck *et al.* (2001b)].

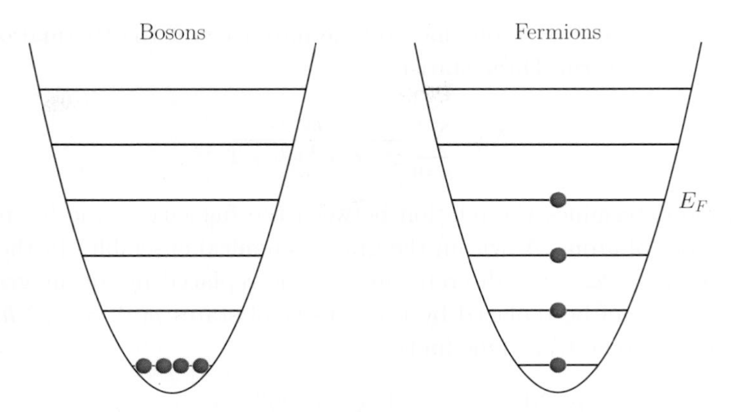

Fig. 21.6 Quantum statistics. Left: bosons at $T = 0$. All atoms of an ideal Bose gas are in the ground state and form a Bose-Einstein condensate. Right: polarized Fermions. Because of the Pauli exclusion principle, each energy level can accomodate at maximum only one polarized fermion. At $T = 0$, the ground state for a system of N fermions is obtained by occupying the N lowest-energy single particle states, thereby forming the so-called Fermi sea. The highest-energy occupied state defines the Fermi level, whose energy is the Fermi energy, E_F.

21.7.1 *Ideal Fermi gas in a three-dimensional harmonic trap*

As with bosons,[25] it is instructive to work out the quantum statistical calculations for the ideal polarized Fermi gas confined in a harmonic trap. The total number of single-particle states with an energy smaller than a given energy $E_{\tilde{n}}$ is given by

$$S[\tilde{n}] = \sum_{n=0}^{\tilde{n}} g_n = \frac{(\tilde{n}+1)(\tilde{n}+2)(\tilde{n}+3)}{6}. \tag{21.12}$$

Let N be the total number of polarized fermions in the trap. If $N \gg 1$, the largest value $\tilde{n}$ of the vibration quantum number is, according to (21.12), given by:

$$\tilde{n} \simeq (6N)^{1/3}. \tag{21.13}$$

and the largest energy E_F of the atoms, also called the *Fermi energy* is equal to:

$$E_F = [\tilde{n} + (1/2)]\,\hbar\omega_0 \simeq \hbar\omega_0(6N)^{1/3}. \tag{21.14}$$

From the Fermi energy E_F, one can define the *Fermi temperature* T_F

$$T_F = E_F/k_B \simeq (\hbar\omega_0/k_B)(6N)^{1/3}. \tag{21.15}$$

The Fermi gas enters the degenerate regime when T is lower than T_F.

It is also interesting to evaluate the order of magnitude of the spatial extension of the atomic cloud, which is on the order of the spatial extent of the wave function of the highest occupied state:

$$(\Delta x)_{\tilde{n}} \simeq (\Delta x)_0\sqrt{\tilde{n}} \simeq (\Delta x)_0(6N)^{1/6} \tag{21.16}$$

where $(\Delta x)_0$ is the spatial extent of the ground state.

[25] See Sec. 21.3.2.

At a temperature $T \neq 0$, one has to take into account the thermal occupation of each state using Fermi-Dirac statistics

$$N = \sum_{n=0}^{\infty} \frac{g_n}{z^{-1} e^{\varepsilon_n / k_B T} + 1}. \tag{21.17}$$

This equation determines the relation between the fugacity z, the temperature T and the number of atoms N within the grand canonical ensemble. In the semiclassical limit, $k_B T \gg \hbar\omega_0$, the discrete sum can be replaced by an integral and the degeneracy term can be replaced by the density of states $\rho(\varepsilon) = \varepsilon^2 / 2(\hbar\omega_0)^3$ for a 3D isotropic harmonic trap. One finds

$$N = \int_0^{\infty} \frac{\rho(\varepsilon) \mathrm{d}\varepsilon}{z^{-1} e^{\varepsilon / k_B T} + 1} = -\frac{1}{x^3} \sum_{n=1}^{\infty} \frac{(-z)^n}{n^3} = -\frac{1}{x^3} \mathrm{Li}_3(-z), \tag{21.18}$$

where Li_s is the polylogarithm function of index s.

21.7.2 *Cooling fermions*

Pioneering work on low temperature dilute Fermi gases in view of attaining the degenerate regime were initiated with deuterium [Silvera and Walraven (1980b)]. Experiments devoted to the study of degenerate fermionic gases re-started in 1997 taking advantage of the powerful laser and evaporative cooling techniques. As already emphasized, some care had to be taken to implement evaporative cooling with fermions. Indeed, collisions between identical fermions are suppressed at low energy since quantum statistics forbids s-wave collisions between identical fermions. Since evaporative cooling relies on binary elastic collisions to rethermalize the gas after selective removal of the most energetic atoms, it cannot be applied directly to a gas of cold polarized fermions. To circumvent this limitation, two species interacting through a sufficiently large interspecies s-wave cross-section were used instead. The first realization of an ultracold Fermi degenerate gas by Brian DeMarco and Deborah Jin at JILA involved thermal contact between potassium-40 atoms in two different hyperfine states [DeMarco and Jin (1999)].

Sympathetic cooling, described in the previous section, has been essential for reaching quantum degeneracy with fermionic atoms. For instance, the group of Massimo Inguscio at LENS reached quantum degeneracy in a mixed gas of fermionic potassium-40 and bosonic rubidium-87 using evaporative cooling on rubidium atoms only [Roati *et al.* (2002)]. Degenerate Fermi gases of lithium-6 have been realized using various schemes: with lithium-7 as a coolant [Truscott *et al.* (2001); Schreck *et al.* (2001b)], with sodium-23 as a coolant [Hadzibabic *et al.* (2002)], and with a mixture of optically trapped atoms in the two lowest hyperfine states [Granade *et al.* (2002)]. Degenerate Bose-Fermi mixture of metastable helium atoms has also been produced by sympathetic cooling of helium-3 with helium-4 [McNamara *et al.* (2006)].

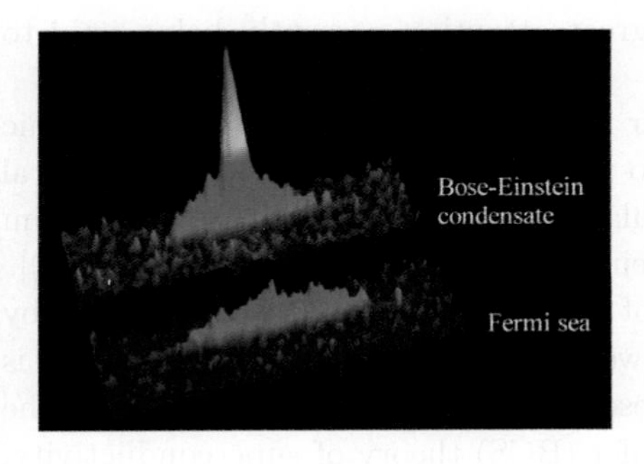

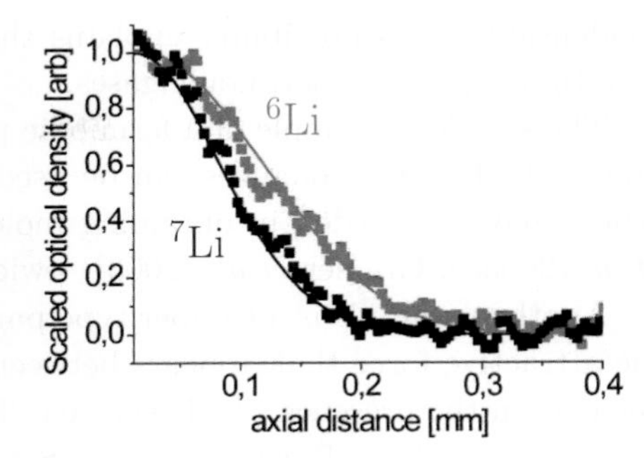

Fig. 21.7 Left: picture of a Bose-Einstein condensate of lithium-7 immersed in a Fermi sea of lithium-6 atoms that have been imaged separately. Courtesy of C. Salomon. Right: observation of Fermi pressure. Spatial distributions integrated over the imaging axis of lithium Fermi-Bose mixture. The temperature is $1.4\ \mu\text{K} = 1.1\ T_C$ for the bosons and $0.33\ T_F$ for the fermions [Schreck *et al.* (2001b)]. Copyright: American Physical Society.

21.7.3 *Spatial distribution and Fermi pressure*

The first studies on degenerate Fermi gases were devoted to the observation of the Fermi pressure as a signature of the degenerate regime [Truscott *et al.* (2001); Schreck *et al.* (2001b)]. Indeed, for polarized fermions, the Pauli exclusion principle restricts the occupation number of a given single particle state to one. At $T = 0$, atoms completely fill the lowest energy single particle states up to the Fermi energy E_F (see Fig. 21.6). As a result, the size of a Fermi degenerate gas at $T = 0$ in the presence of the confinement increases with the number of atoms. For a 3D isotropic harmonic trapping, the quadratic size of the Fermi gas, $(\Delta x)_F$, at $T = 0$ is on the order of $\sim N^{1/6}\Delta x_0$ where Δx_0 is the quadratic size of the ground state wave function (see Eq. (21.16)). This effect has been clearly observed experimentally [Truscott *et al.* (2001); Schreck *et al.* (2001b)]. Figure 21.7 shows the spatial distribution obtained by absorption imaging of a mixture of the bosonic isotope of lithium-7 and of the fermionic isotope lithium-6 . The two distributions can be imaged separately using the isotope shift of the resonance lines of the two isotopes which allows each of them to be selectively excited. The temperature is such that the bosonic gas is just above the critical temperature for BEC while the fermionic gas is in the degenerate regime $T = 0.33T_F < T_F = E_F/k_B$. However, the size of the Fermi gas is larger than that of the bosonic gas, a direct consequence of Fermi pressure.

21.7.4 *Pairs of fermionic atoms*

Consider a pair of two fermionic atoms. The total number of fermions of such a system (electrons, protons, neutrons) is even. A bound pair of fermions is therefore a composite boson and obeys Bose-Einstein statistics. At ultralow temperatures and high enough densities, an ensemble of such pairs can condense and form a condensate. This possibility explains the great attention presently being paid to quantum degenerate fermionic gases.

The simplest example of a fermionic pair is a molecule formed by two fermionic atoms. Feshbach resonances can be used to produce such molecules[26] and several groups have succeeded in obtaining molecular Bose-Einstein condensates [Jochim *et al.* (2003a); Greiner *et al.* (2003); Zwierlein *et al.* (2003); Bourdel *et al.* (2004)].

Another example is a Cooper type pair of two fermionic atoms weakly bound by the attractive force that appears between two atoms when the scattering length is negative in the presence of a Fermi sea. These pairs are somewhat analogous to the Cooper pairs of the Bardeen-Cooper-Schrieffer (BCS) theory of superconductivity: in a metal, two electrons can feel a weak long range attraction due to the exchange of a phonon, and form, in the presence of the Fermi sea of the other electrons, weakly bound pairs that can condense into a superfluid phase, thereby explaining the vanishing of the resistance. Is it possible to observe analogous phenomena with quantum degenerate fermionic gases? Important results have been obtained in this research field and fruitful connections between ultracold atom physics and many body physics have been established. We shall return to these topics and analogies in Chap. 26.

21.8 Conclusion

The road from the initial proposal of Einstein in 1925 to the first observation of gaseous BEC in 1995 has been long. The pioneering work on hydrogen by the groups of Daniel Kleppner and Thomas Greytak at MIT and Isaac Silvera and Jook Walraven at Amsterdam played a crucial role for the success of this ambitious enterprise, even if BEC was first observed on alkalis and not on hydrogen. Several ingenious schemes, such as wall free confinement of neutral atoms and evaporative cooling, were implemented along this road and turned out to be very useful, not only for BEC, but for other investigations in ultracold atom physics.

One could argue that the observation of gaseous BEC is a spectacular feat, but that it does not really bring new ideas, since it is fully predictable from Bose-Einstein statistics whose foundations were established several decades ago. In fact, this conclusion is completely erroneous, as demonstrated by the spectacular developments in this field since the first observation of BEC. The key point is that gaseous condensates are an ideal paradigm for simulating more complex situations found in other fields of physics to gain a deeper understanding of them. The primary reason is the high degree of control that one has over the various physical parameters that characterize the condensate: shape, time dependence and dimensionality of the trapping potential, especially with the use of optical potentials; strength of the atom-atom interactions with the use of Feshbach resonances; control of the in-

[26]See Sec. 16.6 of Chap. 16.

ternal degrees of freedom with the use of optical pumping; precise and sensitive optical diagnosis of the spatial and momentum distributions. In this way, one can observe and study several interesting phenomena, such as the competition between tunneling and interactions in a double well potential (situation analogous to the one occurring in a Josephson junction), in a periodic array of potential wells provided by optical lattices (situation occurring in the superfluid-Mott insulator transition). One can study phase transitions like the Berezinskii-Kosterlitz-Thouless transition in two-dimensional systems, or investigate the effect of disorder on Anderson localization. With pairs of fermions, one can study the crossover between molecular condensates and superfluid phases of Cooper type pairs of fermionic atoms (equivalent of the BEC-BCS crossover). Of particular interest is the possibility to study regimes in which the atom-atom interactions are very strong (at the center of a Feshbach resonance) or where the populations of the two-spin states of the fermionic atoms are unequal. The number of experimental and theoretical publications dealing with these problems has increased in a spectacular way, as has the number of meetings gathering atomic and condensed matter physicists. In Chap. 26, we will review a few of these recent developments showing the great vitality of the field. Beforehand, we will briefly describe in the next chapters a few basic properties of gaseous condensates, such as mean field interactions, coherence and superfluidity.

Chapter 22

Mean field description of a Bose-Einstein condensate

22.1 Introduction

In the previous two chapters, Bose-Einstein condensation was presented for the case of an ideal gas. We now explain how it is possible to include the effect of atom-atom interactions in a regime where the mean field approximation is valid.

In the absence of interactions, a condensate of N atoms trapped in an external potential is a product of N identical three-dimensional wave functions describing each atom in the ground state φ_0 of the trapping potential. In the presence of interactions, the ground state Ψ of the system of N atoms is no longer a product of three-dimensional wave functions, but a $3N$-D wave function. In Sec. 22.2, we look for the best product of N identical wave functions φ to approximate the true ground state Ψ. This amounts to using a variational approximation to describe the condensate. We show that this approach leads to a nonlinear equation called the *Gross-Pitaevskii equation* (GP equation) that describes how the wave function of each atom is determined by the combined effect of the kinetic energy, the trapping potential and the mean field exerted on a given atom by the $N-1$ other atoms. This explains the name *mean field approximation* given to this treatment. In Chap. 24, we will discuss the first corrections to this approximation and we will show that, for an homogeneous condensate, they scale as $(na^3)^{1/2}$, where n is the spatial density and a is the scattering length. The GP equation is therefore valid for dilute gases that satisfy $na^3 \ll 1$.

In Sec. 22.3, we show that interactions introduce a new length scale associated with the many-body wave function, and called the *healing length* ξ which is determined by a balance between the confinement and interaction energies. The physical meaning of ξ is analyzed for the simple case of a condensate in a box, and it is shown that ξ characterizes the typical distance beyond which the condensate "forgets" a local perturbation.

Section 22.4 is devoted to the important case of a condensate in a harmonic trap. We show that simple scaling arguments provide an estimate of the size of the condensate as a function of the number N of atoms, the width a_{ho} of the ground state of the harmonic potential and the scattering length a. The important dimensionless parameter $\chi = Na/a_{\rm ho}$ is introduced to characterize the importance of interactions. When $\chi \gg 1$ and $a > 0$, a regime called the *Thomas-Fermi* regime, the kinetic energy term can be neglected in the GP equation which becomes simple algebraic equations for the shape of the condensate and various physical quantities.

Another important issue is the role played by the sign of the scattering length, which turns out to have dramatic consequences on the stability of the condensate, as discussed in Sec. 22.5.

In Sec. 22.6, we work out the solution of the Gross-Pitaevskii equation with a non-zero angular momentum in an axially symmetric trap. This will allow us to introduce new states of the condensate called *quantum vortices* which will be useful for the discussion of the rotating properties of weakly interacting Bose-Einstein condensates presented in Chap. 24.

In several experiments performed with Bose-Einstein condensates, the trapping potential is varied in time. To describe the condensate dynamics in this context, we use the time-dependent Gross-Pitaevskii equation that can be derived from a least-action principle (Sec. 22.7). We also recast this equation as a set of hydrodynamic equations (Sec. 22.7.2). This formalism is used to describe the ballistic expansion of the condensate.The collective modes are finally discussed for a harmonically trapped Bose-Einstein condensate in the appendix (Sec. 22.9).

22.2 Mean field description of the condensate

In this section, we consider N identical bosons trapped in an external potential $V_{\rm trap}(\vec{r})$ in equilibrium at a temperature $T = 0$. In the absence of interactions, all atoms are in the ground state $|\varphi_0\rangle$ of the trap, and the N-body wave function of the condensate is $|\psi\rangle = |\varphi_0(1)\rangle \otimes |\varphi_0(2)\rangle \otimes \cdots \otimes |\varphi_0(N)\rangle$.

In the presence of interactions, the wave function that describes the structure of the condensate is the ground state wave function of the N-body Hamiltonian:

$$H = \sum_{i=1}^{N} \left[\frac{\vec{p_i}^{\,2}}{2m} + V_{\rm trap}(\vec{r_i}) \right] + \frac{1}{2} \sum_{i} \sum_{j \neq i} V(\vec{r_i} - \vec{r_j}), \qquad (22.1)$$

where the terms $V(\vec{r_i} - \vec{r_j})$ account for two-body interactions. In most cases, it is impossible to determine the exact expression for the ground state energy of H and the associated N-body wave function. Alternatively, one can resort to an approximate determination of the ground state using a variational approach.

22.2.1 *Variational calculation of the condensate wave function*

Such a calculation is performed within a given family of functions. By extension of the exact N-body wave function in the absence of interactions, we restrict ourselves to the family of tensor products of N single-particle identical states:

$$|\psi\rangle = |\varphi(1)\rangle \otimes |\varphi(2)\rangle \otimes \cdots \otimes |\varphi(N)\rangle \,. \qquad (22.2)$$

These states are by definition symmetric in particle permutations, as required for identical bosons. Being tensor products of N states, they cannot describe the quantum correlations between the N atoms.

In the subspace generated by the vectors (22.2), the best state to approximate the ground state of H minimizes the energy functional $E_{\text{tot}}[\varphi]$ defined by:

$$E_{\text{tot}}[\varphi, N] = \langle H \rangle = \frac{\langle \psi \,|H|\, \psi \rangle}{\langle \psi | \psi \rangle}, \tag{22.3}$$

with the constraint $\langle \psi | \psi \rangle = 1$, or equivalently $\langle \varphi | \varphi \rangle = 1$.

The method of Lagrange multipliers permits one to recast the problem as a minimization of $\langle \psi | H | \psi \rangle - \mu \langle \psi | \psi \rangle$, where μ is the Lagrange multiplier associated with the conservation of the norm of the wave function. The functional differentiation $\delta(\langle \psi |\, H \,| \psi \rangle - \mu \, \langle \psi \,| \psi \rangle) = 0$ gives

$$\begin{aligned} N \int \mathrm{d}^3 r \delta\varphi^*(\vec{r}) \Big\{ & -\frac{\hbar^2}{2m} \Delta\varphi(\vec{r}) + V_{\text{ext}}(\vec{r})\varphi(\vec{r}) \\ & + (N-1) \left[\int \mathrm{d}^3 r' V(\vec{r} - \vec{r}\,') |\varphi(\vec{r}\,')|^2 \right] \varphi(\vec{r}) - \mu\varphi(\vec{r}) \Big\} + \text{c.c.} = 0. \end{aligned} \tag{22.4}$$

Since the variations of $\delta\varphi^*$ and $\delta\varphi$ can be considered to be independent, the coefficient of $\delta\varphi^*$ must vanish, yielding:

$$\begin{aligned} & -\frac{\hbar^2}{2m} \Delta\varphi(\vec{r}) + V_{\text{trap}}(\vec{r})\varphi(\vec{r}) \\ & + (N-1) \left[\int \mathrm{d}^3 r' \, V\left(\vec{r} - \vec{r}\,'\right) \left|\varphi\left(\vec{r}\,'\right)\right|^2 \right] \varphi(\vec{r}) = \mu\varphi(\vec{r}). \end{aligned} \tag{22.5}$$

This equation, which resembles the Schrödinger equation, gives the evolution of each atom in the trapping potential and in the mean-field created at its position by $(N-1)$ other atoms.[1]

22.2.2 *Stationary Gross-Pitaevskii equation*

The variational method neglects the correlations between atoms at short distances, and therefore the gas is supposed to be dilute ($n|a|^3 \ll 1$). In this approximation, atoms are essentially far away from each other, and the interactions are governed by the large distance asymptotic behavior of the wave function. Under this assumption, one can replace the true interaction potential with the corresponding pseudo-potential, introduced in Chap. 15, that has the same scattering length a as the real potential [Dalibard (1999); Castin (2001)]. In Eq. (23.4), the wave functions upon which $V\left(\vec{r} - \vec{r}\,'\right)$ acts are not singular for $\vec{r} = \vec{r}\,'$. We can thus use the delta potential form of the pseudo-potential without regularization:

[1]The method used here is analogous to the Hartree-Fock approach used in atomic physics to describe poly-electronic atoms. In this approach, the evolution of each electron is described by an effective potential that takes into account the attraction of the nucleus and the average repulsive effect of the other electrons (except for the fact that in that case, one must obviously use antisymmetric functions).

$V_{\rm pseudo}(\vec{r}-\vec{r}\,') = g\,\delta(\vec{r}-\vec{r}\,') = (4\pi\hbar^2 a/m)\,\delta(\vec{r}-\vec{r}\,')$. With such a contact potential, Eq. (23.4) takes the simple form

$$-\frac{\hbar^2}{2m}\Delta\varphi(\vec{r}) + V_{\rm trap}(\vec{r})\,\varphi(\vec{r}) + (N-1)\,g\,|\varphi(\vec{r})|^2\,\varphi(\vec{r}) = \mu\,\varphi(\vec{r}). \qquad (22.6)$$

Usually, we deal with a sufficiently large number of atoms ($N \gg 1$) that we can replace in the previous equation $N-1$ by N. Equation (22.6), referred to as the stationary Gross-Pitaevskii equation, plays a central role in the study of the static properties of Bose-Einstein condensates in the dilute limit [Dalfovo *et al.* (1999)].

In order to relate the Lagrange multiplier μ to a known physical quantity, we replace the real interaction potential into the energy functional (22.3) with $V_{\rm pseudo}$, and get:

$$E_{\rm tot}[\varphi, N] = N\int d^3r\varphi^*(\vec{r})\left[-\frac{\hbar^2}{2m}\Delta + V_{\rm trap}(\vec{r}) + \frac{(N-1)g}{2}|\varphi(\vec{r})|^2\right]\varphi(\vec{r}). \qquad (22.7)$$

$E_{\rm tot}[\varphi, N]$ depends explicitly on the number of atoms N, and also implicitly through the N-dependence of φ so that:

$$\frac{dE_{\rm tot}[\varphi,N]}{dN} = \frac{\partial E_{\rm tot}[\varphi,N]}{\partial N} + \frac{\delta E_{\rm tot}[\varphi,N]}{\delta\varphi}\frac{\partial\varphi}{\partial N} = \frac{\partial E_{\rm tot}[\varphi,N]}{\partial N} + 0 \qquad (22.8)$$

$$= \int d^3r\varphi^*(\vec{r})\left[-\frac{\hbar^2}{2m}\Delta + V_{\rm trap}(\vec{r}) + \left(N-\frac{1}{2}\right)g\,|\varphi(\vec{r})|^2\right]\varphi(\vec{r}),$$

where we have explicitly used the fact that the functional derivative $\delta E_{\rm tot}[\varphi,N]/\delta\varphi$ vanishes since φ is such that $E_{\rm tot}[\varphi, N]$ is extremal for any variation of φ. The Gross-Pitaevskii equation (22.6) gives an integral expression for the Lagrangian multiplier μ:

$$\mu = \int d^3r\varphi^*(\vec{r})\left[-\frac{\hbar^2}{2m}\Delta + V_{\rm trap}(\vec{r}) + (N-1)g\,|\varphi(\vec{r})|^2\right]\varphi(\vec{r}), \qquad (22.9)$$

where we have used the normalization property $\langle\varphi|\varphi\rangle = 1$. If we compare Eq. (22.8) with Eq. (22.9), in the limit of large N, we deduce that

$$\mu(N) = \frac{\partial E_{\rm tot}[\varphi]}{\partial N} = E_{\rm tot}[\varphi, N] - E_{\rm tot}[\varphi, N-1]. \qquad (22.10)$$

The Lagrange multiplier μ therefore corresponds to the variation of the total mean energy when N varies by one unit, which is nothing but the definition of the chemical potential.

22.2.3 *Expression of the various quantities in terms of the spatial density*

From Eq. (22.7), we can write the total energy $E_{\rm tot}$ as a sum of three terms $E_{\rm tot} = E_{\rm kin} + E_{\rm trap} + E_{\rm int}$ that can be expressed in terms of the spatial density $n(\vec{r}) = N\,|\varphi(\vec{r})|^2$:

- the kinetic energy due to the confinement:[2]

$$E_{\text{kin}} = N\frac{\hbar^2}{2m}\int \mathrm{d}^3r \left|\vec{\nabla}\varphi\left(\vec{r}\right)\right|^2, \tag{22.11}$$

- the trapping energy:

$$E_{\text{trap}} = N\int \mathrm{d}^3r V_{\text{trap}}\left(\vec{r}\right)\left|\varphi\left(\vec{r}\right)\right|^2 = \int \mathrm{d}^3r\, V_{\text{trap}}\left(\vec{r}\right) n\left(\vec{r}\right), \tag{22.12}$$

- and the interaction energy:

$$E_{\text{int}} = \frac{N\left(N-1\right)}{2}\, g\int \mathrm{d}^3r\, \left|\varphi\left(\vec{r}\right)\right|^4 \simeq \frac{g}{2}\int \mathrm{d}^3r \left[n\left(\vec{r}\right)\right]^2. \tag{22.13}$$

It is instructive to rewrite the chemical potential in terms of these three energies. Multiplying the Gross-Pitaevskii equation (22.6) by $\varphi^*(\vec{r})$ and integrating over r gives

$$\mu = \frac{E_{\text{kin}} + E_{\text{trap}} + 2E_{\text{int}}}{N} = \frac{E_{\text{tot}} + E_{\text{int}}}{N}. \tag{22.14}$$

We conclude, contrary to the ideal case, that the chemical potential is not equal to the mean total energy per atom ($\mu \neq E_{\text{tot}}/N$). This is due to the fact that, contrary to E_{trap} and E_{kin}, E_{int} does not increase linearly with N.

Finally, we give an extra relation between the three energies that enter the expression of the total energy and which is valid for a harmonic trapping potential:

$$2E_{\text{kin}} - 2E_{\text{trap}} + 3E_{\text{int}} = 0. \tag{22.15}$$

This equation results from the virial theorem [Dalfovo *et al.* (1999)].

22.3 Condensate in a box and healing length

22.3.1 *Condensate in a one-dimensional box*

The atom-atom interactions yield a new characteristic length associated with the many-body wave function, the healing length. Its physical meaning becomes apparent by considering a three-dimensional condensate in a box of volume L^3 with periodic boundary conditions along two axes and strict boundary conditions along the planes $z = 0$ and $z = L$. In the absence of interactions, all the atoms are in the ground state of the trap. Their wave function is thus given by

$$\Psi(x,y,z) = \frac{1}{L}\varphi_0\left(z\right) \;\text{ with }\; \varphi_0\left(z\right) = \frac{2}{\sqrt{L}}\sin\left(\frac{\pi\, z}{L}\right), \tag{22.16}$$

and the corresponding atomic linear density $N|\varphi_0|^2$ is inhomogeneous. In the presence of interactions, the wave function $\varphi(z)$ still has to vanish at $z = 0$ and $z = L$, but tends to be homogeneous far from the walls since this minimizes the interaction energy for repulsive interactions (see Fig. 22.1(b)).

[2] We have used the relation $\vec{\nabla}(\varphi^*\vec{\nabla}\varphi) = \varphi^*\Delta\varphi + |\vec{\nabla}\varphi|^2$ and the divergence theorem.

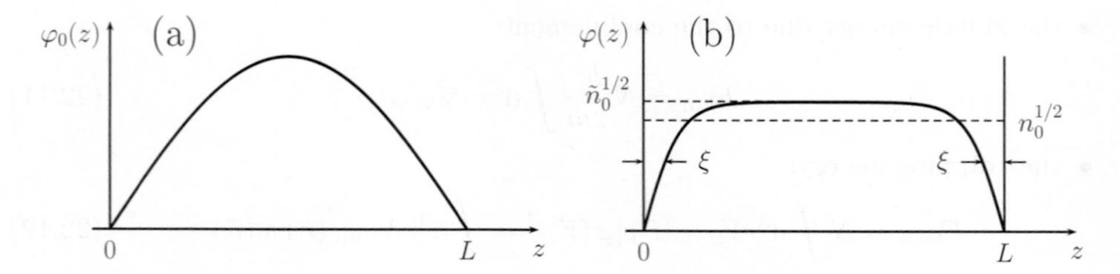

Fig. 22.1 (a) Ground state wave function in a one-dimensional box of size L with strict boundary conditions. (b) Ground state wave function with the same confinement but in the presence of interactions that tend to flatten the wave function. The healing length ξ is the characteristic distance beyond which the boundary condition does not affect the wave function.

22.3.2 *Healing length*

One may wonder what is the characteristic length scale ξ over which the wave function, in the presence of interactions, varies from 0 at the wall position to its constant value $\tilde{n}_0$ (see Fig. 22.1(b)). With the total number of atoms N fixed, the depletion of atoms near the walls increases the spatial density $\tilde{n}_0$ far from the walls and thus the interaction energy E_{int}. We deduce that when ξ increases, the interaction energy E_{int} increases, while simultaneously, the kinetic energy E_{kin} decreases since the gradient of density near the walls is weaker. The equilibrium shape of the condensate corresponds to the value of ξ for which the sum of these two energies is minimum. This characteristic length ξ is called the healing length, and more generally, represents the length after which the condensate "recovers from" a local perturbation (which, here, is due to the walls).

The order of magnitude of the healing length is readily obtained by the scaling of both the kinetic and interaction energies. The order of magnitude of E_{kin} is given by

$$E_{\text{kin}}(\xi) = \frac{\hbar^2}{2m}\int \mathrm{d}z \left(\frac{\partial\sqrt{n}}{\partial z}\right)^2 \simeq \frac{\hbar^2}{2m}2\xi\frac{n}{\xi^2} \simeq \frac{\hbar^2 n_0}{m\xi}. \tag{22.17}$$

In order to estimate the interaction energy, one needs the expression for the density plateau $\tilde{n}_0$. From Fig. 22.1, one finds approximately $\tilde{n}_0 - n_0 \simeq (2\xi/L)\, n_0$, which implies $\tilde{n}_0 \simeq (1 + 2\xi/L)\, n_0$ assuming $\xi \ll L$. The order of magnitude of E_{int} is given by

$$E_{\text{int}}(\xi) = \frac{g}{2}\int \mathrm{d}z\, n^2 \simeq \frac{g}{2}\,(L - 2\xi)\, n_0^2\, (1 + 2\xi/L)^2 \simeq \frac{g n_0^2 L}{2} + g\, n_0^2\, \xi. \tag{22.18}$$

As expected intuitively, $E_{\text{kin}}(\xi)$ is a decreasing function of ξ and $E_{\text{int}}(\xi)$ an increasing function of ξ. The order of magnitude of the healing length is obtained by minimizing $E_{\text{kin}}(\xi) + E_{\text{int}}(\xi)$:

$$\frac{\partial}{\partial\xi}\left(\frac{\hbar^2}{m\xi} + g\, n_0 \xi\right) = 0 \qquad \Rightarrow \qquad \xi \simeq \frac{\hbar}{\sqrt{m\, g\, n_0}} = \frac{1}{\sqrt{4\,\pi\, a\, n_0}}. \tag{22.19}$$

Alternatively, one can solve the Gross-Pitaevskii equation for a one-dimensional box:

$$-\frac{\hbar^2}{2m}\frac{\mathrm{d}^2\varphi}{\mathrm{d}z^2} + Ng\varphi^3(z) = \mu\varphi(z), \tag{22.20}$$

with $\varphi(z)$ a real function that obeys the boundary conditions $\varphi(z=0)=\varphi(z=L)=0$. Far from the walls ($z \sim L/2$), one can neglect $\mathrm{d}^2\varphi/\mathrm{d}z^2$ and deduce the approximate value for the chemical potential $\mu \simeq Ng\varphi^2(z) = gn(z) \simeq g\tilde{n}_0$. Let us introduce the *standard definition of the healing length*,

$$\xi_0 = \left(\frac{\hbar^2}{2mg\tilde{n}_0}\right)^{1/2} = (8\pi a\tilde{n}_0)^{-1/2}. \tag{22.21}$$

Using the dimensionless variable $\zeta = z/\xi_0$, the stationary Gross-Pitaevskii equation (22.20) can be rewritten in the form:

$$\frac{\mathrm{d}^2\varphi(\zeta)}{\mathrm{d}\zeta^2} - \frac{N}{n_0}\varphi^3(\zeta) + \varphi(\zeta) = 0. \tag{22.22}$$

The solution of Eq. (22.22) reads $\varphi(\zeta) = \sqrt{n_0/N}\,\mathrm{th}\left(\zeta/\sqrt{2}\right)$. Starting from 0 at $z=0$, the wave function reaches a constant value after a few healing lengths ξ_0.

22.4 Condensate in a harmonic trap

Condensates are usually formed in a magnetic trap or a far-off resonance dipole trap[3] where they experience harmonic confinement. Let us first consider an isotropic harmonic trap of angular frequency ω_0. In the absence of interactions, the ground state is a Gaussian whose characteristic size is the oscillator length $a_{\mathrm{ho}} = (\hbar/m\omega_0)^{1/2}$. Atom-atom interactions modify the condensate wave function. In the following, we shall use a Gaussian ansatz for the wave function of a weakly interacting condensate. The width of this Gaussian is denoted wa_{ho}, where w is a dimensionless parameter that is estimated by minimizing the total energy. The width estimated in this manner results from a balance between the kinetic, harmonic potential and interaction energies. This approach is useful to classify the different interacting regimes.

22.4.1 *Total energy and the different interaction regimes*

The Gaussian ansatz for the one-body wave function reads:

$$\varphi(\vec{r}) = \frac{1}{\pi^{3/4}\left(w^3 a_{\mathrm{ho}}^3\right)^{1/2}}\exp\left[-\frac{r^2}{2w^2a_{\mathrm{ho}}^2}\right]. \tag{22.23}$$

This ansatz combined with Eqs. (22.11)–(22.13) yields

$$\begin{aligned} E_{\mathrm{tot}}[w] &= E_{\mathrm{kin}}[w] + E_{\mathrm{trap}}[w] + E_{\mathrm{int}}[w], \\ &= N\hbar\omega_0\left[\frac{3}{4}\frac{1}{w^2} + \frac{3}{4}w^2 + \frac{1}{\sqrt{2\pi}}\frac{aN}{a_{\mathrm{ho}}}\frac{1}{w^3}\right]. \end{aligned} \tag{22.24}$$

[3] See Chap. 14.

In the absence of interactions ($a = 0$), the last term of Eq. (22.24) vanishes, and the minimum of the sum of kinetic and trapping energies is obtained for $w = 1$. In this limit, we recover the well-known expression for the ground state wave function of an harmonic oscillator. If the scattering length is non-zero, the order of magnitude of the interaction energy compared to the kinetic and trapping energies for $w = 1$ is determined by the dimensionless parameter:

$$\chi = Na/a_{\text{ho}}. \tag{22.25}$$

If $\chi \ll 1$, interactions can be ignored. In the opposite limit ($\chi \gg 1$), referred to as the Thomas-Fermi limit, the last term of Eq. (22.24) plays a crucial role and one must determine the new value of w that minimizes the total energy. The result depends on the sign of the scattering length: if $a > 0$, the effective interactions are repulsive and w scales as $N^{1/5}$, i.e. the size of the ground state wave function increases with the number of condensed atoms; if $a < 0$, the effective interactions are attractive and one finds $w < 1$ when a solution does exist (see Sec. 22.5).

22.4.2 *Condensate with a positive scattering length and the Thomas-Fermi limit*

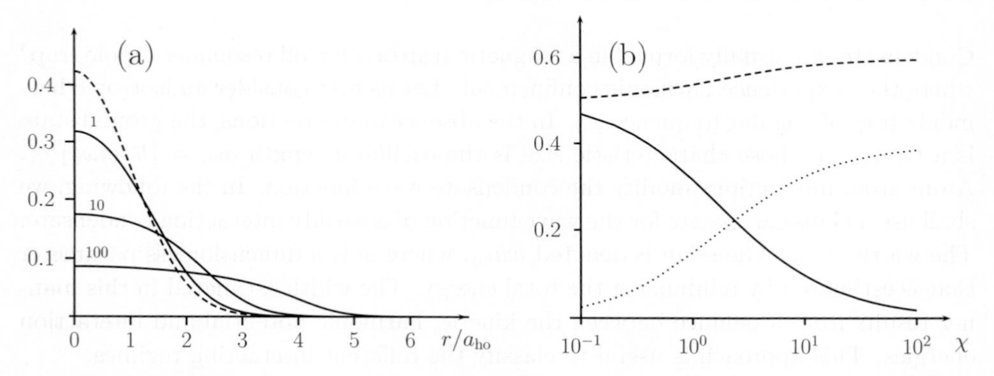

Fig. 22.2 Condensate wave function, at $T = 0$, obtained by solving numerically the stationary Gross-Pitaevskii equation Eq. (22.6) in a spherical trap and with repulsive interactions. The dashed line corresponds to the ideal gas ($a = 0$); the solid lines to $\chi = 1, 10, 100$. Figure courtesy of S. Giorgini. (b) Relative contributions of the different energy terms to the total energy as a function of the dimensionless parameter $\chi = Na/a_{\text{ho}}$ with $a > 0$: $E_{\text{trap}}/E_{\text{tot}}$ (dashed line), $E_{\text{kin}}/E_{\text{tot}}$ (solid line), $E_{\text{int}}/E_{\text{tot}}$ (dotted line).

For a positive scattering length, both the kinetic and interaction energies are decreasing functions of w, whereas the trapping energy increases with w. There is always a value of w that minimizes the total energy and that corresponds to a stable condensate. The radius of the condensate increases when the strength of the repulsive interactions increases, as illustrated in Fig. 22.2(a), which gives different numerical solutions of the stationary GP equation for increasing values of χ. In

Fig. 22.2(b), we have plotted the relative contribution to the total energy of the trapping, kinetic and interaction energies as a function of χ for $a > 0$.

In the limit $\chi \gg 1$, one can neglect the kinetic energy term, and the Gross-Pitaevskii equation becomes a simple algebraic equation (Thomas-Fermi limit)

$$V_{\text{trap}}(\vec{r}) + Ng|\varphi(\vec{r})|^2 = V_{\text{trap}} + gn_0(\vec{r}) = \mu. \tag{22.26}$$

For harmonic confinement, $V_{\text{trap}}(\vec{r}) = m\omega_0^2 r^2/2$, and the spatial density $n_0(\vec{r}) = N|\varphi(\vec{r})|^2$ has the shape of an inverted parabola,

$$n(\vec{r}) = [\mu - m\,\omega_0^2\, r^2/2]/g \tag{22.27}$$

which starts from the value μ/g for $r = 0$ and vanishes for $r \geq r_{\text{max}} = (2\mu/m\omega_0^2)^{1/2}$. The expression for the chemical potential is obtained from the normalization condition. Integrating the density profile over r, we find

$$\mu(N) = \frac{\hbar\omega_0}{2}\left(15\frac{Na}{a_{\text{ho}}}\right)^{2/5}. \tag{22.28}$$

The total energy is obtained by integrating Eq. (22.10) using (22.28):

$$E_{\text{tot}}(N) = \int_0^N \mu(N')\,\mathrm{d}N' = \frac{\hbar\omega_0}{2}\left(15\frac{a}{a_{\text{ho}}}\right)^{2/5}\frac{5\,N^{7/5}}{7}, \tag{22.29}$$

and the total energy per particle is equal to $E_{\text{tot}}(N)/N = 5\mu/7$. Using the relations (22.14) and (22.15), one deduces the expression for the interaction energy per particle in the Thomas-Fermi limit, for which the kinetic energy is negligible:[4]

$$\frac{E_{\text{int}}(N)}{N} = \frac{2}{7}\mu(N) = \frac{\hbar\omega_0}{7}\left(15\frac{Na}{a_{\text{ho}}}\right)^{2/5}. \tag{22.30}$$

The size of the condensate is given by the value of r at which $n(\vec{r})$ given in Eq. (22.27) vanishes. It depends on the number of atoms through the chemical potential:[5]

$$r_{\text{max}}(N) = \sqrt{\frac{2\mu}{m\omega_0^2}} = a_{\text{ho}}\left(\frac{15Na}{a_{\text{ho}}}\right)^{1/5}. \tag{22.31}$$

This atom number dependence of the size is illustrated in Fig. 22.3(a) where the fit proportional to $N^{1/5}$ is in good agreement with the size observed experimentally for different condensed atom numbers.

The interaction energy can be measured by abruptly removing the confinement [Mewes *et al.* (1996)]. Just after the switch off, the total energy is equal to the interaction energy (which has not changed) plus the kinetic energy which is negligible

[4] Actually, the calculation of the kinetic energy using Eq. (22.11) and the Thomas-Fermi density profile gives rise to a logarithmic divergence since, close to the boundary, the Thomas-Fermi approximation breaks down and the kinetic energy term can no longer be ignored. A more refined analysis can be carried out to circumvent this artefact [Dalfovo *et al.* (1996)].

[5] It is interesting to note that the three lengths a_{ho}, r_{max} and ξ_0 are related through the relation $r_{\text{max}}\xi_0 = a_{\text{ho}}^2$.

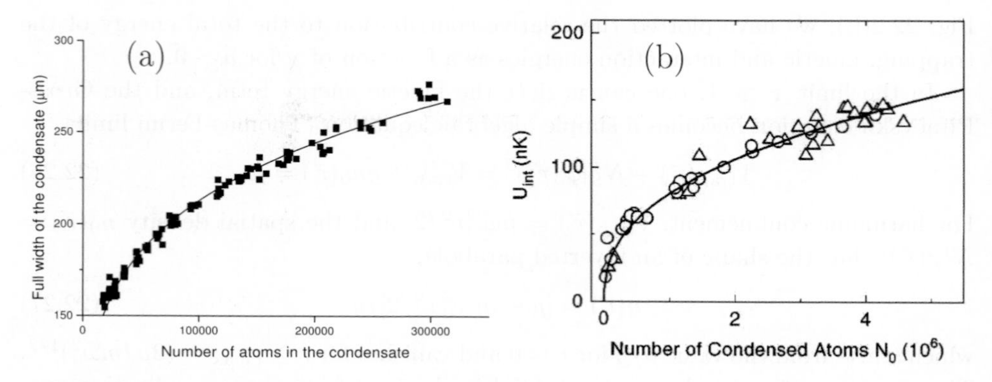

Fig. 22.3 (a) Variations of the full width of the condensate along one axis, as a function of the number of condensed atoms N. The solid line is a fit proportional to $N^{1/5}$. Reprinted with kind permission from Springer Science+Business Media: [Söding *et al.* (1998)]. (b) Mean-field energy per condensed atom versus the number of atoms in the condensate. The solid line is a fit proportional to $N^{2/5}$. Figure from [Mewes *et al.* (1996)]. Copyright: American Physical Society.

in the Thomas-Fermi limit. The interaction energy is converted into kinetic energy during the expansion of the condensate (see Sec. 22.7.4.2). The measurement of this energy shows that it varies with the number of condensed atoms as $N^{2/5}$ (see Fig. 22.3(b)), as expected from the Thomas-Fermi limit result (22.30).

Most experiments use cylindrically-symmetric harmonic traps:

$$V(z,r) = \frac{1}{2}m\left[\omega_z^2 z^2 + \omega_\perp^2 r^2\right].$$

The inverted parabola shape of the density profile is limited axially to $\pm z_{\max}$ and radially to $\pm r_{\max}$ defined respectively by $m\omega_z^2 z_{\max}^2 = 2\mu$ and $m\omega_\perp^2 r_{\max}^2 = 2\mu$. The aspect ratio of the condensate is therefore given by $z_{\max}/r_{\max} = \omega_\perp/\omega_z$. In the absence of interactions, this aspect ratio is equal to the ratio of the oscillator lengths of each oscillator:

$$\frac{z_{\max}}{r_{\max}} = \frac{a_{\text{ho}}^z}{a_{\text{ho}}^\perp} = \sqrt{\frac{\omega_\perp}{\omega_z}} < \frac{\omega_\perp}{\omega_z} \quad \text{if} \quad \frac{\omega_\perp}{\omega_z} > 1. \tag{22.32}$$

We deduce that interactions tend to magnify the aspect ratio of the condensate with respect to an ideal Bose gas.

A remarkable feature of Bose-Einstein condensates in the Thomas Fermi limit is that interactions can be important even though the gas is dilute. Consider a condensate contained in a volume R^3. It can be considered to be dilute as soon as the mean distance $d = (N/R^3)^{-1/3}$ between atoms is large compared to the scattering length a.[6] Since R is always larger than a_{ho}, one has:

$$\frac{a}{d} < \frac{a}{a_{\text{ho}}}N^{1/3} = \chi^{1/3}\left(\frac{a}{a_{\text{ho}}}\right)^{2/3}. \tag{22.33}$$

[6] See Sec. 24.2 of Chap. 24.

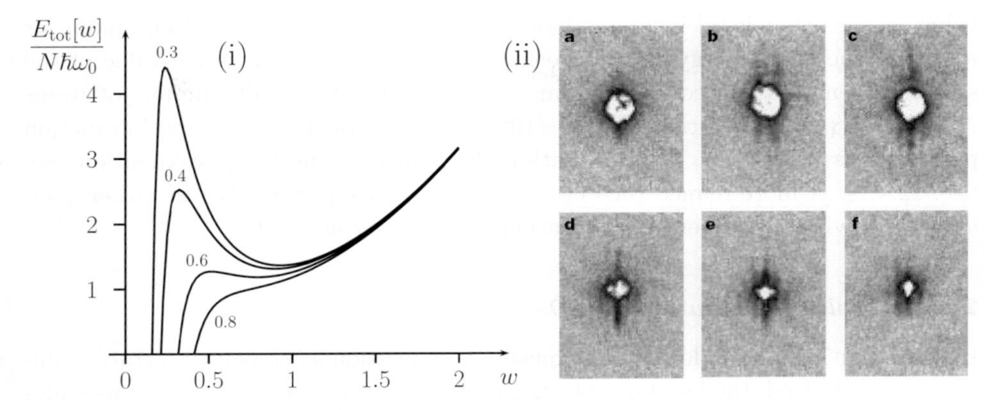

Fig. 22.4 (i) Total energy normalized to $N\hbar\omega_0$, as a function of the width parameter w for the Gaussian model of Sec. 22.4.1 for different values of the interaction parameter $|\chi| =$ 0.3, 0.4, 0.6, 0.8. (ii) Pictures from a to f correspond to the atomic density evolution of a Bose-Einstein condensate with a time step of 2 ms between each images after the sudden switch of the interaction to a large negative value using a Feshbach resonance. One observes the dynamics of the collapsing and exploding of the BEC. Reprinted by permission from Macmillan Publishers Ltd: Nature [Donley *et al.* (2001)].

The ratio a/d increases as $N^{1/3}$, whereas $\chi = Na/a_{\rm ho}$ which characterizes the importance of interactions increases as N. One can therefore have a condensate in the Thomas-Fermi regime ($\chi \gg 1$) while remaining dilute with $a \ll d$. For example, consider the case of rubidium-87 atoms, whose scattering length is $a = 5\,$nm, in a harmonic trap of frequency $\omega/2\pi = 250\,$Hz. This yields an oscillator length $a_{\rm ho} = 0.68\,\mu$m. For a condensate of $N = 10^6$ atoms, one finds that the interaction parameter is large, $\chi = Na/a_{\rm ho} \simeq 7400 \gg 1$. The Thomas-fermi radius is then $r_{\rm max} \simeq 6.9\ \mu$m and the parameter that characterizes the diluteness of the gas remains small $a/d \sim (a/r_{\rm max})\, N^{1/3} \simeq 7.2 \times 10^{-2} \ll 1$.

22.5 Condensate with a negative scattering length

22.5.1 *Condition of stability in 3D*

Theoretical studies predict that a homogeneous Bose-Einstein condensate with attractive interactions is unstable. In the presence of a confinement, a condensate may form if the atom number is not too large. Physically, the stability originates from the tradeoff between the attractive interaction energy, which tends to contract the cloud, and the kinetic energy term resulting from the position-momentum uncertainty in the presence of confinement.

This balance is illustrated for the case of an isotropic harmonic confinement in Fig. 22.4(i), where we have plotted the total energy per particle normalized to $\hbar\omega_0$ as a function of the effective width w for the Gaussian model of Eq. (22.24), for several values of the interaction parameter $|\chi| = N|a|/a_{\rm ho}$. There exists a critical value χ_c such that if $|\chi|$ is larger than χ_c, there is no longer a local minimum for $w \neq 0$ of the total energy $E_{\rm tot}$. In other words, for a given negative value of the scattering length a, the condensate can accommodate only a finite number of atoms $N < N_c = \chi_c a_{\rm ho}/|a|$ [Bradley *et al.* (1997)]. The critical value of the interaction parameter is equal to $\chi_c \simeq 0.671$ within the Gaussian ansatz approximation (see Sec. 22.4.1). A more refined theoretical analysis yields $\chi_c \simeq 0.574$,[7] in rather good agreement with experimental observations [Roberts *et al.* (2001)].

22.5.2 *Solitonic solution in 1D*

By constraining the condensate to a quasi-one dimensional geometry,[8] a stable solution exists. Indeed, the last term of (22.24) that accounts for the interactions scales as w^{-1} in 1D, and, when w decreases, the interactions cannot over-compensate the kinetic energy that scales as w^{-2}. One finds a new class of solutions, called matter-wave solitons, in which the wave-packet dispersion exactly compensates attractive forces. The first experimental realizations of matter-wave solitons were performed in 2002 with lithium-7 atoms by two groups. A condensate was produced with a positive scattering length, and then effective interactions between atoms were tuned from repulsive to attractive with a Feshbach resonance. An optical guide was used to ensure the quasi-1D geometry. In the group of Christophe Salomon at ENS, the condensate had an initial number of atoms of 2×10^4, and a single soliton made of few thousand atoms was observed propagating without expansion [Khaykovich *et al.* (2002)]. In the group of Randall Hulet at Rice University, the condensate initially had 3×10^5 atoms, and the same procedure yielded the observation of multiple solitons [Strecker *et al.* (2002)].

Let us finally emphasize the similarity between bright matter-wave solitons and optical solitons in optical fibers where the attractive forces are replaced by the self-focussing (Kerr) nonlinearities.

22.5.3 *Collapse and explosion of a condensate in 3D with a negative scattering length*

In three dimensions and beyond the stability limit, self-attraction overwhelms the repulsion due to the quantum pressure and causes the condensate to collapse. During the collapse, the density rises yielding a dramatic increase of inelastic collision rates, such as three-body recombination, thereby inducing atom losses. One may wonder what happens after the collapse.

The group of Randall Hulet was able to observe a lithium-7 atom condensate regrowing after the collapse [Gerton *et al.* (2000)]. Indeed, just after the collapse,

[7] See [Dalfovo *et al.* (1999)] and references therein.

[8] The requirement on the geometry means that the mean-field energy gn is below the energy spacing for the transverse degrees of freedom. In this way, radial excitations are inhibited and the motion is effectively 1D.

the gas is out of equilibrium and a condensate may form filled through elastic collisions between thermal atoms in the gas. The condensate was found to undergo many cycles of growth and collapses [Sackett *et al.* (1998)].

The JILA group has explored the dynamics of collapse and its subsequent explosion when the balance of forces governing the Bose-Einstein condensate size and shape is suddenly altered. The collapse is induced by abruptly tuning the interactions from repulsive to attractive using an externally applied magnetic field close to a Feshbach resonance. This technique has allowed the observation of an exploding atomic ejection from the collapsing Bose-Einstein condensate [Donley *et al.* (2001)] (see Fig. 22.4(ii)).

22.6 Quantum vortex in an homogeneous condensate

In view of its importance for the superfluidity properties of weakly interacting Bose-Einstein condensates, we consider in this section a new kind of solution of the stationary Gross-Pitaevskii equation for which all atoms carry one unit of angular momentum $\hbar$. As explained below, the solutions for effective repulsive interaction ($a > 0$) exhibit a vortex line. We shall discuss in more detail in Chap. 24 how vortices can be produced and observed in practice.

22.6.1 *Effective Gross-Pitaevskii equation*

For the sake of simplicity, we consider here an homogeneous condensate and use the cylindrical coordinates $(r_\perp, \varphi, z)$ around the z-axis where $r_\perp$ is the radial distance from the z-axis, φ the azimuthal angle, and z the vertical coordinate. The wave function is of the form

$$\psi(r_\perp, z, \varphi) = \chi(r_\perp, z)e^{i\varphi}, \tag{22.34}$$

where $\chi(r_\perp, z)$ is a real function. The angular momentum is therefore readily calculated

$$L_z\psi = \frac{\hbar}{i}\frac{\partial\psi}{\partial\varphi} = \hbar\psi, \tag{22.35}$$

which explicitly shows that the wave function (22.34) carries a quantum $\hbar$ of angular momentum along the z-axis.

Using the expression of the Laplacian in cylindrical coordinates

$$\Delta = \frac{1}{r_\perp}\frac{\partial}{\partial r_\perp}\left(r_\perp\frac{\partial}{\partial r_\perp}\right) + \frac{\partial^2}{\partial z^2} + \frac{1}{r_\perp^2}\frac{\partial^2}{\partial\varphi^2} = \frac{1}{r_\perp}\frac{\partial}{\partial r_\perp}\left(r_\perp\frac{\partial}{\partial r_\perp}\right) + \frac{\partial^2}{\partial z^2} - \frac{1}{r_\perp^2}\frac{L_z^2}{\hbar^2}, \tag{22.36}$$

Eq. (22.6) yields the following effective Gross-Pitaevskii equation for the wave function $\chi(r_\perp, z)$

$$-\frac{\hbar^2}{2m}\left[\frac{1}{r_\perp}\frac{\partial}{\partial r_\perp}\left(r_\perp\frac{\partial}{\partial r_\perp}\right) - \frac{1}{r_\perp^2}\right]\chi + Ng|\chi|^2\chi = \mu\chi, \tag{22.37}$$

where a centrifugal potential barrier term, $\hbar^2/(2mr_\perp^2)$, appears as a result of the angular momentum. A direct consequence is that the wave function $\chi(r_\perp, z)$ shall necessarily vanish on the z-axis $\chi(r_\perp \to 0, z) \to 0$. Furthermore, the minimizers of the total energy are the uniform functions along the z-axis, $\chi(r_\perp, z) = \chi(r_\perp)$.

At large distances, the centrifugal term becomes negligible and one recovers the homogeneous solution $\mu = gN|\chi(r_\perp \to \infty)|^2 = g\tilde{n}_0$ where $\tilde{n}_0$ is the spatial density which becomes constant at large $r_\perp$. The characteristic size of the central density hole is nothing but the healing length $\xi_0 = (8\pi a n_0)^{-1/2}$. Indeed, using the dimensionless variable $s = r_\perp/\xi_0$, Eq. (22.37) can be recast in the form

$$\left(\frac{\mathrm{d}^2}{\mathrm{d}s^2} + \frac{1}{s}\frac{\mathrm{d}}{\mathrm{d}s}\right)\chi(s) + \left(1 - \frac{1}{s^2}\right)\chi(s) - \chi(s)^3 = 0. \tag{22.38}$$

In other words, the size of the central density hole is determined by the interaction through the healing length.

22.6.2 *Properties of the velocity field*

In quantum mechanics, the velocity field at a location where the density is non-zero is proportional to the gradient of the phase S of the wave function:

$$\vec{v} = \frac{\hbar}{m}\vec{\nabla}S. \tag{22.39}$$

In the absence of phase singularities, as for the condensate without angular momentum studied in the previous sections, the velocity field is necessarily irrotational ($\vec{\nabla} \times \vec{v} = \vec{0}$), and its circulation on any closed contour therefore vanishes.

The situation is radically different when atoms carry an angular momentum. For the wave function (22.34), we get a velocity field whose only non-zero component is orthoradial:

$$v_z = v_{r_\perp} = 0, \qquad \text{and} \qquad v_\varphi = \frac{\hbar}{mr_\perp}. \tag{22.40}$$

The velocity field has thus a singularity at $r_\perp = 0$.

Let us calculate the circulation of the velocity field (22.40) on a circle of radius $r_\perp$ centered on the symmetry axis Oz:

$$\oint_{\mathcal{C}} \vec{v} \cdot \mathrm{d}\vec{\ell} = \int_0^{2\pi} r_\perp v_\varphi \mathrm{d}\varphi = 2\pi\frac{\hbar}{m} = \frac{h}{m}. \tag{22.41}$$

This result is actually valid on any closed contour that contains the symmetry axis.

More generally, the circulation of the velocity field on a closed contour $\mathcal{C}$ that does not intersect a line with zero density is given by

$$\oint_{\mathcal{C}} \vec{v} \cdot \mathrm{d}\vec{\ell} = \frac{\hbar}{m}\oint_{\mathcal{C}} \vec{\nabla}S \cdot \mathrm{d}\vec{\ell} = \frac{\hbar}{m}\delta S. \tag{22.42}$$

As the wave function takes a unique value at each space point, one finds that necessarily $\delta S = 2\pi p$, where p is an integer. The circulation of the velocity field can therefore take only discrete values

$$\oint_{\mathcal{C}} \vec{v} \cdot \mathrm{d}\vec{\ell} = p\frac{h}{m}. \tag{22.43}$$

The ratio h/m appears as the quantum of circulation. Combining the fact that for any $r_\perp \neq 0$, the curl of the velocity field (22.40) vanishes and that the velocity field fulfills Eq. (22.43), we deduce that

$$\vec{\nabla} \times \vec{v} = p\frac{h}{m}\delta^{(2)}(\vec{r}_\perp)\vec{e}_z. \qquad (22.44)$$

Finally, let us point out the analogy with magnetostatics. Indeed, the velocity field for a single vortex line has the same structure as the magnetic field produced by a current running in a straight wire along the z-axis. They obey the same kind of equations:[9]

$$\vec{\nabla} \cdot \vec{B} = 0 \qquad \text{and} \qquad \vec{\nabla} \times \vec{B} = \frac{j}{\varepsilon_0 c^2}\delta^{(2)}(\vec{r}_\perp)\vec{e}_z. \qquad (22.45)$$

22.7 Time-dependent problems

22.7.1 *Time-dependent Gross-Pitaevskii equation*

In several experiments performed with Bose-Einstein condensates, the trapping potential is varied in time. For instance, in time-of-flight experiments, one suddenly switches off the confinement to observe the ballistic expansion of the condensate and deduce its properties. To describe its dynamics in this context, one needs to extend to time-dependent phenomena the Gross-Pitaevskii equation introduced in Sec. 22.2 for analyzing the static properties of condensates.

The hydrodynamic formulation of this equation (Sec. 22.7.2) helps to recover the Thomas-Fermi approximation (Sec. 22.7.3), to describe the ballistic expansion through a scaling factor method (Sec. 22.7.4.1) and as well as the normal collective modes of the trapped condensate which are reviewed as an Appendix at the end of this chapter.

As for the time-independent case, we take the N-particle wave function as a product of N identical time-dependent single-particle functions:

$$\psi(\vec{r}_1, \ldots \vec{r}_N, t) = \varphi(\vec{r}_1, t)\,\varphi(\vec{r}_2, t)\cdots\varphi(\vec{r}_N, t)\,. \qquad (22.46)$$

Such a fully symmetric state describes a situation where all N bosons evolve in the same way and neglects quantum correlations between the atoms.

Inserting the ansatz (22.46) into a least-action principle approach yields the time-dependent Gross-Pitaevskii equation [Cohen-Tannoudji (1998a)]:

$$i\hbar\frac{\partial}{\partial t}\varphi(\vec{r}, t) = -\frac{\hbar^2}{2m}\Delta\varphi(\vec{r}, t) + V_{\text{trap}}(\vec{r}, t)\,\varphi(\vec{r}, t) + N\,g\,|\varphi(\vec{r}, t)|^2\,\varphi(\vec{r}, t), \quad (22.47)$$

where we used the delta potential to describe interactions.[10]

If $V_{\text{trap}}(\vec{r}, t) = V_0(\vec{r})$ does not depend on time, one can look for stationary solutions of (22.47) of the form $\varphi(\vec{r}, t) = \varphi_0(\vec{r})\exp(-i\mu t/\hbar)$. In this way, one exactly recovers the time-independent Gross-Pitaevskii equation (22.6).

[9] One can readily show from Eq. (22.40) that $\vec{\nabla} \cdot \vec{v} = 0$.
[10] See Sec. 22.2.2.

Gray solitons

It is worth emphasizing that Eq. (22.47) in 1D has another class of propagating solutions in addition to the propagating bright matter wave solitons already discussed in Sec. 22.5.2: the so-called gray solitons that correspond to the propagation of a density hole. In an homogeneous system, one can show that a stable gray soliton can propagate at a velocity $v_0 < c$ where $c = (gn_0/m)^{1/2}$ is the sound velocity[11] and that the relative hole depth is proportional to $1 - (v_0/c)^2$ [Tsuzuki (1971)]. The gray soliton thus disappears when its velocity reaches the sound velocity. Experimental observations of gray solitons have been reported by the group of Wolfgang Ertmer at Hannover and the group of William Phillips at Gaithersburg [Burger *et al.* (1999); Denschlag *et al.* (2000)].

22.7.2 *Analogy with hydrodynamic equations*

The time-dependent Gross-Pitaevskii equation can be rewritten in the form of hydrodynamic equations. Their expression is particularly simple in the Thomas-Fermi limit, and turns out to be very useful for the interpretation of several physical effects.

22.7.2.1 *Density field ρ and velocity field $\vec{v}$*

In order to derive the set of hydrodynamic equations that is equivalent to the time-dependent Gross-Pitaevskii equation, it is convenient to normalize the wave function to the number of particles: $\int \mathrm{d}^3r \, |\varphi(\vec{r},t)|^2 = N$. The spatial density is then given by the modulus of the wave function: $n(\vec{r},t) = |\varphi(\vec{r},t)|^2$. With such a choice for the normalization of the wave function, there is no more explicit dependence of the time-dependent Gross-Pitaevskii on the number of atoms N:

$$i\hbar\frac{\partial}{\partial t}\varphi(\vec{r},t) = -\frac{\hbar^2}{2m}\Delta\varphi(\vec{r},t) + V_{\text{trap}}(\vec{r},t)\,\varphi(\vec{r},t) + g\,|\varphi(\vec{r},t)|^2\,\varphi(\vec{r},t). \quad (22.48)$$

It is instructive to rewrite the wave function $\varphi(\vec{r},t)$ in terms of its phase $S(\vec{r},t)$ and its modulus:

$$\varphi(\vec{r},t) = \sqrt{n(\vec{r},t)}\exp\left[i\,S(\vec{r},t)\right]. \quad (22.49)$$

22.7.2.2 *Continuity equation. Evolution equations of $\vec{v}$ and ρ*

Using Eq. (22.48) to derive the equation fulfilled by the modulus $n(\vec{r},t)$, one obtains:

$$\frac{\partial}{\partial t}n(\vec{r},t) + \vec{\nabla}\cdot\left[n(\vec{r},t)\,\vec{v}(\vec{r},t)\right] = 0, \quad (22.50)$$

where $\vec{v}$ denotes the velocity field and is proportional to the gradient of the phase $S(\vec{r},t)$:

$$\vec{v}(\vec{r},t) = \frac{\hbar}{m}\vec{\nabla}S(\vec{r},t). \quad (22.51)$$

[11] The expression for the sound velocity in an homogeneous condensate is determined in Sec. 22.7.3 and also in Chap. 24.

Equation (22.50) is the continuity equation. It shows that the integral of $n(\vec{r},t)$ over space does not change in time, i.e. that if φ is normalized at $t=0$, it remains normalized for all $t>0$. The quantity $n(\vec{r},t)\,\vec{v}(\vec{r},t)$ is the current density. We emphasize that the velocity field obeys the equation $\vec{\nabla}\times\vec{v}(\vec{r},t)=\vec{0}$ and is therefore irrotational.[12]

Similarly, the equation for the evolution of the phase $S(\vec{r},t)$, and thus of the velocity field $\vec{v}(\vec{r},t)$, is readily inferred from Eq. (22.48):

$$m\frac{\partial \vec{v}(\vec{r},t)}{\partial t}=\vec{\nabla}\left[\frac{\hbar^2}{2m}\frac{1}{\sqrt{n(\vec{r},t)}}\Delta\sqrt{n(\vec{r},t)}-\frac{1}{2}m\vec{v}(\vec{r},t)^2-V_{\text{trap}}(\vec{r},t)-gn(\vec{r},t)\right] \tag{22.52}$$

The continuity equation (22.50) for the density evolution and the Euler-like equation (22.52) for the velocity field evolution are often referred to as hydrodynamic equations, and are strictly equivalent to the time-dependent Gross-Pitaevskii equation (22.48) as soon as solutions do not exhibit singularities.

22.7.2.3 *Quantum pressure*

The first term on the right-hand side of the Euler equation (22.52) is the only one that explicitly contains $\hbar$, and is called the quantum pressure term. It originates from the kinetic energy arising from density gradients, and is a direct consequence of the Heisenberg uncertainty principle.

In order to understand under which circumstances this term plays a role, let us denote by R the characteristic length of the spatial variations of the atomic density $n(\vec{r})$. The quantum pressure term scales as $\hbar^2/2mR^2$ and is negligible compared to the interaction term gn when

$$R\gg\left(\frac{\hbar^2}{2mgn}\right)^{1/2}=\xi, \tag{22.53}$$

where ξ is the healing length. The healing length thus appears as the characteristic length ξ such that the energy of confinement in a volume ξ^3 is equal to the interaction energy.

22.7.3 *The two contributions to the kinetic energy: Thomas-Fermi approximation for time-dependent problems*

In the stationary case, the time-independent Gross-Pitaevskii equation can be simplified in the Thomas-Fermi limit ($\chi\gg1$) by neglecting the kinetic energy term.

In problems involving the dynamics of the condensate, it is not correct to neglect the kinetic energy term, $-\hbar^2\Delta\varphi/2m$, in the time-dependent Gross-Pitaevskii equation (22.48) even in the limit $\chi\gg1$. For example, in the ballistic expansion of a condensate that occurs after suddenly switching off the confinement, the kinetic

[12] In the presence of vortices, the phase is not defined at the location of the vortices, and the use of the modulus-phase formalism requires some care to cope with singularities.

energy term is small at the beginning of the expansion, compared to the interaction energy, but the interaction energy is converted into kinetic energy in the course of the expansion which therefore becomes very large and even dominant for long expansion times.

However, the Thomas-Fermi approximation takes a simple form in the Euler hydrodynamic equation where the contributions of the amplitude gradient and phase gradient to the kinetic energy are clearly separated, being represented respectively by the first two terms on the right-hand side of Eq. (22.52). When $\chi \gg 1$, the amplitude gradient (appearing in the quantum pressure term) remains small at all times, whereas the second term, coming from phase gradients, can become very large.

The Thomas-Fermi limit, in the time-dependent case, thus corresponds to a situation where the quantum pressure term can be neglected, so that the set of hydrodynamic equations is, in this limit, equivalent to the following two equations

$$\frac{\partial n}{\partial t} + \vec{\nabla} \cdot [n\vec{v}] = 0,$$

$$m\frac{\partial \vec{v}}{\partial t} = \vec{\nabla}\left[-\frac{1}{2}m\vec{v}^{\,2} - V_{\mathrm{trap}} - gn\right]. \tag{22.54}$$

It is worth noticing that in the Thomas-Fermi regime, $\hbar$ no longer appears in the equation for the velocity field, which consequently appears as a classical Euler equation that describes the motion of a fluid in the trapping potential and in the pressure field arising from the density of other particles. The motion of a condensate, in the Thomas-Fermi limit, and in the mean-field approximation, can thus be described by classical, irrotational hydrodynamics (since by definition the velocity field obeys the relation $\vec{\nabla} \times \vec{v}\,(\vec{r}, t) = \vec{0}\,$).

In the static case, for which there is no global motion of the condensate, the velocity field $\vec{v}$ is equal to $\vec{0}$, and Eq. (22.54) gives:

$$\vec{\nabla}\left[V_{\mathrm{trap}}\,(\vec{r}\,) + gn\,(\vec{r}\,)\right] = \vec{0}. \tag{22.55}$$

We recover here the equilibrium shape in the Thomas-Fermi limit derived in Sec. 22.4.2.

The hydrodynamic equations (22.54) also give access to the motion of the small amplitude oscillations of the condensate in the presence of a confinement ($V_{\mathrm{trap}} \neq 0$). For this purpose, one linearizes the hydrodynamic equations around the equilibrium state defined by

$$n_0\,(\vec{r}\,) = \left[\mu - V_{\mathrm{trap}}\,(\vec{r}\,)\right]/g, \qquad \text{and} \qquad \vec{v}_0\,(\vec{r}\,) = \vec{0}. \tag{22.56}$$

One finds that the first order corrections δn to the density and $\delta\vec{v}$ to the velocity field obey the linear set of equations:

$$\frac{\partial \delta n}{\partial t} = -\vec{\nabla} \cdot (n_0 \delta\vec{v}\,), \tag{22.57}$$

$$m\frac{\partial \delta\vec{v}}{\partial t} = -\vec{\nabla}\left(V_{\mathrm{trap}} + gn_0 + g\,\delta n\right) = -g\vec{\nabla}\delta n. \tag{22.58}$$

Eliminating $\delta\vec{v}$ between the two equations (22.54) leads to the equation of motion of $\delta n(\vec{r},t)$:

$$\frac{\partial^2 \delta n\left(\vec{r},t\right)}{\partial t^2} = \vec{\nabla}\cdot\left[c^2(\vec{r})\vec{\nabla}\delta n\left(\vec{r},t\right)\right], \qquad \text{with} \qquad c^2(\vec{r}) = \frac{g}{m} n_0\left(\vec{r}\right). \tag{22.59}$$

The quantity $c(\vec{r})$ is a local sound velocity. Sound waves can propagate in a non uniform medium. For a cylindrical geometry with a transverse harmonic confinement, the sound velocity in the longitudinal direction[13] is $(\mu/2m)^{1/2}$ [Zaremba (1998); Kavoulakis and Pethick (1998); Stringari (1998)]. The propagation of sound waves in such a geometry has been studied experimentally in [Andrews *et al.* (1997b)].

22.7.4 *Harmonic confinement*

In this section, we suppose that the trapping potential is harmonic but not necessarily isotropic:

$$V_{\text{trap}}\left(\vec{r},t\right) = \frac{1}{2}\sum_{i=x,y,z} m\,\omega_i^2\left(t\right) r_i^2, \tag{22.60}$$

with $(r_1, r_2, r_3) = (x, y, z)$, and that the condensate is in the Thomas-Fermi regime, so that we can use the hydrodynamic equations (22.54) without the quantum pressure term. The time dependence of the trapping frequencies $\omega_i(t)$ will allow us to analyze several problems which can be readily investigated experimentally.

22.7.4.1 *Scaling transformation*

We assume that the condensate is in the Thomas-Fermi regime and at equilibrium at $t = 0$. When the strength of the confining potential is changed as a function of time for $t > 0$, the time-dependent Gross-Pitaevskii can be solved using a scaling transformation. In the following, we denote by $\omega_i(0)$ the angular frequency along the $i = x, y, z$ axis for a time $t \leq 0$, and $\omega_i(t)$ for $t \geq 0$. The time dependent density resulting from the excitation is searched in the form [Castin and Dum (1996); Kagan *et al.* (1997)]:[14]

$$n\left(r_i,t\right) = \frac{1}{b_x\left(t\right) b_y\left(t\right) b_z\left(t\right)} n_0\left(\frac{r_i}{b_i\left(t\right)}\right) = \frac{1}{\Pi_j b_j\left(t\right)} n_0\left(\frac{r_i}{b_i\left(t\right)}\right), \tag{22.61}$$

where n_0 is the initial equilibrium density distribution. The prefactor $\Pi_j b_j^{-1}(t)$ ensures the normalization of the density to the number of atoms. The ansatz (22.61) inserted in the continuity equation gives the expression for the velocity field:

$$v_j\left(\vec{r},t\right) = \frac{\dot{b}_j\left(t\right)}{b_j\left(t\right)} r_j. \tag{22.62}$$

[13] This result differs from the one obtained in a box by a factor of $1/\sqrt{2}$ since it is the average density over the radial direction, and not the peak density, that determines the value of the sound velocity in a such an elongated geometry.

[14] This scaling ansatz yields an exact solution of the time-dependent GP equation in the Thomas-Fermi regime in 3D. In 2D, it gives an exact solution of the time-dependent GP equation in all interaction regimes.

The Euler equation (22.54) yields:

$$m\frac{\partial v_j}{\partial t}+\frac{\partial}{\partial r_j}\left(\frac{1}{2}mv^2+V_{\text{trap}}+gn\right)=m\frac{\ddot{b}_j\left(t\right)}{b_j\left(t\right)}r_j+m\,\omega_j^2\left(t\right)r_j+g\frac{\partial n}{\partial r_j}. \qquad (22.63)$$

The calculation of the last term of (22.63) requires the knowledge of the equilibrium Thomas-Fermi profile (22.27), from which one deduces

$$n\left(r_i,t\right)=\frac{1}{\Pi_j b_j\left(t\right)}\left[\frac{\mu}{g}-\frac{m}{2g}\sum_i\omega_i^2(0)\,\frac{r_i^2}{b_i^2\left(t\right)}\right]. \qquad (22.64)$$

Combining this expression with (22.63), we find the set of nonlinear coupled equations fulfilled by the dilation factors b_j:

$$\ddot{b}_j\left(t\right)+\omega_j^2\left(t\right)b_j\left(t\right)-\frac{\omega_j^2(0)}{b_j\left(t\right)}\frac{1}{\Pi_i b_i\left(t\right)}=0. \qquad (22.65)$$

The scaling transformation thus enables one to account for large-amplitude oscillations and to investigate nonlinear features associated with, for example, the dynamics of the expansion of the gas, by simply solving a set of three nonlinear coupled ordinary differential equations instead of partial differential equations. It is worth noting however, that such an approach is restricted to quadratic potentials.

22.7.4.2 *Ballistic expansion*

To analyze the properties of the condensate, the standard method consists of monitoring the evolution of the shape of the condensate after having suddenly switched off the trapping potential [Anderson *et al.* (1995)]. Such a ballistic expansion is usually necessary for an optical detection, since the in-trap transverse size of the condensate is on the order of few μms, close to the optical resolution of the imaging system. In addition, the optical density is very large. The ballistic expansion is therefore of particular interest since it acts as a probe of the ground state N-body wave function.

For an ideal gas, the *in situ* position dispersion $\Delta x_i(0)$ along a given axis gives rise, through the Heisenberg principle, to a velocity dispersion $\Delta v_i=\hbar/(2m\Delta x_i(0))$. After switching off the trap, the cloud expands and the position dispersion evolves according to $\Delta x_i(t)=[(\Delta x_i(0))^2+(\Delta v_i)^2t^2]^{1/2}$. For long expansion times, the size is dominated by the velocity dispersion term. If the initial harmonic potential is anisotropic, the expansion reflects this anisotropy, since the velocity dispersion along one axis is inversely proportional to the initial position dispersion. The ellipticity of the clouds is reversed in the course of the expansion; this *inversion of ellipticity* is usually considered to be the "smoking-gun" evidence for Bose-Einstein condensation.

Let us consider now a condensate in the presence of atom-atom interactions that is initially well described by the Thomas-Fermi approximation. We can therefore apply the scaling factors formalism of Sec. 22.7.4.1 to the analysis of time-of-flight

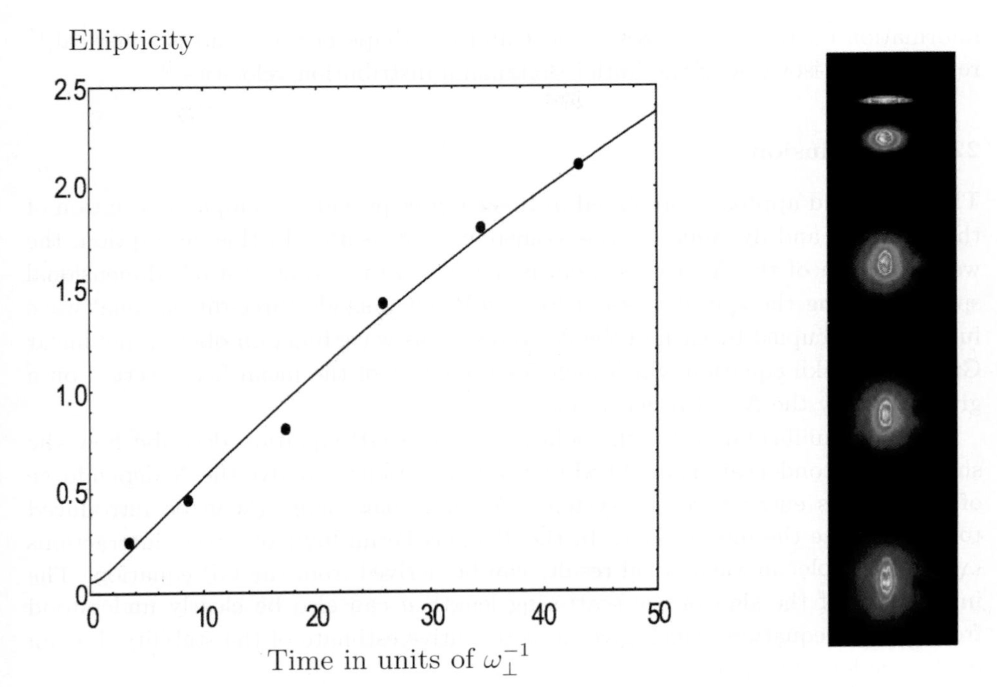

Fig. 22.5 Aspect ratio of an expanding condensate as a function of time. The experimental data were taken with an ellipticity at time $t = 0$ equal to 0.1, the solid line is obtained by integrating the set of Eqs. (22.66). Figure taken from [Guéry-Odelin (1998)].

experiments. The angular frequencies are modified at time $t = 0$ from $\omega_j\,(t < 0) = \omega_j \neq 0$ to $\omega_j\,(t \geq 0) = 0$. Let us suppose that the trap has a cylindrical symmetry around Oz, and a cigar shape: $\omega_x = \omega_y = \omega_\perp \gg \omega_z$. For $t \geq 0$, Eq. (22.65) can be recast in the form

$$\frac{d^2 b_\perp(\tau)}{d\tau^2} = \frac{1}{b_\perp^3(\tau)\, b_z(\tau)}, \quad \text{and} \quad \frac{d^2 b_z(\tau)}{d\tau^2} = \frac{\lambda^2}{b_\perp^2(\tau)\, b_z^2(\tau)}, \tag{22.66}$$

where $\tau = \omega_\perp t$ and $\lambda = \omega_z/\omega_\perp$. The solution of Eqs. (22.66) to the second order in λ when $\lambda \ll 1$ can be worked out analytically [Castin and Dum (1996)].

It is worth mentioning that this equation means that the information on the initial anisotropy of the wave function is not lost during the expansion. Actually, this anisotropy is transferred to the velocity field, and, in the course of the expansion, which acts as a magnifier for the condensate wavefunction, one observes an inversion of the ellipticity of the cloud (see Fig. 22.5). This is to be contrasted with what one would observe with a Boltzmann gas initially at equilibrium in an anisotropic trap. The density profile of a trapped cloud before switching off the confinement reflects the anisotropy of the trap. However, after a sufficiently long expansion time, the information on the initial sizes are lost and the shape of the cloud is spherical,[15] reflecting the isotropy of the initial Boltzmann distribution velocities.[16]

22.8 Conclusion

The mean field approach presented in this chapter provides a simple description of the structure and dynamics of Bose-Einstein condensates. In this description, the wave function of the N-atom system is not a wave function in a $3N$-dimensional space (ignoring the spin degrees of freedom), but a single three-dimensional wave function φ occupied by each of the N atoms. This wave function obeys a nonlinear Gross-Pitaevskii equation which includes the effect of the mean field exerted on a given atom by the $N-1$ other atoms.

In the equilibrium state, the solutions of the GP equation describe how the shape of the condensate is modified by the interactions and give the N-dependence of the various energies of the system. A single parameter χ can be introduced to characterize the interactions. In the Thomas-Fermi limit of strong interactions ($\chi \gg 1$), simpler mathematical results can be derived from the GP equation. The importance of the sign of the scattering length a can also be clearly understood from the GP equation, which gives a quantitative estimate of the stability domain of the condensates with $a < 0$.

Non-equilibrium situations can also be analyzed with the time-dependent GP equation. A remarkable feature is that this equation is strictly equivalent to a set of hydrodynamics equations which become purely classical in the Thomas-Fermi limit. Several phenomena, like ballistic expansion, wave propagation in the condensate, and collective modes of vibration (see the appendix at the end of this chapter), can be quantitatively described with these equations. All the theoretical predictions derived from the time-independent as well as the time-dependent GP equation are in excellent agreement with the observations.

This description of a cloud with a large number N of atoms all occupying the same wave function φ provides new insights into the physical meaning of the wave function in quantum mechanics. Usually, the wave function of a single particle cannot be considered to be a classical wave. The continuity equation satisfied by the probability density ρ and the probability current $\vec{J}$ could suggest that the particle can be described as a classical fluid occupying a certain volume. This is not correct. The wave function is a position dependent probability amplitude from which one can calculate the probability to find the point-like particle in a given point. But, in the situation described in this chapter, we have a huge number of particles all occupying the same wave function. In any small coarsed-grained volume d^3r, we

[15] See Sec. 21.5.2 of Chap. 21.

[16] Actually, this is only true if elastic collisions play a minor role during the expansion. If the classical gas enters the hydrodynamic regime, i.e. if the collision rate is not negligible with respect to the trapping frequencies, the asymptotic expansion does reflect the initial anisotropy of the trap [Pedri *et al.* (2003)].

have many particles $\rho(\vec{r})\mathrm{d}^3r$, so that we can consider that φ describes a classical fluid moving according to the hydrodynamic equations derived in this chapter. In the limit of a very large number of bosonic particles all occupying the same wave function φ, a new interpretation of φ as a classical matter wave can thus be given.[17] For fermions, such an interpretation is not possible, because there is at most one fermion in a given state φ.

A similar situation exists in electromagnetism. The classical Maxwell waves can be considered to be wave functions of the quantum Maxwell field occupied by a macroscopic number of photons, which are bosons. In this limit, we can consider the electric and magnetic fields as classical fields. We know, however, that there are other quantum states of the field, for example single or two photon states, that have no classical interpretation. The correct treatment of electromagnetism requires working in a larger state space (the so-called Fock space), with field operators for creating and destroying photons. Inspired by this well known example of quantum field theory, we can, for a system of bosons, consider a Fock space where other states than φ can contain particles and where atomic field operators can be introduced to create and destroy atoms in a given state. This is in fact the well known second quantization description of a system of identical particles. Such an approach will be followed in the next chapter. It will allow us to describe, in a quantitative way, the coherence properties of Bose-Einstein condensates and to point out the close analogies between quantum optics (with light fields) and atom optics (with atomic fields), as well as the differences between these two fields (photons do not interact whereas atoms interact). This will be also a first step towards a more quantitative description of the system which will be given in Chap. 24, and which will be essential for understanding important effects like superfluidity.

22.9 Appendix: Normal modes of a harmonically trapped condensate

The study of the elementary excitations is one of the standard techniques used to characterize the behavior of quantum many-body systems. These excitations determine the thermodynamic properties and can explain, for instance, the superfluid properties of liquid helium. In the context of dilute Bose-Einstein condensates, they can be derived by linearizing the hydrodynamic equations or, equivalently, the time-dependent Gross-Pitaevskii equation. In the absence of confinement, they correspond to sound waves propagating with the velocity $c = (gn_0/m)^{1/2}$ as it clearly appears from Eq. (22.59). For a harmonically confined Bose-Einstein condensate, an extra time scale given by the inverse of the trap frequency ω_{ho} appears and new kind of oscillatory solutions of Eq. (22.59) are found. These excitations, referred to as low-lying collective modes, have an energy on the order of $\hbar\omega_{\mathrm{ho}}$ and correspond to a periodic motion of the whole condensate. They involve length scales larger

[17] See Volume 3, Sec. 21-4 of Ref. [Feynman (1963)].

than the healing length ξ_0. Sound wave-like solutions are also present for excitations with a wavelength smaller than the size of the condensate but larger than the healing length.

The experimental study of these low energy modes was done soon after the first realization of alkali Bose-Einstein condensates. The measurement of their frequencies can be carried out with a high precision, and therefore allows for precise quantitative comparisons with theory. In principle, such comparison requires numerically solving the time-dependent Gross-Pitevskii equation. However, analytical results provide simple pictures for the evolution of the deformation of the condensate after its excitation and are thus interesting to study.

In this section, we consider, unless otherwise stated, that the condensate is initially in the Thomas-Fermi regime. In this limit, analytical results on both the collective mode frequencies and eigenfunctions can be readily derived from Eq. (22.59). As intuitively expected, one finds that the low-energy frequencies of these modes strongly depend on the geometry of the trap.

22.9.1 *Isotropic trap*

We first consider the case of isotropic harmonic confinement $V_{\text{trap}}(\vec{r}) = m\omega_0^2 r^2/2$. The time dependence of the density fluctuation induced by the excitations is taken in the form $\delta n(\vec{r}t) = \delta n(\vec{r})e^{-i\omega t}$ so that Eq. (22.59) reads:

$$-\omega^2 \delta n(\vec{r}) = \vec{\nabla} \cdot \left[c^2(\vec{r}) \vec{\nabla} \delta n(\vec{r}) \right]. \qquad (22.67)$$

The solutions of this equation are discretized normal modes of the form [Stringari (1996b)]

$$\delta n(\vec{r}) = P_\ell^{(2n)}(r/R)\, r^\ell Y_m^\ell(\theta, \varphi), \qquad (22.68)$$

where $P_\ell^{(2n)}$ are polynomials of degree $2n$, and $Y_m^\ell(\theta,\varphi)$ are the spherical harmonics. The frequencies ω obey the following dispersion law:[18]

$$\omega(n,\ell) = \omega_0 \left(2n^2 + 2n\ell + 3n + \ell\right)^{1/2}. \qquad (22.69)$$

In contrast with the uniform case for which the dispersion law is $\omega = cq$, the dispersion law (22.69) does not depend on the interaction parameter. This surprising feature can be understood as follows.

As the low-lying modes have an energy $\sim \hbar\omega_{\text{ho}}$, their wave vector q is on the order of the inverse of the size of the condensate $r_{\max}$ given in (22.31). The frequency of the mode scales as $\omega = cq$ where $q \simeq 1/r_{\max}$ and $c = \sqrt{gn/m} \simeq \sqrt{gN/mr_{\max}^3}$. The dependence of ω on a and N is thus proportional to $\sqrt{aN/r_{\max}^5}$, which, according to (22.31), turns out to be independent of a and N [Dalfovo *et al.* (1999)]. Note that for large values of n and ℓ the density modulation occurs on small spatial scales, and it is no longer correct to neglect the quantum pressure term.

[18] For an ideal Bose gas, one finds $\omega(n,\ell) = \omega_0(2n+\ell)$.

Monopole mode

The monopole mode $n = 1$, $\ell = 0$ has a frequency $\sqrt{5}\,\omega_0$ and corresponds to the breathing of the radius. It therefore does not conserve the volume and can be shown to be intrinsically linked to the compressibility of the condensate [Singh and Rokhsar (1996)]. For negative scattering length and thus outside the validity domain of the Thomas-Fermi approximation, the compressibility increases all the more when the number of atoms is close to the critical value N_c (defined in Sec.22.5.1) [Ueda and Leggett (1998)]. As intuitively expected, one finds that the larger the compressibility, the lower the monopole mode frequency.[19]

For a Bose gas above the critical temperature and if the mean-field is neglected, or for a Fermi gas with a single spin component, this mode has a frequency $2\,\omega_0$ regardless of the elastic collision rate. This remarkable property is due to the fact that this mode involves only the invariant quantities in elastic collisions [Guéry-Odelin *et al.* (1999)].

Dipole modes

The $n = 0$, $\ell = 1$ modes are the dipole modes. They have the same angular frequency ω_0 as the confinement and describe a global translation of the cloud. This is easily justified by considering, for instance, the $\ell = 1$ and $m = 0$ mode for which $\delta n(\vec{r}) \propto (rY_0^1) \sim z$. The density thus reads:

$$n(\vec{r}, t) = n_0(\vec{r}) + \delta n(\vec{r}, t) = \frac{\mu}{g}\left(1 - \frac{x^2 + y^2}{R^2} - \frac{(z - z_0(t))^2}{R^2}\right), \tag{22.70}$$

where $z_0(t) \propto \cos(\omega_0 t)$ is the position of the center of mass of the condensate. The dipole mode, also referred to as the Kohn's mode, characterizes the center of mass oscillation, and is therefore not affected by two-body interactions,[20] nor by the temperature or the statistics. Therefore, this mode is often used to calibrate the value of the trap frequencies experimentally.

Surface modes

These correspond to $n = 0$ and have eigenfrequencies equal to $\omega_0\sqrt{\ell}$. Their velocity field is of the form $\delta\vec{v}(\vec{r}) \sim \vec{\nabla}\left(r^\ell Y_m^\ell(\theta, \varphi)\right)$. Because of the properties of the spherical harmonics, $\vec{\nabla} \cdot \delta\vec{v} \sim \Delta\left(r^\ell Y_m^\ell(\theta, \varphi)\right) = 0$, which shows that the surface modes are divergence free. As a result, the continuity equation for those modes can be recast in

$$\frac{\partial n}{\partial t} + (\vec{v} \cdot \vec{\nabla})n = \frac{\mathrm{d}n}{\mathrm{d}t} = 0. \tag{22.71}$$

[19] A sum rule approach can be usefully employed to work out the link with compressibility [Zambelli and Stringari (2002)].

[20] This is a general property of the center of mass motion that can easily be shown using the Ehrenfest theorem. Actually, the same is true for the classical counterpart and this mode is therefore unaffected by the temperature of the gas: it remains the same above and below the critical temperature T_c.

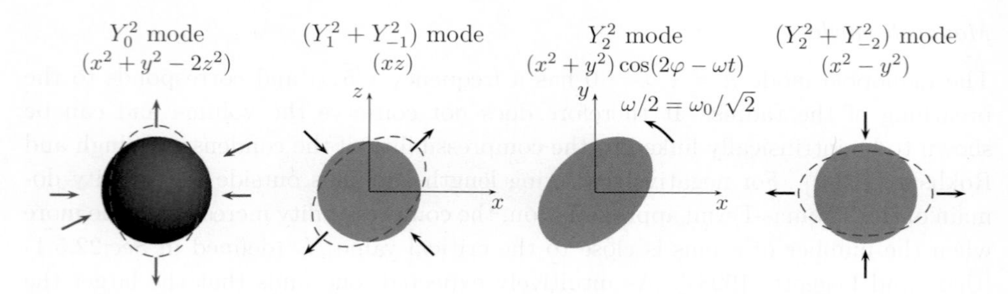

Fig. 22.6 Example of quadrupole modes for a spherical trap.

The cancellation of the total derivative of the density means that in the course of the oscillation the density of the fluid does not change: the "surface" of the condensate is modified while its volume remains constant. This is the reason for the name surface modes. For this class of modes, the compressibility of the condensate does not play a role since the condensate behaves as if it was incompressible.

Quadrupole modes

The surface modes with $\ell = 2$ are referred to as quadrupole modes and have a characteristic frequency equal to $\sqrt{2}\,\omega_0$. Let us detail the kind of density deformation they generate[21] (see also Fig. 22.6):

- The mode $\ell = 2, m = 0$ corresponds to a density perturbation

$$\delta n(\vec{r}, t) \sim \mathrm{Re}(r^2 Y_0^2(\theta, \varphi) e^{-i\omega t}) \sim (2z^2 - x^2 - y^2)\cos(\omega t).$$

  The compression in the transverse plane is accompanied by a dilation along the longitudinal axis, and vice-versa.
- The modes $\ell = 2, m = \pm 1$ generate an elliptical deformation along an axis at 45° with respect to the Ox–Oz axis and Oy–Oz axis, respectively.
- The mode $\ell = 2, m = 2$ corresponds to a density perturbation

$$\delta n(\vec{r}, t) \sim \mathrm{Re}(r^2 Y_2^2(\theta, \varphi) e^{-i\omega t}) \sim (x^2 + y^2)\cos(2\varphi - \omega t).$$

  It corresponds to an elliptic deformation in the plane xOy which rotates at a velocity $\omega/2 = \omega_0/\sqrt{2}$ (see Fig. 22.6). The same conclusion holds for the mode $\ell = 2, m = -2$ but with a reverse rotation velocity. Those modes play an important role in the nucleation of vortices realized by superimposing a small rotating anisotropy to axi-symmetric traps.[22] The linear superposition of those two modes with equal weights gives a density perturbation of the form $\delta n(\vec{r}, t) \sim (x^2 - y^2)\cos\omega t$ that has an opposite sign along the x- and y-axis, and is therefore commonly referred to as the $x^2 - y^2$ mode. The linear superposition

[21] We recall the expression of the corresponding spherical harmonics: $Y_0^2 \propto (3\cos^2\theta - 1)$, $Y_{\pm 1}^2 \propto \sin\theta\cos\theta e^{\pm i\varphi}$ and $Y_{\pm 2}^2 \propto \sin^2\theta e^{\pm 2i\varphi}$.
[22] See Sec. 24.5.4 of Chap. 24.

with opposite sign gives rise to the xy mode. These modes do not affect the z-axis.

A general feature of these surface modes is that they are insensitive to the form of the equation of state $\mu(n)$. This is the reason why they have the same expression for a Bose gas above the critical temperature where it can be described by hydrodynamic equations, i.e. in the regime for which the collision rate is very large in comparison to the trapping angular frequency [Griffin *et al.* (1997)].

In practice, the frequencies of the monopole mode and the quadrupole modes $(x^2 + y^2 - 2z^2)$ and $(x^2 - y^2)$ can be readily obtained using the scaling approach developed in Sec. 22.7.4.1, i.e. by linearizing the set of Eqs. (22.65) around $b_i = 1$. This method provides the modes that have the symmetry adapted to the scaling factors, but can also be used for a non-spherical trap.

22.9.2 *Cylindrically-symmetric trap*

Most experiments are performed using a trapping geometry with cylindrical symmetry. Atoms thus experience a confining potential of the form:

$$V_{\text{trap}}(x, y, z) = \frac{1}{2}m\omega_\perp^2(x^2 + y^2) + \frac{1}{2}m\omega_z^2 z^2. \tag{22.72}$$

For instance, Ioffe Pritchard traps[23] give rise to cigar shaped clouds ($\omega_z \ll \omega_\perp$), and TOP traps to disk shape with $\omega_z = \sqrt{8}\,\omega_\perp$. The normal modes are strongly affected by the geometry. However, the azimuthal quantum number m remains a good quantum number because of the axial symmetry. This is no longer the case for the quantum number ℓ.

One finds that a perturbation density of the form $\delta n(\vec{r}) = r^2 Y_m^2(\theta, \varphi)$ is still a solution of Eq. (22.67) with eigenfrequency $\omega = \omega_\perp\sqrt{2}$ for $m = \pm 2$, and $\omega = (\omega_\perp^2 + \omega_z^2)^{1/2}$ for $m = \pm 1$. A density perturbation of the form $r^2 Y_0^2(\theta, \varphi)$ is no longer an eigenmode. However, two linear superpositions of the $m = 0$ quadrupole $r^2 Y_0^2 \propto x^2 + y^2 - 2z^2$ and monopole[24] $r^2 Y_0^0 \propto x^2 + y^2 + z^2$ give the eigenfrequencies[25] [Stringari (1996b)]:

$$\omega_\pm^2 = 2\omega_\perp^2 + \frac{3}{2}\omega_z^2 \mp \frac{1}{2}\sqrt{9\omega_z^2 - 16\omega_z^2\omega_\perp^2 + 16\omega_\perp^4}. \tag{22.73}$$

This prediction has been successfully confronted to experimental results [Jin *et al.* (1996)].

22.9.3 *Scissors mode for anisotropic traps*

For a cylindrically symmetric trap, the component of the angular momentum $L_z = xp_y - yp_x$ along the symmetry axis is a conserved quantity. This means that if

[23] See Sec. 14.3.3 and 14.3.4 of Chap. 14.

[24] The polynomial $P_0^2 \propto r^2$.

[25] For a spherical trap, one recovers the frequency $\omega_- = \sqrt{2}\omega_0$ of the quadrupole mode $n = 0, \ell = 2, m = 0$ and the frequency $\omega_+ = \sqrt{5}\omega_0$ of the monopole mode $n = 1, \ell = 0, m = 0$.

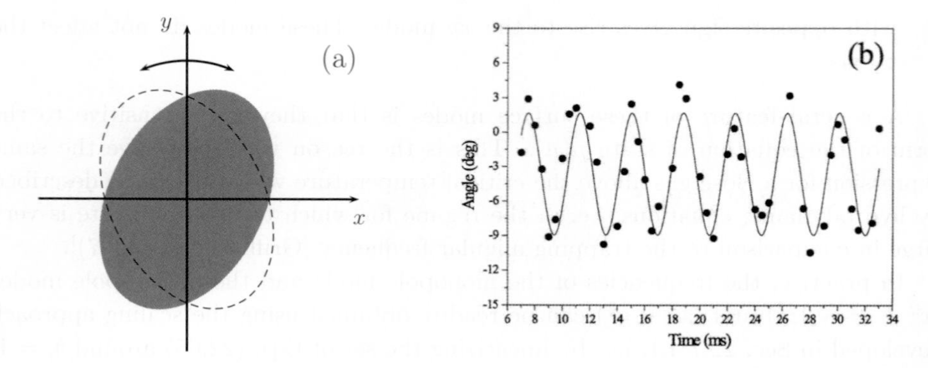

Fig. 22.7 Scissors mode. (a) Angular oscillation of the cloud. This mode is excited by a sudden rotation of the eigenaxis by a small angle. (b) Experimental measurement of the angular oscillation associated with the scissors mode. Figure taken from [Maragò *et al.* (2000)]. Copyright: American Physical Society.

the condensate is set in rotation with an angular velocity along this axis, it will rotate for ever. We consider in the following what happens if we break slightly this symmetry. For this purpose, we consider an anisotropic trap of the form:

$$V_{\rm trap}(\vec{r}) = \frac{1}{2}m\left[\omega_0^2(1+\varepsilon)x^2 + \omega_0^2(1-\varepsilon)y^2 + \omega_z^2 z^2\right]. \tag{22.74}$$

The dimensionless parameter ε accounts for the transverse symmetry breaking. A direct consequence is that the cloud will experience a torque $\Gamma_z = (\vec{r}\times(-\vec{\nabla}V_{\rm trap})/m)_z \propto \varepsilon xy$. As a result, the angular momentum is coupled to the $Y^2_{\pm 1}$ quadrupole modes. In this way, the study of surface modes yields information on the moment of inertia of the condensate [Zambelli and Stringari (2001)]. The scissors mode belongs to such modes and is excited by suddenly rotating the eigenaxis by a small angle around the z-axis. The corresponding density perturbation is of the form $\delta n(\vec{r},t) = \lambda_0 xy\cos(\omega t)$, and is the solution of the linearized hydrodynamic equations with $\omega^2 = \omega_x^2 + \omega_y^2 = 2\omega_0^2$ [Guéry-Odelin ans Stringari (1999)] (see Fig. 22.7). This density deformation can be easily interpreted when the angle of rotation of the trap is small compared to the deformation ε: it corresponds to a pendulum motion of the BEC in the x-y plane. Note that, under the same excitation, a classical gas or a Bose gas above the critical temperature in the collisionless regime exhibits oscillations with two frequencies $|\omega_x \pm \omega_y|$. This difference is intrinsically due to the fact that the velocity field of Bose-Einstein condensates has to be irrotational (in the absence of vortices). As a result, the moment of inertia of the condensate is ε times smaller than in the non-degenerate regime. The first experimental investigation of the scissors mode with a BEC was carried out by the group of Christopher Foot at Oxford University [Maragò *et al.* (2000)].

Let us finally point out that the interaction between the condensate and a thermal cloud leads to damping and frequency shifts of condensate collective modes. These effects have been studied in several experiments [Jin *et al.* (1997); Stamper-Kurn *et al.* (1998); Maragò *et al.* (2001); Chevy *et al.* (2002)].

Chapter 23

Coherence properties of Bose-Einstein condensates

23.1 Introduction

Two years after the observation of BEC in ultracold atomic gases, a spectacular experiment was performed at MIT that demonstrated the interference of two condensates (see Fig. 23.1) [Andrews *et al.* (1997a)]. An elongated condensate was cut in two parts with a blue-detuned laser beam. The trapping potential was switched off, allowing the two condensates to fall under gravity and expand. After a sufficiently long expansion time, an absorption image was taken. In the overlapping zone of the two condensates, a clear interference pattern was observed (right part of Fig. 23.1).

Such an experiment shows that matter waves interfere like electromagnetic waves. This confirms the description, given in the previous chapter, of Bose-Einstein condensates in terms of matter waves. We also know that these condensates are composed of particles, which are identical bosonic atoms (obeying Bose-Einstein statistics), in the same way as light waves are composed of photons. In optics, there are detection signals, such as double counting rates, that give the probability to detect one photon at a given place $\vec{r}$ at a given time t, and another photon at $\vec{r}\,'$ and t'.[1] To correctly interpret this type of signal, and more generally the coherence properties of light beams, Roy Glauber has shown that it is necessary to introduce correlation functions of the field operators that describe the quantum electromagnetic field [Glauber (1965, 2006)]. In other words, the description of optical experiments in terms of classical Maxwell waves is not always sufficient. There are situations that require a quantum description of the field for their interpretation.[2]

This leads us to ask the following questions:

- Is it possible to develop a quantum theory of coherence for matter waves analogous to the one used in quantum optics to describe the coherence properties of light waves?

[1] See Sec. 19.2.1 of Chap. 19.

[2] A simple introduction to quantum optics can be found in [Grynberg *et al.* (2010)].

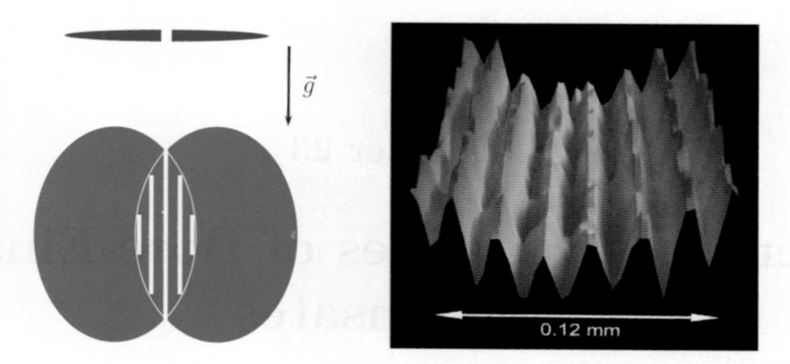

Fig. 23.1 Left: principle of the experiment. An elongated condensate is cut in two by a blue detuned laser beam. After a 40 ms time-of-flight, the two condensates partially overlap. Right: the overlapping zone shows matter wave interference fringes. Courtesy of Wolfgang Ketterle.

- Are there differences between thermal gases and Bose-Einstein condensates analogous to the differences between thermal light and laser light?

In Sec. 23.2 of this chapter, we show that the second quantization formalism used for describing a system of identical particles in quantum mechanics allows a simple introduction of atomic field operators analogous to the quantum electric field operators in quantum optics. Then, in Sec. 23.3, we introduce the correlation functions of these atomic field operators. We show that they are convenient tools for analyzing the coherence and statistical properties of matter waves, not only matter waves involving a very small number of atoms, but also macroscopic matter waves associated with condensates. We calculate these correlation functions for a few simple cases (ideal gas in a box, trapped condensate with interactions) and show that they allow a clear understanding of the differences between the coherence properties of thermal gases and condensates. We also describe a few experimental tests of these predictions.

Another important problem raised by the experiment shown in Fig. 23.1 is the issue of the relative phase of two condensates. How can one obtain interference fringes with two independent condensates if there is no phase relation between them? In Sec. 23.4, we address these problems and we show how a relative phase between two initially independent condensates can emerge as a result of a succession of detection processes.

Finally, one must keep in mind an important difference between light waves and matter waves. In matter waves, atoms interact, which is not the case for photons (except, indirectly, through the coupling with a nonlinear atomic medium). Can these atom-atom interactions give rise to new effects in atom optics, such as dephasing or nonlinear processes? In Sec. 23.6, we will see that these effects exist and will briefly describe a few of them.

23.2 Atomic field operators and correlation functions

23.2.1 *Brief reminder on second quantization*

To describe an ensemble of N identical particles, it is convenient to consider an enlarged space of states, called *Fock space*, in which the number of particles is not fixed. Let $\{|\varphi_\alpha\rangle\}$ be an orthonormal basis of one-particle states. In Fock space, one can introduce an orthonormal basis $\{|\ldots n_\alpha \ldots n_\beta \ldots\rangle\}$, defined by the number n_α of particles in each state $|\varphi_\alpha\rangle$. In the case of bosons, the n_α's can be any non-negative integer. For fermions, n_α is equal to 0 or 1. Note that in this description the different particles are not labelled with numbers. The only relevant indices characterizing the states are the occupation numbers n_α. The Fock space is a direct sum of subspaces corresponding to a well-defined value of the total number $N = \sum_\alpha n_\alpha$ of particles. The subspace with no particles is denoted $|0\rangle$ and called the vacuum.

It is convenient to introduce creation and annihilation operators that create and destroy a particle in each state $|\varphi_\alpha\rangle$ in Fock space. The creation operator of a particle in the state $|\varphi_\alpha\rangle$ is denoted $\hat{a}_\alpha^\dagger$ and the destruction operator by $\hat{a}_\alpha$. These two operators are hermitian adjoints, and when the identical particles are bosons, they obey the commutation relations:

$$\left[\hat{a}_\alpha, \hat{a}_\beta^\dagger\right] = \delta_{\alpha\beta}, \qquad \text{and} \qquad [\hat{a}_\alpha, \hat{a}_\beta] = \left[\hat{a}_\alpha^\dagger, \hat{a}_\beta^\dagger\right] = 0. \tag{23.1}$$

One can show that these commutation relations ensure that the states of the Fock space are fully symmetric under any exchange of particles. In the case of fermions, the commutators of (23.1) have to be replaced by anticommutators that ensure that the states of the Fock space are fully antisymmetric. In this chapter, we will only consider the case of bosons.

The operators $\hat{a}_\alpha$ and $\hat{a}_\alpha^\dagger$ acting on a Fock state $|\ldots n_\alpha \ldots n_\beta \ldots\rangle$ give:

$$\begin{aligned} \hat{a}_\alpha^\dagger |\ldots n_\alpha \ldots n_\beta \ldots\rangle &= \sqrt{n_\alpha + 1}\, |\ldots n_\alpha + 1 \ldots n_\beta \ldots\rangle\,, \\ \hat{a}_\alpha |\ldots n_\alpha \ldots n_\beta \ldots\rangle &= \sqrt{n_\alpha}\, |\ldots n_\alpha - 1 \ldots n_\beta \ldots\rangle\,. \end{aligned} \tag{23.2}$$

From these equations, one can easily show that:

$$|\ldots n_\alpha, \ldots n_\beta, \ldots\rangle = \frac{1}{\sqrt{\ldots n_\alpha! \ldots n_\beta! \ldots}} \ldots \left(\hat{a}_\alpha^+\right)^{n_\alpha} \ldots \left(\hat{a}_\beta^\dagger\right)^{n_\beta} \ldots |0\rangle \tag{23.3}$$

It will be also useful to express the various operators of the system of bosons in terms of the creation and annihilation operators. Let $f(i)$ be a one-particle operator associated with the particle labelled i in the usual *first quantization* formulation. For instance, $f(i)$ can represent the kinetic energy of particle i. The corresponding operator for a system made of N particles reads $F = \sum_{i=1}^N f(i)$. In second quantization, one can show that this operator becomes [Landau and Lifshitz (1996)]

$$\hat{F} = \sum_\alpha \sum_\beta \langle\varphi_\beta| f |\varphi_\alpha\rangle\, \hat{a}_\beta^\dagger \hat{a}_\alpha. \tag{23.4}$$

Similarly, the two-particle operator acting on a system made of N particles is given, in first quantization, by $G = (1/2)\sum_{i=1}^{N}\sum_{j\neq i} g(i,j)$, where $g(i,j) = g(j,i)$. For example, $g(i,j)$ describes an interaction between particles i and j, depending only on their distance $r_{ij} = |\vec{r}_i - \vec{r}_j| = r_{ji}$. The factor (1/2) is introduced in the expression of G, in order to avoid a double counting of the interaction between a pair (i,j) of particles. In second quantization, the expression of G is given by:

$$\hat{G} = \frac{1}{2}\sum_{\alpha}\sum_{\beta}\sum_{\gamma}\sum_{\delta} \hat{a}_{\delta}^{\dagger}\hat{a}_{\gamma}^{\dagger}\hat{a}_{\beta}\hat{a}_{\alpha}\,\langle\varphi_{\delta}(1)\varphi_{\gamma}(2)|\,g(1,2)\,|\varphi_{\alpha}(1)\varphi_{\beta}(2)\rangle\,. \tag{23.5}$$

Note that $\hat{F}$ and $\hat{G}$ both contain an equal number of creation and annihilation operators, and so can only connect states with the same total number of particles.

23.2.2 *Atomic field operators*

The second quantized formulation sketched in the previous subsection is a powerful tool for introducing field operators $\hat{\Psi}^{\dagger}(\vec{r})$ and $\hat{\Psi}(\vec{r})$ which create and destroy an atom at $\vec{r}$. They are in close analogy with the electric field operators $\hat{E}^{(-)}(\vec{r})$ and $\hat{E}^{(+)}(\vec{r})$ introduced in quantum optics to account for the coherence properties of light, and are useful for analyzing the coherence properties of Bose-Einstein condensates, as illustrated in the following with the calculation of the first- and second-order correlation functions.

The field operators are linear combinations of the creation and annihilation operators with coefficients that are the single-particle wave functions:

$$\hat{\Psi}(\vec{r}) = \sum_{\alpha}\varphi_{\alpha}(\vec{r})\,\hat{a}_{\alpha} \quad \text{and} \quad \hat{\Psi}^{\dagger}(\vec{r}) = \sum_{\alpha}\varphi_{\alpha}^{*}(\vec{r})\,\hat{a}_{\alpha}^{\dagger}, \tag{23.6}$$

and where the sum runs over the complete set of single-particle quantum numbers. By construction, these operators are defined at each space point $\vec{r}$. Their interpretation becomes clear when one calculates the action of the operator $\hat{\Psi}^{\dagger}(\vec{r})$ on the vacuum state $|0\rangle$:

$$\hat{\Psi}^{\dagger}(\vec{r})\,|0\rangle = \sum_{\alpha}\varphi_{\alpha}^{*}(\vec{r})\,\hat{a}_{\alpha}^{\dagger}\,|0\rangle \underset{1^{\text{st}}\text{quantiz.}}{\longrightarrow} \sum_{\alpha}\varphi_{\alpha}^{*}(\vec{r})\,|\varphi_{\alpha}\rangle\,. \tag{23.7}$$

Using the property $\varphi_{\alpha}^{*}(\vec{r}) = \langle\varphi_{\alpha}|\vec{r}\rangle$ and the fact that $\{|\varphi_{\alpha}\rangle\}$ forms a complete basis for the single-particle Hilbert space obeying the closure relation $\sum_{\alpha}|\varphi_{\alpha}\rangle\langle\varphi_{\alpha}| = 1$, one finds

$$\hat{\Psi}^{\dagger}(\vec{r})\,|0\rangle \underset{1^{\text{st}}\text{quantiz.}}{\longrightarrow} \sum_{\alpha}\varphi_{\alpha}^{*}(\vec{r})\,|\varphi_{\alpha}\rangle = \sum_{\alpha}|\varphi_{\alpha}\rangle\langle\varphi_{\alpha}\mid\vec{r}\rangle = |\vec{r}\rangle\,. \tag{23.8}$$

The field operator $\hat{\Psi}^{\dagger}(\vec{r})$ is therefore an operator that creates an atom at $\vec{r}$. Similarly, the operator $\hat{\Psi}(\vec{r})$ annihilates an atom at $\vec{r}$. It follows from (23.1) that the field operators for bosons satisfy simple commutation relations

$$\left[\hat{\Psi}(\vec{r}),\hat{\Psi}^{\dagger}(\vec{r}\,')\right] = \delta(\vec{r}-\vec{r}\,') \quad \text{and} \quad \left[\hat{\Psi}(\vec{r}),\hat{\Psi}(\vec{r}\,')\right] = \left[\hat{\Psi}^{\dagger}(\vec{r}),\hat{\Psi}^{\dagger}(\vec{r}\,')\right] = 0. \tag{23.9}$$

The field operators $\hat{\Psi}^{\dagger}(\vec{r})$ and $\hat{\Psi}(\vec{r})$ play the same role in the basis $\{|\vec{r}\rangle\}$ as the operators $\hat{a}_{\alpha}^{\dagger}$ and $\hat{a}_{\alpha}$ in the basis $\{|\varphi_{\alpha}\rangle\}$.

23.2.3 *Examples of physical operators. Field correlation functions*

A physical operator does not change the number of atoms. It must therefore contain the same number of creation and annihilation operators. In this subsection, we give a few examples of such operators.

(i) *Spatial density at a given point*

Consider, in the first quantization formulation, the one-particle operator:

$$\rho_I(\vec{r}_0) = \sum_{i=1}^{N} |i, \vec{r}_0\rangle \langle i, \vec{r}_0| \tag{23.10}$$

Its average value can be written:

$$\langle \rho_I(\vec{r}_0)\rangle = \mathrm{Tr}\left[\sigma_N \rho_I(\vec{r}_0)\right] = \langle \vec{r}_0| \, \sigma_N \, |\vec{r}_0\rangle \tag{23.11}$$

where σ_N is the fully symmetric N-atom density operator and

$$\sigma_1 = \mathrm{Tr}_{2,3,\ldots N}[\sigma_N] \tag{23.12}$$

is the reduced one-atom density operator. This average value represents the probability density to find any particle at $\vec{r}_0$.

Using (23.4), one gets the following expression for the second quantized form of the spatial density operator at $\vec{r}_0$ in terms of the field operators:

$$\hat{\rho}_I\left(\vec{r}_0\right) = \iint \mathrm{d}^3r\mathrm{d}^3r' \hat{\Psi}^\dagger(\vec{r}\,')\hat{\Psi}(\vec{r}) \left\langle \vec{r}\,'\right| \left(\left|\vec{r}_0\right\rangle \left\langle\vec{r}_0\right|\right) \left|\vec{r}\right\rangle = \hat{\Psi}^\dagger(\vec{r}_0)\hat{\Psi}(\vec{r}_0). \tag{23.13}$$

(ii) *Spatial coherence between two points. First-order spatial correlation function*

In the first quantization formulation, the average value of the one-particle operator

$$\rho_I(\vec{r}_0, \vec{r}\,'_0) = \sum_{i=1}^{N} |i, \vec{r}\,'_0\rangle \langle i, \vec{r}_0| \tag{23.14}$$

is equal to

$$\langle \rho_I(\vec{r}_0, \vec{r}\,'_0)\rangle = \mathrm{Tr}\left[\sigma_N \rho_I(\vec{r}_0, \vec{r}\,'_0)\right] = N \langle \vec{r}_0| \, \sigma_1 \, |\vec{r}\,'_0\rangle \tag{23.15}$$

and represents the spatial coherence between the two points $\vec{r}_0$ and $\vec{r}\,'_0$.

The expression for the corresponding operator in second quantization is:

$$\hat{\rho}_I\left(\vec{r}_0, \vec{r}\,'_0\right) = \iint \mathrm{d}^3r\mathrm{d}^3r' \hat{\Psi}^\dagger(\vec{r}\,')\hat{\Psi}(\vec{r}) \left\langle \vec{r}\,'\right| \left(\left|\vec{r}\,'_0\right\rangle \left\langle\vec{r}_0\right|\right) \left|\vec{r}\right\rangle = \hat{\Psi}^\dagger(\vec{r}\,'_0)\hat{\Psi}(\vec{r}_0). \tag{23.16}$$

The average value of $\hat{\rho}_I\left(\vec{r}, \vec{r}\,'\right)$ is called the first-order spatial correlation function:

$$G^{(1)}\left(\vec{r}, \vec{r}\,'\right) = \langle \hat{\Psi}^\dagger(\vec{r})\hat{\Psi}(\vec{r}\,')\rangle. \tag{23.17}$$

Physically, $G^{(1)}\left(\vec{r}, \vec{r}\,'\right)$ is related to the visibility of the interference fringes obtained when two matter waves originating from $\vec{r}$ and $\vec{r}\,'$ are superimposed (see Sec. 23.3.4 (ii)). We will describe below an experiment of this type.

(iii) *Density correlations between two points. Second-order spatial correlation function*

We now consider the two-particle operator whose expression in first quantization is:

$$\rho_{II}(\vec{r}_0, \vec{r}\,'_0) = \sum_{i=1}^{N} \sum_{j \neq i} g(i,j) = \sum_{i=1}^{N} \sum_{j \neq i} |i, \vec{r}_0; j, \vec{r}\,'_0\rangle \langle i, \vec{r}_0; j, \vec{r}\,'_0| \,. \tag{23.18}$$

The average value of this operator is the probability of finding a particle at $\vec{r}_0$ and another one at $\vec{r}\,'_0$. Note that there is no factor $1/2$ in the expression of ρ_{II} because we have here $g(i,j) \neq g(j,i)$, so that there is no risk of double counting.[3] The corresponding expression of this operator in second quantization is:

$$\hat{\rho}_{II}(\vec{r}_0, \vec{r}\,'_0) = \hat{\Psi}^\dagger(\vec{r}_0)\hat{\Psi}^\dagger(\vec{r}\,'_0)\hat{\Psi}(\vec{r}\,'_0)\hat{\Psi}(\vec{r}_0). \tag{23.19}$$

The average value of $\hat{\rho}_{II}(\vec{r}, \vec{r}\,')$ is called the second-order spatial correlation function:

$$G^{(2)}(\vec{r}, \vec{r}\,') = \langle \hat{\Psi}^\dagger(\vec{r})\hat{\Psi}^\dagger(\vec{r}\,')\hat{\Psi}(\vec{r}\,')\hat{\Psi}(\vec{r})\rangle. \tag{23.20}$$

This function characterizes the tendency of the atoms to cluster or to stay apart from each other.

(iv) *Generalizations*

Higher-order correlation functions like

$$G^{(3)}(\vec{r}, \vec{r}\,', \vec{r}\,'') = \left\langle \hat{\Psi}^\dagger(\vec{r})\hat{\Psi}^\dagger(\vec{r}\,')\hat{\Psi}^\dagger(\vec{r}\,'')\hat{\Psi}(\vec{r}\,'')\hat{\Psi}(\vec{r}\,')\hat{\Psi}(\vec{r})\right\rangle, \tag{23.21}$$

which represents the probability to find one atom at $\vec{r}$, a second one at $\vec{r}\,'$ and a third one at $\vec{r}\,''$ can be introduced.

It will be useful to define normalized correlation functions

$$g^{(1)}(\vec{r}, \vec{r}\,') = \frac{G^{(1)}(\vec{r}, \vec{r}\,')}{\sqrt{G^{(1)}(\vec{r}, \vec{r})G^{(1)}(\vec{r}\,', \vec{r}\,')}}, \tag{23.22}$$

$$g^{(2)}(\vec{r}, \vec{r}\,') = \frac{G^{(2)}(\vec{r}, \vec{r}\,')}{G^{(1)}(\vec{r}, \vec{r})G^{(1)}(\vec{r}\,', \vec{r}\,')}, \tag{23.23}$$

$$g^{(3)}(\vec{r}, \vec{r}\,', \vec{r}\,'') = \frac{G^{(3)}(\vec{r}, \vec{r}\,', \vec{r}\,'')}{G^{(1)}(\vec{r}, \vec{r})G^{(1)}(\vec{r}\,', \vec{r}\,')G^{(1)}(\vec{r}\,'', \vec{r}\,'')}. \tag{23.24}$$

Time-dependent correlation functions can be defined by considering field operators $\hat{\Psi}(\vec{r}, t)$ in the Heisenberg picture. For example, the time-dependent first-order correlation function is:

$$G^{(1)}(\vec{r}, t; \vec{r}\,', t') = \langle \hat{\Psi}^\dagger(\vec{r}, t)\hat{\Psi}(\vec{r}\,', t')\rangle. \tag{23.25}$$

[3] Although $g(i,j)$, in this case, is not symmetric, the sum $\sum_{i=1}^{N}\sum_{j\neq i} g(i,j)$ does not change in any permutation of the N particles.

23.2.4 *Heisenberg equation of the field operator*

The time evolution of the field operator $\hat{\Psi}(\vec{r},t)$ is obtained from the Heisenberg equation

$$i\,\hbar\frac{\partial}{\partial t}\hat{\Psi}(\vec{r},t) = \left[\hat{\Psi}(\vec{r},t),\hat{H}\right], \tag{23.26}$$

where the Hamiltonian $\hat{H}$ is the sum of three contributions $\hat{H} = \hat{H}_{\text{kin}}+\hat{H}_{\text{trap}}+\hat{H}_{\text{int}}$. The first two terms, $\hat{H}_{\text{kin}}$ and $\hat{H}_{\text{trap}}$, are one-atom operators, whereas the last one describes the interactions and is a two-atom operator. Using Eqs. (23.4) and (23.5) we readily obtain the second quantized form of these operators in terms of the field operators

$$\begin{aligned}
\hat{H}_{\text{kin}} &= \int \mathrm{d}^3r\,\hat{\Psi}^\dagger(\vec{r})\left(-\frac{\hbar^2}{2m}\Delta\right)\hat{\Psi}(\vec{r}),\\
\hat{H}_{\text{trap}} &= \int \mathrm{d}^3r\,\hat{\Psi}^\dagger(\vec{r})\,V_{\text{trap}}(\vec{r})\,\hat{\Psi}(\vec{r}),\\
\hat{H}_{\text{int}} &= \frac{1}{2}\int \mathrm{d}^3r\,\mathrm{d}^3r'\hat{\Psi}^\dagger(\vec{r})\hat{\Psi}^\dagger(\vec{r}')\,V(\vec{r}-\vec{r}')\,\hat{\Psi}(\vec{r}')\hat{\Psi}(\vec{r}).
\end{aligned} \tag{23.27}$$

We now replace the two-body interaction potential V by the delta function potential:

$$V(\vec{r}-\vec{r}') = g\delta(\vec{r}-\vec{r}')\;\Rightarrow \hat{H}_{\text{int}} = \frac{1}{2}\int \mathrm{d}^3r\,\left(\hat{\Psi}^\dagger(\vec{r})\right)^2\left(\hat{\Psi}(\vec{r})\right)^2. \tag{23.28}$$

Using the bosonic commutation rules (23.9) of the field operators, the Heisenberg equation yields:

$$i\,\hbar\frac{\partial}{\partial t}\hat{\Psi}(\vec{r},t) = \left[-\frac{\hbar^2}{2m}\Delta + V_{\text{trap}}(\vec{r}) + g\,\hat{\Psi}^\dagger(\vec{r},t)\hat{\Psi}(\vec{r},t)\right]\hat{\Psi}(\vec{r},t). \tag{23.29}$$

This equation coincides with the time-dependent Gross-Pitaevskii equation when the field operator $\hat{\Psi}(\vec{r},t)$ is replaced by the wave function $\psi(\vec{r},t)$ that describes the condensate in the mean field approach. More precise treatments can be investigated by trying to take into account the difference $\delta\hat{\Psi}(\vec{r})$ between the field operator $\hat{\Psi}(\vec{r},t)$ and the wave function $\psi(\vec{r},t)$.

23.3 Calculation of correlation functions in a few simple cases

23.3.1 *First-order correlation function for an ideal Bose gas in a box*

We consider an ideal gas of bosons of mass m in a box of size L. An orthonormal basis of single particle states with periodic boundary conditions is made of plane waves

$$\psi_{\vec{k}_i}(\vec{r}) = \frac{e^{i\vec{k}_i.\vec{r}}}{L^{3/2}} \qquad \text{where} \qquad k_i = \frac{2\pi}{L}n_i, \tag{23.30}$$

and n_i is any integer. The field operators are defined by

$$\hat{\Psi}(\vec{r}) = \frac{1}{L^{3/2}} \sum_{\vec{k}} \hat{a}_{\vec{k}} e^{i\vec{k}\cdot\vec{r}} \qquad \text{and} \qquad \hat{\Psi}^\dagger(\vec{r}) = \frac{1}{L^{3/2}} \sum_{\vec{k}} \hat{a}^\dagger_{\vec{k}} e^{-i\vec{k}\cdot\vec{r}}. \tag{23.31}$$

The Hamiltonian of the ideal Bose gas is given by:

$$\hat{H} = \sum_{\vec{k}} \varepsilon_{\vec{k}} \hat{a}^\dagger_{\vec{k}} \hat{a}_{\vec{k}} \qquad \text{with} \qquad \varepsilon_{\vec{k}} = \frac{\hbar^2 k^2}{2m}. \tag{23.32}$$

The calculation of $G^{(1)}(\vec{r},\vec{r}\,')$ will be performed in the grand-canonical formalism where the populations of the various single particle states are independent. The expression of the density matrix is then $\hat{\rho}_{\text{eq}} = e^{-\beta(\hat{H}-\mu\hat{N})}/Z_G$ and we have

$$G^{(1)}(\vec{r},\vec{r}\,') = \frac{1}{L^3} \sum_{\vec{k},\vec{k}\,'} e^{-i\left(\vec{k}\cdot\vec{r}-\vec{k}\,'\cdot\vec{r}\,'\right)} \text{Tr}\left[\hat{\rho}_{\text{eq}} \hat{a}^\dagger_{\vec{k}} \hat{a}_{\vec{k}'}\right]. \tag{23.33}$$

The invariance by translation implies $\langle \hat{a}^\dagger_{\vec{k}} \hat{a}_{\vec{k}\,'} \rangle = \langle n_{\vec{k}} \rangle \delta_{\vec{k},\vec{k}\,'}$. The mean value $\langle n_{\vec{k}} \rangle$ obtained for an ideal Bose gas in the grand-canonical ensemble is given by:

$$\langle n_{\vec{k}} \rangle = \text{Tr}[\hat{\rho}_{\text{eq}} \hat{a}^\dagger_{\vec{k}} \hat{a}_{\vec{k}}] = \frac{z e^{-\beta\varepsilon_k}}{1 - z e^{-\beta\varepsilon_k}} = \sum_{\ell=1}^{\infty} z^\ell e^{-\ell\beta\varepsilon_k} \tag{23.34}$$

where $z = e^{\beta\mu}$ is the fugacity and μ the chemical potential. The first-order correlation function is therefore given by the Fourier transform of the momentum distribution. This gives:

$$\begin{aligned} G^{(1)}(\vec{r},\vec{r}\,') &= \frac{N_0}{L^3} + \frac{1}{(2\pi)^3} \int \mathrm{d}^3\vec{k}\, e^{i\vec{k}\cdot(\vec{r}\,'-\vec{r})} \sum_{\ell=1}^{\infty} z^\ell e^{-\ell\beta\hbar^2k^2/2m} \\ &= \frac{N_0}{L^3} + \frac{1}{\lambda_{\text{T}}^3} \sum_{\ell=1}^{\infty} \frac{z^\ell}{\ell^{3/2}} \exp\left(- \frac{\pi(\vec{r}\,'-\vec{r})^2}{\ell\lambda_{\text{T}}^2}\right), \end{aligned} \tag{23.35}$$

where λ_{T} is the thermal de Broglie wavelength. The fact that $G^{(1)}(\vec{r},\vec{r}\,')$ depends only on the relative distance $|\vec{r}\,'-\vec{r}|$ is a direct consequence of invariance by translation. The first term on the left-hand side of Eq. (23.35) gives the contribution of Bose condensed atoms, and the second term gives that of thermal atoms.

The first order correlation for $\vec{r} = \vec{r}\,'$ is nothing but the atomic density, which is uniform in a box with periodic boundaries:

$$G^{(1)}(\vec{r},\vec{r}) = n(\vec{r}) = \frac{N_0}{L^3} + \frac{1}{\lambda_{\text{T}}^3} \sum_{\ell=1}^{\infty} \frac{z^\ell}{\ell^{3/2}} = \frac{N}{L^3}. \tag{23.36}$$

It is instructive to work out two limits:

- The limit where classical statistics is valid: $n\lambda_{\text{T}}^3 \ll 1$, $N_0 \ll N$, $z \ll 1$,

$$G^{(1)}(\vec{r},\vec{r}\,') \simeq \frac{N}{L^3} \exp^{-\pi(\vec{r}\,'-\vec{r})^2/\lambda_{\text{T}}^2}. \tag{23.37}$$

  We find that for a classical gas the coherence length is $\lambda_{\text{T}}/\sqrt{\pi}$.

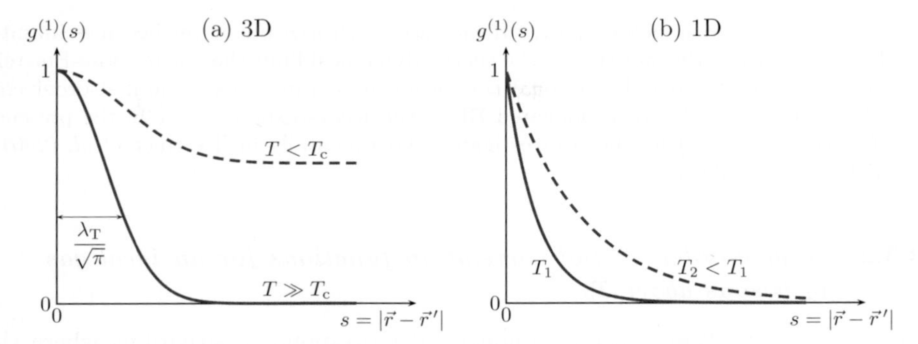

Fig. 23.2 Normalized first order correlation function $g^{(1)}$ for a homogeneous ideal Bose gas in (a) 3D and (b) 1D. In 3D, below T_c, $g^{(1)}(s)$ goes to the constant value N_0/N when $s \to \infty$, clearly displaying off-diagonal long-range order. In contrast, in 1D, the correlation function decays exponentially for any temperature.

- Below the critical temperature, the contribution of the condensed atoms yields an infinite coherence length. $G^{(1)}$ reveals the presence of a long range spatial order due to the condensate (see Fig. 23.2(a)).

Coherence: the role of the dimensionality

For a one-dimensional ideal Bose gas in the degenerate regime within the validity range of the semiclassical description, $k_B T \gg |\mu| \gg h^2/mL^2$, the first-order correlation function reads

$$G^{(1)}(x,0) = \frac{1}{2\pi\hbar}\int \frac{e^{-ipx/\hbar}}{z^{-1}e^{\beta p^2/2m} - 1}\mathrm{d}p. \tag{23.38}$$

Its behavior for large $|x|$ is determined by the behavior at small p of the integrand of (23.38). Expanding the denominator of this integrand in powers of p leads to:

$$G^{(1)}(x,0) \underset{\text{large } |x|}{\simeq} \frac{mk_B T}{\pi\hbar}\int \frac{e^{-ipx/\hbar}}{p^2 + p_c^2}\,\mathrm{d}p \propto e^{-|x|/\xi} \tag{23.39}$$

where $p_c^2 = 2m|\mu|$ and $\xi = \hbar/p_c$. The first-order correlation function exhibits an exponential decay with a coherence length ξ and therefore there is no long range order (see Fig. 23.2(b)).
More insight can be obtained into ξ by re-expressing it in terms of the one-dimensional spatial density $n_{1D} = N/L$ where N is the total number of atoms given by:

$$N = \int G^{(1)}(0,0)\mathrm{d}x \simeq \frac{Lmk_B T}{\pi\hbar}\int \frac{\mathrm{d}p}{p^2 + p_c^2} = \frac{Lmk_B T}{\hbar p_c}. \tag{23.40}$$

From this equation one deduces that $p_c = h/n_{1D}\lambda_T^2$, where λ_T is the thermal de Broglie wavelength, so that:

$$\xi = \frac{\hbar}{p_c} = \frac{n_{1D}\lambda_T^2}{2\pi}. \tag{23.41}$$

This shows that when the mean distance between atoms becomes smaller than the thermal de Broglie wavelength, i.e. when $n_{1D}\lambda_T \gg 1$, the coherence length becomes larger than λ_T by a factor $n_{1D}\lambda_T$. It increases with n_{1D} but never becomes infinite. This is related to the fact that in the thermodynamical limit there is no Bose-Einstein condensation for a one-dimensional Bose gas confined in a box. The first-order correlation function of a very elongated BEC still decays exponentially in the presence of interactions[4] and has been investigated experimentally in [Dettmer *et al.* (2001); Richard *et al.* (2003)].

23.3.2 *Higher-order spatial correlation functions for an ideal gas of bosons above T_c*

Experiments with Bose-Einstein condensates correspond to situations where the total number of atoms in the trap is fixed, and so should be described with the canonical formalism. However, the grand canonical distribution, which leads to simpler calculations, can be used when T is well above T_c since the fluctuations of the occupation numbers remain small in relative value.[5] In the presence of a condensate, the fluctuations of the condensate population is overestimated in the grand canonical description [Politzer (1996); Naraschewski and Glauber (1999)] and the higher-order correlation functions, which are sensitive to these fluctuations, cannot be calculated with the grand canonical distribution. A new method for calculating these higher-order correlation functions in the presence of a condensate is thus needed, and will be presented in the next subsection.

Here, we will suppose that T is sufficiently large compared to T_c so that one can still use a grand canonical distribution, and the Wick's theorem, which applies in this case and states that the average value of a product of creation and annihilation operators in equal numbers can be expressed as sums of products of average values involving only two operators (one creation and one annihilation). Any high-order correlation function can thus be written as a product of a certain number of first-order correlation functions. For example:

$$\begin{aligned} G^{(2)}(\vec{r},\vec{r}\,') &= \langle \hat{\Psi}^\dagger(\vec{r})\hat{\Psi}^\dagger(\vec{r}\,')\hat{\Psi}(\vec{r}\,')\hat{\Psi}(\vec{r})\rangle \\ &= \langle \hat{\Psi}^\dagger(\vec{r})\hat{\Psi}(\vec{r})\rangle\langle \hat{\Psi}^\dagger(\vec{r}\,')\hat{\Psi}(\vec{r}\,')\rangle + \langle \hat{\Psi}^\dagger(\vec{r})\hat{\Psi}(\vec{r}\,')\rangle\langle \hat{\Psi}^\dagger(\vec{r}\,')\hat{\Psi}(\vec{r})\rangle \end{aligned} \tag{23.42}$$

which gives for the normalized correlation function

$$g^{(2)}(\vec{r},\vec{r}\,') = 1 + \left|g^{(1)}(\vec{r},\vec{r}\,')\right|^2 \quad \text{so that} \quad g^{(2)}(\vec{r},\vec{r}) = 1 + 1 = 2. \tag{23.43}$$

Similar calculations can be made for the third-order correlation function, leading to:

$$\begin{aligned} g^{(3)}(\vec{r},\vec{r}\,',\vec{r}\,'') = 1 &+ \left|g^{(1)}(\vec{r},\vec{r}\,')\right|^2 + \left|g^{(1)}(\vec{r}\,',\vec{r}\,'')\right|^2 + \left|g^{(1)}(\vec{r}\,'',\vec{r})\right|^2 \\ &+ 2\mathrm{Re}\left[g^{(1)}(\vec{r},\vec{r}\,')g^{(1)}(\vec{r}\,',\vec{r}\,'')g^{(1)}(\vec{r}\,'',\vec{r})\right] \\ &\text{so that} \quad g^{(3)}(\vec{r},\vec{r},\vec{r}) = 3! = 6. \end{aligned} \tag{23.44}$$

[4] This is valid for a quasi-one dimensional Bose gas that is not in the Tonks regime (see Sec. 26.6 of Chap. 26).

[5] It can be also used for the calculation of the first-order spatial correlation function even if $T < T_c$ because $G^{(1)}$ is not very sensitive to these fluctuations.

We deduce from Eqs. (23.42) or (23.43) that $G^{(2)}(\vec{r},\vec{r}) = 2\left[G^{(1)}(\vec{r},\vec{r})\right]^2$: the probability of finding two bosonic atoms at the same place is twice as large as the product of the two probabilities of finding each of these atoms at this place. There is therefore a bunching effect for bosonic atoms analogous to the bunching effect for photons, also known as the Hanbury Brown and Twiss effect [Hanbury Brown and Twiss (1956)]. This effect is even larger for the higher correlation functions. For example, one deduces from (23.44) that the probability of finding three bosonic atoms at the same place is six times larger than the product of the three probabilities of finding each of these atoms at this place. In Sec. 23.3.4 we will describe a few experiments that demonstrate the existence of these bunching effects.

23.3.3 *Correlation functions for a Bose-Einstein condensate*

When a condensate is present, it is appropriate to write the field operator as:

$$\hat{\Psi}(\vec{r}) = \hat{a}_0\,\varphi_0(\vec{r}) + \sum_{\alpha\neq 0}\hat{a}_\alpha\,\varphi_\alpha(\vec{r}) \tag{23.45}$$

where φ_0 is the wave function of the condensate, solution of the time-independent Gross-Pitaevski equation, forming with the φ_α's an orthonormal basis. Instead of using a grand canonical distribution, we will suppose that all atoms are in the state φ_0, so that the state of the system is $|\psi\rangle = |n_0 = N\,,\;\; n_{\alpha\neq 0} = 0\rangle$. This description of the state of the condensate is more appropriate for a system with a fixed number of atoms well below T_c.

Straightforward calculations then give

$$G^{(1)}(\vec{r}) = \langle n_0 = N|\hat{\Psi}^\dagger(\vec{r})\Psi(\vec{r})|n_0 = N\rangle = N|\varphi_0(\vec{r})|^2. \tag{23.46}$$

$$\begin{aligned} G^{(2)}(\vec{r},\vec{r}\,') &= \langle n_0 = N|\,\hat{\Psi}^\dagger(\vec{r})\hat{\Psi}^\dagger(\vec{r}\,')\hat{\Psi}(\vec{r}\,')\hat{\Psi}(\vec{r})\,|n_0 = N\rangle \\ &= N(N-1)\,|\varphi_0(\vec{r})|^2\,|\varphi_0(\vec{r}\,')|^2 \simeq N^2\,|\varphi_0(\vec{r})|^2\,|\varphi_0(\vec{r}\,')|^2\,. \end{aligned} \tag{23.47}$$

$$\begin{aligned} G^{(3)}(\vec{r},\vec{r}\,',\vec{r}\,'') &= \langle n_0 = N|\,\hat{\Psi}^\dagger(\vec{r})\hat{\Psi}^\dagger(\vec{r}\,')\hat{\Psi}^\dagger(\vec{r}\,'')\hat{\Psi}(\vec{r}\,'')\hat{\Psi}(\vec{r}\,')\hat{\Psi}(\vec{r})\,|n_0 = N\rangle \\ &= N(N-1)(N-2)\,|\varphi_0(\vec{r})|^2\,|\varphi_0(\vec{r}\,')|^2\,|\varphi_0(\vec{r}\,'')|^2 \\ &\simeq N^3\,|\varphi_0(\vec{r})|^2\,|\varphi_0(\vec{r}\,')|^2\,|\varphi_0(\vec{r}\,'')|^2\,. \end{aligned} \tag{23.48}$$

From Eqs. (23.43) and (23.44), one easily deduces that:

$$g^{(2)}(\vec{r},\vec{r}) = 1 \qquad \text{and} \qquad g^{(3)}(\vec{r},\vec{r},\vec{r}) = 1. \tag{23.49}$$

In other words, the bunching effect found in the previous subsection for a thermal Bose gas does not exist for a condensate.

To summarize the results of this section, one can say that there is a strong analogy between

- a thermal Bose gas above T_c and a thermal radiation field. In both cases, all higher-order correlation functions can be expressed in terms of the first-order one and there is a bunching effect.

- a quasi-classical radiation field and a Bose-Einstein condensate. Glauber has shown that for a field in a quasi-classical coherent state, all correlation functions factorize: they appear as products of the classical fields associated with the quantum coherent state. Consequently, there is no bunching effect. Similarly, for a condensate, all correlation functions factorize. They appear as products of classical fields $\sqrt{N}\varphi_0(\vec{r})$ where $\varphi_0(\vec{r})$ is the wave function of the condensate. Here again, there is no bunching effect.

23.3.4 *A few experimental results*

(i) *Atom lasers*

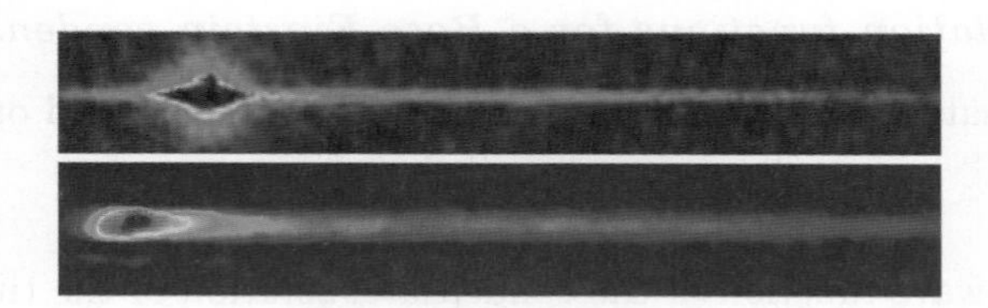

Fig. 23.3 Examples of guided atom lasers in an optical guide realized from trapped Bose-Einstein condensates. Upper part: the condensate is confined in a mixed magnetic and optical trap. Atoms are outcoupled by radio-frequency spin flips. Figure taken from [Guerin *et al.* (2006)]. Copyright: American Physical Society. Lower part: the condensate is trapped in a crossed dipole trap. Atoms are outcoupled by applying an increasing magnetic field gradient. Figure taken from [Couvert *et al.* (2008a)]. Copyright: EDP Sciences.

The term "atom laser" is employed to denote a coherent matter wave beam. So far, atom lasers have been produced by the coherent outcoupling of atoms from a trapped Bose-Einstein condensate. The first demonstrations of BEC output couplers were based on a change of internal state from a magnetically trapped to an untrapped state. In 1997, the MIT group used radiofrequency pulses to induce such transitions and could outcouple packets of atoms, which fell out of the trap under gravity [Mewes *et al.* (1997)]. Later, the Munich group used this radiofrequency method to produce long pulses of matter waves, often referred to as quasicontinuous atom laser beams [Bloch *et al.* (1999)]. Alternatively, the group of William Phillips at NIST produced a well-collimated quasicontinuous beam of atoms using stimulated Raman transitions between magnetic sublevels [Hagley *et al.* (1999b)]. In 2003, a well-controlled spilling of atoms from a dipole trap by reducing the trap depth was demonstrated [Cennini *et al.* (2003)]. In such a scheme the internal state is unchanged.

The realization of horizontally guided atom lasers have been recently reported. Compared to free falling atom lasers, the acceleration experienced by the atoms can be more than 100 times less than that in the gravity field. This allows one to keep the de Broglie wavelength relatively large (few hundreds of nm) over large propagation distances (few mm). In such experiments, the guided atom laser is produced by outcoupling the atoms from a trapped Bose-Einstein condensate (BEC) into a far-off resonance optical guide. In the experiment carried out by the group of Philippe Bouyer and Alain Aspect, the trap is magnetic and atoms are decoupled by radio-frequency spin flips [Guerin *et al.* (2006)]. In the experiment led by David Guéry-Odelin, a BEC in a well-defined Zeeman sublevel m_F is held in a far-off resonance crossed dipole trap. In this scheme, one of the two beams of the crossed dipole trap is nearly horizontal and is used as a guide. Two outcoupling schemes have been demonstrated: (i) magnetic outcoupling by progressively increasing a magnetic field gradient applied along the direction of the guide [Couvert *et al.* (2008a)] and (ii) optical outcoupling by lowering the intensity of the one of the beams of the crossed dipole trap [Gattobigio *et al.* (2009)]. In the latter experiments, the beam could be produced with more than 85% of the flux in the transverse ground state. A large population of the transverse ground state has also been observed in a vertical guiding beam configuration [Dall *et al.* (2010)].

In the previous schemes, the atom laser production relied on the progressive depletion of a reservoir of condensed atoms. A truly cw atom laser producing a continuous, coherent matter wave output has not been achieved yet. However, noticeable advances in this direction have been made:

- Bose-condensed atoms have been periodically transported over large distances using a moving optical dipole trap to feed a reservoir of Bose condensed atoms made by another dipole trap [Chikkatur *et al.* (2002)]. However, there was no outcoupling to remove atoms continuously from the reservoir.
- Another strategy consists in applying the cooling techniques that are successful with a cloud of atoms to an atomic beam. The production of an intense guided beam of cold atoms in the collisional regime was reported in [Lahaye *et al.* (2004)], and a gain of one order of magnitude on the on-axis phase space density of the beam was demonstrated one year later by implementing evaporative cooling [Lahaye *et al.* (2005)].
- An atom laser made from a Bose-Einstein condensate that is simultaneously pumped and output-coupled was recently reported [Robins *et al.* (2008)].

(ii) *Interference between two matter waves extracted from two different points of a condensate*

A very elegant method for determining the coherence length is to measure the visibility of the interference fringes between two matter waves extracted from different points located on two different surfaces of an atomic cloud [Bloch *et al.* (2000)]. The extraction of a matter wave is achieved by a position-selective output-coupler that uses a continuous RF wave to transfer atoms from the magnetic sublevel where they are trapped to another one where they are untrapped. By using two RF waves with different frequencies, one couples out two matter waves from two different surfaces of the condensate. The visibility of the interference fringes between these matter

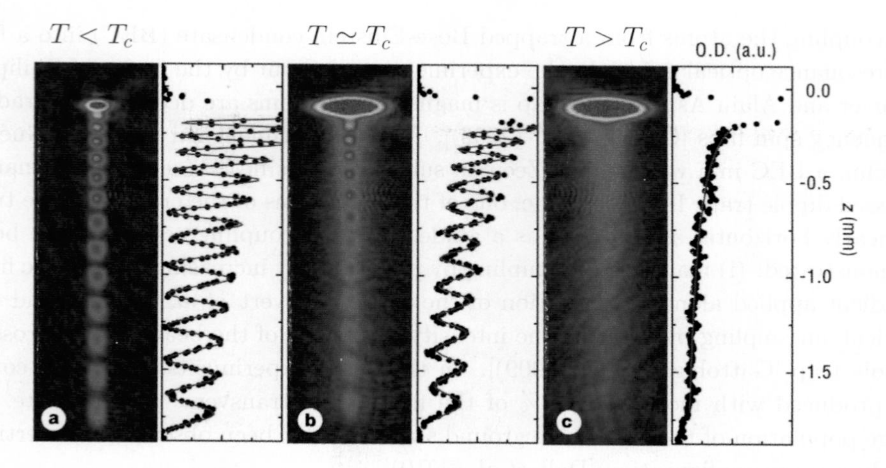

Fig. 23.4 Interference fringes between two matter waves extracted from two different points of a condensate and falling in the gravity field. No fringes are visible above T_c while they clearly appear below T_c. Reprinted by permission from Macmillan Publishers Ltd: Nature [Bloch *et al.* (2000)].

waves is directly related to the first order correlation function $G^{(1)}(\vec{r}, \vec{r}\,')$. The evolution of the visibility of the fringes can be studied across the phase transition when T is decreased from a value above T_c to a value below T_c (see Fig. 23.4), giving clear evidence for the change of the coherence length from the thermal de Broglie wavelength λ_T when $T > T_c$, to a value on the order of the size of the condensate when $T < T_c$.

(iii) *Other methods for measuring the spatial coherence length of a condensate*

For an inhomogeneous system, such as a trapped condensate, $G^{(1)}(\vec{r}, \vec{r}\,')$ does not depend only on $\vec{r} - \vec{r}\,'$ because there is no translation invariance. To characterize the spatial coherence length ξ, it is appropriate to introduce the function $G(\vec{s}) = \int d^3r\, G^{(1)}(\vec{r}, \vec{r}+\vec{s})$ and to define ξ as the width of $G(\vec{s})$. In fact, one can show[6] that $G(\vec{s})$ is nothing but the Fourier transform of the momentum distribution $P(\vec{p})$, so that ξ is also on the order of $\hbar/\Delta p$, where Δp is the width of $P(\vec{p})$. Two experiments have used this connection between ξ and $P(\vec{p})$ or $G(\vec{s})$ to measure ξ.

The first experiment, performed at MIT, consists of measuring $P(\vec{p})$ by Bragg spectroscopy [Stenger *et al.* (1999)]. A velocity-selective two-photon Raman transition couples two states with the same internal quantum numbers but with different momenta. From the spectrum obtained by sweeping the frequency difference between the two counter-propagating beams, one can derive the momentum distribution $P(\vec{p})$ and in particular its width Δp. The measured value of Δp is consistent with the Heisenberg limit $\Delta p \simeq \hbar/\sigma_0$, where σ_0 is the spatial width of the solution φ_0 of the Gross-Pitaevskii equation. This shows that the coherence length of the

[6]See Sec. 17.2.2 of Chap. 17.

condensate is equal to its spatial extent and that there is a single phase throughout the condensate.

In the second experiment [Hagley *et al.* (1999b)], a pair of short pulses of stationary waves creates two copies of the condensate separated by $\vec{s}$. Absorption imaging techniques allow the direct measurement of the overlap integral of these two copies, which is precisely $G(\vec{s})$.

(iv) *Investigation of higher-order correlation functions*

We will mention here three different types of experiments:

(1) In the first one, $g^{(2)}(\vec{r},\vec{r})$ can be inferred from the energy released during the ballistic expansion of a Bose-Einstein condensate [Ketterle and Miesner (1997)]. This can be understood by noting that the mean interaction energy of a condensate gives information on $G^{(2)}(\vec{r},\vec{r})$ since

$$\left\langle \hat{H}_{\text{int}} \right\rangle = \frac{1}{2}\iint \mathrm{d}^3r\, \mathrm{d}^3r'\, V(\vec{r}-\vec{r}')\, G^{(2)}(\vec{r},\vec{r}') \tag{23.50}$$

$$= \frac{g}{2}\int \mathrm{d}^3r\, G^{(2)}(\vec{r},\vec{r}) = \frac{g}{2}\int \mathrm{d}^3r\, \rho^2(\vec{r})\, g^{(2)}(\vec{r},\vec{r})\,. \tag{23.51}$$

We have used $V(\vec{r}-\vec{r}') = g\delta(\vec{r}-\vec{r}')$. If the trap is switched off, the mean interaction energy is converted into kinetic energy which can be measured. The results of the measurements are consistent with $g^{(2)}(\vec{r},\vec{r}) = 1$ for a condensate and exclude $g^{(2)}(\vec{r},\vec{r}) = 2$.

(2) In the second experiment, $g^{(2)}(\vec{r},\vec{r}')$ is directly measured by detecting atoms one by one and correlating their relative distances. This has been successfully realized using multichannel plate detectors for metastable helium atoms [Jeltes *et al.* (2007)]. Figure 23.5 shows the results obtained by releasing a cloud of metastable helium-4 atoms at a temperature above T_c. The bunching effect predicted for bosons clearly appears and differs radically with the antibunching effect observed if one replaces the cloud of helium-4 atoms by a cloud of helium-3 atoms, which are fermions. The same experiment was also performed with a condensate of helium-4 atoms, and the bunching effect disappeared as expected.[7]

(3) Third-order correlations can be investigated through the study of three-body losses [Burt *et al.* (1997)]. The loss rate obeys the equation:

$$\frac{\mathrm{d}N}{\mathrm{d}t} = -\kappa \int n^3(\vec{r},t)\,\mathrm{d}^3r, \tag{23.52}$$

where κ is proportional to $g^{(3)}(\vec{r},\vec{r},\vec{r})$. One expects $g^{(3)}(\vec{r},\vec{r},\vec{r}) = 3! = 6$ for a thermal gas, and 1 for a condensate as a signature of its coherence. The ratio of the loss rates in both situations has been measured to be 7.4 ± 2.6 [Burt *et al.*

[7] Other experiments testing the bunching effect for a thermal cloud of bosonic alkali atoms and its suppression for a condensate have been also performed, using high-finesse optical cavities [Ritter *et al.* (2007)] or single-atom sensitive fluorescence imaging [Manz *et al.* (2010)].

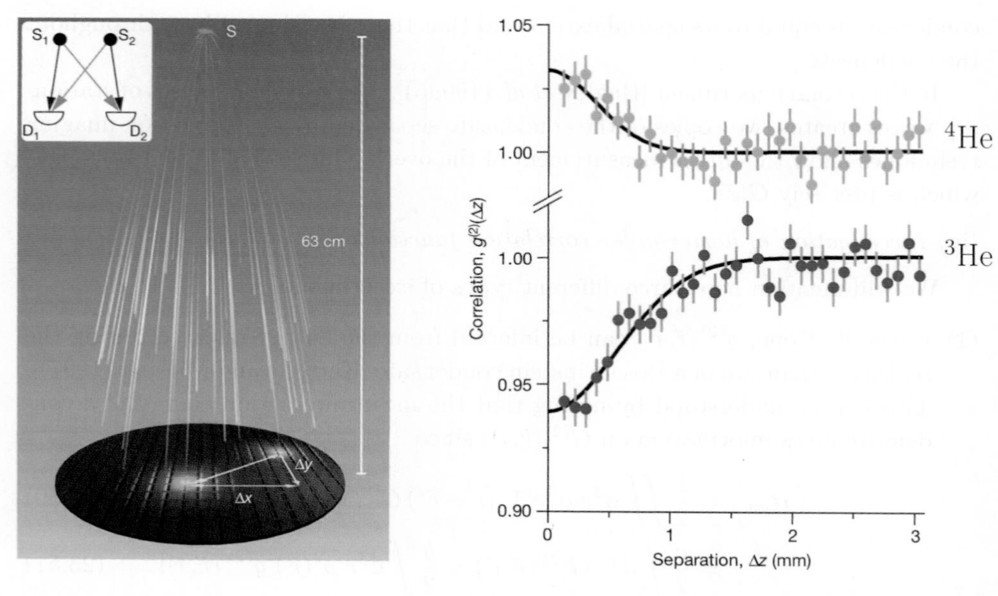

Fig. 23.5 Second-order spatial correlation function observed for a cloud of metastable helium atoms released above a multichannel plate. The bunching effect clearly appears for helium-4 atoms at a temperature above T_c. It transforms into an antibunching effect if helium-4 atoms are replaced by fermionic helium-3 atoms. Reprinted by permission from Macmillan Publishers Ltd: Nature [Jeltes *et al.* (2007)].

(1997)]. This experiment therefore clearly demonstrates the high coherence of a 3D Bose-Einstein condensate (in particular, the reduction of density fluctuations as compared to the thermal gas). Alternatively, one can infer the properties of the third-order spatial correlation functions using the correlation of arrival times of metastable helium atoms [Hodgman *et al.* (2011)].

23.4 Relative phase of two independent condensates

In this section, we investigate the conditions under which a well-defined relative phase can exist between two separate condensates. We identify the quantum variable conjugate to the relative phase and show how it is possible to observe interference fringes when two independent condensates without an initial relative phase are released and overlap.

The two condensates are formed in two different traps centered at $\vec{r}_1$ and $\vec{r}_2$ respectively and do not overlap.[8] The solutions of the Gross-Pitaevskii equation for each trap are denoted ψ_1 and ψ_2 respectively. Our assumptions for the two condensates imply that $\langle\psi_1|\,\psi_2\rangle = 0$ as well as $\psi_1(\vec{r}_2) = \psi_2(\vec{r}_1) = 0$. We take

[8]In Chap. 26, we will consider a condensate in a double well potential with the possibility of tunnelling of atoms from one well to the other.

a basis $\{|\psi_i\rangle\}$ of individual states including $|\psi_1\rangle$ and $|\psi_2\rangle$. In our case, the only relevant basis states in the Fock space are $|n_1, n_2, n_i = 0 \quad \text{for} \quad i \neq 1,2\rangle$. In the following, we simplify the notation by omitting n_i for $i \neq 1,2$. We impose that the total number of bosons $N = n_1 + n_2$ is fixed. Can one find a quantum state describing the two condensates with a well-defined relative phase, i.e. such that the first-order correlation function connecting $\vec{r}_1$ and $\vec{r}_2$ is different from zero:

$$G^{(1)}(\vec{r}_1, \vec{r}_2) = \left\langle \hat{\psi}^\dagger(\vec{r}_1)\hat{\psi}(\vec{r}_2) \right\rangle \neq 0 \ ? \tag{23.53}$$

23.4.1 *Two condensates in Fock states*

Suppose first that each condensate is in a Fock state, so that the state of the whole system is $|n_1, n_2\rangle$. Because all modes $i \neq 1,2$ are empty, and because $\psi_1(\vec{r}_2) = \psi_2(\vec{r}_1) = 0$, we can, in a normally ordered product of field operators, replace $\hat{\psi}(\vec{r})$ and $\hat{\psi}^\dagger(\vec{r})$ by:

$$\hat{\psi}(\vec{r}_1) = \psi_1(\vec{r}_1)\hat{a}_1 \quad \text{and} \quad \hat{\psi}(\vec{r}_2) = \psi_2(\vec{r}_2)\hat{a}_2. \tag{23.54}$$

The first-order correlation function (23.53) is easily found to be equal to:

$$\begin{aligned} G^{(1)}(\vec{r}_1, \vec{r}_2) &= \langle n_1, n_2 | \hat{\psi}^\dagger(\vec{r}_1)\hat{\psi}(\vec{r}_2) | n_1, n_2 \rangle \\ &= \psi_1^*(\vec{r}_1)\psi_2(\vec{r}_2) \langle n_1, n_2 | \hat{a}_1^\dagger \hat{a}_2 | n_1, n_2 \rangle = 0 \end{aligned}$$

This result shows that two condensates in Fock states do not possess a well defined relative phase.

23.4.2 *Phase states*

Another idea consists of taking all N bosons in the same linear combination of ψ_1 and ψ_2:

$$|\eta_1, \eta_2, \theta\rangle = \eta_1 |\psi_1\rangle + \eta_2 e^{i\theta} |\psi_2\rangle \tag{23.55}$$

where η_1, η_2 are real and verify $\eta_1^2 + \eta_2^2 = 1$, and θ is the relative phase between the two condensates.

Let $|N, \theta\rangle$ be the state $|N, \eta_1, \eta_2, \theta\rangle$ corresponding to N bosons in the state $|\eta_1, \eta_2, \theta\rangle$. We call this a *phase state*. To find its expression, we introduce the creation operator $\hat{a}^\dagger_{\eta_1\,\eta_2\,\theta}$ of a boson in the state $|\eta_1, \eta_2, \theta\rangle$:

$$\hat{a}^\dagger_{\eta_1\,\eta_2\,\theta} = \eta_1 \hat{a}_1^\dagger + \eta_2 e^{i\theta} \hat{a}_2^\dagger. \tag{23.56}$$

We have:

$$\begin{aligned} |N, \theta\rangle &= \frac{1}{\sqrt{N!}} \left[\eta_1 \hat{a}_1^\dagger + \eta_2 e^{i\theta} \hat{a}_2^\dagger \right]^N |0\rangle \\ &= \sum_{\substack{n_1=0 \\ n_2=N-n_1}}^{N} \sqrt{\frac{N!}{n_1! n_2!}} \eta_1^{n_1} \eta_2^{n_2} e^{in_2\theta} |n_1, n_2\rangle . \end{aligned} \tag{23.57}$$

Such a phase state appears as a linear superposition of Fock states $|n_1, n_2\rangle$ with $n_1 + n_2 = N$ fixed. Equation (23.57) shows that a phase state is not a product of a state of "mode" 1 by a state of "mode" 2, but exhibits quantum correlations between the two condensates.

The distribution of n_1 is given by:

$$\mathcal{P}(n_1) = \frac{N!}{n_1!n_2!}\eta_1^{2n_1}\eta_2^{2n_2} \quad \text{with} \quad \begin{cases} \eta_2^2 = 1 - \eta_1^2, \\ n_2 = N - n_1. \end{cases} \tag{23.58}$$

The distribution of n_1 is thus a binomial distribution, with the standard results

$$\begin{cases} \overline{n}_1 = N\eta_1^2 \quad \text{and} \quad \overline{n}_2 = N\eta_2^2 = N - \overline{n}_1 \\ (\Delta n_1)^2 = N\eta_1^2\left(1 - \eta_1^2\right) = N\eta_1^2\eta_2^2 = (\Delta n_2)^2 \end{cases} \tag{23.59}$$

To characterize the phase state $|N, \theta\rangle$, it is also interesting to study the distribution of

$$n = n_1 - n_2 \tag{23.60}$$

which can take values from $-N$ to N by steps of 2. Its mean value is $\overline{n} = \overline{n}_1 - \overline{n}_2$, while its dispersion is $\Delta n = 2\Delta n_1 = 2\sqrt{N}\eta_1\eta_2$. The dispersion of n in a phase state is thus large in absolute value, but remains small compared to N.

The entanglement of the two condensates in a phase state appears in the density matrix $\hat{\rho} = |N, \theta\rangle\langle N, \theta|$ as off-diagonal elements that couple states with different values of $n = n_1 - n_2$. The extent of this off-diagonality in n is on the order of $\sqrt{N}$. By contrast, the reduced density matrix $\hat{\rho}^{(1)}$ of condensate 1 contains only diagonal elements. Taking the trace over n_2 of $|N, \theta\rangle\langle N, \theta|$ gives:

$$\hat{\rho}^{(1)} = \sum_{n_1} \mathcal{P}(n_1)|n_1\rangle\langle n_1| \quad \text{with} \quad \mathcal{P}(n_1) = \frac{N!}{n_1!(N - n_1)!}\eta_1^{2n_1}\eta_2^{2(N-n_1)}. \tag{23.61}$$

Correlation functions can be easily calculated in a phase state [Cohen-Tannoudji and Robillard (2001)]. One finds:

$$G^{(1)}(\vec{r}, \vec{r}\,') = \psi^*(\vec{r})\psi(\vec{r}\,'), \tag{23.62}$$

$$G^{(2)}(\vec{r}, \vec{r}\,') = \psi^*(\vec{r})\psi^*(\vec{r}\,')\psi(\vec{r}\,')\psi(\vec{r}), \tag{23.63}$$

where

$$\psi(\vec{r}) = \underbrace{\eta_1\sqrt{N}}_{\sqrt{\overline{n}_1}}\psi_1(\vec{r}) + \underbrace{\eta_2\sqrt{N}}_{\sqrt{\overline{n}_2}}e^{i\theta}\psi_2(\vec{r}). \tag{23.64}$$

More generally, in the presence of interactions, it can be shown that all correlation functions $G^{(k)}$ with $k \ll N$ factorize and can be written as products of $2k$ functions ψ and ψ^*. We can thus conclude that two interfering macroscopic matter waves $\sqrt{\overline{n}_1}\psi_1(\vec{r})$ and $\sqrt{\overline{n}_2}\psi_2(\vec{r})e^{i\theta}$ can be associated with two condensates in a phase state.

Coherent states

Consider a condensate 1 in a coherent state:

$$|\alpha_1\rangle = e^{-|\alpha_1|^2/2} \sum_{n_1=0}^{\infty} \frac{\alpha_1^{n_1}}{\sqrt{n_1!}} |n_1\rangle \quad \text{where} \quad \alpha_1 = \sqrt{\bar{n}_1} e^{i\theta_1}. \tag{23.65}$$

The probability of having n_1 photons in this state obeys a Poisson law with a mean value $\bar{n}_1 = |\alpha_1|^2$ and a variance $\Delta n_1^2 = \bar{n}_1$. Since (23.65) is an eigenstate of $\hat{a}_1$ with eigenvalue α_1, the average value of all normal products of creation and annihilation operators can be easily calculated by just replacing $\hat{a}_1$ and $\hat{a}_1^\dagger$ by α_1 and α_1^* in the expression of the operators. In particular, $\hat{\psi}_1(\vec{r})$ can be replaced by $\sqrt{n_1} e^{i\theta_1} \psi_1(\vec{r})$. This property makes the coherent states very convenient for calculating correlation functions of field operators, though their physical meaning is not clear because they involve linear superpositions of states with different values of the number n_1 of atoms. One can however introduce statistical mixtures of products of two coherent states describing two independent condensates such that the resulting density operator $\hat{\rho}$ of the whole system has no off-diagonal elements between two states with different values of the total number $N = n_1 + n_2$ of atoms. Consider for example the products of coherent states $\left|\alpha_1 = \sqrt{\overline{n}_1} e^{i\theta_1}\right\rangle \otimes \left|\alpha_2 = \sqrt{\overline{n}_2} e^{i\theta_2}\right\rangle$ with $\theta_2 = \theta_1 + \theta$ and θ_1 uniformly distributed in $[0, 2\pi]$. The corresponding density operator is written [Cirac *et al.* (1996)]

$$\hat{\rho} = \frac{1}{2\pi} \int_0^{2\pi} \mathrm{d}\theta_1 \left|\sqrt{\overline{n}_1} e^{i\theta_1}\right\rangle \otimes \left|\sqrt{\overline{n}_2} e^{i(\theta_1+\theta)}\right\rangle \left\langle\sqrt{\overline{n}_1} e^{i\theta_1}\right| \otimes \left\langle\sqrt{\overline{n}_2} e^{i(\theta_1+\theta)}\right| \tag{23.66}$$

with $\overline{n}_1$, $\overline{n}_2$ and θ fixed. One can show that $\hat{\rho}$ is diagonal in N and can be rewritten as a statistical mixture of phase states $|N, \theta\rangle$ with θ fixed and N distributed in an interval of size $\sqrt{\overline{N}}$ around $\overline{N}$.

23.4.3 *Conjugate variable of the relative phase*

We now address the question of the conjugate variable of the relative phase θ. We consider a statistical mixture of phase states $|N, \theta\rangle$ with N fixed and θ distributed according to a distribution $W(\theta)$:

$$\hat{\rho} = \int_0^{2\pi} \mathrm{d}\theta \, W(\theta) \, |N, \theta\rangle \langle N, \theta| \,. \tag{23.67}$$

Using (23.57), one then gets:

$$\hat{\rho} = \sum_{n_1=0}^{N} \sum_{n_1'=0}^{N} \sqrt{\frac{N!}{n_1! n_2!}} \sqrt{\frac{N!}{n_1'! n_2'!}} \eta_1^{n_1+n_1'} \eta_2^{n_2+n_2'} \times \left[\int_0^{2\pi} \mathrm{d}\theta \, W(\theta) e^{-i(n_2'-n_2)\theta}\right] |n_1, n_2\rangle \langle n_1', n_2'| \,. \tag{23.68}$$

If one notices that $N = n_1 + n_2 = n_1' + n_2'$ implies $n_2' - n_2 = n_1 - n_1'$ and thus $n - n' = n_1 - n_2 - (n_1' - n_2') = 2(n_2' - n_2)$, the integral in (23.68) can be rewritten:

$$\mathcal{I} = \int_0^{2\pi} \mathrm{d}\theta \, W(\theta) e^{-i(n_2'-n_2)\theta} = \int_0^{2\pi} \mathrm{d}\theta \, W(\theta) e^{-i(n-n')\theta/2}. \tag{23.69}$$

The broader $W(\theta)$, the smaller the factor multiplying the off-diagonal elements of $\hat{\rho}$ between two states with different values of $n = n_1 - n_2$. If $W(\theta)$ is flat ($W(\theta) = 1/2\pi$), the integral over θ gives $\delta_{n,n'}$, which means that $\hat{\rho}$ is diagonal, not only in N, but also in $n = n_1 - n_2$. The difference $n = n_1 - n_2$ between the number of atoms in the two condensates thus appears as the conjugate variable of the relative phase θ. This relative phase is well defined only if there is a large uncertainty in n.

23.4.4 *Emergence of a relative phase in an interference experiment*

This problem has been addressed by several authors [Javanainen and Yoo (1996); Cirac *et al.* (1996); Wong *et al.* (1996); Castin and Dalibard (1997)]. Here we will give a qualitative discussion similar to that given in Chap. 19 for interpreting the spatial modulations that can be observed in optics when one detects double counting rates in a two-mode field.

Generally, two independent condensates do not find themselves in a state with a well-defined relative phase, but rather in a product of Fock states $|N_1, N_2\rangle = |N_1\rangle \otimes |N_2\rangle$. Such a state corresponds to a zero dispersion on $n = n_1 - n_2$, and thus to a totally undefined relative phase.

However, the detection processes induce a dispersion on n, so that the relative phase of the two condensates becomes increasingly well known. Indeed, the first detected boson can come either from mode 1 or mode 2. After this detection, the state vector becomes

$$|\psi\rangle = \alpha\,|N_1 - 1, N_2\rangle + \beta\,|N_1, N_2 - 1\rangle \tag{23.70}$$

where α and β are coefficients that depend on the position of the first detected boson. This first detection process entangles the two condensates and given rise to a certain relative phase. Similarly, the second detection process changes the state vector into:

$$|\psi\rangle = \lambda\,|N_1 - 2, N_2\rangle + \mu\,|N_1 - 1, N_2 - 1\rangle + \nu\,|N_1, N_2 - 2\rangle\,. \tag{23.71}$$

The off-diagonality of $\hat{\rho} = |\psi\rangle\langle\psi|$ in $n = n_1 - n_2$ thus increases with the number of detected bosons and a more and more well-defined relative phase θ builds up. In this way, one understands how interference fringes can be observed when two independent condensates are released from their traps and made to overlap, as in the experiment described in the introduction of this chapter.

It is important to understand what is the physical quantity measured in this experiment. We do not measure the probability $w_1(\vec{r})$ to detect a single atom at $\vec{r}$ and look for a possible spatial modulation of $w_1(\vec{r})$. Such a measurement would imply making an ensemble average of experiments in which one would detect a single atom in the overlap zone of the two condensates, repeat the same experiment several times with two new condensates prepared each time in the same conditions, and finally make an ensemble average of the results obtained for the position of the single detected atom in each experiment. It is clear that the absence of a relative phase between the two condensates would then imply that the probability of detecting a single atom would not be spatially modulated: $w_1(\vec{r})$ does not depend on $\vec{r}$, and no interference fringes are predicted in the ensemble average. The difference is that in the experiment described in Sec. 23.1, we detect several atoms, and not just one, in a single experimental realization. What we measure is the probability $w_N(\vec{r}_1, \vec{r}_2, \ldots \vec{r}_N)$ to detect one atom at $\vec{r}_1$, one atom at $\vec{r}_2$, ...one atom at $\vec{r}_N$. These detections are correlated and thus give rise to a spatially modulated repartition of the detections. If the same experiment is repeated, it would give a spatial modulation of the detections with the same spatial period, but the center of the system of fringes would not be the same. In other words, an experiment involving several detections in a single run makes a relative phase appear between the two condensates, even if such a phase does not exist in the initial state. But this relative phase varies in a random way from one run to another.

23.5 Long range order and order parameter

23.5.1 *Long range order*

The theoretical tools that we have introduced so far, such as field operators, correlation functions and one-body density operators have been applied in the preceding sections either to an ideal or a weakly interacting condensate within the mean field approximation. In the presence of a condensate, the first-order correlation function reveals the existence of a long range order.[9] For more complex situations, such as for liquid helium-4, for which the interactions cannot be simply described by a mean field approach, one can still introduce the one-body density operator $\hat{\sigma}_1$ and define the condensate fraction through the properties of the eigenvalues of this operator, as originally done by Lars Onsager and Oliver Penrose [Penrose and Onsager (1956)].

Let us recall the properties of the one-atom density operator, $\hat{\sigma}_1$. It is a positive definite operator which has positive eigenvalues π_α whose sum is equal to unity since $\text{Tr}(\hat{\sigma}_1) = 1$. The orthonormal basis formed by the eigenvectors $\{|\varphi_\alpha\rangle\}$ yields the expansion

$$\hat{\sigma}_1 = \sum_\alpha \pi_\alpha |\varphi_\alpha\rangle \langle\varphi_\alpha| \tag{23.72}$$

where the coefficients π_α obey the relations $0 < \pi_\alpha \leq 1$ and $\sum_\alpha \pi_\alpha = 1$. The mean number of atoms in the state $|\varphi_\alpha\rangle$ is therefore equal to $N\pi_\alpha$. The spatial coherence between two points is given by the off-diagonal elements of $\hat{\sigma}_1$:

$$\langle\vec{r}'| \hat{\sigma}_1 |\vec{r}\rangle = \sum_\alpha \pi_\alpha \varphi_\alpha^* (\vec{r}') \varphi_\alpha (\vec{r}) . \tag{23.73}$$

[9] See Sec. 23.3.

In the presence of a Bose-Einstein condensate, one of the states, $|\varphi_0\rangle$, has a macroscopic population, much larger than the population of all other states

$$\pi_0 \gg \sum_{\alpha \neq 0} \pi_\alpha. \tag{23.74}$$

Let us separate in Eq. (23.73) the contribution of $|\varphi_0\rangle$

$$\langle \vec{r}\,' |\, \hat{\sigma}_1 \,| \vec{r}\,\rangle = \pi_0 \, \varphi_0^* \,(\vec{r}\,')\, \varphi_0 \,(\vec{r}\,) + \sum_{\alpha \neq 0} \pi_\alpha \, \varphi_\alpha^* \,(\vec{r}\,')\, \varphi_\alpha \,(\vec{r}\,). \tag{23.75}$$

The inequality (23.74) implies that the contribution of the first term is predominant. For instance, the coherence length associated with this term is on the order of the spatial extent of the wave function $|\varphi_0\rangle$ for an inhomogeneous condensate. The contribution of the last term is much smaller and does not give rise to a large coherence length because of the destructive interference between the large number of terms with $\alpha \neq 0$.

The diagonalization of the one-atom density operator $\hat{\sigma}_1$ of a degenerate Bose gas thus allows one to single out one state $|\varphi_0\rangle$ which corresponds to the largest eigenvalue of $\hat{\sigma}_1$. One can define the state $|\varphi_0\rangle$ as the wave function of the condensate. For a weakly interacting Bose gas, the corresponding wave function coincides with the solution of the Gross-Pitaevskii equation [Castin (2001)]. This approach is also used to extend the Bogolubov approach to inhomogeneous condensates by defining an appropriate basis of one-atom states including the state $|\varphi_0\rangle$ [Pitaevskii and Stringari (2003)].[10]

23.5.2 *Order parameter*

In the presence of a condensate, even at $T = 0$ K, one cannot in principle identify the mean value of the field operator $\langle \hat{\Psi}(\vec{r}\,)\rangle$ to $\sqrt{N}\varphi_0(\vec{r}\,)$. Indeed, the number of atoms associated with the state $|\varphi_0\rangle$ is fixed and the operator $\hat{\Psi}(\vec{r}\,)$, which changes the number of atoms, cannot have a non-zero average value in $|\varphi_0\rangle$. This average value is non-zero only for a condensate described as a linear superposition of states with different atom numbers, in particular with coherent states. As already stressed, a statistical mixture of Fock states, for which the number of atoms has a peak distribution around N, can also be described as a statistical mixture of coherent states $|\alpha\rangle = |\sqrt{N}\exp(i\theta)\rangle$ with random phases θ randomly distributed between 0 and 2π. The description of the condensate by a given coherent state amounts to operating the so-called symmetry breaking prescription which fixes a certain value of θ. Within this formalism, the mean value of the field operator reads $\langle \hat{\Psi}(\vec{r}\,)\rangle = \sqrt{N}\varphi_0(\vec{r}\,)e^{i\theta}$, and is referred to as the *order parameter*.

[10]Similarly, the condensation of Cooper pairs in the Bardeen-Cooper-Schriffer superconductivity theory is analyzed through the properties of the pair correlation function (see Sec. 26.5 of Chap. 26).

23.6 New effects in atom optics due to atom-atom interactions

The results derived in the previous sections clearly show that bosonic quantum gases exhibit many analogies with quantum optical fields. However, there are also some major differences, in particular those related to the interactions which exist between atoms but not between photons. In this section, we first show how atom-atom interactions, which have been neglected in the previous calculations, can give rise to collapses and revivals of the first order correlation function of two condensates in different traps. We then describe a four-wave mixing experiment with matter waves, which is a nonlinear effect facilitated by interactions.

23.6.1 *Collapse and revival of first-order coherence due to interactions*

As in the previous section, we consider two condensates situated at $\vec{r}_1$ and $\vec{r}_2$ and study the time evolution of the spatial coherence between the two condensates under the effect of interactions within each condensate (we neglect the interactions between the two condensates). Using the Heisenberg point of view, one gets:

$$\left\langle \hat{\psi}^\dagger(\vec{r}_1,t)\hat{\psi}(\vec{r}_2,t)\right\rangle = \psi_1^*(\vec{r}_1)\psi_2(\vec{r}_2)\left\langle \hat{a}_1^\dagger(t)\hat{a}_2(t)\right\rangle \tag{23.76}$$

with

$$\hat{a}_1^\dagger(t) = e^{i\hat{H}_1 t/\hbar}\hat{a}_1^\dagger e^{-i\hat{H}_1 t/\hbar} \qquad \text{and} \quad \hat{a}_2(t) = e^{i\hat{H}_2 t/\hbar}\hat{a}_2 e^{-i\hat{H}_2 t/\hbar}. \tag{23.77}$$

In this equation, $\hat{H}_i$ is the Hamiltonian of condensate i, including the interaction potential. Inserting a closure relation in (23.76) yields:

$$\begin{aligned}\hat{a}_1^\dagger(t) &= \sum_{n_1}\sqrt{n_1+1}\,|n_1+1\rangle\langle n_1|\,e^{i[E(n_1+1)-E(n_1)]t/\hbar},\\ \hat{a}_2(t) &= \sum_{n_2}\sqrt{n_2}\,|n_2-1\rangle\langle n_2|\,e^{i[E(n_2-1)-E(n_2)]t/\hbar},\end{aligned} \tag{23.78}$$

which finally gives

$$\begin{aligned}\left\langle \hat{\psi}^\dagger(\vec{r}_1,t)\hat{\psi}(\vec{r}_2,t)\right\rangle = &\sum_{n_1,n_2}\psi_1^*(\vec{r}_1)\psi_2(\vec{r}_2)\sqrt{n_2}\sqrt{n_1+1}\,\langle n_1,n_2|\,\hat{\rho}\,|n_1+1,n_2-1\rangle\\ &\times e^{i[E(n_1+1,n_2-1)-E(n_1,n_2)]t/\hbar}.\end{aligned} \tag{23.79}$$

The spatial coherence between the two condensates thus appears as a sum of terms proportional to the coherence between $|n_1,n_2\rangle$ and $|n_1+1,n_2-1\rangle$ oscillating at the corresponding Bohr frequencies.

Let us consider an initial state of the system with a well-defined spatial coherence, for instance a phase state $|N,\theta\rangle$. In the following, we label the states $|n_1,n_2\rangle$ by $n = n_1 - n_2$, $n_1 + n_2 = N$, which is fixed. One can write $n_{1,2} = \frac{1}{2}(N \pm n)$ and n can take all values from $-N$ to N varying by steps of 2. The spatial coherence $\left\langle \hat{\psi}^\dagger(\vec{r}_1,t)\hat{\psi}(\vec{r}_2,t)\right\rangle$ between the condensates is then proportional to $\sum_n \sqrt{(N-n)(N+n+1)}\,\langle n|\,\hat{\rho}\,|n+2\rangle \exp(i\,[E(n+2)-E(n)]\,t/\hbar)$.

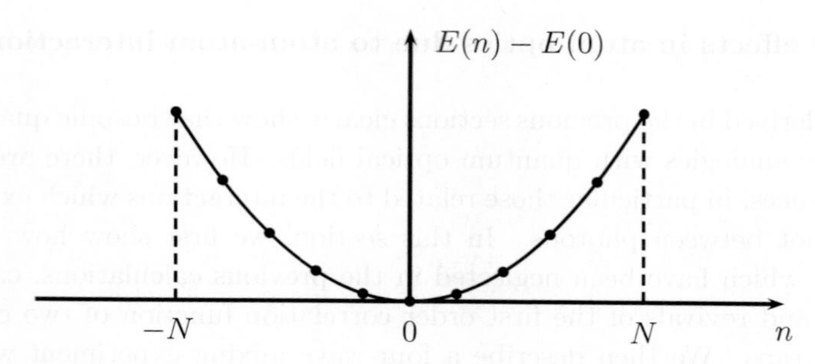

Fig. 23.6 Variation with $n = n_1 - n_2$ of the total energy of two homogeneous condensates including the effect of interactions within each condensate.

To go further in the discussion while keeping the calculations as simple as possible, we consider the case of two homogeneous condensates in boxes of size L, with the same scattering length a; in this case the Hamiltonian is reduced to the interaction Hamiltonian, namely $\hat{H}_1 + \hat{H}_2 = \frac{g}{2L^3}(\hat{N}_1^2 + \hat{N}_2^2)$, with $g = 4\pi\hbar^2 a/m$. This yields

$$E(n) = \frac{g}{8L^3}\left[(N+n)^2 + (N-n)^2\right] = \frac{g}{4L^3}(N^2 + n^2). \tag{23.80}$$

As shown in Fig. 23.6, $E(n)$ varies quadratically with n, so that the relevant quantity

$$E(n+2) - E(n) = \frac{g}{L^3}(n+1) \tag{23.81}$$

is linear in n. As a consequence of the atom-atom interactions within each condensate, the evolution frequencies of the coherences $\langle n|\,\hat{\rho}\,|n+2\rangle$ form a comb of equally spaced frequencies $\frac{g}{\hbar L^3}(n+1)$, where n varies from $-N$ to N. We saw in Sec. 23.4.2 that the dispersion of n in a phase state is on the order of $\sqrt{N}$. Then the spreading $\Delta\omega$ of the frequencies that appears in the evolution of the spatial coherence $\left\langle \hat{\psi}^\dagger(\mathbf{r}_1, t)\hat{\psi}(\mathbf{r}_2, t)\right\rangle$ is on the order of $\Delta\omega \simeq \sqrt{N}\frac{g}{\hbar L^3}$, leading to a damping time

$$T_{\text{damp}} \simeq \frac{\hbar L^3}{g\sqrt{N}} = \frac{\hbar\sqrt{N}}{\mu} \tag{23.82}$$

where $\mu = gN/L^3$ is the chemical potential.

Moreover, since the frequencies $g(n+1)/\hbar L^3$ are equally spaced by $g/\hbar L^3$, one expects a revival of the spatial coherence after a time

$$T_{\text{rev}} \simeq \frac{\hbar L^3}{g} = \frac{\hbar N}{\mu} \gg T_{\text{damp}}. \tag{23.83}$$

Remark

Consider the interference experiment described in Sec. 23.1 in which an elongated condensate is cut into two parts with a laser beam. Just after the cutting, the two condensates have a well-defined relative phase since they originate from the same condensate. After a time long compared to T_{damp} but short compared to T_{rev}, the two condensates can be considered to be independent, without a well-defined relative phase.

23.6.2 *An example of nonlinear effects in atom optics: Four-wave mixing with matter waves*

In Sec. 9.4.4 of Chap. 9, we have described a process in nonlinear optics called degenerate four-wave mixing. Consider three waves 1, 2, 3 with the same frequency ω, propagating in an atomic medium with waves 1 and 2 going in opposite directions (see Fig. 9.20(a) of Chap. 9). The nonlinear susceptibility of the medium in which the waves propagate gives rise to a nonlinear process consisting of the absorption of a photon from wave 1, the stimulated emission of a photon in wave 3, the absorption of a photon from wave 2 and, finally, the emission of a photon in a new wave 4 (see Fig. 9.20(b) of Chap. 9). Conservation of energy implies that wave 4 must have the same frequency as the three other waves. The phase matching condition $\vec{k}_1 + \vec{k}_2 = \vec{k}_3 + \vec{k}_4$, which expresses that the atom emitting wave 4 must not recoil if one wants the waves emitted by the various atoms to interfere (no which-path information), implies that $\vec{k}_4 = -\vec{k}_3$ since $\vec{k}_1 = -\vec{k}_2$. The new wave 4 must therefore propagate in a direction opposite to wave 3. Another interpretation of the condition $\vec{k}_1 + \vec{k}_2 = \vec{k}_3 + \vec{k}_4$ can be given in terms of the Bragg scattering of wave 2 by the intensity grating resulting from the interference of the two waves 1 and 3.

The Gross-Pitaevskii equation describing the evolution of the matter waves contains a nonlinear term proportional to $g|\psi|^2\psi$ that describes the mean field interactions between atoms. This term is analogous to the term $\chi^{(3)}|E|^2E$ in the wave equation of the electric field propagating in a medium whose nonlinearity is described by the nonlinear susceptibility $\chi^{(3)}$. It is this term that gives rise to the four-wave mixing in nonlinear optics. The similarity of the two equations for ψ and E leads us to conclude that atom-atom interactions can give rise to a nonlinear mixing of matter waves. If we start with three matter waves with wave vectors $\vec{k}_1$, $\vec{k}_2 = -\vec{k}_1$, $\vec{k}_3$ and the same modulus, a new wave $\vec{k}_4$ must appear with $\vec{k}_4 = -\vec{k}_3$. Two atoms are transferred from waves $\vec{k}_1$ and $\vec{k}_2$ into waves $\vec{k}_3$ and $\vec{k}_4$ with conservation of the total energy and the total momentum.

This nonlinear mixing of matter waves has been observed by the group of William Phillips at NIST, Gaithersburg [Deng *et al.* (1999)]. Starting from a condensate at rest ($\vec{k}_1 = \vec{0}$), two Bragg pulses are used to produce two matter waves from this condensate with momenta $\vec{k}_3 = \sqrt{2}k\hat{e}_{\pi/4}$, where $\hat{e}_{\pi/4} = (\hat{e}_x + \hat{e}_y)\sqrt{2}$, and $\vec{k}_2 = 2k\hat{e}_x$, respectively. In a frame moving with velocity $(\hbar k/m)\hat{e}_x$, these three momenta become $\vec{k}'_1 = -k\hat{e}_x$, $\vec{k}'_2 = +k\hat{e}_x$, $\vec{k}'_3 = k\hat{e}_y$, respectively. This configuration coincides with that considered above, and one expects the generated wave to appear with a wave vector $\vec{k}'_4 = -k\hat{e}_y$. This is what is observed experimentally (see Fig. 23.7). The three upper wave packets correspond to the matter waves $\vec{k}'_1$, $\vec{k}'_2$, $\vec{k}'_3$. The lower, less intense, wave packet corresponds to the generated wave $\vec{k}'_4$.

23.7 Conclusion

In this chapter, we have shown that Bose-Einstein condensates have interesting coherence properties that make them similar to laser light. A close analogy also exists between thermal light and thermal bosonic gases. There are, however, important differences due to the existence of atom-atom interactions. Nonlinear effects directly appear in atom optics, without requiring the existence of an underlying nonlinear medium like in nonlinear optics.

Atomic field operators and field correlation functions are very important tools for analyzing the coherence properties of Bose-Einstein condensates, for introducing important physical properties like the coherence length and for interpreting the bunching effect in thermal bosonic clouds. They also introduce quantum features in the description of the system that do not exist in the mean field approach given in the previous chapter and using a c-number wave function. The Heisenberg equation for the field operator (23.29) looks like a time-dependent Gross-Pitaevskii equation, but it is an equation between field operators and not between wave functions. Higher order field correlation functions also incorporate the discrete nature of atoms and can describe processes in which several atoms are detected. They can explain how successive detections of atoms can produce an entanglement between two independent condensates and give rise to a relative phase between these condensates even if this phase did not initially exist.

The second quantized description of the bosonic gas introduced in this chapter will also be very useful in the next chapter for introducing quantum corrections, like the quantum depletion, in the description of the ground state of the condensate and for characterizing its first elementary excitations in terms of *quasi-particles*. In certain cases, there is a threshold below which an external perturbation cannot excite such quasi-particles and we will see that this can be used to explain important physical effects such as superfluidity.

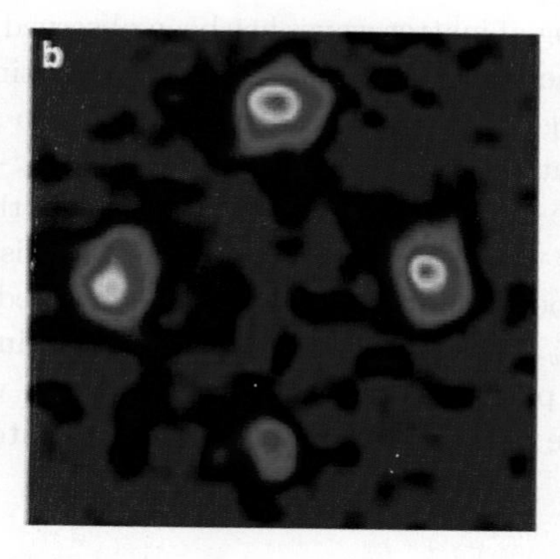

Fig. 23.7 Four-wave mixing with matter waves. Reprinted by permission from Macmillan Publishers Ltd: Nature [Deng *et al.* (1999)].

Chapter 24

Elementary excitations and superfluidity in Bose-Einstein condensates

24.1 Introduction

In the previous chapters, all atoms of the condensate were taken to be in the same one-particle state. This was the starting point of the variational approach followed in Chap. 22 that consisted of looking for the best approximation of the ground state of the system in the family of all possible tensor products of N identical one-particle states. The elementary excitations of the system were described as normal modes of vibration of the macroscopic matter wave occupied by the N bosons. The frequencies ω of these modes were calculated from a linearized time-dependent Gross-Pitaevskii equation, but the modes were considered to be classical waves with an energy that varied continuously, without any quantized structure. In the same way as the energy of a classical Maxwell wave becomes quantized in a quantum description of the electromagnetic field, one expects the energy of the elementary excitations of the condensate to become quantized in a full quantum description of the system of N bosons. A first step towards such a description was made in Chap. 23 with the introduction of atomic field operators and correlation functions, but the energies and the wave functions of the ground state of the system and of its first excited states were not calculated.

In this chapter, we present such a calculation, valid in the limit of dilute systems. It provides new physical insights into the structure of the ground state that cannot be obtained with the treatment in Chap. 22. It also gives the quantized energies of the elementary excitations of the condensate, which can be considered as the first excited states of the system. The knowledge of these energies is important for the understanding of whether or not an external excitation can provide energy to the condensate in its ground state and create elementary excitations. We will show that, in certain conditions, such a transfer of energy from the external excitation to the ground state is impossible. For example, if the external excitation is a microscopic probe particle moving through the condensate with a sufficiently low velocity, the quenching of energy transfer from the particle to the condensate prevents any slowing down of the particle, which thus moves through the fluid without any friction.

Such a situation exhibits superfluidity, as shown for the first time by Lev Landau in the context of superfluid helium [Landau (1941b)].

In Sec. 24.2, we present the Bogolubov approach for calculating the wave function and energy of the condensate in its ground state as well as the first quantized elementary excitations. We restrict ourselves to an homogeneous system leading to simpler calculations. This approach differs from the usual perturbative approach that consists of treating the effect of atom-atom interactions described by the Hamiltonian $\hat{H}_{\text{int}}$ as a perturbation to the ideal gas Hamiltonian containing only the kinetic energy term $\hat{H}_{\text{kin}}$(for an homogeneous system). In a condensate, $\langle \hat{H}_{\text{kin}} \rangle$ is very small, and treating $\langle \hat{H}_{\text{int}} \rangle$ as a perturbation is clearly questionable. We will see in Sec. 24.2 that the Bogolubov approach gives corrections to the mean field result that scale as $(na^3)^{1/2}$, where a is the scattering length and n the atomic density. This shows that the Bogolubov approach is valid only for dilute systems for which $na^3 \ll 1$ (small number of atoms in a volume a^3). In this theory, the condensate wave function Ψ describes a situation where the number N_0 of atoms in the ground state of the box is not equal to the total number N of atoms: Ψ contains admixtures of the excited states of the box and $(N - N_0)/N$ is called the *quantum depletion*, to be distinguished from the *thermal depletion* describing thermal excitations of the system. The quantum depletion is proportional to $(na^3)^{1/2}$, which shows that the small parameter of the Bogolubov approach is $(N - N_0)/N$, and not $\langle \hat{H}_{\text{int}} \rangle / \langle \hat{H}_{\text{kin}} \rangle$ as this would be the case in a usual perturbative approach.

Section 24.3 is devoted to the discussion of the Landau criterion for an homogeneous system. Because of translational invariance, the momenta of the particles are good quantum numbers. We first consider a situation where a probe atom, distinguishable from the atoms of the condensate, is sent with a velocity v through the condensate. A simple analysis, based on a graphic construction, shows that exchanges of energy and momentum between the probe atom and the condensate are impossible if the velocity v of the probe atom is smaller than the sound velocity c in the condensate, calculated in Sec. 24.2. An experimental investigation of this effect in gaseous condensates is briefly described. We also present a more macroscopic analysis of superfluidity, closer to Landau's analysis, by considering a situation where a condensate moves with a velocity v through a capillary.

Superfluidity also occurs in more complex situations. In Sec. 24.4, we consider a condensate contained in a bucket rotating at the angular frequency Ω around the z-axis. We show that there exists a critical value Ω_c of Ω below which the creation of an elementary excitation with a well defined angular momentum $L_z = \ell\hbar$ is not energetically favorable. In other words, if $\Omega < \Omega_c$, the condensate remains at rest and does not rotate with the bucket, though other possible states of the condensate play an important role and must be considered to determine the thermodynamically stable state of the condensate in the rotating bucket. These states are the quantized vortices briefly introduced in Chap. 22, where *all* atoms of the condensate have a well defined angular momentum $\ell\hbar$ along the z-axis.[1] Because of the term $-\Omega L_z$ appearing in the Hamiltonian of the system in the rotating frame (where the bucket is at rest), the energy of a vortex state, for which $\ell \neq 0$, can be smaller than the energy of the condensate without vortex, for which $\ell = 0$. This occurs when Ω is larger than another critical velocity $\Omega_{\rm v}$. In Sec. 24.4, we discuss the various regimes which can be observed depending on the value of Ω, compared to the two critical values $\Omega_{\rm c}$ and $\Omega_{\rm v}$.

Quantum vortices are an important feature of quantum fluids, and gaseous condensates are particularly interesting for the study of vortices because they allow one to study the kinetics of their formation. In Sec. 24.5, we present an example of such an experimental investigation. We first briefly review the various methods that have been used to create, detect and measure the angular momentum of vortices in a condensate. We then describe an experiment in which one starts with a pure condensate (without vortex) in a cylindrical trap that has a small elliptical anisotropy in the xy-plane perpendicular to the symmetry axis. The trap anisotropy is set in rotation with an angular frequency Ω, which is slowly increased from a zero value. One observes vortices appearing when Ω reaches a certain value, which turns out to be the rotation frequency of the surface mode $\ell = 2$ studied in Chap. 22. These experimental results are compared to analytical predictions derived from the solution of the time-dependent Gross-Pitaevskii equation in the rotating frame. They show that, in this experiment, vortices are nucleated through a dynamical instability associated with the excitation of a surface mode of the condensate by the rotating elliptical trap.

24.2 Bogolubov approach for an homogeneous system

In Chap. 22, we investigated the ground state of a weakly interacting N-boson system trapped at zero temperature. The state of the system was determined using the variational approach which describes the motion of each particle in the mean-field created by the $N-1$ other particles. The N-body wavefunction corresponds to a classical field that obeys the Gross-Pitaevskii equation. In such an approximate treatment, the correlations between atoms are neglected.

In this section, we present a more elaborate theoretical framework for the weakly interacting Bose gas first introduced by Nikolay Bogolubov in 1947 [Bogolubov (1947)]. In this approach, one takes into account the quantum fluctuations about the state in which all atoms share the same single quantum state, which is the ground state of the box containing the condensate. As a consequence of quantum fluctuations, the atoms are not all in the same single particle state even at

[1] A clear distinction must be made between this vortex and an elementary excitation of the condensate with angular momentum $\ell\hbar$. In a vortex, all atoms have an angular momentum $\ell\hbar$, so that the total angular momentum is $N\ell\hbar$. When an elementary excitation $\ell\hbar$ is created, the angular momentum of the whole system is just equal to $\ell\hbar$.

$T = 0$ (quantum depletion), and correlations appear between them. This theory also provides more insight into the first excited states of the Bose gas, called *elementary excitations*, with which quasi-particles can be associated.

The Bogolubov approach relies on the description of the N-particle state in Fock space, and is therefore performed within the second quantization formalism. It is valid for a sufficiently dilute Bose gas for which only two-body collisions are taken into account. It is developed here for an homogeneous Bose gas in which particles are assumed to be in a box.[2] The single particle state that has a macroscopic population is always the ground state of the box.[3] In Sec. 24.2.5, we briefly mention the extension of the Bogolubov approach to an inhomogeneous system.

24.2.1 *Second quantized Hamiltonian*

For a gas confined in a box of volume L^3 with periodic boundary conditions, the single particle states are plane waves

$$\langle \vec{r}\,|\varphi_{\vec{k}}\rangle = \varphi_{\vec{k}}(\vec{r}) = \frac{1}{L^{3/2}} \exp(i\,\vec{k}\cdot\vec{r}) \quad \text{with} \quad k_{x,y,z} = \frac{2\pi}{L}\, n_{x,y,z}, \tag{24.1}$$

where $n_{x,y,z}$ is any integer. The operator $\hat{a}_{\vec{k}}$ ($\hat{a}^\dagger_{\vec{k}}$) destroys (creates) an atom in state $|\varphi_{\vec{k}}\rangle$. The second quantized form of the many-body Hamiltonian $\hat{H} = \hat{H}_{\rm kin} + \hat{H}_{\rm int}$ contains (i) a kinetic term $\hat{H}_{\rm kin}$ which is the sum of products of two creation and annihilation operators ($\hat{a}^\dagger_{\vec{k}}$ and $\hat{a}_{\vec{k}}$ respectively) of an atom in state $\vec{k}$ of the box, and (ii) the two-body interaction term $\hat{H}_{\rm int}$ that involves the sum of products of four creation and annihilation operators:

$$\hat{H}_{\rm kin} = \sum_{\vec{k}} \frac{\hbar^2 k^2}{2m}\, \hat{a}^\dagger_{\vec{k}} \hat{a}_{\vec{k}}, \tag{24.2}$$

$$\hat{H}_{\rm int} = \frac{1}{2L^3} \sum_{\vec{k}_1} \sum_{\vec{k}_2} \sum_{\vec{k}} \tilde{V}(\vec{k}\,) \hat{a}^\dagger_{\vec{k}_1} \hat{a}^\dagger_{\vec{k}_2} \hat{a}_{\vec{k}_2+\vec{k}} \hat{a}_{\vec{k}_1-\vec{k}},$$

where we have explicitly used the conservation of the total linear momentum, and introduced the Fourier transform of the two-body interaction potential

$$\tilde{V}(\vec{k}\,) = \int {\rm d}^3 r\, V(\vec{r}\,) e^{-i\vec{k}\cdot\vec{r}}. \tag{24.3}$$

In the following, we assume that $\tilde{V}(\vec{k}\,)$ depends only on the modulus of k because of the spherical symmetry of the interaction potential.

[2] We thus assume a positive scattering length to ensure that the system is stable against collapse.

[3] This is to be contrasted with the case of a trapped weakly interacting condensate, where the ground state is modified as the number of atoms increases.

24.2.2 *Bogolubov quadratic Hamiltonian*

The first step of the Bogolubov approach relies on the fact that the Bose gas is sufficiently dilute, and differs only slightly from an ideal gas, so that the ground state φ_0 of the box with zero momentum remains macroscopically occupied. Mathematically, this assumption is exploited by replacing the operators $\hat{a}_0$ and $\hat{a}_0^\dagger$ by c-numbers equal to $N^{1/2}$ where N is the total number of particles. In a second step, one neglects the products containing less than two $\hat{a}_0$ or $\hat{a}_0^\dagger$ in the expression of $\hat{H}_{\text{int}}$. One is then left with the total Hamiltonian $\hat{H}$ with a quadratic expression in $\hat{a}_k$ and $\hat{a}_k^\dagger$. The Bogolubov theory consists of diagonalizing this quadratic Hamiltonian.

The ground state of an ideal Bose gas in a box is characterized by N particles in the state φ_0:

$$|\psi_0\rangle = |N_0 = N, 0, \ldots, 0, \ldots\rangle = \frac{(a_0^\dagger)^N}{(N!)^{1/2}}|0, 0, \ldots, 0, \ldots\rangle. \tag{24.4}$$

In a dilute Bose gas near $T = 0$, the population N_0 is macroscopic and very large compared to the total population in all other states. It is then justified to make a theory limited to the first-order in

$$\varepsilon = \frac{N - N_0}{N} = \frac{1}{N}\sum_{k\neq 0} N_k \ll 1. \tag{24.5}$$

The action of $\hat{a}_0$ on a Fock state can therefore be approximated by

$$\hat{a}_0 \left|\{N_k\}\right\rangle = N_0^{1/2}|\{N_k - \delta_{k0}\}\rangle \simeq N^{1/2} \left|\{N_k - \delta_{k0}\}\right\rangle. \tag{24.6}$$

The matrix elements of $\hat{a}_0$ and $\hat{a}_0^\dagger$ are on the order of $N^{1/2} \gg 1$. We will neglect the non-commutation of $\hat{a}_0$ and $\hat{a}_0^\dagger$ and use the so-called Bogolubov prescription[4] by replacing them by a c-number $N^{1/2}$: $\hat{a}_0^\dagger \simeq \hat{a}_0 \simeq N^{1/2} \simeq N_0^{1/2}$.

By substituting the Bogolubov prescription into the other terms of $\hat{H}_{\text{int}}$ in Eq. (24.2) and keeping only the leading terms of order N^2 and N, one can finally rewrite the Hamiltonian $\hat{H} = \hat{H}_{\text{kin}} + \hat{H}_{\text{int}}$ in a quadratic form in the operators $\hat{a}_{\vec{k}}$ and $\hat{a}_{\vec{k}}^\dagger$:[5]

$$\begin{aligned}\hat{H} \simeq \hat{H}_Q = {} & \frac{n\tilde{V}(0)}{2}N + \sum_{\vec{k}} \frac{\hbar^2 k^2}{2m}\hat{a}_{\vec{k}}^\dagger \hat{a}_{\vec{k}} \\ & + \frac{n}{2}\sum_{\vec{k}\neq\vec{0}} \tilde{V}(k)\left[\hat{a}_{\vec{k}}^\dagger\hat{a}_{\vec{k}} + \hat{a}_{-\vec{k}}^\dagger\hat{a}_{-\vec{k}} + \hat{a}_{\vec{k}}^\dagger\hat{a}_{-\vec{k}}^\dagger + \hat{a}_{\vec{k}}\hat{a}_{-\vec{k}}\right]\end{aligned} \tag{24.7}$$

[4]This rather crude, but frequently used, treatment of the operators $\hat{a}_0$ and $\hat{a}_0^\dagger$ can be improved in a more refined presentation described for example in [Castin and Dum (1998)].

[5]In the term $\hat{a}_0^\dagger\hat{a}_0^\dagger\hat{a}_0\hat{a}_0$ containing four operators with $i = 0$, we cannot just replace $\hat{a}_0$ and $\hat{a}_0^\dagger$ by $\sqrt{N}$. Using $\hat{a}_0^\dagger\hat{a}_0 = \hat{N} - \sum_{k\neq 0} \hat{a}_{\vec{k}}^\dagger\hat{a}_{\vec{k}}$ and keeping only the linear terms in $\sum_{k\neq 0} \hat{a}_{\vec{k}}^\dagger\hat{a}_{\vec{k}}$, we get

$$\hat{a}_0^\dagger\hat{a}_0^\dagger\hat{a}_0\hat{a}_0 = \hat{a}_0^\dagger\left(\hat{a}_0\hat{a}_0^\dagger - 1\right)\hat{a}_0 \simeq \hat{N}(\hat{N}-1) - 2\hat{N}\sum_{k\neq 0}\hat{a}_{\vec{k}}^\dagger\hat{a}_{\vec{k}}.$$

where $n = N/L^3$ is the spatial density. This truncated form of $\hat{H}$ neglects interactions of particles out of the condensate. Its important feature is that it can be diagonalized exactly since it is a quadratic form in the operators $\hat{a}_{\vec{k}}$ and $\hat{a}^\dagger_{\vec{k}}$.

If we write the Heisenberg equation for the time evolution of the operator $\hat{a}_{\vec{k}}$, we find that $\hat{a}_{\vec{k}}$ is coupled to $\hat{a}_{\vec{k}}$ and $\hat{a}^\dagger_{-\vec{k}}$, and similarly that the time evolution of $\hat{a}^\dagger_{-\vec{k}}$ is coupled to $\hat{a}^\dagger_{-\vec{k}}$ and $\hat{a}_{\vec{k}}$:

$$
\begin{aligned}
i\hbar\frac{\mathrm{d}\hat{a}_{\vec{k}}}{\mathrm{d}t} &= [\hat{a}_{\vec{k}}, \hat{H}_Q] = \left(\frac{\hbar^2k^2}{2m} + n\tilde{V}(k)\right)\hat{a}_{\vec{k}} + n\tilde{V}(k)\hat{a}^\dagger_{-\vec{k}}, \\
i\hbar\frac{\mathrm{d}\hat{a}^\dagger_{-\vec{k}}}{\mathrm{d}t} &= \left[\hat{a}^\dagger_{-\vec{k}}, \hat{H}_Q\right] = -\left(\frac{\hbar^2k^2}{2m} + n\tilde{V}(k)\right)\hat{a}^\dagger_{-\vec{k}} - n\tilde{V}(k)\hat{a}_{\vec{k}}.
\end{aligned}
\tag{24.8}
$$

This suggests to search for the diagonalization of the Hamiltonian (24.7) two linear combinations of these two operators:

$$
\hat{b}_{\vec{k}} = u_k\hat{a}_{\vec{k}} + v_k\hat{a}^\dagger_{-\vec{k}} \quad \text{and} \quad \hat{b}^\dagger_{-\vec{k}} = u_k\hat{a}^\dagger_{-\vec{k}} + v_k\hat{a}_{\vec{k}},
\tag{24.9}
$$

where u_k and v_k are real coefficients depending only on $k = |\vec{k}|$. The corresponding transformation is unitary if the new set of operators $\hat{b}_{\vec{k}}$ and $\hat{b}^\dagger_{-\vec{k}}$ obeys the bosonic commutation relations

$$
\left[\hat{b}_{\vec{k}}, \hat{b}^\dagger_{\vec{k}}\right] = u_k^2\left[\hat{a}_{\vec{k}}, \hat{a}^\dagger_{\vec{k}}\right] + v_k^2\left[\hat{a}^\dagger_{-\vec{k}}, \hat{a}_{-\vec{k}}\right] = u_k^2 - v_k^2 = 1,
\tag{24.10}
$$

for $k \neq 0$. This relation is automatically fulfilled by searching the coefficients u_k and v_k in the form $u_k = \cosh\theta_k$ and $v_k = \sinh\theta_k$. The parameter θ_k is determined by imposing that the expression of the Hamiltonian $\hat{H}$ in terms of the creation and annihilation operators $\hat{b}_{\vec{k}}$ and $\hat{b}^\dagger_{\vec{k}}$ is a sum of independent harmonic oscillator Hamiltonians within a constant. This condition is fulfilled by imposing

$$
\tanh 2\theta_k = \frac{n\tilde{V}(k)}{n\tilde{V}(k) + \hbar^2k^2/2m}.
\tag{24.11}
$$

The final expression for the diagonalized Hamiltonian is found to be

$$
\hat{H}_Q = E_0' + \sum_{\vec{k}\neq\vec{0}} \hbar\omega_k^B\, \hat{b}^\dagger_{\vec{k}}\hat{b}_{\vec{k}} \quad \text{with} \quad \hbar\omega_k^B = \sqrt{\frac{\hbar^2k^2}{2m}\left(\frac{\hbar^2k^2}{2m} + 2n\tilde{V}(k)\right)},
\tag{24.12}
$$

and

$$
E_0' = \frac{1}{2}Nn\tilde{V}(0) + \sum_{\vec{k}\neq\vec{0}}\left\{\frac{\hbar^2k^2}{2m}v_k^2 + n\tilde{V}(k)\left[v_k^2 - u_kv_k\right]\right\}.
\tag{24.13}
$$

24.2.3 *Physical discussion*

For the two problems discussed in this subsection, we can approximate the interaction potential by a delta potential:

$$
V(\vec{r}) = g\delta(\vec{r}) \quad \text{with} \quad g = 4\pi\hbar^2a/m
\tag{24.14}
$$

where a is the scattering length. In this case, we have $\tilde{V}(0) = \tilde{V}(k) = g$. We will discuss the validity of this approximation in the next subsection.

24.2.3.1 *Elementary excitations. Quasi-particles*

The first excited states of the system with momentum $\vec{k}$ and energy $\hbar\omega_k^B$ are given by $\hat{b}^\dagger_{\vec{k}}|\Psi_0\rangle = u_k\,\hat{a}^\dagger_{\vec{k}}|\Psi_0\rangle + v_k\,\hat{a}_{-\vec{k}}|\Psi_0\rangle$. They describe elementary excitations, or quasi-particles, whose dispersion relation is given by $\hbar\omega_k^B(k)$ (see Eq. 24.12). A quasi-particle $\{\vec{k}, \omega_k^B\}$ is therefore obtained by creating a particle of wave vector $\vec{k}$ and by annihilating a particle with an opposite wave vector $-\vec{k}$ from the ground state $|\Psi_0\rangle$, i.e. creating a hole. A quasi-particle thus appears as a linear superposition of a particle and a hole, with amplitudes u_k and v_k.

To discuss the dispersion relation of the elementary excitations, we replace in (24.12) $2n\tilde{V}(k)$ by:

$$2n\tilde{V}(k) = 2gn = 2\mu, \tag{24.15}$$

where μ is the chemical potential, or equivalently by:

$$2n\tilde{V}(k) = 8\pi\hbar^2 na/m = \hbar^2 k_0^2/m, \tag{24.16}$$

where k_0 is the inverse of the healing length ξ

$$k_0 = 1/\xi = \sqrt{8\pi na}. \tag{24.17}$$

This leads to:

$$\hbar\omega_k^B = \sqrt{\frac{\hbar^2k^2}{2m}\left(\frac{\hbar^2k^2}{2m} + 2\mu\right)} = \sqrt{\frac{\hbar^2k^2}{2m}\left(\frac{\hbar^2k^2}{2m} + \frac{\hbar^2k_0^2}{m}\right)} \tag{24.18}$$

In the long-wavelength limit for which $\lambda = 2\pi/k$ is large compared to the healing length $\xi = 1/k_0$, the chemical potential $\mu = gn$ is large compared to the kinetic energy ($\mu \gg \hbar^2k^2/2m$) and the dispersion relation becomes linear with k: $\omega_k^B \simeq ck$, where

$$c = \left(\frac{\mu}{m}\right)^{1/2}. \tag{24.19}$$

This linear dispersion relation describes phonons that propagate in the condensate with the sound velocity c. Furthermore, we have in this limit, according to Eq. (24.11), $\tanh 2\theta_k \simeq 1$, so that $u_k \simeq v_k$. The elementary excitations are linear superpositions with equal amplitudes of a particle and a hole.

In the opposite limit $\hbar^2k^2/2m \gg \mu$, the dispersion relation is given by

$$\hbar\omega_k^B \simeq \frac{\hbar^2k^2}{2m} + \mu. \tag{24.20}$$

One simply finds that the kinetic energy of a free particle is shifted by the chemical potential μ, i.e. the mean-field energy. In this regime, we have, according to Eq. (24.11), $\tanh 2\theta_k \simeq 0$, so that u_k is on the order of 1 and v_k becomes negligible. The elementary excitation is thus very close to an ordinary particle propagating in the condensate with its energy slightly modified by an amount μ due to its interaction with the condensed particles. These results concerning the dispersion law are summarized in Fig. 24.1.

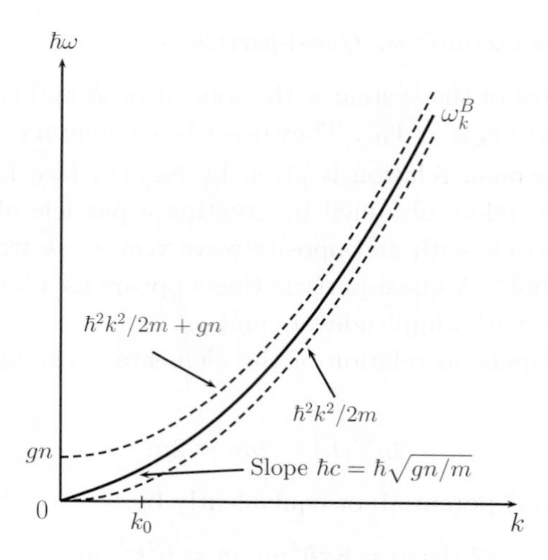

Fig. 24.1 Bogolubov dispersion law for the elementary excitations. The transition between a phonon-like dispersion, $\omega_k^B \simeq ck$, and a free particle one, $\omega_k^B \simeq \hbar k^2/2m + \mu/\hbar$, occurs around $k = k_0 = \xi^{-1}$, where ξ is the healing length.

24.2.3.2 *New ground state and quantum depletion*

The Hamiltonian (24.12) is a sum of Hamiltonians of harmonic oscillators. Its ground state $|\Psi_0\rangle$ is therefore obtained when all these oscillators are in their ground state. The state $|\Psi_0\rangle$ obeys the relations $\hat{b}_{\vec{k}}|\Psi_0\rangle = 0$ for all $\vec{k} \neq \vec{0}$, and does not coincide with the Fock state $|\psi_0\rangle$ of Eq. (24.4) where all the atoms are in the ground state $|\varphi_0\rangle$ of the box. The Bogolubov theory enables one to determine the number of particles out of the state $|\varphi_0\rangle$ when the gas of bosons is in its ground state:

$$N - N_0 = \sum_{\vec{k}\neq\vec{0}} \langle\Psi_0|\, \hat{a}^\dagger_{\vec{k}}\hat{a}_{\vec{k}}\, |\Psi_0\rangle. \tag{24.21}$$

To explicitly calculate this quantity, one expresses the operators $\hat{a}^\dagger_{\vec{k}}$ and $\hat{a}_{\vec{k}}$ in terms of the quasi-particle operators $\hat{b}^\dagger_{\vec{k}}$ and $\hat{b}_{\vec{k}}$:

$$\begin{aligned} N - N_0 &= \sum_{\vec{k}\neq\vec{0}} \langle\Psi_0| \left(u_k\hat{b}^\dagger_{\vec{k}} - v_k\hat{b}_{-\vec{k}}\right)\left(u_k\hat{b}_{\vec{k}} - v_k\hat{b}^\dagger_{-\vec{k}}\right)|\Psi_0\rangle\,, \\ &= \sum_{\vec{k}\neq\vec{0}} v_k^2\,\langle\Psi_0|\,\hat{b}_{-\vec{k}}\hat{b}^\dagger_{-\vec{k}}\,|\Psi_0\rangle = \sum_{\vec{k}\neq\vec{0}} v_k^2. \end{aligned} \tag{24.22}$$

Using Eq. (24.11) with $\tilde{V}(k) = g$ for calculating $v_k = \sinh\theta_k$, we find:

$$v_k^2 = \frac{1}{2}\left[\frac{k^2 + k_0^2}{\sqrt{k^2(k^2 + 2k_0^2)}} - 1\right] \tag{24.23}$$

where k_0 is defined in (24.17). Inserting (24.23) into (24.22), using the dimensionless variable $x = k/k_0$ and replacing the discrete sum by an integral, we finally get

$$\frac{N - N_0}{N} = \sqrt{\frac{32}{\pi}}(na^3)^{1/2}\int_0^\infty x\left[\frac{x^2+1}{\sqrt{x^2+2}} - x\right]\mathrm{d}x = \frac{8}{3\sqrt{\pi}}(na^3)^{1/2}. \tag{24.24}$$

The quantum depletion is therefore small since our calculation is performed under the diluteness assumption $na^3 \ll 1$.

24.2.4 *Energy of the ground state*

In the expression (24.12) of the Bogolubov Hamiltonian, E_0' clearly appears as the energy of the ground state of this Hamiltonian when no elementary excitation is present. The first term of (24.13) is the mean field result and the second is the Bogolubov correction to this result.

First suppose that we continue to use the delta potential, i.e. that we replace in (24.13) $\tilde{V}(0)$ and $\tilde{V}(k)$ by g. The first term gives $Ngn/2$. In the second term, we can replace u_k^2, v_k^2 and $u_k v_k$ by expressions similar to (24.23). One gets a sum over k of a function of k, which behaves, for large k as

$$-\frac{8\pi^2\hbar^2 n^2 a^2}{m}\sum_{\vec{k}\neq\vec{0}}\frac{1}{k^2}. \tag{24.25}$$

This sum diverges. This is an example of the difficulties that can appear when one uses a delta potential. Using the pseudo-potential eliminates these difficulties and leads to the following finite result for the energy of the ground state [Lee and Yang (1957); Lee *et al.* (1957)]:

$$E_0' = \frac{Ngn}{2}\left[1 + \frac{128}{15\sqrt{\pi}}\left(na^3\right)^{1/2}\right]. \tag{24.26}$$

We also find here that the correction to the mean-field result ($Ngn/2$) scales in relative value as $(na^3)^{1/2}$ and is therefore small if the gas is dilute.

Recovering this result with the delta potential truncated in momentum space

We would like to show here that another method for eliminating the divergences in the calculation of E_0' is to use the delta potential with a truncation in momentum space described in Sec. 15.5 of Chap. 15. The Fourier transform of this potential $g_0\delta_{k_c}(r)$ is equal to

$$\tilde{V}(k) = g_0\eta(k) \tag{24.27}$$

where $\eta(k) = 1$ if $k \ll k_c$ and $\eta(k) = 0$ if $k \gg k_c$. The coupling constant g_0 appearing here is not equal to $g = 4\pi\hbar^2 a/m$. It is determined by requiring that the value of the scattering length deduced from this potential is equal to a. In Sec. 15.5 of Chap. 15, we established the relation that must exist between g and g_0:

$$g_0 = \frac{g}{1 - \dfrac{4\pi a}{L^3}\displaystyle\sum_{\vec{k}}\frac{\eta(k)}{k^2}}. \tag{24.28}$$

If, in the first term of (24.13) for the ground state energy, we replace $\tilde{V}(k)$ by $g_0\eta(0) = g_0$, and if we use (24.28), we get for this term

$$\frac{Ng_0}{2} = \frac{2\pi\hbar^2 a/m}{1 - \dfrac{4\pi a}{L^3}\displaystyle\sum_{\vec{k}}\frac{\eta(k)}{k^2}}. \tag{24.29}$$

This expression is not linear in a and therefore does not give the mean field result. We will suppose here that a is not too large [6] and we choose k_c small enough to have $ak_c \ll 1$. By replacing the sum over k in Eq. (24.29) by $(L/2\pi)^3 \int 4\pi k^2 \mathrm{d}k$, we find that the second term in the denominator of (24.29) is on the order of ak_c and we can thus make an expansion of the fraction in powers of ak_c. To order 2 in ak_c, we get:

$$\frac{Ng_0}{2} = \frac{Ng}{2} + \frac{8\pi^2\hbar^2 n^2 a^2}{m} \sum_{\vec{k}} \frac{\eta(k)}{k^2}. \tag{24.30}$$

The second term of (24.13), which is a correction, can be calculated to the lowest order in a at which it appears. We can thus replace $g_0\eta(k)$ by $g\eta(k)$ in this term. The leading term for large k of this term coincides with (24.25), with $1/k^2$ replaced by $\eta(k)/k^2$. It thus cancels out with the last term of (24.30) and eliminates the divergences that appear when one uses a pure delta potential. The remaining part of the second term of (24.13) is a convergent integral of a function of k which takes significant values on a interval of k on the order of k_0. We assume that $k_c \gg k_0$, which is equivalent to $1/k_c \ll \xi$. We can thus replace $\eta(k)$ by 1 in the interval of integration and recover the expression (24.26) for E_0' with no dependence on the value k_c of the cut-off.

Remarks

(i) Condition $1/k_c \ll \xi$ means that the range of the potential is small compared to the healing length, which characterizes the scale of the spatial variations of the condensate. The mean field experienced by an atom at $\vec{r}$ due to the other atoms is proportional to the convolution product $\int V(\vec{r}-\vec{r}\,')|\psi(\vec{r}\,')|^2 \mathrm{d}^3\vec{r}\,'$. It is proportional to $|\psi(\vec{r})|^2$, and the Gross-Pitaevskii equation is a partial differential equation (and not an integral equation), only if the width of $V(\vec{r}-\vec{r}\,')$ is small compared to the scale of variation of $|\psi(\vec{r}\,')|^2$.

(ii) The two conditions that we have supposed to be fulfilled by k_c, $k_c a \ll 1$ and $k_c\xi \gg 1$, can be simultaneously satisfied only if $\xi \gg a$. This condition can be also written $na^3 \ll 1$ and is nothing but the low density approximation at the basis of the Bogolubov approach.

(iii) One can now understand why the use of a delta potential does not give rise to mathematical difficulties for the calculation of the quantum depletion presented in the previous subsection. The quantum depletion does not appear in a mean field treatment and its expression, at the lowest order in a at which it appears, is given by a single term. To calculate this term at the lowest order in a with the truncated delta potential, one can replace g_0 by $g\eta(k)$ if $k_c a \ll 1$. The integrand of the integral appearing in (24.24) takes appreciable values for an interval of $x = k/k_0$ on the order of 1, i.e. for an interval of k on the order of k_0. Since $k_0 \ll k_c$, one can replace $\eta(k)$ by 1, and the calculation is then the same as if one had used a pure delta potential from the beginning.

24.2.5 *Extension to inhomogeneous systems*

The Bogolubov formalism developed above for the homogeneous case gives a clear interpretation of the first excited states of the system in terms of quasi-particles having a well defined energy and momentum, and obtained from the vacuum by the

[6] The system is not placed in the vicinity of the center of a Feshbach resonance (see Chap. 16).

creation of a particle-hole pair. The dispersion relation (24.12) can be obtained by linearizing the time-dependent Gross-Pitaevskii equation for a condensate confined in a box, but the equation for a classical matter wave does not give the quantized energies of the elementary excitations. Neither does it give the quantum depletion (24.24) and the energy of the ground state (24.26), which are new results that cannot be derived from a mean-field approach.

The shapes and the frequencies of the normal modes of vibrations of a trapped Bose-Einstein condensate can be obtained by searching for a solution of the time-dependent Gross-Pitaevskii equation of the form

$$\Psi(\vec{r},t) = \left[\Psi_0(\vec{r}) + \sum_i \left(u_i(\vec{r})e^{-i\omega_i t} + v_i^*(\vec{r})e^{i\omega_i t}\right)\right] e^{-i\mu t/\hbar}. \tag{24.31}$$

If one keeps only the terms linear in u_k and v_k in the time-dependent GP equation, one finds that the functions $u_i(\vec{r})$ and $v_i(\vec{r})$ and the frequencies ω_i obey the following set of equations referred to as the Bogolubov-de Gennes equations:

$$\begin{aligned} \hbar\omega_i u_i(\vec{r}) &= [H - \mu + 2gn(\vec{r})]\, u_i(\vec{r}) + g(\Psi_0(\vec{r}))^2 v_i(\vec{r}) \\ -\hbar\omega_i v_i(\vec{r}) &= [H - \mu + 2gn(\vec{r})]\, v_i(\vec{r}) + g(\Psi_0^*(\vec{r}))^2 u_i(\vec{r}). \end{aligned} \tag{24.32}$$

The normal modes of vibrations are then deduced by numerically solving these equations (see for example [Edwards *et al.* (1996)]).

The elementary excitations of the trapped condensate are the quantized version of the normal modes of vibrations. They are obtained by first writing the atomic field operator $\hat{\Psi}(\vec{r},t)$ as

$$\hat{\Psi}(\vec{r},t) = \hat{\Psi}_0(\vec{r})e^{-i\mu t/\hbar} + \delta\hat{\Psi}(\vec{r},t) \tag{24.33}$$

where $\psi_0(\vec{r})$ is the classical matter wave solution of the time-independent Gross-Pitaevskii equation, and $\delta\hat{\Psi}(\vec{r},t)$ is a field operator describing the quantum fluctuations. One then inserts (24.33) and the corresponding expression for $\hat{\Psi}^\dagger(\vec{r},t)$ into the full quantum Hamiltonian[7] expressed in terms of $\hat{\Psi}$ and $\hat{\Psi}^\dagger$. The equivalent of the Bogolubov approach presented above for the homogeneous case consists of keeping only terms up to order 2 in $\delta\hat{\Psi}$ and $\delta\hat{\Psi}^\dagger$ in this Hamiltonian. In this way, one gets a quadratic Hamiltonian that can be diagonalized: it turns out that the unitary transformation used for this diagonalization involves the same functions u_i and v_i as those appearing in Eq. (24.31) for the solution of the linearized time-dependent Gross-Pitaevskii equation and leads to the same frequencies [Fetter (1972); Castin (2001)]. The generalized Bogolubov approach gives the energies $\hbar\omega_i$ of the elementary excitations, as well as the quantum depletion and the energy of the ground state in the dilute gas limit.

Experimental studies

The excitation spectrum ω_k^B gives important insight into the superfluidity properties as explained below. It was measured in liquid helium using scattering of a monochromatic beam of neutrons [Henshaw and Woods (1961)]. The scattering process can

[7]See Eq. (23.27) of Chap. 23.

be considered as a collision of a neutron with a quasiparticle. By knowing the initial energy of the neutrons and their energy change after the scattering along different directions, it is possible to infer the spectrum of the quasiparticles. For trapped condensates of dilute alkali gases, the Bogolubov spectrum has been measured using Bragg spectroscopy, which consists of determining the maximum probability of excitation from an absorption-stimulated emission elementary process with a well-controlled angular frequency ω and wave vector $\vec{k}$ [Vogels *et al.* (2002); Steinhauer *et al.* (2002)].

24.3 Landau criterion for superfluidity in an homogeneous system

24.3.1 *Microscopic probe*

A major consequence of the Bogolubov dispersion relation (24.12) for the elementary excitations is that according to Landau's criterion, a Bose-Einstein condensate is superfluid, with a critical velocity given by the speed of sound $c = (gn/m)^{1/2}$. In this section, we briefly recall a basic derivation of Landau's criterion for superfluidity.

We consider a point-like[8] test particle of mass M (an impurity) moving in the condensate with a velocity $\vec{v}$. As we consider an homogeneous system, the momentum $\vec{P} = M\vec{v}$ is a well-defined quantity. The damping of this microscopic probe is related to how it can deposit energy in the system during its propagation. The particle will experience a drag force if and only if it interacts with the fluid, i.e if it can create Bogolubov elementary excitations having a momentum $\hbar\vec{k}$ and an energy $\hbar\omega_k^B$; the impurity acquires in the process a new momentum $M\vec{v}'$ and an energy $E' = Mv'^2/2$. Conservation of energy and momentum reads

$$\frac{1}{2}Mv^2 = \frac{1}{2}Mv'^2 + \hbar\omega_k^B \quad \text{and} \quad M\vec{v} = M\vec{v}' + \hbar\vec{k}.$$

Eliminating $\vec{v}'$ between those two equations, we get

$$\frac{1}{2}Mv^2 - \hbar\omega_k^B = \frac{1}{2}M\left(\vec{v} - \frac{\hbar\vec{k}}{M}\right)^2.$$

Therefore, if this process occurs, one must necessarily have

$$\vec{v}\cdot\vec{k} = v\,k\cos\theta = \omega_k^B + \frac{\hbar k^2}{2M} \geq \omega_k^B. \tag{24.34}$$

Suppose that the critical velocity v_c defined as

$$v_c = \min_k \frac{\omega_k^B}{k}$$

is non-zero. Then we see that for $v \leq v_c$, the relation (24.34) cannot be fulfilled; therefore the test particle cannot experience any drag in the condensate. From the

[8] An impurity with a size larger than the healing length, ξ, can generate other types of excitations, different from the Bogolubov quasiparticles (for instance vortex rings), which changes the conclusions of the present paragraph [Frisch *et al* (1992); Raman *et al.* (1999); Aftalion *et al.* (2003)].

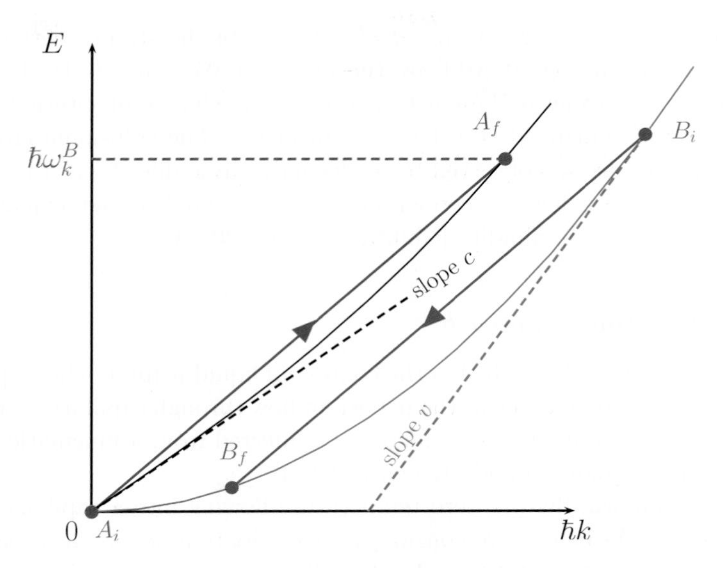

Fig. 24.2 Graphic construction explaining the exchanges of energy and momentum between the probe particle and the condensate when $\theta = 0$. Black curve: dispersion relation (24.12) of the elementary excitations of the condensate, with a slope at $k = 0$ equal to the speed of sound c (black dashed line). Red curve: parabolic dispersion relation of the probe particle. Initial state: the probe particle is in the energy state represented by B_i. The slope of the tangent to the parabola at B_i is equal to the initial velocity v of the probe particle (red dashed line). The condensate is in the state represented by A_i that coincides with $k = 0$, that is, there are no elementary excitations. Final state: The probe particle is at B_f, an elementary excitation represented by A_f is created. The energy and momentum change of the probe particle corresponds to the vector $\overrightarrow{B_iB_f}$ while the energy and momentum change of the condensate corresponds to the vector $\overrightarrow{A_iA_f}$. Conservation of momentum and energy implies that $\overrightarrow{A_iA_f} + \overrightarrow{B_iB_f} = \vec{0}$. The concavity of the probe parabola (red curve) shows that the slope s of $\overrightarrow{B_iB_f}$ is smaller than the slope of the tangent to the parabola at B_i, equal to v. But the slope of $\overrightarrow{B_iB_f}$ is also equal to the slope of $\overrightarrow{A_iA_f}$ which is larger than the slope of the black curve at O, equal to c. These two inequalities $s < v$ and $s > c$ can be simultaneously fulfilled only if $v > c$. If $v < c$, the probe particle cannot deposit energy and momentum to the condensate.

Bogolubov dispersion relation, one finds that $v_c = c$. We thus conclude that in the presence of atom-atom interactions, a condensate at $T = 0$ is superfluid, with a critical velocity equal to the speed of sound (see Fig. 24.2 for a graphic explanation of the superfluid criterion). For the ideal gas, one has the usual quadratic dispersion relation of free particles, which yields a vanishing critical velocity. Therefore, *an ideal BEC is not superfluid.*

The threshold velocity below which no dissipation occurs when a microscopic probe is sent through a superfluid has been investigated experimentally in both superfluid helium-II and dilute Bose-Einstein condensate of alkali atoms. In helium, a detailed study of the breakdown of superfluidity in good agreement with Landau's theory has been performed using, as microscopic probe, negative ions injected into liquid helium at $T = 0.35$ K [Allum *et al.* (1977)]. In the studies performed with Bose-Einstein condensates at MIT by the group of Wolfgang Ketterle, impurity atoms were created using a Raman transition. The change of internal state was accompanied by a controlled transfer of momentum. The collisional cross-section of the impurity atoms was observed to be dramatically reduced when the impurity velocity was reduced below the condensate speed of sound, in agreement with the Landau criterion for superfluidity [Chikkatur *et al.* (2002)].

24.3.2 *Macroscopic approach*

Superfluidity was initially studied in the context of liquid helium, where spectacular effects have been observed such as the persistent flow through capillaries without any dissipation. The Landau criterion can also be inferred from a kinematic argument [Landau (1941a); Pitaevskii and Stringari (2003)].

Consider a uniform fluid at zero temperature flowing into a capillary at a constant velocity $\vec{V}$. As before, we consider that dissipation occurs only through the creation of elementary excitations. In the following, we just consider a single excitation of momentum $\vec{p}$ in the reference frame, K, attached to the fluid. In this frame, the energy of the fluid plus the single excitation is $E = E_0 + \varepsilon(\vec{p})$, where E_0 is the ground state energy of the fluid and $\varepsilon(\vec{p})$ is the energy of the elementary excitations. It is instructive to analyze the same situation in the reference frame, K_0, attached to the capillary. The Galilean transformation from K to K_0 yields [Landau and Lifshitz (1996)]

$$E \longrightarrow E' = E_0 + \varepsilon(\vec{p}) + \vec{p}\cdot\vec{V} + \frac{1}{2}MV^2,$$

$$\vec{0} + \vec{p} \longrightarrow \vec{p}\,' = \vec{p} + M\vec{V}, \tag{24.35}$$

where M denotes the total mass of the fluid. In the absence of any excitation, the energy of the fluid in the frame K_0 is $E'_0 = E_0 + MV^2/2$. There will be dissipation if the energy E' is smaller than the energy E'_0. Conversely, there is no dissipation if $\varepsilon(\vec{p}) + \vec{p}\cdot\vec{V} > 0$, i.e. if the creation of the elementary excitation is not energetically favorable, that is, if the relative velocity between the fluid and the capillary verifies

$$|V| \leq \min_{\vec{p}} \frac{\varepsilon(\vec{p})}{p}. \tag{24.36}$$

24.4 Extension of Landau criterion for a condensate in a rotating bucket

The previous reasoning can be extended to a non-Galilean rotating frame transformation. This provides a new criterion for superfluids subjected to rotation and reveals non-classical signatures such as quantum vortices in the response of the system.

24.4.1 *The rotating bucket*

Consider a cylindrical bucket rotating around the z-axis with an angular velocity Ω with respect to a frame K_0. Suppose that this bucket contains a condensate at rest in K_0 with an energy E_0 and an angular momentum $L_z = 0$ in K_0. Can the rotating bucket communicate an angular momentum to the condensate by exciting an elementary excitation with an angular momentum $L_z = \ell\hbar$ and an energy $\varepsilon(\ell)$?

In the frame K rotating with the bucket, the energy of the condensate is:

$$E = E_0 - \Omega L_z. \tag{24.37}$$

When the condensate has no angular momentum in K_0, this equation gives $E = E_0$. If an elementary excitation with angular momentum $L_z = \ell\hbar$ and energy $\varepsilon(\ell)$ appears in K_0, the energy of the condensate in K becomes:

$$E = E_0 + \varepsilon(\ell) - \ell\Omega. \tag{24.38}$$

If $\varepsilon(\ell) - \ell\Omega > 0$, i.e. if $\Omega < \varepsilon(\ell)/\ell$, the creation of this elementary excitation is not energetically favorable. The conclusion of this analysis is that, if the minimum value of $\varepsilon(\ell)/\ell$ over all possible values of ℓ has a non-zero value Ω_c,[9] and if $\Omega < \Omega_c$, the condensate remains at rest in the rotating bucket and does not acquire any angular momentum.

Remark

When a condensate is moving with a velocity $v < c$ in a capillary, its state is stable with respect to the creation of an elementary excitation of momentum p and energy $\varepsilon(p)$, but does not correspond to the full thermodynamical equilibrium state in the capillary, which is obtained for $v = 0$. By contrast, if the condensate is put in a bucket rotating at an angular frequency $\Omega < \Omega_c$, its state is not only stable with respect to the creation of an elementary excitation $\{\ell\hbar, \varepsilon(\ell)\}$, but is also the full thermodynamical equilibrium state (lowest value of E given in (24.38)).

24.4.2 *Other possible states of the condensate: quantized vortices*

In Chap. 22, we studied another solution of the Gross-Pitaevskii equation which was not the lowest energy state, but a quantized vortex in which each of the N atoms occupied the same single particle state with an angular momentum $\ell\hbar$, with $\ell = 1, 2, \ldots$. A condensate with a vortex $\ell \neq 0$ has an energy E_ℓ higher that the energy E_0 of the usual condensate without vortex ($\ell = 0$). We denote the energy increase per atom of the condensate when it contains one vortex as $\delta\varepsilon = (E_\ell - E_0)/N$.

Consider a condensate with a quantized vortex $\ell\hbar$ in a rotating bucket. How does the energy change $\delta\varepsilon$ per atom in K vary when the rotation angular frequency Ω of the bucket increases? From Eq. (24.37), one deduces:

$$E_\ell(\Omega \neq 0) = E_\ell(\Omega = 0) - \Omega L_z = E_\ell - N\ell\hbar\Omega \tag{24.39}$$

[9] For the case of a hamonically confined weakly interacting Bose-Einstein condensate, the spectrum $\varepsilon(\ell)$ has been calculated numerically in [Dalfovo and Stringari (2000b)]. The quantity $\varepsilon(\ell)/\ell$ exhibits a clear minimum when ℓ varies.

and, consequently

$$\delta\varepsilon(\Omega \neq 0) = \delta\varepsilon(\Omega = 0) - \ell\hbar\Omega. \tag{24.40}$$

It is clear from this equation that for a sufficiently high value of Ω, the condensate with one vortex will have an energy smaller than the state without vortex in the frame K rotating with the bucket. More precisely, if

$$\Omega > \Omega_{\rm v} \quad \text{with} \quad \Omega_{\rm v} = \frac{\delta\varepsilon(\Omega = 0)}{\ell\hbar} \tag{24.41}$$

a state with one vortex can become the thermodynamically stable state.

To get an idea of the value of $\Omega_{\rm v}$, it is useful to first consider an ideal gas confined in a cylindrical harmonic trap that is symmetric with respect to the z-axis. The rotating bucket here corresponds to rotating this cylindrical trap. Let $\omega_\perp$ be the trapping frequency in the plane perpendicular to the symmetry axis. The one-vortex state $\ell = 1$ corresponds to the first excited states of the two-dimensional harmonic oscillator associated with the trap, so that for this state:

$$\delta\varepsilon(\Omega = 0) = \hbar\omega_\perp \tag{24.42}$$

and consequently, according to (24.40), $\Omega_{\rm v} = \omega_\perp$. But, if the rotation frequency Ω of the trap is equal to $\omega_\perp$, the centrifugal force compensates the restoring force of the trap and the gas is no longer confined.

A more realistic model should take into account atom-atom interactions. Figure 24.3 gives qualitatively the energies per atom of the zero and one vortex $\ell = 1$ states versus $\hbar\Omega$ for an ideal gas (dashed lines) and for a weakly interacting condensate (full lines) confined in a two-dimensional isotropic harmonic trap of angular frequency $\omega_\perp$. For an ideal gas and for $\ell = 0$, $\varepsilon_0 = E_0/N$ is independent of Ω (horizontal dotted line). The first excited state (for $\ell = 1$) of an ideal gas is two-fold degenerate, but the rotation lifts this degeneracy and yields two branches and $\varepsilon_1 = E_1/N$ varies as $\varepsilon_0 + \hbar\omega_\perp \pm \hbar\Omega$ (we represent only the state with a slope -1 as a dashed line in Fig. 24.3). The two dashed lines intersect for $\Omega = \omega_\perp$ as explained above. These two dashed lines are shifted upwards when interactions (taken to be repulsive) are included. First suppose that the interactions are weak and can be treated perturbatively. The shifts $\tilde{\varepsilon}_1 - \varepsilon_1$ and $\tilde{\varepsilon}_2 - \varepsilon_2$ are small compared to $\hbar\omega_\perp$. The state with zero vortex is shifted more than the state with one vortex because it is more compact, and thus allows stronger interactions. The two full lines of Fig. 24.3 intersect for a value $\Omega_{\rm v}$ smaller than $\omega_\perp$ as a result of the repulsive interactions. This shows that the first excited state of the transverse harmonic oscillator with a density hole at its center, corresponding to the state with one vortex, becomes thermodynamically more favorable than the state without a vortex for a rotation frequency larger than $\Omega_{\rm v}$ while remaining smaller than $\omega_\perp$ which ensures the stability of the trap.

For stronger interactions, the shifts $\delta\varepsilon_1 = \tilde{\varepsilon}_1 - \varepsilon_1$ and $\delta\varepsilon_0 = \tilde{\varepsilon}_0 - \varepsilon_0$ due to interactions become larger than $\hbar\omega_\perp$. The energy increase per atom $\delta\varepsilon = \tilde{\varepsilon}_1 - \tilde{\varepsilon}_0$ due to the presence of a vortex remains smaller than $\hbar\omega_\perp$, so that the critical angular velocity $\Omega_{\rm v}$ above which a one-vortex state can become the thermodynamical equilibrium state in K remains smaller than $\omega_\perp$.

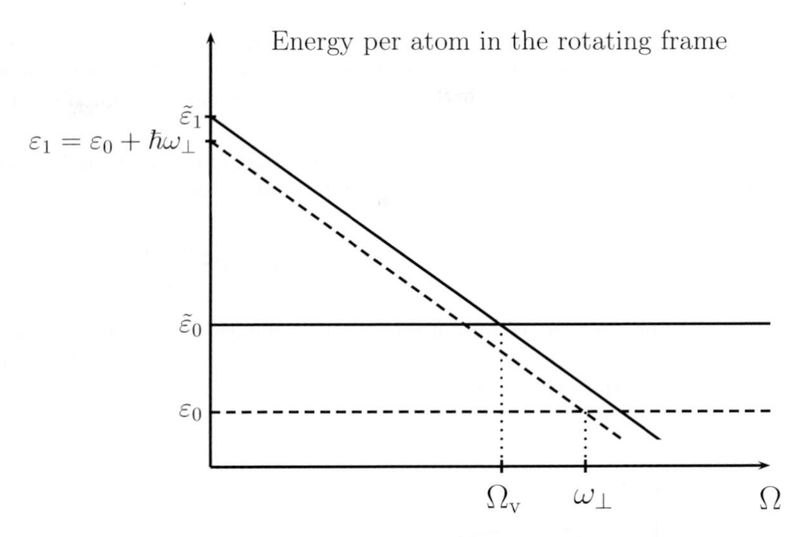

Fig. 24.3 Energies of the zero and one vortex states versus the rotation frequency of the trap for an ideal gas (dashed lines) and for a weakly interacting condensate (full lines).

Energy increase per atom $\delta\varepsilon$ due to the presence of a vortex

In the Thomas-Fermi regime, the differences between the solutions of the Gross-Pitaevskii equation in the presence and in the absence of a vortex are two-fold:

- The wave function in the presence of a centered vortex tends to zero when the distance $r_\perp$ from the symmetry axis of the trap tends to zero in an interval of width ξ_0, where ξ_0 is the healing length[10] (see Fig. 24.4). Outside this region, and up to the Thomas-Fermi radius $R_\perp$, the modulus of the two wave functions are very similar.
- In the presence of a vortex, the wave function is complex, describing an azimuthal velocity field $v_\varphi = h/mr_\perp$.

The main contribution to the increase of energy when a vortex appears in the condensate comes from the azimuthal kinetic energy associated with the azimuthal velocity field. Its order of magnitude is given by:

$$\delta E_{\text{azim}} \simeq \rho_0 L \int_{\xi_0}^{R_\perp} 2\pi r_\perp \frac{M v_\varphi^2}{2} \mathrm{d}\, r_\perp = \pi \rho_0 L \frac{\hbar^2}{m} \mathrm{Log} \frac{R_\perp}{\xi_0} \tag{24.43}$$

where L is the length of the condensate along the z-axis (on the order of twice the Thomas-Fermi radius of the confining potential along the z-axis). The lower bound of the integral is given by the size of the vortex core. Dividing (24.43) by the number of atoms $N \simeq \rho\pi R_\perp^2 L$ gives for the contribution of the azimuthal kinetic energy to the increase of energy per atom:

$$\delta\varepsilon_{\text{azim}} \simeq \frac{\hbar^2}{mR_\perp^2} \mathrm{Log} \frac{R_\perp}{\xi_0}. \tag{24.44}$$

[10]See Sec. 22.3 of Chap. 22.

Other contributions to $\delta\varepsilon$ come from radial confinement in the core of the vortex and from the increase of interaction energy in the region $\xi_0 < r_\perp < R_\perp$ due to the increase of the spatial density coming from the atoms that are removed from the core of the vortex. Their orders of magnitude are $\hbar^2/mR_\perp^2$.

A more precise calculation that takes the parabolic shape of the Thomas-Fermi profiles into account can be found in [Lundh *et al.* (1997)] and gives:

$$\delta\varepsilon = \frac{5}{2}\frac{\hbar^2}{mR_\perp^2}\mathrm{Log}\frac{0.671\,R_\perp}{\xi_0} = \frac{5}{2}\hbar\omega_\perp\left(\frac{a_{\mathrm{ho}}}{R_\perp}\right)^2\mathrm{Log}\frac{0.671\,R_\perp}{\xi_0}, \tag{24.45}$$

in good agreement with the results of numerical calculations. One can check that $\delta\varepsilon$ is smaller than $\hbar\omega_\perp$. When interactions increase, $R_\perp$ increases, but the decrease of $(a_{\mathrm{ho}}/R_\perp)^2$ is faster than the increase of the logarithmic term $\mathrm{Log}(R_\perp/\xi_0)$.

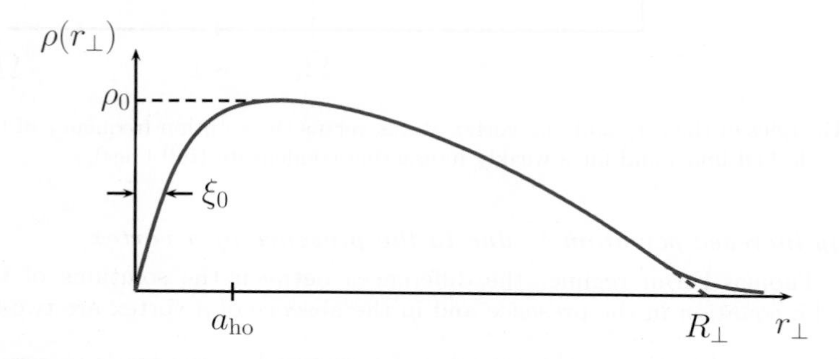

Fig. 24.4 Radial dependence of the solution of the Gross-Pitaevskii equation in the presence of a vortex. ξ_0 is the healing length, $R_\perp$ the Thomas Fermi radius, $a_{\mathrm{ho}} = \sqrt{\hbar/m\omega_\perp}$ the characteristic length of the radial harmonic confinement.

24.4.3 *Various threshold rotation frequencies*

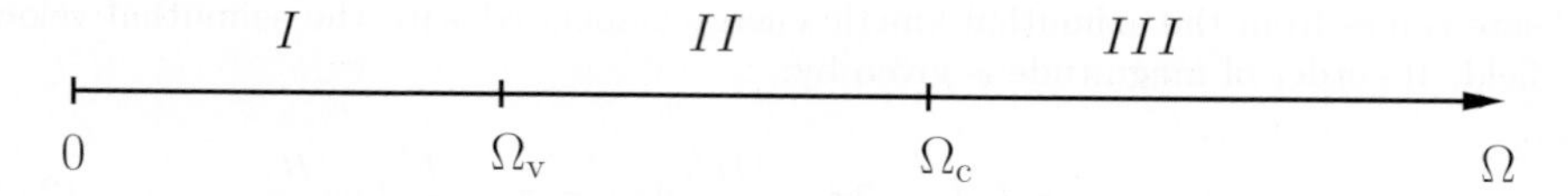

Fig. 24.5 Various threshold rotation frequencies.

In the two previous subsections, we introduced two threshold rotation frequencies Ω_{c} and Ω_{v} which are represented along the Ω axis in Fig. 24.5. Generally, Ω_{c} is larger than Ω_{v}. Three different domains clearly appear in this figure:

- Domain I: $0 < \Omega < \Omega_{\mathrm{v}}$.
  The condensate without vortex is the thermodynamical equilibrium state. The states with one vortex or with one elementary excitation have higher energy in K.

- Domain II: $\Omega_v < \Omega < \Omega_c$.
  In the mean-field approximation, the state with one vortex becomes more favorable. Since it has a completely different topological structure compared to the condensate without vortex, it is not easily formed when Ω is slowly increased from zero. The condensate generally remains without any vortex when Ω reaches the domain II: it is in a metastable state.
- Domain III: $\Omega_c < \Omega$.
  The creation of elementary excitations becomes energetically favorable. We will see in the next section that the creation of these elementary excitations is experimentally observed and that it is an efficient way for nucleating vortices.

24.5 Experimental study of vortices in gaseous condensates

24.5.1 *Introduction*

The wave function of a vortex has a quantized circulation of the velocity as explained, in Chap. 22. A first possibility for generating them starting from a trapped Bose-Einstein condensate consists in printing the corresponding velocity field onto the atomic wave function. Such a phase engineering technique using a combination of a laser and a microwave field was employed by the Boulder group [Williams and Holland (1999); Matthews *et al.* (1999)]. In this first experiment, the vortex was produced in a double component condensate with one component standing still at the center of the trap and the other component in quantized rotation around it. In subsequent experiments, the core at the center was removed [Anderson *et al.* (2000)]. Vortices were also imprinted using a pair of Raman beams [Andersen *et al.* (2006)] or topological phases as demonstrated by the MIT group [Leanhardt *et al.* (2002)].

Another strategy, initially developed in the group of Jean Dalibard at ENS, consists of subjecting a condensate initially confined in a cylindrically symmetric trap to a rotating perturbation by revolving off-resonance laser beams around it (see Fig. 24.6) [Madison *et al.* (2000)]. The potential experienced by the atoms is therefore of the form:

$$V(\vec{r}) = \frac{1}{2}M\omega_\perp^2(x^2+y^2) + \frac{1}{2}M\omega_z^2 z^2 + \frac{1}{2}M\omega_\perp^2(\epsilon_X X^2 + \epsilon_Y Y^2), \qquad (24.46)$$

where M is the atom mass, $\epsilon_{X,Y}$ depends on the characteritics of the stirring beams, and the XY axes rotate in the plane xy at an angular velocity Ω. This approach is a transposition of the rotating bucket experiment performed with liquid helium-4.

24.5.2 *A few experimental results*

In Sec. 22.6 of Chap. 22, we showed that, in a fluid of density n, the radius of the vortex core is determined by the healing length $\xi_0 = (8\pi na)^{-1/2}$ where a is

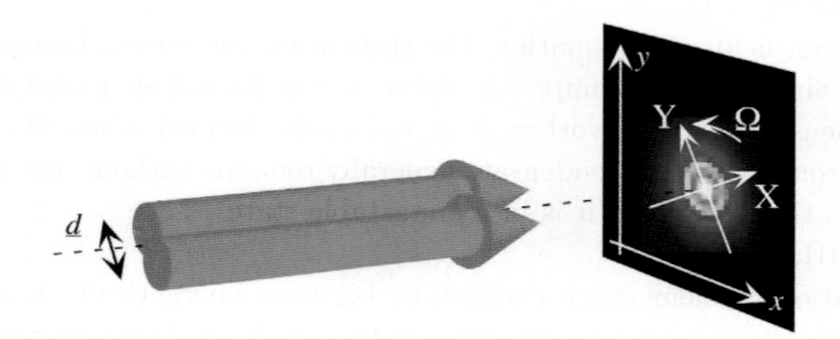

Fig. 24.6 A condensate in an axisymmetric trap is stirred using an off-resonance focussed beam whose motion is controlled using acousto-optic deflectors. This laser propagates along the symmetry axis and toggles back and forth on a very short timescale compared to the oscillation periods along the axial and transverse directions of the trap. As a result, it generates an average dipole potential that is anisotropic in the transverse plane for the atoms. This extra anisotropic potential is then set into rotation at a controlled angular velocity Ω. Courtesy of Frédéric Chevy.

the scattering length. For typical experimental parameters, ξ_0 is of the order of hundreds of nm, which is too small to be observed optically. However, this size can be expanded using a time-of-flight technique. The hole remains in the expansion because of the conservation of the angular momentum [Dalfovo and Modugno (2000a)]. The pictures of Fig. 24.7 are absorption images obtained in this manner. Just above a critical rotation frequency, a density dip appears at the center of the cloud. The energy of the configuration with one vortex is minimum when the vortex is at the center of the trap. At higher frequencies, more vortices are nucleated. A remarkable feature is that multiple vortex configurations most often correspond to symmetric arrangements of the vortex cores. This is a consequence of the balance between the repulsive interaction between vortex lines [11] and the restoring force towards the center that effectively acts on them. The lowest energy configuration is a triangular lattice [Madison *et al.* (2000); Abo-Schaeer *et al.* (2001)].

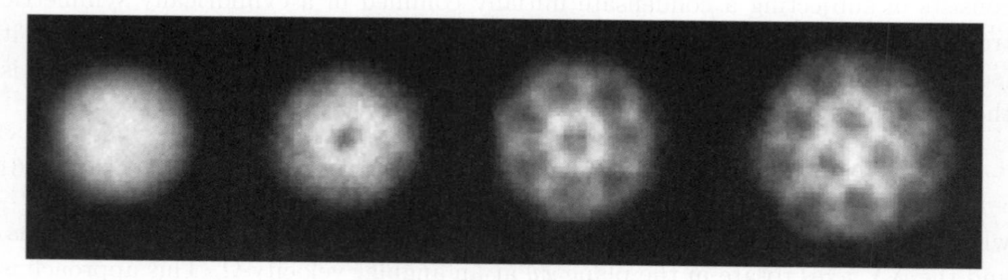

Fig. 24.7 Density profiles after a time-of-flight expansion of a stirred Bose-Einstein condensate. From left to right, the rotating angular frequency Ω increases: below the threshold frequency for nucleating a vortex, no modification of the condensate is observed; increasing the rotating frequency of the stirring ellipticity further, one observes arrays of vortices. Courtesy of Jean Dalibard.

[11] As a result of the Magnus force, two vortices of the same sign (velocity field rotates in the same sense) repel each other.

Regular lattices were originally predicted by Alexei Abrikosov to occur in type-II superconductors in the presence of a magnetic field [Abrikosov (1957)] and were subsequently observed experimentally [Träuble and Essmann (1968); Sarma (1968)]. Evidence of vortex lattice arrangements in rotating superfluid liquid helium-4 was also observed [Yarmchuk *et al.* (1979)]. The size of the core of the vortex is on the order of 0.1 nm in liquid helium that is on the order of an atomic size, while it can reach a few micrometers after expansion in dilute Bose-Einstein gases, thus allowing it to be observed directly by optical means.

Large vortex number limit

In the large vortex number limit, the density of vortices can be inferred from the correspondence principle. In this limit, the coarse grained average of the quantum velocity field on a scale larger than the distance between vortex lines tends to the classical rigid body rotation velocity $\vec{v} = \vec{\Omega} \times \vec{r}$. Indeed, for a uniform surface density of vortices, n_v, the circulation of the velocity field on a circle of radius R centered on the rotation axis is $\mathcal{N}(R)h/M$ where $\mathcal{N}(R) = \pi R^2 n_v$ is the number of vortices contained in the disk of radius R [Feynman (1955); Nozières and Pines (1990)]:

$$\oint_C \vec{v} \cdot \mathrm{d}\vec{r} = 2\pi R(R\Omega) = \pi R^2 n_v \frac{h}{M}. \tag{24.47}$$

The rigid body ciculation is recovered when $n_v = 2M\Omega/h$.

24.5.3 *Measuring the angular momentum per atom in a rotating condensate*

When the condensate accomodates one centered vortex line, the angular momentum per particle is $\hbar$. However, when the position of the vortex core is not well centered, the average angular momentum per particle is less than $\hbar$. One may wonder how to directly measure this angular momentum and how it increases as the number of vortices increases. To answer this question, Francesca Zambelli and Sandro Stringari suggested the study of the frequencies ω_+ and ω_- of the two transverse quadrupole modes of a cylindrically symmetric condensate, that correspond respectively to excitations with angular momentum $m = 2$ and $m = -2$ [Zambelli and Stringari (1998)]. If the condensate does not rotate, these frequencies are equal. On the contrary, if the condensate has a non-zero angular momentum $\langle L_z \rangle$ per particle, the degeneracy is lifted and the difference between the two frequencies for a condensate in the Thomas-Fermi regime is found to be equal to

$$\omega_+ - \omega_- = \frac{2\langle L_z \rangle}{M\langle r_\perp^2 \rangle} \tag{24.48}$$

where $\langle r_\perp^2 \rangle$ is related to the transverse size of the condensate. In practice, one excites a transverse quadrupolar oscillation that is a superposition of both the $m = 2$ and $m = -2$ modes.[12] In the presence of vortices, the axis of the quadrupolar

[12]See Sec. 22.9 of Chap. 22.

oscillation precesses, and the precession frequency is $\dot{\theta} = (\omega_+ - \omega_-)/4$. This method was implemented in the ENS and Boulder groups [Chevy *et al.* (2000); Haljan *et al.* (2001)]. This measurement is analogous to the experiment performed with superfluid liquid helium by William Vinen in which a single quantum of circulation in rotating He II was detected by measuring the frequencies of two opposing circular vibrational modes of a thin wire placed at the center of the rotating fluid [Vinen (1958); Vinen (1961)].

A typical result is shown in Fig. 24.8. At a critical frequency, an average momentum per particle $\langle L_z \rangle$ on the order of $\hbar$ is measured and corresponds to the apparition of a centered vortex. The angular momentum increases regularly as new vortices appear. The larger the ellipticities $\epsilon_{X,Y}$, the wider the region where vortices are nucleated. Interestingly, for very low stirring ellipticity, the curve exhibits a resonant shape centered on the stirring velocity $2\pi \times 125$ Hz. This corresponds to the situation where the rotation frequency Ω of the trap coincides with $\omega(\ell)/\ell$ for $\ell = 2$, i.e. $\Omega = \sqrt{2}\omega_\perp/2 = \omega_\perp/\sqrt{2}$.

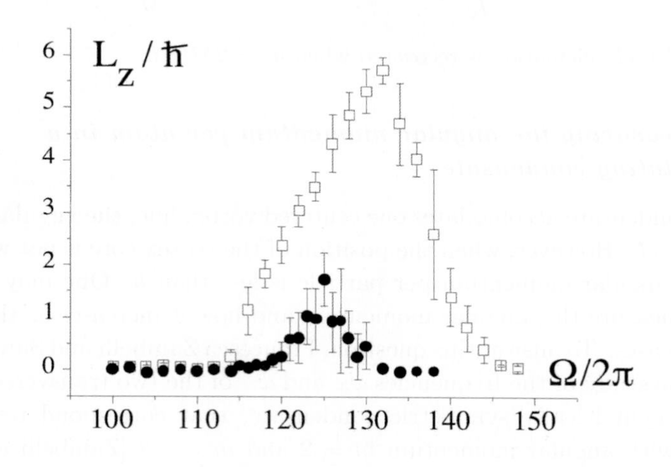

Fig. 24.8 Angular momentum per particle of the condensate, $\langle L_z \rangle$, deduced from the precession of the transverse quadrupole mode as a function of the stirring frequency Ω and ellipticities (black disks: $\epsilon = 0.010 \pm 0.002$; open circles: $\epsilon = 0.019 \pm 0.004$). $\omega_\perp/2\pi = 172$ Hz. Courtesy of Frédéric Chevy.

24.5.4 *Routes to vortex nucleation*

In order to explain this latter observation, we first begin with a reminder of the stationary rotating states of a harmonically trapped Bose-Einstein condensate without any vortex [Recati *et al.* (2001)].

24.5.4.1 *Rotational motion of an interacting Bose-Einstein condensate*

The trapping is taken to be elliptic:

$$V_{\text{trap}}(x,y,z) = \frac{m}{2}\left[\omega_x^2 x^2 + \omega_y^2 y^2 + \omega_z^2 z^2\right] \tag{24.49}$$

with an anisotropy parameter

$$\epsilon = \left(\omega_x^2 - \omega_y^2\right) / \left(\omega_x^2 + \omega_y^2\right). \tag{24.50}$$

It rotates at an angular velocity Ω around the z-axis. In the frame K rotating with the trap, the new coordinates x', y', z', are related to those of the lab frame K_0 by:

$$x' = x\cos\Omega t + y\sin\Omega t, \qquad y' = -x\sin\Omega t + y\cos\Omega t, \quad \text{and} \quad z' = z. \tag{24.51}$$

In the rotating frame, the trapping potential is time-independent and the Gross-Pitaevskii equation takes the form:

$$\begin{aligned} i\hbar\frac{\partial}{\partial t}\phi(\vec{r}\,',t) &= \left[-\frac{\hbar^2}{2m}\Delta' + V_{\text{trap}}(\vec{r}\,') + g\left|\phi(\vec{r}\,',t)\right|^2\right]\phi(\vec{r}\,',t) \\ &\quad - \frac{\hbar}{i}\Omega\left(x'\frac{\partial}{\partial y'} - y'\frac{\partial}{\partial x'}\right)\phi(\vec{r}\,',t) \end{aligned} \tag{24.52}$$

where the last term comes from the $-\Omega L_z$ term appearing in the new Hamiltonian H'. The hydrodynamic equations equivalent to (24.52) for the modulus and the phase of the wave function, $\phi(\vec{r}\,',t) = \sqrt{\rho(\vec{r}\,',t)}\exp[iS(\vec{r}\,',t)]$, simplify when the quantum pressure term can be neglected.[13] They lead to explicit expressions[14] for ρ and S, which are quadratic functions of x', y', z', and which depend only on three parameters, the anisotropy parameter of the trap ϵ given in (24.50), the angular velocity Ω, and a new parameter α given by the following equation:

$$2\alpha^3 + \alpha\left(\omega_x^2 + \omega_y^2 - 4\Omega^2\right) + \Omega\left(\omega_x^2 - \omega_y^2\right) = 0. \tag{24.53}$$

Before discussing the solutions of this equation, we first briefly describe how the shape of the condensate and the velocity field are related to α.

The shape of the condensate in K is determined by $\rho(\vec{r}\,',t)$. In the transverse $x'y'$ plane it has an elliptical shape with the same symmetry axis as the trap. One finds the following relation between the mean values of x'^2 and y'^2, and α:

$$\frac{\langle x'^2\rangle - \langle y'^2\rangle}{\langle x'^2\rangle + \langle y'^2\rangle} = \frac{\alpha}{\Omega}. \tag{24.54}$$

The anisotropy of the condensate in K, characterized by the left-side of this equation, is thus equal to α/Ω. If $\alpha = 0$, the cloud is isotropic in the $x'y'$ plane.

The phase $S(\vec{r}\,',t)$ is found to be equal to $m\alpha x'y'/\hbar$ so that the velocity field is given by:

$$\vec{v}\,' = \frac{\hbar}{m}\vec{\nabla}'S = \alpha\,\vec{\nabla}'(x'y'). \tag{24.55}$$

[13] See Sec. 22.7.3 of Chap. 22.

[14] These solutions are not unique. Other solutions (with quantum vortices) exist and play an important role in the discussion of the stability of the condensate presented in Sec. 24.4.

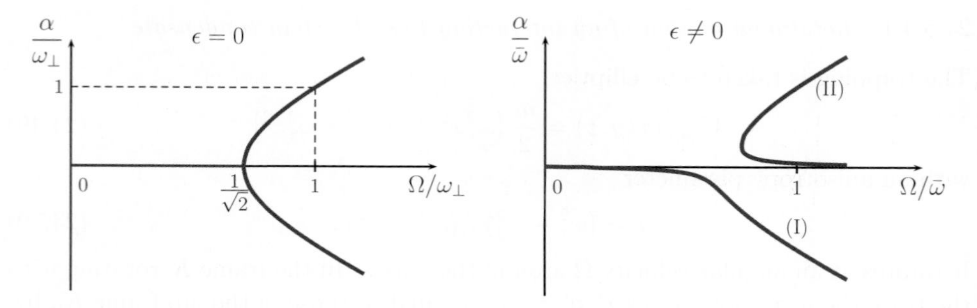

Fig. 24.9 Left: values of the parameter α for an axisymmetric trap ($\epsilon = 0$). Right: values of the parameter α for an elliptical trap ($\epsilon \neq 0$).

The velocity field is proportional to α. If $\alpha = 0$, the velocity field vanishes. One must keep in mind here that the velocity field $\vec{v}\,'$ is the velocity field in the laboratory frame expressed in terms of the coordinates in the rotating frame. This is due to the fact that in the transformation to the rotating frame, the new canonical momentum $\vec{p}\,'$ is equal to the old one $\vec{p} = m\vec{v}$ so that the operator $(\hbar/i)\nabla'$ associated with $\vec{p}\,'$ describes the physical quantity $m\vec{v}$ [Landau and Lifshitz (1982)].

We now come back to Eq. (24.53) and consider the case of an isotropic trap ($\omega_x^2 = \omega_y^2$, or equivalently $\epsilon = 0$). The last term of this equation then vanishes and its solutions are:

$$\begin{aligned} &\alpha = 0, \\ &\alpha = \pm\sqrt{2\Omega^2 - \omega_x^2} \quad \text{if} \;\; \Omega \geq \omega_x/\sqrt{2}. \end{aligned} \tag{24.56}$$

These are represented in Fig. 24.9. As long as Ω remains smaller than $\omega_x/\sqrt{2} = \omega_y/\sqrt{2} = \omega_\perp/\sqrt{2}$, there is a single solution, $\alpha = 0$. The condensate remains isotropic and without any velocity field. Then, when Ω increases, a bifurcation occurs, corresponding to a symmetry breaking, toward two new states where the condensate becomes elliptical with a non-zero velocity field. This bifurcation at $\Omega = \omega_\perp/\sqrt{2}$ corresponds to a resonance with the $\ell = 2$ quadrupole surface mode in an axisymmetric trap [15] and is clearly connected with the Landau criterion: When Ω is larger than $\varepsilon(\ell = 2)/(\ell = 2) = \omega_\perp\sqrt{2}/2 = \omega_\perp/\sqrt{2}$, the creation of an elementary excitation $\ell = 2$ becomes energetically favorable and the condensate can gain angular momentum from the rotating trap.

For a non-symmetric trap ($\epsilon \neq 0$), the various branches (left part of Fig. 24.9) are deformed and transform into those represented in the right part of Fig. 24.9 which can be used for interpreting quantitatively the experiment described in the next subsection.

A stability analysis in the plane of parameters Ω-ϵ has been performed in [Sinha and Castin (2001)]: the lower part of branch II turns out to be stable while branch I and the upper part of branch II become unstable for Ω large enough ($\Omega \geq 0.778\omega_\perp$ for $\epsilon = 0.022$).

[15]See Sec. 22.9 of Chap. 22.

24.5.4.2 *The experiment*

In practice, the condensate is created in a static harmonic potential and then the rotating perturbation is switched on. The final state of the condensate results from its evolution in a time-dependent rotating potential characterized by the two parameters $\epsilon(t)$ and $\Omega(t)$. Let us introduce the convenient dimensionless quantities $\tilde{\alpha} = \alpha/\bar{\omega}$, $\tilde{\Omega} = \Omega/\bar{\omega}$ with $\bar{\omega} = \sqrt{(\omega_X^2 + \omega_Y^2)/2}$ and $\omega_{X,Y} = \omega_\perp(1+\epsilon_{X,Y})$. If the time dependencies are slow enough, the condensate at every instant can be described by a stationary state corresponding to the instantaneous values of ϵ and $\tilde{\Omega}$.

In a first set of experiments, the ellipticity was fixed at a value $\epsilon \sim 0.025$ while the rotating frequency, $\tilde{\Omega}$, was either ramped from a zero value frequency to a final value in the range (0.5, 1.5) or initially fixed at a large value (> 1.5) and decreased to zero [Madison *et al.* (2001)]. For each final frequency studied, the parameter α and the angular momentum per particle were measured. The results are summarized in Fig. 24.10.

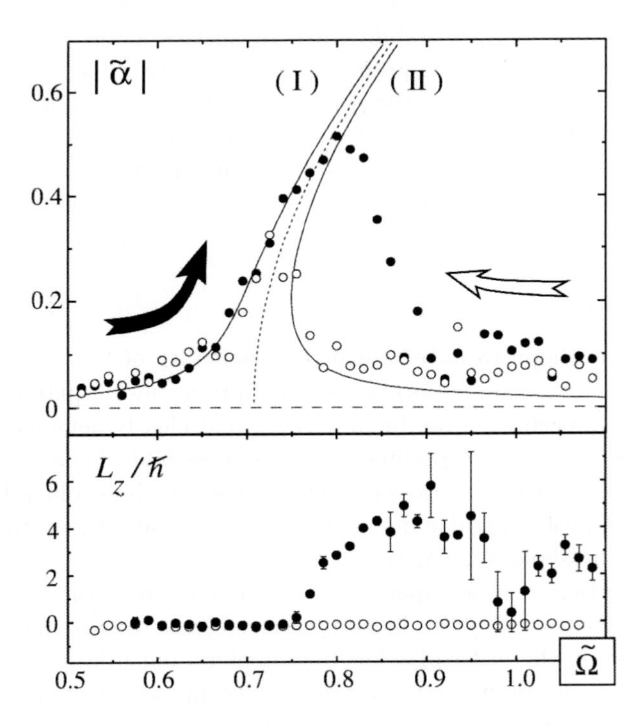

Fig. 24.10 Measurement of the dimensionless parameter $\tilde{\alpha}$ and of the angular momentum per particle as a function of the normalized angular frequency $\tilde{\Omega}$ of the rotating perturbation (see text) in two different situations: in the ascending ramp (branch I, black arrow, black dots) $\tilde{\Omega}$ increases; in the descending ramp (branch II, white arrow, open circles), $\tilde{\Omega}$ decreases. This hysteretic behavior where vortices are nucleated only for the ascending ramp stresses the importance of dynamical instability in the formation of vortices. Figure extracted from [Madison *et al.* (2001)]. Copyright: American Physical Society.

For the descending ramp, the condensate follows the lower part of the stationary solutions of branch II as long as such a solution of Eq. (24.53) exists, then switches to branch I and sticks to it down to $\tilde{\Omega} = 0$. The fact that no vortices are nucleated along this path can be considered as evidence that the lower part of branch II is stable. The situation is radically different for the ascending ramp. Indeed the condensate is observed to start on and to follow branch I until a frequency of $\tilde{\Omega} \simeq 0.75$ is reached. Beyond this point, the measurement of the angular momentum reveals the presence of vortices. The vortex nucleation appears to occur via a dynamic instability that occurs when the rotation resonantly excites surface modes. When these dynamical instabilities occur, the state of the condensate no longer corresponds to the irrotational solution (24.55). New kinds of solutions appear that can accomodate vortices. This resonant nucleation process explains why the frequency range over which vortices are generated is notably smaller than that expected from thermodynamics.

The leading contribution to the rotating potential is the quadratic term included in Eq. (24.46) that couples to quadrupole modes as already explained. The possibility for exciting dynamical instabilities about surface mode frequencies depends on the symmetries of the stirring potential. For instance, using a single beam (i.e. suppressing the toggling motion), the stirring potential contains in addition a linear and a cubic term. The linear term affects only the center of mass motion but the cubic term allows for the resonant excitation of the hexapole mode at a frequency $\omega_\perp/\sqrt{3}$ accompanied with nucleation of vortices as experimentally demonstrated in [Chevy *et al.* (2001)].

24.6 Conclusion

In this chapter, we hope to have shown the originality of the atomic physics approach toward investigating important physical properties of quantum fluids, such as superfluidity. Experiments can be performed that clearly demonstrate the physical mechanisms underlying superfluidity: the impossibility for a probe particle to undergo collisions with the condensate below a certain threshold velocity and the impossibility to transfer angular momentum from a rotating trap to a condensate below a certain threshold angular velocity.

Quantized vortices can be experimentally observed and studied in great detail in gaseous condensates. They can be easily imaged since the core has, after ballistic expansion, a size on the order of a few micrometers, much larger than the size of the vortices which can be observed in other superfluid systems, such as superfluid helium or superconductors. The mechanism of nucleation can also be identified by progressively changing the angular velocity of the trap containing the condensate and measuring the number of vortices which appear and the total angular momentum transferred to the condensate. The importance of resonances with the surface modes of the condensate for triggering dynamical instabilities and giving rise to the generation of vortices was clearly experimentally demonstrated.

Quantized vortices play an important role in the physics of gaseous condensates because their presence is a clear signature of the superfluid character of the condensate. The superfluid velocity is the gradient of the phase of a wave function, and such a wave function only exists if the system contains a condensed part. The superfluid velocity field only depends on the condensed part, whereas the density profile depends on both the condensed and non-condensed parts.

The experimental proof of the superfluid nature of a fluid usually relies on both techniques, the absence of dissipation for a sufficiently slow probe and the observation of quantized vortices. For instance, the superfluidity of ultracold pairs of fermionic atoms in different interacting regimes has been investigated in both ways [Miller *et al.* (2007); Zwierlein *et al.* (2005)]. We will come back to this problem in Chap. 26.

The analysis of superfluidity presented in this chapter requires going beyond the mean field description given in Chap. 22. It is based on the Bogolubov approach which provides several interesting results, such as the quantum depletion and the energies of the quantized elementary excitations. This approach is, however, limited to weakly interacting or dilute systems, in which $na^3 \ll 1$. The possibility of considerably increasing the scattering length a by tuning Feshbach resonances has greatly stimulated the search for new theories that go beyond the Bogolubov approach. This research field is presently expanding very rapidly and its description is beyond the scope of this book. In Chap. 26, we will just give a few examples of recent investigations showing how quantum gases can contribute to a better understanding of complex and strongly interacting many-body systems.

Originally, the relation between superfluidity and Bose-Einstein condensation was extensively discussed in the literature devoted to liquid helium-4 properties. In this chapter, we have presented a simple introduction restricted to weakly interacting Bose gases at zero temperature. Other fundamental problems related to superfluidity are outside the scope of this book. For example, the physics of superfluidity turns out to be much more involved at finite temperature, or/and in low-dimensional systems or/and in liquids. More insight into superfluidity can be also obtained by studying the sensitivity of the many-body wave function to changes in the boundary conditions (see for example [Bloch *et al.* (2008); Leggett (2006)]).

PART 8
Frontiers of atomic physics

Introduction

The spectacular developments that have occurred in atomic physics during the last few decades are at the origin of an increasing dialogue between this discipline and other fields of physics. Atomic physics can bring new insights into difficult problems found elsewhere, provide new results that not only confirm, but also complete those obtained in other experiments that deal with completely different energy domains. The purpose of the last part of this book is to present a few examples of these new frontiers of atomic physics in which fruitful exchanges of ideas and explorations of new approaches are flourishing.

Tests of fundamental theories

The precision of atomic physics measurements is ever increasing due to the development of new ultra-stable laser sources, new techniques such as laser cooling and frequency combs. This opens the possibility of performing new stringent tests of fundamental theories, and determining fundamental constants with improved accuracy. A few examples have already been given in this book and are listed below with references to the chapters in which they are described.

- Tests of the Einstein gravitational redshift with space clocks (Chap. 18)
- Tests of quantum electrodynamics (measurements of the g-2 electron spin anomaly and the Lamb shift of simple atoms,[16] see Chap. 8)
- Tests of Bell's inequalities (Chap. 19)
- Searching for a variation of fundamental constants (Chap. 18)
- Measurement of the fine structure constant and the Rydberg constant (Chap. 18)

In the present part, we have chosen to describe in more detail the atomic physics experiments that test fundamental symmetries. In Chap. 25, we show how these experiments have demonstrated the existence of a parity violation in atoms and how the results obtained in this way on the parameters of the standard model complement the information deduced from high energy experiments. A few indications are also given on the experiments testing time reversal symmetry by looking for electric dipole moments.

Strongly interacting many-body systems

Strongly interacting systems of quantum particles appear in many fields of physics: in neutron stars, in the quark-gluon plasma of the early universe, in atomic nuclei,

[16] In a recent experiment performed by an international collaboration at the Paul Scherrer Institute, the Lamb shift of muonic hydrogen was measured by laser spectroscopy [Pohl *et al.* (2010)]. Because of its larger mass, the muon orbit is much smaller than that of an electron, and the Lamb shift is more sensitive to the size of the proton. The value obtained in this experiment differs by 4% (and 5.0 standard deviations) with other determination deduced from hydrogen spectroscopy or electron scattering. This discrepancy is not yet understood.

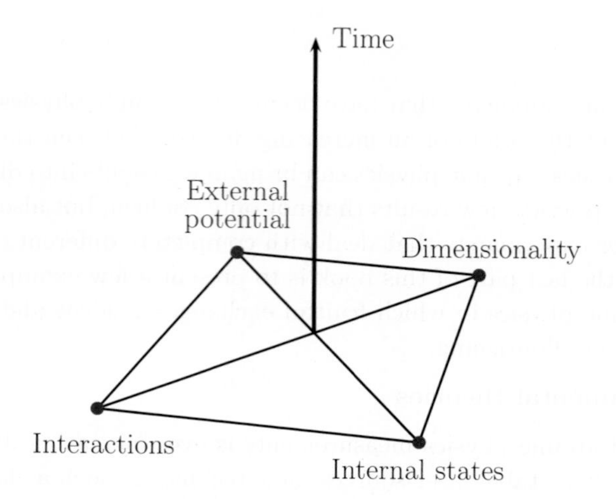

Fig. 24.11 Various parameters that can be varied to reach the strong interaction regime.

and in solid state physics. Understanding their properties is the central problem of many-body physics. In Chap. 26, we describe a few recent experiments showing how quantum gases can provide new interesting guidelines for understanding the physics of strongly interacting systems.

The advantage of quantum gases for investigating these problems is that they offer different routes for reaching the regime of strong correlations (see Fig. 24.11):

- Atoms can be confined in the wells of an optical lattice.[17] The gas may then be described in a certain range of parameters by the Hubbard Hamiltonian, which was initially introduced in solid state physics to study the transition between an insulator and a metal. When the confinement is strong, on-site interaction energies can be larger than the kinetic energy associated with the tunneling between adjacent wells, yielding strong many-body correlations.
- The atom-atom interaction strength can be varied using Feshbach resonances. A particularly interesting case concerns mixtures of fermionic atoms in different spin states. In the $a > 0$ region of the Feshbach resonance, weakly bound molecular states of two fermionic atoms in different spin states can be formed in the vicinity of the resonance.[18] It turns out that the collisional lifetime of these molecules due to inelastic processes can be much longer than for bosonic atoms due to the inhibition of collisions resulting from the Pauli principle. Molecular condensates can then be obtained. If the magnetic field is tuned to the $a < 0$ region of the Feshbach resonance, effective attractive forces exist between atoms that can lead to the formation of weakly bound pairs of fermionic atoms, analogous to the Cooper pairs in the Bardeen-Cooper-Schrieffer (BCS) theory

[17] See 7.5.4 of Chap. 7 and 14.4.3 of Chap. 14.
[18] See Chap. 16.

of superconductivity. A superfluid phase of Cooper pairs can be produced. Varying the magnetic field through the Feshbach resonance also made it possible to achieve, for the first time, a physical realization of the BEC-BCS crossover, a long-standing model of condensed matter physics.
- A full control of internal states is possible. For example, in the investigation of the BEC-BCS crossover, one can vary the imbalance between the two spin state populations and determine for what value of the imbalance the pairing of fermionic atoms is inhibited.
- Atomic motion can be frozen along one or two directions using a tightly confining potential. The reduction of the number of nearest neighbors and the quenching of the kinetic energy along the confined directions enhance the many body correlations. Interesting quantum phase transitions, such as the Berezinskii-Kosterlitz-Thouless transition of 2D quantum gases, or the Tonks-Girardeau regime for 1D quantum gases have been observed.

Several of these parameters can be simultaneously varied. For example, one can tune the interactions for atoms trapped in an optical lattice, or apply an optical lattice to atoms whose motion is confined in two directions. Note finally that these parameters can be easily varied in time, thereby allowing the study of time dependent effects and non-equilibrium phenomena for strongly interacting atoms.

Extreme light

Progress in basic research is enabled by the development of advanced technologies, while reciprocally, the invention of new technologies very often results from a better understanding of basic physical processes. A very clear example of these fruitful interactions is provided by lasers. Lasers could not have been invented without knowledge of the properties of stimulated emission and of non-equilibrium situations in which populations are inverted. Since their first realization in 1960, the performance of lasers has been continuously improved, opening the way to completely new investigations in basic science which quite often suggest new ideas for developing new types of laser sources.

Obviously, we cannot review all the developments in laser technology in this book. In Chap. 27, we have chosen to focus on two examples showing how the interplay between basic and applied research has allowed the realization, in table-top experimental set-ups, of laser sources with unprecedented performances.

We describe the *chirped pulse amplification* (CPA) technique, an ingenious method for amplifying short laser pulses after stretching them to avoid damaging the nonlinear crystals of the amplifier, and then re-compressing them at the end. Laser intensities in the terawatt range can be easily obtained, which has allowed several groups to study the behavior of atoms in intense laser fields and to get a better understanding of several phenomena such as high order harmonic generation or above threshold ionization.[19] New applications of these high intensity

[19]See Chap. 10.

laser pulses are also investigated, such as the acceleration of particles in wake fields and relativistic nonlinear optics.

The ionization of atoms in sufficiently intense laser pulses is no longer a multi-photon process. It involves a tunneling of the electron through the Coulomb barrier created by the nucleus and lowered by the laser electric field. In Chap. 10, we have shown that, once ionized, the photoelectron still interacts with the laser field and can come back in the vicinity of the parent ion and re-collides with it. In the last chapter of this part, we describe an interesting application of these re-collision processes, the generation of ultrashort light pulses in the attosecond domain, that open the way to a wealth of new investigations of electron dynamics in atomic and molecular systems.

Chapter 25

Testing fundamental symmetries. Parity violation in atoms

25.1 Introduction

25.1.1 *Historical perspective*

Until the mid 1950's, it was strongly believed that parity was conserved in all physical interactions. More precisely, consider the mirror image of a physical process obeying the laws of physics. It was considered to be obvious that the mirror image of the physical process also obeyed the laws of physics. Quantum mechanically, the Hamiltonian describing an isolated system was considered to be invariant under space reflection meaning that its eigenstates had a well defined parity. Any transition between these states induced by an interaction invariant under space reflection can only connect states with the same parity. Parity is conserved.

The first anomaly indicating that this rule could be violated came from the observation in the early 1950's that the decay of K mesons can lead to two different final products with opposite parities. To explain this anomaly, Tsung Dao Lee and Chen Ning Yang [Lee and Yang (1956)] suggested two possibilities. The first was that the K meson is a parity doublet, i.e. a superposition of two states with opposite parities. The second was that parity is not conserved by the weak interactions that are responsible for the decay. They also suggested how it could be possible to experimentally check the second possibility. Such an experiment, demonstrating parity violation in weak interactions, was readily performed by Chien-Shiung Wu and her collaborators [Wu *et al* (1957)] who observed that, in the beta decay of polarized ^{60}Co nuclei, electrons are predominantly emitted along the direction opposite to the nuclear spin (see upper part of Fig. 25.1). A reflection through a mirror perpendicular to the nuclear spin does not change the orientation of the nuclear spin, but changes the direction along which the electrons are emitted. The electrons then seem to be emitted along the direction of the nuclear spin (see lower part of Fig. 25.1), a result that contradicts the experimental observation and thus shows that parity is violated. This result stimulated a lot of theoretical and experimental work and it became clear that weak interactions break the left-right symmetry of the physical space.

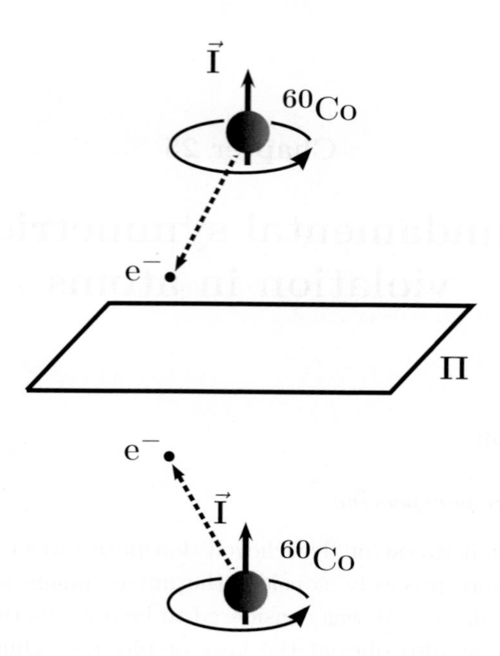

Fig. 25.1 Upper part: in the beta decay of a spin polarized ^{60}Co nucleus, the electrons e^- are preferentially emitted along the direction opposite to the nuclear spin. Lower part: in a mirror reflection through a mirror Π perpendicular to the nuclear spin, the spin direction does not change whereas the direction of the emitted electrons e^- is changed. The reflected case is not what is observed experimentally.

A better description of electromagnetic and weak interactions, the so-called *Electroweak theory*,[1] was developed at the end of the 1960's. This theory provides a unified treatment of these interactions in terms of exchanges of four bosons: the photon, already known for electromagnetic interactions; the charged $W^\pm$ bosons, already known for weak interactions responsible for beta decay, and a new neutral boson, the Z_0 boson. An important step in the development of this theory was the demonstration by Gerard 't Hooft of its renormalizability ['t Hooft (1971)].

Despite these theoretical achievements, the possibility that the neutral Z_0 boson could couple to electrons, neutrinos and nucleons, and thereby give rise to *weak neutral current interactions*, was far from being taken for granted in the early 1970's. All observed weak interaction processes involved *weakly charged current interactions* associated with exchanges of charged $W^\pm$ bosons that thus modified the charge of the interacting particles. No experimental evidence existed for the exchange of Z_0 bosons without a change in the charge of the interacting particles. The first observation of weak neutral current interactions in high-energy neutrino-electron and neutrino-nucleon scattering experiments was in fact realized at CERN in 1973 by the Gargamelle collaboration [Hasert *et al.* (1973)]. These experiments

[1] For a simple presentation of this theory, see for example ['t Hooft (1980)].

were difficult and only one event compatible with weak neutral current interactions in antineutrino-electron scattering was obtained. Other experiments at Fermilab gave negative results before confirming the Gargamelle results one year later. At the same time as the publication of the Gargamelle result, a paper was submitted [Bouchiat and Bouchiat (1974a)] that proposed a detection of a parity violation in atoms as a way to get evidence of the exchange of the Z_0 boson between an electron and the nucleus. It was very important at that time to show that the exchange did exist and that it gave rise to parity violation, since other theories, different from the standard model, predicted no parity violation. Experiments were started in 1973 to try to detect parity violation in atoms, and the first positive result was obtained in 1982 [Bouchiat *et al.* (1982)]. Meanwhile, high energy experiments were also undertaken at SLAC to look for parity violation effects in the scattering of polarized electrons by deuterons. These led to positive results in 1978 [Prescott *et al.* (1978, 1979)]. Finally, the Z_0 was directly observed at CERN in 1983 in high-energy proton-antiproton collisions [Arnison *et al.* (1983)].

Parity violation in electroweak processes is now well established. The purpose of this chapter is to describe the contribution of atomic physics to this research field and to show that, despite the weak energies involved in these experiments, they can bring very valuable information complementary to that obtained in high energy experiments.

25.1.2 *Atomic parity violation (APV)*

Parity violation is expected in atoms because a Z_0 boson can be exchanged between an atomic electron and the nucleus without changing the charge of the particles. Figure 25.2 shows the two possible exchanges of a bosonic particle between the nucleus and an electron: (a) the exchange of a photon described by an electromagnetic amplitude A_E; (b) the exchange of a Z_0 boson described by a weak amplitude A_W. In fact, the standard model predicts that there are two parts to A_W: one part, A_W^{even}, is invariant under mirror reflection and thus conserves parity; the other part, A_W^{odd}, transforms like a pseudoscalar, and is responsible for parity violation in atoms.

The difficulty of APV experiments is that weak interactions in atoms are much weaker than electromagnetic ones. Because of the very large mass of the Z_0 boson ($M_{Z_0} = 80$ GeV/c^2), the range of weak interactions, $\rho = \hbar/M_{Z_0}c$, is extremely short: about 10^{-7} times smaller than the Bohr orbit! By contrast, electromagnetic interactions, which are due to an exchange of zero mass photons, have an infinite range. As shown by Marie-Anne and Claude Bouchiat in their initial proposal [Bouchiat and Bouchiat (1974a,b, 1975)], enhancement mechanisms appear if the experiment is performed with heavy atoms.

Because of the very short range of weak interactions, only s-electrons, which have a non zero probability of being at the position of the nucleus, are expected

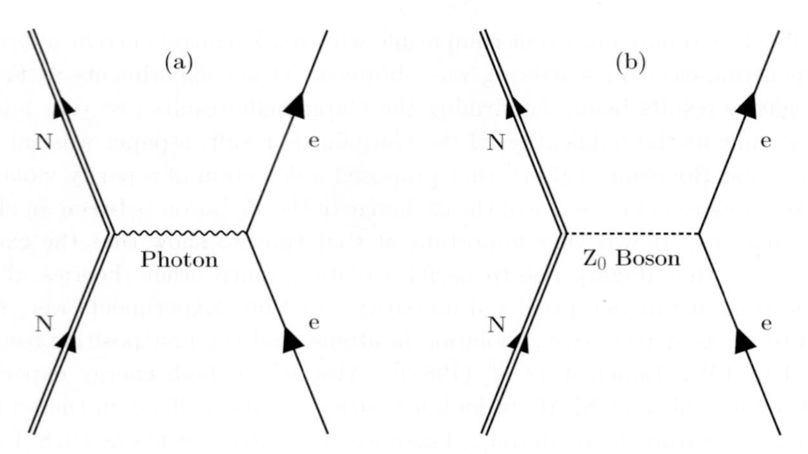

Fig. 25.2 (a) Electromagnetic interactions between an electron e and a nucleus N described by the exchange of a photon. (b) Weak interactions described by the exchange of a neutral boson Z_0.

to contribute to parity violation. One can then understand why it is interesting to look for APV in heavy atoms: Near the nucleus, the charge of the nucleus for a penetrating s-electron is not shielded by the other electrons, and the wave function is more concentrated. The velocity of the electron is also higher. Furthermore, the electron can exchange Z_0 bosons with a large number of nucleons. All these effects can be shown to increase parity violation effects by a factor slightly larger than Z^3, where Z is the atomic number.

Because of its smallness, A_W^{odd} is difficult to detect. The chance of detecting this amplitude can be increased, however, if one uses a highly forbidden transition for which the electromagnetic amplitude A_E is greatly reduced. In this way, the relative contribution of parity violating weak interactions, compared to electromagnetic ones, is increased. In the transition probability $P = |A_E + A_W|^2$, it is the interference term $A_E^* A_W^{\text{odd}} + \text{c.c.}$ between A_E and A_W^{odd} that provides the signature of parity violation. The left-right asymmetry A_{LR}, defined as the normalized difference between the transition probabilities P_L and P_R corresponding to two different left and right experimental configurations that are mirror images of one another, is given by:

$$A_{\text{LR}} = \frac{P_L - P_R}{P_L + P_R} \simeq 2\text{Re}(A_W^{\text{odd}}/A_E). \tag{25.1}$$

Another possibility is to use the interference between A_W^{even} and A_W^{odd}, though this interference term is much smaller than the one involving A_E since $A_W \ll A_E$. Note that the idea of detecting A_W^{odd} by its interference with A_E is similar to the one used for the heterodyne detection of a small signal.[2]

[2]In the beta decay experiment of Ms Wu, it is the interference between A_W^{even} and A_W^{odd} which is detected because $A_E = 0$. Both weak amplitudes are of the same order and the left-right asymmetry is nearly equal to 1. In atoms, A_{LR} rarely exceeds 10^{-6}. This is the origin of all

All the previous arguments clearly show that it is better to use highly forbidden transitions in heavy atoms to detect APV. On the other hand, atomic physics calculations needed for connecting the weak amplitude measured experimentally to the parameters of the electroweak theory are more difficult for heavy atoms. This is why Marie-Anne and Claude Bouchiat suggested studying a heavy alkali atom, like cesium, which has only one valence electron outside closed shells. Atomic calculations in this case are simpler and more reliable than those for heavy atoms with several valence electrons.

25.1.3 *Organization of this chapter*

We begin this chapter by describing the first experiment performed using cesium atoms by the group of Marie-Anne Bouchiat (Sec. 25.2). We focus on this experiment because it is the first one to demonstrate the existence of a parity violation in atoms [Bouchiat *et al.* (1982, 1984)]. This experiment can also be considered as following the tradition initiated by the work of Alfred Kastler and Jean Brossel with the development of optical methods.[3] The general idea is to excite a vapor with a polarized light beam and detect the fluorescence light emitted with a certain polarization. One tries to detect a chiral effect, i.e. an effect non-invariant under mirror reflection. In Sec. 25.3, we connect the measured signal to the parameters of the electroweak theory such as the weak charge of the nucleus. We just give a brief survey of the atomic calculations that have been performed for cesium as well as for other atoms, referring the interested reader to detailed references like [Ginges and Flambaum (2004)]. Section 25.4 gives a brief review of the results obtained up to now as well as indications on new experiments in progress. The significance of these results for the validity of the standard model of particle physics is analyzed in Sec. 25.5. Finally, in an appendix to this chapter (Sec. 25.6), we briefly review experiments that are trying to test time reversal symmetry by looking for permanent electric dipole moments of neutrons, atoms and molecules.

25.2 The first cesium experiment

25.2.1 *Principle of the experiment*

Figure 25.3(a) shows the two transitions that are used to excite the cesium atoms and detect the re-emitted fluorescence light. The excitation transition at 540 nm connects the two states $6s_{1/2}$ and $7s_{1/2}$ that have the same parity. This transition is a magnetic dipole transition with a very small M_1 amplitude (on the order of $4 \times 10^{-5}\mu_B/c$ where μ_B is the Bohr magneton) because of the different radial

experimental difficulties: high sensitivity is necessary along with rigorous control of systematic effects that may mimic the genuine parity violation effect when left and right configurations are interchanged.

[3] See Chap. 3.

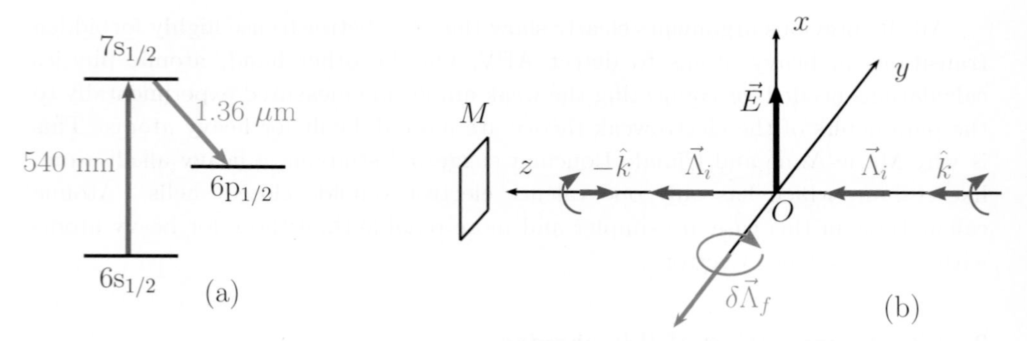

Fig. 25.3 (a) Transitions of the cesium atom used for excitation and detection. (b) Scheme of the experiment.

quantum numbers. In fact the non-zero value of M_1 is only due to higher order effects like core polarization with spin-orbit coupling. The detection uses the allowed electric dipole transition $7s_{1/2}$-$6p_{1/2}$ at 1.36 μm.

Because of the small value of M_1, it is difficult to observe the transition $6s_{1/2} \to 7s_{1/2}$ well above the background due to loosely bound cesium dimers. This is why a static electric field $\vec{E}$ is applied (see Fig. 25.3(b)) whose magnitude allows the strength of the electromagnetic amplitude to be controlled. The cesium vapor is excited by a circularly polarized laser beam at 540 nm with a wave vector $\vec{k}$ perpendicular to $\vec{E}$. The circular polarization is characterized by an helicity $\xi = \pm 1$ that gives the sign of the projection of the angular momentum $\vec{\Lambda}_i$ of the photon along $\hat{k} = \vec{k}/k$: $\vec{\Lambda}_i = \xi\hbar\vec{k}$.

Suppose that the incident beam $\hat{k}, \xi$ is reflected by a mirror M perpendicular to $\hat{k}$ (see Fig. 25.3(b)). This reflection changes the sign of $\hat{k}$, which is a polar vector, but $\vec{\Lambda}_i$, which is an axial vector, remains unchanged. The helicity ξ is a pseudo scalar whose sign changes during the reflection. One gets a new laser beam characterized by $-\hat{k}$ and $-\xi$, while $\vec{\Lambda}_i$ remains unchanged. The new laser configuration formed by the two laser beams thus remains unchanged in a reflection through the plane xOy perpendicular to $\hat{k}$. If parity is conserved, one should expect the properties of the fluorescence light emitted at 1.36 μm to remain unchanged in this reflection. This is not the case. We will see below that the fluorescence light emitted along a direction perpendicular to $\hat{k}$ and $\vec{E}$ is circularly polarized, demonstrating the existence of an angular momentum $\delta\vec{\Lambda}_f$ of the emitted photon along this direction (see Fig. 25.3(b)). Since the sign of $\delta\vec{\Lambda}_f$ changes in a reflection through the plane xOy containing $\delta\vec{\Lambda}_f$, this circularly polarized fluorescence is a chiral effect that gives evidence of an atomic parity violation.

25.2.2 *Transition dipole moment*

The transition dipole moment between the two states $6s_{1/2}$ and $7s_{1/2}$ is the sum of three distinct contributions:

- The magnetic dipole moment $\vec{M}$ (responsible for the magnetic dipole transition M_1).
- The Stark induced dipole moment corresponding to the matrix elements of the electric dipole moment $\vec{D}$ between the states $6s_{1/2}$ and $7s_{1/2}$ contaminated by states of opposite parities via the Stark coupling with the applied electric field $\vec{E}$.
- The parity violating dipole moment corresponding to the matrix elements of the electric dipole moment $\vec{D}$ between the states $6s_{1/2}$ and $7s_{1/2}$ contaminated by states of opposite parities via the parity violating part of the atomic Hamiltonian.

Once the radial integrals of these various matrix elements are performed, one is left with the matrix elements of an *effective dipole operator* $\vec{D}_{\rm eff}$ acting only on the spin quantum numbers $M_S = \pm 1/2$ of the states $6s_{1/2}$ and $7s_{1/2}$. The interaction of the atom with a laser beam with wave vector $\vec{k}$ and polarization $\vec{\varepsilon}$ is described (for the spin variables) by the Hamiltonian

$$H_{\rm int} = -\vec{\varepsilon} \cdot \vec{D}_{\rm eff}. \tag{25.2}$$

Here, we will not calculate $\vec{D}_{\rm eff}$, but instead just give its structure and explain how it can be understood with simple symmetry arguments. One finds:

$$\vec{D}_{\rm eff} = \vec{D}_{\rm mag} + \vec{D}_{\rm el}^{\rm Stark} + \vec{D}_{\rm el}^{\rm pv}, \tag{25.3}$$

where

$$\begin{aligned} \vec{D}_{\rm mag} &= M_1 \vec{\sigma} \times \hat{k}, \\ \vec{D}_{\rm el}^{\rm Stark} &= -\alpha E \mathbb{1}_{\rm op} - i\beta \vec{\sigma} \times \vec{E}, \\ \vec{D}_{\rm el}^{\rm pv} &= -i E_1^{\rm pv} \vec{\sigma}. \end{aligned} \tag{25.4}$$

In these equations, the vector $\vec{\sigma}$ is related to the electron spin operator $\vec{S}$ by $\vec{S} = (\hbar/2)\vec{\sigma}$ (the three matrices $\sigma_x, \sigma_y, \sigma_z$ are the Pauli matrices). The quantities M_1, α, β and $E_1^{\rm pv}$ are chosen to be real, so that the factors i that appear in the equations describe $\pi/2$ phase shifts between the oscillations of the various components of the transition dipole.[4] The first two terms of (25.3), $\vec{D}_{\rm mag}$ and $\vec{D}_{\rm el}^{\rm Stark}$, are associated with purely electromagnetic interactions. The corresponding terms in $H_{\rm int}$ given in (25.2) must be invariant under space reflection as well as time reversal (since electromagnetic interactions do not change in a time reversal). Since the polarization $\vec{\varepsilon}$ of the laser field is, as the laser electric field, a polar vector odd under space reflection and even under time reversal, one concludes that $\vec{D}_{\rm mag}$ and $\vec{D}_{\rm el}^{\rm Stark}$ must also be polar vectors, invariant under time reversal. The last term of (25.3), $\vec{D}_{\rm el}^{\rm pv}$, is due to the parity violating weak interactions, so that its contribution to $H_{\rm int}$ must be odd in a space reflection. However it must be invariant under time reversal since, in the electroweak theory, weak interactions obey this symmetry in a

[4] These factors i come, as will be shown later on, from time reversal symmetry considerations.

lowest order treatment. One concludes that $\vec{D}_{\rm el}^{\rm pv}$ must be an axial vector, invariant under time reversal.

We now explain the three terms of (25.4): The first contains the cross product $\vec{\sigma} \times \hat{k}$ because the magnetic dipole moment, proportional to $\vec{\sigma}$, interacts with the magnetic field of the laser wave parallel to $\hat{k} \times \vec{\varepsilon}$ and not to $\vec{\varepsilon}$, so that $\vec{\sigma} \cdot (\hat{k} \times \vec{\varepsilon})$ is proportional to $\vec{\varepsilon} \cdot (\vec{\sigma} \times \hat{k})$. Let Π and T be the operators describing space reflection and time reversal. We have:

$$\Pi \hat{k} \Pi^\dagger = -\hat{k}, \qquad \Pi \vec{\sigma} \Pi^\dagger = \vec{\sigma}, \qquad T \hat{k} T^\dagger = -\hat{k}, \qquad T \vec{\sigma} T^\dagger = -\vec{\sigma}. \tag{25.5}$$

From these relations, one easily confirms that $\vec{D}_{\rm mag}$ is a polar vector invariant under time reversal.

The second line of (25.4) contains two terms that describe the electric dipole moment induced by $\vec{E}$. The first comes from the scalar polarizability ($\mathbb{1}_{\rm op}$ is the unit operator in the spin space), and is a polar vector since $\vec{E}$ is reversed under space reflection. Because of the spin orbit coupling in the states that are coupled to $6s_{1/2}$ and $7s_{1/2}$ by the Stark Hamiltonian, there is also a vector polarizability[5] described by the second term of the second line of (25.4). Only two vectors appear in the calculation of the vector polarizability: $\vec{\sigma}$ and $\vec{E}$. The only vector that can be formed from these two vectors is their vector product, which explains the appearance of $\vec{\sigma} \times \vec{E}$ in the vector polarizability. The vector product of an axial vector ($\vec{\sigma}$) and a polar vector ($\vec{E}$) is a polar vector. The transformation of the two terms of the second line of (25.4) under time reversal remains to be studied. One must not forget, in this case, that T is an anti-unitary operator, so that:

$$T i T^\dagger = -i. \tag{25.6}$$

Using Eqs. (25.5), (25.6) and $T\vec{E}T^\dagger = \vec{E}$, one easily checks that the two terms of the second line of (25.4) are invariant under time reversal.

Finally, the last term of (25.4) that describes the effect of weak interactions is, like $\vec{\sigma}$, an axial vector contrary to the other terms and give rise to parity violation. However, it remains invariant under time reversal since the signs of both i and $\vec{\sigma}$ are changed under this transformation.

Static electric dipole moment

In the previous discussion, we considered only the off diagonal matrix elements of the electric dipole moment operator between two different manifolds of states $6s_{1/2}$ and $7s_{1/2}$. Can P-violating, T-conserving weak interactions give rise to a non-zero static electric dipole moment in each manifold? Inside a manifold like $6s_{1/2}$ with a well defined angular momentum, the effective dipole operator $\vec{D}_{\rm pv}$ associated with weak interactions is, according to the Wigner-Eckart theorem, necessarily proportional to the angular momentum $\vec{\sigma}$. Because of the time reversal symmetry of weak interactions, the same arguments as above show that one must have $\vec{D}_{\rm pv} = i\gamma\vec{\sigma}$, where γ is real (the change of sign of the factor i compensates the change of sign of $\vec{\sigma}$ in a time

[5] There is no tensor polarizability because a tensor of rank 2 has no matrix elements between two states of angular momentum 1/2.

reversal so that $\vec{D}_{\rm pv}$ remains invariant under this transformation). It follows that $\vec{D}_{\rm pv}$ is anti-hermitian and that its mean value in any state of the manifold $6s_{1/2}$ is purely imaginary. Consequently, it has no physical meaning for a physical quantity like a static electric dipole and the constant γ must be equal to zero. Time reversal symmetry thus forbids the existence of a static electric dipole moment in a well defined non-degenerate angular momentum state. The same argument does not hold for off diagonal elements of $\vec{D}_{\rm pv}$. The anti-hermitian character of $\vec{D}_{\rm pv}$ means that

$$\left\langle 6s_{1/2}M_S\right| \vec{D}_{\rm pv} \left|7s_{1/2}M_S'\right\rangle = -\left\langle 7s_{1/2}M_S'\right| \vec{D}_{\rm pv} \left|6s_{1/2}M_S\right\rangle^* .$$

which does not require that these matrix elements be equal to zero.

25.2.3 *Existence of a chiral signal in the re-emitted light*

From the expressions (25.3) and (25.4) of the transition dipole moment we now calculate the atomic density matrix in the state $7s_{1/2}$ after the absorption of a photon and show how a chiral signal proportional to $E_1^{\rm pv}$ appears in the fluorescence light emitted from $7s_{1/2}$.

After the absorption of a photon $\hat{k}, \vec{\varepsilon}$, an atom in the state $|\psi\rangle$ of $6s_{1/2}$ is brought to the state $\vec{\varepsilon}\cdot\vec{D}_{\rm eff}\,|\psi\rangle$ of $7s_{1/2}$. Initially the atoms are not polarized in the lower state, so their density operator is equal to 1/2 times the unit operator $\mathbb{1}_{\rm op}$. One concludes that the density operator ρ_e in the excited state is given by

$$\frac{1}{2}(\vec{\varepsilon}\cdot\vec{D}_{\rm eff})\mathbb{1}_{\rm op}(\vec{\varepsilon}^{\,*}\cdot\vec{D}_{\rm eff}^{\dagger}). \tag{25.7}$$

Of the four terms appearing on the right-hand side of Eqs. (25.4), the largest is the scalar polarizability proportional to α. If we keep only this term in (25.7), we get a unit matrix in $7s_{1/2}$ equal to $\alpha^2E^2/2$ corresponding to a total population equal to α^2E^2.

The next largest terms result from interference between the term in α and one of the other three terms in β, M_1 and $E_1^{\rm pv}$. Let us calculate the $\alpha E_1^{\rm pv}$ interference term. The laser beam polarization is given by:

$$\vec{\varepsilon} = \frac{1}{\sqrt{1+\xi^2}}\left(\vec{\varepsilon}_x + i\,\xi\,\vec{\varepsilon}_y\right) \tag{25.8}$$

where ξ is the helicity. If $\xi = +1(-1)$, the laser light is right (left) circularly polarized. We can also take $\xi = 0$, in which case the polarization is linear along the x-axis. Using

$$\vec{\varepsilon}\cdot\left(-\alpha\vec{E}\right) = -\frac{1}{\sqrt{1+\xi^2}}\alpha E \quad\text{and}\quad \vec{\varepsilon}\cdot\left(-i\,E_1^{\rm pv}\vec{\sigma}\right) = \frac{1}{\sqrt{1+\xi^2}}E_1^{\rm pv}\left(\xi\sigma_y - i\sigma_x\right), \tag{25.9}$$

we get for this interference term

$$\begin{aligned}[\vec{\varepsilon}\cdot(-\alpha\vec{E})][\vec{\varepsilon}^{\,*}\cdot(+i\,E_1^{\rm pv}\vec{\sigma})] + \text{h.c.} &= -2\alpha\frac{\xi}{1+\xi^2}EE_1^{\rm pv}\sigma_y \\ &= -2\alpha\frac{\xi}{1+\xi^2}EE_1^{\rm pv}\vec{\sigma}\cdot\left(\hat{k}\times\hat{E}\right). \end{aligned} \tag{25.10}$$

If the polarization is circular, $\xi = \pm 1$ and $1 + \xi^2 = 2$. Dividing the αE_1^{pv} interference term by the total population $\alpha^2 E^2$ of e, we get for the contribution of this interference term to the normalized density operator ρ_e in e:

$$\rho_e = -\frac{E_1^{\text{pv}}}{\alpha E}\xi\,\vec{\sigma}.\,\left(\hat{k}\times\hat{E}\right). \tag{25.11}$$

So far, we have considered only the beam $\hat{k}, \xi$ of Fig. 25.3(b). We must add the effect of the second beam $-\hat{k}, -\xi$ obtained by reflection of the first beam on the mirror M. Since, according to Eq. (25.11), ρ_e does not change when $\hat{k}$ and ξ are both reversed, the value of ρ_e remains the same in the presence of the two beams. From Eq. (25.11) one then deduces the spin polarization $\delta\vec{P} = \langle\vec{\sigma}\rangle = \text{Tr}(\rho_e\vec{\sigma})$ appearing in e. Using $\text{Tr}(\sigma_y^2) = 2$, one then gets:

$$\delta\vec{P} = -\frac{2E_1^{\text{pv}}}{\alpha E}\xi\,\left(\hat{k}\times\hat{E}\right). \tag{25.12}$$

This calculation thus predicts the appearance of a chiral polarization, proportional to E_1^{pv}, in the excited state e, that gives rise to a circularly polarized fluorescence light emitted in the direction perpendicular to $\hat{k}$ and $\hat{E}$.

Similar calculations can be done for the other interference terms in $\alpha\beta$ and αM_1, but here we just give the results. When the laser light is circularly polarized ($\xi = \pm 1$), the $\alpha\beta$ interference term gives rise to a polarization $\vec{P}$ parallel to $\hat{k}$:

$$\vec{P} = -\frac{2\beta}{\alpha}\xi\,\hat{k}. \tag{25.13}$$

When the polarization is linear and parallel to $\hat{E}$ and when a single beam $\hat{k}, \xi = 0$ is used, one finds that the αM_1 interference term gives rise to a polarization $\vec{P}^{(1)}$ perpendicular to $\hat{k}$ and $\hat{E}$ given by:

$$\vec{P}^{(1)} = -2\frac{M_1}{\alpha E}\left(\hat{k}\times\hat{E}\right). \tag{25.14}$$

Contrary to $\delta\vec{P}$, $\vec{P}^{(1)}$ does not describe a chiral effect. Here we use a single beam $(\vec{k}, \vec{\varepsilon}_x)$, so that in a reflection through the plane xOy the beam $(\vec{k}, \vec{\varepsilon}_x)$ is transformed into a beam $(-\vec{k}, \vec{\varepsilon}_x)$ and $\vec{P}^{(1)}$ is transformed into $-\vec{P}^{(1)}$. A rotation of angle π around Ox shows that the new configuration is isomorphic to the first.

It thus appears that the interference between the different contributions to the transition dipole moment gives rise to various types of spin polarization in the excited state when one photon is absorbed. These spin polarizations differ by their direction. The important point is that they behave differently when one of the three quantities, helicity ξ, direction $\hat{k}$ of the first laser beam, or direction $\hat{E}$ of the static electric field is reversed. This is very useful for discriminating the parity violating signal, proportional to E_1^{pv}, against those that conserve parity as well as for eliminating systematic effects that could mimic the parity violation effect.

25.2.4 *Calibration of the parity violation amplitude*

It is clear from the above expressions of $\delta\vec{P}$, $\vec{P}$ and $\vec{P}^{(1)}$ that the measured quantities are always related to amplitude ratios: $E_1^{\rm pv}/\alpha E$, β/α, $M_1/\alpha E$. To extract a precise value of $E_1^{\rm pv}$ from the measurements, to be compared to the predictions deduced from the electroweak theory, one needs to calibrate the parity violation amplitude $E_1^{\rm pv}$ with one of the other amplitudes α, β, or M_1, that can be known accurately.

In fact, as shown in [Bouchiat and Piketty (1988)] and [Bouchiat and Guéna (1988)], such a precisely known transition amplitude, $M_1^{\rm hf}$, does exist. It involves the contribution of the off-diagonal hyperfine interaction to the magnetic dipole transition amplitude M_1. To explain its origin we need to discuss the role of the hyperfine interaction between the electronic and the nuclear spins $\vec{S} = \hbar\vec{\sigma}/2$ and $\vec{I}$ in more detail. This hyperfine interaction gives rise to two hyperfine levels $F' = 4, 3$ in $7s_{1/2}$ and to two hyperfine levels $F = 4, 3$ in $6s_{1/2}$. The nuclear spin is not coupled to the laser electric field and plays no role in the Stark and parity violating electric dipole amplitudes. The expressions of $\vec{D}_{\rm el}^{\rm Stark}$ and $\vec{D}_{\rm el}^{\rm pv}$ given in Eq. (25.4) remain valid. Of course, the experimental signal is the fluorescence emitted from one of the two hyperfine levels $F' = 4, 3$ of $7s_{1/2}$. One must therefore extend the previous treatments of the $\alpha\beta$ and $\alpha E_1^{\rm pv}$ interference terms to calculate the density matrix in the hyperfine level F' of $7s_{1/2}$ after the resonant absorption of a photon from the hyperfine level F of $6s_{1/2}$. This can be easily done using the Wigner-Eckart theorem and simple properties of the projection operators $P_F = \sum_{m_F} |F, m_F\rangle \langle F, m_F|$ such as $P_F + P_{F' \neq F} = \mathbb{1}_{\rm op}$. The corresponding calculations can be found in [Bouchiat *et al.* (1985)].

The situation is different for the magnetic dipole amplitude M_1. As already mentioned above, in a single-particle non-relativistic model the radial parts of the wave functions of the two states $6s_{1/2}$ and $7s_{1/2}$ are orthogonal and the matrix elements of $\vec{S}$ between these two states are equal to zero. The non-zero value of M_1 is due to higher order effects. The hyperfine structure Hamiltonian $V_{\rm hf}$ gives rise to one of them. Its non-diagonal element between the two states, $\langle 7s_{1/2}| V_{\rm hf} |6s_{1/2}\rangle$, is at the origin of a contamination of $6s_{1/2}$ by $7s_{1/2}$ and vice versa, on the order of $\langle 7s_{1/2}| V_{\rm hf} |6s_{1/2}\rangle / (E_{7s_{1/2}} - E_{6s_{1/2}})$. This contamination allows $\vec{S}$ to have a non-zero matrix element, $M_1^{\rm hf}$, between the two contaminated states that depends only on $\langle 7s_{1/2}| V_{\rm hf} |6s_{1/2}\rangle$ and on $E_{7s_{1/2}} - E_{6s_{1/2}}$. The important point is that, in the single-particle non-relativistic model, it is easy to show that

$$\langle 7s_{1/2}| V_{\rm hf} |6s_{1/2}\rangle = \left[\langle 7s_{1/2}| V_{\rm hf} |7s_{1/2}\rangle \langle 6s_{1/2}| V_{\rm hf} |6s_{1/2}\rangle\right]^{1/2} \tag{25.15}$$

so that $M_1^{\rm hf}$ can be expressed entirely in terms of quantities that can be measured precisely: the hyperfine splittings in $6s_{1/2}$ and $7s_{1/2}$ and the energy difference between these two states.[6] Another important point is that one can show that $M_1^{\rm hf}$

[6] Corrections to $M_1^{\rm hf}$ beyond the single-particle non-relativistic model as well as the effect of the contamination of $7s_{1/2}$ by $d_{3/2}$ states are evaluated in [Bouchiat and Piketty (1988)]. The correction to $M_1^{\rm hf}$ is smaller than 0.25%, as confirmed by recent many-body relativistic calculations.

has no matrix elements between states of $6s_{1/2}$ and $7s_{1/2}$ with the same value of F ($\Delta F = 0$). It contributes only to the transitions $\Delta F = \pm 1$ with opposite signs and so it can be easily identified and measured.

Finally, by measuring the ratios of the signals proportional to $\delta\vec{P}$ and $\vec{P}$, one can determine E_1^{pv}/β. A precise measurement of M_1^{hf}/β, as explained in [Bouchiat and Guéna (1988)], then provides an absolute determination of E_1^{pv}: $E_1^{\text{pv}} = 0.8376 \times 10^{-11}$ atomic units $|e|a_0$, with a contribution to the theoretical uncertainty less than 0.25% corresponding to $2 \times 10^{-14}|e|a_0$.

25.3 Connection between the parity violation amplitude and the parameters of the electroweak theory

Once E_1^{pv} is measured, the next step is to deduce a value of the parameters of the electroweak theory from it and to compare with the predictions of the standard model and to the corresponding data obtained from high energy experiments. In this subsection, we give a brief outline of these calculations.

25.3.1 *Non-relativistic limit of the weak interaction Hamiltonian*

The largest term of the parity-violating electron-nucleus potential V_{pv} describes processes where the Z_0 boson couples to the electron as an axial vector and to the nucleons as a polar vector. Because of this vector coupling to the nucleons one can show that V_{pv} is proportional to the *weak charge* Q_W of the nucleus given by the sum of the weak charges of the quarks u and d forming the various nucleons. For the weak interactions of the quarks, the weak charges are the analogs of the electric charge for charged particles and they share the same additivity property. The proton is made of two u quarks and one d quark, while the neutron is made of one u quark and two d quarks. It follows that:

$$Q_W = ZQ_W(\text{proton}) + NQ_W(\text{neutron}) = (2Z + N)Q_W(u) + (Z + 2N)Q_W(d) \tag{25.16}$$

where Z and N are the number of protons and neutrons, respectively.

If one first neglects the other smaller terms of V_{pv}, which will be discussed later, one can show that in the non-relativistic limit, V_{pv} takes the following form:

$$V_{\text{pv}}(\vec{r}_e) = \frac{Q_W G_F}{4\sqrt{2}} \left[\delta^3(\vec{r}_e)\,\vec{\sigma}_e.\,\vec{v}_e/c + \text{h. c.}\right] \tag{25.17}$$

where $G_F = 4 \times 10^{-14}$ Rydberg $\times\, a_0^3$ is the Fermi constant and $\vec{r}_e$ and $\vec{v}_e$ give the position and the velocity of the electron, respectively. In a space reflection, $\vec{v}_e$ is changed into $-\vec{v}_e$, while $\vec{\sigma}_e$ does not change. It follows that $\vec{v}_e \cdot \vec{\sigma}_e$ is a pseudo scalar, as expected for a parity violating Hamiltonian. Under time reversal, both $\vec{v}_e$, and $\vec{\sigma}_e$ have their signs changed, which shows that V_{pv} is invariant under time reversal as expected.

25.3.2 *Calculation of the parity violation amplitude*

The parity violation amplitude, $-iE_1^{\rm pv}$, is related to the matrix element of the electric dipole moment operator D_z between the states $|\widetilde{6s_{1/2}}\rangle$ and $|\widetilde{7s_{1/2}}\rangle$ resulting from the contamination of $6s_{1/2}$ and $7s_{1/2}$ by states of opposite parities under the effect of $V_{\rm pv}$:

$$-iE_1^{\rm pv} = \langle\widetilde{7s_{1/2}}|D_z|\widetilde{6s_{1/2}}\rangle \tag{25.18}$$

To first order in $V_{\rm pv}$, one gets:

$$-iE_1^{\rm pv} = \sum_n \frac{\langle 7s_{1/2}|\, D_z \,|np_{1/2}\rangle\langle np_{1/2}|\, V_{\rm pv}\,|6s_{1/2}\rangle}{E(6s_{1/2}) - E(np_{1/2})} + \text{crossed terms} \tag{25.19}$$

Because of the very short range of $V_{\rm pv}$, we neglect in Eq. (25.19) the contribution of the states $np_{3/2}$ which can be shown to be less than 2×10^{-3} (see [Bouchiat and Piketty (1991)]), and thus sufficiently small to be ignored in first approximation. The calculation of the sum appearing in Eq. (25.19) requires knowledge of the wave functions and the energies of the states $6s_{1/2}$, $7s_{1/2}$ and $np_{1/2}$.

Several methods have been used to perform these atomic physics calculations. These can be classified into two broad categories. The first is semi-empirical and incorporates relevant measured quantities in a consistent way: energy of valence states, radiative lifetimes related to dipole matrix elements, hyperfine splittings (see for example [Bouchiat and Piketty (1986)]). The second starts from first principles and uses many-body relativistic calculations inspired by field theory. Spectacular progress has been achieved in these calculations. The interested reader can find an exhaustive presentation of these calculations in the review paper [Ginges and Flambaum (2004)]. The most recent and most precise calculation for cesium reaches an accuracy estimated to be 0.27% [Porsev *et al.* (2009)].

25.3.3 *Nuclear spin-dependent parity violating interactions. Anapole moment*

We now turn to the terms that have been neglected in the expression (25.4) of $D_{\rm el}^{\rm pv}$. Using symmetry arguments similar to those of Sec. 25.2.2, it is possible to show that the nuclear spin-dependent PV interactions induce two new contributions to $D_{\rm el}^{\rm pv}$ that are proportional to $i\vec{I}$ and $\vec{S} \times \vec{I}$, respectively. The corresponding PV interactions have three different origins. A first term describes processes where the Z_0 boson couples to the electron as a polar vector and to the nucleons as an axial vector. The weak charges of the quarks are replaced by their spins, which add like angular momenta, so that the net result does not increase with the number of nucleons like Q_W. A second term describes a modification of the nuclear spin-independent parity violation due to the contamination of the wave functions by the hyperfine Hamiltonian and can be considered to be a parity violating contribution to the hyperfine interaction. Finally there is a third term, the largest one, that describes the effect of weak interactions between the various quarks inside the nucleus. This is the sole term that involves charged as well as neutral currents. In the following, we focus on this third term.

The PV induced modification of the electromagnetic field created by the nucleus was first conjectured by Yakov Zel'dovich [Zel'dovich (1957)] who introduced the concept of the *nuclear anapole moment*. Claude Bouchiat and Claude-Annette Piketty [Bouchiat and Piketty (1991)] proposed a physical interpretation of this APV effect in terms of a chiral spin magnetization inside the nucleus. This magnetization is not homogeneous along the direction of the spin $\vec{I}$ but instead has a helicoidal structure due to a small transverse component $\vec{M}^{\rm pv}$, whose direction has a revolution symmetry around the spin axis and whose magnitude increases linearly with the distance from the axis (see Fig. 25.4). The anapole moment is the Ampere current associated with this chiral magnetization and is a pseudo vector that interacts with the electromagnetic current of the electron. The contribution of this PV Ampere current is shown to be larger than that of the toroidal distribution of currents often invoked to describe the anapole moment [Bouchiat (1999)]. The anapole moment increases with the nuclear radius, proportional to $A^{2/3}$.

The anapole moment can be determined experimentally from measurements performed on different hyperfine lines. For example, for cesium, it can be related to the following quantity,[7] which can be measured experimentally:

$$r_{\rm hf} = \frac{E_1^{\rm pv}(6s - 7s,\, \Delta F = -1)}{E_1^{\rm pv}(6s - 7s,\, \Delta F = +1)} - 1 \tag{25.20}$$

A non-zero value of $r_{\rm hf}$ indicates the existence of an anapole moment. The first evidence for this chiral property of a nucleus was obtained at JILA [Wood *et al.* (1997)]. There are still difficulties for quantitatively interpreting the magnitude of the observed effect.

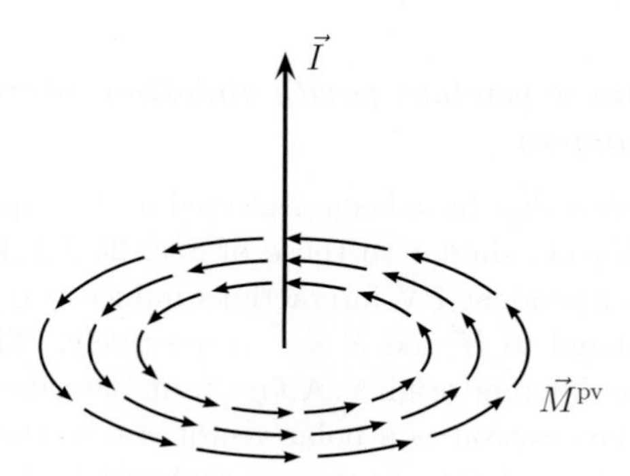

Fig. 25.4 Helicoidal distribution of nuclear magnetization corresponding to an anapole moment. A small rotating transverse magnetization with a revolution symmetry around the spin axis is added to the uniform magnetization parallel to the nuclear spin and represented by the vertical arrow.

[7] See [Ginges and Flambaum (2004)] and [Bouchiat and Piketty (1991)].

25.4 Survey of experimental results

25.4.1 *Cesium experiments*

The experiments performed by the group of Marie-Anne Bouchiat at ENS in Paris in 1982 and 1983 [Bouchiat *et al.* (1982, 1984)] were the first to be carried out in cesium, and the first to give a quantitative test of the electroweak theory in the electron-hadron sector. The principle of the experiment has been explained above. A detailed description of the experiment can be found in [Bouchiat *et al.* (1985, 1986a,b)]. The experimental accuracy was 12% and the theoretical uncertainty was at that time, less than 8%.

In the early 1980's, another cesium experiment was started at JILA by the group of Carl Wieman. The principle of the experiment was different from that of the previous one. The parity violating signal was to appear in the transition rate of single Zeeman component of the optical line rather than on the polarization induced in the excited state. In this case, one must therefore apply (in addition to the static electric field $\vec{E}$) a static magnetic field $\vec{B}$ in order to single out a single Zeeman component. In addition, one must also use an atomic beam perpendicular to the laser beam in order to eliminate, or at least reduce, the Doppler broadening of the optical line so that a single Zeeman component can be resolved. The first result, obtained in 1985 [Gilbert *et al.* (1985)], was in good agreement with the Paris result with a slightly better precision.

Improved versions of this experiment in which a polarized atomic beam was used in order to enhance the signal were later realized at JILA [Noecker *et al.* (1988); Wood *et al.* (1997)]. To provide a much higher laser intensity for the excitation of the $6s_{1/2} - 7s_{1/2}$ transition, the beam passed through a power buildup cavity before reaching a detection zone where the transition rate was measured for different orientations of $\vec{E}$, $\vec{B}$ and for different polarizations of the laser light. The gain in the signal was accompanied by a difficult side effect associated with asymmetric line shapes caused by large spatially inhomogeneous light shifts of the atomic sublevels that were induced by the high intensity laser standing wave in the buildup cavity. Several tests were needed to reduce the systematic errors. The authors of [Wood *et al.* (1997)] estimate their accuracy to 0.35%. They also present measurements performed on the two hyperfine components $\Delta F = \pm 1$ of the optical line which provides a value of the parameter $r_{\rm hf}$ given in (25.20) equal to $(4.8 \pm 0.7) \times 10^{-3}$. This was the first evidence obtained for the existence of an anapole moment.

The value of Q_W deduced from this experiment was found to differ from the standard model prediction by about 2.6 standard deviations [Bennett and Wieman (1999)]. This stimulated theorists to improve their calculations and to look for possible "new physics" beyond the standard model which could explain the deviation of the experimental results from the theoretical prediction. After several investigations described in [Ginges and Flambaum (2004); Guéna *et al.* (2005a)], the discrepancy was reduced to less than one σ. In fact, a very recent atomic physics calculation [Porsev *et al.* (2009)] has now reached an accuracy of 0.27%, estimated from the ability to reproduce precisely known test-parameters (hence better than the experimental accuracy) and gives results in perfect agreement with the Boulder experimental result of 1997.

Meanwhile, the Paris group has mounted a new cesium experiment with the aim of obtaining a much improved detection efficiency on a different observable quantity. It is actually very important to have several independent determinations of the weak charge using different methods. Instead of looking at the fluorescence light emitted on the transition $7s_{1/2} - 6p_{1/2}$, one may probe the spin alignment induced by the laser beam at 540 nm in one hyperfine substate $7s_{1/2}$ by a second laser beam at 1.36 μm resonant with the transition $7s_{1/2} - 6p_{1/2}$. The probe laser is amplified by stimulated emission and the change of its polarization is measured. The advantage of this method is that the left-right asymmetry itself is amplified during the propagation of the probe beam through the cell. A series of improvements described in [Guéna *et al.* (2003, 2005b)] have led to results with an accuracy of 2.6% , in agreement with the Boulder value. Proposals have been made to convert the set up from a longitudinal to a transverse field geometry which, thanks to a larger amplification factor, could allow a 0.1% precision to be reached [Guéna *et al.* (2005c)].

25.4.2 *Experiments using other atoms*

We first mention an experiment performed in Berkeley on thallium.[8] As in the cesium experiment, a highly forbidden M_1 transition is used, the $6p_{1/2} - 7p_{1/2}$ at 293 nm. The experimental accuracy is similar to that obtained with the first cesium experiment, but the atomic physics calculations are much less precise because of the complexity of the atom.

Most other experiments use allowed M_1 transitions of heavy atoms like bismuth and lead. The suppression factor of the electromagnetic transition probability is much smaller than for cesium (10^5 instead of 10^{14}), but one can get a better signal to noise ratio. As with the thallium experiments, the more complicated atomic structure does not allow as precise atomic calculations. Those experiments use the fact that the index of refraction of the vapor is not the same for a σ^+ and a σ^- polarized beam because of the handedness introduced by parity violation. One measures the rotation of the polarization of a linearly polarized beam when it propagates through the vapor. The measured angle of rotation is typically on the order of 10^{-8} to 10^{-6} rad. Experiments of this type have been performed in Oxford, Seattle, Novosibirsk and Moscow (see [Bouchiat and Bouchiat (1997)] and [Ginges and Flambaum (2004)] and references therein).

Very recently, interesting results have been obtained in Berkeley with ytterbium atoms [Tsigutkin *et al.* (2009)]. Parity violation amplitudes about 100 times larger

[8] See [Drell and Commins (1985)] and references therein.

than for cesium have been measured for the forbidden $6s^2\,{}^1S_0 - 5d6s\,{}^3D_1$ transition at 408 nm. The large value of the parity violation amplitude is due to the fact that the state $5d6s\,{}^3D_1$ is very close to the state $6s6p\,{}^1P_1$ which has an opposite parity. The contamination of $5d6s\,{}^3D_1$ by $6s6p\,{}^1P_1$ due to the parity violation Hamiltonian is thus enhanced, leading to a large value of E_1^{pv}. Atomic calculations are of course much more difficult for ytterbium than for cesium, but the interest of ytterbium is that it has a long chain of seven stable isotopes. One can show that the ratio of the PV amplitudes between different isotopes should provide the ratio of the corresponding weak charges, without invoking, to first approximation, any atomic physics calculations. The magnitude of the weak charge is largely dominated by the contribution of the neutrons. Therefore, very accurate measurements in ytterbium could give interesting information on the neutron distribution inside the nucleus.

25.5 Conclusion about the importance of APV experiments

In this last section, we would like to summarize a certain number of important features of APV experiments and to show how they can improve our understanding of the standard model.

(1) The parity violation amplitude E_1^{pv} measured in APV experiments provides a precise determination of the weak charge of the nucleus, Q_W. Equation (25.16) shows that this quantity is a certain linear combination of the weak charges $Q_W(u)$ and $Q_W(d)$ of the quarks u and d. It is interesting to compare this information to that deduced from high energy SLAC experiments [Prescott *et al.* (1978, 1979)]. At such high energies, the nucleus is broken into its fundamental constituents and the contributions of the various quarks add incoherently, contrary to what happens in atomic physics experiments. One can show that the quantity measured in high energy experiments is another linear combination of $Q_W(u)$ and $Q_W(d)$: $(2/3)Q_W(u) - (1/3)Q_W(d)$. Rather than $Q_W(u)$ and $Q_W(d)$, particle physicists use $C_u^1 = (-1/2)Q_W(u)$ and $C_d^1 = (-1/2)Q_W(d)$. Figure 25.5 shows the results deduced from the SLAC experiments and from the most precise cesium experiment which are represented in the (C_u^1, C_d^1) plane by two bands whose width is determined by the experimental precision. It is apparent that the two bands are nearly orthogonal, which means that the APV results *complement*, and not just confirm, those of SLAC. Furthermore, they are more precise. The graduated straight line of the figure gives the predictions of the standard model for $\sin^2\theta$, where θ is the Weinberg angle. Its intersection with the two bands gives a value of θ that is in close agreement with the world average. More recently, a global analysis of PV measurements in polarized electron scattering on nuclear targets has improved the accuracy of the data coming from colliders and further illustrates the complementarity with respect to atomic physics [Young *et al.* (2007)].

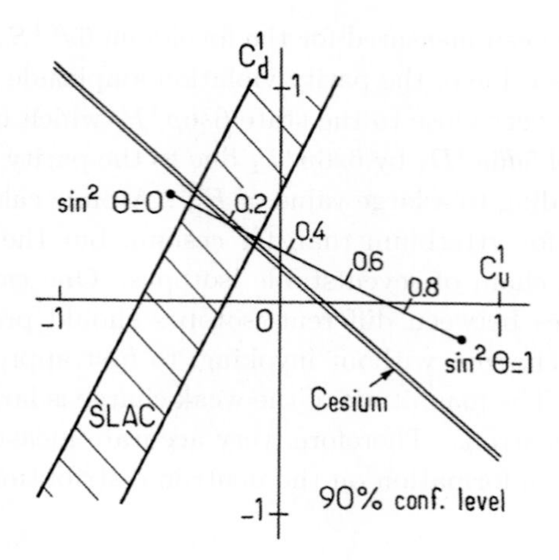

Fig. 25.5 Domains of the values of C_u^1 and C_d^1 deduced form the SLAC and from the APV experiments. The widths of the quasi orthogonal bands is related to the experimental precision. The graduated straight line gives the predictions of the standard model for $\sin^2\theta$ where θ is the Weinberg angle. Figure adapted from [Bouchiat and Bouchiat (1997)]. Copyright: Institute of Physics.

(2) APV and high energy experiments test the electroweak electron-hadron interaction for different ranges of the momentum transfer q: about 1 MeV/c for cesium experiments and about 100 GeV/c for high energy experiments at LEP (even several TeV/c in near future at LHC). This is another feature showing how both types of experiments complement each other. For instance, by combining their results, it is already possible to detect the 3% energy dependence of the electroweak force (so called "running" of the effective coupling strength) predicted by the standard model over the whole range explored experimentally [Porsev *et al.* (2009)].
(3) Increasing the accuracy of both APV experiments and atomic physics calculations could allow an exploration of new physics beyond the standard model. Already now, with the present experimental and theoretical states of the art, interesting constraints can be put on a variety of new physics scenarios beyond the standard model (see for example [Porsev *et al.* (2009)]).
(4) Despite the low energies that are involved, APV experiments have been able, for the first time, to provide new information about static chiral properties of the nucleus, like the anapole moment, that result from weak interactions taking place inside the nucleus. The results obtained by the Boulder group regarding the anapole moment of the cesium nucleus need to be confirmed since they are only in semi qualitative agreement with nuclear physics theoretical predictions. Measurements of the PV amplitude for a long chain of isotopes could also provide information on the distribution of neutrons inside the nucleus.
(5) The interpretation of APV experiments has stimulated the development of new

powerful methods for atomic physics calculations, that will find, undoubtedly, other domains of applications. They have also attracted a lot of attention towards highly forbidden transitions and on a wealth of interesting effects associated with the interference between different transition amplitudes. One can consider that these experiments provide another demonstration of the power of optical methods introduced in Chap. 3 of this book.

25.6 Appendix: Testing time reversal symmetry by looking for electric dipole moments

In this appendix, we briefly discuss other atomic physics experiments that test another fundamental symmetry, the time reversal symmetry T, by looking for a permanent electric dipole moment (EDM) in neutrons, atoms and molecules which would violate both P and T. We will just give the general motivation and the principle of these experiments, and refer the reader to more detailed presentations [Ginges and Flambaum (2004)]. A simple description of the experiments can also be found in [Fortson *et al.* (2003)].

We have already seen (at the end of Sec. 25.2.2) that, if P is violated but T holds, a static EDM cannot exist in a non-degenerate state. More generally, the observation of a permanent EDM of a neutron, atom or molecule would imply a T-violation. If CPT is a good symmetry,[9] which is the case for gauge theories, a T-violation would imply a CP-violation. Up to now, CP-violation has been clearly observed only in the decay of the K_0 meson [Christenson *et al.* (1964)] and T-violation has never been observed in lepton physics. The theoretical problems are not the same in lepton and hadron physics.

In this chapter, we have assumed that, in the standard model, time reversal is a good symmetry. But this assumption is only valid to the lowest order. CP-violation appears in higher-order processes, though the predicted violation is too small to generate EDM of fundamental particles that would be large enough to be seen in any present or planned experiment.[10] The matter-anti-matter asymmetry predicted from the standard model is also too small to explain the asymmetry observed in the universe. By contrast, there are several new theories, beyond the standard model, like supersymmetry, that predict EDM that could be detected with the present state of the art. This clearly shows the appeal of EDM experiments. Observation of a non-zero EDM of the electron would immediately imply new physics beyond the standard model. Even if it is not observed, the upper bound obtained on this EDM eliminates theoretical models predicting an EDM larger than this bound.

The principle of the experiments looking for an electric dipole moment $\vec{D}$ of a particle is to look for a shift $\pm\vec{D}\cdot\vec{E}/\hbar$ of its Larmor frequency depending on the

[9] C means charge conjugation, consisting of replacing each particle by its antiparticle.

[10] For example, the predicted EDM of the neutron and the proton are on the order of 10^{-34} e·cm. The predicted EDM of the electron is even smaller, on the order of 10^{-38} e·cm.

relative orientation of the two applied parallel or antiparallel static magnetic and electric fields, $\vec{B}$ and $\vec{E}$. The Ramsey method using two oscillatory fields that are separated either in space or time is used for reducing the width of the resonance. As in APV experiments, the main difficulty of these experiments is the minimization of the systematic effects that can mimic the EDM shift.

Neutron experiments started in the 1950's, though their accuracy has been considerably improved with the use of ultracold neutron facilities in St Petersburg and Grenoble. The upper bound on the neutron EDM now reaches a few 10^{-28} e·cm, where e is the electron charge (see for example the review paper [Fortson *et al.* (2003)]).

For experiments performed on atoms, it is important to distinguish two situations: atoms with closed electronic shells and atoms with unfilled electronic shells.

Atoms with closed electronic shells, like mercury or xenon, have no net electronic spin. But certain isotopes such as mercury-199 and xenon-129 have a nuclear spin $\vec{I}$ that can exhibit an EDM as a result of T-violating interactions between the quarks. The screening of the applied electric field $\vec{E}$ by the electronic cloud then seems to raise a problem. Consider a neutral system of point charge particles submitted to a static field $\vec{E}$. The local field at the position of each particle must vanish. Otherwise, the particle would be accelerated. This means that the various particles rearrange their configuration in such a way that the electrostatic Coulomb field "seen" by each particle compensates for the external applied field [Schiff (1963)]. In fact, this screening is not perfect for a nucleus because it has a finite size, so that the electric field due to the electrons is not uniform over the nuclear volume. One can show that the EDM of the nucleus can be coupled to the applied field $\vec{E}$. The reduction of the screening is higher for heavy nuclei,[11] which explains why experiments are performed on atoms such as mercury-199 and xenon-129.

Atoms with unfilled electronic shells, like thallium or cesium, have unpaired electrons and can be used to detect an EDM of the electron. Electrons are point particles and the reduction of the screening of the external field by the volume effect mentioned above no longer exists for the electron. For heavy atoms, however, the unpaired electron is relativistic and there are magnetic forces which not only suppress the screening effect, but increase the magnitude of the external electric field at the position of the electron. This enhancement effect, discovered by Pat Sandars [Sandars (1965, 1966)], increases with the atomic number Z. Its mechanism is completely different from the Z^3 enhancement of APV discussed above. The best present limit for the EDM of the electron, $D(e) = (7 \pm 8) \times 10^{-28}$e·cm corresponding to $D(e) < 1.6 \times 10^{-27}$e·cm, has been obtained on thallium [Regan *et al.* (2002)].

Another enhancement mechanism of the external electric field, also discovered by Pat Sandars [Sandars (1967)], appears in polar molecules. This mechanism can be understood by noting that, in the molecules aligned along the static electric field $\vec{E}$, a very large intramolecular electric field parallel to $\vec{E}$ becomes available. This large

[11] See for example [Ginges and Flambaum (2004)] and references therein.

field can be used to try to detect the EDM of nuclei in closed shell molecules such as TlF, or the EDM of the electron in a naturally polarized diatomic polar molecule such as YbF. Recent experiments of this type are in progress at the University of Sussex by the group of Ed Hinds [Hudson *et al.* (2002)].

It is clear that EDM experiments are quite promising. Their sensitivity is in constant progress. If an EDM is detected, either for the electron or for a nucleus, it will be a spectacular result with implications on our understanding of new physics beyond the standard model. Meanwhile, more and more stringent bounds are imposed on existing models, and this is also an important source of progress. To conclude this section, we would like to emphasize the crucial role of optical pumping methods, described in the first chapters of this book, for the realization of the EDM experiments. The precession of the electric dipole moment $\vec{D}$ in the applied electric field is very similar to the precession of the spin magnetic moment $\vec{M}$ in an applied magnetic field and all the optical methods developed for polarizing $\vec{M}$ and detecting its evolution can be directly extended to $\vec{D}$. A clear example of this connection between the two types of situations can be found in the experiment by the group of Larry Hunter looking for an EDM of the electron with cesium atoms [Murthy *et al.* (1989)]. In the experiment with mercury-199 and xenon-129 atoms, performed by the group of Noval Fortson ([Fortson *et al.* (2003)] and references therein), ^{199}Hg atoms are polarized by optical pumping while xenon-129 atoms are polarized by spin exchange with optically pumped rubidium atoms [Vold *et al.* (1984)].[12] Note also that in these EDM experiments the magnetic field is frequently measured *in-situ* by using optically pumped atoms as magnetometers. This clearly shows that the know-how acquired in one research field can stimulate the appearance and development of important advances several decades after.

[12] A recent result obtained by this group for mercury-199 lowers the limit of the EDM of the nucleus by a factor 7 and reaches an accuracy of 1.3×10^{-29} e·cm. [Griffith *et al.* (2009)].

Chapter 26

Quantum gases as simple systems for many-body physics

26.1 Introduction

As shown in Chap. 22, mean-field theories using the Gross-Pitaevskii equation can account for many of the relevant properties of dilute weakly interacting Bose-Einstein condensates such as density profile, stability domain, collective excitations, structure of vortices. Examples of effects beyond mean-field, such as the so-called quantum depletion, have been addressed in Chap. 24 with the Bogolubov theory, though this theory is valid only when the inequality $n|a|^3 \ll 1$ is fulfilled, where n is the spatial density and a the scattering length.

In the research field of ultracold quantum gases, increasing attention is now being focussed on strongly interacting systems where $n|a|^3$ is no longer small compared to 1 and mean-field approaches are no longer sufficient. The motivation of these studies is to try to use ultracold quantum gases as simple model systems, in which all experimental parameters can be fully controlled. They can then provide a better understanding of more complex situations that appear in many-body physics and be used as benchmarks to validate or eliminate theoretical models.

In this chapter, we describe a few important examples that illustrate these trends. Two of them deal with bosonic atoms, and two with fermionic atoms. From the various methods described in the introduction of this part for increasing atom-atom interactions, we will consider here the confinement in a series of potential wells and Feshbach resonances. We will also restrict ourselves to three-dimensional systems.

Strong collisional losses are observed with bosonic atoms near a Feshbach resonance. This explains why the method used for increasing the relative strength of interactions, compared to the kinetic energy, has been essentially the confinement of the bosonic gas in an optical lattice. We begin in Sec. 26.2 with the simple case of a Bose gas in a double well potential. This simple model can be considered to be the equivalent, for condensed matter physics, of the two-level system in atomic physics. It can be easily worked out with simple calculations and provides an introduction to the Josephson effect [Josephson (1974)] through an atomic physicist approach.

The simple calculations performed for the double well problem lead to conclusions that are useful for explaining, in a qualitative way, the so-called superfluid-Mott insulator transition observed with bosonic atoms in optical lattices that will be described in Sec. 26.3.

Fermionic atoms can be also confined in an optical lattice. In Sec. 26.4, we describe how it is possible to observe a fermionic Mott insulator transition with additional features resulting from the Pauli principle, which prevents two atoms in the same internal state from occupying the same potential well.

A very interesting feature of fermionic quantum gases is that collisional losses are very weak near Feshbach resonances for reasons which will be explained in Sec. 26.5 of this chapter. This opens a fascinating territory because it becomes possible to produce a quantum gas in which the scattering length a is very large, even larger that the inter-particle spacing $n^{-1/3}$, where n is the spatial density. Furthermore, two fermionic atoms in different spin states can form *pairs* that behave as composite bosons which, contrary to fermionic atoms, can form a Bose-Einstein condensate. The nature of these pairs changes, depending on the sign of the scattering length. In the $a > 0$ region of the Feshbach resonance, where the two-body system has a bound state, these pairs are Feshbach molecules, or dimers.[1] Qualitatively, in the region $a < 0$ of the Feshbach resonance where effective interactions are attractive, these pairs are analogous to the Cooper pairs that play a central role in the Bardeen-Cooper-Schrieffer (BCS) theory of superconductivity. In Sec. 26.5, we describe a few experimental results that demonstrate the condensation and the superfluidity of pairs of fermionic atoms along the whole crossover region between the BEC and the BCS regimes. This region is explored by scanning the amplitude of a magnetic field around the value corresponding to the Feshbach resonance. A few theoretical models, such as the BCS theory, valid when $k_F|a| \ll 1$, where k_F is the Fermi wave number, and the Leggett theory extending this theory to other values of $k_F a$, are also briefly recalled. Our motivation is to provide to atomic physicists, who are not necessarily familiar with BCS theory, a few guidelines for understanding the recent experiments performed with interacting Fermi gases.

An analytical description of the many-body wave function in the crossover region does not exist up to now, but quantum Monte-Carlo approaches are availaible. A thermodynamic description in terms of an equation of state is an interesting alternative, especially in the unitary limit ($a = \infty$) where the properties of the gas become "universal". A few recent results in this direction are briefly described.

Finally, after the conclusion, a few brief comments and references are given, concerning experiments with quantum gases that are not discussed in this chapter but which are interesting for many-body physics, such as the low dimensional quantum gases and the localization of matter waves in a disordered potential.

[1]See Chap. 16.

26.2 The double well problem for bosonic gases

26.2.1 *Introduction*

Consider the situation depicted in Fig. 26.1 of a symmetric double well potential that is obtained, for instance, by inserting a potential barrier in a harmonic trap.

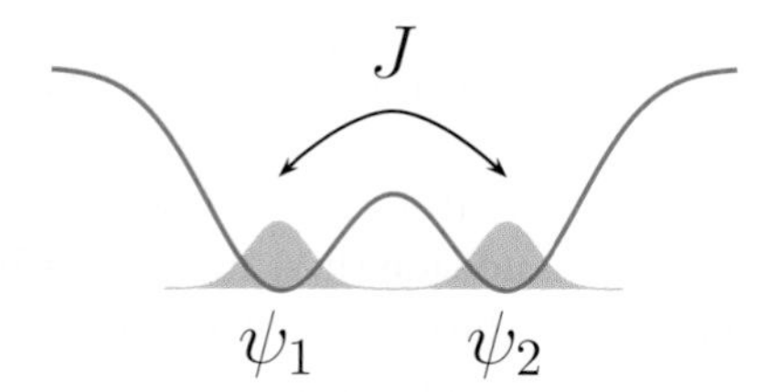

Fig. 26.1 Double well potential. The ground state wave functions in each well are denoted ψ_1 and ψ_2.

In this section,[2] we restrict ourselves to the two-state problem. Let $\psi_1(\vec{r})$ and $\psi_2(\vec{r})$ be the ground state wave functions in each well when the tunnelling and interactions are neglected (see Fig. 26.1).[3] In the absence of interactions, the tunnelling lifts the degeneracy between the two states, and the single particle ground state is the linear combination $\psi_+(\vec{r}) = (\psi_1(\vec{r}) + \psi_2(\vec{r}))/\sqrt{2}$. At sufficiently low temperature, an ideal bosonic gas condenses in this state. Consider an out-of-equilibrium situation in which the two wells contain different atom numbers. Such a situation can be obtained experimentally by smoothly changing the relative depths of the potential wells and then by suddenly re-establishing the symmetric situation. In the following, we study the interplay between tunnelling and interactions in the subsequent evolution of the system. We show how such an initial state gives rise to the existence of a relative phase θ between the two parts of the condensate trapped in each well and to a current of particles between the two wells that is proportional to $\sin\theta$. This effect can be considered to be a Josephson effect for ultracold atoms. We establish the Josephson equations which describe the coupled evolution of θ and the population imbalance $\overline{n} = \overline{n}_1 - \overline{n}_2$ between the two wells and show that these two variables are conjugate. We also answer the question of how this effect depends on the height of the potential barrier between the two wells and to what extent these oscillations are affected by the competition between interactions and tunnelling.

[2] We have chosen here to focus on bosonic gases in a double well to make the connection with Josephson effect in condensed matter. Fermionic gases in a double well are also interesting since they provide an enligthning introduction to magnetic properties of fermionic systems (see [Georges (2010)]).

[3] We assume the temperature of the bosonic gas to be sufficiently low so that all other excited states can be ignored.

26.2.2 The Hubbard Hamiltonian

Let $\hat{a}_1$ and $\hat{a}_2$ be the operators that destroy a particle in the ground states $|\psi_1\rangle$ and $|\psi_2\rangle$ of each well. These states have the same energy E_0 since the double well potential is symmetric (see Fig. 26.1). The Hubbard Hamiltonian of a Bose gas in the double well system can be written in the form:

$$H = E_0(\hat{n}_1 + \hat{n}_2) - \frac{\hbar J}{2}\left(\hat{a}_1^\dagger \hat{a}_2 + \hat{a}_2^\dagger \hat{a}_1\right) + \frac{\hbar U}{2}\left[\hat{n}_1\left(\hat{n}_1 - 1\right) + \hat{n}_2\left(\hat{n}_2 - 1\right)\right]. \quad (26.1)$$

where $\hat{n}_i = \hat{a}_i^\dagger \hat{a}_i$ are the number operators for each well i.

The second term of the right-hand side (first line) describes the tunnel effect: an atom is transferred from one well to the other, with a tunnel amplitude J taken to be positive. The last term describes the interactions between atoms in each well. The parameter U describes the strength of the interactions and is positive if interactions are repulsive and negative if interactions are attractive. The interactions between atoms in different wells can be neglected.

In order to describe the out-of-equilibrium state envisioned in the introduction of this section, it is convenient to use the Heisenberg representation for which the operators $\hat{a}_1$ and $\hat{a}_2$ are time-dependent and obey the following equations:

$$\begin{aligned} i\hbar\frac{\mathrm{d}\hat{a}_1}{\mathrm{d}t} &= \left[\hat{a}_1, \hat{H}\right] = E_0\hat{a}_1 - \frac{\hbar J}{2}\hat{a}_2 + \hbar U\hat{n}_1\hat{a}_1, \\ i\hbar\frac{\mathrm{d}\hat{a}_2}{\mathrm{d}t} &= \left[\hat{a}_2, \hat{H}\right] = E_0\hat{a}_2 - \frac{\hbar J}{2}\hat{a}_1 + \hbar U\hat{n}_2\hat{a}_2. \end{aligned} \quad (26.2)$$

26.2.3 The superfluid regime

In the absence of interactions ($U = 0$), the restriction of the Hamiltonian (26.1) to the single particle manifold has two eigenstates $|\psi_\pm\rangle = (|\psi_1\rangle \pm |\psi_2\rangle)/\sqrt{2}$ with eigenvalues $E_0 \mp \hbar(J/2)$. The ground state $|\psi_+\rangle$ is separated from the first excited state $|\psi_-\rangle$ by an energy $\hbar J$. If the double well potential accomodates N atoms and if the temperature is such that $k_B T \ll \hbar J$, the system will thus condense in the state $|\Psi\rangle = (2^N N!)^{-1/2}(\hat{a}_1^\dagger + \hat{a}_2^\dagger)^N|0\rangle$ corresponding to N atoms in the state $|\psi_+\rangle$. This state is a phase state with a relative phase $\theta = 0$, associated with a macroscopic wave function $\sqrt{N/2}\,[\psi_1(\vec{r}) + \psi_2(\vec{r})]$ with the same mean number of atoms in each well: $\overline{n}_1 = \overline{n}_2 = N/2$.

If we start at $t = 0$ with different mean numbers of atoms in each well, the state of the system at a later time t will correspond to a situation in which all atoms will be in the same linear combination of $|\psi_1\rangle$ and $|\psi_2\rangle$:

$$|\overline{n}_1, \overline{n}_2, \theta\rangle = \eta_1|\psi_1\rangle + \eta_2 e^{i\theta}|\psi_2\rangle, \quad (26.3)$$

where $\eta_1 = (\overline{n}_1/N)^{1/2}$ and $\eta_2 = (\overline{n}_2/N)^{1/2}$. The creation operator for such a state is $a^\dagger_{\overline{n}_1,\overline{n}_2,\theta} = \eta_1 a_1^\dagger + e^{i\theta}\eta_2 a_2^\dagger$. The phase state describing the condensate for which all atoms are in state (26.3) is

$$|N,\theta\rangle = \frac{(a^{\dagger}_{\overline{n}_1,\overline{n}_2,\theta})^N}{\sqrt{N!}}|0\rangle = \sum_{n_1=0,\,n_2=N-n_1}^{N} \sqrt{\frac{N!}{n_1!n_2!}}\eta_1^{n_1}\eta_2^{n_2}e^{in_2\theta}|n_1,n_2\rangle, \qquad (26.4)$$

i.e. a linear superposition of Fock states $|n_1,n_2\rangle$ with $n_1+n_2=N$. The problem is then to find the time evolution of the parameters $\overline{n}_1$, $\overline{n}_2$ and θ appearing in Eq. (26.3), which represent the mean number of atoms in each well and the relative phase between the two parts of the condensate respectively.

Description in terms of coherent states

Mathematically, calculations are not that easy when using the state (26.4). As explained in Sec. 23.4 of Chap. 23, the coherent states $|\alpha\rangle$ are much more convenient for calculations. The unphysical feature of these states is that they are linear superpositions of states with different values of the total number of atoms. This difficulty is solved by taking a statistical mixture of coherent states with the same modulus $|\alpha|$ and a phase uniformly distributed over the interval $[0,2\pi[$. This statistical mixture is identical to a statistical mixture of Fock states $|N\rangle$ with a Poisson distribution of N having a mean value equal to $|\alpha|^2$.

The coherent state associated with $a^{\dagger}_{\overline{n}_1,\overline{n}_2,\theta}$ is equal to $e^{-|\alpha|^2/2}e^{\alpha\hat{a}^{\dagger}_{\overline{n}_1,\overline{n}_2,\theta}}|0\rangle$. Since $\hat{a}^{\dagger}_1$ and $\hat{a}^{\dagger}_2$ commute, this state is also equal to a product of coherent states $|\alpha_1=\sqrt{\overline{n}_1}\rangle\otimes|\alpha_2=\sqrt{\overline{n}_2}e^{i\theta}\rangle$. We can thus describe the state of the condensate by a statistical mixture of products of two coherent states $|\alpha_1=\sqrt{\overline{n}_1}e^{i\theta_1}\rangle\otimes|\alpha_2=\sqrt{\overline{n}_2}e^{i\theta_2}\rangle$ with $\theta_2=\theta_1+\theta$. At the end of the calculation, we must take an average over θ_1 that is uniformly distributed on $[0,2\pi[$.[4]

Evolution equations of θ and $\overline{n}=\overline{n}_1-\overline{n}_2$

If we take the average value of Eqs. (26.2) in this product of coherent states, we obtain a new system of two equations where $\hat{a}_1$ and $\hat{a}_2$ are simply replaced by $\alpha_1=\sqrt{\overline{n}_1}e^{i\theta_1}$ and $\alpha_2=\sqrt{\overline{n}_2}e^{i\theta_2}$. Equating the real and imaginary parts of the two sides of each equation leads to a system of four equations:

$$\frac{\mathrm{d}\overline{n}_1}{\mathrm{d}t} = -J\sqrt{\overline{n}_1\overline{n}_2}\sin\theta, \qquad \frac{\mathrm{d}\theta_1}{\mathrm{d}t} = \frac{J}{2}\sqrt{\frac{\overline{n}_2}{\overline{n}_1}}\cos\theta - \frac{E_0}{\hbar} - U\overline{n}_1,$$

$$\frac{\mathrm{d}\overline{n}_2}{\mathrm{d}t} = +J\sqrt{\overline{n}_1\overline{n}_2}\sin\theta, \qquad \frac{\mathrm{d}\theta_2}{\mathrm{d}t} = \frac{J}{2}\sqrt{\frac{\overline{n}_1}{\overline{n}_2}}\cos\theta - \frac{E_0}{\hbar} - U\overline{n}_2. \qquad (26.5)$$

Taking the difference between the equations for $\overline{n}_1$ and $\overline{n}_2$ and for θ_1 and θ_2 finally leads to the equations of motion of $\overline{n}$ and θ:

$$\frac{\mathrm{d}\overline{n}}{\mathrm{d}t} = -J\sqrt{N^2-\overline{n}^2}\,\sin\theta, \qquad \text{and} \qquad \frac{\mathrm{d}\theta}{\mathrm{d}t} = \frac{J\overline{n}}{\sqrt{N^2-\overline{n}^2}}\cos\theta + U\overline{n}, \qquad (26.6)$$

[4] See Sec. 23.4.2 of Chap. 23.

where we have used $\overline{n}_1 = (N + \overline{n})/2$ and $\overline{n}_2 = (N - \overline{n})/2$. Note that the absolute phase θ_1 has disappeared and that only the relative phase $\theta = \theta_2 - \theta_1$ remains. These two equations can be derived from a classical Hamiltonian

$$H(\overline{n}, \theta) = -J\sqrt{N^2 - \overline{n}^2}\cos\theta + \frac{U\overline{n}^2}{2} \tag{26.7}$$

since it is easy to check that Eqs. (26.6) can be rewritten as a set of Hamiltonian equations:

$$\frac{\mathrm{d}\overline{n}}{\mathrm{d}t} = -\frac{\partial H}{\partial \theta}, \quad \text{and} \quad \frac{\mathrm{d}\theta}{\mathrm{d}t} = \frac{\partial H}{\partial \overline{n}}, \tag{26.8}$$

showing that $\overline{n}$ and θ are conjugate variables.

Josephson effect for ultracold atoms

(i) *Current of particles between the two wells*

The first equation (26.5) gives the rate of decrease of $\overline{n}_1$, which is also the rate of increase of $\overline{n}_2$, describing a current of particles I going from well 1 to well 2:

$$I = -\frac{\mathrm{d}\overline{n}_1}{\mathrm{d}t} = +\frac{\mathrm{d}\overline{n}_2}{\mathrm{d}t} = -J\sqrt{\overline{n}_1\overline{n}_2}\sin\theta. \tag{26.9}$$

This current is proportional to J and to $\sin\theta$. It is non-zero only if a phase difference θ, different from 0 and π, exists between the two parts of the condensate. The two values $\theta = 0$ and $\theta = \pi$ correspond to the two energy states $|\psi_+\rangle$ and $|\psi_-\rangle$ introduced above. These two states are stationary, which explains why they cannot accommodate a current.

(ii) *Josephson oscillations*

The equilibrium situation of the condensate corresponds to $\overline{n} = 0$ and $\theta = 0$. The dynamics of the system for small deviations from the equilibrium correspond to small values of $\overline{n}/N$ and θ. Expanding the Hamiltonian (26.7) to the lowest order in $\overline{n}/N$ and θ leads to an approximate quadratic Hamiltonian equal, within a constant term, to:

$$H = \frac{1}{2}\left(U + \frac{J}{N}\right)\overline{n}^2 + \frac{1}{2}JN\theta^2. \tag{26.10}$$

Since $\overline{n}$ and θ are conjugate variables, this Hamiltonian describes a one-dimensional harmonic oscillator whose angular frequency ω is equal to:[5]

$$\omega_J = \sqrt{JN\left(U + \frac{J}{N}\right)}. \tag{26.11}$$

[5]In condensed matter physics, the term *Josephson oscillations* describes the existence of an ac current flowing between two superconductors separated by an insulating barrier when a (fixed) dc bias voltage is applied to the junction. Strictly speaking, the corresponding situation in a BEC double well junction would correspond to a fixed chemical potential difference (i.e. population imbalance) and an ac superfluid current between the two wells. This is in fact what happens in the nonlinear self-trapping regime described below (due to the finite total number of atoms, the chemical potential difference actually undergoes small oscillations, but around a non-zero value). To keep the terminology that is the most widely used in the cold atom community, we shall call Josephson oscillations the oscillation with frequency ω_J, the other regime being called *nonlinear self-trapping*.

For a non-interacting gas ($U = 0$), or when $U \ll J/N$, ω_J is equal to J, which is the tunnel oscillation frequency in the single atom case. When U increases by positive values, the stiffness of the oscillator increases and ω_J increases. The repulsive interactions between atoms tend to increase the rate of variation of the population difference n more rapidly: the oscillation mode becomes collective. In the limit $U \gg J/N$, ω_J tends to $\sqrt{UJN}$, which is called the *plasma frequency* by analogy with the modification of the oscillation of pairs in the usual Josephson effect due to their Coulomb interaction. When U is negative (attractive gas), ω_J is smaller than J, and the previous treatment is no longer valid when $|U| > J/N$, regime for which all atoms collapse in one well.

The study of the behavior of the system for large deviations from equilibrium requires the use of the nonlinear equations (26.6). In fact, these equations are similar to those describing the motion of a non-rigid pendulum (with a variable length proportional to $\sqrt{N^2 - \overline{n}^2}$), with tilt angle θ, and angular momentum $\overline{n}$. The harmonic motion described by Eq. (26.11) corresponds to small deviations from the vertical equilibrium position. For large values of the initial population difference, the initial angular momentum is large enough to allow the pendulum to reach the top position. The angular momentum, which is also the population difference $\overline{n}$, then exhibits a periodic motion around a non-zero value of $\overline{n}$. One of the two wells is preferentially populated, an effect called *nonlinear self-trapping*. Josephson oscillations and self-trapping have been experimentally observed by the group of Markus Oberthaler in Heidelberg [Albiez *et al.* (2005)] (see Fig. 26.2). Other interesting experiments with rubidium-87 condensates, simulating the application of a constant voltage on two weakly coupled superconductors (a.c. Josephson effect) or a constant current (d.c. Josephson effect) have been performed by the group of Jeff Steinhauer [Levy *et al.* (2007)].

26.2.4 *The insulator regime*

So far, we have supposed that the system remains in a condensed state described by two condensates which are located in each well and have a well defined relative phase. The state of the system is described as a product of two coherent states$|\alpha_1 = \sqrt{\overline{n}_1}e^{i\theta_1}\rangle \otimes |\alpha_2 = \sqrt{\overline{n}_2}e^{i\theta_2}\rangle$ with θ_1 randomly distributed between 0 and 2π and $\theta_2 = \theta_1 + \theta$ with θ fixed. As shown in Chap. 23, the distribution of the number of atoms in each well follows a binomial law, well approximated by a Poisson law, with a variance of n_1 and n_2 proportional to $\bar{n}_1$ and $\bar{n}_2$, respectively.

We will now show that if interactions are repulsive ($U > 0$) and strong enough, the ground state of the system becomes a product of two Fock states containing equal numbers $N/2$ of atoms (we assume that the total number of atoms is even and equal to N): $|\Psi\rangle = |n_1 = N/2\rangle \otimes |n_2 = N/2\rangle$. In other words, the number of atoms in each well is no longer a random variable. To demonstrate this result, we come back to the Hamiltonian (26.1) and suppose first that $J = 0$. It is then

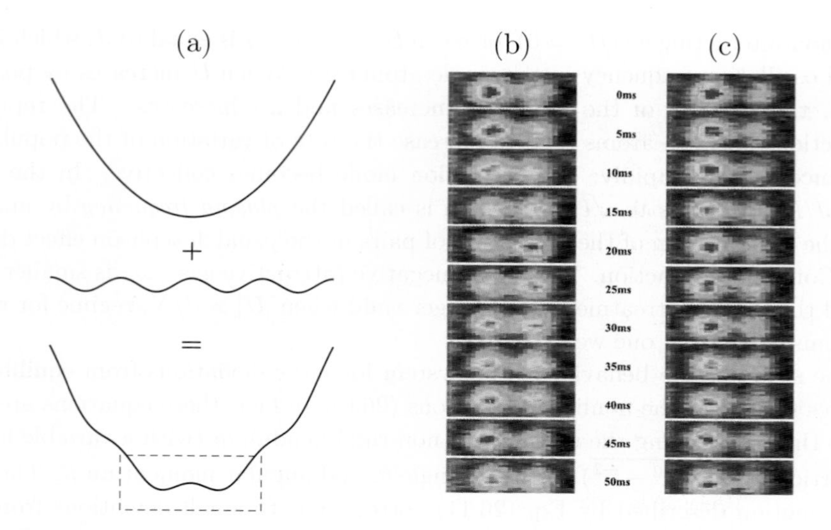

Fig. 26.2 Different regimes for a condensate in a double well potential. (a) Experimental realization of an optical double well potential by superimposing an optical lattice on a trap. By displacing the lattice with respect to the harmonic trapping potential, one can vary the asymmetry of the double well. (b) A small initial population difference n between the two wells; the populations n_1 and n_2 exhibit Josephson oscillations. (c) Self-trapping in one potential well observed for a large initial population difference. Figure adapted from [Albiez *et al.* (2005)]. Copyright: American Physical Society.

easy to find the eigenstates of $\hat{H}$: they are the states $|n_1\rangle \otimes |n_2\rangle$ with $n_1 + n_2 = N$. The corresponding eigenvalues are equal to $NE_0 + \hbar(U/2)(n_1^2 + n_2^2 - N)$. Because of the concavity of the parabolic law, the smallest value of $n_1^2 + n_2^2$ with $n_1 + n_2$ fixed and equal to N is obtained when n_1 and n_2 are equal. We conclude that the ground state of $\hat{H}$ is $|n_1 = N/2\rangle \otimes |n_2 = N/2\rangle$ (see Fig. 26.3). The first excited states form a two-dimensional manifold $\{|n_1 = N/2 + 1\rangle \otimes |n_2 = N/2 - 1\rangle, |n_1 = N/2 - 1\rangle \otimes |n_2 = N/2 + 1\rangle\}$, located at a distance $\hbar U$ above the ground state (see Fig. 26.3).

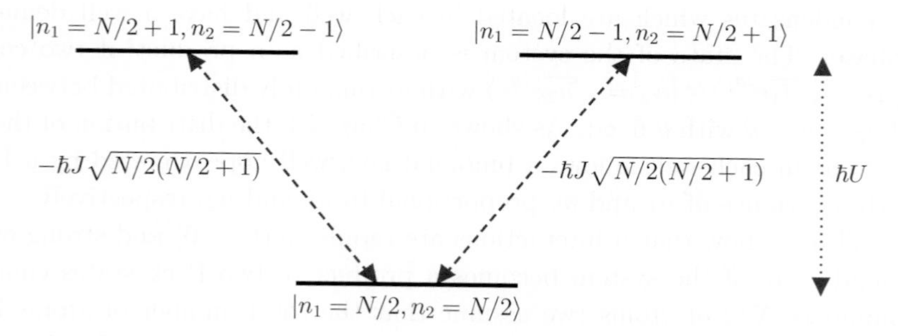

Fig. 26.3 Ground state and first excited manifold of the Hamiltonian $\hat{H}$ when $J = 0$ and $U > 0$. The first excited manifold is at a distance $\hbar U$ above the ground state. The oblique arrows describe the non-zero matrix elements of the tunnelling Hamiltonian.

The effect of the tunnelling between the two wells can now be studied by calculating the matrix elements of the tunnelling Hamiltonian between these states. One finds that the ground state and the two states of the first excited manifold are coupled with matrix elements equal to $-\hbar J\sqrt{N/2(N/2+1)}$. One concludes that, if $U \gg NJ$, this coupling is small compared to the distance between the unperturbed states, so that the Fock state $|n_1 = N/2\rangle \otimes |n_2 = N/2\rangle$ can be considered as the ground state of the system.

26.2.5 *Connection between the superfluid and insulator regimes*

The breakdown of the mean field approach in the large interaction regime suggests to extend the Bogolubov approach to this problem [Pitaevskii and Stringari (2003)]. The Hubbard Hamiltonian (26.1) can be rewritten in terms of the operators $\hat{a}_\pm = (\hat{a}_1 \pm \hat{a}_2)/\sqrt{2}$ and $\hat{a}_\pm^\dagger = (\hat{a}_1^\dagger \pm \hat{a}_2^\dagger)/\sqrt{2}$, and is a polynomial of degree four of these operators. To study the first corrections to the mean field results in the presence of interactions ($U \neq 0$), one implements the Bogolubov prescription[6] applied to the ground state $(a_+^\dagger)^N|0\rangle$. This prescription consists of

- replacing in the Hubbard Hamiltonian $\hat{a}_+$ and $\hat{a}_+^\dagger$ by $\sqrt{N}$,
- neglecting all terms containing less than two operators $\hat{a}_+$ and $\hat{a}_+^\dagger$.

This leads to an Hamiltonian quadratic in $\hat{a}_-$ and $\hat{a}_-^\dagger$ that can be exactly diagonalized. One finds

$$\hat{H} = \tilde{E}_0 + \hbar\omega_B \hat{\beta}^\dagger \hat{\beta}, \qquad \text{where} \qquad \omega_B = \sqrt{J(J+NU)} = \omega_J, \tag{26.12}$$

$\hat{\beta}$ and $\beta^\dagger$ are linear combinations of $\hat{a}_-$ and $\hat{a}_-^\dagger$ that destroy and create elementary excitations of energy $\hbar\omega_J$. In the new vacuum state $|\tilde{\Psi}_0\rangle$, of energy $\tilde{E}_0$, which obeys $\hat{\beta}|\tilde{\Psi}_0\rangle = 0$, one can calculate the population $\langle\tilde{\Psi}_0|\hat{a}_-^\dagger a_-|\tilde{\Psi}_0\rangle$ of atoms which are not condensed (quantum depletion effect). In this way, one recovers that quantum depletion becomes important when $U \gg NJ$. Similarly, one can estimate the coherence between the two wells by calculating the quantity $\langle\tilde{\Psi}_0|\hat{a}_1^\dagger\hat{a}_2|\tilde{\Psi}_0\rangle$.

Alternatively, one can directly numerically diagonalize the Hamiltonian (26.1) represented by a $(N+1)\times(N+1)$ tridiagonal matrix when expressed in the Fock basis [Guéry-Odelin and Lahaye (2011)]. Figure 26.4 shows the result of such a numerical calculation and its comparison with Bogolubov approach in its validity domain. One distinguishes easily three regimes when U/J is varied:

- For $U \ll J/N$, the effect of the interactions are always negligible and the physics is dominated by the tunnel effect. One thus has the same ground state as in the non-interacting case, with full coherence between the wells and Poissonian fluctuations in the number of atoms in each well. This corresponds to the single-particle oscillation regime.

[6]See Sec. 24.2 of Chap. 24.

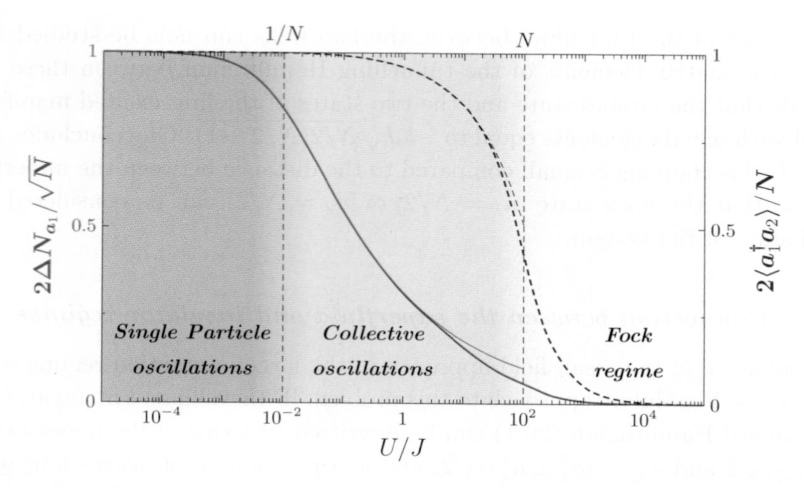

Fig. 26.4 Evolution of the variance of the number of atoms in well 1 (red/solid line) and of the coherence (blue/dashed line) between the two wells as a function of U/J [Guéry-Odelin and Lahaye (2011)]. The thin lines correspond to the results obtained with the Bogolubov approach (valid only as long as $U/J \ll N$), while the thick lines are obtained numerically. One distinguishes three regimes: for $U \ll J/N$, the effect of interactions is negligible, and one observes the single particle behavior, with strong fluctuations of the particle number and full coherence between the wells. For $1/N \ll U/J \ll N$, repulsive interactions tend to reduce the number of fluctuations (giving rise to some number squeezing), but the coherence is still high. In this *Josephson regime*, the mean-field description of the system is thus still valid. When $U \gg NJ$, interactions suppress both fluctuations and coherence: one observes fragmentation of the condensate towards the Fock state $|N/2, N/2\rangle$. This is the so-called *Fock regime*.

- For $1/N \ll U/J \ll N$, the coherence is still very close to its maximum value; however, the atom number fluctuations on a site start to decrease significantly (one observes number squeezing). In this regime, the coherence is still significant. This is the regime where Josephson oscillations and nonlinear self-trapping can be observed, and for this reason is called the *Josephson* regime. Experimentally, number squeezing was observed recently in such a system [Estève *et al.* (2008)].
- Finally, when $U \gg NJ$, the system loses its coherence and number fluctuations are almost completely frozen out. One observes the fragmentation of the condensate into two distinct macroscopically populated modes. In this regime, referred to as the Fock regime, the Bogolubov approach breaks down.

26.2.6 *Production of Schrödinger cat states when interactions are attractive*

If $U < 0$, the interaction energy is negative and its absolute value reaches its maximum when n_1, or n_2, is maximum. We conclude that, in the absence of tunnelling ($J = 0$), the ground state of $\hat{H}$ is a two-dimensional manifold $\{|n_1 =$

$N\rangle \otimes |n_2 = 0\rangle, |n_1 = 0\rangle \otimes |n_2 = N\rangle\}$ well separated from the excited state manifolds. The tunnelling Hamiltonian couples the two states of the lowest manifold via N intermediate states. The new ground state, including the effect of tunnelling, is thus a linear combination of the two states $\{|n_1 = N\rangle \otimes |n_2 = 0\rangle$ and $|n_1 = 0\rangle \otimes |n_2 = N\rangle\}$. If N is macroscopic, such a state is a Schrödinger cat state [Ho and Ciobanu (2004)].

26.2.7 *Controlling the tunnelling rate with a modulation of the difference of the two potential depths*

If a linear potential is added to the double well potential of Fig. 26.1, a difference is introduced between the depths of the two wells. Suppose that the slope of the linear potential is modulated in time at the frequency ω. The difference between the depths of the two potentials is modulated in time at the same frequency. In the absence of interactions ($U = 0$), the new Hamiltonian describing the dynamics of the system becomes

$$\hat{H} = -\hbar\frac{J}{2}\left(\hat{a}_1^\dagger \hat{a}_2 + \hat{a}_2^\dagger \hat{a}_1\right) + \hbar\frac{K}{2}\left(\hat{n}_2 - \hat{n}_1\right)\cos\omega t \tag{26.13}$$

where K is a constant proportional to the maximum slope of the linear potential.

The Hamiltonian (26.13) can be written in a more suggestive way by noting that the three operators

$$\hat{S}_z = \frac{1}{2}\left(\hat{a}_2^\dagger \hat{a}_2 - \hat{a}_1^\dagger \hat{a}_1\right), \qquad \hat{S}_x = \frac{1}{2}\left(\hat{a}_2^\dagger \hat{a}_1 + \hat{a}_1^\dagger \hat{a}_2,\right) \qquad \hat{S}_y = \frac{i}{2}\left(\hat{a}_2^\dagger \hat{a}_1 - \hat{a}_1^\dagger \hat{a}_2\right),$$

obey the commutation relations of a spin angular momentum, so that $\hat{H}$ can be rewritten as

$$\hat{H} = \hbar J\,\hat{S}_x + \hbar K\,\hat{S}_z\cos\omega t \tag{26.14}$$

which with the change $x \leftrightarrows z$, turns out to be identical to the Hamiltonian used in Sec. 8.2 of Chap. 8 for studying the g-factor of an atom dressed by a high frequency RF field. If $\omega \gg J$, the same calculations as those done in Chap. 8 lead to the conclusion that the tunnelling rate J is changed into a dressed tunnelling rate given by

$$J \to J_{\text{dressed}} = J \times J_0\left(\frac{K}{\omega}\right) \tag{26.15}$$

where J_0 is the Bessel function of order 0. The effective tunnelling rate vanishes if the ratio K/ω between the amplitude and the frequency of the modulation of the difference of potential depths coincides with a zero of the Bessel function J_0. This effect has been observed experimentally for both a double well potential [Kierig *et al.* (2008)] and an optical lattice [Lignier *et al.* (2007)].

26.3 Superfluid-Mott insulator transition for a quantum bosonic gas in an optical lattice

In this section, we switch from the double well problem to the multiple well problem realized with an optical lattice, and consider as before the interplay of tunnelling and interactions. Here also we restrict ourselves to the lowest energy level in each well (single band model). We describe a quantum phase transition for ultracold bosonic atoms with repulsive interactions loaded in an optical lattice. Cold atoms provide a realization of the Bose-Hubbard model which predicts a superfluid-Mott insulator transition as first theoretically investigated in [Fisher *et al.* (1989)]. The idea of applying this model to ultracold bosonic atoms in an optical lattice was first suggested in [Jaksch *et al.* (1998)]. This transition for cold atoms was first observed in 2002 by the Munich group [Greiner *et al.* (2002)].

26.3.1 *Bose Hubbard model*

The Bose Hubbard model describes the dynamics of atoms in a periodic potential with a Hamiltonian that generalizes the expression (26.1) given above for a double well potential to an array of potential wells:

$$\hat{H} = -\frac{J}{2}\sum_{i,j}\hat{a}_i^\dagger\hat{a}_j + \frac{U}{2}\sum_i \hat{n}_i(\hat{n}_i - 1). \tag{26.16}$$

The first term describes the tunnelling between two neighboring sites i and j of the lattice. The amplitude J is given by:

$$\frac{J}{2} = -\int \mathrm{d}^3r\, w\left(\vec{r}-\vec{r}_i\right)\left[-\frac{\hbar^2\vec{\nabla}^2}{2m} + V(\vec{r})\right] w\left(\vec{r}-\vec{r}_j\right). \tag{26.17}$$

The second term, where $\hat{n}_i = \hat{a}_i^\dagger\hat{a}_i$, describes the repulsive interactions within each well i. The parameter U, which is positive (repulsive interactions), is given by

$$U = \frac{4\pi\hbar^2 a}{m}\int \mathrm{d}^3r\left|w\left(\vec{r}-\vec{r}_i\right)\right|^4, \tag{26.18}$$

where a is the scattering length. In Eqs. (26.17) and (26.18), $w\left(\vec{r}-\vec{r}_i\right)$ is the Wannier wave function localized in the potential well i.[7]

26.3.2 *Qualitative interpretation of the superfluid-Mott insulator transition*

Suppose first that tunnelling is the dominant process. This can be realized for example by reducing the intensity of the laser beams that produce the optical lattice, so that the depth of the potential wells is very small which allows the tunnelling

[7] We use here a single band model, which is a good approximation if the distance between the lowest band and the first excited one is large compared to all relevant energies.

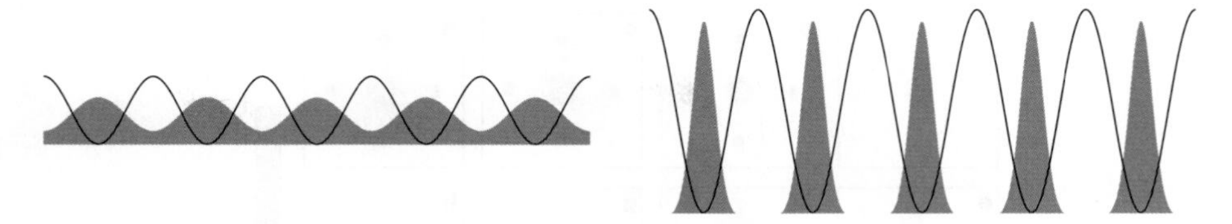

Fig. 26.5 Left: optical lattice with very small potential depths (solid line). Each atom is in the lowest Bloch state of the lattice and the many-body state is a macroscopic coherent wave delocalized over the whole lattice. The gray surface corresponds to the density probability of the wave function. Right: optical lattice with very large potential depths. Each atom is localized at a given site.

rate J to be large compared to the strength U of interactions. The many-body ground state is then well approximated by a product state where each atom is in the lowest Bloch state of the lowest band. This leads to a macroscopic coherent and superfluid wave that extends over the whole lattice (see Fig. 26.5(a)) and is described by the following state:

$$|\Psi\rangle \propto \left(\sum_{i=1}^{M} \hat{a}_i^\dagger \right)^N |\text{vac}\rangle , \tag{26.19}$$

where M is the number of sites.

If the depth of the potential wells is increased to a value so large that the tunnelling rate between adjacent sites becomes negligible, interactions become dominant and each atom is localized at a given site (see Fig. 26.5(b)). In the case of commensurate filling of the lattice described by the integer $\nu = N/M$,[8] the lowest energy state of the system is obtained by minimizing $\sum_i n_i^2$ with $\sum_i n_i$ fixed and equal to N. This is achieved by taking all n_i's equal to $\nu = N/M$. The state of the system is now:

$$|\Psi\rangle \propto \prod_{i=1}^{M} \left(\hat{a}_i^\dagger \right)^\nu |\text{vac}\rangle . \tag{26.20}$$

For example, for $\nu = 1$, each site is occupied by a single atom and this situation is analogous to the Mott insulator state studied in condensed matter.

It is important to emphasize the differences between the two states (26.19) and (26.20). In (26.19), all atoms are in the same state and are not correlated. The system can be described by a mean field theory with the Gross-Pitaevskii equation. The number of atoms at a given site is a random variable described by a Poisson law and this explains the coherence between the matter waves coming from the various sites when the optical lattice is switched off. The state (26.20) is a highly

[8] In practice, experiments are performed by superimposing an optical lattice on a harmonic trap. The density of atoms thus varies in space and there are always regions in the trap where the filling is commensurate. At sufficiently low temperature, superfluid-Mott insulator phases corresponding to different values of ν appear in the trap and can be directly observed through shell structures of the atomic density [Fölling *et al.* (2006); Campbell *et al.* (2006)].

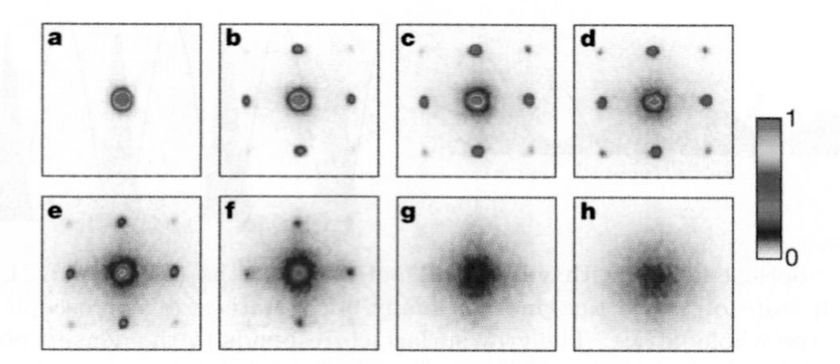

Fig. 26.6 Set of absorption images revealing the evolution of the interference pattern observed after a switching off of the optical lattice for increasing values of the intensity of the laser beams producing the lattice. Reprinted by permission from Macmillan Publishers Ltd: Nature [Greiner *et al.* (2002)].

correlated state. As a consequence of a well defined number of atoms in each site, there is no phase coherence between the various matter waves localized at different sites. For a commensurate filling $n = 1$, each site is occupied by a single atom in the lowest energy state of this site. The first excited state would correspond to a doubly occupied state and to an empty state with all other states occupied by a single atom. The energy of this state is easily calculated to be at an energy $\hbar U$ above the ground state. Contrary to the superfluid state (26.19), the Mott insulator state (26.20) thus exhibits a gap in its excitation spectrum.

26.3.3 *Experimental observation*

When the depth of the potential wells is progressively increased, one expects a transition from the superfluid state (26.19) to the Mott insulator state (26.20). Experimentally, one detects the decrease (when U/J is increase) of the visibility of the interference pattern that is observed after a switching off of the optical lattice and a subsequent time-of-flight (see Fig. 26.6). The interference between the matter waves originating from the various sites gives rise to a clear diffraction profile when the light intensity of the optical lattice laser beams is small, and disappears when it becomes large.[9]

26.4 Quantum fermionic gas in an optical lattice

Ultracold fermionic atoms in a periodic potential are also particularly interesting since they can be used as simple models of electrons in a crystal. As with the bosonic case, the optical lattice enhances the relative importance of interactions compared to the potential and kinetic energies. In addition, new features emerge as a result of the Pauli exclusion principle. As an example, we briefly describe ex-

[9]The evolution of this system has also been studied with single atom imaging [Bakr *et al.* (2009); Sherson *et al.* (2010)].

periments on the investigation of the metallic-Mott insulator transitions performed with an ultracold fermionic gas of potassium-40 atoms in the presence of repulsive interactions [Schneider *et al.* (2008); Jördens *et al.* (2008)].

A Fermi-Hubbard Hamiltonian can be introduced for studying the dynamics of the system. The creation and destruction operators of an atom at a given site now obey anti-commutation relations. We suppose that the gas consists of a mixture of equal proportions of fermionic atoms in two different spin states. Compared to the situation with bosonic atoms, the Pauli exclusion principle introduces new features. First, two atoms in the same spin state do not interact because of the quenching of s-wave scattering for polarized ultracold fermions. Second, in the single band model that we consider here, a given site cannot be occupied by two fermionic atoms in the same spin state. Double occupancy only occurs for two atoms in different spin states.

In addition, we will suppose that a harmonic trapping potential is added to the optical lattice to provide an external confinement of the gas. The observation of the size of the cloud when the confinement is increased provides a measurement of the compressibility, which can be used for detecting the metallic-insulator transitions.[10]

Figure 26.7 sketches the evolution of the system as the external confinement is progressively increased. For a very weak confinement and a small enough number of atoms, most sites are empty. The two spin states are represented by different colors (see upper left part of the figure). Increasing the confinement pushes the atoms to the centre of the trap. Since most sites are empty, atoms can move to occupy the empty sites. The system behaves as a "compressible metal". When all sites are occupied by one atom (end of the compression phase 1 in the figure), we have a Mott insulator (see upper right part of the figure). It costs too much energy to have two atoms (in different spin states) at the same site and the system is incompressible. With a further increase of the trapping potential, the potential energy of the atoms far from the center becomes so high that it becomes energetically favorable to have two atoms (in different spin states) at the same site (end of the compression phase 2 in the figure). As long as all sites are not doubly occupied, one atom in a doubly occupied site can move to another site containing a single atom. We recover a compressible metal (see lower right part of the figure). More and more atoms fill the singly occupied sites until all sites are doubly occupied (end of the compression phase 3 in the figure). We get an incompressible band insulator (see lower left part of the figure) which lasts until the next higher band becomes occupied.

The scenario of Fig. 26.7 has been confirmed experimentally, both by measuring the compressibility of the gas and observing plateaus corresponding to the metallic-Mott insulator and metallic-band insulator transitions [Schneider *et al.* (2008)], and by measuring the number of doubly occupied sites [Jördens *et al.* (2008)].

[10] A similar confinement also exists for the experiments performed on bosonic gases. It gives rise to shell structures in the superfluid-Mott insulator transitions [Fölling *et al.* (2006)].

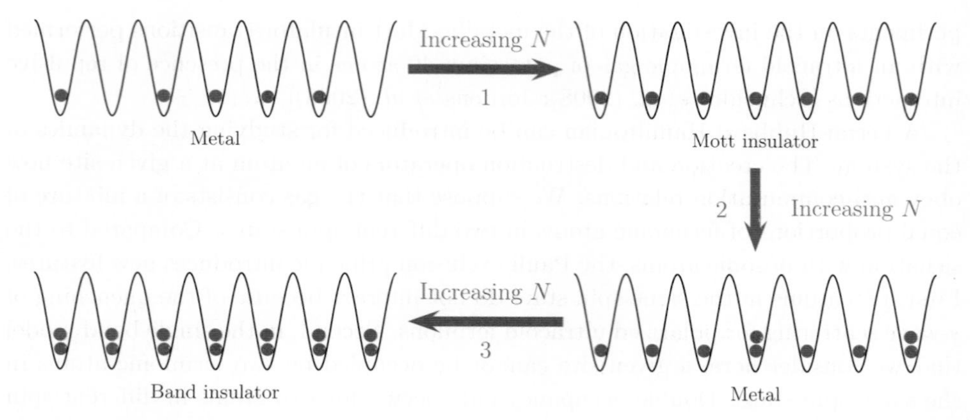

Fig. 26.7 Evolution of the occupation of the various sites when the chemical potential is increased (this is performed by changing the strength of the external confinement). Two types of insulator are expected one resulting from the interactions (Mott-insulator) and the other one from the statistics (band insulator).

26.5 Feshbach resonances and Fermi quantum gases

26.5.1 *Introduction*

As detailed in Chap. 16, Feshbach resonances dramatically change the two-body scattering properties. They give the possibility of achieving situations in which strong interactions, repulsive or attractive, appear in a many-body system, like a quantum degenerate Bose or Fermi gas: the scattering length can even become larger than interparticle distances. It is therefore not surprising that several groups began to look for new physical effects appearing under these conditions soon after the discovery of gaseous BEC in 1995, the observation of Feshbach resonances in 1998 [Inouye *et al.* (1998)], and the first evidence for quantum degeneracy in Fermi gases in 1999 [DeMarco and Jin (1999)].

When Feshbach resonances were explored in Bose gases, strong losses due to inelastic collisions were observed [Stenger *et al.* (1999)], thereby limiting the number of studies of strongly interacting condensates [Papp *et al.* (2008)]. Several groups observed quantum degenerate dimers formed by two bosonic atoms in the region $a > 0$ of the resonance where the two-body system has a bound state [Herbig *et al.* (2003); Xu *et al.* (2003); Dürr *et al.* (2004)], but the lifetime of these dimers was too short to allow thermodynamical equilibrium to be established and thus to speak of a Bose-Einstein condensate.

The situation is radically different if the dimers are formed with two fermionic atoms. It was observed in 2003 by the group of Christophe Salomon in Paris [Cubizolles *et al.* (2003)] that the lifetime of Li_2 dimers formed in the $a > 0$ side of the broad Feshbach resonance[11] from two lithium-6 atoms in different spin states

[11]Broad Feshbach resonances correspond to a small contribution of the closed channel to the paired state on a large range of scattering length ($0 < a^{-1} \ll k_F$) when the magnetic field is

was very long, on the order of 0.5 sec.[12] As already mentioned in Chap. 16, the stability against inelastic atom-dimer and dimer-dimer collisions which could transfer the molecule to a deep bound state in the interatomic potential originates from the combination of two effects: (i) the Pauli exclusion principle which dramatically reduces the probability of three fermions (two of them are necessarily identical) approaching each other at a short distance and (ii) the small wave function overlap between deeply bound molecular states and the giant halo state in which dimers are produced using Feshbach resonances ([Petrov *et al.* (2004, 2007)] and references therein). This long lifetime of fermionic Feshbach molecules, which behave as composite bosons, then allowed several groups to obtain molecular condensates [Jochim *et al.* (2003a); Greiner *et al.* (2003); Zwierlein *et al.* (2003); Bourdel *et al.* (2004); Partridge *et al.* (2005)].

These results are very important because they showed that it is possible to study strongly interacting quantum Fermi gases near a Feshbach resonance. What happens on the negative side of this resonance ($a < 0$), where the two-body system has no bound state? The effective attraction (associated with the negative sign of a) between the two fermions in different spin states could, in the presence of the Fermi sea of the other atoms, give rise to a pairing of fermions analogous to the Cooper pairing of two electrons in a superconductor. Can the Bardeen-Cooper-Schrieffer (BCS) theory of superconductivity [Bardeen *et al.* (1957)] be extended to Fermi gases, not only in the the region of weak interactions ($k_F|a| \ll 1$ where k_F is the Fermi wave number) corresponding to the domain of validity of the BCS theory, but near the resonance where $k_F|a| \gg 1$? Does the Fermi gas collapse like a Bose-Einstein condensate when attractive forces between fermions become too large? Can one observe the crossover between the BEC regime of dimers and the BCS regime of Cooper pairs? Is the Fermi gas superfluid and does it remain superfluid in the crossover? What happens if the populations of the two spin states are unequal? In this section, we answer those questions through a brief review of the developments in this research field.[13]

26.5.2 *Brief survey of BCS theory*

Cooper pairing

For a two-body system with attractive interactions, one can show that a bound state always exist in 1D and 2D. In 3D, it appears only if the strength of the interaction is above a certain threshold. This is related to the differences between the densities of states $\rho(\epsilon)$, varying as $\epsilon^{-1/2}$ in 1D, ϵ^0 in 2D, $\epsilon^{+1/2}$ in 3D. The situation is different in

tuned through the resonance [Partridge *et al.* (2005); Bloch *et al.* (2008)]. The opposite case is referred to as narrow Feshbach resonance.

[12] See also [Strecker *et al.* (2003)] where long-lived molecules were observed in the vicinity of a narrow Feshbach resonance.

[13] More details can be found in [Ketterle and Zwierlein (2008); Giorgini *et al.* (2008)].

a three-dimensional ideal Fermi gas at $T = 0$ K, as shown by an heuristic argument by Leon Cooper [Cooper (1956)]. All states with $k < k_F$ are then filled. If one tries to construct a bound state of two attracting fermions on top of the Fermi sea, only states with $k > k_F$ are available for constructing the bound state. Near the Fermi level, the density of states is nearly constant and equal to $\rho(E_F)$. The situation is thus analogous to a two-dimensional problem and the two attracting fermions on top of the three-dimensional Fermi sea can always form a bound state for arbitrarily weak attractions.[14]

The BCS wave function

One year after Leon Cooper's work, John Bardeen, Leon Cooper and John Schrieffer extended this theory and constructed a ground state in which all electrons form bound pairs [Bardeen *et al.* (1957)].[15] This theory was the first microscopic and non-perturbative many-body theory developed to characterize conventional superconductors.

Consider a balanced mixture of a two-component Fermi gas. In the absence of interactions, the ground state of the gas corresponds to the complete filling of the one particle energy states by two fermions with different internal states up to the Fermi energy E_F. The ground state energy of this normal state is $E_0 = N \times 3E_F/5$. In the presence of attractive interspecies interactions, the new ground state of the Fermi gas is approximated by the BCS ansatz in which all pairs share the same state. We justify the form of this ansatz in the following manner [de Gennes (1966)].

The state $\phi(i,j)$ of a pair i,j of two fermions is written as the tensor product of an orbital part described by a symmetric wave function $|\varphi(i,j)\rangle$ and a spin singlet state $|\chi(i,j)\rangle = (|\uparrow\rangle_i |\downarrow\rangle_j - |\downarrow\rangle_i |\uparrow\rangle_j)/\sqrt{2}$:

$$\phi_\chi(\vec{r}_i, \vec{r}_j) = e^{i\vec{q}\cdot(\vec{r}_i+\vec{r}_j)/2}\varphi(\vec{r}_i - \vec{r}_j)|\chi(i,j)\rangle, \tag{26.21}$$

where $\vec{q}$ is the pair center of mass momentum taken equal to zero in the following because this turns out to be energetically favorable. The many-body state $|\Psi_N\rangle$ corresponds to a situation in which all pairs share the state ϕ_χ which will in turn be determined by a variational principle. The BCS theory is thus a mean-field theory analogous to the Gross-Pitaevskii description of Bose-Einstein condensates, valid for weak attraction.

---

[14]Such an instability of the normal phase yielding pairing phases occurs systematically at sufficiently low temperature when a partial wave component of the interaction gives rise to an effectively attractive interaction. This result is referred to as the Kohn-Luttinger theorem or instability [Kohn and Luttinger (1965)].

[15]The attraction between electrons in a superconductor is an indirect attraction, mediated by the phonons of the crystal lattice.

The many-body state can be written:

$$|\Psi_N\rangle = \mathcal{A}\Big[\Big(|\varphi(1,N'+1)\rangle \otimes |\chi(1,N'+1)\rangle\Big) \otimes \dots \otimes\Big(|\varphi(N',N'+N')\rangle \otimes |\chi(N',N'+N')\rangle\Big)\Big], \tag{26.22}$$

where $N' = N/2$, $\mathcal{A}$ is the antisymmetrization operator for the fermions coordinates. It is convenient to expand the pair wave function on the plane wave basis:

$$\varphi(\vec{r}_i - \vec{r}_{N'+i}) = \sum_{\vec{k}} g_{\vec{k}} e^{i\vec{k}.(\vec{r}_i - \vec{r}_{N'+i})}. \tag{26.23}$$

where $g_{\vec{k}}$ is the Fourier transform of φ.[16] The state (26.22) can thus be rewritten in the form

$$|\Psi_N\rangle = \sum_{\vec{k}_1}\cdots\sum_{\vec{k}_{N'}} g_{\vec{k}_1}\dots g_{\vec{k}_{N'}}\mathcal{A}\Big[\Big(|1:\vec{k}_1; N'+1:-\vec{k}_1\,\rangle \otimes |\chi(1,N'+1)\rangle\Big) \otimes\dots\otimes\Big(|N':\vec{k}_{N'}; N'+N':-\vec{k}_{N'}\,\rangle \otimes |\chi(N',N'+N')\rangle\Big)\Big]. \tag{26.24}$$

and its second quantized form simply reads

$$|\Psi_N\rangle = \mathcal{N}_N (C^\dagger)^{N'}|0\rangle, \qquad \text{where} \qquad C^\dagger = \sum_{\vec{k}} g_{\vec{k}} c^\dagger_{\vec{k},\uparrow} c^\dagger_{-\vec{k},\downarrow} \tag{26.25}$$

and $\mathcal{N}_N$ is a normalization factor so that $\langle\Psi_N|\Psi_N\rangle = 1$. The operator $c^\dagger_{\vec{k},\uparrow}$ ($c^\dagger_{-\vec{k},\downarrow}$) creates a spin up (down) fermion in state $\vec{k}$ ($-\vec{k}$).

The BCS ansatz[17] for the ground state is inspired by the form of the coherent state for bosons associated with the state $|\Psi_N\rangle$:

$$|\Psi_{\rm BCS}\rangle = \mathcal{N} e^{C^\dagger}|0\rangle = \sum \lambda_N |\Psi_N\rangle = \mathcal{N}\prod_k \left(1 + g_k c^\dagger_{\vec{k}\uparrow} c^\dagger_{-\vec{k}\downarrow}\right)|0\rangle, \tag{26.26}$$

where $\lambda_N = \mathcal{N}/(\mathcal{N}_N N!)$. It can be rewritten in the form

$$|\Psi_{\rm BCS}\rangle = \prod_k \left(u_{\vec{k}} + v_{\vec{k}} c^\dagger_{\vec{k}\uparrow} c^\dagger_{-\vec{k}\downarrow}\right)|0\rangle, \tag{26.27}$$

where $\mathcal{N} = \prod_{\vec{k}} u_{\vec{k}}$ and $g_{\vec{k}} = v_{\vec{k}}/u_{\vec{k}}$. We have used the fact that the $c^\dagger_{\vec{k},\uparrow} c^\dagger_{-\vec{k},\downarrow}$'s for different values of $\vec{k}$ commute and that in the expansion of $\exp(g_{\vec{k}} c^\dagger_{\vec{k},\uparrow} c^\dagger_{-\vec{k},\downarrow})$ only the zero and first order terms appear because $(c^\dagger_{-\vec{k}})^2 = 0$. Since $g_{\vec{k}}$ is real here and only depends on k (see footnote 16), $u_{\vec{k}}$ and $v_{\vec{k}}$ are also real and only depend on k.

For a sufficiently large number of atoms, the distribution of the λ_N coefficients is peaked about the mean value $\langle\hat{N}\rangle = \langle\Psi_{\rm BCS}|\hat{N}|\Psi_{\rm BCS}\rangle = 2\sum_{\vec{k}} g_k^2/(1+g_k^2)$ with relative fluctuations that scale as $\langle\hat{N}\rangle^{-1/2}$. The use of coherent states rather than

[16] We restrict here to s-wave pairing for which $g_{\vec{k}}$ is real and isotropic.

[17] We address here the homogeneous case. The BCS ansatz can be generalized to the inhomogenous case [Castin (2008)].

Fock states is analogous to the use of grand canonical ensemble instead of the canonical ensemble. The number of atoms can vary but the chemical potential is fixed.

By construction, the state $|\Psi_{\mathrm{BCS}}\rangle$ involves only pairs with a zero total momentum for which the center of mass wave function is uniform. This approximation made in the original BCS paper thus amounts to neglecting density fluctuations. As a consequence, the BCS theory in its simplest form cannot describe collective excitations such as sound waves.

The determination of the explicit value for the u_k and v_k coefficients is performed by minimizing the free energy of the BCS state $\langle\Psi_{\mathrm{BCS}}|\hat{H} - \mu\hat{N}|\Psi_{\mathrm{BCS}}\rangle$ where μ is the Lagrange multiplier for the number of atoms, i.e. the chemical potential and $\hat{H}$ is the second quantized form of the Hamitonian of the interacting two-component Fermi gas

$$\hat{H} = \sum_{\vec{k},s} \epsilon_k c^\dagger_{\vec{k},s} c_{\vec{k},s} + \frac{V_0}{\Omega} \sum_{\vec{k},\vec{k}} c^\dagger_{\vec{k}\uparrow} c^\dagger_{-\vec{k}\downarrow} c_{-\vec{k}'\downarrow} c_{\vec{k}'\uparrow} \tag{26.28}$$

where $\epsilon_k = \hbar^2 k^2/2m$ is the single particle kinetic energy and Ω the volume of the box containing the gas. The interaction potential has been taken to be an effective contact potential of strength V_0, $V(\vec{r}) = V_0\delta(\vec{r})$. Note that the interaction term in $\hat{H}$ only connects pairs with zero total linear momentum, since we consider only these pairs.

The minimization of $\langle\Psi_{\mathrm{BCS}}|\hat{H} - \mu\hat{N}|\Psi_{\mathrm{BCS}}\rangle$ gives the explicit form of the coefficients of the BCS ansatz:

$$u_k^2 = \frac{1}{2}\left(1 + \frac{\xi_k}{E_k}\right), \qquad \text{and} \qquad v_k^2 = \frac{1}{2}\left(1 - \frac{\xi_k}{E_k}\right), \tag{26.29}$$

where we use the notation

$$\xi_k = \epsilon_k - \mu, \qquad \text{and} \qquad E_k = (\xi_k^2 + \Delta^2)^{1/2}, \tag{26.30}$$

and introduce the energy gap parameter

$$\Delta = -\frac{V_0}{\Omega} \sum_k u_k v_k \tag{26.31}$$

whose physical interpretation will be given later on. Using Eqs. (26.29) and (26.30), one can rewrite (26.31) as:

$$-\frac{1}{V_0} = \frac{1}{\Omega} \sum_{\vec{k}} \frac{1}{2E_k} = \int \frac{\mathrm{d}^3 k}{(2\pi)^3} \frac{1}{2E_k}. \tag{26.32}$$

Since E_k depends on Δ and μ, this is an implicit equation for Δ, referred to as the *gap equation*. In the domain of validity of the BCS theory (weak interaction), the chemical potential μ, which appears in this equation through the ξ_k's, is well approximated by the Fermi energy E_F.

The sum over $\vec{k}$ in Eq. (26.32) leads to a divergent integral if the upper limit of the integral is extended to infinity. In the case of superconductors, this difficulty does not appear because the attractive forces between electrons are mediated by phonons whose frequencies cover an interval of width typically equal to the Debye frequency ω_D. The sum over k in Eq. (26.32) can then be limited to an interval around k_F corresponding to an energy spread on the order of $\hbar\omega_D$. Such a situation does not exist in a Fermi gas of neutral atoms where attractive interactions are direct. The divergence appearing in Eq. (26.32) is however artificial. It is due to the fact that we have used a delta potential $V_0\delta(\vec{r})$, and thus Fourier components extending to infinity. The real potential always has a finite range, so that there is necessarily a cut-off at high momentum. For solving this difficulty, one can use a pseudo-potential, or, a delta potential truncated in momentum space as shown in Sec. 15.5 of Chap. 15. In this way, one introduces the scattering length which, when expressed in terms of V_0, contains the same divergence

$$\frac{m}{4\pi\hbar^2 a} = \int \frac{\mathrm{d}^3k}{(2\pi)^3}\frac{\eta(k)}{2\epsilon_k} + \frac{1}{V_0}, \qquad \text{with} \qquad \epsilon_k = \frac{\hbar^2k^2}{2m}, \tag{26.33}$$

where $\eta(k)$ is the truncating function. Using Eq. (26.33) for replacing V_0 by a in (26.32) leads to a new gap equation

$$-\frac{m}{4\pi\hbar^2 a} = \int \frac{\mathrm{d}^3k}{(2\pi)^3}\left(\frac{1}{2E_k} - \frac{1}{2\epsilon_k}\right). \tag{26.34}$$

which is divergence free (so that $\eta(k)$ can be replaced by 1) and can be integrated in the BCS limit considered in this section ($k_F|a| \ll 1$) to give [Gor'kov and Melik-Barhudarov (1961)]:

$$\Delta \simeq \left(\frac{2}{e}\right)^{7/3} E_F e^{-\pi/2k_F|a|}. \tag{26.35}$$

The gap varies in a non-analytic way with the scattering length. We cannot thus get this result with a perturbation theory.

Physical discussion

(i) *Energy of the ground state*

Using Eqs. (26.28)–(26.31), the energy of the BCS state is found to be equal to

$$E_{\text{BCS}} = \langle\Psi_{\text{BCS}}|\hat{H}|\Psi_{\text{BCS}}\rangle = E_0 - \frac{1}{2}\rho(E_F)\Delta^2 < E_0, \tag{26.36}$$

The energy associated with the BCS state, E_{BCS}, is, as expected, lower than the energy of the normal state, E_0. This clearly shows that the ground state of the ideal Fermi gas is unstable against attractive interactions and that pairing reduces the energy of the system.

(ii) *Order parameter*

For a degenerate Bose gas, the existence of a condensate is characterized by the appearance of off diagonal long range order in the reduced one-body density operator $\hat{\rho}_1$:

$$\langle\hat{\Psi}^\dagger(\vec{r})\hat{\Psi}(\vec{r}\,')\rangle \to \psi(\vec{r})\psi^*(\vec{r}\,') \quad \text{if} \quad |\vec{r} - \vec{r}\,'| \to \infty \tag{26.37}$$

$\psi(\vec{r}) = \langle\hat{\Psi}(\vec{r})\rangle$ is called the *order parameter* and is the wave function of the eigenvector of $\hat{\rho}_1$ (with a normalization $\mathrm{Tr}(\hat{\rho}_1) = N$) with a macroscopic eigenvalue. Equation (26.37) cannot be extended to a Fermi gas since no quantum state can have a macroscopic occupation because of Fermi Dirac statistics: there is no fermionic condensate for a Fermi gas of atoms. But pairs of fermionic atoms, which can be considered to be composite bosons, can condense. Instead of $\hat{\rho}_1$, one has in this case to introduce the reduced two-body density operator

$$\hat{\rho}_2\left(\vec{r}_1, \vec{r}_2, \vec{r}\,'_1, \vec{r}\,'_2\right) = \left\langle \hat{\Psi}^\dagger_\uparrow\left(\vec{r}_1\right) \hat{\Psi}^\dagger_\downarrow\left(\vec{r}_2\right) \hat{\Psi}_\downarrow\left(\vec{r}\,'_2\right) \hat{\Psi}_\uparrow\left(\vec{r}\,'_1\right) \right\rangle. \tag{26.38}$$

If a pair condensate exists, one could have a pair of two atoms located at $\vec{r}_1$ and $\vec{r}_2$ with a center of mass located at $\vec{R} = (\vec{r}_1 + \vec{r}_2)/2$, and another pair of two atoms at $\vec{r}\,'_1$ and $\vec{r}\,'_2$ with a center of mass at $\vec{R}\,' = (\vec{r}\,'_1 + \vec{r}\,'_2)/2$. The existence of a long range order can be now defined by:

$$\left\langle \hat{\Psi}^\dagger_\uparrow\left(\vec{r}_1\right) \hat{\Psi}^\dagger_\downarrow\left(\vec{r}_2\right) \hat{\Psi}_\downarrow\left(\vec{r}\,'_2\right) \hat{\Psi}_\uparrow\left(\vec{r}\,'_1\right) \right\rangle \to \psi\left(\vec{r}_1, \vec{r}_2\right) \psi^*\left(\vec{r}\,'_1, \vec{r}\,'_2\right) \quad \text{if } \left|\vec{R} - \vec{R}'\right| \to \infty \tag{26.39}$$

where

$$\psi\left(\vec{r}_1, \vec{r}_2\right) = \left\langle \hat{\Psi}^\dagger_\uparrow\left(\vec{r}_1\right) \hat{\Psi}^\dagger_\downarrow\left(\vec{r}_2\right) \right\rangle \tag{26.40}$$

can be considered to be the wave function of a pair.[18]

(iii) *Size of the pairs*

Using $\hat{\Psi}^\dagger_\uparrow(\vec{r}) = \sum_{\vec{k}} \hat{c}^\dagger_{\vec{k},\uparrow} \exp(-i\vec{k}.\vec{r})/\sqrt{\Omega}$, Eq. (26.40) can be rewritten:

$$\psi(\vec{r}_1, \vec{r}_2) = \frac{1}{\Omega} \sum_{\vec{k}} u_k v_k \exp\{-i\vec{k}.(\vec{r}_1 - \vec{r}_2)\} \tag{26.41}$$

which means that $\psi(\vec{r}_1, \vec{r}_2)$ is the Fourier transform of $u_k v_k$.

For an ideal Fermi gas, v_k is equal to 1 for $k < k_F$ and to 0 for $k > k_F$, whereas u_k is equal to 0 for $k < k_F$ and to 1 for $k > k_F$. In the presence of interactions, the variations of u_k and v_k between 0 and 1 become smoother and extend over an energy interval $\delta\epsilon$ on the order of Δ, according to Eqs. (26.29) and (26.30). It follows that $u_k v_k$ is non-zero only in an interval of width δk around k_F, with $\delta k = (dk/d\epsilon)\delta\epsilon \sim \Delta/\hbar v_F$ where $v_F = \hbar k_F/m$ is the Fermi velocity. The spatial extent ξ of the wave function φ of the Cooper pairs is thus given by:

$$\xi \simeq \frac{1}{\delta k} \simeq \frac{\hbar v_F}{\Delta}. \tag{26.42}$$

It is interesting to compare ξ to the mean distance d between atoms, which is on the order of $1/k_F$. We get $d/\xi \simeq \Delta/E_F \ll 1$, since, according to Eq. (26.35), $\Delta \ll E_F$

[18] Another wave function φ for the pair has been introduced in Eq. (26.21). Its Fourier transform is equal to v_k/u_k. Depending on the problem to be studied, it is more appropriate to use φ or ψ. Both wave functions have similar behavior and features in Fourier and real space. For interpreting the RF excitation spectrum to a third state, φ is more appropriate, whereas ψ appears in the order parameter. For a more detailed discussion, see [Ketterle and Zwierlein (2008)].

in the BCS limit ($k_F|a| \ll 1$). This inequality means that there is a strong overlap between pairs!

(iv) *Excitation spectrum*

An approach somewhat similar to the Bogolubov approach can be followed for simplifying the expression (26.28) of the Hamiltonian and for finding the first excited states of the system. According to Eq. (26.41), the average value $C_k = u_k v_k = \langle c^\dagger_{\vec{k}\uparrow} c^\dagger_{-\vec{k}\downarrow} \rangle$ is the Fourier transform of the order parameter in BCS theory. One expects that the fluctuations of the operator $c_{\vec{k}\uparrow} c_{-\vec{k}\downarrow}$ (destroying a pair) around its mean value are small. Replacing in Eq. (26.28) the operators destroying or creating a pair by their average values plus the fluctuating term, and keeping only terms containing at most one fluctuating term leads to a quadratic Hamiltonian which can thus be exactly diagonalized.[19] The result of this diagonalization is that the energies of the elementary excitations are just equal to E_k.

To facilitate the comparison of the interacting Fermi gas with the ideal Fermi gas, we first recall the excitation spectrum of an ideal Fermi gas at $T = 0$. In this latter case, two kinds of excitation can be produced (i) by transfer of a particle from the Fermi surface to a state of momentum $p > p_F$, the corresponding excitation energy is $\varepsilon(p) = p^2/2m - p_F^2/2m$. This is referred to as particle-like excitation ; (ii) by transfer of a particle from a state with momentum p inside the Fermi sphere to the surface of the Fermi sphere, associated with an excitation energy $\varepsilon(p) = p_F^2/2m - p^2/2m$. In this kind of excitation, a hole in the Fermi sea is created, and this branch of the excitation spectrum is thus called the hole-like excitation branch. From this point of view, the single particle elementary excitation spectrum of the ideal Fermi gas is $\varepsilon(p) = |p^2 - p_F^2|/2m$, represented as a dashed line in Fig. 26.8.

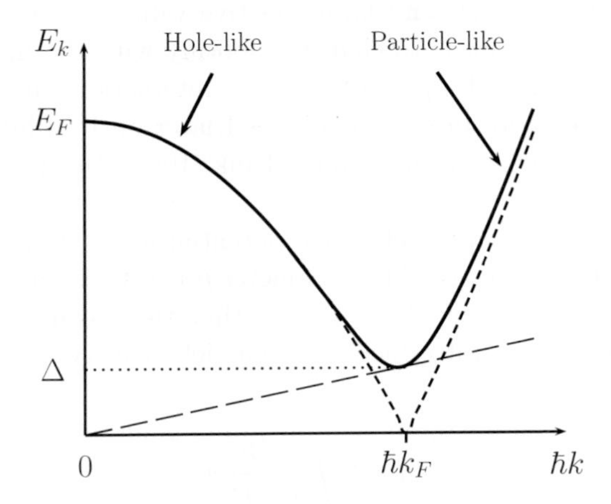

Fig. 26.8 Single particle excitation spectrum for $k_F a < 0$. The minimum energy corresponds to the gap Δ and is obtained for $k = (2m\mu/\hbar^2)^{1/2} \simeq k_F$. The slope of the long dashed line gives the critical velocity for superfluidity (see text).

[19] A more rigorous derivation can be found in [Castin (2008)].

In the presence of attractive interactions, the elementary excitations have an energy $E_k = (\xi_k^2 + \Delta^2)^{1/2}$. Their energy spectrum is represented by the full lines of Fig. 26.8. It clearly appears that the ground state energy is separated from the energies E_k of the first excited states by at least the gap energy Δ. One can produce a single excitation from the BCS state only at the expense of a small but finite amount of energy Δ, called for this reason, the energy gap. One can show that E_k corresponds either to the energy cost to add a single fermion on top of the BCS state or the energy cost to remove a fermion of state $|\vec{k}\uparrow\rangle$ from the BCS state leaving an unpaired fermion in state $|-\vec{k}\downarrow\rangle$ [Leggett (2006)]. In both cases, a pair is broken.

Remark

The excitation spectrum discussed in this section does not contain collective excitations like sound waves because the pairs described by the BCS state have zero total linear momentum. A more precise treatment predicts a sound mode, also referred to as the Bogolubov-Anderson mode. Its dispersion relation is $\omega_k = kc$ where the speed of sound c is related to the Fermi velocity: $c = [(n/m)(\partial\mu/\partial n)_S]^{1/2} = v_F/\sqrt{3}$.

26.5.3 *A simple model for the BEC-BCS crossover*

In 1980, extending early works performed in various contexts such as superconductors [Popov (1966); Keldysh and Kozlov (1968); Eagles (1969)] or helium-3, Anthony Leggett proposed a generalization of the BCS approach to describe the ground state properties of the Fermi gas outside the regime $k_F|a| \ll 1$, where it has been first introduced [Leggett (1980)]. He assumed that the BCS variational wave function can still be used when the interaction parameter $1/k_F a$ is varied between large negative values (BCS regime) and large positive values (molecular BEC regime). The physical properties appear to change smoothly with the interaction strength, in particular in the region $1/k_F|a| \ll 1$ where interactions are very strong. The extension to finite temperature was performed later on by Philippe Nozières and Stefan Schmitt-Rink [Nozières and Schmitt-Rink (1985)] (see also [Sá de Melo *et al.* (1993)]).

Outside the BCS regime, the chemical potential μ can no longer be considered to be equal to the Fermi energy. The parameter μ is actually a Lagrange multiplier and its value can be determined by writing that the mean value of the number operator $\hat{N}$ is equal to N. Introducing the particle density $n = N/\Omega$, one gets an equation

$$n = 2\int \frac{\mathrm{d}^3 k}{(2\pi)^3}\, v_k^2 \tag{26.43}$$

called the *number equation* which, when combined with the gap equation (26.34), allows for the determination of both Δ and μ [Ketterle and Zwierlein (2008)]. Here, we just briefly review a few of their properties.

(i) *Chemical potential*

In the BCS regime ($a < 0$ and $k_F|a| \ll 1$) and at $T = 0$, the chemical potential is very close to the Fermi energy, $\mu \simeq E_F$. Then, when $|a|$ increases, μ decreases, then vanishes when $1/k_F a \simeq 0.55$, and, in the BEC limit ($a > 0$, $1/k_F a \gg 1$) takes the value

$$\mu \simeq -\frac{\hbar^2}{2ma^2} + \frac{\pi\hbar^2 an}{m}. \tag{26.44}$$

The first term accounts for the binding energy per fermion of the dimer made of two fermions with opposite spins. The second term is a mean field term which accounts for half the repulsive interaction between these dimers. It can indeed be written $2\pi\hbar^2 a_{MM} n_M/m_M$, where $n_M = n/2$ is the density of dimers, $m_M = 2m$ their mass and $a_{MM} = 2a$ the scattering length for collisions between two dimers, except that the correct value of a_{MM} is $0.6a$ and not $2a$ [Petrov *et al.* (2004)]. The model presented here for the BEC-BCS crossover describes correctly the wave function of the dimer but does not properly include the correlations between dimers. An exact generalization of the BCS theory in the BEC limit yields the correct mean field term with the proper scattering length [Leyronas and Combescot (2007)].

(ii) *Size of the pairs*

The size ξ of the pairs is larger than the inter-particle distance $1/k_F$ in the BCS regime. It then decreases when $|a|$ increases, becomes on the order of $1/k_F$ at the center of the Feshbach resonance where $1/k_F a$ vanishes, and finally takes a value on the order of the scattering length a in the BEC regime where a characterizes the size of the Feshbach molecules[20] which can be, in this regime, identified with the Cooper pairs.

(iii) *Elementary excitations*

In the BCS regime, the excitation spectrum exhibits an energy gap that reaches its minimum value Δ for $k \simeq k_F$. When $|a|$ increases, the position of the minimum shifts to lower values of k and the energy gap increases. When a reaches a value such that μ becomes negative (in the BEC zone), the minimum of the spectrum is located at $k = 0$. The energy E_k is then related to the binding energy of the dimer (plus mean field corrections) since breaking a pair corresponds in this region to dissociating a dimer. The low energy collective mode in this region are the well known sound modes of a Bose-Einstein condensate.[21]

(iv) *Inadequacies of this model*

The generalization of the BCS theory presented in this section has the advantage of providing a qualitative description of the behavior of the Fermi gas along the whole crossover, including the limit $1/k_F a \to 0$ of strong interactions. Our presentation

[20]See Chap. 16.

[21]One can show that this sound mode connects in the BCS zone to the Anderson-Bogolubov mode mentioned above.

of the BCS theory is however far from being fully satisfactory since it is limited to $T = 0$ K and does not take into account the motion of the pairs. It is actually a mean-field theory, since the variational wave function is a product (26.22) of identical wave functions for the various pairs. A more precise theory, that gives better agreement with the experimental observations, in particular at finite temperatures, should include the motion of the pairs as well as fluctuations around the mean field description. Several theoretical works addressing this point, such as quantum Monte Carlo and diagrammatic techniques, are reviewed in [Giorgini *et al.* (2008)].

26.5.4 *Experimental investigations*

(i) *Fermionic condensates*

The first observations of the condensation of fermionic pairs in the BCS side of the Feshbach resonance ($a < 0$) were performed in 2003, at JILA [Greiner *et al.* (2003)], and shortly after at MIT [Zwierlein *et al.* (2003)]. The observation of the pairs could not be done in the usual way of a ballistic expansion of the cloud, because the weakly bound pairs cannot remain bound during the expansion. Before the expansion, the magnetic field was swept from the region $a < 0$ to the region $a > 0$ in order to transform the Cooper pairs into dimers which do no dissociate during the expansion. The sweep has to be slow enough to allow the adiabatic transformation of Cooper pairs into dimers, but fast enough to prevent the many-body interactions from changing the spatial distribution of the pairs which one wants to measure in the time-of-flight images.[22] Following this procedure, bimodal structures were clearly observed below a certain temperature, revealing the existence of a fermionic condensate.

During the following years, several experiments were performed to explore the physical properties of the system in the crossover region, triggering the creation of many novel techniques to probe strongly interacting Fermi gases. For example, one interesting technique to determine the gap was provided by the radio-frequency excitation of one of the components of the Fermi gas, say $|\uparrow\rangle$, into a third state. However, the analysis may be complicated and thus delicate because of the interaction between this final state and the other spin state $|\downarrow\rangle$ [Chin *et al.* (2004); Kinnunen *et al.* (2004); He *et al.* (2005); Schirotzek *et al.* (2008); Stewart *et al.* (2008)]. Alternatively, photoassociation has been used to measure the pairing amplitude in the BEC-BCS crossover [Partridge *et al.* (2005); Werner *et al.* (2009)].

A first remarkable observation was the stability of the two-component Fermi gas while crossing the resonance. It is not trivial at all that the gas can remain stable in the vicinity of the resonance because of inelastic processes. We know, for instance, that a Bose-Einstein condensate is not stable in the presence of a sufficiently large

[22] A full theoretical description of the effect of this sweep is challenging, and is a clear example of interesting time-dependent problems in many-body physics.

negative scattering length.[23] When the inter-species scattering length changes from a large positive value to a large negative value, the many-body physics, combined with the Fermi pressure, turns out to stabilize the gas.

(ii) *Fermionic superfluidity*

A great challenge of these experimental studies was to unambiguously demonstrate the superfluid character of the Fermi gas and to check if it remains superfluid throughout the whole crossover. This was achieved by two experiments performed by the group of Wolfgang Ketterle at MIT.

In the first one [Zwierlein *et al.* (2005)], the Fermi gas was rotated in the trap above a certain critical angular frequency.[24] The cloud was observed by time-of-flight absorption imaging preceded by a sweep of the magnetic field that transformed the pairs into dimers, while maintaining the information on the velocity field, as explained above. The results shown in Fig. 26.9, clearly demonstrate the existence of lattices of quantized vortices for a wide range of values of the scattering length around the unitary limit. The Fermi gas is thus superfluid along the whole BEC-BCS crossover.

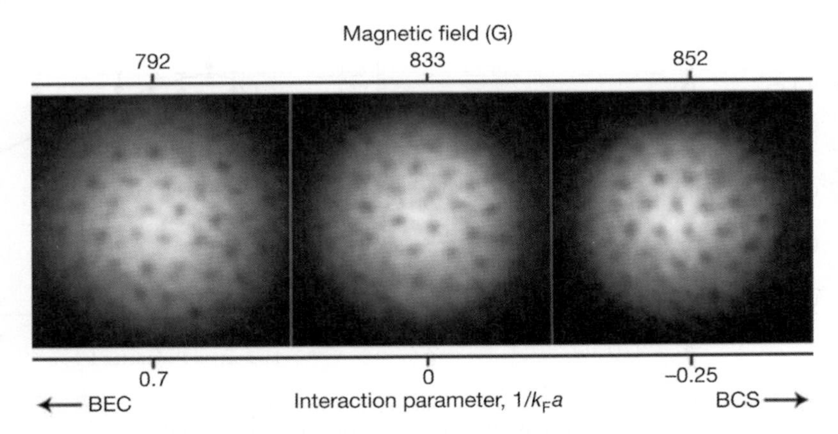

Fig. 26.9 Experimental observation of quantized vortices in a superfluid Fermi gas along the BEC-BCS crossover. Reprinted by permission from Macmillan Publishers Ltd: Nature [Zwierlein *et al.* (2005)].

The second experiment [Miller *et al.* (2007)], inspired by the original formulation of the Landau criterion, was performed by superimposing a moving one-dimensional optical lattice on the cloud. The optical potential acts as a moving roughness with a well-defined velocity. The experiment consists of searching for the threshold critical velocity above which dissipation occurs. This critical velocity v_c is given by

$$v_c = \min_{\vec{k}} \frac{E_k}{\hbar k} \tag{26.45}$$

[23]See Sec. 22.5 of Chap. 22.
[24]See Chap. 24.

where E_k is the energy of the elementary excitations. In the deep BCS regime, v_c is given by the slope of the straight line (long dashed line in Fig. 26.8) starting from the origin and tangent to the elementary excitation curve (solid line in Fig. 26.8). It is thus on the order of $\Delta/\hbar k_F$. There is also the Anderson Bogolubov sound mode, but the corresponding sound velocity, on the order of v_F, is much larger than $\Delta/\hbar k_F$, so that in the BCS region the critical Landau velocity is governed by the single-particle excitation (pair breaking). On the BEC side on the other hand, the usual sound velocity $c = \sqrt{\mu/m}$ is the dominant process for determining the Landau critical velocity. The left part of Fig. 26.10, extracted from [Combescot *et al.* (2006)] summarizes the theoretical predictions for the Landau critical velocity versus the interaction parameter $1/k_Fa$. The right part of this figure shows the experimental results described in [Miller *et al.* (2007)]. They clearly exhibit a maximum around the center of the Feshbach resonance. Among the different quantities that characterize the gas in the BEC-BCS crossover, the Landau critical velocity exhibits the most pronounced changes in the crossover region.[25]

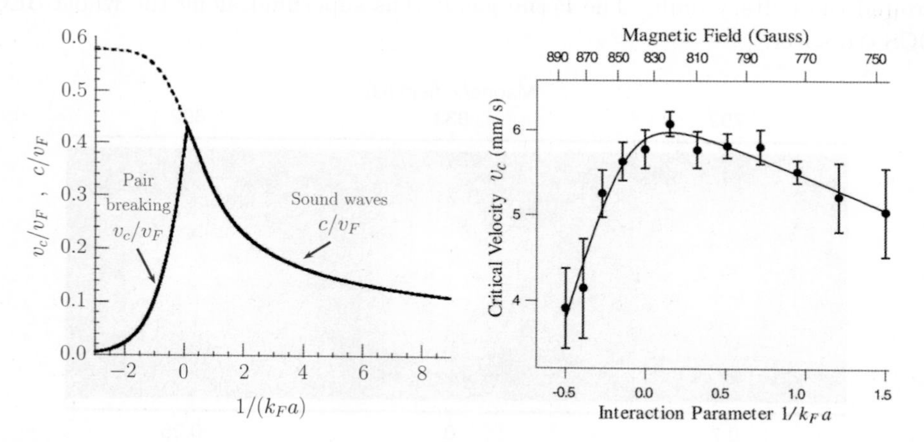

Fig. 26.10 Left: critical velocity v_c (full line) and sound velocity c (dashed line) as a function of $1/(k_Fa)$. The Landau criterion for superfluidity is governed by the Bogolubov sound waves in the BEC regime and by the pair breaking in the BCS regime. The sound mode becomes the Bogolubov Anderson mode on the BCS side. Figure extracted from [Combescot *et al.* (2006)]. Right: measurement of the critical velocities along the BEC-BCS crossover. A pronounced maximum is found at resonance. Figure extracted from [Miller *et al.* (2007)]. Copyright: American Physical Society.

(iii) *Fermi gas with imbalanced spin populations*

Interesting phenomena are expected when the two different spin states are not equally populated since not every particle can find a partner for pairing. In this context, one may wonder to what extent superfluidity can persist. This question

[25] The collective modes of strongly interacting two-species Fermi gases have been also investigated for trapped samples [Kinast *et al.* (2004); Bartenstein *et al.* (2004); Altmeyer *et al.* (2007)].

has been raised in various fields [Casalbuoni and Nardulli (2004)]. In the context of superconductors, one way to realize the imbalance between spin up and spin down electrons consists of applying a magnetic field. In 1962, Albert Clogston and Bellur Chandrasekhar independently established an upper limit of the magnetic field, above which superconductivity is destroyed [Clogston (1962); Chandrasekhar (1962)]. The argument of Clogston simply results from a comparison between Gibb's free energy of the superconducting and normal states. However, the imbalance of the two populations of electrons created by applying a magnetic field is in practice hindered by the Meissner effect.

With neutral two-component fermionic atoms, the imbalance of population can be more easily controlled and the consequences of this imbalance can be investigated along the whole BEC-BCS crossover [Zwierlein *et al.* (2006); Partridge *et al.* (2006)]. In the BCS regime, the breakdown of superfluidity is expected close to the point for which the chemical potential difference $\delta\mu = \mu_\uparrow - \mu_\downarrow$ is on the order of the pairing gap energy Δ. This superfluid to normal transition occurs even at zero temperature. It has been investigated experimentally by studying the number of vortices that could be generated by setting up the sample in rotation as a function of the population imbalance and for various values of the interaction parameter $1/(k_F a)$ [Zwierlein *et al.* (2006)] or by the equation of state (see next paragraph) [Navon *et al.* (2010)]. Surprisingly, superfluidity turns out to be remarkably stable against population imbalance at resonance.

(iv) *Universality and equation of state*

At the center of the interspecies Feshbach resonance, the scattering length diverges. There remains thus only one length scale, the interparticle distance, and one energy scale, the Fermi energy. The physics here no longer depends on the detail of the interaction (provided it is short range[26]) and is called for this reason "universal".

In this regime, also called the unitary regime, the gas is strongly interacting and there are no theoretical tools for giving the expression of the strongly correlated N-body wave function. Experiments are therefore of a crucial importance since they allow the determination of phenomelogical parameters of a thermodynamical description of this strongly interacting gas. As an example, the chemical potential at $T = 0$ of a two-species Fermi gas at unitarity is necessarily related to the energy scale of the problem, the Fermi energy, through a numerical factor, $\mu = \xi E_F$. Experiments have permitted the determination of the value of the dimensionless parameter $\xi \simeq 0.41(1)$ and provide a testbed for Quantum Monte-Carlo simulations (see [Navon *et al.* (2010); Hu *et al.* (2010)] and references therein). The stability of the two-component gas in this universal regime is ensured by the fact that $\xi > 0$. Indeed, the compressibility $\partial\mu/\partial n$ remains positive as expected for a stable phase.

The thermodynamical properties such as the occurrence of phase transitions can

[26]This means that the interparticle distance is large compared to the van der Waals range of the interaction and to the effective range. This is the case for most broad Feshbach resonances.

be inferred from the equation of state. Experiments can give a lot of insight into the equation of state as explained in the following. Such a macroscopic approach gives a direct derivation of important quantities such as the compressibility that would have been difficult to infer from a microscopic analysis.

Consider an homogeneous system (for which most theories have been developed). For a balanced mixture ($\mu = \mu_\uparrow = \mu_\downarrow$), the grand potential $\Xi = -P\Omega$ within the grand canonical ensemble depends on the volume Ω and on two intensive variables μ and T. In addition it also depends on the scattering length. A dimensional analysis then shows that the equation of state can be written as

$$\Xi(\mu, T, \Omega) = \Omega P_0(\mu) g\left(\frac{\mu}{k_B T}; \frac{\mu}{\hbar^2/(ma^2)}\right), \tag{26.46}$$

where $P_0(\mu) = \alpha m^{3/2}\mu^{5/2}/\hbar^3$ is the pressure of a single component ideal Fermi gas and $\alpha = 2^{3/2}/15\pi^2$ is a numerical factor. At zero temperature and in the unitary limit, the dimensionless function g reduces to the numerical factor $g(\infty; \infty)$.

When the gas is held in a trap, absorption imaging gives a priori access only to thermodynamic quantities that are averaged over the trap [Stewart *et al.* (2006); Turlapov *et al.* (2008)]. One may thus wonder to which extent is it possible to determine experimentally the function g that sets the macroscopic properties of an homogeneous system. To probe the local atom density, one can use an Abel transform of in situ images of a trapped gas as demonstrated in [Shin (2008)]. Another strategy consists in using the fact that the doubly integrated density profiles is simply proportional to the local pressure P of the grand canonical ensemble [Cheng and Yip (2007); Ho and Zhou (2009); Nascimbène *et al.* (2010)]. Both methods rely on the validity of the local density approximation. The equation of state of the spin-balanced unitary Fermi gas, that is the function $g(\mu/k_B T, \infty)$, has been determined and compared with various theoretical approaches such as diagrammatic and quantum Monte Carlo calculations, successive virial expansions up to the third order, self-consistent thermodynamics theory [Nascimbène *et al.* (2010)]. Interestingly, none of these different approaches coincides with the experimental data along the nearly three orders of magnitude range of $(\exp(-\mu/k_B T))$ explored experimentally.

The same procedure has been employed to determine the equation of state of a two component ultracold Fermi gas for a wide range of interaction strengths thanks to Feshbach resonance [Navon *et al.* (2010)]. Beyond mean-field corrections to the total energy of the gas could be studied systematically and compared with both Monte-Carlo calculations and Lee-Huang-Yang corrections. In addition, the low-temperature phase diagram of the spin imbalanced gas and the domain of validity of Fermi liquid behavior of the partially polarized normal phase for different interactions strength have been determined. Accurate experimental results obtained with such fermionic systems can therefore be used as a benchmark for many-body theory and are relevant to other fermionic systems such as the outer shell of neutron stars.

26.6 Conclusion

In this chapter, we have described a few experiments that clearly demonstrate how strong interactions can drastically change the properties of a quantum degenerate gas.

In certain cases, the observed phenomena are reminiscent of quantum phase transitions well known in condensed matter physics, such as the superfluid-Mott insulator transition. This was first illustrated in Sec. 26.2 by the analysis of the behavior of a quantum gas in a double potential. Such a system is the atomic physics analogue of the Josephson junction. It allows simple calculations that provide deeper insight into the Josephson effect, and give a quantitative description of the evolution of the system when the strength of the interactions is progressively increased: the frequency of the Josephson oscillations begins to change, then the coherence between the two wells is destroyed and the initial superfluid coherent state is transformed into a Fock state. Similar phenomena appear when the quantum gas is put in an optical lattice, a series of equidistant potential wells. In Secs. 26.3 and 26.4, we have described superfluid-Mott insulator transitions which can be observed with bosonic and fermionic quantum gases when the intensity of the laser beams producing the lattice is progressively increased.

In other situations, such as those described in Sec. 26.5 of this chapter, the observed phenomena are completely new. Scanning a Feshbach resonance gives the possibility of changing, in a continuous way, the scattering length a that describes atom-atom interactions at very low temperatures, and moreover to explore the region around the resonance where a is infinite. In Sec. 26.5, we described experiments studying the superfluidity of a Fermi system in the crossover region connecting the BEC and BCS regimes appearing on both sides of the Feshbach resonance. This crossover region was never explored before. These experiments greatly stimulate the development of new theoretical approaches for interpreting the results and clearly show the appeal of quantum gases for the study of strongly interacting systems:

- Possibility to control atom-atom interactions
- Possibility to control internal states and to easily change the respective populations of the two spin states which can form fermionic pairs
- Possibility to get clear signatures of superfluidity, such as the observation of lattices of quantum vortices
- Possibility to experimentally determine the equation of state of the gas, which can then be applied to the understanding of other systems

....

Other interesting examples of strongly interacting systems could have been given. Here, we just mention a few of them without entering into a detailed description.

Low dimensional gases

Quantum and thermal fluctuations are more important in 1D and 2D than in three dimensions. It is well known, for example, that no BEC can appear in a one or two dimensional ideal homogeneous gas.[27] In the presence of short range interactions, one can show that in 1D and 2D long range order is impossible in the thermodynamic limit at any non-zero temperature (Mermin-Wagner theorem, see [Mermin and Wagner (1966); Hohenberg (1967)] and references therein). The density of states in low dimensions is such that the low frequency, long wavelength, modes have a contribution large enough to destroy the long range order.

First consider the case of two-dimensional Bose gases.[28] Thermal fluctuations are strong enough to prevent the appearance of a long range order, but not strong enough to suppress superfluidity in an interacting Bose gas at low, but non-zero, temperature. There is a phase transition, first described by Vadim Berezinskii, John Kosterlitz and David Thouless (BKT) [Berezinskii (1971); Kosterlitz and Thouless (1973)] that occurrs at a critical temperature T_c. Below T_c, the Bose gas is superfluid and the first order correlation function $g_1(\vec{r})$ decreases algebraically with distance. The gas is said to exhibit a *quasi-long-range order*. Above T_c, the superfluidity disappears and the first order correlation function decreases exponentially with distance.

The fluctuations of the order parameter responsible for the BKT transition are essentially phase fluctuations.[29] The important point is that the source of these phase fluctuations are essentially vortices. Below T_c, vortices can only exist under the form of pairs of vortices with opposite circulations. The velocity field created by these pairs decreases very rapidly with the distance from the pair because the velocity fields of each vortex of the pair have opposite signs. As a result, the perturbation of the phase associated with the superfluid current is small and this current is not blocked. Above T_c, the pairs dissociate and free unbound vortices appear and produce a random phase pattern which inhibits the establishment of a superfluid current.

An experimental proof of this mechanism of the BKT transition has recently been obtained with two-dimensional quantum gases using an interferometric detection scheme (see Fig. 26.11) [Hadzibabic *et al.* (2006); Cladé *et al.* (2009a)]. Another interesting experiment concerns the determination of the equation of state which, as in the three-dimensional case studied in this chapter, provides macro-

[27] See Sec. 20.3.6 of Chap. 20.

[28] For a review of the properties of two-dimensional Bose fluids from an atomic physics perspective, see [Hadzibabic and Dalibard (2009)].

[29] The density fluctuations are reduced by the interactions.

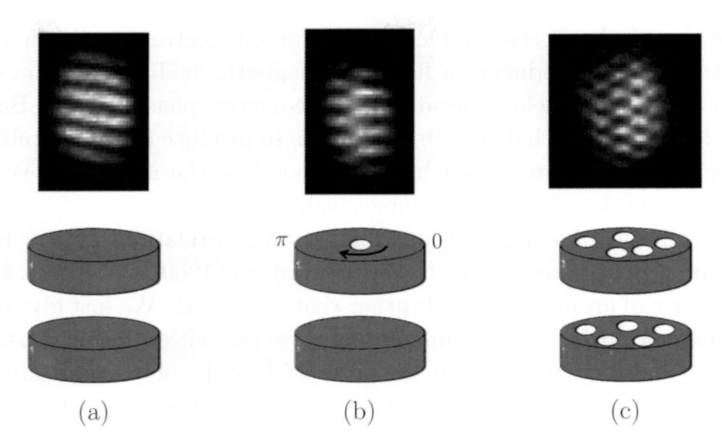

Fig. 26.11 Two two-dimensional Bose gases initially confined in two parallel planar traps are observed through the interference fringes that appear when these two gases expand and overlap after a time-of-flight. (a) If no free vortex exits in either plane, one observes fringes parallel to the planes with a contrast that gives an indication of the degree of spatial coherence. (b) If one free vortex appears in the center of one of the two planes, the matter wave at the two extremities of this plane have a π-phase shift because of the variation of the phase around the core of the vortex. The interference fringes with the matter wave coming from the other plane without vortex will thus exhibit a dislocation between the two extremities of the planes. (c) When the temperature increases, the contrast of the fringes decreases and, at a certain temperature, dislocations appear proving the proliferation of free vortices. Courtesy of J. Dalibard.

scopic information on the system that complements the microscopic analysis [Rath *et al.* (2010); Hung *et al.* (2011)]. The challenge is to reach the strongly interacting regime where the existing theories are still in debate and where experimental results on two-dimensional Bose gases could serve as benchmarks for these theories. It turns out that, in two-dimensional gases, interactions only depend on a dimensionless ratio a/ℓ_0, where a is the scattering length and ℓ_0 a length characterizing the transverse confinement. When this ratio has a small value and neglecting the energy dependence of the scattering length [Petrov and Shlyapnikov (2001)], the equation of state has a scale invariance: it is not an independent function of the chemical potential μ and the temperature T and it depends only on the ratio μ/k_BT. Strong correlations only appear for large values of a/ℓ_0 and the scale invariance of the equation of state is no longer valid.[30]

The physics of rotating two-dimensional Bose gases is also very interesting because it presents close connections with condensed matter problems and offers the possibility to prepare quantum Hall-like states. The Coriolis force in the rotating

[30]Another original feature of two-dimensional systems concerns the consequences of the indistinguishibility of identical particles. Because of topological constraints specific to two-dimensional geometries, the number of possible routes that two identical particles can follow to go from an initial state to a final state is increased. As a result, the Bose and Fermi rules used to combine amplitudes can be generalized to accomodate new particles called anyons [Leinaas and Myrheim (1977); Wilczek (1982); Camino *et al.* (2005)].

frame is similar to the Lorentz force, which acts on electrons in a magnetic field. Another method for introducing a fictitious magnetic field in a system of neutral atoms is to use a gauge field generated by a geometric phase like the Berry phase [Lin *et al.* (2009)].[31] The challenge in this field is to produce highly correlated states where the number of atoms is not large compared to the number of vortices and where a mean-field description is no longer valid.

Another example of a situation in which strong correlations appear is found in one-dimensional Bose gases [Tonks (1936); Girardeau (1960)]. However, this occurs when the density of atoms is reduced rather than increased. We just give qualitative arguments here. Consider a one-dimensional Bose gas with a linear density n. The mean distance between atoms is on the order of $1/n$. If interactions are repulsive, each atom is confined between two neighbors in an interval of length $1/n$. The corresponding kinetic energy of confinement is $E_{\text{kin}} \simeq \hbar^2 n^2/m$. The interaction energy is equal to $E_{\text{int}} \simeq gn$, where g is the coupling constant. Strong correlations appear when the ratio $E_{\text{int}}/E_{\text{kin}} \simeq gm/\hbar^2 n$ is large, i.e. when n is small. In such one-dimensional strongly interacting Bose gases, bosons acquire some properties reminiscent of those of a one-dimensional fermionic ideal gas with, for instance, the same modulus of the N-body wavefunction $|\psi_B(x_1, \ldots, x_N)| = |\psi_F(x_1, \ldots, x_N)|$. The gas in this regime is called a *Tonks-Girardeau gas*. For a description of a few recent experiments performed on Tonks-Girardeau gases, see [Paredes *et al.* (2004); Kinoshita *et al.* (2004); Haller *et al.* (2009)].

Quantum gases in disordered potentials. Anderson localization

We turn now to a one-body problem of importance in condensed matter and that has been recently revisited with cold atoms. We finally discuss how interactions between atoms open interesting perspectives for strongly interacting regimes in the presence of disorder.

In a quantum description of the conductivity of electrons in a metal, the wave nature of electrons plays a crucial role and yields surprising results such as the perfect conductivity of an ideal crystal. The resistivity is then explained as being due to the scattering of the electronic waves from the imperfections of the crystal. The question that is immediately raised by this quantum result is the link between the amount of disorder and the conductivity properties. In 1958, Philip Anderson published a paper suggesting that a slight change of the amount of disorder could trigger a sudden phase transition from conductor to insulator [Anderson (1958)]. This dramatic change of the material properties is an interference phenomenon. For a sufficiently large amount of disorder, the interferences between the many paths associated with coherent multiple scattering from random impurities localize the wavefunction, which gets exponentially decaying tails. Actually, the nature of the Anderson localization depends on the dimensionality: in 1D, the wave functions are

[31] See Sec. 14.5 of Chap. 14.

expected to be exponentially localized in the presence of disorder for any energy, in 2D a similar behavior is expected, but in 3D, a phase transition occurs with states that are exponentially localized and others that are extended, depending on the relative value of the electronic wave vector compared to the so-called mobility edge wave vector.[32]

In most materials, however, this so-called Anderson localization is hardly separable from other mechanism that induce metal-insulator transitions. Later on, it was realized that Anderson localization is ubiquitous in wave physics. It has been observed in a large variety of systems involving classical waves such as sound waves, light in diffusive media, microwaves, etc.[33] In 2008, two experimental groups succeeded in observing the one-dimensional Anderson localization of matter-waves by taking advantage of the versatility and the large degree of control available in ultracold atomic samples. At Institut d'Optique in Palaiseau (France) the disordered potential was obtained by superimposing a laser speckle potential pattern on a matter-wave guide [Billy *et al.* (2008)]. At LENS in Florence (Italy) a one-dimensional quasi-periodic incommensurate lattice was used to realize a 1D Aubry-André Hamiltonian, which exhibits an Anderson localization in a certain range of parameters [Roati *et al.* (2008)]. The experiments were performed by letting the atoms of a non-interacting Bose-Einstein condensate expand along these one-dimensional potentials. The observations realized by absorption or fluorescence imaging reveal the exponential tails of the atomic density distribution as a proof of Anderson localization. In 2008, 3D Anderson localization was observed in momentum space at PhLAM laboratory in Lille (France) with cold atoms in a quasiperiodic kicked rotor potential. The critical exponents of the phase transition deduced from this experiment are consistent with 3D Anderson model [Chabé *et al.* (2008)].

A system of non-interacting particles in disordered potentials is an oversimplistic description of a solid. It is thus important to understand the interplay of interactions and disorder. In 2010, an experiment performed at LENS with potassium atoms has explored the role of weak repulsive interactions using a Feshbach resonance [Deissler *et al.* (2010)]. An Anderson glass phase, for which several exponentially localized states coexist without phase coherence, was identified as a result of this interplay. For stronger interactions, one expects the suppression of single-particle localization, though new concepts such as many-body Anderson-localization [Basko *et al.* (2007)] or Bose glass are involved. The investigation of two-component Fermi gases along the BEC-BCS crossover in the presence of disorder is also a fascinating prospect [Sanchez-Palencia and Lewenstein (2010)].

[32] The destructive interferences responsible for the localization occur when the mean free path between two scattering events is smaller than $\lambda/2\pi$ where λ is the de Broglie wavelength associated with the electrons. This latter criterion is referred to as the Ioffe-Regel criterion.

[33] See [Lagendijk *et al.* (2009)] and references therein.

Chapter 27

Extreme light

27.1 Introduction

Most of the time, advances in atomic physics are determined by technological breakthroughs in the development of light sources with unprecedented performances. In this respect, lasers continue to open a wealth of new research fields that could not be tackled without the coherence, the high intensity and the short pulse duration offered by laser light. In this chapter, we illustrate this idea by describing a new frontier of atomic physics, attosecond science, which is being explored thanks to a new type of extreme light consisting of ultra-short pulses that last a few tens of attoseconds (1 as $= 10^{-18}$s). What are the underlying physical mechanisms giving rise to such short pulses? How can they be measured? What kind of new problems do they allow one to investigate? These are a few questions that we would like to address in this chapter.

Figure 27.1 shows the decrease of the minimum duration of achievable laser pulses over several decades. The continuous decrease from 1964 to 1986 is due to the invention and development of mode locking. At the end of this span of time, ultrashort pulses reached a duration of about 6 fs, corresponding to about 2 or 3 oscillations of the optical or infrared laser field. Going below this value would have required using shorter wavelength lasers and finding adequate nonlinear material for extending mode locking techniques to this frequency domain. But this is not easy to achieve. A completely different approach for breaking the femtosecond barrier was provided by a completely different scheme using the high harmonic generation phenomenon (HHG) described in Chap. 10. This is a beautiful example of how a breakthrough in technology can be achieved, not by a continuous improvement of existing technologies, but by a discontinuous transition involving radically new ideas.

In Sec. 27.2, we give a brief survey of the evolution of this research field and of the new problems that can be investigated with attosecond pulses. More details may be found in recent review papers, such as [Corkum and Krausz (2007); Krausz and Ivanov (2009)]. We first explain the mechanism of production of such ultrashort pulses (Sec. 27.2.1). The first experiments using HHG produced trains of attosecond

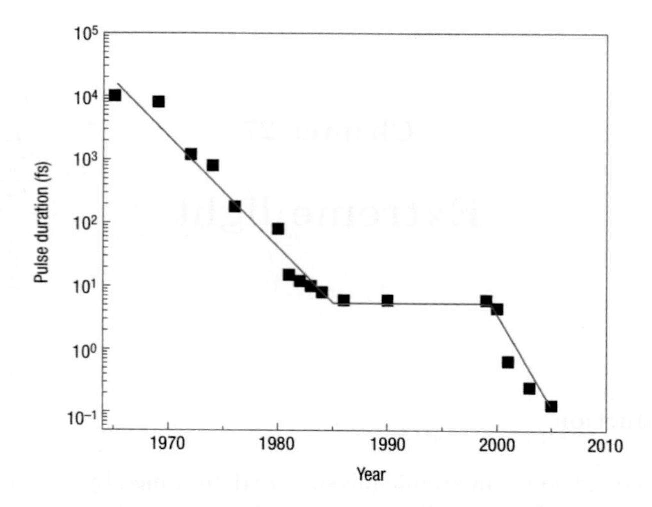

Fig. 27.1 Evolution of the minimum duration of achievable laser pulses during the last few decades. Reprinted by permission from Macmillan Publishers Ltd: Nature Physics [Corkum and Krausz (2007)].

pulses in the UV and X domains (XUV). In Sec. 27.2.2, we describe how the relative phases of the various harmonics that beat to produce attosecond pulses can be measured by the so-called "RABITT" method . Having a train of attosecond pulses is not very convenient for doing time-resolved spectroscopy, so researchers tried to produce single isolated attosecond pulses by using few-cycle optical or near-infrared (IR) laser pulses for producing HHG. The difficulty that then appears is that the time-offset Δt_{peak} between the crests of the carrier laser field and the maximum of the envelope of the few cycle laser pulse generally varies from one pulse to the next, producing large fluctuations in the production of attosecond pulses. In Sec. 27.2.3, we describe how this carrier-envelope phase can be controlled in order to create reproducible attosecond pulses. Once the production of well-controlled and well synchronized laser and XUV pulses is mastered, an important issue is then to measure their characteristics, for example the shape and duration of one XUV attosecond pulse. This is not easy because the usual methods of nonlinear optics developed for femtosecond pulses cannot be extended to this time-scale. In Sec. 27.2.4, we describe a new method that was invented for solving this problem called "attosecond streak imaging" or "atomic transient recorder" [Kienberger *et al.* (2004)]. Finally, in Sec. 27.2.5, we briefly review a few recent applications of attosecond pulses concerning the observation of molecular structures and the real time investigation of multi-electron dynamics.

The XUV attosecond bursts described in Sec. 27.2 appear only if the infrared femtosecond laser pulses used for producing HHG have a high enough intensity, on the order of 10^{14} W/cm^2. Attosecond science is therefore strongly related to the development of high intensity lasers. In Sec. 27.3, we give a very brief review of important advances in this field, which are based essentially on the mode locking and chirped pulsed amplification (CPA) techniques. These have the great advantage of allowing the realization of table-top high intensity lasers that can be used in small scale laboratories. High intensity lasers have, of course, important applications in other fields outside atomic physics, such as plasma physics, beam acceleration or relativistic optics. Here, we will only list a few of them, and refer the reader to recent review papers [Brabec and Krausz (2000); Mourou *et al.* (2006); Krausz and Ivanov (2009)].

27.2 Attosecond science

27.2.1 *Mechanism of production of attosecond pulses*

In Chap. 10, we showed that at high enough laser intensities, the ionization of an atom is no longer a multiphoton process. This occurrs when the laser field becomes on the order of the Coulomb field binding the atomic electron to the ion core. The Coulomb barrier, which prevents the electron from escaping, acquires a small enough width that the electron can tunnel through it. The interesting feature of this mechanism is that once the electron has been released, it is accelerated by the oscillating laser field and can come back near the ion and *re-collide* with it. This produces the ultrashort pulses of XUV radiation responsible for high harmonic generation.[1]

Because of the sensitivity of tunnelling rates to the width and height of the potential barrier that the electron has to cross, tunnel ionization occurs only near the crests of the laser electric field, in a time window about ten times smaller than the period of the laser field, i.e. on the order of 200 or 300 as. This explains how the attosecond timescale appears in the electron bursts that are produced by tunnel ionization and in the subsequent XUV pulses that are emitted when the freed electrons recollide with the parent ion. This research field has known spectacular developments during the last ten years (see Fig. 27.1).

The first clear evidence of attosecond structures in the train of XUV pulses generated by a multiple-cycle laser pulse was obtained in 2001 by the group of Pierre Agostini, which demonstrated the production of trains of XUV pulses with a duration of 250 as each [Paul *et al.* (2001)]. Another similar experiment was also realized in 2001 by the group of Ferenc Krausz, which demonstrated the production of single attosecond pulses with a duration less than 650 as [Hentschel *et al.* (2001)].

27.2.2 *Multiple-cycle laser pulse. Train of attosecond pulses*

If the infrared laser pulse contains many oscillations of the laser field separated by the period $T = 2\pi/\omega_L$ of this field, a train of XUV bursts separated by $T/2$

[1]It is important to note that this recollision process strongly depends on the ellipticity of the polarization of the IR pulse. As shown in Chap. 10, recollision processes are most efficient when the polarization is linear and do not occur when the polarization is circular.

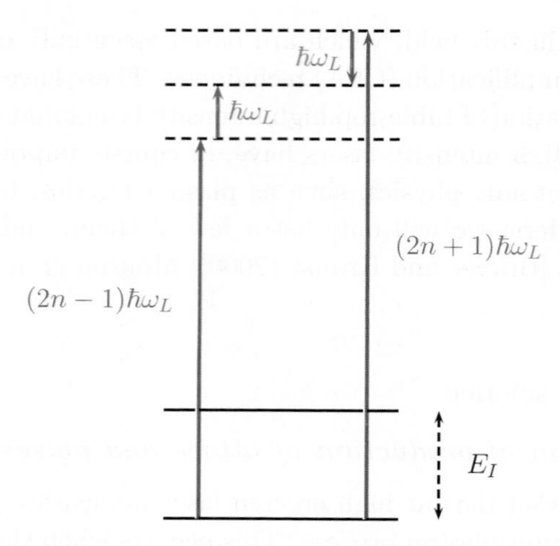

Fig. 27.2 Dominant two-photon processes involving the absorption of one XUV photon and the absorption or stimulated emission of one infrared photon. These processes are responsible for the appearance of peaks at even harmonics of $\hbar\omega_L$ in the photoelectron spectrum of the probe atom.

is produced, whose Fourier decomposition is a series of harmonics $(2n+1)\omega_L$. If one knows the amplitudes and the relative phases of these harmonics, one can reconstruct the XUV field resulting from their superposition and determine the duration of the individual pulses of the train.

The amplitudes of the harmonics is easily deduced from the spectrum of the XUV radiation. To determine the relative phases, one idea is to photoionize another probe atom by the train of XUV pulses in the presence of the infrared pulse. One then looks at the photoelectron spectrum of the probe atom where new structures are expected to appear as a result of two-photon transitions involving one harmonic XUV photon and one infrared laser photon [Véniard *et al.* (1996)]. Figure 27.2 explains the principle of this method, dubbed "reconstruction of attosecond beating by interference of two-photon transitions" (RABITT) [Muller (2002)]. Consider two successive high order harmonics at frequency $(2n-1)\omega_L$ and $(2n+1)\omega_L$. Absorption of one photon of these harmonics by the probe atom brings it to states of the ionization continuum located at energies $(2n-1)\hbar\omega_L$ and $(2n+1)\hbar\omega_L$ above the ground state g (E_I is the ionization energy). Even energies $2n\hbar\omega_L$ appear in the spectrum, and result (to the lowest order) from two distinct processes: (i) absorption of one harmonic photon $(2n-1)\hbar\omega_L$ and absorption of one laser photon $\hbar\omega_L$ (left part of Fig. 27.2) (ii) absorption of one harmonic photon $(2n+1)\hbar\omega_L$ and stimulated emission of one laser photon $\hbar\omega_L$ (right part of Fig. 27.2).[2] The two transition amplitudes describe processes ending at the same final state. They interfere and

[2]Figure 27.2 represents only the dominant processes where the XUV is first absorbed. Other processes could be considered where the absorption of the XUV photon occurs after the absorption or stimulated emission of the infrared photon, but they have a much smaller amplitude because of the large energy denominator in the two-photon matrix element.

the magnitude of the new peak at $2n\hbar\omega_L$ in the photoelectron spectrum depends on the relative phase $\varphi_{2n+1}-\varphi_{2n-1}$ of the two harmonics. If one introduces a time delay τ between the XUV and infrared pulses, one observes oscillations in the magnitude of the peak at $2n\hbar\omega_L$ when τ is varied, from which one deduces $\varphi_{2n+1} - \varphi_{2n-1}$. By repeating this analysis for the other even peaks of the photoelectron spectrum, one can determine the relative phases of the various harmonics and reconstruct the field resulting from their interference. It is in this way that trains of pulses with a duration of 250 as each have been identified (see [Paul *et al.* (2001)] for more detail). Since then, other approaches have been implemented [Corsi *et al.* (2006); Wagner *et al.* (2007)].

27.2.3 *Few-cycle laser pulse. Control of the carrier-envelope phase*

Laser pulses with only two or three optical cycles are preferred if one wants to get a single XUV burst for triggering a physical process and for studying, with another delayed pulse, its subsequent evolution. In such a few-cycle laser pulse, there are only a few crests with different amplitudes within the envelope of the pulse, and the largest crest does not necessarily coincide with the maximum of the pulse envelope. The time-offset Δt_{peak} between the position of the two maxima can be written $\Delta t_{\text{peak}} = \varphi/\omega_L$, where φ is called the "carrier-envelope phase" (CE phase). Figure 27.3 shows the shape of two pulses corresponding to $\varphi = 0$ (called cosine waveform) and $\varphi = \pi/2$ (called sine waveform).

It is clear that the number and the energy of the XUV photons emitted by a few-cycle laser pulse depend on the CE phase. For example, if $\varphi = 0$, one can show that a single XUV photon, with the largest possible energy, is emitted by a electron freed by tunnel-ionization near the negative crest preceding the largest crest and recollides with the parent ion near the first zero of the laser field following the largest

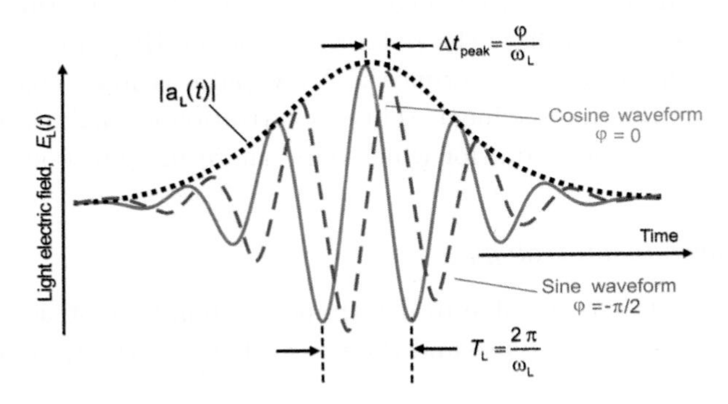

Fig. 27.3 Few-cycle laser pulse with a carrier envelope phase $\varphi = 0$ (full line) and $\varphi = \pi/2$ (dashed line). The envelope of the pulse is represented by the dotted line. Figure extracted from [Krausz and Ivanov (2009)]. Copyright: American Physical Society.

crest.[3] By filtering the XUV radiation near the cut-off of the HHG spectrum, one can isolate this XUV photon. Similarly, if $\varphi = \pi/2$, one can show that there are two photons with the highest possible energy emitted at two times differing by $T/2$. Near the cutoff, this spectrum exhibits modulations at the frequency $2\omega_L$. By looking at the filtered upper part of the XUV spectrum one can thus distinguish a CE phase $\varphi = 0$ (continuous decrease of the spectrum) from a CE phase $\varphi = \pi/2$ (modulation of the spectrum). Other signals, like the upper plateau of the ATI spectrum are also very sensitive to the CE phase and can be used for its estimation.[4]

The important issue is of course to precisely measure φ and stabilize it so that it keeps the same value from one laser shot to the next so that the same XUV pulse is emitted for each laser pulse. In fact, if φ varies from one shot to the next, this means that the group velocity of the laser light (velocity of the maxima of the pulse envelope) differs from the phase velocity (velocity of the crests). One can get a better understanding of the CE phase by considering the Fourier analysis of the train of femtosecond pulses delivered by the mode locked laser, which is a comb of angular frequencies ω_n given by:[5]

$$\omega_n = n\,\omega_r + \omega_{CE} \tag{27.1}$$

where ω_r is the repetition frequency of the pulses and ω_{CE} an offset. The CE phase, and more precisely its change $\Delta\varphi$ from one pulse to the next, is simply related to ω_{CE}: $\Delta\varphi = 2\pi\omega_{CE}/\omega_r$. To get the same CE phase for all pulses, one must therefore measure and control ω_{CE}. This is achieved by using a so-called "$f - 2f$ interferometer". If the frequencies of the comb span an octave, one measures the beat note between the frequency $2\omega_n = 2\omega_{CE} + 2n\omega_r$ of the frequency doubled comb and the frequency $\omega_{2n} = \omega_{CE} + 2n\omega_r$ of the original comb. The frequency of this beat note is nothing but ω_{CE}. For more details about the sequence of operations leading to a full control of the CE phase of few-cycle laser pulses, see [Baltuska *et al.* (2003); Udem and Riehle (2007)] and [Krausz and Ivanov (2009)].

Once the CE phase is stabilized, a convenient method for isolating a single XUV attosecond pulse is to modulate the polarization of the IR pulse and use the fact that the recollision process generating the XUV pulse is most efficient when the polarization of the IR pulse is linear. This "polarization gating" thus provides the possibility of favoring conditions for which only one recollision process is possible.[6]

27.2.4 *Attosecond metrology*

We now explain how it is possible to determine the temporal structure of the two pulses, the XUV attosecond pulse and the few-cycle infrared (IR) laser pulse.

---

[3]See Sec. 10.5.3 of Chap. 10.
[4]See Sec. 10.3 of Chap. 10.
[5]See Sec. 6.7 of Chap. 6.
[6]See for example [Sansone *et al.* (2009)] and references therein.

To introduce the principle of the method it will be simpler to first explain how it is possible to reconstruct the IR pulse when the duration of the attosecond pulse is much shorter than the period of the laser field. When the phase matching conditions are well satisfied for the propagations of the two IR and XUV pulses, the two pulses propagate along the same direction and the XUV pulse is laser-like, with an angular divergence smaller than the IR pulse (because of its smaller wavelength) and consequently a smaller transverse radius. Both pulses are reflected by a two-component mirror composed of a movable central part, which reflects only the XUV pulse because of its smaller radius, and an annular part, which reflects only the IR pulse. By piezo-electrically moving the central part of the mirror, one can introduce a variable delay τ between the two pulses (a displacement of 100 nm gives rise to a delay of 300 as). The two pulses are then focused on a beam of rare gas atoms which are photoionized only by the XUV pulse during a time interval corresponding to the duration of the attosecond pulse, which is negligible compared to the period of the IR field.[7] The energy of the photoelectron is measured with a time-of-flight electron spectrometer for different values of the delay τ between the two pulses.

When the photoelectron is created by the XUV pulse, it is created in the presence of the IR pulse. The electric field of this pulse lasts a longer time than the attosecond XUV pulse. We saw in Sec. 10.5.3 of Chap. 10 (see Eq. 10.17) that the IR pulse can increase the velocity of the electron by an amount $\Delta v = -(|q|/m)A(t_0)$ depending only on the value of the potential vector $A(t_0)$ of the IR pulse at the instant t_0 when the electron appears. By changing τ, one can change the instant of birth of the photoelectron produced by the XUV pulse with respect to the IR pulse, and therefore determine the time variations of the potential vector of the IR field, and after differentiation, those of the electric field. In fact, with the spectrometer one measures the kinetic energy of the photoelectron. If v_{in} is its initial velocity at the instant t_0 of the photoionization, its energy when it is detected is equal to $m[(v_{in} + \Delta v)]^2/2$. If $\Delta v \ll v_{in}$, the energy change ΔE

$$\Delta E \simeq m v_{in} \Delta v \tag{27.2}$$

is proportional to Δv. The variations with τ of the energy of the electron measured by the detector reflect the variations of Δv and thus those of $A(t_0)$. Figure 27.4 shows the reconstruction of the oscillations of the IR electric field obtained in this way.

For measuring the shape of the XUV pulse, one uses another method, called "attosecond streak imaging technique". As in the previous experiment, one measures the energy spectrum of the photoelectrons obtained by illuminating rare gas atoms with the few-cycle laser pulse and the XUV pulse with an adjustable delay. One chooses this delay in such a way that the XUV pulse, whose duration is smaller

[7] We suppose that the intensity of the IR pulse is not high enough for ionizing the atom, but high enough for communicating an appreciable velocity change to the electron that has been freed by the XUV pulse.

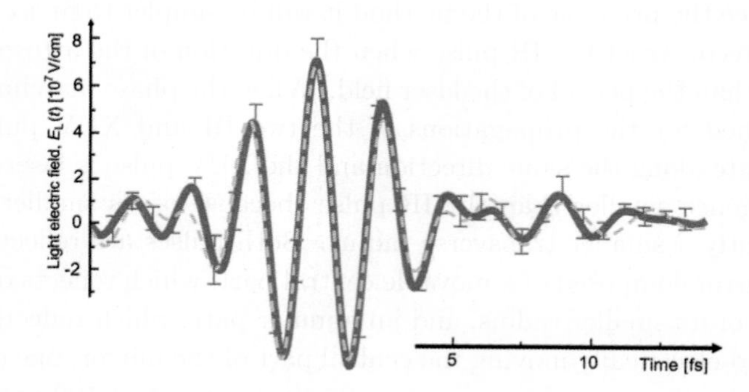

Fig. 27.4 Reconstruction of the time variations of the electric field of a near infrared few-cycle laser pulse [Goulielmakis *et al.* (2004)]. Figure adapted from [Krausz and Ivanov (2009)]. Copyright: American Physical Society.

than a half period of the laser pulse, is contained in a time window where the potential vector $A(t)$ is a monotonic function of t. For example, this time window is located between a minimum of $A(t)$ and its next maximum. As shown in Sec. 10.5.3 of Chap. 10, an electron ionized at time t by the XUV pulse undergoes a velocity change $\Delta v(t)$ proportional to $A(t)$ before reaching the detector. This produces a mapping of the time-dependent intensity profile of the XUV pulse into a corresponding final velocity (or energy) distribution of the photoelectron (see Fig. 27.5). Knowing $A(t)$, one can thus retrieve the temporal dependence of the XUV pulse. This method can be considered to be an optical extension of the streak camera driven by a microwave field used for analyzing time dependent phenomena with a sub-picosecond resolution [Bradley *et al.* (1971); Schelev *et al.* (1971)].

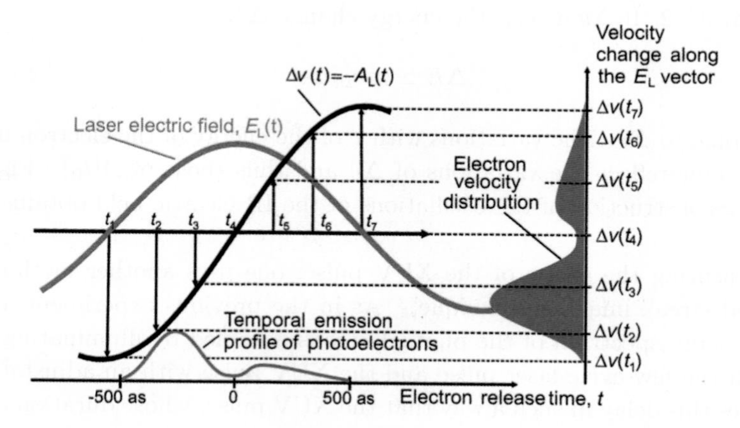

Fig. 27.5 Principle of attosecond streak imaging which maps a time profile into a velocity distribution. Figure adapted from [Krausz and Ivanov (2009)]. Copyright: American Physical Society.

Remark

We have implicitly supposed in the previous discussion that the initial velocity of the photoelectrons produced by the XUV pulse is the same during the whole duration of the attosecond pulse. This is not true in general because the attosecond pulse can be chirped, with a frequency varying in time, resulting in a time-dependent initial energy of the photoelectron. Tomographic techniques are then required to extract the time profile of the attosecond pulse from the velocity distribution of the photoelectrons [Kienberger *et al.* (2004)]. The frequency chirp of the attosecond pulse is in fact due to a lack of synchronization of the high-harmonic XUV photons, which increases the duration of the attosecond pulse (see Ref. [Mairesse *et al.* (2003)] where proposals for reducing this lack of synchronization are presented).

27.2.5 *A few applications of attosecond pulses*

(i) *Laser induced electron diffraction*

Quantum calculations, confirmed by experiments, indicate that the wave packet describing the freed electron expands laterally and has a lateral spread on the order of a few Å when it recollides with the atomic or molecular parent ion. In a typical experiment, the current density seen by the parent ion is on the order of 10^{11} A cm^{-2}, which is quite appreciable [Corkum and Krausz (2007)]. These wave packets can be diffracted by the parent ion to provide an image of the molecular structure [Meckel *et al.* (2008)]. The interesting feature of these experiments is that electron diffraction occurs in a very short time window, on the order of a few hundreds attoseconds. They thus combine attosecond time resolution with angstrom spatial resolution.

(ii) *Attosecond imaging of electronic wave functions*

The high-harmonic emission produced by the recollision process is due to the dipole spatial distribution $\vec{d}(\vec{r})$, proportional to $\psi^*_{\rm el}(\vec{r})\, q\vec{r}\, \psi_{\rm ion}(\vec{r})$, where $q\vec{r}$ is the dipole moment and $\psi_{\rm el}(\vec{r})$ and $\psi_{\rm ion}(\vec{r})$ are the wave functions of the re-colliding electron and parent ion. The attosecond pulse that is emitted during this process thus contains spatial information on the two wave functions, which can be extracted by tomographic reconstruction, as suggested by Jiro Itatani [Itatani *et al.* (2004)]. Experiments of this type have recently been realized [Haessler *et al.* (2010)]: the molecular orbitals of aligned N_2 molecules have been imaged at different times to visualize the motion of a density hole in the molecular ion.

This technique of attosecond imaging can actually be considered to be the reverse of photoelectron spectroscopy where information on the light that produces the photoionization can be obtained by studying the properties of the photoelectrons. Here, the re-colliding electron is captured by the ion and the properties of the light that is emitted in this process gives information on the atomic or molecular system that captures the electron.

(iii) *Electron dynamics*

Laser femtosecond pulses are very useful for investigating the motion of atoms in molecules because the timescale of this motion is on the order of a few tens to a few hundreds of femtoseconds. For example, femtochemistry [Zewail (2000)] has allowed a real-time observation of molecular processes, such as vibrations or chemical reactions. The motion of electrons in atoms, molecules and solids is much faster, in the attosecond range.

For example, the characteristic time for the motion of an electron in the ground state of the hydrogen atom h/E_I, where $E_I = m_e c^2 \alpha^2/2$ is the ionization energy and α the fine structure constant, is on the order of 300 as.[8] Attosecond pulses are thus essential for observing real-time electron dynamics.

Examples of interesting problems that are now being investigated are complex re-arrangement processes resulting from electron-electron interactions following a sudden perturbation, due to the recolliding electron or to the XUV burst. Several non-sequential and sequential mechanisms have been identified in multiple-ionization of rare gas atoms [Rudenko *et al.* (2004)]. Another example concerns Auger processes. Inner shell vacancy can be produced by photoionization of deeply bound electrons by the XUV burst. This vacancy can be filled by a valence electron, the energy loss of this electron being taken by another (Auger) electron that leaves the atomic ion. The lifetime of the Auger recombination has been measured [Drescher *et al.* (2002)] (see also [Uiberacker *et al.* (2007)] where the effect of "shake-up" processes is studied). Attosecond spectroscopy has been applied not only to gas-phase systems but also to condensed matter and to surfaces. Photoelectron emission from single-crystal tungsten has been probed with attosecond resolution to demonstrate a delay of about 100 as between the photoelectrons originating from localized core states of the metal and those that are freed from delocalized conduction-band states [Cavalieri *et al.* (2007)].

27.3 Ultra intense laser pulses

The XUV attosecond bursts described in the previous sections were made possible by the development of high intensity pulsed lasers. In this section, we give a brief overview of the different techniques that have been developed in the course of time to generate such light sources.

A laser pulse is characterized by its duration and energy, or its power (i.e. energy divided by pulse duration). A crucial quantity for atomic physics is the local intensity (i.e. power per unit surface) achievable at the atom position in an atomic gas or jet.

In Chap. 10, we have discussed some nonlinear effects such as the shift of the

[8]The atomic unit of time, equal to 24 as, is defined as the time required to travel over a distance equal to the Bohr radius a_0 with a velocity $c\alpha$ where α is the fine structure constant.

ionization energy and the above threshold ionization spectrum that occur when the kinetic energy associated with the oscillatory motion of the electron in the laser field is on the order of few eV, which corresponds to intensity up to few 10^{13} W/cm^2 typically. Let us recall that an intensity of 10^{14} W/cm^2 corresponds to a laser electric field strength on the order of 3×10^{10} V/m, a value comparable to the field binding the electron to the atom.[9] In this case, tunnel ionization can then be observed.[10] At a macroscopic scale, the forefront of a focussed laser pulse creates a plasma in which the rest of the pulse propagates. For intensity above $\sim 10^{18}$ W/cm^2, the electrons of the plasma generated by the pulse are accelerated to velocity close to the speed of light, this is the domain of relativistic optics.

The quest for high intensity lasers started with the Q-switched lasers during the very early days of laser technology. This technique, described in Sec. 27.3.1, offers the possibility to stock a large amount of energy (tens of J) which is then delivered in a ns pulse typically. Alternatively, the development of the mode-locking technique yielded higher intensities using a much shorter pulse (few tens of fs) with a smaller amount of energy (mJ range).[11] In Sec. 27.3.2, we discuss some key features of this latter technique and underline the importance of the amplifying medium properties. Both techniques are sometimes applied at once.[12]

In Fig. 27.6, we summarize the evolution of the laser peak intensity as a function of time during the last few decades. After the rapid increase due to Q-switching and mode locking in the sixties, this intensity reached a plateau for nearly twenty years. A breakthrough occurred in the mid-eighties with the advent of the chirped pulsed amplification (CPA) technique which enables one to reach output pulses with a very large intensity while avoiding damages in the intermediate amplifying stages. The principle of the CPA technique is presented in Sec. 27.3.3.

High intensity lasers have numerous applications.[13] We briefly discuss a few of them in Sec. 27.3.4: (i) nonlinear relativistic optics with the example of the relativistic self-focussing, (ii) the use of laser-plasma interaction to generate bunches of electrons with GeV energy with possible applications in medicine for instance.

27.3.1 *Q-switched lasers*

Q-switching is realized by introducing a Q-switch device in the optical resonator cavity of the laser. Such a device is used to switch the quality factor (Q) of the laser

[9]The Coulomb field created by a proton at a distance equal to twice the Bohr radius is on the order of 1.5×10^{11}V/m.
[10]See Chap. 10.
[11]A 10 fs pulse with an energy of 1mJ has a peak power of 100 GW (which corresponds to the output power of 100 nuclear reactors, of course for a very short amount of time!). When such a pulse is focussed on a waist of 50 μm, it delivers a local peak intensity as high as 4×10^{15} W/cm^2.
[12]High-energy (>kJ) petawatt (10^{15} W) lasers are currently under development, for instance at Lawrence Livermore National Laboratory, for heating and igniting deuterium and tritium fuel to fusion temperatures. The description of this class of extreme lasers is beyond the scope of this chapter.
[13]See [Malka *et al.* (2008)] and references therein.

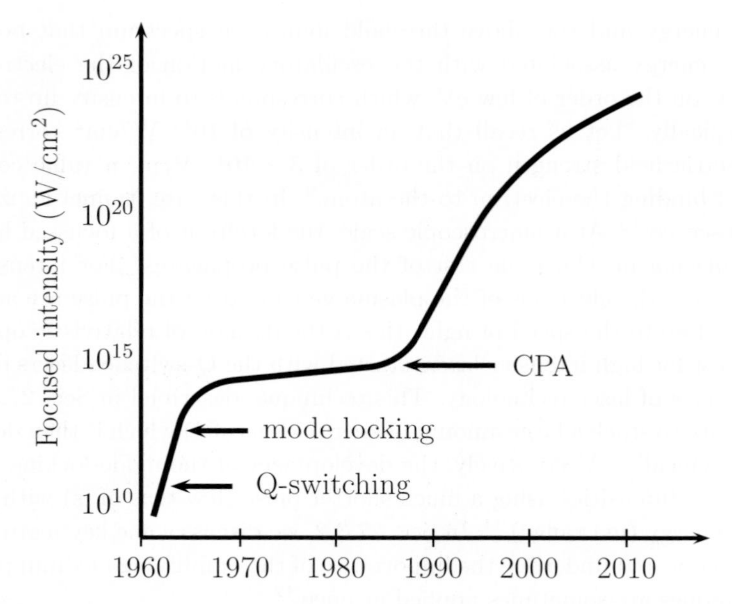

Fig. 27.6 Evolution during the last decade of the laser intensity (left scale) of a 1 cm^2 cross-section laser pulse.

cavity from a low to a high value. When the laser medium starts to be pumped, the Q-switch device degrades the quality factor of the cavity and the laser operation cannot build up. When the amount of energy that can then be stored in the gain medium reaches a maximum value, the Q-switch device is changed to provide a high quality factor. As a result, the light amplification by stimulated emission starts and the fast depletion of the energy stored in the gain medium gives rise to a short and intense light pulse. In practice, the Q-switch device can be either controlled externally, through a variable attenuator (active Q-switching), or benefit from the nonlinear response of a saturable absorber when the intensity is increased (passive Q-switching) [McClung and Hellwarth (1962)]. With this technique, it is possible to produce a ns pulse that incorporates several kilojoules of energy.

27.3.2 *Mode locking techniques*

Ultrashort laser pulses are generated by mode locking the different frequency components of the amplified light in the laser cavity. For this purpose an amplitude modulation with a time period equal to the cavity round trip time T_r is required. Several techniques have been developed to generate such a modulation. They depend on the characteristics of the laser and consist of either active mode locking driven externally or passive mode locking from an intensity dependent loss mechanism. As a result of the interference of the many coherent phase-locked longitudinal modes, the field is non-zero only for very short regularly spaced time intervals. This

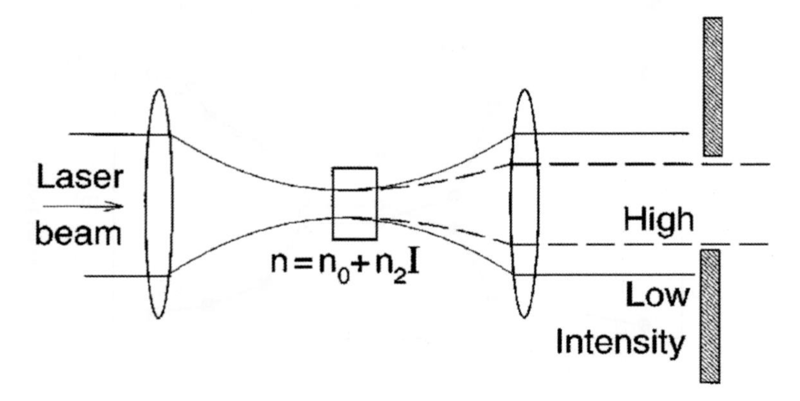

Fig. 27.7 The Kerr effect in a crystal combined with a suitable aperture provides an ultrafast-response saturable effect that is implemented in the laser cavity to generate short pulses. This technique is referred to as the Kerr-lens mode locking technique. Figure extracted from [Brabec and Krausz (2000)]. Copyright: American Physical Society.

interference phenomenon gives rise to a short light pulse that circulates in the resonator. The leaking light from the mirror of the cavity is thus made of short pulses of light repeated periodically with a period T_r, equal to the round trip time in the cavity. From the properties of the Fourier transform, an ultrashort pulse in the time domain must have a broad spectrum in the frequency domain.[14] One thus understands the importance of the performance of the resonator which must have a gain medium with a sufficiently broad spectral bandwidth and optical elements with sufficiently broadband response (in reflection and transmission). Passive mode locking techniques require the introduction of an optical component having a nonlinear response with intensity in the cavity: pioneering experiments have been performed with a saturable absorber, a material for which the absorption of light decreases with increasing light intensity. More recent experiments rely on the Kerr-lens mode locking technique (see Fig. 27.7 and the end of this section). Several other mode locking techniques exist. They will not be detailed here.[15]

The first generation (1965-1970) of mode locked lasers relied on solid state gain media such as ruby, Nd:glass or Nd:YAG and produced pulses of durations less than 100 ps. They played a crucial role for the development of nonlinear optics and time-resolved spectroscopy. The second generation of experiments used organic saturable absorbers. Let us mention for instance the first demonstrations of ps and sub -ps pulses from a mode locked dye laser by the group of Erich Ippen and Charles Shank in the early seventies [Ippen *et al.* (1972); Shank and Ippen (1974)]. Bandwith limited sub-picosecond pulses were generated soon after [Ruddock and

[14]Note the fact that mode locking could generate Fourier limited pulses was originally pointed out by Willis Lamb in 1964 [Lamb (1964)].
[15]See [Brabec and Krausz (2000)] and references therein.

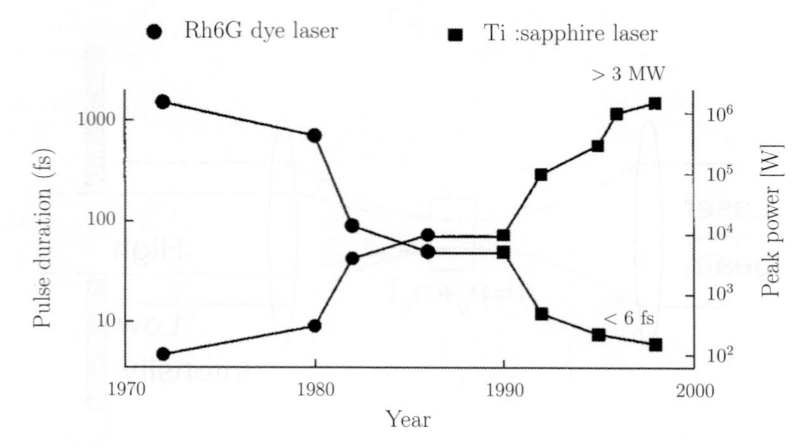

Fig. 27.8 Evolution of passively mode locked Rhodamine 6G dye and Ti:sapphire lasers in terms of pulse duration (decreasing curve) and peak power (increasing curve). Figures adapted from [Brabec and Krausz (2000)]. Copyright: American Physical Society.

Bradley (1976)]. In contrast to the first generation involving solid-state media, the response time of the absorber was no longer a limitation because of the active role of gain saturation in the pulse formation [Brabec and Krausz (2000)]. It is impossible to summarize here all the investigations made for shortening the pulse duration. Let us mention that to generate shorter fs pulses one has to take care of (i) the *group velocity dispersion* i.e. the fact that different frequency components travel at different speeds, and (ii) the *self-phase modulation* that arises from the nonlinear variation with intensity of the refractive index of the intracavity media [French (1995)]. The pioneering ultrashort pulse experiments were performed with organic dye gain media because of their large tunability. Figure 27.8 shows the evolution of passively mode locked dye laser in terms of both pulse duration (decreasing curve) and peak power (increasing curve). A plateau in the performances was observed during the eigthies. Further improvements came in the late eighties with new solid state gain media that exploited vibronic transitions in order to benefit from a large tunability (see Fig. 27.8). A significant advantage of these solid-state gain media is their better conversion efficiencies, thermal characteristics, and peak power. In addition, they are user friendly compared to dye lasers. Among the wide variety of solid state amplifying media that were demonstrated, Titanium-doped sapphire crystal, introduced by Peter Moulton at the MIT Lincoln Laboratory [Moulton (1986)], remains one of the most used. It has the advantage of having a very large fluorescence bandwidth (660 to 1180 nm) and a relatively high surface damage threshold. Its absorption bandwidth is in the blue-green (400-600 nm) which allows it to be pumped by Argon lasers for instance.

The standard mode locking method for generating very short pulses is the so-called Kerr-lens mode-locking technique.[16] It relies on the change of index of refraction with intensity of a Kerr-crystal introduced in the cavity. When a beam with a transverse peak intensity profile propagates into such a crystal, the refractive index experienced by the beam is larger at the center than at its edges. As a result, the beam tends to focus as illustrated in Fig. 27.7. Combined with an aperture of suitable size, the Kerr crystal favors the transmission of a larger fraction of the laser beam at high intensity (see Fig. 27.7). This extra component thus behaves like a fast saturable absorber with reduced losses for large intensity, and can initiate and sustain the formation of short pulses in the cavity. In 2001, the most advanced Ti:sapphire oscillators emitted 5-fs pulses which spanned an octave in wavelength from 600 to 1200 nm (see Fig. 27.8)[Ell *et al.* (2001)]. In these laser systems, the round-trip time in the cavity was on the order of 10 ns and more than one million phase-locked longitudinal resonator modes were oscillating.

27.3.3 *Chirped pulse amplification*

A breakthrough in the realization of tabletop high intensity lasers occurred in 1985 when Gerard Mourou and his coworkers proposed the principle of chirped amplification (CPA) [Strickland and Mourou (1985); Maine *et al.* (1988)] (see also the review papers [Perry and Mourou (1994); Mourou *et al.* (2006)]).

The principle of CPA is easy to understand and is illustrated in Fig. 27.9. Consider a laser pulse of about 10 fs. Before being amplified, this pulse is frequency chirped and stretched by a factor on the order of a few thousand by reflection on two gratings (giving a longer path for the blue components of the light wave packet than for the red ones). The peak intensity of the 1 ns stretched pulse is considerably reduced so as to not damage the glass optical components of the amplifier through which it is sent. The total energy of the pulse is subsequently amplified by a factor on the order of 10^{11}! The pulse is then re-compressed by a reflection on a pair of gratings (giving now a longer path for the red components of the light wave packet than for the blue ones). This technique is used on small micro-joule fiber la- sers all the way up to large-scale kilo-joule systems. The important point is that it is now possible to obtain femtosecond pulses with powers on the order of 10 terawatt (10^{13} W) by this method, in tabletop experiments (in university laboratories). In larger facilities, the power can reach the petawatt range (10^{15} W). When focused, these lasers give intensities larger than 10^{21} W/cm^2.

27.3.4 *A few applications of high intensity table-top lasers*

At laser intensities on the order of 10^{19} W/cm^2 the medium is totally ionized and becomes a plasma. The electrons oscillate in the laser field E_L with a velocity close to the speed of light and with a vibration energy that can reach 1 MeV, larger than

[16]See [Brabec and Krausz (2000)] and references therein.

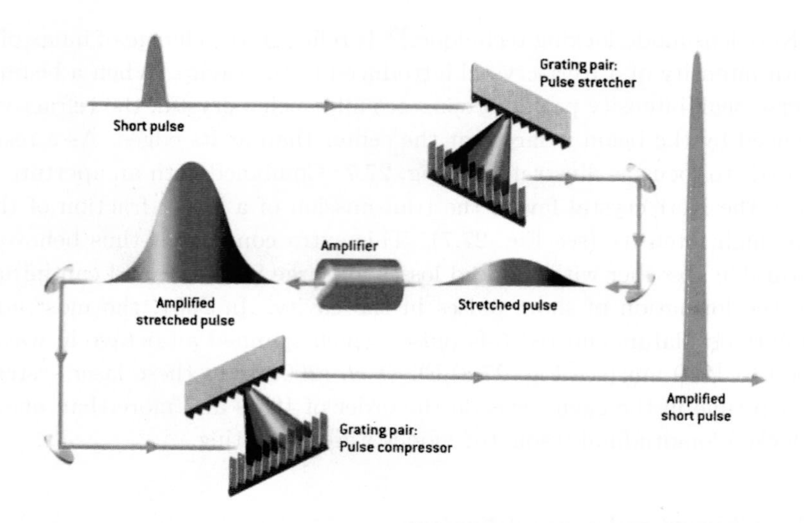

Fig. 27.9 Principle of chirped pulse amplification. Courtesy of G. Mourou.

the rest mass energy of the electron. The laser field E reaches values on the order of 10^{13} V/m, much larger than the Coulomb field of the nucleus.[17] The Lorentz force due to the magnetic field of the laser wave becomes equal to the force due to the electric field, a situation giving rise to extremely large radiation pressure, on the order of 0.3 Gbar. It is clear that, with these orders of magnitude, several new physical effects appear. We just mention two of them.

(i) *Relativistic self-focusing*

The relativistic self-focusing of a laser pulse propagating in an underdense plasma is a spectacular nonlinear optical phenomenon that originates from the nonlinear change of the local index of refraction in the high intensity region [Litvak (1970); Max *et al.* (1974)]. The index of refraction n of a plasma for a laser wave of frequency ω is equal to $n = \left[1 - \left(\omega_P^2/\omega^2\right)\right]^{1/2}$, where $\omega_P = 4\pi n_e q^2/m$ is the plasma frequency (n_e is the density of electrons and q the electron charge). One can show that in the relativistic regime, the mass m which enters the expression of the refraction index is equal to γm_e where γ is the relativistic factor associated with the electron motion transverse to the laser propagation. γ can be expressed in terms of the normalized vector potential of the laser field $a = eA/m_e c^2 \propto \sqrt{I}$ which is thus proportional to the square root of the intensity. The refraction index thus depends on the intensity through the relativistic factor. When I increases, γm_e increases, ω_P decreases and n increases.

As a result of the inhomogeneous transverse intensity profile of the laser, the index of refraction is peaked on the laser beam axis where the intensity has a maximum. This makes the plasma act like a positive lens and produces the self-

[17]See footnote 9 on page 705.

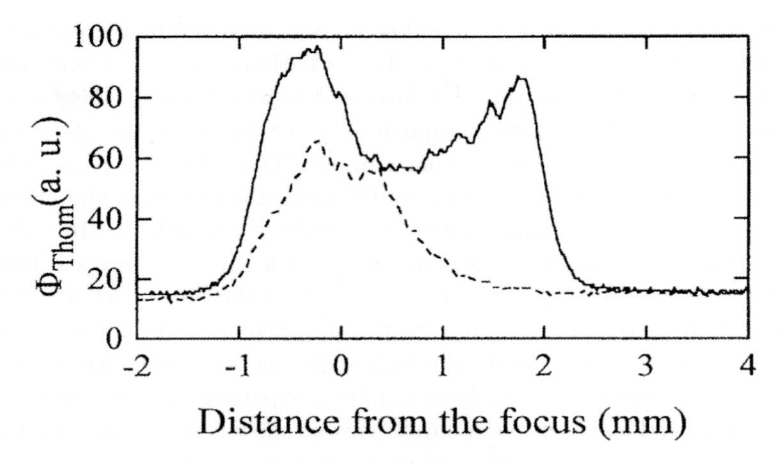

Fig. 27.10 The propagation of the laser in the plasma is studied through the Thomson scattered light at 90°. The critical power is $P_c = 6.8$ W for an electronic density of 2.5×10^{18} cm^{-3}. Dashed line: below the critical power, $P/P_c \simeq 0.15$, the laser beam is focussed over two Rayleigh length ($2z_R = 0.6$ mm). Solid line: $P/P_c \simeq 1.5$, the beam propagates over the plasma length (3 mm). Figure extracted from [Monot *et al.* (1995)]. Copyright: American Physical Society.

focusing of the laser beam. A more careful analysis shows that this effect in an homogeneous medium has a power threshold $P_c \simeq 17(\omega/\omega_P)^2$ GW (see Fig. 27.10)

In the experiments carried out to observe this effect, a focused laser pulse with a peak intensity on the order 10^{19} W/cm^2 interacts with a gas jet. The leading edge of the wavefront ionizes the gas and thus creates a plasma in which the rest of the pulse propagates. As illustrated in Fig 27.10, a propagation of the laser pulse due to relativistic self-focussing over a distance much larger than the Rayleigh length has been clearly observed for power $P > P_c$ [Monot *et al.* (1995); Wagner *et al.* (1997); Krushelnick *et al.* (1997)]. The scattered light suggests that successive foci are formed. This structure is attributed to the competition between diffraction and self-focussing effects.

(ii) *Wakefield acceleration of electrons*

The possibility of accelerating charged particles using the interaction of intense lasers with plasma was first suggested by Toshiki Tajima and John Dawson in 1979 [Tajima and Dawson (1979)]. However, it is only with the advent of CPA laser technology that the necessary requirements in terms of power and pulse durations could be achieved. We briefly discuss here how the interaction between an intense laser pulse with a plasma can create an accelerator through the generation of a wakefield travelling behind the laser pulse. Several other strategies involving light pulses and plasmas have been explored. Their description is beyond the scope of the book.[18]

When an intense laser pulse propagates through an underdense plasma, the mo-

[18]See for instance [Esarey *et al.* (2009)] for more details.

tion of electrons becomes relativistic and is strongly modified by the ponderomotive force associated with the pulse envelope. The radiation pressure exerted by the laser pulse is very large and pushes the electrons much more efficiently than the ions, due to their low mass. This generates charge density oscillations in the plasma that follow the light pulse like a wake following a moving boat. For this reason this effect is called wakefield. The charge separation between the electrons and the ions can generate extremely high longitudinal electric fields of the order 10-100 GV$\cdot$m^{-1}.[19] Indeed plasma as a medium has the advantage of having no electrical breakdown limit. Electrons can be trapped and accelerated by the huge electric fields of the wakefield. In this case, the electrons "surf" the wake over the interaction length. The longer the interaction length, the higher the energy of the electron beam. In practice, different protocols have been explored experimentally to inject electrons in the wakefield and lenghten the interaction zone such as self-relativistic focussing, the use of tailored gas jet targets or of preformed plasma channels.

The first observations of laser-accelerated electrons were reported in 1993-1994 and were associated with the demonstration of the principle of wakefield acceleration [Clayton *et al.* (1993); Everett *et al.* (1994)]. The accelerated electrons were characterized by a large energy spread. With a higher control of the laser and plasma parameters, it turned out to be possible to generate bunches of relativistic electrons with a relative energy spread of a few percent, a low divergence and an energy on the order of 200 MeV for an acceleration performed on few millimeters [Mangles *et al.* (2004); Geddes *et al.* (2004); Faure *et al.* (2004, 2006)]. Each bunch of accelerated electrons contains typically a few 10^8 electrons.[20] As laser-plasma accelerator are usually driven by femtosecond pulses, they produce sources of femtosecond electrons bunches. The ultimate performance of such accelerators depends on the possibility of sustaining ultrahigh accelerating fields over a substantial length to achieve a significant energy gain. For instance, in 2006, a high-quality electron beam with 1 GeV energy was created by first shaping a a 3-cm long channel through the gas with a powerful laser, and then accelerating bunches of electrons through this plasma [Leemans *et al.* (2006)]. Different mechanisms of ion acceleration are also involved in the interaction of a laser with a plasma [Malka *et al.* (2008)].

(iii) *Other applications*

Small low-energy accelerators based on laser-plasma interaction thus provide beams of electrons (with energy of GeV) and protons (with energy of few hundreds MeV). Such beams may facilitate biomedical applications. For instance, in radiotherapy cancer treatment, X-rays with energies of a few MeV are most commonly used. However, energetic electron beams have been proved to provide better clinical results for some types of tumors [Malka *et al.* (2008)]. For radio-resistant cancers and deep

[19]Conventional radiofrequency-based linear accelerators are limited to relatively low accelerating fields (< 100 MV.m^{-1}).

[20]The range of parameters of interest for high energy physicists is $> 10^{11}$ electrons per pulse accelerated to TeV energies.

tumors, proton and hadron therapies represent promising methods. So far these therapies can only be available close to large conventional accelerator facilities. A challenging perspective is thus opened by the use of accelerated protons generated by laser-plasma interaction.

Extreme light infrastructures are under construction in few places all around the world. They will be dedicated to the fundamental study of laser-matter interaction in the ultra-relativistic regime. Such facilities shall favor the connection between optics, high-energy and nuclear physics. A long term and fascinating goal of this field is to reach intensities as high as 10^{29} W/cm^2 (the so-called Schwinger limit) above which the light pulse is expected to generate electron-positron pairs out of the vacuum according to QED [Mourou *et al.* (2006)].

27.4 Conclusion

In this chapter, we have described a few recent advances in atomic physics made possible by the invention of new techniques for producing extreme laser light characterized by extremely short pulse durations, on the order 100 as, and high intensities, on the order or larger than 10^{15} W/cm^2 in table-top set-ups. We have shown that the generation of attosecond pulses is due to the tunnel ionization of the atom by an intense femtosecond infrared laser pulse, followed by a recollision of the electron produced by this ionization when it comes back near the parent ion after having been accelerated by the infrared laser field. Ultrashort pulses of XUV light are then generated during the collision. We have shown how it is possible to obtain reproducible attosecond pulses (by controlling the carrier-envelope phase of the infrared pulse) and to determine their shape and duration by new imaging techniques (such as attosecond streak imaging).

One interesting feature of the tunnel ionization-recollision mechanism is that the de Broglie wavelength of the electron produced by tunnel ionization, when it recollides with the parent ion, is on the order of the typical lengths characterizing the spatial structure of the ion. High spatial resolution is therefore added to the high time resolution provided by the attosecond pulses. This explains the great number of applications of attosecond physics for the determination of atomic and molecular structures as well as for the investigation of electron dynamics. Femtosecond pulses have led to a revolution in chemistry and molecular dynamics by allowing the motion of nuclei in molecules to be followed in real time. A new revolution is now appearing with attosecond pulses, which give access to the observation of the electronic motion.

The intensity of the infrared femtosecond pulses is also a crucial parameter because HHG can appear only above a certain intensity threshold. In this chapter, we have also described the mode locking technique and the chirped pulse amplification which allow very high laser intensities to be obtained in table-top experiments. These techniques have thus been essential for the generation of attosecond pulses.

They also allow the investigation of new domains where relativistic effects play an important role in atom-laser interactions at high intensity.

Finally, we hope to have given in this last chapter of the book a clear example of the intimate connection between fundamental research and technology. It is when trying to understand the non-perturbative features of high harmonic generation and above threshold ionization that physicists identified the tunnel ionization-recollision mechanism. It then appeared that this mechanism was also giving rise to attosecond pulses and this was the birth of attosecond physics which was a breakthrough in the technology of ultrashort pulses. But the tunnel ionization-recollision mechanism could not have been discovered without the availability in university laboratories of intense femtosecond pulses provided by the combination of mode locked lasers and chirped pulse amplification.

Chapter 28

General conclusion

In this book, we have surveyed several domains of atomic physics where spectacular results have been obtained. A few general trends characterize the evolution of this discipline. In this conclusion, we try to point out a few of them. We think that they explain the advances which have been achieved and that they could be at the origin of future developments.

An increasing control of atomic systems
Spectacular progress has been achieved in our ability to manipulate all degrees of freedom of atomic systems (electrons, ions, atoms, molecules). The spin polarization and the internal energy can be controlled by optical pumping. Atomic velocities can be reduced to extremely low values by laser cooling and evaporative cooling. Various types of traps, with different geometries and dimensionalities, have been implemented for charged as well as for neutral particles. Finally, atom-atom interactions can be fully controlled with Feshbach resonances.

Higher frequency resolutions
With the development of ultrastable laser sources and the invention of new techniques such as frequency combs for measuring optical frequencies, the resolution of spectroscopic measurements has been dramatically improved. For example, the relative accuracy of atomic clocks has increased by one order of magnitude every ten years during the last six decades (see Fig. 18.14 of Chap. 18), reaching now, with optical clocks, values in the range of 10^{-17}.

New schemes for overcoming the limits
For reaching these high resolutions, a certain number of apparently fundamental limits had to be overcome, stimulating the emergence of new original ideas.

A first interesting example is related to the recoil communicated to an atom when it absorbs or emits a photon and which gives rise to a Doppler shift. The resulting Doppler broadening of optical lines is a severe limitation for high resolution optical spectroscopy. Several ingenious methods using nonlinear effects, confinement, or the shelving method have been invented to circumvent this limit and to get very narrow optical lines. Another completely different strategy, which was explored later on, was to use the recoil effect itself to produce velocity dependent forces which could cool the atom to temperatures so low that the Doppler shift became completely negligible. A "bad" effect, the recoil, which introduces a recoil shift, is then used in a positive way! New fundamental limits, however, appeared in laser cooling due to the random nature of the photons spontaneously emitted during the cooling process that gave rise to a momentum spread that prevented the cooling of atoms below the single photon recoil limit. Here also, a new method using dark states was invented for cooling atoms below this limit and for reaching the nanokelvin range. Cooling laser pre-cooled atoms by evaporation was another important idea which allowed the observation of BEC in a gas for the first time. The history of trapping is also rich with examples of interplay between apparent limitations and new ideas for overcoming them: purely electrostatic potentials cannot trap charged particles, but Paul and Penning traps provide a solution for trapping electrons and ions; radiation pressure alone cannot be used to trap neutral atoms, but it can be combined with magnetic field gradients and gives rise to the magneto-optical trap (MOT); atom losses due to Majorana spin flips are a severe limitation of purely quadrupole traps, but this problem has been solved with the Ioffe-Pritchard and the TOP traps.

More stringent tests of fundamental laws

Thanks to the better control of atomic systems and the higher precision of the measurements, new tests of fundamental laws have been performed with an unprecedented accuracy: tests of quantum electrodynamics (electron spin anomaly, Lamb shift), general relativity (gravitational redshift), quantum mechanics (violation of Bell's inequalities) and the standard model of particle physics (parity violation in atoms). Fundamental constants, such as the Rydberg constant or the fine structure constant have also been re-measured with a higher precision.

One may wonder how atomic physics tests of fundamental laws can compete with high energy experiments. In fact, the weakness of the energies involved in atomic physics is compensated for by the very high precision of the measurements. Furthermore, the information deduced from atomic physics is not only a confirmation of the high energy results, but complement these results. For example, as shown in Fig. 25.5 of Chap. 25, atomic parity violation experiments, combined with high energy electron scattering, considerably reduce the uncertainty on the weak charges of the quarks.

From microscopic to macroscopic quantum systems

It is now possible to manipulate quantum systems at the single particle level, to make experiments on a single electron, a single atom or a single photon in a cavity. A few decades ago, experiments of this type were considered to be totally unrealistic "gedanken" experiments. They now become real.

A very interesting feature of atomic physics is the possibility of progressively increasing the complexity of the systems that are studied. For example, it has recently become possible to associate ultracold atoms into ultracold molecules and to produce exotic systems such as Efimov trimers and tetramers for the first time. These *few-body systems* presently attract a great interest and one can reasonably expect that a better understanding of their properties will be obtained. Another important research field of atomic physics, which is not described in this book, is the physics of clusters which allows one to study the continuous transition between small individual atoms and bulk condensed matter.

At the other extreme, it is now possible to produce macroscopic and well controlled systems, such as Bose-Einstein condensates and Fermi degenerate gases. Spatially periodic arrays of atoms trapped in optical lattices provide models of perfect crystals and allow the investigation of quantum effects such as Bloch oscillations and phase transitions involving tunnelling between adjacent wells.

Extracting information. A spectacular evolution

Once a quantum atomic system is well controlled, it is important to extract the information. Here also, the progress is spectacular. The development of CCD cameras allows a precise optical imaging and the detection of the fluorescence light emitted by a single atom. Time-of-flight techniques have revolutionized the measurement of extremely low temperatures. Ingenious methods have been invented for detecting a single photon without destroying it.

Another recent trend is the development of new methods of analysis based on the statistical properties of the noise of the signal. It has also become possible to image a single atom embedded in a larger structure such as optical lattice, to manipulate it, to remove it from its position and to displace it to another one.

The progress in time resolution is also impressive. The ultrashort XUV pulses obtained by high order harmonic generation and tunnel ionization now provide time resolution in the attosecond range. For example, new imaging techniques such as attosecond streak imaging have allowed the reconstruction of the time dependence of the electric field of a few cycle femtosecond infrared pulse. It is clear that these developments are revolutionizing the investigation of ultrashort phenomena, and in particular, the electronic dynamics in atoms, molecules and solids.

An increasing and fruitful dialogue with other disciplines

Ultracold quantum gases, especially in the strong interaction regime achievable with Feshbach resonances, stimulate a lot of fruitful interactions with other fields, such as condensed matter. They appear as simple model systems in which all experimental parameters can be fully controlled and which can provide a better understanding of more complex situations that appear in many-body physics. For example, the experimental determination of the equation of state of a quantum fluid as a function of a few universal dimensionless parameters can be applied to other quantum fluids and used as a benchmark to validate or eliminate theoretical models.

It must be kept in mind, however, that the appeal of strongly interacting quantum gases is not just to provide models for other many-body systems. They also allow the investigation of new problems, with new features, such as strong population differences between spin states or out-of-equilibrium situations resulting from strong time-dependent perturbations.

Finally, if the dynamics of a quantum gas is governed by a well defined Hamiltonian and if this Hamiltonian cannot be solved numerically, an experiment performed on the quantum gas can provide the solution. This is the idea of a *quantum simulator* first proposed by Richard Feynman.

Bibliography

Abella, I. D. (1962). *Phys. Rev. Lett.* **9**, 453.

Abo-Schaeer, J. R., Raman, C., Vogels, J. M. and Ketterle, W. (2001). *Science* **292**, 476.

Abragam, A. (1961). *The Principles of Nuclear Magnetism* (Clarendon Press).

Abragam, A. (1964). *L'effet Mössbauer et ses applications à l'étude des champs internes* (Gordon and Breach, New-York).

Abraham, E. R. I., McAlexander, W. I., Gerton, J. M., Hulet, R. G., Côté, R. and Dalgarno, A. (1997). *Phys. Rev. A* **55**, R3299.

Abrikosov, A. A. (1957). *J.E.T.P.* **32**, 1442.

Adams, C. S., Lee, H. J., Davidson, N., Kasevich, M. and Chu, S. (1995). *Phys. Rev. Lett.* **74**, 3577.

Aftalion, A., Du, Q. and Pomeau, Y. (2003). *Phys. Rev. Lett.* **91**, 090407.

Agosta, C. C. and Silvera, I. F. (1989). *Spin Polarized Quantum Systems*, ed. by Stringari, S. (Word Scientific, Singapore), 254.

Agostini, P., Barjot, G., Bonnal, J.-F., Mainfray, G., Manus, C. and Morellec, J. (1968). *IEEE J. Quantum Electron.*, **4**, 667.

Agostini, P., Fabre, F., Mainfray, G., Petite, G. and Rahman N. (1979). *Phys. Rev. Lett.* **42**, 1127.

Agostini, P., Kupersztych, J., Lompré, L., Petite, G. and Yergeau, F. (1987). *Phys. Rev. A* **36**, 4111.

Albert, M. S., Cates, G. D., Driehuys, B., Happer, W., Saam, B., Springer, C. S. and Wishnia, A. (1994). *Nature* **370**, 199.

Albiez, M., Gati, R., Fölling, J., Hunsmann, S., Cristiani, M., and Oberthaler, M. K. (2005). *Phys. Rev. Lett.* **95**, 010402.

Aleksandrov, E. B. (1964). *Optics and Spectrocopy* **17**, 522.

Alekseev, V. A. and Krylova, D. D. (1992). *JETP Lett.* **55**, 321.

Alekseev, V. A. and Krylova, D. D. (1996). *Opt. Commun.* **124**, 568.

Allen, L., Barnett, S. M. and Padgett, M. J. (2003). *Optical Angular Momentum* (IOP Publishing).

Allum, D. R., McClintock, P. V. E. and Phillips, A. (1977). *Phil. Trans. R. Soc. Lond. A*, **284**, 179.

Altmeyer, A., Riedl, S., Wright, M. J., Kohstall, C., Hecker Denschlag, J. and Grimm, R. (2007). *Phys. Rev. A* **76**, 033610.

Altshuler, S., Frantz, L. M. and Braunstein, R. (1966). *Phys. Rev. Lett.* **17**, 231.

Alzetta, G., Gozzini, A., Moi, L. and Orriols, G. (1976). *Il Nuovo Cimento* **36B**, 5.

Aminoff, C., Steane, A., Bouyer, P., Desbiolles, P., Dalibard, J. and Cohen-Tannoudji, C. (1993). *Phys. Rev. Lett.* **71**, 3083.

Ammosov, M. V., Delone, N. B. and Krainov, V. P. (1986). *Sov. Phys. JETP* **64**, 1191.
Anderlini, M. and Guéry-Odelin, D. (2006). *Phys. Rev. A* **73**, 032706.
Andersen, M. F., Ryu, C., Cladé, P., Natarajan, V., Vaziri, A., Helmerson, K. and Phillips, W. D. (2006). *Phys. Rev. Lett.* **97**, 170406.
Anderson, P. W. (1958). *Phys. Rev.* **109**, 1492.
Anderson, M. H., Ensher, J. R., Matthews, M. R., Wieman, C. E. and Cornell, E. A. (1995). *Science* **269**, 198.
Anderson, B. P., Haljan, P. C., Wieman, C. E. and Cornell, E. A. (2000). *Phys. Rev. Lett.* **85**, 2857.
Andrews, M. R., Townsend, C. G., Miesner, H.-J., Durfee, D. S., Kurn, D. M. and Ketterle, W. (1997a). *Science* **275**, 637.
Andrews, M. R., Kurn, D. M., Miesner, H.-J., Durfee, D. S., Townsend, C. G., Inouye, S. and Ketterle, W. (1997b). *Phys. Rev. Lett.* **79**, 553; *ibid* **80** 2967(E).
Aoyama, T., Hayakawa, M., Kinoshita, T. and Nio, M. (2008). *Phys. Rev. D* **77**, 053012.
Apanasevich, P. A. and Kilin, S. J. (1979). *J. Phys.* **B12**, L83.
Appel, J., Windpassinger, P. J., Oblak, D., Hoff, U. B., Kjærgaard, N. and Polzik, E. S. (2009). *Proc. Nat. Acad. Sci.* **106**, 10960.
Arditi, M. and Carver, T. R. (1961). *Phys. Rev.* **124**, 800.
Arimondo, E. and Orriols, G. (1976). *Lett. Nuovo Cimento* **17**, 333.
Arimondo, E., Phillips, W. and Strumia, F. (eds) (1992). *Laser Manipulation of Atoms and Ions*, Proceedings of the International School of Physics "Enrico Fermi", Course CXVIII (Elsevier North-Holland Press, Amsterdam).
Arimondo, E. (1996). *Progress in Optics*, ed. Wolf, E. (Elsevier Science, Amsterdam) pp. 257.
Armstrong, L., Lee Beers, B. and Feneuille, S. (1975). *Phys. Rev. A* **12**, 1903.
Arndt, M., Szriftgiser, P., Dalibard, J. and Steane, A. M. (1996). *Phys. Rev. A* **53**, 3369.
Arndt, M., Nairz, O., Voss-Andreae, J., Keller, C., van der Zouw, G. and Zeilinger, A. (1999). *Nature* **401**, 680.
Arnison, G. *et al.* (UA1 Collab.) (1983). *Phys. Lett.* **122B**, 103; *ibid* **126B**, 398.
Arons, A. B. and Peppard, M. B. (1965). *Am. J. Phys.* **33**, 367.
Ashcroft, N. W. and Mermin, N. D. (1976). *Solid State Physics* (Saunders, Philadelphia).
Ashkin, A. (1970a). *Phys. Rev. Lett.* **24**, 156.
Ashkin, A. (1970b). *Phys. Rev. Lett.* **25**, 1321.
Ashkin, A. (1978). *Phys. Rev. Lett.* **40**, 729.
Ashkin, A., Dziedzic, J. M. ,Bjorkholm, J. E. and Chu, S. (1986). *Opt. Lett.* **11**, 288.
Ashkin, A. and Dziedzic, J. M. (1987). *Science* **235**, 1517.
A. Ashkin, J. M. Dziedzic, T. Yamane (1987). *Nature* **330**, 769.
Askar'yan, G. A. (1962). *Sov. Phys. JETP* **15**, 1088.
Aspect, A., Roger, R., Reynaud, S., Dalibard, J. and Cohen-Tannoudji, C. (1980a). *Phys. Rev. Lett.* **45**, 617.
Aspect, A., Imbert, C. and Roger, R. (1980b). *Opt. Comm.* **34**, 46.
Aspect, A., Grangier, P. and Roger, R. (1981). *Phys. Rev. Lett.* **47**, 460.
Aspect, A., Grangier, P. and Roger, R. (1982a). *Phys. Rev. Lett.* **49**, 91.
Aspect, A., Dalibard, J. and Roger, R. (1982b). *Phys. Rev. Lett.* **49**, 1804.
Aspect, A., Dalibard J., Heidmann, A., Salomon, C. and Cohen-Tannoudji C. (1986). *Phys. Rev. Lett.* **57**, 1688.
Aspect, A., Arimondo, E., Kaiser, R., Vansteenkiste, N. and Cohen-Tannoudji, C. (1988). *Phys. Rev. Lett.* **61**, 826.
Aspect, A., Arimondo, E., Kaiser, R., Vansteenkiste, N. and Cohen-Tannoudji, C. (1989). *J. Opt. Soc. Am. B* **6**, 2112.

Astumian, R. D. and Hänggi, P. (2002). *Phys. Today* **55**, (11) 33.
Autler, S. H. and Townes, C. H. (1955). *Phys. Rev.* **100**, 703.
Avan, P., Cohen-Tannoudji, C., Dupont-Roc, J. and Fabre, C. (1976). *J. Phys. (Paris)* **37**, 993.
Bagnato, V. S., Lafyatis, G. P., Martin, A. G., Raab, E. L., Ahmad-Bitar, R. N. and Pritchard, D. E. (1987a). *Phys. Rev. Lett.* **58**, 2194.
Bagnato, V. S., Pritchard, D. E. and Kleppner, D. (1987b). *Phys. Rev. A* **35**, 4354.
Bagnato, V. S. and Kleppner, D. (1991). *Phys. Rev. A* **44**, 7439.
Baklanov, Y. V. and Chebotayev, V. P. (1972). *Sov. Phys. JETP* **34**, 490.
Baklanov, Y. V., Dubetsky, B. Y. and Chebotayev, V. P. (1976). *Appl. Phys.* **9**, 171.
Bakr, W. S., Peng, A., Folling, S. and Greiner, M. (2009). *Nature* **462**, 74.
Baltuska, A., Udem, Th., Uiberacker, M., Hentschel, M., Goulielmakis, E., Gohle, Ch., Holzwarth, R., Yakovlev, V. S., Scrinzi, A., Hänsch, T. W. and Krausz, F. (2003). *Nature* **421**, 611.
Balykin, V., Letokhov, V. and Mushin, V. (1979). *JETP Lett.* **29**, 560.
Balykin, V. I., (1980). *Opt. Commun.* **33**, 31.
Bardeen, J., Cooper, L. N. and Schrieffer, J. R. (1957). *Phys. Rev.* **108**, 1175.
Bardou, F., Bouchaud, J.-P., Emile, O., Aspect, A. and Cohen-Tannoudji, C. (1994a). *Phys. Rev. Lett.* **72**, 203.
Bardou, F., Saubamea, B., Lawall, J., Shimizu, K., Emile, O., Westbrook, C., Aspect, A. and Cohen-Tannoudji C. (1994b). *C. R. Acad. Sci. Paris* **318**, 877.
Bardou, F., Bouchaud, J.-P., Aspect, A. and Cohen-Tannoudji, C. (2002). *How Rare Events Bring Atoms to Rest?* (Cambridge University Press).
Barontini G., Weber C., Rabatti F., Catani J., Thalhammer G., Inguscio M. and Minardi F. (2009). *Phys. Rev. Lett.* **103**, 043201; *ibid.* **104**, 059901.
Barrat, J. P. (1959a). *J. Phys. Radium* **20**, 541.
Barrat, J. P. (1959b). *J. Phys. Radium* **20**, 633.
Barrat, J. P. (1959c). *J. Phys. Radium* **20**, 657.
Barrat, J. P. (1961). *Proc. Roy. Soc.* **A263**, 371.
Barrat, J. P. and Cohen-Tannoudji, C. (1961). *J. Phys. Radium* **22**, 329 and 443.
Barrett, M. D., Sauer, J. A. and Chapman, M. S. (2001). *Phys. Rev. Lett.* **87**, 010404.
Barry Bebb, H. (1966). *Phys. Rev.* **149**, 25.
Bartenstein, M., Altmeyer, A., Riedl, S., Jochim, S., Chin, C., Hecker Denschlag, J. and Grimm, R. (2004). *Phys. Rev. Lett.* **92**, 203201.
Basko, D. M., Aleiner, I. L. and Altshuler, B. L. (2007). *Phys. Rev. B* **76**, 052203.
Becker, W., Grasbon, F., Kopold, R., Milosevic, D. B., Paulus, G. G. and Walther, H. (2002). *Adv. At. Mol. Opt. Phys.* **48**, 35.
Bell, J. S. (1964). *Physics*, **1**, 195.
Bell, W. E. and Bloom, A. L. (1957). *Phys. Rev.* **107**, 1559.
Bell, D. A., Hess, H. F., Kochanski, G. P., Buchman, S., Pollack, L., Xiao, Y. M., Kleppner, D. and Greytak, T. J. (1986). *Phys. Rev. B* **34**, 7670.
Bell M.T. and Softley T.P. (2009). *Molecular Physics* **107**, 99.
Ben Dahan, M., Peik, E., Reichel, J., Castin, Y. and Salomon, C. (1996). *Phys. Rev. Lett.* **76**, 4508.
Bender, P. (1956). Thesis (Princeton University).
Bennett, S. C. and Wieman, C. E. (1999). *Phys. Rev. Lett.* **82**, 2484.
Berezinskii, V. L. (1971). *Sov. Phys. JETP* **34**, 610.
Bergeman, T., Erez, G. and Metcalf, H. J. (1987). *Phys. Rev. A* **35**, 1535.
Bergmann, K., Theuer, H. and Shore, B. W. (1998). *Rev. Mod. Phys.* **70**, 1003.
Bergeman, T., Moore, M. G. and Olshanii, M. (2003). *Phys. Rev. Lett.* **91**, 163201.

Bergquist, J. C., Lee, S. A. and Hall, J. L. (1977). *Phys. Rev. Lett.* **38**, 159.

Berkeland, D. J., Miller, J. D., Bergquist, J. C., Itano, W. M. and Wineland, D. J. (1998). *Phys. Rev. Lett.* **80**, 2089.

Bethe, H. A., Brown, L. M. and Stehn, J. R. (1950). Phys. Rev. **77**, 370.

Bethlem, H. L., Berden, G. and Meijer, G. (1999). *Phys. Rev. Lett.* **83**, 1558.

Bigelow, N. P. and Prentiss, M. G. (1990). *Phys. Rev. Lett.* **65**, 555.

Billy, J., Josse, V., Zuo, Z., Bernard, A., Hambrecht, B., Lugan, P., Clément, D., Sanchez-Palencia, L., Bouyer, P. and Aspect, A. (2008). *Nature* **453**, 891.

Biraben F., Cagnac, B. and Grynberg, G. (1974). *Phys. Rev. Lett.* **32**, 643.

Bitter, F. (1949). *Phys. Rev.* **76**, 833.

Bjorkholm, J. E., Freeman, R. R., Ashkin, A. and Pearson, D. B. (1978). *Phys. Rev. Lett.* **41**, 1361.

Bloch, F. and Siegert, A. (1940). *Phys. Rev.* **57**, 522.

Bloch, F. , Hansen, W. W. and Packard M. (1946). *Phys. Rev.* **70**, 474.

Bloch, I., Hänsch, T. W. and Esslinger, T. (1999). *Phys. Rev. Lett.* **82**, 3008.

Bloch, I., Hänsch, T. W., and Esslinger, T. (2000). *Nature* **403**, 166.

Bloch, I., Greiner, M., Mandel, O., Hänsch, T. W., and Esslinger, T. (2001). *Phys. Rev. A* **64**, 021402(R).

Bloch, I., Dalibard, J. and Zwerger, W. (2008). *Rev. Mod. Phys.* **80**, 885.

Bloembergen, N. (1982). *Rev. Mod. Phys.* **54**, 685.

Bloembergen, N. (1996). *Nonlinear Optics* (4th edition, World Scientific, Singapore).

Bloom, A. L. (1962). *Applied Optics*, **1**, 61.

Bogolubov, N. N. (1947). *J. Phys. (USSR)* **11**, 23.

Bohm, D. (1951). *Quantum Theory* (Prentice-Hall, Englewood Cliffs, NJ).

Bohn J. L. and Julienne P. S. (1997). *Phys. Rev. A* **56**, 1486.

Bohr, N. (1935). *Phys. Rev.* **48**, 696.

Bollen, G., Hartmann, H., Kluge, H.-J., König, Otto, M. T., Savard, G., Stolzenberg, H. and The ISOLDE Collaboration (1992). *Phys. Scr.* **46**, 581.

Bollinger, J. J., Prestage, J. D., Itano, W. M. and Wineland, D. J. (1985). *Phys. Rev. Lett.* **54**, 1000.

Bongs, K., Burger, S., Dettmer, S., Hellweg, D., Arlt, J., Ertmer, W. and Sengstock, K. (2001). *Phys. Rev. A* **63**, 031602(R).

Bonse, U. and Rauch, H. (eds.) (1979). *Neutron Interferometry* (Clarendon, Oxford).

Bordé, Ch. J. (1970). *C. R. Acad. Sci. Paris*, **271B**, 371.

Bordé, Ch. J. (1976). *C. R. Acad. Sci. Paris*, **283B**, 181.

Bordé, Ch. J. (1977). *Laser Spectroscopy III*, eds. Hall, J. L. and Carlsten, J. L. (Springer-Verlag, Berlin), pp. 121.

Bordé, Ch. J., Salomon, C., Avrillier, S., Van Lerberghe, A., Bréant, C., Bassi, D. and Scoles, G. (1984). *Phys. Rev. A*, **30**, 1836.

Bordé, Ch. J. (1989). *Phys. Lett.*, **A140**, 10.

Bose, S. N. (1924). *Z. Phys.* **26**, 178.

Bose, S. N., Theimer, O. (Translator), and Ram, B. (Translator) (1977). *Am. J. Phys.* **45**, 242.

Bouchendira, R., Cladé, P., Guellati-Khélifa, S., Nez, F. and Biraben, F. (2011). *Phys. Rev. Lett.* **106**, 080801.

Bouchiat, M. A., Carver, T. R. and Varnum C. M. (1960). *Phys. Rev. Lett.* **5**, 373.

Bouchiat, M. A. and Bouchiat, C. (1974a). *Phys. Lett.* **48B**, 111

Bouchiat, M. A. and Bouchiat, C. (1974b). *J. Phys. (Paris)* **35**, 899.

Bouchiat, M. A. and Bouchiat, C. (1975). *J. Phys. (Paris)* **36**, 493.

Bouchiat, M. A., Guéna, J., Hunter, L. and Pottier, L. (1982). *Phys. Lett.* **117B**, 358.

Bouchiat, M.A. (1982). *Parity Violation in Atoms*, in Les Houches Summer School XXXVIII, 1982, eds. Grynberg, G. and Stora, R. (Elsevier Sci. Publ., 1984).
Bouchiat, M. A., Guéna, J., Hunter, L. and Pottier, L. (1984). *Phys. Lett.* **134B**, 463.
Bouchiat, M. A. (1985). *Ann. Phys. Fr.* **10**, 923.
Bouchiat, M. A., Guéna, J. and Pottier, L. (1985). *J. Phys. (Paris)* **46**, 1897.
Bouchiat, M. A., Guéna, J. and Pottier, L. (1986a). *J. Phys. (Paris)* **47**, 1175.
Bouchiat, M. A., Guéna, J., Pottier, L. and Hunter, L. (1986b). *J. Phys. (Paris)* **47**, 1709.
Bouchiat, C. and Piketty, C. A. (1986). *Europhys. Lett.* **2**, 511.
Bouchiat, C. and Guéna, J. (1988). *J. Phys. (Paris)* **49**, 2037.
Bouchiat, C. and Piketty, C. A. (1988). *J. Phys. (Paris)* **49**, 1851.
Bouchiat, C. and Piketty, C. A. (1991) *Z. Phys. C* **49**, 49.
Bouchiat, M. A. and Bouchiat, C. (1997). *Rep. Prog. Phys.* **60**, 1351.
Bouchiat, C. (1999). *The Nuclear Anapole Moment*, in Parity Violation in Atoms and Polarized Electron Scattering, eds. Frois, B. and Bouchiat, M. A. (World Scientific).
Bourdel, T., Khaykovich, L., Cubizolles, J., Zhang, J., Chevy, F., Teichmann, M., Tarruell, L., Kokkelmans, S. J. J. M. F. and Salomon, C. (2004). *Phys. Rev. Lett.* **93**, 050401.
Bouwmeester, D. , Ekert, A. K. and Zeilinger, A (2000). *The Physics of Quantum Information: Quantum Cryptography, Quantum Teleportation, Quantum Computation* (Springer-Verlag, Heidelberg).
Brabec, T. and Krausz, F. (2000). *Rev. Mod. Phys.* **72**, 545.
Bradley, D. J., Liddy, B., Sleat, W. E. (1971). *Opt. Commun.* **2**, 391.
Bradley, C. C., Sackett, C. A., Tollett, J. J. and Hulet, R. G. (1995). *Phys. Rev. Lett.* **75**, 1687.
Bradley, C. C., Sackett, C. A. and Hulet, R. G. (1997). *Phys. Rev. Lett.* **78**, 985.
Bradley, M. P., Porto, J. V., Rainville, S., Thompson, J. K. and Pritchard, D. E. (1999). *Phys. Rev. Lett.* **83**, 4510.
Bretin, V., Stock, S., Seurin, Y. and Dalibard, J. (2004). *Phys. Rev. Lett.* **92**, 050403.
Brewer, R. G., Kelly, M. J. and Javan A. (1969). *Phys. Rev. Lett.* **23**, 559.
Brossel, J. and Kastler, A. (1949). *C. R. Acad. Sci.* **229**, 1213.
Brossel, J. and Bitter, F. (1952). *Phys. Rev.* **86**, 308.
Brossel, J., Kastler A. and Winter, J. M. (1952). *J. Phys. Radium* **13**, 668.
Brossel, J., Cagnac, B. and Kastler A. (1953). *C. R. Acad. Sci.* **237**, 984.
Brossel, J., Cagnac, B. and Kastler A. (1954). *J. Phys. Radium* **15**, 6.
Brossel J., Margerie, J. and Kastler, A. (1955). *C. R. Acad. Sci.* **241**, 865.
Brown, L. S. and Gabrielse, G. (1986). *Rev. Mod. Phys.* **58**, 233.
Brumer, P. and Shapiro, M. (1986). *J. Chem. Phys.* **84**, 410.
Brune, M., Schmidt-Kaler, F. , Maali, A. , Dreyer, J. , Hagley, E., Raimond, J.-M. and Haroche, S. (1996a). *Phys. Rev. Lett.* **76**, 1800.
Brune, M., Hagley, E., Dreyer, J., Maître, X., Maali, A., Wunderlich, C., Raimond, J.-M. and Haroche, S. (1996b). *Phys. Rev. Lett.* **77**, 4887.
Budil, K. S., Salières, P., Perry, M. D. and L'Huillier, A. (1993). *Phys. Rev. A* **48**, R3437.
Buggle, Ch., Léonard, J., von Klitzing, W. and Walraven, J. T. (2004). *Phys. Rev. Lett.* **93**, 173202.
Burger, S., Bongs, K., Dettmer, S., Ertmer, W., Sengstock, K., Sanpera, A., Shlyapnikov, G. V. and Lewenstein, M. (1999). *Phys. Rev. Lett.* **83**, 5198.
Burke, J. P. Jr., Bohn, J. L., Esry, B. D. and Greene, C. H. (1998). *Phys. Rev. Lett.* **80**, 2097.
Burt, E. A., Ghrist, R. W., Myatt, C. J., Holland, M. J., Cornell, E. A. and Wieman, C. E. (1997). *Phys. Rev. Lett.* **79**, 337.
Cacciapuoti, L. and Salomon, C. (2009). *Eur. Phys. J. Special Topics* **172**, 57.

Cadoret, M., de Mirandes, E., Cladé, P., Guellati-Khlifa, S., Schwob, C., Nez, N., Julien, L. and Biraben, F. (2008). *Phys. Rev. Lett.* **101**, 230801.

Cagnac, B. (1961). *Ann. de Phys.* **6**, 467.

Cagnac, B., Grynberg, G. and Biraben, F. (1973). *J. Phys. (Paris)* **34**, 845.

Cagnac, B. (1975). *Lecture Notes in Physics*, Laser Spectroscopy, Proceedings of the Second International Conference, Megève, June 22-27 1975, eds. Haroche, S., Pebay-Peyroula, J. C., Hänsch, T. W. and Harris, S. E. (Springer-Verlag, Berlin), pp. 165.

Camino, F. E., Zhou, W. and Goldman, V. J. (2005). *Phys. Rev. B* **72**, 075342.

Campbell, R. (1955). *Théorie générale de l'équation de Mathieu*, ed. Masson.

Campbell, G. K., Mun, J., Boyd, M., Medley, P., Leanhardt, A. E., Marcassa, L. G., Pritchard, D. E. and Ketterle, W. (2006). *Science* **313**, 649.

Carmichael, H. J. (1993). *An Open Systems Approach to Quantum Optics, Lecture Notes in Physics* (Springer, Berlin).

Carnal, O. and Mlynek, J. (1991). *Phys. Rev. Lett.* **66**, 2689.

Carnal, O., Faulstich, A. and Mlynek, J. (1991a). *Appl. Phys. B* **53**, 88.

Carnal, O., Sigel, M., Sleator, T., Takuma, H. and Mlynek, J. (1991b). *Phys. Rev. Lett.* **67**, 3231.

Carr, L. D. and Ye, J. (2009). *New. J. Phys.*, **11**, 055409.

Casalbuoni, R. and Nardulli, G. (2004). *Rev. Mod. Phys.* **76**, 263.

Casati, C., Chirikov, B. V., Ford, J. and Izrailev, F. M. (1979). *Lect. Notes Phys.* **93**, 334.

Castin, Y., Wallis, H. and Dalibard, J. (1989). *J. Opt. Soc. Am. B* **6**, 2046.

Castin, Y., Dalibard, J. and Cohen-Tannoudji, C. (1991). *Light Induced Effects of Atoms, Ions and Molecules*, eds. Moi, L., Gozzini, S., Gabanini, S., Arimondo, E. and Strumia, F. eds editrice (Pisa).

Castin, Y. and Dalibard, J. (1991). *Europhys. Lett.* **14**, 761.

Castin, Y. and Dum, R. (1996). *Phys. Rev. Lett.* **77**, 5315.

Castin, Y. and Dalibard, J. (1997). *Phys. Rev. A* **55**, 4330.

Castin, Y. and Dum, R. (1998). *Phys. Rev. A* **57**, 3008.

Castin, Y. (2001). *Coherent atomic matter waves*, Lecture Notes of Les Houches Summer School, p.1-136, eds. Kaiser, R., Westbrook, C. and David, F. (EDP Sciences and Springer-Verlag).

Castin, Y. (2008). *Basic tools for degenerate Fermi gases* In Ultra-cold Fermi gases, Proceedings of the International School of Physics "Enrico Fermi", Course CLXIV, eds. Inguscio, M., Ketterle, W. and Salomon, C. (IOS Press, Amsterdam).

Cavalieri, A. L., Müller, N., Uphues, Th., Yakovlev, V. S., Baltuska, A., Horvath, B., Schmidt, B., Blümel, L., Holzwarth, R., Hendel, S., Drescher, M., Kleineberg, U., Echenique, P. M., Kienberger, R., Krausz, F. and Heinzmann, U. (2007). *Nature* **449**, 1029.

Cennini, G., Ritt, G., Geckeler, C. and Weitz, M. (2003). *Phys. Rev. Lett.* **91**, 240408.

Ceperley, D. M. (1995). *Rev. Mod. Phys.* **67**, 279.

Cesar, C. L., Fried, D. G., Killian, T. C., Polcyn, A. D., Sandberg, J. C., Yu, I. E., Greytak, T. J., Kleppner, D. and Doyle, J. M. (1996). *Phys. Rev. Lett.* **77**, 255.

Chabé, J., Lemarié, G., Grémaud, B., Delande, D., Szriftgiser, P. and Garreau, F. C. (2008). *Phys. Rev. Lett.* **101**, 255702.

Chandrasekhar, B. S. (1962). *Appl. Phys. Lett.* **1**, 7.

Chapman, M. S., Ekstrom, C. R., Hammond, T. D., Rubenstein, R. A., Schmiedmayer, J., Wehinger, S. and Pritchard, D. E. (1995). *Phys. Rev. Lett.* **74**, 4783.

Chen, C., Yin, Y.-Y. and Elliott, D. S. (1990). *Phys. Rev. Lett.* **64**, 507.

Cheneau, M., Rath, S. P., Yefsah, T., Günter, K. J., Juzeliunas, G. and Dalibard, J. (2008). *Europhysics Lett.* **83**, 60001.

Cheng,C.-H. and Yip, S.-K. (2007). *Phys. Rev. B* **75**, 014526.

Chevy, F., Madison, K. X. and Dalibard, J. (2000). *Phys. Rev. Lett.* **85**, 2223.

Chevy, F., Madison, K. W., Bretin, V. and Dalibard, J. (2001). Proceedings of the workshop Trapped particles and fundamental physics (Les Houches 2001), organized by S. Atutov, K. Calabrese and L. Moi.

Chevy, F., Bretin, V., Rosenbusch, P., Madison, K. W. and Dalibard, J. (2002). *Phys. Rev. Lett.* **88**, 250402.

Chikkatur, A. P., Görlitz, A., Stamper-Kurn, D. M., Inouye, S., Gupta, S. and Ketterle, W. (2000). *Phys. Rev. Lett.* **85**, 483.

Chikkatur, A. P., Shin, Y., Leanhardt, A. E., Kielpinski, D., Tsikata, E., Gustavson, T. L., Pritchard, D. E. and Ketterle, W. (2002). *Science* **296**, 2193.

Chin, C., Bartenstein, M., Altmeyer, A., Riedl, S., Jochim, S., Hecker Denschlag, J. and Grimm, R. (2004). *Science* **305**, 1128.

Chin, C., Grimm, R., Julienne, P. and Tiesinga, E. (2010). *Rev. Mod. Phys.* **82**, 1225.

Chou, C. W., Hume, D. B., Koelemeij, J. C. J., Wineland, D. J. and Rosenband, T. (2010a). *Phys. Rev. Lett.* **104**, 070802.

Chou, C. W., Hume, D. B., Rosenband, T. and Wineland, D. J. (2010b). *Science* **329**, 1630.

Christenson, J. H., Cronin, J. W., Fitch, V. L. and Turlay, R. (1964). *Phys. Rev. Lett.* **13**, 138.

Chu, S., Hollberg, L., Bjorkholm, J. E., Cable, A. and Ashkin, A. (1985). *Phys. Rev. Lett.* **55**, 48.

Chu, S., Bjorkholm, J., Ashkin, A., and Cable, A. (1986). *Phys. Rev. Lett.* **57**, 314.

Chu, S. (1998). *Rev. Mod. Phys.* **70**, 685.

Church, D.A. and Dehmelt, H. G. (1969). *J. Appl. Phys.* **40**, 3421.

Cirac, J. I. and Zoller, P. (1995). *Phys. Rev. Lett.* **74**, 4091.

Cirac, J. I., Gardiner, C. W., Naraschewski, M. and Zoller, P. (1996). *Phys. Rev. A* **54**, R3714.

Cittert, P. V. (1934). *Physica* **1**, 201.

Cladé, P., de Mirandes, E., Cadoret, M., Guellati-Khélifa, S., Schwob, C., Nez, F., Julien, L., and Biraben, F. (2006a). *Phys. Rev. Lett.* **96**, 033001.

Cladé, P., de Mirandes, E., Cadoret, M., Guellati-Khélifa, S., Schwob, C., Nez, F., Julien, L., and Biraben, F. (2006b). *Phys. Rev. A* **74**, 052109.

Cladé, P., Ryu, C., Ramanathan, A., Helmerson, K. and Phillips, W. D. (2009a). *Phys. Rev. Lett.* **102**, 170401.

Cladé, P., Guellati-Khélifa, S., Nez, F. and Biraben, F. (2009b). *Phys. Rev. Lett.* **102**, 240402.

Clairon, A., Salomon, C., Guellati, S. and Phillips, W. D. (1991). *Europhys. Lett.* **16**, 165.

Clauser, J. F., Horne, M. A., Shimony, A. and Holt, R. A. (1969). *Phys. Rev. Lett.* **23**, 880.

Clauser, J. F. (1976). *Phys. Rev. Lett.* **36**, 1223.

Clauser, J. F. and Shimony, A. (1978). *Rep. Prog. Phys.* **41**, 1881.

Clayton, C. E., Marsh, K. A., Dyson, A., Everett, M., Lal, A., Leemans, W. P., Williams, R. and Joshi, C. (1993). *Phys. Rev. Lett.* **70**, 37.

Clogston, A. M. (1962). *Phys. Rev. Lett.* **9**, 266.

Cohen-Tannoudji, C. (1961a). *C. R. Acad. Sci.* **252**, 394.

Cohen-Tannoudji, C. (1961b). *C. R. Acad. Sci.* **253**, 2899.

Cohen-Tannoudji, C. (1962). *Ann. Phys.* **7**, 423; *ibid.* **7**, 469.

Cohen-Tannoudji, C. and Haroche, S. (1965). *C. R. Acad. Sci.* **261**, 5400.

Cohen-Tannoudji, C. and Haroche, S. (1966a). *C. R. Acad. Sci.* **262B**, 37.

Cohen-Tannoudji, C. and Haroche, S. (1966b). *C. R. Acad. Sci.* **262B**, 268.

Cohen-Tannoudji, C. and Kastler, A. (1966). *Progress in Optics* **V**, 1.

Cohen-Tannoudji, C., Dupont-Roc, J., Haroche, S. and Laloë, F. (1969). *Phys. Rev. Lett.* **22**, 758.

Cohen-Tannoudji, C., Dupont-Roc, J., Haroche, S. and Laloë, F. (1970). *Rev. de Phys. Appl.* **5**, 102.

Cohen-Tannoudji, C. and Dupont-Roc, J. (1972). *Phys. Rev. A* **5**, 968.

Cohen-Tannoudji, C. (1977). Atoms in strong resonant fields, *Frontiers in Laser spectroscopy*, Volume 1, Les Houches, Session XXVII, eds. Balian, R., Haroche, S. and Liberman, S. (North-Holland Publishing Company), pp. 1.

Cohen-Tannoudji, C., Diu, B. and Laloë (1977). Quantum mechanics, John Wiley, Wiley-Interscience.

Cohen-Tannoudji, C. and Reynaud S. (1977). *J. Phys. B* **10**, 345.

Cohen-Tannoudji, C. and Reynaud S. (1979). *Phil. Trans. R. Soc. Lond. A* **293**, 223.

Cohen-Tannoudji, C., Dupont-Roc, J. and Grynberg, G. (1989). *Photons and Atoms: Introduction to Quantum Electrodynamics*, (John Wiley & Sons).

Cohen-Tannoudji, C. (1989). Lectures at college de France 1989-1990, unpublished (see Web site: http://www.phys.ens.fr/cours/college-de-france/).

Cohen-Tannoudji, C (1992a). Atomic Motion in Laser Light, in Fundamental Systems in Quantum Optics, Les Houches, Session LIII, 1990, eds. Dalibard, J., Raimond, J. M. and Zinn Justin, J. (Elsevier Science Publisher B.V.) pp. 1-164.

Cohen-Tannoudji, C., Dupont-Roc, J. and Grynberg, G. (1992b). *Atom-Photon Interactions* (John Wiley & Sons).

Cohen-Tannoudji, C., Bardou, F. and Aspect, A. (1992c). Review on fundamental processes in laser cooling, in *Laser Spectroscopy X*, eds. Ducloy, M., Giacobino, E. and Camy, G. (World Scientific, 1992) p. 3.

Cohen-Tannoudji, C. (1997). Lectures at Collège de France 1996-1997, unpublished (see Web site: http://www.phys.ens.fr/cours/college-de-france/).

Cohen-Tannoudji, C. (1998a). Lectures at Collège de France, unpublished (see Web site: http://www.phys.ens.fr/cours/college-de-france/).

Cohen-Tannoudji, C. (1998b). *Rev. Mod. Phys.* **70**, 707.

Cohen-Tannoudji, C. and Robillard, C. (2001). *C. R. Acad. Sci. Paris* **2**, 445.

Colegrove, F. ,Franken, P., Lewis, R. and Sands, R. (1959). *Phys. Rev. Lett.* **3**, 420.

Colegrove, F. D., Schearer L. D., and Walters, G. K. (1963). *Phys. Rev.* **132**, 2561.

Colombe Y., Knyazchyan E., Morizot O., Mercier B., Lorent V., Perrin H. (2004). *Europhys. Lett.* **67**, 593.

Colombe, Y., Steinmetz, T., Dubois, G., Linke, F., Hunger, D. and Reichel, J. (2007). *Nature* **450**, 272.

Combescot, R., Kagan, M. Yu, and Stringari, S. (2006). *Phys. Rev. A* **74**, 042717.

Cook, R. J. (1979). *Phys. Rev. A* **20**, 224.

Cook, R. J. and Hill, R. K. (1982). *Opt. Commun.* **43**, 258.

Cook , R. J. and Kimble H. J. (1985). *Phys. Rev. Lett.* **54**, 1023.

Cooper, L. N. (1956). *Phys. Rev.* **104**, 1189.

Cooper, N. R. (2008). *Advances in Physics* **57**, 539.

Corkum, P. B. (1993). *Phys. Rev. Lett.* **71**, 1994.

Corkum, P. B. and Krausz, F. (2007). *Nature Phys.* **3**, 381.

Cornell, E. A. and Wieman, C. E. (2002). *Rev. Mod. Phys.* **74**, 875.

Corney A. and Series, G. (1964). *Proc. Phys. Soc.* **83**, 207.

Cornish, S. L., Claussen, N. R., Roberts, J. L., Cornell, E. A. and Wieman, C. E. (2000). *Phys. Rev. Lett.* **85**, 1795.

Cornish, S. L., Thompson, S. T. and Wieman, C. E. (2006). *Phys. Rev. Lett.* **96**, 170401.

Corsi, C., Pirri, A., Sali, E., Tortora, A. and Bellini, M. (2006). *Phys. Rev. Lett.* **97**, 023901.

Courteille, Ph., Freeland, R. S., Heinzen, D. J., van Abeelen, F. A. and Verhaar, B. J. (1998). *Phys. Rev. Lett.* **81**, 69.

Couvert, A., Kawalec, T., Reinaudi, G. and Guéry-Odelin, D. (2008a). *Europhys. Lett.* **83**, 13001.

Couvert, A., Jeppesen, M., Kawalec, T., Reinaudi, G., Mathevet, R. and Guéry-Odelin, D. (2008b). *Europhys. Lett.* **83**, 50001.

Cren, P., Roos, C. F., Aclan, A., Dalibard, J. and Guéry-Odelin, D. (2002). *Eur. Phys. J. D.* **20**, 107.

Cronin, A. D., Schmiedmayer, F. and Pritchard, D. E. (2009). *Rev. Mod. Phys.* **81**, 1051.

Cubizolles, J., Bourdel, T., Kokkelmans, S. J. J. M. F., Shlyapnikov, G. V. and Salomon, C. (2003). *Phys. Rev. Lett.* **91**, 240401.

Dalfovo, F., Pitaevskii, L. P. and Stringari, S. (1996). *Phys. Rev. A* **54**, 4213.

Dalfovo, F., Giorgini, S., Pitaevskii, L. P. and Stringari, S. (1999). *Rev. Mod. Phys.* **71**, 463.

Dalfovo, F. and Modugno, M. (2000a). *Phys. Rev. A* **61**, 023605.

Dalfovo, F. and Stringari, S. (2000b). *Phys. Rev. A* **63**, 011601(R).

Dalibard, J., Dupont-Roc, J., Cohen-Tannoudji, C. (1982). *J. Phys. (Paris)* **43**, 1617.

Dalibard, J. and Cohen-Tannoudji, C. (1985a). *J. Phys. B* **18**, 1661.

Dalibard, J. and Cohen-Tannoudji, C. (1985b). *J. Opt. Soc. Am. B* **2**, 1707.

Dalibard, J. and Cohen-Tannoudji, C. (1986). *Europhys. Lett.* **1**, 441.

Dalibard, J., Reynaud, S. and Cohen-Tannoudji, C. (1987). *"Interaction of radiation with matter"* volume in honor of Adriano Gozzini (Scuola Normale Superiore, Pisa, Italy, 1987) pp. 29.

Dalibard, J. and Cohen-Tannoudji, C. (1989). *J. Opt. Soc. Am. B* **6**, 2023.

Dalibard, J., Castin Y. and Mølmer, K. (1992). *Phys. Rev. Lett.* **68**, 580.

Dalibard, J., (1999). Proceedings of the International School of Physics "Enrico Fermi", Course CXL, Bose-Einstein Condensation in Atomic Gases.

Dalibard, J., Gerbier, F., Juzeliūnas, J. and Öhberg, P. (2010). arXiv:1008.537.

Dall, R. G., Hodgman, S. S., Johnsson, M. T., Baldwin, K. G. H. and Truscott, A. G. (2010). *Phys. Rev. A* **81**, 011602(R).

Davidson, N., Lee, H. J., Adams, C. S., Kasevich, M. and Chu, S. (1995). *Phys. Rev. Lett.* **74**, 1311.

Davis, K. B., Mewes, M. -O., Andrews, M. R., van Druten, N. J., Durfee, D. S., Kurn, D. M. and Ketterle, W. (1995a). *Phys. Rev. Lett.* **75**, 3969.

Davis, K. B., Mewes, M.-O. and Ketterle, W. (1995b). *Appl. Phys. B* **60**, 155.

Davisson, C. and Germer, L. H. (1927). *Phys. Rev.* **30**, 705.

de Beauvoir, B., Nez, F., Julien, L., Cagnac, B., Biraben, F., Touahri, D., Hilico, L., Acef, O., Clairon, A. and Zondy, J. J. (1997). *Phys. Rev. Lett.* **78**, 440.

de Broglie, L. (1923). *Nature* **112**, 540.

de Gennes, P. G. (1966). *Superconductivity of Metals and Alloys* (W. A. Benjamin, Inc).

Dehmelt, H. G. and Walls, F. L. (1968). *Phys. Rev. Lett.* **21**, 127.

Dehmelt, H. G.(1975). *Bulletin of the American Physical Society* **20**, 60.

Dehmelt, H. G.(1981). *Proc. of the int. conf. on Atomic Physics* **7**, Cambridge, Masschusetts, 1980, eds. Kleppner, D. and Pipkin, F. M. (Plenum Press, New York London 1981) pp. 337.

Dehmelt, H. (1990). *Rev. Mod. Phys.* **62**, 525.

Deissler, B., Zaccanti, M., Roati, G., D'Errico, C., Fattori, M., Modugno, M., Modugno, G. and Inguscio, M. (2010). *Nature Phys.* **6**, 354.
DeMarco, B. and Jin, D. S. (1999). *Science* **285**, 1703.
Deng, L., Hagley, E. W., Wen, J., Trippenbach, M., Band, Y., Julienne, P. S., Simsarian, J. E., Helmerson, K., Rolston, S. L. and Phillips, W. D. (1999). *Nature* **398**, 218.
Denschlag, J., Simsarian, J. E., Feder, D. L., Clark, C. W., Collins, L. A., Cubizolles, J., Deng, L., Hagley, E. W., Helmerson, K., Reinhardt, W. P., Rolston, S. L., Schneider, B. I. and Phillips, W. D. (2000). *Science* **287**, 97.
Desruelle, B., Boyer, V., Bouyer, P., Birkl, G., Lécrivain, M., Alves, F., Westbrook, C. I. and Aspect, A. (1998). *Eur. Phys. J. D* **1**, 255.
Dettmer, S., Hellweg, D., Ryytty, P., Arlt, J. J., Ertmer, W., Sengstock, K., Petrov, D. S., Shlyapnikov, G. V., Kreutzmann, H., Santos, L. and Lewenstein, M. (2001). *Phys. Rev. Lett.* **87**, 160406.
Deutsch Ch., Ramirez-Martinez F., Lacroûte C., Reinhard F., Schneider T., Fuchs J.N., Piéchon F., Laloë F., Reichel J. and Rosenbuch P. (2010). *Phys. Rev. Lett.* **105**, 020401.
Dicke, R. H. (1953). *Phys. Rev.* **89**, 472.
Dicke, R. H. (1954). *Phys. Rev.* **93**, 99.
Diedrich, F., Peik, E., Chen, J. M., Quint, W. and Walter, H. (1987). *Phys. Rev. Lett.* **59**, 2931.
Dietrich, P., Burnett, N. H., Ivanov, M. and Corkum, P. B. (1994). *Phys. Rev. A* **50**, R3585.
Dirac P. A. M. (1926). *Proc. Roy. Soc.* **A 112**, 661.
Dixit, S. N. and P. Lambropoulos, P. (1979). *Phys. Rev. A* **19**, 1576.
Dodd, J. N., Fox, W. N., Series, G. W. and Taylor, M. J. (1959). *Proc. Phys. Soc.* **74**, 789.
Dodd, J. N., Kaul, R. D. and Warrington, D. M. (1964). *Proc. Phys. Soc.* **84**, 176.
Dodd, J. N., Sandle, W. J. and Zissermann, D. (1967). *Proc. Phys. Soc.* **92**, 497.
Donley, E. A., Claussen, N. R., Cornish, S. L., Roberts, J. L., Cornell, E. A. and Wieman, C. E. (2001). *Nature* **412**, 295.
Doyle, J. M., Sandberg, J. C., Yu, I. A., Cesar, C. L., Kleppner, D. and Greytak, T. J. (1991). *Phys. Rev. Lett.* **67**, 603.
Drell, P. S. and Commins, E. D. (1985). *Phys. Rev. A* **32**, 2196.
Drescher, M., Hentschel, M., Kienberger, R., Uiberacker, M., Yakovlev, V., Scrinzi, A., Westerwalbesloh, Th., Kleineberg, U., Heinzmann, U. and Krausz, F. (2002). *Nature* **419**, 803.
Drullinger, R. E., Wineland, D. J. and Bergquist, J. C. (1980). *Appl. Phys.* **22**, 365.
Dulieu O. and Gabbabini C. (2009). *Rep. Prog. Phys.* **72**, 086401.
Dum, R., Zoller P. and Ritsch, H. (1992). *Phys. Rev. A* **45**, 4879.
Dupont-Roc, J., Haroche, S. and Cohen-Tannoudji, C. (1969). *Phys. Lett.* **A28**, 638.
Dupont-Roc, J., Fabre, C., Cohen-Tannoudji, C. (1978). *J. Phys. B* **11**, 563.
Dupont-Roc, J., Cohen-Tannoudji, C. (1984). New trends in AtomicPhysics, Les Houches, Session XXXVIII 1982, eds. Grynberg, G. and Stora, R. (Elsevier) pp. 156.
Durfee, D. S., Shaham, Y. K. and Kasevich, M. A. (2006). *Phys. Rev. Lett.* **97**, 240801.
Dürr, S., Volz, T., Marte, A. and Rempe, G. (2004). *Phys. Rev. Lett.* **92**, 020406.
Eagles, D. M. (1969). *Phys. Rev.* **186**, 456.
Eberhard, P. (1978). *Nuovo Cimento B* **46**, 392.
Edwards, M., Ruprecht, P. A., Burnett, K., Dodd, R. J. and Clark, C. W. (1996). *Phys. Rev. Lett.* **77**, 1671.
Efimov V. (1970). *Phys. Lett. B* **33**, 563.
Efimov V. (1971). *Sov. J. Nucl. Phys.* **12**, 589.

Eibl, M., Kiesel, N., Bourennane, M., Kurtsiefer, C. and Weinfurter, H. (2004). *Phys. Rev. Lett.* **92**, 077901.
Eides, M. I., Grotch, H. and Shelyuto, V. A. (2001). *Phys. Reports* **342**, 63.
Einstein, A. (1905). *Annalen der Physik* **17**, 132.
Einstein, A. (1909). *Phys.Z.* **10**, 817
Einstein, A. (1924). *Sitzungsber. Preuss. Akad. Wiss.*, 261.
Einstein, A. (1925). *Sitzungsber. Preuss. Akad. Wiss.*, **1** 3.
Einstein, A., Podolsky, B. and Rosen, N. (1935). *Phys. Rev.* **47**, 777.
Ekstrom, C. R., Keith, D. W. and Pritchard, D. E. (1992). *Appl. Phys. B* **54**, 369.
Ekstrom, C. R., Schmiedmayer, J., Chapman, M. S., Hammond, T. D. and Pritchard, D. E. (1995). *Phys. Rev. A* **51**, 3883.
Ell, R., Morgner, U., Kärtner, F. X., Fujimoto, J. G., Ippen, E. P., Scheuer, V., Angelow, G., Tschudi, T., Lederer, M. J., Boiko, A. and Luther-Davies, B (2001). *Opt. Lett.* **26**, 373.
Ertmer, W., Blatt, R., Hall, J. and Zhu, M. (1985). *Phys. Rev. Lett.* **54**, 996.
Esarey, E., Schroeder, C. B. and Leemans, W. P. (2009). *Rev. Mod. Phys.* **81**, 1229.
Esslinger, T., Bloch, I. and Hänsch, T. W. (1998). *Phys. Rev. A* **58**, R2664.
Esterman, I. and Stern, O. (1930). *Z. Phys.* **61**, 95.
Estève, J., Gross, C., Weller, A., Giovanazzi, S. and Oberthaler, M. K. (2008). *Nature* **455**, 1216.
Everett, M., Lal, A., Gordon, D., Clayton, C. E., Marsh, K. A. and Joshi, C. (1994). *Nature* **368**, 527.
Fano, U. (1961). *Am. J. Phys.* **29**, 539.
Fattori, M., Koch, T., Goetz, S., Griesmaier, A., Hensler, S., Stuhler, J. and Pfau, T. (2006). *Nature Phys.* **2**, 765.
Faure, J., Glinec, Y., Pukhov, A., Kiselev, S., Gordienko, S., Lefebvre, E., Rousseau, J.-P., Burgy, F. and Malka, V. (2004). *Nature* **431**, 541.
Faure, J., Rechatin, C., Norlin, A., Lifschitz, A., Glinec, Y. and Malka, V. (2006). *Nature* **444**, 737.
Fedichev, P. O., Kagan Yu., Shlyapnikov, G. V. and Walraven, J. T. M. (1996a). *Phys. Rev. Lett.* **77**, 2913.
Fedichev, P. O., Reynolds, M. W., Rahmanov, U. M. and Shlyapnikov, G. V. (1996b). *Phys. Rev. A* **53**, 1447.
Felber, J., Müller, G., Ghäler, R. and Golub, R. (1990). *Physica B* **162**, 191.
Ferlaino F., Knoop S. and Grimm R. (2009). *Cold Molecules: Theory, Experiment, Applications* eds. Krems, R. V., Friedrich, B. and Stwalley, W. C. (CRC Press).
Ferlaino F. and Grimm R. (2010). *Physics* **3**, 9.
Ferray, M., L'Huillier, Li, X. F., Lompré, A., Mainfray, G. and Manus, C. (1988). *J. Phys. B* **21**, L31.
Fermi E. (1926). *Rend. Lincei* **3**, 145.
Fetter, A. L. (1972). *Ann. Phys.* **70**, 67.
Fetter, A. L. (2009). *Rev. Mod. Phys.* **81**, 647.
Feynman, R. P. (1948). *Rev. Mod. Phys.* **20**, 367.
Feynman, R. P. (1955). *Progress in Low Temperature Physics* ed. Gorter, C. J. (North-Holland, Amsterdam) Vol. 1, Chap. 2.
Feynman, R. P. (1962). The present status of quantum electrodynamics, in *La théorie quantique des champs* (Interscience, New-York), pp. 61.
Feynman, R. P., Leighton, R. B. and Sands, M. (1963). *The Feynman Lectures on Physics* (Addison-Wesley).

Feynman, R. P. and Hibbs, A. (1965). *Quantum Mechanics and Path Integrals* (Mc Graw Hill, New York).

Fisher, M. P. A., Weichman, P. B., Grinstein, G. and Fisher, D. S. (1989). *Phys. Rev. B* **40**, 546.

Fixler, J. B., Foster, G. T., McGuirk, J. M. and Kasevich, M. A. (2007). *Science* **315**, 74.

Flambaum, V. V., Gribakin, G. F. and Harabati, C. (1999). *Phys. Rev. A* **59**, 1998.

Fölling, S., Widera, A., Müller, T., Gerbier, F. and Bloch I. (2006). *Phys. Rev. Lett.* **97**, 060403.

Foot, C. J. (2005). *Atomic Physics*, Oxford master series in physics (Oxford University Press, New York).

Fortson, N., Sandars, P. and Barr, S. (2003). *Phys. Today* **56**, (6)33.

Franken, P. A., Hill, A. E., Peters, C. W. and Weinreich, G. (1961). *Phys. Rev. Lett.* **7**, 118.

Freedman, S. J. and Clauser, J. F. (1972). *Phys. Rev. Lett.* **28**, 938.

Freeman, R., Bucksbaum, P., Milchberg, H., Darack, S., Schumacher, D. and Geusic, M. (1987). *Phys. Rev. Lett.* **59**, 1092.

French, P. M. W. (1995). *Rep. Prog. Phys.* **58**, 169.

Fried, D. G., Killian, T. C., Willmann, L., Landhuis, D., Moss, S. C., Kleppner, D. and Greytak, T. J. (1998). *Phys. Rev. Lett.* **81**, 3811.

Friedman, N., Kaplan, A., Carasso, D. and Davidson, N. (2001). *Phys. Rev. Lett.* **86**, 1518.

Frisch, R. (1933). *Z. Phys.* **86**, 42.

Frisch, T., Pomeau, Y. and Rica, S. (1992). *Phys. Rev. Lett.* **69**, 1644.

Fry, E. S. and Thompson, R. C. (1976). *Phys. Rev. Lett.* **37**, 465.

Fuchs, J. N., Gangardt, D. M. and Laloë, F. (2002). *Phys. Rev. Lett.* **88**, 230404.

Fuchs, J. N., Gangardt, D. M. and Laloë, F. (2003). *Eur. Phys. J. D* **25**, 57.

Fujita, J., Mitake, S. and Shimizu, F. (2000). *Phys. Rev. Lett.* **84**, 4027.

Fukuhara, T., Sugawa, S. and Takahashi (2007). *Phys. Rev. A* **76**, 051604.

Fukuhara, T., Sugawa, S., Takasu, Y. and Takahashi (2009). *Phys. Rev. A* **79**, 021601(R).

Gabrielse, G., Dehmelt, H. and Kells, W. (1985). *Phys. Rev. Lett.* **54**, 537.

Gabrielse, G. and Dehmelt, H. (1985). *Phys. Rev. Lett.* **55**, 67.

Gabrielse, G., Hanneke, D., Kinoshita, T., Nio, M. and Odom, B. (2006). *Phys. Rev. Lett.* **97**, 030802; *ibid.* **99**, 039902.

Gaebler, J. P., Stewart, J. T., Bohn, J. L. and Jin, D. S. (2007). *Phys. Rev. Lett.* **98**, 200403.

Gattobigio, G. L., Couvert, A., Jeppesen, M., Mathevet, R. and Guéry-Odelin, D. (2009). *Phys. Rev. A* **80**, 041605(R).

Geddes, C. G. R., Toth, Cs., van Tilborg, J., Esarey, E., Schroeder, C. B., Bruhwiler, D., Nieter, C., Cary, J. and Leemans, W. P. (2004). *Nature* **431**, 538.

Gemelke, N., Sarajlic, E. and Chu, S. (2010). arXiv:1007.2677v1

Georges, A. (2010). Lectures notes at Collège de France.

Gerasimov, A. S. and Kazarnovskii, M. V. (1976). *Sov. Phys. JETP*, **44**, 892.

Gerlach, W. and Stern, O., (1924). *Ann. Physik* **74**, 673.

Gerton, J. M., Strekalov, D., Prodan, I. and Hulet, R. D.(2000). *Nature* **408**, 692.

Gibble, K. and Chu, S. (1993). *Phys. Rev. Lett.* **70**, 1771.

Gilbert, S. L., Noecker, M. C., Watts, R. N. and Wieman, C. E. (1985). *Phys. Rev. Lett.* **55**, 2680.

Giltner, D. M., McGowan, R. W. and Lee, S. A. (1995). *Phys. Rev. A* **52**, 3966.

Ginges, J. S. M. and Flambaum, V. V. (2004). *Phys. Rep.* **397**, 63.

Giorgini, S., Pitaevskii, L. P. and Stringari, S. (2008). *Rev. Mod. Phys.* **80**, 1215.

Girardeau, M. (1960). *J. Math. Phys.* **1**, 516.

Gisin, N., and I. C. Percival (1992). *J. Phys. A* **25**, 5677.

Glauber, R. J., (1965). *Quantum Optics and Electronics* (Les Houches 1964), eds. de Witt, C., Blandin, A. and Cohen-Tannoudji, C. (Gordon and Breach, New York).

Glauber, R. J., (2006). *Rev. Mod. Phys.* **78**, 1267.

Gleyzes, S., Kuhr, S., Guerlin, C., Bernu, J., Deléglise, S., Busk Hoff, U., Brune, M., Raimond, J.-M. and Haroche, S. (2007). *Nature* **446**, 297.

Göppert Mayer, M. (1931). *Ann. Phys. (Leipzig)* **9**, 273.

Gordon, J. P. and Ashkin, A. (1980). *Phys. Rev A* **21**, 1606.

Gordon, J. P., Zeiger, H. J. and Townes, C. H. (1954). *Phys. Rev.* **95**, 282.

Gor'kov, L. P. and Melik-Barhudarov, T. K. (1961). *Sov.Phys. JETP* **13**, 1018.

Gott, Y.V., Ioffe, M.S. and Telkovskii, V.G. (1962). *Nuclear Fusion Supplement*, **3**, 1045.

Gould, P. L., Ruff, G. A. and Pritchard, D. E. (1986). *Phys. Rev. Lett.* **56**, 827.

Goulielmakis, E., Uiberacker, M., Kienberger, R., Baltuska, A., Yakovlev, V., Scrinzi, A., Westerwalbesloh, Th., Kleineberg, U., Heinzmann, U., Drescher, M. and Krausz, F. (2004). *Science* **305**, 1267.

Granade, S. R., Gehm, M. E., O'Hara, K. M. and Thomas, J. E. (2002). *Phys. Rev. Lett.*, **88**, 120405.

Greenberger, D. M., Horne, M. A. and Zeilinger, A. (1989). *Bell's Theorem, Quantum Theory, and Conceptions of the Universe*, ed. Kafatos Kluwer, M. (Dordrecht), pp. 69–72.

Greenberger, D. M., Horne, M. A., Shimony, A. and Zeilinger, A. (1990). *Am. J. Phys.* **58**, 1131.

Greene, C. H. (2010). *Phys. Today*, **63**, (3)40.

Greiner, M., Bloch, I., Hänsch, T. W. and Esslinger, T. (2001). *Phys. Rev. A* **63**, 031401.

Greiner, M., Mandel, O., Esslinger, T., Hänsch, T. W. and Bloch, I. (2002). *Nature* **415**, 39.

Greiner, M., Regal, C. A. and Jin, D. S. (2003). *Nature* **426**, 537.

Gribakin, G. F. and Flambaum, V. V. (1993). *Phys. Rev. A* **48**, 546.

Griesmaier, A., Werner, J., Hensler, S., Stuhler, J. and Pfau, T. (2005). *Phys. Rev. Lett.* **94**, 160401.

Griffin, A., Wu, W.-C. and Stringari, S. (1997). *Phys. Rev. Lett.* **78**, 1838.

Griffith, W. C., Swallows, M. D., Loftus, T. H., Romalis, M. V., Heckel, B. R. and Fortson, E. N. (2009). *Phys. Rev. Lett.* **102**, 101601.

Grimm, R., Weidemller, M. and Ovchinnikov, Yu. B. (2000). *Adv. At. Mol. Opt. Phys.* **42**, 95.

Grisenti, R. E., Schöllkopf, W., Toennies, J. P., Hegerfeldt, G. C. and Köhler, T. (1999). *Phys. Rev. Lett.* **83**, 1755.

Gross, N., Shotan, Z., Kokkelmans, S. and Khaykovich, L. (2009). *Phys. Rev. Lett.* **103**, 163202.

Grotch, H. and Kazes, E. (1977). *Am. J. Phys.* **45**, 618.

Grynberg, G., Lounis, B., Verkerk, P., Courtois, J.-Y. and Salomon, C. (1993). *Phys. Rev. Lett.* **70**, 2249.

Grynberg, G., Aspect, A. and Fabre C. (2010). *Introduction to Quantum Optics* (Cambridge University Press, Cambridge).

Guéna, J., Chauvat, D., Jacquier, Ph., Jahier, E., Lintz, M., Sanguinetti, S., Wasan, A., Bouchiat, M. A., Papoyan, A. V. and Sarkisyan, D. (2003). *Phys. Rev. Lett.* **90**, 143001.

Guéna, J., Lintz, M. and Bouchiat, M. A. (2005a). *Mod. Phys. Lett.* **A 20**, 6, 375.

Guéna, J., Lintz, M. and Bouchiat, M. A. (2005b). *Phys. Rev. A* **71**, 042108.

Guéna, J., Lintz, M. and Bouchiat, M. A. (2005c). *J. Opt. Soc. Am. B* **22**, 21.

Guerin, W., Riou, J.-F., Gaebler, J. P., Josse, V., Bouyer, P. and Aspect, A. (2006). *Phys. Rev. Lett.* **97**, 200402.

Guerlin, C., Bernu, J., Deléglise, S., Sayrin, C., Gleyzes, S., Kuhr, S., Brune, M., Raimond, J.-M. and Haroche (2007). *Nature* **448**, 889.

Guéry-Odelin, D., (1998). Thesis, University Paris VI, available on http://tel.archives-ouvertes.fr/.

Guéry-Odelin, D. and Stringari, S. (1999). *Phys. Rev. Lett.* **83**, 4452.

Guéry-Odelin, D., Zambelli, F., Dalibard, J. and Stringari, S. (1999). *Phys. Rev. A* **60**, 4851.

Guéry-Odelin, D. and Lahaye, T. (2011). Proceedings of the Les Houches Summer School of Physics in Singapore, Ultracold gases and Quantum Information, eds. Chuan, K. L., Christian, M., Berthold-Georg, E. and Benoît, G. (Oxford Press).

Guiochon-Bouchiat, M.-A., Blamont, J.-E. and Brossel, J. (1956). *C. R. Acad. Sci.* **243**, 1859.

Guiochon-Bouchiat, M.-A., Blamont, J.-E. and Brossel, J. (1957). *J. Phys. Radium* **18**, 99.

Gupta, S., Kokorowski, D., Rubenstein, R. and Smith, W. (2001). *Adv. At. Mol. Phys.* **46**, 243.

Gustavson, T. L., Bouyer, P. and Kasevich, M. A. (1997). *Phys. Rev. Lett.* **78**, 2046.

Gustavson, T. L., Landragin, A. and Kasevich, M. A. (2000). *Class. Quantum Grav.* **17**, 2385.

Gustavson, T. L., Chikkatur, A. P., Leanhardt, A. E., Görlitz, A., Gupta, S., Pritchard, D. E. and Ketterle, W. (2001). *Phys. Rev. Lett.* **88**, 020401.

Hackermüller, L., Uttenthaler, S., Hornberger, K., Reiger, E., Brezger, B., Zeilinger, A. and Arndt, M. (2003). *Phys. Rev. Lett.* **91**, 090408.

Hadzibabic, Z., Stan, C. A., Dieckmann, K., Gupta, S., Zwierlein, M. W., Görlitz, A. and Ketterle, W. (2002). *Phys. Rev. Lett.* **88**, 160401.

Hadzibabic, Z., Krüger, P., Cheneau, M., Battelier, B. and Dalibard, J. (2006). *Nature* **441**, 1118.

Hadzibabic, Z. and Dalibard, J. (2009). Two-dimensional Bose fluids: An atomic physics perspective, in *Nano Optics and Atomics: Transport of Light and Matter Waves*, Proceedings of the International School of Physics "Enrico Fermi", Course CLXXIII, eds by Kaiser, R. and Wiersma, D. (IOS Press, Amsterdam).

Haessler, S., Caillat, J., Boutu, W., Giovanetti-Teixeira, C., Ruchon, T., Auguste, T., Diveki, Z., Breger, P., Maquet, A., Carré, B., Taïeb, R. and Salières, P. (2010). *Nature Phys.* **6**, 200.

Häffner, H., Hänsel, W., Roos, C. F., Benhelm, J., Chek-al-kar, D., Chwalla, M., Körber, T., Rapol, U. D., Riebe, M., Schmidt, P. O., Becher, C., Gühne, O., Dür, W. and Blatt, R. (2005). *Nature* **438**, 643.

Hagley, E. W., Maître, X., Nogues, G., Wunderlich, C., Brune, M., Raimond, J.-M. and Haroche, S. (1997). *Phys. Rev. Lett.* **79**, 1.

Hagley, E. W., Deng, L., Kozuma, M., Wen, J., Helmerson, K., Rolston, S. L. and Phillips, W. D. (1999a). *Science* **283**, 1706.

Hagley, E. W., Deng, L., Kozuma, M., Trippenbach, M., Band, Y. B., Edwards, M., Doery, M., Julienne, P. S., Helmerson, K., Rolston, S. L. and Phillips, W. D. (1999b). *Phys. Rev. Lett.* **83**, 3112.

Haljan, P. C., Coddington, I., Engels, P. and Cornell, E. A. (2001). *Phys. Rev. Lett.* **87**, 210403.

Haljan, P. C., Anderson, B. P., Coddington, I. and Cornell, E. A. (2001). *Phys. Rev. Lett.* **86**, 2922.

Hall, J. L. (1973). *Atomic Physics 3* (Plenum Press, New-York), pp. 615.
Hall, J. L., Bordé, C. J. and Uehara K. (1976). *Phys. Rev. Lett.* **37**, 1339.
Hall, J. L. (2006). *Rev. Mod. Phys.* **78**, 1279.
Haller, E., Gustavsson, M., Mark, M. J., Danzl, J. G., Hart, R., Pupillo, G. and Nägerl, H.-C. (2009). *Science* **325**, 1224.
Hamilton, W. A., Klein, A. G., Opat, G. I. and Timmins, P. A. (1987). *Phys. Rev. Lett.* **58**, 2770.
Hanbury Brown, E. and Twiss, R. Q. (1956). *Nature*, **178**, 1046.
Hanle, W. (1924). *Z. Phys.* **30**, 93.
Hanle, W. (1926). *Z. Phys.* **35**, 346.
Hanneke, D., Fogwell, S. and Gabrielse, G. (2008). *Phys. Rev. Lett.* **100**, 120801.
Hänsch, T. W., Levenson, M. D. and Schawlow, A. L. (1971a). *Phys. Rev. Lett.* **26**, 946.
Hänsch, T. W., Shahin, I. S. and Schawlow, A. L. (1971b). *Phys. Rev. Lett.* **27**, 707.
Hänsch, T. W., Shahin, I. S. and Schawlow, A. L. (1972). *Nature* **235**, 63.
Hänsch, T. W., Lee, S. A., Wallenstein, R. and Wieman, C. (1975). *Phys. Rev. Lett.* **34**, 307.
Hänsch, T. W. and Schawlow, A. (1975). *Opt. Commun.* **13**, 68.
Hänsch, T. W. (2006). *Rev. Mod. Phys.* **78**, 1297.
Harber, D. M., McGuirk, J. M., Obrecht, J. M. and Cornell, E. A. (2003). *J. Low Temp. Phys.* **133**, 229.
Haroche, S. (1966). Thesis, Paris University, available on http://tel.archives-ouvertes.fr/.
Haroche, S., Cohen-Tannoudji, C., Audoin, C. and Schermann, J.-P. (1970). *Phys. Rev. Lett.* **24**, 861.
Haroche, S. and Cohen-Tannoudji, C. (1970). *Phys. Rev. Lett.* **24**, 974.
Haroche, S. (1971). *Ann. Phys. (Paris)* **6**, 189 ; *ibid.* **6**, 327.
Haroche, S. and Hartmann, F. (1972). *Phys. Rev. A* **6**, 1280.
Haroche, S. and Raimond, J.-M. (2006). *Exploring the Quantum* (Oxford University Press).
Harris, S. (1997). *Phys. Today* **50**, (7)36.
Hartmann, F., Rambosson, M., Brossel, J. and Kastler, A. (1958). *C. R. Acad. Sci.*, **246**, 1522.
Hasert, F. J. *et al.* (1973). *Phys. Lett.* **46B**, 121; *ibid.* **46B**, 138.
Hau, L. V., Busch, B. D., Liu, C., Dutton, Z., Burns, M. M. and Golovchenko, J. A. (1998). *Phys. Rev. A* **58**, R54.
Hau, L. V., Harris, S. E., Dutton, Z. and Behroozi, C. H. (1999). *Nature* **397**, 594.
He, Y., Chen, Q. and Levin, K. (2005). *Phys. Rev. A* **72**, 011602.
Hecht, C. E. (1959). *Physica* **25**, 1159.
Heinzen, D. J., (1999). Proceedings of the International School of Physics "Enrico Fermi", Course CXL, Bose-Einstein Condensation in Atomic Gases.
Henkel, C., Steane, A. M., Kaiser, R. and Dalibard, J. (1994). *J. Physique II* **4**, 1877.
Henshaw, D. G. and Woods, A. D. (1961). *Phys. Rev.* **121**, 1266.
Hentschel, M., Kienberger, R., Spielmann, Ch., Reider, G. A., Milosevic, N., Brabec, T., Corkum, P., Heinzmann, U., Drescher, M. and Krausz, F. (2001). *Nature* **414**, 509.
Herbig, J., Kraemer, T., Mark, M., Weber, T., Chin, C., Nägerl, H.-N. and Grimm, R. (2003). *Science* **301**, 1510.
Hess, H. F. (1986). *Phys. Rev. B* **34**, 3476.
Hess, H. F., Kochanski, G. P., Doyle, J. M., Masuhara, N., Kleppner, D. and Greytak, T. J. (1987). *Phys. Rev. Lett.* **59**, 672.
Hijmans, T. W., Luiten, O. J., Setija, I. D. and Walraven, J. T. M. (1989). *J. Opt. Soc. Am. B* **6**, 2235.
Ho, T.-L. and Ciobanu, C. V. (2004). *J. Low Temp. Phys.* **135**, 257.

Ho, T.-L. and Zhou, Q. (2009). *Nature Phys.* **6**, 131.
Hodapp, T. W., Gerz, C., Furtlehner, C., Westbrook, C. I., Phillips, W. D. and Dalibard, J. (1995). *Appl. Phys. B* **60**, 135.
Hodby, E., Hechenblaikner, G., Marago, O.M., Arlt, J., Hopkins, S. and Foot, C.J. (2000). *J. Phys. B* **33**, 4087.
Hodgman, S. S., Dall, R. G., Manning, A. G., Baldwin, K. G. H. and Truscott, A. G. (2011). *Science* **331**, 1046.
Hofferberth, S., Lesanovsky, I., Fischer, B., Verdu, J. and Schmiedmayer, J. (2006). *Nature Phys.* **2**, 710.
Höffges, J. T., Baldauf, H. W., Lange, W. and Walther, H. (1997). *J. Mod. Optics* **44**, 1999.
Hohenberg, P. C. (1967). *Phys. Rev.* **158**, 383.
Horodecki, R., Horodecki, P. and Horodecki, M. (1995). *Phys. Lett.*, **A200**, 340.
Hu, H., Liu, X.-J., and Drummond, P. D. (2010). *New J. Phys.* **12**, 063038.
Huang, K. (1952). *Am. J. Phys.* **20**, 479.
Huang, K. (1988). *Statistical Mechanics*, (2nd Revised edition, John Wiley and Sons).
Hudson, J. J., Sauer, B. E., Tarbutt, M. R. and Hinds, E.A. (2002). *Phys. Rev. Lett.* **89**, 023003.
Hughes, V. and Grabner, L. (1950a). *Phys. Rev.* **79**, 314.
Hughes, V. and Grabner, L. (1950b). *Phys. Rev.* **79**, 829.
Hulet, R. G., Hilfer, E. S. and Kleppner D. (1985). *Phys. Rev. Lett.* **55**, 2137.
Hung, C.-L., Zhang, X., Gemelke and N., Ching, C. (2010). *Nature* **470**, 236.
Inouye, S., Andrews, M. R., Stenger, J., Miesner, H.-J., Stamper-Kurn, D. M. and Ketterle, W. (1998). *Nature* **392**, 151.
Ippen, E. P., Shank, C. V. and Dienes, A. (1972). *Appl. Phys. Lett.* **21**, 348.
Itano, W. M., Bergquist, J. C., Bollinger, J. J., Gilligan, J. M., Heinzen, D. J., Moore, F. L., Raizen, M. G. and Wineland, D. J. (1993). *Phys. Rev. A* **47**, 3554.
Itano, W. M., Bergquist, J. C., Bollinger, J. J. and Wineland, D. J. (1995). *Phys. Scripta* **T59**, 106.
Itatani, J., Levesque, J., Zeidler, D., Niikura, H., Pépin, H., Kieffer, J. C., Corkum, P. B. and Villeneuve, D. M. (2004). *Nature* **432**, 867.
Itzykson, C. and Zuber J.-B. (1986). *Quantum Field Theory* (McGraw Hill).
Jackson, J. D. (1975). *Classical Electrodynamics* (2nd edition, Wiley, New-York).
Jaksch, D., Bruder, C., Cirac, J. I., Gardiner, C. W. and Zoller, P. (1998). *Phys. Rev. Lett.* **81**, 3108.
Jamieson, M. J., Dalgarno, A. and Yukich, J. N. (1992). *Phys. Rev. A* **46**, 6956.
Jamieson, M. J., Dalgarno, A. and Kimura, M. (1995). *Phys. Rev. A* **51**, 2626.
Janik, G. Nagourney, W. and Dehmelt, H. (1985). *J. Opt. Soc. Am. B* **2**, 1251.
Javanainen, J. and Yoo, S. M. (1996). *Phys. Rev. Lett.* **76**, 161.
Jaynes, E. T. and Cummings, F. W. (1963). *Proc. IEEE* **51**, 89.
Jeffery, A., Elmquist, R. E., Shields, J. Q., Lee, L. H., Cage, M. E., Shields, S. H. and Dziuba, R. F. (1998). *Metrologia* **35**, 83.
Jeltes, T., McNamara, J. M., Hogervorst, W., Vassen, W., Krachmalnico, V., Schellekens, M., Perrin, A., Chang, H., Boiron, D., Aspect, A., and Westbrook, C. I. (2007). *Nature* **445**, 402.
Jensen A.S., Riisager K., Fedorov D.V. and Garrido E. (2004). *Rev. Mod. Phys.* **76**, 215.
Jessen, P. S., Gerz, C., Lett, P. D., Phillips, W. D., Rolston, S. L., Spreeuw, R. J. C. and Westbrook, C. I. (1992). *Phys. Rev. Lett.* **69**, 49.
Jhe, W., Anderson, A., Hinds, E. A., Meschede, D., Moi L. and Haroche, S. (1987). *Phys. Rev. Lett.* **58**, 666.

Jin, D. S., Ensher, J. R., Matthews, M. R., Wieman, C. E. and Cornell, E. A. (1996). *Phys. Rev. Lett.* **77**, 420.
Jin, D. S. , Matthews, M. R., Ensher, J. R., Wieman, C. E. and Cornell, E. A. (1997). *Phys. Rev. Lett.* **78**, 764.
Joachain C. J. (1983). *Quantum Collision Theory* (North Holland, Amsterdam).
Jochim, S., Bartenstein, M., Altmeyer, A., Hendl, G., Riedl, S., Chin, C., Hecker-Denschlag, J. and Grimm, R. (2003a). *Science* **302**, 2101.
Jochim, S., Bartenstein, M., Altmeyer, A., Hendl, G., Chin, C., Hecker-Denschlag, , J. and Grimm, R. (2003b). *Phys. Rev. Lett.* **91**, 240402.
Johnson, B. R., Denker, J. S., Bigelow, N., Lévy, L. P., Freed, J. H. and Lee, D. M. (1984). *Phys. Rev. Lett.* **52**, 1508.
Johnson, G. A., Hedlund, L. and MacFall, J. (1998). *Physics World* **11**, 35.
Jönsson, C. (1961). *Z. Phys.* **161**, 454.
Jönsson, C. (1974). *Am. J. Phys.* **42**, 4.
Jördens, R., Strohmaier, N., Günter, K., Moritz, H. and Esslinger, T. (2008). *Nature* **455**, 204.
Josephson, B. D. (1974). *Rev. Mod. Phys.* **46**, 251.
Kagan, Y., Vartanyantz, I. A. and Shlyapnikov, G. (1980). *Sov. Phys. JETP* **54**, 590.
Kagan, Y, Surkov, E. L. and Shlyapnikov, G. V. (1997). *Phys. Rev. A* **55**, R18.
Kaiser, W. and Garrett, C. (1961). *Phys. Rev. Lett.* **7**, 229.
Kapitza, P. L. and Dirac, P. A. M. (1933). *Proc. Cambridge Phil. Soc.* **29**, 297.
Kapteyn, H. C., Murnane, M. M. and Christov, I. P. (2005). *Phys. Today* **58**, (3)39.
Kasevich, M. and Chu, S. (1991). *Phys. Rev. Lett.* **67**, 181.
Kasevich, M. and Chu, S. (1992a). *Phys. Rev. Lett.* **69**, 1741.
Kasevich, M. and Chu, S. (1992b). *Appl. Phys. B* **54**, 321.
Kastler, A. (1950). *J. Phys. Radium* **11**, 255.
Katori, H. (2002). Sixth Symposium on Frequency Standards and Metrology, ed. Gill, P. (World Scientic, Singapore).
Kavoulakis, G. M. and Pethick, C. J. (1998). *Phys. Rev. A* **58**, 1563.
Kazantsev, A. P. (1972). *Sov. Phys. JETP* **36**, 861.
Keith, D. W., Schattenburg, M. L., Smith, H. I. and Pritchard, D. E. (1988). *Phys. Rev. Lett.* **61**, 1580.
Keith, D. W., Ekstrom, C. R., Turchette, Q. A. and Pritchard, D. E. (1991). *Phys. Rev. Lett.* **66**, 2693.
Keldysh, L. V. (1965). *Sov. Phys. JETP* **20**, 1307.
Keldysh, L. V. and Kozlov, A. N. (1968). *Sov. Phys. JETP* **27**, 521.
Ketterle, W., Davis, K. B., Joffe, M. A., Martin, A. and Pritchard, D. E. (1993). *Phys. Rev. Lett.* **70**, 2253.
Ketterle, W. and van Druten, N. J. (1996). In *Advances in Atomic, Molecular and Optical Physics* Vol. 37, 181 eds. Bederson, B. and Walther, H. (Academic Press, San Diego).
Ketterle, W. and Miesner, H.-J. (1997). *Phys. Rev. A* **56**, 3291.
Ketterle, W., Durfee, D. S. and Stamper-Kurn, D. M. (1999). Making, probing and understanding Bose-Einstein condensates, in *Bose-Einstein Condensation in Atomic Gases*, Proceedings of the International School of Physics "Enrico Fermi", Course CXL, eds. Inguscio, M., Stringari, S. and Wieman, C. E. (IOS Press, Amsterdam) pp. 67.
Ketterle, W. (2002). *Rev. Mod. Phys.* **74**, 1131.
Ketterle, W. and Zwierlein, M. W. (2008). Making, probing and understanding ultracold Fermi gases, in *Ultra-cold Fermi Gases*, Proceedings of the International School of Physics "Enrico Fermi", Course CLXIV, eds. Inguscio, M., Ketterle, W. and Salomon, C. (IOS Press, Amsterdam) pp. 95.

Khaykovich, L., Schreck, F., Ferrari, G., Bourdel, T., Cubizolles, J., Carr, L., Castin, Y. and Salomon, C. (2002). *Science* **296**, 1290.
Kienberger, R., Goulielmakis, E., Uiberacker, M., Baltuska, A., Yakovlev, V., Bammer, F., Scrinzi, A., Westerwalbesloh, Th., Kleineberg, U., Heinzmann, U., Drescher, M. and Krausz, F. (2004). *Nature* **427**, 817.
Kierig, E., Schnorrberger, U., Schietinger, A., Tomkovic, J. and Oberthaler, M. K. (2008). *Phys. Rev. Lett.* **100**, 190405.
Kim, M. S., Knight, P. L. and Wodkiewicz, K. (1987). *Opt. Comm.* **62**, 385.
Kimble, H. J., Dagenais, M. and Mandel (1977). *Phys. Rev. Lett.* **39**, 691.
Kinast, J., Hemmer, S. L., Gehm, M. E., Turlapov, A. and Thomas, J. E. (2004). *Phys. Rev. Lett.* **92**, 150402.
Kinast, J., Turlapov, A., Thomas, J. E., Chen, Q., Stajic, J. and Levin, K. (2005). *Science* **307**, 1296.
Kinnunen, J., Rodriguez, M. and Törmä, P. (2004). *Science* **305**, 1131.
Kinoshita, T., Wenger, T. and Weiss, D. S. (2004). *Science* **305**, 1125.
Kitagawa, M. and Ueda, M. (1993). *Phys. Rev. A* **47**, 5138.
Klaers J., Schmitt, J., Vewinger, F. and Weitz, M. (2011). *Nature* **468**, 545.
Klarsfeld, S. and Maquet, A. (1972). *Phys. Rev. Lett.* **29**, 79.
Kleppner, D. (1981). *Phys. Rev. Lett.* **47**, 233.
Kleppner, D. and Haroche, S. (1989). *Phys. Today* **42**, (1)24.
Koba Z. (1949). *Prog. Theor. Phys.* **3**, 319.
Kocher, C. A. and Commins, E. D. (1967). *Phys. Rev. Lett.* **18**, 575.
Köhler, T., Góral, K. and Julienne, P. S. (2006). *Rev. Mod. Phys.* **78**, 1311.
Kohn, W. and Luttinger, J. M. (1965). *Phys. Rev. Lett.* **15**, 524.
Kosterlitz, J. M. and Thouless, D. J. (1973). *J. Phys. C: Solid State Phys.* **6**, 1181.
Kraemer, T., Mark, M., Waldburger, P., Danzl, J. G., Chin, C., Engeser, B., Lange, A. D., Pilch K., Jaakola, A., Nägerl, H. C. and Grimm, R. (2006). *Nature* **440**, 315.
Kraft, S., Vogt, F., Appel, O., Riehle, F. and Sterr, U. (2009). *Phys. Rev. Lett.* **103**, 130401.
Krausz, F. and Ivanov, M. (2009). *Rev. Mod. Phys.* **81**, 163.
Kuga, T., Torii, Y., Shiokawa, N., Hirano, T., Shimizu Y. and Sasada, H. (1997). *Phys. Rev. Lett.* **78**, 4713.
Krushelnick, K., Ting, A., Moore, C. I., Burris, H. R., Esarey, E., Sprangle, P. and Baine, M. (1997). *Phys. Rev. Lett.* **78**, 4047.
Kuklinski, J. R., Gaubatz, U., Hioe, F. T. and Bergmann, K. (1989). *Phys. Rev. A* **40**, 6741.
Kulin, S., Saubamea, B., Peik, E., Lawall, J., Hijmans, T. W., Leduc, M. and Cohen-Tannoudji, C. (1997). *Phys. Rev. Lett.* **78**, 4185.
Kulinski, J. R., Ganbatz, U., Hioe, F.T. and Bergmann, K. (1989). *Phys. Rev. A* **40**, 6741.
Kuppens, S. J. M., Corwin, K. L., Miller, K. W., Chupp, T. E. and Wieman, C. E. (2000). *Phys. Rev. A* **62**, 013406.
Kush, P. and Foley, H. M. (1948). *Phys. Rev.* **74**, 250.
Kwek, L. C., Miniatura, C., Berthold-Georg, E. and Benoît, G. (2011). Proceedings of the Les Houches Summer School of Physics in Singapore, Ultracold Gases and Quantum Information (Oxford Press).
Kwiat, P., Mattle, K., Weinfurter, W., Zeilinger, A., Sergienko, A. and Shih, Y. (1995). *Phys. Rev. Lett.* **75**, 4337.
Kyröla, E. and Stenholm, S. (1977). *Opt. Commun.* **22**, 123.
Labeyrie, G., Michaud, F. and Kaiser, R. (2006). *Phys. Rev. Lett.* **96**, 023003.
Lagendijk, A., van Tiggelen, B. and Wiersma, D. S. (2009). *Phys. Today* **62**, (8)24.

Lahaye, T., Vogels, J. M., Günter, K., Wang, Z., Dalibard, J. and Guéry-Odelin, D. (2004). *Phys. Rev. Lett.* **93**, 093003.

Lahaye, T., Wang, Z., Reinaudi, G., Rath, S. P., Dalibard, J. and Guéry-Odelin, D. (2005). *Phys. Rev. A.* **72**, 033411.

Lahaye, T., Koch, T., Fröhlich, B. , Fattori, M., Metz, J., Griesmaier, A., Giovanazzi, S. and Pfau, T. (2007). *Nature* **448**, 672.

Lahaye, T., Menotti, C., Santos, L., Lewenstein, M. and Pfau, T. (2009). *Rep. Prog. Phys.* **72**, 126401.

Lai, S. B., Knight, P. L. and Eberly, J. H. (1974). *Phys. Rev. Lett.* **32**, 494.

Laloë, F. (2001). *Am. J. Phys.* **69**, 655.

Lamb, W. E. (1939). *Phys. Rev.* **55**, 190.

Lamb, W. E. and Retherford, R. C. (1947). *Phys. Rev.* **72**, 241.

Lamb, W. E. (1964). *Phys. Rev.* **134**, A1429.

Lamporesi, G., Bertoldi, A., Cacciapuoti, L., Prevedelli, M. and Tino, G. M. (2008). *Phys. Rev. Lett.* **100**, 050801.

Landau, L.D. (1941a). *J. Phys.* **5**, 71.

Landau, L.D. (1941b). *Phys. Rev.* **60**, 356.

Landau, L.D., and Lifshitz, E.M. (1982). *Mechanics, Course of Theoretical Physics* (3rd revised edition, Butterworth-Heinemann).

Landau, L.D. and Lifshitz, E.M. (1996). *Quantum Mechanics, Course of Theoretical Physics* (3rd revised edition, Butterworth-Heinemann).

Landre, C., Cohen-Tannoudji, C., Dupont-Roc, J. and Haroche, S. (1970). *C.R. Acad. Sci.* **270**, 73.

Lawall, J., Bardou, F., Saubamea, B., Shimizu, K., Leduc, M., Aspect, A. and Cohen-Tannoudji, C. (1994). *Phys. Rev. Lett.* **73**, 1915.

Lawall, J., Kulin, S., Saubamea, B., Bigelow, N., Leduc, M. and Cohen-Tannoudji, C. (1995). *Phys. Rev. Lett.* **75**, 4194.

Lea, S. N., Clairon, A., Salomon, C., Laurent, P., Lounis, B., Reichel, J., Nadir, A. and Santarelli, G. (1994). *Phys. Scr.* **T51**, 78.

Leanhardt, A. E., Görlitz, A., Chikkatur, A. P., Kielpinski, D., Shin, Y., Pritchard, D. E. and Ketterle, W. (2002). *Phys. Rev. Lett.* **89**, 190403.

Leduc, M., Nacher, P. J., Betts, D. S., Daniels, J. M., Tastevin, G. and Laloë, F. (1987). *Europhys. Lett.* **4**, 59.

Lee, T. D. and Yang, C. N. (1956). *Phys. Rev.* **104**, 254.

Lee, T. D., and Yang, C. N. (1957). *Phys. Rev.* **105**, 1119.

Lee, T. D., Huang, K. and Yang, C. N. (1957). *Phys. Rev.* **106**, 1135.

Leemans, W. P., Nagler, B., Gonsalves, A. J., Toth, Cs., Nakamura, K., Geddes, C. G. R., Esarey, E., Schroeder, C. B. and Hooker, S. M. (2006). *Nature Phys.* **2**, 696.

Leggett, A. J. (1980). In *Modern Trends in the Theory of Condensed Matter* eds. Pekalski, A. and Przystawa, R. (Springer-Verlag, Berlin).

Leggett, A. J. (2006). *Quantum Liquids: Bose Condensation and Cooper Pairing in Condensed-Matter Systems* (Oxford Graduate Texts).

Lehmann, J.-C. (1967). Thesis, Ecole Normale supérieure (Paris), available on http://tel.archives-ouvertes.fr/.

Leibfried, D., Blatt, R., Monroe, C. and Wineland, D. (2003). *Rev. Mod. Phys.* **75**, 281.

Leibfried, D., Knill, E., Seidelin, S., Britton, J., Blakestad, R. B., Chiaverini, J., Hume, D. B., Itano, W. M. Jost, J. D., Langer, C., Ozeri, R., Reichle, R. and Wineland, D. J. (2005). *Nature* **438**, 639.

Leinaas, J. M. and Myrheim, J. (1977). *Il Nuovo Cimento* **37**, 1.

Lemonde, P., Petit, P., Audoin, C., Salomon, C., Laurent, Ph., Simon, E., Santarelli, G., Clairon, A. and Dimarcq, N. (1998). *Eur. Phys. J. D* **3**, 201.
Lemonde, P., Laurent, P. , Santarelli, G., Abgrall, M., Sortais, Y., Bize, S., Nicolas, C., Zhang, S., Schehr, G., Clairon, A., Dimarcq, N., Petit, P., Mann, A., Luiten, A., Chang, S. and Salomon, C. (2001). *Topics Appl. Phys.* **79**, 131 In Frequency measurement and control, ed. Luiten, A. N. (Springer-Verlag, Berlin).
Leo, P. J., Williams, C. J. and Julienne, P. S. (2000). *Phys. Rev. Lett.* **85**, 2721.
Léonard, J. (2003). Thesis, University Paris VI, available on http://tel.archives-ouvertes.fr/.
Letokhov, V. S. (1968). *JETP Lett.* **7**, 272.
Letokhov, V. S., Minogin, V. G. and Pavlik, B. D. (1976). *Opt.Commun.* **19**, 72.
Letokhov, V. S. (2007). *Laser Control of Atoms and Molecules* (Oxford University Press).
Lett, P. D., Watts, R. N., Westbrook, C. I., Phillips, W. D., Gould, P. L. and Metcalf, H. J. (1988). *Phys. Rev. Lett.* **61**, 169.
Lett, P. D., Phillips, W. D., Rolston, S. L., Tanner, C. E., Watts, R. N. and Westbrook, C. I. (1989). *J. Opt. Soc. Am. B* **6**, 2084.
Levenson, M. D. and Bloembergen, N. (1974). *Phys. Rev. Lett.* **32**, 645.
Levy, S., Lahoud, E., Shomroni, I. and Steinhauer, J. (2007). *Nature* **449**, 579.
Lewandowski, H. J., Harber, D. M., Whitaker, D. L. and Cornell, E. A. (2002). *Phys. Rev. Lett.* **88**, 070403.
Lewandowski, H. J., Harber, D. M., Whitaker, D. L. and Cornell, E. A. (2003). *J. Low. Temp. Phys.* **132**, 309.
Lewenstein, M., Salières, P. and L'Huillier, A. (1995). *Phys. Rev. A* **52**, 4747.
Lewis, G. N. (1926). *Nature* **118**, 874.
Leyronas, X. and Combescot, R. (2007). *Phys. Rev. Lett.* **99**, 170402.
Lhuillier, C. and Laloë, F. (1979). *J. Phys. (Paris)*, **40**, 239.
Lhuillier, C. and Laloë, F. (1982a). *J. Phys. (Paris)*, **43**, 197.
Lhuillier, C. and Laloë, F. (1982b). *J. Phys. (Paris)*, **43**, 225.
Li, X. F., L'Huillier, A., Ferray, M., Lompré, L. A. and Mainfray, G. (1989). *Phys. Rev. A* **39**, 5751.
Lignier, H., Sias, C., Ciampini, D., Singh, Y., Zenesini, A., Morsch, O. and Arimondo, E. (2007). *Phys. Rev. Lett.* **99**, 220403.
Lin, Y.-J., Compton, R. L., Jiménez-García, K., Porto, J. V. and Spielman, I. B. (2009). *Nature* **462**, 628.
Lin, Y.-J., Compton, R. L., Jiménez-García, K., Phillips, W. D., Porto, J. V. and Spielman, I. B. (2011). arXiv:1008.4864, to appear in *Nature Physics.*
Lindblad G. (1976). *Commun. Math. Phys.*, **48**, 119.
Litvak, A. G. (1970). *Sov. Phys. JETP* **30**, 344.
Liu, W., Boshier, M. G., Dhawan, S., van Dyck, O., Egan, P., Fei, X., Grosse Perdekamp, M., Hughes, V. W., Janousch, M. , Jungmann, K., Kawall, D., Mariam, F. G., Pillai, C., Prigl, R., zu Putlitz, G., Reinhard, I., Schwarz, W., Thompson, P. A. and Woodle, K. A. (1999). *Phys. Rev. Lett.* **82**, 711.
London, F. (1938). *Nature* **141**, 643.
Luiten, O. J., Reynolds, M. W. and Walraven, J. T. M. (1996). *Phys. Rev. A* **53**, 381.
Lundh, E., Pethick, C., Smith, H. (1997). *Phys. Rev. A* **55**, 2126.
Luo, F., Giese, C. F. and Gentry W. R. (1996). *J. Chem. Phys.* **104**, 1151.
Mackenzie, A. P. and Maeno, Y. (2003). *Rev. Mod. Phys.* **75**, 657.
Madison, K. W., Chevy, F., Wohlleben, W. and Dalibard, J. (2000). *Phys. Rev. Lett.* **84**, 806.
Madison, K. W., Chevy, F., Bretin, V. and Dalibard, J. (2001). *Phys. Rev. Lett.* **86**, 4443.

Maine, P., Strickland, D., Bado, P., Pessot, M., and Mourou, G. (1988). *IEEE J. Quantum Electron.* **24**, 398.

Mainfray, G. and Manus, C. (1978). *J. Phys. Colloques* **39**, C1–1.

Mairesse, Y., de Bohan, A., Frasinski, L. J., Merdji, H., Dinu, L. C., Monchicourt, P., Breger, P., Kovacev, M., Taïeb, R., Carré, B., Muller, H. G., Agostini, P. and Salières, P. (2003). *Science* **302**, 1540.

Major, F. G. and Dehmelt, H. G. (1968). *Phys. Rev.* **170**, 91.

Major, F. G., Gheorghe, V. N. and Werth, G. (2005). *Charged Particles Trap*, Springer series on Atomic, Optical, and Plasma physics (Springer-Verlag Berlin Heidelgerg).

Majorana, E. (1932). *Nuovo Cimento* **9**, 43.

Malka, V., Faure, J., Gauduel, Y. A., Lefebvre, E., Rousse, A. and Ta Phuoc, K. (2008). *Nature Phys.* **4**, 447.

Malnar, L. and Mosnier, J. P. (1961). *Annales de Radioélectricité* **16**, 3.

Mandel, O., Greiner, M., Widera, A., Rom, T., Hänsch, T. and Bloch, I. (2003). *Phys. Rev. Lett.* **91**, 010407.

Mangles, S. P. D., Murphy, C. D., Najmudin, Z., Thomas, A. G. R., Collier, J. L., Dangor, A. E., Divall, E. J., Foster, P. S., Gallacher, J. G., Hooker, C. J., Jaroszynski, D. A., Langley, A. J., Mori, W. B., Norreys, P. A., Tsung, F. S., Viskup, R., Walton, B. R. and Krushelnick, T. (2004). *Nature* **431**, 535.

Manz, S., Bücker, R., Betz, T., Koller, Ch., Hofferberth, S., Mazets, I. E., Imambekov, A., Demler, E., Perrin, A., Schmiedmayer, J. and Schumm, T. (2010). *Phys. Rev. A* **81**, 031610.

Maragò, O. M., Hopkins, S. A., Arlt, J., Hodby, E., Hechenblaikner, G. and Foot, C. J. (2000). *Phys. Rev. Lett.* **84**, 2056.

Maragò, O., Hechenblaikner, G., Hodby, E. and Foot, C. (2001). *Phys. Rev. Lett.* **86**, 3938.

Margerie, J., and Brossel, J. (1955). *C. R. Acad. Sci.*, **241**, 373.

Marte, P., Dum, R., Taï eb, R., Zoller, P., Shahriar, M. S. and Prentiss, M. (1994). *Phys. Rev. A* **49**, 4826.

Marte, A., Volz, T., Schuster, J., Dürr, S., Rempe, G., van Kempen, E. G. M. and Verhaar, B. J. (2002). *Phys. Rev. Lett.* **89**, 283202.

Martin, P. J., Oldaker, B. G., Miklich, A.H. and Pritchard, D. E. (1988). *Phys.Rev. Lett.* **60**, 515.

Martinez de Escobar, Y. N., Mickelson, P. G., Yan, M., DeSalvo, B. J., Nagel, S. B. and Killian, T. C. (2009). *Phys. Rev. Lett.* **103**, 200402.

Marton, L., Arol Simpson, J. and Suddeth, J. A. (1953). *Phys. Rev.* **90**, 490.

Masuhara, N., Doyle, J. M., Sandberg, J. C., Kleppner, D., Greytak, T. J., Hess, H. F. and Kochanski, G. P. (1988). *Phys. Rev. Lett.* **61**, 935.

Matthews, M. R., Anderson, B. P., Haljan, P. C., Hall, D. S., Wieman, C. E. and Cornell, E. A. (1999). *Phys. Rev. Lett.* **83**, 2498.

Max, C. E., Arons, J. and Langdon, A. B. (1974). *Phys. Rev. Lett.* **33**, 209.

McClung, F.J. and Hellwarth, R.W (1962). *J. Appl. Phys.* **33**, 828.

McGuirk, J. M., Snadden, M. J. and Kasevich, M. A. (2000). *Phys. Rev. Lett.* **85**, 4498.

McGuirk J.M., Lewandowski H.J., Harber D.M., Nikuni T., Williams J.E. and Cornell E.A. (2002). *Phys. Rev. Lett.* **89**, 090402.

McNamara, J. M., Jeltes, T., Tychkov, A. S., Hogervorst, W. and Vassen, W. (2006). *Phys. Rev. Lett.* **97**, 080404.

McPherson, A., Gibson, G., Jara, H., Johann, U., Luk, T. S., McIntyre, I. A., Boyer, K. and Rhodes, C. K. (1987). *J. Opt. Soc. Am. B* **4**, 595.

Meckel, M., Comtois, D., Zeidler, D., Staudte, A., Pavicic, D., Bandulet, H. C., Pépin,

H., Kieffer, J. C., Dörner, R., Villeneuve, D. M. and Corkum, P. B. (2008). *Science* **320**, 1478.

Mehlstäubler, T. E., LeGouët, J., Merlet, S., Holleville, D., Clairon, A., Landragin, A. and Pereira Dos Santos, F. (2007). *Proceedings of the XLII-th Rencontres de Moriond*, Gravitational Waves and Experimental Gravity, eds. Dumarchez, J. and Van, J.T.T. (The Gioi Publishers), p. 323.

Mennerat-Robilliard, C., Lucas, D., Guibal, S., Tabosa, J., Jurczak, C., Courtois, J.-Y. and Grynberg, G. (1999). *Phys. Rev. Lett.* **82**, 851.

Mermin, N. D. and Wagner, H. (1966). *Phys. Rev. Lett.* **17**, 1133.

Mermin, N. D. (1990). *Am. J. Phys.* **58**, 731.

Messiah, A. (2003). *Quantum Mechanics: Two Volumes Bound As One* (Dover Publications Inc.).

Metcalf, H. J. and Van Der Straten, P. (2001). *Laser Cooling and Trapping* (revised edition, Springer-Verlag New York Inc.)

Mewes, M.-O., Andrews, M. R., van Druten, N. J., Kurn, D. M., Durfee, D. S. and Ketterle, W. (1996). *Phys. Rev. Lett.* **77**, 416.

Mewes, M.-O., Andrews, M. R., Kurn, D. M., Durfee, D. S., Townsend, C. G. and Ketterle, W. (1997). *Phys. Rev. Lett.* **78**, 582

Miffre, A., Jacquey, M., Büchner, M., Trénec, G. and Vigué (2006). *Phys. Rev. A* **73**, 011603(R).

Migdall, A. L., Prodan, J. V., Phillips, W. D., Bergeman, T. H. and Metcalf, H. J. (1985). *Phys. Rev. Lett.* **54**, 2596.

Miller, D. E., Chin, J. K., Stan, C. A., Liu, Y., Setiawan, W., Sanner, C. and Ketterle, W. (2007). *Phys. Rev. Lett.* **99**, 070402.

Milner, V., Hanssen, J. L., Campbell, W. C. and Raizen, M. G. (2001). *Phys. Rev. Lett.* **86**, 1514.

Miniatura, C., Perales, F., Vassilev, G., Reinhardt, J., Robert, J. and Baudon, J. (1991). *J. Physique II* **1**, 425.

Minogin, V. G. and Serimaa, O. T. (1979). *Opt. Commun.* **30**, 373.

Moal, S., Portier, M., Kim, J., Dugué, J., Rapol, U. D., Leduc, M. and Cohen-Tannoudji, C. (2006). *Phys. Rev. Lett.* **96**, 023203.

Modugno, G., Ferrari, G., Roati, G., Brecha, R. J., Simoni, A. and Inguscio, M. (2001). *Science* **294**, 1320.

Moerdijk A.J., Verhaar B.J. and Axelsson A. (1995). *Phys. Rev. A* **51**, 4852.

Mohr, P. J., Taylor, B. N. and Newell, D. B. (2008). *Rev. Mod. Phys.* **80**, 633.

Möller, H. E., Chen, X. J, Saam, B., Hagspiel, K. D., Johnson, G. A, Altes, T.A., de Lange, E. E. and Kauczor, H. U. (2002). *Magnetic Resonance in Medicine* **47**, 1029.

Mollow, B. R. (1969). *Phys. Rev.* **188**, 1969.

Monot, P., Auguste, T., Gibbon, P., Jakober, F., Mainfray, G., Dulieu, A., Louis-Jacquet, M., Malka, G. and Miquel, J. L. (1995). *Phys. Rev. Lett.* **74**, 2953.

Monroe, C., Swann, W., Robinson, H. and Wieman, C. (1990). *Phys. Rev. Lett.* **65**, 1571.

Monroe, C. R., Cornell, E. A., Sackett, C. A., Myatt, C. J. and Wieman,C. E. (1993). *Phys. Rev. Lett.* **70**, 414.

Monroe, C., Meekhof, D. M., King, B. E., Jefferts, S. R., Itano, W. M., Wineland, D. J. and Gould, P (1995). *Phys. Rev. Lett.* **75**, 4011.

Moore, F. L., Robinson, J. C., Bharucha, C., Williams, P. E. and Raizen, M. G. (1994). *Phys. Rev. Lett.* **73**, 2974.

Morellec, J., Normand, D., Mainfray, G. and Manus C. (1980). *Phys. Rev. Lett.* **44**, 1394.

Morinaga, M., Yasuda, M., Kishimoto, T., Shimizu, F., Fujita, J.-i. and Matsui, S. (1996). *Phys. Rev. Lett.* **77**, 802.

Morizot, O., Colombe, Y., Lorent, V. and Perrin, H. (2006). *Phys. Rev. A* **74**, 023617.
Mortensen, A., Nielsen, E., Matthey, T. and Drewsen, M. (2006). *Phys. Rev. Lett.* **96**, 103001.
Moshinsky, M. (1952). *Phys. Rev.* **88**, 625.
Moskowitz, P. E., Gould, P. L., Atlas, S. R. and Pritchard, D. E. (1983). *Phys. Rev. Lett.* **51**, 370.
Mössbauer, R. L. (1958a). *Z. Phys.* **151**, 124.
Mössbauer, R. L. (1958b). *Naturwiss.* **45**, 538.
Mössbauer, R. L. (1959). *Z. Naturforch.* **14a**, 211.
Mottelson, Ben (1976). *Rev. Mod. Phys.* **48**, 375.
Moulton, P. F. (1986). *J. Opt. Soc. Am. B* **3**, 125.
Mourou, G., Tajima, T. and Bulanov S. (2006). *Rev. Mod. Phys.* **78**, 309.
Muller, H. G. (2002). *Appl. Phys. B* **74**, S17.
Müller, H., Chiow, S.-W., Long, Q., Herrmann, S. and Chu, S. (2008). *Phys. Rev. Lett.*, **100**, 180405.
Murphy, S. A., Krause Jr, D., Li, Z. L. and Hunter, L.R. (1989). *Phys. Rev. Lett.* **63**, 965.
Myatt, C. J., Burt, E. A., Ghrist, R. W., Cornell, E. A. and Wieman, C. E. (1997). *Phys. Rev. Lett.* **78**, 586.
Nacher, P. J., Tastevin, G., Leduc, M., Crampton, S. B. and Laloë, F. (1984). *J. Physique Lett.* **45**, L-441.
Nagourney, W., Janij, G. and Dehmelt, H. (1983). *Proc. Nat. Acad. Sci.* **80**, 643.
Nagourney, W., Sandberg, J. and Dehmelt, H. (1986). *Phys. Rev. Lett.* **56**, 2797.
Nakagawa, K., Suzuki, Y., Horikoshi, M. and Kim, J. B. (2005). *Appl. Phys. B* **81**, 791.
Naraschewski, M. and Glauber, R. J. (1999). *Phys. Rev. A* **59**, 4595.
Narevicius, E., Libson, A., Parthey, C. G., Chavez, I., Narevicius, J., Even, U. and Raizen, M. G. (2008). *Phys. Rev. Lett.* **100**, 093003.
Nascimbène, S., Navon, N., Jiang, K. J., Chevy, F. and Salomon, C. (2010). *Nature* **463**, 1057.
Navon, N., Nascimbène, S., Chevy, F. and Salomon, C. (2010). *Science* **328**, 729.
Nesvizhevsky, V. V., Börner, H. G., Petukhov, A. K., Abele, H., Baeβler, S., Rueβ, F. J., Stöferle, T., Westphal, A., Gagarski, A. M., Petrov, G. A. and Strelkov, A. V. (2002). *Nature* **415**, 297.
Neuhauser, W., Hohenstaff, M., Toschek, P. and Dehmelt, H. (1978). *Phys. Rev. Lett.* **41**, 233.
Neuhauser, W., Hohenstatt, M., Toschek, P. E. and Dehmelt, H. (1980). *Phys. Rev. A* **22**, 1137.
New, G. H. C. and Ward, J. F. (1967). *Phys. Rev. Lett.* **19**, 556.
Ni, K.K., Ospelkaus, S., de Miranda, M. H. G., Pe'er, B., Neyenhuis, B., Zirbel, J. J., Kotochigova, S., Julienne, P. S., Jin, D. S. and Ye, J. (2008). *Science* **322**, 231.
Niu, Q., Zhao, X.-G., Georgakis, G. A. and Raizen, M. G. (1996). *Phys. Rev. Lett.* **76**, 4504.
Noecker, M. C., Masterson, B. P. and Wieman, C. E. (1988). *Phys. Rev. Lett.* **61**, 310.
Nozières, P., and Schmitt-Rink, S., (1985). *J. Low Temp. Phys.* **59**, 195.
Nozières, P. and Pines, D. (1990). *The Theory of Quantum Liquids* (Addison-Wesley, Redwood City).
Odom, B., Hanneke, D., D'Urso, B. and Gabrielse, G. (2006). *Phys. Rev. Lett.* **97**, 030801.
Oktel, M. Ö. and Levitov, L. S. (2002). *Phys. Rev. Lett.* **88**, 230403.
Olshanii, M. A. and Minogin, V. G. (1991). *Quant. Opt.* **3**, 317.
Olshanii, M. A. and Minogin, V. G. (1992). *Opt. Comm.* **89**, 393.
Omont, A. (1964). *C. R. Acad. Sci.* **258**, 1193.

Oskay, W. H., Diddams, S. A., Donley, E. A., Fortier, T. M., Heavner, T. P., Hollberg, L., Itano, W. M., Jefferts, S. R., Delaney, M. J., Kim, K., Levi, F., Parker, T. E. and Bergquist, J. C. (2006). *Phys. Rev. Lett.* **97**, 020801.

Ouhayoun, M. and Bordé, C. J. (1972). *C. R. Acad. Sci.* **274B**, 411.

Pancharatnam, S. (1966). *J. Opt. Soc. Am.* **56**, 1636.

Papp, S. B. and Wieman, C. E. (2006). *Phys. Rev. Lett.* **97**, 180404.

Papp, S. B., Pino, J. M., Wild, R. J., Ronen, S., Wieman, C. E., Jin, D. S. and Cornell, E. A. (2008). *Phys. Rev. Lett.* **101**, 135301.

Paredes, B., Widera, A., Murg, V., Mandel, O., Fölling, S., Cirac, I., Shlyapnikov, G. V., Hänsch, T. W. and Bloch, I. (2004). *Nature* **429**, 277.

Partridge, G. B., Strecker, K. E., Kamar, R. I., Jack, M. W. and Hulet, R. G. (2005). *Phys. Rev. Lett.* **95**, 020404.

Partridge, G. B., Li, W., Kumar, L., Liao, Y. and Hulet, R. G. (2006). *Science* **311**, 503.

Paul, W. (1990). *Rev. Mod. Phys.* **62**, 531.

Paul, P. M., Toma, E. S., Breger, P., Mullot, G., Augé, F., Balcou, Ph., Muller, H. G. and Agostini, P. (2001). *Science* **292**, 1689.

Paulus, G. G., Nicklich, W., Huale Xu, Lambropoulos, P. and Walther, H. (1994). *Phys. Rev. Lett.* **72**, 2851.

Pedri, P., Guéry-Odelin, D. and Stringari, S. (2003). *Phys. Rev. A* **68**, 043608.

Peik, E., Ben Dahan, M., Bouchoule, I., Castin, Y. and Salomon, C. (1997). *Phys. Rev. A* **55**, 2989.

Penrose, O. and Onsager, L. (1956). *Phys. Rev.* **104**, 576.

Pereira Dos Santos, F., Léonard, J., Wang, J., Barrelet, C. J., Perales, F., Rasel, E., Unnikrishnan, C. S., Leduc, M. and Cohen-Tannoudji, C. (2001). *Phys. Rev. Lett.* **86**, 3459.

Perelomov A.M., Popov V.S. and Terent'ev M.V. (1966). *Sov. Phys. JETP* **23**, 924.

Perelomov A.M., Popov V.S. and Terent'ev M.V. (1967). *Sov. Phys. JETP* **24**, 207.

Peres, A. (1996). *Phys. Rev. Lett.* **77**, 1413.

Perkins, T. T., Smith D. E. and Chu S. (1994a). *Science* **264**, 819.

Perkins, T. T., Quake, S. R., Smith D. E. and Chu S. (1994b). *Science* **264**, 822.

Perry, M. D. and Mourou, G. (1994). *Science* **264**, 917.

Peters, A., Chung, K. Y. and Chu, S. (2001). *Metrologia* **38**, 25.

Petrich, W., Anderson, M. H., Ensher, J. R. and Cornell, E. A. (1995). *Phys. Rev. Lett.* **74**, 3352.

Petrov, D. S. and Shlyapnikov, G. V. (2001). *Phys. Rev. A* **64**, 012706.

Petrov, D. S., Salomon, C. and Shlyapnikov, G. V. (2004). *Phys. Rev. Lett.* **93**, 090404.

Petrov, D. S., Salomon, C. and Shlyapnikov, G. V. (2005). *Phys. Rev. A* **71**, 012708.

Petrov, D. S., Salomon, C. and Shlyapnikov, G. V. (2007). Proceedings of the international school of physics "Enrico Fermi", Course CLXIV, eds. Inguscio, M., Ketterle, W. and Salomon, C. (IOS press, Amsterdam) pp. 385-412.

Phillips, W. D. (1998). *Rev. Mod. Phys.* **70**, 721.

Phillips, W. D., and Metcalf, H. (1982). *Phys. Rev. Lett.* **48**, 596.

Phillips, W. D., and Prodan, J. V. (1983). Laser-Cooled and Trapped Atoms, ed. Phillips, W. D. (Natl. Bur. Stand., Washington, DC), Spec. Publ. **653**, 137.

Picqué , J.-L., and Vialle, J.-L. (1972). *Opt. Commun.* **5**, 402.

Pillet, P., Vanhaecke, N., Lisdat, C., Comparat D., Dulieu, O., Crubellier, A. and Masnou-Seeuws F. (2003). *Phys. Scr.* **T105**, 7.

Pinkse, P. W. H., Mosk, A., Weidemller, M., Reynolds, M. W., Hijmans, T. W., Walraven, J. T. M. and Zimmermann, C. (1997). *Phys. Rev. Lett.* **79**, 2423.

Pitaevskii, L. and Stringari, S. (2003). *Bose-Einstein Condensation* (Oxford University Press).
Pohl, R., Antognini, A., Nez, F., Amaro, F. D., Biraben, F., Cardoso, J. M. R., Covita, D. S., Dax, A., Dhawan, S., Fernandes, L. M. P., Giesen, A., Graf, T., Hänsch, T.W., Indelicato, P., Julien, L., Kao, C.-Y., Knowles, P., Bigot, E.-O.L., Liu, Y.-W., Lopes, J. A. M., Ludhova, L., Monteiro, C. M. B., Mulhauser, F., Nebel, T., Rabinowitz, P., dos Santos, J. M. F., Schaller, L. A., Schuhmann, K., Schwob, C., Taqqu, D., Veloso, J. F. C. A. and Kottmann. F. (2010). *Nature* **466**, 213.
Poli, N., Wang, F.-Y., Tarallo, M. G., Alberti, A., Prevedelli, M. and Tino, G. M. (2011). *Phys. Rev. Lett.* **106**, 038501.
Politzer, H. D. (1996). *Phys. Rev. A* **54**, 5048.
Pollack, S. E., Dries, D. and Hulet, R. G. (2009). *Science* **326**, 1683.
Popov, V. N. (1966). *Sov. Phys. JETP* **23**, 1034.
Porsev, S. G., Beloy, P. and Derevianko, A. (2009). *Phys. Rev. Lett.* **102**, 181601.
Potter, E. D., Herek, J. L., Pedersen, S., Liu, Q. and Zewail, A. H. (1992). *Nature* **355**, 66.
Pound, R. V. and Rebka, G. A. (1960). *Phys. Rev. Lett.* **4**, 337.
Prescott, C. Y. *et al.* (1978). *Phys. Lett.* **77B**, 347.
Prescott, C. Y. *et al.* (1979). *Phys. Lett.* **84B**, 524.
Prestage, J. D., Tjoelker, R. L. and Maleki, L. (2001). *Recent Developments in Microwave Ion Clocks, Topics in Applied Physics: Frequency Measurement and Control* (Springer Berlin, Heidelberg).
Price, G. N., Bannerman, S. T., Viering, K., Narevicius, E. and Raizen, M. G. (2008). *Phys. Rev. Lett.* **100**, 093004.
Pricoupenko, L., Perrin, H. and Olshanii, M. (Eds.) (2003). Quantum gases in low dimensions, *Journal de Physique IV* **116**.
Pritchard, D. E., (1983). *Phys. Rev. Lett.* **51**, 1336.
Pritchard, D. E., Helmerson K., and Martin A. G. (1989). in *Atomic Physics 11* eds. Haroche, S., Gay, J. C. and Grynberg, G. (World Scientific, Singapore).
Prodan, J. V., Phillips, W. D. and Metcalf, H. (1982). *Phys. Rev. Lett.* **49**, 1149.
Pryce, M. H. L. (1950). *Phys. Rev.* **77**, 136.
Purcell, E. M., Torrey, H. C. and Pound, R. V. (1946). *Phys. Rev.* **69**, 37.
Raab, E. L., Prentiss, M., Cable, A., Chu, S. and Pritchard, D. E. (1987). *Phys. Rev. Lett.* **59**, 2631.
Rabi, I. I. and Cohen, V. W. (1934). *Phys. Rev.* **46**, 707.
Raizen, M. G. (2009). *Science* **324**, 1403.
Raman, C. V. and Nagendra Nath, N. S.(1936). *Proc. Ind. Acad. Sci.* **3**, 459.
Raman, C., Köhl, M., Onofrio, R., Durfee, D. S., Kuklewicz, C. E., Hadzibabic, Z. and Ketterle, W. (1999). *Phys. Rev. Lett.* **83**, 2502.
Ramsey, N. F. (1949). *Phys. Rev.* **76**, 996.
Ramsey, N. F. (1950). *Phys. Rev.* **78**, 695.
Ramsey, N. F. (1956). *Molecular Beams* (Oxford University Press).
Ramsey, N. F. (1990). *Rev. Mod. Phys.* **62**, 541.
Rath, S. P., Yefsah, T., Günter, K. J., Cheneau, M., Desbuquois, R., Holzmann, M., Krauth, W. and Dalibard, J. (2010). *Phys. Rev. A* **82**, 013609.
Rauch, H., Treimer, W. and Bonse, U. (1974). *Phys. Lett. A*, **47**, 369.
Rauch, H., and Werner, S. (2000). *Neutron Interferometry* (Clarendon Press, Oxford).
Regan, B. C., Commins E. D., Schmidt, C. J. and DeMille, D. (2002). *Phys. Rev. Lett.* **88**, 071805.
Recati, A., Zambelli, F. and Stringari, S. (2001). *Phys. Rev. Lett.* **86**, 377.

Reichel, J., Bardou, F., Ben Dahan, M., Peik, E., Rand, S., Salomon, C. and Cohen-Tannoudji, C. (1995). *Phys. Rev. Lett.* **75**, 4575.
Reichel, J., Hänsel, W. and Hänsch, T. W. (1999). *Phys. Rev. Lett.* **83**, 3398.
Reichel, J. (2002). *Appl. Phys. B* **74**, 469.
Renn, M. J., Donley, E. A., Cornell, E. A., Wieman, C. E. and Anderson, D. N. (1996). *Phys. Rev. A* **53**, R648.
Reynaud, S., Dalibard, J. and Cohen-Tannoudji, C. (1988). *IEEE J. Quant. Electron.* **24**, 1395.
Richard, S., Gerbier, F., Thywissen, J. H., Hugbart, M., Bouyer, P. and Aspect, A. (2003). *Phys. Rev. Lett.* **91**, 010405.
Riedel, M. F., Böhi, P., Li, Y., Hänsch, T. W., Sinatra, A. and Treutlein, P. (2010). *Nature* **464**, 1170.
Riehle, F., Kisters, Th., Witte, A., Helmcke, J. and Bordé, Ch. J. (1991). *Phys. Rev. Lett.* **67**, 177.
Riis, E., Weiss, D., Moler, K. and Chu, S. (1990). *Phys. Rev. Lett.* **64**, 1658.
Ringot, J., Szriftgiser, P., Garreau, J. C. and Delande, D. (2000). *Phys. Rev. Lett.* **85**, 2741.
Ritter, S., Ottl, A., Donner, T., Bourdel, T., Köhl, M., and Esslinger, T. (2007). *Phys. Rev. Lett.* **98**, 090402.
Roati, G., Riboli, F., Modugno, G. and Inguscio, M. (2002). *Phys. Rev. Lett.* **89**, 150403.
Roati, G., D'Errico, C., Fallani, L., Fattori, M., Fort, C., Zaccanti, M., Modugno, G., Modugno, M. and Inguscio, M. (2008). *Nature* **453**, 895.
Robert, A., Sirjean, O., Browaeys, A., Poupard, J., Nowak, S., Boiron, D., Westbrook, C. I. and Aspect A. (2001). *Science* **292**, 461.
Roberts, J. L., Claussen, N. R., Burke, J. P., Greene, C. H., Cornell, E. A. and Wieman, C. E. (1998). *Phys. Rev. Lett.* **81**, 5109.
Roberts, J. L., Claussen, N. R., Cornish, S. L., Donley, E. A., Cornell, E. A. and Wieman, C. E. (2001). *Phys. Rev. Lett.* **86**, 4211.
Robins, N. P., Figl, C., Jeppesen, M., Dennis, G. R. and Close, J. D. (2008). *Nature Phys.* **4**, 731.
Robinson, H. G. ,Ensberg, E. S. and Dehmelt, H. G. (1958). *Bull. Am. Phys. Soc.* **3**, 9.
Roman P. (1965). *Advanced Quantum Theory* (Addison Wesley).
Roos, C. F., Cren, P., Guéry-Odelin, D. and Dalibard, J. (2003). *Europhys. Lett.* **61**, 187 (2003).
Roos, C. F., Riebe, M., Häffner, H., Hänsel, W., Benhelm, J., Lancaster, G. P. T., Becher, C., Schmidt-Kaler, F. and Blatt, R. (2004). *Science* **304**, 1478.
Rosenband, T., Hume, D. B., Schmidt, P. O., Chou, C. W., Brusch, A., Lorini, L., Oskay, W. H., Drullinger, R. E., Fortier, T. M., Stalnaker, J. E., Diddams, S. A., Swann, W. C., Newbury, N. R., Itano, W. M., Wineland, D. J. and Bergquist, J. C. (2008). *Science* **319**, 1808.
Rudenko, A., Zrost, K., Feuerstein, B., de Jesus, V. L. B., Schröter, C. D. Moshammer, R. and Ullrich, J. (2004). *Phys. Rev. Lett.* **93**, 253001.
Ruddock, I. S. and Bradley, D. J. (1976). *Appl. Phys. Lett.* **29**, 296 (1976).
Sá de Melo, C. A. R., Randeria, M. and Engelbrecht, J. R. (1993). *Phys. Rev. Lett.* **71**, 3202.
Sackett, C. A., Stoof, H. T. C. and Hulet, R. D. (1998). *Phys. Rev. Lett.* **80**, 2031.
Sagnac, G. (1913). *C. R. Acad. Sci.* **157**, 708; *ibid.* 1410.
Sakurai, J. J. (1994). *Modern Quantum Mechanics*, (revised edition, Addison-Wesley, New-York).

Salières, P., L'Huillier, A., Antoine, P. and Lewenstein, M. (1999). *Adv. At. Mol. Opt. Phys.* **41**, 83.
Salières, P., Carré, B., Le Déroff, L., Grasbon, F., Paulus, G. G., Walther, H., Kopold, R., Becker, W., Milosevic, D. B., Sanpera, A. and Lewenstein, M. (2001). *Science* **292**, 902.
Salomon, C., Dalibard, J., Aspect, A., Metcalf, H. and Cohen-Tannoudji, C. (1987). *Phys. Rev. Lett.* **59**, 1659.
Salomon, C., Dalibard, J., Phillips, W. D., Clairon, A. and Guelatti, S. (1990). *Europhys. Lett.* **12**, 683.
Salour, M. and Cohen-Tannoudji, C. (1977). *Phys. Rev. Lett.* **38**, 757.
Sanchez-Palencia, L. and Lewenstein, M. (2010). *Nature Phys.* **6**, 87.
Sandars, P. G. H. (1965). *Phys. Lett.* **14**, 194.
Sandars, P. G. H. (1966). *Phys. Lett.* **22**, 290.
Sandars, P. G. H. (1967). *Phys. Rev. Lett.* **19**, 1396.
Sansone, G., Benedetti, E., Caumes, J. P., Stagira, S., Vozzi, C., Nisoli, M., Poletto, L., Villoresi, P., Strelkov, V., Sola, I., Elouga, L. B., Zaïr, A., Mével, E. and Constant, E. (2009). *Phys. Rev. A* **80**, 063837.
Santarelli, G., Laurent, P., Lemonde, P., Clairon, A., Mann, A. G., Chang, S., Luiten, A. N. and Salomon, C. (1999). *Phys. Rev. Lett.* **82**, 4619.
Sarma, N. V. (1968). *Philos. Mag.* **17**, 1233.
Saubaméa, B., Hijmans, T. W., Kulin, S., Rasel, E., Peik, E., Leduc, M. and Cohen-Tannoudji, C. (1997). *Phys. Rev. Lett.* **79**, 3146.
Saubaméa, B., Leduc, M. and Cohen-Tannoudji, C. (1999). *Phys. Rev. Lett.* **83**, 3796.
Sauter, Th., Neuhauser, W., Blatt, R. and Toschek, P. E. (1986). *Phys. Rev. Lett.* **57**, 1696.
Schafer, K. J., Baorui Yang, DiMauro, L. F. and Kulander, K. C. (1993). *Phys. Rev. Lett.* **70**, 1599.
Schelev, M. Ya., Richardson, M. C. and Alcock, A. J. (1971). *Appl. Phys. Lett.* **18**, 354.
Schieder, R., Walther, H. and Wöste, L. (1972). *Opt. Commun.* **5**, 337.
Schiff, L.I. (1963). *Phys. Rev.* **132**, 2194.
Schiff, L. I. (1969). *Quantum Mechanics* (3rd revised edition, McGraw-Hill).
Schirotzek, A., Shin, Y., Schunck, C. H. and Ketterle, W. (2008). *Phys. Rev. Lett.* **101**, 140403.
Schmidt, P. O., Hensler, S., Werner, J., Griesmaier, A., Görlitz, A., Pfau, T. and Simoni, A. (2003). *Phys. Rev. Lett.* **91**, 193201.
Schmidt, P. O., Rosenband, T., Langer, C., Itano, W. M., Bergquist, J. C. and Wineland, D. J. (2005). *Science* **309**, 749.
Schneider, U., Hackermüller, L., Will, S., Best, Th., Bloch, I., Costi, T. A., Helmes, R. W., Rasch, D. and Rosch, A. (2008). *Science* **322**, 1520.
Schöllkopf, W. and Toennies, J. P. (1996). *J. Chem. Phys.* **104**, 1155.
Schreck, F., Ferrari, G., Corwin, K. L., Cubizolles, J., Khaykovich, L., Mewes, M.-O. and Salomon, C. (2001a). *Phys. Rev. A* **64**, 011402(R).
Schreck, F., Khaykovich, L., Corwin, K. L., Ferrari, G., Bourdel, T., Cubizolles, J. and Salomon, C. (2001b). *Phys. Rev. Lett.* **87**, 080403.
Schreck, F (2003). *Ann. de Phys.* **28**, 1.
Schweikhard, V., Coddington, I., Engels, P., Mogendorff, V. P. and Cornell, E. A. (2004). *Phys. Rev. Lett.* **92**, 040404.
Schwinger, J. (1948). *Phys. Rev.* **73**, 416.
Schwob, C., Jozefowski, L., de Beauvoir, B., Hilico, L., Nez, F., Julien, L., Biraben, F., Acef, O., Zondy, J.-J. and Clairon, A. (1999). *Phys. Rev. Lett.* **82**, 4960.

Scully, M. O. and Zubairy, M. S. (1997). *Quantum Optics*, Cambridge University Press.
Sesko, D., Walker, T., Monroe, C., Gallagher, A. and Wieman, C. (1989). *Phys. Rev. Lett.* **63**, 961.
Seres, E., Seres, J., Krausz, F. and Spielmann, C. (2004). *Phys. Rev. Lett.* **92**, 163002.
Series, G., Schawlow, A. and Hänsch, T. (1979). *Scientific American* **240**, 94.
Setija, I. D., Werij, H. G. C., Luiten, O. J., Reynolds, M. W., Hijmans, T. W. and Walraven, J. T. M. (1993). *Phys. Rev. Lett.* **70**, 2257.
Shank, C. V. and Ippen, E. P. (1974). *Appl. Phys. Lett.* **24**, 373.
Shahriar, M. S., Hemmer, P. R., Prentiss, M. G., Marte, P., Mervis, J., Katz, D. P., Bigelow, N. P. and Cai, T. (1993). *Phys. Rev. A* **48**, R4035.
Shen, Y. R. (1984). *Principles of Nonlinear Optics* (John Wiley and Sons Inc.).
Sherson, J. F., Weitenberg, C., Endres, M., Cheneau, M., Bloch, I. and Kuhr, S. (2010). *Nature* **467**, 68.
Shimizu, F., Shimizu, K. and Takuma, H. (1992). *Phys. Rev. A* **46**, R17.
Shimizu, F. and Fujita, J.-i. (2002). *Phys. Rev. Lett.* **88**, 123201.
Shin, Y. (2008). *Phys. Rev. A* **77**, 041603(R).
Shlyapnikov, G. V., Walraven, J. T. M., Rahmanov, U. M. and Reynolds, M. W. (1994). *Phys. Rev. Lett.* **73**, 3247.
Short, R. and Mandel, L. (1983). *Phys. Rev. Lett.* **51**, 384.
Silvera, I. F. and Walraven, J. T. M.(1980a). *Phys. Rev. Lett.* **44**, 164.
Silvera, I. F. and Walraven, J. T. M.(1980b). *Phys. Rev. Lett.* **45**, 1268.
Singh, K. G. and Rokhsar, D. S. (1996). *Phys. Rev. Lett.* **77**, 1667.
Sinha, S. and Castin, Y. (2001). *Phys. Rev. Lett.* **87**, 190402.
Smith, S. B., Cui, Y. and Bustamente, C. (1996). *Science* **271**, 795.
Söding, J., Guéry-Odelin, D., Desbiolles, P., Ferrari, G. and Dalibard, J.(1998). *Phys. Rev. Lett.* **80**, 1869.
Söding, J., Guéry-Odelin, D., Desbiolles, P., Chevy, F., Inamori, H. and Dalibard, J.(1999). *Appl. Phys. B* **69**, 257.
Sokol, P. (1995).*Bose-Einstein condensation*, eds. Griffin, A., Snoke, D. W. and Stringari, S. (Cambridge University Press, Cambridge).
Sørensen, A. S. and Mølmer, K. (2001). *Phys. Rev. Lett.* **86**, 4431.
Sortais, Y., Bize, S., Nicolas, C., Clairon, A., Salomon, C. and Williams, C. (2000). *Phys. Rev. Lett.* **85**, 3117.
Spoden, P., Zinner, M., Herschbach, N., van Drunen, W. J., Ertmer, W. and Birkl, G. (2005). *Phys. Rev. Lett.* **94**, 223201.
Spreeuw, R. J. C., Gerz, C., Goldner, L. S., Phillips, W. D., Rolston, S. L., Westbrook, C. I., Reynolds, M. W. and Silvera, I. F. (1994). *Phys. Rev. Lett.* **72**, 3162.
Sprik, R., Walraven, J. T. M. and Silvera, I. F. (1985). *Phys. Rev. B* **32**, 5668.
Steinhauer, J., Ozeri, R., Katz, N. and Davidson, N. (2002). *Phys. Rev. Lett.* **88**, 120407.
Stewart, J. T., Gaebler, J. P. and Jin, D. S. (2008). *Nature* **454**, 744.
Stamper-Kurn, D. M., Miesner, H.-J., Inouye, S., Andrews, M. R. and Ketterle, W. (1998). *Phys. Rev. Lett.* **81**, 500.
Stärck, J. and Meyer, W. (1994). *Chem. Phys. Lett.* **225**, 229.
Steane, A. M., Szriftgiser, P., Desbiolles, P. and Dalibard, J. (1995). *Phys. Rev. Lett.* **74**, 4972.
Steinhauer, J., Ozeri, R., Katz, N., and Davidson, N. (2002). *Phys. Rev. Lett.* **88**, 120407.
Stellmer, S., Tey, M. K., Huang, B., Grimm, R. and Schreck, F. (2009). *Phys. Rev. Lett.* **103**, 200401.
Stenger, J., Inouye, S., Andrews, M. R., Miesner, H.-J., Stamper-Kurn, D. M. and Ketterle, W. (1999). *Phys. Rev. Lett.* **82**, 2422.

Stenger, J., Inouye, S., Chikkatur, A., Stamper-Kurn, D. M. and Ketterle, W. (1999). *Phys. Rev. Lett.* **82**, 4569.

Stenholm, S. (1986). *Rev. Mod. Phys.* **58**, 699.

Stewart, J. T., Gaebler, J. P., Regal, C. A. and Jin, D. S. (2006). *Phys. Rev. Lett.* **97**, 220406.

Storey, P. and Cohen-Tannoudji, C. (1994). *J. Physique II* **4**, 1999.

Strecker, K. E., Partridge, G. B., Truscott, A. G. and Hulet, R. G. (2002). *Nature* **417**, 150.

Strecker, K. E., Partridge, G. B. and Hulet, R. G. (2003). *Phys. Rev. Lett.* **91**, 080406.

Strickland, D. and Mourou, G. (1985). *Opt. Commun.* **56**, 219.

Stringari, S. (1996a). *Phys. Rev. Lett.* **76**, 1405.

Stringari, S. (1996b). *Phys. Rev. Lett.* **77**, 2360.

Stringari, S. (1998). *Phys. Rev. A* **58**, 2385.

Strecker, K. E., Partridge, G. B., Truscott, A. G. and Hulet, R. G. (2002). *Nature* **417**, 150.

Stuhler, J., Griesmaier, A., Koch, T., Fattori, M., Pfau, T., Giovanazzi, S., Pedri, P. and Santos, L. (2005). *Phys. Rev. Lett.* **95**, 150406.

Stwalley, W. C. and Nosanow, L. H. (1976). *Phys. Rev. Lett.* **36**, 910.

Szriftgiser, P., Guéry-Odelin, D., Arndt, M. and Dalibard, J. (1996). *Phys. Rev. Lett.* **77**, 4.

't Hooft, G. (1971). *Nucl. Phys.* **B33**, 173 and **B35**, 167.

't Hooft, G. (1980). *Scientific American* **242**, 90.

Tajima, T. and Dawson, J. M. (1979). *Phys. Rev. Lett.* **43**, 267.

Takasu, Y., Maki, K., Komori, K., Takano, T., Honda, K., Kumakura, M., Yabuzaki, T. and Takahashi, Y. (2003). *Phys. Rev. Lett.* **91**, 040404.

Tanguy, C., Reynaud, S. and Cohen-Tannoudji, C. (1984). *J. Phys. B* **17**, 4623.

Tannor, D. J., Kosloff, R. and Rice, S. A. (1986). *J. Chem. Phys.* **85**, 5805.

Tapster, P. R., Rarity, J. G., and Owens, P. C. M. (1994). *Phys. Rev. Lett.* **73**, 1923.

Tastevin G., Nacher P.J., Leduc M. and Laloë F. (1985). *J. Physique Lett.* **46**, L-249.

Taylor, G. I. (1909). *Proc. Cam. Phil. Soc.* **15**, 114.

Teklemariam, G., Fortunato, E. M., Pravia, M. A., Sharf, Y., Havel, T. F., Cory, D. G., Bhattaharyya, A. and Hou, J. (2002). *Phys. Rev. A* **66**, 012309.

Thalhammer G., Theis M., Winkler K., Grimm R. and Hecker Denschlag J. (2005). *Phys. Rev. A* **71**, 033403.

Theis M., Thalhammer G., Winkler K., Hellwig H., Ruff G., Grimm R. and Hecker Denschlag J. (2004). *Phys. Rev. Lett.* **93**, 123001.

Thomas, N. R., Kjærgaard, N., Julienne, P. S. and Wilson, A. S. (2004). *Phys. Rev. Lett.* **93**, 173201.

Thompson, S. T., Hodby, E. and Wieman, C. E. (2005). *Phys. Rev. Lett.* **95**, 190404.

Tisza, L. (1938). *Nature*, **141**, 913.

Tittel, W., Brendel, J., Zbinden, H. and Gisin, N. (1998). *Phys. Rev. Lett.* **81**, 3563.

Tollett, J. J., Chen, J., Story, J. G., Ritchie, N. W. M., Bradley, C. C. and Hulet, R. G. (1990). *Phys. Rev. Lett.* **65**, 559.

Tollett, J. J., Bradley, C. C., Sackett, C. A. and Hulet, R. G. (1995). *Phys. Rev. A* **51**, R22.

Tonks, L. (1936). *Phys. Rev.* **50**, 955.

Tonomura, A., Endo, J., Matsuda, T., Kawasaki, T. and Ezawa, H. (1989). *Am. J. Phys* **57**, 117.

Townsend, C. G., Edwards, N. H., Cooper, C. J., Zetie, K. P., Foot, C. J., Steane, A. M., Szriftgiser, P., Perrin, H. and Dalibard, J. (1995). *Phys. Rev. A* **52**, 1423.

Träuble, H. and Essmann, U. (1968). *J. Appl. Phys.* **25**, 273.

Truscott, A. G., Strecker, K. E., McAlexander, W. I., Partridge, G. B. and Hulet, R. G. (2001). *Science* **291**, 2570.

Tsigutkin, K., Dounas-Frazer, D., Family, A., Stalnaker, J. E., Yashchuk, V. V. and Budker, D. (2009). *Phys. Rev. Lett.* **103**, 071601.

Tsuzuki, T. (1971). *J. Low Temp. Phys.* **4**, 441.

Turlapov, A., Kinast, J., Clancy, B., Luo, L., Joseph, J. and Thomas, J. E. (2008). *J. Low Temp. Phys.* **150**, 567.

Udem, Th., Huber, A., Gross, B., Reichert, J., Prevedelli, M., Weitz, M. and Hänsch, T. W. (1997). *Phys. Rev. Lett.* **79**, 2646.

Udem, Th., and Riehle, F. (2007). *La Rivista del Nuovo Cimento* **30**, 563.

Ueda, M. and Leggett, A. J. (1998). *Phys. Rev. Lett.* **80**, 1576.

Uiberacker, M., Uphues, Th., Schultze, M., Verhoef, A. J., Yakovlev, V., Kling, M. F., Rauschenberger, J., Kabachnik, N. M., Schröder, H., Lezius, M., Kompa, K. L., Muller, H.-G., Vrakking, M. J. J., Hendel, S., Kleineberg, U., Heinzmann, U., Drescher, M. and Krausz, F. (2007). *Nature* **446**, 627.

Ungar, P. J., Weiss, D. S., Riis, E., and Chu, S. (1989). *J. Opt. Soc. Am. B* **6**, 2058.

Unruh, W. G. and Zurek, W. H. (1989). *Phys. Rev. D* **40**, 1071.

Vanhaecke, N., Meier, U., Andrist, M, Meier, B. H. and Merkt, F. (2007). *Phys. Rev. A* **75**, 031402.

van Abeelen, F. A. and Verhaar, B. J. (1999). *Phys. Rev. A* **59**, 578.

Van Dyck, Jr R. S., Schwinger, P. B. and Dehmelt, H. G. (1986). *Phys. Rev. D* **34**, 722.

Van Dyck, R. S. Jr., Schwinberg, P. B. and Dehmelt, H. G. (1987). *Phys. Rev. Lett.* **59**, 26.

van Enk, S. J. and G. Nienhuis, G. (1992). *Phys. Rev. A* **46**, 1438.

van Enk, S. J. (1994). *Quant. Opt.* **6**, 445.

van Roijen, R., Berkhout, J. J., Jaakkola, S. and Walraven, J. T. M. (1988). *Phys. Rev. Lett.* **61**, 931.

Vasilenko, L. S., Chebotayev, V. P. and Shishaev, A. V. (1970). *JETP Lett.* **12**, 113.

Véniard, V., Taïeb, R. and Maquet, A. (1996). *Phys. Rev. A* **54**, 721.

Venturi, V., Whittingham, I. B., Leo, P. J. and Peach, G. (1999). *Phys. Rev. A* **60**, 4635.

Verkerk, P., Lounis, B., Salomon, C., Cohen-Tannoudji, C., Courtois, J.-Y. and Grynberg, G. (1992). *Phys. Rev. Lett.* **68**, 3861.

Vessot, R. and Levine, M. (1979). *J. Gen. Rel. Grav.* **10**, 181.

Vinen, W. F. (1958). *Nature* **181**, 1524.

Vinen, W. F. (1961). *Proc. Roy. Soc. A* **260**, 218.

Vitanov, N. V., Fleischauer, M., Shore, B.W. and Bergmann, K. (2001). *Adv. At. Mol. Opt. Phys.* **46**, 55.

Vogels, J. M., Xu, K., Raman, C., Abo-Shaeer, J. R. and Ketterle, W. (2002). *Phys. Rev. Lett.* **88**, 060402.

Vold, T. G., Raab, F. J., Heckel, B. R. and Fortson, N. (1984). *Phys. Rev. Lett.* **52**, 2229.

Vollhardt, D. and Wölfle, P. (1990). *The Superfluid Phases of Helium 3* (Taylor and Francis, New York).

Voronov, G. and Delone, N. B.(1966). *Sov. Phys. JETP* **23**, 54.

Wagner, R., Chen, S.-Y., Maksimchuk, A. and Umstadter, D. (1997). *Phys. Rev. Lett.* **78**, 3125.

Wagner, N., Wuest, A., Christov, I., Popmintchev, T., Zhou, X., Murnane, M. and Kapteyn, H. (2006). *Proc. Nat. Acad. Sci.* **103**, 13279.

Wagner, N., Zhou, X., Lock, R., Li, W., Wüest, A., Murnane, M. and Kapteyn, H. (2007). *Phys. Rev. A* **76**, 061403(R).

Walraven, J. T. M., (1996). In *Quantum Dynamics of Simple Systems*, eds. Oppo, G. L., Barnett, S. M., Riis, E. and Wilkinson, M. (Institute of Physics, London), pp. 315.

Walker, T., Sesko, D. and Wieman, C. (1990). *Phys. Rev. Lett.* **64**, 408.

Walker, B. Sheehy, B., Kulander, K. C. and DiMauro, L. F. (1996). *Phys. Rev. Lett.* **77**, 5031.

Wallis, H., Dalibard, J. and Cohen-Tannoudji, C. (1992). *Appl. Phys. B* **54**, 407.

Waschke, C., Roskos, H. G., Schwedler, R., Leo, K., Kurz, H. and Köhler, K. (1993). *Phys. Rev. Lett.* **70**, 3319.

Weber, T., Herbig, J., Mark, M., Nägerl, H.-A. and Grimm, R. (2003). *Science* **299**, 232.

Weber, C., Barontini, G., Catani, J., Thalhammer, G., Inguscio, M. and Minardi, F. (2008). *Phys. Rev. A* **78**, 061601(R).

Weihs, G., Jennewein, T., Simon, C., Weinfurter, H. and Zeilinger, A. (1998). *Phys. Rev. Lett.* **81**, 5039.

Weiner, J., Bagnato, V. S., Zilio, S. and Julienne, P. S. (1999). *Rev. Mod. Phys.* **71**, 1.

Weisskopf, V. F. (1975). Intuitive approaches to field theory *Lepton and hadron structure, Subnuclear Physics Series*, Vol. 12, ed. Zichichi, A. (New York, Academic Press), p. 307.

Welton, T. A. (1948). *Phys. Rev.* **74**, 1157.

Werner, F., Tarruel, L. and Castin, Y. (2009). *Eur. Phys. J. B* **68**, 401.

Weyers, S., Aucouturier, E., Valentin, C. and Dimarcq, N. (1997). *Opt. Comm.* **143**, 30.

Wilczek, F. (1982). *Phys. Rev. Lett.* **49**, 957.

Wilkinson, S. R., Bharucha, C. F., Madison, K. W., Niu, Q. and Raizen, M. G. (1996). *Phys. Rev. Lett.* **76**, 4512.

Williams, J. and Holland, M. (1999). *Nature* **401**, 568.

Williams, J. E., Nikuni, T. and Clark, C. W. (2002). *Phys. Rev. Lett.* **88**, 230405.

Wilson, M. (2011). *Phys. Today* **64**, 10.

Wineland, D. J., Ekstrom, P. and Dehmelt, H. (1973). *Phys. Rev. Lett.* **31**, 1279.

Wineland D. J. and Dehmelt, H. (1975). *Bull. Am. Phys. Soc.* **20**, 637.

Wineland, D. J., Drullinger, R. E. and Walls, F. L. (1978). *Phys. Rev. Lett.* **40**, 1639.

Wineland D. J. and Itano W. M. (1979). *Phys. Rev. A* **20**, 1521.

Wineland D. J. and Itano W. M. (1987). *Phys. Today* **40**, (6)34.

Wineland, D. J., Bergquist, J. C., Itano, W. M., Bollinger, J. J. and Manney, C. H. (1987). *Phys. Rev. Lett.* **59**, 2935.

Wineland, D. J., Dalibard, J. and Cohen-Tannoudji, C. (1992a). *J. Opt. Soc. Am. B* **9**, 32.

Wineland, D. J., Bollinger, J. J., Itano, W. M., Moore, F. L. and Heinzen, D. J. (1992b). *Phys. Rev. A* **46**, R6797.

Wing, W. H. (1984). *Prog. Quant. Electr.* **8**, 181.

Winter, J. (1955). *C. R. Acad. Sci.*, **241**, 375.

Wong, T., Collett, M. J. and Walls D. F. (1996). *Phys. Rev. A* **54**, R3718.

Wood, C. S., Bennett, S. C., Cho, D., Masterson, B. P., Roberts, J. L., Tanner, C. E. and Wieman C. E. (1997). *Science* **275**, 1759.

Wu, C. S., Ambler, E., Hayward, R. W., Hoppes, D. D. and Hudson, R. P. (1957). *Phys. Rev.* **105**, 1413.

Xu, K., Mukaiyama, T., Abo-Shaeer, J. R., Chin, J. K., Miller, D. E. and Ketterle, W. (2003). *Phys. Rev. Lett.* **91**, 210402.

Yamazaki, R., Taie, S., Sugawa, S. and Takahashi, Y. (2010). *Phys. Rev. Lett.* **105**, 050405.

Yarmchuk, E. J., Gordon, M. J. V. and Packard, R. E. (1979). *Phys. Rev. Lett.* **43**, 214.

Young, R. D., Carlini, R. D., Thomas, A. W. and Roche, J. (2007). *Phys. Rev. Lett.* **99**, 122003.

Zaccanti M., Deissler B., D'Errico C., Fattori M., Josa-Lasinio M., Mller S., Roati G., Inguscio M. and Modugno G. (2009). *Nature Phys.* **5**, 586.
Zambelli, F. and Stringari, S. (1998). *Phys. Rev. Lett.*, **81**, 1754.
Zambelli, F. and Stringari, S. (2001). *Phys. Rev. A*, **63**, 033602.
Zambelli, F. and Stringari, S. (2002). *Laser Phys.*, **12**, 240.
Zaremba, E. (1998). *Phys. Rev. A* **57**, 518.
Zel'dovich Ya. B. (1957). *Sov. Phys. JETP* **6**, 1184.
Zernike, F. (1938). *Physica* **5**, 785.
Zewail, A. H. (2000). *J. Phys. Chem. A*, **104**, 5660.
Zewail, A. H. (2003). Femtochemistry: Atomic-Scale Dynamics of the Chemical Bond Using Ultrafast Lasers, Nobel Lectures in Chemistry 1996-2000, ed. Grenthe, I. (World Scientific, Singapore) pp. 262
Zhu, L., Kleiman, V., Li, X., Lu, S. P., Trentelman, K. and Gordon, R. J. (1995). *Science* **270**, 77.
Zobay, O. and Garraway, B. M. (2001). *Phys. Rev. Lett.* **86**, 1195.
Zoller, P., Marte, M. and Walls, D. F. (1987). *Phys. Rev. A* **35**, 198.
Zurek, W. H. (2003). *Rev. Mod. Phys.* **75**, 715.
Zwierlein, M. W., Stan, C. A., Schunck, C. H., Raupach, S. M. F., Gupta, S., Hadzibabic, Z. and Ketterle, W. (2003). *Phys. Rev. Lett.* **91**, 250401.
Zwierlein, M.W., Abo-Shaeer, J. R., Schirotzek, A., Schunck, C. H. and Ketterle, W. (2005). *Nature* **435**, 1047.
Zwierlein, M.W., Schirotzek, A., Schunck, C. H. and Ketterle, W. (2006). *Science* **311**, 492.

Index